# 16th International Conference on X-ray Absorption Fine Structure (XAFS16)

AA006468

Journal of Physics: Conference Series
Volume 712

Karlsruhe, Germany
23 – 28 August 2015

**Editors:**

**J.-D. Grunwaldt**
**M. Hagelstein**
**J. Rothe**

ISBN: 978-1-5108-2451-5
ISSN: 1742-6588

**Printed from e-media with permission by:**

Curran Associates, Inc.
57 Morehouse Lane
Red Hook, NY 12571

proceedings.com

**Some format issues inherent in the e-media version may also appear in this print version.**



Printed by Curran Associates, Inc. (2016)

For permission requests, please contact the Institute of Physics
at the address below.

Institute of Physics
Dirac House, Temple Back
Bristol BS1 6BE UK

Phone: 44 1 17 929 7481
Fax: 44 1 17 920 0979

techtracking@iop.org

**Additional copies of this publication are available from:**

Curran Associates, Inc.
57 Morehouse Lane
Red Hook, NY 12571 USA
Phone: 845-758-0400
Fax: 845-758-2633
Email: curran@proceedings.com
Web: www.proceedings.com

# Table of contents

## Volume 712

**16th International Conference on X-ray Absorption Fine Structure (XAFS16)**
**23-28 August 2015, Karlsruhe, Germany**

**Accepted papers received: 15 April 2016**
**Published online: 6 June 2016**

**New sources and new instrumentation**

**Advanced methods**

**Chemistry, catalysis, operando and time-resolved studies**

**Radionuclides, actinides, earth and environmental**

**Materials science**

**Energy-related materials**

**Soft matter and biology**

**Microscopy, beamlines, applications, cultural heritage**

**Industrial session**



**Workshop "Data acquisition, treatment, storage-quality assurance in XAFS spectroscopy"**

Journal of Physics: Conference Series **712** (2016) 011001 doi:10.1088/1742-6596/712/1/011001

# The 16th International Conference on X-ray Absorption Fine Structure (XAFS16)

**J-D Grunwaldt[1,2], M Hagelstein[3], J Rothe[4]**

[1] Institute for Chemical Technology and Polymer Chemistry (ITCP), KIT, Kaiserstr. 12, D-76131 Karlsruhe, Germany
[2] Institute of Catalysis Research and Technology (IKFT), KIT, Hermann-von-Helmholtz-Platz 1, D-76344 Eggenstein-Leopoldshafen, Germany
[3] Institute for Beam Physics and Technology (IBPT), KIT, Hermann-von-Helmholtz-Platz 1, D-76344 Eggenstein-Leopoldshafen, Germany
[4] Institute for Nuclear Waste Disposal (INE), KIT, Hermann-von-Helmholtz-Platz 1, D-76344 Eggenstein-Leopoldshafen, Germany

E-mail: grunwaldt@kit.edu

**Abstract**. This preface of the proceedings volume of the 16th International Conference on X-ray Absorption Fine Structure (XAFS16) gives a glance on the five days of cutting-edge X-ray science which were held in Karlsruhe, Germany, August 23 – 28, 2015. In addition, several satellite meetings took place in Hamburg, Berlin and Stuttgart, a Sino-German workshop, three data analysis tutorials as well as special symposia on industrial catalysis and XFELs were held at the conference venue.

## 1. Preface

The 16th International Conference on X-ray Absorption Fine Structure (XAFS16 - www.xafs16.org) was held at the Karlsruhe Institute of Technology (KIT) in Karlsruhe, Germany, from 23rd to 28th August 2015. More than 550 participants - including more than 160 early career scientists - from 37 countries all around the world attended this important event which took place, appropriately, in the UNESCO "International Year of Light and Light-based Technologies" (http://www.light2015.org/). At the same time, the XAFS16 host city Karlsruhe celebrated the 300th anniversary of its foundation by Margrave Karl Wilhelm of Baden-Durlach in 1715, a birthday commemorated with spectacular events like the "Schlosslichtspiele" throughout the town center (Figure 1), which many of the participants attended.

Since the first XAFS meeting in 1981 in Daresbury, UK, the XAFS conference series has become the foremost international meeting covering developments and applications in X-ray absorption spectroscopy and related techniques. The preceding conferences took place in Stanford in 2006 [1], Camerino in 2009 [2] and Bejing in 2012 [3]. The last and up to now only XAFS conference in Germany was XAFS8, held in Berlin in 1994. Jointly organized by KIT (Karlsruhe), DESY (Hamburg), HZB (Berlin) and the European XFEL (Hamburg) and under the auspices of the International X-ray Absorption Society (IXAS), XAFS16 attracted a new record number of participants (Figure 2). The local organization was provided by the KIT institutes IBPT, INE, IKFT, ITCP and the ANKA synchrotron facility, with Professor Jan-Dierk Grunwaldt chairing the local organizing committee.

**Figure 1.** Illumination of the façade of Karlsruhe's palace at the heart of the town center during the "Schlosslichtspiele" in summer 2015. The installation symbolizes the discovery of electromagnetic waves (1886) by Heinrich Hertz in Karlsruhe at the Technische Hochschule, a predecessor of today's KIT (Copyright: Karlsruhe | Tourismus).

The symposia and accompanying poster sessions addressed virtually all aspects of X-ray absorption spectroscopy – from the latest theoretical developments to instrumentation and applications – covering topics from the in-situ characterization of catalytic reaction processes, environmental contaminant and radionuclide speciation, to novel developments in medical applications and forensic analysis of cultural and historical objects. Altogether 9 plenary lectures, 26 keynotes and almost 200 contributed talks were presented at the conference in 18 different topics (cf. section 3) grouped in 9 scientific fields - starting with field I '*Theory, modelling and data analysis*'. In the development of XAFS theory a gradual transition from semi-empirical codes to first principle codes is being observed, also encompassing the metal L-edges. Data analysis is at the heart of all XAFS techniques and, apart from the hands-on courses and the satellite meeting '*Data acquisition, treatment, storage - quality assurance in XAFS spectroscopy*', XAFS data treatment and modelling was the topic of several invited and contributed talks at the conference. In field II '*New sources and new instrumentation*' the status of numerous existing and newly established synchrotron beamlines was covered (cf. several contributions in these proceedings) - the trend is towards faster measurements, specialized beamlines (e.g., for catalysis, energy or radionuclide research), new detectors or detectors that operate at room temperature, to name a few examples. A further prominent trend is towards spatial resolution, even into the 5 – 10 nm regime, and unprecedented energy resolution with high energy fluorescence detected XAS and X-ray emission spectroscopy. While XFELs definitely develop towards an interesting source, pushing the frontiers of the XAFS technique at large scale accelerator facilities (as demonstrated in a dedicated workshop at the conference), an equally challenging and promising trend was also observed: the development of competitive laboratory XAS and XES instrumentation based on X-ray tubes or plasma sources. The trends in field III '*Advanced methods*' were strongly connected to the previous fields: X-ray emission (XES) and valence-to-core (VTC) emission spectroscopy was already one topic highlighted in the first plenary talk by Serena de Beer on the nitrogenase enzyme, and continued through many talks focusing on resonant and non-resonant XES applications. In particular the recent improvements of the underlying theory have made XES a powerful hard X-ray probe with impact in many areas, like field IV '*Chemistry, catalysis, operando and time-resolved studies*', where in situ and operando studies were reported in vast broadness. Sophisticated high energy resolution methods such as high energy resolved fluorescence detected (HERFD) or partial fluorescence yield (PFY) XANES are meanwhile applied in other fields such as radionuclide and earth and environmental research (field V), e.g., for the study of transient or minority oxidation states in mixed valence actinide systems, or for the investigation of the mobility and bioaccumulation of heavy metal contaminants such as mercury, as illustrated in the plenary lecture given by Alain Manceau.

There was a substantial number of materials science studies submitted to XAFS16, which were split into the separate fields related to the '*Science of materials*' (VI) and '*Energy-related materials*' (VII). In this context, not only studies highlighting structure-property relationships were presented, but also exciting operando studies on batteries and energy storage materials. The broadness of the XAFS application field was further demonstrated with plenary lectures, keynote and contributed talks in the field of '*Biology and soft matter*' (VIII). Last but not least, a special session on spatially resolved XAS and X-ray microscopy (field IX: '*Microscopy, beamlines, applications, cultural heritage*') gave an excellent overview of the dynamic developments in this area. Microscopy is now conducted on all length scales from the μm (or even mm/cm) to the 50 nanometer range. Although ptychography offers new opportunities, achieving X-ray absorption spectroscopy contrast at the ultimate 5 nm scale remains a wish not yet fulfilled. Finally, the important role that X-ray absorption spectroscopy has traditionally played in the commercial use of synchrotrons and for industrial research was reflected in a special symposium dedicated to current and potential applications in industrial XAFS research.

**Figure 2.** The official conference photo (Copyright: Carina Braun | KIT) taken at the site of the old stadium, KIT Campus South. The inset shows the conference logo designed by Wilfrid Schroeder, KIT († 2015).

In addition to the main conference a number of satellite meetings covering specialist topics were held at HZB (20 - 21 August 2015, '*Time and energy-resolved soft X-ray spectroscopy*', organized by Emad Flear Aziz), at DESY (21 - 22 August 2015, '*Data acquisition, treatment, storage and quality assurance in XAFS spectroscopy*', organized by Edmund Welter), and at the MPI-IS in Stuttgart (30 August – 2 September 2015, '*Application of XAFS to the study of magnetic materials*', organized by Eberhard Goering). Moreover, a Sino-German Workshop on '*In situ Spectroscopy on Catalysis and Membranes*' at KIT (August 28 – 29, 2015, organized by Ying Zhou, Andreas Goldbach and Jan-Dierk Grunwaldt), and three data analysis tutorials coordinated by Federico Benzi at KIT on FEFF9 (August 23, 2015, John Rehr), GNXAS (August 23, 2015, Andrea Di Cicco) and CTM4XAS (August 29, 2015, Frank de Groot) were held.

The International X-ray Absorption Society (IXAS) used the occasion of the XAFS16 conference to honour a number of scientists whose activities have contributed to the continued development of X-ray absorption spectroscopy: the Edward Stern Prize was awarded jointly to Majed Chergui, Professor of Physics and Chemistry at EPFL Lausanne, Switzerland (for his pioneering work in the field of time-resolved X-ray absorption spectroscopy) and Andrea Di Cicco, Professor of Physics at the Università di Camerino, Italy (for outstanding achievements and pioneering work in the application of XAS to disordered systems). Dr. Amélie Juhin of the French National Centre for Scientific Research in Paris received the Farrel Lytle Prize for her contribution to the understanding of X-ray dichroism and resonant inelastic X-ray scattering, and the Dale Sayers Prize for outstanding young scientists went to Dr. Narcizo Souza-Neto, beamline scientist at the Brazilian Synchrotron Light Laboratory in Campinas, for the development of instrumentation for the use of synchrotron techniques under extreme conditions, such as high pressures (Figure 3).

**Figure 3.** The prize-winners together with the conference organizers - from left to right: Pieter Glatzel, new IXAS chairman, Bruce Bunker, former IXAS chairman, Andrea Di Cicco, Majed Chergui, Amélie Juhin, Narcizo Souza-Neto, and XAFS16 chairman Jan-Dierk Grunwaldt (Copyright: Carina Braun | KIT).

The social program of XAFS16 encompassed excursions to attractive locations near Karlsruhe - a guided walking tour of Heidelberg's old town and castle, a wine tasting tour of the Affentaler cellars in Bühl near Baden-Baden and a visit of the medieval Maulbronn Monastery, one of Germany's UNESCO world heritage sites. The conference banquet was held at the Center of Art and Media (ZKM) in Karlsruhe, a worldwide renowned institution for computer based art and media design. The dinner was preceded by a tour of the '*Global Games*' exhibition at ZKM, a hands-on exhibition of the developments in the area of computer games during the past decades.

At the conclusion of the conference the participants had the opportunity to visit the ANKA synchrotron radiation facility at KIT Campus North, with currently three beamlines dedicated to X-ray spectroscopy and where the CAT-ACT beamline for catalyst and actinide research is currently being commissioned. In the closing presentations the hosts of the next conference, XAFS17 (www.xafs2018.com), which will take place in Krakow, Poland in 2018 were introduced. We look forward to meet many of the XAFS16 attendees at this occasion!

## 2. Nationality of XAFS16 conference attendees

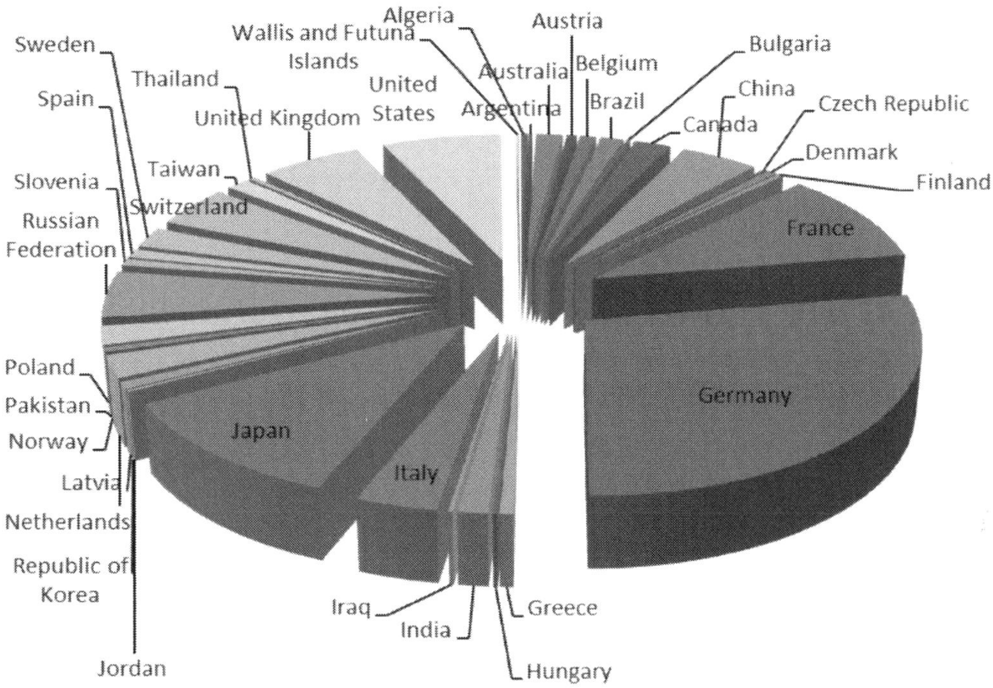


## 3. Original list of topics offered for abstract submission

The following topics were grouped by nine program committees into scientific fields I - IX for refereeing and setting up the time schedule of the conference:

1. X-ray facilities, instrumentation, optics
2. Materials science and energy-related materials
3. Catalysis
4. Radioactive and nuclear materials
5. Life science, biology and biochemistry
6. Earth and environmental sciences
7. Cultural heritage
8. Atoms and molecules
9. Nanotechnology
10. Surfaces and interfaces
11. Magnetism
12. Soft matter
13. X-ray scattering and electron spectroscopy (XES, RIXS, NIXS, HERFD, and XPS)
14. Theoretical methods
15. Software, data analysis
16. New methods and combination of techniques
17. Spectroscopies at XFEL sources, time-resolved and ultrafast techniques
18. X-ray microscopy

## 4. Committees, scientific sessions and staff

We would like to thank the members of the International Advisory Committee, the Program Committee and all reviewers evaluating more than 600 original abstracts submitted to the conference and the manuscripts collected in this *J. Phys.: Conf. Ser.* proceedings volume from the XAFS16 conference and the satellite meeting in Hamburg for their dedication and invaluable effort! Many of them have served in chairing the scientific sessions of the many different topics and setting up the program in 9 fields. Melissa A. Denecke is especially acknowledged for organizing and chairing the poster award committee.

*Chairman of the Conference:*
Jan-Dierk Grunwaldt

*International Advisory Committee:*
Kiyotaka Asakura (Japan)
Valérie Briois (France)
Chris Chantler (Australia)
Stephen Cramer (United States)
Melissa A. Denecke (Great Britain)
Pieter Glatzel (France)
Frank de Groot (Netherlands)
Hugh Harris (Australia)
Krystyna Ławniczak-Jabłońska (Poland)
John Rehr (United States)
Gerrit van der Laan (Great Britain)
Gustavo Azevedo (Brazil)
Bruce Bunker (United States), IXAS chair
Andrea Di Cicco (Italy)
Paola D'Angelo (Italy)
Ronald Frahm (Germany)
Robert Gordon (Canada)
Jan-Dierk Grunwaldt (Germany)
Daniel Haskel (United States)
Stefano Della Longa (Italy)
Philippe Sainctavit (France)
Ziyu Wu (China), † 2014

*Local Organizing Committee:*
Jan-Dierk Grunwaldt
Federico Benzi
Maria Casapu
Michael Drees
Michael Hagelstein
Clemens Heske
Stefan Mangold
Ralph Steininger
Tonya Vitova
David Batchelor
Carina Braun
Kathy Dardenne
Jörg Göttlicher
Marc Häming
Henning Lichtenberg
Jörg Rothe
Marika Vespa
Anna Zimina

*Secretaries:*
Margit Costarelli
Evangelia Mina
Elke Kalina (C & M)
Ogarit Uhlmann (C & M)
Birgit Stahl (C & M)

*Technical Support:*
Elke Bohnert
Tim Prüßmann
Rony Sobiesiak

*Graphics and Photos:*
Carina Braun

*Conference registration and commercial exhibition:*
congress & more Klaus Link GmbH, Karlsruhe, Germany

## 6. Acknowledgments
The conference was generously supported by Karlsruhe Institute of Technology (KIT), IXAS, IUCr, Deutsche Forschungsgemeinschaft (DFG), DESY (Hamburg), XFEL (Hamburg), Helmholtz-Zentrum Berlin (HZB) and the "Komitee Forschung mit Synchrotronstrahlung" (KFS). Financial contributions by all companies participating in the commercial exhibition are acknowledged with gratitude. Finally, we wish to thank IOP for publishing the conference proceedings, the participants for an exciting meeting and the authors for submitting the manuscripts to these conference proceedings which has allowed us to give the present overview on the research conducted in the field of X-ray absorption fine structure.

## References
[1] B. Hedman and P. Pianetta 2007 *AIP Conf. Proc.* **882**
[2] A Di Cicco and A Filipponi 2010 *J. Phys. Conf. Ser.* **190** 011001
[3] Z Y Wu 2013 *J. Phys. Conf. Ser.* **430** 011001

# Peer review statement

All papers published in this volume of *Journal of Physics: Conference Series* have been peer reviewed through processes administered by the proceedings Editors. Reviews were conducted by expert referees to the professional and scientific standards expected of a proceedings journal published by IOP Publishing.

# *Quanty* for core level spectroscopy - excitons, resonances and band excitations in time and frequency domain

**Maurits W. Haverkort**
Max Planck Institute for Chemical Physics of Solids, Nöthnitzer Straße 40, 01187 Dresden, Germany

E-mail: Maurits.Haverkort@cpfs.mpg.de

**Abstract.** Depending on the material and edge under consideration, core level spectra manifest themselves as local excitons with multiplets, edge singularities, resonances, or the local projected density of states. Both extremes, i.e., local excitons and non-interacting delocalized excitations are theoretically well under control. Describing the intermediate regime, where local many body interactions and band-formation are equally important is a challenge. Here we discuss how ***Quanty***, a versatile quantum many body script language, can be used to calculate a variety of different core level spectroscopy types on solids and molecules, both in the frequency as well as the time domain. The flexible nature of ***Quanty*** allows one to choose different approximations for different edges and materials. For example, using a newly developed method merging ideas from density renormalization group and quantum chemistry [1–3], ***Quanty*** can calculate excitons, resonances and band-excitations in x-ray absorption, photoemission, x-ray emission, fluorescence yield, non-resonant inelastic x-ray scattering, resonant inelastic x-ray scattering and many more spectroscopy types. ***Quanty*** can be obtained from: http://www.quanty.org.

Core level spectroscopies have long been valuable tools in the field of material research for a huge range of compounds with different degree of correlations. The theoretical interpretation can be roughly divided into two different approaches. The first approach is based on density functional theory. These methods include the full band formation, either using a one electron basis [4–6] or using multiple scattering theory [7, 8]. At this level of approximation, x-ray absorption is identified as the unoccupied density of states, which works quite well for several edges [9–12]. The core-valence interaction can be modeled as an additional potential resulting in an edge singularity in the spectral function [13–15]. The inclusion of the full band-structure comes at a price and in order to make the calculations tractable approximations have to be made on the local Coulomb interaction and multiplets. The second approach starts from a local ansatz. The method approximates the solid by a small cluster in which all local Coulomb interactions and multiplets can be accounted for. Very good agreement with experiment exists for many transition metal $L_{2,3}$ and rare earth $M_{4,5}$ edges, which are strongly influenced by multiplet effects [16, 17]. The resulting spectra are intrinsically many-body and must not be confused with the unoccupied density of states. In order to make the calculations tractable, the full multiplet methods neglect band formation and non-local dispersions. The later approach used to be parameter based, but several groups recently managed to calculate *ab intio* spectra including full multiplet theory [1, 18–21].

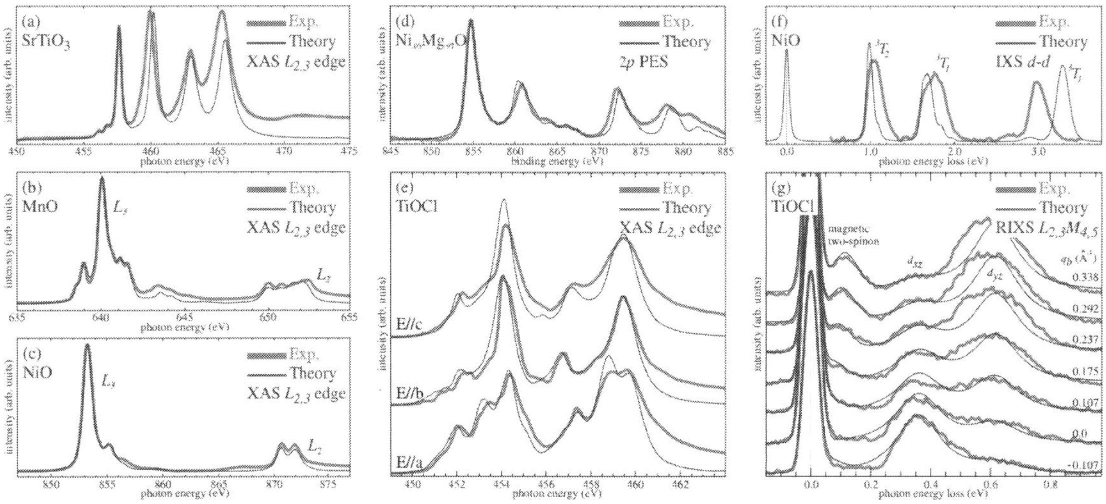


**Figure 1.** *Ab initio* theoretical and experimental core level spectra for several materials and experimental techniques. (a-c) $2p$ to $3d$ x-ray absorption in transition metal compounds. (d) $2p$ core level photoemmission of Ni impurities in MgO. (e) $2p$ to $3d$ x-ray absorption in TiOCl, including the natural linear dichroism induced due to the low crystal symmetry. (f) $d - d$ excitations in NiO induced by inelastic x-ray scattering. (g) $d - d$ and dispersive magnetic excitations in TiOCl induced by resonant inelastic x-ray scattering at the Ti $L$ edge. Experimental spectra reproduced from: (a) Uehara *et al.* [22], (b) Csiszar *et al.* [23], (c) Alders *et al.* [24], (d) Altieri *et al.* [25], (e) Glawion *et al.* [19, 26], (f) Verbeni *et al.* [27, 28], (g) Glawion *et al.* [19, 26]. Theory reproduced from: (a-d,f) Haverkort *et al.* [1], (e,g) Glawion *et al.* [19, 26].

There is a range of materials, for example correlated metals and a set of experiments, including excitations involving resonances with multiplets, where neither of the two approaches give a satisfactory result. For insulators with a large core-valence interaction a bound state of the core-hole and additional valence electron can form. These bound states are called excitons and show in the spectrum as sharp peaks outside the continuum. If the exciton is degenerate with a set of continuum excitations the core hole and valence electron do not form a perfect bound state. One still can find an enhancement in the absorption, but these peaks are broader than excitonic excitations and have an asymmetric lineshape [29]. These virtually bound states are called (Fano) resonances. Resonances are generally difficult to describe as both local multiplets and non-local band-formation is important. This problem is not unique to core level spectroscopy, but also exists at low photon energies related to visible light. Describing the absorption of light into a local resonance with multiplets and the subsequent decay into a delocalized hole and electron state is a true theoretical challenge. The exponential scaling of the computational effort with system size makes the calculation intractable without further approximations. Despite the difficulty of the problem, progress is made into merging the two different approaches for the calculation of core level spectroscopy. On the side of the band-structure approaches several groups implemented the Bethe Salpeter equations, which allows one to include core-valence interactions in the final state for band-insulators, i.e. $Ti^{4+}$ and $Ca^{2+}$ compounds [30, 31]. The local cluster models can be extended to include the band width of the material at the level of an Anderson impurity model [32–36]. Using a newly developed method to solve the Anderson impurity model [2], this has recently been applied to metals and correlated metals [3].

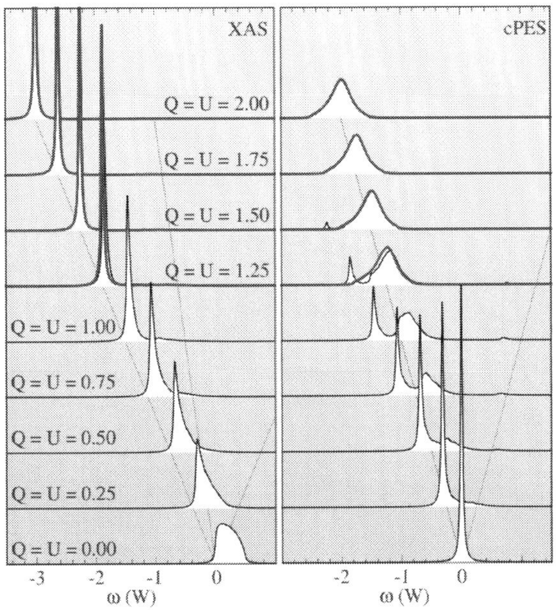


**Figure 2.** Theoretical model calculations of x-ray absorption (left) and core level photoemission (right) spectra of a band-metal (bottom) to a correlated Mott-insulator (top). The calculations include core-valence interactions ($Q$) and local valence-valence interactions ($U$). The infinite correlated lattice is exactly solved locally and non-locally approximating the many body interactions on a (dynamical) mean-field level. For small Coulomb interaction ($U \leq 1.50$) the ground-state is a metal (thin lines) for large Coulomb interactions ($U \geq 1.25$) the ground-state is insulating (thick blue lines). For intermediate correlation strength ($1.25 \leq U \leq 1.50$) we show results for both a metallic and an insulating ground-state. Reproduced from Haverkort *et al.* [3].

***Quanty*** is a program, or script language that can be used to calculate several core level spectroscopies within different approximations. In the first part of the paper we will show *ab initio* results obtained with ***Quanty***. We show multiplet ligand field theory calculations, where the band dispersion is neglected and also show calculations that include the non-local interactions on a (dynamical) mean-field level, governing the complete range of possible spectra. The second part of the paper shows how these calculations can be implemented in ***Quanty***.

In multiplet ligand field theory, the solid is approximated by a single (transition-metal) ion surrounded by an effective set of ligand orbitals mimicking the effect of the solid. For transition metal oxides one can think of these as the O-$2p$ orbitals. Traditionally, multiplet ligand field theory was parameter based [16, 17]. One can implement this method as a post density functional theory method which removes the need to fit parameters [1]. In Fig. 1 we show the result for a large variety of different spectroscopy techniques and transition metal compounds. As one can see, there is good agreement between theory and experiment.

Multiplet ligand field theory is, by the construction of its local basis, not able to describe band excitations or resonances well. The spectra in Fig. 1 are all excitonic excitations in insulators. An extension, which can deal with metals, can be made using the dynamical mean field approximation. Within this approximation locally all correlations are included, but non-locally these are approximated on a mean field level. In Fig. 2 we show the x-ray absorption and core-level photoemission spectra of a one band model for different interaction strengths. The entire range of spectral features, from density of states to resonances to edge singularities to excitons is captured on this level of theory.

Being able to handle both excitons with multiplets as well as resonances in one theory, is for example, important in the description of shallow core level spectra [38]. In Fig. 3 we show the $3p$ to $3d$ excitation in MnO as measured with non-resonant inelastic x-ray scattering. For small momentum transfer (panel c and f) one measures a dipole transition. For large momentum transfer (panel a,b,d,e) one measures an octupole transition [39]. Note that although a dipole transition does not possess linear dichroism in a cubic crystal, an octupole transition does, thereby making it possible to distinguish between transitions into the $t_{2g}$ and $e_g$ orbitals by

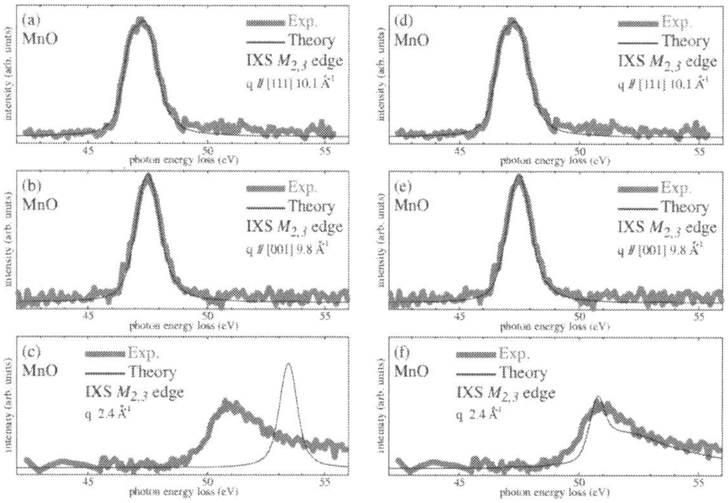


**Figure 3.** $3p$ to $3d$ IXS excitations in MnO. (a,b,d,e) large momentum transfer, octupole transitions into excitonic states. (c,f) small momentum transfer, dipole transitions into resonances. (a-c) including only a local basis, (d-f) including local correlations as well as the full (dynamical) mean-field approximated band-structure in the theory. Experimental data reproduced from Gordon *et al.* [37]

symmetry[37, 40]. In panel a to c we compare the experiment to the theoretical performed calculations on an *ab initio* multiplet ligand field theory level. The excitonic bound transitions below the continuum edge at $\approx$ 50 eV are well described. The dipole transition above the continuum edge is in the ligand field theory a sharp peak, whereas it is a broad continuum, showing resonant like features in the experiment. Including the continuum in the theory on a dynamical mean field level does not change the excitonic bound states, but does allow one to also capture the resonances correctly as can be seen in panel d to f.

***Quanty*** is a script language which enables its users to calculate these spectra and allows one to solve many other problems in quantum mechanics with a minimal amount of coding. As a first step ***Quanty*** needs to know the basis set used. Although ***Quanty*** is a many body code, the basis set is defined by one particle modes. These are the boxes (basis states) one can place a particle in. The minimal information ***Quanty*** needs to define a basis is the number of Fermionic and Bosonic modes. Fermionic modes can be different orbitals with different spins (spin-orbitals) or different lattice sites. For calculations in solid state physics or chemistry these could refer to the Wannier orbitals in a solid or the molecular orbitals for a molecule. We generally would obtain those with the use of either DFT or Hartree Fock calculations.

```
NF = 6 -- Number of spin-orbitals
NB = 0 -- Number of phonon-modes
```

Once a basis set is defined one can create operators using second quantization. Any operator can be written as:

$$\alpha^{(0,0)} + \sum_i \alpha_i^{(1,0)} a_i^\dagger + \alpha_i^{(0,1)} a_i + \sum_{i,j} \alpha_{i,j}^{(2,0)} a_i^\dagger a_j^\dagger + \alpha_{i,j}^{(1,1)} a_i^\dagger a_j + \alpha_{i,j}^{(0,2)} a_i a_j + \sum_{i,j,k} \cdots, \qquad (1)$$

with $\alpha$ numerical prefactors and $a_i^\dagger$ an operator creating an electron in spin-orbital $i$. In ***Quanty*** one can create a creation operator on spin-orbital $i = 0$ ($a_0^\dagger$) with the function:

```
OppC0 = NewOperator("Cr",NF,0)
```

Annihilation operators are created by replacing the string "Cr" by "An":

```
OppA0 = NewOperator("An",NF,0)
```

One is allowed to multiply operators, either with (complex) numbers or by other operators. Furthermore, one can add operators. In this way one is able to create all possible operators needed.

```
OppNO = OppCO * OppAO             -- Number operator
```

Several standard operators for atomic physics, ligand field theory and different model calculations are pre-defined. It is possible to read the results of several density functional theory codes and obtain a tight-binding Hamiltonian from these codes. This tight binding Hamiltonian can be used as a starting point for many-body calculations and spectroscopy [1, 41].

Wavefunctions are defined by the command:

```
psi = NewWavefunction(NF, NB, {{"100000",sqrt(1/2)}, {"000010",sqrt(1/2)}})
```

where the string defines a Slater determinant and the total function is a sum over different determinants. Expectation values can be obtained by multiplying a wavefunction with an operator with a wavefunction.

```
NO = psi * OppNO * psi
```

For a given operator ($H$) one can calculate the $N_{psi}$ lowest eigenstates with the function "Eigensystem()". ***Quanty*** does not need to keep the number of electrons fixed. In order to restrict the solutions to a fixed number of electrons one can specify restrictions. This also allows one to do restrictive active space calculations when needed.

```
StartRestrictions = {NF, NB, {"111111",2,2}}
Npsi = 3
psiList = Eigensystem(H, StartRestrictions, Npsi)
```

Spectra are implemented by calculating the Green's function. We calculate the complex energy dependent quantity:

$$G(\omega) = \left\langle \psi_i \middle| T^\dagger \frac{1}{\omega - H + \imath\Gamma/2} T \middle| \psi_i \right\rangle, \tag{2}$$

with $T$ and $H$ operators given in second quantization and $\psi_i$ a many particle wavefunction.

```
G = CreateSpectra(H,T,psi)
```

For photoemission, the transition operator $T$ would be an annihilation operator, for absorption it is the product of a creation and annihilation operator and for inverse photoemission $T$ becomes a creation operator. ***Quanty*** has many transition operators for standard types of spectroscopy predefined, but you are always able to create your own transition operator.

Resonant spectra are implemented by calculating a third order Green's function or susceptibility ($\chi_3$):

$$G^3(\omega_1,\omega_2) = \left\langle \psi_i \middle| T_1^\dagger \frac{1}{\omega_1 - H_1 - \imath\Gamma/2} T_2^\dagger \frac{1}{\omega_2 - H_2 + \imath\Gamma/2} T_2 \frac{1}{\omega_1 - H_1 + \imath\Gamma/2} T_1 \middle| \psi_i \right\rangle. \tag{3}$$

For $2p$ core level resonant inelastic x-ray scattering measuring magnons or $d-d$ excitations $T_1$ would excite a $2p$ core electron into the $3d$ valence orbitals and $T_2$ would de-excite a $3d$ electron into the $2p$ core hole. For core-core excitations $T_2$ would de-excite for example a $3s$ core electron into the $2p$ core hole. ***Quanty*** can calculate resonant spectra with the function:

```
G3 = CreateResonantSpectra(H1, H2, T1, T2, psi)
```

The Green's functions calculated do not only describe the spectra one can measure in the frequency domain, but also spectra one can measure in the time domain. These Green's functions are on a fundamental level related to how an excitation propagates through the sample. The Green's function $G(t) = \langle\psi_i|T^\dagger(t)T(0)|\psi_i\rangle$ describes the probability where an excitation made at $t=0$ can be found at $t=t$. The relation between $G(\omega)$ and $G(t)$ is given by a Fourier transform: $G(\omega) = \imath\int_0^\infty e^{\imath\omega t}G(t)$. Pump probe spectroscopy where one investigates the propagation of one excitation by a second excitation is expressed as the Fourier transform of $G^3$.

In conclusion, we here present several calculations of different forms of x-ray spectroscopy (XAS, cPES, RIXS, nIXS) including excitons, resonances and continuum excitations obtained with the script language ***Quanty***. A basic introduction to the script language is presented.

## References

[1] Haverkort M W, Zwierzycki M and Andersen O K 2012 *Phys. Rev. B* **85** 165113
[2] Lu Y, Höppner M, Gunnarsson O and Haverkort M W 2014 *Phys. Rev. B* **90** 085102
[3] Haverkort M W, Sangiovanni G, Hansmann P, Toschi A, Lu Y and Macke S 2014 *Europhys. Lett.* **108** 57004
[4] Hehre W J, Stewart R F and Pople J A 1969 *J. Chem. Phys.* **51** 2657
[5] Andersen O K 1975 *Phys. Rev. B* **12** 3060
[6] Blaha P, Schwarz K, Sorantin P and Trickey S B 1990 *Comput. Phys. Commun.* **59** 399
[7] Rehr J J, de Leon J M, Zabinsky S I and Albers R C 1991 *J. Am. Chem. Soc.* **113** 5135
[8] Ebert H, Ködderitzsch D and Minár J 2011 *Rep. Prog. Phys.* **74** 096501
[9] Zaanen J, Alouani M and Jepsen O 1989 *Phys. Rev. B* **40** 837
[10] de Groot F M F, Faber J, Michiels J, Czyzyk M T, Abbate M and Fuggle J C 1993 *Phys. Rev. B* **48** 2074
[11] Ankudinov A L, Ravel B, Rehr J J and Conradson S D 1998 *Phys. Rev. B* **58** 7565
[12] Soininen J A, Hämäläinen K, Caliebe W A, Kao C C and Shirley E L 2001 *J. Phys.: Condens. Matter* **13** 8039
[13] Roulet B, Gavoret J and Nozières P 1969 *Phys. Rev.* **178** 1072
[14] Nozières P, Gavoret J and Roulet B 1969 *Phys. Rev.* **178** 1084
[15] Doniach S and Sunjic M 1970 *J. Phys. C: Solid State Phys.* **3** 285
[16] Thole B T, van der Laan G, Fuggle J C, Sawatzky G A, Karnatak R C and Esteva J M 1985 *Phys. Rev. B* **32** 5107
[17] de Groot F M F, Fuggle J C, Thole B T and Sawatzky G A 1990 *Phys. Rev. B* **42** 5459
[18] Ikeno H, Tanaka I, Koyama Y, Mizoguchi T and Ogasawara K 2005 *Phys. Rev. B* **72** 075123
[19] Glawion S, Heidler J, Haverkort M W, Duda L C, Schmitt T, Strocov V N, Monney C, Zhou K J, Ruff A, Sing M and Claessen R 2011 *Phys. Rev. Lett.* **107** 107402
[20] Roemelt M, Maganas D, DeBeer S and Neese F 2013 *J. Chem. Phys.* **138** 204101
[21] Pinjari R V, Delcey M G, Guo M, Odelius M and Lundberg M 2014 *J. Chem. Phys.* **141** 124116
[22] Uehara Y, Lindle D W, Callcott T A, Terminello L T, Himpsel F J, Ederer D L, Underwood J H, Gullikson E M and Perera R C C 1997 *Appl. Phys. A* **65** 179
[23] Csiszar S, Haverkort M W, Hu Z, Tanaka A, Hsieh H, Lin H J, Chen C T, Hibma T and Tjeng L H 2005 *Phys. Rev. Lett.* **95** 187205
[24] Alders D, Tjeng L H, Voogt F C, Hibma T, Sawatzky G A, Chen C T, Vogel J, Sacchi M and Iacobucci S 1998 *Phys. Rev. B* **57** 11623
[25] Altieri S, Tjeng L H, Tanaka A and Sawatzky G A 2000 *Phys. Rev. B* **61** 13403
[26] Glawion S, Haverkort M W, Berner G, Hoinkis M, Gavrila G, Kraus R, Knupfer M, Sing M and Claessen R 2012 *J. Phys.: Condens. Matter* **24** 255602
[27] Huotari S, Pylkkänen T, Vankó G, Verbeni R, Glatzel P and Monaco G 2008 *Phys. Rev. B* **78** 041102
[28] Verbeni R, Pylkkänen T, Huotari S, Simonelli L, Vankó G, Martel K, Henriquet C and Monaco G 2009 *J. Synchrotron Rad.* **16** 469
[29] Fano U 1961 *Phys. Rev.* **124** 1866
[30] Laskowski R and Blaha P 2010 *Phys. Rev. B* **82** 205104
[31] Vinson J, Rehr J J, Kas J J and Shirley E L 2011 *Phys. Rev. B* **83** 115106
[32] Gunnarsson O and Schönhammer K 1983 *Phys. Rev. B* **28** 4315
[33] Zaanen J, Sawatzky G A, Fink J, Speier W and Fuggle J C 1985 *Phys. Rev. B* **32** 4905
[34] van der Laan G, Zaanen J, Sawatzky G A, Karnatak R and Esteva J M 1986 *Phys. Rev. B* **33** 4253
[35] Fuggle J C, Gunnarsson O, Sawatzky G A and Schönhammer K 1988 *Phys. Rev. B* **37** 1103
[36] Cornaglia P S and Georges A 2007 *Phys. Rev. B* **75** 115112
[37] Gordon R A, Haverkort M W, Gupta S S and Sawatzky G A 2009 *J. Phys.: Conf. Ser.* **190** 012047
[38] Sen Gupta S, Bradley J A, Haverkort M W, Seidler G T, Tanaka A and Sawatzky G A 2011 *Phys. Rev. B* **84** 075134
[39] Haverkort M W, Tanaka A, Tjeng L H and Sawatzky G A 2007 *Phys. Rev. Lett.* **99** 257401
[40] van Veenendaal M A and Haverkort M W 2008 *Phys. Rev. B* **77** 224107
[41] Gunnarsson O, Andersen O K, Jepsen O and Zaanen J 1989 *Phys. Rev. B* **39** 1708

# New Graphical User Interface for EXAFS analysis with the GNXAS suite of programs

**Keisuke Hatada[1,2], Fabio Iesari[2], Leonardo Properzi[2] , M. Minicucci[2], and Andrea Di Cicco[2]**

[1]Groupe Théorie, Département Matériaux-Nanosciences, Institut de Physique de Rennes, UMR UR1-CNRS 6251, Université de Rennes1, France

[2]Physics Division, School of Science and Technology, University of Camerino, Camerino, 62032, Italy

E-mail: keisuke.hatada.gm@gmail.com

**Abstract.** GNXAS is a suite of programs based on multiple scattering calculations which performs a structural refinement of EXAFS spectra. It can be used for any system although it has been mainly developed to determine the local structure of disordered substances. We developed a user-friendly graphical user interface (GUI) to facilitate use of the codes by using wxPython. The developed GUI and the codes are multiplatform running on Windows, Macintosh and Linux systems, and are free shareware (http://gnxas.unicam.it). In this work we illustrate features and potentials of this newly developed version of GNXAS ($w$-GNXAS).

## 1. Introduction

The GNXAS suite of programs [1, 2, 3] is widely used for EXAFS refinements of the local structure of disordered systems by means of multiple scattering calculations through the n-body expansion of the x-ray absorption cross section. [2] Due to the n-body expansion, disorder effects are correctly included in the multiple-scattering signals of any order. Different modeling of the pair, triplet and higher-order distributions, based on Gaussian and non-Gaussian shapes, are possible. This approach has been applied to several interesting cases ranging from free molecules, crystalline and nanocrystalline solids, amorphous/glassy systems, liquids and solutions. Extension to Reverse Monte Carlo (RMC) modeling was also carried out and applied successfully to molecular [4] and liquid systems [5, 6].

GNXAS was originally developed for VMS, Unix, and Linux. Due to the use of obsolete Fortran expressions, the code was only able to be compiled by g77 compiler. We have recently revised the codes to suppress those old statements, thus now the code can be compiled by modern gfortran compilers. [7] The code has been compiled in static mode, so that it is fully portable, namely the users do not need to install additional libraries and dependencies. Furthermore, GNXAS was conceived to run with command lines. When it was developed, the text based input and output files were standard, so that the users were used to the operations. Nowadays, most of computer users are familiar only with the window based operations with a mouse, therefore this old style operation creates difficulty for novice users to start using the codes. For example another popular EXAFS analysis code FEFF [8] has very useful graphical interfaces [9, 10], which contributed to increase their users. Not only that but it helps to reduces human errors

Journal of Physics: Conference Series **712** (2016) 012002 doi:10.1088/1742-6596/712/1/012002

```
../pha/Ge6pha.tl
../pha/Ge6pha.s0
cGe
3,6                     ! mode + Xn limit
7.                      ! RpdT
1                       ! calculates derivatives (1)
 3,1,3
    2.44479    4.00    1  2     1   2
    3.99232   12.00    1  2     1   6
    4.68142   12.00    1  2     1  19
 1,1,1
    2.44479   2.44479 109.47122    6.00  2 1 1.0  1  2  2    1   2
3
                                           2 2.0  2  1  2    2   1
6
                         12.000
0,1,0
0                        ! L=0    K-edge
.000001,6,.000001,150 ! toll_lmax,mmax and Eps (Number of
iterations)
3                        ! IOUT output flag
0.                       ! weight k**wh of the plot for090 and for033
0.5,12.,.05              ! k_min, k_max, and dk of the interpolated
signal
0.2                      ! corresponding energy of the theoretical zero
```

**Figure 1.** A typical example of text based input file for GNXAS.

of input parameters. Thus this kind of software development is becoming very important for progress of theoretical codes. In this communication we present the newly developed graphical interface for GNXAS, which facilitates the use of the package for various operating systems.

## 2. Discussion

The original GNXAS is manipulated in terms of traditional command line. For example, to run `gnxas` which calculates $n$-body distribution functions, $g_n$, for successive processes of EXAFS analysis, one should type a command in a console window as

```
# ./gnxas < c-Ge_xas.in
```

where `./gnxas` is a command for its execution and `c-Ge_xas.in` is a name of input file. The symbol "<" is for directing the input file to the executable file, which is used only for Linux and Macintosh versions of GNXAS, while Windows version does not use this symbol. For plotting output data, there are scripts provided for TopDrawer [11] and gnuplot [12]. The executable of TopDrawer is provided at the site of GNXAS.

Fig. 1 shows the input file `c-Ge_xas.in` from the data-analysis examples [13] as a typical example of text based input file. All the parameters are explained in so called "red book" [1]. We see that the text type input file is not intuitive, and one should spend some time to be familiar with it. Moreover, one should learn command line manipulations. Thus, these text based operations discourage novice users.

A human friendly interface called $w$-GNXAS which is based on a Graphical User Interface (GUI) tool through the Python library "wxPython" [14] controles these text based operations of GNXAS programs. Since the Python code is platform independent, it can be executed on any modern operating systems, such as Linux, Windows and Macintosh. With this newly developed GUI, the usage of GNXAS is intended to be more intuitive and facilitated. In this release we use Python 2.7, which is well tested and very stable. Users can control it directly typing commands on the appropriate panels and windows, on any operating systems. (See the image of Fig. 2) The input parameters and the data are read via the $w$-GNXAS interface, and they are directly transferred to GNXAS calculations. The plotting scripts are also prepared automatically. The results may be plotted by TopDrawer (or gnuplot, or other graphical software).

The left panel of Fig. 3 shows a typical image for the new $w$-GNXAS interface where the same input data in Fig. 1 are used. The $w$-GNXAS interface is constructed with several tabs in its main frame, such as `crymol`, `gnxas`, etc., which correspond to the programs of GNXAS suite.

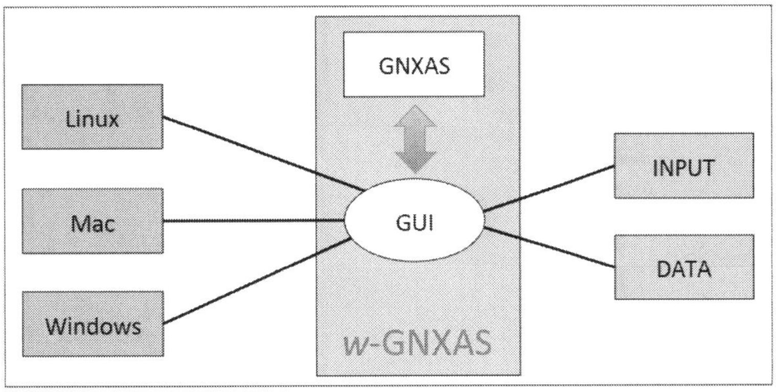


**Figure 2.** Flow diagram for the *w*-GNXAS program. GNXAS programs are controlled by the GUI of *w*-GNXAS. It is a cross platform toolkit, namely it can be used on Linux, Mac and Windows systems. The *w*-GNXAS types input commands and reads accessing to text based input and output files behind of its graphical windows. These files are compatible among different platforms as well. The output can be plotted by the TopDrawer or other graphical software, and script files for plotting are generated by GNXAS.

Hence users click a tab and choose a project to work on. The details of projects are described in the "red book".

After inserting the requested inputs inside the dialog panels (or prepare an input file and load it into the panel), users can run the code by clicking the "run" button at the bottom of each window. (Fig. 4) Below that, there is a space to show the messages appearing in the console window during execution (log file). Users can check all the log messages (warning/errors) there. In the right side of Fig. 4 we show a typical output for EXAFS refinement of crystalline germanium at 300 K, using the data-analysis examples provided with the software [13]. The plot was generated by TopDrawer with a script prepared by GNXAS.

## 3. Summary

We realized a new user friendly graphical interface (*w*-GNXAS) for the GNXAS suite of programs, exploiting all of the potential of this advanced multiple-scattering data-analysis package. The new interface allows users to overcome the problem of inserting large amount of data with the old fashion text-style for input and output command files. All the codes of package including graphical interface are cross platform and statically compiled. Thus presently EXAFS structural refinements using GNXAS can be performed using this newly developed *w*-GNXAS adapted to Windows, Mac and Linux systems. After finishing tests, we will upload the program at the beginning of 2016 at the GNXAS website [15].

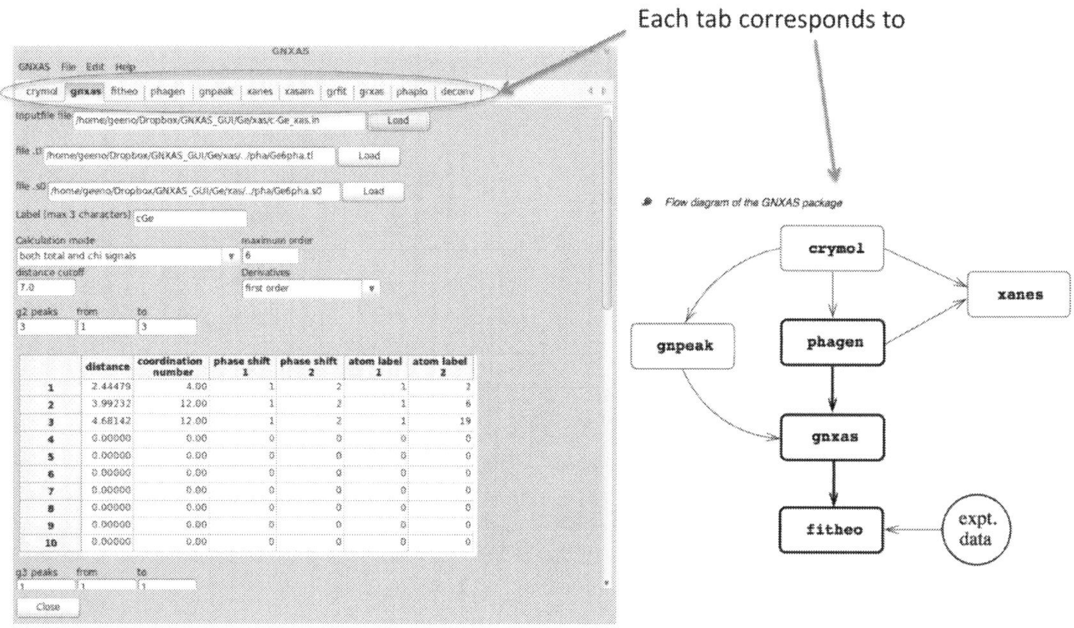


**Figure 3.** *w*-GNXAS has several tabs in the main frame (left panel) which correspond to the different programs in the GNXAS suite (shown in the right side of the figure). Users can select a specific calculation or project by clicking on the corresponding tabs.

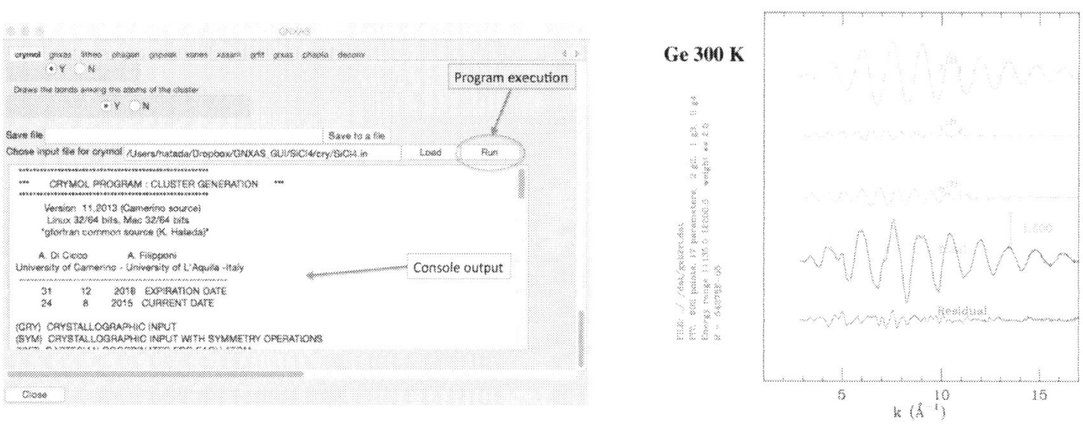


**Figure 4.** Left panel: typical graphical window of *w*-GNXAS when selecting the program `crymol` to run. The console output can be checked and information about the selected model structure and possible warnings and error messages can be verified visually. Right panel: typical graphical output for the EXAFS refinement of crystalline Ge obtained by GNXAS (see data-analysis examples [13])

## 4. Acknowledgments

Parts of this work have been funded by European FP7 MSNano network under Grant Agreement No. PIRSES-GA-2012-317554, by FP7 MS-BEEM under Grant Agreement No. FP7-PEOPLE-2013-IEF-625388 and by COST Action MP1306 EUSpec.

## References

[1] *Extended suite of programs for advanced x-ray absorption data-analysis: methodology and practice*, edited by A. Di Cicco (TASK publishing, Gdansk, Poland, ADDRESS, 2009).
[2] A. Filipponi, A. Di Cicco, and C. R. Natoli, Phys. Rev. B **52**, 15122 (1995).
[3] A. Filipponi and A. Di Cicco, Phys. Rev. B **52**, 15135 (1995).
[4] A. Di Cicco and A. Trapananti, J. Phys.: Condens. Matter **17**, S135 (2005).
[5] A. Di Cicco, A. Trapananti, S. Faggioni, and A. Filipponi, PRL **91**, 135505 (2003).
[6] A. Di Cicco *et al.*, Phys. Rev. B **89**, 060102 (2014).
[7] GFORTRAN, http://www.gnuplot.info/.
[8] J. J. Rehr *et al.*, Phys. Chem. Chem. Phys. **12**, 5503 (2010).
[9] B. Ravel and M. Newvillei, JSR **12**, 537 (2005).
[10] S. Webb, SixPACK, http://www.sams-xrays.com/#!sixpack/rovht.
[11] J. Clement, TopDrawer, http://www.rcnp.osaka-u.ac.jp/okamura/ftp/pub/index.php?dir=topdrawer/.
[12] T. Williams, C. Kelley, and many others, Gnuplot 5.0: an interactive plotting program, http://gnuplot.sourceforge.net/.
[13] GNXAS data-analysis examples, http://gnxas.unicam.it/XASLABwww/pag_gnxas/gnxas_examp.html.
[14] R. Dunn and H. Pasanen, wxPython, http://www.wxpython.org/.
[15] GNXAS downloading site, http://gnxas.unicam.it/XASLABwww/pag_gnxas/gnxas_download.html.

# Disappearance of correlations in the atom motion upon hydrogen intercalation into $ReO_3$ lattice

**Janis Timoshenko, Alexei Kuzmin and Juris Purans**

Institute of Solid State Physics, University of Latvia, Kengaraga street 8, Riga, LV-1063, Latvia

E-mail: janis.timoshenko@gmail.com

**Abstract.** The influence of hydrogen intercalation on the local structure of rhenium trioxide is studied in-situ by the Re $L_3$-edge EXAFS spectroscopy and analysed using a novel approach, based on the use of evolutionary algorithm and wavelet transform. The proposed method allows us to perform accurate EXAFS analysis within the multiple-scattering approach taking into account contributions from outer coordination shells and to access the information on correlations in atomic thermal motion.

## 1. Introduction

The analysis of extended X-ray absorption fine structure (EXAFS) data from the first coordination shell around the absorbing atom is nowadays a common tool to study the local structure of materials [1]. At the same time, EXAFS spectra, acquired at modern brilliant X-ray sources, often contain much more information, especially for crystalline materials, where a contribution of distant coordination shells (up to 10 Å and more) is frequently present in the total EXAFS spectrum. The precise analysis of EXAFS spectra beyond the first coordination shell using conventional methods is, however, often impossible since the contributions of outer shells overlap, and the total number of parameters, required to describe the local structure, growths exponentially with the increase of the number of coordination shells, included in the analysis [2]. Moreover, conventional EXAFS analysis even for the first coordination shell can be imprecise for significantly distorted or disordered systems, where the distributions of interatomic distances are often far from the Gaussian one [3].

This problem may be treated by applying simulation-based techniques as molecular dynamics (MD) [3, 4] and reverse Monte Carlo (RMC) method [5]. For our purposes the latter is especially interesting, since the RMC method does not require the knowledge and calculation of interatomic forces and relies solely on the available experimental EXAFS data. In RMC approach a 3D structure model of the material is constructed using an iterative random process, aimed to minimize the difference between experimental EXAFS and configuration-averaged EXAFS spectra, calculated for the given structure model. Once a good agreement between simulated and experimental spectra is obtained, one may expect that also the constructed model provides a good approximation of the material structure. The existing RMC implementations for EXAFS analysis, however, usually neglect the influence of multiple-scattering (MS) effects or treat them in an approximate way due to concomitant significant computational costs. As a consequence, the information amount, obtainable from EXAFS spectrum, again, is rather limited.

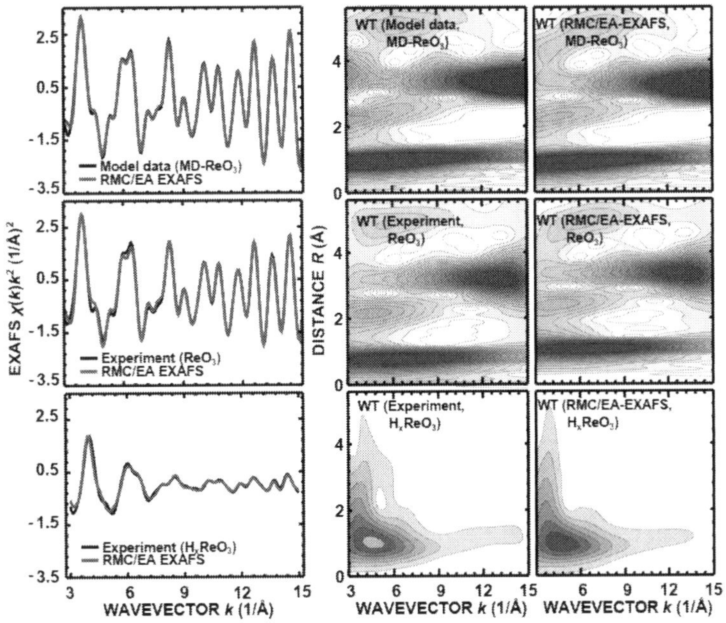


**Figure 1.** Comparison of the EXAFS spectra (MD model for $ReO_3$, experimental data for pure $ReO_3$ and $H_xReO_3$) with the results of RMC/EA-EXAFS calculations. Left column - Re $L_3$ edge EXAFS, two right columns - corresponding Morlet wavelet transforms (WT).

In the presented study we treat this problem and complement the conventional RMC scheme with (i) a powerful evolutionary algorithm (EA) for efficient structure model optimization [6] and (ii) the use of wavelet transform (WT) [7], which allows us to represent EXAFS spectra in real and frequency spaces simultaneously and, hence, to obtain more information from the same experimental data [8]. This advanced approach allows us to probe more distant coordination shells and to reveal the anisotropy and correlations of atom thermal motion and static displacements.

## 2. Experimental and simulation details

The experimental Re $L_3$-edge EXAFS spectra of pure polycrystalline rhenium trioxide ($ReO_3$) and of rhenium trioxide after in-situ intercalation with hydrogen ions ($H_xReO_3$) were taken from [9], and for both spectra RMC/EA simulations were performed. Additionally, to validate the proposed approach, RMC/EA simulations were performed also for the model EXAFS spectrum, obtained in MD simulations of cubic $ReO_3$ in [10] (MD-$ReO_3$).

Detailed description of our RMC/EA scheme is given in [6]. RMC/EA calculations were performed using a 4×4×4 supercell (256 atoms). Positions of atoms were optimized by minimizing the mismatch between WTs for experimental EXAFS (or MD-EXAFS) and configuration-averaged EXAFS, obtained in ab-initio FEFF8 calculations within multiple-scattering approach [11]. As a result, a reasonable agreement was obtained for the final atomic configuration for all three cases (Fig. 1). These final structure models are analyzed below.

## 3. Results and discussion

Radial distribution function (RDF) around absorbing rhenium atom can be obtained straightforwardly from the results of RMC/EA calculations. To demonstrate the accuracy of the

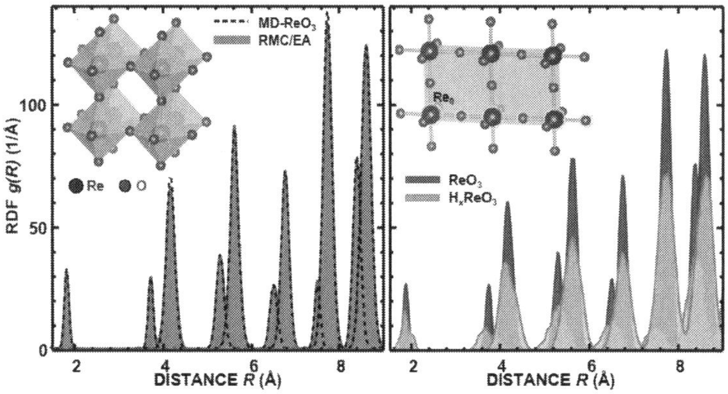


**Figure 2.** Original and reconstructed by RMC/EA approach RDFs around Re atom for MD model (left panel) and RDFs obtained by RMC/EA calculations for experimental EXAFS for crystalline $ReO_3$ and $H_xReO_3$ (right panel).

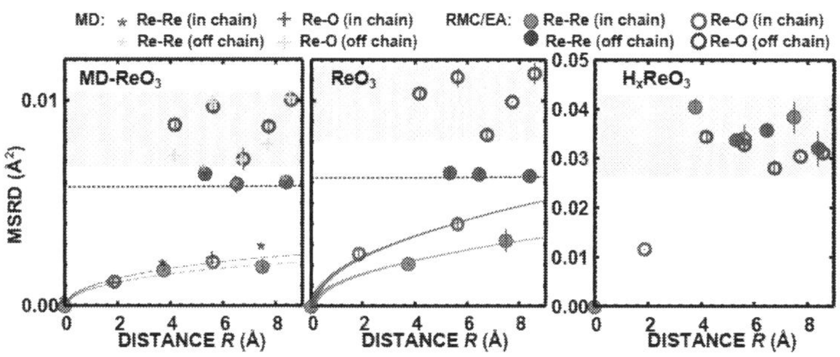


**Figure 3.** MSRD factors obtained by RMC/EA analysis for Re–Re and Re–O atom pairs as a function of interatomic distance for MD model (left panel; MSRDs obtained directly from MD simulations are also shown), pure $ReO_3$ (middle panel) and $H_xReO_3$ (right panel). In-chain and off-chain atoms are shown separately; dashed lines are guides for the eye; vertical lines are estimated error bars of RMC/EA analysis results.

proposed approach, the RDFs, obtained in RMC/EA simulation for the model EXAFS data and directly from MD simulation, are compared in Fig. 2. The agreement between initial structure model and the results of RMC/EA reconstruction is remarkable both for close and more distant coordination shells up to 8-9 Å. Note that the longest interatomic distance, probed by RMC/EA, is limited by the size of the used supercell.

The RDFs, obtained in RMC/EA simulations of experimental data for $ReO_3$ and $H_xReO_3$ compounds (right panel in Fig. 2), indicate that the intercalation by hydrogen ions leads to the significant broadening of all RDF peaks due to the presence of strong disorder in the system.

The widths of RDF peaks are characterized by mean square relative displacement (MSRD) factors $\sigma^2$ [1]. MSRD for a pair of atoms A and B can be expressed as $\sigma^2_{AB} = \langle u^2_A \rangle + \langle u^2_B \rangle - 2\langle u_A u_B \rangle$, where $\langle u^2_A \rangle$ and $\langle u^2_A \rangle$ are mean square displacement factors, related to the amplitudes of atom oscillations in the direction of interatomic bond. The last term $\langle u_A u_B \rangle$ accounts for the correlation in atom motion. The MSRD factors for several Re–Re and Re–O pairs are shown in Fig. 3 as a function of interatomic distance. Note that the results of RMC/EA calculations for the model $ReO_3$ system are in a good agreement with the true values of MSRD factors, obtained directly from the atom coordinates, which suggests a good reliability of the method.

The results, obtained from the experimental EXAFS of pure $ReO_3$, are shown in the middle panel of Fig. 3. Here the values of the MSRD factors for Re–Re and Re–O atom pairs suggest the difference in the correlations of atom motion along linear –Re–O–Re– chains (see the inset in the right panel of Fig. 2) with respect to the correlations in the off-chain directions.

The MSRD factors for atom pairs along the chains are small, reflecting strong interatomic interactions, and grow slowly with the increase of interatomic distance. Such behaviour can be explained by the presence of strong correlation in the atom motion, which naturally decreases at large distances. Nevertheless, our result suggests that correlations in the atom motion for the in-chain atoms are important even for atoms that are located as far as 9 Å from the absorbing rhenium atom. Note also that such behaviour of the MSRD factors for crystalline $ReO_3$ is in a qualitative agreement with the results, obtained from the MD simulations in [10], which are able to reproduce the dynamics of $ReO_3$ lattice.

The values of the MSRD factors for the off-chain directions are significantly larger, suggesting weaker interactions, and show almost no distance dependence, thus indicating that a correlation in the atom motion is negligible. Note that fluctuations in the values of MSRD factors for Re–O pairs for the off-chain atoms are a consequence of the anisotropy of oxygen thermal motion in the $ReO_3$ lattice, hence the MSRD factors depend on the angle between corresponding interatomic bond and lattice vectors.

The intercalation of $ReO_3$ by hydrogen ions affects the correlations in the atom motion dramatically (right panel in Fig. 3). First, the values of the Re–Re and Re–O MSRD factors in $H_xReO_3$ are several times larger than those in pure $ReO_3$. Moreover, our results suggest complete disappearance of interatomic correlations beyond the first coordination shell, since there is almost no difference in the MSRD factors for in-chain and off-chain atom pairs in $H_xReO_3$. Thus one may conclude that the intercalation of hydrogen ions leads not only to tilting and distortion of the $ReO_6$ octahedra, evidenced by diffraction studies [12], but also to significant changes in the atomic bonding, hence – electronic structure of the material, observed previously at the Re $L_{1,3}$-edge XANES as a shift of the absorption edges to smaller energies and attributed to the lowering of the valence state of rhenium ions [9].

## 4. Conclusions

The ability to accurately analyze contributions to EXAFS spectra from atoms located at distances up to 8 Å from absorbing atom is demonstrated for the first time in $ReO_3$ and $H_xReO_3$. For this purpose we employed a complex approach, based on the use of reverse Monte Carlo method, evolutionary algorithm and wavelet transform. This approach allowed us to reveal the presence of very strong and relatively long-range correlations in the motion of Re and O atoms within linear –Re–O–Re– chains in pure $ReO_3$. It was also demonstrated that these correlations are destroyed by the intercalation of hydrogen ions into crystalline $ReO_3$ lattice.

## Acknowledgments

This work was supported by the Latvian Science Council grants no. 187/2012 and 402/2012.

## References

[1] Rehr J J and Albers R C 2000 *Rev. Mod. Phys.* **72** 621
[2] Provost K, Beret E, Muller D, Marcos E S and Michalowicz A 2013 *J. Phys.: Conf. Ser.* **430** 012015
[3] Chill S T, Anderson R M, Yancey D F, Frenkel A I, Crooks R M and Henkelman G 2015 *ACS Nano* **9** 4036
[4] Kuzmin A and Chaboy J 2014 *IUCrJ* **1** 571
[5] McGreevy R L and Pusztai L 1988 *Mol. Simul.* **1** 359
[6] Timoshenko J, Kuzmin A and Purans J 2014 *J. Phys.: Condens. Matter* **26** 055401
[7] Timoshenko J and Kuzmin A 2009 *Comp. Phys. Commun.* **180** 920
[8] Munoz M, Argoul P and Farges F 2003 *Am. Mineral.* **88** 694–700
[9] Gaidelene J, Kuzmin A, Purans J and Guéry C 2005 *Phys. Status Solidi (c)* **2** 149
[10] Kalinko A, Evarestov R A, Kuzmin A and Purans J 2009 *J. Phys.: Conf. Ser.* **190** 012080
[11] Ankudinov A L, Ravel B, Rehr J J and Conradson S D 1998 *Phys. Rev. B* **58** 7565
[12] Dickens P G and Weller M T 1983 *J. Solid State Chem.* **48** 407

# Finite difference method accelerated with sparse solvers for structural analysis of the metal-organic complexes

**A A Guda[1], S A Guda[2], M A Soldatov[1], K A Lomachenko[1,3], A L Bugaev[1,3], C Lamberti[1,3], W Gawelda[4], C Bressler[4,5], G Smolentsev[1,6], A V Soldatov[1], Y Joly[7,8]**

[1] International Research Center "Smart Materials", Southern Federal University, 344090 Rostov-on-Don, Russia
[2] Faculty of Mathematics, Mechanics and Computer Science Southern Federal University, 344090 Rostov-on-Don, Russia
[3] Department of Chemistry, NIS and CrisDi Centers, Turin University and INSTM Reference Center, 10125 Turin, Italy
[4] European XFEL, Albert-Einstein-Ring 19, 22761 Hamburg, Germany
[5] The Hamburg Centre for Ultrafast Imaging (CUI), Universität Hamburg, Luruper Chaussee 149, 22761 Hamburg, Germany
[6] Paul Scherer Institute, Villigen, 5232, Switzerland
[7] Inst NEEL, Université Grenoble Alpes, 38042 Grenoble, France
[8] Institut NEEL, CNRS, 38042 Grenoble, France

E-mail: guda@sfedu.ru

**Abstract**. Finite difference method (FDM) implemented in the FDMNES software [*Phys. Rev. B*, **2001**, *63*, 125120] was revised. Thorough analysis shows, that the calculated diagonal in the FDM matrix consists of about 96% zero elements. Thus a sparse solver would be more suitable for the problem instead of traditional Gaussian elimination for the diagonal neighbourhood. We have tried several iterative sparse solvers and the direct one MUMPS solver with METIS ordering turned out to be the best. Compared to the Gaussian solver present method is up to 40 times faster and allows XANES simulations for complex systems already on personal computers. We show applicability of the software for metal-organic $[Fe(bpy)_3]^{2+}$ complex both for low spin and high spin states populated after laser excitation.

## 1. Introduction

The main working time of the full-potential finite difference calculation of XANES spectra in FDMNES [1] is occupied by the solution of a linear system. Nonzero elements of the matrix concentrate near the diagonal and the solver performs traditional gaussian elimination for the diagonal neighborhood. Thorough analysis shows, however, that the calculated diagonal consists of about 96% zeros, thus a sparse solver would be more suitable for the problem. We tried several iterative sparse solvers in Matlab and the direct one (Matlab uses the solver UMFPACK of Timothy A. Davis [2]) as well as massively parallel MUMPS solver [3,4]. The comparison of different solvers for FDMNES matrix was published in [5]. Direct solvers can use a variety of ordering packages. MUMPS solver with METIS [6] ordering turned out to be the best and we chose it for latest FDMNES release build.

Highly accelerated version of the software makes possible now using accurate finite difference modelling for structural analysis on personal computers. We show applicability of the method for metal-organic molecule iron tris bipyridine. Visible light absorption can trigger transition of $Fe(bipy)_3$ complex from singlet low-spin (LS) state $^1A_1$ to quintet high-spin (HS) state $^5T_2$. The transition occurs via singlet metal-to-ligand-charge-transfer process [7,8]. Theoretical simulations within contemporary quantum chemistry methods provide accurate information about the geometries and electron structure for both LS and HS states [9-12]. However results of DFT simulations strongly depend on the choice of the exchange-correlation functional, which in turn require experimental verification. In contrast to EXAFS data XANES can provide quantitative information about both distances and bond angles [13]. In this work we have extracted the information about bond distances along with rotation angle of the bipyridine ligands using pump-probe x-ray absorption spectroscopy above Fe K-edge [14].

## 2. Compilation details for the sparse release

Each new release of the accelerated FDMNES version is available on the official web-site (http://neel.cnrs.fr/spip.php?rubrique1007). It contains serial executables. The history of builds which includes also parallel executables is available on GitHub repository (https://github.com/gudasergey/fdmnes). We have implemented FDMNES build system as a web service. The figure 1 depicts its structure. The online build system is available to FDMNES developers, and the request for access can be sent to corresponding author. When source code is uploaded the building process starts automatically. The Linux FDMNES version is built locally on the server, while windows – on the remote windows machine. In the first version of build system the FDMNES for Windows was built by MINGW compiler and was linked with BLAS and LAPACK. But then we discovered, that the Windows FDMNES version is 5 times slower than Linux version built by Intel compiler with MKL on the same computer. Thus we decided to setup building of FDMNES for Windows by Intel compiler on Windows machine. Some difficulties appeared because MUMPS is distributed with make-files for Linux. To overcome this we have installed on windows machine MSYS and mingw-make, configured the MUMPS make-file to use Intel compiler and Microsoft Library Manager.

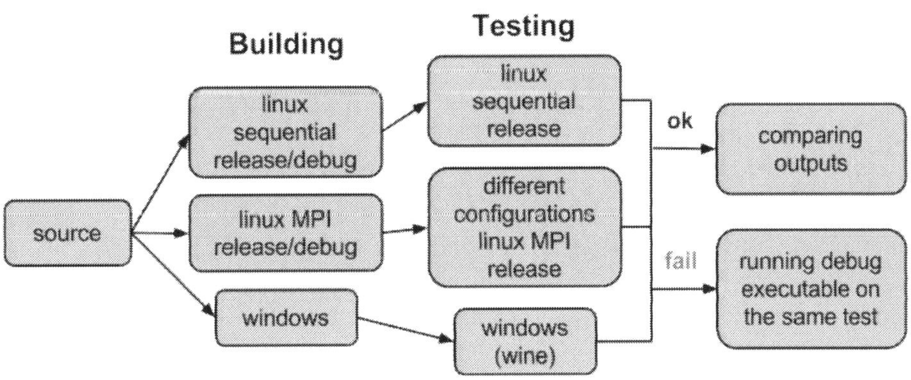


**Figure 1. Structure of FDMNES build system.**

The authors of the ordering package METIS increased portability of the sources by using cmake in the last 5th version. It can generate projects for Visual Studio. FDMNES code successfully links with the libraries if option -NODEFAULTLIB:LIBCMT was passed to the linker. When compiling by Intel compiler FDMNES also requires options -heap-arrays and -zero (-Qzero for windows). To generate stack traces in case of errors we have also added the options -g1 and -traceback. If building procedure was finished successfully the testing process starts. We run sequential and several parallel FDMNES release configurations with different MPI overall and MUMPS host number parameters. When some of tests fail, procedure is passed to the same parallel configuration of debug FDMNES. If all the tests finished successfully, their outputs are compared with the output of sequential FDMNES.

## 3. Iron tris(bipyridine) deformations upon spin transition

A set of structural parameters for the bipyridines of the $Fe(bipy)_3^{2+}$ that strongly influence on the shape of x-ray absorption spectra was determined. These include the Fe-N distance changes and rotation of bipyridine ligands. The procedure of the multidimensional interpolation approximation, as described in [15] and implemented in FitIt software was used to find the accurate values of these parameters. Simulations of the XANES spectra were performed by means of finite-difference method implemented in the FDMnes program.

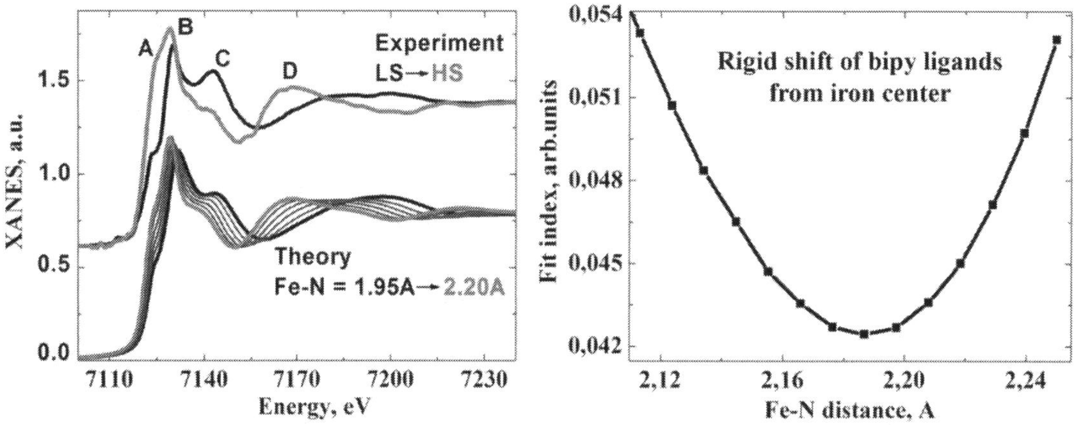


**Figure 2.** Left panel shows the experimental Fe K-edge XANES spectra for LS and HS states of $[Fe(bipy)_3]^{2+}$ and simulation for the series of Fe-N distances obtained by the rigid shift of bipyridine ligands. Right panel shows the variation of the fit index for the experimental HS spectrum along with the change of Fe-N distance.

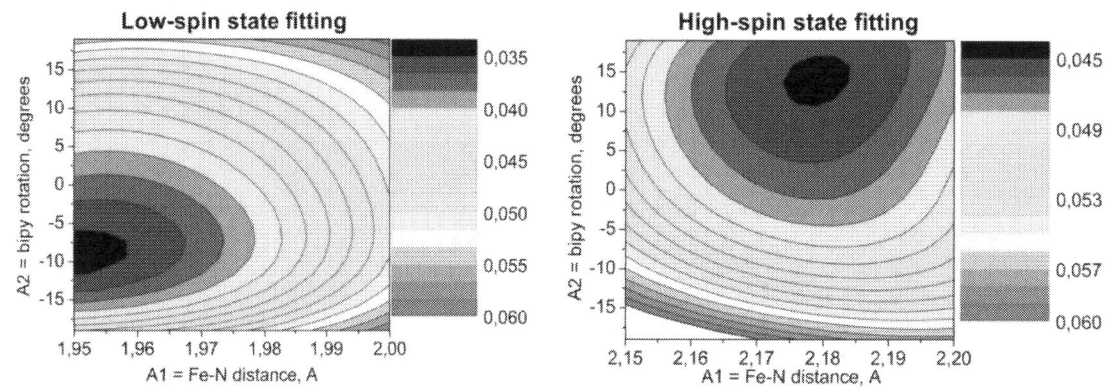


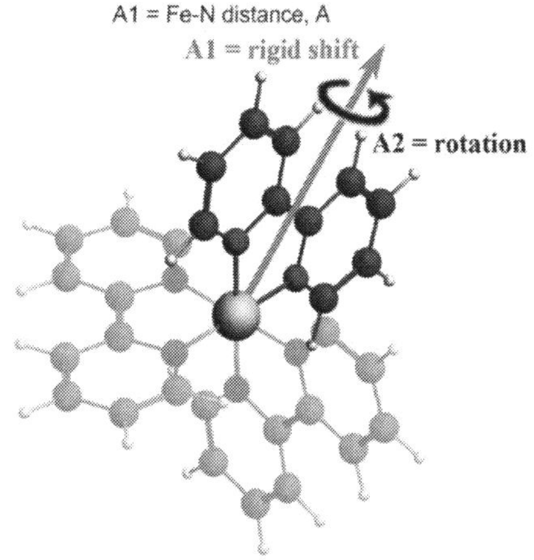


**Figure 3.** Calculated fit indices for a series of structural parameters for the low spin and high spin states of the $[Fe(bipy)_3]^{2+}$. Deformations were performed starting from ideal octahedral coordination of the Fe by nitrogens from bipyridine ligands. Then whole ligands were allowed to move and rotate (three simultaneously by same values) along the axes shown in the left panel.

In the LS state six nitrogens around iron form a slightly distorted octahedron with the Fe-N distances 1.967 Å [16]. When the molecule undergoes a transition to the high-spin state the electrons are transferred from highest occupied the nonbonding molecular orbitals to the antibonding ones with significant admixture of ligand contribution. Such electronic transitions change the energetically favorable positions of the ligands around Fe. DFT simulations show that Fe-N bonds are elongated by 0.2 Å [11]. We have calculated Fe K-edge XANES for a set of Fe-N distances, obtained by a rigid shift of bipyridine ligands from iron center. Figure 2 shows the evolution of the fit index for an experimental high spin spectrum as a function of Fe-N distance. The minimum value was obtained for 2.185 Å in a good agreement with DFT simulations.

Figure 3c shows the axes for the distortion of the iron tris-bipyridine structure in order to perform the XANES fitting for two structural parameters. During the fitting procedure we have started from the ideal octahedral configuration of the molecule. Two parameters were chose for deformations in order to represent the structure of the LS and HS states. They are: A1 – rigid shift of the bipyridine (figure 3c); A2 – rotation of the bipyridine around same axis (positive value of A2 corresponds to the counterclockwise rotation). Fitting was performed both for the low spin and high spin states of the molecule using data before and after laser excitation. The obtained equilibrium distances for Fe-N bond are in a good agreement with previous EXAFS analysis [14]. Additionally we have obtained the information about the rotation of the bipyridine ligands.

Present experimental XAS data for the high spin state does not have enough time resolution to see intermediate states of the complex. Further generation of the XFELs would provide femtosecond time resolution for x-ray absorption spectroscopy. The structural analysis of these states can be performed by means of the presented above methods.

**Acknowledgements**
C.L., A.B., and K.L. would like to thank the Russian Ministry of Education and Science for the support (megagrant of the Russian Federation Government no. 14.Y26.31.0001). A.G., M.S., and A.S. would like to thank the Russian Ministry of Education and Science for the support (project RFMEFI 58714X0002, agreement no. 14.587.21.0002). W.G. and C.B. would like to acknowledge the Deutsche Forschungsgemeinschaft (via SFB925, TPA4), the European XFEL GmbH, and the Hamburg Center for Ultrafast Imaging (University of Hamburg) for funding. G.S. would like to thank Swiss National Science Foundation (grant nr 200021-135226) and European Commission's Seventh Framework Programme (FP7/2007-2013) under grant agreement no. 290605 (COFUND: PSIFELLOW)

**References**
[1] Joly Y 2001 *Phys. Rev. B* **63** 125120
[2] Davis T A 2004 *ACM Trans. Math. Softw.* **30** 196-99
[3] Amestoy P R, Duff I S, L'Excellent J and Koster J 2001 *SIAM J. Matrix Anal. Appl.* **23** 15-41
[4] Amestoy P R, Guermouche A and L'Excellent J-Y 2006 *Parallel Comput.* **32** 136-56
[5] Guda S A et.al. 2015 *J. Chem. Theory Comput.* **11** 4512–21
[6] Karypis G and Kumar V 1998 *SIAM J. Sci. Comput.* **20** 359-92
[7] Gütlich P 2004 Spin Crossover in Transition Metal Compounds *Springer Berlin* pp.1
[8] Hauser A 2004 Spin Crossover in Transition Metal Compounds *Springer Berlin* pp.786
[9] Alexander B D, Dines T J and Longhurst R W 2008 *Chem. Phys.* **352** 19-27
[10] Hauser A, Enachescu C, Daku M L et al. 2006 *Coord. Chem. Rev.* **250** 1642-52
[11] Daku M L, Vargas A, Hauser A et. al. 2005 *ChemPhysChem* **6** 1393-410
[12] Pierloot K and Vancoillie S 2006 *J. Chem. Phys.* **125** 124303
[13] Smolentsev G, Guilera G, Tromp M et.al. 2009 *J. Chem. Phys.* **130** 174508
[14] Gawelda W, Pham V T, van der Veen R M, et.al. 2009 *J. Chem. Phys.* **130** 124520
[15] Smolentsev G, Soldatov A V and Feiters M C 2007 *Phys. Rev. B* **75** 144106
[16] Dick S 1998 *Z. Kristallogr. - New Cryst. Struct.* **213** 356

Journal of Physics: Conference Series **712** (2016) 012005 doi:10.1088/1742-6596/712/1/012005

# Tracking the signature of low symmetry environments in the XAS *K* pre-edge

**M Hunault[1,*], V Vercamer[1,2], M W Haverkort[3], M-A Arrio[1], C Brouder[1], G Calas[1] and A Juhin[1]**

[1] Sorbonne Universités, UPMC Univ Paris 06, CNRS, IRD, MNHN, Institut de Minéralogie, de Physique des Matériaux et de Cosmochimie, 4 place Jussieu, 75252 Paris cedex 05, France
[2]AGC France SAS, 100 rue Léon Gambetta, 59168 Boussois, France
[3]Max-Planck-Institut für Chemische Physik fester Stoffe Nöthnitzer Str. 40, 01187 Dresden, Germany

*E-mail: myrtille.hunault@impmc.upmc.fr

**Abstract**. For Co-containing crystalline compounds measured by XAS, the influence of the local symmetry on the *K* pre-edge features is studied using advanced Ligand Field Multiplet (LFM) calculations, which accurately take into account the *p-d* hybridization. The LFM theory is used to calculate the *K* pre-edge spectra of $Co^{2+}$ in various environments and the absolute intensities of electric-quadrupole and -dipole transitions involved in the pre-edge. This enables to reproduce the spectra for cubic ($O_h$, $T_d$) and lower symmetries ($C_{4v}$, $D_{3h}$), and allows quantitative derivation of the local *p-d* mixing.

## 1. Introduction

Site geometry of transition metal (TM) ions substituted in crystals is a crucial characteristic, which drives their properties and related applications, from solid-state physics to coordination chemistry. In particular, exotic geometries like 5-fold coordinated sites are often predicted or suspected in coordination complexes, surfaces and interfaces, glasses and amorphous compounds. However their presence is difficult to assess because the spectral signature is largely unknown. X-ray absorption spectroscopy and in particular, *K* pre-edge features are powerful tools to determine the local structure of TM. In the case of cobalt(II), a widely used coloring agent, its coordination number (CN) plays a major role in the coloration of crystals and glasses [1,2]. Understanding how the local structure controls the properties in crystals is a motivation for applying core level spectroscopy to investigate the local geometry of $Co^{2+}$.

$Co^{2+}$ can be found in 4-, 5- or 6-fold coordination with a large diversity of site symmetries, which can be distinguished according to the intensity, and shape of the pre-edge [3]. The variation of the intensity of the pre-edge depends on the *3d-4p* hybridization that, when it is allowed by the absence of an inversion center, enables an electric dipole contribution in addition to the electric quadrupole transitions. This effect is well-known for the cases of $T_d$ site geometry and has been reproduced for iron ions by calculations in the Ligand Field Multiplet model [4]. In order to describe exotic sites, it is necessary to consider lower symmetries though it is challenging. In this paper we compare the influence of $Co^{2+}$ symmetry ($O_h$, $T_d$, $C_{4v}$, $D_{3h}$) and calculate the *p-d* hybridization, which is expected to vary for 6-, 5- and 4-fold coordinated $Co^{2+}$. A good agreement is obtained between calculations and experimental Co K-edge XANES spectra.

## 2. Experimental

Our work focuses on a benchmark set of four crystalline samples bearing $Co^{2+}$ in various symmetries for which we expected the *p-d* hybridization to vary. We chose two different spinels: $CoCr_2O_4$ (4-fold coordinated $Co^{2+}$ in $T_d$) and $Co_2GeO_4$ (6-fold coordinated $Co^{2+}$ in $D_{3d}$ that we approximate to $O_h$) and two compounds containing 5-fold coordinated Co sites: $BaCoP_2O_7$ ($C_{4v}$) and $LiCoBO_3$ ($D_{3h}$). $CoCr_2O_4$ was synthesized by coprecipitation [5], $BaCoP_2O_7$ by precipitation-calcination [3] and $LiCoBO_3$ by a traditional ceramic route [6]. All samples were characterized by x-ray diffraction. Samples were prepared as powders sandwiched between two Kapton® tapes.

Co *K*-edge (7709 eV) XAS spectra were collected at beamlines 13-BM-D and 13-ID-E at the Advanced Photon Source (Chicago IL, USA – 7 GeV and 102 mA injected positron). The incident beam was monochromated using Si(111) or Si(311) double crystal monochromators. XANES spectra were recorded in transmission mode at room temperature. The energy calibration was achieved using a Co foil as reference. All XANES spectra presented here result from the average of four successively obtained spectra. The spectra were normalized to an absorption step of 1 using the "ATHENA" software [7]. The pre-edge features presented in this paper are given after background removal, using an arctangent function.

## 3. Theory and calculations

In order to extract quantitative information from *p-d* hybridization of the pre-edge, we performed LFM calculations using the method developed by Thole et al. [8] in the framework established by Cowan [9] and Butler [10]. Here, this approach is implemented in the code Quanty [11], in which operators are defined using second quantization and spectra are calculated using Green functions [12]. It takes into account all the *3d-3d* and *1s-3d* electronic Coulomb interactions, as well as the spin-orbit coupling on every open shell of the absorbing atom. For all sites, we applied a reduction factor of $\beta = 0.6$ to the Slater integrals calculated for an isolated ion and spin orbit coupling at 80% of its free ion value [13]. $Co^{2+}$ is considered as an isolated ion and its geometrical environment is treated through a parameterized crystal-field potential defined by the point group symmetry of the absorbing site. These parameters were fitted in agreement with the optical spectra of the compounds. For non-centrosymmetric sites, hybridization is allowed between *3d* and *4p* orbitals of $Co^{2+}$ and is described by an hybridization Hamiltonian built from the $U^{(1-)}$ and $U^{(3-)}$ unit tensor operators as derived in [4], which mixes the electronic configurations $1s^2 3d^7 4p^0$ and $1s^2 3d^6 4p^1$ that describes the initial state. The *p-d* hybridization Hamiltonian depends on the difference $\Delta$ between the average energies of these two electronic configurations. For $Co^{2+}$, the *ab initio* value is $\Delta = 13.5$ eV. Depending on the symmetry, the hybridization Hamiltonian is described by one ($T_d$ and $D_{3h}$) or two ($C_{4v}$) parameters [14]. The hybridization parameters are adjusted in order to fit the experimental intensity of the *K* pre-edge. The absolute intensities are calculated at T = 300 K and the population of the initial state levels is given by the Boltzmann law. The spectra are convoluted by a Lorentzian (full width half maximum, FWHM: $2\gamma = 1.33$ eV) and a Gaussian (FWHM: 0.4 eV), which respectively accounts for the lifetime of the *1s* core hole of Co and the instrumental resolution. Finally, the transitions are normalized by the edge jump at the Co *K*-edge, calculated for a Co atom as $3.0\ 10^{-4}$ Å$^2$ [15]. Hence, the calculated spectra can be directly compared to the normalized experimental ones.

## 4. Results and discussion

The normalized background-corrected resulting spectra are presented in Fig.1. They reveal the influence of the site symmetry on the intensity and shape of the pre-edge transitions. As expected, the lowest intensity is observed for the 6-fold coordinated site in $Co_2GeO_4$. We used the centrosymmetric

$O_h$ approximation to model for the symmetry of the slightly distorted $D_{3d}$ site. In this case, hybridization is symmetry-forbidden and only electric quadrupole transitions contribute to the pre-edge feature. For the other point groups, we observe an increase in the intensity from $C_{4v}$ to $D_{3h}$ to $T_d$, which is in line with the effect of the *p-d* hybridization allowed for non-centrosymmetric cases.

On Fig.1, we compare the experimental results with the calculated pre-edge for each case. The hybridization parameters are given in Table I. These calculations enable to reproduce the measured intensities and shapes of pre-edge features as shown in Fig.1. This validates the use of ideal point group symmetry to model the symmetry of the slightly distorted sites. For $O_h$ and $T_d$ symmetries, the range of the fitted parameters agrees with that found for Fe K-edge XANES spectra [4]. Furthermore we observe that the fraction of the dipole contribution to the total pre-edge area for Co and Fe in $T_d$ are similar (68%). For the cases of $C_{4v}$ and $D_{3h}$, the fraction of electric dipole contribution to the pre-edge is intermediate between $O_h$ and $T_d$ cases. The symmetry of the ground state (Table I) agrees with the energy level diagrams [3].

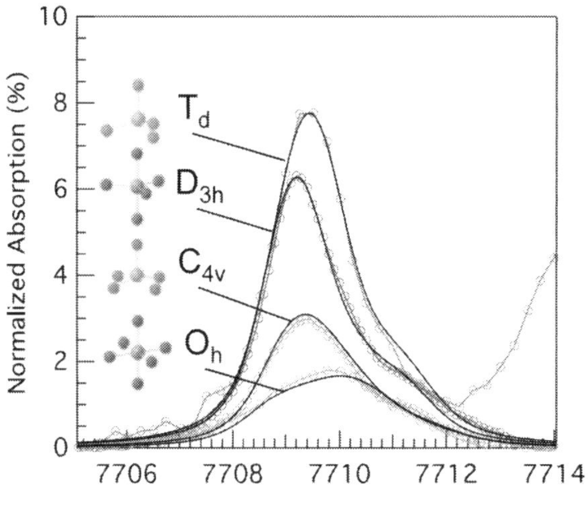


**Fig.1** Normalized Co *K* pre-edge spectra for each of the four symmetries of this study: experimental (colored line and markers) and respective calculation (plain black line). Green: $CoCr_2O_4$ ($T_d$), dark blue: $LiCoBO_3$ ($D_{3h}$), light blue: $BaCoP_2O_7$ ($C_{4v}$) and pink: $Co_2GeO_4$ ($O_h$).

Table I: Parameters used for the calculation of the Co *K* pre-edge features.

| Sample | Symmetry (CN) | Crystal field parameters (eV) | Hybridization parameters (eV) | Ground state symmetry | Dipole contribution (%) |
|---|---|---|---|---|---|
| $Co_2GeO_4$ | $O_h$(6) | $D_q = 0.11$ | - | $^4T_{1g}$ | 0 |
| $BaCoP_2O_7$ | $C_{4v}$(5) | $D_q = 0.148$<br>$D_s = -0.087$<br>$D_t = 0.055$ | $V_{p\text{-}d}(a_1) = 0.25$<br>$V_{p\text{-}d}(e) = 0.84$ | $^4E$ | 26 |
| $LiCoBO_3$ | $D_{3h}$(5) | $D_\mu = 0.0$<br>$D_\nu = -0.096$ | $V_{p\text{-}d} = 5.5$ | $^4A_2'$ | 61 |
| $CoCr_2O_4$ | $T_d$(4) | $D_q = -0.055$ | $V_{p\text{-}d} = 8.5$ | $^4A_2$ | 68 |

## 5. Conclusion

In this work, we calculated and quantified the influence of hybridization on the *K* pre-edge spectral features. The good agreement obtained between calculations and experiment demonstrates the validity of our model and of the Ligand Field Multiplet approach. It offers the possibility to interpret quantitatively the pre-edge features of $Co^{2+}$ or other TM ions in terms of spectral characteristics and of degree of admixture between *4p* and *3d* levels in the ground state. This information can be further used to interpret the optical absorption spectra of Co-bearing materials, in which the energy and intensity of the *3d*-transitions strongly rely on the *p-d* hybridization. This aspect will be further discussed in an upcoming paper. The rationalization of the relation between the local structure and the spectroscopic properties offers a powerful tool to help determining the speciation of TM in amorphous materials as for instance in glasses, which show a strong influence of glass composition on its color.

**Acknowledgements**
This work is part of the « VITRAUX » project supported by a Sorbonne University Convergence grant (SU-14-R-ScPC-15-2). This research used resources of the Advanced Photon Source, a U.S. Department of Energy (DOE) Office of Science User Facility operated for the DOE Office of Science by Argonne National Laboratory under Contract No. DE-AC02-06CH11357. We acknowledge the support of GeoSoilEnviroCARS (Sector 13), which is supported by the National Science Foundation - Earth Sciences (EAR-1128799), and the Department of Energy, Geosciences (DE-FG02-94ER14466)

**References**
[1] Hunault M, Calas G, Galoisy L, Lelong G and Newville M 2014 Local Ordering Around Tetrahedral $Co^{2+}$ in Silicate Glasses *J. Am. Ceram. Soc.* **97** 60–2
[2] Hunault M, Bauchau F, Loisel C, Hérold M, Galoisy L, Newville M and Calas G 2015 Spectroscopic investigation of the coloration and fabrication conditions of medieval blue glasse *J. Am. Ceram. Soc.* doi:10.1111/jace.13783
[3] Hunault M, Robert J-L, Newville M, Galoisy L and Calas G 2014 Spectroscopic properties of five-coordinated $Co^{2+}$ in phosphates *Spectrochim. Acta. A. Mol. Biomol. Spectrosc.* **117** 406–12
[4] Arrio M-A, Rossano S, Brouder C, Galoisy L and Calas G 2000 Calculation of multipole transitions at the Fe K pre-edge through *p-d* hybridization in the Ligand Field Multiplet model *Europhys. Lett. EPL* **51** 454–60
[5] Wang Y, Zhou Z, Jia M, Zhu X, Zhang W and Jiang D 2005 Spinel-Type Cobalt Chromites as Novel and Highly ortho-selective Catalysts for Phenol Alkylation *Catal. Lett.* **104** 67–71
[6] Tao L, Neilson JR, Melot BC, McQueen TM, Masquelier C and Rousse G 2013 Magnetic structures of $LiMBO_3$ (M = Mn, Fe, Co) lithiated transition metal borates *Inorg. Chem.* **52** 11966–74
[7] Ravel B and Newville M 2005 ATHENA, ARTEMIS, HEPHAESTUS: data analysis for X-ray absorption spectroscopy using IFEFFIT *J. Synchrotron Radiat.* **12** 537–41
[8] Thole B T, Cowan R D, Sawatzky G A, Fink J and Fuggle J C 1985 New probe for the ground-state electronic structure of narrow-band and impurity systems *Phys. Rev. B* **31** 6856–8
[9] Cowan R D 1981 *The Theory of Atomic Structure and Spectra* (University of California Press)
[10] Butler P H 1981 *Point group symmetry applications: methods and tables* (Plenum Press)
[11] Haverkort M W Quanty - a quantum many body script language, http://www.quanty.org/
[12] Haverkort M W, Zwierzycki M and Andersen O K 2012 Multiplet ligand-field theory using Wannier orbitals *Phys. Rev. B* **85** 165113-22
[13] Slater integrals for Co free ion: $F^2_{3d3d}$ = 11.60483 eV, $F^4_{3d3d}$ = 7.20942 eV, $\zeta_{3d}$ = 0.066 eV;
[14] The hybridization Hamiltonian mixes *d* and *p* functions of same symmetry: $t_2$ for $T_d$, e' for $D_{3h}$, and $a_1$ and e for $C_{4v}$. For each point group symmetry, one parameter corresponds to one symmetry.
[15] Gullikson E M 2010 X-ray interactions with matter, CXRO, http://www.cxro.lbl.gov

# First-principles study of phonon effects in x-ray absorption near-edge structure spectroscopy

**R Nemausat,[1,2] Ch Brouder,[1] Ch Gervais,[2] D Cabaret[1]**

[1] Sorbonne Universités, UPMC Univ Paris 06, IMPMC, UMR CNRS 7590, 4 place Jussieu, F-75005, Paris, France

[2] Sorbonne Universités, UPMC Univ Paris 06, LCMCP, UMR CNRS 7574, Collège de France, 11 place Marcelin Berthelot, F-75005 Paris, France

E-mail: ruidy.nemausat@impmc.upmc.fr

**Abstract.** Usually first-principles x-ray absorption near-edge structure (XANES) calculations are performed in the Born-Oppenheimer approximation assuming a static lattice, whereas the nuclear motion undoubtedly impacts XANES spectra notably at the $K$ pre-edge of light elements in oxides. Here, an efficient method based on density-functional theory to account for quantum thermal fluctuations of nuclei is developed and is successfully applied to the $K$ edge of corundum for temperatures up to 930 K. The zero-point motion influence is estimated. Comparison is made with previous theoretical approaches also developed to account for vibrations in XANES.

## 1. Introduction

Standard x-ray absorption near-edge structure (XANES) calculations rest upon the Born-Oppenheimer (BO) approximation, considering nuclei fixed at their equilibrium positions defined by the set of nuclear coordinates $\overline{\mathbf{R}}_{\mathrm{eq}}$. Yet, significant temperature-induced effects are noticed, particularly at the Al $K$-edge of corundum ($\alpha$-$Al_2O_3$): the pre-edge peak can only be explained by $1s \rightarrow 3s$ forbidden transitions induced by lattice vibrations [1, 2, 3]. In this paper, the influence of quantum thermal fluctuations of nuclei on XANES is modeled within the BO approximation and quasiharmonic approximation (QHA) using a density-functional theory (DFT) framework. The dynamical matrix is determined to generate temperature-dependent sets of nonequilibrium nuclear configurations $\overline{\mathbf{R}}$ for which individual XANES spectra are calculated. Their average at a given temperature may be compared to the corresponding experimental spectrum. The method is applied at the Al $K$ edge in corundum, which has been measured for temperatures up to 930 K [2]. The calculated spectra are compared with two previous theoretical approaches developed to account for thermal motion, either by small absorbing-atom displacements (AAD) [2] or by moving the $1s$ initial wave function in the crude Born-Oppenheimer (CBO) approximation [3].

## 2. Method

In the single-electron and electric-dipole approximations, the $K$-edge XANES cross section with x-rays of energy $\hbar\omega$, polarized along the $\hat{\mathbf{e}}$ vector is given by [4]

$$\sigma(\hbar\omega) = 4\pi^2\alpha\hbar\omega \sum_n |\langle\psi_n| \hat{\mathbf{e}} \cdot \mathbf{r} |\psi_{1s}\rangle|^2 \, \delta(\varepsilon_n - \varepsilon_{1s} - \hbar\omega), \tag{1}$$

where $\alpha$ is the fine structure constant, and $|\psi_n\rangle$ and $|\psi_{1s}\rangle$ the final and initial monoelectronic states of energy $\varepsilon_n$ and $\varepsilon_{1s}$, respectively. In Eq. (1) the nuclei are fixed in a given configuration, $\overline{\mathbf{R}}$, which is a parameter of the calculation: $\sigma(\hbar\omega) = \sigma(\hbar\omega; \overline{\mathbf{R}})$.

An expression accounting for the total system of nuclei and electrons is

$$\sigma_{\text{tot}}(\hbar\omega) = 4\pi^2\alpha\hbar\omega \sum_{n,j} \left|\langle\Psi_n^j|\,\hat{\mathbf{e}}\cdot\mathbf{r}\,|\Psi_{1s}^0\rangle\right|^2 \delta(E_n^j - E_{1s}^0 - \hbar\omega), \tag{2}$$

where the total $\Psi_{1s}^0$ initial and $\Psi_n^j$ final states have energies $E_{1s}^0$ and $E_n^j$, respectively. Using the BO approximation and the first-order expansion of the Green's function of the total system as derived in Ref. [5], $\sigma_{\text{tot}}$ can be written as

$$\sigma_{\text{tot}}(\hbar\omega) = \int d\overline{\mathbf{R}}\; \rho(\overline{\mathbf{R}})\; \sigma(\hbar\omega; \overline{\mathbf{R}}), \tag{3}$$

with $\rho(\overline{\mathbf{R}})$ the QHA weighting displacement distribution. In Eq. (3), $\sigma_{\text{tot}}$ is the average of individual configuration cross sections $\sigma(\hbar\omega; \overline{\mathbf{R}})$ using a probability distribution that takes into account temperature and quantum nuclear motion (see Ref. [5] for details).

In this paper, temperature-dependent XANES cross sections [Eq.(3)] are calculated within DFT in the PBE approximation [6] using the Quantum ESPRESSO suite of codes [7]. The calculations are performed using ultrasoft pseudopotentials, with plane-wave energy cutoff of 60 Ry and 720 Ry for the wave-functions and charge density, respectively. Two Al pseudopotentials (one for the absorbing atom and one for the others) are constructed starting from a $3s3p3d$ valence configuration with a maximum cutoff radius of 2.10 Bohr. The O pseudopotential is the one of Ref. [5]. The experimental lattice parameters of corundum at 300, 650 and 930 K are used, while the 0 K calculation is carried out using the 100 K experimental parameters [8]. The dynamical matrices are obtained on a $3 \times 3 \times 3$ **k**-point grid. Fifty $2 \times 2 \times 2$ trigonal-supercell configurations of 80 atoms including one $1s$ core-hole are generated, upon which self-consistent charge densities are calculated at the $\Gamma$ point, and XANES calculations are performed using the XSpectra package [9, 10] on a $4 \times 4 \times 4$ **k** point grid with a broadening parameter of 0.4 eV. The 50 calculated XANES spectra are averaged by taking into account the core-level shift and the averaged spectra are shifted by 1565 eV so that the experimental and calculated main-edge peak energies match at 300 K.

## 3. Results and discussion

### *3.1. Comparison with experiments*

The experimental and calculated temperature-dependent Al $K$-edge XANES spectra of corundum are presented in Fig. 1(a). As the temperature increases, the P pre-edge emerges and the A peak decreases, both peaks move down to lower energy. The $B_1$ and $B_2$ features are smoothed by temperature. The QHA calculation reproduces the experimental trends, as highlighted by the similarity of the difference of each spectrum with the room-temperature one shown in Fig. 1(b). From 300 to 930 K, the experimental and calculated A peak intensities decrease by 20 and 31%, respectively and the energy shift agreement between experiment and calculation is satisfactory. However, in the calculations, the P peak is less resolved and slightly underestimated. This discrepancy is expected to come from the fact that only the first order of the Green's function is considered in the calculation. An equivalent experiment-calculation matching is observed at the Mg $K$-edge in MgO [5].

It has been demonstrated that the P pre-edge emerges from forbidden $1s \rightarrow 3s$ electronic transitions induced by vibrations [1, 2]. The observation of the pre-edge in the 0 K calculation, where only residual zero-point motion subsists, emphasizes the importance of accounting for quantum vibrations in theoretical XANES.

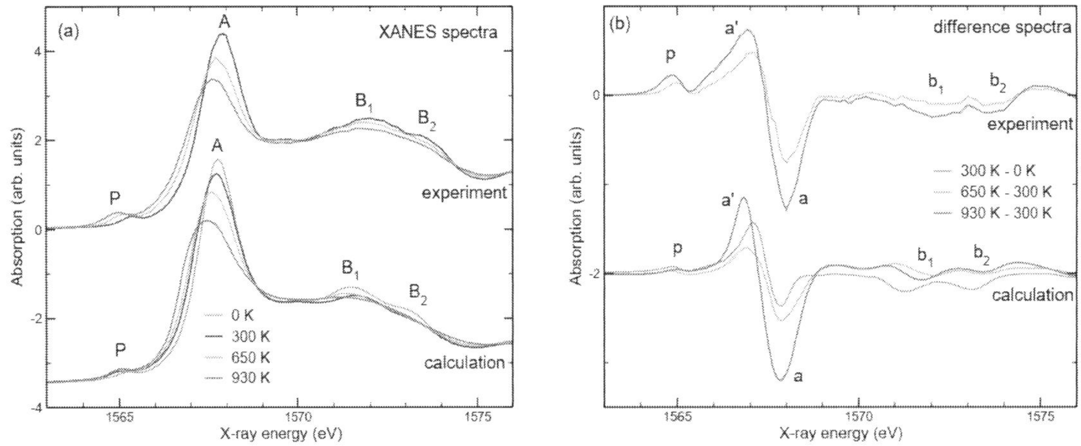


**Figure 1.** (a) Temperature-dependent experimental [2] and calculated Al $K$-edge XANES spectra of corundum. (b) Experimental and calculated difference spectra with the 300 K ones.

### *3.2. Comparison with previous theoretical approaches*

Figure 2 compares the 300 K experimental XANES spectrum and four theoretical spectra, which are obtained using the $\overline{\mathbf{R}}_{\mathrm{eq}}$ equilibrium configuration, the QHA method (Sec. 2) and two previous theoretical works, here called absorbing-atom displacement (AAD) [2] and crude Born-Oppenheimer (CBO) [3]. The calculated spectra are here all performed with the norm-conserving Troullier-Martins pseudopotentials used in Ref. [2]. The AAD method only considers the absorbing-atom displacement in the equilibrium configuration. In this approximation, the XANES spectra calculated for each off-center absorbing-atom displacement $\mathbf{R}_a$ are averaged with a weighting core harmonic displacement distribution function $\varrho(\mathbf{R}_a)$ (see Ref. [2] for details). In the CBO approximation, the electronic wave function of the final state is supposed to not significantly vary for small atomic motions. Therefore, the final states are calculated for the $\overline{\mathbf{R}}_{\mathrm{eq}}$ configuration, and only the initial state is displaced. Then, with $\gamma$ including both the core-hole lifetime and instrumental resolution, the CBO XANES cross section is expressed as [3]

$$\sigma_{\mathrm{CBO}}(\hbar\omega) = 4\pi\alpha \int d\mathbf{R}_a\ \varrho(\mathbf{R}_a) \sum_n \left| \int d\mathbf{r}\ \psi_n^*(\mathbf{r})\ \hat{\mathbf{e}}\cdot\mathbf{r}\ \psi_{1s}(\mathbf{r}-\mathbf{R}_a) \right|^2 \frac{(\varepsilon_n - \varepsilon_{1s})}{(\varepsilon_n - \varepsilon_{1s} - \hbar\omega)^2 + \gamma^2}. \quad (4)$$

All methods produce spectra in better agreement with the experiments than the equilibrium spectrum. Unlike the equilibrium calculation, each method induces a P pre-edge. The QHA and AAD spectra show a similar weak P peak. The displacement of either the absorbing atom or the whole nuclei configuration creates hybridizations between the Al $s$ and $p$ empty states, allowing to probe the absorbing Al $3s$ empty states through the electric-dipole approximation. The CBO approximation produces a strong P peak due to electric-dipole transitions towards absorbing Al $s$ states from the $\psi_{1s}(\mathbf{r}-\mathbf{R}_a)$ displaced $1s$ initial state, which includes $\ell = 1$ components [3]. As observed at the Ti $K$-edge in rutile ($TiO_2$) [3], the CBO method significantly overestimates the pre-edge intensity.

Peak A varies according to the method used. Since the CBO method calculates the final states at $\overline{\mathbf{R}}_{\mathrm{eq}}$, peak A almost matches the equilibrium one. Using AAD, peak A is slightly shifted to lower energy and this shift is even more pronounced using QHA. Such an energy shift results from the influence of the collective quantum dynamics on the band gap [5, 11].

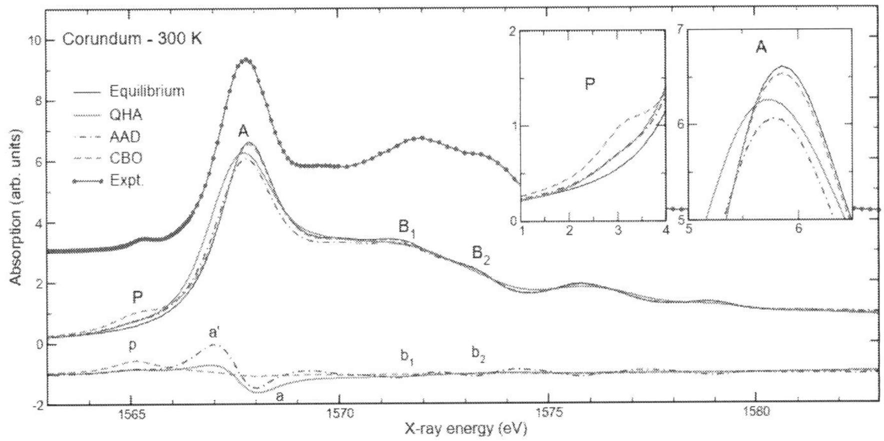


**Figure 2.** Calculated Al $K$-edge XANES spectra of corundum, at the 300 K volume, in the equilibrium case, using QHA method and previous approaches [2, 3]. The differences of theoretical spectra with respect to the equilibrium one are also displayed, as the well as the experimental 300 K XANES spectrum. The insets focus on the pre-edge and main peaks.

## 4. Conclusion

An efficient framework to account for temperature in XANES spectra is proposed. The good agreement between calculated and measured XANES spectra demonstrates that this approach successfully explains the temperature effects experimentally observed in XANES. Unlike standard equilibrium calculations, temperature-dependent theoretical spectra show the emergence of an additional P pre-edge peak, in good agreement with experiments. Our approach also reproduces other experimental trends, which are observed with increasing the temperature, such as the growth of the pre-edge peak, the intensity decrease of the main-edge peak and the shift towards lower energy of both peaks. The comparative analysis with former studies highlights the importance of accounting for the whole nuclear dynamics using a quantum treatment.

## Acknowledgments

This work was supported by French state funds managed by the ANR within the Investissements d'Avenir programme under reference ANR-11-IDEX-0004-02, within the Cluster of Excellence MATISSE led by Sorbonne Universités. Computer facilities were provided by GENCI-IDRIS (Project No. 2015-100172). The authors acknowledge Y. Joly for fruitful discussions.

## References

[1] Cabaret D and Brouder C 2009 *J. Phys.: Conf. Ser.* **190** 012003
[2] Manuel D, Cabaret D, Brouder C, Sainctavit P, Bordage A and Trcera N 2012 *Phys. Rev. B* **85** 224108
[3] Brouder C, Cabaret D, Juhin A and Sainctavit P 2010 *Phys. Rev. B* **81** 115125
[4] Brouder C 1990 *J. Phys.: Condens. Matter* **2** 701–738
[5] Nemausat R, Cabaret D, Gervais C, Brouder C, Trcera N, Bordage A, Errea I and Mauri F 2015 *Phys. Rev. B* **92** 144310
[6] Perdew J P, Burke K and Ernzerhof M 1996 *Phys. Rev. Lett.* **77** 3865
[7] Giannozzi P, *et al.* 2009 *J. Phys.: Condens. Matter* **21** 395502
[8] Fiquet G, Richet P and Montagnac G 1999 *Phys. Chem. Miner.* **27** 103–111; Lucht M, Lerche M, Wille H C, Shvyd'ko Y V, Rüter H D, Gerdau E and Becker P 2003 *J. Appl. Crystallogr.* **36** 1075–1081
[9] Taillefumier M, Cabaret D, Flank A M and Mauri F 2002 *Phys. Rev. B* **66** 195107
[10] Gougoussis C, Calandra M, Seitsonen A P and Mauri F 2009 *Phys. Rev. B* **80** 075102
[11] O'Brien W L, Jia J, Dong Q Y, Callcott T A, Mueller D R and Ederer D L 1992 *Phys. Rev. B* **45** 3882

Journal of Physics: Conference Series **712** (2016) 012007 doi:10.1088/1742-6596/712/1/012007

# Theory of x-ray absorption and linear dichroism at the Ca $L_{23}$-edge of $CaCO_3$

**Peter Krüger**[1], **Calogero R. Natoli**[2]

[1] Graduate School of Advanced Integration Science, Chiba University, Chiba 263-8522 Japan
[2] INFN Laboratori Nazionali di Frascati, Via E Fermi 40, c.p. 13, I-00044 Frascati, Italy

E-mail: pkruger@chiba-u.jp

**Abstract.** X-ray absorption calculations of Ca $L_{23}$-edge spectra of calcium carbonate in its two main crystal phases, calcite and aragonite, are reported. The multichannel multiple scattering theory with a correlated particle-hole wave function and a partially screened core-hole potential is used. Very good agreement with experiment for both $CaCO_3$ phases is obtained, while the independent particle approximation completely fails. For aragonite, appreciable linear dichroism is predicted in agreement with recent observations.

## 1. Introduction

Transition element $L_{23}$-edge spectra are a major tool for probing the local structure and electronic properties around a metal site in crystals, molecules and nanomaterials. Systems with high point symmetry can be successfully modeled using the crystal field multiplet (CFM) model [1]. In cubic symmetry there is only one crystal field parameter which can be meaningfully fitted to experiment. For lower symmetry, however, the number of free parameters increases quickly and the usefulness of the CFM model becomes questionable. Nanoparticles and composite systems often have very low symmetry. Through X-ray absorption spectroscopy at the $L_{23}$-edge small structural and electronic variations at the metal site can be detected [2]. However, without predictive computational methods the structural information cannot be extracted from the data. The development of first principles methods for transition element $L_{23}$-edge spectra is thus highly desirable, but has been a long standing theoretical challenge mainly because of the strong atomic-multiplet-like configuration mixing induced by the core-hole in the final state. In the last decade, good progress has been made by a number of new approaches which may be divided into two categories. The first ones are full multiplet approaches where the basis states are Slater determinants of all electrons in $d$-like orbitals. These methods include small cluster quantum-chemistry schemes [3] such as the complete active space method [4] and the multiplet ligand-field theory using Wannier orbitals [5]. The second class are particle-hole schemes, where configuration mixing is essentially restricted to the core-hole and the excited electron. This effect dominates for the light transition elements, especially for those with a formal ($3d^0$) configuration, where the spectral weight redistribution with respect to the independent particle approximation is largest. Particle-hole schemes include time-dependent density functional theory [6], the Bethe-Salpeter-Equation (BSE) [7] approach and the multichannel multiple scattering (MCMS) theory developed by the present authors [8, 9]. Note that the general MCMS theory [10] is not limited to particle-hole wave functions, and we have recently devised a scheme for full multiplet coupling in arbitrary open shell configurations [11]. MCMS inherits the flexibility and physical intuitiveness of standard multiple scattering, the most popular method for X-ray absorption calculations. Previously, we have shown that for light transition elements, MCMS gives excellent results [8, 9] equivalent to those of the BSE [7] but at a much lower computational cost. So far we have focused on bulk systems with rather high symmetry (cubic or tetragonal). Here we apply the MCMS method to systems of very low point symmetry, namely the calcium carbonate phases calcite and aragonite. We investigate the polarization dependence of the absorption spectra, and compare the results with recent experimental data [12, 13].

## 2. Computational

We use the MCMS method with particle-hole wave function [8, 9]. First, a standard multiple scattering calculation is carried out for the system without absorber which yields the reflectivity of the "environment". In this real space calculation, the cluster size is increased until convergence of the spectra. For calcite (aragonite) we have used a cluster of 177 (176) atoms and 286 (68) empty spheres which were found largely converged. The atomic and empty sphere potentials have been generated using the local density approximation to density functional theory and the linear-muffin-tin-orbital code [14]. The ground state bulk potential is used for all atoms except the absorber. For the absorber atom, a partially screened core-hole potential is employed. It is a linear mixture $V = (1-\alpha)V_s + \alpha V_u$ between the statically screened ($V_s$) and the unscreened ($V_u$) core-hole potential. $V_s$ is obtained from a self-consistent supercell calculation with a core-hole on a single Ca atom, i.e. it corresponds to the fully relaxed core-hole. The use of the one-electron Fermi's golden rule and the potential $V_s$ (i.e. $\alpha = 0$) shall be called "final state rule" (FSR) below. $V_u$ is the sum of the ground state potential and the Hartree potential of the core-hole. $V_u$ is related to a sudden limit, in which the excited electron would feel the bare potential of the core-hole, before relaxation and screening set in [10]. The ratio $\alpha$ is the only empirical parameter of the present theory, and it turns out to be very robust, since we could describe all Ca and Ti compounds with the same value (0.1) [8, 9]. To take account of multiplet effects, i.e. configuration mixing, the $T$-matrix of the absorber atom is calculated using multichannel scattering theory with the variational R-matrix approach. By combining the reflectivity of the environment and the multichannel T-matrix, an exact embedding of the configuration-mixed particle-hole wave function into the first-principles electronic structure of the large cluster is achieved. Further theoretical details can be found in our previous articles [8, 9]. The calculated spectra have been shifted to match the experimental energy scale and are broadened with a Lorentzian and a Gaussian, both of 0.2 eV FWHM.

## 3. Results and discussion

Figure 1 (a,b) shows the local atomic structure around the Ca atom in calcite and aragonite. A central Ca atom and its $CO_3$ ligand units are shown. The point symmetry on the Ca sites is obviously very low. If only the first nearest oxygen atoms are considered, calcite has a weakly distorted octahedral coordination. Rez et al. [12] have modeled the calcite spectra using the CFM model assuming an octahedral crystal field and obtained good results for the unpolarized spectra. Recently, the linear dichroism was measured by the same group and found negligible for calcite [13]. In aragonite, Ca has a 9-fold O coordination, which was approximated by a tetragonal ligand field in the CFM calculations of Refs. [12, 13]. Good agreement with experiment was obtained for the unpolarized spectra, but the linear dichroism, clearly evident in the experimental data, could not be explained with those CFM model calculations [13].

Figure 1 (c) shows the Ca $L_{23}$-edge spectra computed in the independent particle approximation using either the ground state (GS) potential or the statically screened potential $V_s$ corresponding to the FSR, along with the experimental data [12]. With the GS potential, the spectral features are very broad, which reflects the large Ca $3d$ conduction band width. In

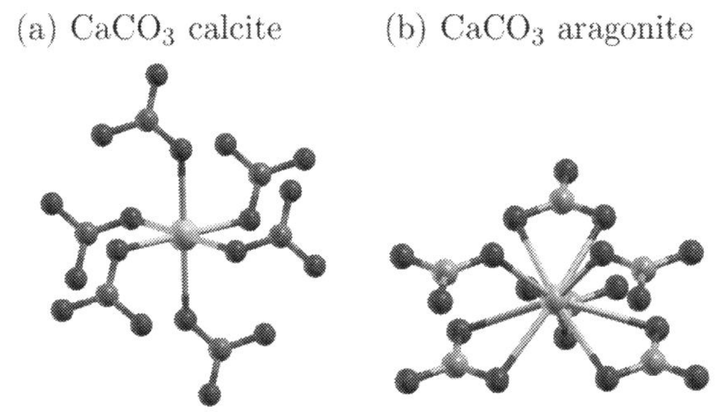


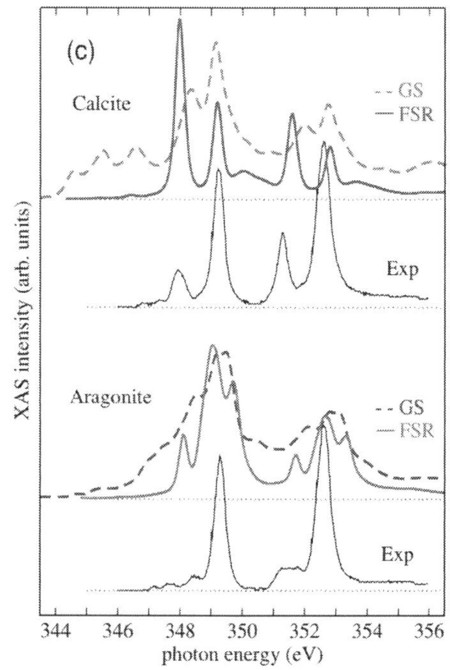


**Figure 1.** Top panels (a,b): Local structure around Ca of $CaCO_3$ phases calcite (a) and aragonite (b). Ca in blue, C in grey, O in red. Right panel (c): Ca $L_{23}$-edge spectra. Comparison between experiment (Exp) [12] and calculations in the independent particle approximation using either the ground state (GS) or the final state rule (FSR) potential.

contrast, the experimental peaks are very sharp, which shows that these spectra are highly excitonic, i.e. the core-hole potential pulls the $3d$ resonance into the band gap, creating a fairly localized final state. This effect can partially be described by adding a fully relaxed core-hole according to the FSR as can be seen in Fig. 1 (c). The peaks in the FSR spectra are narrower than those of the GS spectra and thus the spectral shapes agree better with experiment. While this effect is weak for aragonite, it leads to a strong improvement in the case of calcite, where the FSR peaks are sharp and the relative peak positions agree quite well with experiment. The relative intensities of the four main peaks are, however, totally wrong. This is the case for all spectra obtained in the independent particle approximation, where the L3:L2 branching ratio is necessarily close to statistical (2:1), and thus more than twice larger than the experimental one. In conclusion, the highly excitonic character of the Ca-$L_{23}$-edge spectra can partially be accounted for by using FSR potentials, but all spectra obtained in the independent particle approximation have completely wrong peak intensities.

Figure 2 shows the multichannel calculations with a partially screened core-hole potential, using $\alpha = 0.1$ as in our previous calculations [8, 9]. The L3-L2 splitting (a–b) and in calcite also the T2g-Eg splittings (a1–a2 and b1–b2) are slightly overestimated as compared to experiment. Moreover, the theoretical spectra contain a few weak features not present in the data, namely a peak at 354.5 eV in calcite and shoulders around 350 and 354 eV in aragonite. Apart from these small differences, the agreement with the experimental data is very good in both phases.

Figure 3 shows the calculated spectra for linearly polarized X-rays along Cartesian axes x, y, z. Calcite (space group No. 167) has trigonal symmetry and so x and y polarizations are equivalent. In orthorhombic aragonite (space group No. 62), all three polarizations are different. In calcite the dichroism is negligible in agreement with experiment [13]. This reflects the fact that Ca has an almost perfect octahedral coordination, where all Ca-O bond lengths are identical within 0.1% and the bond angles differ from 90° by less than 3°. In aragonite, however, marked linear dichroism is found. First, the main peak intensities are larger for x- than for y or z polarized light in agreement with the crystal field multiplet analysis in Ref. [13]. Second, peak "d" is clearly maximum for z-polarized light. This finding nicely agrees with experiment [13],

Journal of Physics: Conference Series **712** (2016) 012007 doi:10.1088/1742-6596/712/1/012007

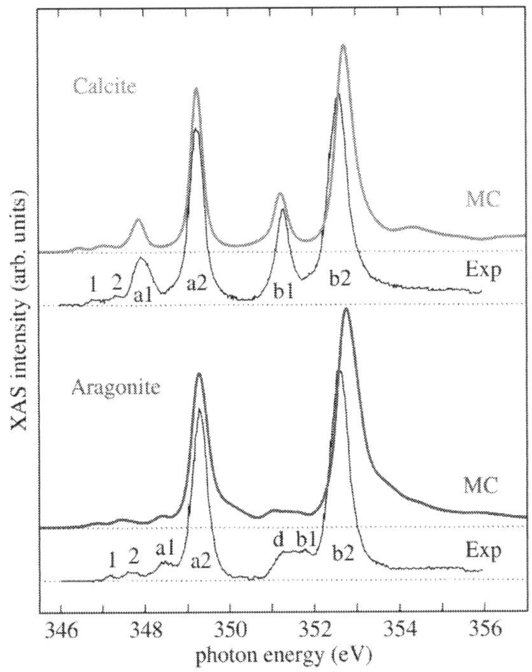


**Figure 2.** Ca $L_{23}$-edge spectra of $CaCO_3$ computed in multichannel theory (MC) with $\alpha = 0.1$ vs experimental data (Exp) [12]. Peaks are labeled as in [13].

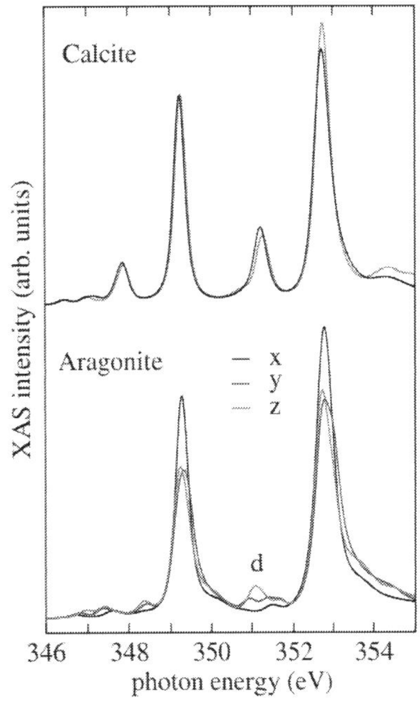


**Figure 3.** Ca $L_{23}$-edge spectra for linearly polarized X-rays along crystal axes x, y and z, computed in multichannel theory with $\alpha = 0.1$.

where an increase of the "d" peak was observed when the polarization was turned from the xy-plane to the z-axis

## 4. Conclusions

In summary, we have theoretically studied Ca $L_{23}$-edge spectra of $CaCO_3$ in its main crystal phases calcite and aragonite. Using the multichannel multiple scattering method with a partially screened core-hole potential, we have obtained good agreement with experiment in all cases while the independent particle approximation completely fails. Appreciable linear dichroism is predicted for aragonite which qualitatively agrees with the available experimental data.

## References

[1] de Groot F M F and Kotani A 2008 *Core Level Spectroscopy of Solids* (New York: Taylor & Francis)
[2] Bittencourt C *et al* 2012 *Beilstein J Nanotech* **3** 789
[3] Roemelt M, Manganas D, DeBeer S and Neese F 2013 *J. Chem. Phys.* **138** 204101
[4] Ikeno H, Mizoguchi T and Tanaka I 2011 *Phys. Rev.* B **83** 155107
[5] Haverkort M W, Zwierzycki M and Andersen O K 2012 *Phys. Rev.* B **85** 165113
[6] Schwitalla J and Ebert H 1998 *Phys. Rev. Lett.* **80** 4586
[7] Laskowski R and Blaha P 2010 *Phys. Rev.* B **82** 205104
[8] Krüger P and Natoli C R 2004 *Phys. Rev.* B **70** 245120
[9] Krüger P 2010 *Phys. Rev.* B **81** 125121
[10] Natoli C R, Benfatto M, Brouder C, Lopez M F R and Foulis D 1990 *Phys. Rev.* B **42** 1944
[11] Natoli C R *et al* 2012 *J. Phys. Condens. Matter* **24** 365501
[12] Rez P and Blackwell A 2011 *J. Phys. Chem.* B **115** 11193
[13] Metzler R A and Rez P 2014 *J. Phys. Chem.* B **118** 6758
[14] Andersen O K and Jepson O 1984, *Phys. Rev. Lett.* **53** 2571

# XAFSmass: a program for calculating the optimal mass of XAFS samples

**K Klementiev[1] and R Chernikov[2]**

[1] MAX IV Laboratory, 22363 Lund, Sweden
[2] DESY Photon Science, 22607 Hamburg, Germany

konstantin.klementiev@maxlab.lu.se

**Abstract**. We present a new implementation of the XAFSmass program that calculates the optimal mass of XAFS samples. It has several improvements as compared to the old Windows based program XAFSmass: 1) it is truly platform independent, as provided by Python language, 2) it has an improved parser of chemical formulas that enables parentheses and nested inclusion-to-matrix weight percentages. The program calculates the absorption edge height given the total optical thickness, operates with differently determined sample amounts (mass, pressure, density or sample area) depending on the aggregate state of the sample and solves the inverse problem of finding the elemental composition given the experimental absorption edge jump and the chemical formula.

## 1. Introduction

XAFS samples for transmission measurements can be optimized in quantity based on purely statistical arguments [1]. The XAFS signal-to-noise ratio

$$\frac{\mathrm{S}}{\mathrm{N}} \propto \mu_T d \left(\frac{1}{\mathrm{att}I_0} + \frac{1}{(1-\mathrm{att}I_0)\mathrm{att}I_t \exp(-\mu_T d)}\right)^{-1/2}$$

is maximized with respect to the attenuation by the 1st ionization chamber (IC), att$I_0$, and the optical thickness of the sample, $\mu_T d$. The attenuation by the 2nd IC ($I_t$ detector), att$I_t$, is taken as a selectable parameter. If a reference foil is placed between the 2nd and the 3rd IC, the fraction of x-rays absorbed by the 2nd IC is usually set to 50%. If the reference foil is not needed, one can select the total absorption (100%). For these two cases the optimal absorption of the 1st IC and $\mu_T d$ are found from figure 1 showing the levels of signal-to-noise ratio.

The elemental composition of the sample is typically known to the experimentalist. The remaining problem is to find the sample quantity given its optimal optical thickness $\mu_T d$ and the atomic absorption cross-sections.

## 2. Parts of XAFSmass

XAFSmass is a free open source software (MIT License). It consists of three Python modules – one for material properties (essentially, the atomic absorption cross-sections calculated via tabulated scattering factors *f''*), one for a chemical formula parser and the actual calculations and one for a graphical user interface (GUI) built with the platform independent application framework Qt. XAFSmass requires Python interpreter installed (any of its two branches).

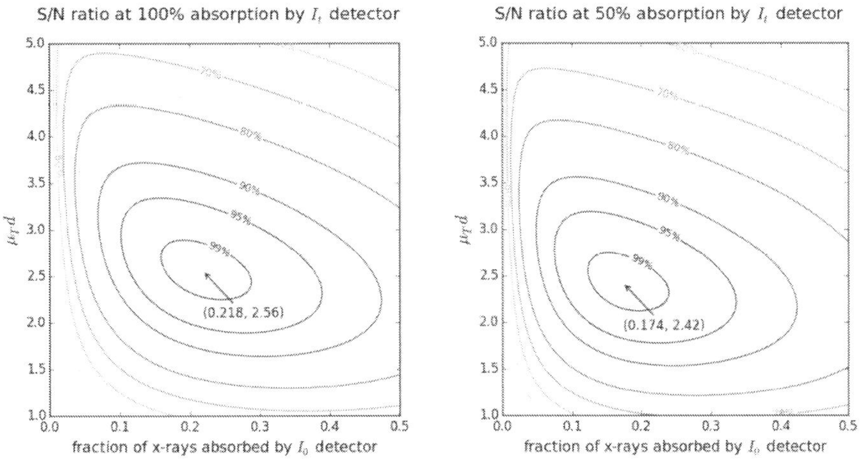


**Figure 1.** The levels of XAFS S/N at 100% absorption by the 2[nd] IC (left) and at 50% of it (right).

### *2.1. Tabulated scattering factors*

The tabulations are taken from Henke et al. (10 eV < $E$ < 30 keV) [2], Chantler (11 eV < $E$ < 405 keV) [3] and Brennan & Cowan (30 eV < $E$ < 509 keV) [4]. XAFSmass provides a plotting window for visualizing the scattering factors of the selected chemical elements.

### *2.2. Chemical formula parser*

The parser understands usual chemical formulas as "YBa2Cu3O6.93" and "Fe2(SO4)3" but also inclusion-to-matrix weight percentages as "Cu%1Zn%1((Al2O3)%10SiO2)". The latter formula is interpreted as 1 wt% of Cu and 1 wt% of Zn in a matrix composed of 10 wt% of alumina in silica.

## 3. Usage of XAFSmass

There are four operation modes depending on the sample nature. Samples can be: 'powder', 'foil, film, glass etc.' (those with known density), 'gas' or 'has unknown concentration'. Figure 2 shows the corresponding GUI views. At the top part of them one can see the underlying calculation formulas.

### *3.1. Powder samples*

The optimal optical sample thickness $\mu_T d$ depends on the absorption levels selected for the ionization chambers (see above). Typically, $\mu_T d$ is taken between 2 and 3 (e.g. for a 17.4% absorption level for the 1[st] chamber and a 50% level for the 2[nd] chamber, the optimal thickness is 2.42). However, if you get the absorption step more that 1.5 (reported by the drop-down list "absorptance step"), it is recommended to reduce the sample mass to avoid the potential thickness effect [5] due to possible inhomogeneity in the wafer. One can see in figure 1 that the S/N ratio has a wide maximum: $\mu_T d$ can be taken two times greater or two times less with a loss in S/N of only ~30%.

### *3.2. Samples with known density*

In this mode one can calculate the thickness of the sample with known density (usually, a foil). Commercial foils are highly homogeneous in thickness, so that one may ignore large step jumps with paying attention to $\mu_T d$ only. By putting $\mu_T d$ equal to 1, one gets the absorption length.

### *3.3. Gas pressure in ionization chambers*

If a reference foil is placed between the 2[nd] and the 3[rd] IC, the fraction of x-rays absorbed by the 2[nd] IC is usually set to 50%. If the reference foil is not needed, one can select the total absorption (exact 100%, however, is never possible). For these two cases the optimal absorption of the 1[st] IC is found from Figure 1.

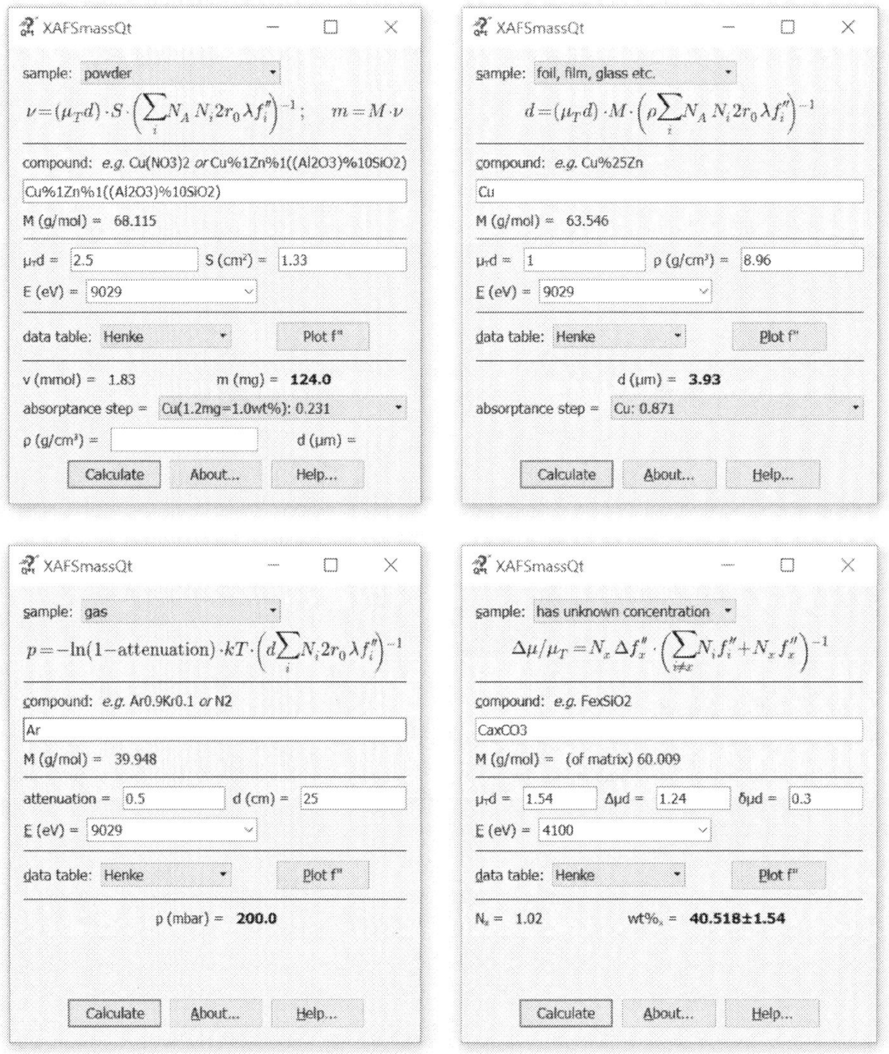


**Figure 2.** The GUI views for the four operation modes of XAFSmass.

### *3.4. Calculation of an unknown elemental concentration*

*Method 1: The matrix composition is known.* An absorption spectrum taken without the sample (*empty* spectrum) but with the same state of the ionization chambers is required. It must be then subtracted from the spectrum of the sample, e.g. in VIPER program, and thus a true (i.e. not vertically shifted) absorption coefficient is obtained. Determine the value of $\mu_T d$ above the edge, the edge jump ($\Delta\mu d$) and its uncertainty ($\delta\mu d$). Specify the chemical formula with x. The unknown molar concentration $N_x$ is then found from the formula visible in the lower-right part of Figure 2. Notice that the experimental ratio $\Delta\mu/\mu_T$ is determined much better than the step height alone since the errors in the numerator and the denominator are equal and of the same sign: $\frac{\Delta\mu\pm\delta\mu}{\mu_T\pm\delta\mu} \approx \frac{\Delta\mu}{\mu_T} \pm \frac{\mu_T-\Delta\mu}{\mu_T^2}\delta\mu$. This method, therefore, is more precise than the next one that relies only on the absorption step height.

*3.4.1. Method 2: The sample mass and area are known.* Determine the edge jump ($\Delta\mu d$). For the pure element find such a value for $\mu_T d$ that the absorption step in the pull-down list is equal to the experimental $\Delta\mu d$. This gives the mass of the element of interest that needs to be divided by the total sample mass to get the weight percentage.

*3.4.2. Experimental example.* Fine powder of $CaCO_3$ (4.3 mg) was mixed with powder polyethylene and pressed into a 13-mm-diameter pellet. The XAFS spectra are shown in figure 3. After subtracting the empty spectrum (here, of pure polyethylene pellet), the true XAFS spectrum gives: $\mu_T d$ = 1.54, $\Delta\mu d$ = 1.24, $\delta\mu d$ = 0.3. This results in wt%(Ca) = 40.5 ± 1.5 by the 1st method and wt%(Ca) = 40 ± 10 by the 2nd one. The true value is 40.04 wt%.

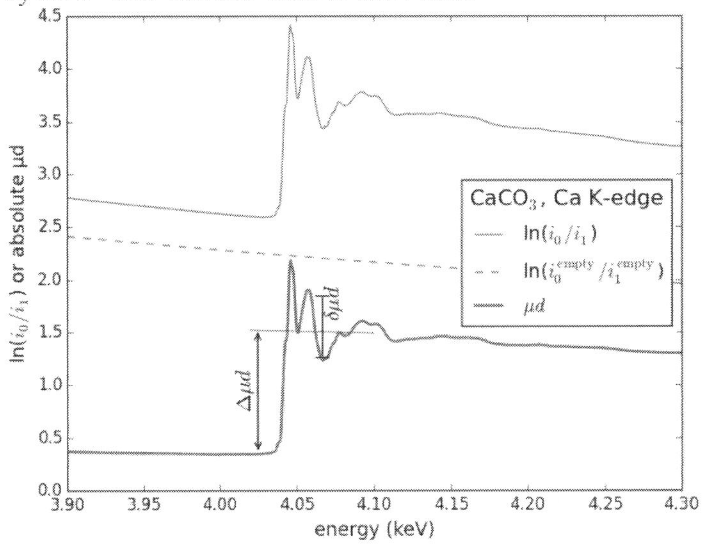


**Figure 3.** Experimental XAFS spectra of $CaCO_3$, from top to bottom: as measured, empty and true (= measured minus empty).

## 4. Conclusion

The presented program XAFSmass facilitates routine calculations of sample masses by combining a versatile chemical formula parser and several tabulations of the scattering factors. The calculated absorption edge height and its experimental value are an important hint to the experimentalist on the sample uniformity (presence of pin holes). Inversely, the edge height can be used to determine an elemental concentration assuming no holes in the sample.

The program is available at pypi.python.org. The version used in this paper is 1.2.0. XAFSmass is independent of Python version: it works with both branches, 2.x and 3.x, without modification.

## References

[1] Lee P A, Citrin P H, Eisenberger P and Kincaid B M 1981 Extended x-ray absorption fine structure—its strengths and limitations as a structural tool *Rev. Mod. Phys.* **53**, 769 – 806
[2] Henke B L, Gullikson E M and Davis J C 1993 X-ray interactions: photoabsorption, scattering, transmission, and reflection at E = 50–30000 eV, Z = 1–92 *Atomic Data and Nuclear Data Tables* **54** (no.2) 181–342
[3] Chantler C T 1995 Theoretical Form Factor, Attenuation, and Scattering Tabulation for Z = 1–92 from E = 1–10 eV to E = 0.4–1.0 MeV *J. Phys. Chem. Ref. Data* **24** 71–643
[4] Brennan S and Cowan P L 1992 A suite of programs for calculating x-ray absorption, reflection and diffraction performance for a variety of materials at arbitrary wavelengths *Rev. Sci. Instrum.* **63** 850–3
[5] Stern E A and Kim K 1981 Thickness effect on the extended-x-ray-absorption-fine-structure amplitude *Phys. Rev.* B **23** 3781–7

# Ab initio molecular dynamics simulations of the Sc K-edge EXAFS of scandium trifluoride

**Dmitry Bocharov[1,2], Matthias Krack[1], Aleksandr Kalinko[2], Juris Purans[2], Francesco Rocca[3], Shehab E. Ali[4], Alexei Kuzmin[2]**

[1]Paul Scherrer Institute, Laboratory for Reactor Physics and Systems Behaviour, Villigen, CH-5232, Switzerland
[2]Institute of Solid State Physics, University of Latvia, Kengaraga street 8, LV-1063, Riga, Latvia
[3]IFN-CNR, Institute for Photonics and Nanotechnologies, Unit FBK-Photonics of Trento, Povo, Trento, Italy
[4]Physics Department, Faculty of Science Suez Canal University, Ismailia, Egypt

E-mail: bocharov@latnet.lv

**Abstract.** Scandium fluoride $ScF_3$ has a simple cubic structure and attracts attention due to its large negative thermal expansion (NTE) over a wide range of temperatures (0-1100 K). In this study we present ab initio molecular dynamics (AIMD) simulations of $ScF_3$ and their validation using the Sc K-edge EXAFS spectra in the temperature range from 300 K to 1000 K measured at the XAFS beamline of ELETTRA. The obtained results allow an assessement of the employed AIMD model and provide insight into the local structure and the lattice dynamics of $ScF_3$ beyond the harmonic approximation. A strong anisotropy of the fluorine atom vibrations in the direction orthogonal to the –Sc–F–Sc– chain is observed. An expansion of the $ScF_6$ octahedra is found upon increasing temperature despite of the overall NTE of the crystal.

## 1. Introduction

At atmospheric pressure $ScF_3$ has a cubic perovskite-type structure (space group $Pm\overline{3}m$) down to at least 10 K [1] and undergoes a strong negative thermal expansion (NTE) below 1100 K [1]. In spite of the experimental and theoretical efforts [1, 2, 3] that have been performed to understand the large NTE observed for $ScF_3$, its behaviour in respect with similar isostructural crystals is still under debate. Most often it is explained in terms of the so-called rigid unit modes (RUMs) model [4, 5], which involves coupled vibrations of the rigid $ScF_6$ octahedra. When neighboring rigid octahedra librate in opposite directions, the distance between the centers of the octahedra decreases leading to the lattice contraction. Thus, to elucidate the nature of the NTE phenomena in details, one needs to follow temperature dependent changes of the lattice dynamics and structure of $ScF_3$ at the atomic scale.

Molecular dynamics (MD) simulations are a native approach to take into account vibrational and other dynamical properties that could be important for the observed NTE effect [2, 6, 7, 8]. In this study, ab initio molecular dynamics (AIMD) simulations were performed to address the problem beyond the harmonic approximation. The AIMD simulations were validated based on the Sc K-edge EXAFS spectra, previously measured at the XAFS beamline of ELETTRA from 10 K up to 1100 K.

## 2. Computational details

AIMD simulations based on Kohn-Sham density functional theory (DFT) [9] were performed at three temperatures (300 K, 600 K, and 1000 K) using the CP2K code [10]. A spin polarized implementation of the exchange-correlation functional PBEsol [11] was selected for this study, because it performs better for equilibrium properties of densely packed solids than the original PBE functional proposed by Perdew, Burke, Ernzerhof [12], in particular for lattice constants.

CP2K employs a localized basis set of Gaussian-type orbital functions for the description of the Kohn-Sham matrix within the framework of the Gaussian Plane Waves method [13, 14]. An auxiliary basis set of plane waves is employed to expand the electronic density using an density cutoff of 600 Hartree. Scalar-relativistic norm-conserving Goedecker-Teter-Hutter pseudopotentials [15, 16, 17] were employed for Sc and F atoms including 11 ([Ne] $3s^2\,3p^6\,4s^2\,3d^1$) and 7 ([He] $2s^2\,2p^5$) valence electrons, respectively. All calculation were performed at the Gamma point only using MOLOPT basis sets [18] optimized for these pseudopotentials. In this way fully unconstrained cell optimizations were performed for a $ScF_3$ model system consisting of $4 \times 4 \times 4$ unit cells, i.e. no symmetry was employed neither for the atomic position nor for the cell shape during the optimization procedure. A relaxed lattice constant of $a_0 = 4.027$ Å for $T = 0$ K was obtained using the PBEsol functional whereas PBE resulted in a value of $a_0 = 4.065$ Å.

A set of atomic configurations was generated within the canonical ensemble (NVT) for a supercell $4a_0{\times}4a_0{\times}4a_0$ ($a_0$ is the lattice parameter of $ScF_3$ taken from [1] for the respective temperature: 4.013 Å for 300 K, 4.0075 Å for 600 K, and 4.0055 Å for 1000 K temperature, correspondingly) containing 256 atoms. After an equilibration run of 5 ps, an AIMD sampling run was performed for 10 ps. The required temperature was maintained during the simulation runs using the Nosé-Hoover thermostat. The generated atomic configurations were used in the next step to calculate the Sc K-edge configuration-averaged EXAFS spectra using the *ab initio* real-space multiple-scattering (MS) FEFF9.64 code [19] within the MD-EXAFS scheme [20, 21].

The MS contributions into EXAFS were accounted up to the sixth order within the radius of 8 Å around the absorbing Sc atom (7 coordination shells composed of 110 atoms). The calculation of the cluster potential was performed only once for the average crystallographic $ScF_3$ structure [1], thus neglecting its small variation due to thermal disorder. The complex exchange-correlation Hedin–Lundqvist potential and default values of muffin-tin radii ($R_{\mathrm{mt}}$(Sc)=1.31 Å and $R_{\mathrm{mt}}$(F)=1.01 Å), as provided within the FEFF9.64 code [19], were employed. The amplitude of the theoretical EXAFS spectrum was multiplied by the $S_0^2 = 0.9$ factor.

## 3. Results and discussions

The Sc K-edge EXAFS spectra, obtained from the AIMD simulations, are compared with the experimental data at $T$ = 300 K, 600 K and 1000 K in Fig. 1. The agreement between the theoretical and experimental data is quite good, but not ideal. Note that no fitting parameters were employed, and all MS contributions were accounted. Some overestimation of the amplitude of the first two peaks, located at 1.6 Å and 3.6 Å, is visible at $T$ = 300 K which can be caused by the neglect of quantum effects in AIMD simulations, leading to underestimation of thermal disorder. The best agreement is observed for the 600 K spectra up to 4.5 Å, i.e. for the first three coordination shells around the absorbing Sc atoms. The poorer agreement for the next peak at 5.5 Å can be tentatively attributed to the beginning of the border effect, since the supercell half size is about 8 Å. One can also note that in spite of an octahedral coordination of Sc, the MS contribution from its first coordination shell, expected at ~2.5 Å between the first and second main peaks, is significantly reduced due to the dependence of the MS terms on the final $l$ state ($l = 1$ for the Sc K-edge) [22].

The atomic coordinates obtained in the AIMD simulations were used to calculate a number of structural properties as the radial distribution functions (RDFs) $G$(Sc–F) and $G$(Sc–Sc) and the inter-octahedral Sc–F–Sc bond angle distribution function (BADF) (Fig. 2). Note that for the

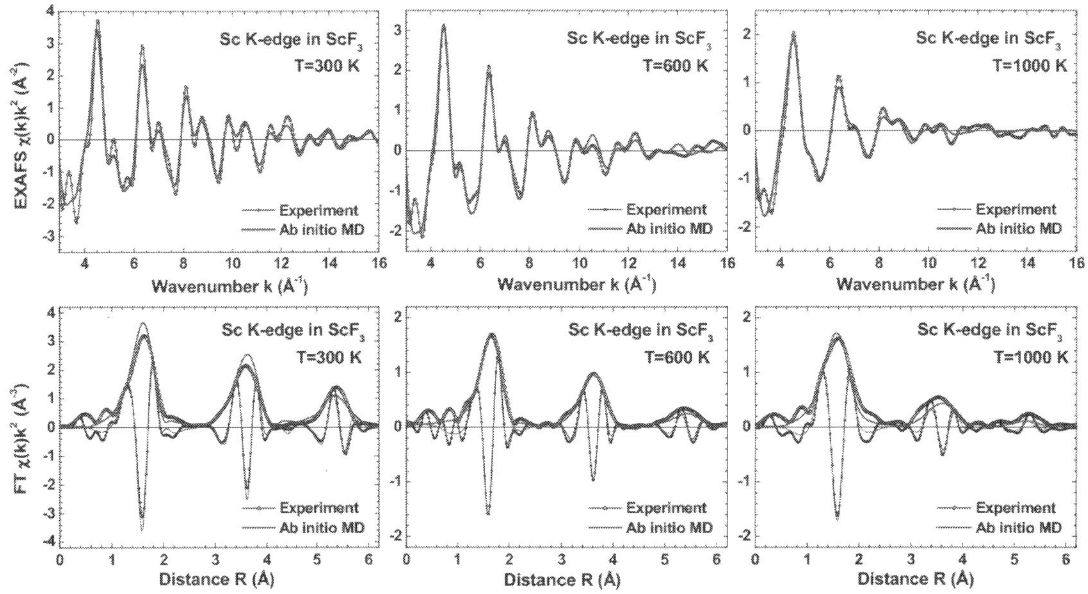


**Figure 1.** Experimental (open circles) and calculated (solid lines) Sc K-edge EXAFS and their Fourier transforms (modulus and imaginary parts are shown by thick and thin lines, correspondingly) at $T = 300$ K, 600 K and 1000 K.

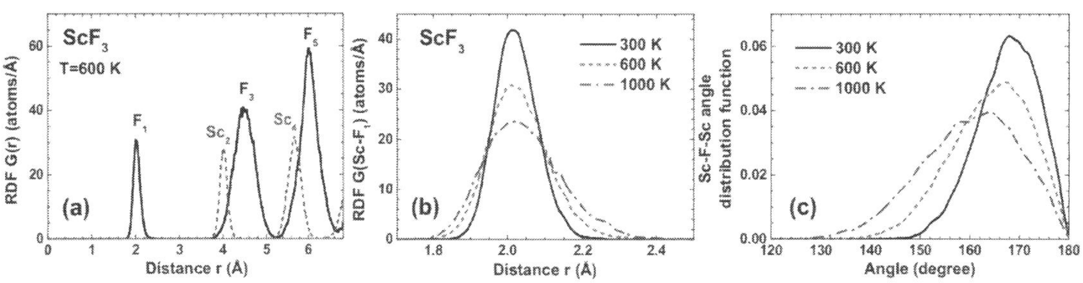


**Figure 2.** Radial distribution functions (RDFs) $G(r)$ for Sc–F and Sc–Sc atom pairs at $T = 600$ K (a). RDFs for the first coordination shell Sc–$F_1$ (b) and the inter-octahedral Sc–F–Sc bond angle distribution functions (c) at $T = 300$ K, 600 K and 1000 K.

atoms located in the crystallographic positions the Sc–F–Sc bond angle is equal to 180 degree. This value is reduced significantly when evaluated over instantaneous atomic positions with a broad distribution. Of course, the average value in the 3D space is still 180 degree, but with very low probability, as was well explained previously for $SrTiO_3$ in [23]. Note that similar but much narrower BADF for the inter-octahedral angle Re–O–Re was found at temperatures below 573 K in cubic $ReO_3$ [24], which has very weak NTE [25]. In $ReO_3$ [24] the smallest value of the inter-octahedral angle at $T < 600$ K is larger than $\sim 165°$ against $\sim 140°$ in $ScF_3$, suggesting stronger tilting of $ScF_6$ octahedra than that of $ReO_6$ units in $ReO_3$ [26].

According to AIMD, the vibration of the F atoms is strongly anisotropic: the amplitude of their displacements along the –Sc–F–Sc– atomic chains is about 2-2.5 times smaller than in the orthogonal direction, and both amplitudes increase about 3-4 times upon increasing temperature from 300 K to 1000 K. The vibration amplitude of the Sc atoms also increases, but only about 2 times in this temperature range, and its distribution remains isotropic. Another interesting result is that while the lattice parameter $a_0$ decreases by 0.01 Å from 300 K to 1000 K [1], the calculated average Sc–F distance increases by 0.02 Å, and the shape of the first coordination shell peak in RDF $G$(Sc–F) becomes asymmetric (Fig. 2(c)). This means that the $ScF_6$ octahedra expand slightly upon increasing temperature due to an anharmonicity of the Sc–F bonding.

## 4. Conclusions

We have demonstrated the efficiency of the MD-EXAFS approach [20] in combination with ab initio molecular dynamics method for theory validation and temperature dependent structural properties description in $ScF_3$. In the framework of this approach configuration-averaged Sc K-edge EXAFS spectra were produced for three different temperatures $T = 300$ K, 600 K and 1000 K and compared with the experimental data, suggesting good qualitative agreement. It was found that upon temperature increase the $ScF_6$ octahedra do not behave as rigid units: the strong rotation of $ScF_6$ octahedra is accompanied by their expansion caused by the anharmonic Sc–F bonding, despite of a negative thermal expansion of the crystal lattice.

## Acknowledgments

The calculations were performed on the Paul Scherrer Institute cluster Merlin4, HPC resources of the Swiss National Supercomputing Centre (CSCS) in Lugano as well as at the Latvian SuperCluster (LASC). We gratefully acknowledge the assistance of Giuliana Aquilanti and of the staff members of XAFS beamline at Elettra during the EXAFS experiment (proposal no. 20120215). Sergei Piskunov gave us important suggestions, in the initial stage of the MD simulations. The work was partially supported by the Latvian Science Council grant no. 187/2012.

## References

[1] Greve B K, Martin K L, Lee P L, Chupas P J, Chapman K W and Wilkinson A P 2010 *J. Am. Chem. Soc.* **132** 15496
[2] Li C, Tang X, Muñoz J, Keith J, Tracy S, Abernathy D and Fultz B 2011 *Phys. Rev. Lett.* **107** 195504
[3] Liu Y, Wang Z, Wu M, Sun Q, Chao M and Jia Y 2015 *Comp. Mater. Sci.* **107** 157–162
[4] Barrera G D, Bruno J A O, Barron T H K and Allan N L 2005 *J. Phys.: Condens. Matter* **17** R217
[5] Takenaka K 2012 *Sci. Tech. Adv. Mater.* **13** 013001
[6] Zhou W, Wu H, Yildirim T, Simpson J R and Walker A R H 2008 *Phys. Rev. B* **78** 054114
[7] Fang H, Dove M T, Rimmer L H N and Misquitta A J 2013 *Phys. Rev. B* **88** 104306
[8] Sanson A 2014 *Chem. Mater.* **26** 3716
[9] Kohn W and Sham L J 1965 *Phys. Rev.* **140** A1133
[10] CP2K developers group 2000–2015 (http://www.cp2k.org)
[11] Perdew J P, Ruzsinszky A, Csonka G I, Vydrov O A, Scuseria G E, Constantin L A, Zhou X and Burke K 2008 *Phys. Rev. Lett.* **100** 136406
[12] Perdew J P, Burke K and Ernzerhof M 1996 *Phys. Rev. Lett.* **77** 3865
[13] Lippert G, Hutter J and Parrinello M 1997 *Mol. Phys.* **92** 477
[14] VandeVondele J, Krack M, Mohamed F, Parrinello M, Chassaing T and Hutter J 2005 *Comput. Phys. Commun.* **167** 103
[15] Goedecker S, Teter M and Hutter J 1996 *Phys. Rev. B* **54** 1703
[16] Hartwigsen C, Goedecker S and Hutter J 1998 *Phys. Rev. B* **58** 3641
[17] Krack M 2005 *Theor. Chem. Acc.* **114** 145
[18] VandeVondele J and Hutter J 2007 *J. Chem. Phys.* **127** 114105
[19] Rehr J J, Kas J J, Vila F D, Prange M P and Jorissen K 2010 *Phys. Chem. Chem. Phys.* **12** 5503
[20] Kuzmin A and Evarestov R A 2009 *J. Phys: Condens. Matter* **21** 055401
[21] Kuzmin A and Evarestov R A 2009 *J. Phys.: Conf. Series* **190** 012024
[22] Kuzmin A, Purans J, Benfatto M and Natoli C R 1993 *Phys. Rev. B* **47** 2480
[23] Hui Q, Tucker M G, Dove M T, Wells S A and Keen D A 2005 *J. Phys.: Condens. Matter* **17** S111
[24] Kalinko A, Evarestov R A, Kuzmin A and Purans J 2009 *J. Phys.: Conf. Series* **190** 012080
[25] Wdowik U D, Parlinski K, Chatterji T, Rols S and Schober H 2010 *Phys. Rev. B* **82** 104301
[26] Purans J, Fornasini P, Ali S E, Dalba G, Kuzmin A and Rocca F 2015 *Phys. Rev. B* **92** 014302

# Real-space multiple-scattering theory of XMCD including spin-orbit interaction in scattering process

**Akihiro Koide**[1,2], **Kaori Niki**[1], **Seiji Sakai**[2] **and Takashi Fujikawa**[1]

[1]Graduate School of Advanced Integration Science, Chiba University, 1-33 Yayoi-cho, Inage-ku, Chiba, 263-8522, Japan

[2]Advanced Science Research Center, Japan Atomic Energy Agency, 2-4 Shirakata-Shirane, Tokai, Naka, 319-1195, Japan

E-mail: koide-a@chiba-u.jp

**Abstract.** The effects of the spin-orbit interaction on surrounding atoms for XMCD spectra are studied by a real-space multiple-scattering theory. The present numerical calculation for Fe *K*-edge XMCD spectra from BCC iron demonstrates the importance of the spin-orbit interaction on scattering atoms, which has been disregarded in previous works. These effects will be inevitable for *K*-edge XMCD analyses of light elements surrounded by heavy magnetic atoms.

## 1. Introduction

XMCD spectra of light elements have been measured to study spintronic materials and nanostructures such as graphene/Ni(111) [1]. For a bulk system, O *K*-edge XMCD intensity was also obtained in $CrO_2$ [2]. To observe XMCD signals, both the spin-orbit interaction (SOI) and spin polarization are inevitable. In case of the X-ray absorption by light elements like carbon, XMCD are still observed, which implies the importance of electron scattering by heavy magnetic element like Fe. Fujikawa et al. [3, 4] developed a real-space multiple-scattering (MS) XMCD theory with relativistic corrections: they neglected the SOI except that on the X-ray absorbing atom. This approximation does not work well for the XMCD spectrum of light elements. Although MS codes already include relativistic effects on all sites [5, 6, 7], infuluence of the SOI on scattering sites was not studied intensively for XMCD spectra. Before the investigation of complicated systems including light elements, we carefully study the SOI on surrounding atoms, and numerically calculate Fe *K*-edge XMCD specra for BCC iron.

## 2. Theory

The absorption intensity $I(\omega, m_p)$ for the incident photon energy $\omega$ with the helicity ($m_p = \pm 1$) of circularly polarized X-rays is written as [4]

$$I(\omega, m_p) = -2\mathrm{Im}\,\langle c|\, H_{ep}^{\dagger}(m_p) g_D(\varepsilon) H_{ep}(m_p)\, |c\rangle \tag{1}$$

$$= T_{11}(\omega, m_p) + T_{12}(\omega, m_p) + T_{21}(\omega, m_p) + U_{11}(\omega, m_p) + \cdots \tag{2}$$

$$U_{11}(\omega, m_p) \sim -2\mathrm{Im}\,\langle \varphi_c|\, \Delta^*(m_p) g(\varepsilon) \delta V g(\varepsilon) \Delta(m_p)\, |\varphi_c\rangle \tag{3}$$

where $g_D(\varepsilon)$ is the one-electron Dirac Green's function with the photoelectron energy $\varepsilon$, $H_{ep}(m_p)$ is the electron-photon interaction operator with the photon helicity $m_p$, and $|c\rangle$ is the 4-spinor core orbital. As demonstrated by Gesztesy et al. [8], $g_D(\varepsilon)$ is expanded by the correlated non-relativistic Green's function $g(\varepsilon)$, which yields $T_{11}$, $T_{12}$, $T_{21}$ and $U_{11}$. The intensity $T_{11}$ corresponds to the term in the non-relativistic limit. It does not contribute to the $K$-edge XMCD intensity because the $1s_{1/2}$ core state, and the final state (phoelectron) in $T_{11}$ have no SOI. The intensities $T_{12}$ and $T_{21}$ reflect some relativistic effects only on the absorbing atom. The intensity $U_{11}$ includes SOI $\delta V$ for the photoelectron state as shown (3). The large component $|\varphi_c\rangle$ of the 4-spinor $|c\rangle$ is influenced by $\Delta_{m_p}$, the electron-photon interaction operator in the 2-spinor form. The total SOI $\delta V$ can be separated into each atomic site contribution as $\delta V \sim \sum_\alpha \delta v_\alpha$, where $\delta v_\alpha$ is the SOI on the site $\alpha$. Then, we derive MS expansion for $U_{11}$. If we go beyond this perturbation prescription, the SOI should be renormalized and embedded in full MS paths.

The non-relativistic Green's function $g(\varepsilon)$ is expanded in terms of the site-$t$ matrix $t_\alpha(= v_\alpha + v_\alpha g_0 t_\alpha : v_\alpha$ is the atomic potential at the $\alpha$ site), the free propagator $g_0$ and the inner and inter site propagator $g_\alpha(= g_0 + g_0 t_\alpha g_0)$:

$$g = (1-\delta_{ab})g_b v_b g_a + g_a + g_b \sum_{\alpha(\neq a,b)} t_\alpha g_a + g_b \sum_{\beta(\neq \alpha,b)} t_\beta g_0 \sum_{\alpha(\neq a)} t_\alpha g_a + \cdots \tag{4}$$

The site-$t$ matrix $t_\alpha$ describes the full inner atomic scattering at the $\alpha$ site without the SOI. The first term vanishs when the $a$ and $b$ sites represent the same site, for instance, the absorbing atom in the case of the X-ray abosorption intensity calculation. The intensity $U_{11}$ is separated into the non-scattering and scattering terms as $U_{11} = U_{11,A} + U_{11,sc}$, where $A$ is the absorbing site. The intensity $U_{11,A}$ includes the SOI on $A$ in non-scattering processes: the photoelectron is affected by the SOI when it initially goes out from or finally comes back to $A$. This term has already been derived in references [3, 4]. On the other hand, $U_{11,sc}$ involves the SOI on scattering sites which include $A$. The scattering term $U_{11,sc}$ is written as

$$U_{11,sc}(m_p) = -\frac{2}{\pi}\mathrm{Im}\sum_\sigma \rho_\sigma{}^2 (u^{GG}_{m_p,\sigma} + u^{Z'G}_{m_p,\sigma} + u^{Z'Z'}_{m_p,\sigma}) \tag{5}$$

$$u^{GG}_{m_p,\sigma} = \sum_{B(\neq A)} \sum_L G^{AB,\sigma}_{1m_p,L} \left( m\sigma\{[\xi^\sigma_B]^{jj}_l - 4ik[(\xi v)^\sigma_B]^{jj}_l - 4k^2[(v\xi v)^\sigma_B]^{jj}_l\}\right) G^{BA,\sigma}_{L,1m_p} \tag{6}$$

$$u^{Z'G}_{m_p,\sigma} = \sum_{B(\neq A)} \sum_L Z'^{AB,\sigma}_{1m_p,L} \left( m\sigma\{2[\xi^\sigma_B]^{Rj}_l - 4ik[(\xi v)^\sigma_B]^{Rj}_l\}\right) G^{BA,\sigma}_{L,1m_p} \tag{7}$$

$$u^{Z'Z'}_{m_p,\sigma} = \sum_\alpha \sum_L Z'^{A\alpha,\sigma}_{1m_p,L} \left( m\sigma[\xi^\sigma_\alpha]^{RR}_l\right) Z'^{\alpha A,\sigma}_{L,1m_p} \tag{8}$$

The spin index $\sigma$ is $+$ $(-)$ for the up (down) spin of the photoelectron, $\rho$ is the dipole radial integral and $k$ is the momentum of the photoelectron. The propagater $G^{BA}_{L,1m_p}$ corresponds to the photoelectron migration from $A$ with angular momentum $(1, m_p)$ to $B$ with $L$, while $Z'^{A\alpha}_{1m_p,L}$ describes the migration from $A$ to $\alpha$ with MS from double scattering up. We note that $G$ is arisen from the first and second terms of the expansion described by eq. (4), which are the zeroth order of $t$. In addition, the atomic potential $v$ in eqs. (6) and (7) is not replacement of site-t matrix, but comes from $g_B v_B g_A$ in eq. (4). The sandwiched terms $(\cdots)$ by $G$ or $Z'$ describe overlaps between the photoelectron wavefunction and the SOI on a scattering site. The radial integrals $[\cdots]$ are written as

$$[\xi^\sigma_\alpha]^{\varphi_1\varphi_2}_l = \left[\int dr r^2 \varphi_{1,l}(kr)\xi(r)\varphi_{2,l}(kr)\right]_{\alpha,\sigma} \tag{9}$$

$$[(\xi v)^\sigma_\alpha]^{\varphi_1\varphi_2}_l = \left[\int dr dr' r^2 r'^2 \varphi_{1,l}(kr)\xi(r)R_l(kr_<)f_l(kr_>)v(r')\varphi_{2,l}(kr')\right]_{\alpha,\sigma} \tag{10}$$

$$[(v\xi v)^{\sigma}_{\alpha}]^{\varphi_1\varphi_2}_{l} = \left[\int dr dr' dr'' r^2 r'^2 r''^2 \right.$$
$$\left. \times\, \varphi_{1,l}(kr)v(r)R_l(kr_<)f_l(kr_>)\xi(r')R_l(kr'_<)f_l(kr'_>)v(r'')\varphi_{2,l}(kr'')\right]_{\alpha,\sigma} \quad (11)$$

where $\xi$ is the radial part of the SOI. The wavefunctions $\varphi_1$ and $\varphi_2$ are the spherical Bessel function $j$ or the numerical regular solution $R$. The $R$ and numerical irregular solutions $f$ in $[(\xi v)]$ and $[(v\xi v)]$ describe the intra atomic propagation between $\xi$ and the atomic potential $v$. The integral variables $r_{<(>)}$ and $r'_{<(>)}$ represent the $max(min)(r, r')$ and $max(min)(r', r'')$, respectively. We numerically check the effect of the scattering site SOI in the next section.

## 3. Results & Discussion

We calculate Fe $K$-edge XMCD for BCC iron to investigate the contribution of the SOI on scattering sites. We adopt the muffin-tin potential for phaseshift calculations. The cluster to calculate MS matrices includes 113 atoms within a 7 Å distance from the absorbing atom. XMCD spectra are obtained by $\Delta I = I(+) - I(-)$ under the antiparallel magnetization to the incident X-rays. XMCD intensity is obtained in percentage of a main peak one of a corresponding XAS spectrum (not shown).

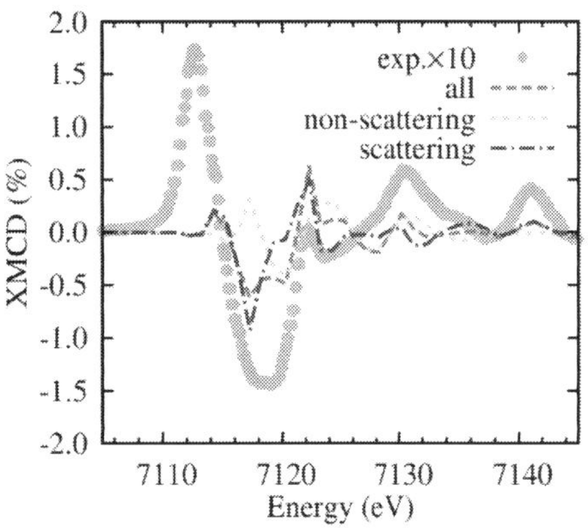


**Figure 1.** Calculated Fe $K$-edge XMCD spectra for BCC iron. Three different approximations are used: the SOI in all processes (red solid), in only non-scattering processes (green dashed) and in only scattering processes (blue dotted) are considered. The experimental result (gray dot) is shown for comparison [9].

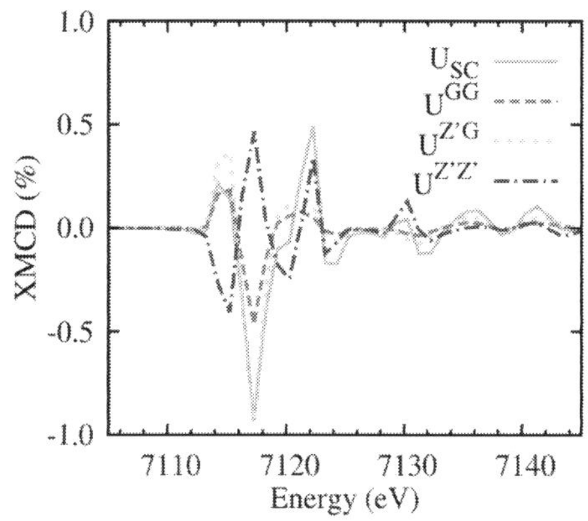


**Figure 2.** Calculated XMCD spectra for the components of $U_{sc}$; $U^{GG}$ (red dashed), $U^{Z'G}$ (green dots) and $U^{Z'Z'}$ (blue chain). The scattering term $U_{sc}$ (gray solid) is also shown for comparison.

Figure 1 shows the calculated Fe $K$-edge XMCD spectra for three different approximations: the SOI in all processes, in only non-scattering processes and in only scattering processes are considered for the XMCD calculations. The scattering term corresponds $\Delta U_{sc} = U_{sc}(+) - U_{sc}(-)$. The non-scattering term is obtained by $\Delta U_A + 2\Delta T_{12}$, which includes the usual non-SOI MS [4]. In Fig. 1, the scattering term is comparable in intensity to the non-scattering term. Although the calculated spectra need to a scale factor to be compared with the experimental one [9], the all proscesses spectrum is improved by the scattering one. These indicate that the SOI on scattering sites is still important for $K$-edge XMCD spectra of transition metals. Figure 2 shows XMCD spectra for the components of $\Delta U_{sc}$. The component $U^{GG}$ is obtained by $U^{GG} = -\frac{2}{\pi}\mathrm{Im}\sum_{\sigma}{\rho_\sigma}^2 u^{GG}_{\sigma}$. The components $U^{Z'G}$ and $U^{Z'Z'}$ are also obtained similarly.

All components of $\Delta U_{sc}$ have comparable in intensity to $U_{sc}$. The previous calculation of the single SOI scattering showed negligibly weak XMCD, and suggested that the weak XMCD is attributed to the absence of exchange (site-$t$ matrix) scattering [3]. However, the single SOI scattering $U^{GG}$ is not negligible, and comparable in intensity to $U^{Z'G}$ and $U^{Z'Z'}$ which include the exchange scattering in $Z'$ as shown in (7) and (8). To understand this discrepancy, we investigate the components of $U^{GG}$ in (6).

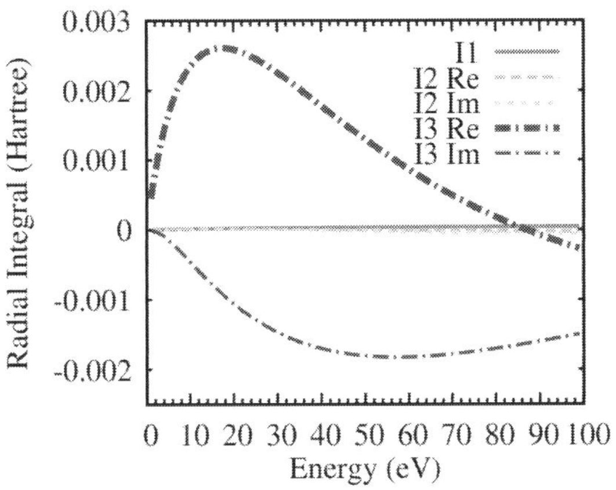


**Figure 3.** Radial integrals $I1$, $I2$ and $I3$, which compose $U^{GG}$, represent $[\xi_B^\sigma]_l^{jj}$, $-4ik[(\xi v)_B^\sigma]_l^{jj}$ and $-4k^2[(v\xi v)_B^\sigma]_l^{jj}$, respectively. The calculated results for $\sigma$ = (up), $l$ = 1 and $B$ = Fe are shown. The integral $I1$ (red solid) is real, while the $I2$ (real: orange dahsed, imaginary: green dotted) and $I3$ (real: thick blue chain, imaginary: thin blue chain) have real parts and imaginary parts. The horizontal axis shows photoelectron kinetic energies.

Figure 3 shows radial integrals which compose $U^{GG}$. In Fig. 3, $I3$ ($[(v\xi v)]^{jj}$) is considerably large compared with the other terms. Because $\xi$ is the gradient of the atomic site potential $v$, $\xi$ is strong only near the nucleus. The free photoelectron described by the $j$ function has small amplitude near the nucleus. Then, $[\xi]^{jj}$ and $[\xi v]^{jj}$, which include the direct overlap between $\xi$ and $j$, are negligible. The result for $[\xi]^{jj}$ is the same as the previous result [3]. On the other hand, the potential $v$ can intermediate and enhance the overlap between the $j$ function and $\xi$ in $[(v\xi v)]^{jj}$. Thus, $U^{GG}$ gives the comparable XMCD intensity. Similarly, $U^{Z'G}$ ($[(\xi v)]^{Rj}$) and $U^{Z'Z'}$ ($[(\xi)]^{RR}$) are also enhanced by $v$: the $R$ function already includes the effect of $v$, and has relatively larger amplitude near the nucleus than the $j$ function. This indicates that the photoelectron should be affected by the attractive potential $v$ and the SOI $\xi$ simultaneously.

The present numerical calculations for Fe $K$-edge XMCD clearly show the importance of the SOI on scattering sites, which was not specificaly studied before. This result should be important for the XMCD analyses in case that a light absorber is surrounded by rather heavy atoms.

## Acknowledgments

This work was supported by Grant-in-Aid for JSPS Fellows Grant Number 15J07459.

## References

[1] Matsumoto Y, Entani S, Koide A, Ohtomo M, Avramov P V, Naramoto H, Amemiya K, Fujikawa T and Sakai S 2013 *J. Mater. Chem.* C **1** 5533
[2] Goering E, Bayer A, Gold S, Schütz G, Rabe M, Rüdiger U and Güntherodt 2002 *Europhys. Lett.* **6** 906.
[3] Fujikawa T and Nagamtsu S 2002 *Jpn. J. Appl. Phys.* **41** 2005
[4] Fujikawa T and Nagamtsu S 2003 *J. Electron Spectro. Relat. Phenom.* **129** 55
[5] Rehr J J, Kas J J, Prange M P and Jorissen K 2010 *Phys. Chem. Chem. Phys.* **12** 5503.
[6] Joly Y, Martteo S D and Natoli C R 2004 *Phys. Rev. B* **69** 224401.
[7] Ebert H, Ködderitzsch D and Minár J 2011 *Rep. Prog. Phys.* **74** 096501.
[8] Gesztesy F, Grosse H and Thaller B 1984 *Ann. Inst. Henri Poincaré* **40** 159
[9] Mathon O, Baudelet F, Itié J-P, Pasternak S, Polian A and Pascarelli S 2004 *J. Synchrotron Rad.* **11** 423

# DFT+$U$ calculation of the Cr $K$ pre-edge structures in $ZnCr_2O_4$

**E Rozsályi[1], L Verger[2,3], D Cabaret[2] and A Juhin[2]**

[1] Wigner Research Centre for Physics, Hungarian Academy of Sciences, Budapest, Hungary
[2] Sorbonne Universités, UPMC Univ Paris 6, UMR CNRS 7590, IMPMC, Paris, France
[3] Cité de la Céramique, Sèvres et Limoges, Sèvres, France

E-mail: rozsalyi.emese@wigner.mta.hu

**Abstract.** Zinc chromite $ZnCr_2O_4$ has been investigated by X-ray absorption near edge structure (XANES) spectroscopy at the Cr K-edge, with the objective of better understanding the origin of the electronic excitations giving rise to the pre-edge features. In particular, the origin of a third feature visible only in Cr-rich compounds of the $ZnAl_2O_4$-$ZnCr_2O_4$ solid solution is studied. Standard density functional theory (DFT) XANES calculations were first performed, then a Hubbard $U$ parameter was included in an attempt to improve the relative peak positions by increasing the localization of the Cr $3d$ states. The origin of the pre-edge features in $ZnCr_2O_4$ has been analyzed using DFT+$U$ calculations, local projected densities of states and the assignment of the spectral features for the dilute $ZnAl_2O_4$:$Cr^{3+}$ compound. We show that DFT+$U$ does not overcome the limitations of standard DFT in modeling the relative peak positions, and that approaches including advanced description of the electron-hole interaction are needed.

## 1. Introduction

Spinels exhibit a wide range of solid solutions, the color of which has been studied extensively, since the great variety of coloration make them the basis of most ceramic pigments [1, 2]. In the gahnite-zincochromite $ZnAl_{2-x}Cr_xO_4$ the change of color from pink to green through brownish pink upon increasing chromium content has been the subject of several studies [2, 3, 4]. X-ray absorption near edge structure spectroscopy (XANES) has been used in our previous work to interpret the influence of chromium on the color and structural changes for two compounds of the $ZnAl_{2-x}Cr_xO_4$ series [5]. In this work the isotropic XANES spectra of a compound diluted in Cr, with composition $ZnAl_{1.875}Cr_{0.125}O_4$, and of the end-member $ZnCr_2O_4$ were calculated using density functional theory (DFT) and compared with the experimental spectra measured for $ZnAl_{1.8}Cr_{0.2}O_4$ and $ZnCr_2O_4$, respectively. The agreement with experiment has been considered fairly good for the dilute sample but only satisfactory for the concentrated one regarding the overall spectral features. To have a better agreement with experiment, a promising approach is to improve the description of electron correlation effects present in the $ZnCr_2O_4$ spinel by the inclusion of the Hubbard $U$ term [6]. This method often improves the description of the electronic structure for systems in which standard DFT fails [7, 8]. Therefore, this paper reports the influence of the $U$ parameter when simulating XANES within DFT in the case of $ZnCr_2O_4$.

## 2. Computational methods

The first-principles analysis has been conducted within DFT using the PWscf and Xspectra [10, 11] codes included in the QUANTUM-ESPRESSO package [9], utilizing spin-polarized PBE generalized gradient approximation [12], and norm-conserving pseudopotentials. The parameters used for the pseudopotential generation are given in Ref. [13]. A plane-wave kinetic energy cut-off of 90 Ry was found to be optimal from convergence tests. At room temperature, $ZnCr_2O_4$ has a cubic structure ($Fd\bar{3}m$) with $a = 8.33$ Å [5]. $Zn^{2+}$ cations occupy tetrahedral sites while $Cr^{3+}$ with S=3/2 populate octahedral sites. We considered the cubic cell (56 atoms) and the collinear antiferromagnetic structure proposed by Chan *et al.* [14]. The atomic positions were allowed to vary to minimize the total energy and the interatomic forces while keeping the lattice parameter fixed to its experimental value. A Hubbard $U$ correction term, which is applied to the $3d$ states of the Cr atoms to increase their localization, was added to the conventional DFT calculation. Singh *et al.* [15], for example, used $U$=3.0 eV for Cr $d$ orbitals in $ZnCr_2O_4$ with the local spin-density approximation (LSDA). However, the value of $U$ has not been reported relevant to this study and, consequently, different values have been chosen for $U$ in the 0-3 eV range. The Cr $K$-edge electric dipole (E1) and electric quadrupole (E2) absorption cross sections were then calculated including the Hubbard $U$ term. The charge density of the system was calculated self-consistently, with a $1s$ core hole on one Cr among the 16 in $ZnCr_2O_4$. A single $k$-point was used to perform self-consistent spin-polarized charge-density calculation including the core-hole, and a Monkhorst-Pack grid of $2 \times 2 \times 2$ $k$-points in the Brillouin zone for the absorption cross section calculation. The energy dependent broadening parameter used for the XANES cross section calculation is identical to that used in Ref. [16]. The isotropic XANES spectrum is the sum of the E1 and E2 contributions. Local projected density of states (DOS) were calculated on the system with a core-hole, using Löwdin projections, $2 \times 2 \times 2$ Monkhorst-Pack $k$-points grid and a Gaussian broadening of 0.00368 eV.

## 3. Results and discussion

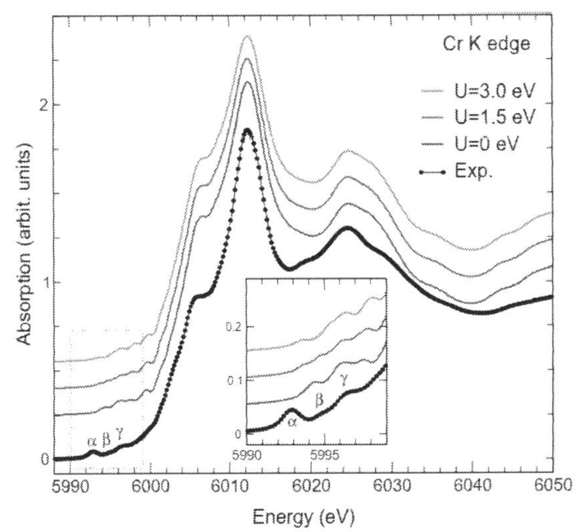


**Figure 1.** Comparison between experimental and calculated XANES spectra at the Cr $K$-edge of the spinel $ZnCr_2O_4$ using DFT with and without Hubbard $U$ parameter, namely $U$=0, 1.5 and 3 eV. The standard DFT result corresponds to $U$=0 eV. The inset is a focus on the pre-edge region. The isotropic XANES spectrum calculated in standard DFT, and the experimental isotropic spectrum have been replotted from Ref. [5].

Figure 1 displays the experimental and calculated Cr $K$-edge spectra in $ZnCr_2O_4$. The theoretical spectra have been aligned with the experimental one with respect to the energy position of the main peak. The experimental pre-edge spectrum exhibits three main features at 5993, 5995 and 5996.5 eV, labeled $\alpha$, $\beta$ and $\gamma$, respectively. Based on Ref. [5] (Figs. 6 and 9), the assignment of these peaks is straightforward when comparing the calculated spectrum of $ZnCr_2O_4$ with that of $ZnAl_2O_4$:$Cr^{3+}$. Accordingly, peaks $\alpha$ and $\beta$ are due to E2 transitions of the core electron to unoccupied $3d$ states of the absorber. The $\gamma$ feature is due to non-local E1 transitions to empty $4p$ states of the absorber, hybridized with empty $3d$ states of the next-nearest Cr neighbors through empty $p$ states of O. As can be seen and also reported in Ref. [5], the calculated pre-edge peaks lie higher in energy than the analogous peaks in the experimental spectrum. Furthermore, two peaks can be well-distinguished in the calculated spectrum belonging to the feature $\gamma$, which are not as distinct in the experimental spectrum.

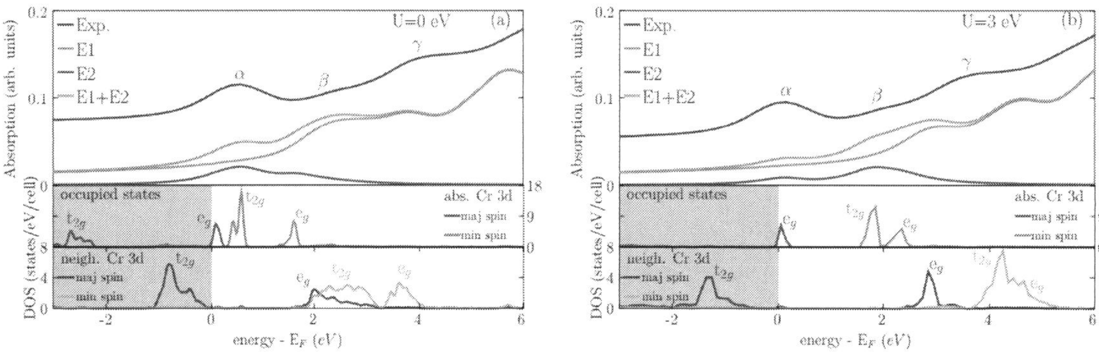

**Figure 2.** Analysis of the pre-edge region of $ZnCr_2O_4$ calculated with (a) standard DFT (replotted from Ref. [5]), and (b) DFT+$U$ for U=3 eV. Upper panel: experimental and calculated pre-edge spectra and its decomposition into E1 and E2 transitions. Lower panels: $3d$ density of states projected on the Cr absorber and the Cr next-nearest neighbors. The experimental spectrum has been shifted to the theoretical data to match the position of peak $\alpha$. $E_F$ is the Fermi energy.

The identification of the pre-edge peaks observed in Cr $K$-edge calculated XANES spectrum of $ZnCr_2O_4$ seems not to be straightforward in the present case, and both local and non-local transitions noticed in the pre-edge regions are at too high energies compared to experiment. To assign these transitions, the pre-edge regions of the experimental and calculated spectra as well as their decompositions into E1 and E2 transitions are shown in the top panels of Figure 2 while the density of states (DOS) projected on the $3d$ absorbing Cr and the $3d$ neighboring Cr are plotted in the middle and bottom panels at (a) $U$=0 and (b) 3 eV. As can be seen from the local $d$ DOS on the absorbing Cr at $U$=0 (Figure 2 (a), middle panel), peak $\alpha$ is the result of two E2 transitions to minority spin $t_{2g}$ and majority spin $e_g$ orbitals while peak $\beta$ is due to E2 transitions to minority spin $e_g$ states. Also shown in Figure 2 (a), peak $\gamma$ is due to E1 transitions to the $p$ states of the absorbing Cr, orbitally mixed with the $3d$ orbitals of the neighboring Cr. These observations are also presented briefly in Table 1.

The lower panels of Figure 2 (b) illustrate the impact of the Hubbard $U$ on the electronic DOS. When the $U$ term is added, the minority spin $t_{2g}$ and $e_g$ orbitals are found closer to each other. Furthermore, the majority spin states are separated from the minority spin ones leading to a decrease in intensity of peak $\alpha$, in disagreement with experiment. Similarly as in the previous case, the observed peaks in the pre-edge region at $U$=3 eV, shown in the top panel of Figure 2 (b), are assigned to specific transitions and listed in Table 1. Comparing the transitions at $U$=0 eV to the transitions at $U$=3 eV, it can be observed that the Cr $3d$ $t_{2g}^{min}$ contribution is moved from peak $\alpha$ to peak $\beta$ as the Hubbard parameter $U$ is changed from 0 to 3 eV.

In summary, it appears that DFT+$U$ does not overcome the two main issues arising in the standard DFT calculation of the Cr $K$-edge spectrum of $ZnCr_2O_4$. First, the E1 peaks in the pre-edge are too close to the main edge peak, that is, the $3d$ states of Cr are too close to the $p$ ones. Second, the E2 peaks are also found at too high energies, as the core-hole interaction is not well modeled by our calculations. As the distance between neighboring core-holes is 8.33 Å,

**Table 1.** Tentative assignment of $\alpha$, $\beta$ and $\gamma$ peaks in terms of local (E2) and non-local (E1) transitions.

| peak | Assign. for $U$=0 eV | Assign. for $U$=3 eV |
|---|---|---|
| $\alpha$ | E2: $1s \rightarrow 3d(e_g^{maj} + t_{2g}^{min})$ of abs. Cr | E2: $1s \rightarrow 3d\ e_g^{min}$ of abs. Cr |
| $\beta$ | E2: $1s \rightarrow 3d\ e_g^{maj}$ of abs. Cr | E2: $1s \rightarrow 3d(t_{2g}^{min} + e_g^{min})$ of abs. Cr |
| $\gamma$ | E1: $1s \rightarrow p$ of abs. Cr orbitally mixed with $3d$ of neigh. Cr | E1: $1s \rightarrow p$ of abs. Cr orbitally mixed with $3d$ of neigh. Cr |

no improvement is expected by increasing the supercell size. Vibrational effects could affect the peak intensities rather than their splitting and are not responsible for the observed failure of the current calculation. Beyond DFT approaches could improve the modeling of the core-hole-electron interaction but the Bethe-Salpeter equation remains very challenging for such open-shell system, while the efficiency of TD-DFT relies on the choice of the used kernel. Besides, ligand field multiplet calculations could better reproduce the local $\alpha$ and $\beta$ peaks but is unable to account from the non-local $\gamma$ peak.

## 4. Conclusion

The evolution of the Cr $K$ pre-edge features along the $ZnAl_2O_4$-$ZnCr_2O_4$ was first reported by Verger *et al.* [5] and interpreted using XANES calculations performed in standard DFT. In the present study, in an attempt to improve the agreement with the experimental spectrum for $ZnCr_2O_4$, DFT+$U$ calculations were carried out with different values chosen for $U$ on the Cr $3d$ states. Our findings show that the pre-edge features, and in particular their relative splitting, are however not better reproduced when taking into account the Hubbard parameter. Approaches including an improved modeling of the electron-hole interaction are expected to provide a better overall agreement with experiment, although correlated systems such as $ZnCr_2O_4$ represent a significant computational challenge. From the point of view of experiment, high energy resolution fluorescence detection could be used to better resolve the pre-edge features.

## Acknowledgement

This work was partly granted access to the HPC resources of IDRIS under the allocation 2015-100172 made by GENCI (Grand Équipement National de Calcul Intensif). E.R. gratefully acknowledges the support of the Postdoctoral Fellowship of the Hungarian Academy of Sciences.

## References

[1] Martos M, Martínez M, Cordoncillo E and Escribano P 2007 *J. Eur. Ceram. Soc.* **27** 4561-67
[2] Prim S, García A, Galindo R, Cerro S, Llusar M, Folgueras M and Monrós G 2013 *Ceram. Int.* **39** 6981-89
[3] Halenius U, Andreozzi G B and Skogby H 2010 *Am. Mineral.* **95** 456-62
[4] Bosi F, Andreozzi G B, Halenius U and Skogby H 2011 *Am. Mineral.* **96** 594-8
[5] Verger L *et al.* 2015 *Phys. Chem. Miner.*, DOI 10.1007/s00269-015-0771-8
[6] Anisimov V I, Aryasetiawan F and Lichtenstein A I 1997 *J. Phys. Condens. Matter.* **9** 767
[7] Panchmatia P M, Sanyal B and Oppeneer P M 2008 *Chem. Phys.* **343** 47-60
[8] Rivero P, Loschen C, Moreira I D P R and Illas F 2009 *J. Comput. Chem.* **30** 2316-26
[9] Giannozzi P *et al.* 2009 *J. Phys. Condens. Matter* **21** 395502
[10] Gougoussis C, Calandra M, Seitsonen A and Mauri F 2009 *Phys. Rev. B* **80** 075102
[11] Taillefumier M, Cabaret D, Flank A M and Mauri F 2002 *Phys. Rev. B* **66** 195107
[12] Perdew J P, Burke K and Ernzerhof M 1996 *Phys. Rev. Lett.* **77** 3865-8
[13] Juhin A, Calas G, Cabaret D, Galoisy L and Hazemann J L 2007 *Phys. Rev. B* **76** 054105
[14] Chan K, Sau J, Zhang P and Cohen M 2007 *Phys. Rev. B* **75** 054304
[15] Singh N and Rhee J Y 2010 *J. Korean Phys. Soc.* **57** 1233-7
[16] Bordage A, Rossano S, Horn A H and Fuchs Y 2012 *J. Phys. Condens. Matter* **22** 225401

Journal of Physics: Conference Series **712** (2016) 012012 doi:10.1088/1742-6596/712/1/012012

# PrestoPronto: a code devoted to handling large data sets

**S J A Figueroa**[1], **C Prestipino**[2]

[1] Centro Nacional de Pesquisa em Energia e Materiais (CNPEM), Laboratório Nacional de Luz Síncrotron (LNLS), Rua Giuseppe Máximo Scolfaro, 10.000, Polo II de Alta Tecnologia, Caixa Postal 6192, 13083- 970 Campinas, SP, Brasil.
[2] Institut des Sciences Chimiques de Rennes (ISCR), UMR CNRS N°6226, 263 Avenue du Général Leclerc, 35042 RENNES, France.

E-mail: carmelo.prestipino@univ-rennes1.fr

**Abstract**. The software PrestoPronto consist to a full graphical user interface (GUI) program aimed to execute the analysis of large X-ray Absorption Spectroscopy data sets. Written in Python is free and open source. The code is able to read large datasets, apply calibration, alignment corrections and perform classical data analysis, from the extraction of the signal to EXAFS fit. The package includes also programs with GUIs] to perform, Principal Component Analysis and Linear Combination Fits. The main benefit of this program is allow to follow quickly the evolution of time resolved experiments coming from Quick-EXAFS (QEXAFS) and dispersive EXAFS beamlines.

## 1. Introduction

Starting on early eighties with the first developments on fast X-ray Absorption Spectroscopy as dispersive EXAFS [1,2] and QEXAFS [3,4] beamlines , the use of time resolved experiments become widespread. From the early technical developments to the later ones [5] the amount of data that experimentalists must treat easily go from some tens of spectra to several hundred till some thousand. Here we present a new program developed in Python, a full graphic tool for the analysis of large datasets measured by dispersive or QEXAFS beamlines. In order to analyze large datasets generated in time or space resolved experiments several codes have been already wrote [6–8] and analysis macro languages as Ifeffit [9] and Larch were developed capable to satisfies all analysis strategy. Nevertheless the learning curve for analysis macro languages is pretty steep for users without coding experience and writing script is relatively time consuming. The main aims of PrestoPronto package is to cover the need to rapidly screen data during experiments and perform basic data analysis, dealing with an important set of spectra in a quick, intuitive and easy way, using the algorithm that are the most close to the user experience. The code has been used already in few study [10–12] it is free and open source so could be adapted to specific requirements as different data formats, peculiar setup or different type of analysis directly by users or in a collaborative way.

## 2. Structure of the code

The code is written in Python 2.7 using the standard Tk/Tkinter graphical toolkit for user interfaces, and mathplotlib [13] package for plotting data. In the software the sequence spectra is stored in a python list where each element is a class containing a spectrum with bound methods and properties. The EXAFS specific functions are performed using the Ifeffit engine [9].

In order to improve the interaction with the user and the data processing the software charge into the memory the entire sequence of spectra and define new matrix at each step. Consequently the software is limited to handling set of data that are smaller than the available memory. To our knowledge the program has been tested up to 1300x1050 data matrix, however counting the memory needed to process and store intermediate data, a modern computer with one Giga of available memory should be able to treat a set of 25000 spectra of 4000 data points or an equivalent size matrix.

The package is composed by three programs: *PrestoPronto*, *Linearcomb_GUI* and *PCA_GUI*.

### 2.1. *PrestoPronto*

The main goal of PrestoPronto (soon ready, from Italian) is to allow a fast analysis of a sequence spectra that come as a result of QEXAFS and Dispersive XAS experiments, principally the ones that involve *in-situ* and *operando* studies. The program is designed for rapidly calibrate, average, extract and model data in order to follow with a minimum time delay the progression of an experiment. The software shows a single notebook window (figure 1) and a terminal were information for user are printed. The notebook tabs must be used from left to right, where each tab correspond to a step of the analysis that could be plotted or saved in a multicolumn ASCII file.

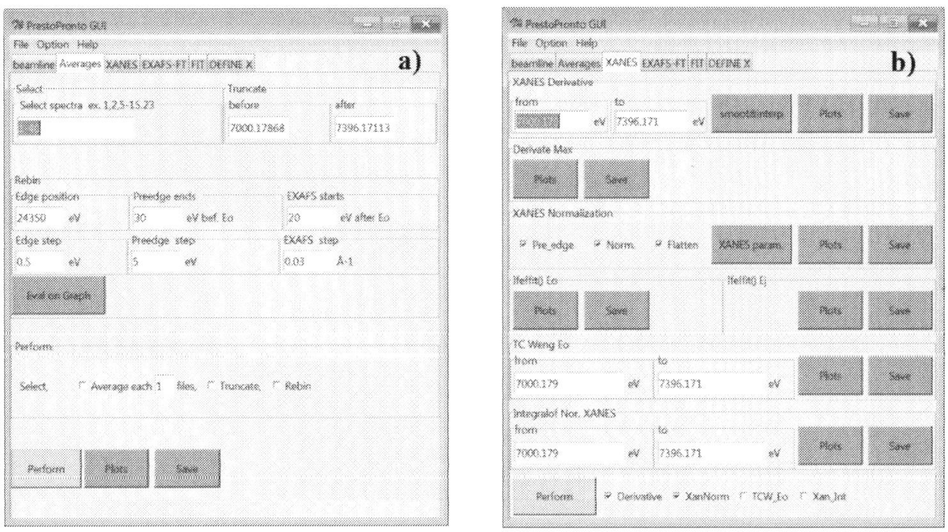


**Figure 1.** Main window of Prestopronto Gui, Averages and XANES pages part a) and b) respectively

The first tab "beamlines" is used to import data from different formats and corrects energy shift using a calibration sample (a reference spectra collected simultaneously with the data). The tab "Averages" could rebinning and truncate the spectra, as well could select a subset of data and cyclically average (every N spectra). The tab XANES perform a XANES normalization and plot the edge position, the edge jump and the integral of a selected region of the spectra. Tab EXAFS-FT is devoted to EXAFS signal extraction and Fourier transformation. Finally the tab Fit is devoted to EXAFS analysis using theoretical path calculated by FEFF code [14], the spectra are fitted individually (nowadays the model is s limited to only two shell). In order to simplify the choice of parameters and visualization of the full set of data during the normalization in the plot interface a slider has been added helping to stack plot or to select the relevant parameters (figure 2).

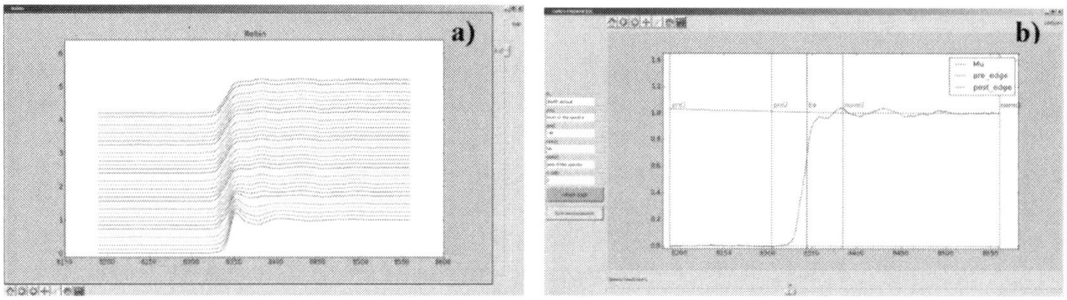


**Figure 2.** Plot windows for stacked plot and selection of parameters part a) and b) respectively.

### *2.2. PCA_GUI*

As additional resources to the standard data treatment the Principal Component Analysis (PCA) algorithm is added with a dedicated GUI. This module is showed by a notebook with three tab: Setup, PCA and ITTFA. The "Setup" tab is used to input a multicolumn ASCII file and select the interested subset and the spectral range to analyse. The "PCA" tab visualizes the results of PCA analysis [15] showing several parameters to determine the number of factors: IND and MAD functions; residual variance and Reduced eigenvalue F-test (REV) (see figure 3).

Finally, the "ITTFA" tab is based on the approach used in the program Factor (developed by Marcos Fernandez Garcia) where the Iterative Target Transformation Factor Analysis is performed. This procedure allows to align theoretical components obtained from PCA analysis to physical meaningfully spectra [16,17].

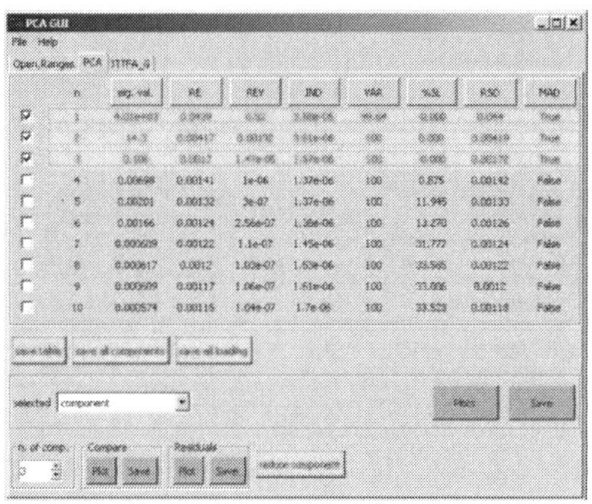

**Figure 3**. PCA_GUI, PCA windows showing parameters calculated to decide the number of components

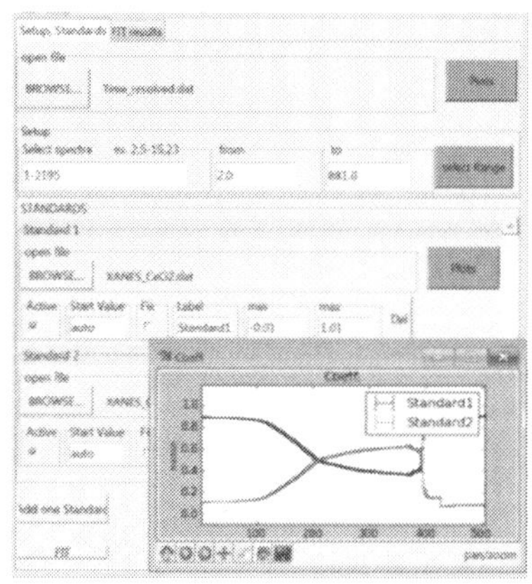

**Figure 4**. Linearcomb_GUI, setup windows and plot showing refined coefficients

### *2.3. Linearcomb_GUI*

Finally to complete this package we include the Linearcomb_GUI figure 4. This part is a front end for the python library LMFIT [18] which allows to fit a set of spectra in a multicolumn file by linear combination of a series of standards. The code use the Levenberg-Marquardt algorithm to preform minimization and standard error calculate at 1σ interval of confidence.

## 3. Resource

A project page for PrestoPronto is accessible at http://soonready.github.io/PrestoPronto/. The self-installer and source are available. The screen shots in this article were made on a Windows 7 computer. The programs may be different in appearance on other systems.

Journal of Physics: Conference Series **712** (2016) 012012 doi:10.1088/1742-6596/712/1/012012

**Acknowledgments**
The authors acknowledge the European Synchrotron Radiation Facility for beam time provision at ID24, BM23 (ex-BM29) beamlines and the support of Sakura Pascarelli and Mark A. Newton in helping to develop this program. We also thanks Matt Newville and Marcos Fernández-García for gently open their codes.

**References**
[1] Kaminaga U, Matsushita T and Kohra K 1981 *Jpn. J. Appl. Phys.* **20** 355
[2] Matsushita T and Phizackerley R P 1981 *Jpn. J. Appl. Phys.* **20** 2223
[3] Frahm R 1988 *Nucl. Instruments Meth. A* **270** 578
[4] Stötzel J, Lützenkirchen-Hecht D and Frahm R 2010 *Rev. Sci. Instrum.* **81** 073109
[5] Rodríguez J A, Hanson J C and Chupas P J 2013 *In-situ characterization of heterogeneous catalysts* (New York: Wiley)
[6] Del Rio M S and Dejus R J *Proc. SPIE 8141, Advances in Computational Methods for X-Ray Optics II, 814115*
[7] Ressler T 1998 *J. Synchrotron Radiat.* **5** 118
[8] San-Miguel A 1995 *Phys. B: Condens. Matter* **208** 277
[9] Newville M 2001 *J. Synchrotron Radiat.* **8** 322
[10] Moog I, Prestipino C, Figueroa S, Majimel J and Demourgues A 2014 *J. Phys. Chem. C* **118** 22746
[11] Achilli E, Minguzzi A, Lugaresi O, Locatelli C, Rondinini S, Spinolo G and Ghigna P 2014 *J. Spectrosc.* **2014** 480102
[12] Nilsson J, Carlsson P-A, Fouladvand S, Martin N M, Gustafson J, Newton M A, Lundgren E, Grönbeck H and Skoglundh M 2015 *ACS Catal.* **5** 2481
[13] Hunter J D 2007 *Comput. Sci. Eng.* **9** 90
[14] Rehr J J and Albers R C 1990 *Phys. Rev. B* **41** 8139
[15] Malinowski E R 2002 *Factor analysis in chemistry* (New York: Wiley)
[16] Fernandez-Garcia M, Marquez Alvarez C and Haller G L 1995 *J. Phys. Chem.* **99** 12565
[17] Rossberg A, Reich T and Bernhard G 2003 *Anal. Bioanal. Chem.* **376** 631
[18] Newville M, Stensitzki T, Allen D B, Rawlik M, Ingargiola A and Nelson A R J 2014 LMFIT *Zenodo.* 10.5281/zenodo.11813

# Local structure of perovskites $ReO_3$ and $ScF_3$ with negative thermal expansion: interpretation beyond the quasiharmonic approximation

**Juris Purans[1], Sergei Piskunov[1], Dmitry Bocharov[1], Aleksandr Kalinko[1], Alexei Kuzmin[1], Shehab E. Ali[2,3], Francesco Rocca[2]**

[1]Institute of Solid State Physics, University of Latvia, Kengaraga Str. 8, LV-1063, Riga, Latvia
[2]IFN-CNR, Institute for Photonics and Nanotechnologies, Unit FBK-Photonics of Trento, Povo, Trento, Italy
[3]Physics Department, Faculty of Science Suez Canal University, Ismailia, Egypt

E-mail: purans@cfi.lu.lv

**Abstract.** We propose an approach beyond the quasiharmonic approximation for interpretation of EXAFS and XRD data and for *ab initio* calculations of electronic and vibration properties of materials with negative thermal expansion. *Ab initio* electronic structure and lattice dynamics calculations for cubic and distorted $ScF_3$ were performed using the linear combination of atomic orbitals (LCAO) method. The band gap obtained in calculations for $ScF_3$ is equal to 10.54 eV and agree well with the expected value. The calculated infrared spectra of F displaced (FD) cubic $ScF_3$ allow us to predict that its mean Sc–F–Sc angle within NTE deviates from 180 degree.

## 1. Introduction

The interest towards $ReO_3$-type perovskites has been recently renewed by the discovery of large negative thermal expansion (NTE) in cubic $ScF_3$ structure in a wide temperature range [1], to be compared with a much less intense effect measured for the cubic $ReO_3$ [2, 3, 4]. At present, the great difference between the isostructural compounds is not yet explained, but it is found that the materials represent a case of strong anharmonic effects [4, 5]. Anharmonicity is of general importance in condensed matter in relation to thermal expansion, structural phase transitions, soft modes in ferroelectrics, melting and related phenomena. Usually anharmonicity in crystals is weak enough and thus so called implicit anharmonicity can be described in the framework of quasiharmonic approximation (QHA) as a variation of the phonon frequency due to the change in volume with temperature or pressure. However, this might be not the case for strongly explicit anharmonic systems, like perovskites with NTE. The explicit anharmonic effect links the phonon frequencies to the amplitude of the atomic vibrations.

Synchrotron radiation EXAFS studies of local structure with femtometer accuracy [6] offer the possibility to study implicit and explicit anharmonic effects for perovskites with NTE, since the temperature dependent EXAFS spectra include both anharmonic effects. The relative contributions of these two effects can be estimated by the extent to which quasiharmonic calculations of amplitude of the atomic vibrations reproduce the experimental EXAFS data. In spite of the experimental and theoretical efforts that have been performed to understand the

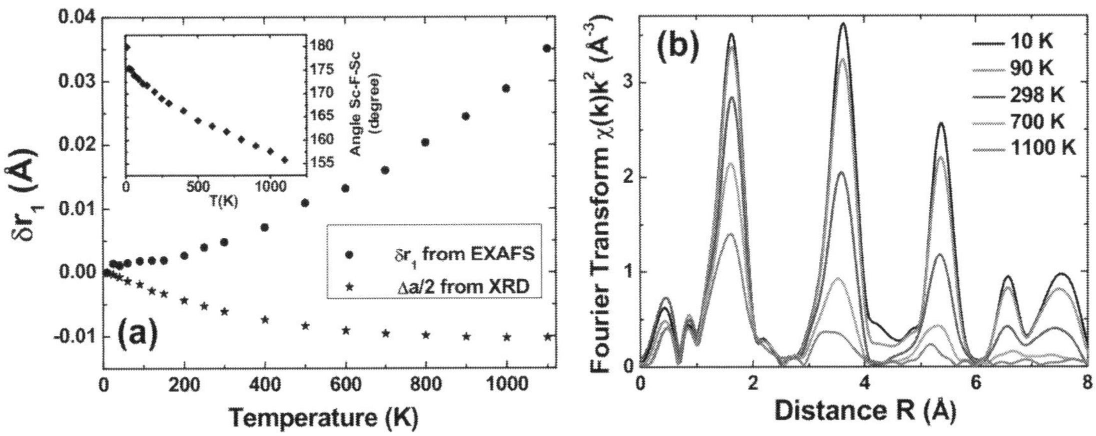


**Figure 1.** (Color online) (a) The variation of the Sc–F bond length measured by EXAFS and crystallographic distance measured by XRD (in house work). The angle Sc–F–Sc is calculated from EXAFS (Sc–F) and XRD (Sc–Sc) data. (b) Temperature dependence of the experimental Sc K-edge $\chi(k)k^2$ Fourier transforms (FT) for polycrystalline $ScF_3$.

NTE in $ReO_3$ and $ScF_3$, its origin is still under debate. Most often it is explained in terms of so-called rigid unit modes (RUMs) model, which involves coupled vibrations of rigid $ReO_6$ ($ScF_6$) octahedra. When neighboring octahedra librate in opposite directions the distance between the centers of octahedra decreases leading to the lattice contraction. Obviously, the distance Sc–F must decrease in the QHA according to the coefficient of NTE. On the other hand, our EXAFS data (Fig. 1) demonstrate the positive thermal expansion of the distance in the first coordination shell. Thus, to elucidate the nature of the NTE phenomena in details, one needs to follow temperature dependent changes of lattice dynamics and structure at the atomic scale beyond the QHA.

In this study, we propose an approach beyond the QHA for the interpretation and *ab initio* calculation of electronic and vibration properties of materials with NTE on the basis of local structure reconstructed from EXAFS analysis. *Ab initio* electronic structure and lattice dynamics calculations for cubic (space group $Pm\bar{3}m$) and fluorine displaced (FD) cubic (space group $Im\bar{3}$) $ScF_3$ were performed using the linear combination of atomic orbitals (LCAO) method.

## 2. Computational details

Both perfect and FD cubic $ScF_3$ crystals were simulated by means of LCAO within the framework of the hybrid HF/DFT approach. In order to perform hybrid LCAO calculations, we used the periodic crystal code [7], which employs Gaussian-type functions (GTFs) centered on atomic nuclei as the basis sets (BSs). In the current study, the triple-zeta all-valence BSs with polarization functions were used for both fluorine and scandium atoms as suggested in Ref. [8]. The revised version of the PBE0 hybrid exchange-correlation DFT functional with 33.3333% of Hartree-Fock exchange [9] was used in our calculations. Chosen exchange-correlation functional yields $ScF_3$ lattice constant $a_0 = 4.0257$ Å, which is in excellent agreement with the experimental value of 4.026 Å [1]. Until now the band gap of $ScF_3$ was not precisely measured experimentally, but it is estimated to be larger than 7–8 eV [10]. Our calculations for the equilibrium structure of perfect cubic $ScF_3$ ($a_0 = 4.0257$ Å) suggest that $ScF_3$ is an insulator with relatively large band gap $E_g = 10.54$ eV that correlates very well with expected experimental value.

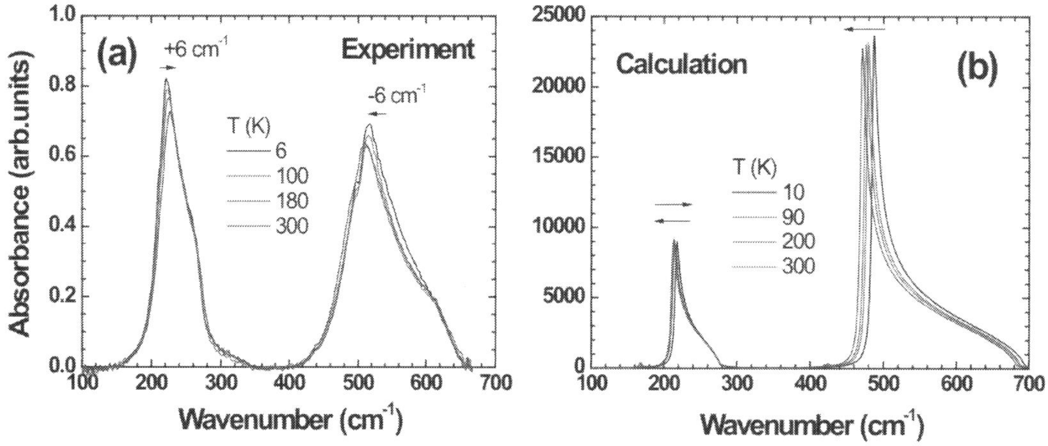


**Figure 2.** (Color online) (a) Temperature dependence of the experimental infrared spectra of $ScF_3$. Two bands due to the F–Sc–F bending mode (at 220 $cm^{-1}$) and Sc–F stretching (at 520 $cm^{-1}$) are visible. A shift of the band maxima by about 6 $cm^{-1}$ in the opposite directions occurs upon increasing temperature. (b) Calculated infrared absorption spectra of F-displaced cubic $ScF_3$ for several values of contracting lattice parameter $a_0$ and fluorine atoms displacements. The tilt angle Sc–F–Sc changes from 180° at 10 K to 158° at 300 K.

## 3. Results and discussions

The perovskite-type $ScF_3$ structure consists of corner-linked $ScF_6$ regular octahedra with Sc atoms at the centers and linear Sc–F–Sc chains. The space between eight octahedra forming the cube is fully vacant allowing for a significant rotation of the $ScF_6$ octahedra and, thus, may result in the anisotropy of fluorine atom thermal vibration. Our recent EXAFS measurements confirm the expansion of the first shell Sc–F (Fig. 1) and that the amplitude of fluorine thermal vibrations is larger in the direction perpendicular to the Sc–F–Sc chain, while, according to XRD, the vibrations of scandium are almost isotropic. In fact, this may lead to the mean displacement of fluorine atom forming Sc–F–Sc bond with the mean angle Sc–F–Sc reported in the insert in Fig. 1. To simulate the $ScF_3$ behavior at elevated temperatures with expanded Sc–F bond length, as measured by EXAFS, and contracted lattice constant (mean length of the linear Sc–F–Sc chain, as measured by x-ray diffraction (XRD), we elaborated a model allowing for fluorine atom displacement in the direction perpendicular to the Sc–F–Sc chain, but keeping its original cubic structure (FD model). For this purpose larger supercell consisting of 32 atoms was adopted. The FD model was also used to describe temperature dependence of the far-infrared (far-IR) absorption spectra of $ScF_3$, obtained by Fourier transform infrared (FTIR) spectroscopy.

Fig. 2a shows the experimental temperature dependence of the far-IR absorption in $ScF_3$. Two absorption bands at 220 and 520 $cm^{-1}$ correspond to the F–Sc–F bending and Sc–F stretching modes, respectively. As temperature increases from 6 K to 300 K, the two bands shift toward each others of $\sim$6 $cm^{-1}$. Note that infrared spectra calculated for perfect cubic $ScF_3$ show opposite directions of band shifted with respect to contracted lattice constant. At the same time, the FD cubic $ScF_3$ model (Fig. 2b) yield good agreement with measured data: the band near 500 $cm^{-1}$ shifts to the left as $a_0$ contracts, while the band near 220 $cm^{-1}$ starts shifting to the right when the Sc–F–Sc angle is smaller than $\sim$172°.

## 4. Conclusions

In this study three experimental techniques (EXAFS, XRD and FTIR) were used to probe the structure of polycrystalline $ScF_3$ in the wide temperature range. We found that the temperature dependence of the Sc–F distance, reconstructed from the analysis of the first coordination shell of scandium, shows opposite behaviour to that of the lattice parameter due to negative thermal expansion. Based on this structural information, we propose an approach beyond the quasiharmonic approximation for interpretation of the electronic and vibration properties of cubic $ScF_3$. The electronic structure and infrared spectra of locally distorted $ScF_3$ were studied within the framework of the first-principles hybrid density functional theory calculations. The calculated band gap of 10.5 eV agrees with that experimentally observed. It has been found that the position of scandium atoms in the middle of regular $ScF_6$ octahedra is relatively stable, while the mean Sc–F–Sc angle deviates from 180° due to the anisotropy of thermal vibrations of fluorine atoms. Good agreement between calculated and experimentally obtained far-IR spectra was found, that additionally supports the accuracy of our predictions.

## Acknowledgments

This work was supported by Latvian Government Research grant no. 402/2012. We gratefully acknowledge the assistance of Giuliana Aquilanti and of ELETTRA beamline staff members during the EXAFS experiment (proposal no. 20120215)

## References

[1] Greve B K, Martin K L, Lee P L, Chupas P J, Chapman K W and Wilkinson A P 2010 *J. Am. Chem. Soc.* **132** 15496
[2] Purans J, Fornasini P, Ali S E, Dalba G, Kuzmin A and Rocca F 2015 *Phys. Rev. B* **92** 014302
[3] Timoshenko J, Kuzmin A and Purans J 2014 *Journal of Physics: Condens. Matter* **26** 055401
[4] Chatterji T, Henry P, Mittal R and Chaplot S 2008 *Phys. Rev. B* **78** 134105
[5] Li C, Tang X, Muñoz J, Keith J, Tracy S, Abernathy D and Fultz B 2011 *Phys. Rev. Lett.* **107** 195504
[6] Purans J, Afify N D, Dalba G, Grisenti R, De Panfilis S, Kuzmin A, Ozhogin V I, Rocca F, Sanson A, Tiutiunnikov S I and Fornasini P 2008 *Phys. Rev. Lett.* **100** 055901
[7] Dovesi R, Saunders V R, Roetti C, Orlando R, Zicovich-Wilson C M, Pascale F, Civalleri B, Doll K, Harrison N M, Bush I J, DArco P, Llunell M, Causá M and Noël Y 2014 *CRYSTAL14 User's Manual* University of Torino Torino
[8] Peintinger M F, Oliveira D V and Bredow T 2013 *J. Comp. Chem.* **34** 451–459
[9] Guido C A, Brémond E, Adamo C and Cortona P 2013 *J. Chem. Phys.* **138** 021104
[10] Umeda M, Tezuka Y, Shin S and Yagishita A 1996 *Phys. Rev. B* **53** 1783–1789

# Improvement of the efficient referencing and sample positioning system for micro focused synchrotron X-ray techniques

**T Spangenberg[1], J Göttlicher and R Steininger**

Karlsruhe Institute of Technology, ANKA Synchrotron Radiation Facility, Hermann-von-Helmholtz-Platz 1, 76344 Eggenstein-Leopoldshafen, Germany

E-mail: thomas.spangenberg@kit.edu

**Abstract**. An efficient referencing and sample positioning system is a basic tool for a micro focus beamline at a synchrotron. The seven years ago introduced command line based system was upgraded at SUL-X beamline at ANKA [1]. A new combination of current server client techniques offers direct control and facilitates unexperienced users the handling of this frequently used tool.

## 1. Motivation

Experiments performed using a micro focus beamline at synchrotrons offer a set of methods to investigate small sized regions on samples. The beam sizes are normally much less than a square millimeter. But mounting conditions, multiple samples on a single mounting device or limited options for the sample preparation lead to much larger sample areas on which the region of interest (ROI) for micro focus experiments must be located. Typically the ROI represents a region much smaller in size (2-3 orders on magnitude) compared to the entire sample.

The localization on large samples becomes crucial for saving preparation- and experiment time. It becomes essential if the same sample is investigated with different experimental setups or methods and each method should be applied on the same ROI.

Furthermore the experimental setup of sample, detectors, supporting equipment, remote access, and synchrotron radiation limits the applicable options for aligning the sample.

## 2. Current solution

At ANKA's SUL-X beamline a sample positioning system has been implemented that turned out as an essential tool for setting up the experiments. The transportable sample mount, shown in figure 1, was designed with an extension of about 90mm x 60mm and offers approximately 6 orders larger sample area than a typical beam size (~100 to 30μm) of a micro focus experiment at this experimental station. An offline localization of ROIs with a wide zoom optical microscope (Zeiss Axio Imager.Z1m) is used as a first orientation on the sample. Four reference marks on the outermost edges of a sample mount are always included into the set of coordinates.

[1] To whom any correspondence should be addressed.

The prepared sample mount is inserted to the experimental stage and the reference marks are localized. Based on these data the ROIs are transformed into the experiment coordinate system. This was done solely by a command line (CL) interface and a standalone image visualization. The solution is stable and saves experiment time. Crucial to that CL based method were the training curve of unexperienced users and the finding of the right focus position.

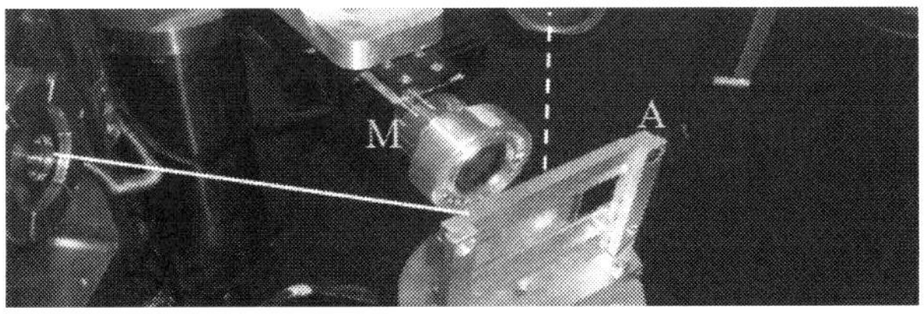


**Figure 1.** The real experimental situation at SULX with standard sample holder. The large area sample holder (A) indicates the possible area of interest. The focused beam might be $10^6$ times smaller. The microscope (M) is used to figure out the area of investigation (ROI).

### 2.1. *User and focus problematic*
The experimental setup forces the user to deal with a Windows based proprietary software for the offline characterization and a CL interface for aligning the sample at the experimental station. Despite the reduced and partially automated options of the SPEC [2] CL interface it still introduces a high threshold for unexperienced users.

An optical microscope with cross hair is used for the offline sample preparation as well as for the experiment position control (custom product, IEEE1394 camera with 1392x1032 pixel). In two dimensions a position is marked reliably in a range of a micron (see figure 2). The third dimension is indicated by the focus position. The focus indicators are the observed image and the evaluation of the image sharpness influenced by not quantifiable factors such as illumination or region selection. An estimation is around ±10µm correctness for that dimension for a flat sample. If the sample roughness is increased the uncertainty for that dimension increases, especially for unexperienced users. In figure 2 the problematic is illustrated with a test sample.

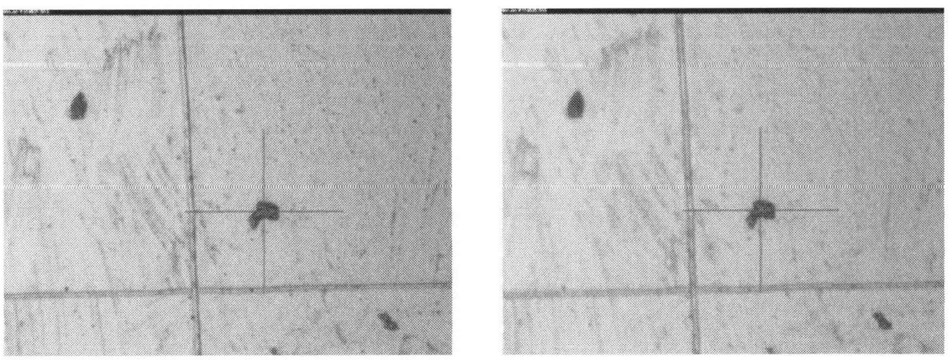

**Figure 2.** Two images of the same sample region but with a shift of $\Delta y$=30µm in the sample distance towards the microscope. The image sharpness is decreased in the right image. Pixel size of the image is about 0.5µm, allowing positioning in the related plain of about two µm. The crosshair indicates 200µm.

The normally used 45° geometry between beam and sample surface leads to a mismatch of the optically selected region (crosshair) and the by X-rays investigated region. Figure 3 shows this in a sketch. The focus distance mismatch is in this geometry equal to the location mismatch of the X-ray exposed sample region. In case of a micro focus experiment and the assumed uncertainty the two regions might not overlap at all.

## 3. Improvements

Mainly an automatic focus search algorithm and the user interface were revised. The focus search makes use of the fact that the image memory size for JPEG or PNG depends on the number of details in it. A maximum size might be expected for the sharpest state. As shown in figure 4 the idea is correct for flat samples. A rough structured surface broads and distorts the peak while flat and weak textured surface let the peak disappear. The manual adjustment remains important in these cases.

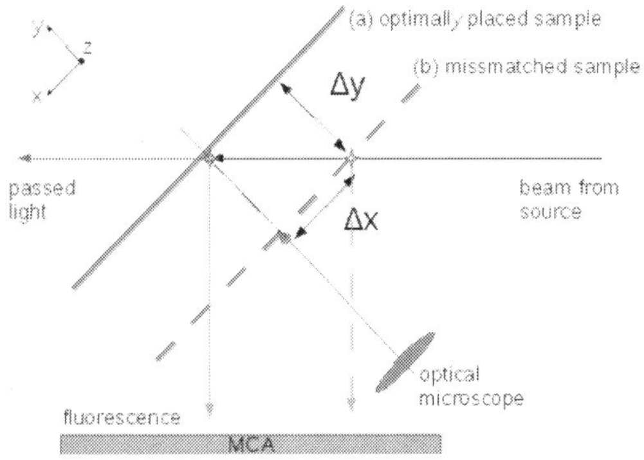


**Figure 3.** A typical fluorescence setup at SUL-X microfocus beamline at ANKA. The optically selected sample position is marked with (a). The sample position (b) is typical for a Δy focus mismatch. The point of investigation (✧) differs from the optically selected position by Δx.

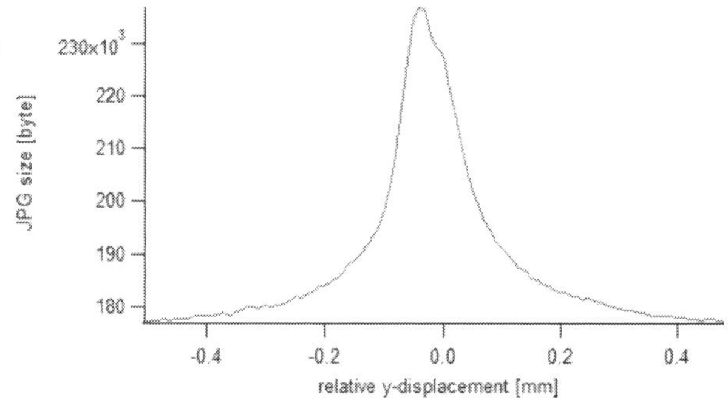


**Figure 4.** The image sharpness correlates to the storage size of a JPEG (norm ISO/IEC 10918-1). The shown storage size was taken from images each taken in 10 µm steps +/- 0.5mm around an approximately focused sample. The distorted peak shape structure indicates a rough sample.

The second improvement was focused on the user interface (figure 5). WEB techniques based applications are platform independent for the user and appropriate for remote support. Therefore the central point of the development was to adopt the position control into a WEB service with respect of the introduced Tango [3] environment.

The interface offers now a direct distance control for all three axis and additionally a direct position selection for points of interest which are already visible in the microscope image. Any position can be memorized and recalled by its given name. This can be done either by scripting in SPEC or by selecting it in the interface.

Camera shutter time, light control for the microscope and a snapshot option complete the improved user interface based on a combination of a few Tango servers.

Tango is a network based environment which mainly focuses onto the connection of different devices and organizing a fluent data exchange. The visualization of data is normally realized by external client applications. The developed approach overcomes the strict division and implements an interface to WEB servers as an additional cooperation option into the Tango server itself. This method offers an essential change in the human interaction with Tango servers.

### *3.1. The HTTP interface*

A WEB server provided common gateway interface (CGI) is used to link the HTTP based WEB interface temporarily to a Tango server. The protocol overhead is limited to 3 additional Tango commands without an impact to other Tango provided functions.

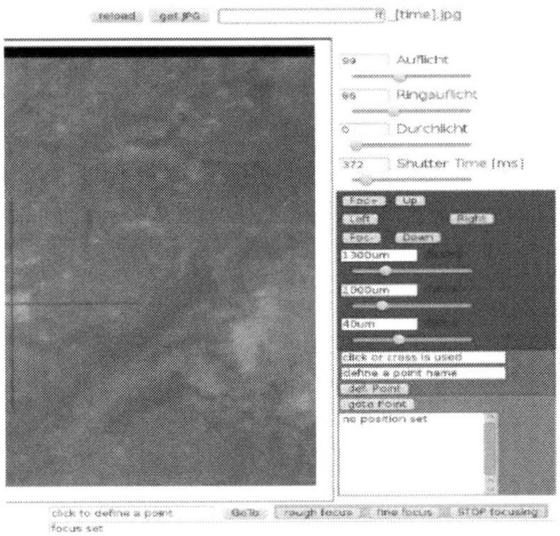


**Figure 5.** A section of the user interface for the microscope positioning and operating assistance system. The beam position is marked by the cross. The vertical and horizontal size of the cross is 200µm.

The developed solution maps incoming HTTP requests by a combined Perl and C++ CGI software into a sequence of Tango commands. The Tango server provided response is transferred back to the HTTP server. The developed CGI software retrieves the information about the chosen Tango server from the URL parameters and from a currently config-file maintained address database. The service might be extended for an unlimited number of cooperating Tango servers.

Though the processing is not limited to a fixed programed response, e.g. a camera image, the Tango server can also be used to process scripts.

The presented solution is constructed as a demonstrator for further implementations. Anything was programed up to the limit where a real script interpreter would be involved into to the process. The assumed macro functions are simulated by fixed programed C++ code.

Technically the approach is similar a server side include (SSI) scripting known for HTML processing of HTTP servers.

## 4. Summary

A semi automatically software assisted focus determination was implemented. The portable scheme requires mostly an easy and fast image capturing without an autofocusing camera and fixed capture time and a reliable remote control of the focus distance.

Surfaces with e.g. weak contrast, or granular surfaces result in uncertainty for the right focal distance. A manual intervention is strictly advised in these cases.

The presented software relies on HTTP/Tango and a custom camera driver and auxiliary Tango servers (light, motion system). But the presented focus scheme is highly portable and does not depend on a special environment.

Due to the WEB technology an assistance from the beamline scientists can be realized by remote control.

## References

[1] Spangenberg T, Göttlicher J and Steininger R 2009 An Efficient Referencing And Sample Positioning System To Investigate Heterogeneous Substances With Combined Microfocused Synchrotron X-ray Techniques *AIP Conference Proceedings* **1092** 93.
[2] SPEC, http://www.certif.com/. [15 July 2015].
[3] TANGO, http://www.tango-controls.org/. [15 July 2015].

Journal of Physics: Conference Series **712** (2016) 012015 doi:10.1088/1742-6596/712/1/012015

# A Modern Laboratory XAFS Cookbook

**GT Seidler(*), DR Mortensen, AS Ditter, NA Ball, AJ Remesnik**

Department of Physics, University of Washington
Seattle, Washington, United States 98195-1560

(*)E-mail: seidler@uw.edu

**Abstract**. We have recently demonstrated a very favorable, inexpensive modernization of lab-based x-ray absorption fine structure (XAFS) and high-resolution x-ray emission spectroscopy (XES) using only commercially-available optics and x-ray tube sources. Here, we survey several proven instrument designs that can be readily implemented in any laboratory setting to achieve synchrotron-quality XAFS and XES for many systems in the 5 keV to 10 keV energy range. These approaches are based on our immediate experience with the development of: (1) an inexpensive, low-powered monochromator capable of performing either XAFS or XES, (2) a mid-scale XAFS user facility having $10^6$/sec flux with sub-eV bandwidth on each of two independent beamlines, and (3) multiple XES spectrometers having outstanding performance that can complement synchrotron XAFS studies in battery and actinide research.

## 1. Introduction

The advent of synchrotron light sources has led to the steady growth and impressive scientific reach of XAFS. While the synchrotron light sources provide a gold-standard of technical capability and user support, the fact remains that the nearly total localization of XAFS to the synchrotron facilities is an anomaly in current scientific practice. By contrast, none of laser spectroscopies, NMR, electrical transport, or x-ray diffraction would be so broadly important across the sciences, engineering, and technology if students and future users could train and use only a few dozen, heavily subscribed facilities worldwide. For most techniques, there is a range of instrumentation spanning the reciprocal continuum of access to cost and capability. Inexpensive, benchtop systems are widely distributed and can be used for routine studies, education, and initial sample characterization, while only the most advanced systems exist as limited-access world-class shared-user facilities or as unique, expensive research instruments.

Building on the above observations [1], we have argued that the access limitations inherent to synchrotron facilities adversely impacting active XAFS research programs and the education of the next generation of XAFS users, but also preemptively exclude a large body of scientific studies or industrial research and development. This includes: work by scientists in developing countries who lack the travel funding needed for frequent synchrotron visits, studies requiring complex *in situ* apparatus that are not easily transported to the light sources, and system-specific considerations that inhibit sample transport to the light source, *e.g.*, extreme air sensitivity or radiological safety considerations. The same concerns have existed from the earliest days of XAFS synchrotron beamline development [2].

For these reasons we have begun a long-term research program [1] aimed at rejuvenating laboratory-based XAFS and also enabling laboratory-based high resolution x-ray emission spectroscopy (XES); note that we discuss XES in another entry to the present proceedings [3]. This paper continues as follows. In section 2, we review some basic issues involved in the operation of Rowland circle monochromators. In section 3, we discuss the design and performance of our low-powered XANES and

Journal of Physics: Conference Series **712** (2016) 012015 doi:10.1088/1742-6596/712/1/012015

XES system that has now been operating for over a year. In section 4, we present an overview of the design of a high-powered two-beamline user facility that is now well into commissioning at the University of Washington. In section 5, we briefly discuss a new single-motor drive system and some future directions. Finally, in section 6 we summarize and conclude.

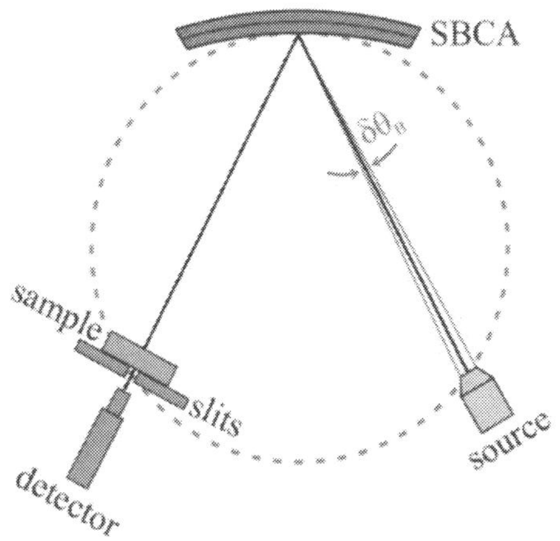


**Figure 1**: Schematic of a Rowland circle monochromator using a spherically bent crystal analyzer (SBCA). Polychromatic bremsstrahlung from a conventional x-ray tube is incident on the SBCA. In the plane of the figure, excluding the small Johann error from the optic, the radiation is refocused to the symmetrically-located exit point on the Rowland circle but is also wavelength-selected via Bragg scattering. Energy scanning uses symmetric motion of the source and detector on the Rowland geometry. For transmission-mode XAFS, as shown in the figure, it is convenient to place the sample in front of the exit aperture (slits) placed on the Rowland circle in front of the detector.

## 2. The Rowland Circle Monochromator

The problem of point-to-point monochromatization of hard x-rays is widespread, and decades of instrument development across numerous platforms support the excellent function and versatility of the Rowland circle configuration, as depicted in figure 1. In such a system the source, detector and optic are all placed as shown on a circle whose diameter must match the radius of curvature of the optic. Cylindrical optics were used in the earlier generation of laboratory XAFS systems and still see some use in high-resolution applications [4], but the overwhelming majority of present-day, high-resolution Rowland circle spectrometers are instead based on spherically-bent crystal analyzers (SBCA).

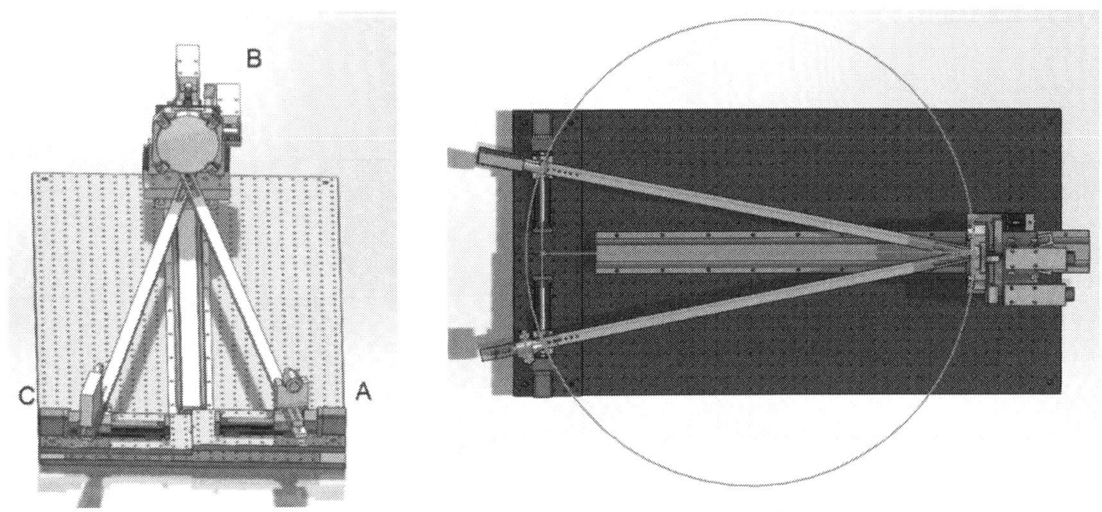


**Figure 2**: *Left*: A perspective rendering of the low-powered lab XANES system. (A) The source positioner with the x-ray tube mounted on a V-block on the source steering bar; (B) SBCA mounted on a 2-axis tilt stage which is in turn mounted on a passive linear slide; (C) The detector positioner with the detector mounted onto a hold on the detector steering bar. For clarity of presentation the helium flight path and radiation enclosure have been suppressed. *Right*: A top-view rendering of the same instrument with the 1-m diameter Rowland circle superimposed.

There are many ways to implement the scanning motion on the Rowland circle in the laboratory frame, with the main consideration being whether some benefit accrues from having one of the major components held stationary in the laboratory frame. For example, the low-powered instrument described in section 3 holds no component fixed in the lab frame while the XANES user-facility and high-powered XES systems in sections 4 and 5 keep the (fairly heavy) source fixed. No matter the choice of drive mechanism, when working not too far from backscatter the energy resolution is still dominated by the source size in the plane of the Rowland circle [1]. Consequently it is the proliferation of available SBCA crystal orientations that helps enable the present effort; SBCA can be fabricated or purchased having Bragg angles of 70-85 degrees for almost any absorption edge in the hard x-ray range.[5] This allows ~1-eV energy resolution in the hard x-ray range when using ~0.5 – 1 mm source spot sizes [1], greatly expanding the useful selection of inexpensive commercial x-ray sources.

## 3. Low-powered Monochromator/Spectrometer for XAFS and XES

Our entry into laboratory-based spectroscopy was somewhat inadvertent. We sought to have in-house capability to test x-ray optics, either in support of mass-scale manufacture of optics for a next-generation multi-analyzer system for inelastic x-ray scattering [6] or of a project on time-resolved XAFS and XES of luminescent lanthanide compounds [7]. A broad survey of commercial x-ray sources found a family of inexpensive low-powered air-cooled x-ray sources that appealed on the basis of convenience and source spot size. To our surprise, an order of magnitude calculation of flux for a 1-m Rowland circle predicted excellent performance for XES and useful capability for transmission-mode x-ray absorption near edge structure (XANES). The decision was quickly made to develop such an instrument and in the subsequent 60 days all components were acquired and the instrument was assembled, computer-interfaced, and safety certified. The construction of a laboratory-based, high-resolution Rowland-circle monochromator is not a major project in cost, time, or manpower.

In figure 2 we present CAD renderings of this instrument, somewhat modified from the original report.[1] Relative to the earlier iteration, the present system instead mounts the SBCA 2-axis tilt unit onto a passive (non-motorized) linear slide. Positioning of the SBCA along this slide is then set by a system of three conjoined steel wires (not shown), each of length 0.5 m, anchored to the pins under the

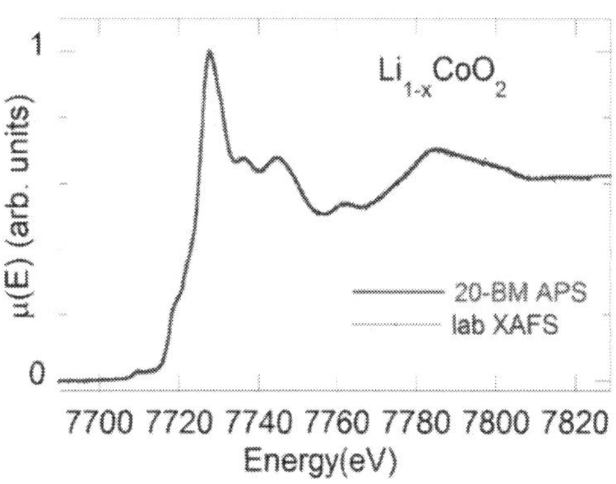


**Figure 3**: Transmission-mode XANES for an *ex situ* battery electrode material. The data indicated by the red curve was taken at beamline 20-BM of the Advanced Photon Source while the blue data (small dots) was taken with the lab XANES instrument. The optic was Ge (444) with a 1-m radius of curvature. No source slit was used.

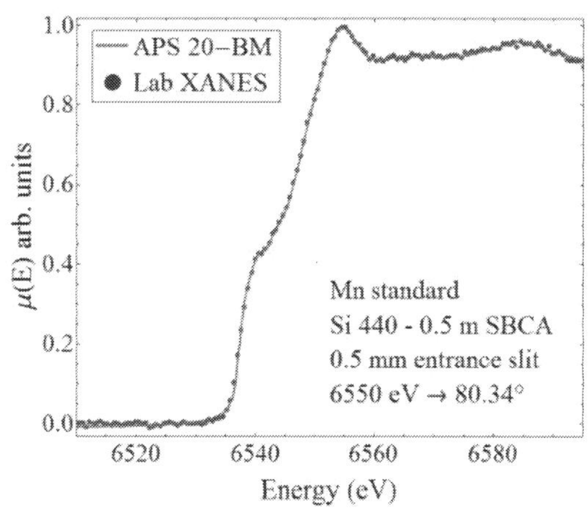


**Figure 4**: Transmission-mode XANES for an Mn reference standard. The data indicated by the red curve was taken at beamline 20-BM of the Advanced Photon Source while the blue data (small dots) was taken with the lab XANES instrument. The optic was Si (440) with a 0.5-m radius of curvature and a 0.5-mm wide source slit was used.

source, detector, and SBCA locations. The wire triad's raw geometrical constraint forces the three components to be on a 1-m diameter circle, and the alignment of the system establishes the requisite symmetry on the circle. Energy scanning involves synchronous scanning of the source and detector location and is accompanied by a translation of the Rowland circle in the lab frame.

As shown in figures 3 and 4, the performance of this system for transmission-mode XANES of concentrated samples is very favorable. The measurement time for the lab XANES data in each case was about 8 hours. However, note that the results in figure 4 were taken with one of the 10-cm wafer diameter, 0.5-m radius of curvature SBCA (XRS Tech) that are only now becoming available; the obvious modifications where made to achieve the tighter Rowland circle but we had no helium flight path for the smaller radius Rowland circle so measurements were taken entirely in air, which more than counteracted the increased solid angle of the SBCA. That said, if such optics become more common, the overall instrument footprint could be correspondingly decreased for a more truly "benchtop" system, in addition to having a larger collection solid-angle and improved measurement times.

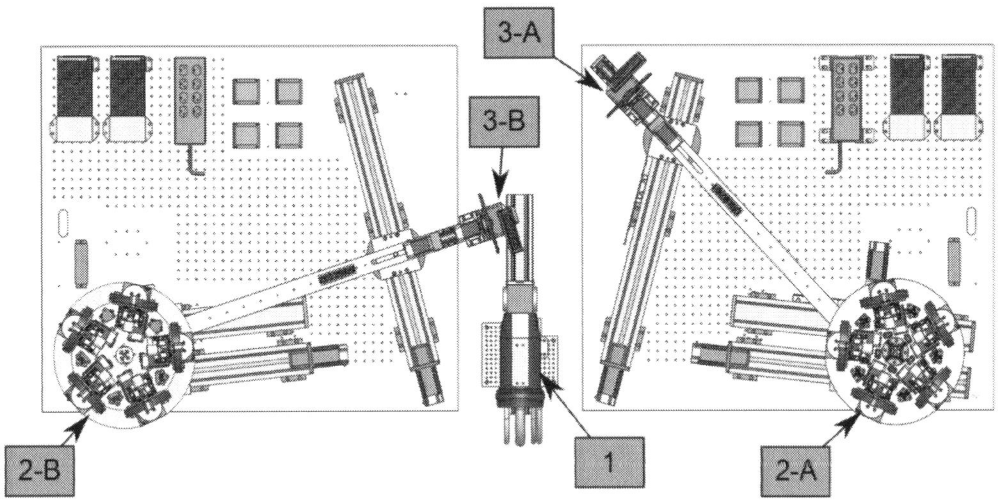


**Figure 5**: A top-view rendering of the CEI-XANES instrument. (1) X-ray tube housing with shutters in both horizontal directions, 6-degree take-off angle. (2-A) and (2-B) analyser optic turret. (3-A) and (3-B) detector sub-assemblies including sample turret. The radiation enclosure, helium flight path, and additional support structure have been supressed for clarity of presentation.

## 4. The University of Washington Clean Energy Institute XANES User Facility

XAFS has proven to be a powerful tool in the world-wide effort to improve electrical energy storage for vehicle electrification and grid-level storage. This situation presents an important example of how laboratory-based XAFS can have high scientific and social impact. The ongoing research work has the common character of materials chemistry: rapid iteration or hypothesis, synthesis, and evaluation is needed, but such rapid turn-around is typically problematic, albeit not impossible, at the synchrotron light sources. However, the best research battery cell, the so-called 'pouch cell', can have outstanding agreement with the technical demands of transmission-mode XAFS studies and consequently with lab-based XAFS applications, thus meeting the demand for rapid feedback and ultra-high access.

It is this rationale that motivated the ongoing development of a two-beamline lab XAFS user facility at the University of Washington, funded by the UW Clean Energy Institute. This instrument, referred to as CEI-XANES although its available energy scanning range will often extend into the EXAFS regime, is now almost fully assembled and one of the two monochromators is very near to completion of commissioning. CAD renderings of this instrument are presented in figures 5 and 6. For the purpose of the present manuscript, i.e., the discussion of different designs of lab XAFS instruments, the most

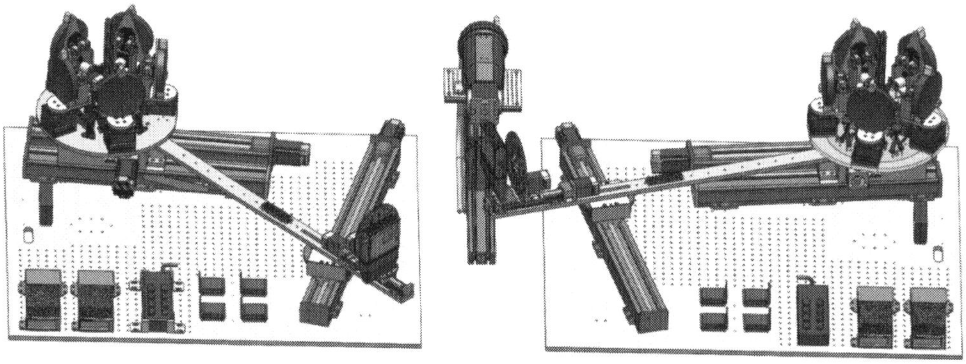

**Figure 6**: CEI perspective: A perspective rendering of the CEI-XANES instrument. The two monochromators operate fully independently. The radiation enclosure, helium flight path, and additional support structure have been supressed for clarity of presentation.

important technical point is that this system is almost entirely 'modern', in that rather than using mechanical linkages to enforce the Rowland circle constraints upon energy scanning the CEI-XANES instrument must instead synchronize multiple non-symmetry-related motor drives. This design choice followed from the need to have an SBCA turret to enable easy selection of energy range and thus simplify user operation and staff support. In this case, it is difficult, e.g., to passively track the Bragg angle of the SBCA to half-angle the detector arm motion.

CEI-XANES commissioning has been more challenging than that for the simpler instrument of the preceding section. Considerable effort must be placed on coordinating the motors for a systematic-free calibration and also the demands on smoothness of motion of the Bragg-angle stage under each SBCA are more strict than originally anticipated. At the time of submission of this proceeding, these challenges are almost completely overcome. Good motor tracking and the flux milestone of $10^6$/sec has been reached in the 5-10 keV range and *in situ* studies of pouch cell batteries will being shortly. A more complete report on this instrument will follow, but we find the effort already to be a significant success.

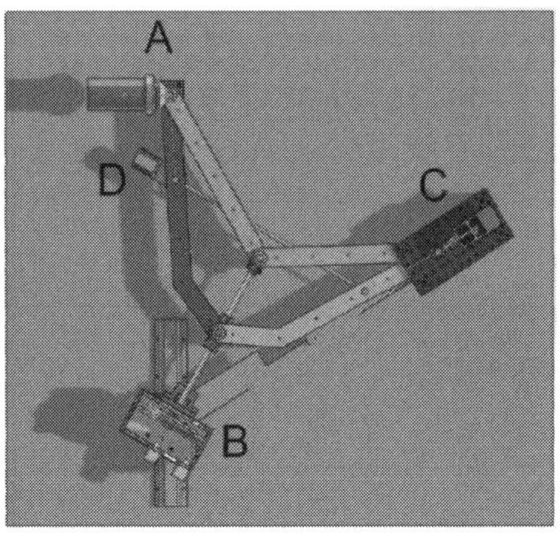


**Figure 7**: A top-view CAD rendering of a 'linear spectrometer' design under construction. (A) X-ray source and sample for XES; (B) SBCA 2-axis tilt on passive linear slide; (C) Detector platform; (D) Stepper motor drive. The stepper-driven leadscrew pushes the detector and source arms apart, with the shown mechanical couplings enforcing both placement and $\Theta$-2$\Theta$ symmetry on the Rowland circle.

## 5. Revisiting the Linear XAFS Spectrometer

Given the above *caveats* on the full nominal modernization to motorization and computer control over multiple degrees of freedom to drive energy scanning on the Rowland circle for systems where the x-ray source must be fixed in the lab frame, it is interesting to revisit the classic 'linear spectrometer' design from the 1970's and 1980's [8-10], updated only to operate closer to backscatter and use SBCA rather than the older cylindrical optics, thus allowing high energy resolution at all absorption edges. The key point is that energy scanning of these instruments can be achieved by a single motorized lead-screw drive with smooth tracking of the SBCA orientation driven by high-quality mechanical linkage. Three such systems are now under construction in our group; one for in-house use, one that will become an XES instrument for the study of batteries in the user facility of the Joint Center for Energy Storage Research at Argonne National Labs, and one that is destined for XES of *f*-electron materials at Los Alamos National Labs. A CAD rendering of these systems in the XES configuration is presented in Figure 7. Switching to a generic line-focus few-kW tube in line-of-sight to the optic, in analogy to the CEI-XANES instrument, converts this XES spectrometer to an XAFS monochromator. Full details of this design will be published after commissioning in the autumn of 2015.

Before concluding, in the context of our 'modern cookbook' one important component has not been discussed: the ready availability of high spatial resolution x-ray cameras. As investigated by other speakers at this conference, this allows numerous wavelength-dispersive designs having their own unique performance advantages and disadvantages. We particularly endorse the use of such systems at energies lower (few keV) and much higher (15-20 keV) than considered here. In the former case, this approach has led to outstanding performance for XES [4] and the switch to a bremsstrahlung source at the sample location would convert such an instrument to an interesting XAFS system. For the latter case, early reports of high-quality lab-based EXAFS at energies of actinide $L_3$-edge studies would benefit from this approach and also a possible change of the Laue optic to a curved geometry [11].

## 6. Conclusions

We have reported some of our experience in the development of point-focusing, Rowland-circle monochromators for laboratory-based XAFS and XES instruments. Each of the source, optic, and detector characteristics have diversified in economically- and technically-favorable ways in the few decades since lab-based XAFS fell out of favor. We propose that the high demand for even 'routine', transmission-mode XAFS of concentrated samples supports the rebirth and rapid spread of lab-based systems and user facilities. Our simple, low-powered system is particularly favorable for this purpose.

## 7. Acknowledgements

This work has been supported by: (1) the U.S. Department of Energy, Basic Energy Sciences under Grant No. DE-FG02-09ER16106 and also by Fusion Energy Sciences and the National Nuclear Security Administration thought Grant No. DE-SC0008580; (2) the State of Washington through the University of Washington Clean Energy Institute; (3) The Joint Center for Energy Storage Research (JCESR), an Energy Innovation Hub funded by the U.S. Department of Energy, Office of Science, Basic Energy Science; and (4) The U.S. Department of Defense through a subcontract from Los Alamos National Laboratories. APS Sector 20 operations are supported by the US Department of Energy and the Canadian Light Source, with additional support from the University of Washington.

## References

[1] Seidler G T, Mortensen D R, *et al.*, 2014 *Rev. Sci. Instrum.* **85** 113906
[2] X-RayAbsorption: Principles,Applications,Techniques of EXAFS, SEXAFS and XANES, edited by D. C. Koningsberger and R. Prins (John Wiley & Sons, 1988), pp. 163–210.
[3] Mortensen D R, Seidler G T, Ditter A S, in these proceedings.
[4] Kavcic M, *et al.*, 2012 *Rev. Sci. Instrum.* **83** 033113
[5] Gog T, Casa D M, *et al.*, 2013 *J. Synchrotron Rad.* **20** 74
[6] Fister T T, Seidler G T, *et al.*, 2006 *Rev. Sci. Instrum.* **77** 063901
[7] Pacold J I, Tatum D S, *et al.*, 2014 *J. Amer. Chem. Soc.* **136** 4186
[8] Georgopoulos P, Knapp G S, 1981, *J. Appl. Crystallogr.* **14**, 3
[9] Knapp G S, Chen H, Klippert T E, 1978 *Rev. Sci. Instrum* **49** 1658
[10] Laboratory EXAFS Facilities, edited by Stern E A (AIP, New York, 1980)
[11] Lecante P, Jaud A, *et* al., 1994 *Rev. Sci. Instrum.* **65** 845

# Data reduction for XAS experiments with the 100 element Ge Detector

**L A Martín-Montoya[1,2], A Rothkirch[1] and W Caliebe[1]**

[1]Deutsches Elektron Synchrotron (DESY), Notkestr. 85, 22607 Hamburg, Germany
[2]Universität Paderborn, Warburger Str. 100, 33098 Paderborn, Germany

E-mail: ligia.andrea.martin.montoya@desy.de

**Abstract.** X-ray absorption spectroscopy on dilute samples is still a challenge nowadays. By means of the fluorescence processes that follow the absorption it is possible to obtain the XAS spectrum; however, in order to gain sufficient statistics a multi-pixel fluorescence detector is required. A 100 pixel Ge detector will be available at beamline P64 at PETRA III for ultra dilute sample systems. Here, we propose a method of data reduction that resolves the line of interest from the scattering peak, improving the signal-to-noise ratio of the final XAS spectrum compared with the one obtained with a simple region of interest (ROI).

## 1. Introduction

Chemical processes on biological samples, like metalloproteins, are of great interest. X-ray Absorption Spectroscopy (XAS) allows to get structural and chemical information and does not require crystallinity; therefore, it is potentially an adequate tool for the investigation of disordered materials like metalloproteins in solution [1]. Those samples are typically in ultra dilute concentrations, meaning that the absorption is too low to be detected with the standard XAS in transmission mode [2]. This drawback can be overcome by measuring XAS in fluorescence mode and employing a multi-pixel fluorescence detector to improve the signal-to-noise ratio by a factor of $\sqrt{n}$, where $n$ is the number of pixels [3, 4].

The beamline P64 at PETRA III is planned for BioXAS experiments among others. Those studies will be possible with the 100 pixel Germanium detector, that is a segmented-pixellated planar fluorescence detector where each pixel performs like a single detector, aiming to improve the signal-to-noise ratio of the final XAS spectrum by a factor of 10 [5]. The data collection works as follows: a full fluorescence spectrum is collected at a determined incident energy and per pixel, leading to a final 3D data array of dimensions $n \times k \times m$, where $k$ is the number of incident beam energies and $m$ is the length of each fluorescence spectra (see Figure 1). The 3D data array is to be reduced efficiently into a final XAS spectrum of length $k$.

## 2. Data Reduction

Reducing $n$ fluorescence spectra at a single incident energy could be done in principle with an average. Nevertheless, this approach is not completely accurate since there can be several artifacts coming from some pixels that should not be taken into account e.g. glitches [6]. In

**Figure 1.** 3D data array of the sample $[Cu_2(RSSR)_2](OTf)_2$ measured at PETRA III P11. The image contains 100 pixels with incident energy vs. fluorescence energy where 78 pixels were turned on.

order to reject those pixels with low performance, we propose a method that at a single incident energy calculates a quantity $\text{Diff}_i$ for each pixel $i$, as:

$$\text{Diff}_i = \hat{S}^j - S_i^j \tag{1}$$

where $\hat{S}^j$ is the entry $j$ of the mean fluorescence spectrum and $S_i^j$ is the entry $j$ of the fluorescence spectrum of the pixel $i$.
Only if $\text{Diff}_i$ is equal or lower than the shot noise $\sqrt{\hat{S}^j}$ [4], it is taken into account for the final fluorescence spectrum. From this procedure 20 pixels were deleted from the mean. As an alternative, the median can be used instead of the mean spectrum in order to delete artifacts like glitches. A comparison between the three methods is exhibited in Figure 2 and shows that the corrected mean and the median decrease slightly the overlapping between the line of interest and the scattering.

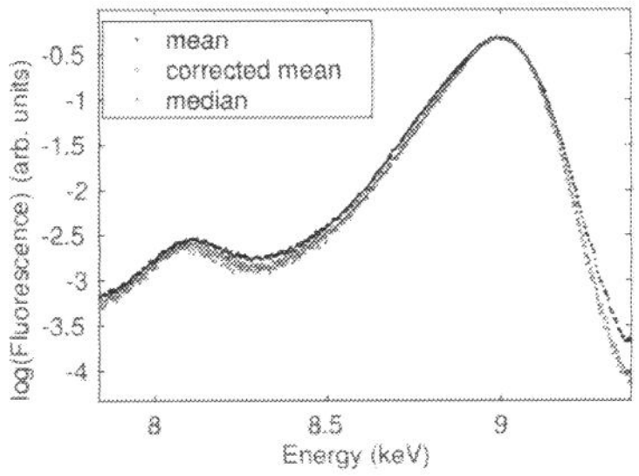


**Figure 2.** Comparison between the mean spectrum, the corrected mean without 20 rejected pixels and the median spectrum at the incident energy of 8.98 keV.

Since the position of the Compton scattering exposes an angular dependence and its intensity is comparable to the Thomson scattering at angles close to $\pi/2$, the width of the scattering line (Compton + Thomson) is different depending on the pixel's column at energies near the Cu absorption edge. Therefore, for accurateness the spectra should be analyzed by columns instead of an overall treatment of all the pixels.
Once the final fluorescence spectrum per incident energy is obtained, it is necessary to extract

the total number of counts under the fluorescence peak of interest. This value per incident energy would build up the final XAS spectrum. One of the procedures is made by a selection of a region of interest ROI around the peak of interest and a further integration of the number of counts in this region. However, the strong overlapping of the line of interest and the tail of the scattering line at energies close to the absorption edge makes this method inaccurate [7] (see Figure 3).

Aiming to extract the number of counts that correspond to the fluorescence line of interest, an automatic fitting procedure of a region between the line of interest and the scattering line is proposed. The line of interest as well as every fluorescence line in this region is fitted as a Gaussian function and the scattering line as an Exponentially Modified Gaussian Function $EMG$ [8, 9, 10], described as:

$$\mathrm{EMG} = A \exp\left[\frac{1}{2}\left(\frac{\sigma}{t_0}\right)^2 + \frac{x-\beta}{t_0}\right]\left[\mathrm{erf}\left(\frac{1}{2}\left[\frac{\beta}{\sigma} + \frac{\sigma}{t_0}\right]\right) + \mathrm{erf}\left(-\frac{x-\beta}{\sqrt{2}\sigma} - \frac{\sigma}{\sqrt{2}t_0}\right)\right], \quad (2)$$

where A is the amplitude, $\beta$ the center, $\sigma$ the FWHM and $t_0$ the skewness. This model allows to fit the strong asymmetry of the scattering line by means of the parameter $t_0$.

The data from the sample $[Cu_2(RSSR)_2](OTf)_2$ measured at PETRA III beamline P11 was evaluated with the proposed fitting algorithm. At low energies (XANES) the fit includes a Gaussian, corresponding to $CuK_\alpha$, plus an EMG (Figure 3).

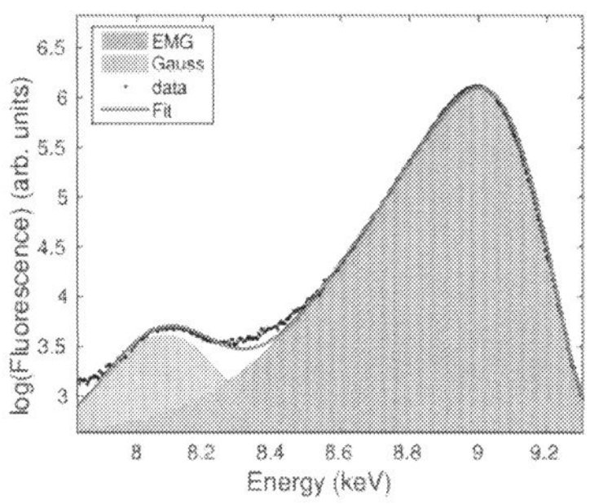


**Figure 3.** Data and fit result at an incident energy of 9 keV. Magenta denotes the Gaussian fit to the $CuK_\alpha$ line and blue the EMG fit to the scattering line. The final fit is indicated by the red curve.

The fitting algorithm was automatized to fit the amplitude of the Gaussian $A_{gauss}$ correspondent to the $CuK_\alpha$ line at each incident energy and to determine the area under the Gaussian ($\sqrt{2\pi}A_{gauss}\sigma_{gauss}$) in order to build up the XANES spectrum (see Figure 4). To avoid as much as possible errors in the fitting procedure only the amplitudes of the Gaussians were fitted and other parameters were calculated based on the peak positions, typical skewness and FWHM. The fitting algorithm results in a smoother XAS curve compared to the ROI curve and the quality of the fit is always above the $R^2 = 0.998$.

The data under analysis corresponds to the first test beamtime of the 100 pixel Ge detector; therefore, due to set-up issues, it is too noisy to be analyzed as normal XAFS. Nevertheless, the XANES region on Figure 4 shows that the fitting algorithm gives a smoother curve than the ROI. Furthermore, the pre-edge region of the ROI curve is strongly contaminated by the scattering as shown in Figure 4. In this sense, the fitting algorithm is more reliable in the pre-edge and the XANES regions.

At higher energies, e.g. at the Mo absorption edge, the elastic and inelastic scattering peaks are discerned. In such a case, the data reduction should include a model that treats the Thomson and Compton lines independently.

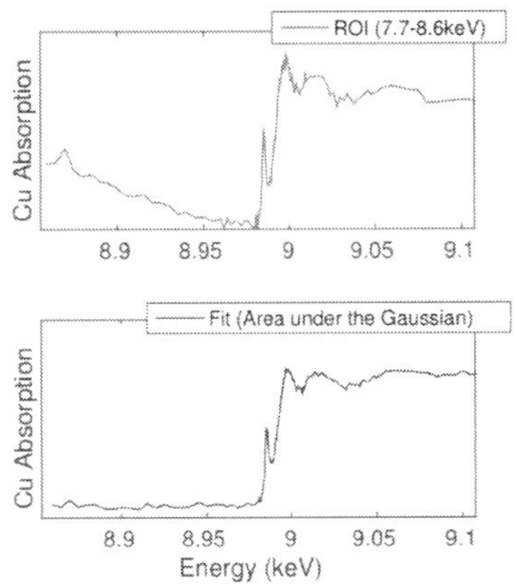


**Figure 4.** Final XANES spectra obtained by integration over a ROI between 7.7–8.6 keV (in red) and with the Area under the Gaussian, calculated with the proposed fitting algorithm (in blue).

## 3. Conclusions

Here we present an automatic routine of data reduction for the 100 pixel Ge detector with XAS experiments. The routine reduces 100 fluorescence spectra at each incident energy into one, by comparing each spectra to the average and dismissing those which difference is above the shot noise. A further fitting procedure that takes into account the asymmetry of the scattering peak is presented and compared to the ROI procedure to extract the XANES curve. The fitting algorithm proposed here gives satisfactory results in terms of goodness ($R^2$) and signal-to-noise ratio. The pre-edge region is better described with the fitting algorithm because the contamination from the scattering is reduced compared to the ROI algorithm.

## Acknowledgments

We would like to thank the staff from PETRA III beamline P11 for help during the beamtime, our colleagues R. Chernikov, A. Kalinko, V. Murzin, M. Naumova, M. Tolkiehn, B. Walz and E. Welter, as well as the professors G. Henkel and J. Lindner for valuable discussions and encouragement, and an unknown reviewer for enlightening thoughts concerning Compton scattering. This project is funded by DESY and the BMBF project "BioXAS at PETRA III".

## References

[1] Shi W *et al* 2011 Characterization of metalloproteins by high-throughput X-ray absorption spectroscopy *Cold Spring Harbor Laboratory Press* **21** 898–907
[2] Ascone I, Meyer-Klaucke W and Murphy L 2003 Experimental aspects of biological X-ray absorption spectroscopy *J. Synchrotron Rad.* **10** 16–22
[3] Cramer S P, Tench O, Yocum M and George G N 1988 A 13-elment Ge detector for fluorescence EXAFS *Nuclear Instruments and Methods in Physics Research* A **266** 586–91
[4] Janesick J R 2007 *Photon Transfer: DN → λ, Chapter 3: Photon Transfer Noise Sources* Bellingham, Washington 98227-0010 USA SPIE 21–8
[5] Camberra Industries, High-purity Germanium (HPGe) Detectors
[6] Korbas M, Fulla Marsa D and Meyer-Klaucke W 2006 KEMP: A program script for automated biological x-ray absorption spectroscopy data reduction *Review of Scientific Instruments* **77** 063105
[7] Tamenori Y, Moritab M and Nakamuraa T 2011 Two-dimensional approach to fluorescence yield XANES measurement using a silicon drift detector *J. Synchrotron Rad.* **18** 747–52
[8] Hangi D and Carr P W 1985 Errors in Exponentially Modified Gaussian Equations in the Literature *Anal. Chem.* **57** 2394–95
[9] Walz B 2011 Untersuchung von Fremdatomen in kristallinen Materialien mit kinematischen stehenden Röntgenwellen *Deutsches Elektronen-Synchrotron* 52–6
[10] Eggert T, Boslau O, Kemmer J, Pahlke A and Wiest F 2006 The spectral response of silicon X-ray detectors *Nuclear Instruments and Methods in Physics Reearch A* **568** 1–11

# XAFS data acquisition with 2D-detectors: Transmission mode XAFS and grazing incidence EXAFS spectroscopy

**D Lützenkirchen-Hecht*, J-C Gasse, R Bögel, R Wagner, R Frahm**
Bergische Universität Wuppertal, Gaußstr. 20, 42097 Wuppertal, Germany

*E-mail: dirklh@uni-wuppertal.de

**Abstract.** XAFS-experiments in transmission and reflection modes have been performed using a Pilatus 100K pixel detector. Transmission mode XAFS spectra from a Co metal foil and $Co_3O_4$ were recorded to evaluate the data quality offered by this 2D-detector. Furthermore, the pixel detector was also used to measure reflection mode grazing incidence EXAFS data. Using different regions of interest in the collected scattering patterns, we will show that the diffuse scattering can be separated for the different contributing surfaces and interfaces, allowing simultaneous investigations of surfaces and buried interfaces within multi-layered samples.

## 1. Introduction

The quality of transmission mode EXAFS measurements rely on the homogeneity of the investigated samples in the region transmitted by the X-ray beam. Lateral thickness variations, holes or inhomogeneous sample compositions may lead to severe distortions of the EXAFS spectra. So the amplitude of the EXAFS oscillations is reduced compared to a perfect sample, and the determined coordination numbers and Debye-Waller factors are erroneous [1]. A careful sample preparation may prevent such artifacts in the case of ex-situ experiments. For in-situ studies, however, there are many situations where the sample considerably changes, e.g., in the case of catalysts in a reacting environment [2] or a battery material during charge [3]. While ionization chambers measure the average absorption through the sample, a 2D-area pixel detector may provide local absorption spectra for each position within the sample. Recent experiments have demonstrated the capabilities of such a setup for in-situ experiments of battery materials, where local charging / discharging occurs [3]. In this contribution, we will present transmission mode EXAFS data of model samples measured with a large, unfocussed beam of a wiggler source in order to evaluate the data quality offered by pixel detectors also for real samples.

In a second series of experiments, we addressed grazing incidence X-ray absorption spectroscopy measurements with the 2D detector, where the X-rays impinge on the surface of the sample under a small incident angle $\phi_i$. Besides the specular reflected beam (incident angle $\phi_i$ = exit angle $\phi_e$), a second distinct scattering peak (Yoneda-peak) occurs if $\phi_e = \phi_c$, the critical angle of the scattering material [4], with $\phi_c(Au) \approx 0.35°$ and $\phi_c(Bi) \approx 0.25°$. From related XAFS experiments, a depth profiling of the sample is feasible, and structural information of the surface and buried layers may be obtained [5, 6]. If ionization chambers are used as detectors, the EXAFS has to be measured sequentially for the different scattering (exit) angles, which makes Yoneda-XAFS experiments extremely time-consuming. Using a 2D-area detector, the entire scattering pattern is acquired simultaneously for each energy in the EXAFS scan, and a wealth of information is available in a moderate time. Here we will discuss the feasibility of Yoneda-XAFS measurements for the investigation of multilayered samples.

## 2. Experimental details

A Pilatus 100K detector was used for all experiments [7]. Transmission mode studies were conducted at beamline 10 at the DELTA storage ring [8], where incident and transmitted intensities from the sample were monitored by $N_2$-filled ionization chambers, and the 2D-detector was placed behind the second ionization chamber in order to measure the lateral distribution of the transmitted X-rays. The unfocussed incident beam from the Si(111) channel-cut monochromator was defined by two slit systems in front of the sample. Kapton foils downstream the second ionization chamber were used to adjust the photon flux on the Pilatus detector in order to avoid counter overflows and not to exceed the maximum countrate accepted by the detector ($10^6$ counts $s^{-1}$ $pixel^{-1}$). Co metal foils and $Co_3O_4$ powder samples were investigated, with about 0.5 s – 1 s integration time per data point (ca. 25 min/spectrum).

For the conducted grazing incidence X-ray scattering experiments, the 2D-detector was directly placed in the scattered radiation fan without a second ionization chamber. The measurements were done at the MRH endstation of the ROBL beamline BM20 at the ESRF using a Si(111) monochromator and a mirrored, focussed beam. Test sample was a Bi (20 nm)/Au (200 nm)-bilayer on a float glass substrate prepared by vacuum deposition. The entire scattering pattern was evaluated for each energy while scanning across the Au $L_3$ (11919 eV) and the Bi $L_3$ (13419 eV) absorption edges.

## 3. XAFS experiments

### *3.1. Transmission mode EXAFS*

In order to study the feasibility of laterally resolved EXAFS experiments and to evaluate the data quality that can be achieved, first tests were performed using a Co-metal foil. In figure 1(a), a map of the beam behind the sample is shown in the right inset for an acquisition time of 1 s. The illuminated region of the sample consists of ca. 6 pixels vertically, and 40 pixels horizontally. With a pixel size of 172 μm ([7], see inset), this corresponds to a beam size of 1 x 7 $mm^2$. The raw absorption data measured using the ionization chambers are compared to different areas of the pixel detector. Obviously, the data quality of a spectrum calculated from a single pixel is excellent with respect to the reproduction of sharp absorption features close to the edge as well as to the low noise of the $k^3$-weighted $\chi(k)$ data depicted in the inset. If a vertical cross section of the beam, i.e. a line of 6 pixels, is used for the calculation of an absorption spectrum, the data quality slightly improves, and if the entire region of the sample that is illuminated from the beam is taken into consideration (40 lines of 6 pixels = 240 pixels), the obtained spectrum is almost indistinguishable to that from the ionization chamber.

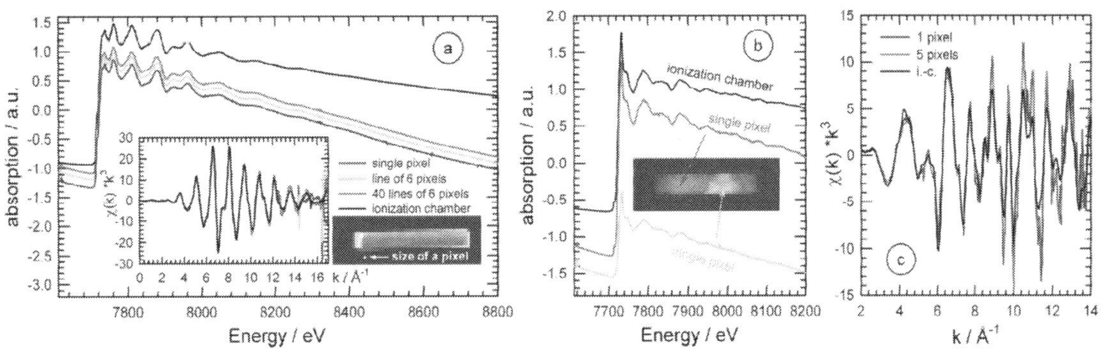


**Figure 1.** (a) Comparison of the K-edge X-ray absorption spectrum of a Co metal foil measured in transmission at room temperature using ionization chambers as detectors (—) and using the intensities on the 2D-detector (1 s / point). Different regions of this detector were used for spectrum calculation: a single pixel (—), a vertical cross section of 6 pixels (—) and the entire illuminated area (40x6 pixel,—). The insets show the $k^3$-weighted EXAFS oscillations $\chi(k)*k^3$ for the different spectra depicted, and the intensity distribution on the Pilatus detector measured at E = 7600 eV. (b) Raw absorption data (0.5 s / point) for an inhomogeneous $Co_3O_4$ powder sample with an intensity map. (c) $\chi(k)*k^3$ data extracted from a single pixel (—) and 5 neighboring pixels (—) compared to the ionization chamber data (—).

While no substantial differences can be seen in the extracted $\chi(k)*k^3$ data up to ca. 12 Å$^{-1}$ for all the different measurements, uncompensated glitches are visible at about 14.4 Å$^{-1}$ and 16 Å$^{-1}$ in the data of the pixel detector, limiting the usable k-range here. The EXAFS of a $Co_3O_4$ powder sample is shown in figure 1(b). The inhomogeneous intensity distribution on the 2D-detector is directly reflected in the variation of the edge jump on two selected positions (arrows in figure 1(b)). Moreover, the EXAFS oscillations measured using the ionization chamber have substantially smaller amplitudes compared to the data of single pixels, especially at larger k-values, see figure 1(c). Thus, it should be promising to measure laterally resolved EXAFS spectra of inhomogeneous samples containing variations of thickness, composition, etc., with a lateral resolution of the pixel size, i.e. ca. 200 μm in the present case.

### *3.2. Grazing incidence XAFS*

Some selected scattering patterns obtained from the Bi/Au-bilayer sample are shown in figure 2 for three different energies in the vicinity of the Au $L_3$ edge and $\phi_i$=0.5°, placing the 2D-detector 0.6 m behind the sample. As can clearly be seen, the intensity of the diffuse scattering strongly depends on the X-ray energy and the scattering angle, i.e. the diffuse scattering from the smaller exit angle, i.e. the outer Bi-air surface with smaller density (ROI 2), is not dramatically affected when the energy is scanned through the Au $L_3$-edge. In contrast, the pattern related to the larger exit angle (ROI 2), the inner Bi-Au interface with the larger Au density, dramatically drops down in this energy range (figure 2(c)).

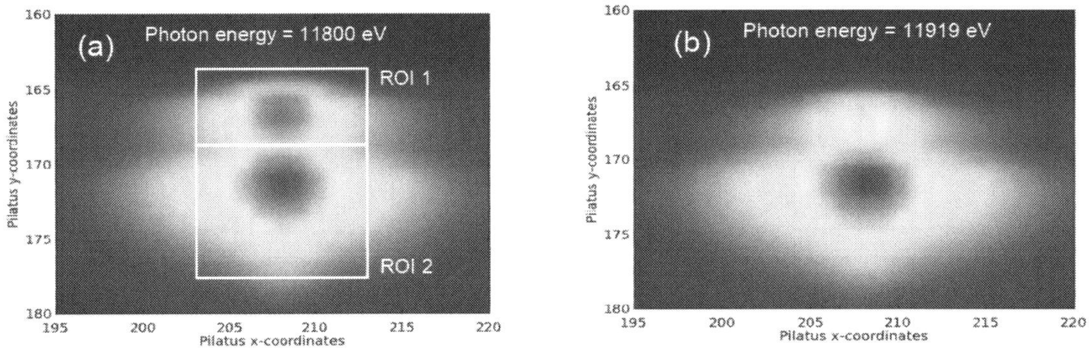

**Figure 2.** Scattering pattern from a Bi / Au / glass sample for different photon energies in the vicinity of the Au $L_3$-edge and an incidence angle $\phi_i$=0.5°: (a) 11800 eV, (b) 11919 eV, (c) 12000 eV. The intense feature at the y-coordinate of about 173 ± 3 pixel ($\phi_e \approx 0.25°$, ROI 2) corresponds to the Yoneda peak of the outer Bi-air surface, and the less intense, smaller spot at y = 167 ± 2 pixel ($\phi_e \approx 0.35°$, ROI 1) belongs to the inner Bi-Au interface, respectively. Regions of interest (ROIs 1 and 2) were defined around these two spots for the collection of representative diffuse EXAFS data.

This behaviour is also observed in the EXAFS spectra calculated from these diffuse X-ray scattering profiles. As can be seen in figure 3, the intensity measured for the Yoneda-peak with the larger $\phi_e$ (smaller y-coordinate in figure 2) is always smaller compared to that related to the smaller $\phi_e$ (larger y-coordinate). While a distinct Au $L_3$-edge can be seen in the first Yoneda peak, no edge can be seen in the scattering related to the second feature. Furthermore, the Bi $L_3$-edge spectrum related to the first Yoneda peak shows a fine structure similar to that of metallic Bi, while that of the second peak resembles that of Bi-oxide $Bi_2O_3$. The qualitative interpretation of these observations is clear: The diffuse scattering related to the larger $\phi_e$ is related to the inner Au-Bi interface, where both elements are present in a metallic form. On the other hand, no gold is present at the outer surface of the bilayer, and thus no Au $L_3$-edge is observed for the smaller $\phi_e$. Due to the chemical reactivity of Bi metal, the latter metal is oxidized at its air-side surface, forming a thin oxide layer of about 1-2 nm thickness [9]. Although the incidence angle is larger than $\phi_c$, the technique appears to be extremely surface sensitive [5]. Thus, the X-ray absorption fine structure in the diffuse scattering reveals features similar to that of $Bi_2O_3$ (see e.g. [10]) due to strong contributions of the outermost surface to the scattered signals [5].

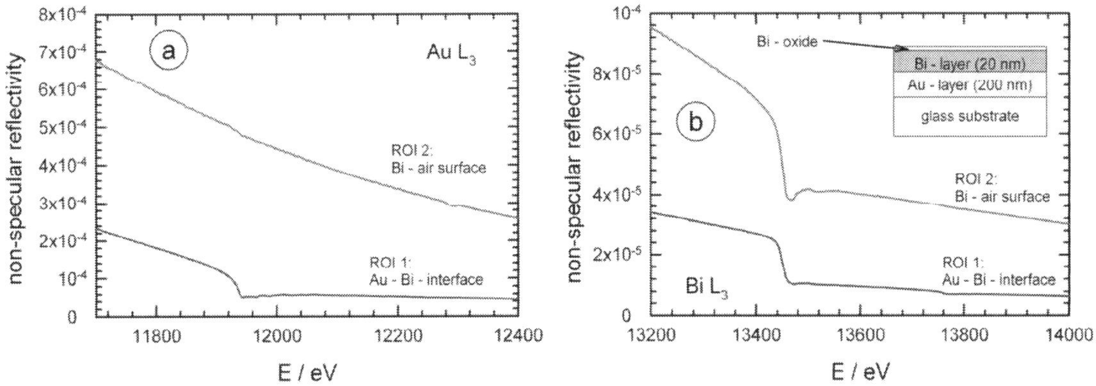


**Figure 3.** Non-specular reflection mode EXAFS experiments at the (a) Au $L_3$- and (b) Bi $L_3$-edges for an incidence angle $\phi_i = 0.5°$ and exit angles according to the Yoneda peaks shown in figure 2, i.e. $\phi_e \approx 0.25°$ (—) and $\phi_e \approx 0.35°$ (—). The inset depicts a model of the layer structure of the sample.

A detailed modelling of the measured data (for calculation details see [5]) is currently in progress in order to determine the thickness of the Bi oxide layer at the surface. Furthermore, the formation of a gold-bismuth alloy at the inner interface can be initiated by an annealing process of the bilayer. Such an alloy formation may result in substantial modifications of the EXAFS detected at the Yoneda peak at larger exit angles, and, thus, a careful analysis may identify these interfacial phases, their thickness and their changes with temperature and time, which will be subject of a forthcoming publication.

## 4. Conclusions

We have investigated the feasibility of transmission and reflection mode EXAFS experiments using a 2D-pixel detector. The results show that acceptable data quality can be achieved even if a short acquisition time and a single detector pixel is used for the evaluation, and the data quality improves if the signals from several pixels are averaged. For surface sensitive grazing incidence X-ray absorption spectroscopy, the pixel detector allows the simultaneous measurement of the entire scattering pattern, thus the structure of different surfaces and interfaces can be investigated in parallel.

## References

[1] Bridges F, Li GG and Wang X 1992 *Nucl. Instrum. Meth. A* **320** 548
[2] Schroer CG and Grunwaldt J-D 2013 *In: In-situ characterization of heterogeneous catalysts. Eds.: J.A. Rodriguez, J.C. Hanson, and P.J. Chupas.* John Wiley & Sons pp. 49 - 73.
[3] Tanida H, Yamashige H, Orikasa Y, Oishi M, Takanashi Y, Fujimoto T, Sato K, Takamatsu D, Murayama H, Arai H, et. al. 2011 *J. Synchrotron Rad.* **18** 919
[4] Yoneda Y 1963 *Phys. Rev.* **131** 2010
[5] Keil P, Frahm R and Lützenkirchen-Hecht D 2010 *Corrosion Sci.* **52** 1305
[6] Lützenkirchen-Hecht D, Keil P and Frahm R 2007 *Surf. Sci.* **601** 4232
[7] Kraft P, Bergamaschi A, Broennimann C, Dinapoli R, Eikenberry EF, Henrich B, Johnson I, Mozzanica A, Schlepütz CM, Willmott PR and Schmitt B 2009 *J. Synchrotron Rad.* **16** 368
[8] Lützenkirchen-Hecht D, Wagner R, Szillat S, Hüsecken AK, Istomin K, Pietsch U and Frahm R 2014 *J. Synchrotron Rad.* **21** 819
[9] Atkinson R and Curran E 1985 *Thin Solid Films* **128** 333
[10] Jiang N and Spence JCH 2006 *J. Phys.: Condens. Matter* **18** 8029

# High performance emission spectrometer at Balder/MAX IV beamline

**K Klementiev, I Preda, S Carlson, K Sigfridsson and K Norén**

MAX IV Laboratory, 22363 Lund, Sweden

E-mail: konstantin.klementiev@maxlab.lu.se

**Abstract**. The emission spectrometer at Balder/MAX IV beamline is presented. Its unique features are described. Comparison is given with other types of curved crystals analyzers.

## 1. Introduction

Balder is a hard x-ray XAFS beamline at the brand new Swedish synchrotron MAX IV. It was designed for a broad energy range (2.4 – ~50 keV), high flux (> $10^{13}$ ph/s), quick scanning (>1 Hz repetition rate for full EXAFS [Extended X-ray Absorption Fine Structure]) and simultaneous high-resolution emission detection compatible with various sample set-ups [1].

High flux beamlines frequently experience overload problems in detecting fluorescence signals. Even with the last generation of signal processors, solid state detectors can hardly handle high count rates of >~1 Mcps/element. The repetition rate of spectra is thus limited by detectors, not by the primary flux. On the other hand, high energy resolution XES [X-ray Emission Spectroscopy] paves the way towards *site-selective* XAFS [X-ray Absorption Fine Structure] — disentangled spectra of individual species in a mixture. In such applications, high incoming flux is necessary to detect weak satellite lines and/or to work with low elemental concentrations. Balder will use its high flux source for XES but will have to tackle the overload problem.

In the next section we show that both tasks, high flux fluorescence detection and high resolution emission spectroscopy, can be solved with one instrument. The following sections describe the spectrometer at Balder beamline with focusing at its most crucial components — the crystals.

## 2. Comparison of 1D-bent crystal analyzers

In this section we compare Rowland circle based spectrometers utilizing simply bent (Johann [2]) and ground-bent (Johansson [3]) crystals and a von Hamos spectrometer [4]. The schematics of the three spectrometers can be found on the web page of xrt program [5] and are briefly outlined as follows. The Johann crystal is simply bent from a flat crystal to a cylinder of radius $2R$; the Johansson crystal is first cylindrically ground to a radius $2R$ and then bent down to a radius $R$. In both cases, the source (sample), the crystal and the detector lie on a circle (Rowland circle) of radius $R$. The grinding and/or bending occurs in the plane of the circle. The von Hamos crystal has *sagittal* bending, being flat in the diffraction plane. When a Johann crystal is rotated by 90° around the sample-to-crystal line, it becomes a von Hamos crystal that has to be put at a correct distance. This case is labeled as "Johann as von Hamos" and has a larger bending radius and thus a larger working distance than the other von Hamos crystals compared in this study. The Johann and Johansson crystals are 100(meridional)×20(sagittal) mm² with the

Rowland circle diameter 1 m. The von Hamos crystals are 100(sagittal)×50(meridional) mm² with the sagittal radius 250 mm, diced and continuously bent. All the crystals are Si(444).

The ray tracing study was done with the program xrt [5]. The resulted beam "images" are reported on the program's web page; here we only summarize the end results — energy resolution and efficiency.

In comparing with a von Hamos spectrometer, one should realize its strongest advantage — inherent energy dispersive operation without a need for energy scan. This advantage is especially important for broad emission lines. Below, the comparison is made for two cases: (1) a narrow energy band (figure 1, left), which is more interesting for valence band RIXS and which assumes a high resolution monochromator in the primary beam and (2) a wide energy band (figure 1, right), which is more interesting for core state RIXS and normal fluorescence detection. The desired position on the charts is in the upper left corner. As seen in the figures, the efficiency of the von Hamos crystals (i) is independent of the energy band (equal for the left and right charts), which demonstrates truly energy-dispersive behavior of the crystals but (ii) is significantly lower as compared to the Johann and Johansson crystals. A way to increase efficiency is to place the crystal closer to the source, which obviously decreases energy resolution because of the increased angular source size. Inversely, if the crystal is put at a further distance, the energy resolution is improved (see the von Hamos case with larger radius and distance, "Johann as von Hamos", square symbols) but the efficiency is low because of a smaller solid angle collected.

Finally, among the compared 1D-bent spectrometers the Johansson type is the best in the combination of good energy resolution and high efficiency and therefore it can function both as a high resolution spectrometer and a fluorescence detector. One should bear in mind, however, two very strong advantages of von Hamos spectrometers: (1) they do not need alignment: a crystal and a detector positioned approximately will probably immediately work and (2) the image is inherently energy dispersive with a flat (energy independent) detector response. The low efficiency and mediocre energy resolution are a price for the alignment-free energy dispersive operation. Rowland circle based spectrometers will always require good alignment.

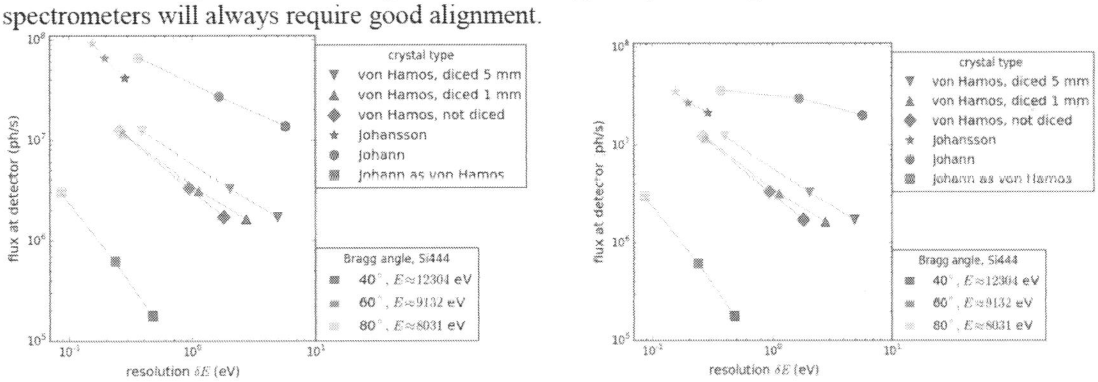


**Figure 1.** Resolution-efficiency charts of 1D-bent Si444 crystal analyzers at $10^{13}$ph/s incoming flux and 100×100 μm² source size. The left chart is with a narrow source energy band equal to the 6-fold energy resolution; the right one is with a wide energy band equal to $8 \cdot 10^{-4} \cdot E$ (approx. width of $K\beta$ lines [6]).

## 3. Spectrometer design

The main design objectives are: 1) to be compatible with set-ups used for transmission XAFS detection in order to eliminate the need for a large *ad hoc* fluorescence window, 2) to have the emission energy axis dispersive, i.e. recorded without scanning, 3) to implement 2D detection for quicker and more visual alignment, 4) to serve both as a high resolution spectrometer and a standard fluorescence detector and 5) to leave ample space around the sample, also above it, for bulky sample holders as a cryostat or an in-situ chemical cell.

These objectives will be achieved by the back-scattering escape direction – not the usual 90°, variable in-Rowland-circle position – by the translation of the spectrometer towards the sample, large 2D pixel detectors, large ground-bent crystals and the detector stage *under* the sample position.

The overall design of the Balder end station, including the spectrometer, is sketched in figure 2. The crystals of 300×40 mm² size are grouped in two branches, tunable for two, possibly different energies. The primary beam passes between the two branches. Two 2D detectors (one of them is under the table and not visible in figure 2) are coupled to the two crystal branches. The detectors are outside the vessel, behind a large curved window. The beam height at the end station enables a Rowland circle of ~1.5 m diameter. A light-weight removable vessel (several Viton-sealed covers) will be added at a later stage.

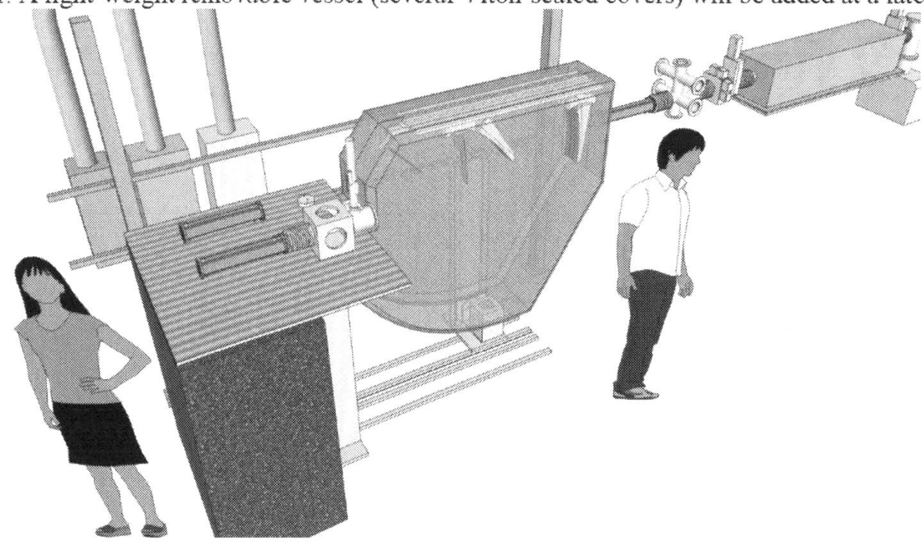

**Figure 2.** Balder end station. The monochromatic beam enters from the right. Shown are, in downstream direction, a vessel with harmonic rejecting mirrors (one of them will additionally focalize in vertical), beam conditioning and diagnostics elements, the spectrometer (with the semitransparent vessel), an adjustable table with two switchable experiments and a gas distribution system with various gas types (behind the table).

## 4. Crystals

The central part of the spectrometer – several crystals of various orientations – will be manufactured in-house. This project includes grinding, lapping, polishing, metrology measurements, special surface treatment, bending, assembling and testing with x-rays. The crystals will be of the Johansson type, i.e. ground to a radius equal to the Rowland circle diameter and bent down to one half of this radius. In order to bend a crystal, it has to be sufficiently thin. A safe thickness-to-radius ratio for silicon crystals is usually taken as 5000. We therefore aim at a thickness of ~300 μm.

We have established a silicon processing workshop at MAX IV Laboratory. For a test ground and polished crystal, so far not bent, we have achieved a high radius precision $R = 1503.6 \pm 0.6$ mm. The resulted silicon arc is shown in figure 3.

**Figure 3.** A ground and polished test silicon crystal. Notice that this is not a bent piece; the crystal planes are still flat and parallel to the table.

The next processing stage is the surface treatment. The amorphized surface layer is usually removed by chemical etching. Having the crystal thin and large, we believe a safer and more determined way is to subject the crystal to recrystallization in an ultra-high vacuum oven. A small test piece has exhibited re-appearance of a LEED pattern measured *in situ* in the oven. We are building a larger volume oven, also with an *in situ* LEED instrument, to work on the production crystals.

Meanwhile, the first batch of six Si(110) 310-mm-long crystals has been cut and ground (not lapped yet) to the radius 1500 mm, see figure 4.

**Figure 4.** The first production batch of crystal analyzers. The upper surface is ground to 1.5 m radius.

## 5. Outlook

The spectrometer will be implemented in stages, starting with a smaller number of branches and crystal types and limited to a moderate energy range. Later on a lightweight vacuum chamber and a specialized low-energy detector will enable the usage of the spectrometer at low energies. A 2D on-the-fly image analysis will be developed for converting detector images into XES spectra.

The mechanical design is in progress. The spectrometer is planned for the first tests by end 2016.

## 6. Conclusion

Among various curved crystal analyzers, the Johansson type is the best in the combination of good energy resolution and high efficiency. This property is due to the exact point-to-point imaging provided by each point of the crystal along the Rowland circle. The future spectrometer at the Balder beamline is therefore selected to be of the Johansson type.

The back scattering (relatively to the sample) geometry makes the usual side windows in a cryostat or an *in situ* cell unnecessary and requires only the entrance window.

Large ground-bent crystals are being produced at the MAX IV Laboratory. The test pieces have demonstrated maturity of the developed technological steps.

**Acknowledgments**
The Balder beamline is funded by the Knut and Alice Wallenberg Foundation and Swedish universities.

**References**

[1] Klementiev K, Norén K, Carlson S, Sigfridsson Clauss K and Persson I 2015 The BALDER Beamline at the MAX IV Laboratory, this volume
[2] Johann H H 1931 *Z. Phys.* **69** 185–206
[3] Johansson T 1933 *Z. Phys.* **82** 507–28
[4] von Hámos L 1934 *Ann. Phys.* **411** 252–260
[5] Klementiev K and Chernikov R 2014 Powerful scriptable ray tracing package xrt *Advances in Computational Methods for X-Ray Optics III* (Proc. SPIE 9209, 2014) ed Sanchez del Rio M and Chubar O 92090A pypi.python.org/pypi/xrt, pythonhosted.org/xrt
[6] Henke B L, Gullikson E M and Davis J C 1993 *At. Data Nucl. Data Tables* **54** 181–342

Journal of Physics: Conference Series **712** (2016) 012019 doi:10.1088/1742-6596/712/1/012019

# The CAT-ACT Beamline at ANKA: A new high energy X-ray spectroscopy facility for CATalysis and ACTinide research

**A Zimina[1,2], K Dardenne[3], M A Denecke[4], J D Grunwaldt[1,2], E Huttel[5], H Lichtenberg[1,2], S Mangold[5], T Pruessmann[1,2], J Rothe[3], R Steininger[5] and T Vitova[3]**

[1]Institute of Catalysis Research and Technology (IKFT), Karlsruhe Institute of Technology, Germany
[2]Institute for Chemical Technology and Polymer Chemistry (ITCP), Karlsruhe Institute of Technology, Germany
[3]Institute for Nuclear Waste Disposal (INE), Karlsruhe Institute of Technology, Germany
[4]The University of Manchester, Dalton Nuclear Institute, Manchester, United Kingdom,
[5]ANKA Synchrotron Radiation Facility, Karlsruhe Institute of Technology, Germany

E-mail: anna.zimina@kit.edu

**Abstract**. A new hard X-ray beamline for CATalysis and ACTinide research has been built at the synchrotron radiation facility ANKA. The beamline design is dedicated to X-ray spectroscopy, including 'flux hungry' photon-in/photon-out and correlative techniques with a special infrastructure for radionuclide and catalysis research. The CAT-ACT beamline will help serve the growing need for high flux/hard X-ray spectroscopy in these communities. The design, the first spectra and the current status of this project are reported.

## 1. Introduction

Synchrotron radiation based techniques have become key speciation methods in actinide [1] and catalysis research [2,3]. In the field of actinide research reliable radionuclide speciation techniques are important to characterize, e.g. geochemical processes such as mobilization/immobilization of long-lived actinide and fission product nuclides to assess long-term safety of the disposal of nuclear waste. The increasing need to understand degradation processes of nuclear waste forms such as discharged reactor fuel and vitrified reprocessing residues during extended interim storage demands the development of speciation methods capable of the direct characterization of highly radioactive samples with a minimum of pre-treatment. In this context standard X-ray absorption spectroscopy (XAS) and advanced high resolution X-ray emission techniques have been recently shown to be well suited to study the chemical state and electronic structure of radionuclides in various matrices - a prerequisite for predicting their reactivity [1,4,5]. For catalysis research in-situ and operando X-ray based characterization methods emerged as key elements for the rational design of heterogeneous catalysts [2,6-8]. This approach was not only appreciated by the scientific community, but also applied by industry, e.g. Haldor Topsøe A/S as one of the pioneering companies in Europe ([6] and refs therein). Knowledge of oxidation state, coordination environment, and nanoparticle morphology contributed significantly to deeper understanding of heterogeneously catalyzed reactions and serves today as a basis for the design of new catalysts and for first-principles kinetic modeling. At the Karlsruhe Institute of Technology a new hard X-ray beamline for CATalysis and ACTinide research at the ANKA synchrotron has been installed and commissioned and will be operated in close collaboration with ANKA scientists by the two KIT communities having essential expertise in these two application fields. The beamline design places emphasis on XAS in terms of X-ray absorption near edge structure (XANES) and extended X-ray absorption fine structure (EXAFS) combined with 'flux hungry' photon-in / photon-out techniques like X-ray emission spectroscopy (XRES), X-ray diffraction (XRD) and further correlative spectroscopies (infrared spectroscopy, IR, UV-vis spectroscopy), including gas analysis (mass spectrometry, on line IR, GC-analysis) in catalytic studies. Special infrastructure for experiments with radioactive samples and catalytic studies under reaction conditions is currently underway. The CAT-ACT beamline will help serve the growing need for high flux / high energy spectroscopy, complementing strongly overbooked existing beamlines in these fields.

## 2. Beamline layout and characteristics

Figure 1 shows the main components of the CAT-ACT beamline with a superconducting multipole wiggler (2.5 T, K=11.2, 15 mm vacuum gap, 10.2 keV critical photon energy, 4.7 kW radiated power), two mirrors for beam collimation, focusing and rejection of higher harmonics (vertically collimating Si mirror coated with Rh and Pt stripes and toroidal focusing mirror with Rh and Ir stripes) and a double crystal monochromator with Si(111) and Si(311) crystal pairs (fixed-exit, $LN_2$ cooled crystals, direct drive goniometer with maximal speed of 4°/s). The components are designed to provide high intensity (flux ~ $10^{13}$-$10^{14}$ photons/s/100 mA) in a wide photon energy range (3.1-60 keV) at fixed positions (beam size with mirrors ~1x1 mm, mirrorless operation above about 35 keV) at two end stations (CAT and ACT) in two separate experimental hutches, built in series.

The wiggler source gives access to K absorption edges of heavy elements like Rh, Pd, Ag and Ce (of particular interest for catalysis research) and f-element actinide $L_1$-edges and lanthanide K-edges up to Gd, while the low energy limit of the beamline still allows investigations at the K-edges of transition metals and actinides $M_{4,5}$-edges. Compared to ANKA bending magnet radiation, the photon flux at 20 keV will be about two orders of magnitude higher, which is especially beneficial with respect to the spectroscopic sensitivity for dilute sample systems, e.g., for studying catalyst promoters or poisons or trace concentrations in environmental radionuclide studies.

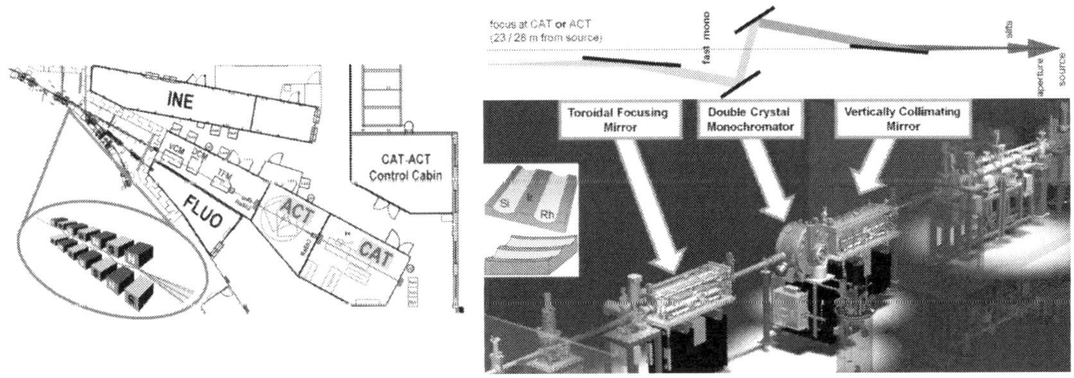


**Figure 1.** Layout of the CAT-ACT beamline at ANKA with two in-line experimental stations (left). CAT-ACT optical components designed and manufactured by FMB Oxford (Oxford, United Kingdom). From right to left: frontend components (intensity absorber, X-ray beam position monitor, white beam slit), collimating mirror, double crystal monochromator, diagnostics module, toroidal focusing mirror, beam shutter separating ACT hutch from optics section. Inset: focusing mirror coating scheme.

## 3. Experimental hutches

The CAT-ACT beamline comprises two alternately operated experimental stations. The ACT experimental hutch will be operated by KIT-INE as a temporary controlled area with the necessary infrastructure for handling radioactive samples, adhering to the double containment concept established at the ANKA INE beamline [9] to safely enclose all radioactive materials by two independent layers of protection. Measurements of non-fissile radionuclides with activities up to $10^6$ times the European exemption limit and 200 mg of fissile U-235 or Pu-239 will be possible. The experimental infrastructure includes standard equipment for XAS measurements in transmission and fluorescence detection mode, as well as a multi analyzer crystal high resolution X-ray emission spectrometer as central component. The latter has already been designed and commissioned at the INE beamline. Equipment for diffraction techniques (high energy X-ray scattering and single crystal X-ray diffraction) are envisaged as possible future upgrades.

The CAT experimental station will be operated by KIT-ITCP and KIT-IKFT and equipped with a special infrastructure for in-situ and in-operando catalytic studies, including supply of reactive gases. A number of specially designed sample environments for catalytic experiments under realistic reaction conditions in gas phase, at high pressure, high temperature or in liquid phase have been set up. They are presently complemented by further, more demanding set-ups for combining XAS with complementary characterization techniques (e.g., infrared spectroscopy and X-ray diffraction). They will allow to analyze the local molecular environment and the long range order structures of catalytically active materials and to study adsorbed reaction species simultaneously.

## 4. Project milestones

In 2011 the ANKA experimental hall was extended to accommodate the CAT-ACT experimental hutches and the control cabin. The superconducting multipole wiggler, designed and built by Budker Institute of Nuclear Physics (Novosibirsk, Russia), was installed into a short straight section of the ANKA storage ring in July 2014. The frontend beamline section built by FMB Berlin (Berlin, Germany), containing diagnostic elements and white beam slits, the radiation protection hutches comprising three separate sections (one for optics and the two end stations, ACT and CAT) and the lock room providing controlled access to the ACT station were installed shortly afterwards. The optical components were designed and installed inside the optics hutch in 2015 by FMB Oxford (Oxford, United Kingdom). Currently the commissioning of the optics components, implementation of the control system and integration of the CAT-ACT beamline into the personal safety and radiation protection system at ANKA are in progress. Installation of special infrastructure for actinide and catalysis research will be finished in the middle of 2016.

## 5. First results

Pilot experiments with synchrotron light were successfully conducted after the first commissioning phase in 2015 and first XAS spectra of samples relevant for both scientific fields were measured.

Research at ITCP and IKFT is focused on catalysts for energy related applications (solar fuels, conversion of biomass platform molecules), selective oxidation, sustainable fine chemical processes and exhaust gas purification. Figure 2 shows Pt $L_3$–edge XAS spectra of two differently treated diesel oxidation catalysts which have been subject of intensive research [10,11], along with the spectrum of a Pt foil. The catalysts were prepared by incipient wetness impregnation of $\gamma$-$Al_2O_3$, dried, calcined and reduced in 5% $H_2$/He ("fresh" catalyst) and then aged by mild hydrothermal aging procedure ("aged" catalyst). Spectra were measured in transmission mode using ionization chambers. The quality of the spectra shows already at this stage the endstation's potential for future studies in catalysis. They furthermore reveal the Pt particles are more reduced after ageing as a result of particle sintering during aging (reoxidation of small Pt-paricles in air is faster for the small freshly reduced Pt-particles).

One of the major aspects of research performed at INE is the incorporation of nuclear fuel reprocessing residues (i.e., highly radioactive waste concentrates, 'HAWC') in specially designed glass matrices - a key step in immobilizing and conditioning separated minor actinides and fission products prior to their transport and final disposal. Nuclear waste glass simulates are investigated to optimize miscibility and radionuclide loading capacities of glass formulations used in a real industrial process. Figure 3 depicts the first U $L_3$-edge XAS spectra recorded at CAT-ACT. The red curve shows the spectrum of a borosilicate glass doped with 4.7 wt.% $UO_2$, prepared under oxidizing conditions at 1200 °C in a Pt crucible [12]. Comparison to the spectrum of uranyl ($(O{=}U^{VI}{=}O)^{2+}$) species present in the mineral meta-schoepite (blue curve) unambiguously proves that uranium in the glass is oxidized to the hexavalent state during the melting process.

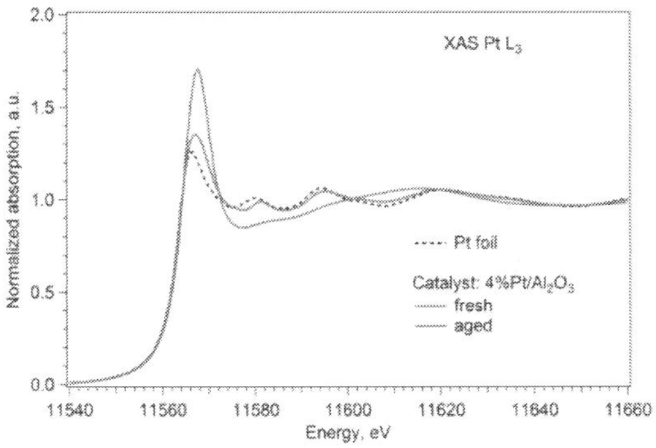


**Figure 2.** Pt $L_3$-edge XAS spectra of freshly prepared and aged 4%Pt/$Al_2O_3$ diesel oxidation catalysts measured in transmission mode at CAT-ACT. The spectrum of a Pt foil was measured as reference and for energy calibration and is shown for comparison.

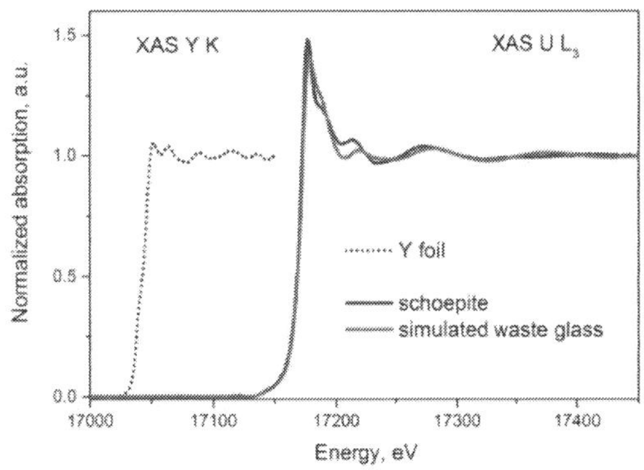


**Figure 3.** U $L_3$-edge XAS spectra of U(VI) or 'uranyl' species in the mineral meta-schoepite ($UO_3$ • $nH_2O$) and a simulated U-borosilicate nuclear waste glass (4.7 wt.% $UO_2$ loading). Y metal was measured as energy calibration standard.

**6. Acknowledgement**
The authors thank KIT, BMBF (within the project "Materials in Action"), Helmholtz programs Renewable Energies (EE), Science and Technology of Nanosystems (STN), Storage Systems and Cross-linked Infrastructure (SCI), from Matter to Materials and Life (MML), Nuclear Waste Management and Safety (NUSAFE) and the European Commission Joint Research Centre – Institute for Transuranium Elements (JRC-ITU) for financial support. T.V. received funding through the Helmholtz Young Investigator Group VH-NG-734. The simulated glass was kindly provided by S. Bahl (INE), the catalyst - by A. Gänzler (ITCP).

**References**
1. Kvashnina K O, Kvashnin Y O, Butorin S M, 2014 *J. Electron Spec. Relat. Phenom.* **194** 27
2. Grunwaldt J-D, Caravati M, Hannemann S, Baiker, A 2004 *Phys. Chem. Chem. Phys.* **6** 3037
3. Bordiga S, Groppo E, Agostini G, van Bokhoven J A, Lamberti C 2013, *Chem. Rev.* **113** 1736
4. Dardenne K, González-Robles E, Rothe J, Müller N, Christill G, Lemmer D, Praetorius R, Kienzler B, Metz V, Roth G, Geckeis H 2015 *J. Nucl. Mater.* **460** 209
5. Vitova T, Green J C, Denning R G, Löble M, Kvashnina, K, Kas J J, Jorissen K, Rehr J J, Malcherek T, Denecke M A 2014 *Inorg. Chem.* **54** 174
6. Topsøe H 2003 *J. Catal.* **216** 155
7. Weckhuysen B M 2003 *Phys. Chem. Chem. Phys.* **5** 4351
8. Grunwaldt J-D and Clausen B S 2002 *Topis Catal.*,**18** 37
9. Rothe J, Butorin S, Dardenne K, Denecke M A, Kienzler B, Löble M, Metz V, Seibert A, Steppert M, Vitova T, Walther C, Geckeis H 2012 *Rev. Sci. Instrum.* **83** 043105
10. Boubnov A, Gänzler A, Conrad S, Casapu M. and Grunwaldt J-D 2013 *Topics Catal.* **56** 333
11. Gänzler A M, Casapu M, Boubnov A, Müller O, Conrad S, Lichtenberg H, Frahm R, Grunwaldt J-D 2015 *J. Catal.* **328**, 216
12. Ollier N 2003 *Opt. Mater.* **24**, 63

# Fast EXAFS in synchronous scanning mode at PETRA P06

**Roman Chernikov[1], Edmund Welter[1], Wolfgang Caliebe[1], Gerd Wellenreuther[2] and Gerald Falkenberg[1]**

[1] DESY Photon Science, Notkestrasse 85, 22607 Hamburg, Germany
[2] European XFEL, Albert-Einstein-Ring 19, 22607 Hamburg, Germany

E-mail: roman.chernikov@desy.de

**Abstract.** First fast EXAFS spectra have been successfully measured at the Hard X-ray Micro-probe Beamline P06 (PETRA III) synchronously scanning the U32 undulator gap and the monochromator Bragg axis. Stability and spatial homogeneity of the beam are proved to be the limiting factors for the quality of the EXAFS spectra, whereas the performance of the data acquisition electronics limits the energy resolution when total scan time is reduced to 30 seconds. Results of the test measurements give us the estimate of the utmost performance of 5-10 seconds per full EXAFS scan considering certain hardware modifications.

## 1. Introduction

One of the goals in an EXAFS experiment is to measure the highest quality spectra in the shortest time; in practice a typical measurement at a bending magnet source could last up to 20-30 minutes in order to accumulate decent signal-to-noise level. Nowadays the flux created by insertion devices is high enough to provide sufficient photon counts even at millisecond timescales. The efficiency of conventional step scans therefore tends to zero, as a significant portion of the time is wasted for monochromator movement and stabilization. Moreover long exposure of the samples to high intensity beam raises the problem of radiation damage. The general solution in this case is operation in continuous mode, when the energy is scanned non-stop at fixed velocity and the signal is read out on the fly. This approach works flawlessly at bending magnets or wigglers that generate a smooth spectrum, however its implementation at an undulator source is non-trivial – one must either scan the undulator gap together with the Bragg angle [1,2] or taper the undulator gap [3.4]. The former guarantees higher flux and predictable spatial profile of the beam for the price of lower mechanical reliability and the need for precise monochromator-undulator movement synchronization. The latter suffers from the limited energy bandwidth and complicated spatial structure of the beam changing along the scan. Both approaches are feasible at the Hard X-Ray Micro-probe Beamline P06 (PETRA III). Here we present the results of our attempts to apply the synchronous scanning for the fast EXAFS measurement on powder samples.

## 2. Synchronization of the undulator and monochromator movement

Details of the beamline layout are published elsewhere [5], the most important are the PETRA III spectroscopy undulator U32 ($\lambda_u$=31.4mm) and the FMB-Oxford High Heatload Monochromator. All beamline components including the measurement system are controlled via TANGO, the measurement routine was written in Python invoking the pyTango module according to the flowchart at figure 1.

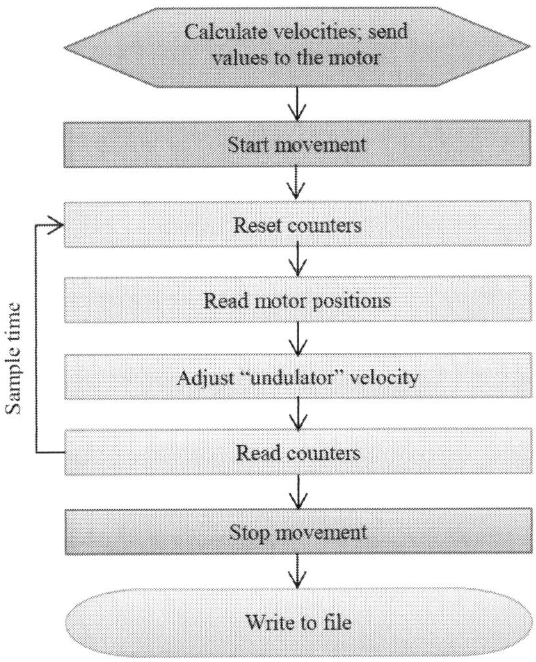


**Figure 1.** Flowchart of the control procedure

Our main goal was to synchronize the continuous Bragg axis movement with the undulator gap scan in order to match the same photon energy. The task is complicated by a nonlinear relation between the motor- and energy axes; for the undulator the photon energy is related to the gap width via the magnetic field:

$$E_{und} = \frac{9.5\ E_e^2}{(1 + K^2/2)\lambda_u}$$

$$K = 0.0934\ B_0\lambda_u$$

$$B_0 = a_u \exp(b_u\left(\frac{g}{\lambda_u}\right) + c_u\left(\frac{g}{\lambda_u}\right)^2)$$

Here $E_e$ is the particle energy in the storage ring in GeV, $\lambda_u$- undulator period in mm, $K$ – deflection parameter, $B_0$ – maximum magnetic field in T, g – full magnetic gap of the undulator in mm; $a_u$, $b_u$ and $c_u$ – undulator-specific characteristic parameters [6]. Rotation of the Bragg axis of the monochromator is converted to the energy units according to the well-known Bragg law $E_{mono} = c\hbar/2d \sin\Theta$, where $c$ denotes the speed of light in vacuum, $\hbar$ - the Planck constant, $d$ –spacing between lattice planes, Θ – Bragg angle.

In the operational range of the undulator the shape of $E_{und}(g)$ is almost linear while $E_{mono}(\Theta)$ is cosecant and in case of the simultaneous linear movement of the motors along $g$ and Θ axes the misalignment between $E_{und}$ and $E_{mono}$ will reach 20 eV over the 1000 eV range. For the narrow spectral peak this energy difference can result in 25% variation of flux intensity and introduce unwanted variation in the spatial profile of the beam.

Therefore, the control routine includes the negative feedback loop which reads the positions of the motors defining the Bragg angle and undulator gap width from corresponding encoders, estimates the misalignment and regulates the motors movement. Due to limitations of the control system it was not possible to adjust the Bragg axis rotation speed once the movement was started, so the monochromator was set as the master device while the undulator operated in the slave mode making the gap width follow the rotation of the crystals. The desired energy range and total measurement time were set as initial parameters for the calculation of motor positions and movement velocities.

Resulting differences in the energy positions (synchronization precision) of undulator and monochromator did not exceed 2 eV over the range of 1000 eV, so that the beam was always matching the top of the undulator 3-rd harmonic peak.

The energy resolution was limited by the readout performance (sample time), that in turn is determined by the delay in communication between the beamline computer and Tango servers.

$$dE = t_{sample}\frac{E_{max} - E_{min}}{t_{total}}$$

Distribution of the readout time calculated from the timestamp is given at the figure 3, average sample time amounted ~ 45ms. Thus, for the 120s scan over a 1000 eV range one could estimate the energy step to be $dE \sim 0.375$ eV and reaching up to 0.75 eV in certain points.

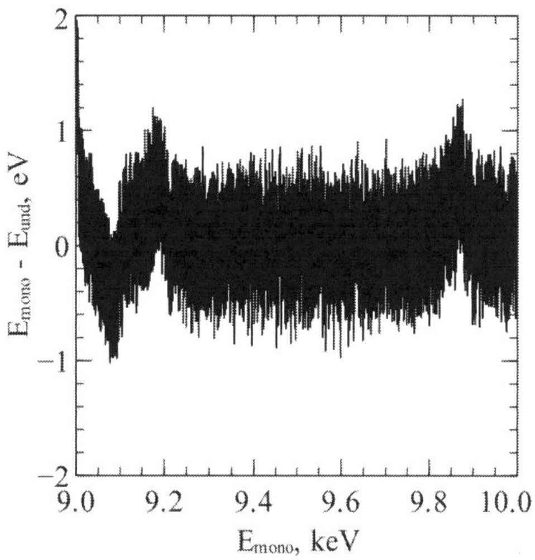


**Figure 2.** Energy misalignment in a 120s scan

In a 30s scan the XANES region is already distorted due to insufficient energy resolution, therefore, in order to have both long EXAFS and high quality XANES one needs a faster communication interface and faster electronics.

The other problem we address is the beam quality – both spatial and temporal. The beam is highly coherent in the vertical plane, so propagation through the entrance slit creates a pronounced Fresnel-like spatial pattern (figure 4). At the same time the long distance between the monochromator and the sample position converts small vibrations of the crystals into a perceptible beam movement across the sample. The combination of these factors drastically reduces the quality of the spectra of inhomogeneous samples (in contrast to the reference metal foils real powder samples are always inhomogeneous). In order to minimize these effects in our experiment we collimated the beam to 0.5x0.5 $mm^2$ with the slits located 4 m upstream from the sample.

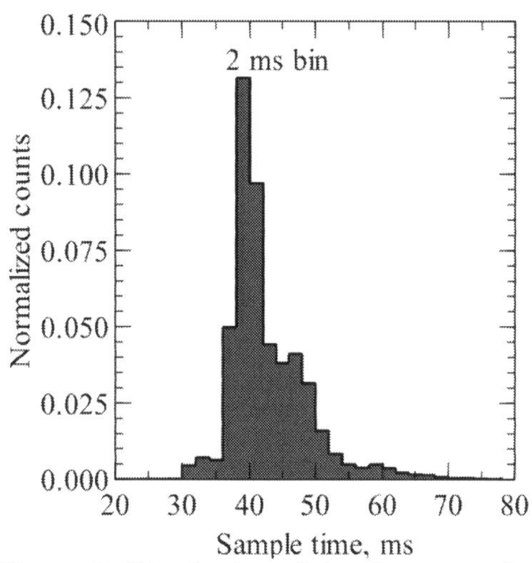


**Figure 3.** Distribution of the time-per-point in a 120s scan

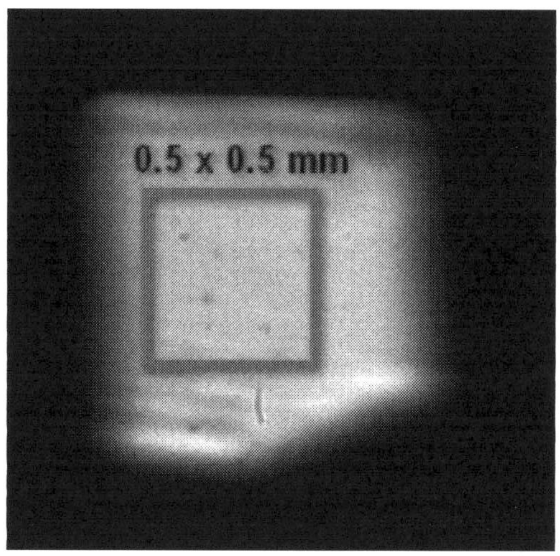


**Figure 4.** Beam profile at the sample

## 3. XAFS measurements

The sample was prepared according to standard procedures by pressing the grinded powder in a pellet with cellulose. X-ray absorption spectra previously measured at beamline C (DORIS) for this very sample were taken for comparison [7], identical EXAFS extraction parameters were used in both cases. The substance we used for the tests, $LuB_{12}$, allows measuring rather long EXAFS and features a sharp white line at the Lu $L_{III}$ edge.

Custom built ionization chambers optimized for the high flux applications were used to measure the incident and transmitted beam intensity [8]. Both chambers were filled with nitrogen at optimal pressure and operated at high (2 kV) bias-voltage. The currents from the ionization chambers were amplified by Keithley 428 current amplifiers and digitized with 2 MHz VFC ADC [9].

The XANES spectrum and the $k^2$ weighted EXAFS function are compared in figures 5 and 6 with the results from a conventional step scan (accumulation time increasing proportionally to $k^2$ in EXAFS region). A slight deformation of the white line due to insufficient energy resolution is easily noticeable in the 30s scan.

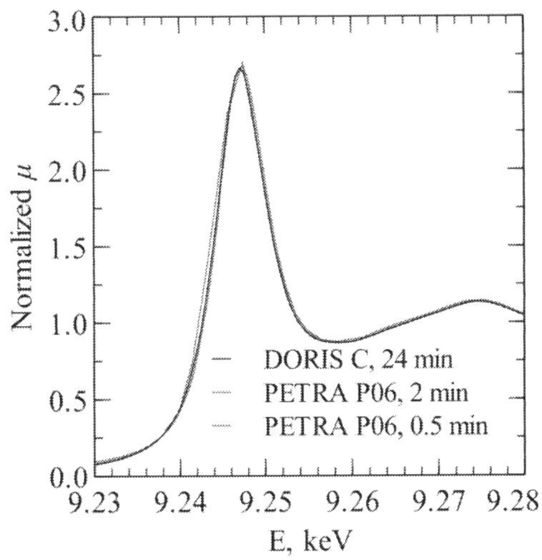


**Figure 5.** $LuB_{12}$ $L_{III}$-Lu XANES

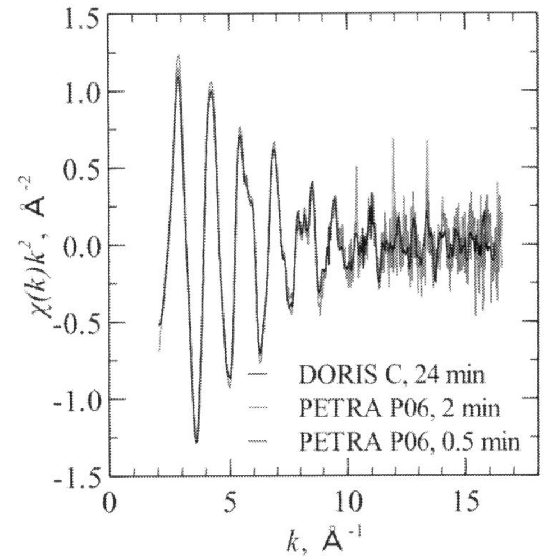


**Figure 6.** $LuB_{12}$ $L_{III}$-Lu EXAFS function

High frequency oscillations in the high-*k* region of EXAFS spectra do not distort the Fourier transform (not shown here) and allow accurate data reduction. Here the quality of the spectra is more likely limited by the inhomogeneity of the sample than by the insufficiency of the photon counts.

## 4. Conclusion

We have proved the feasibility of fast EXAFS experiment in synchronous scan mode at the PETRA III undulator beamlines. We expect that this mode can be implemented at other beamlines built on the similar hardware and relying on TANGO as the control framework.

[1] Rogalev A, Gotte V, Goulon J, Gauthier C, Chavanne J and Elleaume P 1998 *J. Synchrotron Rad* **5** 989
[2] Oyanagi H, Ishii M, Lee C-H, Saini N, Kuwahara Yu, Saito A, Izumi Ya and Hashimoto H 1999 *J. Synchrotron Rad* **6** 155
[3] Curtin M, Bhowmik A, Brown J, McMullin W, Metty P, Benson S V and Madey J M J 1990 *Nucl. Instrum. Methods Phys. Res. A* **296** 69
[4] Boyanov B, Bunker G, Lee J M and Morrison T I 1994 *Nucl. Instrum. Methods Phys. Res. A* **339** 596
[5] Schroer C, Boye P, Feldkamp J, Patommel J, Samberg D, Schropp A, Schwab A, Stephan S, Falkenberg G, Wellenreuther G and Reimers N 2010 *Nucl. Instrum. Methods Phys. Res. A* **616** 93
[6] Kim K J 1989 AIP Conf. Proc. **184** 565
[7] Menushenkov A P, Yaroslavtsev A A, Zaluzhnyy I A, Kuznetsov A V, Chernikov R V, Shitsevalova N Yu and Filippov V B 2013 *JETP Letters* **98** 165
[8] Müller O, Stötzel J, Lützenkirchen-Hecht D and Frahm R 2013 *J. Phys.: Conf. Ser.* **425** 092010
[9] Kracht Th and Zink H 2005 *HASYLAB Annual Report* **1** 119

Journal of Physics: Conference Series **712** (2016) 012021 doi:10.1088/1742-6596/712/1/012021

# LISA: the Italian CRG beamline for x-ray Absorption Spectroscopy at ESRF

**F. d'Acapito*, A. Trapananti, A. Puri**

CNR-IOM-OGG c/o ESRF, 71 Avenue des Martyrs, Grenoble (France)

E-mail: *dacapito@esrf.fr

**Abstract.** LISA is the acronym of **L**inea **I**taliana per la **S**pettroscopia di **A**ssorbimento di raggi X (Italian beamline for X-ray Absorption Spectroscopy) and is the upgrade of the former GILDA beamline installed on the BM08 bending magnet port of European Synchrotron Radiation Facility (ESRF). Within this contribution a full description of the project is provided.

## 1. Introduction

LISA is the name of the new beamline that will replace GILDA [1] at the BM08 bending magnet port of ESRF. The beamline will be dedicated to X-ray Absorption Spectroscopy and its main characteristics will be:

- high sensitivity for the analysis of diluted samples;
- high spectral and mechanical stability for the collection of low noise XAS data in transmission mode;
- use of grazing incidence/total reflection data collection in linear dichroic mode for surface analysis;
- data collection in differential mode;
- combined multitechnique *in situ* data collection for extreme conditions or *in operando* experiments.

Achieving these targets requires particular care about the quality of the beam spot in terms of homogeneity, spatial and energy stability, reduced size and divergence. To carry out the outlined research programs, the beamline has to fulfill several requirements. First, it has to cover a wide energy range (ex. $5 < E < 70$ KeV) to carry out structural analysis (using XAS at the K absorption edge) of technologically crucial materials like catalysts (containing Mo, Pd), hydrogen reservoirs (Nb, Pd), protonic conductors (In, Ba, Ce, Zr), solar cells and transparent conductors (In,Sn), luminescent and advanced magnetic materials (Rare Earths). Furthermore, it has to provide a beam with reduced divergence, in order to carry out experiments based on X-ray reflectivity, Grazing Incidence and linear dichroic mode. This will allow the structural analysis of thin films, interfaces or the interaction between adsorbed species and model surfaces, as needed for example in anisotropic magnetic materials or environmental science. The upgraded instrument has to provide an intense sub-mm beam in the whole energy range for studies under (moderate) extreme conditions of high temperature or pressure and for pump-probe experiments with electric field or light excitation. And, finally, the optics should be compatible with the new

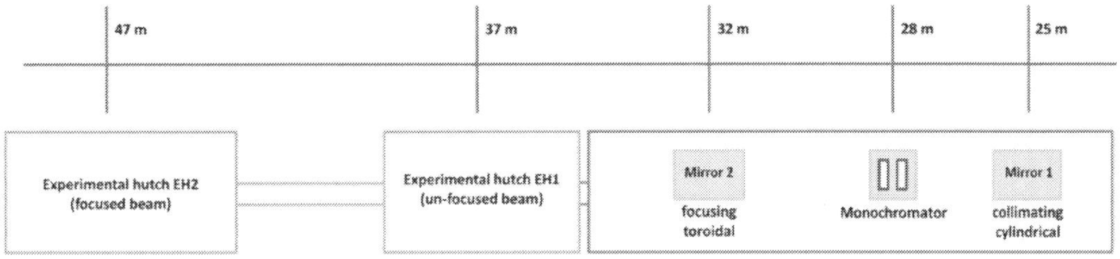


**Figure 1.** Sketch of the proposed beamline: the Optic Hutch (OH) contains all the main optical elements whereas the experiments are carried out in the Experimental Hutches EH2 (focused beam) and EH1 (unfocused beam). In the upper part of the picture the distance of the various elements from the source is indicated.

source that will be available after the upgrade of the ESRF ring that will consist in a 2 poles wiggler with a limited horizontal divergence (1.6 mrad) [2].

## 2. Technical design

### *2.1. Layout*

The layout of the beamline is shown in Fig.1 and will consist in three cabins: an Optical Hutch (OH) and two Experimental Hutches (EH1 and EH2). The OH will contain the optical elements to monochromatize and focus the beam in the middle of the second experimental hutch EH2. Here, there will be the main experimental setup consisting in two sample stations mounted on a long granite bench. A first experimemtal hutch (Experimental Hutch 1, EH1) will be used for experiments with a non focused beam such as EXAFS in transmission mode.

The beamline optical scheme will consists in a collimating first mirror, a double crystal monochromator and a focusing second mirror. The beamline will use mirrors in the energy range 5-40 keV while above 40 keV the monochromator will be the only optical element.

### *2.2. X-ray source*

The beamline will be built on the high field part (0.85 T) of the BM8 bending magnet of ESRF. At this point the X-ray source has size $75\mu m$ (H) $\times$ $32\mu m$ (V) FWHM. The source is located at 25 m from the first optical element. The realization of the new lattice of ESRF [2] will lead to a two pole wiggler source with smaller size ($13 \times 4\mu m$) and an increased distance (+2.5 m) from the Front End. The design is such that it will be possible to cope with these changes and to exploit at maximum the foreseen new features.

### *2.3. Mirrors*

The problem of beam focusing is a major issue for the present project. To meet the requirements stated in the scientific case it is necessary to have at the same time high brilliance, flux and beam homogeneity, conditions that are difficult to meet at the same time. Although proposed long agon in literature [3] only recently a considerable interest for toroidal mirrors has emerged driven by the achromaticity of these elements and the strong technical improvement of these devices emerged in the latest years. A mirror-based focusing scheme has been adopted for recently designed beamlines at APS [4], B18 at DIAMOND [5], the ROBL beamline at ESRF [6], the CLAESS beamline at ALBA [7]. This solution has been retained for the present project as it ensures at the same time a high intensity and homogeneous beam with a reduced size.
The first mirror will have a cylindrical shape and it will be used for collimating the beam to

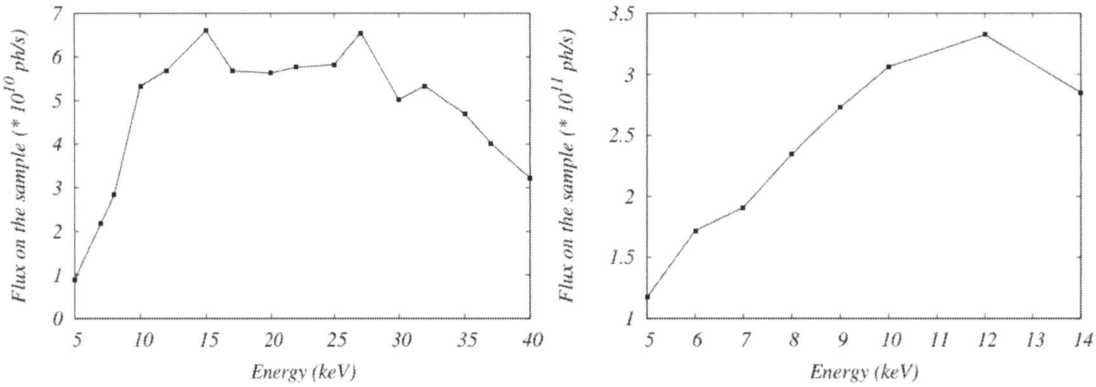


**Figure 2.** Flux foreseen on the focal spot in the case of two possible crystal planes for the monochromator: Si(311) left, Si(11), right. For the EH1 the flux will be about an order of magnitude lower due to the lack of focalization. The electron beam energy was fixed at 6 GeV and the current at 200mA. The input acceptance was 1mrad hor. and 43 $\mu$rad vert. On the focal spot the beam size is calculated to be 103 $\mu$m hor. and 143 $\mu$m vert. FWHM and a divergence of 50 hor. $\times$ 4 vert. mdeg.

achieve an instrumental energy resolution well below the core-hole width of the K edges in the energy range of interest. This mirror will have 2 stripes: Pt and Si in order to cover the whole energy range harmonic free. Si will cover the region 5-15 keV whereas Pt will cover 15-40 keV. The second mirror will have a toroidal shape and will focalize the beam in a 2:1 condition. It will consist in two cylindrical channels dug in a single bendable silicon substrate. This solution has been shown to minimize the impact of chromatic aberrations in the focal spot [4] and keep at a moderate value the overall length of the beamline. The part between the channels section of the second mirror will be left flat to make available to users a non focused beam. Both mirrors will work at 2 mrad incidence angle. A pair of Pt-coated plane mirrors (already present on the beamline) working at 10mrad will be used to reject harmonics in the lowest part of the energy range.

### *2.4. Monochromator*

The monochromator will be a fixed-exit device with a single rotation axis and will use flat crystals cooled with liquid nitrogen. Two pairs of crystals will be mounted to access the whole energy range: Si(311) will cover the interval 5-40 keV ensuring high energy resolution, whereas Si(111) crystal will be used to obtain high flux in the region 5-15 keV. The crystals will be permanently mounted inside the monochromator and the change will be realized by simple horizontal translation of the instrument.

### *2.5. Ray Tracing*

A detailed Ray-Tracing calculation of the parameters of the beamline has been carried out using the SHADOW code [8]. The new beamline is shown to provide a high photon flux ($10^{10}$ – $10^{11}$ ph/s) in a wide energy range and a small beam size in compliance with the experimental requirements cited in the previous sections (See Fig. 2). The constraint of having a source with a low horizontal divergence in the future limits the flux presently available but an increase of about a factor 2 is foreseen by calculations with the 2 pole wiggler parameters.

### *2.6. Endstations*
The beamline will have two experimental hutches (EH) with the main hutch (EH2) be designed for applications requiring a focused beam. The setup will consist in a long granite bench carrying detectors (ion chambers and Ge multi-element fluorescence detectors) and two sample stations. The first experimental station will host a vacuum chamber containing two manipulators: one for standard XAS experiments in fluorescence mode and the other for ReflEXAFS experiments. The second block will be a versatile table with basic translation/rotation stages to be used to mount bulky experimental setups and for combined multi-technique experiments. The first experimental hutch EH1 will be designed for the collection of high quality XAS spectra in transmission mode with un-focused beam. The standard beamline sample environment will include a liquid helium/nitrogen cryostat, a cell for chemical gas-solid reactions and an oven for high temperature measurements up to 1500 K.

## 3. Conclusion
The design of the new Italian beamline at ESRF "LISA" has been presented. A description of the X-ray optics scheme as well as the layout of the experimental stations has been given. The full project is foreseen to be operative in about two years.

## References
[1] D'Acapito F, Colonna S, Pascarelli S, Antonioli G, Balerna A, Bazzini A, Boscherini F, Campolungo F, Chini G, Dalba G, Davoli I, Fornasini P, Graziola R, Licheri G, Meneghini C, Rocca F, Sangiorgio L, Sciarra V, Tullio V and Mobilio S 1998 *ESRF Newsletter* 42
[2] 2015 The ESRF Orange Book URL `http://www.esrf.eu/home/about/upgrade/documentation/orange-book.html`
[3] Cowan P, Brennan S, Deslattes R, Henins A, Jach T and Kessler E 1986 *Nucl. Instrum. and Meth. A* **246** 154 – 158
[4] MacDowell A A, Celestre R S, Howells M, McKinney W, Krupnick J, Cambie D, Domning E E, Duarte R M, Kelez N, Plate D W, Cork C W, Earnest T N, Dickert J, Meigs G, Ralston C, Holton J M, Alber T, Berger J M, Agarde D A and Padmorea H A 2004 *J. Synchrotron Rad.* **11** 447
[5] Dent A J, Cibin G, Ramos S, Smith A D, Scott S M, Varandas L, Pearson M R, Krumpa N A, Jones C P and Robbins P E 2009 *J. Phys. Conf. Ser.* **190** 012039
[6] Scheinhost A the ROBL beamline at ESRF URL `http://www.hzdr.de/db/Cms?pNid=247`
[7] The CLAESS beamline at ALBA URL `https://www.cells.es/en/beamlines/bl22-claess`
[8] Lai B and Cerrina F 1986 *Nucl. Instrum. Methods, A* **246** 337

Journal of Physics: Conference Series **712** (2016) 012022 doi:10.1088/1742-6596/712/1/012022

# Upgrades to the XAFS2 beamline control system and to the endstation at the LNLS

**S J A Figueroa[1,*], J C Mauricio[1], J Murari[1], D B Beniz[1], J R Piton[1], H H Slepicka[1], M Falcão de Sousa[2], A M Espíndola[1], A P S Levinsky[1]**

[1] Centro Nacional de Pesquisa em Energia e Materiais (CNPEM), Laboratório Nacional de Luz Síncrotron (LNLS), Rua Giuseppe Máximo Scolfaro, 10.000, Polo II de Alta Tecnologia, Caixa Postal 6192, 13083- 970 Campinas, SP, Brasil.

[2] Universidade Federal do Ceará, Departamento de Física, Laboratório de Raios X, Caixa Postal 6030, Campus do Pici, 60455-970 Fortaleza, CE, Brasil.
[*]E-mail: santiago.figueroa@lnls.br

**Abstract**. The XAFS2 is a general-purpose X-ray absorption beamline. It is the second one built at the LNLS. After approximately 7 years in operation this beamline has been substantially updated in order to improve its experimental possibilities. Recently arrived, a 4-circle Huber diffractometer has been incorporated to perform combined experiments. This collects XRD patterns with the XAFS. Through the development of a new sampling environment it is now also possible to perform these measurements in situ/operando conditions. Other upgrades include a complete remodelling of the beamline software and its control system. The following new systems are crucial for the next steps that are currently underway at the beamline, namely, (i) enabling remote access for users and (ii) the testing of QEXAFS measurements.

## 1. Introduction

Upon the completion, in 2007, of the commissioned works on the XAFS2 [1], a large number of users have been using this experimental facility in order to perform several kinds of experiments [2-5]. Since 2007, and with the development of combined simultaneous characterizations [6], there are a plethora of new possibilities additional to the usual XAFS experiments. Among these numerous possibilities; a combined XAFS and XRD experiment would be especially useful in order to obtain structural information of the short range order (around a selected absorber atom) and of the long range order, thus unveiling information on the material as a whole. Here one is probing different properties of the same system in a unique experiment under the same conditions, thus bridging the complex gaps in the structure. Therefore, in order to improve those abilities of the beamline, a 4-circle Huber Diffractometer was added to the experimental hutch in order to explore both techniques [7] through user demand. Here below an example of these XAFS and XRD measurements is presented. The example shows the performance of the beamline at work while combining both techniques.

Importantly, the control system of the beamline has been renovated by the installation of a PXI. The PXI is from National Instruments (PXI-NI) and communicates with Galil/Parker controllers on an EPICS platform [8]. Note some parts of the motors were changed in order to improve performance with the upgrade and there were also important changes to the control hardware. The Windows operational system was replaced with Red Hat Linux and the *3-WinDCM* control system [9] with EPICS. Furthermore, a new python based script (*Py4Syn*) was added. This provides high-level abstraction for device manipulation, scan routines, real-time plots and more [10]. This package was created with the aim of providing a simple yet powerful tool to allow scientists and users to develop their own scripts for data acquisition. For user-friendly interface builds a Control System Studio (CS-Studio) is used.

The next step with the XAFS2 upgrade, namely, towards a high-throughput XAFS beamline, will involve testing the viability of performing *QEXAFS* and testing the remote control at the endstation [11]. This will allow access for users to collect their data without having to travel to the LNLS. This is a work in progress and normalized operation is foreseen to commence by the end of 2015.

## 2. Brief overview of the beamline

The design, construction and performance testing of the majority of the elements of the XAFS2 beamline have been completed *in house*, save the mirrors that were purchased from SESO (France). The beamline is positioned at the 15° port of the D08B bending magnet of the LNLS storage ring. It operates with a 1.37 GeV source with a maximum current of 250 mA, a critical energy of 2.08keV and offers a beam life time close to 17 h without coupling correction [12]. The XAFS2 beamline covers the high-energy x-ray range and delivers photons from 3.5 up to 17 keV. The schematic representation of the main components of the beamline and other useful information can be obtained from the web-site (http://lnls.cnpem.br/beamlines/xafs/beamlines/xafs2/).

The optical hutch shapes the white beam, collimating it vertically via the First Rh-coated cylindrical mirror, which sends a parallel synchrotron beam onto the two flat Si(111) Double Crystal Monochromator (DCM). The DCM has fixed exit geometry and is the only optical element with thermal stabilization (i.e. water cooled). A second Rh coated toroidal bendable mirror refocuses the monochromatic beam, in the sample position, to a low eccentricity ellipse of approximately 450μm in diameter. The typical photon flux, in the sample position, for standard beam size and current ring is among the $10^9$ photon/s range. Harmonic content for energies higher than 6 keV is lower than $3x10^{-5}$. For lower energies, harmonic rejection is performed by detuning the second crystal. The standard beamline setup allows for transmission and fluorescence modes. For transmission experiments three ionization chambers, (for beam intensity measurements) are filled with optimal He, $N_2$ and/or Ar gas mixtures at a total pressure of 1.5 bars.

The signals of the ionization chambers are amplified by Stanford picoammeters and digitalized by a voltage to frequency converter before being finally read by the counters at the PXI-NI. A Germanium 15 elements fluorescence detector, from Canberra Inc., and an electron detector complete the detection equipment. Typical absorption data measured in transmission mode at room temperature is presented in figure 1. The single scan data were processed in Athena [13] without any filtering. Almost noise-free data are visible even for k > 16 Å$^{-1}$, which allows for the securing of detailed and precise information of the short range structural order.

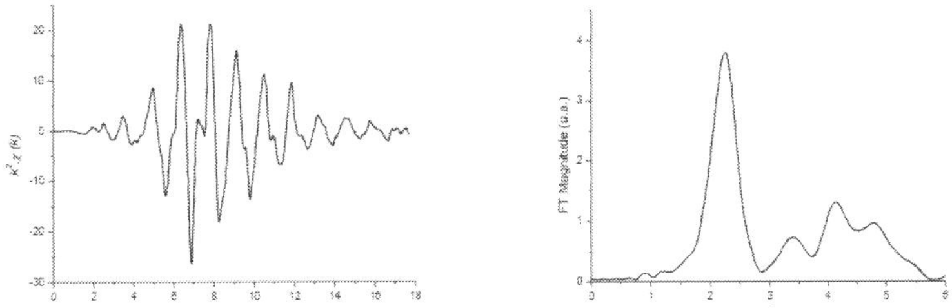


**Figure.1** Cu foil measured at XAFS2: here is shown a $k^2.\chi(k)$ vs. k and the Fourier Transform in R space. Data were collected at 1s per point from 8850 eV up to 9500 eV and then in 3 steps on k space (0.07, 0.05 and 0.03) up to 10500 eV reaching 3s per point on the last step.

Journal of Physics: Conference Series **712** (2016) 012022 doi:10.1088/1742-6596/712/1/012022

## 3. Advances in the experimental set-up and the beamline control

During the summer school program (http://lnls.cnpem.br/education/student-program/bolsas-de-verao/) at the Brazilian Synchrotron Light Laboratory, the XAFS2 beamline was adapted to perform diffraction experiments. This instrumentation development allowed us to open up a new window on the usual XAFS measurements, which is especially useful for combining the techniques to obtain information on short and long range order. Both techniques were explored by performing *in situ* studies with a capillary micro-reactor [14], showing the hydrogen desorption on a $Mg_2CoH_5$ sample (MgCo hydride) during a thermal treatment and under a helium atmosphere.

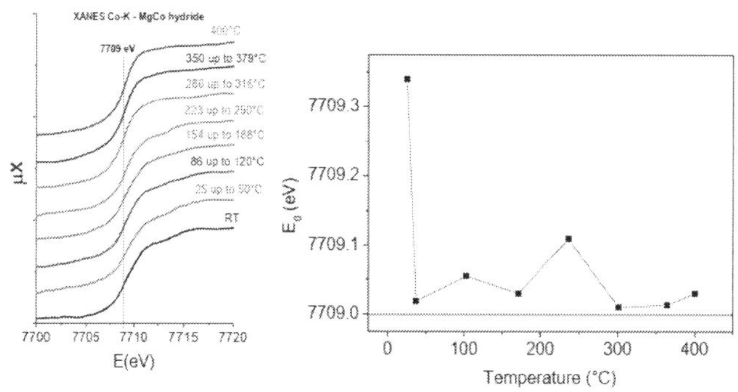


**Figure.2** XANES spectra at Co-K edge on sample $Mg_2CoH_5$ under thermal treatment following the $E_0$ shift. The acquisition time was close to 300s per spectra. Sample measurements are done simultaneously with a Co foil in between $I_1$ and $I_2$ in order to have a good energy calibration (black line at 7709 eV).

The possibility to change between XAFS and XRD experiments (in a few minutes and following the behavior with a mass spectrometer) gives a complete picture of the changes in the material under *in situ* conditions. This is a very promising combination to explore catalysts or functional materials, thus increasing the research potential of the beamline. In figure 2 and 3 it is possible to show the performance of the beamline with this setup. Results in the XANES region, shows slight shifts in the spectrum upon increasing the treatment temperature and characterizes changes in the oxidation state. In this scenario, we use the DCM in a pseudo-channel cut crystal configuration in order to reach an accuracy level of 0.1eV (being the mean of fluctuations at the edge energy when measuring 20 Co foil scans). With the XRD data it is possible to follow the change in the initial tetragonal phase hydride ($\beta$-$Mg_2CoH_5$) to the cubic ($\gamma$-$Mg_2CoH_5$) phase. With the displacement of the diffraction peaks this phase becomes unstable at higher temperatures and the complex hydride decomposes into Mg, Co and $H_2$.

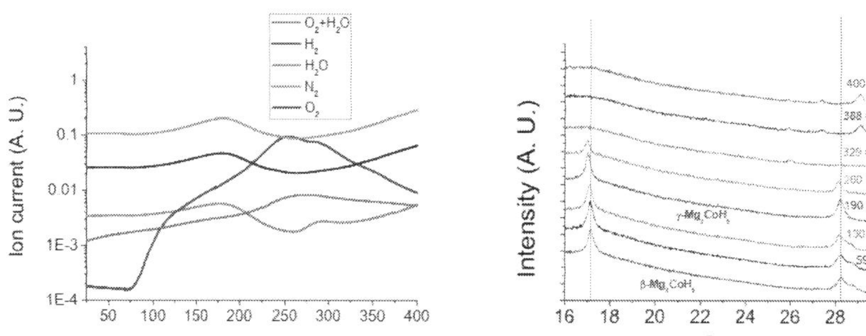


**Figure.3** MS data and XRD patterns obtained on the hydride $Mg_2CoH_5$ under the thermal treatment. The acquisition time for XRD measures was 300s. The energy used to measure XRD data was 11104 eV. Lines in diffractograms guide the eye onto the shifts of the identified phases ($\beta$-$Mg_2CoH_5$ and $\gamma$-$Mg_2CoH_5$).

The results obtained with mass spectrometry indicated a maximum release of $H_2$ at a temperature close to 250°C, an increase in energy is observed at the same temperature when plotted on a graph of $E_0$ *vs.* Temperature (figure 2) indicates a possible increase in the oxidation state, probably due to the loss of electrons from the cobalt upon the $H_2$ release. Correlating the results of diffraction, absorption and Mass Spectrometry (MS), it was noted that the initial release of $H_2$ is not related to change in the structure of the hydride. Rather, it has to do with the departure of the hydrogen interstitial positions in the crystal structure and without it changing per se. More detailed and careful experiments of this kind of system can be shown in the work of Zepon et al [15], differences in temperatures related to the paper are due to their air exposure.

Other changes on the beamline include the integration of motors and the control system in EPICS/Linux replacing the old Delphi/Windows platform. A PXI from the National Instruments(NI) chassis and its modules were made available to EPICS through the NI Real-Time hypervisor virtualization system. In sum, this allows the running simultaneously of the EPICS/Linux and the LabVIEW Real-Time in the same PXI controller, sharing a common memory block as their communication interface. A control software for the beamline developed in Python at the LNLS (*Py4Syn*) [10] completes the picture with a practical and intuitive user interface in CS-Studio. These changes were essential preparations for the next upgrades on the beamline. The first one is to implement the LabWeb project [11] aimed at allowing the remote operation of the beamlines for the user community. The implementation of LabWeb will allow researchers to use the laboratory structure through a web browser, sending the samples to the LNLS without leaving their research centres, reducing time and travel costs in a continental country like Brazil. Thereafter, is the work to develop a fast scan acquisition (QEXAFS) in order to reduce times and in tandem to test the best software and hardware solutions for our future beamlines in the next Brazilian Light Source (Sirius-http://lnls.cnpem.br/sirius/).

**Acknowledgments**
The authors acknowledge the hard work achieved by the staff of GAE (LNLS) and for their invaluable help during the upgrade. And to M. S. Eleotério, A. Barros, F. Zambello and V. Fonseca for their kind support during the experiments, throughout the sampling environment and in-house development. S.J.A.F. thanks H. Westfahl Jr for fruitful comments on the manuscript. Thanks also to D. Leiva and G. Zepon (DEMa-UFSCar) for the samples used in the tests and their help during the experiments.

**References**

[1] Medeiros Azevedo G et al 2009 *14th Inter. Conf. on X ray Absorption Fine Structure (Camerino-Italy)* A new XAFS beamline at LNLS Poster Session 1 n°154
[2] Moreira L M et al 2009 *Acta Materialia* **57** 5174-81
[3] Terra J et al 2009 *Physical Chemistry Chemical Physics* **11** 568-77
[4] Ferreira A P et al 2009 *Journal of Catalysis* **263** 335-44
[5] Duarte G et al 2012 *Geochimica et Cosmochimica Acta* **83** 205-16
[6] Rodriguez J A, Hanson J C and Chupas P J 2013 *In Situ Characterization of Heterogeneus Catalysts* Chapters 13 14 15 and 16 pp 345-474 (New York: J John Wiley & Sons Inc)
[7] Sousa M F et al 2015 *Proc. of the Int. Symp. on Crystallography* Vol **1** 3 p 61
[8] Piton J R et al 2013 *ICALEPCS (San Francisco-USA)* TUPPC036
[9] Piton J R and Duarte L F 1998 *3-WinDCM Internal Technical Manual MT* 01/1998 Laboratório Nacional de Luz Síncrotron Campinas SP Brazil
[10] Slepicka H H, Canova H F, Beniz D B and Piton J R 2015 *J Synchrotron Rad* **22** 1182-89
[11] Slepicka H H et al 2013 *ICALEPCS (San Francisco-USA)* TUPPC037
[12] Rodrigues A R D, Craievich A F and Gonçalves da Silva C E T 1998 *J Synchrotron Rad* **5** 1157-61
[13] Ravel B and Newville M 2005 *J Synch Rad* **12**, 537-41
[14] Figueroa S J A et al 2013 *J Appl Crystallogr* **46** 1523-27
[15] Zepon G et al 2015 *IJHE* **40** 3 1504-15

# The BALDER Beamline at the MAX IV Laboratory

**K Klementiev[1], K Norén[1], S Carlson[1], K G V Sigfridsson Clauss[1*] and I Persson[2]**

[1]MAX IV Laboratory, Lund University, 22592 Lund, Sweden,
[2]Swedish University of Agricultural Science, Dept. of Chemistry and Biotechnology, 75007 Uppsala, Sweden

E-mail: kajsa.sigfridsson_clauss@maxlab.lu.se

**Abstract**. X-ray absorption spectroscopy (XAS) includes well-established methods to study the local structure around the absorbing element – extended X-ray absorption fine structure (EXAFS), and the effective oxidation number or to quantitatively determine the speciation of an element in a complex matrix – X-ray absorption near-edge structure (XANES). The increased brilliance and intensities available at the new generation of synchrotron light sources makes it possible to study, *in-situ* and *in-operando,* much more dilute systems with relevance for natural systems, as well as the micro-scale variability and dynamics of chemical reactions on the millisecond time-scale. The design of the BALDER beamline at the MAX IV Laboratory 3 GeV ring has focused on a high flux of photons in a wide energy range, 2.4–40 keV, where the $K$-edge is covered for the elements S to La, and the $L_3$-edge for all elements heavier than Sb. The overall design of the beamline will allow large flexibility in energy range, beam size and data collection time. The other focus of the beamline design is the possibility to perform multi-technique analyses on samples. Development of sample environment requires focus on implementation of auxiliary methods in such a way that techniques like Fourier transform infrared (FTIR) spectroscopy, UV-Raman spectroscopy, X-ray diffraction and/or mass spectrometry can be performed simultaneously as the XAS study. It will be a flexible system where different instruments can be plugged in and out depending on the needs for the particular investigation. Many research areas will benefit from the properties of the wiggler based light source and the capabilities to perform *in-situ* and *in-operando* measurements, for example environmental and geochemical sciences, nuclear chemistry, catalysis, materials sciences, and cultural heritage.

## 1. Introduction

The low emittance 3 GeV ring at the MAX IV Laboratory in Sweden is undergoing commissioning and will be taken in operation in 2016. The BALDER beamline, dedicated to X-ray absorption spectroscopy (XAS) is one of the first phase beamlines under construction and is expected to initiate user operation in second half of 2016. XAS is one of few techniques that will provide detailed molecular information on elements in complex matrices. Very often regarded as a non-destructive technique where a minimum of pre-treatment is required, only very small amounts of material are needed, and concentrations of the studied elements are often low, XAS has become an indispensable tool in research areas such as life sciences, environmental and geochemical sciences and nuclear chemistry, catalysis and materials sciences, and cultural heritage. The design of the BALDER beamline has focused on a high flux of photons in a wide energy range, 2.4–40 keV, with flexibility in

Journal of Physics: Conference Series **712** (2016) 012023 doi:10.1088/1742-6596/712/1/012023

beam size and data collection time, and with focus towards implementation of multi-technique analyses on samples.

## 2. Beamline design

The BALDER beamline requires high flux, broad energy range, high energy resolution, high beam stability and fast energy scanning capability. These requirements are incorporated into the beamline design, which is based on an in-vacuum wiggler source and includes four main optical elements, see Figure 1:

- Vertically collimating mirror with Si and Ir stripes, water cooled
- Double crystal monochromator with Si(111) and Si(311) LN2-cooled crystals
- Toroid focusing mirror with Si and Ir torus
- Harmonic rejection mirrors with Si and Ir stripes (in experimental hutch, not shown in Figure 1)

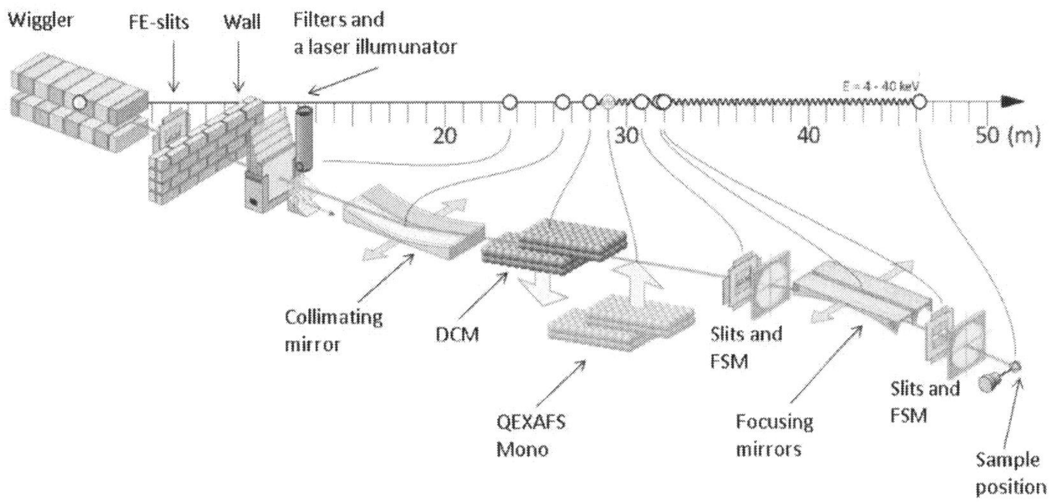


**Figure 1**. Conceptual optical layout of the BALDER beamline. The "QEXAFS Mono" (quick monochromator) is a possible device for future upgrades.

### *2.1. The Source*

The design concept for the BALDER beamline is based on an in-vacuum wiggler source. The wiggler design of choice is an in-vacuum wiggler developed at SOLEIL. The wiggler is composed of 38 periods of 50 mm producing a magnetic peak field of 2.1 T at a minimum gap of 5.5 mm. The magnetic system consists of a sequence of NdFeB permanent magnets with a magnetic remanence of 1.2 T at ambient temperature [1].

As pointed out by E. Welter [2], a multi-pole wiggler installed at a low-emittance storage ring produces a non-uniform beam. We confirm this statement by using two calculation programs – SPECTRA [3] and xrt [4], see Figure 2.

In order to smoothen the radiation distribution, one can perform tapering – varying the magnetic gap along the wiggler. While SPECTRA is capable of calculating tapered insertion devices, it cannot represent both spatial and energy domains at the same time. We therefore use xrt to study the result of tapering. Low energy radiation is more prone to create the spatial and energy fringes. In Figure 3 we present the most difficult case – the lowest beamline energy, 2.5 keV. For an energy band of 200 eV, one can see the chromatic fringes in the spatial distribution and a non-uniform energy distribution. Both features are gone as one gradually increases the tapering up to 2% (in $\Delta$gap/gap). At higher energies, tapering can be smaller, e.g. down to ~0.5% at 9 keV. Moreover, inevitable random errors in the alignment of magnetic poles facilitate the smoothing both in spatial and energy domains.

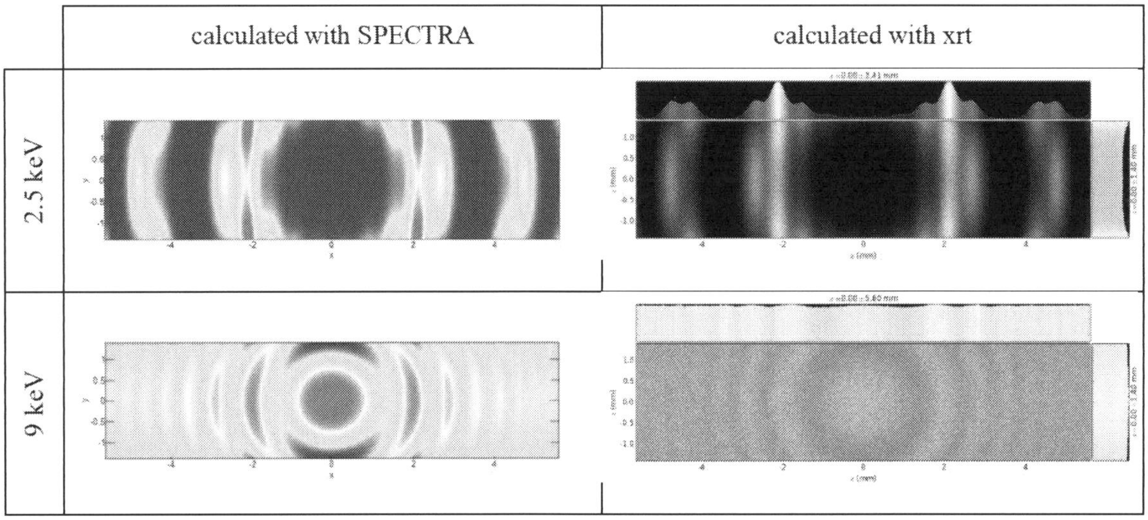


**Figure 2**. Beam cross-sections calculated with SPECTRA [3] and xrt [4] at two photon energies. SPECTRA represents intensity by color mapping; xrt represents intensity by brightness while colors represent energy. Xrt additionally shows two 1D positional histograms. Apart from these two different ways of representation, the results look very similar.

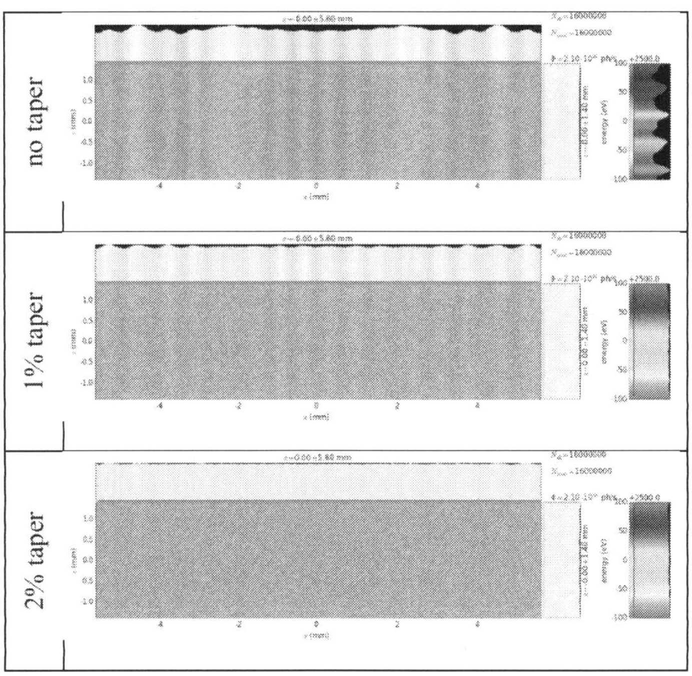


**Figure 3.** Beam cross-sections and energy distributions calculated with xrt at different gap tapering levels.

*2.2. The Optics*

The front end accepts 0.4h×0.1v mrad$^2$ of the wiggler radiation with up to 1.2 kW power in the beam. This power is distributed among three filters, a collimating mirror and the first crystal of a double crystal monochromator (DCM). The latter typically absorbs up to 500 W, in one particular regime – up to 750 W. Therefore, the monochromator is cryogenically cooled. The collimating mirror absorbs up to 800 W and is cooled via grooves filled with liquid metal.

The beam at the sample is focalized into a ~100h×100v μm$^2$ spot by a toroid mirror. There are two toroids with different coatings to serve different energy ranges. In order to minimize the realignment time, the two toroids have equal sagittal radii. The beam at the sample can be unfocused up to ~4.5h×2v mm$^2$ by going off the nominal 2 mrad pitch angle of the mirrors. In order to minimize the realignment time and simplify the end station, the height of the focusing mirror and all the downstream elements is kept invariant under changes of the mirror pitch angle. This can be achieved by a relatively long range of the fixed exit offset of the DCM within 10–32 mm. The DCM is foreseen to be operated in fixed exit mode, but with the possibility to be operated in pseudo channel-cut mode as well, for faster data acquisition time. The DCM has direct drive mechanism in order to operate the Bragg angle at a speed of up to $5^0$/s. The transmitted flux onto the sample by the two DCM crystal pairs, Si 111 and Si 311, is $10^{12}$–$10^{13}$ ph/s as estimated by ray tracing using xrt [4] in various assumptions about optical quality of the mirrors.

## 3. The experimental station

The experimental station requires an easy exchange between different experimental set-ups for both transmission and fluorescence XAS measurements, as well as auxiliary instrumentation and sample preparation laboratory. It is required to have a variety of detectors for fluorescence measurements accessible to the users. An ongoing project is in-house development of an X-ray emission spectrometer, which will be used as the standard florescence detector, for performing site selective XAFS spectroscopy measurements and also as an X-ray Raman spectrometer for probing soft X-ray absorption edges. The spectrometer is also presented in this proceedings volume [5]. Additional detection system that will be available to the user community is a set of standard fluorescence detectors, ranging from Lytle and PIPS detectors to a four-element energy dispersive silicon drift detector. The silicon drift detector will be equipped with an Xpress 3 pulse processor that gives count rates of $3 \cdot 10^6$ counts/element.

Additional instrumentation available for user experiment is a closed cycle cryo-cooler that operates with exchange gas in the sample compartment for temperatures down to 4 K, and a gas delivery system, for pressures ranging from atmospheric pressure to up to 30 bar, equipped with a mass spectrometer. The experimental station is developed to incorporate auxiliary methods in such a way that techniques like Fourier transform infrared (FTIR) spectroscopy, UV-Raman spectroscopy, X-ray diffraction and/or mass spectrometry can be performed simultaneously as the XAS study. It will be a flexible system where different instruments can be plugged in and out depending on the needs for the particular investigation.

## Acknowledgments

The BALDER beamline is funded by the Knut and Alice Wallenberg Foundation and Swedish universities.

## References

[1] Marcouillié O. *et al.* 2010 *AIP Conf. Proc.* **1234** 556
[2] Welter E 2012 *J. Synchrotron Rad.* **19** 905
[3] Tanaka T and Kitamura H 2001 *J. Synchrotron Rad.* **8** 1221
[4] Klementiev K and Chernikov R 2014 *SPIE* Advances in Computational Methods for X-Ray Optics III, 92090A; doi:10.1117/12.2061400
[5] Klementiev K, Preda I, Carlson S, Sigfridsson K and Norén K 2015 *High performance emission spectrometer at Balder/MAX IV beamline*, this volume.

# X-ray Absorption Spectroscopy and Coherent X-ray Diffraction Imaging for Time-Resolved Investigation of the Biological Complexes: Computer Modelling towards the XFEL Experiment

**A L Bugaev**[*, 1, 2], **A A Guda**[1], **O M Yefanov**[3,4], **U Lorenz**[3,5],
**A V Soldatov**[1], **I A Vartanyants**[3,6]

[1]International Research Centre "Smart Materials", Southern Federal University, Zorge street 5, 344090, Rostov-on-Don, Russia
[2]Department of Chemistry, University of Turin, Via P. Giuria 7, 10125 Turin, Italy
[3]Deutsches Elektronen-Synchrotron DESY, Notkestraße 85, D-22607 Hamburg, Germany
[4]Center for Free-Electron Laser Science, DESY, Notkestraße 85, D-22607 Hamburg, Germany
[5]Institute of Chemistry, University of Potsdam, D-14476 Potsdam, Germany
[6]National Research Nuclear University MEPhI (Moscow Engineering Physics Institute), Kashirskoe shosse 31, 115409 Moscow, Russia

E-mail: abugaev@sfedu.ru

**Abstract.** The development of the next generation synchrotron radiation sources – free electron lasers – is approaching to become an effective tool for the time-resolved experiments aimed to solve actual problems in various fields such as chemistry, biology, medicine, etc. In order to demonstrate, how these experiments may be performed for the real systems to obtain information at the atomic and macromolecular levels, we have performed a molecular dynamics computer simulation combined with quantum chemistry calculations for the human phosphoglycerate kinase enzyme with Mg containing substrate. The simulated structures were used to calculate coherent X-ray diffraction patterns, reflecting the conformational state of the enzyme, and Mg K-edge X-ray absorption spectra, which depend on the local structure of the substrate. These two techniques give complementary information making such an approach highly effective for time-resolved investigation of various biological complexes, such as metalloproteins or enzymes with metal-containing substrate, to obtain information about both metal-containing active site or substrate and the atomic structure of each conformation.

## 1. Introduction

Experimental investigation of protein dynamics is a problem of great importance because the conformational changes of the biological complexes in most cases determine their functionality. The solution of this problem becomes possible with the development of next generation synchrotron radiation sources – free electron lasers, providing extremely short pulses with ultrahigh brilliance sufficient for a single molecule imaging. A technique for collecting and evaluating scattering data from such kind of X-ray sources is called Coherent X-ray Diffraction Imaging (CXDI). CXDI


Published under licence by IOP Publishing Ltd

measurements can provide information on the 3D structure of the sample [1-3], e.g. virus, protein or enzyme, however the detailed information about a substrate or an active site cannot be obtained. These tiny changes in the active site may be a key point in understanding of the chemical processes, as in case of iron oxidation in hemoglobin [4] and changes in Fe-porphyrin geometry [5]. For metal containing proteins, such as hemoglobin, X-ray absorption spectroscopy can be applied to investigate atomic and electronic structures of the local environment around the absorbing atom. Being successfully applied to metalloproteins [6] X-ray absorption methods were recognized as an important tool in biophysics to investigate metal binding sites in macromolecules. In this work we suggest a complex technique which combines CXDI and XANES study to show how the complementary structural information can be obtained.

## 2. Materials and methods

A human phosphoglycerate kinase (PGK) [7], an enzyme which plays an important role in the glycolysis, was taken as a sample object. The initial structure of human PGK (Figure 1a) was extracted from the RSBS Protein Data Bank, reference number 2XE7 [8]. The two domains of this enzyme host 1,3-phospoglyceric acid (3PG) and Mg containing adenosine diphosphate (MgADP) and are associated with large scale conformational changes.

Molecular dynamics simulations were performed in Fujitsu Scigress Explorer Ultra version 7.6 using Alligner's MM3 augmented force field [9]. Further refinement of the enzyme active site was carried out using the PM7 Hamiltonian of the Molecular Orbitals Package [10].

CXDI patterns were calculated by means of the MOLTRANS program developed by Dr. E. Weckert in DESY. The Mg K-edge XANES spectra were simulated using the full multiple scattering approach of FEFF8.4 code [11] with the Hedin-Lundqvist exchange correlation potential. All calculations were performed on the Intel Core i7 personal computer with 8 GB RAM.

## 3. Results and discussion

### *3.1. Molecular dynamics simulations*

The total energy of the system was constructed from a number of terms corresponding to bond stretches, bond angles, dihedral angles, improper torsions, van der Waals interactions, electrostatic potentials, and hydrogen bonding energies. The van der Waals cut-off distance was set to 9 Å. The enzyme was surrounded by a 9 nm sphere of water. Prior to the molecular dynamics simulation, the initial geometry had been optimized by molecular mechanics using the same force fields with the parameters described above. The time step was set to 1 fs with data recording every 50 fs.

Calculations were performed at different simulated temperatures from 200 to 450 °K, affecting the kinetic energy per degree of freedom, and indicated that the process is temperature-dependent with PGK being more stable in close conformation at low temperatures, and remaining in the open conformation when heated above 370 °K. The resulting structure obtained by MD simulation at 300 °K is shown in the Figure 1b.

### *3.2. CXDI*

CXDI patterns were calculated for the wavelength $\lambda = 1.24$ Å and the range of the scattering angles $2\theta$ from 0 to 7.125° which resulted in the resolution of 10 Å which was shown to be sufficient for determining different conformations of the enzyme (see figure 2). If one does a reconstruction of such a pattern the projection of the electron density in the chosen direction can be obtained. In order to focus on the process of the enzyme transformation from open to close state, the plane of the enzyme closure angle was oriented perpendicularly to the incoming beam. The projections of the enzyme shown in figure 2b are obtained by Fast Fourier Transformation of the calculated patterns with given intensities and phases.

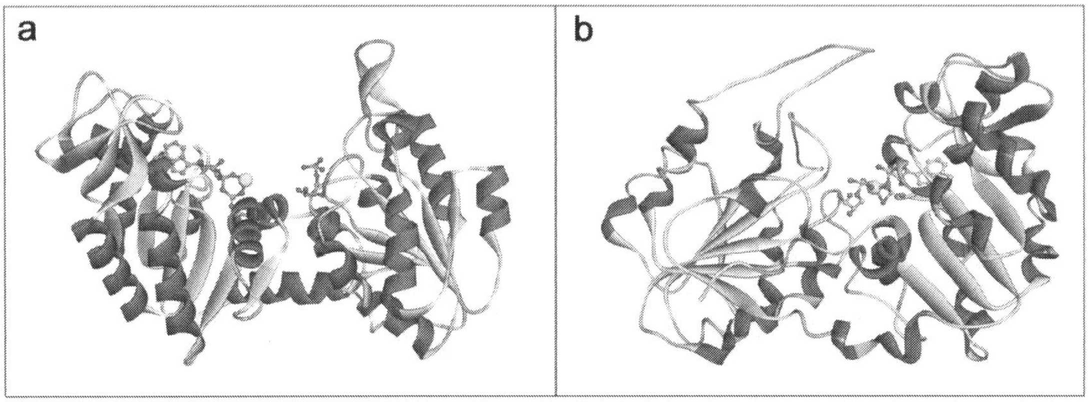


**Figure 1.** Stable open (a) and close (b) conformations of PGK enzyme obtained from MD. Protein structure shown as a solid ribbon and substrate molecules shown in ball and stick style with green Mg atom scaled for better visualization. Water molecules and hydrogen atoms are not shown.

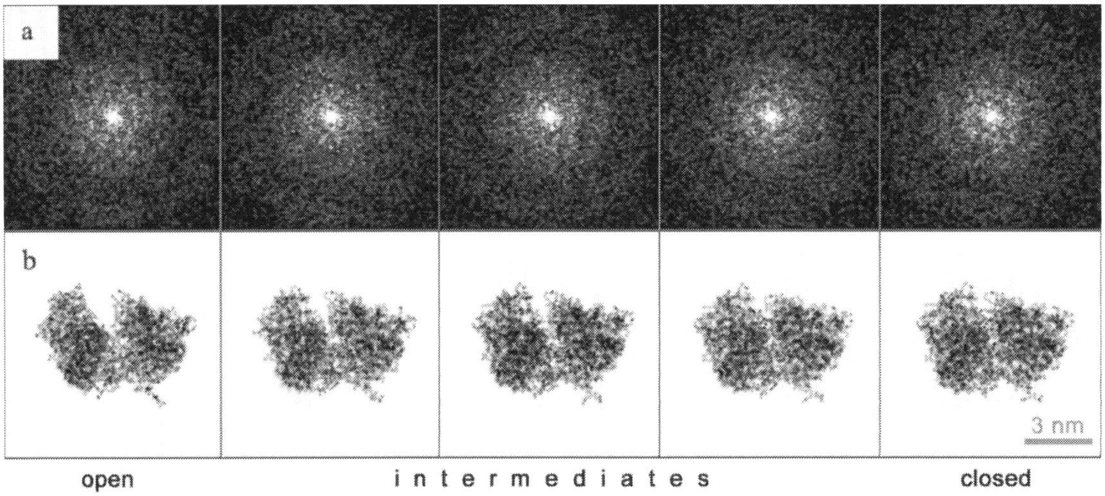


**Figure 2.** (a) Evolution of CXDI patterns in logarithmic scale for different conformation stages of the enzyme starting from the open stage (left pattern) and finishing with the closed stage (right pattern) and (b) corresponding projections in real space obtained by Fourier Transform of the patterns.

*3.3. XANES*

In order to obtain accurate structures for XANES calculations, the geometry of active site with MgADP and 3PG and nearest water molecules was optimized using the molecular orbital package MOPAC2012 [10]. According to the optimization results the Mg atom always remains in the octahedral coordination with water and phosphate group's oxygen atoms in the first coordination shell, which is consistent with previous studies of $Mn^{2+}$ATP complexes [12], analogous to $Mg^{2+}$ATP.

It is important to note that while Mg atom remains to be octo-coordinated by 6 oxygen atoms, these atoms may belong either to the phosphate groups of the substrate or to the surrounding water molecules. Due to the fact that the displacement of the Mg-O distance in case of Mg-water bonds will be much higher than that of rigid Mg-ADP (or Mg-ATP) bonds, leading to larger broadening of the experimental spectra. As obtained from the MOPAC optimization, in the open conformation Mg atom has 4 water oxygens in the first coordination shell and 2 oxygens of phosphate groups, while in the close conformation 2 water molecules are superseded by additional phosphate group, coming from 1,3-phospoglyceric acid. These results correlate with the experimental studies of ATP complexes with $Mn^{2+}$ ion, having similar to $Mg^{2+}$ chemical properties [12].

XANES spectra (Figure 3) were calculated for the obtained structures using FEFF9 code for spherical clusters with a radius of 13 Å around the absorbing Mg atom extracted from the entire enzyme structure. Full multiple scattering calculations were performed within a sphere of 8 Å, the radius of self-consistent field calculation was set to 5 Å.

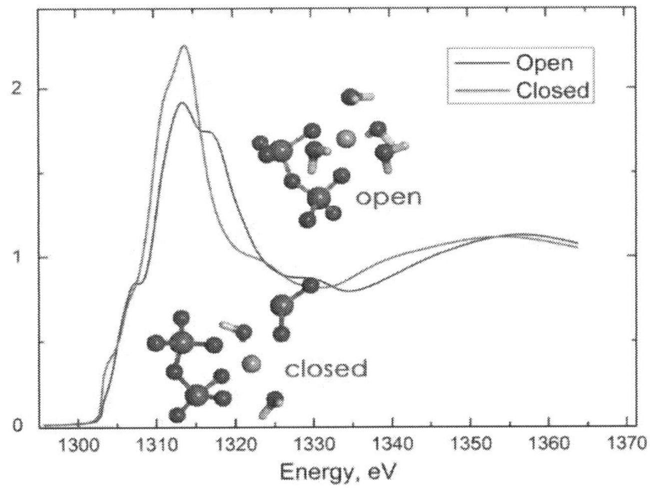


**Figure 3.** XANES spectra calculated for open (blue) and close (red) conformations of PGK. Atomic structures represent the surrounding of Mg atom (green) by oxygens (red), of phosphates (phosphorus atoms are orange) and water (hydrogens are white).

## 4. Conclusions

We suggest an approach of measuring combined CXDI and XANES, to follow both the overall state of an enzyme and evolution of local structure around Mg atom. This approach might be applied to various biological samples, like metal containing proteins, which are associated with conformational changes. Simulations of conformational processes may give additional knowledge about the probability of different conformational structures, aimed to help in experimental data evaluation.

## 5. Acknowledgments

ALB, AAG and AVS would like to thank the Ministry of Education and Science of the Russia for the financial support (agreement number 14.587.21.0002, project identifier RFMEFI 58714X0002).

## 6. References

[1] Nugent K 2010 *Adv. Phys.* **59**(1) 1–99
[2] Mancuso A, Yefanov O and Vartanyants I 2010 *Journal of Biotechnology* **149**(4) 229-237
[3] Vartanyants I and Yefanov O 2015 *X-ray Diffraction. Modern Experimental Techniques* O H Seeck and B M Murphy (Singapore: Pan Stanford Publishing) pp 341-384
[4] Bianconi A, Congiu-Castellano A, Dell'Ariccia M, Giovannelli A, Burattini E and Durham P 1985 *Biochem. and biophys. research commun.* **131** 98-102
[5] Bianconi A, Congiu-Castellano A, Dell'Ariccia M, Giovannelli A, Burattini E, Castagnola M and Durham P 1985 *Biochimica et Biophysica Acta* **831** 120-124
[6] Eisenberger P, Shulman R, Kincaid B, Brown G and Ogawa S 1978 *Nature* **274** 30-34
[7] Auerbach G, Huber R, Grättinger M, Zaiss K, Schurig H, Jaenicke R and Jacob 1997 *Structure* **5** 1475-1483
[8] Zerrad L, Merli A, Schröder G F, Varga A, Gráczer E, Pernot P, Round A, Vas M and Bowler M W 2011 *J. of Biol. Chem.* **286** 14040-14048
[9] Allinger N, Yuh Y and Lii J H 1989 *JACS* **111** (23) 8551–8582
[10] Stewart J P 2007 *J. of Mol. Mod.* **13** 1173-1213
[11] Rehr J and Albers R 2000 *Rev. Mod. Phys* **72** 621-654
[12] Belli M, Scafati A, Bianconi A, Burattini E, Mobilio S, Natoli C R, Palladino L and Reale A 1983 Il Nuovo Cimento 5 1281-1304

# Development of highly stable Bragg polychromator for energy dispersive XAFS

**Kazuo Kato*[1], Takaki Irie[1], Tomoya Uruga[1, 2], Kei Uera[1], Morihiro Kawase[1]**

[1]Japan Synchrotron Radiation Research Institute, SPring-8, 1-1-1, Kouto, Sayo-cho, Sayo-gun, Hyogo, Japan
[2]The University of Electro-Communications, Innovation Research Center for Fuel Cells, Chofu, Tokyo, Japan

E-mail:kkato@spring8.or.jp

**Abstract.** A highly stable Bragg polychromator for energy dispersive XAFS (DXAFS) systems was developed at the bending magnet beamline BL28B2 at SPring-8. The bending device is a four–point-type bender. Additional stoppers for the crystal are attached at both ends of the bender to maintain the position and curvature of the crystal in the bender. The crystal was indirectly cooled in a copper grain bath with good thermal contact, the temperature of which was precisely controlled using water cooling. A performance test of stability of the new Bragg polychromator showed that the energy drift in the DXAFS spectra was reduced to less than 0.01 eV for 2 hrs.

## 1. Introduction

In energy dispersive XAFS (DXAFS) measurements, the temporal stability of a polychromator is an important factor for measuring high-quality spectra. Recently, Mairs et al. developed the Bragg polychromator which have the high stability of the crystal bending shape and the high repeatability of the adjustment of the focal spot [1]. At the DXAFS station at BL28B2, SPring-8, a Bragg-type polychromator was developed and has been used in the energy region from 6 to 12 keV [2]. It consists of a crystal-bending device to obtain a variable curvature but has no cooling devices. However, an energy shift and/or fluctuation in the DXAFS spectra were observed during long-term measurements due to mechanical and thermal instability of the bent crystal in the polychromator. We therefore newly developed a Bragg-type polychromator having a highly stable bending and cooling mechanism.

## 2. Design of the Bragg polychromator

The newly developed Bragg polychromator consists of a four-point bender as a bending mechanism and a cooling bath for a crystal (Fig. 1), which are housed in a helium gas flow chamber under atmospheric pressure. The crystal is a symmetrically cut Si (111) with dimensions of 200(w) x 40(h) x 1(t) mm. The bender has an adjusting mechanism to correct the twist of the crystal. We first developed a 4 point bender made of 4 cylinders, but curvature of the crystal could not be maintained due to the slip of the crystal in the bender. To solve this problem, we adopted the design of the bender made of 2 cylinders and 2 plane surfaces with stoppers. Two stoppers for the crystal are installed at both ends of the bender; one is movable and the other is fixed. The position of the movable stopper is adjusted to fix the end of the crystal by hand after adjusting the bending of the crystal.

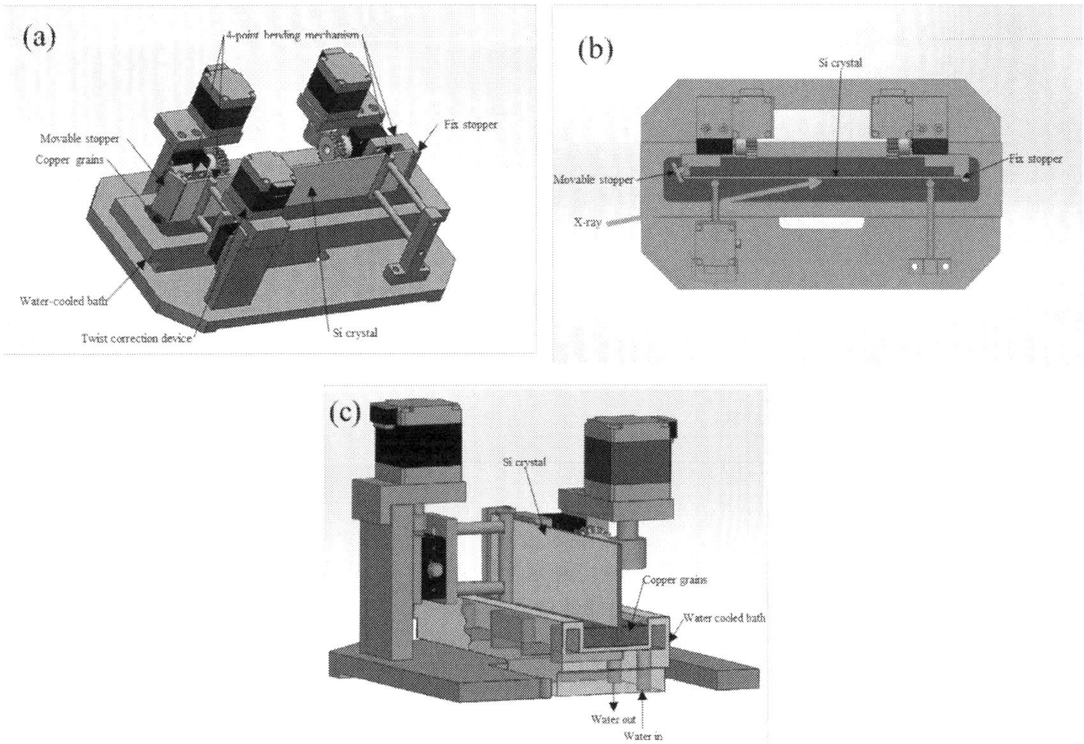


**Figure 1.** New Bragg polychromator: (a) schematic perspective view, (b) top view and (c) cross-section view of the water-cooling bath.

The crystal was indirectly cooled with a newly developed bath. A liquid InGa eutectic alloy is often used as a heating medium of indirect-cooling bending devices due to its excellent thermal conductivity [3, 4]. This alloy, however, is highly corrosive and difficult to be removed once adhered. If the liquid InGa eutectic alloy spills out from the bath, it will inflict damage on the surrounding surface. We used copper grains with an average grain diameter of 0.5 mm (Kojundo Chemical Laboratory Co.) as a substitute for the liquid InGa eutectic alloy, which are much easier to handle and achieve good thermal contact between the crystal and the water-cooling bath. The cooling water is circulated in flow channel in the bath (Fig. 1(c)). The lower part of the crystal (ca. 5 mm wide) was immersed in the copper grain bath. The crystal temperature could be controlled between 26 and 30°C within an accuracy of 0.01°C by using water circulated in the bath.

## 3. Performance of the Bragg polychromator

### *3.1. Long-term mechanical stability of the crystal bending shape*

Figure 2 shows the effectiveness of the stoppers for the bent crystal of the polychromator bender on the long-term stability of the focused beam profiles. In Fig. 2(b), the focused beam size in the horizontal direction of 22 μm (FWHM) is achieved with energy bandwidth of 450 eV at 8 keV. The focused beam profile with no stoppers significantly expanded with 24-hour continuous bending even under no heat load by X-ray irradiation on the crystal (Fig. 2(a)). This indicates that the curvature of the bent crystal gradually decreased due to the slip of the crystal in the bender. The bent shape of the crystal was significantly stabilized by installing stoppers on the polychromator bender (Fig. 2(b)).

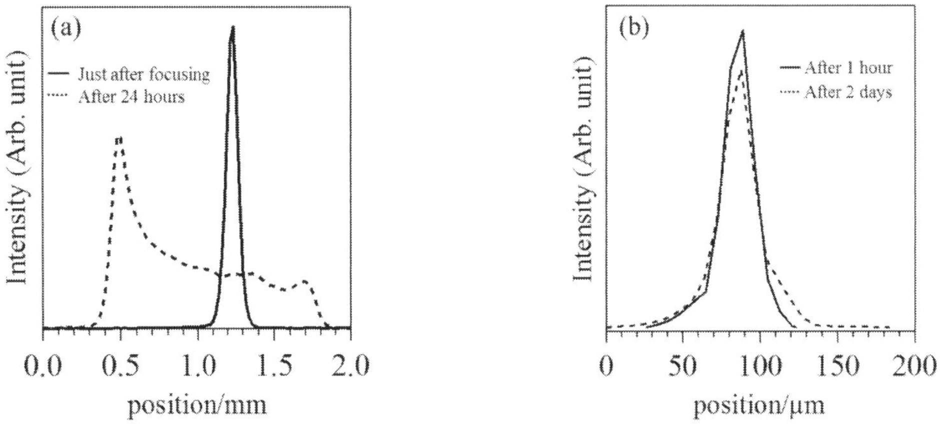


**Figure 2.** Temporal change in focused beam profiles in horizontal direction: (a) without stoppers and (b) with stoppers.

### *3.2. Stability of the crystal temperature*

Figure 3 shows the effects of crystal cooling on the stability of the crystal temperature during X-rays irradiation. The temperature of the crystal was measured using thermistor elements attached to the back surface of the crystal. The temperature of the crystal without cooling (without the copper grains in the cooling bath) increased to 31.05°C after 20 min then gradually increased until 31.1°C (dotted line). In contrast, by cooling the crystal, the crystal temperature reached a steady state at 29.04°C within 20 min and its fluctuation was suppressed within ± 0.01°C.

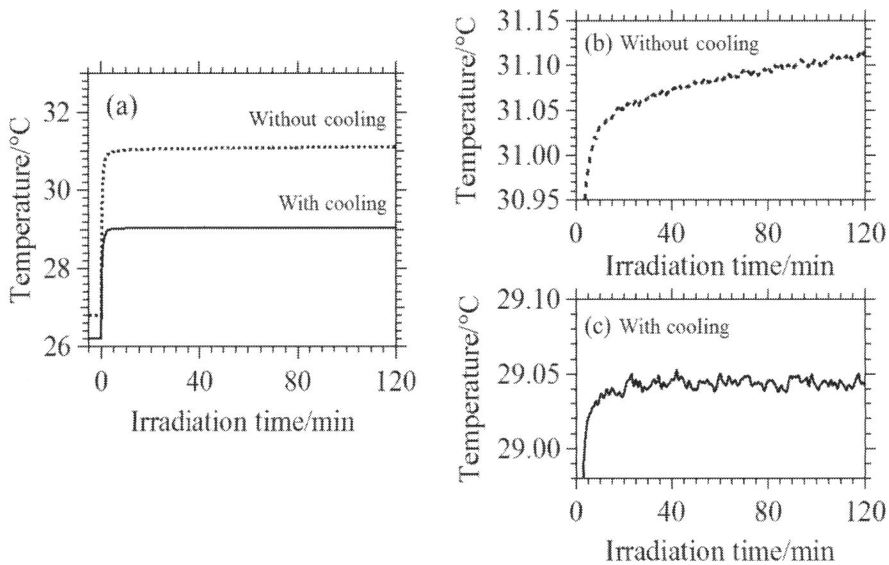


**Figure 3.** Temperature of crystal (a), and magnification of (a) without cooling (b) and with cooling (c) during X-rays irradiation.

### *3.3. Stability of the absorption edge position in DXAFS spectra*

Figure 4 shows the fluctuation of the Ni K-edge position in the DXAFS spectra of the Ni foil during X-ray irradiation. Without cooling, the Ni K-edge position gradually shifted toward high energy after

20 min. With cooling, the DXAFS spectral shape and curvature of the crystal reached a steady state within 20 min, which is the same tendency observed in the stability of the crystal temperature. As a result, the energy fluctuation in the DXAFS spectra decreased to 0.007 eV (FWHM).

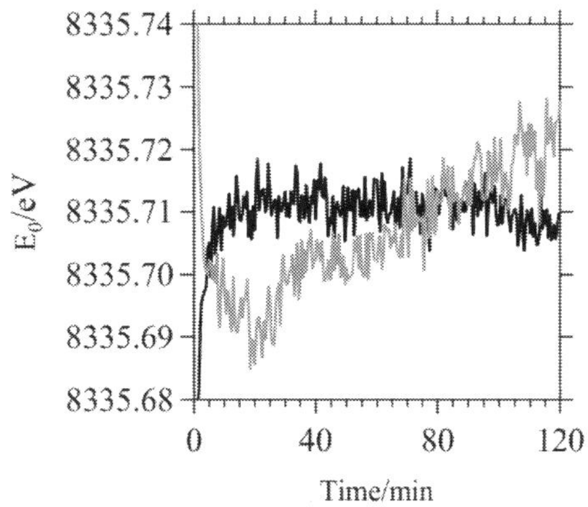


**Figure 4.** Ni K-edge energy of Ni foil in DXAFS spectra with (black) and without cooling of crystal during X-rays irradiation (gray)

## 4. Conclusion and further improvement plan

We developed a highly stable Bragg polychromator for DXAFS measurements at the bending magnet beamline BL28B2 at SPring-8. Long-term stability of the curvature of the crystal and reduction in the energy drift below 0.01 eV were achieved by introducing newly developed equipment, such as a stopper for the crystal and cooling device with copper grain bath.

The DXAFS station at BL28B2 has a beam shutter upstream of the polychromator. The polychromator is irradiated intermittently with intense white X-rays for each measurement. Thus, it is quite important to cut the time needed for stabilizing the X-ray beams incident on a sample for efficient use of the beam time. We plan to develop a high-speed stabilizing system for reaching steady state within a few minutes regarding crystal temperature. We also plan to develop a remote control device to adjust the stopper position without opening the helium gas flow chamber.

## References

[1] T. R. Mairs J. Borrel and O. Mathon Diamond Light Source Proceedings 1 MEDSI 6 2011 1-4
[2] K. Kato, T. Uruga, H. Tanida, S. Yokota, K. Okumura, Y. Imai, T. Irie, Y. Yamakata, AIP Conf. Proc. 879 2007 1214-1217
[3] A. San-Miguel M. Hagelstein J. Borrel G. Marot M. Renier J. Synchrotron Radiattion 5 1998 1396-1397
[4] F. Baudelet Q. Kong L. Nataf J.D. Cafun, A. Monza S.Chagnot and J.P. Itié High Pressure Research 31 2011 136-139

# XAFS at the new materials science beamline 10 at the DELTA storage ring

**D Lützenkirchen-Hecht*, R Wagner, R Frahm**
Bergische Universität Wuppertal, Gaußstr. 20, 42097 Wuppertal, Germany

*E-mail: dirklh@uni-wuppertal.de

**Abstract.** The layout and the characteristics of the hard X-ray beamline BL 10 at the superconducting asymmetric wiggler at the 1.5 GeV Dortmund Electron Accelerator DELTA are described. Equipped with a stable and robust Si(111) channel-cut monochromator, this beamline is suited for XAFS studies in the spectral range from about 4 keV to ca. 16 keV photon energy. We will illustrate the performance of the beamline, and present EXAFS data obtained from several reference compounds. XANES data measured for dilute sample systems as well as surface sensitive grazing incidence EXAFS obtained from thin film samples will also be discussed.

## 1. Introduction

The DELTA synchrotron radiation facility (1.5 GeV electron energy, injection currents of ~120 mA, lifetime 10-15 h) is equipped with several beamlines providing radiation from the UV-visible spectral range to hard X-rays [1]. In order to increase X-ray related research activities, and to meet the user needs for high energy photons especially for spectroscopic studies, a new hard X-ray beamline was installed, making use of the intense radiation emitted by a superconducting asymmetric wiggler (SAW) with a critical energy of about 7.9 keV. The broad radiation fan of this insertion device covers about ± 25 mrad horizontal beam divergence, so that the installation of a third beamline (BL10) at this insertion device in parallel to the two existing ones [2, 3] is feasible, however with some restrictions in terms of the space available for the newly installed instrumentation. In this contribution, we will describe the optical layout and the performance of this new beamline, including some experimental data from user experiments.

## 2. Beamline overview

Due to the angular offset of 15 mrad between BL 10 and the adjacent beamline 9, there is only a limited space available for the added instrumentation of BL 10, especially in the front-end and the optics hutches. Therefore, the installation of X-ray mirrors for focusing, suppression of higher harmonics and for a reduction of the heat load on the monochromator was not possible, and BL10 has a very simple X-ray optical layout. It consists of a fixed, water-cooled absorber mask ca. 12 m from the center of the wiggler source that delivers a radiation fan of 3 mrad x 0.7 mrad (horiz. x vert.) into the beamline, a tilt absorber in a distance of 23.8 m, which is used for the definition of the vertical beam size impinging on the compact, Si(111) channel-cut monochromator that is installed in the experimental hutch 30 m from the source [4]. As proved by finite element calculations, the thermal distortions and slope errors of the first monochromator crystal surface induced by the heat load of the wiggler of about 1.59 W mrad$^{-2}$ mA$^{-1}$ (corresponding to about 0.2 W mm$^{-2}$ for 100 mA of stored electron current) are significantly smaller compared to the Darwin width of the employed Si(111) crystal [4]. Furthermore, changes of the heat load on the monochromator due to the decrease of the electron current in the storage ring have no measurable influence on the monochromator resolution, and observed shifts of the energy scale are smaller than 0.1 eV. Downstream the monochromator, two slit systems finally define the size of the monochromatic beam on the sample. Two different setups are permanently installed in the experimental hutch: (a) a table for standard transmission and fluorescence mode EXAFS experiments using gas-filled ionization chambers and different photodiodes with related detector electronics as detectors, and (b) a 6-axis Kappa-diffractometer, that can be employed for all kinds of diffraction studies, and grazing incidence EXAFS experiments [4, 5]. A typical photon flux at the position of the sample amounts to $5*10^9$ photons s$^{-1}$ mm$^{-2}$ at 8 keV and 100 mA of stored current.

The use of a channel-cut monochromator and the absence of mirrors make harmonic rejection a critical issue for the present beamline. However, due to the critical energy of the wiggler source of only 7.9 keV, the flux decreases substantially for photon energies above 10 keV. Furthermore, if the filling gases of the ionization chambers are chosen properly, the sensitivities for the higher order harmonics are further reduced, so that a harmonic suppression of at least $5*10^{-3}$ or better can be achieved [4]. XAFS experiments of different reference materials have shown that the harmonic suppression achieved at BL 10 appears sufficient for high quality XANES and EXAFS studies.

Due to the simple layout of the beamline, the different detector systems available and the flexibility of the equipment for employing e.g. cryostats (down to 77 K), high temperature reaction cells (300 K – ca. 1500 K) or vacuum chambers ($p < 10^{-8}$ mbar), it is ideally suited for educational purposes, and several projects with students have been performed in the first few months of operation.

## 3. XAFS experiments

### *3.1. Transmission mode EXAFS*

In the commissioning phase of BL 10, several different metal foils in the energy range from the Ti K-edge (4966 eV) to the Au $L_1$-edge (14353 eV) have been examined by transmission mode EXAFS at room temperature for reference purposes. In general, the obtained data quality is excellent with respect to the reproduction of sharp absorption features close to the edges as well as to the low noise of the $k^3$-weighted $\chi(k)$ data as can be seen in figure 1, where a representative example of a Ta-metal foil measured at room temperature within about 35 minutes is shown. The extracted $k^3$-weighted $\chi(k)$ data are almost noise-free up to wavenumbers of 18 Å$^{-1}$, i.e. more than 1200 eV above the edge. It should be noted that the measured k-range is only limited by the onset of the Ta $L_2$-absorption edge at 11136 eV. The experimental results show that the energy resolution of the channel-cut monochromator is close to the values expected from the values for the natural width of the absorption edges of interest, (see ref. [4]) and thus high resolution XANES and EXAFS experiments in the energy range between ca. 4 keV and 16 keV are feasible at BL 10.

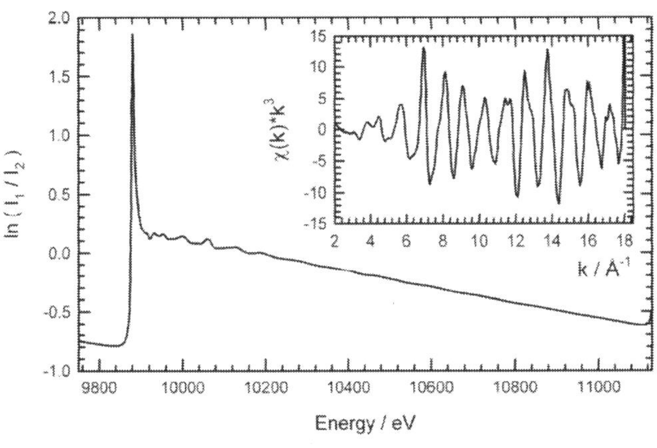


**Figure 1.** $L_3$-edge X-ray absorption spectrum of a Ta metal foil measured in transmission at room temperature. In the inset, the $k^3$-weighted fine structure oscillations $\chi(k) * k^3$ are depicted. The onset of the Ta $L_2$-edge can be seen in the absorption spectrum at about 11130 eV, and at ca. 18 Å$^{-1}$ in the $k^3$-weighted $\chi(k) * k^3$ data, respectively.

### *3.2. Fluorescence mode XANES*

Besides transmission mode experiments, fluorescence detection is feasible at BL10 using large area PIPS (passivated implanted planar silicon) detectors or an energy dispersive Peltier-cooled Si drift-diode with digital pulse processing electronics. As an example, we have recently investigated doped barium aluminate ($BaAl_2O_4$) samples. These materials are of particular interest because of their luminescence properties when they are doped with rare earth or transition metal cations [6, 7]. However, in the case of Cr-doping, the measurement of the K-edge of the dopant is extremely difficult, because the $L_1$-edge of the Ba host has an identical energy of 5989 eV, making transmission mode experiments impossible. As can be seen in figure 2(a), also the fluorescence emissions from the Cr dopant and the $BaAl_2O_4$ host lattice are heavily overlapping, with intense Ba L-lines at about 4460, 4830, 5160 and 5530 eV, and relatively weak Cr K-emissions at 5410 and 5950 eV. However, with a careful setting of the region in the fluorescence spectrum used for the calculation of an EXAFS spectrum, the overlap can be minimized, and absorption data of reasonable data quality can be achieved (see figure 2(b)). The comparison of the measured data with those of reference compounds ($BaCrO_4$, $Cr_2O_3$) shows that the Cr incorporated in the doped samples seems to be in a 6+ state, and the coordination is mainly tetrahedral, as can be anticipated from the intense pre-edge peak at 5993 eV and the edge position at about 6009 eV, which is further supported by a linear combination XANES fit of the edge region (not shown).

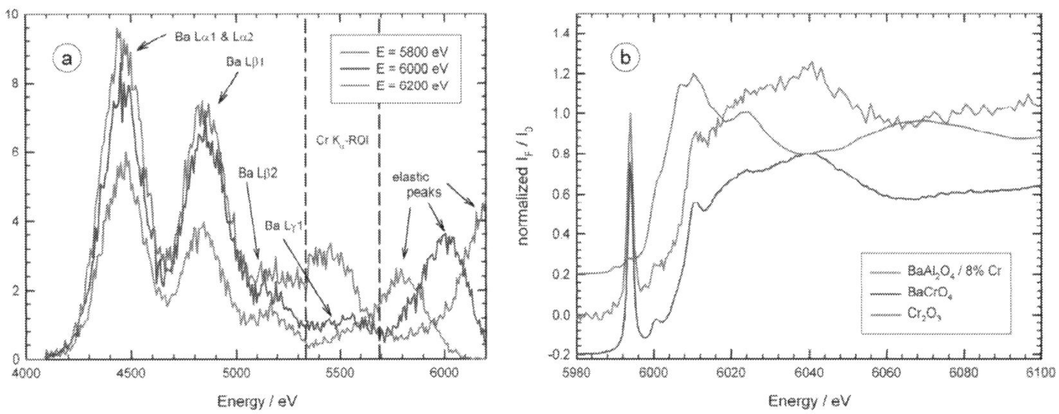


**Figure 2.** (a) X-ray fluorescence spectra from a $BaAl_2O_4$ sample doped with 8 at.% Cr for different excitation energies in the vicinity of the Cr K-edge / Ba $L_1$-edge. Emissions between ca. 5350 eV and 5700 eV are mainly caused by the Cr-fluorescence. (b) Cr K-edge near edge spectrum of $BaAl_2O_4$ sample doped with 8 at.% Cr in comparison to K-edge spectra of barium chromate ($BaCrO_4$) and $Cr_2O_3$.

### *3.3. Grazing incidence XAFS*

Grazing incidence X-ray absorption spectroscopy is an excellent method to study the near surface structure of thin films, surfaces, adsorbates etc. (e.g. refs. [5, 8-11]) and it has also been used in the past for the investigation of intermetallic layers [8]. Here we have *in-situ* explored the formation of NiAl induced by a heat treatment of Al-Ni-bilayers in vacuum using Ni K-edge XAFS. The samples were prepared by thermal evaporation of Al and Ni on float glass substrates under high-vacuum. According to X-ray reflectivity measurements, the thickness of the Al and Ni layers amount to about 80 nm and 72 nm, respectively. For temperatures below 200°C, the measured XAFS data closely resembles the spectrum of metallic Ni metal, while NiAl formation starts at about 210°C. In figure 3, in-situ measured XAFS spectra for different incidence angles Θ are displayed for a substrate temperature of 230°C, and spectra of a NiAl reference sample and Ni metal are shown for comparison. As can be seen in figure 3, the XAFS data from the bilayer sample closely resemble that of the NiAl reference for incidence angles below the critical angle of total reflection of Ni, while they are dominated by XAFS fine structure of Ni metal for larger incidence angles $\Theta > 0.35°$. The present results suggest that a thin NiAl-film has formed at the interface between the metallic Al- and Ni-sublayers. From a fit of the experimental data (not shown), the thickness of this layer can be estimated to about 10 nm. Experiments at more elevated temperatures show that the NiAl film thickness is not increasing further, and it should also be noted that no evidence for other Ni-Al compounds such as $NiAl_2$, $Ni_2Al_3$ etc. could be obtained in the present study.

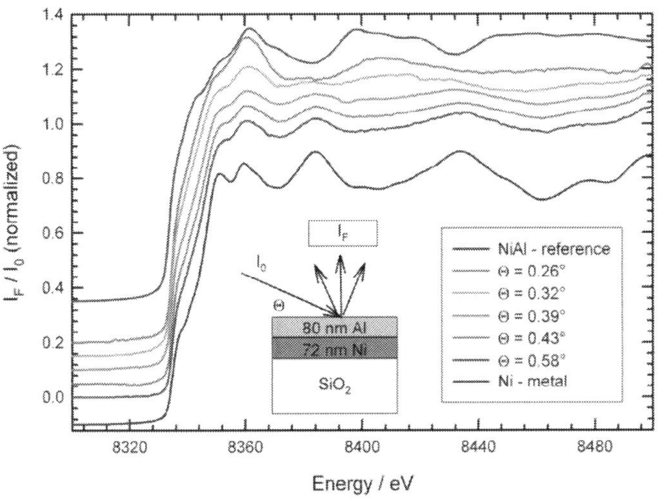


**Figure 3.** Grazing incidence XAFS at the Ni K-edge measured *in-situ* during the annealing of a Al/Ni-bilayer on a glass substrate at T = 230°C. X-ray absorption spectra were measured in fluorescence mode using a large area PIPS diode for different incidence angles Θ as indicated in the inset. For comparison, the spectrum of a NiAl intermetallic compound and a Ni-metal reference are also shown.

## 4. Conclusions

In conclusion, the instrumentation of BL 10 at the DELTA storage ring provides the conditions for a wide range of XAFS investigations. The robust and stable Si(111) channel-cut monochromator covers the energy range from ca. 4 – 16 keV, allowing XAFS experiments at the K- or L-edges of most of the technologically important elements. Besides transmission and fluorescence detection modes, grazing incidence EXAFS experiments are feasible making use of the multi-purpose diffractometer, also for studies in different non-ambient environments (vacuum, gas and liquid phase, cryogenic and high temperature). The results presented here demonstrate that high quality EXAFS data can be achieved at beamline 10 in various different sample environments, making BL 10 thus an attractive instrument for EXAFS studies with hard X-rays.

## References

[1] Tolan M, Weis T, Wille K and Westphal C 2003 *Synchrotron Rad News* **16** 9
[2] Paulus M, Fendt R, Sternemann C, Gutt C, Hövel H, Volmer M, Tolan M and Wille K 2005 *J. Synchrotron Rad.* **12** 246
[3] Lützenkirchen-Hecht D, Wagner R, Haake U, Watenphul A and Frahm R 2009 *J. Synchrotron Rad.* **16** 264
[4] Lützenkirchen-Hecht D, Wagner R, Szillat S, Hüsecken AK, Istomin K, Pietsch U and Frahm R 2014 *J. Synchrotron Rad.* **21** 819
[5] Lützenkirchen-Hecht D, Wulff D, Wagner R, Frahm R, Holländer U and Maier HJ 2014 *J. Mater. Sci.* **49** 5454
[6] Fukuda K, Iwata T and Orito T 2005 *J. Solid State Chem.* **178** 3662
[7] Lin Y, Zhang Z, Tang Z, Zhang J, Zheng Z and Lu X 2001 *Mater. Chem. Phys.* **70** 156
[8] Heald SM, Chen H and Tranquada JM 1988 *Phys. Rev. B* **38** 1016
[9] Borthen P and Strehblow H-H 1995 *J. Phys.: Condens. Matter* **7** 3779
[10] Jiang DT and Crozier ED 1998 *Can. J. Phys.* **76** 621
[11] d'Acapito F, Emelianov I, Relini A, Cavatorta P, Gliozzi A, Minicozzi V, Morante S, Solari PL and Rolandi R 2002 *Langmuir* **18** 5277

# Hard X-ray XAFS beamline, BL5S1, at AichiSR

**M Tabuchi[1], H Asakura[1], H Morimoto[1], N Watanabe[1], Y. Takeda[2]**

[1]Nagoya University, Synchrotron Radiation Research Center, Furocho, Chikusaku, Nagoya 464-8603, Japan
[2]Aichi Science and Technology Foundation, Aichi Synchrotron Radiation Center, 250-3, Minami Yamaguchicho, Seto 489-0965, Japan

m.tabuchi@nusr.nagoya-u.ac.jp

**Abstract**. A XAFS beamline, BL5S1, had been operated at Aichi Synchrotron Radiation Center, Japan since March 2013. The beamline was designed for the measurements in the energy range from 5 to 20 keV. The photon flux of 6 x $10^{10}$ at around 9 keV and beam spot size of 0.5 x 0.3 mm at sample position are as good as designed. For the standard transmission XAFS measurement, both of the step- and quick- scan modes are available. Energy resolution at around 9keV is good enough to discuss the energy shift of the order of 0.1 eV or higher even when the measurements are conducted in the quick-scan mode. With several kinds of detectors for fluorescence and/or CEY detection mode measurements, and various kinds of sample holders which are supported by the XAFS measurement software, users easily obtain spectra for their samples. Such a standard, well operated and easy to access XAFS beamline must be very important to broaden the base of the XAFS society further.

## 1. Introduction

Aichi Synchrotron Radiation Center (AichiSR) was built in Aichi, Japan in 2012 as a new facility to utilize the synchrotron radiation. After a short set-up and commissioning periods of the accelerators and the beamlines for about one year, it has been openly operated for synchrotron-radiation users since March 2013. A beamline, BL5S1 at AichiSR, was designed and built as a hard X-ray XAFS beamline to satisfy the potential demands both of the industrial and the academic users in the local region. In this paper, we introduce the design and initial performance of the BL5S1 with some information on the accelerator as a light source and show recent status after the two years operation for the users.

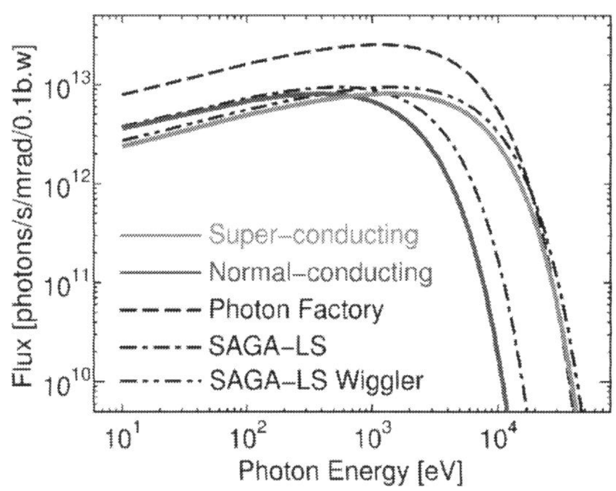


Fig. 1 Theoretically calculated flux for the AichiSR super-(red) and normal-(blue) conducting bending magnets comparing with those of bending magnets and wiggler of SAGA-LS and bending magnets of Photon Factory.

## 2. Aichi Synchrotron Radiation Center

The light source at AichiSR consists of three main components, a LINAC, a booster ring, and a storage ring[1]. The electron beams of 50 MeV

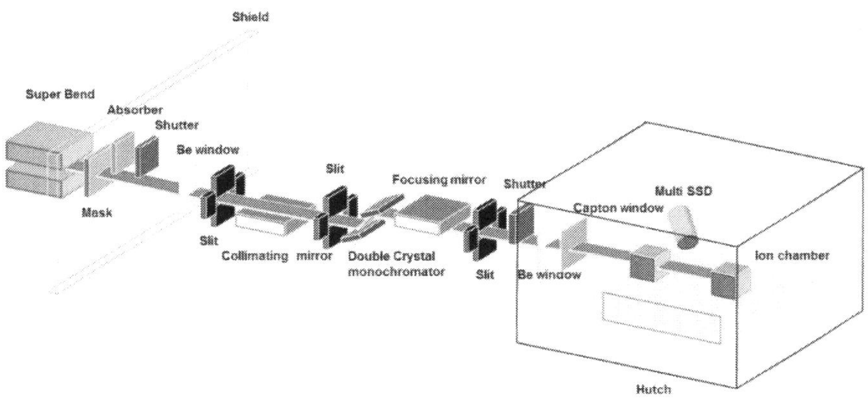


Fig. 2 Schematically drawn optics layout of BL5S1. Optics, the 1st slit, the 1st mirror, the 2nd slit, double crystal monochromator, the 2nd mirror, and the 3rd slit, locate at 8220, 9450, 10910, 11820, 13180, and 15345mm away from the light source, respectively. The focal point in hatch is 23017mm from the light source, since the 2nd mirror is 2:1 focusing mirror in horizontal,.

from the LINAC are accelerated to 1.2 GeV by the booster ring and finally injected to the storage ring. The storage ring is usually operated in top-up mode at 300mA. The storage ring consists of eight normal-conducting and four super-conducting bending magnets with an undulator in a linear section. The four super-conducting bending magnets of 5 T realize to provide hard X-rays of the critical energy at 4.8 keV while the normal-conducting bending magnets of 1.4 T provide soft X-rays of the critical energy at 1.3 keV. Figure 1 shows theoretically calculated photon fluxes as a function of photon energy for the AichiSR super- and normal-conducting bending magnets, comparing with those for the bending magnets and wiggler of SAGA-LS and the bending magnets of Photon Factory. As shown in Fig. 1, for the superconducting bending magnets of AichiSR, the critical energy and thus the shape of the spectrum is similar to those of SAGA-LS wiggler and Photon Factory.

In the energy range from 5.0 to 20.0 keV, where the hard X-ray XAFS beamline BL5S1 was designed to be operated, photon fluxes of several $10^{10}$ photons/s were expected at the sample position from the calculation shown in Fig. 1.

## 3. Hard X-ray XAFS Beamline at AichiSR, BL5S1

Figure 2 schematically shows the optics of BL5S1 including experimental hatch. The main components of the optics are a couple of 1 m long mirrors coated with Rh and a Si (111) double-crystal monochromator. Through the mask and the first slit, the acceptance of the beam line is 2mrad in horizontal. The mirrors are located before and after the monochromator for collimating and focusing. The grazing incidence angle of the mirrors are usually set at 3.36 mrad to realize the cut off energy of 20keV. The first mirror is collimating the incident beam in vertical, and the second bend-cylindrical mirror is focusing the beam at sample position both in horizontal (2:1) and vertical. For the XAFS measurements around the absorption edges less than about 7keV, the grazing angles of mirrors are changed, typically 9 mrad, to reduce the 3rd harmonics. The monochromator, provided by Kohzu Precision Co., Ltd, keeps the water-cooled first crystal on a set of curved arms with cams and control the position and the angle of the crystal mechanically to fix the out going beam position. The second crystal has no cooling system. The monochromator can be driven as fast as 100 eV per several seconds when a quick-scan mode XAFS measurement is performed.

Figure 3 shows (a) the actually measured photon fluxes at the sample position in the experimental hatch, (b) the beam spot image taken at the focusing point of the focusing mirror, and (c) measured Cu-K XANES spectra in step- and quick-scan modes. The quality of the X-ray beam at BL5S1 can be understood roughly from Fig. 3. The flux is as expected and high enough for most of the experiments.

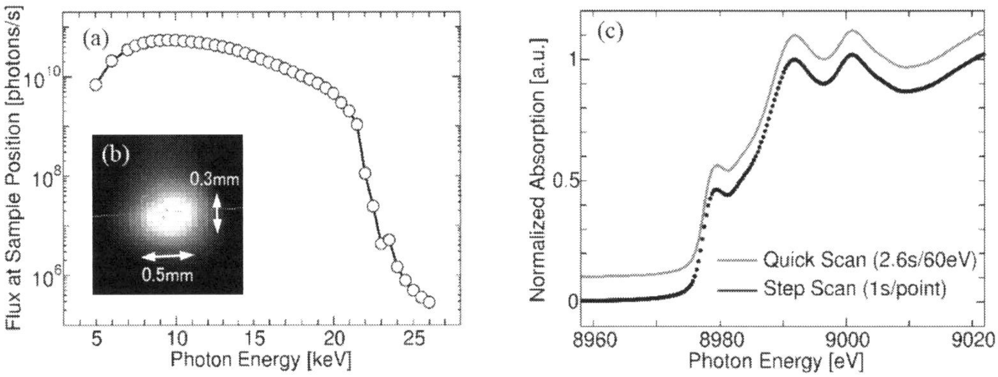


Fig. 3 (a) Measured flux at sample position, (b) Spot size taken at focusing point, (c) XANES spectra of Cu-K edge measured at BL5S1 with step- and quick-scan modes. The spectrum of 'Quick Scan' shifted by 0.1 to upper for visibility.

The beam spot size is achieved by the bend-cylindrical mirror. The vertical spot size can be changed by changing the degree of the bend. However, the horizontal size is determined by the cylinder shape of the mirror. The performance of the focusing mirror is as good as designed. The position of the focused spot moves less than a few 10 μm while usual XAFS measurements in the energy range of several 100 eV. There is no problem to discuss the energy shifts of the order of 0.1 eV or higher even in the quick-scan mode in the energy resolution at around 9 keV.

In the experimental hatch, following a 200 μm-thick Be window at the end of the vacuum pipe, X-Z slits to and 3 ion chambers (S1194B1 and S1196B1, Ohken) are set on the X-ray path for the typical transmission experiments. A set of electronics consists of pre-amplifiers (S2341A, Ohken) connected to the ion chambers, V/F converters (S2341 or N2VF-1H, Tsujicon) and a counter (CT08-01E, Tsujicon) is utilized both for the standard step-scan and quick-scan mode XAFS measurement. Digital multi-meters (34410a, Agilent) to measure the outputs of the pre-amplifiers are also available as alternative choice for users. A 19-elements Ge solid state detector (CJ-9-11-116, Canberra with DXP-XMAP, XIA, LLC), a 4-elements silicon drift detector (XSSD50-07-4 with APV8004X(DSP), TechnoAP) and a 3 grid Lytle detector (EXAFS company) for the fluorescence yield measurements and a CEY(conversion electron yield) measurement chamber are available.

In order to satisfy the demands on various kinds of experiments, three types of sample holders are always stand by. The first one is a position fixed single sample holder which can keep a sample mounted

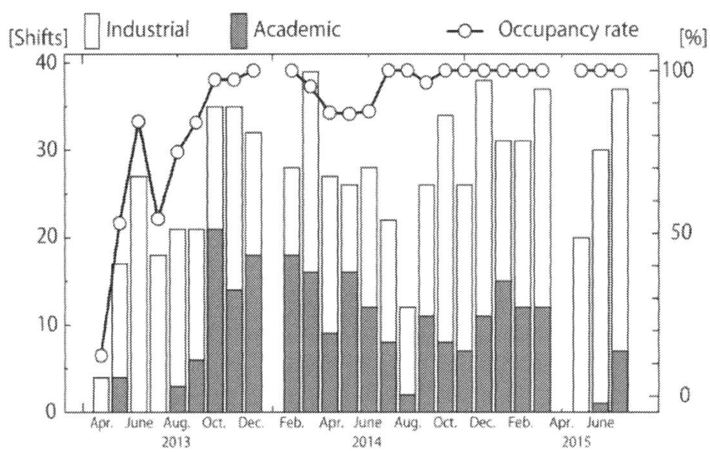


Fig. 4 The number of the shifts spent by users at BL5S1 and the occupancy rate to the available shifts for each month. AichiSR was shut down for about one month in Jan. 2014 and Apr. 2015.

on a standard slide-film frame. The second one is a remote-controllable X-Z stage of which motion range is 10cm x 10cm wide with 1μm resolution. For the X-Z stage, a few kinds of sample-holder blades are preliminary prepared, i.e., 9 samples (3x3) holder blade, 49 samples (7x7) holder blade, and a single sample holder for rather large samples. The third one is a remote controllable four axes (X, Z, ω, θ) stage which can precisely control the alignment of a sample for some experiments like as total reflection mode measurements. The controls of the remote stages are integrated to the XAFS measurement system to make complex measurements easy.

$H_2$, $O_2$, $N_2$, and He gas lines controlled by digital mass flow meters are available in the experimental hatch for the users who conduct in-situ XAFS measurements with the gases. An in-situ cell for transmission measurement which can be heated up to 600°C is ready to use on the XAFS measurement stage with the gases.

The main target of the beamline BL5S1 are users who would like to use the XAFS measurement as one of the tools for their daily research and it is not expected that all the users of BL5S1 are the specialists of the XAFS measurement. Therefore, a large part of the efforts of the staffs working at BL5S1 is spent to make the beamline user-friendly. Most of the instruments at the beamline are preliminarily set up and tuned for the experiments of the day before the users' beam time starts. The XAFS measurement software is so simplified as the users need only a few clicks on the user-interface to start standard XAFS measurements. As a result, BL5S1 is easy to use even for the light users also for the specialists. Such a beamline, with the policy and easy to access for users, must be very important to broaden the base of the XAFS society further.

AichiSR provides beam time to the users in the unit of shift. 1 shift is 4 hours. Figure 4 shows transitions of the shifts spent by users and the occupancy rate to the available shifts for each month. As shown in Fig. 4, the number of the used shifts steeply increased just after the AichiSR started to be operated for public users, and it has been almost saturated at maximum for about two years. The result shows that BL5S1, which is not special but rather standard hard X-ray beamline, has been accepted by the users as one of the very important tools for their research and development.

## 4. Conclusions

A new hard X-ray XAFS beamline, BL5S1, built at Aichi Synchrotron Radiation Center which has been operated since March 2013 was introduced. The beamline, BL5S1, was designed for the measurements using SR in the energy range from 5 to 20 keV from a super-conducting bending magnet of 5 T. Measured photon flux at the sample position was 6 x $10^{10}$ at around 9 keV and the beam spot size of 0.5 x 0.3 mm are as good as designed. For the standard transmission XAFS measurement, both of the step- and quick- scan modes are available. Energy resolution at around 9 keV is good enough to discuss the energy shift of the order of 0.1 eV or higher even when the measurement is conducted in the quick-scan mode. With several kinds of detectors for fluorescence and/or CEY detection mode measurements, and various kinds of sample holders which are supported by the XAFS measurement software, users easily obtain spectra for their samples. Such a standard, well operated and easy to access XAFS beamline must be very important to broaden the base of the XAFS society further.

## References

[1] http://www.astf-kha.jp/synchrotron/en/userguide/kougen/

# Yoneda-XAFS with Area X-Ray Detectors

**J-C Gasse*, D Lützenkirchen-Hecht, R Wagner, R Frahm**

Bergische Universität Wuppertal, Gaußstr. 20, 42097 Wuppertal, Germany

*E-mail: gasse@uni-wuppertal.de

**Abstract**. ReflXAFS is a well-known, surface sensitive method to investigate the structure of e.g. thin films. By a proper choice of incident and exit angles the Yoneda-peak, a distinct intensity maximum in the diffuse scattering, becomes visible. For multilayered samples, several Yoneda-peaks related to surface and interface regions may be observed. It is possible to utilize these peaks to measure the fine structure of an element inside the related surface and interface regions of a multi-layered sample. We will show that the use of 2D-area detectors allow the simultaneous measurement of the entire scattering pattern for each energy in an EXAFS spectrum, providing access to detailed in-situ structure information of the surface and interface regions in multilayered samples.

## 1. Introduction

Reflection mode X-ray absorption fine structure (ReflXAFS) is a well-known method for the determination of the chemical valence and the atomic short-range order around a chosen element of thin films and surfaces. A variation of the incidence angle allows a depth profiling of the sample. For incidence angles below the critical angle of total reflection this method is notably surface sensitive, whereas higher incidence angles lead to bulk sensitivity. In addition to the specular reflected beam, a second intensity maximum is present in the diffuse scattering pattern, firstly discovered by Yoneda [1]. The related scattering angle of this so called Yoneda-peak correlates with the electron density of the scattering materials and can also be utilized for XAFS experiments [2, 3]: By a proper choice of incidence and exit angles (see figure 1) multiple peaks related to the surface and interface regions of the Au/Bi bilayer sample used in this contribution can be found. This provides the possibility to measure the absorption fine structure of a chosen element in a chosen depth position inside a multi-layered sample [4, 5].

In general, a XAFS signal has to be measured with an accuracy of at least 0.1 % to provide reliable results. Due to the extremely low diffuse reflectivities of typically less than $10^{-4}$ even in the maximum of the Yoneda-peak a fine structure requires to be measured with an absolute precision of better than $10^{-5} - 10^{-6}$, which is very demanding from an experimental point of view. Successful experiments so far have been performed using ionization chambers as detectors and incidence as well as exit angles were defined by slit systems [2 - 5]. As a consequence, the different scattering peaks need to be measured sequentially, so that the data collection of a single sample may exceed several hours for a near edge structure (XANES) or even days for the extended fine structure, even if high intensity insertion device beamlines are used.

In this contribution, we will show the feasibility of Yoneda-XAFS experiments with improved area X-ray detectors to collect the entire scattering pattern including the specular reflection peak in one step (see figure 1), and therefore reduce measurement durations substantially with a comparable data quality.

We will demonstrate that Yoneda-XAFS measurements in a wide energy range of about 1000 eV are feasible in less than one hour this way. Furthermore, due to the reduced acquisition times with the new experimental setup, real in-situ experiments of surface layers and buried interfaces become feasible by the combination of Yoneda-XAFS experiments with an in-situ heat treatment of the samples.

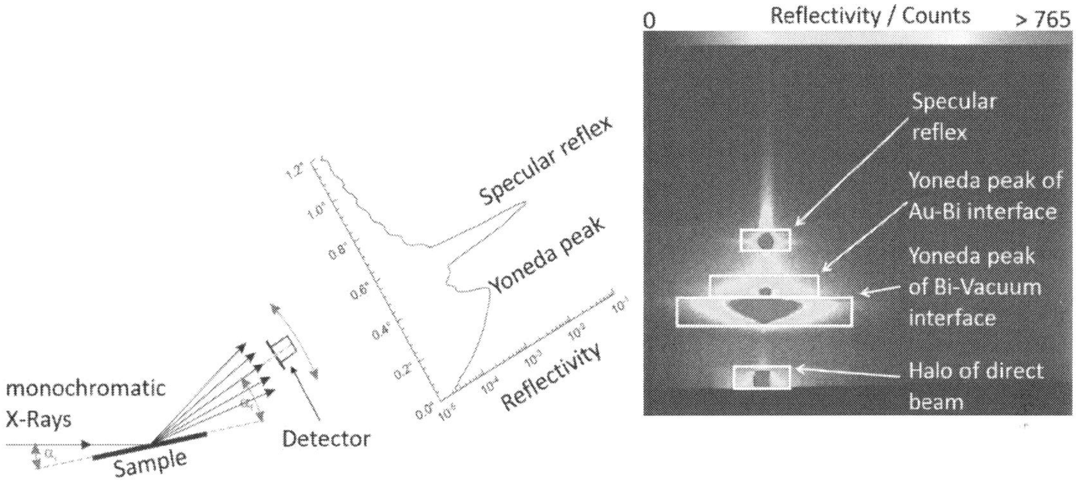


**Figure 1.** Left: Schematic layout of a Yoneda-XAFS experiment with ionization chambers as detectors. The different contributions in the scattering pattern need to be measured sequentially, extending the measurement durations by several hours. Right: The whole scattering pattern measured with an area X-ray detector for a single energy. The different peaks corresponding to the different interfaces inside a multi-layered sample (here: glass/Au/Bi) can be distinguished with a subsequent data analysis. The upper three white boxes show exemplary the regions of interest for a further data evaluation. Measurement durations were reduced substantially.

## 2. Experimental details

The Yoneda-XAFS experiments were conducted at the MRH endstation of the bending magnet Rossendorf beamline BM20 at the ESRF [6]. The synchrotron radiation beam was focused by two mirrors and monochromatized by a double-crystal Si(111) monochromator. Incident intensities were measured with a fluorescence screen and a nitrogen filled ionization chamber respectively, whereas reflected intensities were measured with the Pilatus 100 K 2D-area X-ray detector from Dectris [7]. A vacuum chamber with a sample heater was used as sample stage. The measurements and the in-situ sample heating were performed under vacuum conditions of about $10^{-7}$ mbar. Test sample was a Bi (20 nm)/Au (200 nm)-bilayer on a float glass substrate prepared by physical vapor deposition. The entire scattering pattern was evaluated for each energy while scanning across the Au $L_3$ (11919 eV) and the Bi $L_3$ (13419 eV) absorption edges for different temperatures and incidence angles.

## 3. Yoneda-XAFS experiments

For the bilayer sample we used in this experiment, an incident angle of $\Phi = 0.7\,°$ was chosen. The resulting scattering pattern is shown on the right side of figure 1. From top to bottom four peaks are observable and can be related to the specular reflection, the Au-Bi interface Yoneda-peak, the Bi-air interface Yoneda-peak and the remaining halo of the direct beam, which got mitigated by a lead plate, respectively. The diffuse scattering intensities were evaluated from these two Yoneda-peaks for room temperature, and during a heat treatment at 120 °C. The regions of interest (ROIs) for the data evaluation were carefully set after completing the measurement taking all scattering pattern of a single EXAFS scan into consideration. This way small shifts of the Yoneda-peak with the energy were properly compensated. The intensities in the different ROIs (see figure 1) were integrated for each pattern and plotted for the corresponding energies. The resulting EXAFS spectra at the Au $L_3$- and the Bi $L_3$ absorption edges are shown in figure 2. Comparing the spectra at the Au $L_3$-edge at room temperature it can clearly be seen, that there is a distinct absorption edge in the case of the Au-Bi interface. This feature is absent in the Bi-air surface, due to the lack of gold atoms at the outer surface in its as-deposited state. The bismuth $L_3$-absorption edge at room temperature on the other hand is present in the Bi-air surface as well as in the Au-Bi interface spectrum. This is well expected, because the bismuth layer forms the surface as well as the interface.

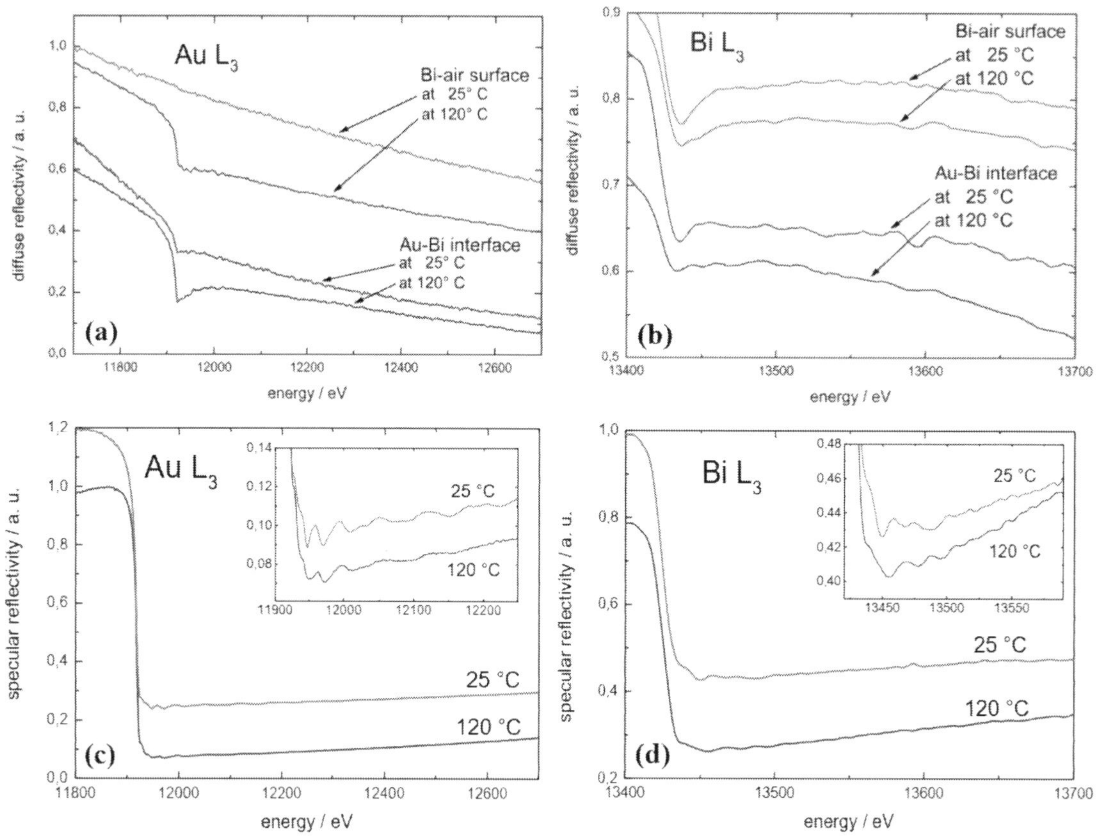


**Figure 2.** In-situ specular and diffuse reflection mode EXAFS experiments at the Au $L_3$- and Bi $L_3$-edges for an incident angle of $\Phi = 0.7$ °. The intensities in (a) and (b) were calculated from the corresponding Yoneda-peaks in the scattering pattern shown in figure 1 at room temperature, and during the heat treatment at 120 °C, respectively. The Au edge related to the outer Bi-air surface forms during the heat treatment due to diffusion of Au through the Bi layer, whereas an alteration in the XANES features at the Bi edge is visible during the heating process. The intensities in (c) and (d) are simultaneously calculated from the corresponding specular reflection peaks in the same scattering pattern as in (a) and (b). An alteration in the fine structure during the heating process is clearly observable and may give additional information about the alloy forming processes.

Heating the sample to 120 °C alters the absorption spectra significantly. As can be seen in the Au-Bi interface spectrum, the Au absorption edge gets more pronounced at 120 °C, which can be interpreted by an increased Au concentration in this sample region. An even more obvious alteration is observable in the spectrum related to the scattering at the outer vacuum-side surface of the bilayer, which now is also showing a distinct gold absorption edge at 11919 eV and XAFS features as well. This alteration clearly indicates that gold atoms are present at the outer vacuum-side surface of the bilayer sample. It can be assumed that the Au got to the surface due to diffusion processes. A comparison of the fine structure to calculated spectra suggests certain differences to a simple metallic gold layer, i.e. the presence of an intermetallic AuBi-species is very likely [8]. The Bi absorption edges on the other hand do not show such a simple behavior during the heating process at 120 °C. The Bi-air surface does not seem to show any alteration, although a small change in the fine structure of the inner Au-Bi interface is observable. This may indicate that the alloy formation process already started in the Au-Bi interface region, but the concentration of gold atoms in the outer surface is not yet sufficient to observe an alteration in the Bi structure.

A more detailed analysis of the collected data is very demanding and currently in progress. Especially the data processing is getting optimized to enhance the data quality and thus clearly differentiate the fine structure from the noise. A subsequent, careful analysis of the EXAFS may give additional information on the formation of a gold-bismuth alloy in the interface or surface region of the sample. Due to the simultaneous measurement of the entire scattering pattern, a further source of information may also be the measured specular reflection mode EXAFS (see figure 2 (c) and (d)) and thus it should be included in the modelling for a more detailed description of the sample and the alloy formation process.

## 4. Conclusion

Due to the utilization of an area X-ray detector like the Pilatus 100 K and the resulting, significant time saving, we were able to perform in-situ Yoneda-XAFS measurements for the first time. The resulting data clearly show an alteration in the EXAFS spectra determined from the interface and surface regions of the sample during the heating process. This indicates the diffusion of gold atoms from the inner gold layer to the outer bismuth layer surface and it can be assumed that an alloy formation process is taking place at the Au-Bi interface region.

## Acknowledgments

The authors gratefully acknowledge the help of Carsten Baehtz and Joerg Exner with the high temperature cell and the experiments at the ESRF. This work was financially supported by the BMBF under project no. 05K13PX1.

## References

[1] Y. Yoneda, Physical Review 131 (1963) 2010.
[2] P. Keil, D. Lützenkirchen-Hecht, R. Frahm. Europhys. Lett. 71 (2005) 77.
[3] P. Keil, D. Lützenkirchen-hecht, R. Frahm. Physica B 357 (2005) 1.
[4] D. Lützenkirchen-Hecht, P. Keil, R. Frahm. Surface Sci. 601 (2007) 4232.
[5] P. Keil, R. Frahm, D. Lützenkirchen-Hecht. Corrosion Sci. 52 (2010) 1305.
[6] W. Matz, N. Schell, G. Bernhard et al. J. Syn. Rad. 6 (1999) 1076.
[7] P. Kraft, A. Bergamaschi, Ch. Broennimann, et.al. J. Syn. Rad. 16 (2009) 368
[8] http://www.crct.polymtl.ca/FACT/phase_diagram.php?file=Au-Bi.jpg&dir=SGTE

# The binuclear nickel center in the A-cluster of acetyl-CoA synthase (ACS) and two biomimetic dinickel complexes studied by X-ray absorption and emission spectroscopy

**P Schrapers[1], S Mebs[1], Y Ilina[2], D S Warner[3], C Wörmann[2], N Schuth[1], R Kositzki[1], H Dau[1], C Limberg[3], H Dobbek[2] and M Haumann[1]***

[1] Department of Physics, Freie Universität Berlin, 14195 Berlin, Germany
[2] Department of Biology, Humboldt-Universität zu Berlin, 10115 Berlin, Germany
[3] Department of Chemistry, Humboldt-Universität zu Berlin, 12489 Berlin, Germany

*E-mail: michael.haumann@fu-berlin.de

**Abstract**. Acetyl-CoA synthase (ACS) is involved in the bacterial carbon oxide conversion pathway. The binuclear nickel sites in ACS enzyme and two biomimetic synthetic compounds containing a Ni(II)Ni(II) unit (**1** and **2**) were compared using XAS/XES. EXAFS analysis of ACS proteins revealed similar Ni-N/O/S bond lengths and Ni-Ni/Fe distances as in the crystal structure in oxidized ACS, but elongated Ni-ligand bonds in reduced ACS, suggesting more reduced nickel species. The XANES spectra of ACS and the dinickel complexes showed overall similar shapes, but less resolved pre-edge and edge features in ACS, attributed to more distorted square-planar nickel sites in particular in reduced ACS. DFT calculation of pre-edge absorption and $K\beta_{2,5}$ emission features reproduced the experimental spectra of the synthetic complexes, was sensitive even to the small geometry differences in **1** and **2**, and indicated low-spin Ni(II) sites. Comparison of nickel sites in proteins and biomimetic compounds is valuable for deducing structural and electronic differences in response to ligation and redox changes.

## 1. Introduction

Carbon oxide ($CO_x$) conversion reactions in biological enzymes are of high interest for potential renewable energy and climate engineering applications. Nickel-containing proteins such as the acetyl-CoA synthase (ACS) are prominently involved in the bacterial $CO_x$ conversion pathway [1]. The active site of ACS denoted A-cluster features a cubane-type [4Fe4S] cluster linked by a cysteine thiolate to a dinickel unit (figure 1A) [2]. The proximal nickel ion ($Ni_p$) is bridged by two cysteine thiolates to the distal nickel ($Ni_d$) ligated by two nitrogen atoms from the protein backbone [3]. Spectroscopic studies have suggested that CO-, methyl- and acetyl groups are bound to $Ni_p$ in the catalytic cycle and that an acetyl species is transferred from ACS to coenzyme-A [4, 5]. $Ni_d$ is believed to remain in the Ni(II) state, whereas $Ni_p$ alternates between Ni(0) or Ni(I) and Ni(II) states in the reaction cycle [1-5]. The nickel valence states, as well as the binding motifs of the small molecule ligands in ACS have been investigated by protein crystallography and spectroscopy [1-5], but are not fully established, partly because of resolution limitations or difficult accessibility of intermediates.

Synthetic model chemistry is a valuable approach for exploring possible coordination motifs and electronic configurations of the biological metal sites. Here, two crystallized dinickel complexes, which mimic the first-sphere coordination of the Ni sites in the ACS by mixed N/S-ligation and thiolate bridging, were studied (figure 1B,C). Both compounds comprise a $Ni^{II}(\mu\text{-}S)_2Ni^{II}$ site as revealed in earlier studies [6]. X-ray absorption and emission spectroscopy (XAS/XES) facilitates molecular fine structure and electronic configuration determination and is sensitive to all valence and spin states of nickel sites [7]. The pre-edge absorption region of the K-edge due to resonant core-level (1s) electron excitation into valence orbitals (core-to-valence transitions, ctv), and the Kß emission region, due to electronic 3p→1s decay (Kß main lines, $K\beta_{1,3}$ and Kß´) or decay from valence levels in the $K\beta_{2,5}$ region (valence-to-core transitions, vtc), bear valuable information on the electronic structure [8-10]. The ctv and vtc features can be calculated using density functional theory (DFT) [8-10]. Here, we combined Ni XAS, Kß XES, and DFT to study the configurations of the nickel sites.

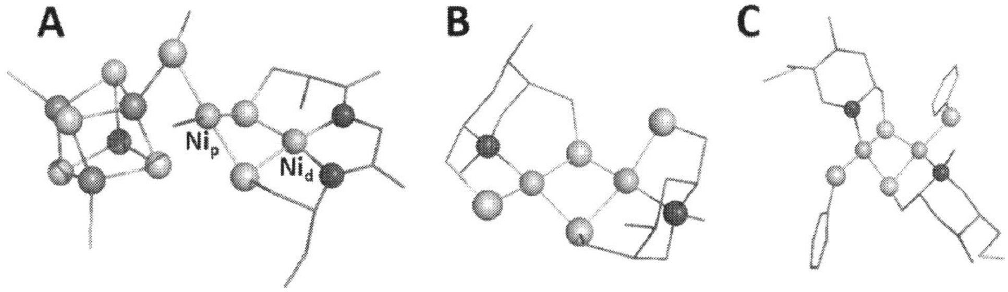


**Figure 1.** Crystal structures of metal centers in (A) acetyl-CoA synthase (ACS) protein (PDB entry 1RU3, 2.20 Å resolution [3]) and in (B, C) synthetic dinickel complexes (**1**, B [6]; **2**, C, synthesis and crystal data for **2** will be reported in a forthcoming publication). Color code: Ni, green; Fe, orange; N, blue; O, red; S, yellow; C, grey. H-atoms are omitted. Proximal and distal Ni ions in ACS are labelled.

## 2. Materials and Methods

ACS was purified according to a modified protocol of Svetlichnyi [3, 11]. The ACS gene from *Carboxydothermus hydrogenoformans* was cloned into pET28 with an N-terminal tween-strep tag, which was cleaved with TEV protease over night at RT. Ni-reconstitution was achieved at 45 °C for 48 h (addition of 5 mM ß-ME, 1:2.5 ACS:$NiCl_2$, unbound nickel was removed by gel filtration). As-isolated (oxidized, $ACS^{ox}$) and titanium(III)-citrate treated (reduced, $ACS^{red}$) protein samples (~1 mM) were prepared. The nickel complex **1** was synthesized as described in [4], the synthesis of complex **2** will be reported in a forthcoming publication. Metal contents in ACS samples were determined by total-reflection X-ray fluorescence analysis. For ACS, XAS at the Ni K-edge was performed at beamline KMC-1 of the BESSY (Berlin, Germany) using a Si[111] double-crystal monochromator and an energy-resolving 13-element Ge detector for Kα-fluorescence detection. For the Ni complexes, XAS/XES spectra were collected at beamline ID26 of the ESRF (Grenoble, France) using a Si[311] double-crystal monochromator and a vertical-plane Rowland spectrometer with 5 Si[551] analyzer crystals and an avalanche photodiode detector for emission detection. Samples were held in a liquid-He cryostat at 20 K. Single-point DFT calculations on crystal structures of **1** and **2** were performed with the ORCA program [12] (TPSSh functional, TZVP basis set). Calculated spectra were 183 eV shifted on the energy axis, scaled, and broadened by Gaussian (ctv) or Lorentzian (vtc) functions.

## 3. Results

*XAS on ACS protein.* Ni XANES and EXAFS spectra of oxidized and reduced ACS samples are shown in figure 2. The XANES spectra exhibit overall similar shapes, excluding major differences in the nickel ligation. A more pronounced shoulder at ~8337 eV in $ACS^{ox}$ and a ~1 eV lowered edge energy in $ACS^{red}$ were observed, the edge downshift presumably reflecting more reduced nickel ions in $ACS^{red}$. EXAFS fit analysis of $ACS^{ox}$ revealed about 2 Ni-S bonds of ~2.20 Å, 1 Ni-S bond of ~2.60 Å, and 1-2 Ni-N/O bonds of ~1.98 Å. Metal-metal interactions of ~2.91 Å (Ni-Ni) and ~2.69 Å (Ni-Fe) contributed relatively weakly to the EXAFS (~0.5 Ni-Ni/Fe distances per Ni ion). For $ACS^{red}$, similar coordination numbers, but overall slightly longer distances were found (Ni-S, ~2.24 Å/~2.62 Å; Ni-N/O, ~1.99 Å; Ni-Ni/Fe, ~2.98 Å and ~2.64 Å). The Fe/Ni ratio of ~3 for the ACS samples was larger than the ratio of 2 for a [4Fe-4S]-Ni-Ni cluster, suggesting sub-stoichiometric Ni contents.

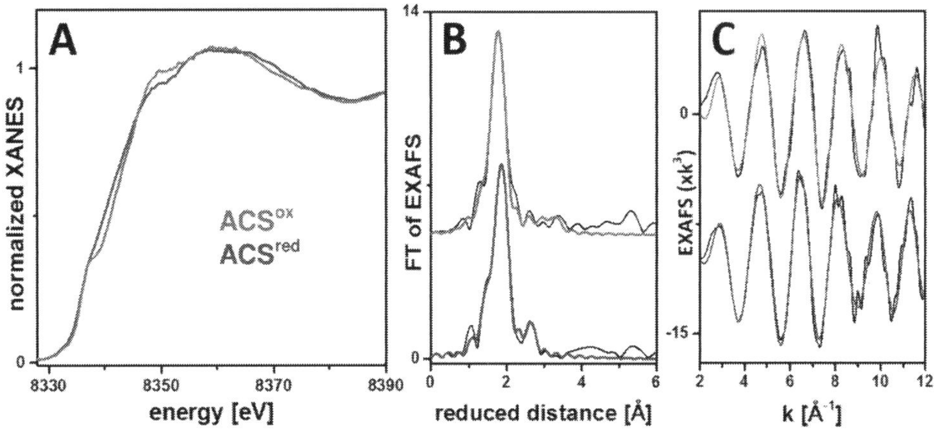


**Figure 2.** Ni XAS spectra of ACS proteins. (A) XANES spectra, (B) Fourier-transforms of EXAFS spectra in (C) (black lines, experimental data; colored lines, simulations with parameters given in the text). Spectra are vertically shifted in (B) and (C) for comparison.

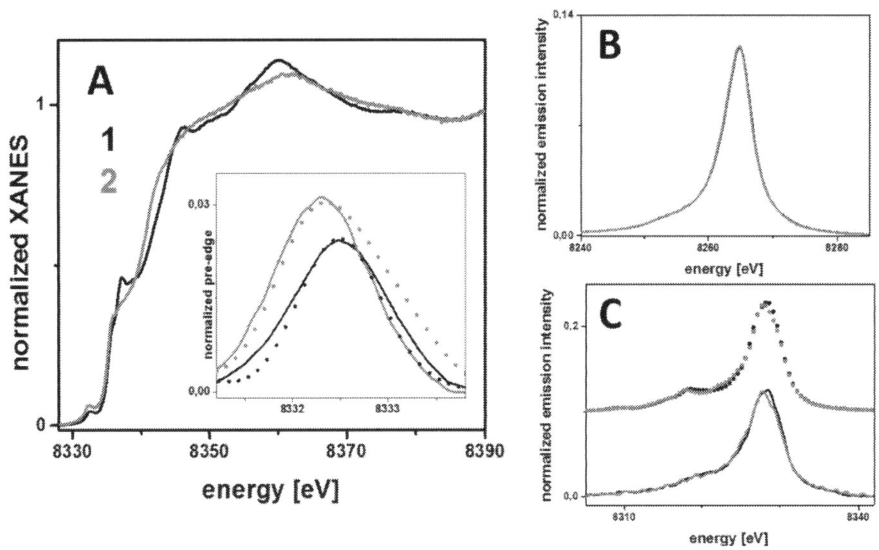


**Figure 3.** XAS/XES spectra of two synthetic dinickel complexes. (A) XANES spectra and isolated pre-edge features (ctv) in the inset (lines, experimental data; dots, DFT-calculated spectra), (B) Kß main line emission, (C) Kß$_{2,5}$ emission lines (vtc; bottom, experimental data; top, spectra from DFT).

*XAS/XES on dinickel complexes.* Both synthetic compounds show XANES shapes resembling the ACS spectra, but revealing better resolved features due to less distorted square-planar Ni sites in the complexes (figure 3). The more pronounced shoulder (at ~8337 eV) in the XANES of **1** is associated with a ~0.4 Å larger Ni-Ni distance and more regular site symmetries compared to **2**. This suggested a more distorted site symmetry also in ACS$^{red}$ compared to ACS$^{ox}$. Complexes **1** and **2** showed a discernable pre-edge feature (~8332.5 eV), as explained by considerable admixtures of molecular orbitals with ligand character to the Ni(d) valence levels, whereas a pre-edge feature was hardly resolved in the ACS spectra. Both complexes showed almost identical Kß main line spectra with dominant Kß$_{1,3}$ (~8265 eV) and small Kß′ (~8255 eV) features due to similar low-spin Ni(II)Ni(II) states ([4], manuscript in preparation). Also the Kß$_{2,5}$ emission lines were alike, reflecting the similar coordination environment and electronic structure of the Ni sites in both complexes. Calculation by DFT of the pre-edge absorption (ctv) and Kß$_{2,5}$ emission (vtc) spectra based on the crystal structures of **1** and **2** accordingly yielded good agreement with the experimental data only for low-spin Ni(II) sites.

## 4. Discussion

We compared the dinickel units in ACS protein and two biomimetic complexes. The nickel-ligand bond lengths and metal-metal distances of ACS in solution were similar to the single-crystal structure, suggesting similar site configurations. Significant reduction of the nickel in ACS manifested in a K-edge energy down-shift and considerable Ni-N/O/S bond elongation, possibly due to formation of mixed-valence Ni(I)Ni(II) species in the reduced enzyme compared to a Ni(II)Ni(II) unit in oxidized ACS. The apparent relatively small contributions to the EXAFS of Ni-Ni/Fe interactions, which are expected according to the crystal structure [3], presumably are explained by incomplete nickel reconstitution (sub-stoichiometric Ni contents) in the ACS protein, by admixtures of (inactive) Ni-Zn sites, and by destructive interference of counterphasic EXAFS oscillations of Ni-Ni/Fe and long Ni-S interactions as also observed for other nickel enzymes [13]. The XANES spectral similarities of ACS and the models suggest overall similar square-planar Ni(II)Ni(II) site geometries, which are considerably more distorted in the ACS. In particular, the pre-edge absorption feature reflecting resonant core-to-valence excitations was less resolved in ACS, as explained in part by monochromator resolution differences, but also site symmetry variations and/or the presence of different active site species in the oxidized and reduced protein samples may be involved. Kß XES and DFT calculations of ctv and vtc spectra confirmed the low-spin Ni(II) centers in **1** and **2**. The spin-state of the nickel sites in ACS remains to be addressed using Kß XES. These results show that comparison of XAS/XES data of biological and synthetic dinickel centers provides access to molecular and electronic structure differences. Further work on the ACS protein along this line is in progress in our laboratories.

**Acknowledgments**
M.H., H.Do., H.Da., and C.L. gratefully acknowledge support within Unicat (Cluster of Excellence Berlin). M.H. thanks the Deutsche Forschungsgemeinschaft for a Heisenberg Fellowship and for funding (grant Ha3265/6-1) and the Bundesministerium für Bildung und Forschung for support within the Röntgen-Angström Cluster (grant 05K14KE1). We thank the groups of P. Glatzel at ID26 of ESRF and of F. Schäfers at KMC-1 of BESSY (Helmholtz Center Berlin) for excellent technical support.

**References**

[1] Appel A M, Bercaw J E, Bocarsly A B, Dobbek H, DuBois D L, Dupuis M, Ferry J G, Fujita E, Hille R, Kenis P J, Kerfeld C A, Morris R H, Peden C H, Portis A R, Ragsdale S W, Rauchfuss T B, Reek J N, Seefeldt L C, Thauer R K and Waldrop G L 2013 *Chem. Rev.* **113** 6621
[2] Jeoung J H, Goetzl S, Hennig S E, Fesseler J, Wörmann C, Dendra J and Dobbek H 2014 *Biol. Chem.* **395** 545
[3] Svetlitchnyi V, Dobbek H, Meyer-Klaucke W, Meins T, Thiele B, Römer P, Huber R and Meyer O 2004 *Proc. Natl. Acad. Sci. U. S. A.* **101** 446
[4] Barondeau D P and Lindahl P 1997 *J. Am. Chem. Soc.* **119** 3959
[5] Seravalli J, Kumar M and Ragsdale S W 2002 *Biochemistry* **41** 1807
[6] Warner D S, Limberg C and Mebs S 2013 *Z. Anorg. Allg. Chem.* **639** 1577
[7] Haumann M (2015) In *Biohydrogen* (Rögner M, Ed), DeGruyter, Berlin, 97
[8] Lambertz C, Chernev P, Klingan K, Leidel N, Siegfridsson K G V, Happe T and Haumann M 2014 *Chem. Sci.* **5** 1187
[9] Chernev P, Lambertz C, Brünje A, Leidel N, Sigfridsson K G V, Kositzki R, Hsieh C H, Yao S, Schiwon R, Driess M, Limberg C, Happe T and Haumann M 2014 *Inorg. Chem.* **53** 12164
[10] Leidel N, Chernev P, Havelius K G V, Schwartz L, Ott S and Haumann M 2012 *J. Am. Chem. Soc.* **134** 14142
[11] George S J, Seravalli J and Ragsdale S W 2005 *J. Am. Chem. Soc.* **127** 13500
[12] Neese F 2012 *Wiley Interdisc. Rev. Comput. Mol. Sci.* **2** 73
[13] Sigfridsson K G V, Leidel N, Sanganas O, Chernev P, Lenz O, Yoon K S, Nishihara H, Parkin A, Armstrong F A, Dementin S, Rousset M, De Lacey A L and Haumann M 2015 *Biochim. Biophys. Acta* **1847** 162

# Lithographically fabricated silicon microreactor for *operando* QEXAFS studies in exhaust gas catalysis during simulation of a standard driving cycle

**D E Doronkin[1,2], S Baier[1], T Sheppard[1,2], F Benzi[1], J-D Grunwaldt[1,2]**

[1] Institute for Chemical Technology and Polymer Chemistry, KIT, Kaiserstr. 12, D-76131 Karlsruhe, Germany
[2] Institute of Catalysis Research and Technology, KIT, Hermann-von-Helmholtz-Platz 1, D-76344 Eggenstein-Leopoldshafen, Germany

E-mail: grunwaldt@kit.edu

**Abstract**. Selective catalytic reduction of $NO_x$ by ammonia over Cu-ZSM-5 was monitored by *operando* QEXAFS during simulation of the New European Driving Cycle. The required fast temperature transients were realized using a novel silicon microreactor, enabling simultaneous spectroscopic and kinetic analysis by X-ray absorption spectroscopy (XAS) and mass spectrometry (MS). Periods of high temperature were correlated to an increase in both $N_2$ production and change of coordination of Cu sites. This *operando* approach using Si microreactors can be applied to other heterogeneous catalytic systems involving fast temperature transients.

## 1. Introduction

The development of more efficient technologies for exhaust gas aftertreatment is a major driver for automotive catalysis research. In particular, increasingly strict legislation in Europe in recent years renders it important to proactively reduce vehicular pollution such as nitrogen oxide ($NO_x$) emissions. Selective Catalytic Reduction (SCR) of $NO_x$ by ammonia ($NH_3$) over iron- or copper-based zeolite catalysts is currently one of the most efficient methods of $NO_x$ abatement. However, the exact mechanism and structure of the active site is still the subject of strong discussion [1,2].

X-ray Absorption Spectroscopy (XAS) is a powerful tool for catalyst characterisation, as it can be applied to a range of realistic samples, allowing determination of metal oxidation state, coordination environment, phase and structural changes. To better understand catalyst performance and to further design future catalysts, it is crucial to perform spectroscopic analysis *in situ* or *operando*, and with high spatial- and time-resolution. However, laboratory conditions for catalytic test measurements are often over-simplified, limiting the validity of such measurements in describing realistic driving conditions. For testing exhaust gas catalysts under close to realistic conditions, so-called driving cycles [3] were developed which mimic conditions found on board a vehicle during driving. One such set of realistic model conditions is the New European Driving Cycle (NEDC), widely used to test exhaust aftertreatment systems in Europe. The cycle determines vehicle speed, which in turn defines the catalyst working conditions (temperature, concentration of exhaust gas components, space velocity). Due to abrupt speed changes (acceleration and braking) and high gas velocities, the catalyst temperature also changes rapidly. To successfully model temperature transients during NEDC and correlate it to the catalyst structure, it is necessary to use a system capable of high heating and cooling rates, and to apply *operando* measurements with high time resolution. Until now only some special lab test benches designed for studies of monolithic catalysts were described [3] but none to our knowledge modelled driving cycles with powder catalysts while allowing simultaneous spectroscopic studies.

Recently our group described a novel silicon microreactor designed for *in situ* XAS measurements [4], allowing particularly rapid temperature transients (approx. 300 °C/s) in combination with product gas analysis by MS. The microreactors are constructed by anodic bonding of two channel-cut silicon (or silicon and glass) wafers, and are most suited for fluorescence measurements due to the presence of minor diffraction peaks observed in transmission, the latter as a consequence of the manufacturing process. The window between diffraction peaks is enough to measure quick XANES in transmission geometry. Here we demonstrate time-resolved *operando* Quick X-ray Absorption Spectroscopy (QEXAFS) measurements during fast temperature transients from an NEDC. As a first example we show structural changes of the active metal sites in zeolite catalysts during $NH_3$-SCR, while following the NEDC temperature profile.

## 2. Experimental Methods

Cu-ZSM-5 zeolite was prepared by aqueous ion-exchange of $NH_4$-ZSM-5 zeolite (Si/Al ratio = 11, Clariant). 5 g of zeolite was stirred in 250 ml of 0.05M Cu(II) acetate solution (Merck) for 24 h at 20 °C. The solid was then filtered, washed with 1 L of de-ionized water, dried overnight at 80 °C and calcined for 4 h at 500 °C in static air. Cu concentration in the final solid was estimated from X-ray absorption as ~2.9 wt.%. Sieve fraction of 100-200 μm was used for the measurements.

The newly designed *in situ* cell (Fig. 1) for catalytic testing is a lithographically etched Si-based microreactor [4] with U-shaped gas channel passing through a catalyst bed (500 μm width, 250 μm depth, approx. 4-5 mm length). Two types of microreactors are available with different composition: a "Si-Si type" and a "Si-Glass type". The "Si-Glass" type can be used at temperatures up to 450-500 °C, allowing combination with light microscopy, UV-Vis and Raman techniques. Temperatures of at least 600 °C can be reached using "Si-Si" reactors which can be combined with IR spectroscopy. The microreactors are capable of performing fast temperature transients with simultaneous collection of catalytic data. A "Si-Si" microreactor was used in the current study. For temperature measurements a Pt100 resistance thermometer was glued below the catalyst channel and connected to a Eurotherm 2216 controller, an integrated Pt layer on the chip surface provided resistive heating.

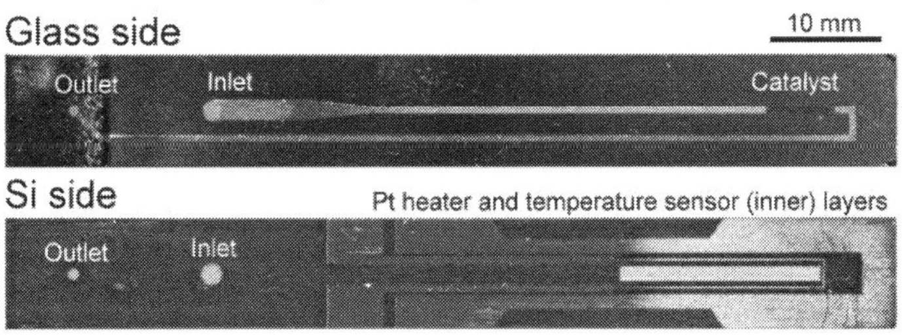


**Figure 1.** Exemplary photo of an Si-Glass type microreactor cell with a loaded catalyst. A sieved catalyst (grain size > 50 μm) is loaded via the inlet hole. A Si stopper holds the catalyst in place.

*Operando* XAS measurements were carried out at the Cu *K*-edge (8979 eV) at the SuperXAS beamline (Swiss Light Source, SLS) equipped with a Si(111) channel cut monochomator to allow data collection in QEXAFS mode at 10 Hz [5]. The storage ring was operated at 2.4 GeV in top-up mode with a ring current of around 400 mA. X-ray absorption spectra were collected in transmission mode using ionization chambers specially developed for quick data collection [6]. QEXAFS spectra were recorded at the beginning of the catalyst bed with a focused X-ray beam of about 100 x 100 μm. Spectra were averaged, normalized and Linear Combination Fitting (LCF) analysis was performed using JAQ 2 software (by O. Müller, P. Becker and R. Frahm). Spectra of the catalyst under SCR gas feed were recorded at 50 °C and 350 °C (Fig. 2, middle) and used as references for LCF. An example of the fitting procedure is shown in Fig. 2, right.

The gas feed contained 1000 ppm NO, 1000 ppm $NH_3$, 5% $O_2$ and approx. 2% $H_2O$ in He (10 ml/min total flow and GHSV ~ 1,000,000 $h^{-1}$ respectively). $N_2$, NO, $NO_2$, $N_2O$ and $NH_3$ signals were monitored using a Hiden Analytical QGA mass spectrometer. The applied temperature cycles for mimicking NEDC temperature profiles were based on ref. [7]. Several repetitions of NEDC were conducted to ensure that the catalyst response was reproducible.

## 3. Results and discussion

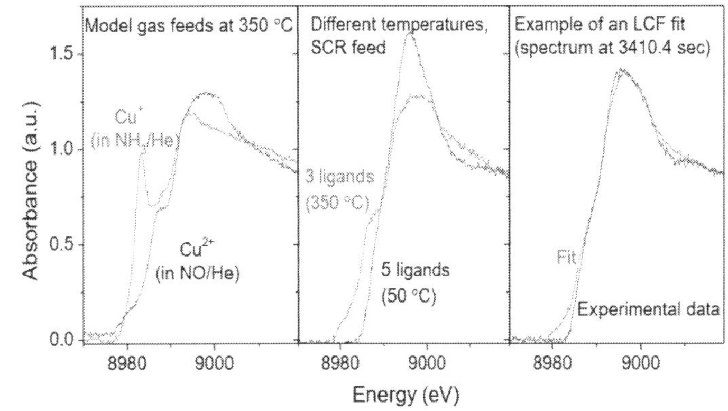


**Figure 2.** XANES spectra at Cu *K*-edge of Cu-ZSM-5 catalyst under: (left) model conditions typically used to determine Cu oxidation state; (middle) SCR feed at low and high temperatures; and (right) an exemplary LCF fit of a spectrum obtained during the NEDC test (at 3410.4 sec, Fig. 3).

First, spectra of Cu sites under steady-state conditions in pure $NH_3$/He and NO+$O_2$/He were recorded (Fig. 2, left) to serve as model spectra with pure oxidized (e.g. $Cu^{2+}$) and reduced (e.g. $Cu^{+}$) species (cf. [2,8]). These differ significantly from spectra obtained in the full SCR feed at different temperatures (Fig. 2, middle; 350 °C with an SCR-active catalyst and 50 °C where no SCR activity is observed). Comparison of the experimental data in SCR feed gas atmosphere at different temperatures with literature [2,9] and also with our previous results obtained in a capillary reactor at steady state or "slow" transient conditions (5-10°C/min vs. ~300°C/s) [8] shows that in this case changes in the coordination geometry of Cu sites were observed on heating/cooling, rather than changes in oxidation state. Although the catalytic activity varies as expected upon changing the temperature, the pre-edge peak ascribed to $NH_3$-$Cu^{+}$ [2,9] was observed in the *operando* spectra only during first heating of the catalyst but disappeared for the consequent heating / cooling cycles. This effect can be due to mass transfer limitations which result from the rather large sieve fraction and possible channeling effects [10]. Furthermore, at the low temperatures used in this study (<100°C) water adsorption could compete with $NH_3$ in the Cu coordination sphere [11] and also be retained in the microreactor, probably compacting the catalyst particles. In contrast, the adverse effects of high water concentration are also suggested by recent results obtained in our group for a Pt/$Al_2O_3$ diesel oxidation catalyst during NEDC cycling under dry conditions. In this case, typical variations of the "white line" in the Pt $L_3$-edge spectra were observed together with conversion of CO, similarly to that shown in ref. [12].

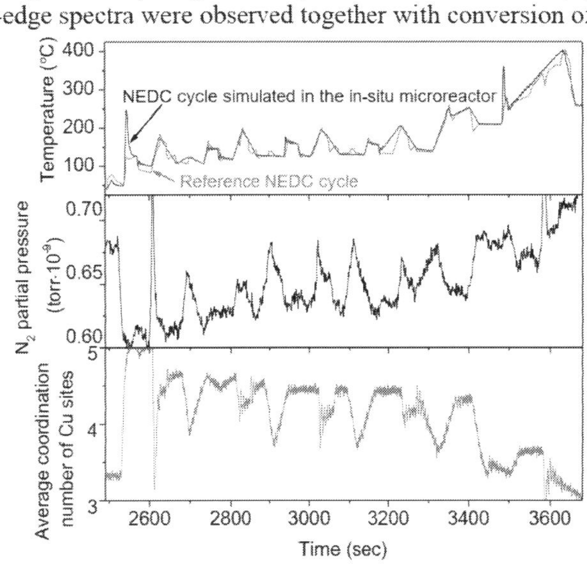


**Figure 3.** Temperature (top), amount of produced $N_2$ (middle), and (bottom) average coordination number of Cu sites in Cu-ZSM-5 catalyst during a simulated NEDC cycle (simulating inlet temperature of the first catalyst after the engine [7]). Conditions: 1000 ppm NO, 1000 ppm $NH_3$, 5% $O_2$, and approx. 2% $H_2O$ in He, 10 ml/min total flow (GHSV ~1,000,000 $h^{-1}$).

During application of rapid heating/cooling cycles for the Cu-ZSM-5 sample, LCF analysis was used to monitor changes in Cu coordination (Fig. 3, bottom). Fig. 3 (top) demonstrates how the simulated NEDC cycle can successfully mimic a real cycle, involving rapid temperature changes with sufficient accuracy. By comparing the temperature profiles (Fig. 3, top) to the catalytic (middle) and spectroscopic data (bottom) a direct correlation between the Cu catalyst performance, the operating temperature and Cu coordination can be observed. Heating of the catalyst led e.g. to an increase in $N_2$ production accompanied by a decreased number of ligands around Cu. A similar variation of the XAS spectra and thus of the coordination sphere was shown by other Cu-zeolite samples such as 1.2 wt.% Cu-SSZ-13 and 3.4wt.% Cu-SAPO-34.

## 4. Conclusions

A silicon based microreactor in combination with *operando* QEXAFS provides new opportunities for catalysis research as it opens up the possibility to understand the structure-activity relationship of heterogeneous catalysts during very fast temperature transients, i.e. under realistic driving conditions. Changes in the structure of SCR catalysts occurred simultaneously to the catalytic performance during fast temperature transients from the NEDC. Collecting catalytic and QEXAFS data on different Cu-zeolites with different pore sizes, sieve fractions and space velocities will help to determine whether differences in catalytic activity result from different diffusivity of gases and mobility of Cu ions or different intrinsic activity of the Cu sites. In addition, the versatility of the microreactor system will permit further studies on a range of exhaust gas aftertreatment processes.

## 5. Acknowledgments

This work was supported by the Federal Ministry of Education and Research (BMBF) projects "X-ray microscopy" (05K10VK1), "*In situ* Nanoskop" (05K10VK2) and "ZeitKatMat" (05K13VK3) as well as the virtual Helmholtz-Institute VI-403. In addition, we thank the "Stiftung der Deutschen Wirtschaft" (S.B) for grants and the Paul Scherrer Institute (Villigen, Switzerland) for providing beamtime at the SLS SuperXAS (X10DA) beamline. We gratefully acknowledge O. Müller and M. Nachtegaal for assistance during the XAS measurements, and M. Casapu (KIT) for fruitful discussions. We also thank GeSiM (Gesellschaft fuer Silizium-Mikrosysteme GmbH) for discussions on the custom-designed microreactor and Clariant for providing the parent zeolite.

## References

[1] Brandenberger S, Kröcher O, Tissler A and Althoff R 2008 Catal. Rev. Sci. Eng. **50** 492
[2] Günter T, Carvalho H W P, Doronkin D E, Sheppard T, Glatzel P, Atkins A, Rudolph J, Jacob C R, Casapu M and Grunwaldt J-D 2015 Chem. Commun. **51** 9227
[3] Weisweiler W and Wunsch R 1998 Chem. Eng. Process. **37** 229
[4] Baier S, Rochet A, Hofmann G, Kraut M and Grunwaldt J-D 2015 Rev. Sci. Instrum. **86** 065101
[5] Frahm R, Nachtegaal M, Stötzel J, Harfouche M, van Bokhoven J A, Grunwaldt J-D, Garrett R, Gentle I, Nugent K and Wilkins S 2010 AIP Conf. Proc. **1234** 251
[6] Müller O, Stötzel J, Lützenkirchen-Hecht D and Frahm R 2013 J. Phys. Conf. Series **425** 092010
[7] Robinson K, Ye S, Yap Y and Kolaczkowski S T 2013 Chem. Eng. Res. Design **91** 1292
[8] Doronkin D E, Casapu M, Günter T, Müller O, Frahm R and Grrunwaldt J-D J. Phys. Chem. C **118** 10204
[9] Borfecchia E, Lomachenko K A, Giordanino F, Falsig H, Beato P, Soldatov A V, Bordiga S and Lamberti C 2015 Chem. Sci. **6** 548
[10] Grunwaldt J-D, Caravati M, Hannemann S and Baiker A 2004 Phys. Chem. Chem. Phys. **6** 3037
[11] Bagnasco G 1996 J. Catal. **159** 249
[12] Gänzler A M, Casapu M, Boubnov A, Müller O, Conrad S, Lichtenberg H, Frahm R, Grunwaldt J-D 2015 J. Catal. **328** 216

# X-ray natural linear dichroism of graphitic materials across the carbon K-edge: Correction for perturbing high-order harmonics

**Jansing C[1], Mertins H C[1], Gaupp A[1], Sokolov A[2], Gilbert M C[1], Wahab H[3] and Timmers H[3]**

[1] Münster University of Applied Sciences, Physical Engineering, Stegerwaldstraße 39, 48565 Steinfurt, Germany, [2] Helmholtz-Zentrum Berlin für Materialien und Energie, Albert Einstein Str. 15, D-12489 Berlin, [3] School of Physical, Environmental and Mathematical Sciences, The University of New South Wales, PO Box 7916, Canberra BC ACT 2610 Australia

E-mail: c.jansing@fh-muenster.de

**Abstract.** Reflectivity measurements on graphitic materials such as graphene at energies across the carbon K-edge are frustrated by significant intensity loss due to adventitious carbon on beamline mirrors. Such intensity reduction enhances effects due to perturbing high-order harmonics in the beam. These effects distort the actual structure of the reflectance curve. In order to overcome this limitation, a correction technique has been developed and demonstrated first with measurements for highly ordered pyrolytic graphite. The same approach may be applied to other graphitic materials such as graphene and it may be used with other synchrotron beamlines. The fraction of high-order harmonics was determined by passing the incident beam through a 87 nm thin silicon nitride absorber that can be well modeled. Using the corrected measurements the x-ray natural linear dichroism of the sample has been determined.

## 1. Introduction

The technology perspectives offered by graphene and other graphitic materials have triggered significant research activity [1],[2]. Two different types of molecular bonds are important for electronic and structural properties of such materials. While $\pi$-bonds, which are perpendicular to the basal plane, are important for electronic transport properties, in-plane oriented $\sigma$-bonds are important for structural properties [1]. The orthogonal orientation of both types of bonds can be investigated by resonant excitation across the carbon K-edge between 280 eV-300 eV using linearly polarized synchrotron radiation (SR). P-polarized light, with the electric field vector parallel to the $\pi$-bonds, excites unoccupied $\pi$-orbitals, predominantly at 285 eV. S-polarized light with the electric field vector aligned with the $\sigma$-bonds excites unoccupied $\sigma$-orbitals above 291 eV.

Measurements at soft x-ray beamlines suffer from the fact that optical components become contaminated with adventitious carbon [3]. This contamination is unavoidable due to residual organic molecules being adsorbed on beamlines optical elements [4]. At certain energies near the carbon K-edge the adventitious carbon on optical elements can lead to huge decreases in intensity of up to two orders of magnitude [3] and a drastic increase of the relative amount of high-order radiation with respect to first order radiation. High-order harmonics are produced in the monochromator and cannot be avoided due to the optical laws of grating diffraction

[5],[6]. They thus act as an unwanted background that creates artefacts that are superimposed on the actual reflection and transmission spectra and obscure resonance structures. Moreover, the unwanted absorption of incident light $I_0$ by contaminated beamline optical elements is polarization sensitive.

Measurements were performed at BESSY II, Berlin, at the elliptical undulator beamline UE56-2 with a plane grating monochromator PGM-2 [5] using the BESSY polarimeter [7]. Incoming, linearly polarized light with vertical polarization vector (Fig 1: $I_{0ver}$, closed symbols) predominantly excites unoccupied $\sigma$-orbitals of adventitious carbon on contaminated optical elements at 290 eV, thus causing a strong intensity dip at this energy. Linear polarized light with horizontal polarization vector (Fig 1: $I_{0hor}$, open symbols) excites mainly unoccupied $\pi$-orbitals at 284 eV leading to an intensity minimum at this energy [3].

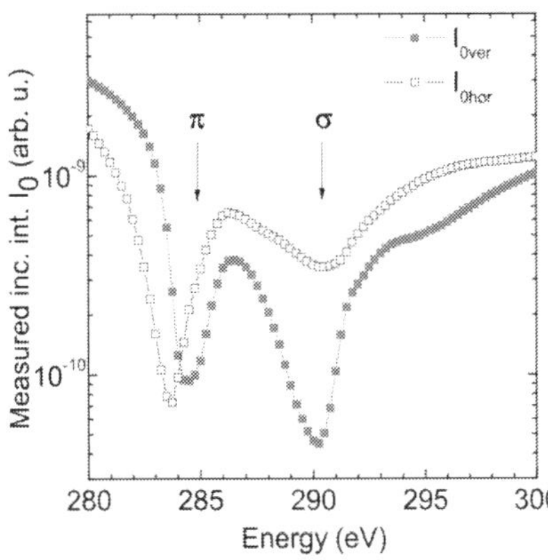


**Figure 1.** Intensity of incoming light for vertically and horizontally polarization with respect to the laboratory system.

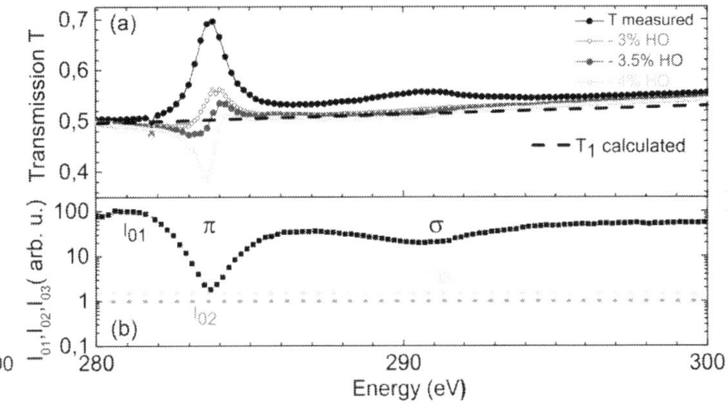


**Figure 2.** (a) Measured transmission $T$ for a 87 nm $Si_3N_4$-absorber, using $I_{0hor}$ and corrections with different amounts of high-order SR. (b) Amount of first order SR $I_{01}$ according to Eq. 1 and high-order harmonics $I_{02}$, $I_{03}$.

## 2. Estimating high-order contributions

In order to determine the high-order fraction of $I_0$ across the carbon K-edge, we recorded transmission spectra for a 87 nm thin $Si_3N_4$-foil that acts as a differential discriminator of different high-order harmonics to determine the high-order fractions in the resonance region and outside at 300 eV.

The high-order fraction of light at soft x-ray PGM beamlines outside the resonances can be determined as suggested by Follath [6]. The author utilized the line structure of undulator radiation at a PGM beamline similar to the arrangement used here with UE56-2 PGM-2 for different fix-focus constants ($c_{ff}$), which is the ratio of the angle of mirror and grating. Follath worked in high resolution mode ($c_{ff}$=2.25) and in high-order suppression mode ($c_{ff}$=1.4) [6]. For the high resolution mode he found a ratio of second to first order SR of $I_{02}(270\,\text{eV})/I_{01}(135\,\text{eV}) = 0.0014$ and of third to first order of $I_{03}(405\,\text{eV})/I_{01}(135\,\text{eV}) = 0.0022$. Furthermore, for $c_{ff}$=1.4 high-order SR was completely suppressed. Due to similarities of the set-ups, our analysis was guided by Follath's results.

Incoming light of intensity $I_0$ is transmitted by the filter and is detected as $I_t$ by a GaAs photo diode [7]. The thickness of the $Si_3N_4$-filter was determined as (87±2) nm and its theoretical transmission (Fig. 2(a), dashed line) was simulated with the computer code REFLEC [8] using tabulated data [9]. The measured transmission spectrum $T = I_t/I_0$ (Fig. 2(a)) shows local maxima near 284 eV and 290 eV. These maxima are caused by high-order contributions and result from the energy dependent transmittance of the $Si_3N_4$-filter, which differentially suppresses first order SR with respect to second and third order radiation. The theoretical transmission coefficients have been determined using REFLEC as $T_1(284\,eV) = 0.50$, $T_2(568\,eV) = 0.74$, $T_3(852\,eV) = 0.90$, respectively. Correcting the transmission spectrum from high-order contributions should result in a straight line as indicated in Fig. 2(a), (dashed line), neglecting possibly existent adventitious carbon on the filter. In order to correct for the artificial maxima the incoming light is separated into first order $I_{01}$ and high-order components $I_{02}$, $I_{03}$. The measured transmitted intensity is then given by $I_t = I_0 T = T_1 I_{01} + T_2 I_{02} + T_3 I_{03}$. The amount of first order SR with respect to high-order components is given according to

$$I_{01} = ((-I_t/I_0 + T_2)I_{02} + (-I_t/I_0 + T_3)I_{03})/(I_t/I_0 - T_1). \quad (1)$$

Assuming only weak energy dependence of high-order SR, the ratio at 300 eV was taken as constant with a ratio of $I_{03}/I_{02} = 1.5$, in agreement with Follath [6], gives an $I_{01}$ contribution according to Eq. 1 as shown in Fig. 2(b). One can clearly see the huge loss of first order radiation $I_{01}$ with respect to the amount of high-order harmonics at the resonances down to 42% of the total incoming intensity $I_0$ at 284 eV. At 300 eV the summed fraction of high-order harmonics $I_{02}+I_{03}$ in $I_0$ amounts to 3.5%. That is a magnitude higher than determined by Follath, possibly caused by different amounts of adventitious carbon on optical components in the respective beamlines.

## 3. High-order correction

The measured transmitted intensity $I_t$ has been corrected across the complete spectrum according to:

$$T_1 = \frac{I_t - T_2 I_{02} - T_3 I_{03}}{I_0 - I_{02} - I_{03}}. \quad (2)$$

A correction with a fraction of (3.5±0.5)% high-order radiation gives the best agreement with expectation (Fig. 2(a)).

For reflection spectroscopy we used highly ordered pyrolytic graphite (HOPG) with 0.4° mosaicity and recorded reflection spectra under an angle of incidence $\theta = 20°$ using again the BESSY polarimeter [7] with exactly the same beamline settings as used for the transmission measurements. The corrected reflectance $R_1$ is obtained by subtracting high-order contributions from the measured intensity $I_r$ according to:

$$R_1 = \frac{I_r - R_2 I_{02} - R_3 I_{03}}{I_0 - I_{02} - I_{03}}. \quad (3)$$

The theoretical reflection coefficients for s- and p-polarized light $R_1, R_2, R_3$ for the respective higher order are again calculated by REFLEC as $R_{p2}(568\,eV) = 2.5 \cdot 10^{-5}, R_{p3}(852\,eV) = 4.9 \cdot 10^{-6}$, $R_{s2}(568\,eV) = 4.3 \cdot 10^{-5}, R_{s3}(852\,eV) = 8.3 \cdot 10^{-6}$, respectively. Experimentally deduced reflectances $R_p$, $R_s$ (filled symbols) and their

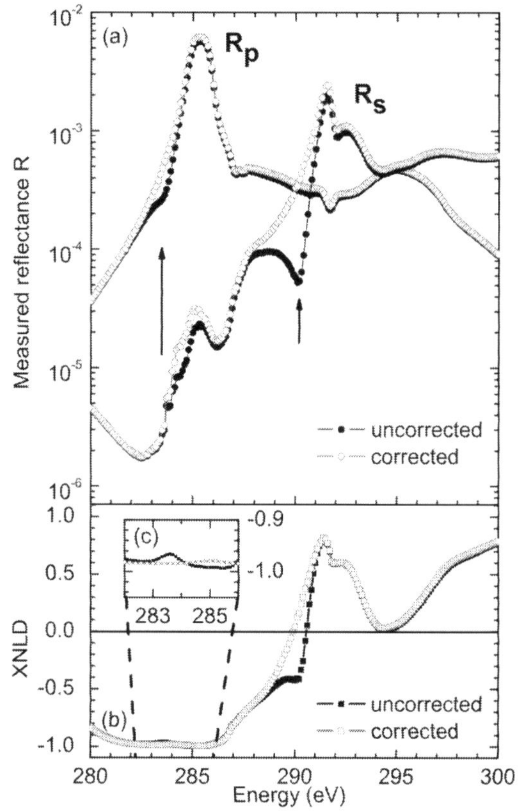


**Figure 3.** (a) Measured reflection spectra of HOPG 0.4° for p-polarized light (using $I_{0hor}$) and s-polarized light (using $I_{0ver}$) with $c_{ff}$=1.4.
(b), (c) XNLD of HOPG 0.4°.

corrected spectra (open symbols) are shown for the high-order suppression mode with $c_{ff}$=1.4 in Fig. 3(a). While reflection measurements performed with p-polarized light ($R_p$) show a huge peak at 285 eV, originating from the excited $\pi$-states of HOPG, measurements with s-polarized light ($R_s$) show a feature around 291.5 eV at the $\sigma$-states of HOPG. One can clearly see remarkable differences between uncorrected and corrected spectra in Fig. 3(a). In the resonances of the adventitious carbon on the beamline optical elements around 284 eV and 290 eV this difference is especially huge. At these energies, $I_0$ shows minima (Fig. 1) with a large amount of high-order SR with respect to first order SR. This leads to minima in uncorrected reflection spectra. Those artificial structures could easily be misinterpreted as resonant excitations.

Although working in the high-order suppression mode with $c_{ff}$=1.4, artificial minima still occur around 284 eV and 290 eV, showing that even a small $c_{ff}$-value is not able to reduce high-order contributions completely in this energy range. In order to avoid noise-introduced, non-physical negative reflectances between 282 eV-285 eV, the reflection coefficients $R_2, R_3$ for $R_s$ in Eq. 3 were set to zero only in this energy range.

Reflection spectra of s- and p-polarized light ($R_s$,$R_p$) play a crucial role in the x-ray natural linear dichroism (XNLD) [2] according to

$$XNLD = (R_s - R_p)/(R_s + R_p).$$

These spectra can directly be compared to polarization spectra to demonstrate the correlation of XNLD and birefringence [2]. Uncorrected reflection spectra lead to distorted XNLD signals as shown in Fig. 3(b),(c). In the minima of $I_0$ (Fig. 1) around 284 eV and 290 eV two artefacts occur. After correcting the reflection spectra $R_s$ and $R_p$, the feature at 290 eV in the XNLD signal completely vanishes and the peak's intensity at 284 eV is strongly reduced.

## 4. Conclusions

We determined the fraction of high-order harmonics in the energy range 280 eV- 300 eV across the carbon K-edge for linearly polarized light in beamline UE56-2 PGM-2 at BESSY II. The technique we have demonstrated using transmission measurements with a $Si_3N_4$-absorber foil gives a fraction of high-order harmonics in $I_0$ of up to 58% in the 284 eV resonance, of about 10% at the 290 eV resonance and of 3.5% out of the resonance region at 300 eV. The high-order fraction has been shown to affect the measured transmission, reflectance and XNLD spectra, since artefacts are introduced. Such artefacts may result in the misinterpretation of resonance structures in the spectra. Through the energy-dependent measurement of the high-order fraction of incident light, a correction of spectra is possible, as it has been demonstrated. The importance of correcting reflection spectra to determine the correct x-ray natural linear dichroism of HOPG has become evident. The procedure may be applied similarly to other graphitic materials including graphene.

## References

[1] Geim A K *et al.* 2007 *Nat. Mater.* **6** 183–91
[2] Mertins H C *et al.* 2004 *Phys. Rev.* B **70** 235106
[3] Gaupp A *et al.* 2013 *J. Phys. Conf. Ser.* **425** 122013
[4] Singh A *et al.* 2013 *J. Phys. Conf. Ser.* **425** 152020
[5] Weiss M R *et al.* 2001 *Nucl. Instrum. Meth.* A **467-468** 449–52
[6] Follath R 2001 *Nucl. Instrum. Meth.* A **467-468** 418–25
[7] Schäfers F *et al.* 1999 *Appl. Opt.* **38** 4074–88
[8] Schäfers F *et al.* 1996 *REFLEC - A program to calculate VUV and soft x-ray optical elements and synchrotron radiation beamlines* (*TB* vol 201) (Berlin, Germany: BESSY) pp 1–17
[9] Henke B *et al.* 1993 *At. Data Nucl. Data Tables* **54** 181 – 342

# Hydride phase formation in carbon supported palladium hydride nanoparticles by *in situ* EXAFS and XRD

**A L Bugaev*[,1,2], A A Guda[1], K A Lomachenko[1,2], A Lazzarini[2], V V Srabionyan[1], J G Vitillo[2,8], A Piovano[3], E Groppo[2], L A Bugaev[1], A V Soldatov[1], V P Dmitriev[1,4], R Pellegrini[5], J A van Bokhoven[6,7], C Lamberti[1,2]**

[1] Southern Federal University, Zorge Street 5, 344090 Rostov-on-Don, Russia
[2] Department of Chemistry, University of Turin, Via P. Giuria 7, 10125 Turin, Italy
[3] Institut Laue-Langevin (ILL), 6 Rue Jules Horowitz, F-38042, Grenoble, France
[4] SNBL at ESRF, 6 Rue Jules Horowitz, F-38042 Grenoble, France
[5] Chimet SpA - Catalyst Division, Via di Pescaiola 74, I-52041 Arezzoб Italy
[6] ETH Zurich, Institute for Chemical and Bioengineering, Zurich, Switzerland
[7] Laboratory for Catalysis and Sustainable Chemistry, SLS, PSI, Villigen, Switzerland
[8] Dipartimento di Scienza e Alta Tecnologia, Università dell'Insubria, Via Valleggio 11, 22100 Como, Italy

E-mail: abugaev@sfedu.ru

**Abstract.** In the current work we present a detailed analysis of the hydride phase formation in industrial Pd/C nanocatalysts by means of combined *in situ* X-ray absorption spectroscopy (EXAFS), X-ray diffraction (XRD) and volumetric measurements for the temperatures from -10 to 50 ℃ in the hydrogen pressure range from 0 to 1000 mbar. α- and β- hydride phases are clearly distinguished in XRD. For the first time, H/Pd atomic ratio were obtained by theoretical fitting of the near-edge region of the absorption spectra (XANES) and compared with volumetric measurements.

## 1. Introduction

Palladium-based nanomaterials play an important role in various fields such as medicine, chemistry, biology and catalysis [1]. In a number of industrial application, supported palladium nanoparticles catalyse hydrogenation reactions (e.g. hydrogenation of alkenes). In the presence of hydrogen, Pd catalyst may undergo one of its hydride phases which nature affects its catalytic performance [2]. Determination of the hydride phase in the palladium Pd nanoparticles is an important step in order to investigate the active phase of the nanocatalyst. Being element specific, EXAFS technique reflects Pd-Pd interatomic distance increase which is proportional to the number of hydrogen absorbed in the nanoparticle and is not affected by the molecular hydrogen adsorbed on the support surface. At the same time, phase specific XRD allows to discriminate between α- and β- hydride phases even in the mixed phase region [3], while EXAFS is able to detect the increase of the average Pd-Pd distance upon H intercalation [4]. In the current work we present a detailed investigation by means of *in situ* EXAFS and XRD of the hydride phase formation in industrial catalysts (Chimet S.p.A.) consisting of palladium deposited on an amorphous carbon support. The X-ray data is complemented by volumetric measurements.

## 2. Materials and methods

5 wt.% Pd on carbon catalyst were supplied by Chimet S.p.A. and it has been prepared by deposition-precipitation method on a wood-based activated carbon (surface area = 980 $m^2g^{-1}$; pore volume = 0.62 $cm^3g^{-1}$). Transmission electron microscopy measurements indicated the averaged size of the nanoparticles to be in the range of 3 – 4 nm. Pd black powder was used as a reference sample.

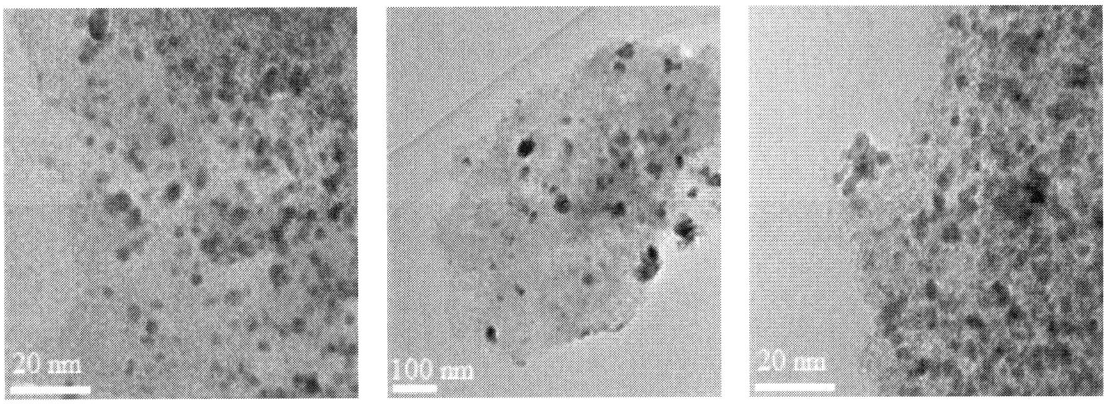


**Figure 1.** TEM images of the Pd/C nanocatalyst.

EXAFS and XRD data were collected at the BM01B (Swiss-Norwegian Beamline) of the European Synchrotron Radiation Facility (ESRF), Grenoble, France. Sample was loaded inside 1 mm glass capillary connected to a remotely controlled gas rig enabling to control the hydrogen pressure during the experiment. Nitrogen blower positioned above the sample was used to control the temperature. EXAFS spectra were collected in the transmission mode in the energy range around the Pd K-edge from 24.1 to 25.4 keV. Pd foil was measured simultaneously for energy calibration. Analysis of EXAFS spectra was performed with the Demeter software [5]. XRD patterns were measured using a $\lambda = 0.50544$ Å radiation and collected by a 2D Image Plate detector. Rietveld analysis of the diffraction profiles was performed by means of Jana2006 code [6].
Synchrotron measurements was complemented with volumetric adsorption isothermal data measured on a Micromeritics ASAP 2020 sorption analyser connected to a H2 cylinder (99.9999vol% purity, Rivoira) in the NIS centre of the University of Turin.

## 3. Results and discussion

### *3.1. EXAFS analysis*

Palladium K-edge absorption spectra and diffraction patterns were collected during the formation of the palladium hydride induced by stepwise increasing of the hydrogen pressure from 0 to 1000 mbar. The evolution of absorption and diffraction data are shown in Figure 2 (left and righ part, respectively). The increase of Pd-Pd interatomic distances leads to the shift of the EXAFS oscillations to the lower energies. At the same time, indicative features of hydride phase are observed in the near-edge (XANES) region, which is in agreement with our previous results [7, 8]. Single-shell analysis of EXAFS indicates Pd-Pd distance increase from 2.74 to 2.82. The obtained coordination number N = 9.3 is consistent with the particle size of about 3 nm [9]. Debye-Waller parameter increase from 0.005 to 0.008 is caused by the coexistence of α- and β- phases with slightly different interatomic parameters. As determined from XRD, the difference $R_\beta$ - $R_\alpha$ is about 0.05 Å: this difference is beyond the limit of standard EXAFS resolution of local structure distortion [10] with relatively short Δk interval (up to 12 Å$^{-1}$).

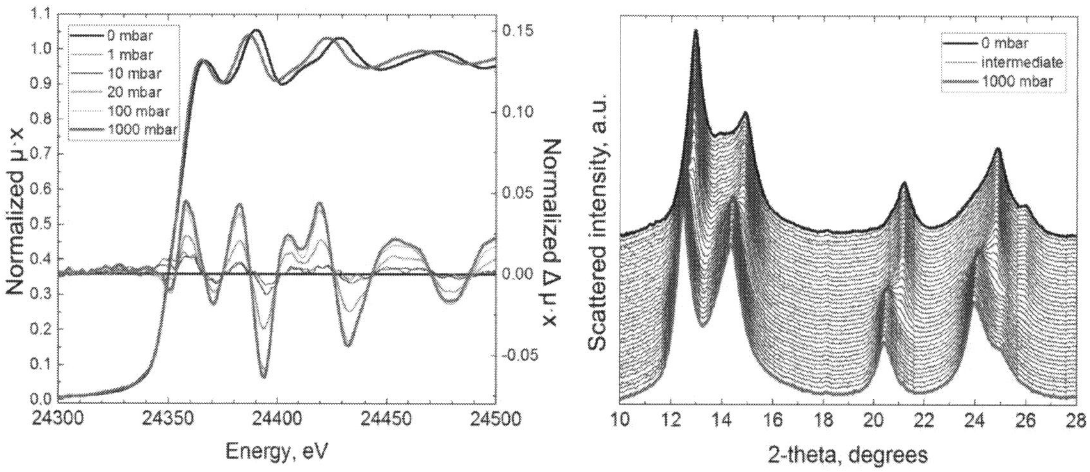


**Figure 2.** (left) Pd K-edge XANES and difference XANES taken at 20 °C. Black and red lines correspond to bare and hydride (final) palladium nanoparticles respectively, coloured lines represent intermediate states. (b) X-ray powder diffraction patterns from bare (black) to hydride (red) palladium nanoparticles taken at 20 °C. For better visualization, the patterns are shifted in vertical direction.

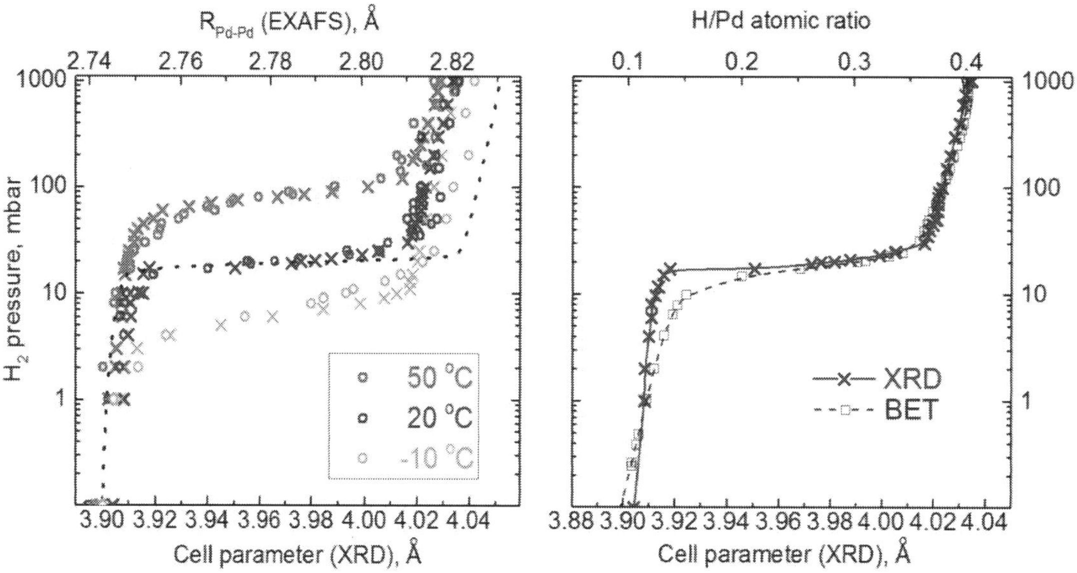


**Figure 3.** (left) Comparison of Pd-Pd interatomic distances obtained from EXAFS (circles) with cell parameter determined by XRD (crosses). Black dashed curved corresponds to the Pd black isotherm. (right) XRD measurements (crosses, blue line) vs volumetric data (squares. dashed navy line).

### *3.2. XRD analysis*

In order to reduce the number of variables, Rietveld refinement was performed in several steps. At the first step, initial and final XRD profiles (palladium in the metallic and hydride form, respectively) were used to obtain profile shape, asymmetry and background. Then, cell parameters (for α- and β-phases), zero angle shift and phase concentrations were refined for each pattern. At the last step zero shift values were averaged and fixed, and only cell parameters of α- and β- phases and their concentrations were optimized.

For comparison with EXAFS averaged cell parameters were calculated. Palladium hydride isotherms obtained by single-shell Fourier analysis of EXAFS and Rietveld analysis of XRD data are shown in Figure 3 (left). To quantify the hydrogen present in the palladium at each pressure, volumetric measurements were performed. Assuming that the amount of hydrogen adsorbed in the carbon support is linearly proportional to the hydrogen pressure, this contribution was subtracted resulting in a similar isotherm as obtained from XRD and EXAFS (Figure 3 right).

### *3.3. Volumetric measurements*

Volumetric data shown in figure 3 (right) provide quantitative information on the amount of hydrogen absorbed by palladium nanoparticles which allows to derive how interatomic distance depends on the hydrogen loading. However, several difficulties in conducting volumetric measurements exists in the case of the nanoparticles. First, the method is not element selective, and one has to perform separate measurement for the support itself, which is needed to be subtracted in order to obtain proper Pd-H isotherm. Second, experimental data reflects also adsorption of hydrogen on the surface of nanoparticles, and on the non-palladium species in the sample. This results in the non-zero H/Pd ratio for hydrogen pressures even less than 0.1 mbar. Third, the method cannot be applied in reaction conditions to estimate the H/Pd loading in the working nanocatalyst.

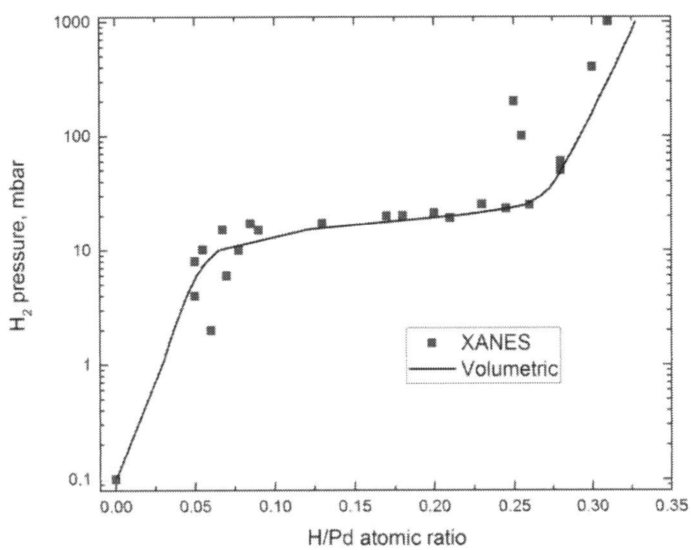


**Figure 4.** Comparison of H/Pd atomic ratio during hydrogen absorption by the palladium nanoparticles obtained by volumetric method (black line) and theoretical fitting of XANES (blue squares). Volumetric curve is shifted by 0.1 to lower H/Pd ratio.

### *3.4. XANES fitting*

As was shown in figure 2 (left) the presence of hydrogen can be monitored by difference XANES. In order to obtain hydrogen concentration we used multidimensional interpolation approach of FitIt-3 [11]. Following the procedure described in our previous work [8] the hydrogen concentration were fitted to obtain the minimum of root-mean-square difference between experimental and theoretical [12] difference spectra. The interatomic distances were set as obtained from EXAFS, while hydrogen loading x in theoretical models of $PdH_x$ clusters was changing from 0 to 0.5. The results of the fit are presented in Figure 4 together with volumetric data. In order to overcome the effect of non-zero hydrogen absorption in the low-pressure region, discussed in section 3.3, volumetric curve was shifted by 0.1 to lower H/Pd ratio.

## 4. Conclusions

We have performed a systematic study of hydride formation in the industrial palladium nanocatalysts by means of element-specific EXAFS and XANES, phase-specific XRD performed simultaneously on the same sample, measured under the same temperature and hydrogen pressure and independently complemented by volumetric measurements. The results obtained from these three techniques are in excellent agreement. We illustrate the advantage of diffraction measurements to get detailed information on the phase fractions which cannot be obtained from EXAFS.

We underline the importance of XANES analysis and show the indicative features of hydride formation region in the near edge region. Quantitative analysis of XANES using theoretically calculated spectra was shown to be in a good agreement with volumetric data. This shows for the first time, that XANES data can give quantitative information on H/Pd ratio in the conditions where volumetric measurements cannot be performed, highlighting the relevance of these method for in situ catalytic experiments.

The obtained pressure-composition isotherms provide deeper understanding of Pd-H phase transitions in the palladium nanoparticles and can be used as a reference for more complex catalytic experiments.

## 5. Acknowledgments

The research was supported by the Mega-Grant of the Ministry of Education and Science of the Russian Federation (14.Y26.31.0001). ALB, AAG acknowledge the Grant of President of Russian Federation for Young Scientists support MK3206.2014.2.

## 6. References

[1] Singh J, Lamberti C and van Bokhoven J A 2010 Chem. Soc. Rev. 39(12) 4754-4766

[2] Teschner D, Borsodi J, Wootsch A, Revay Z, Havecker M, Knop-Gericke A, Jackson S D, R. Schlögl R 2008 Science **320**(5872) 86-89.

Teschner D, Borsodi J, Kis Z, Szentmiklósi L, Révay Z, Knop-Gericke A, Schlögl R, Torres D, Sautet P 2010 *J. Phys. Chem. C* **114**(5) 2293-2299.

[3] Narehood D, Kishore S, Goto H, Adair J, Nelson J, Gutierrez H and Eklund P 2009 Int. *J. Hydrogen Energ.* **34**(2) 952-960

Lewis F A 1982 *Platinum Metals Rev.* **26**(1) 20-27

Jobic H, Renouprez A 1987 *J. Less-Common Met.* **129** 311-316

Davis R, Landry S, Horsley J, Boudart M 1989 *Phys. Rev. B* **39**(15) 10580

[4] McCaulley J A 1993 *J. Phys. Chem.* **97**(40) 10372-10379

Soldatov A, Della Longa S, Bianconi A 1993 *Solid State Commun.* **85**(10) 863-868

Scarano D, Bordiga S, Lamberti C, Ricchiardi G, Bertarione S, Spoto G 2006 *Appl. Catal. A* **307**(1) 3-12

Tew M W, Janousch M, Huthwelker T, van Bokhoven J A 2011 *J. Catal.* **283**(1) 45-54

[5] Ravel B and Newville M 2005 *J. of Sync. Rad.* **12**(4) 537–541

[6] Petricek V, Dusek M and Palatinus L 2014 *Z. Kristallogr.* **229**(5) 345-35

[7] Bugaev A, Srabionyan V, Soldatov A, Bugaev L and van Bokhoven J 2013 *J. Phys. Conf. Ser* **430**(1) 012028-012033

[8] Bugaev A L, Guda A A, Lomachenko K A, Srabionyan V V, Bugaev L A, Soldatov A V, Lamberti C, Dmitriev V P and van Bokhoven J A 2014 *J. Phys. Chem. C* **118**(19) 10416-10423

[9] Srabionyan V V, Bugaev A L, Pryadchenko V V, Avakyan L A, van Bokhoven J A and Bugaev L A 2014 *J. Phys. Chem. Sol.* **75**(4) 470-476

[10] Bugaev L A, Avakyan L A, Srabionyan V V and Bugaev A L 2010 *Phys. Rev. B* **82**(6) 064204

[11] Smolentsev G, Soldatov A V 2007 *Comp. Mater. Sci.* **39**(3) 569-574.

[12] Bugaev A, Guda A, A, Lomachenko K A, Bugaev L A, Soldatov A V 2015 Bull. Russ. Acad. Sci., Phys. **79**(4) 10416-10423

[13] Guda S A, Guda A A, Soldatov M A, Lomachenko K A, Bugaev A L, Lamberti C, Gawelda W, Bressler C, Smolentsev G, Soldatov A V and Joly Y 2015 *J. Chem. Theory Comput.* **11**(9) 4512-4521

Journal of Physics: Conference Series **712** (2016) 012033 doi:10.1088/1742-6596/712/1/012033

# Depth-resolved X-ray magnetic circular dichroism measurement by a multi-anode microchannel plate detector combined with polarization switching

**K Amemiya[1,2,5], M Sakamaki[1,2], S Kishimoto[1,2], T Kosuge[1,2], K Nigorikawa[2,3], M Tanaka[2,4], T Uchida[2,4], M Saito[2,4], M Ikeno[2,4] and K Nakayoshi[2,4]**

[1]Institute of Materials Structure Science, High Energy Accelerator Research Organization, Oho 1-1, Tsukuba, Ibaraki 305-0801, Japan

[2]Open Source Consortium of Instrumentation (Open-It)

[3]Accelerator Laboratory, High Energy Accelerator Research Organization, Oho 1-1, Tsukuba, Ibaraki 305-0801, Japan

[4]Institute of Particle and Nuclear Studies, High Energy Accelerator Research Organization, Oho 1-1, Tsukuba, Ibaraki 305-0801, Japan

E-mail: kenta.amemiya@kek.jp

**Abstract.** A depth-resolved X-ray magnetic circular dichroism (XMCD) technique has been developed in the soft X-ray region by combining a multi-anode microchannel plate detector with polarization switching between right and left circular polarizations. A simultaneous measurement for 30-channel anodes has been achieved with a continuous data acquisition rate of 1 kHz, which is fast enough for 10 Hz polarization switching. This enables us to realize the layer-resolved XMCD analysis with a high signal-to-noise ratio.

## 1. Introduction

In the past decade, the depth-resolved X-ray magnetic circular dichroism (XMCD) technique has been developed in the soft X-ray region, which realizes the atomic-layer resolved XMCD analysis for magnetic ultrathin films [1-3]. The technique is based on the fact that the effective escape depth of the Auger electrons, which are emitted after X-ray absorption, depends on the emission angle of the electrons. Therefore, a set of XMCD data at different probing depths is simultaneously obtained at different detection angles. In addition, it has been demonstrated that the signal-to-noise ratio of the XMCD data is significantly improved by using polarization switching between right and left circular polarizations (RCP and LCP) [4-6]. In this paper, we apply the polarization switching to the depth-resolved XMCD technique in order to realize the high-sensitivity depth-resolved XMCD analysis. To simultaneously record the Auger electron signals at different detection angles with much higher data acquisition rate than the polarization switching frequency (10 Hz), we adopt a multi-anode microchannel plate (MCP) detector. The XMCD spectra are obtained for a naturally-oxidized Co thin film, in which the surface of the film is preferentially oxidized, and it is demonstrated that the spectral features depend on the electron detection angles as expected.

[5] To whom any correspondence should be addressed.

## 2. Detector and data acquisition system

Up until now depth-resolved XMCD measurements have been performed by using an imaging-type detector, consisting of a MCP, a phosphor screen and a CCD camera as shown in Fig. 1(a) [1-3]. A retarding voltage is applied to the second grid in order to preferentially observe the Auger electrons, while the first grid is grounded. To apply the polarization switching technique to the depth-resolved XMCD measurement, it is necessary to simultaneously record the XMCD data at different detection angles, $\theta_d$, at a much higher data acquisition rate than the polarization switching frequency. Although this can be realized by using a high-speed CCD camera, we adopt a multi-anode MCP system, as illustrated in Fig. 1(b), for higher detection efficiency. Figure 1(c) shows the 30-channel multi-anode MCP detector, which was fabricated by Hamamatsu Photonics K.K. The detector was mounted on a vacuum flange, and the 30-channel anodes were separately connected to electric feedthroughs.

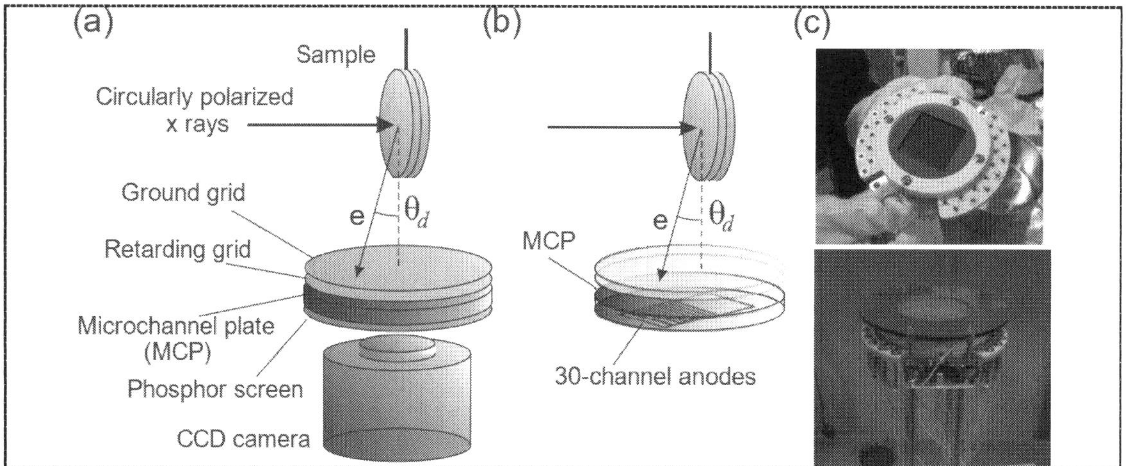


**Figure 1.** Experimental setup for the depth-resolved XMCD system consisting of (a) imaging-type MCP detector with phosphor screen and CCD camera and (b) multi-anode MCP detector, together with (c) images of the multi-anode MCP detector (top), which is mounted on a vacuum flange with HV cables and retarding grids (bottom).

Figure 2 schematically illustrate the data acquisition system. A high voltage (HV) of 2.1 kV is applied to the anodes via electric feedthroughs, by using a HV divider placed out of vacuum, while a bias voltage of 1.85 kV is applied to the bottom of the MCP. A typical pulse height for the obtained electron signal is 5-10 mV. The signals are separately amplified by a 32-channel application specific integrated circuit (ASIC), and are counted by a field-programmable gate array (FPGA). The X-ray intensity signal ($I_0$) is also connected to the ASIC, together with the undulator status signal, which represents the polarization of the X rays at each moment. The pulse counts from the anodes in each 1 ms are continuously recorded at a data acquisition rate of 1 kHz. The recorded data are transferred via SiTCP [7].

## 3. Demonstration of depth-resolved XMCD measurement

The depth-resolved XMCD measurement was performed at the soft X-ray beamline, BL-16A, of the Photon Factory, Japan, by using 10 Hz polarization switching [8,9]. Two APPLE-type undulators, ID1 and ID2, installed in the tandem configuration, are set to opposite circular polarizations (*e.g.* RCP and LCP), and the electron orbit through the undulators is modulated by kicker magnets in order to alternate the undulators which provide light for the beamline [4-6,9]. A naturally-oxidized Co thin film with ~5 nm thickness was used as a test sample, after being magnetized by a pulse magnetic field of ~100 Oe. Figure 3(a) shows time dependence of the electron pulse count, I, for channel (Ch) 18, together with that of the X-ray intensity, $I_0$. The $I_0$ signal exhibits two peaks in each 100 ms, which correspond to RCP and LCP. It is recognized that the pulse count changes in accordance with the change in $I_0$. Moreover, $I/I_0$ for LCP is higher than that for RCP, which directly indicates a negative XMCD signal at the Co $L_3$ edge. The XMCD spectrum at each detection angle, $\theta_d$, is obtained by separately integrating the pulse count for RCP and LCP, and taking the difference of them after the normalization by the X-ray intensity. Here, the I and $I_0$ signals in each 30 ms are used to obtain the XMCD data for RCP and LCP, $\mu_+$ and $\mu_-$, respectively, as indicated in Fig. 3.

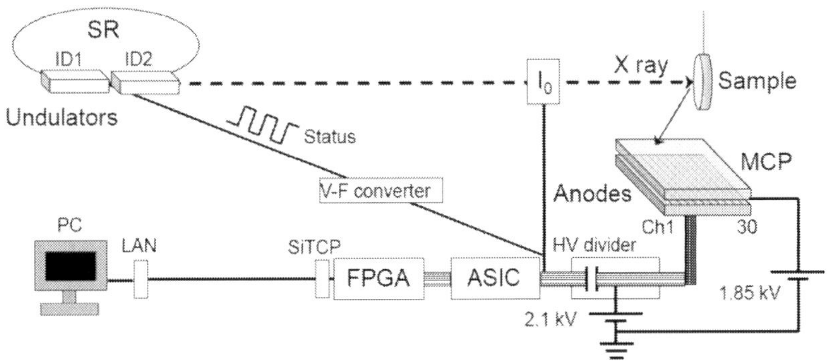


**Figure 2.** Schematic diagram for the data acquisition system.

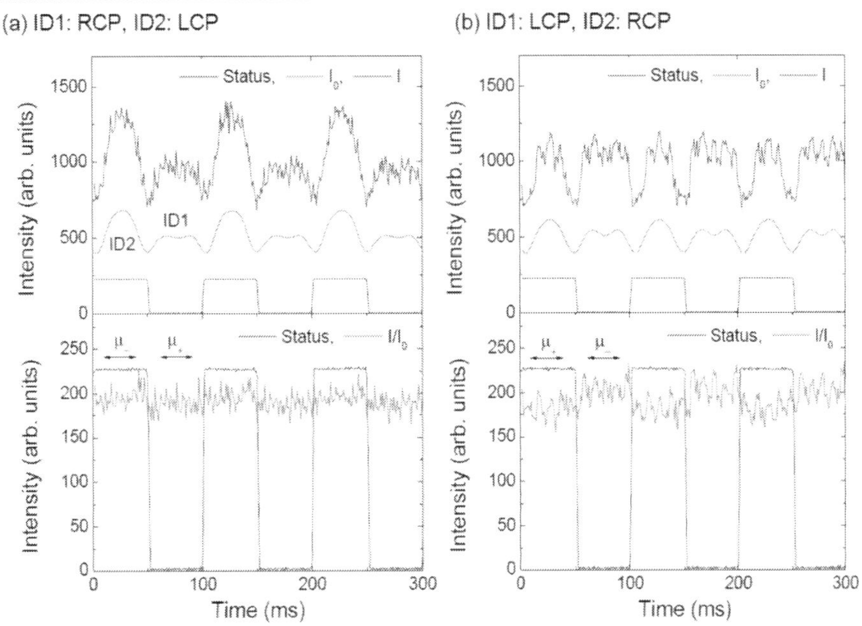


**Figure 3.** Time dependence of electron pulse count (I) for Ch 18, together with that of incident X-ray intensity ($I_0$) and $I/I_0$, taken at the Co $L_3$ peak-top energy (779 eV) with 10 Hz polarization switching. The X-ray polarizations of ID1 and ID2 were set to (a) RCP and LCP, respectively, and (b) *vice versa*. The status signal represents which undulator (ID1 or ID2) is active at each moment.

The obtained XMCD spectra are shown in Fig. 4. Some shoulder features are found in the absorption spectra, which are characteristic of Co oxide. Moreover, the shoulder features are more prominent and the XMCD difference signal is weaker at Ch 29 compared with those at Ch 18. Since the probing depth is smaller for Ch 29 than Ch 18, owing to more grazing electron emission angle, this indicates that the surface of the Co film is preferentially oxidized. Feasibility of the depth-resolved XMCD technique with the multi-anode MCP system and polarization switching has been thus demonstrated.

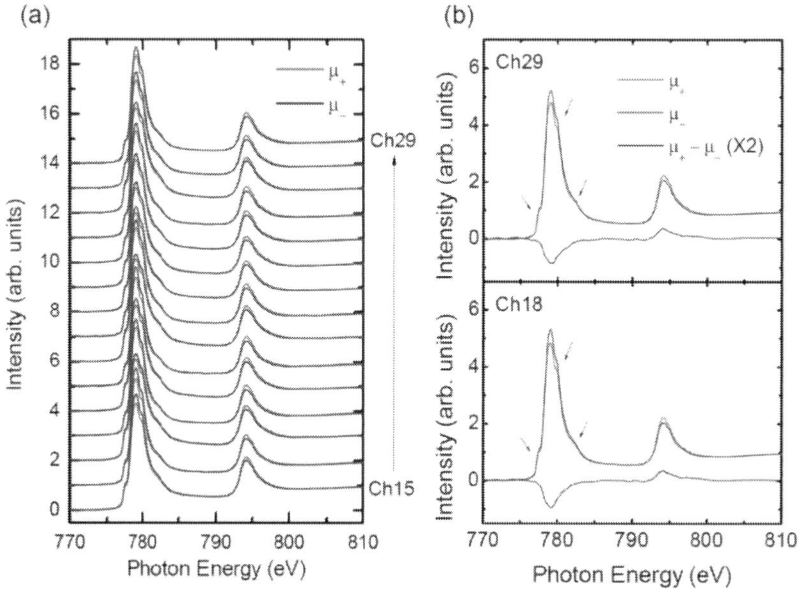


**Figure 4.** (a) Co L-edge X-ray absorption spectra for RCP and LCP, $\mu_+$ and $\mu_-$, respectively, for Ch 15-29, and (b) XMCD spectra for Ch 18 and 29. Arrows indicates spectral features for Co oxide.

## 4. Summary

We have developed the depth-resolved XMCD technique by combining the multi-anode MCP system with polarization switching between RCP and LCP. Feasibility of the technique has been demonstrated by measuring a naturally-oxidized Co thin film, which shows rather weak XMCD signal and depth-dependent chemical and magnetic structures.

## References

[1] Amemiya K, Kitagawa S, Matsumura D, Yokoyama T and Ohta T 2003 *J. Phys.: Condens. Matter* **15** S537.
[2] Amemiya K and Sakamaki M 2011 *Appl. Phys. Lett.* **98** 012501.
[3] Amemiya K 2012 *Phys. Chem. Chem. Phys.* **14** 10477.
[4] Muro T, Saitoh Y, Kimura H, Matsushita T, Nakatani T, Takeuchi M, Hirono T, Kudo T, Nakamura T, Wakita T, Kobayashi K, Hara T, Shirasawa K and Kitamura H 2004 *AIP Conf. Proc.* **705** 1051.
[5] Muro T, Nakamura T, Matsushita T, Kimura H, Nakatani T, Hirono T, Kudo T, Kobayashi K, Saitoh Y, Takeuchi M, Hara T, Shirasawa K and Kitamura H 2005 *J. Electron Spectrosc. Relat. Phenom.* **144-147** 1101.
[6] Muro T, Nakamura T, Matsushita T, Wakita T, Fukumoto K, Kimura H, Hirono T, Kinoshita T, Hara T, Shirasawa K, Takeuchi M and Kitamura H 2007 *AIP Conf. Proc.* **879** 571.
[7] Uchida T 2008, *IEEE Trans. Nucl. Sci.* **55** 1631.
[8] Amemiya K, Toyoshima A, Kikuchi T, Kosuge T, Nigorikawa K, Sumii R and Ito K 2010 *AIP Conf. Proc.* **1234** 295.
[9] Amemiya K, Sakamaki M, Koide T, Ito K, Tsuchiya K, Harada K, Aoto T, Shioya T, Obina T, Yamamoto S and Kobayashi Y 2013 *J. Phys.: Conf. Ser.* **425** 152015.

# The Structure of p-Aminobenzoic Acid in Water: Studies Combining UV-Vis, NEXAFS and RIXS Spectroscopies

**A Gainar[1], J S Stevens[1], E Suljoti[2], J Xiao[2], R Golnak[2], E F Aziz[2,3], S L M Schroeder[1,4,5]**

[1]School of Chemical Engineering and Analytical Science, The University of Manchester, Oxford Road, Manchester, M13 9PL, UK
[2]Joint Ultrafast Dynamics Lab in Solutions and at Interfaces (JULiq), Helmholtz-Zentrum Berlin für Materialien und Energie Institution Albert Einstein Strasse 15, 12489 Berlin, Germany
[3]Freie Universität Berlin, Arnimallee 14, 14195 Berlin, Germany
[4]School of Chemical and Process Engineering, University of Leeds, Leeds, LS2 9JT, UK
[5]DIAMOND Light Source Ltd, Chilton, Didcot, OX11 0QX, UK

E-mail: [1]adrian.gainar@postgrad.manchester.ac.uk; [1,4,5]s.l.m.schroeder@leeds.ac.uk

**Abstract.** NEXAFS-RIXS and home laboratory-based UV-Vis absorption spectroscopy are combined to examine the speciation and electronic structure of *para*-aminobenzoic acid (PABA) in aqueous solution as a function of pH. DFT and TD-DFT electronic structure calculations reproduce the experimental trends and provide a correlation between the experimental HOMO↔LUMO gap as well as the electronic transitions between molecular orbitals in the non-ionic, anionic and cationic forms of PABA.

## 1. Introduction

*Para*-aminobenzoic acid (PABA) has therapeutic effects and sunscreen properties due to its capability to filtering out UV radiation through the electronic conjugation of the *para* orientated amino and carboxyl substituents through the aromatic ring. Core-level spectroscopies such as near-edge X-ray absorption fine structure (NEXAFS) and resonant inelastic X-ray scattering (RIXS) have given incisive insight on the electronic state of PABA species in solution and in the solid state including the anionic and cationic forms in water [1]. While NEXAFS involves excitation of electrons from 1s core orbitals to unoccupied molecular orbital (MO) levels (*e.g.* lowest unoccupied MO, LUMO), RIXS probes transitions from occupied MOs (*e.g.* highest occupied MO, HOMO) to 1s core orbitals. UV-Vis spectroscopy is quite similar to these core level techniques in that it probes electronic transitions, but from occupied MOs such as σ, n (non-participant) or π type orbitals to σ* or π* unoccupied MOs. As NEXAFS, RIXS and UV-Vis often probe the same final-state MOs, their combined measurement supported by electronic structure calculations (particularly density functional theory, DFT, and time-dependent density functional theory, TD-DFT) provides an opportunity to arrive at a deeper understanding of the local electronic structure landscape.

## 2. Experimental

### *2.1. Materials*
PABA crystalline powder (99%, Sigma-Aldrich, UK) was dissolved in distilled, deionized water. pH of the solution was adjusted by dropwise addition of 1.2 M aqueous solution of NaOH or of HCl.

### *2.2. Transmission UV-Vis*
A saturated stock solution of 0.03 M PABA in water was prepared and successively diluted to reach the detection range of the Beckman DU 520 spectrophotometer.

### *2.3. Gaussian calculations*
Non-ionic, anionic and cationic PABA monomers were ground-state optimized with B3LYP/6-31G* in Gaussian09 to obtain the DFT molecular orbitals. Using these optimized monomers in a water continuum, TD-DFT polarizable continuum model (PCM) calculations were employed to generate the simulated spectra and the MOs involved in these transitions.

### *2.4. NEXAFS/RIXS*
NEXAFS and RIXS were recorded at the N K-edge (resolution $\leq$ 0.1 eV) using the liquid microjet of the LiXEdrom endstation, U41-PGM beamline, BESSY II synchrotron. The common energy scale was calibrated using $N_2$ gas, measuring the 1s→2p π transitions in total electron yield mode.

## 3. Results and discussion
Depending on the pH of the aqueous solution, PABA is present in cationic, neutral or anionic form [1,2]. For low (acidic) pH solution (where PABA is almost entirely present as the cationic species), the main absorption band is centered at a wavelength of ~ 225 nm. For high pH, PABA is anionic and the main UV absorption peak occurs at ~ 265 nm, while the neutral species, at intermediate pH, absorbs at ~ 275 nm (Table 1 and Figure 1). TD-DFT calculations reproduce the observed trend, with the main absorption shifted to shorter wavelengths compared to non-ionic PABA (cationic < anionic < non-ionic, Table 1). Within this trend, it is noticeable that the calculated value for the anionic species is significantly lower than the experimentally observed value. It has been suggested previously that calculated charge transfer excitations can be less accurate for ionic π systems due to underestimated excitation energies of ionic components [3]. Perhaps it also plays a role here that the carboxylate group is less electron withdrawing and thus electrons are more consolidated around this group rather than being delocalized over the ring [4].

**Table 1.** Comparison of main absorption wavelengths in experimental UV-Vis and TD-DFT simulated spectra of PABA species in water, with primary transitions from HOMOs to LUMOs.

| PABA species | Wavelength (experimental) / nm | Wavelength (simulated) / nm | Primary transition |
|---|---|---|---|
| Non-ionic | ~ 275 | 266.96 | HOMO→LUMO |
| Anionic | ~ 265 | 228.26 | HOMO–1→LUMO |
| Cationic | ~ 225 | 218.95 | HOMO–1→LUMO |

Collating together the snapshots of the MOs generated from the TD-DFT calculations for the non-ionic, anionic and cationic monomers in water, Figure 1 reveals that the primary transition in the non-ionic PABA takes place between the orbital with annotation A as HOMO and the 1π* orbital as LUMO (~ 267 nm, 4.64 eV), whereas the shift to shorter wavelength in the anionic PABA is due to widening energy separation between A and 1π* (~ 228 nm, 5.43 eV), with A lowering in energy and becoming HOMO–1. Similarly, the noted shift to lower wavelength for cationic PABA (~ 219 nm, 5.66 eV) can be explained by the primary transition between A, which is now the HOMO–1, and 1π*.

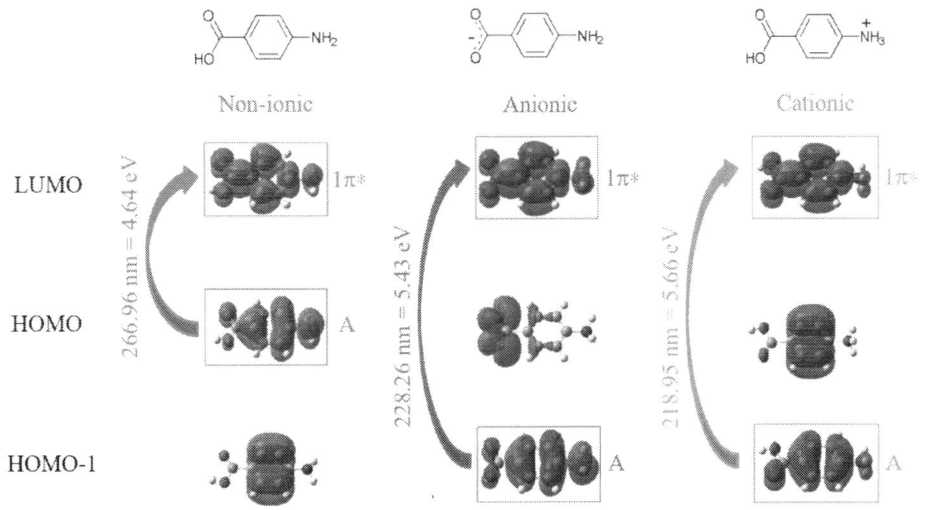


**Figure 1.** Transitions (arrows) between MOs of interest A and 1π* (marked with boxes) in TD-DFT calculations for non-ionic, anionic and cationic PABA monomers in a water continuum.

We can now correlate these observations with the MO information for these systems that we have previously obtained by RIXS and NEXAFS [1]. Figure 2 shows that the experimentally determined HOMO↔LUMO gap for nitrogen (which is not applicable to cationic PABA due to the protonation of nitrogen) is correlated with the calculated energy gap between orbitals A and 1π* from the DFT calculations for the non-ionic and anionic species. In the case of non-ionic PABA, orbital A corresponds to the HOMO and orbital 1π* to the LUMO, whilst for the anionic species, orbital A lowers down in energy becoming HOMO–3 and orbital 1π* raises up in energy as LUMO+1.

Although core hole relaxation effects are expected to affect the energetic position of features in the NEXAFS and RIXS data, the impact on taking the experimental RIXS-NEXAFS energy gap as a measure of the HOMO-LUMO is believed to be of the order of at most a few 100 meV; additionally, when such effects were included in CASTEP calculations for crystalline PABA [5] the inclusion of final-state effects did not change the MO interpretation significantly. For the DFT calculations, omitting relaxation effects may lead to a slight overestimation of transition energies, and thus of the RIXS-NEXAFS energy gap. The reason for the weak effects of relaxation are the strong localized core-level excitations and the weak nature of the intermolecular interactions (*i.e.* hydrogen bonding) compared to intramolecular interactions (*i.e.* covalent bonds and protonation effects).

**Table 2.** Values for the energy gap of interest A↔1π* reflected by the experimental UV-Vis, RIXS-NEXAFS, and by TD-DFT and DFT calculations.

| Energy gap A↔1π* | Non-ionic PABA | Anionic PABA |
|---|---|---|
| UV-Vis (experimental) | ~ 4.51 eV | ~ 4.68 eV |
| TD-DFT (calculated) | 4.64 eV | 5.43 eV |
| RIXS-NEXAFS (experimental) | 5.06 eV | 5.89 eV |
| DFT (calculated) | 5.01 eV | 5.86 eV |

The differences noted in the A↔1π* gap presented in Table 2 between UV-Vis and RIXS-NEXAFS spectroscopies may be explained through the presence of a hole in the valence band in the UV-Vis, that exerts a smaller effect over the final state than the core hole from RIXS-NEXAFS. The TD-DFT PCM calculations also involve a water continuum (thus PABA surface-polarized monomers with slightly different MO energy levels), yet this rather simplistic model is sufficient to picture the increase of the A↔1π* energy gap when moving from non-ionic to anionic PABA species.

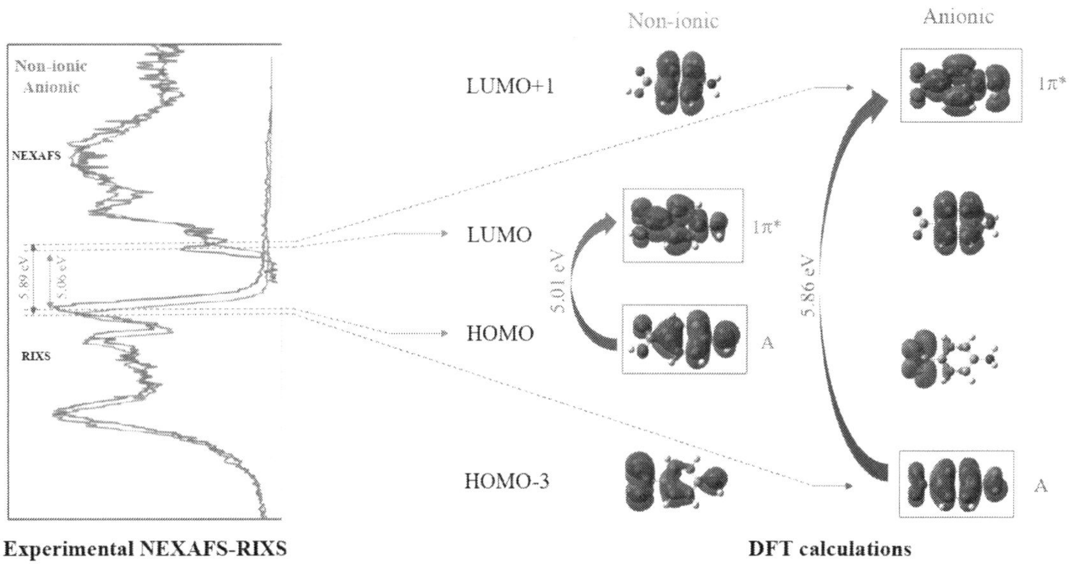


**Figure 2.** Nitrogen RIXS and NEXAFS showing the HOMO↔LUMO gap for non-ionic and anionic species, correlated with transitions (arrows) between DFT MOs of interest labelled as A and 1π*.

## 4. Conclusions

Monitoring the electronic transitions of PABA in the UV range in aqueous solution at intermediate, high and low pH, and interpreting the results with TD-DFT calculations provides a spectroscopic identification of PABA species that is complementary to the fine detail arising from the previously reported NEXAFS-RIXS combination. The effect of pH can be translated in the UV-Vis by the shifts to shorter wavelengths in the main absorption peaks encountered in the spectra of high and low pH solutions with respect to the intermediate pH solutions, which are reflected by the TD-DFT simulations. These shifts are explained by A↔1π* energy gap alterations with the changes in electronic structure accompanying the pH variation; the DFT calculations associated with the experimental RIXS-NEXAFS identify well with the orbital gap observed in the TD-DFT simulations that reflects the main absorption peaks in the UV-Vis, thus bridging the synchrotron X-ray core-level spectroscopies NEXAFS and RIXS with the laboratory UV-Vis technique.

**Acknowledgements**

This work was supported by the EPSRC Critical Mass Grant EP/I013563/1 and use of synchrotron at BESSY II (HZB) by European Community's Seventh Framework Programme, grant no. 312284.

**References**

[1] Stevens J S, Gainar A, Suljoti E, Xiao J, Golnak R, Aziz E F and Schroeder S L M 2015 *Chem. Eur. J.* **21** 7256–63.
[2] van de Graaf B, Hoefnagel A J and Wepster B M 1981 *J. Org. Chem.* **46** 653–57.
[3] Grimme S and Parac M 2003 *Chem. Phys. Chem.* **3** 292-95.
[4] Lampman G M, Pavia D L, Kriz G S and Vyvyan J R 2001 *Spectroscopy* (Belmont: Brooks/Cole Cengage Learning).
[5] Stevens J S, Seabourne C R, Jaye C, Fischer D A, Scott A J and Schroeder S L M 2014 *J. Phys. Chem. B* **118** 12121–28.

# XAFS studies of monodisperse Au nanoclusters formation in the etching process

**Lina Yang, Ting Huang, Wei Liu, Jie Bao, Yuanyuan Huang, Yuanjie Cao, Tao Yao, Zhihu Sun*, and Shiqiang Wei***

National Synchrotron Radiation Laboratory, University of Science and Technology of China, Hefei, Anhui 230029, P. R. China

Corresponding authors. E-mail: zhsun@ustc.edu.cn, sqwei@ustc.edu.cn

**Abstract.** Understanding the formation mechanism of gold nanoclusters is essential to the development of their synthetic chemistry. Here, by using x-ray absorption fine-structure (XAFS) spectroscopy, UV-Vis and MS spectra, the formation process of monodisperse $Au_{13}$ nanoclusters is investigated. We find that a critical step involving the formation of smaller $Au_8$–$Au_{11}$ metastable intermediate clusters induced by the HCl + HSR etching of the polydisperse $Au_n$ precursor clusters occurs firstly. Then these intermediate species undergo a size-growth to $Au_{13}$ cores, followed by a slow structure rearrangement to reach the final stable structure. This work enriches the understanding of cluster formation chemistry and may guide the way towards the design and the controllable synthesis of nanoclusters.

## 1. Introduction

Gold nanoclusters possess size-dependent molecular-like structure and discrete electronic energy levels, which render them diverse applications in fields such as catalysis, optical devices and imaging [1-5]. Therefore, the development of synthetic strategies for monodisperse Au nanoclusters is crucial. Currently, etching polydisperse Au clusters has been widely employed as an useful top-down method and obtained remarkable advances. However, the mechanistic understanding on the etching-induced synthesis of monodisperse nanoclusters is still to be resolved [6]. Therefore, it is important to study etching reactions for developing more routes to fine-tune nanoclusters with desired functionalization.

Currently, direct liberation of the same-sized particle as the target cluster in a single step was hypothesized but not experimentally verified [7]. What happens in the convergence process and how this process is related to the structure and property of the end product is scarcely known. In order to solve these problems, in-situ methods by a combination of various probing techniques with sensibility to transient intermediate species in solution are specially required.

Herein, we report an investigation on the kinetic formation process of an unprecedented monodisperse $Au_{13}(L_3)_2(SR)_4Cl_4$ nanocluster ($L_3$: 1,3-Bis(diphenylphosphino)propane) by in-situ x-ray absorption fine structure (XAFS), UV-vis absorption and MS spectra. The $Au_{13}$ nanocluster is obtained via a facile one-phase reaction of polydisperse $Au_n$ clusters with hydrochloric acid (HCl) and dodecanethiol (HSR). It is found that the formation of the monodisperse $Au_{13}(L_3)_2(SR)_4Cl_4$ nanoclusters is realized in an etching/growth/rearrangement manner including three reaction stages.

## 2. Experimental Section

Sample preparations: The monodisperse Au nanoclusters were synthesized by reaction of HCl and HSR mixture with preformed $L_3$-capped $Au_n$ clusters. The starting material of this reaction, $L_3$-capped $Au_n$ clusters, was synthesized by reduction of $Au_2(L_3)Cl_2$ (70 mg, 0.08 mmol) with $NaBH_4$ (15 mg, 0.04 mmol) in 60 ml dichloromethane solvent at room temperature for 3 h under stirring [6]. The solvent was evaporated to dryness and then redissolved in 12 ml ethanol, to which a mixture of dodecanethiol (50 μL, 0.2 mmol) and HCl (200 μL, 2.4 mmol) was added. The process of the reaction was monitored by in-situ XAFS spectra, UV-Vis, as well as mass spectrometry.

In-situ XAFS measurement: The Au $L_3$-edge XAFS spectra were measured at the BL14W1 beamline of the Shanghai Synchrotron Radiation Facility (SSRF). The storage ring of SSRF worked at 3.5 GeV with a maximum current of 210 mA. The reactive solution was continuously circulated along a microtube by peristaltic pump and flowed into an in-situ cell for XAFS measurement.

## 3. Results and discussion

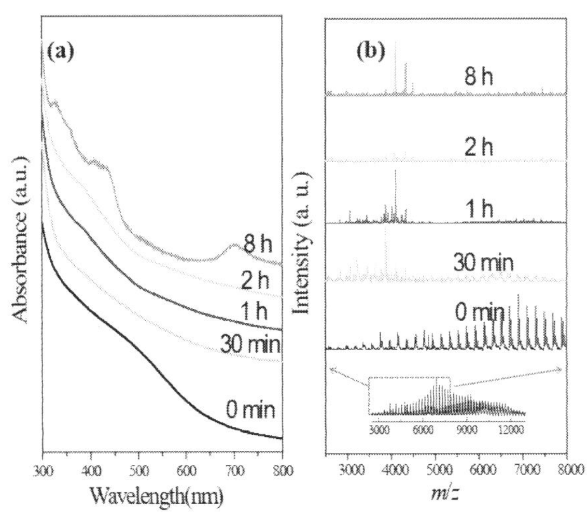


**Figure 1.** *In situ* UV-vis absorption spectra and MALDI mass spectra at different reaction times. The inset shows the MALDI-MS of the starting material at a wide mass range.

The time-dependent UV-vis absorption and MALDI-MS spectra for the formation process of monodisperse $Au_{13}(L_3)_2(SR)_4Cl_4$ nanoclusters were shown in Fig. 1. At first, the UV-Vis spectrum demonstrates a wide hump at around 500 nm, suggesting the polydispersity of the precursor clusters.

At 30 min the spectral profile shows a featureless decay. From 2 h on, absorption bands peaked at 327, 410, 433, and 700 nm become gradually pronounced and reach saturation. The MS spectra for the starting material covers a wide size range of 3000–13000 Da, corresponding to a mixture of $Au_{15}$–$Au_{60}$. Within the 30 min, the wide size distribution is divided into two regions covering roughly 6000–8000 and 3000–4000 Da, which could be assigned to $Au_{31}$–$Au_{39}$ and $Au_8$–$Au_{11}$ clusters respectively. At 2 h, only the peaks at 4101 and 4335 Da are present. These two peaks are corresponding to $Au_{13}(L_3)_2(SR)_4Cl_4$ cluster (a fragment of and intact, respectively), indicating complete size-convergence into $Au_{13}$ clusters.

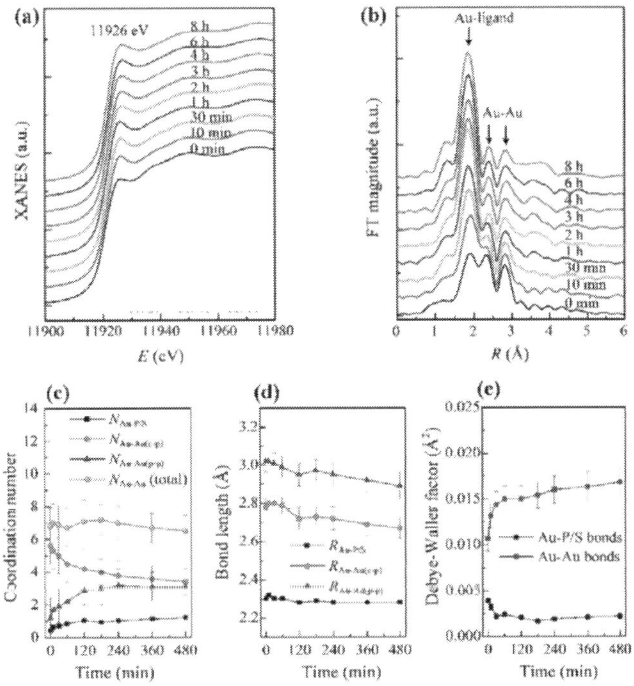


**Figure 2**. (a) Time-dependent XANES spectra, and (b) EXAFS spectra. (c) Coordination number CN, (d) bond distance $R$, and (e) Debye-Waller factor $\sigma^2$ against reaction time extracted from EXAFS curve-fitting. Au–Au(c-p) and Au–Au(p-p) represent the central-peripheral and peripheral-peripheral Au–Au bonds in an complete or incomplete icosahedron, respectively.

In situ XAFS measurements at Au $L_3$-edge were performed to detect the evolutions of the size-conversion process. The XANES spectra in Fig. 2(a) display that the spectrum of the starting material exhibits a peak at 11936 eV (labeled by an arrow), which is characteristic of fcc structured Au. After addition of HCl + HSR, the intensity of this peak is remarkably smeared out at 30 min, and disappears completely at 2h. The peak shape is similar to nanoclusters. The Fourier transformed (FT) EXAFS $k^2\chi(k)$ curves in Fig. 2(b) also show that immediately after the start of reaction (within 30 min), the Au-ligand (Au–S/P/Cl) peak at 1.90 Å is intensified rapidly, while the Au-Au peaks (2.36 and 2.88 Å) are significantly damped, suggesting the decomposition of the larger clusters. Moreover, the white-line peak at around 11926 eV in the XANES spectrum, corresponding to the $2p_{3/2}\rightarrow5d_{5/2,3/2}$

electronic transition of Au atoms, is significantly intensified immediately after the addition of HCl + HSR. With continued reaction, the white-line peak is reduced slightly in intensity, along with the gradually weakened Au-Au peaks in the FT curve. After 2 h, the XANES spectra do not display any remarkable changes, while subtle continuing increase of the Au-ligand EXAFS peak intensity could be discerned .Quantitative structural parameters have been obtained through a least-squares curve-fitting using the ARTEMIS [8] module of IFEFFIT package. We will discuss the details in the following.

Within the first stage of 30 min, $Au_{15}$–$Au_{60}$ clusters with Au-ligand CN of 0.44, are prominently etched into smaller $Au_8$–$Au_{11}$ metastable intermediates. Subsequently, these active small $Au_8$–$Au_{11}$ clusters are immediately stabilized by the absorbed SR, $L_3$ and $Cl^-$ ligands to form metastable intermediates as evidenced by the quickly increased Au-ligand CN (from 0.44 at 0 min to 0.72 at 30 min). These $Au_8$–$Au_{11}$ cores are fully protected by the ligands, forming closed geometrical shells that make the clusters highly resistant against etching. In the second stage (30 min-2 h), these metastable $Au_8$–$Au_{11}$ intermediates focus into $Au_{13}$ cores, which is possibly achieved by incorporating the Au(I) ions or Au(I)-Cl oligomers pre-existing in the solution. The obtained Au-ligand CN (1.02) at 2h is between the nominal values in $Au_{13}(L_3)_2(SR)_4Cl_4$ (0.92) and Au(I)-SR polymers (2.0) [10], suggesting the existence of both species. The proportion $p$ of the Au atoms in $Au_{13}(L_3)_2(SR)_4Cl_4$ could be estimated by the equation $0.92\times p+2\times(1-p)=1.02$, which gives $p\sim0.90$. This result demonstrate that the majority (~90%) of Au atoms at 2 h is in the form of $Au_{13}$ cluster, with a small part (~10%) forming Au(I)-SR polymers or analogues. In the third stage (after 2 h), the atomic structure undergoes a rearrangement process toward the energetically stable structure. The EXAFS curve-fitting show that during the structure rearrangement process, the Au–Au bond displays a trend of contraction. At the beginning of forming $Au_{13}$ skeletons, both the $R_{Au\text{-}Au(c\text{-}p)}$ and $R_{Au\text{-}Au(p\text{-}p)}$ (2.72 and 2.95 Å) of the $Au_{13}$ icosahedron are within the typical range (2.71–2.79 and 2.85–2.95 Å) of $Au_{13}$ clusters as previously reported [9]. After the structure rearrangement, the $R_{Au\text{-}Au(c\text{-}p)}$ and $R_{Au\text{-}Au(p\text{-}p)}$ decrease to 2.67 and 2.89 Å, respectively, suggesting a considerable structural distortion away from the original icosahedral $Au_{13}$ skeleton. The gradually enlarged distortion is also reflected by the slightly increased disorder degree $\sigma^2$ of the Au-Au bonds, from 0.015 to 0.017 $Å^2$. This implies the structural rearrangement of $Au_{13}$ clusters.

## 4. Conclusion

In summary, in-situ XAFS combined with MS and UV-Vis have been used to probe the formation process of the monodisperse $Au_{13}(L_3)_2(SR)_4Cl_4$ nanocluster. It is found that the cluster formation is achieved in an etching/growth/rearrangement manner including three distinct reaction steps. (1) The initial polydisperse $Au_n$ clusters are etched to form much smaller $Au_8$–$Au_{11}$ intermediate clusters. (2) These intermediate species undergo a size growth to $Au_{13}$ cores, by incorporating the pre-existing Au(I) ions or Au(I)-Cl oligomers in the solution. (3) Finally, a slow structure rearrangement of the $Au_{13}$ skeleton and the covering ligand shells occurs. These findings enrich our understanding on the etching mechanism and can guide our way towards the synthesis of nanomaterials in a controllable manner.

**Acknowledgments**

This work was supported by National Natural Science Foundation of China (Grant No. 11475176, 11135008, U1332131, and 11305172). The authors are grateful to SSRF for the valuable beamtime.

**References**

[1] Jin, R. C. *Nanoscale* 2010, 2, 343.

[2] Li, G.; Jin, R. C. *Acc. Chem. Res.* 2013, 46, 1749.

[3] Qian, H. F.; Zhu, M. Z.; Wu, Z. K.; Jin, R. C. *Acc. Chem. Res.* 2012, 45, 1470.

[4] Yau, S. H.; Varnavski, O.; Goodson, T. *Acc. Chem. Res.* 2013, 46, 1506.

[5] Lu, Y. Z.; Chen, W. *Chem. Soc. Rev.* 2012, 41, 3594.

[6] Shichibu, Y.; Konishi, K. *Small* 2010, *6*, 1216.

[7] Duan, H. W.; Nie, S. M. *J. Am. Chem. Soc.* 2007, *129*, 2412.

[8] Ravel, B.; Newville, M. *J. Synchrotron Rad.* 2005, *12*, 537.

[9] Hall, K. P.; Mingos, D. M. P. *Prog. Inorg. Chem.* 1984, *32*, 237.

[10] Menard, L. D.; Frenkel, A. I.; Murray, R. W.; Nuzzo, R. G. *J. Phys. Chem. B* 2006, *110*, 14564.

Journal of Physics: Conference Series **712** (2016) 012036 doi:10.1088/1742-6596/712/1/012036

# Benchtop Nonresonant X-ray Emission Spectroscopy: Coming Soon to Laboratories and XAS Beamlines Near You?

**Devon R. Mortensen[1], Gerald T. Seidler*[1], Alexander S. Ditter[1] and Pieter Glatzel[2]**

[1]Deparment of Physics, University of Washington, Seattle, WA 98195, USA
[2]European Synchrtron Radiation Facility, CS40220, Grenoble 38043 Cedex 9, France

*seidler@uw.edu

**Abstract.** Recently developed instrumentation at the University of Washington has allowed for nonresonant x-ray emission spectra (XES) to be measured in a laboratory-setting with an inexpensive, easily operated system. We present a critical evaluation of this equipment by means of Kβ and valence-level XES measurements for several Co compounds. We find peak count rates of ~5000/s for concentrated samples and a robust relative energy scale with reproducibility of 25 meV or better. We furthermore find excellent agreement with synchrotron measurements with only modest loss in energy resolution. Instruments such as ours, based on only conventional sources that are widely sold for elemental analysis by x-ray fluorescence, can fill an important role to diversify the research applications of XES both by their presence in non-synchrotron laboratories and by their use in conjunction with XAFS beamlines where the complementarity of XAFS and XES holds high scientific potential.

## 1. Introduction

Nonresonant x-ray emission spectroscopy (XES), while far less common as ubiquitous as x-ray absorption spectroscopy (XAS), has seen steady growth in a number of fields over the past several decades. In particular, Kβ (3p→1s) and valence-level XES has seen extensive use in determining spin state, oxidation state, and ligand speciation in 3d transition-metal materials [1-3]. Because this technique probes the emissions resulting from the filling of coreholes, it provides complementary information to XAS and combination of the two spectroscopies is a natural and powerful extension.

Unfortunately the infrastructure for XES has lagged considerably behind that for XAS with only a handful of synchrotron end stations dedicated to such measurements, see for example [4-7]. This access scarcity has significantly hampered the further development of XES as a routine complement to XAS. We have recently demonstrated, however, that benchtop instrumentation based on high-resolution spherically-bent crystal analyzers (SBCAs) can in many cases provide synchrotron-quality XES measurements [8]. Here we explore the capability and limitations of this laboratory-based equipment via a direct comparison to synchrotron data and we also report on an instrumental improvement that helps to finely anchor the accuracy of the energy scale.

## 2. Experimental Details

Laboratory spectra were recorded using a Rowland-circle monochromator recently developed at the University of Washington [8], see figure 1. X rays are generated by a conventional tube source (Moxtek) with an Au anode operated at 40 kV bias and 200 μA current, i.e., only 8 W total electron beam power.

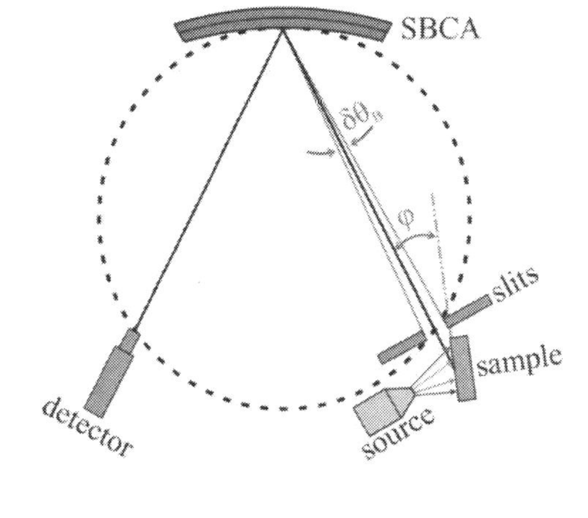


**Figure 1.** The lab-based Rowland-circle spectrometer.[4] The slits in front of the sample constrain the angular width ($\delta\theta_B$) probed by the analyzer which improves energy resolution and reproducibility. The sample is slightly rotated ($\phi$) to allow improved line-of-sight to the SBCA.

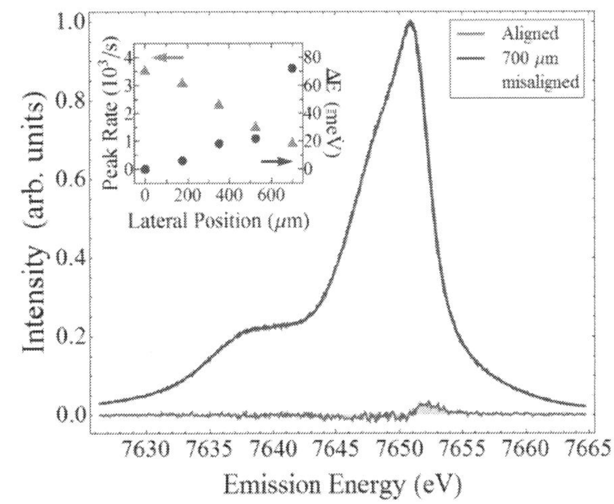


**Figure 2.** Peak-intensity normalized CoO Kβ spectra measured at the aligned and the extreme misaligned location. The gray curve shows the residual between the two spectra. Inset: the peak count rate and energy shift as a function of lateral sample position.

As has been emphasized elsewhere [8], the small anode-to-sample distance possible with this type of source allows the sample to subtend a very large solid angle, giving highly efficient excitation that can reach rates intermediate between those typical for a monochromatized bending magnet and insertion device at a third-generation synchrotron. The sample fluorescence was analyzed and refocused at the detector position by a 1-m radius of curvature Ge (444) SBCA (XRS Tech). A geometric correction, based on ray-tracing calculations [9], is applied to compensate for the Bragg angle dependence of the height of the refocused spot at the finite-height (5-mm) detector.

The sample is rotated slightly (~15°) to provide improved line-of-sight to the analyzer for the fluorescence while still allowing for minimal offset from the x-ray tube anode (~3 mm). Most recently, instead of placing the sample directly on the Rowland geometry we use a narrow entrance slit on the Rowland circle and move the sample + source subassembly ~10mm behind the slit. The purpose of this slit is two-fold. First, the dominant limit on energy resolution is the apparent angular width of the illuminated sample area as seen by the SBCA. The entrance slit gives some ability to tune the resolution by adjusting its width. Higher resolution of course comes at the cost of lower count rates and the optimal tradeoff is determined on a case-by-case basis. For this study a slit width of 0.5 mm was selected. This corresponds to an angular width of 0.03° (0.48 eV) at 82.93° Bragg angle (7650 eV).

The second purpose of the entrance slit is more subtle and involves the difficulty of reproducing exact sample position. Since the relative energy scale is determined geometrically, a small lateral misalignment of the sample will register as an incorrect energy shift in the resulting spectrum. This ambiguity presents a significant barrier to cross-comparisons between different samples, which is especially pressing in the case of Kβ XES from 3d transition metal compounds where shifts in the $K\beta_{1,3}$ peak can indicate changes in spin state and other local electronic properties.[1] With the slit constraining the angular range being sampled by the SBCA, however, this becomes a minor issue; a misalignment large enough to meaningfully shift the energy scale will fall outside of the sampled range, resulting in an obviously reduced signal. For means of example, we measured CoO $K\beta_{1,3}$ at several lateral sample positions as well as a spectrum where the sample was removed, replaced, and realigned – see figure 2. These results demonstrate a highly robust energy scale that is insensitive to exact sample position. The inset of figure 2 shows that shifts in the energy scale as small as ~25 meV are accompanied by factor of 2 decreases in overall intensity. Simple optimization of signal intensity with respect to the sample position behind the entrance slit therefore ensures outstanding consistency of energy scale between different measurements.

For the study below, synchrotron measurements were performed at beamline ID26 of the European Synchrotron Radiation Facility (ESRF). The incident energy was tuned to 7.8 keV using a double Si (111) monochromator. The total incident flux was ~ 2 x $10^{13}$ photons/s. Samples were oriented at 45° in the scattering plane, resulting in a beamsize of 0.5 mm in-plane and 1 mm out-of-plane. The resulting fluorescence was analyzed by an array of five Ge (444) SBCAs in a 1-m Rowland circle geometry. The energy resolution, which varies slightly between SBCAs, is ~1.1 eV as determined from the FWHM of elastic scatter lines.

Samples were prepared from high purity powder (99.99% Co3O4; 99.5% LiCoO2) from Alfa Aesar. The same samples were measured at both the University of Washington lab and the ESRF, thus eliminating possible systematic effects due to variations in sample preparation.

## 3. Results and Discussion

In figure 3 we present a laboratory/synchrotron comparison of Kβ XES for $Co_3O_4$ and $LiCoO_2$. To better overlay with the laboratory data the ESRF spectra have been additionally broadened by convolving with a Gaussian of FWHM = 0.8 eV. This is larger broadening than expected from the finite size of the entrance slit discussed above and is mostly likely due to the SBCA being slightly off circle in the laboratory system. Nonetheless the net effective energy resolution of ~1.4 eV is more than sufficient to resolve the key XES features without loss in scientific merit.

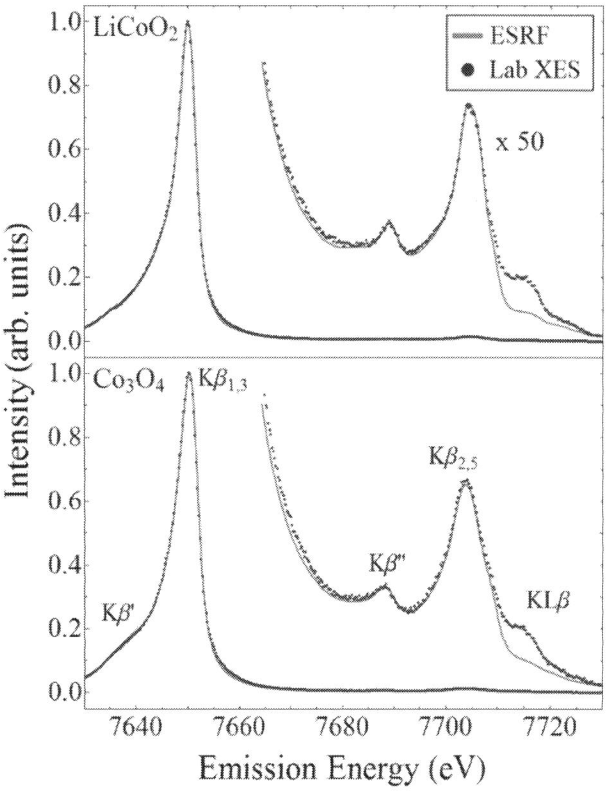


**Figure 3.** Kβ XES for $LiCoO_2$ and $Co_3O_4$ powder samples. The red curve shows spectra collected as ESRF; blue dots show data taken using the laboratory-based instrument described in the text. All spectra have been normalized to peak intensity. For ease of presentation, the valence-to-core emission is shown amplified 50x. The ESRF results have been broadened by 0.8 eV to match the slightly poorer energy resolution of the lab XES results; see the text for discussion.

The two datasets are in excellent agreement for both the main Kβ ($K\beta_{1,3}$ and Kβ') and valence-to-core ($K\beta_{2,5}$ and Kβ'') emission lines. The only discrepancy is the KLβ feature present only in the laboratory data. This emission is the result of a multi-electron excitation [10] and therefore exhibits a threshold behavior at roughly the K + L-edge binding energies. The excitation energy for the ESRF data (7.8 keV) is well below this threshold. The broadband tube source in the laboratory setup, however, emits x rays with energies up to 40 keV resulting in sizeable KLβ emission.

Despite the low-powered x-ray tube used here, the peak count rates at $K\beta_{1,3}$ are ~5 x $10^3$/s, allowing high quality spectra in very useful measurement times. The measurement time for the main Kβ XES was less than one hour while the valence XES required ~6 hours of data collection for cleanest statistics. Count-rate improvement of up to 100x could be realized by upgrading the x-ray source to a kW-level commercial tube of the type used commonly in fluorescence studies for elemental analysis. For comparison the ESRF measurements averaged ~2 x $10^6$/s/SBCA at peak.

## 4. Conclusion

We have demonstrated the effectiveness of inexpensive, easily assembled and operated benchtop equipment using only low-powered, conventional x-ray sources for high-resolution XES measurements. We find easy measurement of even the lowest-intensity valence-level emission lines for concentrated samples and also find excellent agreement with synchrotron-based studies. This versatile instrument design could have significant application in many fields where transition-metal and lanthanide compounds are of central importance, including electrical energy storage, f-electron chemistry, catalysis, and environmental chemistry. In addition to extending routine XES beyond the confines of synchrotron light sources, we propose that this type of user-friendly spectrometer could be used as an on-site supplement to XAFS beamlines allowing XES measurements without using synchrotron beamtime.

## 5. Acknowledgements

This work has been supported by: (1) the U.S. Department of Energy, Basic Energy Sciences under Grant No. DE-FG02-09ER16106 and also by the Office of Science, Fusion Energy Sciences and the National Nuclear Security Administration thought Grant No. DE-SC0008580; (2) the State of Washington through the University of Washington Clean Energy Institute; (3) The Joint Center for Energy Storage Research (JCESR), an Energy Innovation Hub funded by the U.S. Department of Energy, Office of Science, Basic Energy Science; and (4) The U.S. Department of Defense through a subcontract from Los Alamos National Laboratories.

## References

[1] Glatzel P and Bergmann U 2005 *Coord. Chem. Rev.* **249** 65-95
[2] Gallo E and Glatzel P 2014 *Adv. Mater.* **26** 7730-46
[3] Bergmann U and Glatzel P 2009 *Photosynth Res.* **102** 255-66
[4] Llorens I, *et al.*, 2012 *Rev. Sci. Instrum.* **83** 063104
[5] Sokaris D, *et al.*, 2013 *Rev. Sci. Instrum.* **84** 053102
[6] Kavcic M, *et al.*, 2012 *Rev. Sci. Instrum.* **83** 033113
[7] Kleymenov E, *et al.*, 2011 *Rev. Sci. Instrum.* **82** 065107
[8] Seidler G T, Mortensen D R, Remesnik A J, Pacold J I, Ball N A, Barry N, Styczinski M and Hoidn O R 2014 *Rev. Sci. Instrum.* **85** 113906
[9] Mortensen D R, Ph. D dissertation 2016 University of Washington
[10] Glatzel P, Bergmann U, deGroot F and Cramer S 2003 *AIP Conference Proceedings* **652** 250-55

# Separation of Different Atomic Shells in EXAFS Spectrum Using the Regularization Method

**Dong-Seok Yang**

Department of Physics Education,Chungbuk National University, Cheongju, 362-763, Korea

E-mail: dosyang@Chungbuk.ac.kr

**Abstract.** It is shown that the regularization method with two kernels can resolve the double shells of different kinds of atomic species. The method was tested for simulated and experimental spectra. In both works, it provided successful results in separation of the double shells of different atoms.

## 1. Introduction

The extended X-ray absorption fine structure (EXAFS) technique has been used to examine the local structure of liquid and solid matters for several decades since the Fourier transformed skill was utilized in this field in 1971.[1] It has been applied to the analysis of the local structure of catalysis, amorphous and various functional materials. Recently, this technique has been used for various advanced materials containing many kinds of atomic species. However, it was very difficult to separate the double shells or the different atomic shells because it requires severe constraints among fit parameters. It has been shown that the regularization method provided higher R-space resolution for the double shell of monatomic materials,[2, 3], but there were still limitations on the separation of different atomic shells because of the difficulties of construction of the kernels for different atomic species. Nickel-titanium (so called nitinol) and related compounds are such materials. Nitinol is widely used for dental and medical applications because it exhibits an excellent shape memory effect.[4, 5]

Nitinol wire has a special nature with the super elastic and shape memory effect. It has the same atomic ratio of Ni and Ti elements, and the crystalline structure is monoclinic at the low temperature Martensite phase and cubic at the high temperature Austenite phase. The space group of the structure at the Martensite phase is $P_{21/m}$ and the cell parameters are $a = 2.898\AA$, $b = 4.108\AA$ and $c = 4.646\AA$ with the clinic angle $\beta = 97.78$. From the cell parameters, the atomic configuration was obtained by atomic bonds of $R_{Ni-Ti} = 2.53(3), 2.59(1), 2.60(3)$ and $3.28(1)\AA$ and $R_{Ni-Ni} = 2.61(2)$ and $2.90(2)\AA$ with the coordinates in parenthesis. It is very difficult to analyze the local structure with the conventional fit method for this material because mixed atomic shells with two different atomic species are included in its EXAFS spectrum.

In this work, we applied the regularization method with two kernels containing different atoms to the spectra of mixed shells to extract the shells in pair distribution functions. This was done by a simulation test and experiments.

## 2. Experiments

We performed EXAFS experiments for nitinol wire at 8C EXAFS beamline, Pohang light source (PLS), with the electron storage performance of 3.0 GeV and 300 mA electron beam current. The EXAFS experiments were conducted with the focused beam at Ni K-edge (Eo=7112eV). The proper fluorescence rejection filters were employed with the Mn filter. The solid-state detector (SSD) was installed at the detection site for the fluorescent measurements. The EXAFS data were analyzed by the conventional method using IFEFFIT [6, 7] and the regularization method.[2, 8, 9, 10]

## 3. Regularization method with two kernels of mixed shells

The EXAFS equation for multiple shells can be expressed by

$$\chi(k) = \sum_j \int \frac{N_j F_j(k, r_j)}{k r_j^2} e^{-\frac{2r_j}{\lambda(k)}} g_j(r_j)(\sin(2\mathrm{kr_j} + \phi_\mathrm{j}(\mathrm{k}))\mathrm{dr_j} \quad (1)$$

where $j$ and $g_j(r)$ indicate each atomic shell and distribution of the shell, respectively and $N_j$, $F_j(k, r_j)$, $\phi_j(k)$ represent the number of coordination, backscattering amplitude and phase of each neighboring scatterer, respectively.

The equation is transformed to matrix form as

$$\chi = Ag \quad (2)$$

where $A$ is given by the matrix obtained from the EXAFS components in Eq.1. The element of matrix is calculated by the FEFF [11].

$$A = (A_1, A_2) \quad (3)$$

The $g$ vector consists of two vectors, $g_1$ and $g_2$,

$$q = \begin{pmatrix} g_1 \\ g_2 \end{pmatrix} \quad (4)$$

where $A$ is (n,n) matrix, and $A_1$, $A_2$ are (n, $\frac{n}{2}$) matrixes which are the kernels of shell 1 and 2, respectively, and $g_1$, $g_2$ are $\frac{n}{2}$ vectors which are the distribution functions of shell 1 and 2, respectively.

The pair distribution function of $g_1$ and $g_2$ can be obtained by the inverse procedure with regularization from Eq.2. The process of regularization in EXAFS analysis was described in detail elsewhere [3, 8]

## 4. Analysis of Simulated Spectrum

EXAFS spectra shown in Figure 1-(a) for Ni-Ti (green dash line), Ni-Ni (red dot line) and Ni-(Ti,Ni) (black line) shells have been calculated by FEFF with the pair distribution functions (indicated by "model") shown in Figure 1-(b) and (c), and the two spectra were merged to make a double shell spectrum of different atomic species. Each EXAFS spectrum and the related equations were prepared as a vector or a matrix form for the regularization procedure. Here, $A$ was (200 × 200) matrix, and $A_1$ and $A_2$ was (200 × 100) matrices.

The pair distribution functions of Ni-Ti and Ni-Ni pairs extracted from the simulated EXAFS spectra are shown in Figure 1-(b) and (c). As shown in 1-(b), the pair distribution function (green dash line) extracted from the Ni-Ti single shell spectrum almost fits the model (black line) as expected. The pair distribution function (red dot line) obtained from the merged spectrum deviated slightly from the model one but fits fairly. As shown in Figure 1-(c), the pair distribution functions of Ni-Ni shells also fit well to the model one within an acceptable error

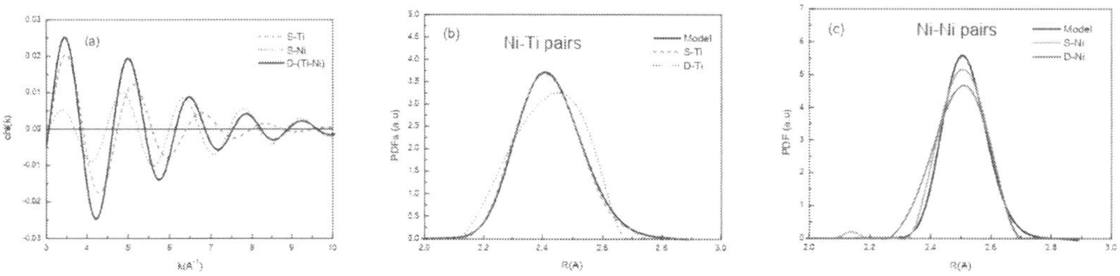


**Figure 1.** Comparison of simulated EXAFS spectra, and the modeled and reconstructed pair distribution functions. (a) EXAFS spectra for Ti shell only (green dash line), Ni shell only (red dot line), mixed shells (black line). (b) The pair distribution functions of Ni-Ti pairs for Model (black line), extracted from Ti single shell (green dash line), extracted from mixed shells (red dot line), (c) The pair distribution function of Ni-Ni pair for model (black line), extracted from Ni single shell (green line), extracted from mixed shells (red line).

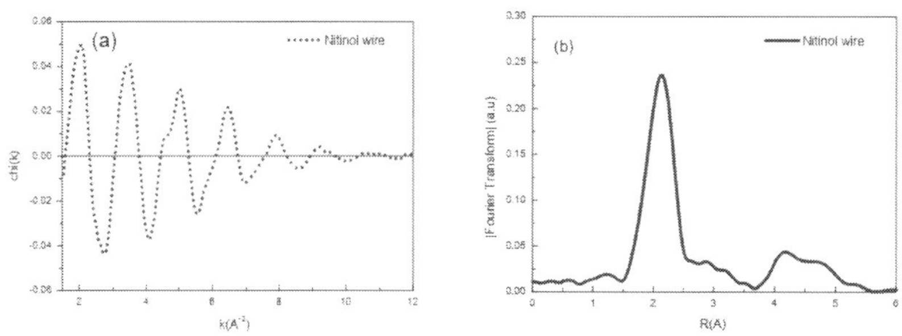


**Figure 2.** (a) Experimental EXAFS spectrum of nitinol wire and (b) its Fourier transform.

margin. From this simulation test, we see that the pair distribution functions of different kinds of atomic shells can be obtained separately from the merge spectrum without any constraint.

## 5. Analysis of Experimental Spectrum

Nitinol is very suitable to test the separation of double shells of different atomic species. Figure 2-(a) and (b) show the experimental chi(k) spectrum and the Fourier transformed spectra obtained by experiment for nitinol wire.

As shown in Figure 2-(a) and (b), the EXAFS spectrum of nitinol wire looks almost like single shell features. However, we know that the atomic arrangement around Ni central atoms in nitinol consists of multiple shells of different atomic species. The experimental EXAFS spectrum shown in Fig. 2-(a) was analyzed by the regularization method to extract the pair distribution functions. The main peak in Figure 2-(b) within the window $R = (1.73 - 2.65)\mathring{A}$ was filtered by the inverse Fourier transform to extract chi(q) spectrum to be processed by the regularization method.

Fig. 3-(a) and (b) show the obtained pair distributions of Ni-Ti and Ni-Ni shells, respectively, and Fig. 3-(c) shows the fit of the filtered spectrum to the calculated spectrum by the regularization method. The peak at $R = 2.51\mathring{A}$ in Fig. 3-(a) corresponds to the Ni-Ti bond at $R = 2.53\mathring{A}$ in the crystallographic data. Also, the peak at $R = 2.96\mathring{A}$ in Fig. 3-(b) corresponds to the Ni-Ni bond at $R = 2.90\mathring{A}$ in the crystallographic data. The other peaks near $R = 3.25\mathring{A}$

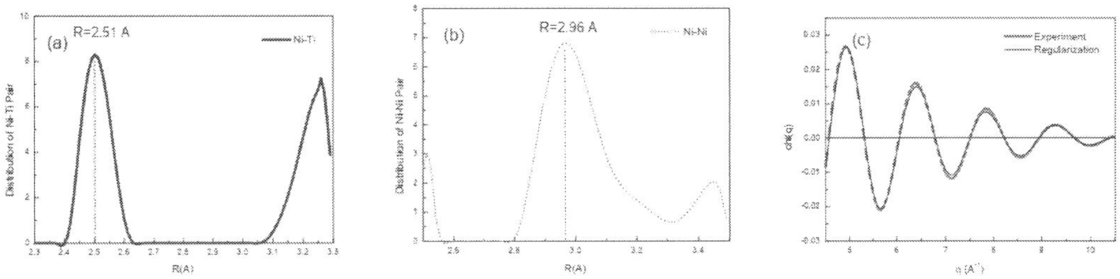


**Figure 3.** Pair distribution functions obtained by regularization method for Ni-Ti (a) and Ni-Ni (b), and EXAFS spectra (c) filtered from the experimental spectrum (red dash) by inverse Fourier transform and calculated (black solid line) with the pair distribution functions obtained by the regularization method.

and $R = 3.43\mathring{A}$ are related to other bond lengths.

Fig. 3-(c) shows the chi(q) spectra obtained by the experiment (red dash) and calculated with the pair distribution functions. As shown in Fig. 3-(c), the reconstructed spectrum fits accordingly to the experimental one. This means that the pair distributions and the bond length were estimated properly with the regularization method.

As pointed out, it was very difficult to separate the merged shells of different kinds of atomic shells with the conventional fitting technique because the two shell fitting requires many fit parameters exceeding the limitation of fit variables. However, as shown in Fig.3, the regularization method provided an adequate separation of pair distribution functions of merged shells of the two different kinds of atoms, but the bond length was not matched exactly to the crystallographic data. This means that this method needs more sophisticated skills to provide clearer structural information of the mixed shells of different atomic species.

## 6. Summary

In this work, we showed that the regularization method with two kernels can resolve the merged shells of two different kinds of atoms. It was tested by simulation and experimental works. In both works, the regularization method with two kernels provided successful results in separation of the double shells.

## References

[1] D. E. Sayers, E. A. Stern, and F. W. Lytle, Phys. Rev. Lett. 27 (1971) 1204.
[2] D.S. Yang, G. Bunker, Phys. Rev. B, 54, (1996) 3169-3172.
[3] Dong-Seok Yang and Jay Min Lee, Physica Scripta. Vol. T115, (2005) 200-201,
[4] M. Dawood, M.W. El-Tahan, B. Zheng, Composites Part B 77 (2015) 238-247.
[5] Q. Zheng, Y. Pivak, L.P.A. Mooij, A.M.J. van der Eerden, H. Schreuders, P.E. de Jongh, J.H. Bitter, B. Dama, international journal of hydrogen energy 37 (2012)4161-4169.
[6] B. Ravel and M. Newville, J. Synchrotron. Rad. 12 (2005) 537.
[7] M. Newville, J. Synchrotron Rad. 8 (2001) 322.
[8] Babanov, Yu. A., Vasin, V. V., Ageev, A. L., Ershov, N. V. Phys. Stat. Sol., 105 (1981) 747-754
[9] Babanov, Yu.A., Vasin, V.V., Ageev, A.L., Ershov, N.V. ,Phys. Stat. Sol., 108, (1981) 103-111.
[10] Yu. Babanov, S. Kiryanov, A. Sidorenkol, L. Romashev, D. Vyalikh, S. Molodtsov, G. Guentherodt, U. Ruediger, Yu. Dedkov, M. Fonine, K. Baberschke, H.Wende and Y. U. Idzerda, Physica Scripta, T115 (2005) 194-196.
[11] J. J. Rehr and R. C. Albers, Rev. Mod. Phys. 72 (2000) 621.

# Local symmetry in liquid metals probed by x-ray absorption spectroscopy

**Fabio Iesari and Andrea Di Cicco**

Physics Division, School of Science and Technology, University of Camerino, Italy

E-mail: fabio.iesari@unicam.it

**Abstract.** The nature of local point symmetry in simple monoatomic liquids has been a fundamental open question for almost 40 years of computational and experimental studies. We present original results of local ordering in different monoatomic liquids obtained by x-ray absorption spectroscopy (XAS), exploiting its high sensitivity to short-range ordering. We used Reverse Monte Carlo (RMC) analysis (RMC-GNXAS package) for obtaining structural models compatible with XAS experimental data (and diffraction techniques for long-range ordering). Pair, bond-angle distributions and suitable indicators for point symmetry are calculated for different liquid metals.

## 1. Introduction

The study of local point symmetry in simple monoatomic liquids has been a fundamental and debated topic for almost 40 years of computational and experimental studies, following the initial Frank's hypothesis [1] about the presence of icosahedral short-range ordering (ISRO) in liquids. The presence of such close-packed five-fold configurations, incompatible with translational symmetry but favored by energetic considerations, was conjectured as a possible explanation for the peculiar undercooling properties of small metal droplets studied by Turnbull in the early fifties [2]. Crystal nucleation [3] is hampered by the positive liquid/crystal inter-facial energy resulting in the possibility to bring a liquid sample into a metastable (undercooled) molten state below the melting temperature $T_m$. The debate about presence and amount of ISRO in undercooled liquids is still open, also because of the experimental difficulties to access deeply undercooled states and the limited structural information supplied by x-ray diffraction experiments. In the last decades, several computational and experimental studies were devoted to investigate locally preferred structures in simple atomic liquids (see for example [4, 5, 6, 7, 8]), and most of these works support the existence and importance of ISRO. In this work, the local order in Ni [9], Cu [10] and Cd [11] is probed by x-ray absorption spectroscopy (XAS), suitable to investigate liquid matter [12] and sensitive to higher order distribution functions. The EXAFS signals are analyzed using a Reverse Monte Carlo (RMC) [13, 14] approach yielding structure models compatible with the measurements and reliable information about presence and amount of ISRO in RMC models is extracted using a suitable geometrical analysis.

## 2. XAS experiment

Two different sample-environment setup (see [15] and refs. therein) have been used for the present experiments: a furnace for high-temperature measurements (Fig. 1) and a large volume

Paris-Edinburgh (PE) press for high-pressure and high-temperature measurements (Fig. 2).

**Figure 1.** The furnace used for high-temperature measurements. In the inset: the crucible where the sample is inserted.

**Figure 2.** The Paris-Edinburgh press for high-pressure and high-temperature measurements.

Using the first setup, samples are prepared from powder mixture of sub-micrometric metal sample and a filler ($Al_2O_3$ for Ni, C for Cu) with a suitable ratio to obtain an optimal XAS signal, pressed into 100 $\mu$m thick 13 mm diameter pellets. The filler powder represents also a container for the metal droplets after melting. The pellets are placed inside a crucible of a furnace, which consist in a 130 mm diameter cylindrical Pyrex glass vessel with a suitable window for simultaneous XRD (x-ray diffraction) and XAS data collection. The heat treatments are performed under high vacuum conditions P $\simeq 10^{-5}$ mbar and temperature measurements are carried out using a high-temperature pyrometer (estimated uncertainty is 20 K).

Instead, using the PE press setup, we use a 10 mm boron/epoxy toroidal gasket as a container for our powder mixture, together with some other element used as a pressure marker (Si or NaCl). The powder obtained is pressed into a cylindrical shape of about 1.6 $mm^3$, insert into a graphite heater and then loaded in the toroidal gasket. The temperature is controlled by using a K type thermocouple inserted through the gasket. Cd [11] was studied using this setup and some extra precautions were used due to its toxicity.

The temperature behavior of the samples and their phase transformations have been monitored throughout the experiments by x-ray absorption temperature scans [16] and x-ray diffraction collected with an area detector. X-ray absorption experiments at variable temperatures using both sample environment setups were performed at the BM29 beamline (now BM23) of the ESRF (Grenoble) and we were able to obtain data with a good signal to noise ratio (Fig. 3). Pressure behavior of solid Cd was also studied [11], but the data are not shown here as goes beyond the purpose of this paper.

## 3. Reverse Monte Carlo XAS analysis

RMC is a data-analysis method based on structural modeling, it simulates a cell that reproduces our experimental data. Its basic algorithm is a variation of the Metropolis Monte Carlo method (MMC), but instead of using some model to obtain configurations and then compare them with experimental data, in RMC we directly use the experimental data as a constraint to drive our structures.

RMC modeling was applied to the measured XAS data [14] using the RMC-GNXAS software [17], simulating a 1372-atoms cubic cell with periodic boundary conditions and fixed density. Due to the short-range nature of the XAS signal, the RMC refinement of the liquid systems was carried out using also a measured pair distribution function $g(r)$ ([18, 19]) mainly as a constraint for long-range order. Technical details about the program can be found in Refs. [14, 17].

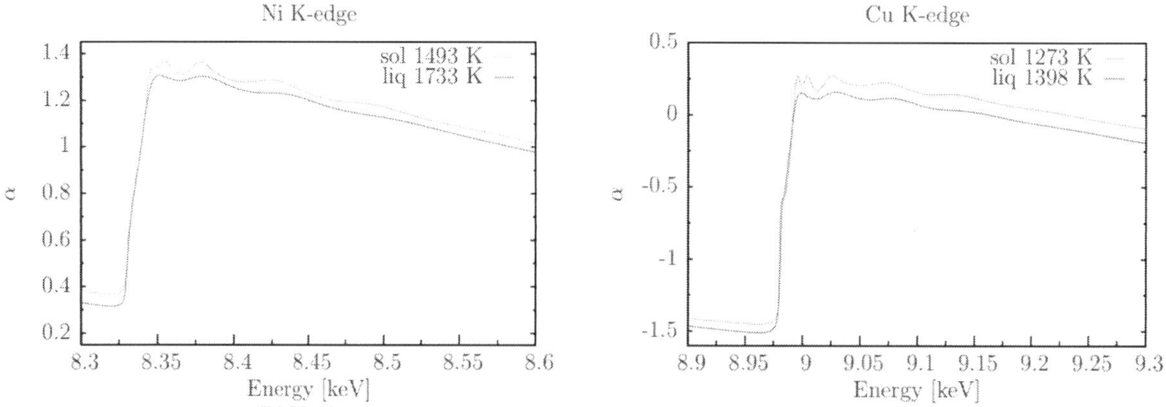


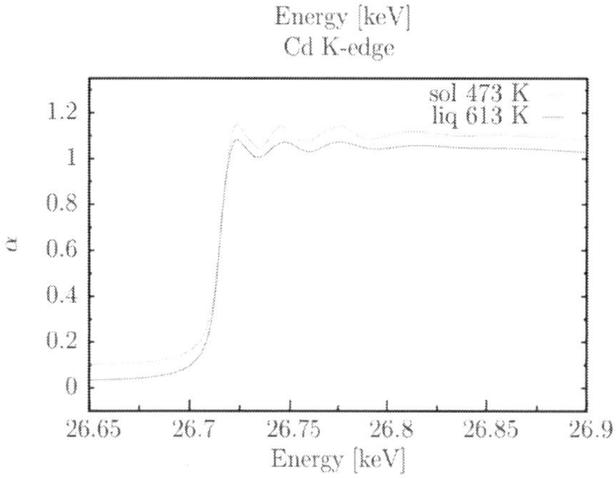


**Figure 3.** Raw XAS spectra for Ni, Cu and Cd in their solid and liquid phase.

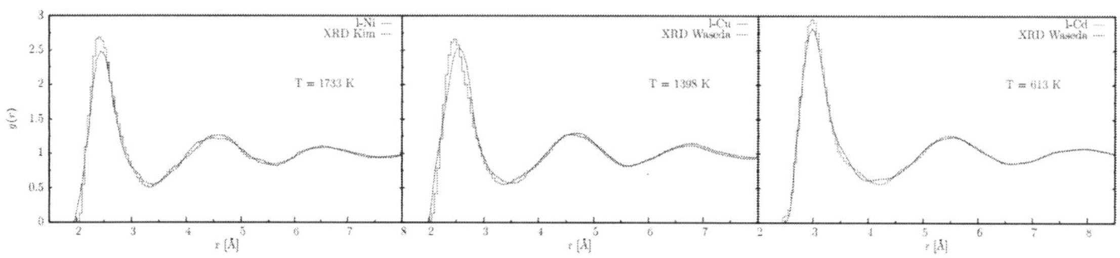


**Figure 4.** Pair distribution functions obtained from averaging over 100 RMC configurations.

A total of 100 different box configurations, obtained after convergence to a minimum and reproducing our experimental data, were finally considered.

## 4. Data analysis and results

Pair distribution functions for the three liquids were calculated using the equilibrated RMC configurations (Fig. 4) and found to be in good agreement with the ones obtained from XRD and used as a constraint for long-range order. The first minimum in the $g(r)$ was used to define the distance to consider first near-neighbor atoms.

To analyze the local geometry we used common-neighbor analysis (CNA) [20]: each pair of nearest-neighbor atoms is classified by using a set of three indexes $jkl$: $j$ is the number of nearest neighbors common to both atoms, $k$ is the number of bonds between the common neighbors, and $l$ is the number of bonds in the longest continuous chain formed by the $k$ bonds between common neighbors. This method is able to distinguish between various local structures:

555 pairs are characteristic of icosahedral structures (ICOS), 544 and 433 are characteristic of strongly distorted or defective icosahedra (DICOS), 666 are found in bcc structures, and 421 and 422 are characteristic of fcc and hcp ordering [18]. These geometrical structures are the most abundant in our models, and their relative weights are reported in table 1, normalized to the total number of nearest-neighbor pairs.

**Table 1.** Percentage of pairs from CNA obtained from RMC configurations

| Pairs | Ni | Cu | Cd |
|---|---|---|---|
| 421 + 422 | 16% | 15% | 19% |
| 433 + 544 | 40% | 38% | 34% |
| 555 | 11% | 10% | 7% |
| 666 | 3% | 3% | 2% |

The results reported in table 1 show that defective and distorted icosahedra are the most present structures in our model, as well as non-negligible percentages of nearly-perfect icosahedra.

On the other hand, structures reminiscent of the crystalline phase are present. It is interesting to notice that in the Cd case, the ISRO is less abundant. This is probably connected to the fact that Cd is not exactly close-packed in its crystalline phase and this is also reflected in its liquid phase. These findings support and quantify the presence of ISRO in close-packing liquid metals, for which a complete analysis of short-range structure and local symmetry will be presented elsewhere.

## References

[1] F. C. Frank, Proc. Royal Society London A **215**, 43 (1952).
[2] D. Turnbull and R. E. Cech, J. Appl. Phys. **21**, 804 (1950).
[3] K. F. Kelton and A. L. Greer, *Nucleation in Condensed Matter* (Elsevier, Amsterdam, 2010).
[4] P. J. Steinhardt, D. R. Nelson, and M. Ronchetti, Phys. Rev. Lett. **47**, 1297 (1981).
[5] S. Nosé and F. Yonezawa, J. Chem. Phys. **84**, 1803 (1986).
[6] H. Jónsson and H. C. Andersen, Phys. Rev. Lett. **60**, 2295 (1988).
[7] T. Kondo and K. Tsumuraya, J. Chem. Phys. **94**, 8220 (1991).
[8] S. Mossa and G. Tarjus, J. Chem. Phys. **119**, 8069 (2003).
[9] A. Di Cicco *et al.*, Phys. Rev. B **89**, 060102(R) (2014).
[10] A. Di Cicco, A. Trapananti, S. Faggioni, and A. Filipponi, Phys. Rev. Lett. **91**, 135505 (2003).
[11] M. Minicucci *et al.*, Physica Scripta. **T115**, 1056 (2005).
[12] A. Filipponi, J. Phys.: Condens. Matter **13**, R23 (2001).
[13] R. L. McGreevy and L. Pusztai, Molec. Simul. **1**, 359 (1988).
[14] A. Di Cicco and A. Trapananti, J. Phys.: Condens. Matter **17**, S135 (2005).
[15] A. Di Cicco *et al.*, Journal of Non-Crystalline Solids **352**, 4155 (2006).
[16] A. Filipponi *et al.*, Rev. Sci. Instrum. **71**, 2422 (2000).
[17] A. Di Cicco *et al.*, *GNXAS, extended suite of programs for advanced x-ray absorption data-analysis: methodology and practice* (TASK Publishing, Gdansk, 2009).
[18] T. Kim and K. Kelton, J. Chem. Phys. **126**, 054513 (2007).
[19] Y. Waseda, *The Structure of Non-Crystalline Materials* (McGraw-Hill, New York, 1980).
[20] J. D. Honeycutt and H. C. Andersen, J. Phys. Chem. **94305**, 4950 (1987).

# Pd nanoparticles formation inside porous polymeric scaffolds followed by *in situ* XANES/SAXS

**A Longo[*,1,2], C Lamberti[*,3,4], G Agostini[5,6], E Borfecchia[5], A Lazzarini[5], W Liu[5,7], F Giannici[8], G Portale[1], E Groppo[5]**

[1]Netherlands Organization for Scientific Research at ESRF, Grenoble (France); [2]CNR, Istituto per lo Studio dei Materiali Nanostrutturati, Palermo (Italy); [3]Southern Federal University, Rostov-on-Don (Russia); [4]Dept. of Chemistry, CrisDi Centre for Crystallography, University of Turin, Turin (Italy); [5]Dept. of Chemistry, NIS and INSTM Reference Centers, University of Turin, Turin (Italy); [6]European Synchrotron Radiation Facility (ESRF), Grenoble (France); [7]School of Environmental and Chemical Engineering Tianjin Polytechnic University, Tianjin (People's Republic of China); [8]Dipartimento di Fisica e Chimica Università di Palermo (Italy).

E-mail: alessandro.longo@esrf.fr; carlo.lamberti@unito.it

**Abstract.** Simultaneous time-resolved SAXS and XANES techniques were employed to follow *in situ* the formation of Pd nanoparticles from palladium acetate precursor in two porous polymeric supports: polystyrene (PS) and poly(4-vinyl-pyridine) (P4VP). In this study we have investigated the effect of the use of different reducing agents ($H_2$ and CO) from the gas phase. These results, in conjunction with data obtained by diffuse reflectance IR (DRIFT) spectroscopy and TEM measurements, allowed us to unravel the different roles played by gaseous $H_2$ and CO in the formation of the Pd nanoparticles for both PS and P4VP hosting scaffolds.

## 1. Introduction

In recent years, Pd nanoparticles (NPs) supported on porous polymers have become a promising class of catalysts because of their unique physical–chemical properties and attractive catalytic performances [1]. The polymer acts simultaneously as support and stabilizer, thanks to the dual effect of porosity (steric stabilization) and charged functional groups (electrostatic stabilization) [2,3]. For this reason, Pd catalysts supported on polymers are more stable than the homogeneous counterparts; moreover, they are cheaper, easy to handle, and easily recovered. XAFS techniques have been largely employed to characterize metal NPs [4,5], sometimes in combination with other techniques such as XRPD, IR, and SAXS [6]. In this contribution we present a simultaneous XANES/SAXS study, performed on the BM26A beamline of the ESRF [7], to follow *in situ* the formation of Pd nanoparticles in PS and P4VP porous polymers. Parallel laboratory IR and TEM study complement the XANES/SAXS investigation.

## 2. Experimental

The starting Pd(II)–polymer composite was prepared by impregnation, stirring the highly cross-linked polymers polymer (both a poly(4-ethylstyrene-co-divinylbenzene), hereafter PS and a poly(4-vinylpyridine-co-divinylbenzene), hereafter P4VP from Aldrich) with an appropriate amount of palladium acetate in acetonitrile, resulting in a final Pd loading of 10 wt % [8,9]. *In situ* reduction of

the Pd(II)–polymer to get Pd(0)-containing polymer was carried out in mild conditions in flow of $H_2$ or CO reducing agents in the gas phase (both diluted at 5% in He).
FT-IR spectra were collected in reflectance mode (DRIFT) on a Nicolet 6700 instrument, equipped with a MCT detector. A Thermo Fisher Environmental Chamber was adopted to record FT-IR spectra in reaction conditions. Transmission electron micrographs were obtained using a JEOL 3010-UHR instrument operating at 300 kV, equipped with a 2k × 2k pixel Gatan US1000 CCD camera.
The SAXS/XAS set up used at BM26A @ ESRF was described elsewhere [7]. The energy delivered by the double crystal Si(111) monochromator was calibrated measuring the XANES spectrum of a Pd foil in transmission mode. Fluorescence XANES spectra at the Pd K-edge (24.3 keV) were collected with a nine-element Ge detector. The SAXS patterns were collected using a 2D Mar detector. The sample–detector distance was calibrated with standard Ag behenate. The energy change between the start and the end of the XANES spectrum (about 120 eV) is irrelevant to SAXS at that high energy, so the incident beam wavelength can be treated as constant, $\lambda = 0.509(1)$ Å. The resulting $q$-range was 0.5–7.0 nm$^{-1}$ ($q = 4\pi \sin(\theta)/\lambda$, where $\theta$ is the X-ray scattering angle), allowing investigation of the 12.5–0.9 nm $d$-spacing interval. Each XANES spectrum was collected in about 300 s. As the readout and erasing time of the 2D Mar detector was 180 s, each SAXS pattern was collected for 120 s resulting in a 1 to 1 correspondence between XANES spectra and SAXS patterns.

## 3. Results and discussion

A summary of the data and of the analysed results is reported in Figures 1 and 2. We have demonstrated that the nature of the polymeric matrix has a strong influence on the formation of the Pd NPs and affects their final properties, in terms of particle size, electronic properties, and type and fraction of accessible surface sites. In particular, the presence of nitrogen-based ligands allows the formation of extremely small, subnanometric Pd clusters in P4VP, as demonstrated by the observation of a specific interaction between Pd nanoparticles and the nitrogen of the pyridine ligands [8]. For both Pd/PS and Pd/P4VP systems, the reduction process is strongly dependent on the nature of the reduction agent ($H_2$ or CO) [9].
For the Pd/PS systems the following conclusions have been obtained: (i) Reduction of palladium acetate starts at lower temperature (already at room temperature) and proceeds faster in the presence of CO than with $H_2$. (ii) The Pd NPs are bigger when formed in CO and show peculiar morphologies (triangles and hexagons, see inset in Figure 1a), whereas those formed in $H_2$ are smaller and spherical in shape (Figure 2b). (iii) Pd NPs formed in CO have unclean surfaces, probably as a consequence of residual by-products (such as acetic anhydride) that are not desorbed at the reduction temperature, as evidenced by DRIFT [9]. The reason for the different properties of the formed Pd NPs has to be related to the different mechanism of palladium acetate reduction. It is likely that, in the presence of CO, Pd acetate carbonyl intermediates are formed, where CO acts as templating agent, promoting the crystal growth in specific directions. This is relevant as the catalytic reactivity of metal NPs depends not only on the size (and size distribution) control but also on the exposure of the peculiar (hkl) planes.
For the Pd/P4VP system (Figure 2f-l), we have demonstrated that the presence of the basic pyridine functional groups inside P4VP facilitates the dispersion of palladium acetate in a monomeric form, in which the acetate ligands assume a mono-dentate configuration [8,9]. This peculiar structure of Pd-acetate in P4VP is at the basis of its different reactivity towards reducing agents with respect to the observations made for Pd/PS, where functional groups are absent. In particular, $H_2$ efficiently reduces palladium acetate in P4VP, to give palladium nanoparticles and acetic acid. The latter is stabilized by the basic functional groups in the polymer, which make possible its spectroscopic observation. The consequence is that reduction occurs at a higher temperature than for palladium acetate dispersed in PS (that has no functional groups), as determined by DRIFT and XANES spectroscopy. In addition, the Pd NPs are smaller (as determined by TEM and SAXS), as they are stabilized by direct covalent interaction with the nitrogen-containing ligands, in good agreement with previous findings [8]. On the contrary, CO does not reduce palladium acetate (data not shown). Isolated $Pd^{2+}$ carbonyl adducts are immediately formed, with the simultaneous displacement of the acetate ligands, as detected by DRIFT

spectroscopy. These $Pd^{2+}$ carbonyl adducts easily migrate and aggregate in units of about 5 nm that, having a higher electronic density with respect to the hosting matrix, are easily revealed by SAXS.

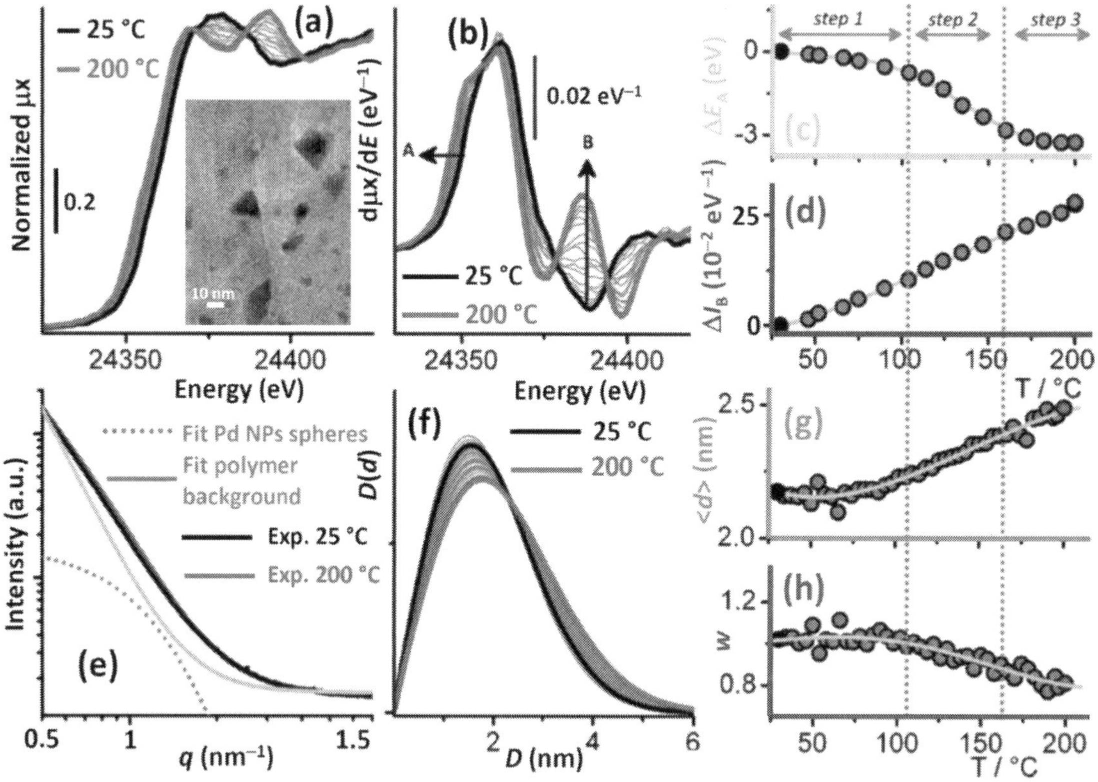


**Figure 1.** SAXS–XAS data collected simultaneously during reduction of Pd/PS in CO/He flow from 25 °C (black) to 200 °C (red); heating rate 2 °C/min. Parts (a, b): evolution of the normalized XANES spectra and of the derivative signal. Parts (c, d): evolution as a function of temperature of features A (shift in energy) and B (change in intensity) of the derivative XANES spectra, see arrows in part (b). Part (e): SAXS pattern at the beginning (black) and the end (red) of the temperature ramp. Also examples of the theoretical signals obtained by considering a distribution of spherical particles (dotted gray) and the background (gray) are reported. Part (f): evolution of the particle size distribution as a function of the temperature. Parts (g,h): average diameter of the Pd particles and relative weight as a function of temperature, respectively, as obtained from the spherical model fitting of the Pd NPs contribution to the SAXS data. The inset in part (a) reports a representative TEM micrograph of the CO-reduced Pd/PS system. Adapted with permission from Ref. [9a], copyright American Chemical Society 2014.

## 4. Conclusions and acknowledgements

These findings have important potential implications in the field of catalysis. For example, it is expected that reactions catalyzed by palladium acetate and involving carbon monoxide would have a different outcome when palladium acetate is hosted in porous P4VP than in other supports without pores and/or nitrogen containing ligands. Finally, the importance of a multi-technique approach in following the whole process of metal nanoparticles formation clearly emerges.

We are grateful to R Pellegrini (Chimet SpA) and G Leofanti for constructive discussions. C. L. acknowledges the megagrant of the Russian Federation Government to support scientific research at the Southern Federal University, no. 14.Y26.31.0001.

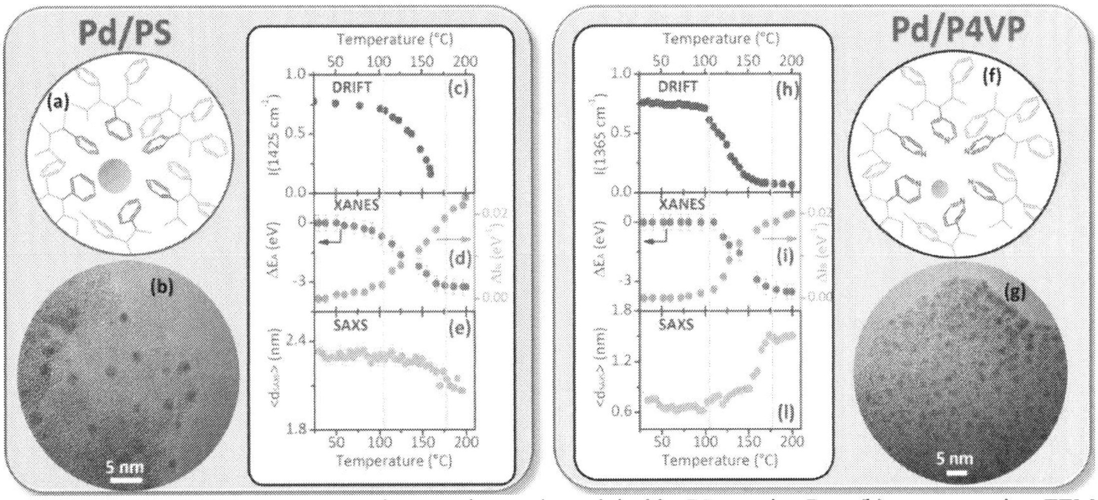


**Figure 2.** Part (a): cartoon representing a Pd NPs hosted inside PS matrix. Part (b) representative TEM micrographs of the $H_2$-reduced Pd/PS system. Parts (c-e): Summary of the DRIFT and of the SAXS−XAS data collected simultaneously during reduction of Pd/PS in $H_2$/He flow. Evolution as a function of temperature of: part (c) the intensity of the IR absorption band, which is characteristic of the acetate ligand [9]; part (d) features A (shift in energy, left axis) and B (change in intensity, right axis) of the derivative XANES spectra (as marked in Figure 1b) and part (e) average Pd particle size as obtained from analysis of the SAXS data (same approach as shown in Figure 1e. (f-l): as parts (a-e) for the DRIFT and the SAXS−XAS data collected simultaneously during reduction of Pd/P4VP in $H_2$/He flow. Previously unpublished figure comparing data discussed in Refs. [9a] and [9b].

## References

[1] (a) Hodge P 1997 *Chem. Soc. Rev.* **26** 417; (b) Akiyama R and Kobayashi, S 2009 *Chem. Rev.* **109** 594

[2] Roucoux A Schulz, J and Patin H 2002 *Chem. Rev.* **102** 3757

[3] Ott L S and Finke R G 2007 *Coord. Chem. Rev.* **251** 1075

[4] (a) Newton M A 2008 *Chem. Soc. Rev.* **37** 2644; (b) Frenkel A I 2012 *Chem. Soc. Rev.* **41** 8163

[5] (a) Bordiga S, Groppo E, Agostini G, van Bokhoven J A and Lamberti C. 2013 *Chem. Rev.* **113** 1736; (b) Mino L, Agostini G, Borfecchia E, Gianolio D, Piovano A, Gallo E and Lamberti C. 2013 *J. Phys. D: Appl. Phys.* **46** 423001; (c) Garino C, Borfecchia E, Gobetto R, van Bokhoven J A, Lamberti C. 2014 *Coord. Chem. Rev.* **277–278** 130.

[6] (a) Lamberti C, Borfecchia E, van Bokhoven J A, and Fernández García M 2016 in *X-Ray Absorption and X-ray Emission Spectroscopy: Theory and Applications*, Eds. van Bokhoven J A, and Lamberti C, Chapter. 12, Vol. 1 (Chichester: John Wiley & Sons) p 303; (b) Groppo E, Agostini G, Piovano A, Muddada NB, Leofanti G, Pellegrini R, Portale G, Longo A, Lamberti C 2012 *J. Catal*, **287** 44; (c) Agostini G, Lamberti C, Pellegrini R, Leofanti G, Giannici F, Longo A, Groppo E, 2014 *ACS Catal.* **4**, 187

[7] (a) Longo A, Portale G, Bras W, Giannici F, Ruggirello A M and Turco Liveri V 2007 *Langmuir* **23** 11482; (b) Nikitenko S, Beale A M, van der Eerden A M J, Jacques S D M, Leynaud O, O'Brien M G, Detollenaere D, Kaptein R, Weckhuysen B M, and Bras W 2008 *J. Synchrotron Radiat.* **15** 632

[8] Groppo E, Liu W, Zavorotynska O, Agostini G, Spoto G, Bordiga S, Lamberti C and Zecchina A, 2010 *Chem. Mater.* **22** 2297

[9] (a) Groppo E, Agostini G, Borfecchia E, W Liu, Giannici F, Portale G, Longo A and Lamberti C, 2014 *J. Phys. Chem. C*, **118** 8406; (b) Groppo E, Agostini G, Borfecchia E, Lazzarini A, W Liu, Lamberti C, Giannici F, Portale G and Longo A 2015 *ChemCatChem*, **7** 2188

# XAS and RIXS study of acetic acid and methyl formate in liquid

**O Takahashi[1,*], N Nishida[2], S Kanai[2], Y Horikawa[3,4], T Tokushima[4]**

[1]Institute for Sustainable Sciences and Development, Hiroshima University, Japan.
[2]Department of Chemistry, Hiroshima University, Japan.
[3]Department of Physics and Information Science, Yamaguchi University, Japan.
[4]RIKEN SPring-8 Center, Japan.

E-mail: shu@hiroshima-u.ac.jp

**Abstract**. Structure of acetic acid (AA) and methyl formate (MF) in the liquid phase is studied using X-ray absorption spectroscopy (XAS) and X-ray emission spectroscopy (XES) experimentally and theoretically. Two distinct XES spectra are observed by tuning photon energy for both molecules. Model structure in the liquid phase is constructed using the classical and first principle molecular dynamics simulations, and XES spectra are calculated using density functional theory. Calculated XES spectra are consistent with experimental ones. The effect of core-hole induced excited state molecular dynamics are discussed.

## 1. Introduction

Soft X-ray spectroscopy is now one of powerful tools to investigate a local structure of various materials in the gas or condensed phases [1, 2]. Experimental technique to observe spectra under ambient conditions has been developed for more than fifteen years. To interpret the observed spectra theoretical analysis with the model structure is needed.

Previously Tokushima and his coworkers investigated structure of acetic acid (AA) in the liquid phase using resonant X-ray emission spectroscopy (resonant XES) [3], a technique that can also be called resonant inelastic X-ray scattering (RIXS). XES spectra of the two oxygen atoms at different sites in the carboxyl group can be distinct by tuning incident photon energy. In the previous study theoretical calculations performed for a single molecule and these results supported their experiments. Calculations using more applicable models is challenging work at the liquid phase.

In the present study, we investigate methyl formate (MF) in the liquid phase using X-ray absorption spectroscopy (XAS) and XES in comparison with AA [4]. MF and AA are isomers of $C_2H_4O_2$. And, it is expected that molecular interaction through a hydrogen bond (H-bond) is different between MF and AA because OH functional group is not included in MF. For AA, strong intermolecular interaction such as OH···O H-bond may be operated. On the other hand, for MF, weak interaction such as CH···O H-bond may be operated only. To elucidate feature of XES spectra of these liquids theoretical model structure was constructed using the first principle molecular dynamics simulations, and XES spectra were produced within the framework of density functional theory (DFT) using the cluster model. Note that Robin et al. [5] reported studies for AA in the gas phase by electron energy loss spectroscopy (EELS).

## 2. Experimental measurements

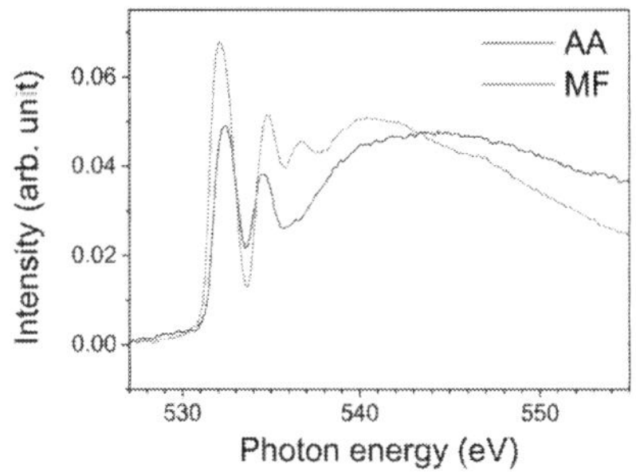


**Figure 1**. Experimental XAS spectra of AA and MF.

XAS and XES spectra of AA and MF were recorded at synchrotron radiation facility SPring-8 at the beamline BL17SU using an X-ray emission spectrometer. All the spectra were measured using a liquid flow cell with a 150 nm-thick Au-coated thin-film SiC window to separate the liquid flowing at atmospheric pressure from the high-vacuum region. Circularly polarized X-rays were used to eliminate any polarization dependence effects. The energy resolution of the $O_{1s}$ XES measurements is typically about 0.35 eV. The background signals from oxygen impurities in the SiC window, i.e., natural surface oxides, were estimated by measuring the XES spectra of pure acetonitrile and these were subtracted from the sample spectra.

## 3. Theoretical methods

Molecular dynamics simulations of AA and MF were performed under periodic boundary conditions with OPLS force field using GROMACS version 4.5.5. An NVT ensemble MD simulation comprising 40 molecules modeled using the OPLS force field was performed. The temperature during the simulation was 300 K, which was controlled using the Nosé-Hoover thermostat. The simulation cell was a cubic cell with a cell length of 15.6 Å. The density of the simulation cell was 1.049 g/cm$^3$ which is in good agreement with the experimental density (1.044 g/cm$^3$ at atmospheric pressure and at 298.15 K). The simulation was performed with a time step of 0.1 fs and a total time of MD was 200 ps. One snapshot was obtained as a final structure, and this was used as the starting configuration for a DFT ab initio NVT ensemble MD simulation using VASP. This was performed to model the liquid more accurately. A plane wave basis set with an energy cutoff of 400 eV was used, and PAW potentials were applied to all atoms. The Perdew, Burke, and Ernzerhof (PBE) exchange correlation functional was used. The simulation temperature was 300 K, controlled by the Nosé-Hoover thermostat. The time step was 0.2 fs and the total time was 4.0 ps. The first half of the simulation was used as the equilibrium time (2.0 ps).
Theoretical XES spectra were calculated using deMon2k code with the extension of X-ray electron spectroscopy parts. Initial geometries of 16 molecules cluster were sampled and 40 geometries were studied. Core-hole induced excited state (CHES) MD calculations were propagated within 20 fs with the time step of 0.25 fs. XES spectra were calculated at every step and weighted together according to the exponential lifetime decay constant of 4.1 fs.

## 4. Results and Discussion

Experimental XAS spectra of AA and MF are shown in Figure 1. The first and second resonant peaks at 532.3 eV and 534.6 eV are assigned as O(C=O) → $\pi^*$ and O(OH) → $\pi^*$ excitations, respectively. In the same way, two resonant peaks at 532.2 eV and 534.7 eV for MF are assigned as O(C=O) → $\pi^*$ and O($OCH_3$) → $\pi^*$ excitations. Thus by tuning photon energy, site-selective XES spectra can be also observed for MF.
In Figure 2, XES spectra of AA and MF in the liquid phase are illustrated. In a similar manner to AA, observed spectra of MF are different depending on excitation energy. In addition a behavior of peak broadening for AA and MF are quite similar although peak positions are slightly different depending on

molecules. This means that difference of H-bond behavior may not be the reason of the peak broadening of O(OH, $OCH_3$) excited XES spectra, and other driving force may work in such situations.

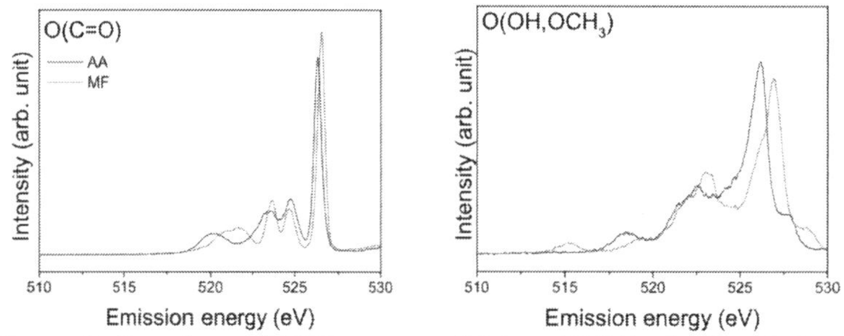


**Figure 2**. Experimental XES spectra of AA and MF in the liquid phase.

To understand the reason of the peak broadening of AA and MF in the XES spectra in detail, CHES dynamics calculations were performed. We measured the variations in the C=O, C-O, and O-H bond lengths ($d_{C=O}$, $d_{C-O}$, $d_{O-H}$) of AA and C=O, C-O, and O-$CH_3$ bond lengths ($d_{C=O}$, $d_{C-O}$, $d_{O-CH3}$) of MF during the 20 fs simulation period, as depicted in Figure 3. In the cluster model, average bond lengths of the different dynamics generated from the various initial cluster configurations are shown. The error bars in these figures indicate the standard deviation about the averaged bond lengths at each time step. First we discuss about AA. In the case of O(C=O) excitation for a single molecule, the displacement of these three bonds is small. In contrast, at the O(OH) excitation (Figure 3(b), dashed lines), an elongation of the C-O bond is observed. This dynamics is specific at the O(OH) excitation. In the case of both O(C=O) and O(OH) excitations for the cluster model, the enhancement of O-H stretching was observed (Figures 3(a) and (b), solid lines). These dynamics are due to the intermolecular OH···O hydrogen bonds. The O-H stretching at the O(OH) excitation is larger than that at the O(C=O) excitation. And period of vibration of the O-H bond at the O(OH) excitation is shorter than that at the O(C=O) excitation. These results suggest that a chemical bond which is near from a core-excited site is excited more aggressively. Furthermore, the elongation of C-O bond occurs at the O(OH) excitation for both the single molecule and cluster models. The speed of this elongation for the cluster model is slower than that for the single molecule model due to the condensed configuration. A shape of XES spectra depends on chemical

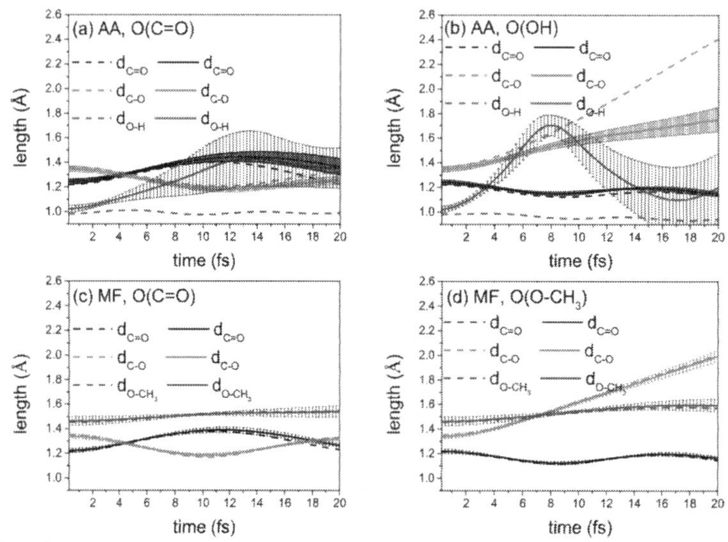


**Figure 3**. (a), (b) Time propagation within 20 fs of three bonds during two oxygen 1s excitations of AA. Dotted and solid lines correspond to monomer and cluster models. The standard deviations of trajectories by cluster models due to different initial geometry samplings are also included. (c), (d) Those of MF.

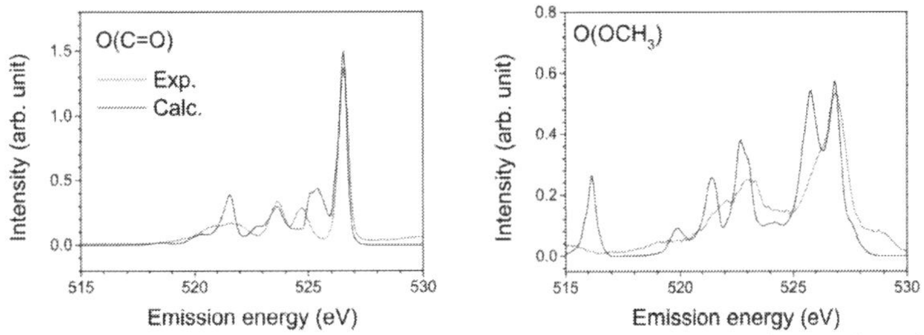


**Figure 4**. XES spectra of MF. Red and black lines correspond to experimental and theoretical spectra.

environment around a core-excited molecule and its CHES dynamics. Contrary to AA, bond deviations of MF are small in general. Common feature which was observed at the O(OH) excitations of AA and the $O(OCH_3)$ excitations of MF is the elongation of C-O bond. We conclude that the feature of spectra of AA at the O(OH) excitation and MF at the $O(OCH_3)$ excitation is originated with the elongation of C-O bond following core-excitation.

Direct comparisons of experimental and calculated XES spectra of MF are depicted at O(C=O) $\rightarrow\pi^*$ and $O(OCH_3)$ $\rightarrow\pi^*$ excitations in Figure 4 [4]. At O(C=O) $\rightarrow\pi^*$ excitation, the sharp features due to slow CHES dynamics are well reproduced. At $O(OCH_3)$ $\rightarrow\pi^*$ excitation, the elongation of C-O bond contributes to the broad feature of XES spectra. It should be noted that such trends can be observed on AA experimentally and theoretically. Theoretical XES spectra were produced the above geometries with initial geometry samplings from MD calculations.

## 5. Conclusion

We explained the features of XES spectra of AA and MF in the liquid phase by dynamical geometry samplings combined with DFT. The strength of H-bond is reflected on the CHES dynamics for both molecules. The broad features of XES spectra at O(OH) and $O(OCH_3)$ $\rightarrow\pi^*$ excitation of AA and MF can be explained by the elongation of C-O bond, which takes place in both molecules.

**Acknowledgements**

This work was supported by a Grant-in-Aid for Scientific Research from JSPS. The synchrotron radiation experiments were performed at the BL17SU of SPring-8 with the approval of the Japan Synchrotron Radiation Research Institute (JASRI) (Proposal No.2014B1738). We thank Research Center for Computational Science, Okazaki, Japan. N.N., S.K., and O.T. thank Prof. Ebata for his helpful support.

**References**

[1] Guo J H, Luo Y, Augustsson A, Rubensson J E, Såthe C, Ågren H, Siegbahn H and Nordgren J 2002 X-Ray Emission Spectroscopy of Hydrogen Bonding and Electronic Structure of Liquid Water *Phys. Rev. Lett.* **89** 137402

[2] Sellberg J A, McQueen T A, Laksmono H, Schreck S, Beye M, DePonte D P, Kennedy B, Nordlund D, Sierra R G, Schlesinger D, Tokushima T, Zhovtobriukh I, Eckert S, Segtnan V H, Ogasawara H, Kubicek K, Techert S, Bergmann U, Dakovski G L, Schlotter W F, Harada Y, Bogan M J, Wernet P, Föhlisch A, Pettersson L G M and Nilsson A 2015 X-ray emission spectroscopy of bulk liquid water in "no-man's land" *J. Chem. Phys.* **142** 044505

[3] Tokushima T, Horikawa Y, Harada Y, Takahashi O, Hiraya A and Shin S 2009 Selective observation of the two oxygen atoms at different sites in the carboxyl group (-COOH) of liquid acetic acid *Phys. Chem. Chem. Phys.* **11** 1679

[4] Nishida N, Kanai S, Tokushima T, Horikawa Y and Takahashi O 2015 Intermolecular interaction dependence of acetic acid and methyl formate in liquid *J. Comput. Chem. Jpn.* **14** 60 (in Japanese)

[5] Robin M B, Ishii I, McLaren R and Hitchcock A P 1988 Fluorination effects on the inner-shell spectra of unsaturated molecules *J. Electron Spectrosc. Relat. Phenom.* **47** 53.

# Active sites in Cu-SSZ-13 deNO$_x$ catalyst under reaction conditions: a XAS/XES perspective

**Kirill A Lomachenko[1,2,*], Elisa Borfecchia[2], Silvia Bordiga[2], Alexander V Soldatov[1], Pablo Beato[3] and Carlo Lamberti[1,2,*]**

[1] Southern Federal University, Zorge street 5, 344090 Rostov-on-Don, Russia
[2] Department of Chemistry and INSTM Reference Center, University of Turin, via P. Giuria 7, 10125 Turin, Italy
[3] Haldor Topsøe A/S, Nymøllevej 55, 2800 Kgs. Lyngby, Denmark

E-mail: carlo.lamberti@unito.it, kirlom@gmail.com

**Abstract.** Cu-SSZ-13 is a highly active catalyst for the $NH_3$-assisted selective catalytic reduction (SCR) of the harmful nitrogen oxides ($NO_x$, x=1, 2). Since the catalytically active sites for this reaction are mainly represented by isolated Cu ions incorporated into the zeolitic framework, element-selective studies of Cu local environment are crucial to fully understand the enhanced catalytic properties of this material. Herein, we highlight the recent advances in the characterization of the most abundant Cu-sites in Cu-SSZ-13 upon different reaction-relevant conditions made employing XAS and XES spectroscopies, complemented by computational analysis. A concise review of the most relevant literature is also presented.

## 1. Introduction

The Cu-exchanged form of the novel SSZ-13 zeolite, with CHA framework and high Si/Al ratio, is attracting a lot of attention due to its outstanding performance in $NH_3$-assisted selective catalytic reduction (SCR) of $NO_x$ gases contained in the exhaust fumes from cars and industrial plants. Although H-SSZ-13 was first synthesized in 1985 [1] it was demonstrated that Cu-SSZ-13 outperforms other SCR catalysts (namely Cu-ZSM-5 and Cu-β zeolites) in terms of activity and hydrothermal stability only in the late 2000s [2, 3]. In order to explain such behavior and to develop the $NH_3$-based SCR reaction mechanism many studies were performed by research groups all over the world involving a very broad range of characterization techniques, such as FTIR, UV-Vis, EPR, NMR, XRD and others, as recently reviewed in [4]. Among them, X-ray spectroscopy (both absorption and emission) is particularly useful due to its element selectivity, since the active sites of SCR reaction in Cu-SSZ-13 are highly diluted Cu ions hosted in the cavities of the zeolite framework. Careful analysis of EXAFS, XANES and XES spectra, often assisted by advanced DFT calculations, yields detailed information on the local environment and oxidation state of Cu centers in different reaction conditions [4-9]. *In situ* and *operando* studies, possible due to the high penetration depth of hard X-rays (around 9 keV, corresponding to the Cu K-edge), provide the means to answer "chemical questions" concerning the reactivity of the Cu species towards particular gases at given temperatures, which is crucial for determining different steps of the reaction mechanism [10-12].

However, most of XAS and XES studies are feasible only exploiting synchrotron sources, which obviously makes these techniques much less common compared to the standard laboratory characterization methods. Therefore, the following review of the advances made in the understanding of the Cu-SSZ-13 chemistry by means of XAS and XES will be useful not only for the X-ray spectroscopy community, but also for the broad chemical audience, as a demonstration of the power of these techniques applied to the study of metal-exchanged porous catalysts under reaction conditions.

## 2. Exploring the activation process

The structure of natural copper-exchanged CHA framework has been known for a long time from single crystal diffraction studies [13]. The structure of Cu-SSZ-13 is similar, but with a lower Al and Cu content. It is composed of double six-membered rings (6-rings) connected in an AABBCC sequence forming cavities with eight-membered windows (8-rings) (figure 1).

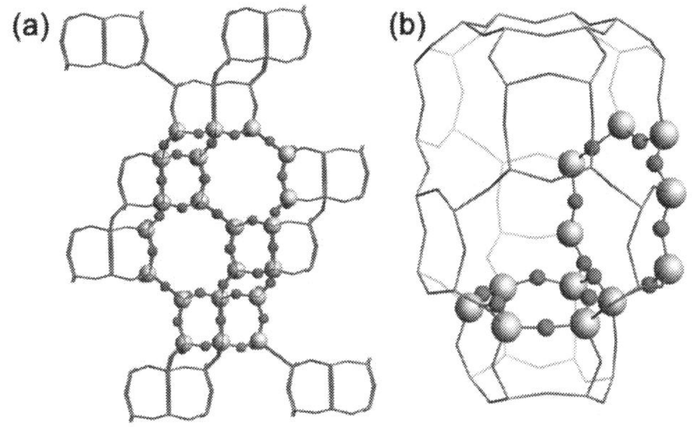


**Figure 1.** (a) Fragment of CHA framework with the atoms forming the large cavity highlighted; (b) close-up of the large CHA cavity showing 6- and 8-membered rings, which are reported to be the most probable hosts for Cu ions. Color code: Si – gray, O – red.

After the discovery of the outstanding catalytic properties of synthetic Cu-SSZ-13 [14], the focus of the structural investigations was gradually shifted from the framework to the location of the isolated Cu ions in the hydrated and thermally activated material, which were suggested to be the active sites of the SCR reaction. One of the first studies, carried out by Fickel and Lobo demonstrated that the Cu ions tend to occupy the centers of the 6-rings [15]. Since every 6-ring is likely to have one negatively charged tetrahedral unit with Al substituting Si, the authors admitted that Cu cations are expected to be shifted towards it from the center of the ring. However, XRD refinement was always converging to central position due to the lack of structural contrast. Nonetheless, it was possible to detect that, being always "projected" on the center of the 6-ring, the Cu ions were gradually approaching its plane upon calcination. The authors suggested that this effect may take place due to the dehydration of the Cu ions.

In a subsequent work Korhonen *et al.* presented Cu K-edge EXAFS data for both calcined and as-synthesized (hydrated) Cu-SSZ-13 [16] and performed a fit using as a model a theoretical structure of Cu in the 6-ring reported earlier (figure 2a) [17].

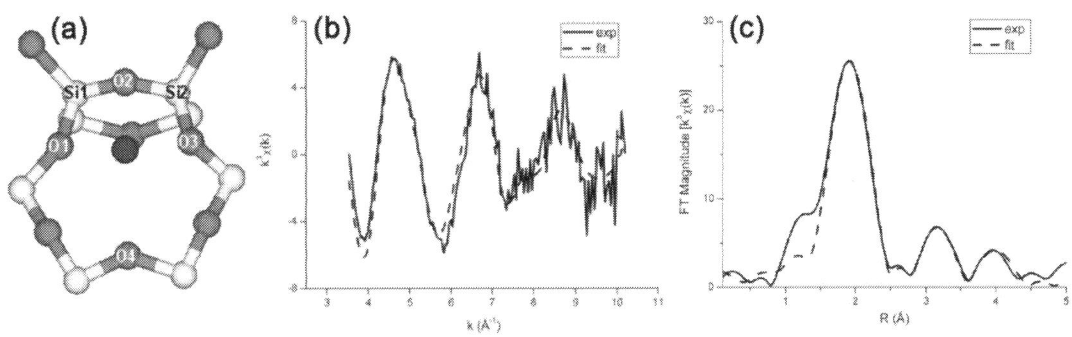


**Figure 2.** (a) DFT model of Cu in SSZ-13 6-ring employed for the fitting; (b) experimental and fitted $k^3$-weighted EXAFS signal of the Cu-SSZ-13 calcined at 500 °C and (c) the modulus of its Fourier transform corrected by Cu-O phase. Adapted with permission from Korhonen *et al.* [16]. Copyright (2011) The Royal Society of Chemistry.

The fit resulted in adequate agreement with the experiment, confirming that Cu is indeed shifted from the center of the 6-ring (figure 2b,c). The very substantial displacement of one of the oxygens in the 6-ring from the initial position during fitting was explained by the lower degree of deformation of the ring compared to the one predicted by DFT. First shell coordination numbers were determined as 4 for hydrated material and 3.2 for the calcined one. Presented XANES results also indicated the decrease of coordination upon dehydration due to the decrease of the main maximum (the white line). Later, the shift of the Cu ion from the center of the 6-ring was also confirmed by Deka *et al.*, by means of XANES simulations [18]. Calculated spectrum of the model with Cu placed off-axis resulted in a much better agreement with the experimental data compared to the centered model.

Hydration issue was raised in more detail in the work by McEwen *et al.* where it was demonstrated that Cu K-edge XANES spectrum of the as-synthesized Cu-SSZ-13 zeolite is very similar to the one of aqueous solution of Cu ions [19]. This was further confirmed by Borfecchia *et al.*, who presented also a comparison of the EXAFS data for these two cases [6]. High degree of similarity of both XANES and EXAFS data suggests that as long as the zeolite is exposed to air at room temperature, copper is covered by a shell of water molecules. Such shielding explains the lack of the framework contribution to the EXAFS data of the hydrated material and confirms the initial hypothesis of Fickel and Lobo [15]. At the same time, a substitution of one $H_2O$ molecule by $OH^-$ group cannot be excluded since these species are hardly distinguishable in EXAFS.

One of the most extensive spectroscopic studies of the activated Cu-SSZ-13 to date was carried out by Borfecchia *et al.* [6]. EXAFS-optimized experimental setup and rather high copper content in the sample (Cu/Al=0.44, Si/Al=13.1) allowed to achieve a very good data quality paving the way to a detailed quantitative analysis. Comparison of both XANES and EXAFS spectra of hydrated material with the data for the zeolite activated in oxygen and in helium revealed the significant differences of these three states (figure 3).

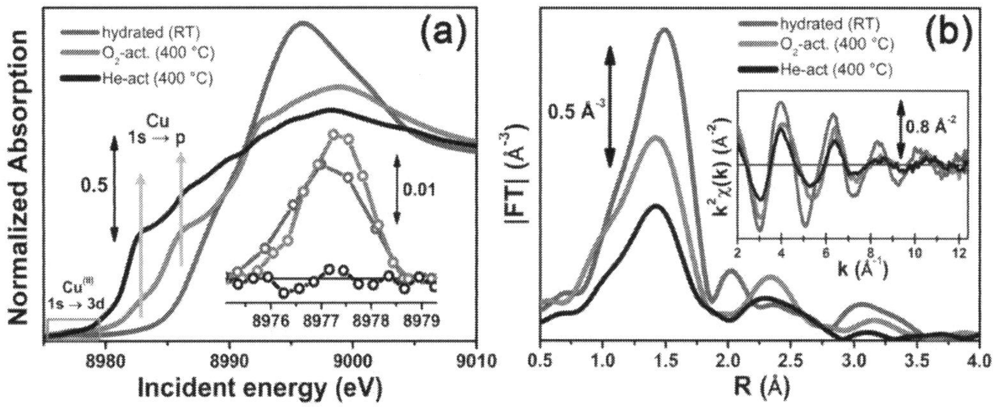


**Figure 3.** *In situ* static XAS data of the hydrated (RT), $O_2$-activated and He-activated (400 °C) Cu-SSZ-13 catalyst. (a) Cu K-edge XANES spectra; the inset shows the background-subtracted pre-edge peak highlighted by the orange box in the main panel. (b) Magnitude of the phase-uncorrected FT EXAFS spectra, obtained by transforming in the (2.4–12.4) Å$^{-1}$ range the k$^2$-weighted $\chi(k)$ curves reported in the inset. Adapted with permission from Borfecchia *et al.* [6]. Copyright (2015) The Royal Society of Chemistry.

It was confirmed that, upon heating in $O_2$, Cu(II) centers undergo progressive dehydration, while interacting more closely with the framework, with no significant modification of their oxidation state. Features typical for Cu(II) in low-symmetry environment were observed in XANES (figure 3a), while EXAFS witnessed the marked decrease of the first shell intensity due to the loss of the coordinated water molecules (figure 3b). Similar evolution of the XANES spectra upon activation in $O_2$ was reported also by Kwak *et al.* [20]. Conversely, as already observed for other zeolitic frameworks [21-28], upon activation in vacuum or in inert atmosphere (e.g. helium) the Cu oxidation state did change to +1, as

Journal of Physics: Conference Series **712** (2016) 012041 doi:10.1088/1742-6596/712/1/012041

evidenced by the disappearance of 1s→3d transition (figure 3a, inset) and by the additional redshift of the edge. Most interestingly, high-quality EXAFS data reveal that the coordination of Cu upon He-activation was further decreased compared to the activation in $O_2$. Coupled with the observation that the reduction in He flow appears only at high temperature (T>250 °C), while at lower T the evolution of the spectra is identical to the $O_2$-activation case, it indicates that a charged extra-ligand is still coordinated to Cu even at high temperature in case of $O_2$-activation.

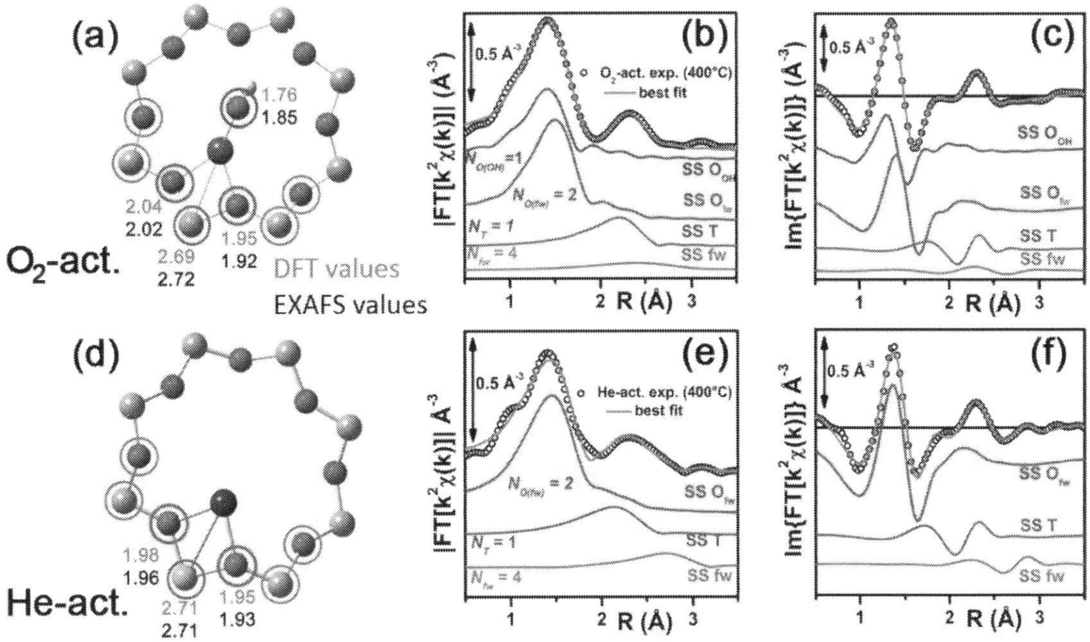


**Figure 4.** (a) DFT-optimized model of Cu in 8-ring that yielded the best agreement with the EXAFS data for the $O_2$-activated Cu-SSZ-13. Atom color code: Cu – green; H – white; O – red; Al – orange, Si – gray; Distances from Cu to the neighboring atoms are given in Å in grey (DFT values) and black (EXAFS values). Colored circles highlight the different coordination shells included in the fitting model; (b and c) experimental EXAFS spectrum of the $O_2$-activated Cu-SSZ-13 and corresponding best fit performed with the model from panel (a); modulus and imaginary part of the FT are shown in panels (b) and (c) respectively, together with the contributions of different shells, drawn with the same color code as in part (a) and translated vertically for the sake of clarity. (d, e, f) theoretical and experimental data for the case of He-activated Cu-SSZ-13, analogous to (a, b, c), respectively. Adapted with permission from Borfecchia *et al.* [6]. Copyright (2015) The Royal Society of Chemistry.

The authors have further tested this hypothesis by performing a set of DFT simulations of Cu ions in different locations in the framework and using the resulting structures as input for EXAFS fits and XANES and XES simulations. The best overall agreement with the experimental data was obtained for the models of Cu in the 8-ring. While in case of He-activation it was a bare Cu(I) cation (figure 4 d-f), after $O_2$-activation the $OH^-$ ligand was found to be coordinated to the Cu(II) ion (figure 4 a-c), confirming the first assignment of the ν(OH) stretching mode at 3657 $cm^{-1}$, done by Giordanino *et al.* [29]. Interestingly, the seeming mismatch between the DFT and EXAFS data observed in the pioneering XAS work by Korhonen *et al.* might be explained. It appeared that the dramatic elongation of the distance between Cu and the "opposite" oxygen in the 6-ring observed by EXAFS led to an antiphase effect, which made the results of the fits in the 8-ring and in 6-ring very similar. Although the presence of the destructive interference between EXAFS paths cannot be automatically considered as a sign of fitting problem being sometimes a characteristic feature of the material, more careful analysis allowed to state that in this case it was a purely mathematical effect and the fit of the $Cu^{2+}(OH)^-$ structure in 8-

ring was much more robust. The presence of the $OH^-$ groups coordinated to Cu was further supported by the FTIR results presented in the same publication [6] and confirmed by the EPR analysis of Godiksen *et al.* [30], while the Cu ions in the 8-ring position were detected also by high-resolution X-ray diffraction by Andersen *et al.* [31].

One of the highly debated questions in the Cu-zeolites community is the possibility to form small Cu oxo-clusters, usually dimers, inside the zeolitic pores under certain conditions. Detection of such species was reported in many studies [32-39]. However, generally the authors clearly state, that these are minority species. Such conclusion was reached, e.g. by Woertnik *et al.*, by means of resonant Raman for ZSM-5 [38] and by Godiksen *et al.* by EPR for SSZ-13 [30]. Since EXAFS always delivers the average picture, such low-abundance species are beyond the detection limit of this technique and the possibility of their existence cannot be neither ruled out nor confirmed from EXAFS data. Conversely, having Cu oxo-clusters as dominant species in Cu-SSZ-13 is very unlikely. In principle, it may be possible to imagine a hypothetic structure of a dimer o trimer, where several scattering paths cancel each other due to the antiphase effect and yield an EXAFS pattern with Cu-Cu contribution at around 2.3 Å (phase uncorrected) compatible with experimental data. However for Cu-SSZ-13, to the best of our knowledge, no comprehensive proof constituting a thorough EXAFS analysis complemented by other spectroscopic techniques (XANES, XES, UV-Vis) and theoretical calculations to support this argument has been presented so far. Contrarily, it was demonstrated that the feature at around 2.3 Å could be fitted very well considering the scattering from the framework Al in much simpler models with isolated Cu ions. DFT calculations based on these structural models were able to reproduce also the Cu K-edge XANES and Cu valence-to-core XES spectra. Being in agreement also with UV-Vis data, these structural models seem much more convincing as dominant species in Cu-SSZ-13 [6, 16, 18, 29].

## 3. Understanding the SCR mechanism

Since one of the most important goals of the Cu-SSZ-13 characterization is the understanding of the SCR reaction mechanism, a lot of efforts were invested into measuring X-ray absorption and emission spectra of this material in different reaction-relevant conditions. In the whole variety of the performed experiments, two major approaches can be distinguished. One comprises the experiments in truly *operando* conditions, when the sample is exposed to the complete reaction mixture (including NO, $NH_3$, $O_2$, He, $H_2O$, in some studies also $NO_2$ and $CO_2$) under controlled temperature. Most often such studies are performed in plug-flow reactors, such as the one described in the work of Kispersky *et al.* [40].

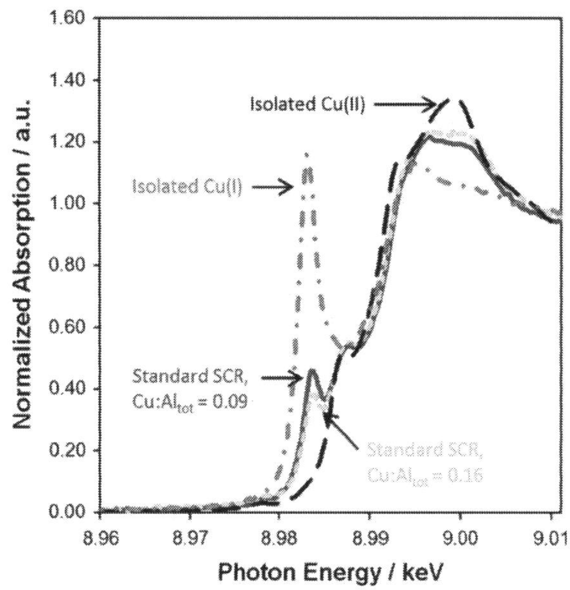


**Figure 5.** *Operando* Cu K-edge XANES spectra of Cu-SSZ-13 samples with Cu:Al ratio of 0.09 and 0.16 under standard SCR reaction conditions compared with the spectra of Cu-SSZ-13 with Cu:Al ratio of 0.02 collected upon the reaction with 1000 ppm $NH_3$ + 1000 ppm NO (isolated Cu(I)) and calcination in 10% $O_2$ (isolated Cu(II) references), both at 200 °C. The standard SCR conditions used were 320 ppm NO, 320 ppm $NH_3$, 10% $O_2$, 8% $CO_2$, 6% $H_2O$, and balance helium at 180 °C. Reproduced with permission from Bates *et al.* [41]. Copyright (2014) Elsevier.

Interpretation of such data is intrinsically complicated, since several different Cu species forming on the different stages of the reaction contribute to the signal. The common approach is to measure also the spectra of relevant reference materials, and, performing linear combination analysis, determine the relative contribution of each species to the spectrum of the catalyst under SCR conditions. Such strategy was adapted in the works of McEwen *et al.* [19] and Bates *et al.* [41], where samples with different Cu loadings were tested. Fitting the data with the Cu(I) and Cu(II) standards allowed to determine the ratio between the amounts of Cu species of these oxidation states during the reaction (figure 5).

The second strategy is to probe separately different stages of SCR reaction thus testing the independently developed hypotheses regarding the behavior of the active sites in particular conditions. Such approach allowed Janssens *et al.* to decouple the oxidation and reduction stages of the suggested SCR cycle [7]. In particular, the authors have tested the reducing capability of different agents, namely NO, $NH_3$ and their mixture (figure 6a). The results clearly indicate that almost complete reduction occurs only after exposure to the mixture of NO and $NH_3$. The resulting spectrum is characteristic for Cu(I) species in linear geometry [5]. Conversely, NO alone is not able to reduce Cu(II). Exposure to $NH_3$ leads only to a partial reduction accompanied by the formation of a linear Cu complex, while around 75% of Cu species remain oxidized, presumably forming tetrammine complexes (figure 6b,c), as observed by Janssens *et al.* [7]. Series of such *in situ* experiments with the components of SCR mixture allowed the authors to develop a feasible mechanism of the SCR reaction (scheme 1), where a key role is played by the nitrite-nitrate equilibrium, which has to be investigated in more detail in future spectroscopic studies.

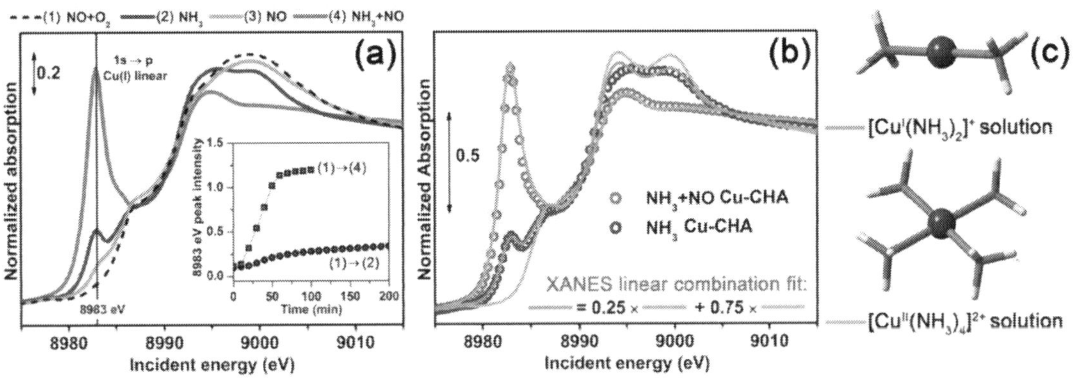


**Figure 6.** (a) *In situ* Cu K-edge XANES demonstrating the reducing capability at 200 °C of 1200 ppm $NH_3$ (2: blue curve), 1000 ppm NO (3: orange curve), and a mixture of 1200 ppm $NH_3$ + 1000 ppm NO (4: red curve) on the Cu(II) state obtained after initial oxidation in a mixture of 1000 ppm NO + 10% $O_2$ (a: dashed black curve). Inset: development of the intensity at 8983 eV with time with $NH_3$ only (1 → 2) and with a mixture of $NH_3$ + NO (1 → 4), visualizing the different reduction behavior with time in these cases. (b) Cu K-edge XANES spectra of the $[Cu^I(NH_3)_2]^+$ and $[Cu^{II}(NH_3)_4]^{2+}$ amino complexes (orange and green lines) compared with the data for Cu-SSZ-13 exposed to 1200 ppm $NH_3$ and 1200 ppm $NH_3$ + 1000 ppm NO at 200 °C (empty circles). Light-blue solid line corresponds to a linear combination of 25% $[Cu^I(NH_3)_2]^+$ and 75% $[Cu^{II}(NH_3)_4]^{2+}$. (c) Models of $[Cu^I(NH_3)_2]^+$ and $[Cu^{II}(NH_3)_4]^{2+}$ amino complexes. Color code: Cu – green; N – blue; H – white. Adapted with permission from Janssens *et al.* [7]. Copyright (2015) American Chemical Society.

Similar strategy of probing particular intermediate states of the reaction was adapted by Günter *et al.* who used XES and high-energy resolution fluorescence detected (HERFD) XANES for characterizing different stages of SCR reaction [42]. It was possible to detect the formation of Cu-N bond upon coordination of $NH_3$ monitoring the shift of Kβ" satellite from 8956.9 to 8958.3 eV (figure 7). Notably, no such shift was observed in the cases when $NH_3$ was absent from the feed, indicating thus the very weak coordination of NO to Cu accompanied by minor changes in Cu local environment.

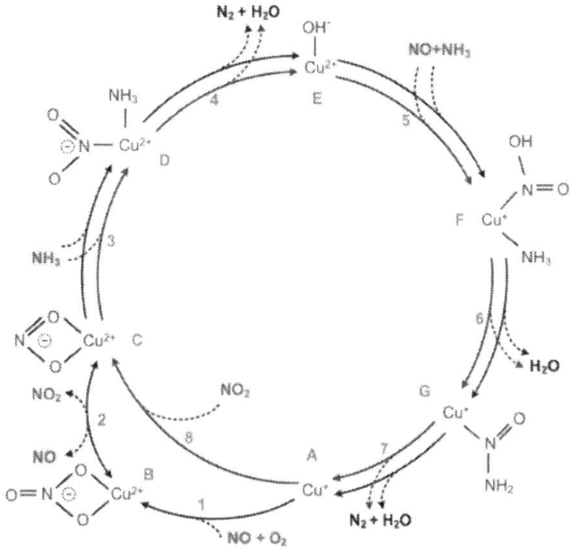

**Scheme 1.** Possible mechanism of SCR reaction. The fast SCR cycle is shown in blue, the NO activation cycle - in black. Reactants are indicated in red, reaction products - in black, and the $NO_2$ intermediate - in green. Reproduced with permission from Janssens et al. [7]. Copyright (2015) American Chemical Society.

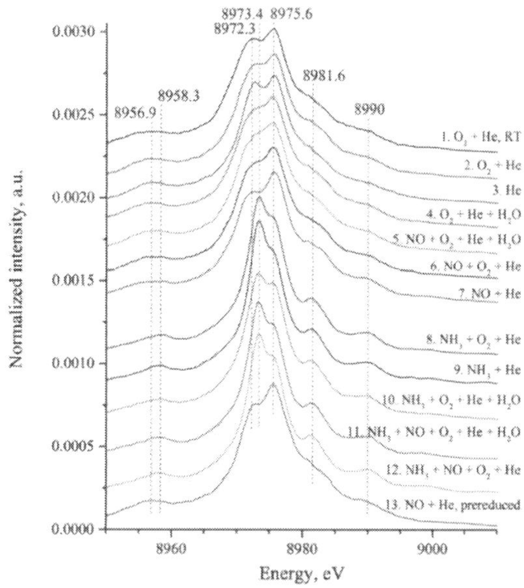

**Figure 7.** Cu valence-to-core ($K\beta_{2,5}$ and $K\beta''$) XES spectra of Cu-SSZ-13 catalyst in the course of the $NH_3$-SCR reaction and under exposure to the related model gas mixtures at 200 °C. Reproduced with permission from Günter et al. [42]. Copyright (2015) The Royal Society of Chemistry.

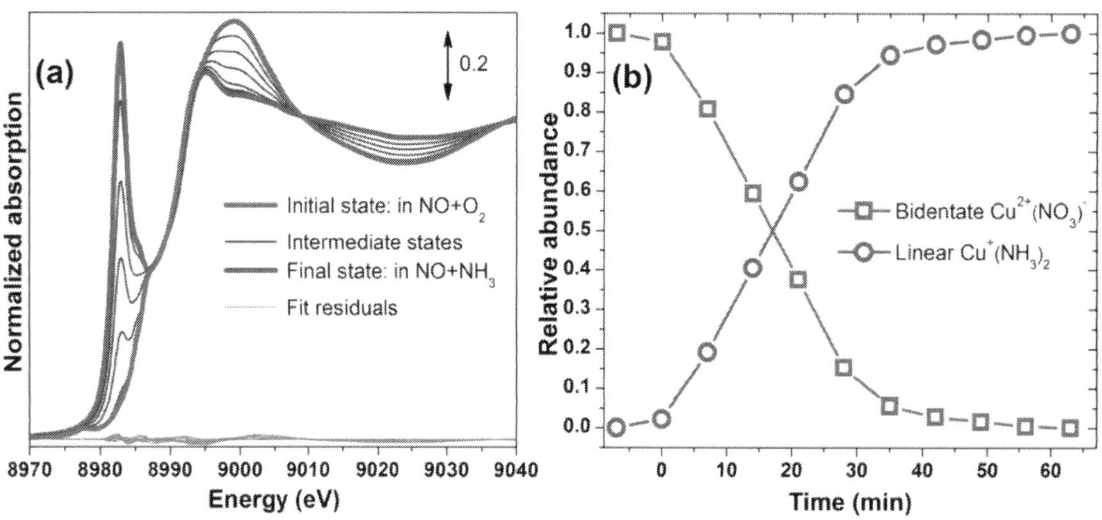


**Figure 8.** (a) Evolution of Cu K-edge XANES spectra of Cu-SSZ-13 after switching from 1000 ppm NO + 10% $O_2$ flux (initial state) to 1000 ppm NO + 1200 ppm $NH_3$ flux (intermediate states and final state) at 200 °C (b) Relative abundance of $Cu^{2+}(NO_3)^-$ bidentate nitrate and linear $Cu^+(NH_3)_2$ complex after changing the flux composition from NO+$O_2$ to NO+$NH_3$. The moment when the fluxes were switched and the measurement of the first "intermediate" spectrum was initiated corresponds to time 0. Each pair of points in panel (b) was obtained performing a linear combination fit of the corresponding intermediate spectrum presented in panel (a) considering the spectra of the initial and final states as standards. Unscaled residuals of all fits are presented in the bottom panel (a).

Two above mentioned approaches can certainly be combined in order to reach a better understanding of the SCR process. In particular, XANES linear combination studies at SCR conditions can be performed using as standards not the spectra of reference materials, but those of Cu-SSZ-13, collected at stable conditions, where the forming species are well-characterized. In such way more accurate quantitative data on the structure of active sites under real working conditions can be obtained. An *in situ* study of stored Cu-nitrates [7] reduction by $NH_3$ and NO in Cu-SSZ-13 presented in figure 8 serves to illustrate the possibilities of such approach. To the best of our knowledge these kind of studies in SCR conditions have not been reported yet, but surely will appear in the literature in the nearest future.

Notably, with the development of specialized computer codes, it has become possible to analyze the XANES data not only comparing the spectra with those of model compounds, but also simulating them directly for a given structural model. Due to the high porosity of the structure and significant anisotropy of Cu local environment, in case of Cu-SSZ-13 full potential codes are likely to be preferable to those that use the muffin-tin approximation and show excellent results for densely-packed systems [43]. The main drawback of full-potential codes is that usually they are up to two orders of magnitude slower than muffin-tin analogues, such as the most widely used FEFF code [44], and therefore require considerable computational resources. However, recently one of the mature full potential codes, FDMNES [45], was considerably sped-up due to the use of sparse matrices and dedicated solvers [46]. Decrease of calculation time by an order of 40 allowed to carry out studies which before were hardly feasible without huge computational clusters. To illustrate this, figure 9 summarizes the results of a structural investigation of the $[Cu(NH_3)_2]^+$ formed in CHA cavity upon the exposure of the material to the mixture of $NH_3$ and NO (figure 9a). The simulated spectrum of the DFT model showed an underestimated gap between the first two XANES maxima and therefore the Cu–N distances were decreased stepwise from the DFT value of 1.92 Å (figure 9b). The best agreement with the experiment was reached at Cu–N distance of 1.82 Å, which is significantly shorter compared to the DFT value (figure 9c). Total calculation time for each of the reported spectra was around 24 hours, which would mean more than a month and twice as much memory with the old version of FDMNES, clearly driving the current task out of the range of feasibility.

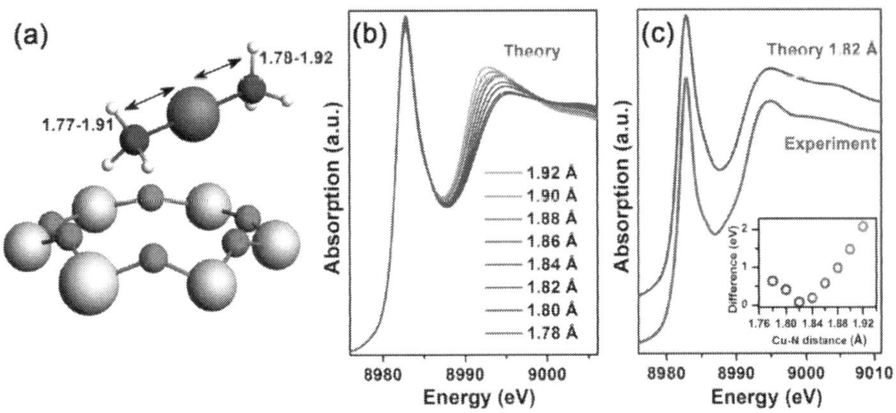


**Figure 9.** (a) Fragment of the structural model of $[Cu(NH_3)_2]^+$ complex in the large CHA cavity above the 6-ring. The variation range of the distances is shown in Å; (b) Theoretical spectra for different Cu–N distances; (c) Comparison of the best simulated spectrum to the experimental data collected at 200 °C in 1200 ppm $NH_3$ + 1000 ppm NO atmosphere. The inset shows the discrepancy of the experimental and theoretical values of the gap between the first two XANES maxima as a function of Cu–N distance.

## 4. Conclusion

Due to the efforts of many research groups, in the past five years there has been a marked advance in the understanding of the structure and reactivity of the active sites in Cu-SSZ-13 catalyst. As illustrated in this review, significant part of the insights were gained by the element-selective XAS/XES studies, complemented by laboratory techniques and theoretical calculations. Since the first mechanisms of SCR reaction catalyzed by Cu-SSZ-13 were proposed [47-49], new XAS insights allowed to discard several seemingly plausible paths, such as $Cu^{2+}$ reduction by NO alone, to characterize the active sites in the activated material and to shed light on the structure of the intermediate species. It allowed to develop a consistent scheme of the catalytic cycle respecting the stoichiometry and charge conservation, avoiding at the same time adsorption and desorption of molecular fragments and isolated ions [7]. Nonetheless, there is undoubtedly still a lot to discover in the further studies, particularly (but not exclusively) exploring the formation of Cu nitrates and nitrites, checking the influence of Cu/Al and Si/Al ratios on the structures that form in Cu-SSZ-13 at different conditions, quantifying the Cu speciation in SCR conditions at different temperatures. We foresee that the role of XAS and XES spectroscopy in solving these open questions will be of particular importance.

**Acknowledgments**
The authors are strongly indebted to the Haldor Topsøe team involved in the SCR catalysis (T.V.W. Janssens, H. Falsig, L.F. Lundegaard, P.N.R. Vennestrøm, S.B. Rasmussen, P.G. Moses), to A. Godiksen and S. Mossin (from DTU) and to F. Giordanino (Turin University), for an outstanding 4-years collaboration on the subject. K.A.L., C.L. and A.V.S. acknowledge the megagrant of the Russian Federation Government to support scientific research at the Southern Federal University, no. 14.Y26.31.0001.

**References**
[1] Zones S I 1985 U.S. Patent 4544538
[2] Kwak J H, Tonkyn R G, Kim D H, Szanyi J and Peden C H F 2010 *J. Catal.* **275** 187-90
[3] Kwak J H, Tran D, Burton S D, Szanyi J, Lee J H and Peden C H F 2012 *J. Catal.* **287** 203-9
[4] Beale A M, Gao F, Lezcano-Gonzalez I, Peden C H F and Szanyi J 2015 *Chem. Soc. Rev.* **44** 7371-405
[5] Giordanino F, Borfecchia E, Lomachenko K A, Lazzarini A, Agostini G, Gallo E, Soldatov A V, Beato P, Bordiga S and Lamberti C 2014 *J. Phys. Chem. Lett.* **5** 1552-9
[6] Borfecchia E, Lomachenko K A, Giordanino F, Falsig H, Beato P, Soldatov A V, Bordiga S and Lamberti C 2015 *Chem. Sci.* **6** 548-63
[7] Janssens T V W *et al.* 2015 *ACS Catal.* **5** 2832-45
[8] Doronkin D E, Casapu M, Günter T, Müller O, Frahm R and Grunwaldt J D 2014 *J. Phys. Chem.* C **118** 10204-12
[9] Boubnov A, Carvalho H W P, Doronkin D E, Günter T, Gallo E, Atkins A J, Jacob C R and Grunwaldt J D 2014 *J. Am. Chem. Soc.* **136** 13006-15
[10] Bordiga S, Groppo E, Agostini G, van Bokhoven J A and Lamberti C 2013 *Chem. Rev.* **113** 1736-850
[11] Van Bohkoven J A and Lamberti C 2016 X-ray absorption and emission spectroscopy for catalysis *X-Ray Absorption and X-ray Emission Spectroscopy: Theory and Applications* vol 2 eds van Bokhoven J A and Lamberti C (New York: John Wiley & Sons) pp 353-83.
[12] Soldatov A V and Lomachenko K A 2016 Nanostructured Materials *X-Ray Absorption and X-ray Emission Spectroscopy: Theory and Applications* vol 2 eds van Bokhoven J A and Lamberti C (New York: John Wiley & Sons) pp 809-27.
[13] Pluth J J, Smith J V and Mortier W J 1977 *Mat. Res. Bull.* **12** 1001-7
[14] Bull I, Boorse R S, Jaglowski W M, Koermer G S, Moini A, Patchett J A, Xue W M, Burk P, Dettling J C and Caudle M T 2008 U.S. Patent 20080226545
[15] Fickel D W and Lobo R F 2010 *J. Phys. Chem.* C **114** 1633-40
[16] Korhonen S T, Fickel D W, Lobo R F, Weckhuysen B M and Beale A M 2011 *Chem. Commun.* **47** 800-2
[17] Pierloot K, Delabie A, Groothaert M H and Schoonheydt R A 2001 *Phys. Chem. Chem. Phys.* **3** 2174-83

[18] Deka U, Juhin A, Eilertsen E A, Emerich H, Green M A, Korhonen S T, Weckhuysen B M and Beale A M 2012 *J. Phys. Chem.* C **116** 4809-18
[19] McEwen J S, Anggara T, Schneider W F, Kispersky V F, Miller J T, Delgass W N and Ribeiro F H 2012 *Catal. Today* **184** 129-44
[20] Kwak J H, Varga T, Peden C H F, Gao F, Hanson J C and Szanyi J 2014 *J. Catal.* **314** 83-93
[21] Cheung T, Bhargava S K, Hobday M and Foger K 1996 *J. Catal.* **158** 301-10
[22] Keith Hall W and Valyon J 1992 *Catal. Lett.* **15** 311-5
[23] Jang H J, Keith Hall W and d'Itri J L 1996 *J. Phys. Chem.* **100** 9416-20
[24] Li Y and Keith Hall W 1991 *J. Catal.* **129** 202-15
[25] Sarkany J, d'Itri J L and Sachtler W M H 1992 *Catal. Lett.* **16** 241-9
[26] Valyon J and Keith Hall W 1993 *J. Phys. Chem.* **97** 1204-12
[27] Turnes Palomino G, Fisicaro P, Bordiga S, Zecchina A, Giamello E and Lamberti C 2000 *J. Phys. Chem.* B **104** 4064-73
[28] Llabrés i Xamena F X, Fisicaro P, Berlier G, Zecchina A, Turnes Palomino G, Prestipino C, Bordiga S, Giamello E and Lamberti C 2003 *J. Phys. Chem.* B **107** 7036-44
[29] Giordanino F, Vennestrom P N R, Lundegaard L F, Stappen F N, Mossin S, Beato P, Bordiga S and Lamberti C 2013 *Dalton Trans.* **42** 12741-61
[30] Godiksen A, Stappen F N, Vennestrom P N R, Giordanino F, Rasmussen S B, Lundegaard L F and Mossin S 2014 *J. Phys. Chem.* C **118** 23126-38
[31] Andersen C W, Bremholm M, Vennestrom P N R, Blichfeld A B, Lundegaard L F and Iversen B B 2014 *IUCrJ* **1** 382-6
[32] Alayon E M C, Nachtegaal M, Bodi A and van Bokhoven J A 2014 *ACS Catal.* **4** 16-22
[33] Ene A B, Bauer M, Archipov T and Roduner E 2010 *Phys. Chem. Chem. Phys.* **12** 6520-31
[34] Groothaert M H, Smeets P J, Sels B F, Jacobs P A and Schoonheydt R A 2005 *J. Am. Chem. Soc.* **127** 1394-5
[35] Groothaert M H, van Bokhoven J A, Battiston A A, Weckhuysen B M and Schoonheydt R A 2003 *J. Am. Chem. Soc.* **125** 7629-40
[36] Lamberti C, Spoto G, Scarano D, Paze C, Salvalaggio M, Bordiga S, Zecchina A, Palomino G T and Dacapito F 1997 *Chem. Phys. Lett.* **269** 500-8
[37] Grundner S, Markovits M A C, Li G, Tromp M, Pidko E A, Hensen E J M, Jentys A, Sanchez-Sanchez M and Lercher J A 2015 *Nat. Commun.* **6**
[38] Woertink J S, Smeets P J, Groothaert M H, Vance M A, Sels B F, Schoonheydt R A and Solomon E I 2009 *Proc. Natl. Acad. Sci. U. S. A.* **106** 18908-13
[39] Alayon E M C, Nachtegaal M, Bodi A, Ranocchiari M and van Bokhoven J A 2015 *Phys. Chem. Chem. Phys.* **17** 7681-93
[40] Kispersky V F, Kropf A J, Ribeiro F H and Miller J T 2012 *Phys. Chem. Chem. Phys.* **14** 2229-38
[41] Bates S A, Verma A A, Paolucci C, Parekh A A, Anggara T, Yezerets A, Schneider W F, Miller J T, Delgass W N and Ribeiro F H 2014 *J. Catal.* **312** 87-97
[42] Günter T, Carvalho H W P, Doronkin D E, Sheppard T, Glatzel P, Atkins A J, Rudolph J, Jacob C R, Casapu M and Grunwaldt J D 2015 *Chem. Commun.* **51** 9227-30
[43] Rehr J J and Albers R C 2000 *Rev. Mod. Phys.* **72** 621-54
[44] Rehr J J, Kas J J, Vila F D, Prange M P and Jorissen K 2010 *Phys. Chem. Chem. Phys.* **12** 5503-13
[45] Joly Y 2001 *Phys. Rev.* B **63** 125120
[46] Guda S A *et al.* 2015 *J. Chem. Theory Comput.* **11** 4512-21
[47] Gao F, Kwak J H, Szanyi J and Peden C H F 2013 *Top. Catal.* **56** 1441-59
[48] Kwak J H, Lee J H, Burton S D, Lipton A S, Peden C H F and Szanyi J 2013 *Angew. Chem. Int. Ed.* **52** 9985-9
[49] Paolucci C, Verma A A, Bates S A, Kispersky V F, Miller J T, Gounder R, Delgass W N, Ribeiro F H and Schneider W F 2014 *Angew. Chem. Int. Ed.* **53** 11828-33

Journal of Physics: Conference Series **712** (2016) 012042 doi:10.1088/1742-6596/712/1/012042

# CO adsorption and decomposition on Pd/$Al_2O_3$ studied by time-resolved XAFS using dispersive optics

**D Matsumura*[1], Y Okajima[2] and Y Nishihata[1]**

[1] Quantum Beam Science Center, Japan Atomic Energy Agency, 1-1-1 Koto, Sayo, Hyogo 679-5165, Japan

[2] SPring-8 Service Co. Ltd., 1-20-1 Shingu-cho Koto, Tatsuno, Hyogo 679-0077, Japan

E-mail: daiju@spring8.or.jp

**Abstract.**
We have studied surface structural change of supported Pd metal nanoparticles caused by CO adsorption and following temperature increase by using *in situ* and time-resolved X-ray absorption fine structure technique with dispersive optics. CO adsorption on Pd metal nanoparticles at room temperature shows two types of adsorption behavior and the less stable CO which brings about surface expansion of Pd metal nanoparticles is desorbed just after temperature increase. Pd carbide layer is created at around 350 °C by CO decomposition, which was clearly detected by the elongation of Pd-Pd interatomic distance and the increase of Debye-Waller factor and third cumulant.

## 1. Introduction

Supported Pd metal nanoparticles have been widely used as practical catalytic reactions such as hydrogenation reaction, cross-coupling reaction, and redox reaction of exhaust gas. Catalysis at surface often accompanies changes of surface structure of Pd metal nanoparticles [1, 2]. Because the surface structural change due to adsorption is relatively small against the bulk structural change, it is needed to build up high-stable observation system for obtaining information of change of surface structure when we adopt useful bulk-sensitive approach such as X-ray absorption fine structure (XAFS).

XAFS spectroscopy with dispersive optics enables us to observe XAFS spectra without mechanical movement at all [3, 4], which means that this technique has a potential of high-stable XAFS observation especially in the high energy region due to the low X-ray band width. We have applied dispersive XAFS system to the structure and shape changes of supported Pd metal nanoparticles [5, 6, 7]. Continuous observation with high sampling rate brings about large data sets, therefore the precise change of structural parameters such as interatomic distance, Debye-Waller factor ($C_2$) and third cumulant ($C_3$) can be revealed by the consistent analysis of large XAFS data sets [8].

In this study, we employed the *in situ* and time-resolved XAFS technique with dispersive optics for understandings of CO surface adsorption on Pd nanoparticles at room temperature and following temperature raising. Time-resolved XAFS observation by dispersive optics were performed for estimation of surface structural change of Pd metal nanoparticles. We have succeeded to reveal the CO surface adsorption and decomposition mechanism on Pd metal nanoparticles.

## 2. Experiments

Pd $K$-edge XAFS spectra were measured by the dispersive mode with Laue configuration at the bending magnet beamline BL14B1 and BL28B2 of SPring-8, Japan [5, 6]. The optics of both beamlines are similar. Dispersed X rays were obtained by Si(422) reflection plane. From the curvature with a radius of 2000 mm, X rays with an energy range of 1000 eV were generated. Samples were set to the focal point of X rays. $Gd_2O_2S$(Tb) was exposed to re-dispersed X-ray and emitted lights were collected using a charge coupled device (CCD) camera. The intensities in the vertical direction were summed up to produce a one-dimensional spectroscopy. All XAFS spectra were collected under *in situ* and time-resolved mode with a sampling rate of 2 Hz.

Supported Pd metal nanoparticles were prepared by the impregnation method with powdered $\gamma$-$Al_2O_3$ and dilute aqueous solution of palladium nitrate, $Pd(NO_3)_2$. Following drying and calcination at 500 °C, Pd(4 wt%)/$Al_2O_3$ sample was obtained.

100 mg of powdered Pd sample was set into a cylindrical sample holder ($\phi 10 \times \phi 7 \times 10$ mm$^3$) and pressed by hand in order to make a disk pellet ($\phi = 7$ mm). The sample holder with the pellet was placed in an *in situ* flow type XAFS cell [9]. Gas flow was checked by a quadrupole mass spectrometer whose capillary was directly attached to the *in situ* XAFS cell. Samples were reduced to complete metal state by $H_2$(10 %)/He of 50 cc/min flow at 400 °C for 10 min. After cooling to room temperature under He flow at 50 cc/min, CO(5 %)/He gas was flowed at 50 cc/min as an adsorption gas.

## 3. Results and Discussion

We show the result of time-resolved extended X-ray absorption fine structure (EXAFS) spectra as Fig. 1. 100 spectra of Pd(4 wt%)/$Al_2O_3$ taken in 50 s at room temperature are displayed. It is understood that there are no distinct distortions and scatters in the spectra set and all spectra were recorded under high precision with clear fine structures. Pd metal nanoparticles were kept metal state before flow of CO(5 %)/He gas. After CO flow, slight change in EXAFS functions is observed in the difference spectra (Fig. 1 (b)). Structure change of Pd metal nanoparticles is brought about by CO surface adsorption on Pd surface.

The differences in XAFS spectra between before and after CO adsorption at room temperature on Pd metal nanoparticles are summarized in Fig. 2. In these data sets, we can see the small difference in near edge structure, EXAFS function and Fourier transform intensity. CO adsorption is thought to influence surface atomic and electronic structure for Pd metal nanoparticles. However, in normal sense for spectra comparison, it is not easy to judge that CO

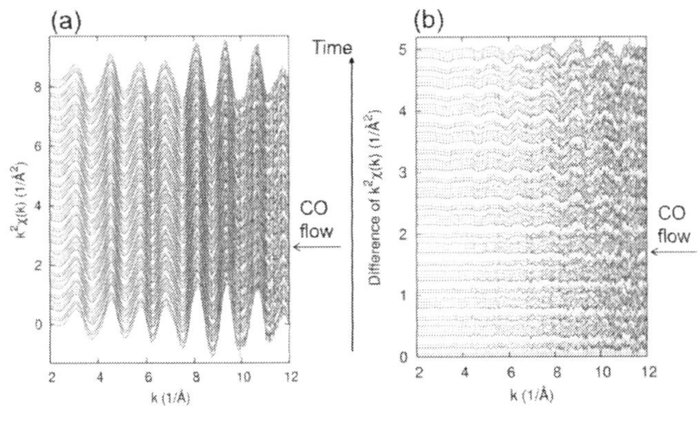


**Figure 1.** (a) Normal and (b) differential $k^2$-weighted EXAFS functions of Pd(4 wt%)/$Al_2O_3$ under 2 Hz observation mode at room temperature. 100 spectra taken in 50 s are displayed. The functions are shifted to vertical direction. CO(5 %)/He gas was started to dose at the position of arrows.

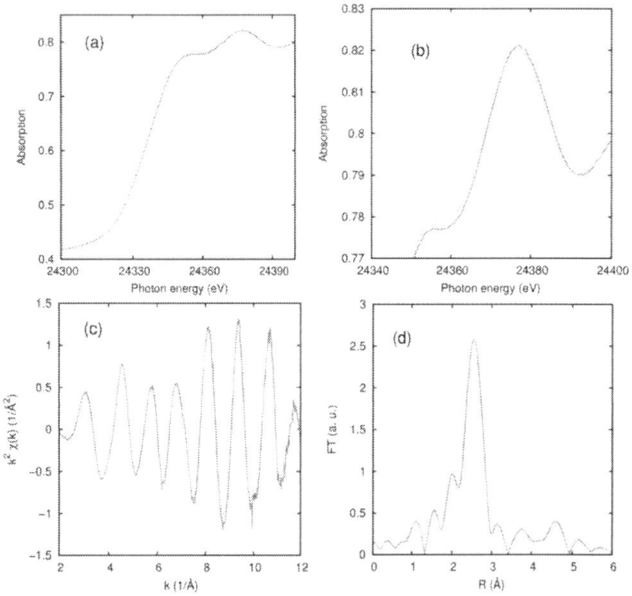


**Figure 2.** Pd $K$-edge (a) X-ray absorptoin near edge structure (XANES) spectra, (b) enlarged XANES spectra just after the absorption edge, (c) EXAFS functions and (d) Fourier transform intensities for Pd(4 wt%)/$Al_2O_3$ metal nanoparticles before (red solid) and after (green dashed) CO adsorption at room temperature. These data correspond to the first and last spectra in Fig. 1.

surface adsorption brings about surface structure change of Pd metal nanoparticles by taking only two spectra before and after CO adsorption into consideration because the difference in the spectra is so small that normal error value hinders the justification of the effect of CO surface adsorption. On the other hand, clear change in continuous XAFS spectra by CO surface adsorption at room temperature is recognized in Fig. 1. Continuous observation by time-resolved XAFS spectra enables us to detect small changes in metal nanoparticles.

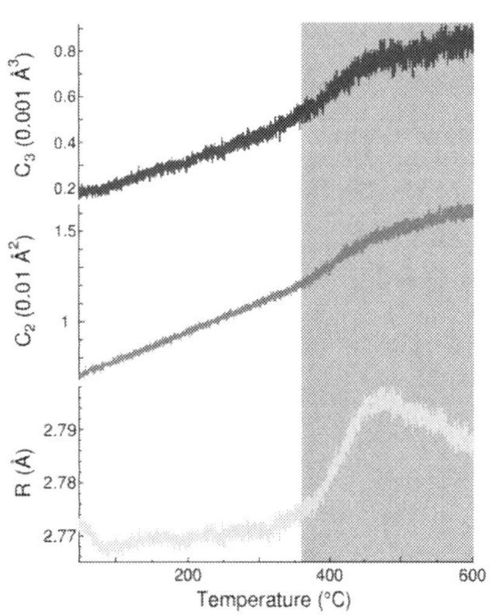


**Figure 3.** Changes of interatomic distance, $C_2$ and $C_3$ for Pd metal nanoparticles during temperature increase at 20 K/min under CO(5 %) flow.

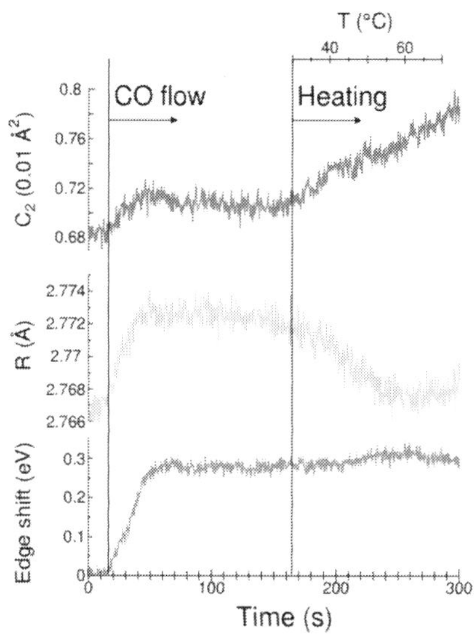


**Figure 4.** Temporal behavior of edge shift, interatomic distance and $C_2$ for Pd nanoparticles during CO adsorption at room temperature and following temperature increase.

In order to understand the details of CO adsorption on Pd nanoparticles, we employed time-resolved XAFS measurement with constant temperature ramp. The sample was heated at a rate of 20 K/min, and XAFS was continuously observed at 2 Hz. Interatomic distance, $C_2$ and $C_3$ for Pd-Pd nearest neighbor shell of Pd metal nanoparticles are plotted in Fig. 3. The $C_2$ and $C_3$ increase linearly by heating, indicating the increase of the thermal vibration and anharmonicity. Over 350 °C, which is shown in the hatched area of Fig. 3, small shifts are recognized in the $C_2$ and $C_3$, and at the same temperature region, the interatomic distance starts to increase. These increases of the interatomic distance, $C_2$ and $C_3$ indicate that the adsorbed CO molecules are decomposed and Pd carbide layer are created by interstitial solution of C atoms [5, 10].

Figure 4 shows the variations of XAFS parameters during CO surface adsorption and following temperature raising for the Pd metal nanoparticles with a sampling rate of 2 Hz. The relative precisions of the interatomic distance and the edge shift are about 0.001 Å and 0.02 eV, respectively. CO surface adsorption reaction at room temperature brings about elongation of interatomic distance and positive chemical shift of Pd nanoparticles. Only expanded interatomic distance recovers its original value by heating treatment. This means that there are two types of CO surface adsorption behavior at least. The less stable CO molecule causes surface expansion of Pd metal nanoparticles, which results increase of mean interatomic distance by about 0.007 Å. The CO is desorbed from surface of Pd metal nanoparticles by heating treatment.

## 4. Conclusions

Dispersive optics for time-resolved X-ray absorption fine structure spectroscopy are powerful tool for the observation of small structure change. The relative precisions of the interatomic distance and the edge shift at room temperature under 2 Hz monitoring reach 0.001 Å and 0.02 eV, respectively. There are at least two CO adsorption behaviors on Pd metal nanoparticles at room temperature and the less stable CO brings about surface expansion of Pd metal nanoparticles. Adsorbed CO shows decomposition and creates Pd carbide layer over 350 °C, which is clearly detected by the elongation of Pd-Pd interatomic distance and the increase of $C_2$ and $C_3$.

## Acknowledgments

We are grateful to Mr. M. Taniguchi of Daihatsu Motor Co., Ltd. for the sample preparation. We also thank Mr. H. Ikushima of Kwansei Gakuin University for experimental support. This work was supported by the Elements Science and Technology Project. This work was also supported by a Grant-In-Aid for Young Scientists B (25790083) from the MEXT. The present experiment was performed under the approval of JASRI (Proposal No. 2010B1205, 2011A3609, 2011A1369, 2011B3609, 2011B3610, 2012B1393, 2013A1432).

## References

[1] Newton M A, Belver-Colderia C, Martínez-Arias A and Fernández-García M 2007 *Nature Materials* **6** 528
[2] Iglesias-Juez A, Kubacka A, Fernández-García M, Michiel M D and Newton M A 2011 *J. Am. Chem. Soc.* **133** 4484
[3] Matsushita T and Phizackerley R P 1981 *Jpn. J. Appl. Phys.* **20** 2223
[4] Kaminaga U, Matsushita T and Kohra K 1981 *Jpn. J. Appl. Phys.* **20** 355
[5] Matsumura D, Okajima Y, Nishihata Y, Mizuki J, Taniguchi M, Uenishi M and Tanaka H 2009 *J. Phys.: Conf. Ser.* **190** 012154
[6] Matsumura D, Nishihata Y, Mizuki J, Taniguchi M, Uenishi M and Tanaka H 2010 *J. Appl. Phys.* **107** 124319
[7] Matsumura D, Okajima Y, Nishihata Y and Mizuki J 2011 *J. Alloys Compd.* **2011** S849
[8] Yokoyama T, Hamamatsu H and Ohta T 1998 *EXAFSH* The University of Tokyo 2nd ed
[9] Suzuki A, Yamaguchi A, Chihara T, Inada Y, Yuasa M, Abe M, Nomura M and Iwasawa Y 2004 *J. Phys. Chem. B* **108** 5616
[10] Newton M A, Michiel M D, Kubacka A and Fernández-García M 2010 *J. Am. Chem. Soc.* **132** 4540

# Reduction process of Pd-containing La-Fe perovskite-type oxides by in-situ Dispersive X-ray absorption spectroscopy

**T Uchiyama[1], K Kamitani[2], K Kato[3] and M Nishibori[4]**

[1]Department of Molecular and Material Sciences, Interdisciplinary Graduate School of Engineering Sciences, Kyushu University, Fukuoka, Japan

[2]Institute for Materials Chemistry and Engineering, Kyushu University, Fukuoka, Japan

[3]Japan Synchrotron radiation Research Institute, SPring-8, Hyogo, Japan

[4]Department of Energy and Materials, Faculty of Engineering Sciences, Kyushu University, Fukuoka, Japan

Email: nishibori.maiko.511@m.kyushu-u.ac.jp

**Abstract**. Reduction process of Pd-containing La-Fe perovskites was investigated by in-situ Pd K-edge dispersive X-ray absorption fine structure as well as mass spectroscopy. The prepared perovskite was characterized by a conventional X-ray absorption spectra to confirm the incorporation of cationic Pd into perovskite matrix. Under the reductive atmosphere (5 vol%$H_2$/He), we found the presence of three reduction processes of Pd cations in perovskite structure. The segregation of Pd metal particles was observed from 200-400 °C although the cationic Pd species remained at 700 °C due to the strong metal-support interaction.

## 1. Introduction

Over the past years, many researchers have investigated the catalytic properties of Palladium-supported/incorporated perovskite-type oxides for three-way catalytic reaction to remove pollutants in automobile emission [1]. Nishihata et al. reported $LaFe_{0.95}Pd_{0.05}O_3$ perovskite, which achieved incorporation of Pd into the perovskite matrix (B-site) as ionic state of $Pd^{n+}$($n \geq 2$) [2]. In this catalyst, Pd can segregate out of and reintegrate in the perovskite lattice under reductive and oxidative atmospheres, respectively. This self-regenerating property prevents Pd particles from sintering and enables to maintain their high catalytic activity with high metal dispersion. This research clearly indicates that the redox properties are one of the most important factors to determine the activity of three-way catalyst. Temperature-programmed reduction (TPR) techniques are commonly used to investigate the redox properties of metal oxides. $H_2$-TPR can be easily applied for a simple system like $MO_x/Al_2O_3$ (M = metal cation) because only one metal cation contributes to the reduction behavior. However, the results always become more difficult in a mixed metal oxide system since we cannot distinguish which cations were reduced based on the conventional TPR. Therefore, it is of great interest to investigate the redox properties for a particular element in the sample. X-ray absorption fine structure (XAFS) technique is one of the candidates for solving the problem as it can obtain information on local structure and electronic states of the target element at the same time. Here, we investigated the reduction process of Pd-containing LaFe perovskites by simultaneously conducting Pd K-edge XAFS and $H_2$-TPR with mass spectrometer. Dispersive XAFS (DXAFS) mode was applied for high speed observation.

## 2. Experimental

Pd-containing La-Fe perovskites ($LaFe_{0.95}Pd_{0.05}O_3$) were prepared by the amorphous citrate method. The aqueous solution of the metal nitrates and citric acid (chemicals and purity; $La(NO_3)_3 6H_2O$ > 95.0 %, $Fe(NO_3)_3 9H_2O$ > 99.0 %, 5%$Pd(NO_3)_2$ aq.solution, citric acid > 99 %) in a molar ratio of La: Fe: Pd: citric acid =1:0.95:0.05:3 were evaporated to dryness under stirring, followed by the calcination in air at 600 °C for 2 h. XRD patterns of the as-prepared perovskites (labeled as $LaFePdO_3$) were indexed with orthorhombic structure and no impurities were detected. Incorporation of Pd was also preliminarily confirmed by a conventional XAFS (Quick XAFS) at BL01B1 of SPring-8 using Si(311) for Pd, La K-edge and BL06 of SAGA-LS using Si(111) for Fe K-edge. In-situ Pd K-edge DXAFS experiments were carried out at BL28B2 of SPring-8[6]. DXAFS spectra were collected by a Si(422) bent crystal with the Laue-configuration polychromator and a position sensitive CCD detector (Hamamatsu Photonics K.K., 656(X) x 178(Y) pixels were used for our experiment). The intensities of the vertical (Y) direction were summed to produce a spectrum and those of the horizontal (X) direction were used for calibration plots to convert pixel into energy. The calibration plots were produced by comparing a conventional XAFS and DXAFS spectrum using Pd foil. The energy region that we have acquired was from 24.10 keV to 25.24 keV. A single frame was acquired at the exposure time of 240 ms. A spectrum used for analysis was obtained by accumulating 10 frames to improve counting statistics (2.4 s/spectrum). The samples were preliminarily treated at 350 °C for 10 min in 20 vol.%$O_2$/He and cooled to 60 °C under the same atmosphere. $H_2$-TPR were carried out from 60 to 700 °C at 10 °C/min using 5 vol.%$H_2$/He at 30 mL/min. The data analysis was processed by using ATHENA and ARTEMIS, a suite of IFEFFIT software programs[7].

## 3. Results and discussion

### *3.1. Incorporation of Pd into perovskite structure*

The local structure of Pd in the prepared material was analysed by a conventional Q-XAFS before starting DXAFS measurement. **Figure 1.** shows normalized Pd K-edge XANES spectra of $LaFePdO_3$ perovskites. The shoulder (24.340-24.345 keV) was clearly observed for PdO, confirming that $Pd^{2+}$ ions are in square planar coordination. For $LaFePdO_3$, XANES spectra couldn't be expressed on the basis of both Pd foil and PdO contributions, indicating the disappearance of PdO on $LaFePdO_3$. In contrast, peak in 24.367 keV (post-edge region) exhibited the formation of a Pd-containing oxide phase because the peak could also be observed for $LaPdO_3$[3]. In addition, the XANES spectra of $LaPdO_3$ and $LaFePdO_3$ were very similar reflecting the similar structure of both materials[3]. These results suggested that Pd-$LaFeO_3$ solid solutions were successfully achieved in $LaFePdO_3$ and Pd occupied distorted octahedral sites (Fe site) of $LaFeO_3$.

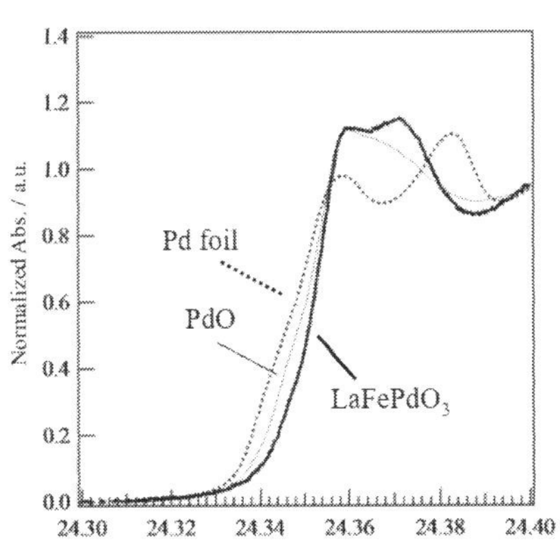

**Figure 1.** Pd K-edge XANES of $LaFePdO_3$

Furthermore, we compared EXAFS oscillations among the three cationic ions (La, Fe, Pd) in $LaFePdO_3$. The non-phase shift corrected $k^3$-weighted EXAFS spectra and their Fourier transformation for $LaFePdO_3$ were presented in **Figure 2**(A) and (B), respectively. EXAFS oscillations of PdO and $LaFePdO_3$ (@La K-edge, (c)) were completely different in both their amplitude and period from that of $LaFePdO_3$ (@Pd K-edge, (a)). On the other hand, $LaFePdO_3$ (@Pd K-edge, (a)) shared several oscillation features in $LaFePdO_3$ (@Fe K-

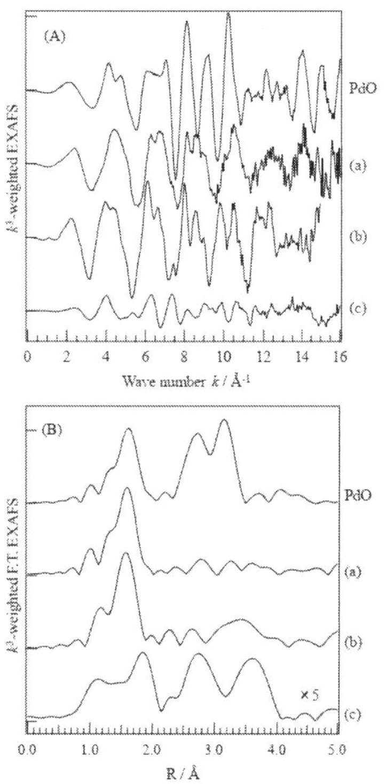


**Figure 2.** $k^3$-weighted Pd K-edge EXAFS (A) and their Fourier transformation (B) for PdO@Pd K-edge, (a) $LaFePdO_3$@Pd K-edge, (b) $LaFePdO_3$@Fe K-edge and (c) $LaFePdO_3$@La K-edge. F.T. range was fixed to 3-15 Å for all EXAFS.

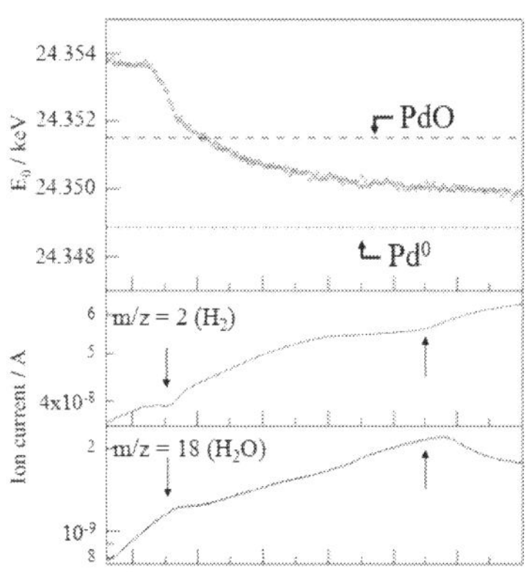


**Figure 3.** Temperature-dependent change of $E_0$ and mass spectrometer signal profile of $LaFePdO_3$ during reduction by hydrogen from 60 °C to 700 °C

edge, (b)) even though each oscillation period was slightly different due to the difference of phase shift and bond length.

As a result, $LaFePdO_3$ showed similar F.T. EXAFS in Pd K-edge (a) and Fe K-edge (b), reflecting the same local structure of these materials. As a consequence, we confirmed that cationic Pd of $LaFePdO_3$ occupies B-site of La-Fe perovskites.

### *3.2. Reduction behaviour of Pd under 5 vol.%$H_2$/He*

The reduction process of $LaFePdO_3$ was investigated by dispersive XAFS measurements. The dispersive Pd K-edge XANES spectra of $LaFePdO_3$ didn't show any shift after changing atmosphere from pretreatment gas ($O_2$) into $H_2$ at 60 °C. Therefore, most of cationic Pd species maintained the state and local structure even in this condition. **Figure 3** shows temperature dependence of XANES spectra and the absorption edge energy ($E_0$) change on $LaFePdO_3$ under 5 vol.%$H_2$/He. The ion current of mass spectrometer ($H_2$ (m/z = 2), $H_2O$ (m/z = 18)) against the temperature were also included. The figure indicated that $E_0$ was stable under 120 °C while a remarkable decrease was observed from 120-200 °C. Then, the position of $E_0$ exhibited between that of PdO and $Pd^0$ the entire temperature from 200 °C to 700 °C. This result means that a part of cationic Pd is still in perovskite lattice, and $Pd^0$ particles, which strongly interacting with support, were formed on the perovskite[4]. In terms of mass signals ($H_2$, $H_2O$), they showed the $H_2$ consumption (negative) and $H_2O$ evolution (positive) peaks at around 120-200 °C and 400-550 °C. The former reduction was coincided with the behavior of $E_0$ while the latter one was not. Even though the slow reduction was confirmed at around 550 °C in $E_0$, it might be difficult to observe as a peak, considering the variation of $E_0$. These results suggest that the reduction at around 400-550 °C is originated from the reduction of $Fe^{3+}$ in perovskite[3]. In particular, the reduction rate both cationic Fe and Pd became quite small and the total reaction seemed to be stopped over 550 °C. This caused the decease of water production and the absence of negative peak in $H_2$.

**Figure 4** shows non-phase shift corrected $k^2$-weighted DXAFS (a) and their Fourier transformation ($k$ range 3-10 Å) (b) for $LaFePdO_3$ under $H_2$-TPR. The EXAFS oscillations exhibited shift of the peak and its amplitude became small as the temperature increased due to the structural change and thermal factor. The peak at 1.5 Å in F.T. EXAFS corresponds to Pd-O shell. It remained over the whole temperature range, which had an agreement with XANES results. On the contrary, Pd-Pd shell at 2.5 Å was observed above 180 °C. In order to understand the structural changes, the local structure of Pd was estimated from EXAFS by curve fitting. We conducted $R$-space fitting using F.T. EXAFS (R range of 1-3.5 Å) with two shell model (Pd-O, Pd-Pd). The reaction with the host material such as alloying of Pd with Fe could not be observed in La-Fe-Pd perovskite system[4]. The fitting results were shown in Figure 4. for Pd-O and Pd-Pd shells. Due to the strong correlation between coordination number (C.N.) and thermal factor, Pd-Pd shell slightly under estimate the values. However, it was observed from 200 °C and became almost constant over 400 °C, whose temperature coincided with the end of a remarkable decrease $E_0$. This suggested that the supply of Pd from the bulk has stopped and particle growth was suppressed by strong metal-support interaction[4]. Besides, C.N. of Pd-O shell showed the constant values until Pd-Pd shell was appeared at 200 °C. This temperature was just under the reduction peak in mass signal at 120-200 °C and the temperature at the intersection of $E_0$ for $LaFePdO_3$ and PdO ($Pd^{2+}$) is just around 200 °C. Thus, the reduction at this temperature range was not the reduction of cationic Pd into Pd metal particles but into stable $Pd^{2+}$ species. Moreover, the constant C.N. in Pd-O indicated that $LaFePdO_3$ perovskite structure around Pd is stable below 200 °C in this condition even if the cationic Pd was reduced or segregated out from the lattice.

Considering these results, the reduction process of Pd species in Pd-incorporated La-Fe perovskites can be expected to show three reduction reactions; I (120-200 °C) Reduction of $Pd^{n+}$ (n = 3 or 4) into $Pd^{2+}$ species in perovskite matrix or PdO particles, II (200-400 °C) the simultaneous reduction of cationic Pd and Fe in perovskite structure (decomposition), III (550-700 °C) the reduction of $Pd^{2+}$ species and Fe in perovskite were almost ended.

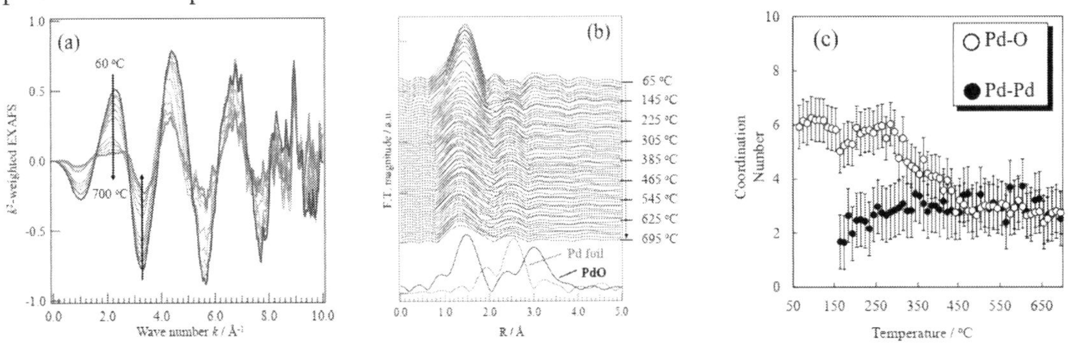


**Figure 4.** Temperature-dependent change of $k^2$-weighted Pd K-edge EXAFS, their Fourier transformation (B) and coordination number of Pd-O and Pd-Pd.

## 4. Conclusion

In-situ Pd K-edge DXAFS and mass spectrometry techniques enabled the simultaneous investigation on XAFS and TPR study. We identified the presence of three reduction processes of Pd on Pd-containing LaFe perovskites. Under 5 vol.%$H_2$/He, the segregation of Pd metal particles observed and Pd couldn't be completely reduced due to the strong metal-support interaction.


## Acknowledgements

The synchrotron radiation experiments were performed with the approval of the Japan Synchrotron Radiation Research Institute (JASRI) (Proposal No. 2014B1861, 2014B1768, 2015A1814). This work was partially supported by the Photon and Quantum Basic Research Coordinated Development Program from the Ministry of Education, Culture, Sports, Science and Technology, Japan.


## References


[1] Li X, Chen C, Liu C, Xian H, Guo L, Lv J, Jiang Z, Vernoux P 2013 *ACS Catalysis* **3**(6) 1071

[2] Nishihata Y, Mizuki J, Akao T, Tanaka H, Uenishi M, Kimura M, Okamoto T and Hamada 2002 Nature **418** 164

[3] Eyssler A, Mandaliev P, Winkler A, Hug P, Safonova O, Figi R, Weidenkaff A, Ferri D 2010 *Journal of Physical Chemistry C* **114**(10) 4584

[4] Katz M B, Graham G W, Duan Y, Liu H, Adamo C, Schlom D G, PAN X 2011 *Journal of the American Chemical Society* **133**(45) 18090

[5] Kato K, Uruga T, Tanida H, Yokota S, Okumura K, Imai Y, Irie T and Yamakata Y 2007 *AIP Conference Proceedings* **879** 1214

[6] Ravel B, Newville M 2005 *Journal of Synchrotron Radiation* **12** 537

# Geometric and electronic structure of Au on Au/$CeO_2$ catalysts during the CO oxidation: Deactivation by reaction induced particle growth

**Ali M. Abdel-Mageed[1], Gabriela Kučerová[1], Ayman Abd El-Moemen[2], Joachim Bansmann[1], Daniel Widmann[1], and R. Jürgen Behm[1,]***

[1]Institute of Surface Chemistry and Catalysis, Ulm University, D-89069 Ulm, Germany
[2]Chemistry Department, Faculty of Science, Suez Canal University, Ismailia, Egypt
*Corresponding author: juergen.behm@uni-ulm.de

**Abstract.** Changes of the geometric and electronic structure of gold on Au/$CeO_2$ catalysts induced by different pre-treatments (oxidative and reductive) and by the CO oxidation reaction at 80°C were followed by *operando* XANES / EXAFS measurements. The results showed that i) oxidative pre-treatment ($O_2$) leads to larger Au nanoparticles than reductive pre-treatment (CO), that ii) Au is predominantly metallic during CO oxidation, irrespective of the preceding pre-treatment, and that iii) there is a reaction induced Au particle growth. Correlations with the activity of the respective catalysts and its temporal evolution give insights into the origin of deactivation of these catalysts under reaction conditions, in particular on reaction induced changes in the Au particle size.

## 1. Introduction

Supported Au catalysts have attracted enormous interest in recent years because of their exceptionally high activity in a number of oxidation and reduction reactions [1]. One of the most prominent examples is the (preferential) CO oxidation reaction. Despite extensive efforts many important aspects of this reaction are still under debate, such as the active Au species or the physical origin of the pronounced deactivation during the reaction [1], which is relevant also for practical applications.

In the meantime there has been increasing evidence that the exact nature and working principle of Au catalysts depends strongly both on the nature of the support and the reaction and catalyst pre-treatment conditions [2,3]. While for CO oxidation on Au/$TiO_2$ at temperatures around 80 °C and higher Au is exclusively present in form of $Au^0$ nanoparticles (NPs), which are also the active Au species [3,4], this may be different for other supports and reaction conditions. Furthermore, detailed information on the geometric structure and on the electronic structure and state of the gold species in these catalysts has so far mainly been provided by *ex situ* techniques such as transmission electron microscopy (TEM) and X-ray photoelectron spectroscopy (XPS) measurements. Considering that these properties may be different under reaction conditions, *in situ* information is highly desirable.

For the CO oxidation on highly active Au/$CeO_2$ catalysts, previous results led to contradicting conclusions on the nature of the active Au species. While some groups claimed that gold in ionic form ($Au^+$, $Au^{3+}$) is predominantly responsible for the activity of the Au catalysts [5], other groups suggested that metallic $Au^0$ NPs represent the active species during reaction [6]. Likewise, deactivation of Au/$CeO_2$ catalysts during CO oxidation had been proposed to mainly originate from (i) effects associated with changes of the formal oxidation state of Au [5,7], (ii) the build-up of surface poisoning carbon containing species at the perimeter of the Au-support interface [8], and/or (iii) sintering of Au NPs during reaction [7,9]. Due to the considerable differences in the pre-treatment and reaction conditions in the different studies, a more general understanding of the active Au species and its role in the reaction and deactivation process can hardly be derived from these data.

We here present results of a combined kinetic and *in situ* XAS (XANES / EXAFS) study where we investigated the structural and electronic properties of the Au species in a Au/$CeO_2$ catalyst upon pre-treatment in either oxidative ($O_2/N_2$) or reductive ($CO/N_2$) atmosphere and during subsequent CO oxidation at 80°C, and their correlation with the catalyst activity and its evolution with time, to unravel the role of changes in the Au particle size and in its electronic state in the deactivation of the catalyst.

## 2. Experimental Section

The *operando* XANES / EXAFS measurements were performed on a highly dispersed 4.5 wt.% Au/$CeO_2$ catalyst with a surface area of 188 $m^2 \cdot g^{-1}$, at the XAFS beamline at the Elettra Synchrotron (Trieste) using a Si(111) double crystal monochromator [10] and a specially designed reaction cell [11]. The catalyst was prepared by a deposition-precipitation procedure described elsewhere [12]. EXAFS/XANES data were recorded before and after the pre-treatment as well as during reaction at the Au $L_3$-edge (11919 eV) in the fluorescence mode, using a Si drift diode detector (KETEK GmbH). The pre-edge region was measured from 11616 - 11885 eV, followed by the XANES (11885 - 11970 eV) and the EXAFS region (11970 - 12500 eV). The EXAFS spectra were evaluated using XDAP software. The theoretical references were calculated by FEFF 8.0, and calibrated by measured reference spectra (Au foil and $Au_2O_3$ powder)[13]. About 20 mg of the diluted catalyst (1:3 with $SiO_2$) were placed in the reaction cell (catalyst bed size: 2 x 10 mm). Next, the catalyst was pre-treated either in oxidative (O400: 10% $O_2$ in $N_2$) or in reductive (CO400: 10% CO in $N_2$) atmosphere. This included first heating up to 400°C in $N_2$, then holding the catalyst for 30 min in the respective gas mixture at 400°C, and afterwards cool-down in $N_2$ to the reaction temperature of 80°C (all at 1.0 atm). Subsequently, the CO oxidation was performed in an idealized gas mixture (1% CO, 1% $O_2$, 98% $N_2$) at 1 atm and 80°C. Kinetic measurements were performed under identical conditions.

## 3. Results and discussion

### *3.1. Influence of pre-treatment on the nature and dispersion of Au nanoparticles*

The influence of the pre-treatment on the nature of the Au species as well as on the initial metal dispersion was determined by recording XANES/EXAFS at the Au $L_3$-edge in $N_2$ at 80°C after the respective pre-treatments (see above) and in $N_2$ at 25°C for the fresh sample (N25). A comparison of the normalized XANE spectra (see Fig.1) clearly demonstrates that the fresh Au/$CeO_2$ catalyst shows almost identical features as the $Au_2O_3$ reference (inflection point: 11918 eV; peak: 11921 eV). Hence, it largely consists of ionic $Au^{3+}$ species. The strong oxide peak visible on the fresh sample, however, disappeared almost completely after CO400 pre-treatment, and instead a fully metallic $Au^0$ species is formed. After the O400 pre-treatment, in contrast, a significant fraction of oxidic species is still present, although to a much lower extent compared to the fresh sample (N25). These results agree well with EXAFS data obtained after the different pre-treatments. After CO400 pre-treatment, the spectrum can be fitted with only metallic Au-Au scatterers, while for O400 pre-treated catalyst it has to be fitted with a model containing both metallic (Au-Au) and oxidic (Au-O) shells (see Table 1 and Fig. 2) [14].

**Table 1.** Structural parameters for Au NPs obtained from EXAFS after O400 and CO400 pre-treatments at 80°C. CN: coordination number, DW: Debye-Waller factor, R: distance, $E_0$: energy shift.

| Pre-treatment | Scatterer | CN | DW / $10^{-3}$Å$^2$ | R / Å | $E_0$ / eV | Dispersion / % |
|---|---|---|---|---|---|---|
| O400 | Au-Au / Au-O | 6.3 / 0.5 | 9.7 / 1.0 | 2.84 / 1.96 | 0.65 / 19.7 | 54 |
| CO400 | Au-Au | 3.9 | 12.0 | 2.80 | -1.65 | 100 |

These findings confirm the presence of oxidic Au species after O400. Hence, the ionic gold species present on the fresh $Au/CeO_2$ catalyst sample are reduced to $Au^0$, either partly (O400) or completely (CO400). Furthermore, the pre-treatment procedure results in remarkable differences in the Au-Au CN and, hence, in the Au NP size, with the higher dispersion / smaller Au particle size for CO400. While the Au-Au CN is 6.3 after O400, it is only 3.9 after CO400. Assuming hemispherical Au NPs one can also calculate the corresponding dispersion, which is much higher after CO400 (100%) as compared to O400 (54%), and the Au particle size, which is much smaller after a CO400 pre-treatment (table 1).

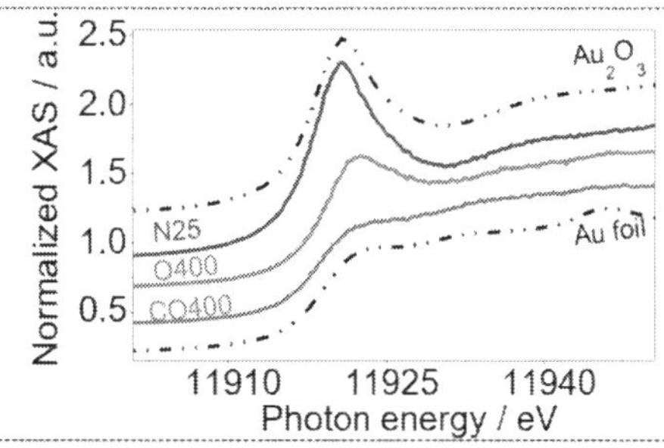


**Figure 1.** XANE spectra of a fresh $Au/CeO_2$ catalyst in $N_2$ at 25°C (N25), after different pre-treatments, and reference spectra of $Au_2O_3$, (top) and a metallic Au foil (bottom).

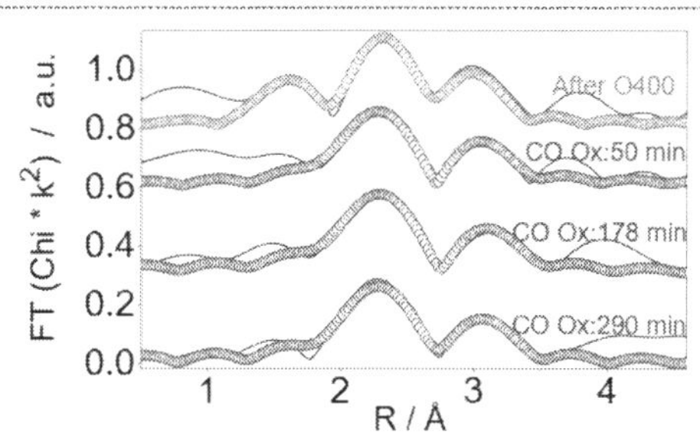


**Figure 2.** $k^2$-weighted FT converted EXAFS data (solid lines) and corresponding fits (open circles) for O400 pre-treatment (red) and subsequent CO oxidation at different reaction times (blue).

*3.2. Reaction induced changes: Au oxidation state and Au particle size*

During kinetic measurements (see Fig. 3) the CO oxidation activity over $Au/CeO_2$ catalysts was found to be i) lower after CO400 than after O400 and ii) strongly decreasing with time on stream for both pre-treatments (deactivation after 1000 min on stream: 36% for O400 and 46% for CO400).

To obtain more information on the Au oxidation state of the $Au/CeO_2$ catalysts, XANES data were recorded at different times on stream during the CO oxidation. While for the CO400 pre-treated catalyst (Fig. S1) they did not show significant changes during the reaction (only metallic Au species), the respective data after O400 pre-treatment (Fig. S2) revealed a fast decrease of the oxidic peak upon changing from $N_2$ to the reaction gas mixture, leading to complete disappearance in less than 5 min. Thus, independent of the pre-treatment, we always find mainly $Au^0$ species after a few minutes on stream. Similar observations of $Au^0$ NPs as dominant Au species during CO oxidation had been reported earlier for other Au catalysts, e.g., for $Au/TiO_2$ [4]. This strong decrease of ionic Au species in the O400 pre-treated sample occurs on a rather short time scale compared to the catalyst deactivation, which is less than 5% within the first 10 min. In contradiction to previous reports [5,7], who claimed that cationic gold species are responsible for the high CO oxidation activity of $Au/CeO_2$, we conclude that the decrease in activity under present reaction conditions is not mainly related to changes in the Au oxidation state, e.g., from $Au^{3+}$ to $Au^0$. This is also supported by the respective data for the CO400 pre-treated catalyst, which equally showed a significant decrease in activity.

Additional information on the Au-Au CN and Au NP size under reaction conditions were derived from EXAFS measurements (see Table S1). These data show a pronounced increase of the Au-Au CN, from 6.3 to 8.0 and from 4.1 to 5.9 for O400 and CO400 pre-treated samples, respectively, within 175 min of reaction. For the first series, the corresponding $k^2$-weighted Fourier Transformations of EXAFS data are shown in Fig. 2 after different time intervals. This increase in CN reflects a decrease in the dispersion of the Au NPs from 54% to 34% (O400) and from 100% to 60% (CO400), indicating a reaction-induced particle growth (see Fig. 4). Although a direct comparison of the activities after O400 and CO400 is not possible due to the rather different catalyst state after both pre-treatments, it is obvious that for both samples the activity decreases with increasing Au particle size, similar to other Au catalysts [8]. Moreover, also on a quantitative scale the loss of activity and the loss of Au dispersion are of similar order of magnitude. The dispersion decreases by a factor of ~0.6 for both pre-treatments within 175 min on stream, while the decrease in activity (within the same time) is 0.81 and 0.57 for O400 and CO400, respectively. Considering the (comparable) time scales of deactivation and Au particle sintering (in contrast to changes in the Au oxidation state) and their similar magnitudes, we conclude that under present reaction conditions reaction-induced Au particle size growth contributes significantly to the catalyst deactivation. Nevertheless, especially in the beginning of the reaction other factors such as changes in the catalyst oxidation state or site blocking by adsorbed surface species e.g., by carbonates or formate species, may contribute to the initial fast deactivation.

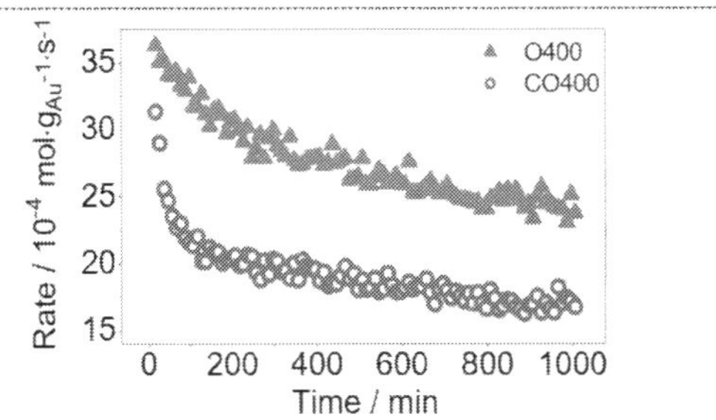


**Figure 3** Au mass-normalized CO oxidation activity in 1% CO, 1% $O_2$, balance $N_2$ at 80 °C (O400: red; CO400: green) with time on stream.

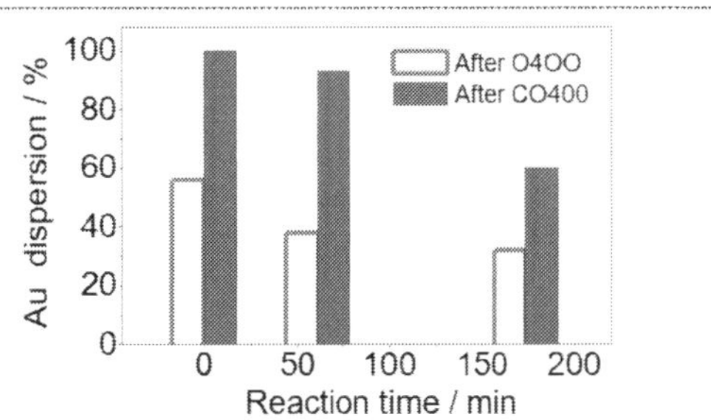


**Figure 4.** Time evolution of the Au dispersion during CO oxidation (O400: open red column; CO400: green filled column).

In summary, different pre-treatments lead to different initial states of the Au/$CeO_2$ catalyst. While Au NPs after O400 pre-treatment contain a significant amount of oxidic species, only metallic Au species could be detected after CO400 pre-treatment. During CO oxidation only metallic Au NPs are present, which grow in size for both pre-treatments. Accordingly, the relatively strong deactivation of Au/$CeO_2$ catalysts with time on stream is mainly attributed to reaction-induced Au particle size growth.

**Acknowledgements**
We acknowledge continuous support by Dr. L. Olivi and G. Aquilanti at the XAFS beamline (Elettra Synchrotrone Trieste) as well as the financial support by Elettra via the European project CALIPSO.

**References**
[1] Hashmi, A S K and Hutchings, G J 2006 *Angew.Chem.Int.Ed.* **45** 7896-7936.
[2] Widmann, D and Behm, R J 2014 *Acc.Chem.Res.* **47** 740-749.
[3] Kast, P, Kucèrová, G and Behm, R J 2015 *Catal.Today* **244** 146-160.
[4] Weiher, N, Bus, E, Delannoy, L, Louis, C, Ramaker, D E, Miller, J T and van Bokhoven, J A 2006 *J.Catal.* **240** 100-107.
[5] Guzman, J, Carrettin, S and Corma, A 2005 *J.Am.Chem.Soc.* **127** 3286-3287.
[6] Deng, W, Carpenter, C, Yi, N and Flytzani-Stephanopoulos, M 2007 *Top.Catal.* **44** 199-208.
[7] Hernández, J. A., Gómez, S. A., Zepeda, T. A., Fierro-González, J. C., and Fuentes, G. A 2015 *ACS Catalysis* **5** 4003-4012.
[8] Chen, S, Luo, L, Jiang, Z and Huang, W 2015 *ACS Catal.* **5** 1653-1662.
[9] Tang, Z R, Edwards, J K, Bartley, J K, Taylor, S H, Carley, A F, Herzing, A A, Kiely, C J and Hutchings, G J 25-7-2007 *J.Catal.* **249** 208-219.
[10] Di Cicco, A, Aquilanti, G, Minicucci, M, Principi, E, Novello, N, Cognigni, A and Olivi, L 2009 *J.Phys.:Conf.Series* **190** 012043-012043-7.
[11] Abdel-Mageed, A M, Eckle, S, Anfang, H-G and Behm, R J 2013 *J.Catal.* **298** 148-160.
[12] Abd El-Moemen, A, Karpenko, A, Denkwitz, Y and Behm, R J 2009 *J.Power Sources* **190** 64-75.
[13] Koningsberger, D C, Mojet, B L, van Dorssen, G E and Ramaker, D E 1-5-2000 *Top.Catal.* **10** 143-155.
[14] Miller, J T, Kropf, A J, Zha, Y, Regalbuto, J R, Delannoy, L, Louis, C, Bus, E and van Bokhoven, J A 2006 *J.Catal.* **240** 222-234.

Journal of Physics: Conference Series **712** (2016) 012045 doi:10.1088/1742-6596/712/1/012045

# Using combined XAS/DRIFTS to study CO/NO Oxidation over Pt/$Al_2O_3$ catalysts

**A M Gänzler[1], H Lichtenberg[1,2], A I Frenkel[3], M Casapu[1], A Boubnov[1], D Wang[4], J-D Grunwaldt[1,2]**
[1]Institute for Chemical Technology and Polymer Chemistry, Karlsruhe Institute of Technology (KIT), 76131 Karlsruhe, Germany
[2]Institute of Catalysis Research and Technology, Karlsruhe Institute of Technology (KIT), 76131 Karlsruhe, Germany
[3]Physics Department, Yeshiva University, 245 Lexington Avenue, New York, NY 10016, USA
[4]Institute of Nanotechnology, Karlsruhe Institute of Technology (KIT), 76131 Karlsruhe, Germany

Email: grunwaldt@kit.edu

**Abstract**. Combined X-ray absorption spectroscopy (XAS) and diffuse reflectance infrared Fourier transform spectroscopy (DRIFTS) were applied to investigate the interaction between reaction atmosphere, adsorbates and Pt oxidation state of Pt/$Al_2O_3$ model diesel oxidation catalysts under CO/NO oxidation conditions. The Pt oxidation state was correlated to the adsorbates on the catalyst's surface. Even at low temperature the reaction atmosphere had a strong impact on the oxidation state of the catalyst, and the oxidation state in turn strongly affected CO adsorption on the Pt particles.

## 1. Introduction

Understanding catalysts at work is a key challenge for the rational design of new catalysts. In the past decades an arsenal of *in situ* techniques was particularly developed to investigate heterogeneous catalysts whose performance is determined by the interaction of numerous factors, e.g. oxidation state and adsorbate coverage [1-3]. However, in order to gain a deeper understanding of catalytic processes under relevant conditions also the reactor design has to be considered, which demands furthermore spatially resolved examination of the catalytic material [1, 4-6]. By combining characterization techniques which provide complementary information about the catalysts structure on different length scales an in depth understanding of catalytic processes under reaction conditions can be established.

Herein, results on the oxidation of CO and NO over a Pt based diesel oxidation catalyst are reported, which were obtained by combining XAS for oxidation state evaluation and DRIFTS for identification of adsorbed species. This combination of techniques allows to probe simultaneously bulk and surface properties of catalysts [7-9], and they were used to unravel the interdependence of the reaction atmosphere, adsorbates and the Pt oxidation state. DRIFTS can not only detect adsorbates on the surface but also provide information on the catalyst's surface structure. The results illustrate that these factors are closely correlated and sensitively depend on each other. This approach can therefore significantly contribute to understand the role of the noble metal oxidation state during these oxidation reactions.

## 2. Experimental

The model 2 wt.% $Pt/Al_2O_3$ catalysts were prepared by incipient wetness impregnation using hexachloroplatinic acid (HCP) in an aqueous solution or platinum acetylacetonate (PAA) in acetone as precursors. The samples were dried overnight at 80 °C, calcined at 500 °C for 2h and reduced in 5 % $H_2$/He at 400 °C for 4h. A Titan 80-300 (FEI Company) FEG electron microscope was used in STEM mode to study the distribution and the size of the resulting Pt nanoparticles on the alumina support.

Simultaneous XAS/DRIFTS experiments were performed at the X18A beamline at the National Synchrotron Light Source (Brookhaven National Laboratory, Brookhaven, USA). The bending magnet radiation was monochromatized by a Si (111) channel-cut crystal monochromator in step scanning mode, to record XANES at the Pt $L_{III}$-edge (11564 eV, beam with a spot size 0.75mm x 0.75 mm) in transmission mode using ionization chambers. The XANES data was evaluated by linear combination analysis using spectra of a Pt-foil and a $PtO_2$ reference with the IFEFFIT package [10]. DRIFTS data of the samples were simultaneously recorded using a special setup consisting of a DaVinci arm (Harrick) attached to a modified Praying Mantis DRIFTS accessory (Harrick) [11].

About 40 mg of the respective catalyst (*PAA, HCP*) was loaded into the sample cup. After pre-reduction in 5 % $H_2$/He (15mL/min) at 200 °C the catalyst was heated to 450 °C in He and background spectra for DRIFTS were acquired in 50 °C intervals during cooling to room temperature. At room temperature, the catalyst was first exposed to 5000 ppm CO, 10 % $O_2$ and He and then NO was added resulting in a 15 ml/min flow of 5000 ppm CO, 5000 ppm NO, 10 % $O_2$ and He. After 30 min the catalyst was heated stepwise (temperature plateaus in 50 °C intervals) to 450 °C and cooled back to 50 °C. At each temperature plateau XANES and DRIFTS data were acquired using the corresponding, previously recorded DRIFTS background spectra. When comparing with other data one should bear in mind that due to the temperature gradients within the DRIFTS cell [12] the actual temperature was usually lower than the set temperature.

## 3. Results & Discussion

According to the TEM results (Figure 1, left, evaluation of more than 400 particles), impregnation with HCP resulted in small (1.1 nm), finely dispersed Pt particles. Impregnation with acetylacetonate also led to very small Pt particles (1.1 nm), however, several very large particles (~50nm) were also found on this sample (Figure 1, right side).

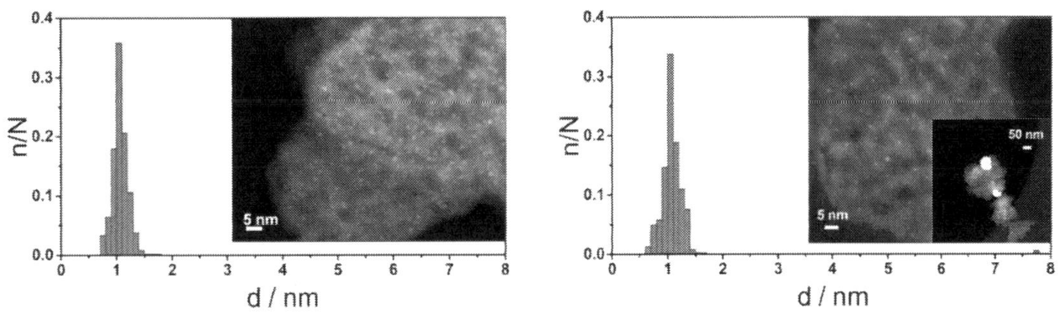


**Figure 1.** TEM images and respective particle size distribution for 2%$Pt/Al_2O_3$ catalyst prepared with HCP (left) and PAA (right).

Figure 2 shows XANES and DRIFTS spectra of a pre-reduced catalyst (PAA), simultaneously acquired during the introduction of the reaction mixture at 50 °C in two steps: (1) only CO along with oxygen (blue spectra), (2) addition of NO (red spectra). During the first step the catalyst remained reduced: the XANES spectra did not change and the only visible peak in the DRIFTS spectra at 2096 $cm^{-1}$ was attributed to CO adsorbed on reduced platinum [13, 14].

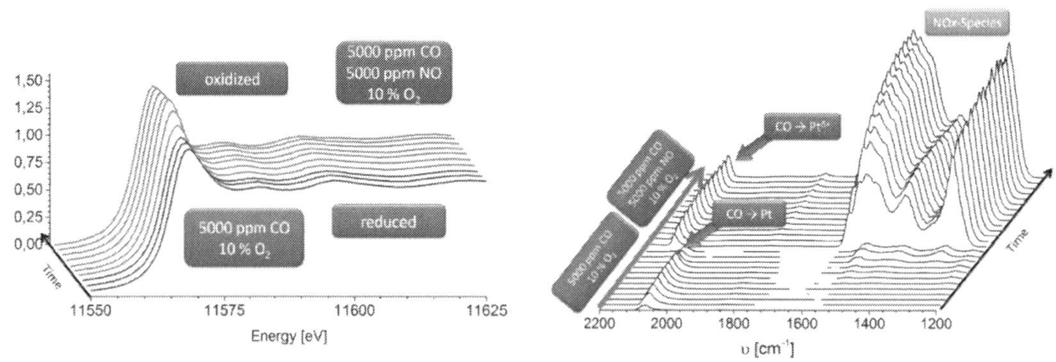


**Figure 2.** Simultaneously measured XANES (left) and DRIFTS (right) during the introduction of 5000 ppm CO, 10 % $O_2$, He (blue) over the pre-reduced PAA catalyst, followed by the addition of 5000 ppm NO (red).

After addition of NO the DRIFTS spectra changed significantly. A broad feature between 1200 $cm^{-1}$ and 1600 $cm^{-1}$, which can be attributed to NO, $NO_2^-$ and $NO_3^-$ species adsorbed on the $Al_2O_3$ support [15, 16], started to dominate the spectra. At 1600 – 1800 $cm^{-1}$ weak bands, corresponding to linear, non-linear or bent $Pt^0$-NO species [17], were observed. The appearance of these features was accompanied by changes in the CO frequency. The peak at 2096 $cm^{-1}$ vanished while a new peak emerged at 2124 $cm^{-1}$ representing CO adsorbed on partially oxidized platinum [18]. The DRIFTS spectra suggest that, unlike the previously dosed $CO/O_2$ feed, the $CO/NO/O_2$ mixture led to partial oxidation of Pt sites already at room temperature. This is supported by the simultaneously recorded XANES spectra: The Pt-$L_3$-edge white line increased significantly when NO was added to the reaction mixture. This change in oxidation state at low temperature is probably due to the oxidation of very small Pt particles, which are present in the sample prepared with the PAA precursor [5, 19].

As a next step this interplay between adsorbates and oxidation state was monitored at different temperatures. In Figure 3 the results of the linear combination fitting (LCF) of the XAS data are presented together with simultaneously acquired DRIFTS spectra at several temperature plateaus in the CO/NO/ $O_2$ reaction mixture for the *HCP* (left) and the *PAA* (right) catalysts.

At room temperature the *HCP* sample (Figure 3, left) was covered by CO and NOx species. CO molecules were adsorbed to Pt sites, which got partially oxidized upon introduction of the reaction mixture, whereas NOx species covered mainly the $Al_2O_3$ support. With increasing temperature Pt was further oxidized, while more CO adsorbed to oxidic platinum sites (2129 $cm^{-1}$). At the same time the DRIFTS bands corresponding to NOx species adsorbed at Pt sites (1600-1800 $cm^{-1}$) disappeared and the NOx coverage on the alumina surface (between 1200 $cm^{-1}$ and 1650 $cm^{-1}$) decreased slightly. At elevated temperature (T > 350°C) all adsorbates desorbed, while the catalyst was reduced again even though exposed to an oxidizing atmosphere. During cooling the Pt oxidation state followed a similar profile as during heating. Again, the catalyst was oxidized accompanied by CO adsorption, before Pt was partially reduced during cooling to 50 °C, however, the reoxidation and reduction was shifted to lower temperatures. On the other side NOx-species on the support adsorbed during heating and cooling in the same manner. According to the LCF results the catalyst ended up slightly more oxidized after the heating-cooling procedure. This was also reflected in the CO adsorption peak, which was slightly shifted to higher wave numbers in the respective DRIFTS spectrum.

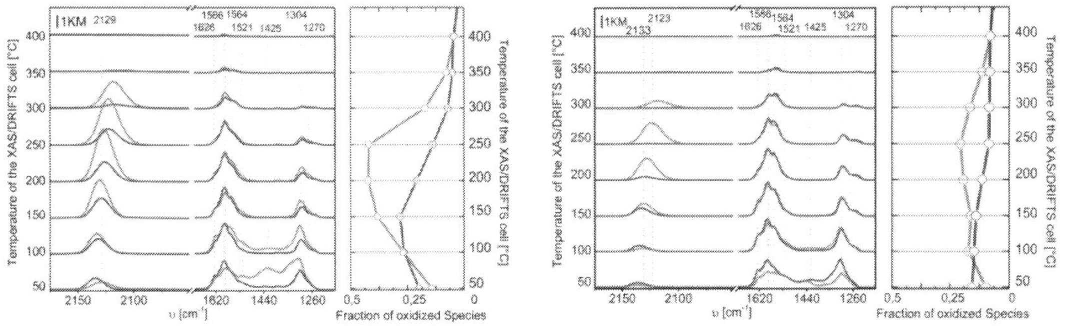


**Figure 3.** DRIFTS and results of LCF from simultaneously measured XANES/DRIFTS during CO/NO over the 2% Pt/$Al_2O_3$ catalyst HCP (left) and PAA (right) during heating (red) and cooling (blue).

In case of the *PAA* catalyst changes in the oxidation state of platinum during the CO/NO oxidation temperature cycle were less pronounced, since this catalyst contains also several larger particles (~50 nm) in addition to very small Pt nanoparticles and, accordingly, the fraction of surface Pt atoms, which easily get oxidized is lower. As shown in Figures 2 and 3 the Pt sites were partially oxidized at room temperature in the CO/NO/$O_2$ mixture. During heating Pt oxidized further also on this catalyst, accompanied by an uptake of adsorbed CO to oxidic Pt sites. At elevated temperatures the Pt oxidation state decreased, as did the CO coverage. When the temperature was decreased again the catalyst remained reduced at elevated temperatures and got partially oxidized at lower temperatures, while CO adsorbed again on oxidic Pt sites. Also on the PAA the NOx adsorption and desorption on the support followed the same trend during cooling and heating. Note that significant concentration and temperature gradients could have evolved along the catalyst bed due to the low gas flow entering the reactor and the rather large diameter of the reactor compared to its length [3, 9]. Therefore, XAS and DRIFTS data might not be any more directly comparable at elevated temperature, even though it was measured in the same cell at the same time.

## 4. Conclusion

According to these experiments the adsorption of CO appears to be strongly dependent on the Pt oxidation state. Changes in the reaction atmosphere significantly influenced the catalyst's oxidation state already at room temperature. On the other side the oxidation state determined the adsorption on the Pt particles. CO adsorption, which poisons Pt sites with respect to oxygen adsorption and therefore inhibits CO oxidation at low temperatures [5, 19], was found to be significantly more pronounced on oxidized Pt. The combination of XAS and DRIFTS shows the highly sensitive interdependence between the composition of the gas atmosphere, adsorbates and oxidation state, which should be furthermore in close relationship to the noble metal morphology and atomic structure [20]. All these factors are known to influence the catalyst performance and therefore have to be studied and considered, also emphasizing the importance of spatially resolved studies [4, 5].

This study illustrates the interdependence of the reaction gas composition, adsorbates and oxidation state of Pt nanoparticles in exhaust gas catalysts. The combined set-up used for these measurements allowed to see correlations between surface coverage and oxidation state. Features in the IR spectra, which are in turn sensitive to the oxidation sate, like the CO band, could be unambiguously assigned. This combination of two powerful characterization techniques can contribute significantly to a better understanding and rational design of catalysts.

## 5. Acknowledgement
Phillip Baumann, Nebojsa Marinkovic and Hudson Carvalho are acknowledged for support during the XAS/DRIFTS experiments, Sabrina Conrad for catalyst preparation and characterization and the BMBF projects "MatAkt" (05K10VKB) and "ZeitKatMat" (05K13VK13) for financial support. AIF gratefully acknowledges funding of his work by the U. S. DOE Grant No. DE-FG02-03ER15476. We acknowledge the facilities support provided at the National Synchrotron Light Source at the Brookhaven National Laboratory (US DOE, Contract No. DE-SC0012704) and the Synchrotron Catalysis Consortium (US DOE, Grant No. DE-SC0012335).

## References
[1] Grunwaldt J-D, Wagner J B, Dunin-Borkowski R E 2013 *ChemCatChem* **5** 1
[2] Weckhuysen B M, In-situ spectroscopy of catalysts, American Scientific Publishers Stevenson Ranch, CA, 2004.
[3] Topsøe H 2003 *J. Catal.* **216** 1–2
[4] Urakawa A, Baiker A 2009 *Top. Catal.* **52** 10
[5] Gänzler A M, Casapu M, Boubnov A, Müller O, Conrad S, Lichtenberg H, Frahm R, Grunwaldt J-D 2015 *J. Catal.* **328**
[6] Hannemann S, Grunwaldt J-D, van Vegten N, Baiker A, Boye P, Schroer C G 2007 *Catal. Today* **126** 1–2
[7] Newton M A, Dent A J, Fiddy S G, Jyoti B, Evans J 2007 *Catal. Today* **126** 1–2
[8] Ferri D, Newton M, Nachtegaal M 2011 *Top Catal* **54** 16-18
[9] Chiarello G L, Nachtegaal M, Marchionni V, Quaroni L, Ferri D 2014 *Rev. Sci. Instrum.* **85** 7
[10] Ravel B, Newville M 2005 *J. Synchrotron. Radiat.* **12** 4
[11] Marinkovic N S, Wang Q, Frenkel A I 2011 *J. Synchrotron. Radiat.* **18** 3
[12] Meunier F C 2010 *Chem. Soc. Rev.* **39** 12
[13] Kappers M, Maas J 1991 *Catal. Let.* **10** 5
[14] Brandt R K, Hughes M R, Bourget L P, Truszkowska K, Greenler R G 1993 *Surf. Sci.* **286** 1–2
[15] Toops T J, Smith D B, Partridge W P 2005 *Appl. Catal., B* **58** 3–4
[16] Toops T J, Smith D B, Epling W S, Parks J E, Partridge W P 2005 *Appl. Catal., B* **58** 3–4
[17] Lévy P J, Pitchon V, Perrichon V, Primet M, Chevrier M, Gauthier C 1998 *J. Catal.* **178** 1
[18] Sheppard N, Nguyen T T, The Vibrational Spectra of Carbon Monoxide chemisorbed in the Surfaces of Metal Catalysts - A suggested Scheme of Interpretation, in: R.G.H. Clark, R.E. Hester (Eds.) Advances in Infrared and Raman Spectroscopy, Heyden, London, 1987, pp. 67-148.
[19] Boubnov A, Gänzler A, Conrad S, Casapu M, Grunwaldt J-D 2013 *Top. Catal.* **56** 1-8
[20] Vendelbo S B, Elkjær C F, Falsig H, Puspitasari I, Dona P, Mele L, Morana B, Nelissen B J, van Rijn R, Creemer J F, Kooyman P J, Helveg S 2014 *Nat. Mater.* **13** 9

# Chemical state of Ag in Conducting Bridge Random Access Memory cells: a depth resolved X-ray Absorption Spectroscopy investigation.

**F d'Acapito*,E Souchier@, P Noe@, P Blaise@, M Bernard@, V Jousseaume@**

* CNR-IOM-OGG c/o ESRF, LISA CRG, 71 Avenue des Martyrs, Grenoble (France), dacapito@esrf.fr

@ Univ. Grenoble Alpes, CEA, LETI, MINATEC Campus, 17 Rue des Martyrs, Grenoble (France)

E-mail: dacapito@esrf.fr

**Abstract.** Conducting Bridge Random Access Memories (CBRAM) are a promising substitute for FLASH technology but problems with limited retention of the low resistance *ON* state still hamper their massive deployment. Depth resolved X-ray Absorption Spectroscopy has been used to describe the chemical state of the atoms of the active electrode (in this case Ag) and to reveal the role of Sb as stabilizer of the metallic state.

## 1. Introduction

Conducting Bridge Random Access Memories (CBRAM) represent one of the possible substitutes for the FLASH technology below the 22 nm node [1]. These memory cells are based on a chalcogenide layer (namely $GeS_2$) sandwiched between two metallic electrodes: one inert (usually W or Pt) and the other active (Ag, Cu). Upon application of an electric field the ions from the active electrode migrate in the chalcogenide layer and eventually form metallic filaments that reduce considerably the resistance of the cell to the Low Resistance State (*ON* state). The application of an opposite electric field leads to the partial dissolution of the wire and to a regain of cell resistivity to the High Resistance State (*OFF* state). CBRAMs offer good scalability but suffer from the instability of the *ON* state due to the spontaneous dissolution of the metallic bridge in the surrounding matrix with time [2] and this represents a severe limitation in the use of these memories in a multilevel configuration. Recently [3] it has been reported that the addition of Sb to the chalcogenide layer has beneficial effects on the retention of the *ON* state. For this study REFLection Extended X-ray Absorption Fine Structure (ReflEXAFS) technique[4, 5, 6, 7, 8] has been used to derive information on the chemical state of Ag below the active electrode and in the chalcogenide layer. These findings shed light on the formation process of the wires and on the role of Sb in stabilizing the metallic phase for Ag.

## 2. Experimental

Two samples, consisting in Ag-doped and Ag+Sb-doped $GeS_2$ chalcogenide thin films, were prepared by co-sputtering from $GeS_2$ and Sb pure targets over a W bottom electrode of 25

nm deposited on Si. The chalcogenide layer had a thickness of 50 nm and Sb was introduced at a 10 % level. Ontop of the layer a 3 nm-thick top Ag electrode was deposited in order to create a real memory cell. One of the cells underwent a serie of 20 resistive switching cycles before the removal of the Ag top electrode by ion beam etching prior to XAS analysis. The lateral dimensions of all samples were 3×80 mm. In the following the samples will be named SW (switched) and NSW (non-switched).
Samples were measured at the Ag K edge (E=25514 eV) in the dedicated experimental station [9] operative at the GILDA beamline at the ESRF [10]. The scattering vector was vertical and the vertical size of the beam was about 50 $\mu$m. The incoming and the reflected beams were measured with ion chambers loaded with Kr gas at 1 atm. The experimental procedure was the same used for investigations on Sb-free samples [11] and for studying the Sb behavior in these samples [10]. For each specimen a reflectivity curve at fixed energy (25600 eV) was first collected (fig.1) and two zones were identified. The *surface* ($\Sigma$) regime at an incidence angle of about 0.07° (where the beam was in total external reflection and the extinction length of the beam was estimated to be 4 nm) and the *layer* ($\Lambda$) regime at about 0.15° (where the probe beam shines the whole chalcogenide layer). Collecting XAS data at these two values for the incidence angle permitted to realize a depth-sensitive investigation in the zones near the active (Ag) electrode and inside the layer. The data were analyzed with the CARD code [12] in its version 04 that permits the treatment of layered samples [13].

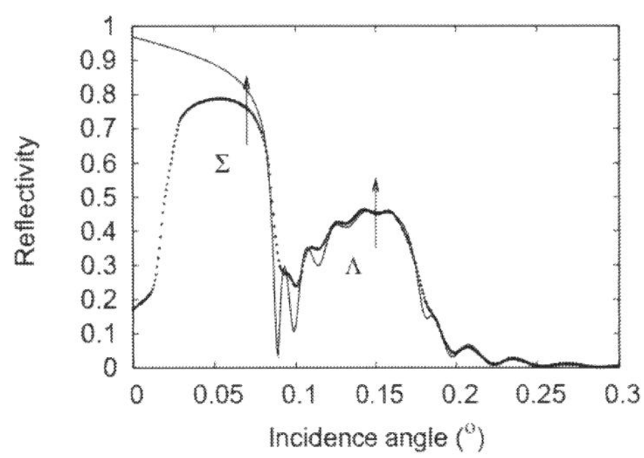


**Figure 1.** Reflectivity of sample NSW collected at 25600 eV (dots) with the best fitting curve (line). The angles for the Surface ($\Sigma$, 0.07°) and Layer ($\Lambda$, 0.15°) datasets are marked by arrows.

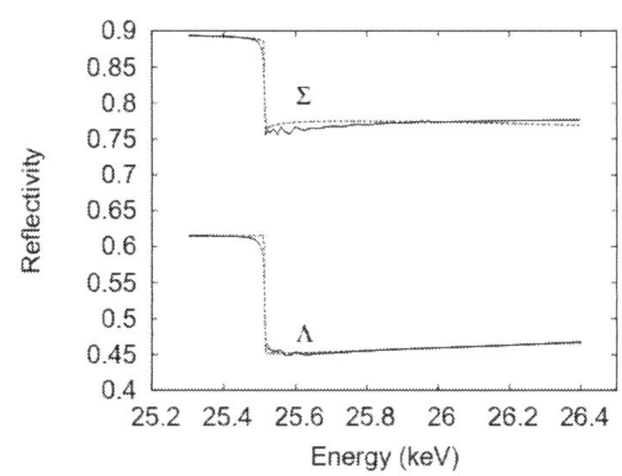


**Figure 2.** Reflectivity curves of sample NSW collected at $\Sigma$ and $\Lambda$ values for the incidence angle (dots) with the best fitting curves (lines).

## 3. Results

For both SW and NSW samples the reflectivity at fixed angle was fitted to a model consisting in two layers Ag:GeS$_x$ and W above a Si substrate. The W thickness was found to be $t_W = 300 \pm 30$ nm with a roughness $\sigma_W = 0.7 \pm 0.1$ nm and the chalcogenide had $t_{chalco} = 500 \pm 50$ nm and $\sigma_{chalco} = 1.9 \pm 0.2$) nm, in good agreement with growth data. From the measured density of the chalcogenide layer ($4.1 \pm 0.3\, g/cm^3$) a composition $(GeS_2)_{0.75}(Ag_2S)_{0.25}$ could be derived. Successively ReflEXAFS spectra were collected at each of the two incidence angle values and the reflectivity fits are shown in fig.2. Once having determined the parameters determining the sample reflectivity a set of suitable modified theoretical EXAFS paths were calculated with CARD and used for the quantitative data fitting using the FEFFIT code [14]. As an example, the EXAFS spectra of the NSW sample are shown in fig. 3 whereas the related Fourier Transforms

| Sample | Metal Fraction $x$ | $R_{AgS}$ (Å) | $R_{AgAg}$ (Å) |
|---|---|---|---|
| NSW-Σ | 0.51(5) | 2.47(3) | 2.86(3) |
| NSW-Λ | 0.37(7) | 2.46(3) | 2.84(4) |
| SW-Σ | 0.70(5) | 2.50(5) | 2.86(3) |
| SW-Λ | 0.40(8) | 2.46(3) | 2.80(5) |

**Table 1.** Main structural parameters derived from the quantitative analysis of the ReflEXAFS data; errors on the last digit are in parentheses. A global amplitude reduction factor of 0.95(5) was used as determined from the spectrum of metallic Ag taken in transmission mode. For comparison, the distance measured on the metal foi was $R_{AgAg}$=2.880(5)Å. The Debye-Waller factors were $\sigma^2_{AgS} = 0.0012(5)$ and $\sigma^2_{AgAg} = 0.0025(5)$

are in fig. 4. The spectra contain two components, a low frequency one due to Ag-S bonds and

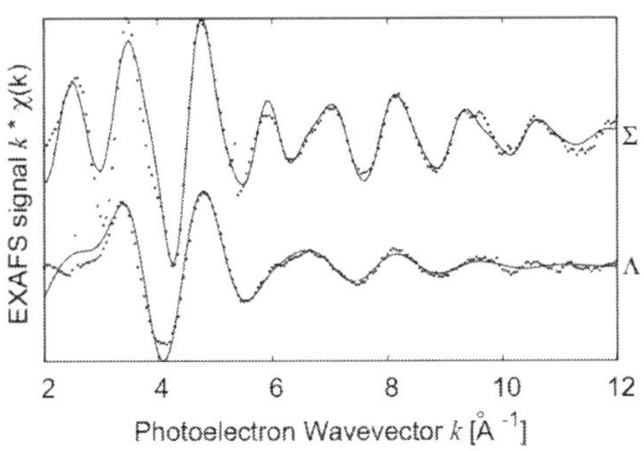


**Figure 3.** ReflEXAFS spectra of the *NSW* sample in Λ and Σ condition obtained from the difference between the reflectivity R and the spline approximant to the atomic background. With this definition these spectra appear to be reversed with respect to standard XAS spectra due to the opposite edge jump.

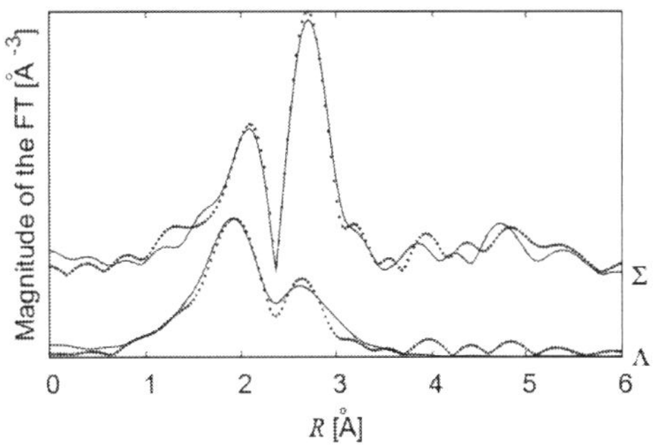


**Figure 4.** FTs of the previous spectra taken in the interval 2-12 Å$^{-1}$. An interpretation of these spectra is not straightforward as the two peaks in the Λ spectrum come from the two AgAg and AgS shells whereas in the case of the Σ spectrum the splitting is dominated by the Ramsauer-Townsend effect [15] on the Ag backscatterer.

a higher frequency one due to Ag-Ag bonds. The data analysis was carried out following the procedure already used for these systems [11, 16] i.e. by mixing a 3-coordinated Ag-sulphide phase plus a 12 coordinated Ag-metal phase with relative weights $x$ and $(1-x)$. Table. 1 resumes the main parameters.

## 4. Discussion

The ReflEXAFS spectra reveal the presence of two kind of bonds for Ag, AgS derived from the Ag ions dispersed in the $GeS_2$ matrix and AgAg coming from the metallic fraction forming the wires that cause the resistance drop in the *ON* state of the cell. Different values for the metallic fraction are observed in surface an in the layer in the *SW* and *NSW* samples. This shows that the metal fraction already before the switching process is higher near to the active electrode and the same situation is retained after the formation of the conducting wires. As a consequence, the wires can be described as cones with the base towards the active electrode. This reveals a low diffusivity of Ag and a reduction process faster than the ion migration as already observed in Ref. [17] in $Cu/Al_2O_3/TiN$ cells. In both cases the SW sample has a higher metal amount than the NSW sample revealing that new metal is formed during the switching process that is not completely dissolved during the switching cycles. It is worth noticing that in all cases the metal fraction is sensibly higher if compared with similar samples grown without Sb [11]. This is due to the action of Sb making stable bonds with the available S ions [16] making more difficult the sulphurization of Ag. This leads to an increased retention of the *ON* state by the cell as pointed out by [3]. The AgAg bond length appears to be slightly shorter in the Λ samples respect to Σ and bulk Ag ($R_{AgAg}$=2.880(5)Å) probably due to the possible presence of Ag oligomers.

## 5. Conclusion

Depth resolved XAS has been used to describe the chemical state of Ag ions Conductive Bridge Random Access Memories in their pristine state or after having undergone several switching cycles. The samples analyzed contained Sb in the chalcogenide layer as recent literature [3] reported a beneficial effect of this element on the retention of the *ON* state. The switched sample presents an overall higher metal content than the non switched one, meaning that this process really leads to the formation of new metal after the cell deposition. The metal content is higher below the active electrode, revealing that the cone-shaped filaments have their thicker part towards this part of the sample. The values of metal fraction observed here are sensibly higher than those found on Sb-free samples [11] : having established that Sb forms stable bonds with S in these samples [16], this finding confirms the Sb role in hampering the sulphurization of Ag.

## References

[1] 2013 *International Technology Roadmap for Semiconductors 2013 edition Process Integration, Devices, and Structures.*
[2] Guy J *et al.* 2014 *Thin Solid Films* **563**
[3] Vianello E *et al.* 2012 *Proceedings of the IEEE International Electron Devices Meeting (IEDM), San Francisco*
[4] Barchewitz R and andG Onori M C V 1978 *J. Phys. C* **11** 4439
[5] Martens G and Rabe P 1980 *Phys. Stat. Sol. A* **58** 415
[6] Heald S 1992 *Rev. Sci. Instrum.* **63** 873
[7] Asakura K, Chun W and Iwasawa Y 2000 *Topics Catal.* **10** 209
[8] Lützenkirchen-Hecht D and Frahm R 2001 *J. Synchrotron Rad.* **8** 478
[9] d'Acapito F, Davoli I, Ghigna P and Mobilio S 2003 *J. Synchrotron Rad.* **10** 260
[10] d'Acapito F, Trapananti A, Torrengo S and Mobilio S 2014 *Notiziario Neutroni e Luce di Sincrotrone* **19** 14
[11] Souchier E, d'Acapito F, Noe P, Blaise P, Bernard M and Jousseaume V 2015 *Phys. Chem. Chem. Phys* **17** 23931
[12] Benzi F, Davoli I, Rovezzi M and d'Acapito F 2008 *Rev. Sci. Instrum.* **79** 103902 URL `http://www.esrf.eu/computing/scientific/CARD/CARD.html`
[13] Costanzo T, Benzi F, Ghigna P, Pin S, Spinolo G and d'Acapito F 2014 *J. Synchrotron Rad.* **21** 395
[14] Ravel B and Newville M 2005 *J. Synchrotron Rad.* **12** 537
[15] McKale A G, Veal B W, Paulikas A P, Chan S K and Knapp G S 1988 *Phys. Rev. B* **38**
[16] d'Acapito F, Souchier E, Noe P, Blaise P, Bernard M and Jousseaume V 2015 *Phys. Status Solidi A* Submitted
[17] Celano U *et al.* 2014 *Nano Lett.* **14** 2401

Journal of Physics: Conference Series **712** (2016) 012047 doi:10.1088/1742-6596/712/1/012047

# Ex-situ and in-situ investigations of thermal anti-oxidation treatments of stainless steels by reflection mode EXAFS

**D Lützenkirchen-Hecht*[1], D Wulff[2], R Wagner[1], U Holländer[2], HJ Maier[2], R Frahm[1]**

[1]Bergische Universität Wuppertal, Gaußstr. 20, 42097 Wuppertal, Germany

[2]Leibniz Universität Hannover, Institut für Werkstoffkunde, An der Universität 2, 30823 Garbsen, Germany

*E-mail: dirklh@uni-wuppertal.de

**Abstract.** Different thermal treatments were performed for the anti-oxidation of steel surfaces, and the processes were investigated ex-situ and in-situ by surface sensitive reflection mode EXAFS experiments at the Cr and Fe K-edges. While the samples for the ex-situ studies were heat-treated in a conveyor belt furnace at temperatures between 600 and 900 °C in inert carrier gases ($N_2$ or Ar) using different additives such as hydrogen ($H_2$) and monosilane ($SiH_4$), the in-situ anti-oxidation treatments have been performed in a high vacuum environment ($p < 10^{-6}$ mbar). While the ex-situ experiments suggest that $SiH_4$-additives are needed for the reduction of the steel to a metallic state, the in-vacuum treatments appear successful for temperatures above approx. 900 °C.

## 1. Introduction

The removal of oxide layers from steel surfaces is of large technological interest, for example, for fluxless brazing processes, where the presence of oxide layers prevent the wetting of the steel surface with the braze at elevated temperatures. Therefore it is required to remove these oxides while heating the base metals in a furnace together with the braze [1]. Appropriate temperatures are in the range of 900-1150 °C using Cu-Ni-alloy brazes for stainless steels, so that an additional oxidation of the steel is very likely during the required high-temperature treatments, even if the oxygen partial pressure is very low. Therefore, the investigation and the understanding of the oxide decomposition processes and their kinetics are of utmost interest in order to optimize fluxless brazing procedures. In this context, the use of additives such as hydrogen ($H_2$) or monosilane ($SiH_4$) in the furnace appears interesting in order to create reducing conditions [1-3]. Especially $SiH_4$ is an interesting additive, since it may quantitatively remove all oxygen and water residues in the gas phase according to the reactions $SiH_4 + 2O_2 \rightarrow SiO_2 + 2H_2O$ and $SiH_4 + 2H_2O \rightarrow SiO_2 + 4H_2$ [1].

The decomposition reaction was studied as a function of the conditions applied during the thermal treatments by a series of surface sensitive, reflection mode EXAFS experiments of different steel samples at the Cr and Fe K-edges in grazing incidence geometry. On one hand, the samples were heat-treated in a conveyor belt furnace, which is especially equipped for experiments with $SiH_4$ containing shielding gases at temperatures between 600 °C and 900 °C using well-defined temperature-time-regimes [3], and the samples were ex-situ investigated with reflection mode XAFS after their removal from the furnace. In addition, a second series of experiments were performed in-situ in order to elucidate the possibility of oxide removal from the steel surfaces under high-vacuum ($p < 10^{-6}$ mbar).

## 2. Experimental details

The ex-situ reflection mode EXAFS studies were performed at the wiggler beamline 10 of the DELTA storage ring operating with 1.5 GeV electrons [4]. A Si(111) channel-cut monochromator was employed, and $N_2$- and Ar-filled ionization chambers were used to monitor the incident and reflected intensities, respectively. The samples were mounted on the multi-purpose diffractometer endstation to enable a precise alignment of the polished samples within the incident X-ray beam, and the measurements were performed in ambient air. A high vacuum furnace with a base pressure below approx. 4 x $10^{-7}$ mbar was used for the in-situ de-oxidation experiments at the MRH endstation of the ROBL beamline BM 20 at the ESRF. The obtained XAFS data from the steel samples were compared to those of reference compounds such as α- and γ-$Fe_2O_3$ (hematite and maghemite), magnetite ($Fe_3O_4$), FeO, goethite (α-FeOOH), $FeCr_2O_4$, $Cr_2O_3$, CrN as well as Fe and Cr metal foils.

The circular stainless steel samples with a nominal composition of 64% Fe, 19.5% Cr, 8.6% Ni, 2% Mn, 0.9% Cu, 0.4% Si, 0.3% Mo in wt.-% and minor traces (< 0.03%) of Ti, V, P, C and S were polished using diamond paste to a final grading of 3 μm to ensure reproducible starting conditions for the thermal treatments. A conveyor belt furnace [3] was used for the sample preparation in different high purity gases in the temperature range from 600 to 900 °C. When monosilane ($SiH_4$) was used for the processing, the resulting partial pressure of oxygen was adjusted to values < $10^{-19}$ mbar. A typical thermal treatment took about 100 minutes, including the heating ramp, a hold time of 5 minutes at the desired maximum temperature, and the cooling phase. These samples were stored in dried $N_2$.

For the in-situ experiments at the ESRF, the samples were heated after the evacuation of the furnace, and the spectra were measured in-situ at temperatures between 600 and 900 °C. In addition to the investigation of steel samples with a native, room temperature oxide, some specimen were oxidized in air at a temperature of 300 °C to prepare a thicker oxide layer on the steel.

## 3. Results and discussion

### *3.1. Ex-situ EXAFS experiments*

A comparison of ex-situ reflection mode Fe K-edge EXAFS data obtained from a sample reduced in an atmosphere containing silane at 900 °C with a sample treated in pure $N_2$ is shown in figure 1. As can already be deduced from these raw data, the sample treated without $SiH_4$ suffers from substantial oxidation as indicated by the intense white line feature directly above the Fe edge, while the heat treatment in silane-containing media efficiently avoids surface oxidations, as indicated by the presence of strong metallic features in the related spectrum.

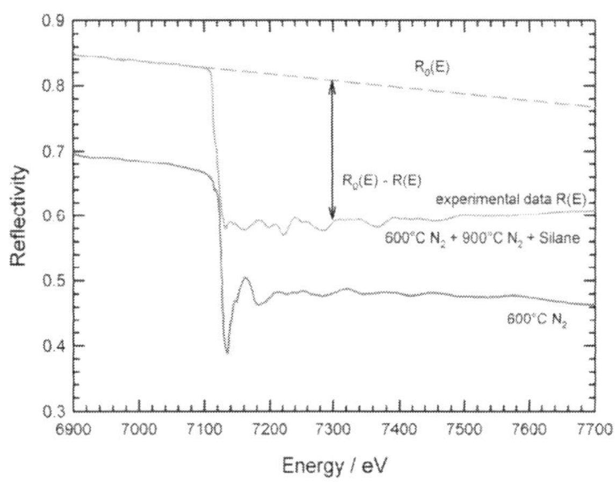


**Figure 1.** Ex-situ reflection mode EXAFS spectra (incidence angle $\phi$ = 0.1°) of two steel samples with different heat treatments at the Fe K-edge: Annealing in pure $N_2$ at 600 °C for 25 min. and an additional heating in a mixture of $N_2$ and addition of 20 ppm $SiH_4$ at 900 °C for 5 min. (——), and for comparison, solely annealing in pure $N_2$ at 600 °C for 5 min. (——). The latter spectrum is vertically shifted by -0.1 for a better comparison of the measured reflection mode fine structures of the two samples.

For a more detailed interpretation of the measured data, a linear pre-edge background $R_0(E)$ was fitted to the experimental data as shown in figure 1, interpolated to the post edge region and the normalized difference $R_0(E)$ - R(E) was calculated in order to obtain a measure of the absorption. For grazing angles smaller than the critical angle of total reflection, the absorption coefficient is approximately proportional to $R_0(E)$ - $R(E)$ as shown in refs. [5, 6], and thus, such an approach is sufficient to allow for a qualitative XANES analysis using the extracted data, i.e. a comparison to reference compounds.

The XANES spectra determined this way are compared to spectra of suited model compounds in figure 2. As can be seen in figure 2(a), the heat treatments in pure noble gas or nitrogen atmospheres generally lead to strong iron oxide contributions, even if $H_2$ is added, as indicated by the white line features, the edge position and the post-edge absorption features. Obviously, traces of oxygen in the process gas lead to oxidation at elevated temperatures. While the native, untreated steel sample shows features near the edge that can be related to metallic iron, the spectra of the heat treated samples are governed by strong oxide features. Since the penetration depth of the X-rays is about 3 nm for the chosen incidence angle of $\phi = 0.1°$, it can be concluded that the thickness of the iron oxide layer is at least of the order of about 5 nm in the absence of $SiH_4$ during the heat treatments. By contrast, for the sample treated in silane containing atmosphere, the measured spectrum at the Fe K-edge closely resembles that of metallic, body centered iron, and no distinct iron-oxide features were found. Compared to the XANES of the Fe metal reference, the amplitude of the EXAFS oscillations are slightly damped due to the grazing incidence geometry and the simple data analysis applied. It should also be noted, that no evidence e.g. for Fe-Si-compounds such as $FeSi_2$ or mixed Fe-Cr-oxides were found in the present experiments. As can be deduced from the spectra obtained at the Cr K-edge (figure 2(b)), there are strong contributions from chromium oxide in all of the treated steel samples. However, only in the Cr K-edge spectrum of the steel sample treated in an atmosphere with $SiH_4$ additions, indications for metallic Cr can be found (see the arrow in figure 2(b)).

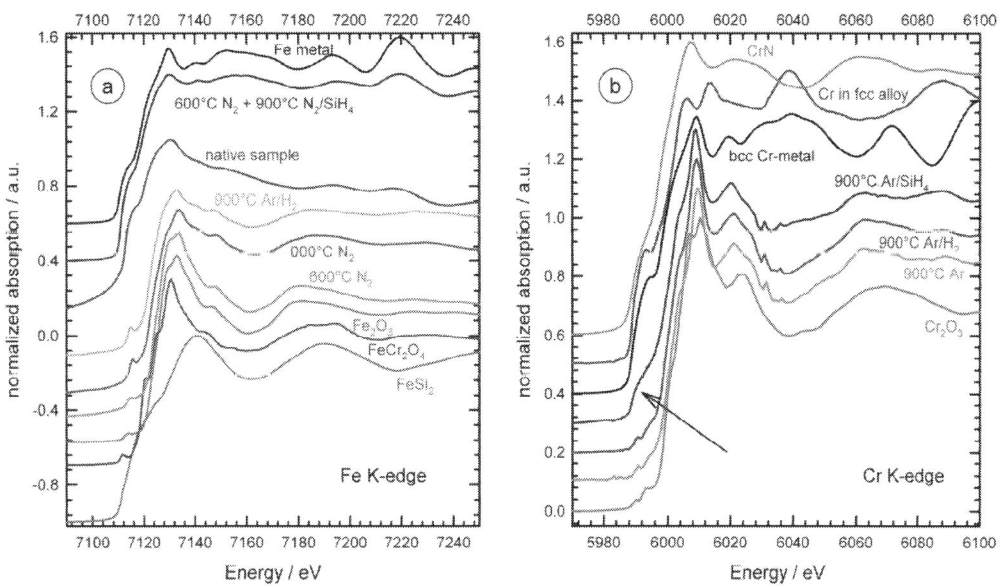


**Figure 2.** Comparison of the XANES spectra determined from ex-situ reflection mode EXAFS spectra (incidence angle $\phi = 0.1°$) of several steel samples with different heat treatments as indicated, with spectra of suited reference compounds: (a) Fe K-edge, (b) Cr K-edge. The arrow in (b) indicates the position of the Cr K-edge of the metal in the sample treated in silane containing atmosphere.

From the present experiments it remains however unclear, at which stage of the heat treatment the Cr-oxides are formed: Due to the ex-situ processing, it cannot be excluded that the oxides have been entirely removed in the hot zone of the conveyor belt furnace. The underlying thermodynamics indicate that even $Cr_2O_3$ should decompose for the applied temperatures and gas compositions [1]. However, as also suggested by the heat treatments conducted at lower temperatures around 600 °C,

$Cr_2O_3$ and also the less stable iron oxides were still present on the surface of the steel after the heat treatments. This suggests that the oxides found after the removal of the samples from the furnace were formed during the cooling phase from the hot central temperature back to room temperature.

### *3.2. In-situ EXAFS studies*

According to the ex-situ measurements presented in the previous section, a de-oxidation was performed under vacuum conditions in order to study the thermal stability of iron and chromium oxides at elevated temperatures. An example of a grazing incidence EXAFS measurement at 900 °C is presented in figure 3. As can be seen, structures of metallic iron can at least be partly recovered in comparison to the air-oxidized surface. However, it is important to keep in mind the iron phase diagram with substantial contributions of cubic face centered Fe (austenite) for $T > 700$ °C, so that a more detailed temperature dependent investigation of the base steel is required here. At the Cr K-edge, no substantial changes of the XANES could be found during the first de-oxidation experiments in vacuum, suggesting that the oxygen partial pressure might still be to large for a Cr-oxide removal.

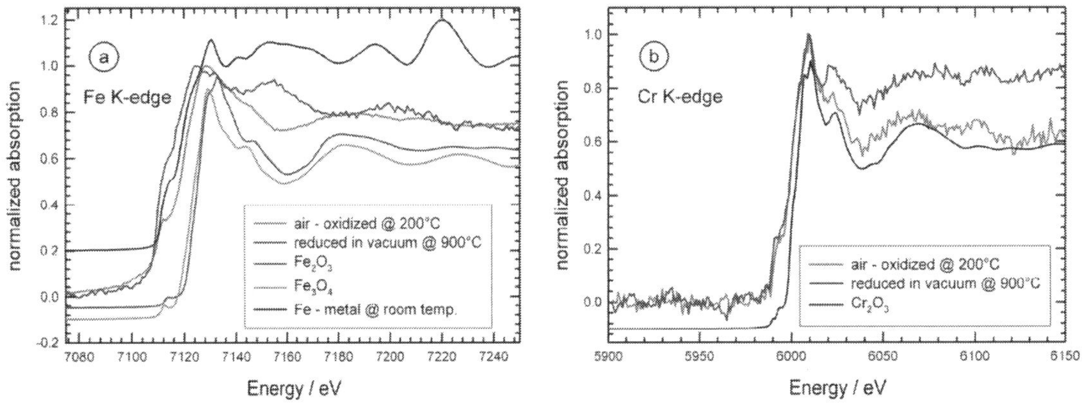


**Figure 3.** Comparison of the XANES from a steel sample measured in-situ in reflection at elevated temperature in vacuum with the XAFS of reference compounds: (a) Fe K-edge and (b) Cr K-edge.

## 4. Conclusions

Ex-situ and in-situ XAFS experiments have been conducted to study the thermal de-oxidation of steel surfaces. The results have shown that the iron oxides can be completely removed by a treatment in $SiH_4$-containing gas atmospheres as well as in vacuum for $T \approx 900$ °C. By contrast, contributions of Cr-oxides are always detectable after the thermal treatments. The decomposition of those oxides as well as in-situ experiments during $SiH_4$-treatments will be subject of future experiments.

## Acknowledgments

The authors like to acknowledge the help of C. Baehtz with the setup at the ESRF. The study was financially supported by the DFG under project no. FR 1388/8-1 and Ma 1175/48-1.

## References

[1] Bach FW, Möhwald K and Holländer U 2010 *Key Eng. Mater.* **438** 73
[2] Jung H and Thompson WJ 1991 *J. Catal.* **128** 218
[3] Bach FW, Möhwald K, Holländer U and Roxlau C 2007 *Tagungsband zum 8. Int. Kolloq. Hart- und Hochtemperaturlöten und Diffusionsschweißen. DVS-Berichte* **243** pp. 235–241
[4] Lützenkirchen-Hecht D, Wagner R, Szillat S, Hüsecken AK, Istomin K, Pietsch U and Frahm R 2014 *J. Synchrotron Rad.* **21** 819
[5] Pizzini S, Roberts KJ, Greaves GN, Harris N, Moore P, Pantos E and Oldman RJ 1989 *Rev. Sci. Instrum.* **60** 2525
[6] Heald SM 1992 *Rev. Sci. Instrum.* **63** 873

# Atomic structure of PtCu nanoparticles in PtCu/C catalysts prepared by simultaneous and sequential deposition of components on carbon support

**L A Bugaev, V V Srabionyan, V V Pryadchenko, A L Bugaev, L A Avakyan, S V Belenov and V E Guterman**
Southern Federal University, Physical and Chemical Faculties, 344090, Zorge str. 5, Rostov-on-Don, Russia

E-mail: bugaev@sfedu.ru

**Abstract.** Nanocatalysts PtCu/C with different distribution of components in bimetallic PtCu nanoparticles (NPs) were synthesized by simultaneous and sequential deposition of Cu and Pt on carbon support. Electrochemical stability of the obtained samples PtCu/C was studied using the cyclic voltammetry. Characterization of atomic structure of as prepared PtCu NPs and obtained after acid treatment was performed by Pt $L_3$- and Cu $K$-edge EXAFS using the technique for determining local structure parameters of the absorbing atom under strong correlations among them. EXAFS derived parameters were used for generation of structural models of PtCu NPs by the method of cluster simulations. Within this approach, the models of atomic structure of PtCu NPs obtained by the two methods of synthesis, before and after post treatment and after two months from their preparation were revealed.

## 1. Introduction

Besides of other valuable applications, the platinum NPs on carbon supports are a basic component of catalytic layers in low-temperature fuel cells. The main reasons that prevent commercial use of these materials are relatively high cost and insufficient corrosive and morphological stability of NPs during the operation. An efficient way to overcome these problems is the synthesis of catalytic layer containing bimetallic NPs of core-shell architecture, with a core of transition metal (Me) atoms and a shell of Pt atoms. For such architecture, the decisive factors determining the catalytic properties of bimetallic NPs are the thickness, composition and continuity of the shell, its atomic structure, shape, surface morphology, stability of the "core-shell" structure in time and during operation.

Despite the progress in synthesis and applications of bimetallic PtMe NPs with the "core-shell" architecture [1,2], the development of reliable and technological methods for the production of catalysts containing large amounts of core-shell NPs on the carbon supports remains a challenging problem. The development of these methods requires in turn the improvement of the methods for diagnostics of NPs atomic structure, which enable to reveal the high concentration of Pt atoms in the near-surface region of NPs – necessary (but not sufficient) condition for high adsorption and catalytic activities of PtMe/C electrocatalysts.

## 2. Experimental results

In this work PtCu/C nanocatalysts were synthesized by two different approaches: simultaneous (obtained samples are referred to as PtCu_sim/C) and sequential (PtCu_seq/C) chemical reduction of

$Cu^{2+}$ ($CuSO_4$) and Pt (IV) ($H_2PtCl_6$) in a carbon (Vulcan XC-72, Cabot) suspension based on a two-component water-organic solvent at pH = 10 (excess of $NH_3$). In the case of sequential reduction, first, nanostructured material Cu/C was obtained by the reduction of $Cu^{2+}$ and, then, without removing of Cu/C from the solution, a certain amount of an aqueous solution of $H_2PtCl_6$ containing an excess of $NH_3$ was added. In the case of simultaneous deposition the metals reduction was performed from a solution containing a mixture of platinum and copper compounds. It is reasonable to assume that the metal component in the resultant PtCu/C catalyst should have the structure of a PtCu solid solution [3]. A part of the resulting PtCu/C catalyst was exposed to post-treatment: the materials were maintained at room temperature (RT) in 1 M nitric acid for 1 hour and then washed and dried. The corresponding materials are referred to as PtCu_sim-at/C and PtCu_seq-at/C respectively.

XRD study of synthesized materials showed, that PtCu_sim/C and PtCu_sim-at/C contain only one phase corresponding to PtCu solid solution, while PtCu_seq/C and PtCu_seq-at/C contain phases of Pt and Cu. These two phases can exist both in the core-shell NPs and as separate Pt and Cu NPs. Copper oxide, which probably exists in all samples according to EXAFS data, is amorphous and therefore is not seen as a separate phase.

Electrochemical measurements were performed at RT using a conventional three-electrode cell. A 0.1 M $HClO_4$ solution saturated with Ar at atmospheric pressure was used as an electrolyte. The glassy carbon electrodes with an applied layer of electrocatalyst powder were used as the working electrodes. The platinum wire was employed as the counter electrode, the Ag/AgCl/KCl saturated electrode – as the reference electrode. The electrochemical surface area of the electrode (ESA) was determined by the quantity of electricity expended for electrochemical adsorption/desorption of atomic hydrogen with input of double layer region [4]. Study of the stability of PtCu NPs was performed using the long-term voltammetric cycling (1000 cycles). Cyclic voltammograms, recorded every 100 cycles (Figure 1), were used to obtain the dependencies of ESA upon the number of cycle for synthesized materials PtCu_seq/C, PtCu_seq-at/C and for the commercial Pt/C material TEC10v30e.

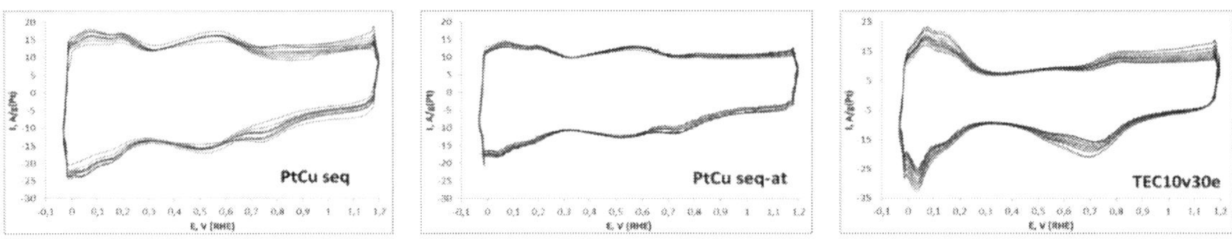


**Figure 1.** Cyclic voltammograms for materials PtCu_seq/C, PtCu_seq-at/C and for the commercial Pt/C material TEC10v30e recorded during 1000 cycles.

Structural characterization of the synthesized PtCu NPs after two months storage was performed by Pt $L_3$- and Cu $K$-edge EXAFS. The spectra were measured at the μSpot beamline of the BESSY-II SRF (Berlin, Germany). Mean electron current of a storage ring was maintained during experiment at 300 mA. The measurements were performed in the transmission mode utilizing a Si(111) channel-cut monochromator, two ionization chambers and a diode for reference channel. The photon energy scanning steps were adjusted to $\delta E$ = 1.0 eV and $\delta k$ = 0.05 $Å^{-1}$ in the XANES and EXAFS respectively, where $E$ is the energy of incident radiation and $k$ is the photoelectron wavenumber.

The processing of EXAFS in bimetallic NPs requires a large number of structural parameters. Because of the correlations among them, the best-fit set of parameters, obtained from the condition of $R$-factor minimum, may differ significantly from the "real" set since the model used for the fit never includes all contributions, which determine the spectrum fine structure. As a result, the effect of unaccounted contributions is redistributed over the values of correlating parameters in the used model, giving instabilities in the fit parameters depending upon the used $\Delta k$ intervals and $k$-weights for $\chi^{exp}(k)$ in FT procedure. To reduce the effect of these correlations, the fitting technique of [5] was implemented, which gives the values of structural parameters using different $\Delta k$ and $k$-weights.

Fit of the Fourier-transforms of EXAFS – $F(R)$, was performed using the expressions for $\chi(k)$ based on the qualitative analysis of $F(R)$ in PtCu/C and XRD data:

$$\chi_{Pt}(k) = N_{Pt-Pt} \cdot \chi_{Pt-Pt} + N_{Pt-Cu} \cdot \chi_{Pt-Cu}; \quad (1)$$

$$\chi_{Cu}(k) = \xi_1 \cdot (N_{Cu-Cu} \cdot \chi_{Cu-Cu} + N_{Cu-Pt} \cdot \chi_{Cu-Pt}) + \xi_2 \cdot N_{Cu-O,C} \cdot \chi_{Cu-O,C}, \quad (2)$$

where $\chi_{A-B}$ –contribution of one atomic pair A-B (A, B denote Pt or Cu); $\chi_{Cu-O,C}$ – contribution of interaction of Cu with light atoms O, C; $\xi_1$, $\xi_2$ – the fractions of Cu atoms interacting with metal atoms and with light atoms respectively ($\xi_1 + \xi_2 > 1$, since a part of $\xi_1$ atoms interact also with O, C atoms and a part of $\xi_2$ atoms interact also with metal ones). Therefore, the independently varied parameters were: amplitudes of each term in (1) and (2), corresponding Debye-Waller parameters ($\sigma^2_{A-B}$), and distances ($R_{A-B}$). The fit gave the coincidence of parameters: $R_{Pt-Cu} \approx R_{Cu-Pt}$ (up to 0.01Å) and $\sigma^2_{Pt-Cu} \approx \sigma^2_{Cu-Pt}$. The last is quite a random, but obtained reasonable for RT values of $\sigma^2$ indicate that the fit models (1), (2) enabled to account for Pt and Cu local structures in the sample. Parameters of the third term in (2) are less informative for PtCu NPs structure, however, the obtained value $R_{Cu-O,C} = 1.93$ Å is in agreement with the Cu-O distance in copper oxide. Therefore, in Table 1 the values of the following parameters are presented: $N_{Pt-Pt}$, $N_{Pt-Cu}$, $\xi_1 \cdot N_{Cu-Pt}$, $\xi_1 \cdot N_{Cu-Cu}$, $R_{Pt-Pt}$, $R_{Pt-Cu}$, $R_{Cu-Cu}$, $\sigma^2_{Pt-Pt}$, $\sigma^2_{Pt-Cu}$, $\sigma^2_{Cu-Cu}$. The fit quality, obtained by this set of parameters, is characterized by the values of $R$-factors $\leq 0.015$ for both Pt $L_3$- and Cu $K$-EXAFS, which were calculated for $\Delta k$ (3.0 – 15.7 Å$^{-1}$) and $k^{n=2}$.

**Table 1.** Parameters of Pt and Cu local atomic structure in materials PtCu/C.

| Material | $N_{Pt-Pt}$ | $N_{Pt-Cu}$ | $\xi_1 \cdot N_{Cu-Cu}$ | $\xi_1 \cdot N_{Cu-Pt}$ | $R_{Pt-Pt}$ | $R_{Pt-Cu}$ | $R_{Cu-Cu}$ | $\sigma^2_{Pt-Pt}$ | $\sigma^2_{Pt-Cu}$ | $\sigma^2_{Cu-Cu}$ |
|---|---|---|---|---|---|---|---|---|---|---|
| PtCu_sim/C | 6.6 | 4.5 | 2.9 | 4.0 | 2.73 | 2.64 | 2.55 | 0.006 | 0.016 | 0.009 |
| PtCu_sim-at/C | 7.4 | 2.6 | 1.5 | 6.6 | 2.73 | 2.66 | 2.61 | 0.007 | 0.016 | 0.010 |
| PtCu_seq/C | 8.4 | 2.5 | 3.5 | 2.2 | 2.75 | 2.62 | 2.54 | 0.006 | 0.015 | 0.010 |
| PtCu_seq-at/C | 8.8 | 2.3 | 3.0 | 4.8 | 2.75 | 2.66 | 2.57 | 0.006 | 0.016 | 0.009 |

## 3. Discussion and conclusions

Broad peaks at 0.5 V (Fig. 1) are associated with the oxidation/reduction of functional groups on the carbon support surface [6]. The absence of anodic peaks corresponding to pure Cu dissolution (0.2 - 0.3 V) or Cu dissolution from the PtCu alloy (0.6 - 0.8 V) on the cyclic voltammograms for the materials PtCu_seq/C and PtCu_seq-at/C indirectly confirms the absence of the Cu/electrolyte interface.

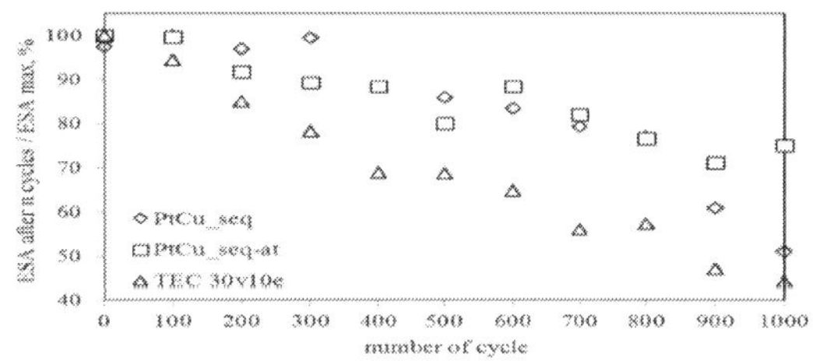


**Figure 2.** The dependencies of ESA upon the number of cycle for materials PtCu_seq/C (diamonds), PtCu_seq-at/C (squares) and for the commercial Pt/C material TEC10v30e (triangles).

The study of changes in the ESA of catalysts in the process of long-term cycling enabled to compare their morphological stability. Figure 2 shows the different decrease of the ESA value in the considered samples with the increase of the number of cycle. The ESA of PtCu_seq/C electrocatalyst decreased after 1000 cycles by 49% from the maximum value of 36 to 18.6 m$^2$/g (Pt). Meanwhile, one can conclude that stability of PtCu NPs increases after acid treatment, since for PtCu_seq-at/C material the ESA decreased after 1000 cycles only by 30% from the maximum value of 28 to 20 m$^2$/g (Pt). For comparison, the stability of the commercial Pt/C material TEC10v30e was also studied: the ESA for this material decreased by 56% from its maximum value of 80 to 35.5 m$^2$/g after a similar test. Thus, one can see that the commercial Pt/C catalyst degraded significantly faster in comparison with prepared PtCu_seq/C and PtCu_seq-at/C catalysts.

### *3.1. Cluster modeling of PtCu nanoparticles*

The study of averaging over the sizes of PtMe NPs and over the Pt:Me ratio in them [5] enabled to conclude that EXAFS derived structural parameters can be considered as the characteristics of the mean PtMe NP, representative for the PtMe/C catalyst, prepared by one of the considered methods. Meanwhile, to determine the relationship between the synthesis conditions – atomic structure –

electrochemical performances of these materials, it is necessary to visualize the character of components distribution over the volume of NP in correspondence with the values of the local structure parameters derived from EXAFS.

To suggest the models of NPs in materials PtCu/C the method of atomic cluster simulations was implemented using the clusters size $D$ = 8 nm according to TEM data. The comparison of $F(R)$ of Pt $L_3$- and Cu $K$- EXAFS of the studied materials with the corresponding $F(R)$ of Pt and Cu foils in the extended $R$-range (up to ~6.0 Å) enables to conclude, that the cluster models of PtCu NPs structure can be generated using the *fcc* cooper cluster, replacing Cu atoms by Pt ones. The placement of atoms in the cluster was performed by fitting the parameters of radial distribution function of Pt and Cu atoms so as to match the values of EXAFS derived parameters: $N_{Pt\text{-}Pt}$, $N_{Pt\text{-}Cu}$ and the ratio $N_{Cu\text{-}Cu}/N_{Cu\text{-}Pt}$ (since $\xi_1$ and $\xi_2$ are undetermined). Cu:Pt ratios obtained by XRF method were used as the boundaries for Cu:Pt ratios in generated clusters: $Cu^{model}$:$Pt^{model}$ < $Cu^{XRF}$:$Pt^{XRF}$, because a part of Cu atoms can exists in oxide states, independently from NPs. By this technique, the models of NPs in materials PtCu/C were generated (Figures 3, 4). These models show that the components distribution in NPs PtCu_sim corresponds to homogeneous solid solution with Cu:Pt ratio ~0.7 (XRF: 0.9) (Figure 3a), while in PtCu_seq the core-shell structure is realized with Cu:Pt ratio ~0.5 (XRF: 0.7) (Figure 4a). After acid treatment, NPs PtCu_sim-at contain a thin (~one-two atomic layers) Pt-shell corresponding to Cu:Pt ratio ~0.35 (XRF: 0.45) (Figure 3b). The acid treatment of PtCu_seq/C removes Cu cores of NPs with discontinuous shells, forming pure Pt NPs. This is reflected in the relative thickening of the Pt shell in the cluster model of the mean PtCu NP with fixed size, corresponding to Cu:Pt ratio ~0.3 (XRF: 0.4) (Figure 4b). The smearing of the boundary of Cu-core in Figure 4b can be caused by the atoms diffusion due to acid treatment and two months storage of materials, leading to the structural changes. The suggested models are only the possible ones, but the obtained character of components distribution is rather stable within the accuracy (≤ 7%) of $N$ determination by the used processing of EXAFS [5]. To solve the problem of stability and unambiguity of the generated models of NPs the used technique of visualization needs further improvement.

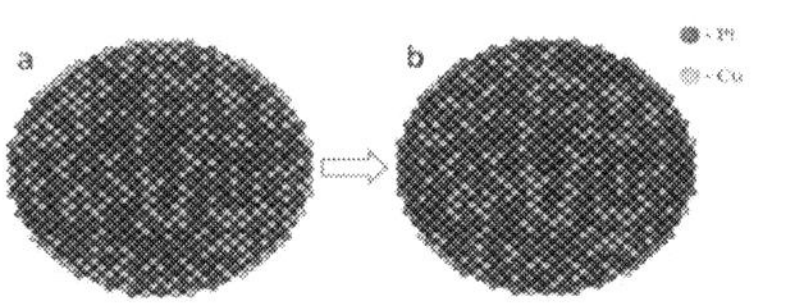


**Figure 3.** Slices of the atomic clusters modeling NPs in PtCu_sim/C (a) and PtCu_sim-at/C (b).

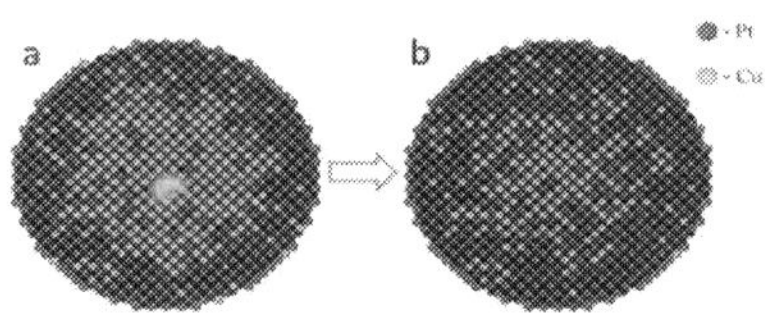


**Figure 4.** Slices of the atomic clusters modeling NPs in PtCu_seq/C (a) and PtCu_seq-at/C (b).

## 4. Acknowledgements

The work was supported by Grant 213.01.-07.2014/10ПЧВГ of Southern federal university. Authors are thankful to I. Zizak (INOT, Helmholtz-Zentrum Berlin) for the help during EXAFS measurements.

## 5. References

[1] Sankar M, Dimitratos N, Miedziak PJ, Wells PP, Kiely CJ and Hutchings GJ. 2012 *Chem. Soc. Rev.* **41** 8099; Russell A E and Rose A. 2004 *Chem. Rev.* **104** 4613–35.

[2] Xu Y, Chen L, Wang X, Yao W and Zhang Q. 2015 *Nanoscale* **7** 10559–83.

[3] Guterman VE, Lastovina TA, Belenov SV, Tabachkova NY, Vlasenko VG, Khodos I and Balakshina EN. 2014 *J. Solid State Electrochem.* **18** 1307-17.

[4] Schulenburg H, Durst J, Muller E, Wokaun A and Scherer GG. 2010 *J. Electroanal. Chem.* **642** 52-60.

[5] Pryadchenko VV, Srabionyan VV, Mikheykina EB, Avakyan LA, Murzin VY, Zubavichus YV, Zizak I, Guterman VE and Bugaev LA. 2015 *J. Phys. Chem. C* **119** 3217-27; Srabionyan VV, Bugaev AL, Pryadchenko VV, Avakyan LA, van Bokhoven JA and Bugaev LA. 2014 *J. Phys. Chem. Solids* **75** 470-6.

[6] Antolini E. 2009 *Appl. Catal. B Environ.* **88** 1–24.

# Structural response of Ni/$ZrO_2$ to feed modulations during $CH_4$ reforming reactions

**M Steib, A Jentys* and J A Lercher**

Department of Chemistry, Catalysis Research Center, Technische Universität München; Lichtenbergstraße 4, 85747 Garching, Germany

E-mail:jentys@tum.de

**Abstract**. Time resolved oxidation and reduction cycles of $ZrO_2$-supported Ni catalysts for dry reforming of $CH_4$ ($CO_2 + CH_4 \leftrightarrow 2\ CO + 2\ H_2$) during feed modulations have been studied. Under reaction conditions (1073 K) Ni remains fully reduced, whereas switching the feed gas to pure $CO_2$ results in a slow (25 sec) formation of a stable NiO phase. When switching back to reaction conditions the NiO phase is rapidly reduced (~ 1 sec) to metallic Ni. In the context of this study, a novel capillary cell has been built, allowing the parallel treatment of 5 catalyst samples with different gas compositions and different pressures. A comparison of the capillary cell to conventional systems regarding the spectral quality and the kinetic data shows that the capillary cell can be used to obtain identical kinetic data and high quality X-ray absorption spectra.

## 1. Introduction

Dry reforming of $CH_4$ ($CO_2 + CH_4 \leftrightarrow 2\ CO + 2\ H_2$) produces a synthesis gas with a $H_2$/CO ratio close to one. To reach acceptable conversion levels, reaction temperatures around 1000 K are required to shift the equilibrium to the side of the products. At this temperature metal sintering and agglomeration of the active phase play an important role. Since the usage of non-noble metals makes the process more economic, metals like Co or Ni were intensively studied in the last decades. Ni shows, in contrast to noble metals like Pt or Ir, a strong deactivation, that depending on the support can result from coking or metal sintering. Since coke deposited on the metal can be removed with gasifying agents such as $O_2$ or $CO_2$, cycle stability of the catalysts becomes an important issue for a successful catalyst development. Takenaka et al. [1] reported that gasifying deposited carbon over a $SiO_2$ supported Ni catalysts leads to a higher rate of particle agglomeration, since $SiO_2$ shows only weak metal-support interaction. Wei et al. [2] have shown that the initial activity of a modified Ni/$ZrO_2$ catalyst for $CH_4$ decomposition can be restored by gasifying the deposited carbon with $CO_2$. The periodic treatment of a Ni/$ZrO_2$ catalyst in $CO_2$ is also applicable for DRM, but there is no study on the influence of this treatment on the structural properties of the active metal published so far.

In this study the structural changes of Ni/$ZrO_2$ during cycles of $CH_4$ dry reforming and periodic gasification of the deposited carbon with $CO_2$ have been investigated. To effectively utilize the synchrotron radiation, a novel five-fold capillary cell has been developed, allowing the pretreatment of five catalysts in parallel. This cell has been tested with respect to spectral quality and comparability of the kinetic data to a lab-scale reactor for dry reforming of $CH_4$.

## 2. Experimental

Ni supported on $ZrO_2$ (Mel chemicals, XZO1501/03) was prepared via incipient wetness impregnation of an aqueous solution of Ni (II) nitrate hexahydrate (Acros) as described previously.[3] After impregnation, the catalyst was dried over night at 373 K, calcined in air at 673 K for 2 h (flow rate: 100 $mL \cdot min^{-1}$) and pre reduced in pure $H_2$ at 873 K for 12 h (heating rate 0.5 $K \cdot min^{-1}$).

The activity tests were carried out in a 4 mm plug flow reactor made of quartz glass. The catalyst was fixed between two plugs of quartz wool. Prior to reaction at 1073 K the catalyst was reduced *in situ* at 873 K in 10 % $H_2$ in He for 90 minutes. The reaction was carried out at ambient pressure with a reactant gas consisting of 15 vol.% $CH_4$ (99.995 vol.%), 15 vol.% $CO_2$ (99.995 vol.%) and 70 vol.% He (99.999 vol.%) with a total flow of 80 $mL \cdot min^{-1}$. The catalytic test in the capillary was carried out at identical conditions. To obtain the same space velocity the total flow was adjusted to 10 $mL \cdot min^{-1}$. The product gas was analyzed with an online mass spectrometer (Pfeiffer Omistar GSD 320).

The X-ray absorption spectra were obtained at the SuperXAS beamline at the SLS synchrotron in Villigen, Switzerland. All data were recorded in transmission mode at the Ni K-edge at 8333 eV using a QEXAFS channel-cut Si(111) monochromator operated at an oscillation frequency of 2 Hz. All *in situ* X-ray absorption spectra were measured in a capillary made of quartz glass (1 mm diameter, 20 μm wall thickness, Hilgenberg), were the pressed catalyst was fixed with two plugs of quartz wool. After reduction at 873 K in 10 % $H_2$ in He for 90 minutes, the catalyst was heated to 1073 K in He with a hot air blower from below before exposing the catalyst to the reactant gas. After 10 minutes, the reactor was flushed with He before introducing $CO_2$ to the catalysts. The capillaries were sealed in a U-shaped sample holder allowing the parallel treatment of 5 samples (Figure 1).

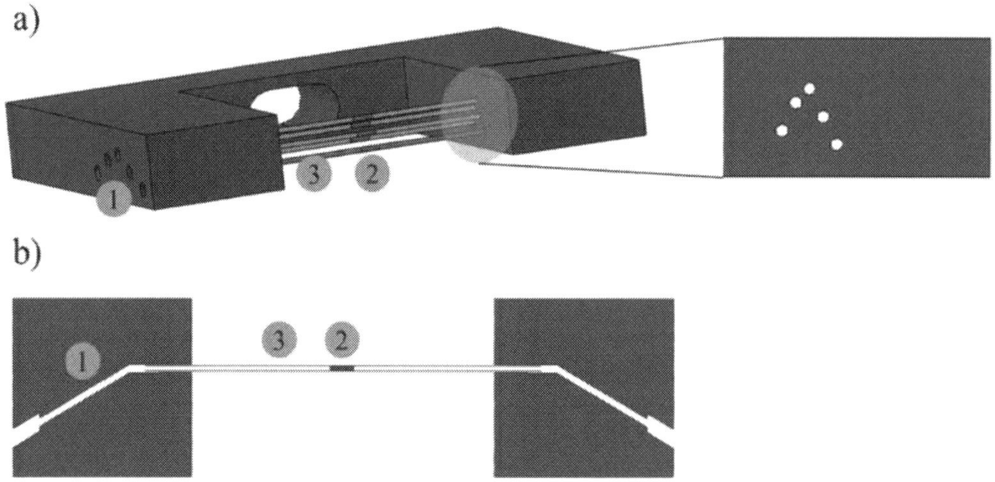


Figure 1: 3D model of the in situ cell (a) and horizontal cut to see the gas supply lines (b), 1: holes for gas supply lines; 2: catalyst bed; 3: capillaries.

The capillaries were located in a vertical distance of 4 mm (center to center) and in a horizontal distance of 4.5 mm (Figure 1 a). For gas supply, 1/8" tubes were mounted at the outside of the cell (Figure 1 b). To compare the signal quality obtained from the capillaries to a conventional pellet cell, a self-supporting wafer of the 10 wt.% $Ni/ZrO_2$ catalyst was prepared and spectra were recorded before and after reduction in 10 % $H_2$ in He.

## 3. Results and discussion

*3.1. Comparison of the capillary cell with conventional reactors with respect to XAS and kinetics*

Figure 2 a) and b) show the $k^2$-weighted EXAFS obtained in a conventional pellet cell [4] in comparison to the capillary of 1 mm outer diameter. The EXAFS are identical up to k = 9 $Å^{-1}$, for higher values of k the noise level in the capillary cell becomes higher.

Since operando studies should be carried out under identical conditions as the kinetic studies in the lab, reactions in a capillary should show the same rates compared to a lab-scale reactor. Figures 2 c) and d) show a comparison of the conversions obtained in a 1 mm capillary and in a 4 mm plug flow reactor used for kinetic studies of the dry reforming of $CH_4$ at the same space velocity. Both reactor types showed similar $CO_2$ and $CH_4$ conversion levels as well as equal $H_2$/CO ratios in the product gas, which proves that both reactors have an identical kinetic behaviour.

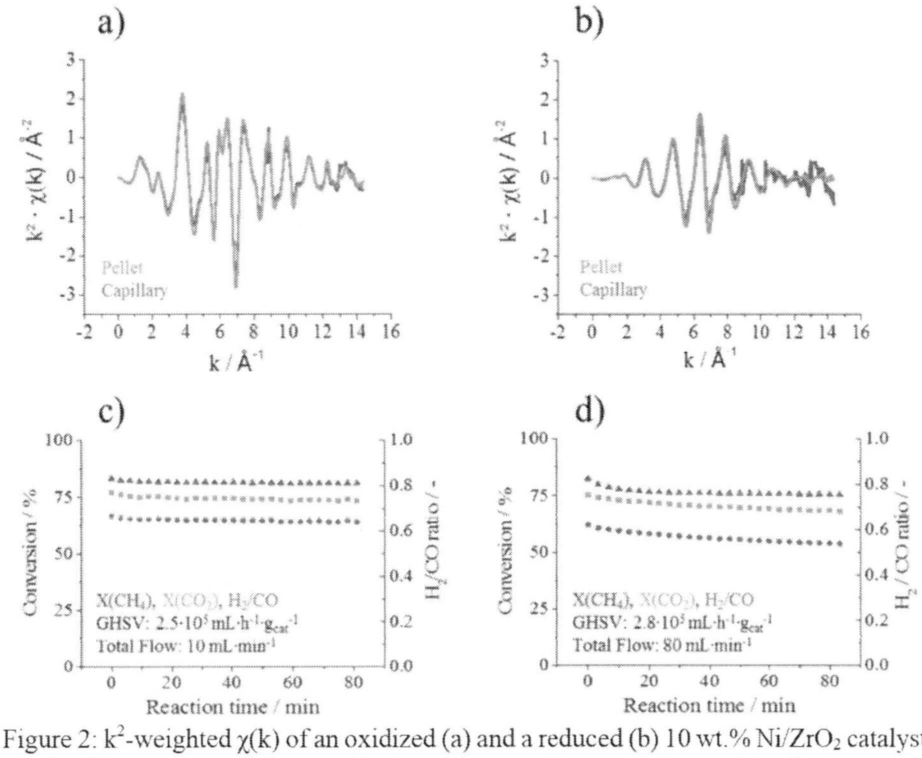


Figure 2: $k^2$-weighted $\chi(k)$ of an oxidized (a) and a reduced (b) 10 wt.% Ni/$ZrO_2$ catalyst obtained in a pellet cell (orange line) and in a capillary cell (1 mm, blue line) at room temperature and comparison of kinetic data obtained with a 2 wt.% Ni/$ZrO_2$ catalyst in a 1 mm capillary (c) and a 4 mm plug flow reactor (d); $CH_4$ conversion (blue dots), $CO_2$ conversion (orange squares) and $H_2$/CO ratio (red triangles).

*3.2. Structural changes of the Ni particles during feed modulations*

The changes in the oxidation state of the Ni/$ZrO_2$ catalyst during feed modulations between 15 % $CH_4$, 15 % $CO_2$ in He (reactant gas for $CH_4$ DRM) and pure $CO_2$ are shown in Figure 3. After the reduction at 873 K and the subsequent heating in He to 1073 K the catalyst was fully reduced. During the following cycle in $CO_2$ atmosphere, Ni was completely oxidized to NiO in ~ 25 sec. During the first reaction cycle (shaded area in figure 3 b), NiO was reduced again to metallic Ni (~ 3 sec), consuming $H_2$ formed in the dry reforming reaction of $CH_4$. Please note that in contrast to Ni/$SiO_2$ [5], Ni/$ZrO_2$ shows no sintering after $CO_2$ treatment (observed from the EXAFS, data not shown), which can be attributed to the stronger metal-support interaction between Ni and $ZrO_2$.

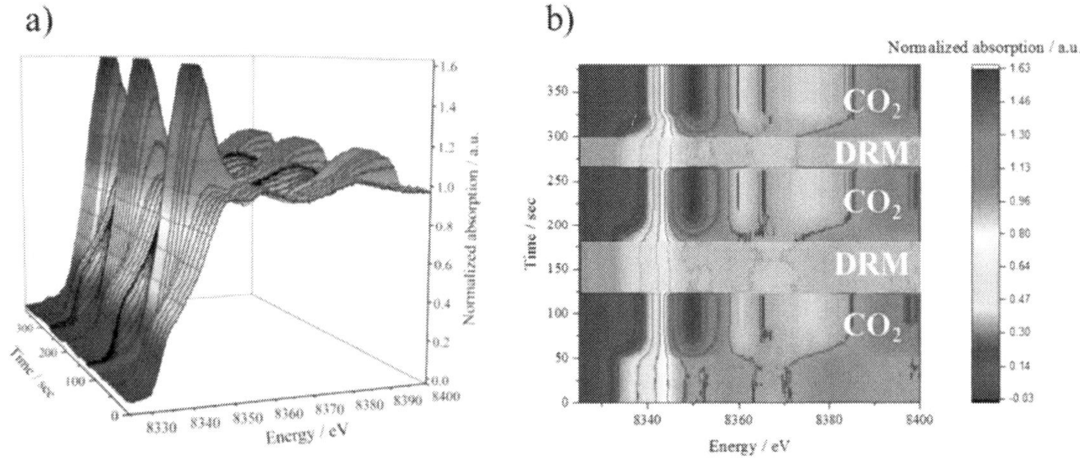


Figure 3: QEXAFS taken every second during feed modulations from reaction gas (grey area) to pure $CO_2$ (a) and color map of the QEXAFS (b); each cycle had a duration of 10 minutes, the time axis was adjusted for better visualization, T = 800 °C.

During the subsequent $CO_2$ treatment of the catalyst, deposited carbon was removed via the reverse Boudouard reaction ($C + CO_2 \leftrightarrow 2\ CO$, CO evolution detected via mass spectrometry, data not shown) and the Ni particles were oxidized to NiO. In the following reaction cycle, NiO was reduced to metallic Ni. Note that Ni shows the same degree of oxidation and the same coordination number after every cycle, which indicites that this procedure can be used to regenerate spent Ni catalysts, if the metal-support interaction is sufficiently strong to prevent agglomeration.

## 4. Conclusions

We have successfully shown the structural changes of a Ni/$ZrO_2$ catalyst during dynamic feed modulations using X-Ray absorption spectroscopy with very high time resolution. Switching the feed gas from dry reforming conditions ($CH_4 + CO_2$ in He) to pure $CO_2$ resulted in a slow oxidation (~ 25 sec) of the metallic Ni to NiO, whereas applying $CH_4$ and $CO_2$ resulted in a fast reduction of the NiO phase (~ 3 sec).

**Acknowledgements**

This work was financially supported by the BMBF under the project ZeitKatMat. The X-Ray absorption experiments were performed on the SuperXAS beamline at the Swiss Light Source, Paul Scherrer Institut, Villigen, Switzerland. The authors gratefully acknowledge the beamline scientist Maarten Nachtegaal and his team at the SuperXAS beamline for help during the beamtime.

## References

[1] S. Takenaka, E. Kato, Y. Tomikubo, K. Otsuka, Structural change of Ni species during the methane decomposition and the subsequent gasification of deposited carbon with $CO_2$ over supported Ni catalysts, J Catal, 219 (2003) 176-185.
[2] B.Q. Xu, J.M. Wei, Y.T. Yu, J.L. Li, Q.M. Zhu, Carbon dioxide reforming of methane over nanocomposite Ni/$ZrO_2$ catalysts, Top Catal, 22 (2003) 77-85.
[3] J.A. Lercher, J.H. Bitter, W. Hally, W. Niessen, K. Seshan, Design of stable catalysts for methane-carbon dioxide reforming, 11th International Congress on Catalysis - 40th Anniversary, Pts A and B, 101 (1996) 463-472.
[4] S. Simson, A. Jentys, J.A. Lercher, Dynamic Phase Separation in Supported Pd-Au Catalysts, J Phys Chem C, 119 (2015) 2471-2482.
[5] S. Takenaka, H. Ogihara, K. Otsuka, Structural change of Ni species in Ni/$SiO_2$ catalyst during decomposition of methane, J Catal, 208 (2002) 54-63.

# Dynamic transformation of small Ni particles during methanation of $CO_2$ under fluctuating reaction conditions monitored by *operando* X-ray absorption spectroscopy

**B Mutz[1,2], H W P Carvalho[1], W Kleist[1,2] and J-D Grunwaldt[1,2]***

[1]Institute for Chemical Technology and Polymer Chemistry (ITCP), Karlsruhe Institute of Technology (KIT), D-76131 Karlsruhe, Germany
[2]Institute of Catalysis Research and Technology (IKFT), Karlsruhe Institute of Technology (KIT), D-76344 Eggenstein-Leopoldshafen, Germany

Email: grunwaldt@kit.edu

**Abstract**. A 10 wt.-% $Ni/Al_2O_3$ catalyst with Ni particles of about 4 nm was prepared and applied in the methanation of $CO_2$ under dynamic reaction conditions. Fast phase transformations between metallic Ni, NiO and $NiCO_3$ were observed under changing reaction atmospheres using *operando* X-ray absorption spectroscopy (XAS). Removing $H_2$ from the feed gas and, thus, simulating a $H_2$ dropout during the methanation reaction led to oxidation of the active sites. The initial reduced state of the Ni particles could not be recovered under methanation atmosphere ($H_2/CO_2 = 4$); this was only possible with an effective reactivation step applying $H_2$ at increased temperatures. Furthermore, the cycling of the gas atmospheres resulted in a steady deactivation of the catalyst. *Operando* XAS is a powerful tool to monitor these changes and the behavior of the catalyst under working conditions to improve the understanding of the catalytic processes and deactivation phenomena.

## 1. Introduction

Triggered by the increasing importance of renewable energies e.g. from wind and solar plants long term energy storage becomes more and more important to balance the seasonal fluctuations in the power grid. A process which addresses this problem is the "power to gas" concept, in which $CO_2$ is hydrogenated with sustainable $H_2$ (e.g. from water splitting) according to the Sabatier reaction. The resulting synthetic natural gas is a diversely applicable energy carrier.[1]

Supported nickel catalysts are well-established in the methanation of $CO_2$ under stationary conditions.[2-4] However, the behavior and deactivation of the catalyst under dynamic conditions have not yet been studied in detail. For example, fast load changes will most probably occur under realistic conditions due to a fluctuating supply of renewable $H_2$. In previous studies we used *operando* X-ray absorption spectroscopy (XAS) to investigate the methanation of $CO_2$ under changing reaction atmospheres and observed a fast oxidation after removal of $H_2$ from the feed gas.[5] In this work we present an extended experiment revealing the effects of cycling and a reactivation step on the catalyst. In this study, a catalyst with more defined Ni nanoparticles supported on $\gamma$-$Al_2O_3$ was used.

## 2. Experimental

X-ray absorption spectroscopy experiments were conducted at the ANKA synchrotron radiation facility in Karlsruhe, Germany, (2.5 GeV ring, 100-150 mA) at the XAS beamline. The measurements were performed at the Ni K edge (8333 eV) in transmission mode using an X-ray beam of 1 mm in height and 7 mm in width. A time resolution of 90 s for X-ray absorption near edge structure (XANES) and 300 s for extended X-ray absorption fine structure (EXAFS) spectra was used. For the *operando* experiments a quartz glass capillary (1 mm diameter, 20 μm wall thickness) was filled with a sieved fraction of the catalyst powder (100-200 μm, diluted with $\gamma$-$Al_2O_3$ 1:1) to get a catalyst bed in 10 mm length. The capillary was heated to the desired temperatures of 30-500 °C (10 K/min) using an Oxford GSB-1300 hot air blower.[6] Individual mass flow controllers (Bronkhorst) were used for gas dosing and online gas analysis was conducted using a Pfeiffer Vacuum mass spectrometer (ThermoStar GSD 320 T1). Gas mixtures of 15 % $H_2$/He, 3.75 % $CO_2$/He and 15 % $H_2$ + 3.75 % $CO_2$/He with constant gas flow of 50 mL/min were used resulting in a weight hourly space velocity (WHSV) of 37 $h^{-1}$ (concerning $CO_2$ flow and catalyst mass). Data analysis was performed using the Athena and Artemis software of the IFEFFIT package.[7] Spectra of Ni and NiO nanoparticles as well as bulk $NiCO_3$ were used as references for linear combination analysis (LCA).

## 3. Results and discussion

Homogeneous deposition-precipitation using urea as precipitating agent [8, 9] led to a 10 wt.-% Ni/$Al_2O_3$ catalyst which was X-ray amorphous, i.e. it did not show any reflections of NiO or Ni after reduction but only the reflections of the $Al_2O_3$ support in X-ray diffraction. Evaluation of the scanning transmission electron microscopy (STEM) images resulted in a mean Ni particle size of 4 nm, a narrow size distribution (3.7±1.2 nm) and a dispersion of 21 %. According to the small particle size, the coordination number obtained from the EXAFS was calculated as 6.1±1.4 for the metallic Ni and is much lower compared to the coordination number in bulk Ni which is 12.0. The Ni-Ni bond distance of 2.46 Å is shortened (2.49 Å in bulk), showing a contraction of the bond length due to the smaller particle size. The specific surface area of the catalyst was 200 $m^2$/g. The catalyst was also tested in the laboratory with a comparable setup yielding in a turnover frequency (TOF) of 38 $h^{-1}$ (measurements analogous to the previous work [5]) and CO as the only byproduct.

Figure 1(a) visualizes the results of the *operando* XAS experiment which was performed based on earlier works in our group.[5] The corresponding XANES spectra in each atmosphere at 400°C can be seen in Figure 2(a). A pre-reduced catalyst was activated in a flow of 15 % $H_2$/He at an oven temperature of 500 °C (10 K/min, temperature 460-470 °C measured with a thermocouple at the location of the capillary) for about 40 min. Obviously, the catalyst was not fully reduced which might be due to the strong interaction of the small Ni particles with the $Al_2O_3$ support and the low $H_2$ concentration used for the activation. On the other hand, no $NiAl_2O_4$ spinel species could be identified in linear combination (LC) fitting. The capillary was cooled to room temperature to receive a spectrum of the reduced catalyst in good quality as reference.

Afterwards the gas atmosphere was switched to methanation conditions ($H_2$/$CO_2$ = 4, 81 % He) and the reactor was heated to 400 °C (10 K/min, real temperature 350-360 °C). Even at room temperature the catalyst was oxidized to some extent but was reduced again when the production of $CH_4$ started at about 200 °C. 80 % of the Ni nanoparticles remained in the reduced state during the 40 min methanation period at 400 °C. After removing $H_2$ from the gas stream 75 % of the Ni fraction were immediately oxidized in the less reducing atmosphere and the formation of $CH_4$ stopped as well in the absence of $H_2$. After another 40 min, $H_2$ was introduced again to restore the previous methanation conditions. In the first few minutes Ni was reduced fast and afterwards more slowly resulting in about 50 % metallic Ni. The oxidation of the catalyst during the next switch to the less reducing atmosphere took longer than during the first switch. In the third and fourth methanation cycle a behavior similar to the second cycle was observed which means that the fast reduction in the first few minutes was decelerating over time.

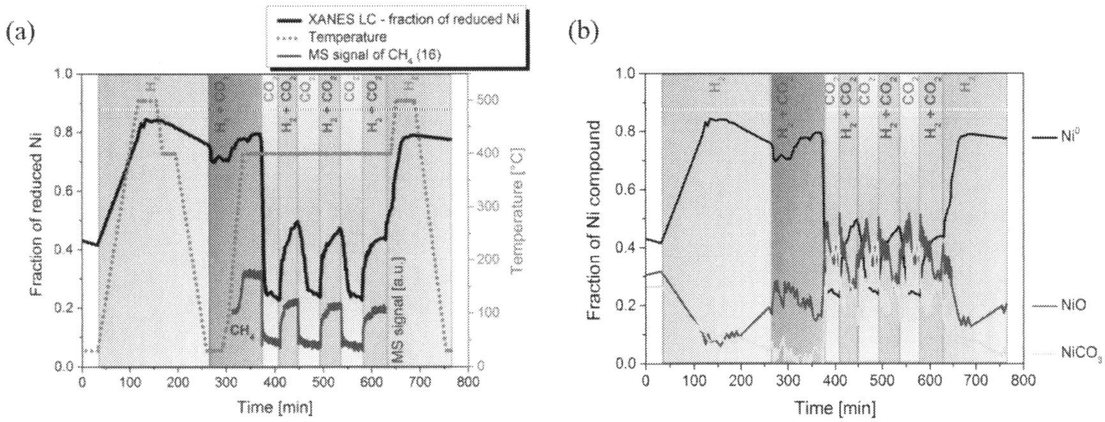


**Figure 1.** (a) *Operando* XAS experiment for the methanation of $CO_2$ with a 10 wt.-% $Ni/Al_2O_3$ catalyst at ANKA synchrotron Karlsruhe. The fraction of the Ni compounds was calculated using linear combination analysis (LCA) of the spectra. (b) Fraction of Ni, NiO and $NiCO_3$ during the experiment.

Overall, the maximum percentage of metallic Ni decreased with the number of cycles and also the produced amount of $CH_4$ decreased slightly. The first $H_2$ dropout caused a faster oxidation than the following cycles which might be due to oxygen contamination in the system. Therefore, a deep bulk oxidation occurred which was not fully reversible under methanation atmosphere. In the following cycles dynamic oxidation and reduction of the Ni particles with a small but continuous deactivation was observed.

Figure 1(b) shows the development of each Ni phase during the experiment calculated from linear combination analysis. After exposure to air the pre-reduced catalyst showed a Ni and NiO phase as well as $NiCO_3$ formation. After cooling to room temperature in $H_2$ atmosphere and the start of the methanation, NiO was formed and only traces of $NiCO_3$ were present in the catalyst. Removal of $H_2$ from the feed gas always led to the formation of NiO in the first step and its subsequent conversion into $NiCO_3$ (best seen in $CO_2$ cycle 2 and 3). In the following methanation atmosphere first a fast reduction of $NiCO_3$ occurred followed by a delayed slower reduction of NiO to metallic Ni. Roughly 30 % of each oxidized compound remained on the catalyst during the methanation sequences after the first $H_2$ dropout. In addition, a band around 1448 $cm^{-1}$ [10] in diffuse reflectance infrared Fourier transform spectroscopy (DRIFTS) experiments could be assigned to $NiCO_3$.

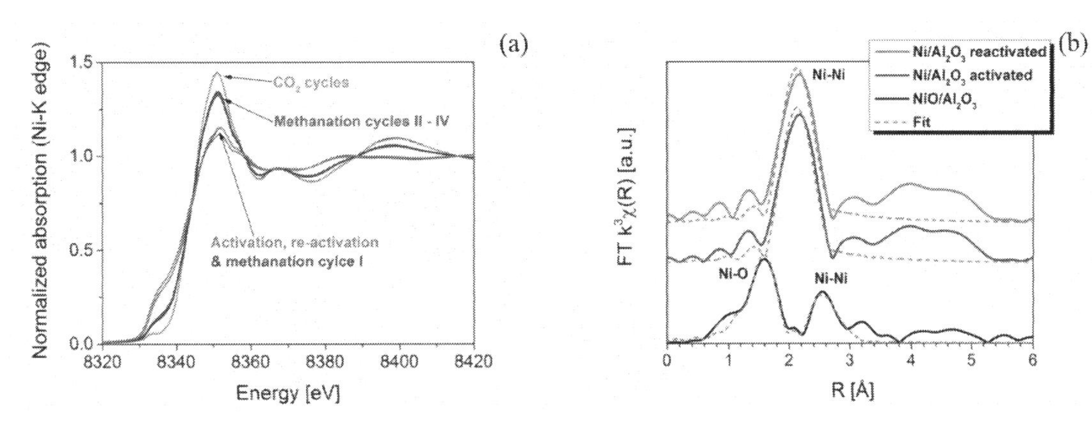


**Figure 2.** (a) Normalized XANES spectra at 400°C for each atmosphere according to Figure 1. (b) Magnitude of Fourier transformed $k^3$-weighted EXAFS spectra recorded at room temperature along with the corresponding fits.

As a result of the Ni oxidation during $H_2$ dropout a reactivation step was conducted at the end of the experiment recovering the same fraction of reduced Ni (78 % at room temperature, see Figure 1(a)) as after the activation in the beginning. Furthermore, an increase in the production of $CH_4$ nearly to the initial yield was observed in laboratory experiments (not shown here). As depicted in Figure 1(b) bulk NiO and $NiCO_3$ could only be reduced in a reactivation step flushing the reactor with $H_2$ at 500 °C. A reactivation step with higher $H_2$ concentrations might be even more effective.

Comparing the EXAFS spectra of the activated catalyst and the catalyst after the reactivation step at room temperature (Figure 2(b)), no significant changes could be observed and the Ni coordination number was calculated to the similar value of 6.0±1.4, the same for the Ni-Ni bond distance of 2.46 Å. Therefore, it can be concluded that the size of the particles did not change during this short experiment. Repeating the experiment several times in the laboratory led to Ni particle sintering up to a mean diameter of 5 nm which was determined from STEM images. Thus, further studies on the dynamic behavior of supported nanoparticles under fluctuating reaction conditions are currently performed in our group.

## 4. Conclusion

A 10 % $Ni/Al_2O_3$ catalyst with an average particle size of 4 nm and a narrow size distribution was prepared and applied in the methanation of $CO_2$ ($H_2/CO_2 = 4$) under dynamic reaction conditions. In a less reducing atmosphere after $H_2$ dropout metallic Ni was oxidized to NiO and $NiCO_3$. Accordingly a slight but steady deactivation of the catalyst with each cycle was observed. After adding $H_2$ to restore methanation conditions, $NiCO_3$ was first converted to NiO which was then reduced to metallic Ni. However, the initial fraction of reduced Ni could not be reached. An effective reactivation step with $H_2$ at 500 °C was able to recover the initial state of the catalyst. This experiment illustrates that *operando* XAS is a powerful tool to analyze catalysts under working conditions and to monitor changes of the active centers. This is crucial for a better understanding of the catalytic processes and the reasons for deactivation, especially in the methanation of $CO_2$ under dynamic reaction conditions.

## 5. Acknowledgement

This work was funded and supported by KIT and the Helmholtz Research School "Energy-Related Catalysis". The ANKA synchrotron radiation source and Dr. S. Mangold as well as the Karlsruhe Nano Micro Facility (KNMF) are gratefully acknowledged for providing beamtime and technical support. The BMBF project "MatAkt" (05K10VKB) is acknowledged for providing the infrastructure for the *in situ* studies. The authors thank Dr. A. Gaur, Dr. D. Doronkin and K. Hengst (XAS), Dr. D. Wang and S. Baier (STEM), H. Köhler (ICP-OES), and A. Beilmann ($N_2$ physisorption).

## References

[1] Sterner M 2009 *Erneuerbare Energien und Energieeffizienz - Renewable Energies and Energy Efficiency* 14 (Kassel: Kassel University Press GmbH)
[2] Du G, Lim S, Yang Y, Wang C, Pfefferle L and Haller G L 2007 *J. Catal.* **249** 370-9
[3] Ocampo F, Louis B and Roger A-C 2009 *Appl. Catal. A* **369** 90-6
[4] Garbarino G, Riani P, Magistri L and Busca G 2014 *Int. J. Hydrogen Energy* **39** 11557-65
[5] Mutz B, Carvalho H W P, Mangold S, Kleist W and Grunwaldt J-D 2015 *J. Catal.* **327** 48-53
[6] Grunwaldt J-D, van Vegten N and Baiker A 2007 *Chem. Commun.* 4635-7
[7] Ravel B and Newville M 2005 *J. Synchrotron. Radiat.* **12** 537-41
[8] Geus J W and van Dillen A J 2008 Preparation of Supported Catalysts by Deposition–Precipitation *Handbook of Heterogeneous Catalysis* (Weinheim: Wiley-VCH) pp 428-67
[9] Bitter J H, van der Lee M K, Slotboom A G T, van Dillen A J and de Jong K P 2003 *Catal. Lett.* **89** 139-42
[10] Westermann A, Azambre B, Bacariza M C, Graça I, Ribeiro M F, Lopes J M and Henriques C 2015 *Appl. Catal. B* **174–175** 120-5

Journal of Physics: Conference Series **712** (2016) 012051 doi:10.1088/1742-6596/712/1/012051

# Operando XAS study of the influence of CO and NO on methane oxidation by $Pd/Al_2O_3$

**V Marchionni[1], M Nachtegaal[1], A Petrov[1], O Kröcher[1,2] and D Ferri[1]**

[1] Paul Scherrer Institut, CH-5232 Villigen PSI, Switzerland

[2] Institute of Chemical Science and Engineering, École polytechnique fédérale de Lausanne (EPFL), CH-1015 Lausanne, Switzerland

E-mail: davide.ferri@psi.ch

**Abstract.** Methane oxidation on $Pd/Al_2O_3$ has been investigated using operando X-ray absorption spectroscopy (XAS) at the Pd K-edge. The influence of CO and NO on methane abatement has been addressed performing temperature programmed reaction runs while recording simultaneously XANES spectra. During CO oxidation Pd is reduced while methane conversion is shifted to higher temperature. NO strongly inhibits both CO and $CH_4$ oxidation, despite the higher fraction of oxidized palladium.

## 1. Introduction

Natural gas vehicles are a clean alternative to gasoline and diesel engines. Despite the numerous advantages of natural gas, unburned methane ($CH_4$) is a potent greenhouse gas; therefore, catalytic after-treatment of the exhaust gases is necessary to reduce its emissions. Typically, Pd-based three-way catalysts (TWC) are employed [1]. The oxidation state of Pd is a key parameter in the abatement of $CH_4$ under lean and stoichiometric conditions [2]. X-ray absorption spectroscopy (XAS) is a powerful tool to probe the oxidation state of Pd under relevant reaction conditions. Despite the reported positive effect of pre-reduction that produced the most active catalyst [3], highly dispersed oxidized Pd has been identified as the most active phase at low temperatures [4, 5]. In the case of $CH_4$ oxidation over three-way catalysts (TWC), the presence of CO and NO can modify the oxidation state of Pd, thus strongly influencing its reactivity. Only few spectroscopic studies considered the influence of other pollutants on $CH_4$ abatement on Pd-based catalysts. For example under simulated operation conditions of a TWC, Pd remained reduced at high temperature irrespective of gas feed composition and thus oxygen concentration, unless heavy lean ($O_2$ excess) conditions were applied [6]. $Pd^0$ was reduced at ca. 400°C under stoichiometric conditions of a $CH_4$-CO-NO feed but the effect of each single pollutant on $CH_4$ abatement was not unveiled [7]. Here, we study $CH_4$ oxidation on $Pd/Al_2O_3$ by operando XAS and DRIFT spectroscopy with the aim to follow changes of Pd oxidation state during reaction as function of the presence of CO and NO in the feed.

## 2. Experimental

The sample (1.6% $Pd/\gamma$-$Al_2O_3$, 40 mg, 50-100 $\mu$m sieve fraction, kindly provided by Umicore) was placed in a homemade plug-flow reactor cell [8] equipped with two graphite windows and was subjected to temperature programmed reaction runs (TPRR) under various reaction conditions while collecting XAS spectra in transmission at the Pd K-edge (24.35 keV). The outlet of the cell was connected to a mass spectrometer and to an FT-IR spectrometer for gas analysis. Prior to all experiments the sample was subjected to in situ pretreatment under stoichiometric conditions (0.25 vol% $CH_4$/1 vol% CO/1 vol% $O_2$ at 350°C for 1 h, GHSV = 120'000 $h^{-1}$) in order to equilibrate the catalyst. X-ray absorption near edge spectra (XANES) were collected every 10°C in the middle of the catalyst bed during consecutive TPRR under stoichiometric (0.25 vol% $CH_4$/1 vol% CO/1 vol% $O_2$ and 0.25 vol% $CH_4$/0.5 vol% $O_2$) and lean (0.25 vol% $CH_4$/1 vol% $O_2$) conditions at 2.5°C/min at the SuperXAS beamline of the SLS to investigate the influence of CO on $CH_4$ oxidation. XANES spectra (ca. 5 s/spectrum) were continuously collected during TPRR under stoichiometric conditions (0.25 vol% $CH_4$/1 vol% CO/1 vol% $O_2$ and 0.25 vol% $CH_4$/1 vol% CO/ 0.22 vol% NO/ 0.89 vol% $O_2$) at 10°C/min at beamline ID24 of the ESRF to analyse the effect of the addition of NO. Linear combination fitting was performed to quantify the amount of $Pd^{2+}$; the spectrum of the fresh sample, and the spectrum at the end of the $CH_4$/CO/$O_2$ TPRR, were used as references for $Pd^{2+}$ and $Pd^0$, respectively. All spectra were analysed with the IFEFFIT software package [9]. An identical stoichiometric $CH_4$/CO/$O_2$ TPRR (50-210°C, 10°C/min) was performed while collecting diffuse reflectance infrared spectra (DRIFTS) at 2 $cm^{-1}$ resolution using the same reactor cell equipped with a $CaF_2$ window. The background spectrum was taken in Ar at 50°C before the pretreatment.

## 3. Results and discussion

### *3.1. $CH_4$ abatement*

Fig. 1 shows the transmission XANES spectra collected during the $CH_4$/$O_2$ TPRR in lean conditions. Initially, Pd is fully reduced.During the lean TPRR, the spectra exhibit a progressive increase in the edge energy and in the white line intensity with increasing reaction temperature that reflect the gradual oxidation of Pd. Fig. 2 displays the corresponding temperature evolution of the fraction of $Pd^{2+}$ and the activity expressed in terms of $CH_4$ conversion. $CH_4$ conversion begins at about 230°C in presence of a mixed Pd/PdO phase (ca. 30% $Pd^{2+}$); the fraction of $Pd^{2+}$ increases with the reaction progress but full oxidation is not attained when full $CH_4$ conversion is reached at ca. 430°C. In an identical experiment under stoichiometric conditions, the fraction of $Pd^{2+}$ increases as well until about 380°C before Pd fully reduces as a consequence of the consumption of all available oxygen. PdO decomposition is a spontaneous thermodynamic process, which is strongly affected by the $O_2$ partial pressure [10]. The average oxidation state of Pd in stoichiometric conditions is lower than under lean conditions and $CH_4$ conversion occurs at higher temperature. The reaction mechanism for methane oxidation was found to be different for highly oxidized Pd nanoparticles [2, 11], possibly explaining the improvement in the catalytic activity in excess of $O_2$. The levels of $Pd^{2+}$ in both TPRR are different when $CH_4$ oxidation sets in and probably as a consequence the light-off is shifted to higher temperature in stoichiometric conditions.

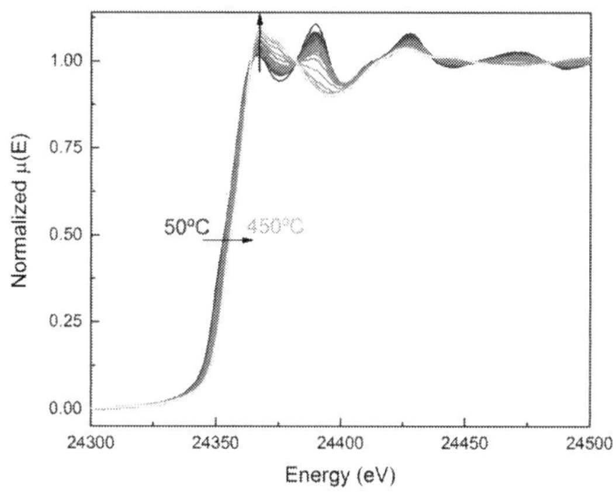


**Figure 1.** Representative operando XANES spectra collected during $CH_4$/$O_2$ TPRR in lean conditions (2.5°C/min).

### *3.2. Influence of CO*

In presence of CO in the feed, the initial $Pd^{2+}$ fraction is higher (ca. 30%) because the experiment was directly subsequent to the pre-treatment. The gradual oxidation of Pd observed in lean conditions is perturbed by an initial slow reduction to ca. 15% $Pd^{2+}$ until 175°C. The quick partial re-oxidation to ca. 30% $Pd^{2+}$ coincides with the sudden increase in CO conversion, which proceeds to completeness at ca. 220°C. The $O_2$ concentration decreases simultaneously to 50% when full CO conversion is reached in agreement with the stoichiometry of the CO oxidation reaction. An identical DRIFTS experiment (Fig. 3 a) demonstrates that the initial slow reduction is caused by CO adsorption on reduced Pd as monitored by the signal of CO bridged bonded to twofold metal sites (ca. 1986 $cm^{-1}$) [12]. Above 175°C the signal quickly vanishes concomitantly to the sudden but partial Pd re-oxidation. CO adsorption inhibits $O_2$ dissociation until the temperature is attained where the oxygen surface coverage is sufficient to boost CO oxidation. CO also inhibits methane oxidation because $CH_4$ conversion is retarded by ca. 20°C. Under these conditions, there is a higher fraction of $Pd^{2+}$ during $CH_4$ oxidation compared to the TPRR without CO, probably due to the lower $CH_4$ conversion and the consequent higher concentration of available oxygen. On the contrary, during the stoichiometric TPRR in presence of CO the same extent of $Pd^{2+}$ to the lean TPRR is attained. However, $CH_4$ oxidation is even more inhibited. Probably, both oxidation reactions can occur on similar sites, but the ease of coordination of CO to Pd favours CO oxidation even when $CH_4$ can be activated. CO does not affect the second reduction temperature (380°C) but induces full Pd reduction compared to the stoichiometric TPRR without CO.

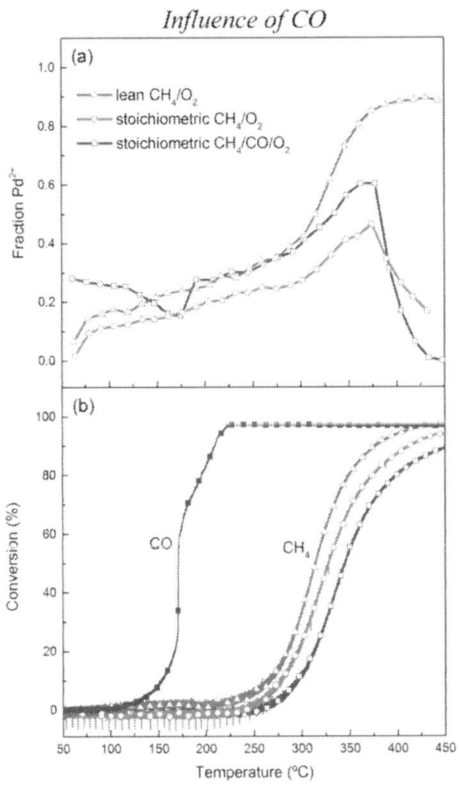


**Figure 2.** Temperature profiles of (a) the fraction of $Pd^{2+}$ and (b) $CH_4$ (open symbols) and CO (full symbols) conversions under various reaction conditions at 2.5°C/min. $CO_2$ and $H_2O$ were the only detected products.

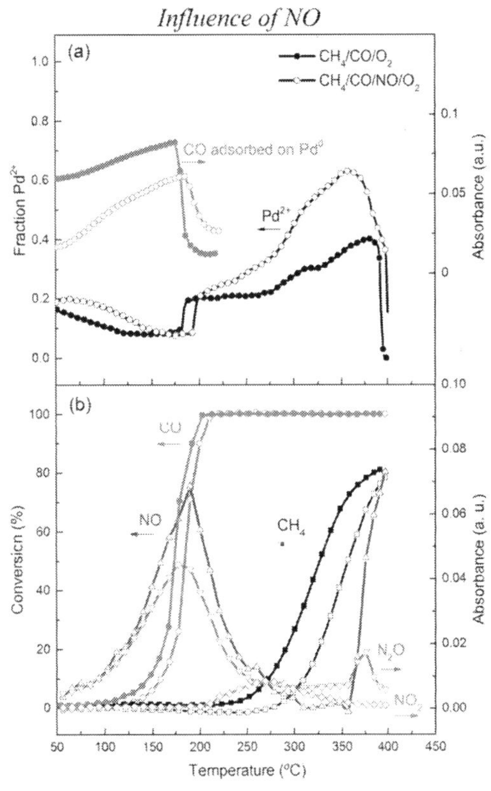


**Figure 3.** (a) Fraction of $Pd^{2+}$ and DRIFT signal of adsorbed CO (ca. 1986 $cm^{-1}$). (b) Conversion of $CH_4$, CO, and NO and FTIR signals of gaseous $N_2O$ and $NO_2$ under various reaction conditions at 10°C/min.

The evolution of $Pd^{2+}$ obtained during the $CH_4/CO/O_2$ TPRR at 10°C/min (Fig. 3) is very similar to that obtained at 2.5°C/min (Fig. 2). The extent of Pd re-oxidation during $CH_4$

conversion is lower when using a faster heating ramp but the temperature of Pd reduction is identical (175 and 380$^o$C). Hence, we assume that the data obtained during $CH_4/CO/O_2$ TPRR are comparable despite the different temperature ramp.

*3.3. Influence of NO*
The influence of NO on $CH_4$ abatement is shown in Fig. 3. In the presence of NO, the initial Pd reduction is accompanied by the undesired NO reduction to $N_2O$. The onset of Pd re-oxidation after the low temperature reduction is shifted by ca. 10$^o$C to higher temperature, which reflects the correspondingly later onset of CO oxidation. After attaining full CO conversion at ca. 220$^o$C, the fraction of $Pd^{2+}$ increases gradually to a higher extent than in the absence of NO. This is concomitant to the development of $NO_2$ from the NO+$O_2$ reaction. $CH_4$ conversion is also shifted to higher temperature by ca. 40$^o$C and sets in at ca. 280$^o$C when the $NO_2$ concentration starts decreasing. Therefore, NO inhibits both CO and $CH_4$ oxidation reactions. The presence of $NO_2$ and the increased fraction of $Pd^{2+}$ are tentatively associated with the poorer performance in $CH_4$ oxidation. Finally, from ca. 350$^o$C NO reduction produces $N_2O$ and $N_2$ while Pd reduces to metal because of the lack of oxygen in the feed.

## 4. Conclusions
Methane oxidation on $Pd/Al_2O_3$ has been examined using operando XAS and DRIFT spectroscopy to correlate the oxidation state of palladium with its catalytic activity in the presence of CO and NO in the feed. The data indicate that a mixed Pd/PdO phase is present at the onset of methane oxidation irrespective of reaction conditions and that both CO and NO inhibit methane oxidation. In presence of CO, Pd is reduced until CO is fully oxidized, then Pd suddenly re-oxidizes before $CH_4$ oxidation can start. We ascribe the CO inhibition to the competition of CO and $CH_4$ oxidation reactions for similar if not identical reaction sites. In the case of NO, the inhibition of $CH_4$ oxidation is tentatively associated with the constantly higher oxidation state of Pd due to NO interaction with Pd.

## Acknowledgments
We are grateful to the Swiss National Science Foundation (SNF, project number 138068/1) and to the Competence Center for Materials Science and Technology (CCMX) for financial support, to SLS and ESRF for beam-time allocation, and to Dr. G. Agostini, Dr. A. Kambolis, and Mrs. M. Sridhar for help during measurements.

## References
[1] Gélin P and Primet M 2002 *Appl. Catal. B* **39** 1
[2] Czekaj I, Kacprzak K A and Mantzaras J 2013 *Phys. Chem. Chem. Phys* **15** 11368
[3] Grunwaldt J D, Maciejewski M and Baiker A 2003 *Phys. Chem. Chem. Phys.* **5** 1481
[4] Grunwaldt J D et al 2009 *J. Phys.: Conf. Ser.* **190** 012160
[5] Matam S K et al 2010 *The Journal of Physical Chemistry C* **114** 9439
[6] Lu Y et al 2013 *Catal. Commun.* **39** 55
[7] Matam S, Chiarello G, Lu Y, Weidenkaff A and Ferri D 2013 *Top. Catal.* **56** 239
[8] Chiarello G L et al 2014 *Rev. Sci. Instrum.* **85** 074102
[9] Ravel B and Newville M 2005 *J. Synch. Rad.* **12** 537
[10] Colussi S et al 2010 *Appl. Catal. A* **390** 1
[11] Wolf M M, Zhu H, Green W H and Jackson G S 2003 *Appl. Catal. A* **244** 323
[12] Lear T et al 2005 *J. Chem. Phys.* **123** 174706

# Local structure of Iridium organometallic catalysts covalently bonded to carbon nanotubes.

**J Blasco[1], V Cuartero[2], G Subías[1], M V Jiménez[3], J J Pérez-Torrente[3], L A Oro[3], M Blanco[4], P Álvarez[4], C Blanco[4] and R Menéndez[4]**

[1]Instituto de Ciencia de Materiales de Aragón, Departamento de Física de la Materia Condensada, CSIC-Universidad de Zaragoza, C/ Pedro Cerbuna 12, 50009 Zaragoza (Spain)
[2]ESRF-The European Synchrotron, 71, Avenue des Martyrs, Grenoble (France)
[3]Departamento de Química Inorgánica, Instituto de Síntesis Química y Catálisis Homogénea-ISQCH, Universidad de Zaragoza-CSIC, C/ Pedro Cerbuna 12, 50009 Zaragoza (Spain)
[4]Instituto Nacional del Carbón, CSIC, P.O. Box 73, 33080 Oviedo (Spain)

E-mail: jbc@unizar.es

**Abstract.** Hybrid catalysts based on Iridium N-heterocyclic carbenes anchored to carbon nanotubes (CNT) have been studied by XAFS spectroscopy. Oxidation of CNT yields a large amount of functional groups, mainly hydroxyl groups at the walls and carboxylic groups at the tips, defects and edges. Different kinds of esterification reactions were performed to functionalize oxidized CNT with imidazolium salts. Then, the resulting products were reacted with an Ir organometallic compound to form hybrid catalysts efficient in hydrogen transfer processes. XANES spectroscopy agree with the presence of Ir(I) in these catalysts and the EXAFS spectra detected differences in the local structure of Ir atoms between the initial Ir organometallic compound and the Ir complexes anchored to the CNT. Our results confirm that the halide atom, present in the Ir precursor, was replaced by oxygen from –OH groups at the CNT wall in the first coordination shell of Ir. The lability of this group accounts for the good recyclability and the good efficiency shown by these hybrid catalysts.

## 1. Introduction

Carbon nanotubes (CNT) can be considered as a versatile matrix to develop heterogeneous catalysts with improved activity [1]. Here we report a thorough characterization by X-ray absorption fine structure spectroscopy (XAFS) of a set of catalysts based on N-heterocyclic carbene (NHC) iridium complexes covalently bonded to CNT properly functionalized with imidazolium salts. These materials showed catalytic activity in hydrogen transfer processes and they were tested in the hydrogenation of cyclohexanone with 2 propanol showing both good efficiency and good recyclability [2].
Chemical oxidation of CNT yields different oxo-groups that can be functionalized. Hydroxyl groups are located at the basal planes of CNT walls whereas carboxylic acids mainly lie at the tips, edges and defects of CNT (see Fig. 1a). These oxidized CNT materials can react directly with some methoxo-cylooctadiene iridium dimer complexes, [{Ir(μ-OMe)(cod)}$_2$] (cod=1,5-cyclooctadiene), resulting poor catalytic activity [2]. Better results were obtained when oxidized CNT were functionalized with the imidazolium salts (Im) like [MeImH(CH$_2$)$_3$OH]Cl (chloride of 3-methyl-1-(3-hydroxypropil)-

imidazolium, hereafter denoted as Im-1) or [MeImH(1-cyclohexyl-2-ol]I (iodide of 3-methyl-1-(1-cyclohexyl-2-ol)-imidazolium (denoted as Im-2) rendering functionalized *f*-CNT supports for NHC iridium complexes synthesis by simple reaction with $[\{Ir(\mu\text{-}OMe)(cod)\}_2]$. Two ways to functionalize CNT were tested, either esterification reactions of –COOH groups with Im-1 or Im-2 (denoted as CNT-1 and CNT-2, respectively) and activation of –OH groups with *p*-nitrophenylchloroformate followed by the reaction with Im-1 (denoted as CNT-pNPh-1). Both functionalized CNT were then reacted with $[\{Ir(\mu\text{-}OMe)(cod)\}_2]$ to yield the hybrid catalysts: CNT-1-Ir, CNT-2-Ir and CNT-pNPh-1-Ir. In the latter case, the Ir complex can be also anchored to the reactive –COOH groups. In order to discriminate activities obtained from those different anchoring points, some CNT were thermally reduced at 400ºC in inert atmosphere to remove –COOH groups. The resulting product was functionalized in the same way as CNT-pNPh-1-Ir and then reacted with the Ir precursor, resulting in a supported NHC complex without secondary products, which was named CNT-T400-pNPh-1-Ir. Figures 1b-e show the proposed final structure for the hybrid catalysts synthesized.
The local iridium structure is unknown but the halide ligand is proposed to remain in the first coordination shell of the Ir atom and the output release of this anion from the coordination sphere of the metal is suggested as the initiation of the catalysis mechanism [2]. In order to examine this possibility, XAFS spectroscopy has been used with the aim of studying both the electronic state of Ir atoms and its local environment in these hybrid catalysts.

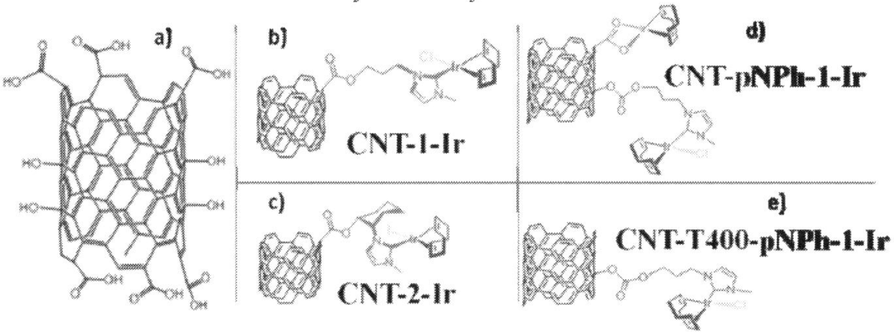


***Figure 1.*** *(a) Oxidized CNT and (b-e) presumed structures for the hybrid catalysts.*

## 2. Experimental section.

XAFS spectra at the Ir $L_3$-edge were recorded at the BL22-Claess (Alba) and at the BM23 (ESRF) beamlines on fresh samples. The measurements were carried out in transmission mode at room temperature on pellets of the materials mixed with cellulose if necessary. The beam was monochromatized by a fixed-exit Si (111) double crystal and harmonic rejection better than $10^{-5}$ was achieved by using the Si mirror coating of the double flat mirror installed after the monochromator. Energy resolution was estimated to be $\sim 8\times 10^{-5}$ at the Ir $L_3$-edge and a pellet of Ir metal with cellulose was simultaneously measured for energy calibration. The XANES spectra were normalized at ~100 eV beyond the absorption edge. We have also recorded spectra of Ir-metal and $IrCl(cod)[MeIm(CH_2)_3OH]$ as references for $Ir^0$ and $Ir^+$ respectively. The second compound is labelled as Ir(I)-reference and its structure can be seen in the inset of the figure 2a.
EXAFS signal extraction and background removal were carried out using Athena software [3] and the EXAFS structural analysis was performed using the ARTEMIS program [3] which makes use of theoretical scattering amplitudes and phases calculated from FEFF6 code [4] for the several single and multiple scattering paths. The fits were carried out in the R-space using a Hanning window in the region 1.1- 2.3 Å which only include contributions from the first coordination shell as it is correlated to the main differences in the local environment of Ir atoms.

## 3. Results and conclusions.

Figure 2a shows the derivative of the absorption coefficient vs energy about the Ir $L_3$-edge for representative catalysts compared to the ones for $Ir^0$ and the Ir(I)-reference. All catalysts exhibit the first inflection point at the same energy that concurs with Ir(I)-reference suggesting a similar oxidation state for the Ir atoms at CNT-catalysts.

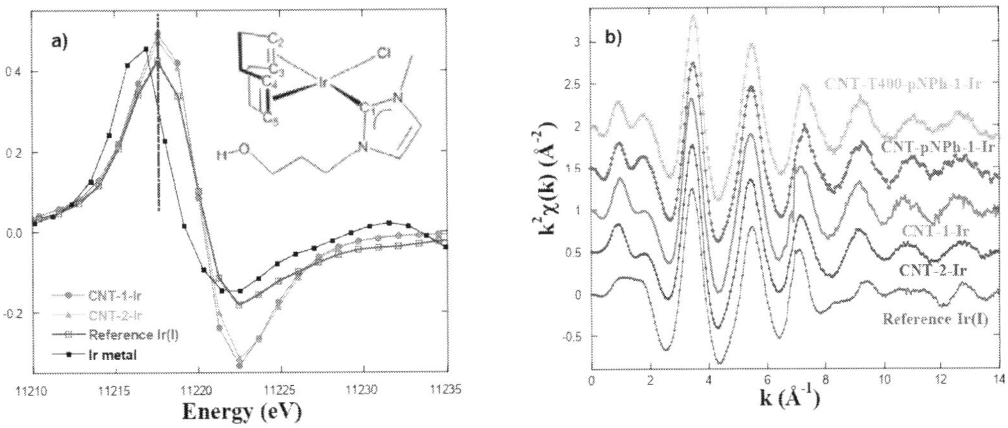


***Figure 2.*** *(a)Derivative of the absorption coefficient vs. energy al the Ir $L_3$-edge for CNT-1-Ir, CNT-2-Ir, Ir(I)-reference and $Ir^0$. Vertical line shows the inflection point for the Ir(I) compounds. Inset: Structure of the Ir(I)-reference compound. (b) EXAFS spectra, $k^2 \times \chi(k)$, at the Ir $L_3$-edge for the indicated CNT-based catalysts and the Ir(I)-reference at room temperature.*

The extracted EXAFS signals at room temperature for all studied samples are compared in Figure 2b for the sake of comparison. CNT-based catalysts exhibit similar EXAFS oscillations but the Ir(I)-reference has noticeable differences in the $8Å^{-1} \leq k \leq 12Å^{-1}$ range. Figure 3a shows the Fourier transform (FT) of the $k^2$-weighted EXAFS spectra, calculated between 2.75 and 14 $Å^{-1}$ using a Hanning window. All CNT-based catalysts show a strong peak at 1.67 Å (without phase shift correction) corresponding to the first coordination shell. This distance agrees with the expected one for Ir-C bond length. The Ir(I)-reference shows the same peak but with less intensity which can be ascribed either to a higher disorder or fewer Ir-C paths respect to the Ir in hybrid catalysts. In addition, the FT of Ir(I)-reference displays a shoulder at ~2.0 Å ascribed to the Ir-Cl bond length that is lacked in the CNT-based catalysts.

The structural analysis was carried out as indicated in the experimental section. We have considered contributions from single scattering paths, Ir-C, Ir-O and Ir-Cl. The representative best fits for the Ir(I)-reference and a representative catalyst are shown in the figure 3b. The Ir(I)-reference was analyzed using the local environment indicated in the figure 2a and the fit agrees with a first coordination shell composed by 5 Ir-C paths and an additional Ir-Cl contribution with the following bond lengths: Ir-$C_1$= 2.03(1) Å, Ir-$C_2$($C_3$)= 2.11(1) Å, Ir-$C_4$($C_5$)= 2.19(1) Å and Ir-Cl= 2.43(1) Å in agreement with reported structural data in related compounds [5]. However, this environment failed to fit the EXAFS signal of CNT-based catalysts. Preliminary fits clearly showed the absence of Cl paths in the CNT-1-based catalysts (or I in the CNT-2-compound) as the resulting bond length was unphysical long and the fit was very poor. Hence, the logical hypothesis is that chloride (or iodide) has been displaced by another nucleophile and the most logical candidate was the –OH groups from the CNT wall. Therefore, we replaced the Ir-Cl path by an Ir-O one and the fit converged to yield an Ir-O distance similar to the previous Ir-$C_1$. Therefore, we performed the structural analysis using this model and some constrains were imposed to avoid over-parametrization. We only refined 5 parameters: an inner potential correction ($\Delta E_0$), and average Debye-Waller factor ($\sigma^2$) and three distances $R_1$, $R_2$ and $R_3$ accounting for $C_1$+O, $C_2$+$C_3$ and $C_4$+$C_5$, respectively. The best fit parameters are summarized in the table 1 for the four catalysts. The amplitude reduction factor $s_0^2$ was set at 1.0, following the previous fit of the reference compound and in agreement with previous calculations [6]. The data from table 1 indicate a similar local environment for Ir atoms in all hybrid catalysts no matter the kind of functionalization performed. In all cases, the first coordination sphere around Ir atoms is composed by six light elements (C and O) without sign of halides.

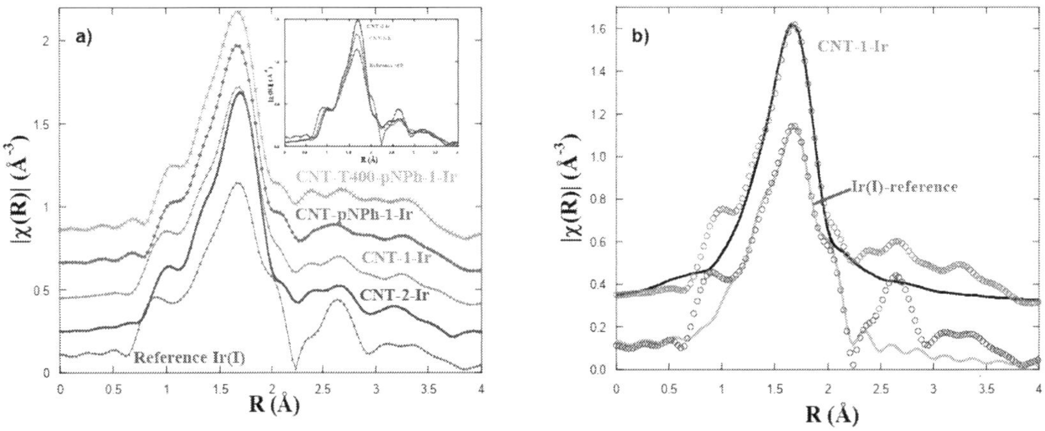


***Figure 3.*** *(a) Modulus of the Fourier transforms (FT) extracted from the curves of Fig. 2b. Inset: Direct comparison of representative samples. (b) Comparison of the FT (circles) to the best fits (lines) for two representative samples. The fit was limited to the first coordination shell.*

**Table 1.** Best fit parameters (inner potential, bond lengths, average Debye-Waller factor and reliability factors [3]) obtained from the fitting of EXAFS data in the R-space mode. Numbers in parentheses refer to standard deviation of the last significant digits.

| Sample | $\Delta E_0$ *(eV)* | $R_1$*(Å)* | $R_2$*(Å)* | $R_3$*(Å)* | $\sigma^2$(Å$^2$) $\times 10^{-3}$ | $R_F$ |
|---|---|---|---|---|---|---|
| **CNT-1-Ir** | 3.9(9) | 2.035(19) | 2.107(26) | 2.186(24) | 4.3(19) | 0.0043 |
| **CNT-2-Ir** | 3.9(7) | 2.029(21) | 2.101(14) | 2.186(20) | 1.7(14) | 0.0028 |
| **CNT-pNph-1-Ir** | 3.6(8) | 2.029(21) | 2.083(34) | 2.172(32) | 3.1(19) | 0.0038 |
| **CNT-pNph-1-T400-Ir** | 3.3(9) | 2.024(37) | 2.090(28) | 2.158(37) | 2.9(22) | 0.0049 |

Summarizing, XAFS spectroscopy resulted very useful to discern the local structure around Ir atoms in CNT-based hybrid catalysts. Halide ions (Cl in CNT-1 or I in CNT-2) are lost during the synthesis process to obtain the hybrid catalysts. They are likely replaced by –OH groups from the CNT wall. The lability of these –OH groups explains the good recyclability shown by these catalysts [2,7].

**Acknowledgments.**

Authors acknowledge the financial support from Mineco/FEDER CTQ2010-15221, MAT2012-38213-C02-01, and DGA E07 and E69 projects. They also thank Alba for granting beam time and the kind assistance of Claess staff. M. Blanco thanks MECD for his FPU grant (AP2010-0025).

**References**

[1] Schaetz A, Zeltner M and Stark W J 2012 *ACS Catal.* **2** 1267.
[2] Blanco M *et al.* 2013 *ACS Catal.* **3** 1307.
[3] Ravel B and Newville M 2005 *J. Synchrotron Radiat.* **12** 537.
[4] Rehr J J and Albers R C 2000 *Rev. Mod. Phys.* **72** 621.
[5] Jiménez M V *et al.* 2011 *Organometallics* **30** 5493.
[6] Roy M, Gurman S J and van Dorssen G 1997 *J. Phys. IV* **7** 151.
[7] Blanco M *et al.* 2016 *Carbon* **96** 66.

# XAS on Rh and Ir metal sites in post synthetically functionalized UiO-67 Zirconium MOFs

**L. Braglia[1,2], E. Borfecchia[1], K. A. Lomachenko[1,2], S. Øien[3], K. P. Lillerud[3], C. Lamberti[1,2]**

[1]University of Turin, Department of Chemistry, NIS and CrisDi interdepartmental centers and INSTM reference center, Via Pietro Giuria 7, 10125 Turin (Italy)

[2]Southern Federal University, Faculty of Physics, Zorge 5, 344090 Rostov on Don (Russia)

[3]inGAP Centre for Research Based Innovation, Department of Chemistry, University of Oslo, P.O. Box 1033, N-0315 Oslo, Norway

luca.braglia@unito.it

**Abstract**. We synthesized UiO-67 metal-organic-frameworks (MOFs) functionalized with different transition metals (Rh, Ir). Using EXAFS we verified that the synthesis has been successful. Furthermore, we observed the change of local environment while varying of metal site. XAS spectroscopy is the most informative technique to characterize these kind of materials and to study the local environment around the metal site.

## 1. Introduction

Metal-organic frameworks (MOFs) are crystalline, porous solids consisting of metal ions or clusters, coordinated with organic molecules. The large number of combinations for inorganic and organic building units offers an infinite variety of structural solutions with a wide range of properties [1]. The recently discovered UiO-66 and UiO-67 classes of iso-structural MOFs are obtained connecting $Zr_6O_4(OH)_4$ inorganic cornerstones with 1,4-benzene-dicarboxylate or 4,4'biphenyl-dicarboxylate linkers, for the UiO-66 and UiO-67 MOFs, respectively [2]. Due to their outstanding stability at high temperatures, high pressures and in presence of different solvents, these materials are among the few MOFs already commercialized for applications in the fields of catalysis, $H_2$ storage, and gas purification. We are currently exploring the possibility to enhance the capabilities of the UiO-67 MOF by grafting to the framework an additional catalytically-active metal center, by chelating bipyridine-dicarboxylate (bpydc) linkers. The resulting metal-functionalized MOFs are attractive candidates for industrial applications aiming to heterogenization of homogeneous catalytic reactions [1]. In a recent study [3] we have characterized Pt-functionalized UiO-67 using three types of syntheses and followed the reactivity of this system towards molecules of small- (e.g. $H_2$), medium- (e.g. $Br_2$), and large-size (e.g. thiol). After these interesting results obtained with Pt, UiO-67 derivatives functionalized with different transition metals were prepared, targeting different kinds of catalytic reactions. Herein we report a preliminary XAS-based study on UiO-67 functionalized with Rh, Ir and Au in *in situ* conditions to demonstrate the effectiveness of synthesis procedure.

## 2. Experimental and Methods

### *2.1. Sample preparation*
The samples were prepared by post-synthesis functionalization (PSF) approach, where premade UiO-67 MOF was suspended in a solution of precursor metal chloride salts. 4,4'-biphenydicarboxylic acid ($H_2$bpdc) 2,2'-bipyridine-5,5'-dicarboxylic acid ($H_2$bpydc) and 2-phenylpyridine-5,4'-dicarboxylic acid ($H_2$ppydc) were used as linkers in the MOF synthesis; the latter is fundamental for providing the chelating site where the metal functionalization occurs. The employed functionalizing metals were Rh, Ir and Au, using $RhCl_3$, $IrCl_3$ and $HAuCl_4$ as precursors, respectively using the same strategies detailed elsewhere for Pt [3]. Thereby the whole sample set included: UiO-67 bpydc-Rh, UiO-67 bpydc-Ir, UiO-67 ppydc-Ir and UiO-67 ppydc-Au.

### *2.2. X-ray Absorption Spectroscopy measurements*
Due to the local character of the functionalization process, XAS spectroscopy has played a key role in clarifying the local structural and electronic properties of the grafted metal centre [4]. We performed our experiments in Max Lab II (Lund, Sweden) at I811 beamline and in ESRF (Grenoble, France) at BM23 beamline. The white beam in I811 is produced by a liquid He-cooled superconducting wiggler operating with a current around 200mA. We did not use mirrors, and the radiation was selected by a horizontally sagitally focused double-crystal Si(111) monochromator detuned to 20% to minimize the third harmonic. The energy beam was detected by 30 cm ionization chambers for $I_0$ and $I_1$ and by photodiode for $I_2$. We operated in transmission mode above the Ir-$L_3$ edge (11215 eV) and Au-$L_3$ edge (11920 eV). The white beam at BM23 was focused in a vertical aperture of 0.3 mm, then the energy beam was selected by a double-crystal monochromator Si(111) and the higher harmonics were suppressed by a Pt-coated mirror at 2 mrad. We collected the data in transmission mode at the Rh-K edge (23220 eV). Three ionizations chambers detected the energy for $I_0$, $I_1$ and $I_2$. An ad hoc cell was used to host the samples [5], in the form of self-supporting pellets of optimized thickness. The XANES and EXAFS spectra were collected in flux of He at room temperature. The data were extracted and analyzed with Athena and Arthemis codes [6].

## 3. Results and Discussion
The strategy used to fit our data has been to use the amplitude and the Debye Waller factor of chlorine ($\sigma_{Cl}^2$) obtained from the fit of reference metal salt samples, while the $\sigma_N^2$ values have been estimated on the basis of the analysis previously performed on the Pt-UiO-67 system [3]. The metal in the $RhCl_3$ and $IrCl_3$ reference compounds is assumed to be six-coordinated with six Cl ligands, in octahedral geometry (Table 1). Once the fit of the salt was acceptable, we set its amplitude ($S_0^2$) in the fits performed on functionalized MOFs. The EXAFS analysis has been performed in the first shell with $k^3$ weight extraction, the $k_{max}$ was 14 Å$^{-1}$ for all the samples. Figure 1 reports the XANES spectra for UiO-67 functionalized with Rh (a), Ir (b) and Au (c); the MOFs are indicated with orange lines (bpydc linkers) and red lines (ppydc), while the salts are reported as grey lines. XAS has allowed us the ineffective of Au grafting in the UiO-67 ppydc-Au. Indeed, in this case the XANES of the precursor (grey) and after the reaction with the MOF (orange) are exactly overlapped (see Figure 1c).

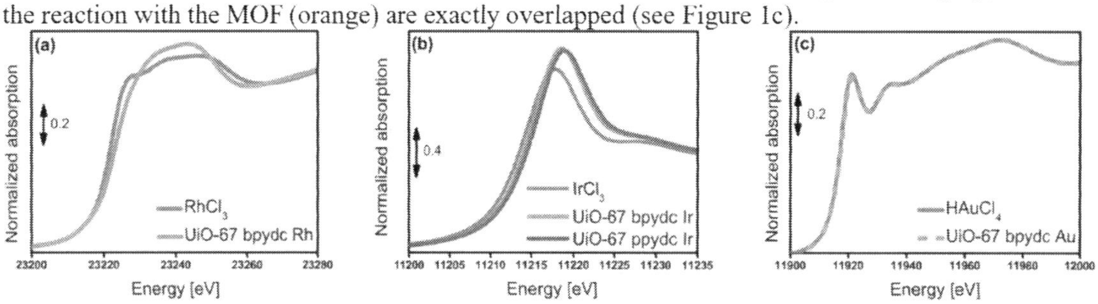


**Figure 1.** XANES spectra at Rh K-edge (a), Ir $L_3$-edge (b) and Au $L_3$-edge (c) for the respective metal-functionalized UiO-67 MOFs, the spectra collected on MOFs are reported as orange and red colored lines (ppydc linkers), while the precursors are reported in grey.

*3.1. Determination of the Rh Local Environment in Rh-UiO-67 MOF*

XANES does not reveal a significant energy shift of the absorption edge for the Rh-UiO-67 sample with respect to the one of the precursor salt. This evidence suggests that the Rh-complex grafted in the MOF shows the same Rh oxidation state of the precursor salt, i.e. 3+. With respect to EXAFS analysis, the fits achieved for modulus and imaginary part of the Fourier Transform are reported in Fig. 2a. The fits are in a good agreement with the experimental spectra. In the UiO-67-Rh, the contributions of Rh-Cl and Rh-N single scattering paths are indicated with green and cyan lines, respectively, and they evidence the presence at shorter distances of two N atoms of the bpydc unit, while three Cl neighbours are located at longer bond distance from the Rh centre. However, it is well known that Rh(III) complexes prefer octahedral coordination. In the present case, the presence of an additional low-atomic-number atom (e.g. from a water or DMF solvent molecule) in the first coordination shell of Rh also results in satisfactory fit of the EXAFS spectra, within the available data quality (fit not reported). All the parameters refined in the fits, are reported in Table 1, where the underlined values have been fixed in the fitting procedure.

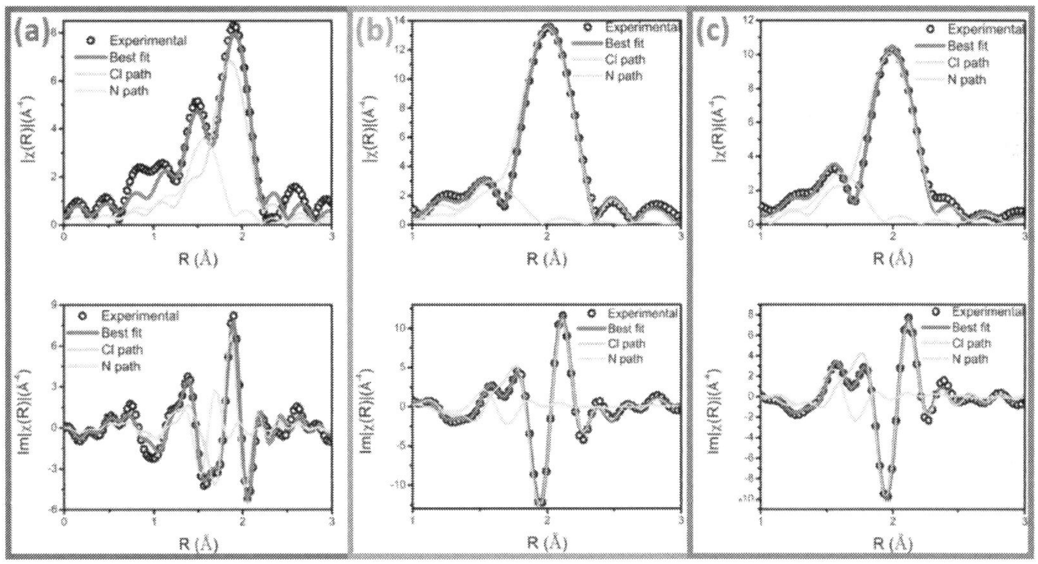


**Figure 2.** (a) Fourier transform of the experimental EXAFS spectra (black dots) and its best fit (red) of UiO-67 bpydc-Rh in modulus (top) and imaginary part (bottom); (b) and (c): same as (a) for UiO-67-Ir MOFs with bpydc (b) and ppydc (c) chelating linkers respectively. The SS paths involving Cl and N shells are shown as green and cyan solid lines, respectively.

*3.2. Determination of the Ir Local Environment in Ir-UiO-67 MOF*

The Ir-$L_3$ edge XANES spectra (Figure. 1b) of the two Ir-UiO-67 MOFs show a slight shift at higher energy (ca. 1 eV) and they show a significant increase of the white line intensity, compared the one of the $IrCl_3$ precursor. This evidence suggests a decrease in the electron density in the d orbitals on the metal center with respect the $IrCl_3$ precursor [3,7]. A similar behaviour has been observed for the UiO-67-bpy-Pt [3]. In Table 1 we observe that the experimentally refined coordination number for the Ir-Cl shell are (4.8 ± 0.3) and (3.5 ± 0.2) for MOFs with bpydc and ppydc linkers, respectively. In the case of Ir bpydc, we observe that the coordination number refined in the fit for N is less than two, suggesting the presence of some Ir sites not grafted to the MOF framework. The simplest alternative, supported also by the rather high Ir-Cl coordination number found, is the presence of a certain fraction of unreacted precursor salt $IrCl_3$ ($N_{Cl}$ = 6) trapped in the MOF cavities. With this respect, we estimated roughly the percentage of grafted Ir-sites over the total number of Ir absorbers, and thus the percentage of salt that did not react. We used the system showed below Eq. (1), where $N_{exp}$ is the coordination number obtained from the fit, $N_{the}$ is the coordination number expected for Cl or N, x is the weight factor of $N_{the}$, y is the weight factor of the precursor and $N_{salt}$ is the coordination number of the salt ($N_{salt}$ (Cl) = 6, $N_{salt}$ (N) = 0 octahedral coordination).

$$\begin{cases} N_{exp} = \frac{(N_{the} \cdot x)+(N_{salt} \cdot y)}{x+y} \\ x + y = 1 \end{cases} \quad (1)$$

Using the Ir-Cl coordination number, the estimated fraction of unreacted $IrCl_3$ is 0.4 ± 0.1 (0.25±0.08 using the Ir-N coordination number) for the MOF with bpydc linkers.
With respect to the UiO-67-Ir-ppydc, Ir(III) is well-known to form N,C-chelating complexes with ppydc type ligands by orthometalation process [8]. The experimentally refinement coordination number for the Ir-N/C shell is compatible with 2 within experimental error while for Ir-Cl shell is (3.5 ± 0.2). This suggest that the orthometalation process occurred in a majority of Ir sites. In addition, the Ir coordination sphere most likely include also three chlorine neighbours and possibly one solvent ligand (minorly contributing to the EXAFS signal as previously discussed for Rh) to complete the octahedral configuration. The charge in these metal complexes immobilized on the linker could be balanced by possible counter-ions present in the cavities and located at distances typically higher than 5 Å in a disordered configuration. Thus their identification by EXAFS is very difficult: future experiments using a third generation source and high k data collection would be required to further clarify this point, as well as to precisely locate possible solvent molecule coordinated to the metal center.

**Table 1.** Best fit parameters for $RhCl_3$ and $IrCl_3$ precursor salts and M-UiO-67 MOFs (M = Rh, Ir and Pt from [3]). The parameters fixed in the fit are underlined. The $S_0^2$ and $\sigma_{Cl}^2$ values have been chosen from the EXAFS fit of metal salt references, while the $\sigma_N^2$ values have been estimated on the basis of the analysis previously performed on the Pt-UiO-67 system [3].

| Sample | Shell | N | $\sigma^2$(Å$^2$) | R(Å) | $E_0$(eV) | $S_0^2$ | R-factor |
|---|---|---|---|---|---|---|---|
| $RhCl_3$ | Rh-Cl | <u>6</u> | 0.0038±0.0004 | 2.331±0.003 | 0±1 | 0.97±0.06 | 0.008 |
| UiO-67-Rh | Rh-N | 2.1±0.7 | 0.003 | 2.02±0.02 | 2±4 | <u>0.97</u> | 0.02 |
| | Rh-Cl | 2.5±0.2 | <u>0.0038</u> | 2.32±0.01 | | | |
| $IrCl_3$ | Ir-Cl | <u>6</u> | 0.0038±0.0005 | 2.35±0.01 | 12±1 | 0.75±0.06 | 0.01 |
| UiO-67-Ir (bpydc) | Ir-N | 1.5±0.5 | 0.003 | 1.98±0.01 | 12±2 | <u>0.75</u> | 0.008 |
| | Ir-Cl | 4.8±0.3 | <u>0.0038</u> | 2.36±0.01 | | | |
| UiO-67-Ir (ppydc) | Ir-N/C | 1.7±0.4 | 0.003 | 1.99±0.01 | 9±1 | <u>0.75</u> | 0.01 |
| | Ir-Cl | 3.5±0.2 | <u>0.0038</u> | 2.35±0.01 | | | |
| UiO-67-Pt | Pt-N | 2.0±0.3 | 0.002±0.001 | 2.02±0.01 | 1±2 | <u>0.92</u> | 0.02 |
| | Pt-Cl | 1.8±0.2 | 0.0025±0.0006 | 2.295±0.008 | | | |

## 4. Conclusions and acknowledgements

XAS has proved PSF as a good method for chelating some metals, a part from Au, to the MOFs linkers, even though the efficiency can still be improved. The EXAFS experiments performed at third generation synchrotron could lead to a more precise comprehension of the presence of possible trapped solvent or other molecules in the MOFs and to solve the local structure around metal grafted to the linker with better accuracy.

C L, K A L and L B acknowledge the megagrant of the Russian Federation Government to support scientific research at the Southern Federal University, no. 14.Y26.31.0001.

## References

[1] Ferey G 2008 *Chem. Soc. Rev.* **37** 191; Long J.R and Yaghi O M 2009 *Chem. Soc. Rev.* **38**, 1213; Bordiga S, Bonino F, Lillerud K-P and Lamberti C 2010 *Chem. Soc. Rev.* **39** 4885; Butova V V, Soldatov M A, Guda A A, Lomachenko K A and Lamberti C, 2015 *Russ. Chem. Rev.*, **84**

[2] Cavka J H, Jakobsen S, Olsbye U, Guillou N, Lamberti C, Bordiga S, Lillerud K-P 2008 *J. Am. Chem. Soc.* **130** 13850; Valenzano L, Civalleri B, Chavan S, Bordiga S, Nilsen M H, Jakobsen S, Lillerud K-P, Lamberti C. 2011 *Chem. Mater.* **23** 1700; Chavan S, Vitillo J G, Gianolio D, Zavorotynska O, Civalleri B, Jakobsen S, Nilsen M H, Valenzano L, Lamberti C, Lillerud K-P, Bordiga S. 2012 *Phys. Chem. Chem. Phys.* **14** 1614.

[3] Øien S, Agostini G, Svelle S, Borfecchia E, Lomachenko K A, Mino L, Gallo E, Bordiga S, Olsbye U, Lillerud K-P and Lamberti C 2015 *Chem. Mater.* **27** 1042.

[4] Bordiga S, Groppo E, Agostini G, van Bokhoven J A and Lamberti C. 2013 *Chem. Rev.* **113** 1736; Mino L, Agostini G, Borfecchia E, Gianolio D, Piovano A, Gallo E, Lamberti C. 2013 *J. Phys. D: Appl. Phys.* **46** 423001; Garino C, Borfecchia E, Gobetto R, van Bokhoven J A, Lamberti C. 2014 *Coord. Chem. Rev.* **277–278** 130.

[5] Lamberti C, Prestipino C, Bordiga S, Berlier G., et al. 2003 *Nucl. Instr. Meth. B,* **200** 196

[6] Ravel B and Newville M 2005 *J. Synchrotron Radiat.* **12** 537.

[7] Lomachenko K, Garino C, Gallo E, Gianolio D, Gobetto R, Glatzel P, Smolentsev N, Smolentsev G, Soldatov A, Lamberti C, Salassa L. 2013 *Phys. Chem. Chem. Phys*, **15** 38.

[8] Sprouse S, King K A, Spellane P J, Watts R J 1984 J. Am. Chem. Soc. 106 6647; Lamansky S, Djurovich P, Murphy D, Abdel-Razzaq F, Kwong R, Tsyba I, Bortz M, Mui B, et al. 2001 *Inorg. Chem.* **40** 1704

# *In situ* characterization of catalysts and membranes in a microchannel under high-temperature water gas shift reaction conditions

**G Cavusoglu[1,2], F Dallmann[3], H Lichtenberg[1,2], A Goldbach[4], R Dittmeyer[3], J-D Grunwaldt[1,2]**

[1]Institute of Catalysis Research and Technology (IKFT), Karlsruhe Institute of Technology (KIT), Germany
[2]Institute for Chemical Technology and Polymer Chemistry (ITCP), KIT, Germany
[3]Institute for Micro Process Engineering (IMVT), KIT, Germany
[4]Dalian Institute of Chemical Physics (DICP), Dalian National Laboratory for Clean Energy (DNL), China

E-mail: grunwaldt@kit.edu

**Abstract.** *Microreactor technology with high heat transfer in combination with stable catalysts is a very attractive approach for reactions involving major heat effects such as methane steam reforming and to some extent, also the high temperature water gas shift (WGS) reaction. For this study Rh/ceria catalysts and an ultrathin hydrogen selective membrane were characterized in situ in a microreactor specially designed for x-ray absorption spectroscopic measurements under WGS conditions. The results of these experiments can serve as a basis for further development of the catalysts and membranes.*

## 1. Introduction

The worldwide growing demand for energy, increasing levels of harmful emissions and limited stocks of fossil fuels have led to increased efforts in the search for environmentally clean, sustainable and renewable energy sources and to a growing interest in efficient hydrogen production. Hydrogen, a feedstock used for diverse industrial applications, can be used as a clean fuel for fuel cells and is a key element for chemical storage of renewable energy. $H_2$ can be produced by the water gas shift (WGS) reaction:

$$CO + H_2O \rightleftharpoons CO_2 + H_2 \qquad \Delta H_R = -41.1 \text{ kJ mol}^{-1}$$

Advantages of microreactor technology such as high surface to volume ratio and high heat and mass transfer coefficients have shown a high potential to intensify production efficiency [1]. Membrane reactors can shift the equilibrium limitation by continuously removing one of the reaction products from the reaction area [2]. Therefore, integration of $H_2$ selective membranes into microreactors will be highly beneficial for hydrogen production processes [3]. In order to gain a deeper understanding of such systems, both the catalysts and the membranes need to be characterized, preferentially under operating conditions [4, 5] and in a spatially resolved manner [6, 7]. For this purpose, a special microreactor, which allows to characterize catalysts and membranes under WGS conditions using X-ray absorption spectroscopy (XAS), was designed. The first results are presented here on a $Rh/CeO_2$ catalyst with an ultrathin hydrogen selective PdAg membrane.

## 2. Experimental

10.5 wt.% Rh/ceria nanoparticles were prepared by flame spray pyrolysis (FSP) [8], using solutions of the corresponding initial precursor in xylene (VWR chemicals). Precursor solutions were prepared by dissolving specified amounts of metal acetylacetonate (Rh(acac)$_3$, 99% (25% Rh), abcr GmBH & Co. KG), and cerium (III) 2-ethylhexanoate 49% 2-ethylhexanoic acid (12%Ce, Alfa Aesar). The precursor solutions (1 mol/L of the noble metal) were sprayed through a capillary into a methane/oxygen flame (750 mL/min $CH_4$, 1600 mL/min $O_2$) at 5 mL/min using a syringe pump and dispersed by oxygen (5 L/min). A cylindrical steel vessel with a glass fiber filter (75 cm diameter, Whatman GF6) was placed above the flame. The produced particles were collected on this filter with aid of a vacuum pump using a recently established setup at KIT [8].

The new catalytic microreactor with hydrogen selective membranes designed for the present study is shown in Figure 1. Two channel plates, one for the catalytic bed and reactant gas flow and the other one for the produced $H_2$, are separated by a commercial hydrogen selective PdAg alloyed membrane. Heating cartridges in the reactor shell and gas connections allow XAS measurements at the Rh and Pd K-edge under reaction conditions. Each of the nine reactor channels is 0.1 cm wide and 20 cm long (total area 18 cm$^2$). The channels were coated with 720 mg (total amount) of the catalyst mixed with 1 mol/L NaOH and dried overnight. The gasses were pre-mixed and the water was added using Control Evaporator Mixer (CEM) (not calibrated). Note that both the coating and the gas dosing system were preliminary and are presently improved.

X-ray Absorption Near Edge Structure (XANES) and Extended X-ray Absorption Fine Structure (EXAFS, ca. 25 min/scan) spectra at the Rh K-edge (23220 eV) along the central reactor channel were recorded at the Swiss Norwegian Beamline (SNBL) at the European Synchrotron Radiation Facility (ESRF) in transmission mode using a Si(111) double crystal monochromator (typical beam size 0.4 x 5 mm$^2$). Two ionization chambers were used to measure the intensity of the incident and transmitted X-ray beam, and a third for energy calibration with a rhodium foil. XAS data analysis was conducted using the Athena and Artemis interfaces of the IFEFFIT software package (version 0.8.056) [9].

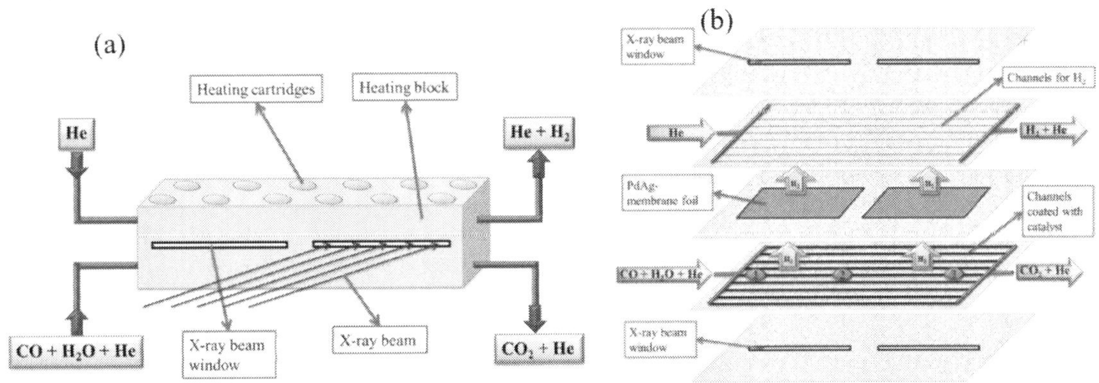


**Figure 1.** (a) Schematic drawing of the microreactor prototype designed by IMVT and ITCP/IKFT (KIT) and built at IMVT for *in situ* spectroscopic studies of catalysts and membranes under WGS conditions (b) Reactor window, channel plates for reactants and the produced hydrogen, and hydrogen selective PdAg membrane foil. The EXAFS analysis positions close to the reactor inlet, middle and reactor exit are indicated.

## 3. Results and Discussion

Figure 2 shows Pd K-edge EXAFS data of the PdAg foil in the microreactor at room temperature before the channels were coated, after coating with 10.5 wt.% Rh/$CeO_2$ and after applying WGS reaction conditions (gas mixture 8% CO, 32% $H_2O$ and 60% He, total flow 25 mL/min) and cooling down at position 1 close to the reactor inlet. According to EXAFS fitting results obtained from this data the distances between absorbing and neighbor atoms are 2.76 Å and 3.87 Å in the first and second shell and the numbers of nearest neighbor atoms are 12 and 6 in the first and second shell. Obviously, neither heating nor WGS reaction conditions have significant influence on the EXAFS data.

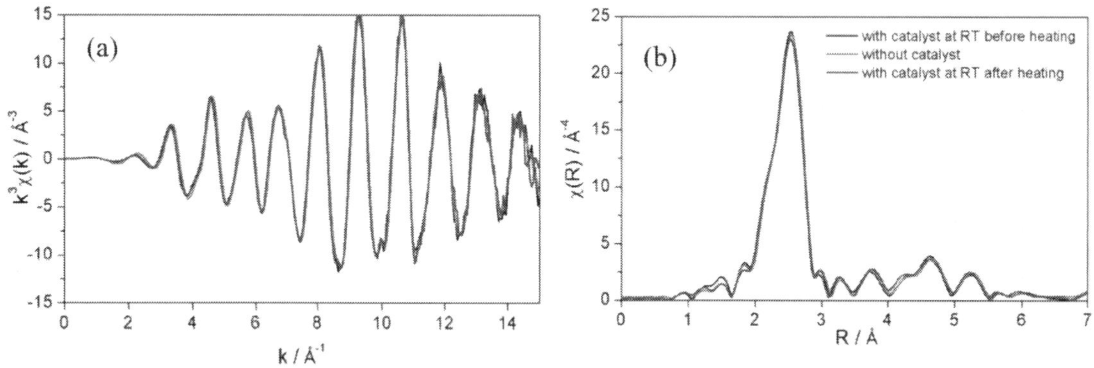


**Figure 2.** (a) $k^3$–weighted Pd K-edge EXAFS data (b) Fourier transformed data of the PdAg foil in the reactor without catalyst, with catalyst at room temperature and after WGS reaction at position 1.

Figure 3(a) shows Rh K edge XANES spectra of the catalyst recorded at the same position (close to the reactor inlet) at different temperatures and in different gas atmospheres: The first 2 spectra in Figure 3(a) were measured at room temperature and after heating to 150 °C in Helium. At this temperature, the gas feed was switched from He to a WGS reaction gas mixture (8% CO, 32% $H_2O$ and 60% He, total flow 25 mL/min). Up to this point, no changes in the XANES spectra occurred; the catalyst was obviously in an oxidized state. Afterwards, the reactor was heated further up to 170 °C. This leads to a decrease in white line intensity which indicates reduction of Rh oxide to metallic particles.

In Figure 3(b) Fourier transformed $k^2$–weighted EXAFS data of the catalyst recorded at two different positions (close to the reactor inlet and close to the middle) are compared. Data measured at a third position near the reactor exit show a high noise level at $k > 5$ Å$^{-1}$ (probably due to sample inhomogeneities) and were therefore not included in the EXAFS analysis. The Fourier transformed EXAFS results in Figure 3(b) show that the Rh particles are in an oxidized state at room temperature, and were reduced at 170 °C under WGS reaction conditions (in accordance with the XANES results in Figure 3(a)) at both positions along the reactor channel. They stayed reduced after heating further up to 290 °C and finally cooling down to room temperature and switching the feed gas back to Helium. The similarity of the EXAFS data recorded at different positions under the same conditions (temperature, feed gas composition) indicates that the structure of the catalyst does not change along the channel. Therefore, further results are only shown for one position (close to the reactor inlet).

Structural parameters were derived from curve fitting of the $k^3$–weighted EXAFS data in R-space. The experimental data that aligned the model implicate two main shells, a Rh – O shell and a Rh – Rh shell. This model was used in a previous study that indicates rhodium (III) oxide ($Rh_2O_3$) at a distance of 2.01 Å for Rh – O and 2.69 Å for Rh – Rh [10]. The first peak is observed at about 2.03 Å, that is same as the Rh – O bond length in $Rh_2O_3$. The coordination number obtained for oxygen is 4.5 – 5 for this catalyst, Rh coordinates with 6 oxygen atoms in $Rh_2O_3$. According to Gayen et al. [11] the Rh ions may substitute such $Ce^{4+}$ sites in the surface layer of the $CeO_2$ crystallites, as surface $Ce^{4+}$ ions have only four $O^{2-}$ ions. The second backscattering peak about 2.70 Å can be attributed to Rh – Rh (second nearest neighbor distance in $Rh_2O_3$ cf. ref [10]). After *in situ* reaction under WGS reaction conditions, the first shell Rh – Rh coordination number increases and no oxygen was observed anymore, indicating the formation of Rh metal particles. Rh exhibited an average coordination number of 9 in the reduced phase of the catalyst, indicating a particle size of 2 – 3 nm.

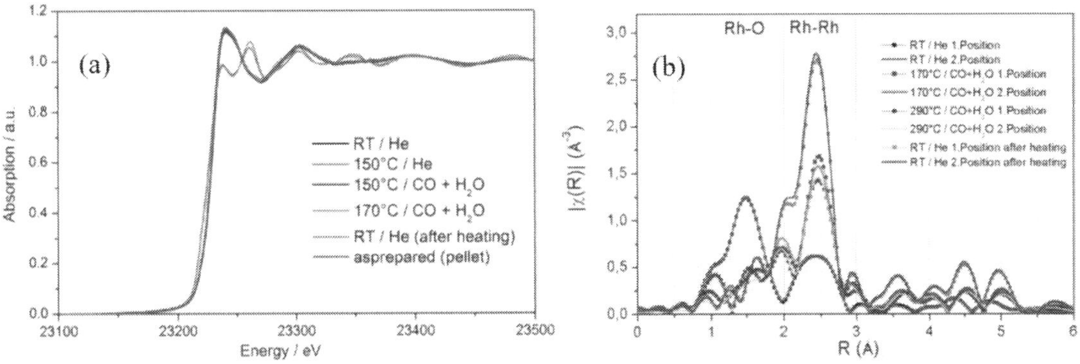


**Figure 3.** (a) X-ray absorption near edge structure data at the Rh K-edge of 10.5 wt.% Rh/$CeO_2$ in the microreactor under WGS reaction conditions, *ex situ* measurements of 10.5 wt% Rh/$CeO_2$ as prepared (pellet). (b) $k^2$–weighted Fourier transformed EXAFS data of the catalyst at RT under He, at 170 °C, and 290 °C under 8% CO, 32% $H_2O$, 60% He and at RT after heating under He in microreactor at position 1 and position 2 (close to the reactor inlet and close to the middle).

## 4. Conclusion

This feasibility study shows that the microreactor specially developed for this project is suitable for X-ray absorption spectroscopic characterization of Rh based WGS catalysts and hydrogen selective Pd alloy membranes under reaction conditions. At both absorption edges high quality XANES and EXAFS data (in a k-range up to 15 $Å^{-1}$) could be acquired, and the reactor remained stable and gas tight at high temperature and in CO/$H_2O$ reaction atmosphere.

## 5. Acknowledgement

Financial support by the Helmholtz Association and the Chinese Academy of Sciences (CAS) for funding of the Helmholtz-CAS Joint Research Group 118 "Integrated Catalytic Technologies for Efficient Hydrogen Production" is gratefully acknowledged. We acknowledge the European Synchrotron Radiation Facility (Grenoble) for provision of synchrotron radiation facilities, and we would like to thank Herman Emerich for assistance in operating SNBL and Hudson W.P. Carvalho, Abhijeet Gaur and Xiangang Ma for assistance in the experiments.

## 6. References

[1] Pfeifer P 2012 *Application of Catalysts to Metal Microreactor Systems* (*Chemical Kinetics*) ed V Patel (Publisher: InTech) chapter 15 pp 325-344
[2] Bi Y, Xu H, Li W and Goldbach A 2009 *Int. J. Hydrogen Energy* **34** 2965
[3] Dittmeyer R, Höllein V, Daub K and Mol J 2001 *Catal. A: Chem.* **173** 135
[4] Weckhuysen B M 2004, *In-situ Spectroscopy of Catalysts*, (California, USA/ American Scientific Publishers)
[5] Grunwaldt J-D, Caravati M, Hannemann S and Baiker A 2004 *Phys. Chem. Chem. Phys.* **6** 3037
[6] Grunwaldt J-D, Wagner J B and Dunin-Borkowski R E *ChemCatChem* 2013 **5** 62
[7] Grunwaldt J-D and Schroer C G 2010 *Chem. Soc. Rev.* **39** 4741
[8] Tepluchin M, Casapu M, Boubnov A, Lichtenberg H, Wang D, Kureti S and Grunwaldt J-D 2014 *ChemCatChem* **6** 1763
[9] Newville M 2001 *J. Synchrotron Radiat.* **8** 322
[10] Cavusoglu G, Miao D, Lichtenberg H, Carvalho H W P, Xu H, Goldbach A and Grunwaldt J-D 2015 *Appl. Catal.* **504** 381
[11] Gayan A, Priolkar K R, Sarode P R, Jayaram V, Hegde M S, Subbanna G N and Emura S 2004 *Chem. Mater.* **16** 2317

# Copper-oxide metalorganic nanocomposite: morphological and X-ray spectroscopy studies

**G Yalovega[1], A Funik[1], T Myasoedova[1] and M Brzhezinskaya[2]**

[1] Southern Federal University, Faculty of Physics, Rostov-on-Don, Russia

[2] Helmholtz Zentrum Berlin fur Materialien und Energie, Berlin, Germany

E-mail: yalovega1968@mail.ru

**Abstract**. Oxidation state, local and electronic structure as well as the surface morphology of a $CuO_x$ nanocomposites were investigated. The research is focused on studying the relationships between surface morphology and structure of the inorganic component of the nanocomposites. The physico-chemical mechanisms of change in local and electronic structure, oxidation state and the surface morphology of nanocomposites in the interaction of metal oxide nanoparticles with different types of matrices were determined.

## 1. Introduction

Composite materials based on metal oxide nano- microcrystals in/on organic and nonorganic matrices, have attracted increasing attention thanks to the possibility of tailoring their chemico-physical properties as a function of particle size, shape, composition and structure. As well known, the size, morphology and structure of nano- microcrystals as well as composition and morphological details of matrix have great effects on electrical and adsorption properties of the resulting nanocomposites. Besides the influence of the kind of matrix to directing the nano-unit nucleation and subsequent organization of metal oxide nanoparticles/ microcrystals is observed.

In the present research we investigated morphology and local atomic and electronic structure of copper oxide based nanocomposites by means of SEM and XAS. The aim of this study is to examine the relationship between local atomic and fractal structure of the copper-based metalorganic nanocomposite as well as to determine the mechanisms of these structures formation by means of XANES.

## 2. Experiment

Copper oxide films were prepared by sol–gel method on $Si/SiO_2$ substrates. In the citrate sol-gel process the alcohol solution of $CuCl_2$, ethylene glycol and citric acid were selected as starting materials. The details of the synthesis are published in [1]. The prepared films were dried at 150°C for 1 hour and then annealed at temperatures 250, 350 and 500$^0$C with a fixed time of 2 hour in an ambient air. The total thickness of films equal 4.2 μm including silicon substrate with thickness order 3 μm, SiO sublayer with thickness equal 1 μm and Cu layer in order 0.2 μm.

Surface morphological study of the deposited film samples was carried out using a scanning electron microscopy (LEO 1560, ZEISS). The SEM operating voltage was employed at 5 kV.

The Cu K-edge XANES spectra of the $CuO_x$ thin films as well as referenced samples were measured in fluorescence mode at the bending-magnet beamline KMC-2 (BESSY II, Berlin) using double-crystal $Si_{1-x}Ge_x$ (111) monochromator [2]. To increase energy resolution, the beam was focused into an area smaller than 400 μm × 600 μm. A copper metal foil (first inflection point at 8979 eV) was used for energy calibration. Pre-edge features were collected using 0.3 eV steps whereas the main edge jump was collected with 0.7 eV till 240 eV above the absorption edge. Samples were set up at grazing incidence (~3° to the X-ray beam) in order to enhance the detection efficiency.

Collected spectra were corrected, energy calibrated, pre-edge subtracted and post-edge normalized using Athena package [3].

## 3. Results and discussion

The X-ray absorption near-edge structure spectroscopy (XANES) is successfully applied to provide information about oxidation states and symmetries as well as chemical nature of bonding investigating elements in thin films [4, 5]. Figure 1 shows the experimental XANES spectra of $CuO_x$ thin films annealed at 150, 250, 350 and 500 °C (we marked these samples as I, II, III and IV respectively) and few other copper-based materials (i.e. CuO, $Cu_2O$, Cu(OH)Cl etc.).

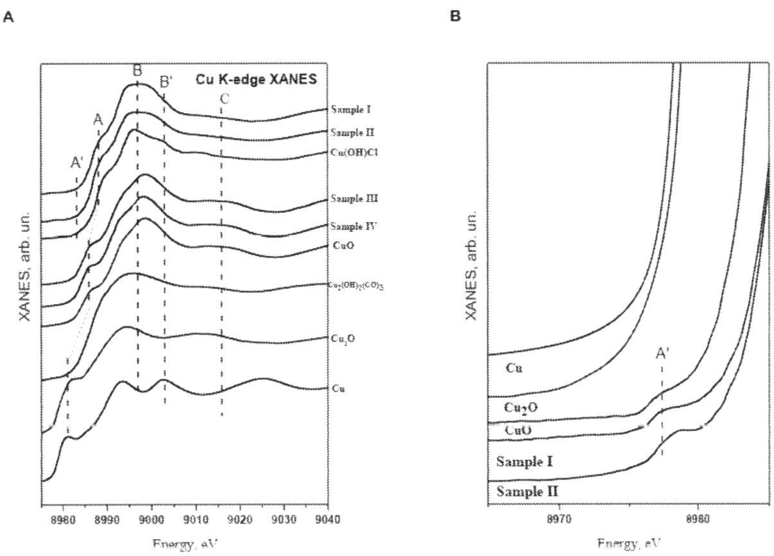


**Figure 1. a** XANES Cu K-edge spectra of the $CuO_x$ thin films in comparison with referenced compounds; **b** pre-edge region XANES Cu K-edge spectra of the $CuO_x$ thin films in comparison with referenced compounds.

It is clearly seen that there are two groups of spectra. Spectra for sample I and II are essentially identical and spectra III and IV are very similar. All spectra under study include main peaks A, B and C with average value of energy ~8987 eV, ~8998 eV and ~9015 eV respectively. One can see that between these groups the main differences in the spectra shape and the energy positions of the features are observed. It indicates that between 250° and 350°C there are some significant changes in local environment of copper. At the same time, this suggests no change in the copper valence and local structure of both samples annealed at 150° and 250° as well as samples annealed at 350° and 500° (fig.1a). We consider in more detail the differences in the spectra of these groups. The edge positions for both groups of samples and the Cu references have been determined from the first derivative function. The shift toward higher energy relative Cu foil (8979 eV) on ~5 eV is observed in the case of samples III and IV, that corresponds to energy position of CuO spectrum (8 983.6 eV) [6]. From comparison of these spectra and measured reference compounds clear formation of CuO oxide after annealed at 350°C and 500°C, that could be explained by formation of CuO crystallites. The spectra I,II are characterized by the pre-edge peak A'(~8977 eV) and broad peak B compared with the spectra of the CuO, shift of energy position of the absorption edge to high energy on ~3 eV (8986.8 eV). The organic component of the film begins to decompose at temperatures above 250° C. The main reason of this process is high concentration of $CuCl_2$ in starting solution compared with citric acid concentration. Therefore, we assume the XANES signal from a thin film I,II can be considered resulting in from a linear combination of signals from Cu atoms with neighborhood similar to that found in Cu(OH)Cl, CuO and some intermediate metalloorganic compounds that results to spectra broadening. The polymer resin (reaction of polycondensation) is formed at 100-110 °C. At any rate, three carboxyl groups of lemon acid take part in the formation of either chelates or the bond with ethylene glycol. As a result, a "nucleus" is formed. The following polymerization runs around the "nucleus". The organic substances decompose only at temperatures above 250 °C and the CuO phase forms that is in good agreement with the SEM results and XANES results. Comparing the spectra of film annealed at 250° and reference compounds, one can observe that peaks A', A, B, B' and C can be used as a fingerprint for the formation of the Cu(OH)Cl phase during annealed. As was shown in [7] a weak pre edge peak around 8976–8978 eV like that of $Cu^{+2}$ reference, indicating that the oxidation state of as-synthesized material is divalent because the reason of this peak is 1s – 3d transition and there is no 3d hole neither in Cu nor in $Cu_2O$ (fig. 1b). At the same time, the K-edge peak A is associated with the main 1s – 4p transition for each of compared compounds [8].

To define the structures resulting from the synthesis of the composite materials at temperatures up to 250 degrees more precisely we carried out a computational modeling of X-ray absorption spectra by the finite difference method with FDM software [9] since in this code there is not muffin-tin approximation formalism of the Green's potential that provides more accurate results for a number of compounds. The simulation results in comparison with experimental spectrum are shown in Figure 2. To compare we have chosen complex hydroxides $Cu_2(OH)_2CO_3$ and Cu(OH)Cl as well as chelate compound $C_6H_{10}CuO_{10}$ [10]. All of the structures were collected thanks to CCDC (Cambridge Crystallographic Data Center).

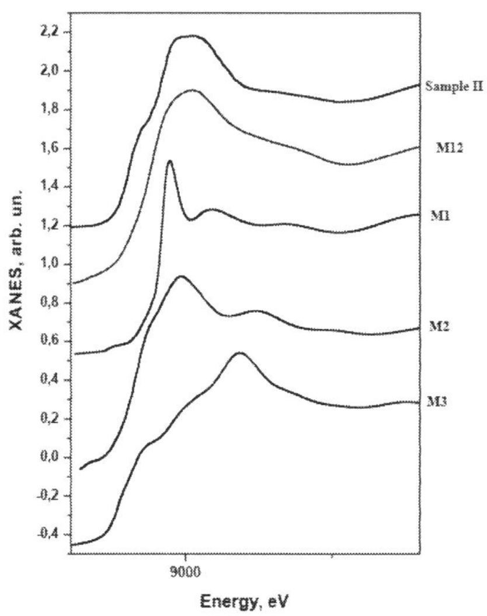


**Figure 2**. Theoretically calculated K-edge XANES spectra of the different compounds in the comparison with Sample II spectra: M1 – Cu K-edge XANES for Cu(OH)Cl, M2 – Cu K-edge

XANES for $Cu_2(OH)_2CO_3$, M12 – Linear combination of the M1 and M2 spectra, M3 - Cu K-edge XANES for $C_6H_{10}CuO_{10}$.

This comparison confirms the assumption that nanocomposite thin film Sample II is formed by several inorganic phases. The main features of the linear combination of two XANES spectra of different complex hydroxides $Cu_2(OH)_2CO_3$ and Cu(OH)Cl are almost similar to those in spectrum for Sample II (energy position for the edge structure, peaks at 9010 and 9025 eV respectively).

XANES results and computational calculations are confirmed by scanning electron microscopy results (Fig. 3). Gradual decomposition of the organic component of the film leads to evolution of the final morphological structure as a function of annealing temperature. As a result, at temperatures above 550 °C crystals of copper oxides are dominated and are converted into fractals with another level of organization.

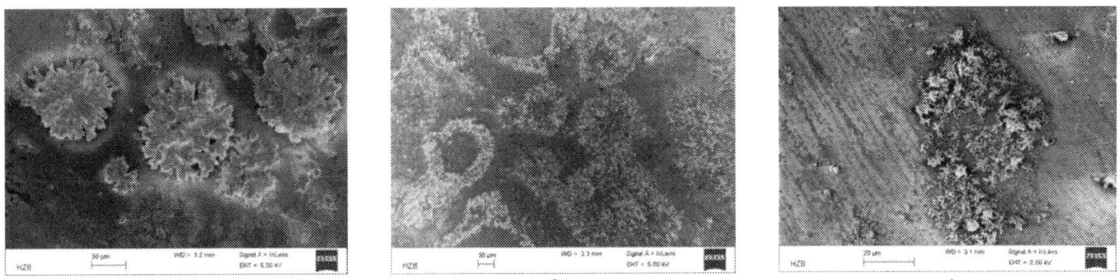

a) b) c)

**Figure 3.** SEM-images of the thin films annealed at: a) 250°C; b) 350°C; c) 500°C.

## 4. Conclusions

We have carried out a comprehensive both experimental and theoretical study of the organometallic nanocomposite thin films based on copper oxides. It was found that at the synthesis temperature below 250 degrees there is complex inorganic component in the structure of the film. With increasing annealing temperature up to 500 formation of CuO crystallites with fractal structure was observed.
While temperature increases changes in the local atomic and electronic structure as well as in the morphology are happened. This phenomenon is explained by the decomposition of the initial organic phases at high temperatures. The copper (II) oxide crystallites are formed on the film surface that is confirmed by results of XANES and SEM investigation.

## 5. Acknowledgments

The research leading to these results received funding from the Ministry of education and science of the Russian Federation under grant agreement № 11.2432.2014/K. We thank HZB (Berlin, Germany) for the allocation of synchrotron radiation beamtime (Beam Line KMC-2 at BESSY II) and Institute for Nanometre Optics and Technology (HZB) for the provided opportunity SEM measurements.

## References

[1] Myasoedova T.N *et al* 2014 *Appl. Mech. and Mat.* V **481** 133
[2] A. Erko *et al* 2001 *Nucl. Instr. and Meth. in Phys. Res.* A **467–468**
[3] B. Ravel and M. Newville 2005 *J. Synchr. Radiat.* **12** 537–541
[4] A. Gaur and B.D. Shrivastava 2012 *A Comp. St. of the M. of Spec. Us. XAFS* V **121** 647-652
[5] G.X. Shen *et al* 2005 *Thin Solid Films* **489** 130
[6] R. Gago *et al* 2013 *J. of Alloys and Comp.* **561** 87–94
[7] M. Vaseem *et al* 2008 *Cat. Comm.* **10** 11–16
[8] J. Fulton *et al* 2000 *J. Phys. Chem. A* ***104*** **(49)** 11651–11663
[9] O. Bunau and Y. Joly 2009 *J. Phys.: Condens. Matter* **21** 345501
[10] A. Lenstra and O.N.Kataeva 2001 *Acta Crystallogr.,Sect.B:Struct.Sci.* **57** 497

# XAFS analysis for quantification of the gallium coordinations in $Al_2O_3$-supported $Ga_2O_3$ photocatalysts

**M. Akatsuka[1], T. Yoshida[2], N. Yamamoto[1], M. Yamamoto[1], S. Ogawa[1], S. Yagi[2]**
[1] Graduate School of Engineering, Nagoya University, Nagoya 464-8603, Japan
[2] Eco Topia Science Institute, Nagoya University, Nagoya464-8603, Japan

E-mail: akatsuka.masato@d.mbox.nagoya-u.ac.jp, tyoshida@ocarina.osaka-cu.ac.jp

**Abstract**. $Ga_2O_3$ loaded $Al_2O_3$ samples ($Ga_2O_3/Al_2O_3$) were prepared to change coordination structures around Ga atoms. Ga K-edge XANES spectra of the $Ga_2O_3/Al_2O_3$ samples showed two peaks assigned to Ga atoms having tetrahedral coordination structure (Ga(t)) and octahedral one (Ga(o)). Curve-fitting analysis of XANES spectra was carried out with a set of pseudo-Voight and arctangent functions, and the fractions of Ga(t) and Ga(o) were quantitatively estimated from the ratio of the peak areas. EXAFS curve-fitting analysis also evaluated the fractions of Ga(t) and Ga(o) and they were in good agreement with those obtained by XANES analysis. It was revealed that the fraction of Ga(t) increased with the decrease in the loading amount of $Ga_2O_3$ due to the interaction of Ga species with $Al_2O_3$. The fractions of Ga(t) and Ga(o) might relate to the photocatalytic activity for $CO_2$ reduction with $H_2O$ over the $Ga_2O_3/Al_2O_3$ samples.

## 1. Introduction

Gallium oxide ($Ga_2O_3$) photocatalysts can reduce $CO_2$ to CO with $H_2O$, although the efficiency of CO production is very low [1]. Generally, photocatalytic activity depends on the coordination structure of the surface active site [2]. However, there are few works investigating the influence of $Ga_2O_3$ coordination structure on the $CO_2$ photoreduction activity. $Ga_2O_3$ has two coordination structures, tetrahedral coordination site (Ga(t)) and octahedral one (Ga(o)). In $\beta$-$Ga_2O_3$ crystal, for example, the ratio of Ga(t) to Ga(o) is known as 1:1 [3], and the other crystal phases of $Ga_2O_3$ have different ratios. In addition, it has been reported that GaAl-oxide has only Ga(t) [4]. Thus, we think that the ratio of Ga(t) to Ga(o) can be controlled by loading $Ga_2O_3$ on the $Al_2O_3$-support to improve photocatalytic activity of $Ga_2O_3$.

In this study, we prepared $Ga_2O_3$ loaded $Al_2O_3$ samples ($Ga_2O_3/Al_2O_3$) and measured Ga K-edge XAFS spectra of these samples. The fractions of Ga(t) and Ga(o) were calculated by curve-fitting analysis of XANES spectra and were confirmed by curve-fitting analysis on EXAFS of first neighboring oxygen atoms (Ga-O shell). In addition, the EXAFS of the second neighboring atoms was investigated in detail to obtain the information about interaction between $Ga_2O_3$ and $Al_2O_3$. We also carried out the photocatalytic $CO_2$ reduction over $Ga_2O_3/Al_2O_3$ samples to reveal the correlation between the photocatalytic activity and the ratio of Ga(t) to Ga(o).

## 2. Experiment

$Ga_2O_3/Al_2O_3$ samples were prepared by impregnation of $\gamma$-$Al_2O_3$ with aqueous solution of gallium nitrate followed by dry and calcination in the air at 773 K for 4 h, as described elsewhere [5]. The loading amounts of $Ga_2O_3$ were 1, 5, 10 and 20 wt%. Unsupported $Ga_2O_3$ sample was also prepared by calcination of gallium nitrate similarly to $Ga_2O_3/Al_2O_3$. The $Ga_2O_3$ sample was employed for XRD measurement to identify the crystal phase of $Ga_2O_3$. Surface areas of all the prepared samples were measured by BET method.

The Ga K-edge XAFS measurement of the samples were conducted at the beamline 5S1 in Aichi Synchrotron Radiation Center. The beamline was equipped with Si(111) double-crystal monochromators. The data were recorded in a transmission mode or a fluorescent X-ray yield mode with a 19 channel silicon semiconductor detector.

XANES spectra were curve-fitted with pseudo-Voight function (1) and arctangent function (2).

$$f(x) = \frac{A}{[1+M(x-x_0)^2/\beta^2]\exp\{(1-M)[(ln2)(x-x_0)^2/\beta^2]\}} \quad (1)$$

$$g(x) = \mathrm{B}\left\{(1/\pi)\arctan\left(\frac{x-x_0'}{\Gamma/2}\right) + (1/2)\right\} \quad (2)$$

In these formula, A, B are heights and β, γ are full width at half maximum. $x_0$ and $x_0'$ are position of the peak of pseudo-Voight curve and the inflection point of arctangent curve, respectively. M is the ratio of contribution of Lorentzian in pseudo-Voigt curve. Nishi *et al.* carried out curve-fitting with Gaussian and arctangent curves on Ga K-edge XANES spectra of $Ga_2O_3$ loaded $Al_2O_3$ samples although the Gaussian fitting arose some deviation from a XANES spectrum especially around the edge [5]. On the other hand, pseudo-Voigt curve used in this study was more suitable for the curve-fitting because XANES spectrum includes natural width due to electron excitation as well as that of the measurement system. Therefore, it is expected that the present curve-fitting leads to better accuracy in the quantitative estimation of Ga(t) and Ga(o). The fractions of Ga(t) and Ga(o) were also evaluated by the non-linear least squares curve-fitting analysis of EXAFS spectra [6].

Photocatalytic $CO_2$ reduction with $H_2O$ was carried out over $Ga_2O_3/Al_2O_3$ and $Ga_2O_3$ samples. $Ga_2O_3/Al_2O_3$ sample was put into the fixed-bed flow reactor cell under $CO_2$ gas with a flow rate at 3.0 mL/min and irradiated by UV-light in the range of 254 ± 10 nm was 29.5 mW/cm$^2$. The reaction gases were measured with the GC-TCD (SHIMADZU, GC-8A).

## 3. Results and discussion

Unsupported $Ga_2O_3$ was identified as γ-$Ga_2O_3$ by the XRD pattern. On the other hand, very weak and broad lines due to Ga oxide phase were observed for $Ga_2O_3/Al_2O_3$ samples but could not be identified clearly. In the EDS measurements of $Ga_2O_3/Al_2O_3$ samples, no remarkable Ga oxide particles were observed. These results suggested that the loaded $Ga_2O_3$ species were highly dispersed on $Al_2O_3$. Based on the BET measurements for a bare $Al_2O_3$ and $Ga_2O_3/Al_2O_3$ samples, we roughly estimated the surface area of Ga oxide phase in each $Ga_2O_3/Al_2O_3$ sample by subtracting the component of $Al_2O_3$ support. We regarded that the surface areas of Ga oxide phase are almost the same each other.

Ga K-edge XANES spectra of $Ga_2O_3/Al_2O_3$ samples are shown in Figure 1 together with those of γ-$Ga_2O_3$ and β-$Ga_2O_3$. Two peaks shown at 10375 eV and 10379 eV are attributed to Ga(t) and Ga(o), respectively [5]. In addition, a small peak is observed around 10384 eV. This peak would be influenced by a multiple scattering of photoelectrons by neighboring atoms. The XANES spectra of the $Ga_2O_3/Al_2O_3$ samples showed both peaks at 10375 eV and 10379 eV, and the former peak reduced and the latter peak grew with increased amount of $Ga_2O_3$ supported on $Al_2O_3$. It is likely that the $Ga_2O_3/Al_2O_3$ samples contain two kinds of Ga species, Ga(t) and Ga(o), and the fraction of each species differs with the loading amount of $Ga_2O_3$.

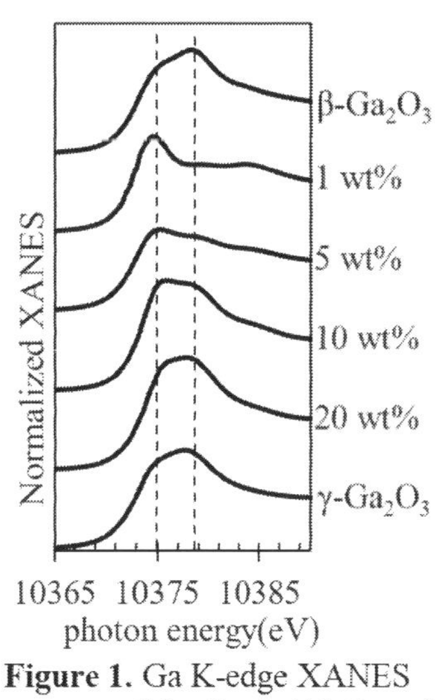


**Figure 1.** Ga K-edge XANES spectra of the $Ga_2O_3/Al_2O_3$ and $Ga_2O_3$ samples.

Here, we calculate the fractions of Ga(t) and Ga(o) in each $Ga_2O_3/Al_2O_3$ sample by curve-fitting on each XANES spectrum with two sets of pseudo-Voigt and arctangent curves. The simulation of the spectrum was made on the basis of the following assumptions: [7] 1) the spectrum is composed of two components; 2) each component consists of an arctangent curve for continuum absorption and a pseudo-Voigt curve for the white line; 3) the ratio of the height of each arctangent curve is equal to the ratio of the peak area of each pseudo-Voigt curve; 4) The small peak around 10384 eV is not employed for the curve-fitting analysis. Figure 2 shows the curve-fitting on XANES spectrum of β-$Ga_2O_3$. The fractions of Ga(t) and Ga(o) in β-$Ga_2O_3$ were evaluated as 51% and 49 %, and the fractions were well accorded with the crystal data of β-$Ga_2O_3$ (fractions of Ga(t) and Ga(o) are 50% and 50 %)

Journal of Physics: Conference Series **712** (2016) 012056 doi:10.1088/1742-6596/712/1/012056

[8]. This result suggests that the fractions of Ga(t) and Ga(o) in $Ga_2O_3/Al_2O_3$ samples can be estimated quantitatively by our curve-fitting analysis, and we also decided the parameter M in the pseudo-Voigt curve as 0.7. Table 1 summarizes the other parameters used for the curve-fitting analysis on XANES spectra of $Ga_2O_3/Al_2O_3$ samples, and the calculated fractions of Ga(t) in the $Ga_2O_3/Al_2O_3$ samples are also shown in Table 2. It is revealed that the coordination structure around a Ga atom was controlled by loading $Ga_2O_3$ on the $Al_2O_3$-support and the Ga(t) was formed preferentially in the sample with lower loading amount of $Ga_2O_3$.

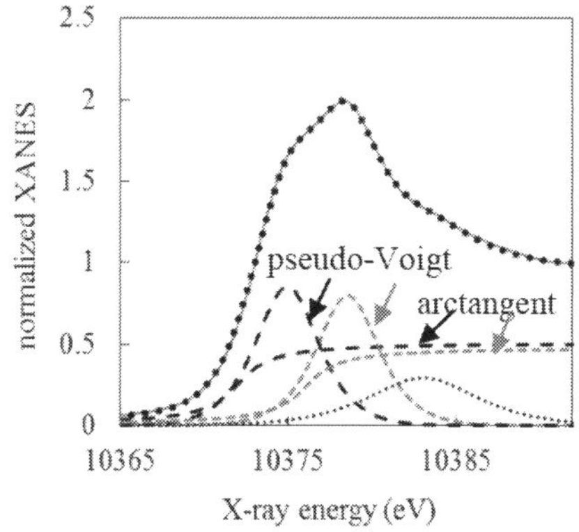


**Figure 2.** Ga K-edge XANES spectrum of $\beta$-$Ga_2O_3$ (solid line) and the simulated spectrum (dotted), which consists of two sets of pseudo-Voigt and arctangent curves (broken).

**Table 1.** Curve-Fitting parameters

| Samples | Ga(t) species (eV) | | | | Ga(o) species (eV) | | | |
|---|---|---|---|---|---|---|---|---|
| | position $X_0$ | position $X_0'$ | fwhm $\beta$ | fwhm $\Gamma$ | position $X_0$ | position $X_0'$ | fwhm $\beta$ | fwhm $\Gamma$ |
| $\beta$-$Ga_2O_3$ | 10374.9 | 10372 | 2.3 | 3 | 10378.5 | 10376 | 2.3 | 3 |
| $\gamma$-$Ga_2O_3$ | 10374.3 | 10371.7 | 2 | 4 | 10378 | 10375 | 3 | 2.5 |
| $Ga_2O_3/Al_2O_3$ | | | | | | | | |
| 1 wt% | 10374.8 | 10372.5 | 2.1 | 2 | 10379.5 | 10378 | 3 | 4 |
| 5 wt% | 10374.6 | 10373 | 2.5 | 4 | 10378.8 | 10377 | 3.4 | 4 |
| 10 wt% | 10375.1 | 10373 | 2.1 | 4 | 10378.6 | 10377 | 3 | 4 |
| 20 wt% | 10375.2 | 10373 | 2.1 | 4 | 10378.5 | 10376 | 3 | 4 |

Fitting error of XANES spectrum with these parameters is ±5%.

**Table 2.** The fraction of Ga(t) in $Ga_2O_3/Al_2O_3$ estimated by XANES and EXAFS analyses

| sample | Loading amount of $Ga_2O_3$ (wt%) | | | | |
|---|---|---|---|---|---|
| | 1 | 5 | 10 | 20 | $\gamma$-$Ga_2O_3$ |
| Ga(t) (%) | 75 | 60 | 46 | 39 | 36 |
| Ga(o) (%) | 25 | 40 | 54 | 61 | 64 |

Errors of Ga(t) and Ga(o) fractions estimated by XANES and EXAFS are ±2% and ±8%, respectively.

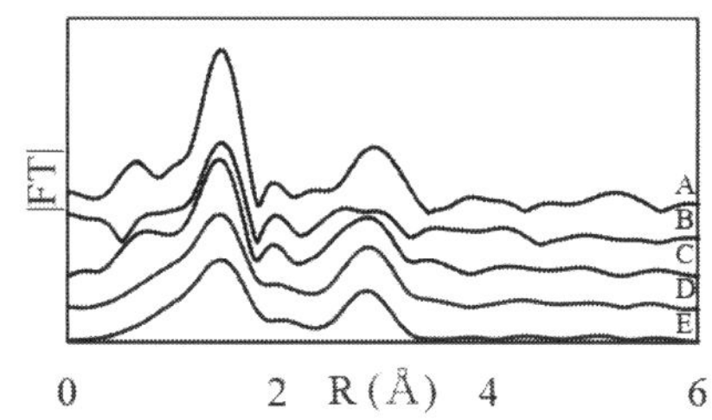


**Figure 3.** Radial structure functions of $Ga_2O_3/Al_2O_3$ and $\gamma$-$Ga_2O_3$. A) 1 wt% $Ga_2O_3/Al_2O_3$, B) 5 wt% $Ga_2O_3/Al_2O_3$, C) 10 wt% $Ga_2O_3/Al_2O_3$, D) 20 wt% $Ga_2O_3/Al_2O_3$, E) $\gamma$-$Ga_2O_3$

We also measured $k^3$-weighted Ga K-edge EXAFS spectra of $Ga_2O_3/Al_2O_3$ and $\gamma$-$Ga_2O_3$ samples as shown in Supporting Information with Figure S1. The Fourier transform was performed on each EXAFS spectrum in the 3-13.7 $Å^{-1}$ region and the radial structure function (RSF) was obtained as shown in Figure 3. Peak appearing at 1-2 Å is assigned to the backscattering from the adjacent oxygen atoms, and peaks around 3 Å show the presence of the second-neighboring metal atoms (Ga or Al). Note that the RSFs of the 10 and 20 wt% $Ga_2O_3/Al_2O_3$ samples were similar to that of $\gamma$-$Ga_2O_3$ rather than $\beta$-$Ga_2O_3$.

The first peak at 1-2 Å was sharp and high for 1 wt% $Ga_2O_3/Al_2O_3$ while it became broader and lower for the other $Ga_2O_3/Al_2O_3$ samples. We performed curve-fitting analysis for the inverse Fourier transform of the first peak at 1-2 Å with two Ga-O shells. For the reference sample, $\beta$-$Ga_2O_3$, the curve-fitting results indicate coordination shells with 2.0 oxygen atoms at 1.85 Å and 3.0 oxygen atoms at 2.00Å, which is in good agreement with the crystallographic data for Ga(t)-O and Ga(o)-O shells, respectively. Similar short Ga(t)-O and long Ga(o)-O shells were found for all $Ga_2O_3/Al_2O_3$ samples, then we estimated the fractions of Ga(t) and Ga(o) by using of the coordination numbers for Ga(t)-O and Ga(o)-O, and the results are also summarized in Table 2. The fractions of Ga(t) and Ga(o) obtained by XANES and EXAFS analyses are good agreement with each other, supporting that the fraction of Ga(t) is higher for the sample with lower loading amount of $Ga_2O_3$.

Here, we investigated the second-neighboring metal atoms in the low and high loaded $Ga_2O_3/Al_2O_3$ samples. Figure 4 shows the envelope functions of inverse Fourier transform of the second peak around 3 Å in RSFs of 1, 20 wt% $Ga_2O_3/Al_2O_3$ samples and $\gamma$-$Ga_2O_3$. The maximum in the envelope appeared at a low k value for the 1 wt% $Ga_2O_3/Al_2O_3$ sample while that in the other envelopes was observed at 8-9 $Å^{-1}$. This result clearly indicates that the envelope of 1wt% $Ga_2O_3/Al_2O_3$ sample is due to backscatterer lighter than Ga atoms, i.e., Al is very likely. Therefore, in the low loaded $Ga_2O_3/Al_2O_3$ sample, highly dispersed Ga species would interact with $Al_2O_3$ support to form Ga(t) structure, while in the high loaded

$Ga_2O_3/Al_2O_3$ sample, the aggregation of Ga species would produce similar structure as $\gamma$-$Ga_2O_3$ to decrease in the fraction of Ga(t).

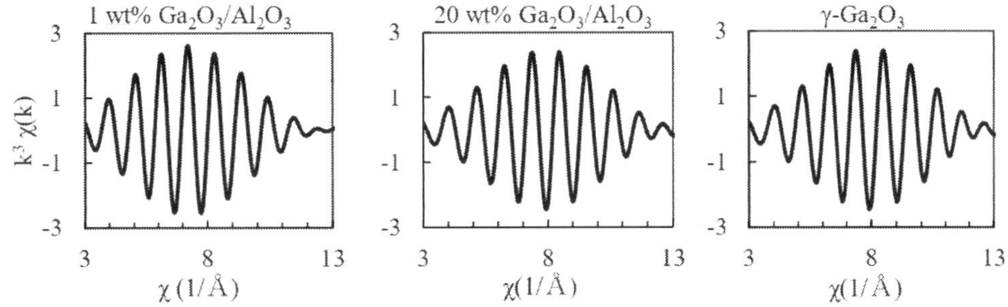


**Figure 4.** The envelope functions of inverse Fourier transform of the second peak around 3 Å in radial structure functions of 1, 20 wt% $Ga_2O_3/Al_2O_3$ and $\gamma$-$Ga_2O_3$.

Photocatalytic $CO_2$ reduction with $H_2O$ was carried out over the $Ga_2O_3/Al_2O_3$ sample. CO production rates over 5, 10 and 20 wt% $Ga_2O_3/Al_2O_3$ samples were higher than those over $\beta$-$Ga_2O_3$ and $Al_2O_3$ samples, while over 1 wt% $Ga_2O_3/Al_2O_3$, it was almost the same as that over $Al_2O_3$ sample. At present, the reason for a small amount of CO production over 1 wt% $Ga_2O_3/Al_2O_3$ has not been clarified yet. This sample probably forms a GaAl-oxide and such a solid state sample might show low activity for $CO_2$ photoreduction.

Figure 5 shows the variation of CO production rate per a Ga atom with the fraction of Ga(t) in 5, 10 and 20 wt% $Ga_2O_3/Al_2O_3$ samples. As mentioned above, since $Ga_2O_3$ species would be highly dispersed on $Al_2O_3$, we assumed that the Ga(t) fraction evaluated by XANES and EXAFS corresponds to the Ga(t) fraction at the surface. In this figure, CO production rate per a Ga atom seemed to increase with the fraction of Ga(t). However, it might be required to estimate carefully the numbers of the surface Ga(t) and Ga(o) really contributing to the photocatalytic reaction by $CO_2$ and $H_2O$ adsorption experiments in our future work.

## 4. Conclusion

Ga K-edge XAFS spectra were recorded on the $Ga_2O_3$ loaded $Al_2O_3$ photocatalysts. Curve-fitting analysis of XANES spectra could estimate the fractions of tetrahedral and octahedral Ga species (Ga(t) and Ga(o)) in these samples, which was confirmed by EXAFS curve-fitting analysis. We concluded that the fraction of Ga(t) increased with lower loading amount of $Ga_2O_3$. CO production rate per a Ga atom seemed to increase with the fraction of Ga(t), however, the further quantitative analysis for the surface Ga(t) and Ga(o) will be required.

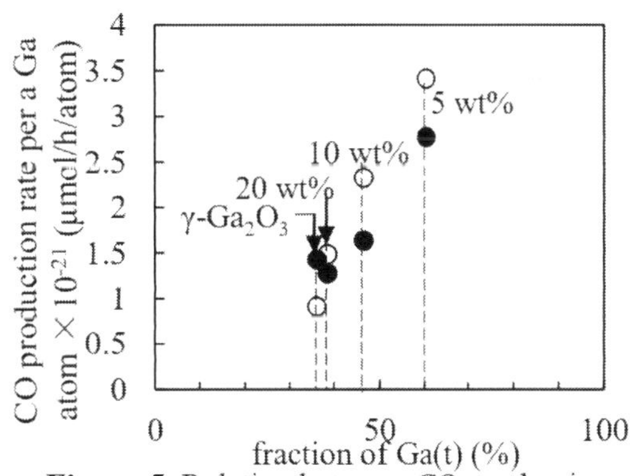


**Figure 5.** Relation between CO production rate per a Ga atom and the fraction of Ga(t). (○ : after 1 h, ● : after 4 h )

## 5. Acknowledgment

This work was supported by a Grant-in-Aid for Scientific Research (No.24360332 and 24656489) from the Japan Society for the Promotion of Science (JSPS). Synchrotron experiments were carried out at Aichi Synchrotron Radiation Center which is supported by No.201505044.

## References

[1] Yoshida H, Zhang L, Sato M, Morikawa T, Kajino T *et al.* 2015 *Catal. Today* **251** 132.
[2] Yoshida S *et al.* 1992 *Catal. Lett.* **12** 277.
[3] Cramer S *et al.* 1976 *J. Am. Chem. Soc.* **98** 1287.
[4] Takahashi M *et al.* 2005 *SPring-8 User Experiment Report* **No. 15 A** 261.
[5] Nishi K, Shimizu K, Takamatsu M, Yoshida H *et al.* 1998 *J. Phys. Chem.* B **102** 10190.
[6] Yoshida T, Tanaka T, Yoshida H, Funabiki T and Yoshida S 1996 *J. Phys. Chem.* **100** 2302.
[7] Kato Y *et al.* 2001 *Phys. Chem. Chem. Phys.* **3** 1925
[8] Geller S 1960 *J. Chem. Phys.* **33** 676.

# Structural characterization of bimetallic Pd-Cu vapor derived catalysts

**Antonella Balerna[1], Claudio Evangelisti[2], Rinaldo Psaro[2], Graziano Fusini[2,3], and Adriano Carpita [3]**

[1] INFN, Frascati National Laboratory, 00044 Frascati, Italy
[2] CNR, Institute of Molecular Science and Technologies (ISTM), 20138 Milano, Italy
[3] Department of Chemistry and Industrial Chemistry, University of Pisa, 56126 Pisa, Italy

E-mail: antonella.balerna@lnf.infn.it

**Abstract.** Pd-Cu bimetallic Solvated Metal Atoms (SMA) were synthesized by metal vapor synthesis technique and supported on PVPy resin. Since the catalytic activity, of the Pd-Cu system turned out to be quite high also compared to the corresponding monometallic system, a structural characterization, using electron microscopy techniques and X-ray Absorption Fine Structure spectroscopy, was performed. HRTEM analysis showed the presence of Pd particles distributed in a narrow range with a mean diameter of about 2.5 nm while the XAFS analysis, confirmed the presence of the Pd nanoparticles but revealed also some alloying with Cu atoms.

## 1. Introduction

Bimetallic catalysts are of great interest from both a scientific and an industrial point of view thanks to their catalytic, electronic and optical properties often different from their corresponding mono-metal systems. Among the numerous structural features which can influence their catalytic efficiency, such as particles size and shape like in the monometallic systems, the chemical states and the distribution of the two metals within the individual nanoparticles (e.g., alloying, core-shell or segregation) play a crucial role [1-3]. Bimetallic catalysts which include a combination of a group VIII, such as Pd, and a group IB, such as Cu, have been extensively studied in many processes such as hydrogenation or oxidation reactions but only a smaller number of studies concern their use in Sonogashira coupling reaction that is a very powerful method for the formation of $C_{sp^2}$ - $C_{sp}$ bonds [4-6]. The Pd/Cu co-catalyzed $C_{sp^2}$ - $C_{sp}$ coupling reaction of an aryl or vinyl halide and an alkyne(Sonogashira reaction), is a powerful tool to prepare important precursors of natural products, agrochemicals and pharmaceuticals, as well as molecular materials (optical or electronic) [7]. The development of supported catalysts is a promising way to ensure an easy separation and reusability of the catalytic systems avoiding the environmental pollution from heavy metallic ions [7]. Recently, Pd-Cu bimetallic nanoparticles also embedded in an ion-exchange resin have been reported to be active in Sonogashira coupling reactions performed in acetonitrile as solvent [6]. In the last years many efforts have been made to develop new Pd-Cu bimetallic systems. In this study the structural characterization of a Pd-Cu bimetallic catalyst derived from Pd and Cu vapors, according to metal vapor synthesis (MVS) technique and supported on a commercially available polyvinylpyridine crosslinked with divinylbenzene (PVPy) resin, that showed high catalytic activity in Sonogashira carbon-carbon coupling reactions, is reported.

## 2. Sample preparation

The synthesis of the Pd-Cu bimetallic system was achieved by the simultaneous co-condensation of Pd and Cu vapors with a mixture of 1–hexene and mesitylene vapors, according to the Metal Vapor Synthesis (MVS) procedure [8]. The MVS approach allowed to easily deposit the metal nanoparticles onto the PYPy polymeric support avoiding further additional treatments (e.g. high temperature calcinations or reducing steps). A quantitative analysis (ICP-OES) performed on the Pd-Cu/PVPy sample revealed a 1.0 wt.% of Pd and 0.3 wt.% of Cu.

## 3. Results and discussion

The structural features, the metal particle size distribution and the metal dispersion on the PVPy support, of all samples were investigated by high resolution transmission electron microscopy (HRTEM) (Fig. 1) and X-ray absorption fine structure (XAFS) spectroscopy.

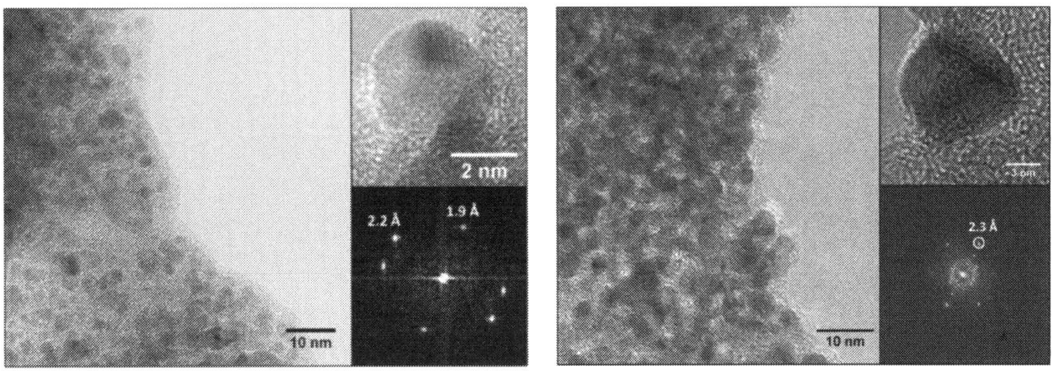


**Figure 1.** HRTEM and lattice fringe analysis of the PYPy-supported bimetallic Pd-Cu (left) and monometallic Pd (right) samples.

The HRTEM analysis was performed using a 200 KV Zeiss LIBRA 200FE analytical transmission electron microscope, equipped with a STEM facility and an EDS probe for chemical analysis. The local atomic structure around the Pd and Cu atoms was investigated by XAFS spectroscopy. Here we report the results achieved at the Pd K absorption edges of the Pd monometallic and Pd-Cu bimetallic samples at the Italian GILDA beamline [9] of the European Synchrotron Radiation Facility (ESRF, Grenoble - France). XAFS data analysis was performed using the IFEFFIT program [10] to best fit experimental data to simulated spectra with theoretical amplitudes and phases provided by the FEFF code [11]. In the analysis, the palladium foil data, where used to calibrated in energy the different spectra and determine and fix to 0.82 the $S_0^2$ parameter.

### *3.1. Catalytic activity*

The high catalytic performances of the bimetallic Pd-Cu/PVPy system, compared to the ones of the Pd/PVPy monometallic system in Sonogashira-type carbon-carbon coupling of iodobenzene with phenylacetylene, are reported in Table 1. The reaction conditions were: 2 mmols of aryl iodide, 2.4 mmols of phenyl acetylene, 4 mmols of pyrrolidine, 21.2 mg of catalyst (molar ratio iodide/Pd = 1000, molar ratio iodide/Cu = 2000), 5 mL of $H_2O$, 1 mmol of TBAB, 95°C, under air. The catalytic system was effectively used for three catalytic runs showing only a slight decrease of the catalytic activity (second run: TOF = 496 $h^{-1}$, third run: TOF = 481 $h^{-1}$).

**Table 1.** Catalytic performances of the bimetallic and monometallic samples; a) Evaluated by gas-chromatographic analyses; b) Calculated as moles of aryl iodide converted/moles of Pd per hour; c) wt.% of metal leached respect to the available starting amount.

| Catalyst | Time (h) | Conversion[a)] (%) | Turnover Frequency (TOF) ($h^{-1}$)[b)] | Metal leaching[c)] (wt. %) |
|---|---|---|---|---|
| Pd-Cu/PVPy | 0.5 | 26.5 | 520 | - |
| (1.0 Pd wt.%, 0.3 Cu wt.%) | 2.5 | 100 | - | <0.1 (Pd) <1.8 (Cu) |
| Pd/PVPy | 0.5 | 14.5 | 289 | - |
| (1.0 wt.%) | 4 | 100 | - | 0.5 (Pd) |

*3.2. Structural characterization using HRTEM and XAFS*

HRTEM analysis (Fig. 1) has shown that in the bimetallic Pd-Cu sample, the PVPy resin, appears densely populated by nanostructured metal particles mainly distributed in a narrow range (1.0 nm - 4.5 nm) with a mean diameter of about 2.5 nm. STEM-EDS analysis performed on different grains of the bimetallic Pd-Cu system revealed a constant Pd/Cu molar ratio which was in agreement with above reported quantitative analysis. Fourier transforms analysis (FFT) of the high-resolution images of individual particles of the bimetallic system showed spots at 2.2 Å and 1.9 Å (see Fig. 1). This values are compatible (within the experimental errors) with the presence of small Pd nanoparticles [12], having a face cubic centered (fcc) structure, showing a contraction of the spacings of the {111} and {200} lattice planes with respect to the monometallic Pd system that gives a value of 2.3 Å, typical of pure metallic Pd. Comparing the bimetallic system with the monometallic Pd/PVPy one, synthesized using the same procedure, the latter contains metal particles with larger mean diameters of about 2.9 nm and a broader size distribution (1.0 nm - 7.5 nm) showing that the presence of Cu has an effect on the size and on the distribution of the metal particles in the bimetallic catalyst. HRTEM lattice fringe analysis and EXAFS analysis (Fig. 2) indicated, for the monometallic and bimetallic samples, the presence of fcc Pd nanoparticles. The monometallic sample, showed the presence of metallic Pd nanoparticles ($N_{Pd-Pd}$= 8, $R_{Pd-Pd}$= 2.74 Å and $\sigma^2{}_{Pd-Pd}$= 0.0049 Å$^2$) and of a little Pd-O ($N_{Pd-O}$= 0.7, $R_{Pd-O}$ = 2.02 Å and $\sigma^2{}_{Pd-O}$= 0.0024 Å$^2$) contribution coming from the superficial oxidation of the Pd nanoparticles [13]. Moving to the bimetallic, Pd-Cu sample the EXAFS data analysis showed the presence of three different Pd-O, Pd-Cu and Pd-Pd contributions. Compared to the monometallic sample, the coordination number of the Pd-O contribution in the bimetallic catalyst is higher but has the same interatomic distance ($N_{Pd-O}$= 1.3, $R_{Pd-O}$ = 2.02 Å and $\sigma^2{}_{Pd-O}$= 0.002 Å$^2$). The higher coordination number probably comes from the smaller nanoparticle dimensions and consequently to the higher surface to volume ratio of the bimetallic catalyst. A small amount of hetero- atomic Pd-Cu bonds ($N_{Pd-Cu}$= 0.7, $R_{Pd-Cu}$= 2.68 Å and $\sigma^2{}_{Pd-Cu}$= 0.008 Å$^2$) and a shortening of the Pd-Pd distance ($N_{Pd-Pd}$= 5, $R_{Pd-Pd}$= 2.72 Å and $\sigma^2{}_{Pd-Pd}$= 0.0068 Å$^2$) were also observed. The shorter Pd-Pd interatomic distance of the bimetallic sample, compared to monometallic one, can be due to the smaller nanoparticles dimension [14] (fcc structure is still observable). Considering that the monometallic and bimetallic samples were prepared using the same Metal Vapor Synthesis (MVS) procedure [8], the main structural effects of the presence of Cu, seems to be some alloying probably present only in the immediate subsurface layers and the reduction of the mean dimension and of the size distribution of the produced nanoparticles as seen by HRTEM. Taking into account the catalytic results these findings suggest that, the presence of

Cu (Pd-Cu heteroatomic bonds) and of smaller Pd nanoparticle dimensions, could be responsible of the enhanced catalytic efficiency.

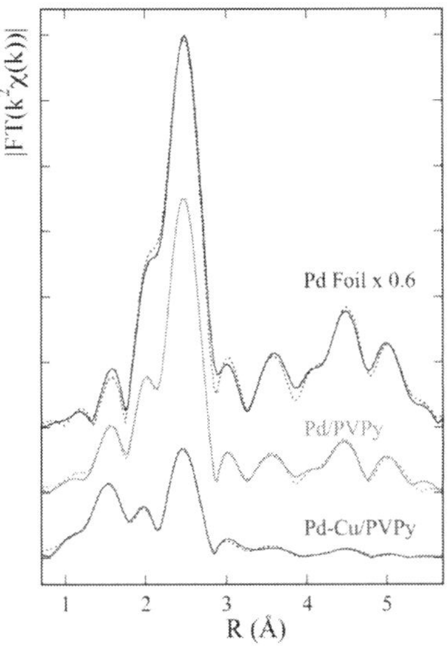


**Figure 2.** FT of the EXAFS signals and their best fit (dotted lines).

## 4. Conclusions

A Pd-Cu/PVPy catalyst, used in Sonogashira type carbon-carbon coupling reactions, resulted largely more active than the corresponding monometallic Pd catalyst. In the bimetallic Pd-Cu/PVPy sample, HRTEM analysis showed the presence of metal particles distributed in a narrow range with a mean diameter of about 2.5 nm. XAFS analysis indicated the presence of small Pd nanoparticles with contracted interatomic distances and of heteroatomic Pd-Cu bonds. The structural effects due to presence of Cu were related to the higher catalytic activity of the Pd-Cu sample.

## 5. References

[1] Toshima N, Yan H and Shiraishi Y 2008 *Metal Nanoclusters in Catalysis and Materials Science. The Issue of Size-Control* ed B. Corain, G. Schmid and N. Toshima (Amsterdam: Elsevier) chapter 3 pp 49–75

[2] Evangelisti C, Panziera N, Pertici P, Vitulli G, Salvadori P, Battocchio C and Polzonetti G 2009 *J. Catal.* **262** 287

[3] Evangelisti C, Schiavi E, Aronica L A, Caporusso A M, Vitulli G, Bertinetti L, Martra G, Balerna A and Mobilio S 2012 *J. Catal.* **286** 224

[4] Karak M, Barbosa L C A and Hargaden G C 2014 *RSC Adv.* **4** 53442

[5] Korzec M, Bartczak P, Niemczyk A, Szade J, Kapkowski M, Zenderowska P, Balin K, Lelatko J and Polanski J 2014, *J. Catal.* **313** 1

[6] Sengupta D, Saha J, De G and Basu B 2014 *J. Mat. Chem. A* **2** 3986

[7] Chinchilla J R and Najera C 2011 *Chem. Soc. Rev.* **40** 5084

[8] Vitulli G, Evangelisti C, Caporusso A M, Pertici P, Panziera N, Bertozzi S and Salvadori P 2008 *Metal Nanoclusters in Catalysis and Materials Science. The Issue of Size-Control* ed B. Corain, G. Schmid and N. Toshima (Amsterdam: Elsevier) chapter 32 pp 437–451

[9] Pascarelli S, D'Acapito F, Antonioli G, Balerna A, Boscherini F, Cimino R, Dalba G, Fornasini P, Licheri G, Meneghini C, Rocca F and Mobilio S 1995 *ESRF Newslett.* **23** 17

[10] Ravel B and Newville M 2005 *J. Synch. Rad.* **12** 537

[11] Ankudinov A L, Ravel B, Rehr J J and Conradson S D 1998 *Phys. Rev. B* **58** 7565

[12] Lee A F, Ellis P J, Fairlam I J S, Wilsona K, 2010 *Dalton Trans.* **39** 10473.

[13] Oberhauser W, Evangelisti C, Jumde R P, Petrucci G, Bartoli M, Frediani M, Mannini M, Capozzoli L, Passaglia E, Rosi L, 2015 *J. Catal.* **330** 187-196

[14] Seriani N, Mittendorfer F and Kresse G 2012 *J. Chem. Phys.* **132** 024711

# Energy dispersive-EXAFS of Pd nucleation at a liquid/liquid interface

**S-Y Chang,[1] S G Booth,[2] A Uehara,[2,5] J F W Mosselmans,[3] G Cibin,[3] V-T Pham,[6,7] L Nataf,[6] R A W Dryfe[2] and S L M Schroeder[1-4]**

[1] School of Chemistry and School of Chemical Engineering and Analytical Science, University of Manchester, Manchester M13 9PL, UK.
[3] Diamond Light Source Ltd., Diamond House, Harwell Science and Innovation Campus, Fermi Ave, Didcot, Oxfordshire OX11 0DE, UK.
[4] Present address: School of Chemical and Process Engineering, Faculty of Engineering, University of Leeds, Leeds LS2 9JT, UK.
[5] Present address: Division of Nuclear Engineering Science, Research Reactor Institute, Kyoto University, Asashironishi, Kumatori, Osaka, 590-0494, Japan.
[6] Synchrotron SOLEIL, L'Orme des Merisiers, Saint-Aubin, BP48, 91192, Gif-sur-Yvette, France.
[7] Center for Quantum Electronics, Institute of Physics, Vietnam Academy of Science and Technology, P.O. Box 429, Boho, 10000 Hanoi, Viet Nam.

Email: s.l.m.schroeder@leeds.ac.uk

**Abstract.** Energy dispersive extended X-ray absorption fine structure (EDE) has been applied to Pd nanoparticle nucleation at a liquid/liquid interface under control over the interfacial potential and thereby the driving force for nucleation. Preliminary analysis focusing on Pd K edge-step height determination shows that under supersaturated conditions the concentration of Pd near the interface fluctuate over a period of several hours, likely due to the continuous formation and dissolution of sub-critical nuclei. Open circuit potential measurements conducted *ex-situ* in a liquid/liquid electrochemical cell support this view, showing that the fluctuations in Pd concentration are also visible as variations in potential across the liquid/liquid interface. By decreasing the interfacial potential through inclusion of a common ion (tetraethylammonium, $TEA^+$) the Pd nanoparticle growth rate could be slowed down, resulting in a smooth nucleation process. Eventually, when the $TEA^+$ ions reached an equilibrium potential, Pd nucleation and particle growth were inhibited.

## 1. Introduction

Nucleation is currently the focus of considerable research activity in crystallisation science and crystal engineering and has been the subject of a number of reviews including [1, 2]. Particle formation from homogeneous solution involves nucleus (or nuclei) formation by self-association of solute molecules and subsequent phase-separation from the bulk solution. By following the solute and solution structure from the under-saturated through the supersaturated state to the regime of spontaneous particle formation, relationships to the structure of the crystallised final products may be established. However, there are considerable experimental challenges arising from the size of the nuclei (sub-nm range), their unknown composition, diffusive or convective movement, as well as the stochastic nature of

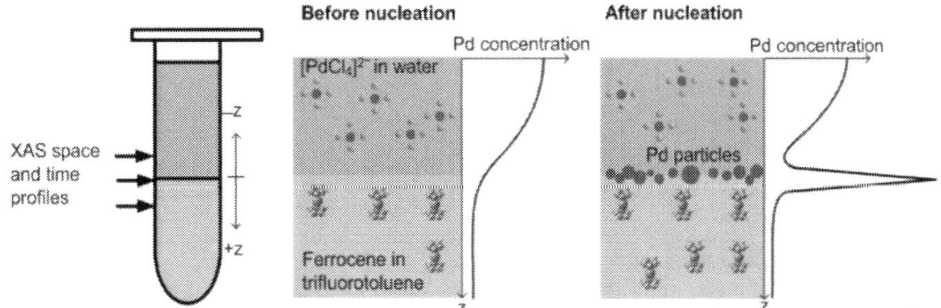


Figure 1. The liquid-liquid system used for studying nucleation. The aqueous layer contains $[PdCl_4]^{2-}$ and the organic layer (trifluorotoluene) contains the reducing agent ferrocene. The build-up of metallic Pd nanoparticles is evident from the absorption coefficient profile along the vertical axis, through the aqueous, interfacial and organic phases. Reproduced from Ref. [14] with permission from the Royal Society of Chemistry.

nucleation events and the possibility that nuclei may be short-lived transient species present in very low concentrations [3, 4]. Techniques previously used include coupled solid and liquid state NMR, liquid-cell TEM, analytical ultracentrifugation with UV-Visible spectroscopy, small angle X-ray scattering (SAXS) and X-ray absorption fine-structure (XAFS) [5-8]. We monitor nucleation at a liquid-liquid interface by XAFS (figure 1). The liquid-liquid interface was between an aqueous phase containing $[PdCl_4]^{2-}$ and an organic (ααα-trifluorotoluene, TFT) solution containing ferrocene ($Fe(cp)_2$). $[PdCl_4]^{2-}$ is reduced by $Fe(cp)_2$ at the liquid/liquid interface as per equation (1) [9], leading to the nucleation of Pd nanoparticles.

$$[PdCl_4]^{2-}{}_{(aq)} + 2\,Fe(cp)_{2\,(org)} \rightleftharpoons Pd_{\,(interface)} + 2\,[Fe(cp)_2]^{+}{}_{(org)} + 4\,Cl^{-}{}_{(aq)} \quad (1)$$

Such an interface between two immiscible liquids provides an elegant avenue to confining homogeneous nanoparticle nucleation and growth to a defined spatial region [9, 10]. This enables targeted analysis by specialist techniques such as XAFS [11-13] while permitting control of the thermodynamic driving force for nucleation by varying the concentration ratio of the redox couple [14]. Nucleation at the liquid/liquid interface results in a high local concentration of particles, providing spectroscopic contrast to enable observation of transient species in the early stages of nucleation [15, 16].

The energy dispersive EXAFS (EDE) work described here used the same experimental system that we employed in [14] where we reported evidence for a randomly fluctuating equilibrium at the liquid/liquid interface by quick EXAFS (QEXAFS) spectroscopy. The use of EDE at SOLEIL's ODE beamline allowed us to characterize the system with higher time resolution than in the previous QEXAFS study at B18/Diamond [14]. In this contribution we will focus on how inclusion of a common ion (tetraethylammonium, $TEA^+$) in both phases fixes the interfacial potential and thereby achieves additional control over the growth process [17].

## 2. Experimental

Solutions were added to 2 mL Eppendorf safe-lock tubes (diameter 1 cm). The concentrations were 10 mM $[PdCl_4]^-$and 100 mM LiCl in the aqueous phase, in conjunction with 20 mM ferrocene and 10 mM [BTPPA][TFPB] in the TFT (organic) phase. In the system where we included $TEA^+$, the concentrations are 100 mM TEACl in the aqueous phase and 1 mM [TEA][TFPB] in the organic phase in addition to the reactants and background electrolytes mentioned above. XAFS data were obtained in transmission mode at the energy dispersive XAFS beamline, ODE at Synchrotron SOLEIL, Gif-sur-Yvette, France [18]. To follow the variation in palladium density the Pd K edge heights were estimated by taking the differences in absorption at 24,331.2 and 24,375.3 eV. We note that the use of a Pd-coated mirror in the ODE beamline optics resulted in a complicated, non-linear transmission function of the spectrometer around the Pd K-edge, which led to a noticeable distortion of the absorption spectrum in the XANES region. However, as the samples were all collected at the ODE beamline the data remains internally consistent. Spectra were collected at different vertical $z$-distances to the interface. Lowest *z*-values (in mm) indicate measurements in the aqueous phase (*above* the interface), while highest values stem from the organic phase *below*. Further details on chemicals and EDE experiments are given in the Supporting Information.

## 3. Results and Discussion

An example of a time series of EDE spectra measured at a fixed position at the interface ($z$ = 0.05 mm) under conditions of supersaturation, but below the onset of spontaneous nucleation is shown in figure 2a and S1. The edge heights varied significantly with time, indicating Pd concentration fluctuations across the width of the beam (300 μm), as expected for a stochastic nucleation equilibrium. A further example of the metastable state formation is shown in figure S2.

The time series profiles were monitored at several other distances relative to the interface ($z$ values of 0.00, 0.05, 0.10, 0.15 and 0.20 mm). Edge height profiles as a function of time are plotted in figure 2b. Near the bulk of the aqueous phase ($z$ = 0.00 mm), there is a brief induction period of ~40 minutes followed by density fluctuations over a period of 5-7 h. Nearer the interface ($z$ = 0.05 to 0.15 mm) these fluctuations were also visible and became more pronounced closer to the interface. The organic phase is at $z$ = 0.20 mm, just 0.05 mm below the region with the highest Pd concentration and strongest fluctuations. At $z$ = 0.20 mm the edge height is near zero due to the absence of Pd. No transfer of Pd from the aqueous phase took place. Our previous observations by QEXAFS [14] indicated stochastic Pd reduction and oxidation taking place near the interface, and not simply adsorption and desorption of Pd nanoparticles.

The density fluctuations observed in the XAFS data should be associated with potential variations at the liquid/liquid interface because $[PdCl_4]^{2-}$ or ferrocene may become locally depleted through the formation of $Pd^0$ and a ferricinium oxidation product. To examine this we applied the use of a common ion ($TEA^+$) in both liquid phases. As the $TEA^+$ is able to transfer between the two phases, an equilibrium potential is formed across the liquid/liquid interface and fixes the interfacial potential. Starting with a large excess of $TEA^+$ in the aqueous phase, $TEA^+$ is then able to partition into the organic phase, resulting in a negative biasing of the aqueous phase relative to the organic phase, slowing the electron transfer from ferrocene to Pd.

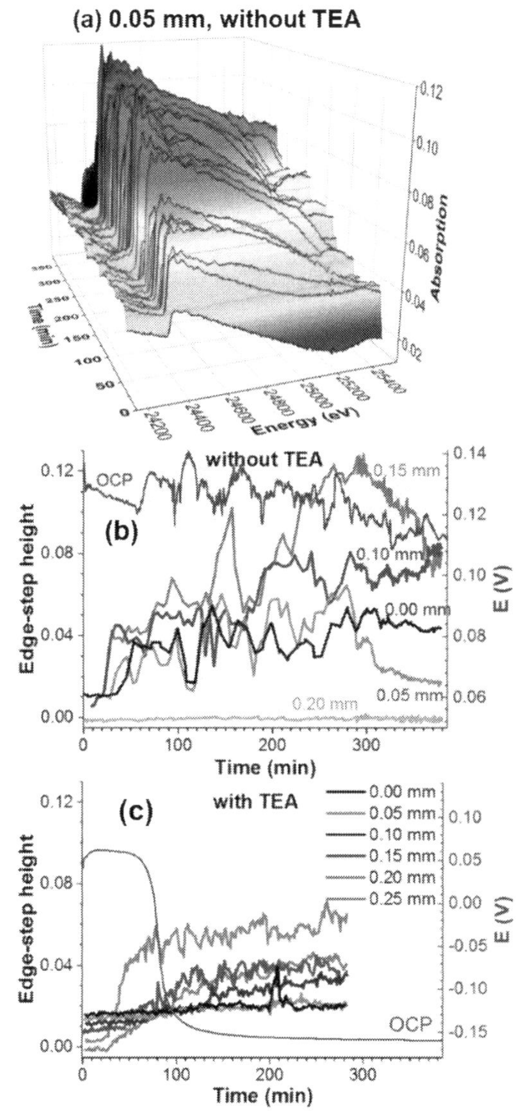


Figure 2. (a) EDE spectra collected at $z$ = 0.05 mm in the sample without $TEA^+$. (b) and (c) show the edge-step heights at different scan depths at the liquid/liquid interface for the sample (b) without added $TEA^+$ and (c) with added $TEA^+$. Overlaid in (b) and (c) are the open circuit potentials.

The EDE experiment under conditions of supersaturation was therefore repeated with inclusion of $TEA^+$. By comparing the experiments with (figure 2c) and without $TEA^+$ (figure 2b) it can be seen that the addition of $TEA^+$ resulted in a slight increase in the induction time before the onset of nucleation, followed by smoother Pd nucleation as can be seen from the supressed fluctuations in edge-step heights. In order to confirm the potential dependence of the growth process, we also measured the open circuit potentials (OCP). Without added $TEA^+$, the interfacial potential fluctuated with a similar frequency to the EDE-detected fluctuations (figure 2b). With $TEA^+$ added a relatively large decrease in potential ~100 minutes after the phases were connected took place before the potential stabilised (figure 2c). OCP stabilisation fitted well with the point at which Pd growth slowed down significantly. Although the potential scales were different for the two samples it is evident that electron transfer was slowed down by negatively biasing the aqueous phase, in accordance with previous electrochemical studies [9, 10].

**Conclusions**

EDE monitoring of Pd nanoparticle nucleation in a galvanic liquid/liquid cell revealed a regime of spatial instability compatible with the density fluctuations expected when growth and dissolution of sub-critical Pd nuclei take place in a supersaturated state. Control of the interfacial potential by inclusion of a common ion ($TEA^+$) in both phases slowed down the nucleation process when the aqueous phase had a negative potential bias relative to the organic phase. Open circuit potential measurements in a liquid/liquid electrochemical cell are in line with the EDE data, indicating that density fluctuations are accompanied by chemical potential variations near the liquid/liquid interface.

**Acknowledgements** All data supporting this study are provided either in the results section of this paper or in the supplementary information accompanying it. SOLEIL awarded ODE beamtime under proposal 20141060. RAWD/SLMS acknowledge financial support through an EPSRC-NSF "Materials World Network" grant (EP/H047786/1). SLMS thanks EPSRC for financial support under the critical mass grant EP/I013563/1. AU thanks Kyoto University foundation for sabbatical support at UoM. SYC thanks UoM as well as Mr and Mrs Clews for PhD sponsorship, and the Royal Society of Chemistry and XAFS16 for conference travel grants. The authors are grateful to the two anonymous reviewers for their constructive feedback.

**References**

[1] Gebauer, D and Cölfen, H 2011, *Nano Today*. **6**, 564-84.
[2] Finney, E E and Finke, R G 2008, *J. Colloid Interf. Sci.* **317**, 351-74.
[3] Davey, R J; Schroeder, S L M, and ter Horst, J H 2013, *Angew. Chem. Int. Ed.* **52**, 2166-79.
[4] Anwar, J and Zahn, D 2011, *Angew. Chem. Int. Ed.* **50**, 1996-2013.
[5] Hughes, C E; Williams, A; Keast, V, et al. 2014, *Faraday Discuss.* **179**, 115-40.
[6] Liao, H-G; Niu, K, and Zheng, H 2013, *Chem. Comm.* **49**, 11720-27.
[7] Voelkle, C; Gebauer, D, and Coelfen, H 2015, *Faraday Discuss.* **179**, 59-77.
[8] Sun, Y and Ren, Y 2013, *Part. Part. Syst. Char.* **30**, 399-419.
[9] Dryfe, R A W; Simm, A O, and Kralj, B 2003, *J. Am. Chem. Soc.* **125**, 13014-15.
[10] Johans, C; Lahtinen, R; Kontturi, K, et al. 2000, *J. Electroanal. Chem.* **488**, 99-109.
[11] Gründer, Y; Ho, H L T; Mosselmans, J F W, et al. 2011, *Phys. Chem. Chem. Phys.* **13**, 15681-89.
[12] Gründer, Y; Mosselmans, J F W; Schroeder, S L M, et al. 2013, *J. Phys. Chem. C.* **117**, 5765-73.
[13] Booth, S G; Uehara, A; Chang, S-Y, et al. 2015, *J. Phys. Chem. C.* **119**, 16785-92.
[14] Chang, S-Y; Gründer, Y; Molleta, L B, et al. 2016, *CrystEngComm.* **Accepted**, DOI: 10.1039/C5CE01883H.
[15] Cheng, Y and Schiffrin, D J 1996, *J. Chem. Soc. Faraday Trans.* **92**, 3865-71.
[16] Platt, M; Dryfe, R A W, and Roberts, E P L 2003, *Electrochim. Acta.* **48**, 3037-46.
[17] Dryfe, R A W, *The Electrified Liquid-Liquid Interface*, in *Advances in Chemical Physics*2009, John Wiley & Sons, Inc. p. 153-215.
[18] Abécassis, B; Testard, F; Kong, Q, et al. 2010, *Langmuir*. **26**, 13847-54.

Journal of Physics: Conference Series **712** (2016) 012059 doi:10.1088/1742-6596/712/1/012059

# Validation of EXAFS Analysis of Iridium Compounds

**M. C. Feiters,[1] A. Longo,[2] D. Banerjee,[2] C. J. M. van der Ham,[3] D. G. H. Hetterscheid[3]**

[1] Institute of Molecules and Materials, Radboud University, Heyendaalseweg 135, 6525 AJ Nijmegen, The Netherlands
[2] Dutch-Belgian Beamline (DUBBLE), ESRF – The European Synchrotron, CS40220, 38043 Grenoble Cedex 9, France
[3] Leiden Institute of Chemistry, Leiden University, P.O. Box 9502, 2300 RA Leiden, The Netherlands

m.feiters@science.ru.nl

**Abstract**. Results of iridium $L_3$ edge EXAFS measurements of compounds relevant for water oxidation catalysis are compared to those of other structural techniques. The structural results from EXAFS for the Ir compounds investigated here compare well to those of other structural techniques. Multiple scattering contributions are important in the coordinated Cp* and NHC ligands as well as in the $IrCl_6$ unit and the $IrO_2$ rutile structure. NHC is relatively weak compared to Ir, Cl, and even Cp* and O, and often out of phase with the other contributions.

## 1. Introduction

New water oxidation catalysts that show higher rates of conversion at milder conditions are important in the search for alternatives for fossil fuels [1], in which solar energy is stored as an energy dense chemical fuel, in a process producing oxygen from water along with a reduced fuel such as hydrogen, methane, methanol, or another hydrocarbon. [IrCp*$(OH)_2$($Me_2$-NHC)] (**2**, prepared from the bis-Cl analogue **1**; $Me_2$-NHC = *N*-dimethylimidazolin-2-ylidene, Cp* = pentamethylcyclopentadienyl, see Fig. 1 (right) for structures) is a catalyst for water oxidation at an electrode surface [2], and parallel to the present study we investigate with EXAFS whether the iridium complex is oxidized to Ir oxide, possibly $IrO_2$ **3**, which is also catalytically active [3], and/or whether the Cp* and $Me_2$NHC ligands are still coordinated to Ir. In order to validate our approach, we describe here how we measured and simulated the EXAFS of the aforementioned Ir compounds **1-3** as well as that of the coordination compound $K_2IrCl_6$ **4** and the bridged dimer [{IrCp*$Cl_2$}$_2$] **5** (Fig. 1), of which most structures were already known by other techniques.

## 2. EXAFS measurements and simulations

X-ray absorption spectra were collected at the Ir $L_3$-edge (11215 eV) on the EXAFS station (BM26A) of the Dutch-Belgian beamline (DUBBLE) [4] at the European Synchrotron Radiation Facility (ESRF) in Grenoble, France. The Ir compounds were diluted with boron nitride and measured as pressed pellets in transmission mode at 20 K; at least 2 scans per sample were averaged to improve the signal-to-noise ratio. The EXAFS was extracted with Viper [5] and simulated with EXCURVE [6] by the same approach as in the recent study of Ir complexes with *N*-heterocyclic carbene and cyclooctadiene or pyridine ligands [7]. The geometries (distances and angles) of the carbene and cyclopentadienyl ligands were adopted from crystallographic data for **1** [8] and **5** [9] and expressed in polar coordinates with the absorber Ir atom at the centre. Structural models (without H) were iteratively refined against the experimental data until a minimum in the fit index was reached by floating the threshold energy EF (in eV), the Debye-Waller (DW) factors (as $2\sigma^2$ in $Å^{-2}$), and the distances (R in Å) to Ir. Multiple scattering (MS) within the ligands was calculated up to 3rd order unless otherwise indicated. In order to reduce the number of parameters, the distances between the ring atoms were treated as restraints [10], so that effectively only the distance between the Ir and the ligand-donor atoms was allowed to float, while the distances between Ir and atoms in equivalent positions were averaged, and the corresponding DW factors grouped as much as possible.

## 3. EXAFS results

The spectra of the Ir-Cp* complexes are shown in Fig. 1. The Fourier transform (FT) of **2** (top) shows that the contributions of the atoms directly linked to Ir, 1 C atom of NHC, 5 C of the Cp* ring, and 2 O atoms, are not resolved and give a single shell at R = 2.0-2.2 Å. Compared to this, the contributions of the Me groups in the Cp* and NHC and the remote atoms of the ring of latter are relatively weak; it was expected that NHC-coordination would give a characteristic EXAFS signature similar to that observed for imidazole coordination in metalloproteins, the so-called 'camel back' [11]. Although a weak resolved feature just above 4 Å in the FT is characteristic for the NHC ring for **1** and **2**, its contribution (remotely resembling a camel back, blue trace) is hardly visible in the experimental EXAFS, in line with our earlier results for hydrogenation catalysts [7]. The Cp* contributions (black) are stronger but out of phase with that of NHC due to the slightly longer Ir-C distance.

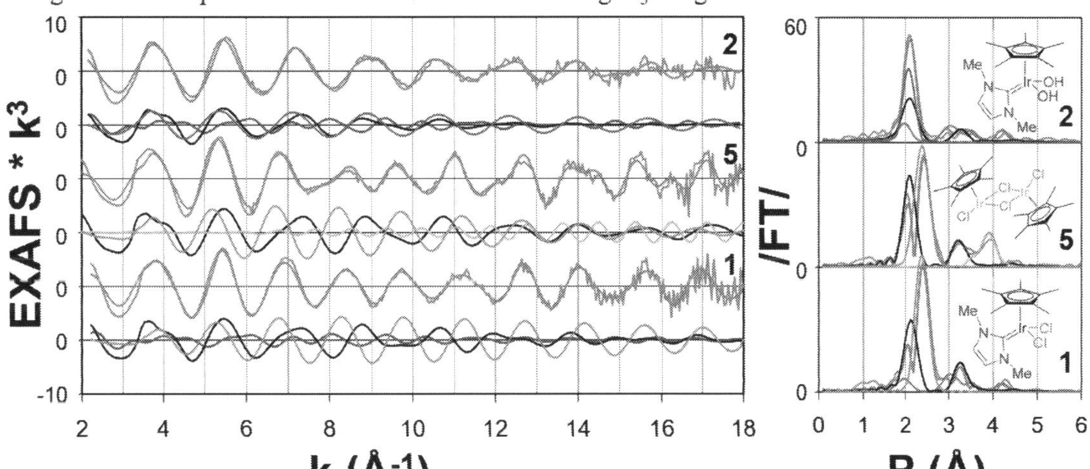


**Figure 1**. $k^3$-weighted Ir $L_3$ edge EXAFS ($\chi*k^3$, left) and phase-shifted modulus of the Fourier transform (/FT/, right) of Ir Cp* compounds **2**, **5**, and **1**. The traces below the experimental (grey) and simulated (red, see Table 1 for parameters) EXAFS and included in the FT panel represent the components: black, Cp*; blue, NHC; green, Cl; orange, Ir.

The FT of **1** (bottom) is dominated by a split major peak at 1.5-3.0 Å which contains contributions of 2 Cl backscatterers (green) at 2.4 Å, the 5 Cp-C atoms at (average) 2.2 Å, and the NHC-C atom at 2.015 Å, which are out of phase with each other for most of the range of the EXAFS. The contribution of the Cp-Me groups at 3.3 Å is relatively strong in this case, whereas that of the NHC outer ring atoms is again weak. The spectrum of **5** (middle) is similar to that of **2**, except that the contribution of NHC is replaced by one of the remote Ir at 3.77 Å, which is also evident from a strong high-frequency contribution (orange) at high k in the EXAFS. It might be expected that the 3 Cl atoms in **5** would give a stronger contribution to the EXAFS than the 2 Cl in **2**, but the crystal structure [9] detects a subtle difference between the distance from Ir to terminal (T) and bridging (B) Cl atoms (Table 1) which is

**Table 1**. Crystallographic and EXAFS simulation results for distances in Ir Cp* compounds

| | IrCp*X$_2$(Me$_2$-NHC); **1**(X=Cl) | | **2** (X=OH) | [{IrCp*Cl$_2$}$_2$] **5** | |
|---|---|---|---|---|---|
| | XRD (average) [8] | EXAFS [a)] | EXAFS [a)] | XRD ave [9] | EXAFS [a)] |
| NHC-C | 2.054 | 2.03(4) /.01(1) | 2.03/.008 | - | - |
| NHC-N | 3.063, 3.066 (3.065) | 3.08(2) /.005(4) | 3.08/.005 | - | - |
| NHC-Me | 4.218, 4.251 (4.235) | 4.27(4)/.01(1) | 4.29/.004 | - | - |
| NHC-C | 3.526, 3.483 (3.505) | 3.50(5)/.01(1) | 3.50/.004 | - | - |
| **Cp-C** | 2.150, 2.155, 2.159, 2.228, 2.261 (2.191) | 2.18(2)/.009(2) | 2.16/.017 | 2.132 | 2.144(6)/.007(1) |
| **Cp-Me** | 3.258, 3.310, 3.308, 3.332, 3.376 (3.319) | 3.31(2)/.006(3) | 3.34/.016 | 3.280 | 3.28(2)/.008(3) |
| OH | - | - | 2.12/.003 | - | - |
| Cl-T [b)] | 2.438, 2.440 (2.439) | 2.426(3)/.0030(5) | - | 2.387 | 2.37(2)/.005(2) |
| Cl-B [b)] | - | - | - | 2.453 | 2.45(2)/.005(1) |
| Ir | - | - | - | 3.769 | 3.77(2)/.005(1) |
| EF (eV) | - | -11.6(6) | -11.1 | - | -10.5(6) |

[a)] Distances in Å/Debye-Waller factors as $2\sigma^2$ in Å$^{-2}$; errors in last decimals for **1** and **5** in parentheses; [b)] Cl-T, Cl-B, terminal and bridging Cl atoms respectively.

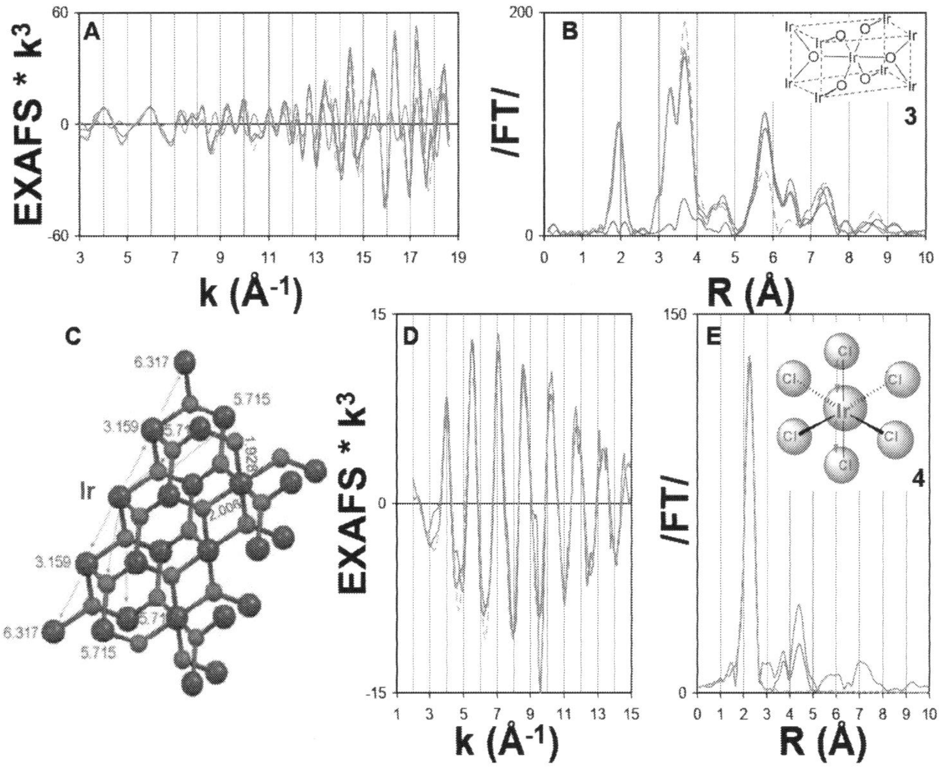


**Figure 2**. EXAFS (A) and FT (B, inset: schematic rutile structure) of $IrO_2$ **3**. (C) Quadrant of the cluster used in the simulation for **3**; central Ir labeled with 'Ir'. EXAFS (D) and FT (E, inset: $IrCl_6$ octahedron with an example of a 4[th] order MS pathway indicated with red arrows) of $K_2IrCl_6$ **4**. Traces: grey, experimental; red, best simulations; dashed orange, single scattering; green, residual MS.

detected by EXAFS as a weakening of their contribution, and adequately simulated with a split shell with same DW-factors but different distances. The good agreement (within 0.02 Å) of the distances in the refined EXAFS simulations (Table 1) with the averages of the relevant crystallographic Ir-ligand atom distances inspires confidence in the phaseshift calculations and the approach for the simulation and the refinement. The EXAFS result for **2** also agrees with that of the crystal structure of **1**, except that the Cl ligands in **1** have to be replaced by O (from OH) in **2**, which are found at 2.12 Å. The proximity of 3 unresolved shells of weak scatterers in the range 2.0-2.2 Å results however in strong correlations between the parameters and high fitting errors for **2**, which are not shown in Table 1.

The EXAFS of **3** (Fig. 2A-B) has sharp and intense features even at high k and the FT displays shells even above 10 Å (not shown). Given the many sets of atoms Ir, A, B in **3** for which the Ir-A-B angle approaches 180°, the influence of MS effects on the spectrum, given in Fig. 2A-B as the difference between the complete simulation and the single scattering component (green trace) is remarkably small, as noted before [12]. In our final simulation a cluster of 126 Ir and O atoms at positions corresponding to a rutile structure (partly shown in Fig. 2C) with shells up to 8.5 Å was refined. There is a split first EXAFS shell due to a subtle tetragonal distortion of the Ir coordination geometry. The results for the nearest atoms compare well with those of neutron diffraction ([13], given in parentheses) 2 O at 1.969 (1.960); 4 O at 2.001 (1.999); 2 Ir at 3.166 (3.159); 4 O at 3.41 (3.423); 8 Ir at 3.555 (3.556-3.564) Å. Interestingly, the spectrum of **4** (Fig. 2D-E) displays an additional shell in the FT at approximately twice the distance of the expected major shell of Cl atoms at 2.32 Å. It was found that both shells could be reasonably well simulated with a unit of 6 Cl arranged in octahedral symmetry around Ir at a distance of 2.326 Å ($2\sigma^2$ = .008) and calculating the MS up to 4th order. The shell in the FT at 4.5 Å does not arise from the presence of atoms at that distance, but from a MS pathway Ir-Cl-Ir-Cl(trans)-Ir with a total path length of twice the Ir-Cl distance, indicated with red arrows in the inset in Fig. 2E. The Ir-Cl distance found corresponds reasonably well with those found in other studies for **4** (2.307 Å, [14]) and its Cs analogue (2.332 Å [15]).

## 4. Acknowledgements

The authors thank the Dutch and Flemish Research Councils (NWO, FWO) for the provision of beam time at the DUBBLE beamline at BM26A of the European Synchrotron Radiation Facility.

## References

[1] Lewis NS, Nocera DC 2006 *Proc. Nat. Acad. Sci. U.S.A.* **103** 15279.
[2] Hetterscheid DGH, Reek JNH 2011 *Chem. Commun.* **47** 2712.
[3] Nakagawa T, Beasley CA, Murray RW 2009 *J. Phys. Chem. C* **113** 12958.
[4] Nikitenko S, Beale AM, van der Eerden AMJ, Jacques SDM, Leynaud O, O'Brien MG, Detollenaere D, Kaptein R, Weckhuysen BM, Bras W 2008 *J. Synchrotron Radiat.* **15** 632.
[5] Klementev KV 2001 *J. Phys. D Appl. Phys.* **34** 209.
[6] Gurman SJ, Binsted N, Ross I 1984 *J. Phys. C Solid State* **17** 143.
[7] van Weerdenburg BJA, Engwerda AHJ, Eshuis N, Longo A, Banerjee D, Tessari M, Fonseca Guerra C, Rutjes FPJT, Bickelhaupt FM, Feiters MC 2015 *Chem. Eur. J.* **21** 10482.
[8] Xiao XQ, Jin GX 2008 *J. Organomet. Chem.* **693** 3363.
[9] Churchill MR, Julis SA 1977 *Inorg. Chem.* **16** 1488.
[10] Binsted N, Strange RW, Hasnain SS 1992 *Biochemistry* **31** 12117.
[11] Feiters MC, Navaratnam S, Al-Hakim M, Allen JC, Spek AL, Veldink GA, Vliegenthart JFG 1988 *J. Am. Chem. Soc.* **110** 7746.
[12] Prouzet E 1995 *J. Phys.-Condens. Mat.* **7** 8027.
[13] Bolzan AA, Fong C, Kennedy BJ, Howard CJ 1997 *Acta Crystallogr. B* **53** 373.
[14] Lindop AJ 1970 *J. Phys. Part C Solid* **3** 1984.
[15] Coll RK, Fergusson JE, Penfold BR, Rankin DA, Robinson WT 1990 *Inorg. Chim. Acta* **177** 107.

# Nb K- and $L_3$-edges XAFS study on the structure of supported Nb carbide catalyst

**N Ichikuni*[1], F Yanagase[1], K Mitsuhara[2], T Hara[1], S Shimazu[1]**

[1] Department of Applied Chemistry and Biotechnology, Graduate School of Engineering, Chiba University, Chiba, Japan
[2] The Research Organization of Science and Technology, Ritsumeikan University, Kusatsu, Japan

E-mail: ichikuni@faculty.chiba-u.jp

**Abstract**. Mesoporous silica SBA-15 supported NbC catalysts were prepared under reduced pressure of $CH_4$-$H_2$ gas in a closed circulating system. Nb *K*- and $L_3$-edges XAFS measurements revealed that small NbC cluster formed on SBA-15 support and gradual carburization process by lengthening the carburization temperature-maintaining period. Carburization degree of Nb species was clearly explained by using threshold energy shift of Nb $L_3$-edge XANES profile.

## 1. Introduction

Early transition metal carbide (ETMC) is one of the most attractive candidates for substituting the noble metal catalysts. In general, high temperature is required for preparation of ETMC, and hence, the particle grows larger and has the narrower surface area, which should be improved as applying to the catalyst. We have already reported the preparation of NbC on oxide support [1] and showed the activity toward ethene hydrogenation. However, the activity was not high enough. Development of smaller NbC particles was thought to improve the catalysis.

In this study, the preparation of the NbC nanocluster on mesoporous silica SBA-15 [2] support and characterization by using Nb *K*- and $L_3$-edges XAFS were reported.

## 2. Experimental

### *2.1. Catalyst preparation*

SBA-15 was prepared in the same manner as presented in the literature [2]. Peroxoniobic acid (PNA) was prepared as in the literature [3] and was used as the Nb precursor for preparation of Nb/SBA catalyst (NbOx/SBA). Carburized catalyst NbC/SBA was obtained in a closed circulating system under $CH_4$ and $H_2$ gas mixture with the pressure of 26.7 kPa and 26.7 kPa, respectively by a temperature programmed reaction (TPR) process; the catalyst was heated up to 1223 K at a linear rate of 10 $K \cdot min^{-1}$ where it was maintained for a certain period (termed as the temperature maintaining period). Nb loading was regulated to 3 wt%. For optimization of temperature maintaining period, the completion of carburization was firstly evaluated by the evolution of CO [4].

For comparison, the silica support (Aerosil, #200) without the mesopore was also employed to the support (NbC/$SiO_2$). In this case, the carburization was completed for 30 min at the temperature-maintaining period.

### 2.2. XAFS measurements

Nb $L_3$-edge XAFS data were collected in partial fluorescence yield (PFY) by silicon drift detector (SDD) and total electron yield (TEY) modes at BL13 of the SR center (Ritsumeikan University, proposal no. R1450) by using InSb(111) double crystal monochromator. Nb *K*-edge XAFS data were taken at PF-AR NW10A (Proposal no. 2014G575) of IMSS-KEK by Si(311) double crystal monochromator in a transmission mode by using two ion-chambers for $I_0$ and I filled with Ar50%($N_2$ 50%) and Ar100%, respectively. Data analysis was conducted using the Athena [5] for Nb $L_3$-edge and EXAFS analysis program REX2000 [6] for Nb *K*-edge. Fourier transform (FT) *k*-range of $k^3$-weighted EXAFS data was 30-150 $nm^{-1}$. Model parameters for Curve-fitting (CF) analysis were extracted from EXAFS oscillations observed for bulk NbC.

### 2. 3. Adsorption measurement and catalytic reaction

$N_2$ adsorption isotherms at 77 K of the samples were measured using BELSORP-MAX (BEL Japan, INC). Samples were degassed at 473 K for 2 h before measurement. The specific surface area and pore volume were obtained by Brunauer–Emmett–Teller (BET) method and by Barrett–Joyner–Halenda (BJH) method, respectively.

Ethene hydrogenation was carried out at 273 K in a closed circulating system. The catalyst was treated with 40.0 kPa of $H_2$ at 1223 K, followed by the evacuation at the same temperature for 15 min, prior to use. The initial pressures of ethene and $H_2$ were 1.3 kPa and 6.7 kPa, respectively. The catalytic activity of the catalyst was determined from the initial rate of ethane formation.

## 3. Results and discussion

### 3.1. Nb K-edge XAFS

Figure 1 showed the Nb K-edge XANES profiles for the carburized catalysts and reference compounds. Peaks at around 18998, 19011 and 19043 eV are distinctly observed in bulk NbC as compared to those in bulk Nb oxide. Those three peaks, designated as α-, β- and γ-peaks, respectively, can be utilized as fingerprint for the existence of NbC species. Nb *K*-edge XANES profiles of the carburized catalyst (Fig. 1 d, e) are similar to that of bulk NbC. NbC composition of NbC/$SiO_2$ and NbC/SBA were 97.9% and 90.7%, respectively, by linear combination analysis.

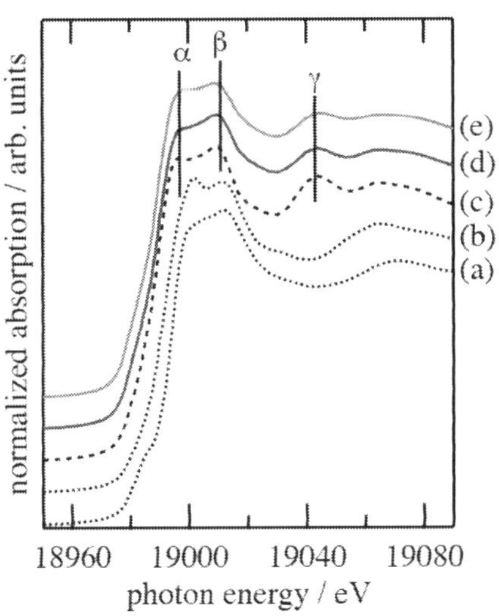


**Figure 1.** Nb *K*-edge XANES spectra for (a) bulk $Nb_2O_5$, (b) bulk $NbO_2$, (c) bulk NbC, (d) NbC/SBA and (e) NbC/$SiO_2$.

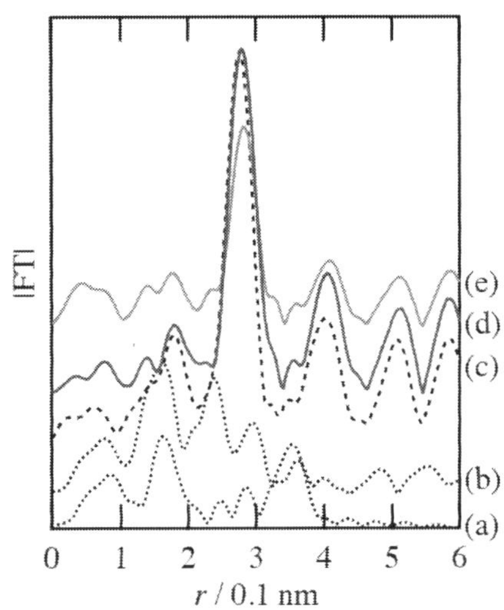


**Figure 2.** FT of $k^3$-weighted Nb *K*-edge EXAFS oscillations for (a) bulk $Nb_2O_5$, (b) bulk $NbO_2$, (c) bulk NbC, (d) NbC/SBA and (e) NbC/$SiO_2$.

Journal of Physics: Conference Series **712** (2016) 012060 doi:10.1088/1742-6596/712/1/012060

To clarify the local structure around Nb atom, Nb *K*-edge EXAFS data were collected and Fourier-transformed as shown in Fig. 2 (phase shifts are not corrected). FT profiles of NbC/SBA and NbC/$SiO_2$ were similar to that of bulk NbC except the intensity. CF analysis was conducted for the main peak at around 0.28 nm, which can be attributed to the nearest Nb-(C)-Nb coordination, besides Nb-C (0.223 nm) and Nb-(C)-Nb (0.446 nm) coordination. CF results were summarized in Table 1.

According to the CF results, the *CN*s of peaks at around 0.2 nm which can be attributed Nb-C was almost 6.0 suggesting the Nb species in both catalysts were converted into NbC. At first, we expected that mesoporous silica SBA-15 may have the templating effect that limiting the NbC cluster size, and hence, the NbC cluster size in SBA-15 would be smaller than in $SiO_2$ support. However, the FT intensity and the *CN* of Nb-(C)-Nb1 of NbC/SBA were larger than those of NbC/$SiO_2$, which means that the larger NbC cluster created on SBA-15. The reason of this reversal cluster size order is still unclear, however, the mesopore size of SBA-15 is thought to be one of the factors.

As estimating the NbC cluster size on SBA-15 from the *CN*s of Nb-(C)-Nb1 and Nb-(C)-Nb2, 1.9 nm in diameter was obtained by assuming the corner cutting cubic structure. Table 2 showed textural properties of NbOx/SBA and NbC/SBA by $N_2$ adsorption measurements. Pore diameter of NbOx/SBA and NbC/SBA were 8.1 and 6.2 nm, respectively. Such large pore size was not able to have the templating effect for such smaller cluster size.

**Table 1.** Curve-fitting results of Nb *K*-edge EXAFS for NbC catalysts.

| | coordination | *CN* | *r* (nm) | d*E* (eV) | $\Delta\sigma^2$ ($10^{-4}$ $nm^2$) |
|---|---|---|---|---|---|
| bulk NbC (model) | Nb-C | 6 | 0.223 | 0.0 | $\sigma$ = 0.006 (nm) |
| | Nb-(C)-Nb1 | 12 | 0.315 | 0.0 | $\sigma$ = 0.006 (nm) |
| | Nb-(C)-Nb2 | 6 | 0.446 | 0.0 | $\sigma$ = 0.006 (nm) |
| NbC/SBA | Nb-C | 6.8±0.9 | 0.224±0.001 | 3.13±1.79 | 0.28±0.16 |
| | Nb-(C)-Nb1 | 8.2±1.5 | 0.318±0.001 | 2.55±2.03 | -0.20±0.11 |
| | Nb-(C)-Nb2 | 3.5±0.7 | 0.449±0.001 | 1.12±1.65 | -0.27±0.14 |
| NbC/$SiO_2$ | Nb-C | 6.6±0.8 | 0.222±0.001 | 1.75±1.68 | 0.45±0.18 |
| | Nb-(C)-Nb1 | 6.5±1.1 | 0.318±0.001 | 7.04±1.86 | 0.00±0.24 |
| | Nb-(C)-Nb2 | 2.2±0.4 | 0.450±0.001 | 3.47±1.60 | -0.20±0.16 |

* *CN*: coordination number, *r*: coordination distance, d*E*: threshold energy differences, $\Delta\sigma^2$: the difference between $\sigma^2$ for the calculated value and $\sigma^2$ for the model parameter.

### *3.2. Nb $L_3$-edge XAFS*

Nb $L_3$-edge XAFS measurements were carried out to clarify the chemical state of Nb species in the NbC/SBA catalysts as shown in Fig. 3. The time presented in catalyst notation means the temperature-maintaining period (Figs. 3e-3h). Peak profile and edge position resembles from bulk $Nb_2O_5$ to bulk NbC as lengthening the period. As prolonging the period, the band about 2383.1 eV gradually grew clearly same as bulk NbC. The completion of the carburization is expected to be in the 180 to 240 min. Obtained NbC/SBA-180 min and NbC/SBA-240 min catalysts showed the ethene hydrogenation activity at 273 K. However, the latter catalyst showed less activity than the former did. And no ethane was detected on NbC/SBA-30 min catalyst. It can be said that the carburization degree must be high to some extent to show the catalysis. However, the excess carburization reduces the catalysis, which may result from the deposited carbon. Indeed, the enough hydrogen treatment at 1223 K before the reaction improved the catalysis. The 15 min hydrogen treated and 90 min hydrogen treated NbC/SBA catalysts showed the initial rate of ethane was 4.8 and 10.6 mmol·$min^{-1}$·$g_{Nb}^{-1}$, respectively. This confirmed that deposited carbon on the carburized catalyst surface affected catalysis. To convert the active carbide catalyst, removal process of deposited carbon is necessary.

On the other hand, the temperature-maintaining period 30 min was enough for the carburization completion for NbC/$SiO_2$ catalyst as estimated from the edge energy position (Fig. 3i). Although NbC species on $SiO_2$ support is smaller than that on SBA support and fully carburized, the initial rate of ethene hydrogenation was 1.6 mmol·min$^{-1}$·$g_{Nb}^{-1}$, which was much smaller than that of NbC/SBA showed. Mesoporous structure might affect the rate of carburization and continuing carbon deposition, and thus it is supposed that the catalytic activity was differed between NbC/SBA and NbC/$SiO_2$. However, the deposited carbon on NbC surface did not affect the Nb $L_3$-edge XAFS data in this condition.

**Table 2.** Textural properties of catalysts

| Catalyst | $S_{BET}$ ($m^2$ $g^{-1}$) | $V^a$ ($cm^3$ $g^{-1}$) | $D^b$ (nm) |
|---|---|---|---|
| NbOx/SBA | 764 | 1.11 | 8.1 |
| NbC/SBA | 389 | 0.67 | 6.2 |

[a] Total pore volume. [b]Pore diameter determined by Barrett-Joyner-Halenda method.

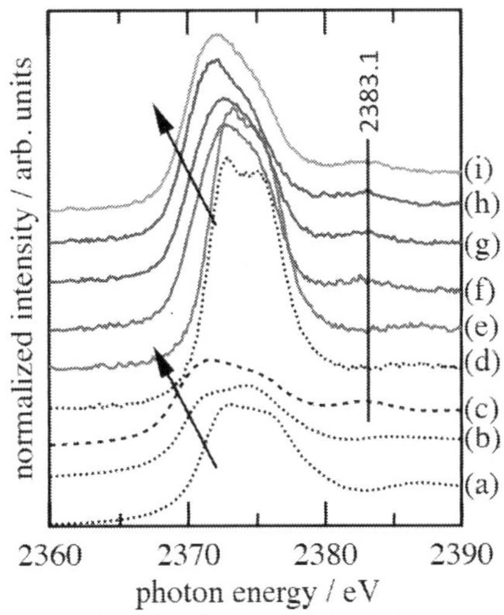


**Figure 3.** Nb $L_3$-edge XANES spectra in PFY mode for (a) bulk $Nb_2O_5$, (b) bulk $NbO_2$, (c) bulk NbC, (d) NbOx/SBA, (e) NbC/SBA-0 min, (f) NbC/SBA-30 min, (g) NbC/SBA-180 min, (h) NbC/SBA-240 min and (i) NbC/$SiO_2$.

## Conclusion

NbC was prepared on the support by using peroxoniobic acid, followed by the carburization in a closed circulating system. Completion of carburization can be visualized from the Nb *K*- and $L_3$-edges XAFS analysis. Also the NbC cluster size could be estimated from Nb *K*-edge XAFS analysis. Although the prolongation of the temperature-maintaining period is seemed to be the effective way to complete the carburization, excess carburization treatment may cause the carbon deposition on the surface and leads to the decrease in the active site.

## Acknowledgment

This study was supported by a Grant-in-Aid for Scientific Research from JSPS (26420784). Nb $L_3$-edge XAFS measurements were carried out by using a Project for Creation of Research Platforms and Sharing of Advanced Research Infrastructure (R1450) at BL13 of the SR Center.

## References

[1] Kodama S, Ichikuni N, Bando K K, Hara T and Shimazu S 2008 *Appl. Catal. A* **343** 25-28
[2] Zhao D, Huo Q, Feng J, Chmelka B F and Stucky G D, 1998 *J. Am. Chem. Soc.* **120** 6024-6036
[3] Wakai Y, Hara T, Bando K K, Ichikuni N and Shimazu S 2009 *Top. Catal.* **52** 1517-1524
[4] Ichikuni N, Sato F, Shimazu S and Uematsu T 2002 *Top. Catal.* **18** 101-104
[5] Ravel B and Newville M 2005 *J. Synchrotron Rad.* **12** 537–541
[6] REX 2000 is a commercially available analytical program for XAFS provided by Rigaku Corp.

# Time-resolved study on dynamic chemical state conversion of $SiO_2$-supported Co species by means of dispersive XAFS technique

**S Chotiwan, H Tomiga, S Yamashita, M Katayama, and Y Inada***

Department of Applied Chemistry, Ritsumeikan University, Kusatsu 525-8577, Japan

E-mail: yinada@fc.ritsumei.ac.jp

**Abstract.** The chemical state conversion of the Co species supported on $SiO_2$ was investigated using the *in-situ* and the time-resolved XAFS techniques. The supported $Co_3O_4$ species was finally reduced to metallic Co with the stable intermediate state of CoO for both the temperature-programmed and time-course processes. The oxidation of $Co^0$ traced the reverse route to $Co_3O_4$, whereas the relative stability of the $Co_3O_4$ species to the CoO intermediate under the oxidative environment diminished the composition of CoO. The time-resolved measurement for the oxidation reaction showed an additional intermediate at the early stage of the CoO intermediate formation suggesting the $Co^0$ particle dispersion.

## 1. Introduction

The supported Co catalysts are well known to have the high activity in various catalytic reactions such as Fischer–Tropsch synthesis, hydrogenation, desulfurization, and ethanol steam reforming [1]. The chemical state conversion of the active Co species causes the degradation of the catalytic performance. Therefore, the direct characterization and its structure determination of such active species under the reaction gas conditions are imperative for the stable utilization of the catalysis process. Besides the applicability of the XAFS spectroscopy to *in-situ* observations of metal species under a real reaction condition, *in-situ* quick XAFS provides quantitative information on supported phases formed during the cobalt catalyst activation, in terms of the structural, as well as temperature and rate of formation and decomposition [2]. The time-resolved XAFS technique is powerful to clarify the reaction mechanism of chemical conversions for catalytically active materials, the time-resolved dispersive XAFS (DXAFS) technique was applied to measure the dynamic process of the chemical state conversion for the supported metal catalysts [3,4]. In this study, the chemical state conversions on a supported Co catalyst have been investigated by means of the *in-situ* XAFS and the time-resolved DXAFS techniques, and the redox reactions of the Co species supported on $SiO_2$ will be discussed.

## 2. Experimental

The Co catalyst was prepared by the impregnation method using $SiO_2$ (192 $m^2/g$) from the Catalysis Society of Japan. The $Co(NO_3)_2{\bullet}6H_2O$ was dissolved in deionized water and the $SiO_2$ powder was suspended in the solution. The mixture was stirred for 1 h and was dried for 24 h at 80 °C. The obtained catalyst precursor was subjected to a calcination treatment at 600 °C for 2 h in air. The Co loading was set to be 10 wt.%.

The *in-situ* XAFS measurements were performed at the BL-12C station of PF (KEK, Japan) using a Si(111) double-crystal monochromator. The sample powder was first exposed to dilute $H_2$ flow (10 vol.% balanced by He, 200 $cm^3$/min) in an *in-situ* observation cell and was heated to 750 °C at 10 °C/min (temperature-programmed reduction: TPR). Once cooling to room temperature by keeping the $H_2$ gas flow, the gas was changed to dilute $O_2$ (10 vol.% balanced by He, 200 $cm^3$/min), and the sample was heated to 600 °C at 10 °C/min (temperature-programmed oxidation: TPO). The XAFS spectra were repeatedly recorded with the interval time of 2 min during both processes.

The dynamic changeover of the chemical state was studied by the time-resolved DXAFS technique at the NW2A station of PF-AR (KEK, Japan) using a Si(111) bent crystal as the polychromator with the bending radius of 2.0 m. The Co catalyst was set in a batch-type DXAFS cell [5], and the sample was heated to 600 °C at 20 °C/min. The reaction gas (pure $H_2$ at 6.5 kPa for reduction of Co oxides and pure $O_2$ at 1.8 kPa for oxidation of metallic Co) was rapidly injected to start the reaction into the cell after the evacuation. The time-resolved spectra were recorded with the time resolution of 10 ms.

## 3. Results and discussion

The XANES spectral changes for the TPR and TPO processes are shown in figure 1. The initial spectrum of TPR was identical to that of $Co_3O_4$, and the XANES spectrum was finally changed to metallic Co. During the conversion from $Co_3O_4$ to $Co^0$, the transient spectrum was almost consistent with that of CoO with the white line peak at 7725 eV. The XANES spectral change for the TPO process showed that the $Co^0$ species was reversibly oxidized to $Co_3O_4$. The composition analysis based on the XANES spectrum was thus carried out by the linear combination of three reference species, $Co^0$, CoO, and $Co_3O_4$, and the calculated values of mole fraction are plotted in figure 2(a) as a function of temperature. Figure 2(a) clarifies that the divalent CoO species quantitatively exists at *ca.* 450 °C under the dilute $H_2$ gas flow. The spectral reproducibility is given in figure 2(b) for a stable CoO intermediate formed during TPR at 436 °C. In contrast, the composition of CoO is diminished to *ca.* 20 % under the oxidative atmosphere during the TPO process. It is evaluated that the CoO species are readily converted to the final product of $Co_3O_4$ (see figure 2(a)) because of the relative stability of $Co_3O_4$ to CoO under the oxidative atmosphere. The *in-situ* XAFS measurements reveal that the reversible redox reactions between $Co_3O_4$ and $Co^0$ take place on supporting $SiO_2$ via the intermediate state of CoO and that the stability of CoO is significantly affected by the gas environment. At room temperature, an obvious composition change was observed when the flow gas was switched from $H_2$ to $O_2$. The composition analysis revealed that the $Co^0$ species was partially oxidized to CoO (20 %) and $Co_3O_4$ (10 %) by the exposure to $O_2$ at room temperature.

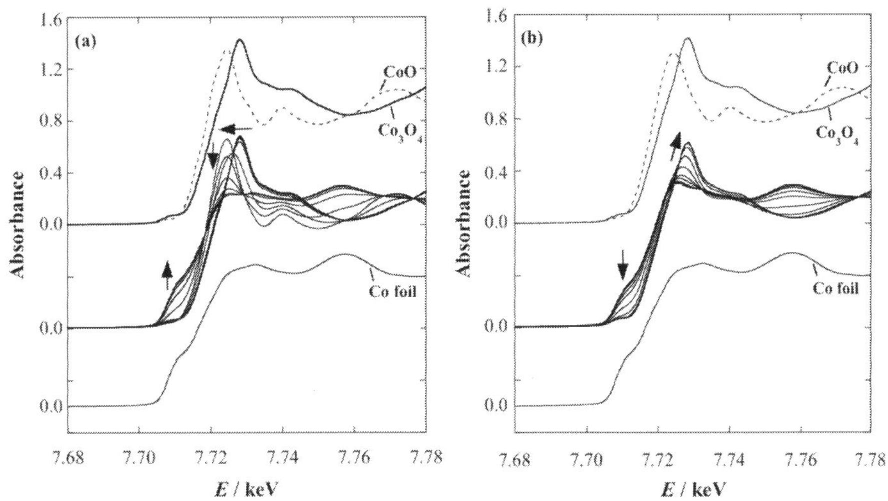


**Figure 1.** XANES spectral change during the TPR (a) and TPO (b) process of $Co/SiO_2$.

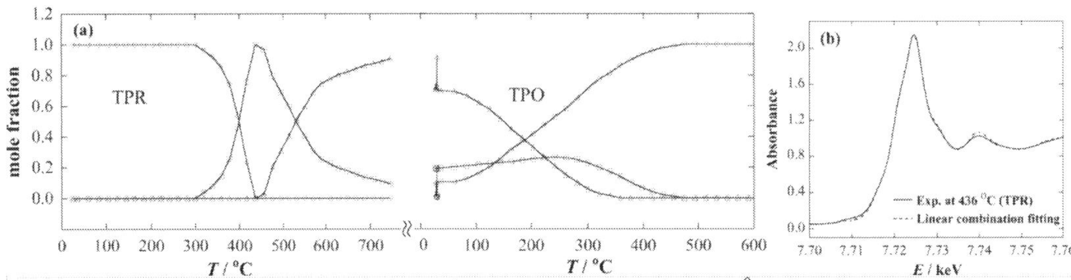


**Figure 2.** The temperature variation of components species, $Co^0$ (triangles), CoO (squares) and $Co_3O_4$ (circles), during the TPR and TPO process (a) and a comparison between the calculated and observed spectra at 436 °C during TPR for a stable CoO intermediate (b).

The time-resolved XANES spectral change is presented in figure 3 for the redox reactions between $Co_3O_4$ and $Co^0$ measured by the DXAFS instrument. For both reactions, the observed spectral changes are quite similar to those for the TPR and TPO processes shown in figure 1. In the case of the reduction of $Co_3O_4$ by $H_2$ (figure 3(a)), the two-phase change is clearly observed and the intermediate spectrum with the white line peak at 7725 eV is in agreement with that of CoO. Thus, it is concluded that the first step corresponds to the reduction from $Co_3O_4$ to CoO, which is followed by the succeeding reduction from CoO to $Co^0$. The stepwise conversions are observed in the dynamic process similar to the TPR process. The intermediate CoO species have the lifetime of a few seconds under the present experimental conditions.

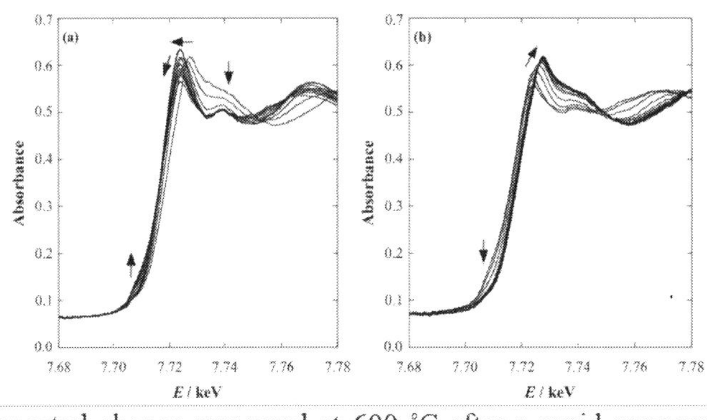


**Figure 3.** XANES spectral change measured at 600 °C after a rapid exposure of $H_2$ (6.5 kPa) to $Co_3O_4/SiO_2$ (a) and of $O_2$ (1.8 kPa) to $Co^0/SiO_2$ (b).

The similar XANES change during the oxidation reaction (figure 3(b)) to that for the TPO process (figure 1(b)) suggests that the reaction proceeds with the unstable intermediate of CoO. The detailed analysis shown in figure 4 for the observed absorbance change at 7721 eV, corresponding to the isosbestic point of the second step, clearly indicates that the oxidation process is composed of three steps. As shown in figure 4(a), a slight enhancement of the absorbance at 7723 eV without the energy shift was observed within 0.5 s after the injection of $O_2$ gas. The energy shift of the white line peak to 7725 eV was observed at the second step from 0.5 s to 1.7 s, as shown in figure 4(b), and it was identical to the difference between metallic cobalt and CoO. The second step is thus assigned to the conversion to CoO, and an additional intermediate species is suggested to form at the first step within 0.5 s. The white line peak was further shifted to the higher energy side after the second step and was finally matched to that of $Co_3O_4$ at 5.0 s as shown in figure 4(c). Only the enhancement of the absorbance without the energy shift at the first step may suggest the $Co^0$ particle dispersion promoted

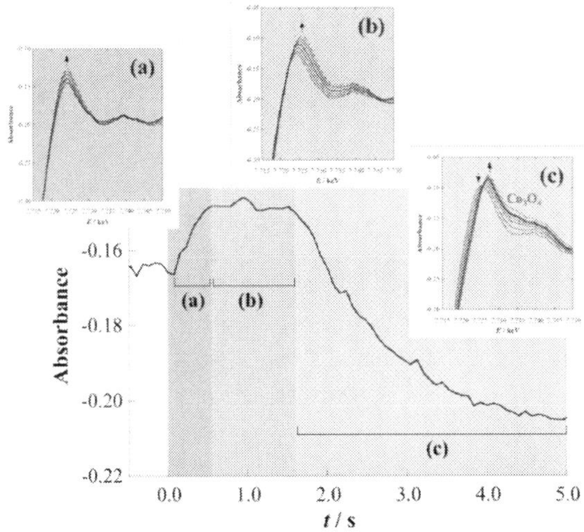


**Figure 4.** The X-ray absorbance change at 7721 eV as a function of time measured at 600 °C under 1.8 kPa of $O_2$. XANES spectral change for the time region after the exposure to $O_2$ from 0 to 0.5 s (a), from 0.5 to 1.7 s (b), and from 1.7 to 5 s (c). In (c), the spectrum of $Co_3O_4$ is shown for comparison.

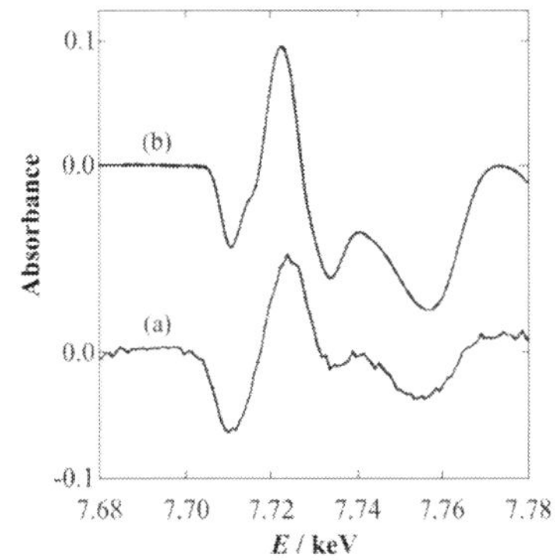


**Figure 5.** The spectral difference at the first step of the oxidation between 0 and 0.5 s (a) and that between Co foil and supported metallic $Co^0$ species measured at 0 s.

by the $O_2$ gas exposure. Figure 5 indicates that the spectral difference observed at the first step between 0 and 0.5 s is consistent with the difference spectrum between Co foil and the supported $Co^0$ particle. This consistency suggests that the first step of the oxidation process (figure 4(a)) corresponds to the dispersion of the supported $Co^0$ particle. It is considered that the additional intermediate state, the dispersed $Co^0$ species, is generated by the adsorption of oxygen at the surface defect of the $Co^0$ particle. The dispersed $Co^0$ species is then oxidized to $Co_3O_4$ via CoO at the succeeding two steps

## 4. Conclusions

The temperature-programmed and time-resolved XAFS measurements for the redox reactions of the supported Co species on $SiO_2$ revealed that the chemical state conversion between $Co_3O_4$ and $Co^0$ proceeded reversibly with the intermediate state of CoO. The transient composition of CoO during the reduction process reached almost 100 %, whereas the diminished composition under the oxidative atmosphere indicated the relative stability of $Co_3O_4$ to CoO. The detail analysis for the oxidation reaction showed an additional intermediate at the early stage suggesting the $Co^0$ particle dispersion.

## Acknowledgement

The XAFS measurements have been performed under the approval of the Photon Factory Program Advisory Committee (Proposal Nos. 2011G531 and 2013G596).

## References

[1] de la Osa A R, de Lucas A, Romero A, Valverde J L and Sanchez P 2011 *Fuel.* **90** 1935
[2] Hong J, Marceau E, Khodakov A Y, Griboval-Constant A, Fontaine C L and Briois V 2012 *Chem. Eur. J.* **18** 2802
[3] Katayama M, Niwa Y, Doi K, Yamashita S and Inada Y 2013 *J. Phys.: Conf. Series* **430** 012053
[4] Yamashita S, Katayama M and Inada Y 2013 *J. Phys.: Conf. Series* **430** 012051
[5] Uemura Y, Inada Y, Bando K K, Sasaki T, Kamiuchi N, Eguchi K, Yagishita A, Nomura M, Tada M, and Iwasawa Y 2011 *Phys. Chem. Chem. Phys.* **13** 15833

Journal of Physics: Conference Series **712** (2016) 012062 doi:10.1088/1742-6596/712/1/012062

# Estimation of the catalytic centre in double metal cyanide catalysts by XAS

**Krystyna Lawniczak-Jablonska*[1] and Arkadiusz Chrusciel[2]**
[1]Institute of Physics, Polish Academy of Sciences, Lotnikow Str 32/46, 02 668 Warsaw, Poland
[2]MEXEO, 47 225 Kędzierzyn-Kozle, Energetykow Str 9, Poland

*E-mail: jablo@ifpan.edu.pl

**Abstract**. Double metal cyanide (DMC) catalysts are commonly applied at industrial ring opening polymerization of epoxides. Nevertheless, the knowledge on the molecular nature of their high activity and selectivity is limited. XAS studies were performed to look for the possible catalytic centre in this family of catalysts. DMC catalysts were synthesized from $ZnCl_2$ and potassium hexacyanocobaltate(III) solution, in the presence of the different organic ligands and show significant fraction of the non-crystalline structure. Two ligands were analysed (*tert*-butanol ($^t$BuOH) or glyme ($CH_3OCH_2CH_2OCH_3$)). EXAFS analysis established that only Zn atoms are the active metallic centers in DMC regardless the used ligand. The coordination around Zn was changed from octahedral in reference non catalytic material to tetrahedral in catalysts, and Cl atoms were detected near some of Zn atoms.

## 1. Introduction

The double metal cyanide (DMC) type catalysts frequently used in commercial production are the complex compounds of zinc(2+) and hexacyanocobaltate(III)(3-) ions containing appropriate kinds of organic ligands. Owing to the heterogeneous nature of DMCs with complex compositions, it is difficult to obtain detailed knowledge on an exact active center. The explanations of DMC catalysts properties described in the literature are based on the hypothesis that their activity is determined by the properties of Zn metal through the formation of the coordination bond with organic co-ligand oxygen atoms. Such bonding is weaker than strong Zn–NC one [1-5]. Some of the papers indicated also the role of Zn–Cl bond in the catalytic properties [6,7]. Unfortunately, no satisfactory explicit experimental data confirming the existing of Zn–O or Zn–Cl bonds in the DMC catalysts have been communicated. Therefore, in order to build up the experimental basis for the molecular justification of the activity of the DMC catalysts, the X-ray absorption spectroscopy (XAS) was applied, due to its sensitivity on the closest atomic neighbourhood of examined elements regardless an amorphous or crystalline surroundings. Moreover, to exam the influence of the used ligand on the Zn and Co local atomic structure the DMC with two different organic co-ligands were studied.

## 2. Experimental

The DMC catalysts and the reference material were synthesized at *MEXEO Kędzierzyn-Koźle* company according to the method described in [8]. The reference material was the ‘pure’, hydrated trizinc *bis*-[hexacyanocobaltate(III)] compound ($Zn_3[Co(CN)_6]_2{\cdot}nH_2O$) of negligibly low catalytic activity, not suitable for practical application. The investigated DMCs were synthesized from $ZnCl_2$ and potassium hexacyanocobaltate(III) solution, in presence of the organic ligands as *tert*-butanol $^t$BuOH) and glyme ($CH_3OCH_2CH_2OCH_3$) and discussed here as DMC BuOH and DMC glyme, respectively. These ligands are frequently used in preparation of this family of DMC catalysts. Due to the quantity of the potential coordination centers of these ligands, the tert-butanol is expected to be the mono-dental ligand but glyme (1,2-dimethoxyethane) should be the bi-dental one. Therefore we suspect the essential difference in their performance to coordinate the active catalytic centre (Zn atoms).The catalytical activity of investigated catalysts was examined on the base of the standard methyloxirane and polypropylene glycol PPG450 polyaddition reaction, according to the method described in [8] and was estimated to be 1600 gPO/gcat*min and 1700 gPO/gcat*min for DMC-BuOH and DMC-glyme, respectively.

The local atomic structure around Zn and Co atoms was determined by EXAFS. Experiments were performed at the beamline SAMBA of the Soleil synchrotron at room temperature in atmospheric pressure and in transmission mode. Appropriate amounts of the samples were mixed with cellulose and pressed into a pellet. The analysis was performed using IFEEFIT package [9].

## 3. Results and discussion

In the figure 1 the Fourier transformed (FT) data of the Zn K-edge EXAFS measurements for the three studied samples are presented. There is clear difference in FT of data for Zn K-edge. Several models of atomic order around Zn were considered, taking into account the models proposed in the literature discussing the role of O and Cl atoms in DMC catalysts [4,6,10,11].

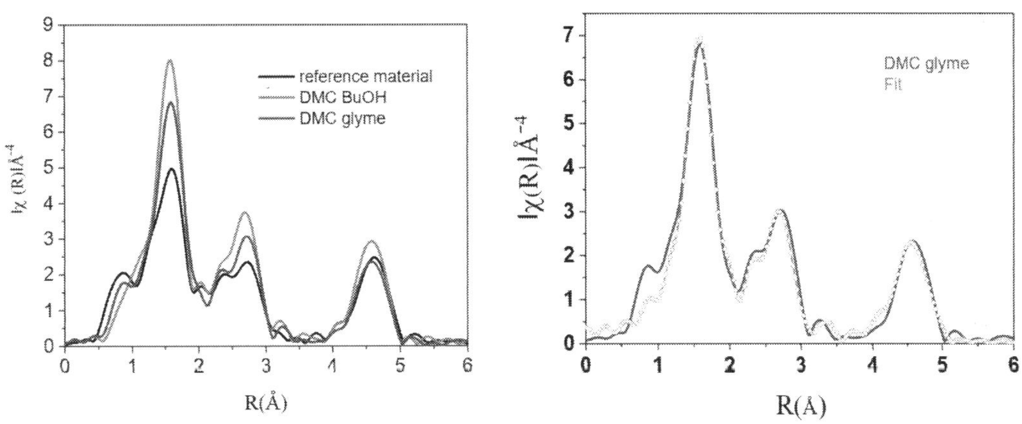


**Figure 1.** K-edges of Zn. The comparison of FT module of the experimental spectra (left); Fit to the EXAFS data of the model for DMC glyme (right).

Similarly as in paper [12] the best fit to the EXAFS data was obtained for the model assuming that considerable fraction of Zn atoms in the catalyst has the local atomic order as should be in the reference anhydrous material with rhombohedral structure (4 cyanide groups) and some of them have Cl in the first coordination sphere (figure 1 (right) and table 1). There is an access of Zn atoms and only 84(3)% and 79(5)% of Zn atoms have 4 NC near neighbour groups for DMC BuOH and DMC glyme, respectively. In the first coordination sphere 0.4 and 0.56 Cl atom per one Zn atom in the distance 2.24 Å, respectively, was identified for investigated catalysts. This Zn–Cl bond length is close to that in $ZnCl_2$ (2.30Å), therefore can be considered as characteristic for Zn–Cl bond. Nevertheless, the existence of $ZnCl_2$ compound was not confirmed, because the atomic order characteristic for $ZnCl_2$ (in first coordination shell 2 Cl in distance 2.3 Å and 2 Cl in distance 3.2 Å and next 4 Zn atoms at distance 3.75 Å and 2 Cl in distance 3.8 Å) could not be fitted. Substituting some of Cl atoms with oxygen leads to an increase of the errors of estimated parameters and to an increase of the $R_f$ parameter which estimate the quality of the fit. Nevertheless, one cannot exclude that oxygen atoms can be bonded to Zn but in the very limited amount. In many reports it is emphasized that the type (coming from water or ligands) and amount of oxygen atoms coordinated to Zn should play an important role in the ring opening polymerization since oxygen atoms coordinated to a zinc ion are believed to be a real active center of DMC catalyst [11,13]. From the performed EXAFS studies, similarly as in ref. [12] in the DMC with the *tert*-butanol ligand, we cannot confirm the importance of Zn–O bonds in the DMC with the glyme ligand. Instead we do confirm the location of Cl close to Zn atoms as well in this DMC. The role of Cl was also discussed in the literature on the base of XPS experiment [6] and molecular calculation [10]. In the paper by Zhang et al., [6] based on overall chemical premises the role of Cl atoms in the catalytic properties of DMC was postulated and reported that the ligand influences on catalytic activity only by helping to form the amorphous structure. Results of performed EXAFS studies are fully in line with this statement.

**Table 1.** The local atomic order around Zn atoms in the reference material, and both catalysts. Number of atoms (No) in the given coordination sphere, distance (R) from Zn atom in Å, and Debye – Waller factors. In the parentheses the confidence interval indicated by IFEFFIT code is given. $R_f$ factor indicate the quality of the fitting.

| Shell | Elem. | Reference material 100% | | | DMC BuOH 84(3)% | | | DMC glyme 79 (5)% | | |
|---|---|---|---|---|---|---|---|---|---|---|
| | | No | R (Å) | Sigma$^2$ | No | R (Å) | Sigma$^2$ | No | R (Å) | Sigma$^2$ |
| 1 | 6 N 2.090 Å | 6 | 2.09(4) | 0.013(1) | N 3.36 | 1.974(3) | 0.003(1) | N 3.16 | 1.98(1) | 0.005(1) |
| 2 | 6 C 3.23 Å | 6 | 3.23(1) | 0.006(2) | C 3.36 | 2.98(2) | 0.013(4) | C 3.16 | 3.00(1) | 0.009(3) |
| 3 | 24 O 4.23 Å | 3(1) | 4.50(6) | 0.008(6) | Co 1.68 | 5.006(4) | 0.004(1) | Co 1.58 | 5.00(1) | 0.007(1) |
| 4 | 6 Co 5.12 Å | 6 | 5.15(8) | 0.018(8) | Co 1.68 | 5.038(4) | 0.004(1) | Co 1.58 | 5.04(1) | 0.007(1) |
| Zn–Cl | | | | | | | | | | |
| 1 | Cl | 0 | | | 0.40(2) | 2.24(1) | 0.003(1) | 0.56(5) | 2.24(1) | 0.005(1) |
| R f | | 0.013 | | | 0.003 | | | 0.018 | | |

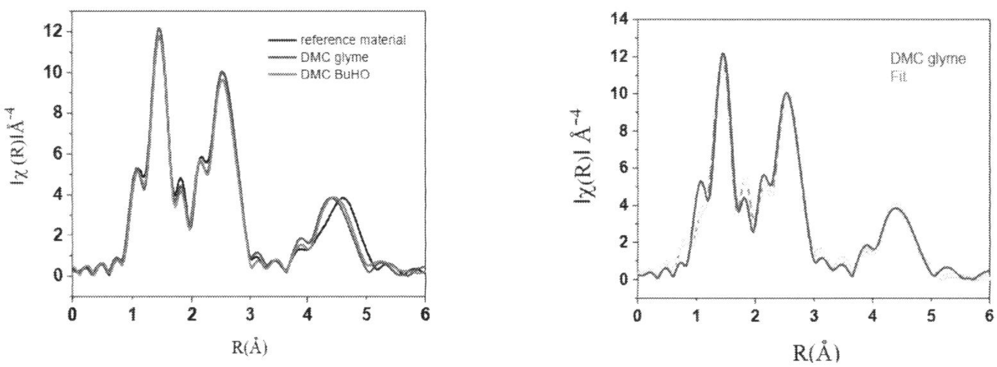


**Figure 2.** K-edges of Co. The comparison of FT module of the experimental spectra (left); Fit to the EXAFS data of the model for DMC glyme (right).

Finally the local atomic structure around Co atom was investigated for DMC glyme and compared with DMC BuHO and reference material. As can be seen in figure 2 (left) practically this structure is not changing for all investigated materials up to about 3 Å. After the fitting of the EXAFS of the reference material to it's model of crystal structure, the atomic order around Co was found like follows: in first shell 6.0(2) carbon atoms at the distance 1.89(1) Å, next 6.0(2) nitrogen at 3.09 Å and 10(4) and 9(3) oxygen atoms in distances of 3.12(1) Å and 4.6(1) Å, respectively, and 6.0(2) Zn atoms at distance of 5.13(3) Å. In best fits of EXAFS data to investigated catalysts the difference in respect to reference material was noticed only in the number of oxygen atoms. All distances of shells and coordination numbers of other than oxygen shells was not changed. In the DMC glyme oxygen shells 12.6(4) and 15.5(3) atoms were found but in DMC BuHO 10(4) and 11(3), respectively. These oxygen atoms can originate from water molecule or ligand. The error indicated by IFEFFIT code for the number of these atoms is quite large indicating that oxygen can be distributed no homogeneously. This confirms that Co metallic center is not active during the catalyst preparation. Interestingly, the atomic order around Co remains like in hydrated form of $Zn_3[Co(CN)_6]_2{\cdot}nH_2O$ with cubic (*Fm*-3*m*) structure. Therefore, in catalyst both metals have different atomic order. Zn atoms like in anhydrous structure (rhombohedral (*R*-3*c*)) bonded with four NC groups and Co atoms like in hydrated structure bonded with six CN groups.

## 4. Conclusions

The results of the EXAFS analysis performed for DMC with glyme ligand, in line with the results reported in [12] for DMC BuHO ligand, support the model proposed by Zhang et al., [6] and the model resulting from molecular calculation performed by Wojdeł et al., in [10]. These models postulate the important role of Zn–Cl bonds in the activity of examined catalysts. The role of Zn–O bond postulated in other papers [1-5,11,13] is not confirmed. We do not detected Zn atoms in the distance 4.38 Å as resulted from calculation performed in the ref. 11 but in distance 5.13 Å. Therefore, we postulate that catalysts form cluster-like complexes with the Co atomic structure not affected, as compared to reference cubic hydrated material. The Zn atoms inside the clusters have changed atomic order which becomes like in anhydrous structure (tetrahedral coordination). This explains that part of Zn atoms having an atomic order similar to that in the reference anhydrous material. Remaining Zn atoms can be bound partially with groups of cyanide and with chlorine atoms. These atoms can be located at the surface of the clusters and be catalyst active centres. EXAFS analysis does not confirmed the existence around Zn atoms the atomic order like in $ZnCl_2$ compounds. Instead the nanoclusters are formed in catalyst with different atomic order around Co and Zn with new, until now not resolved structure, which incorporates Cl atoms.

Financial support from the EU FP7 EAgLE project under the grant agreement REGPOT-CT-2013-316014 is gratefully acknowledged.

## References

[1] Ionescu M, **2005** *Chemistry and Technology of Polyols for Polyurethanes*. R.A.P.R.A. Technology Ltd. Shawbury-Shrewsbury-Shropshire, UK.
[2] Kim I, Ahn J-T, Ha Ch-S, Yang Ch-S, Park I 2003 *Polymer* **44** 3417
[3] Chen S, Zhang P, Chen L 2004 *Prog. Org. Coat.* **50** 269
[4] Kim I, Byun S H, Ha C S 2005 *J. Pol. Sci.* **43** 4393
[5] Lee S, Baek S T, Anas K, Ha Ch-S, Park D-W, Lee J W, Kim I 2007 *Polymer* **48** 4361
[6] Zhang X-H, Hua Z-J, Chen Sh, Liu F, Sun X-K, Qi G-R 2007 *Applied Catalysis* **325** 9
[7] Huang Y-J, Zhang X-H, Hua Z-J, Chen S L, Qi G-R 2010 *Macromol. Chem. Phys.* **11** 1229
[8] Chruściel A *et al.* 2014 *Ind. Eng. Chem. Res.* **53** 6636
[9] Ravel B, Newville M 2005 J. *Synchrotron Rad.* **12** 537
[10] Wojdeł JC, Bromley ST, Illas F, Jansen JC 2007 *J. Mol. Model.* **13** 751
[11] Almorea-Barrios N *et al.* 2015 *ChemCatChem* **7** 928
[12] Lawniczak- Jablonska K *et al.* 2015 *X-Ray Spectrom.* **44** 330
[13] Yoon J H, Lee I K, Choi H Y, Choi E J, Yoon, J H, Shim, S E, Ii K. 2011 *Green. Chem.* **13** 631

# Thermal properties of size-selective nanoparticles: Effect of the particle size on Einstein temperature

**Y Li[1], R M Anderson[2], Z Duan[3], S Chill[3], R M Crooks[2], G Henkelman[3], A I Frenkel[1]**

[1]Department of Physics, Yeshiva University, New York, New York, United States
[2]Department of Chemistry and Biochemistry, Texas Materials Institute, The University of Texas at Austin, 1 University Station, Austin, Texas, United States
[3]Department of Chemistry and Biochemistry, Institute for Computational and Engineering Science, The University of Texas at Austin, 1 University Station, Austin, Texas, United States

E-mail: anatoly.frenkel@yu.edu

**Abstract**. Characterizing size related thermal properties of nanoclusters is challenging due to the requirement to accurately control both their average sizes and the size distributions. In this work, temperature-dependent Extended X-ray Absorption Fine Structure spectroscopy and the phenomenological bond-order-length-strength (BOLS) model were employed to investigate the size-dependent Einstein temperature of Au nanoclusters. Theoretical calculations of Einstein temperature and average bond distance for clusters with different sizes agree quantitatively with experiment. The BOLS model is thus useful for predictive understanding of structure and thermal properties in well-defined metal clusters.

## 1. Introduction

Nanoparticles are characterized by high fraction of uncoordinated surface atoms, asymmetric bond length distributions and non-bulk like electronic structure, all of which affect their properties.[1] Experimental investigations of effects of temperature and pressure on these properties are subjects of continuous research,[2-4] although fundamental theory of thermodynamics of nanoparticles is yet to be developed. The greatest challenge towards that goal is twofold. First, it is difficult to isolate nanoparticles from their environment, such as support and adsorbates, to investigate metal-metal bonding structure and dynamics in a broad temperature range. Second, due to inherent heterogeneities of sizes, shapes, states of crystalline order,[5] and due to the ensemble-averaging nature of most structural probes, it is difficult to interpret the average measurements over an ensemble in terms of the behavior of a "representative" nanoparticle.[6] Computational methods, such as molecular dynamics simulations, are commonly employed to address some of the challenges.[7, 8] Alternative methods, based on analytical approaches and/or phenomenological models, are also sought, as there is a strong need to understand what factors are responsible for the thermal behaviors.

Characteristic temperatures such as melting point and Einstein temperatures are important factors defining thermodynamic properties of materials. From known behavior of these factors, many thermal properties can be deduced.[2, 9, 10] In previous studies, their relationships with the particle size were not easy to understand without accounting for the interaction between particle and environment.[11-13] Here we isolate the effect of particle size on Einstein temperature using dendrimer-encapsulated nanoparticles (DENS). Their Einstein temperatures and structural characteristics were measured by temperature-dependent Extended X-ray Absorption Fine Structure (EXAFS) spectroscopy. The bond-order-length-strength (BOLS) model was used to interpret the data.[14] Agreement between measured and calculated changes in Einstein temperature with particle size demonstrated the usefulness of the BOLS model for prediction of thermodynamic properties of self-supported metal clusters.

## 2. Experimental

Au nanoclusters with two different average sizes were prepared using the wet chemical method as previously reported.[15] The size of particles was controlled by tuning surfactant and $HAuCl_4$ salt ratio. The two samples, denoted S1 and S2 were characterized by transmission electron microscopy (TEM) at the University of Texas, using a JEOL-2010F TEM operating at 200 kV. Temperature-dependent XAFS spectra at Au $L_3$ edge were collected in transmission mode at beamline X18B, NSLS at Brookhaven National Lab. Two Au cluster samples and Au foil were placed side by side onto a closed cycle Displex cryostat sample holder to avoid difference in sensed temperature.

## 3. Results and Discussion

Analysis by TEM showed that for sample S1 the average particle size was 1.6±0.2 nm and for S2 - 1.4±0.2 nm. The temperature-dependent Au $L_3$ XAFS spectra for Au foil are shown in Fig. 1(c). The amplitudes of Fourier transform (R-space) spectra decrease with increasing temperature, as expected, due to the enhanced bond-length disorder at elevated temperature. Similar trend was observed in the temperature-dependent XAFS spectra of samples S1 and S2.

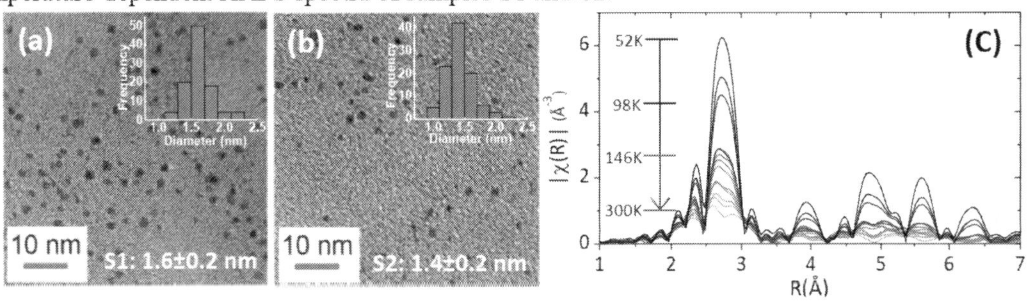


**Figure 1.** (a,b) The image and histogram (inset) of size distribution for samples S1 and S2. (c) The temperature-dependent Fourier transformed XAFS spectra for Au foil.

In the quantitative EXAFS data analysis, we varied the disorder factor and kept the coordination number constant, as no change in particle size or shape is expected in the temperature range below room temperature. In the multiple data-set modeling process, the total bond length disorder (Debye-Waller factor) was separated into the static ($\sigma_s$) and thermal disorder ($\sigma_d$) parameters, that were assumed to be statistically independent from each other:[16]

$$\sigma^2 = \sigma_s^2 + \sigma_d^2. \quad (1)$$

We used Einstein model to model the change of the dynamic disorder with temperature.[17] The coordination number of the first nearest Au-Au bond in S1 was obtained to be 8.8 ± 0.4, slightly larger than that in S2 (8.6 ± 0.4), which agrees with the trend observed in TEM measurements. The first nearest Au-Au bond distances in sample S2 with smaller particle size are smaller than those of S1 with larger particle size (Fig. 2(a)).[18] The average (over the measured temperature range) Au-Au distance in sample S2 and S1 is 2.4% and 1.7% shorter, respectively, than that in Au foil.

Debye-Waller factors for all samples are shown in Fig. 2 (b). The static disorder contribution to the total disorder (Eq. (1)) can be visualized as y-intercept in Fig. 2 (b). It is clearly larger for smaller particles, as shown previously.[5, 19, 20] Einstein temperature follows the opposite trend, increasing from 117.7±9.6 to 124.9±9.4K with particle size and to 134.5±0.8K for Au foil. This effect is expected for small particles, dating back to the works of F. von Lindemann, who demonstrated a monotonic relationship between the Einstein and melting temperatures,[21] and the fact that melting temperature decreases with decreasing size in small particles was reported by many groups.[2, 3, 7]

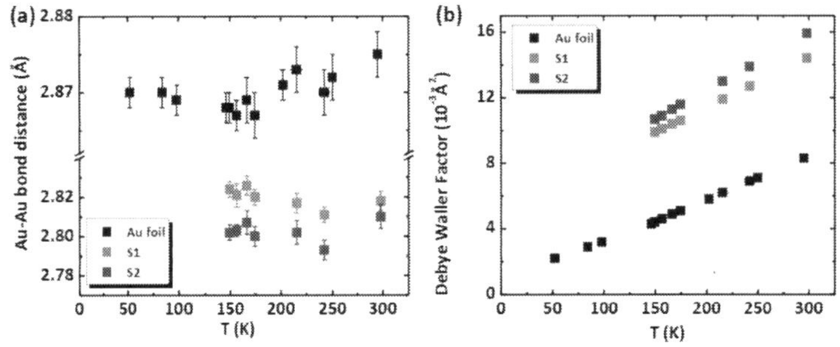


**Figure 2.** The temperature-dependent values of (a) the Au-Au bond distance and (b) Debye Waller factors for samples S1, S2 and Au foil.

These results were compared with those predicted by BOLS model, which calculates average bond distance and melting temperature based on the atomic structure of nanoclusters.[9] The change in melting temperature of nanoclusters relative to that of bulk structures can be expressed as:[9]

$$\frac{\Delta T_m(\mathrm{D})}{T_m(\infty)} = \frac{T_m(\mathrm{D}) - T_m(\infty)}{T_m(\infty)} \cong \frac{\Delta E_T(\mathrm{D})}{E_T(\infty)} = \sum_{i\le 3} \gamma_i (z_{ib} c_i^{-m} - 1);$$
$$\gamma_i = \frac{N_i}{N};\; z_{ib} = \frac{z_i}{z_b};\; c_i(z_i) = \frac{d_i}{d} = 2/[1+\exp\left(\frac{12 - z_i}{8 z_i}\right)]. \tag{2}$$

The average bond distance ($\bar{d}$) over all atoms in the cluster is:[22]

$$\bar{d} = d[1+\Delta d],\; \Delta d = \sum_{i\le 3} \gamma_i (c_i - 1). \tag{3}$$

In Eq. 3, $d$ is the bond distance of bulk. From Eqs. 2 and 3, it is obvious that only the low-coordinated atoms on the surface are responsible for changes in average bond distance and melting temperature of the cluster.

Calculations were performed on cuboctahedral Au clusters containing 55, 147 and 309 atoms because the coordination numbers of these three regular cluster types ($N_{55}$=7.86, $N_{147}$=8.98 and $N_{309}$=9.63)[23] and their diameters of 1.1, 1.7 and 2.3 nm, respectively are comparable to those expected in "average" particles within samples S1 and S2. To directly compare experimental results with calculations, melting temperature was converted to Einstein temperature.[21] The ratio of Einstein temperature of cluster with diameter D to bulk can thus be derived in terms of the corresponding ratio for the melting temperature:

$$\frac{T_E(\mathrm{D})}{T_E(\infty)} = \sqrt{\frac{\Delta T_m(\mathrm{D})}{T_m(\infty)} + 1}\,. \tag{4}$$

Experimental and calculated results on average bond distance and relative Einstein temperatures for Au clusters with different sizes are listed in Table 1.

**Table 1.** Comparison of BOLS model predictions and experimental results.

| BOLS model prediction | | | EXAFS results | | |
|---|---|---|---|---|---|
| | $\bar{d}$ (Å) | $T_E$(D)/$T_E$(∞) | | $\bar{d}$ (Å) | $T_E$(D)/$T_E$(∞) |
| 55 | 2.756 | 0.823 | | | |
| 147 | 2.793 | 0.875 | S2 | 2.802±0.005 | 0.88±0.08 |
| 309 | 2.813 | 0.903 | S1 | 2.820±0.005 | 0.93±0.08 |
| Au foil | 2.870* | 1 | Au foil | 2.870±0.002 | 1 |

* Obtained by EXAFS analysis of Au foil

Similar to the experimental results, BOLS model calculations (Table 1) also predict the decrease of the average bond distance and Einstein temperature with decrease of particle size. The combination of results obtained from TEM, XAFS and BOLS model indicates that cuboctahedral $Au_{147}$ is the appropriate model to describe the size and shape of sample S2 (the smaller sample). The agreement of the experimental and calculated results for that sample validates the BOLS model and its application to unsupported metal clusters. On the other hand, though the average diameter of sample S1 is close to that of S2, the average bond distance and Einstein temperature of sample S1 are closer to those of $Au_{309}$. Taken together, these latter observations suggest that a small portion of large particles dominates the thermal properties of the sample S1, which is a reasonable conclusion, taken into account that XAFS averaging is done according to the volume, not diameter, of nanoparticles. Hence, even a few large particles not accounted for in TEM images, can dominate the XAFS signal. The BOLS model, therefore, should be used with caution in its predictive capacity, and its validity should be tested against the measured particle size distribution whenever possible.

## 4. Conclusions

Temperature dependent EXAFS data in unsupported Au clusters of two different sizes demonstrated that both the average metal-metal bond length and Einstein temperature decreased with particle size. Phenomenological relationship between these two quantities can be also obtained analytically, within BOLS model that utilizes no free parameters. The agreement between the experiment and the calculation validates both the Einstein model in EXAFS data analysis of Au clusters and the use of the BOLS model for predicting their thermal and structural properties. An important prediction validated in this work is the role of surface atoms in tuning thermal properties of nanoclusters.

**Acknowledgment**

We gratefully acknowledge support from the National Science Foundation grant No. CHE-1534184. We acknowledge the facilities support provided at NSLS at BNL (DOE, BES, Contract No. DE-SC0012704) and the Synchrotron Catalysis Consortium (DOE, BES, Grant No. DE-SC0012335).

**References**

[1] Smith A M *et al* 2009 *Nature Nanotech.* **4** 56-63
[2] Zhao Y H and Lu K 1997 *Phys. Rev. B* **56** 14330-14337
[3] Mei Q S and Lu K 2007 *Prog. Mater. Sci.* **52** 1175-1262
[4] Jiang Q *et al* 2000 *Chem. Phys. Lett.* **322** 549-552
[5] Li L *et al* 2013 *J. Am. Chem. Soc.* **135** 13062-13072
[6] Li Y *et al* 2015 *Nature Commun.* **6** 7583
[7] Chushak Y G and Bartell L S 2001 *J. Phys. Chem. B* **105** 11605-11614
[8] Wang L *et al* 2003 *Phys. Lett. A* **310** 197-202
[9] Sun C Q *et al* 2002 *J. Phys. Chem. B* **106** 10701-10705
[10] Avramov I and Michailov M 2008 *J. Phys.: Cond. Matter* **20** 295224
[11] Roldan Cuenya B *et al* 2010 *Phys. Rev. B* **82** 155450
[12] Roldan Cuenya B *et al* 2011 *Phys. Rev. B* **84** 245438
[13] Small M W *et al* 2012 *ACS Nano* **6** 5583-5595
[14] Sun C Q 2007 *Prog. Solid State Chem.* **35** 1-159
[15] Yancey D F *et al* 2013 *Chem. Sci.* **4** 2912-2921
[16] Sayers D E *et al* 1971 *Phys. Rev. Lett.* **27** 1204-1207
[17] Frenkel A I and Rehr J J 1993 *Phys. Rev. B* **48** 585-588
[18] Frenkel A I *et al* 2005 *J. Chem. Phys.* **123** 184701
[19] Witkowska A *et al* 2007 *Phys. Rev. B* **76** 104110
[20] Agostini G *et al* 2013 *J. Phys.: Conf. Ser.* **430** 012031
[21] Lindemann F A 1910 *Phys. Z.* **11** 609-612
[22] Chang Q S *et al* 2002 *J. Phys.: Cond. Matter* **14** 7781
[23] Glasner D and Frenkel A I 2007 *AIP Conf. Proc.* **882** 746-748

# Investigation of oxygen vacancies in $CeO_2$/Pt system with synchrotron light techniques

**L Braglia[1,2], A L Bugaev[2], K A Lomachenko[1,2], A V Soldatov[2], C Lamberti[1,2], A A Guda[2]**

[1] Department of Chemistry, NIS and CrisDi Centers, Turin University and INSTM Reference Center, 10125 Turin, Italy
[2] International Research Center "Smart Materials", Southern Federal University, 344090 Rostov-on-Don, Russia

luca.braglia@unito.it

**Abstract**. A peculiar property of ceria is the ease to form oxygen vacancies, producing reactive sites or facilitating ionic diffusion. For these reasons ceria promotes catalytic activities for a number of important reactions when it is used as a support for transition metals. In our work we study the temporal evolution of oxygen vacancies formation by time-resolved XANES at Ce K-edge and XRD measurements on $CeO_2$/Pt nanoparticles, successfully monitoring the reaction of CO oxidation.

## 1. Introduction

This rare earth oxide claim several applications: in catalysis as a support substrate to improve mechanical and thermal stability and moreover as activity and selectivity of catalysis. Ceria also covers an important role in treatment of toxic emissions, for removal soot from diesel engine exhaust, as integral part of car exhaust three-way automotive catalysts, in low-temperature water-gas shift reaction, in solid oxide fuel lattices (SOFCs), in solar-driven thermochemical $CO_2$ reduction and for biomass reforming [1]. The reason of these wide range of applications is enclosed in the high oxygen storage capacity (OSC) at low temperature, which is the capacity to store oxygen at lean and to release it at rich condition. We focused our attention in the CO oxidation process [2]: $2Ce^{4+}+O^{2-}+CO \rightarrow 2Ce^{3+}+V_o+CO_2$, where $O^{2-}$ is lattice oxygen and $V_o$ is oxygen vacancy. The CO molecules are absorbed on the surface of Pt nanoparticles, where they react with the oxygen provided by non-stoichiometric cerium oxide [3]. Hence, $Ce^{3+}$ concentration, which can be monitored by x-ray absorption spectroscopy [4], is proportionally related with oxygen vacancy concentration. The presence of noble and transition metals facilitate the release of oxygen from ceria, and they actively participate to the catalysis process [5]. However the kinetics of this process that is determined by energy of activation is still unclear. DFT studies predict that energy, required for oxygen atom to migrate from one crystallographic position into another depends on concentration of vacancies and is in the range of 0.46–1.08 eV [6]. *Hui-Ying Li et al.* [7] found out that the most energy favourable mechanism of oxygen vacancy on the surface of $CeO_2$ in 111 plane is the two-step (TS) diffusion [8]. In this work we analyse the process of vacancy formation by means of monitoring $Ce^{3+}$ concentration.

## 2. Experimental and Methods

The change in the oxidation state of $Ce^{4+}$ in $Ce^{3+}$ during the reaction with CO, previously described, permitted the monitoring of the temporal evolution of $Ce^{3+}$ formation with a suitable technique as time-resolved X-ray absorption near-edge spectroscopy (XANES). During the Ce reduction we also observed a change of lattice parameter that it has been studied with powder X-ray diffraction (XRD).
Truncated octahedral $CeO_2$ nanoparticles were prepared by hydrothermal method. The platinum nanoparticles were deposited on ceria by wetness impregnation by tetraammine platinum(II) nitrate (Aldrich, 99.995%) followed by calcination in air at 674K for 4 h and reduction in 5% $H_2$ flow at 573K for 4 h [2]. The diameter of platinum nanoparticles is 1.2 (±0.2) nm, and the concentration is 1.5%wt.
We collected the XANES spectra at BM23 beamline of ESRF (Grenoble, France) at Ce K-edge (40470eV). The measurement has been done in transmission mode using a double-crystal monochromator Si (311). Powder sample was introduced within a glass capillary of 1 mm diameter size and the gas was capable to flow through the two cut edges of the capillary. We measured the steady XANES spectra at 473K in $O_2$, afterwards we switched in reducing flow of CO, following the temporal evolution of $Ce^{3+}$ formation collecting the XANES spectra for 3 h 15 min. The same procedure has been adopted for collecting the two-dimensional XRD patterns ($\lambda$=0.3024 Å) at BM23 beamline. The data were analyzed with Athena software and then the principal component analysis (PCA) [9] has been done with FitIt software [10].

## 3. Results and Discussion

XAS has proved an attractive technique for monitoring the temporal evolution of the $Ce^{3+}$ formation in reductive condition. In figure 1a are reported the Ce K-edge XANES spectra of the sample at the steady state in $O_2$ (blue curve) and in CO (orange curve) after 3 h 15 min of reaction. The changes are slightly small but visible: in CO the absorption threshold of the XANES spectrum is shifted to lower energy and the shape of the second peak is shifted to lower energy indicating increase of Ce-O distances. The energy shift of the absorption threshold confirms the reduction of $Ce^{4+}$ in $Ce^{3+}$ in CO flux.

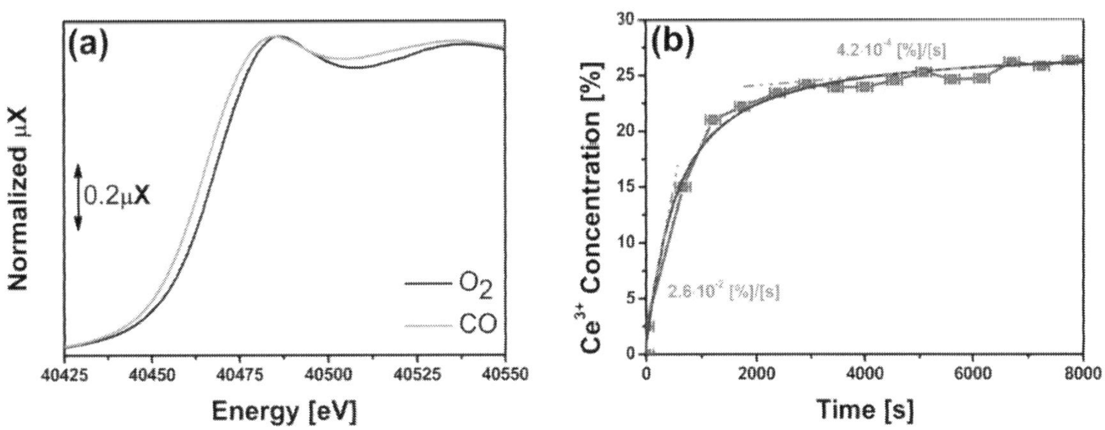


**Figure 1.** (a) XANES spectra at 423K above Ce K-edge of the $CeO_2$/Pt nanoparticles in the steady state in $O_2$ (blue line) and in presence of CO (orange line) after 3 h 15 min of reaction; (b) Temporal evolution of $Ce^{3+}$ formation in reductive condition in function of time at 423K. PCA analysis was performed on the XANES spectra collected at Ce K-edge in CO. We fitted the curve with an exponential function (blue line) and we have drawn the tangents in grey coloured with their respectively slope values.

Principal component analysis (PCA) was performed for series of XANES spectra during the CO fluxing and revealed two independent components. We associated the components to $Ce^{3+}$ and $Ce^{4+}$ state and in

PCA results we observed an exponential time dependent behaviour of the $Ce^{3+}$ formation, shown in figure 1b. The concentration has been figured out considering as references, the XANES spectra of $Ce_2(CO_3)_3$ for 100% $Ce^{3+}$ and the spectrum of $CeO_2$/Pt nanoparticles in $O_2$ for 0% $Ce^{3+}$ ; the XAS measurement was performed in previously experiment [2] and the output of the sample cell was connected to the mass spectrometer allowing to follow the $CO_2$ formation. The trend of the curve in figure 1b presents a drastic rise until 20% corresponding to the percentage of $Ce^{3+}$ formation in the first 20 min that undergoes to a significant decrease of the rate reaching a plateau at around 25% of $Ce^{3+}$%. The plot in figure 1b can be divided in two regions: region I, which occurs during the first 20 min of flow in CO, where the CO molecules absorbed on Pt surface react with the oxygen provided by the $CeO_2$ surface with the following reduction of $Ce^{4+}$ atoms. This mechanism yields an increasing of the oxygen vacancies in the local area around Pt nanoparticles and on the surface of ceria, leading a lack of available O atoms for the reaction with CO absorbed on Pt surface. In region II we observed that the rate of $Ce^{3+}$ formation rapidly decrease, possibly in association with the oxygen migration from the bulk to the surface. We estimated the rate of $Ce^{3+}$% formation in the two regions mentioned before, drawing the two tangents for the respectively parts. The tangent in the region I, reported in figure 1b, presents a slope of $2.6\times10^{-2}$ [%]/[s], conversely the slope achieved in the region II, where the $Ce^{3+}$% formation is much slower, is $4.2\times10^{-4}$ [%]/[s]. The value of the rate obtained in the second region provides information about the oxygen vacancy formation that is directly related with the $Ce^{3+}$ formation.
Cerium forms non-stoichiometric oxides with the fluorite structure ($Fm\bar{3}m$). The effect of $Ce^{3+}$ formation can be also supported with the variation of lattice parameter distances, hence we collected powder XRD patterns with the same setup and procedure used previously for XAS experiment. In figure 2a are shown the XRD patterns for the steady state of the sample in $O_2$ (blue line) and in CO (orange line), and as seen before we collected XRD patterns during the flux of CO in function of time, following the temporal variation of the lattice parameter. The lattice parameter has been derived for each XRD patterns and we distinguished an expansion of the lattice parameter, shown in figure 2b, due to the loss of oxygen atoms, that its behaviour can be described by an exponential curve.

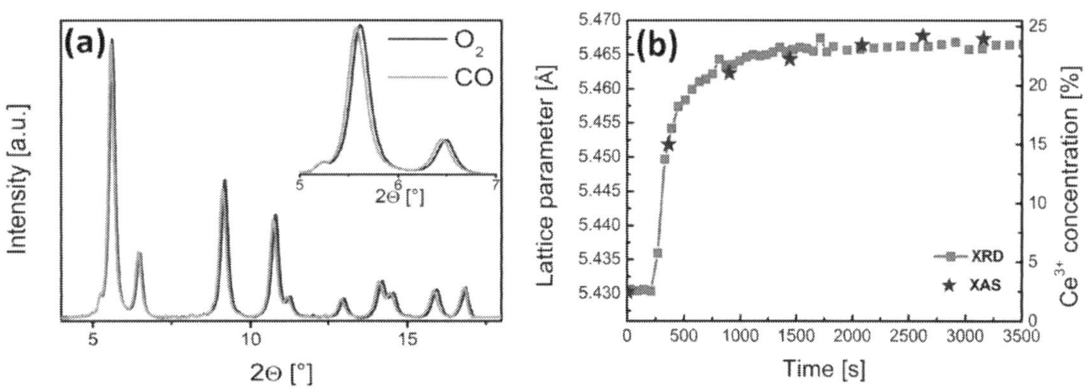


**Figure 2.** (a) The XRD patterns (λ=0.3024 Å) at 423K are shown in completely oxidized state in $O_2$ (blue line) and in reduced state in CO (orange line). (b) Lattice parameter variation in time dependent (red line) after the CO switching derived from the XRD patterns. The blue star scatter indicate the values of the temporal evolution of $Ce^{3+}$ formation described by XANES spectra previously seen in figure 1b.

The curve in figure 2b reproduced the trend studied with XAS measurements, indeed in the first 20 min there is an intense rise of the lattice parameter value, correspondingly the high $Ce^{3+}$ formation, which undergoes to a quick slowdown.

## 4. Conclusions

XANES experiments have proved the formation of $Ce^{3+}$ in the $CeO_2$/Pt nanoparticles in presence of CO. Collecting series of XANES spectra in CO and with a further PCA analysis we have figured out the exponential behaviour of the $Ce^{3+}$ formation and the $Ce^{3+}$% concentrations during the reaction. In 3h 15min in CO, the sample formed ≈ 25% of $Ce^{3+}$ atoms in two main steps, described by the exponential fit in figure 1b: in 20 min the $Ce^{3+}$ formation is rapid until 20% and then bear to a significant slowdown of the rate. The rate of $Ce^{3+}$ and oxygen vacancies formation in the region II is slow and can be related with the migration of oxygen atoms from the bulk to the interface with Pt nanoparticles. The combination of XAS and XRD techniques has been very useful because the $Ce^{3+}$ formation is accompanied by an expansion of the lattice parameter that it has been studied in CO and we observed a similar exponential tendency, shown in figure 2b. X-ray Photoelectron Spectroscopy (XPS) performed at the same environmental condition could be an interesting technique to study more in detail the temporal evolution of oxygen migration, varying the kinetic energy of photoelectrons, the probing depth changes and the oxygen vacancy concentration could be monitored at different depth [11].

**Acknowledgements**

A.B., K.L., A.G. acknowledge the Grant of the President of Russia for Young Scientists MK-3206.2014.2. A.S. and C.L. acknowledge the Mega-grant of the Russian Federation Government to support scientific research under the supervision of the leading scientist at Southern Federal University, No. 14.Y26.31.0001, for the partial funding of the research. L.B. acknowledge Master in Material Science Exploiting Large Scale Facilities (MaMaSELF programme). We also acknowledge O. Safonova from Paul Scherer Institute (PSI, Villigen, Switzerland) who provided the samples.

**References**

[1] Li S L, Wang N L, Yue Y H, Wang G S, Zu Z, and Zhang Y 2015 *Chem. Sci.* **6** 2495-500
[2] Safonova O V, Guda A A, Paun C, Smolentsev N, Abdala P M, Smolentsev G, Nachtegaal M, Szlachetko J, Soldatov M A, Soldatov A V *et al.* 2014 *J. Phys. Chem. C* **118** 1974-82
[3] Gorte R J 2010 *Aiche J.* **56** 1126-35
[4] Rothensteiner M, Sala S, Bonk A, Vogt U, Emerich H, and van Bokhoven J A 2015 *Phys. Chem. Chem. Phys.* **17** 26988-96
[5] Trovarelli A, Catalysis by ceria and related materials, Imperial College Press, London, 2002.
[6] Vayssilov G N, Lykhach Y, Migani A, Staudt T, Petrova G P, Tsud N, Skala T, Bruix A, Illas F, Prince K C *et al.* 2011 *Nat. Mater.* **10** 310-5
[7] Li H Y, Wang H F, Guo Y L, Lu G Z, and Hu P 2011 *Chem. Commun.* **47** 6105-7
[8] Wang H F, Li H Y, Gong X Q, Guo Y L, Lu G Z, and Hu P 2012 *Phys. Chem. Chem. Phys.* **14** 16521-35
[9] Beale A M, Le M T, Hoste S, and Sankar G 2005 *Solid State Sci.* **7** 1141-8
[10] Smolentsev G and Soldatov A V 2007 *Comput. Mater. Sci.* **39** 569-74
[11] Kato S, Ammann M, Huthwelker T, Paun C, Lampimaki M, Lee M T, Rothensteiner M, and van Bokhoven J A 2015 *Phys. Chem. Chem. Phys.* **17** 5078-83

Journal of Physics: Conference Series **712** (2016) 012065 doi:10.1088/1742-6596/712/1/012065

# Flame made ceria supported noble metal catalysts for efficient $H_2$ production via the water gas shift reaction

**G Cavusoglu[1,2], H Lichtenberg[1,2], A Gaur[2], A Goldbach[3], J-D Grunwaldt[1,2,4]**

[1]Institute of Catalysis Research and Technology (IKFT), Karlsruhe Institute of Technology (KIT), Germany
[2]Institute for Chemical Technology and Polymer Chemistry (ITCP), KIT, Germany
[3]Dalian Institute of Chemical Physics (DICP), Dalian National Laboratory for Clean Energy (DNL), China

E-mail: grunwaldt@kit.edu

**Abstract.** *Rh/ceria catalysts were synthesized by flame spray pyrolysis for high temperature water gas shift (WGS) reactions. These catalysts show a high specific surface area due to a high degree of nanocrystallinity. X-ray absorption spectroscopy (XAS) unraveled the formation of small Rh particles under WGS reaction conditions. The catalytic activity was examined at atmospheric pressure by measuring CO conversion as a function of temperature. Some methane formation was observed above 310 °C.*

## 1. Introduction

Hydrogen is a feedstock used for diverse industrial applications, but also a clean fuel for fuel cells and a key element for chemical storage of renewable energy. The WGS reaction is important for hydrogen production and purification and an essential step in many industrial processes. In industry, different catalysts and reaction conditions are applied for this reaction [1].

Hydrogen production over commercial (copper and ferrochrome) catalysts requires two reaction stages for WGS, which need a large volume reactor. Due to their volume, long treatment time, limited durability and susceptibility to reaction conditions, industrial copper (low temperature) and ferrochrome (high temperature) WGS catalysts are not suitable for one stage reactions and fuel cell applications [1, 2]. Furthermore, conventional WGS catalysts are pyrophorous, require time-consuming reduction, and they are sensitive to sulfur and chloride poisoning. In recent years noble metals (mainly Pt, Au and Pd) supported on reducible oxides (ceria, zirconia, titania) have been extensively studied as promising catalysts for high temperature WGS reactions [2, 3]. These catalysts should exhibit high stability, activity and poisoning resistance and convert CO at a single stage at high temperature [2, 4]. Rather little is known about Rh-based catalysts up to now. For this study, $Rh/CeO_2$ catalysts were prepared by flame spray pyrolysis and characterized during WGS by X-ray absorption spectroscopy.

## 2. Experimental

X-ray Absorption Near Edge Structure (XANES) and Extended X-ray Absorption Fine Structure (EXAFS) measurements at the Rh K-edge (23220 eV) were performed at BM25 (the Spanish CRG Beamline) at the European Synchrotron Radiation Facility (ESRF, Grenoble, France) in transmission mode using a Si(111) double crystal monochromator and ionization chambers.

[4]To whom any correspondence should be addressed

4.3 wt.% Rh/$CeO_2$ catalysts were prepared by flame spray pyrolysis (FSP) [5], using solutions of the corresponding initial precursor in xylene (VWR chemicals). Precursor solutions were prepared by dissolving specified amounts of metal acetylacetonate ($Rh(acac)_3$, 99% (25% Rh), abcr GmBH & Co. KG), and cerium (III) 2-ethylhexanoate 49% 2-ethylhexanoic acid (12%Ce, Alfa Aesar). The concentration of the noble metal precursor was 1 mol/L. At first, X-ray diffraction (XRD) patterns of the catalysts were recorded in a 2Θ range of 10°–90° (2 s/step, step size 0.016°) using Cu $K_\alpha$ radiation (λ=0.15406 nm), a nickel filter and a graphite monochromator (D8 Advance Diffractometer by Bruker). For *in situ* XAS measurements, a 100–200 µm sieve fraction of the catalyst was filled into a quartz glass capillary (outer diameter 1.5 mm, wall thickness 20 µm) between quartz wool plugs. The capillary was placed on top of a gas blower (FMB Oxford). First the catalysts were characterized by *in situ* temperature programmed reduction (TPR) during heating from room temperature to 450 °C at 5 °C/min in 5% $H_2$/He (flow 50 mL/min) while recording XANES spectra in Quick-EXAFS mode (recorded at SNBL/ESRF).

**Table 1.** 4.3 wt.% Rh/$CeO_2$ in capillary at BM25 beamline at ESRF. EXAFS spectra were recorded at each gas atmosphere and temperature.

| RT | 200 °C | 300 °C | 400 °C | RT |
|---|---|---|---|---|
| He | CO | CO + $H_2O$ | CO + $H_2O$ | CO |
| CO | $H_2O$ | | | He |
| | CO + $H_2O$ | | | |

Most importantly, the effect of different feed gas compositions and temperature on the catalyst structure was investigated (Table 1). The first series of XAS spectra were recorded at room temperature, first under He, then in 8% CO/He. Afterwards the capillary reactor was heated to 200 °C in 8% CO/He. Further series of XAS spectra were recorded at 200 °C in 8% CO/He, 12% $H_2O$/He, and CO/$H_2O$ (8% CO, 12% $H_2O$ and 80% He), respectively. Afterwards the capillary reactor was heated to 300 °C and 400 °C in the WGS gas mixture. Finally the reactor was cooled to room temperature and further XAS spectra were recorded. Absorption data were analyzed using the IFEFFIT software package [6].

## 3. Results and Discussion

The X-ray diffraction pattern of the 4.3 wt.% Rh/$CeO_2$ catalyst, shown in Figure 1(a), corresponds to crystalline $CeO_2$ in fluorite phase. At 2θ = 41.1° a reflection matching the Rh(111) plane appears. However, it is rather broad, and so weak in intensity that it can hardly be distinguished from the background. Further analysis of ceria reflections at 2θ = 28.6° and 33.0° using the Scherrer equation [7] unraveled $CeO_2$ crystallite sizes of 7.5 nm and 6.3 nm, respectively.

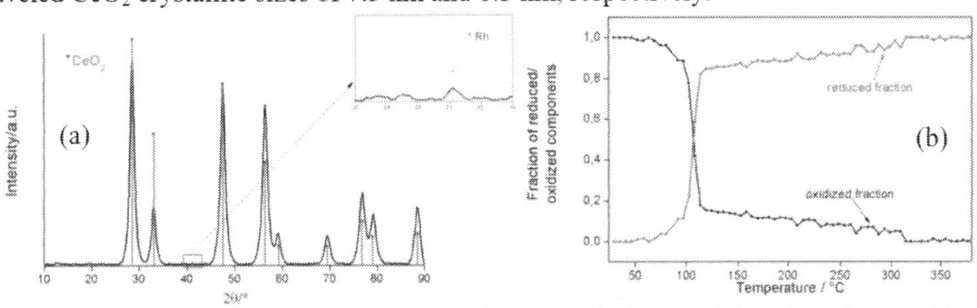


**Figure 1.** (a) XRD pattern of flame made 4.3 wt.% Rh/$CeO_2$ with characteristic ceria reflections (2θ = 28.6°, 33.0°, 47.5°, 56.5°, 59.1°) and an extremely weak Rh peak (2θ = 41.1°). (b) Linear combination fitting results of Rh K-edge XANES spectra recorded during TPR, revealing changes in relative concentrations of oxidized and reduced Rh species.

The specific surface area, determined by BET, amounted to 142 $m^2g^{-1}$, in line with the smaller crystallite size of the support compared to two previously investigated flame made 1.9 wt.% and

3.1wt.% Rh/$CeO_2$ catalysts with crystallite sizes of 11 and 17 nm (reflection at 2θ = 28.6°) and surface areas of 125 and 105 $m^2g^{-1}$, respectively. Next the catalyst was reduced during TPR. The relative concentrations of oxidized and reduced Rh species, determined by linear combination fitting using the initial and final spectrum of the TPR series as reference spectra, are shown in Figure 1(b). The results show that the 4.3 wt.% Rh/$CeO_2$ catalyst is reduced at ca. 114 °C. A previously characterized flame made 3.1 wt.% Rh/$CeO_2$ catalyst was completely reduced at 75 °C [8]. These results show that Rh in contrast to Pt [9] is reduced at rather low temperature.

Figure 2 (a) shows normalized Rh K-edge XANES spectra of the 4.3 wt.% Rh/$CeO_2$ catalyst in different gas atmospheres and at different temperature. At room temperature under He and in CO the catalyst is obviously in an oxidized state. The decrease in white line intensity after *in situ* reaction at higher temperature and in different reaction gas mixtures indicates reduction of the catalyst.

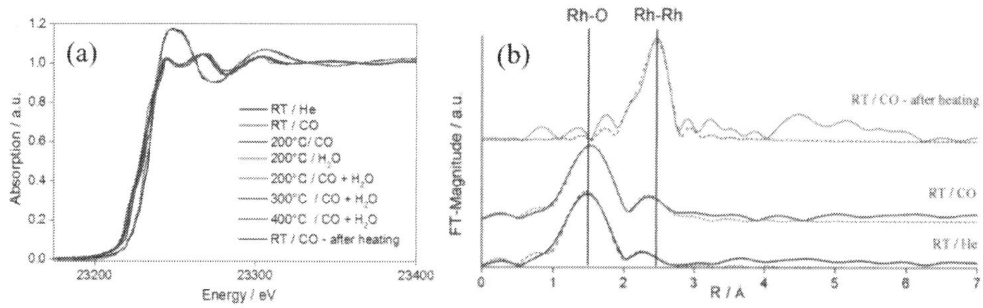


**Figure 2.** (a) Rh K-edge XANES spectra of the 4.3 wt.% Rh/$CeO_2$ catalyst recorded at room temperature under He and in 8% CO/He, at 200 °C in 8% CO/He and in 12% $H_2O$/He, and in CO + $H_2O$ (8% CO, 12% $H_2O$, 80% He) at 200 °C, 300 °C and 400 °C. (b) $k^3$–weighted Fourier transformed EXAFS data of the catalyst at room temperature under He and in CO, and at room temperature in CO after the WGS reaction (solid lines) and fitting results (dashed lines).

Structural parameters were derived from fitting the $k^3$–weighted Fourier transformed EXAFS data in R-space. Representative spectra and the corresponding fits are shown in Figure 2(b). The chemical environment of Rh at room temperature under He and in CO is similar; the absorber atoms are surrounded by ca. 6 oxygen atoms at 2.01 Å, similar to the first coordination sphere (Rh – O) in $Rh_2O_3$. At a larger distance (ca. 2.2 Å in Figure 2(b)) the contribution of a second shell is observed, but this peak is so weak that the coordination number cannot be extracted. After *in situ* WGS reaction the contribution of the Rh – O shell disappears and a Rh – Rh shell clearly dominates. The Rh absorbers are now coordinated by ca. 4 neighboring Rh atoms at 2.71 Å. This low coordination number compared to the coordination number in bulk state (12) indicates a very small Rh particle size, probably due to disruption of the particles, e.g. by Rh carbonyl formation. A similar effect was observed by Grunwaldt et al. [10] for Rh/$Al_2O_3$ catalysts for catalytic partial oxidation of methane. Obviously at higher temperature Rh is reduced to metallic Rh, in accordance with the XANES results (Figure 2(a)). At higher temperature EXAFS analysis was not possible due to the high noise level.

Catalytic activity measurements were performed at atmospheric pressure with a space velocity of 150,000 ml/g $h^{-1}$ for 10.5 wt% Rh/$CeO_2$ (R1) and 50,000 ml/g $h^{-1}$ for 10.5 wt% Rh/$CeO_2$ (R2,R3 and R4) and for 4.3 wt% Rh/$CeO_2$ (R4 and R5). The $H_2O$/CO ratio was also varied from 1.5 (R1, R2 and R4) to 4 (R3 and R5). Figure 3 shows the effect of the space velocity on the conversion of CO. With a $H_2O$/CO ratio of 1.5 feed the CO conversion is ca. 77% at 330 °C and 92% at 350 °C at a space velocity of 150,000 ml/g $h^{-1}$, whereas 50,000 ml/g $h^{-1}$ leads to ca. 93% at 330 °C and 97% at 350 °C. The $H_2O$/CO ratios as well as the partial pressures affect the CO conversion. For 10.5 wt.% Rh/$CeO_2$ CO conversion increased from 15.2% to 26.8%, for 4.3 wt.% Rh/$CeO_2$ from 17.6% to 28.3% at 250°C at a space velocity of 50,000 ml/g $h^{-1}$. The thermodynamic WGS equilibrium was calculated using the Gaseq program [11] and is at 350 °C 92.9% for a S/C ratio of 1.5, and 98.5% for a S/C ratio of 4. Obviously, both catalysts show high activity at high temperature and reach almost equilibrium concentrations, which can be circumvented using membrane reactors [12]. A disadvantage of Rh/$CeO_2$

is the formation of methane ($CH_4$) starting at 310 °C, which has been previously reported [13]. $CH_4$ selectivity ($S_{CH4}$) was calculated by relating the outlet $CH_4$ concentration to the total outlet concentration of carbon containing species [8]. Note that with increasing Rh loading the conversion to $CH_4$ increases.

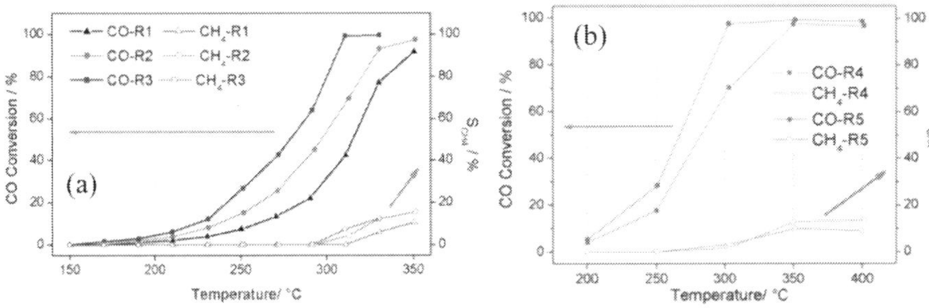


**Figure 3.** Effect of temperature and $H_2O/CO$ ratio on CO conversion over a 10.5 wt.% $Rh/CeO_2$ catalyst (a) and a 4.3 wt.% $Rh/CeO_2$ catalyst (b). For R1, R2 and R4 the $H_2O/CO$ ratio is 1.5, and for R3 and R5 the ratio is 4.

## 4. Conclusion

Nanocrystalline $Rh/CeO_2$ catalysts with high surface area were prepared in one step by flame spray pyrolysis. *In situ* XAS studies combined with linear combination fitting of the XANES data recorded during TPR showed that the flame made 4.3 wt.% $Rh/CeO_2$ catalyst was reduced at rather low temperature (114 °C). Moreover, EXAFS analysis uncovered rather small Rh particles after *in situ* WGS reaction (coordination number around 4), probably forming mobile Rh carbonyl species under reaction conditions. FSP made catalysts are promising candidates for high temperature WGS reactions in membrane reactors.

## 5. Acknowledgement

Financial support by the Helmholtz Association and the Chinese Academy of Sciences (CAS) for funding of the Helmholtz-CAS Joint Research Group 118 "Integrated Catalytic Technologies for Efficient Hydrogen Production" is gratefully acknowledged. We furthermore acknowledge the European Synchrotron Radiation Facility (Grenoble) for providing synchrotron radiation and financial support, and we would like to thank Eduardo Salas-Colera for assistance in operating BM25 and Ying Zhou and Haiyuan Jia for assistance during beamtime as well as Dengyuan Miao for assistance with the catalytic activity tests.

## 6. References

[1] Newsome D S 1980 *Catal. Rev. Sci. Eng.* **21** 275
[2] Zalc J M, Sokolovskii and Löffler D G 2002 *J. Catal.* **206** 169
[3] Liu B, Goldbach A and Xu H 2011 *Catal. Today* **171** 304
[4] Ratnasamy C, Wagner J P 2009 *Catal. Rev. Sci. Eng.* **51** 325
[5] Tepluchin M, Casapu M, Boubnov A, Lichtenberg H, Wang D, Kureti S, Grunwaldt J-D 2014 *ChemCatChem* **6** 1763
[6] Newville M 2001 *J. Synchrotron Rad.* **8** 322
[7] Weidenthaler C 2011 *Nanoscale* **3** 792
[8] Cavusoglu G, Miao D, Lichtenberg H, Carvalho H W P, Xu H, Goldbach A, Grunwaldt J-D 2015 *Appl. Catal. A* **504** 381
[9] Stark W J, Grunwaldt J-D, Maciejewski M, Pratsinis S E, Baiker A. 2005 *Chem Mat.***17** 3352
[10] Grunwaldt J-D, Basini L, Clausen B S 2001 *J. Catal.* **200** 321
[11] http://www.gaseq.co.uk (Version 0.79, C. Morley, 15.08.2015)
[12] Bi Y, Xu H, Li W and Goldbach A 2009 *Int. J. Hydrogen Energy* **34** 2965
[13] Wheeler C, Jhalani A, Klein E J, Tummala S, Schmidt L D 2004 *J. Catal.* **223** 191

# XAFS study on structural order in highly monodispersed thiol-stabilized Au nanoparticles

**Y Huang, W Liu, L Yang, T Huang, Y Jiang*, T Yao* and S Wei**

National Synchrotron Radiation Laboratory, University of Science and Technology of China, Hefei, Anhui 230029, P. R. China

* Email: yaot@ustc.edu.cn; hanyanjy@ustc.edu.cn

**Abstract.** Understanding the influence of thiol on nanoparticle size and structure is essential for the fundamental and applied researches. Here, using x-ray absorption fine structure (XAFS) spectroscopy, we investigate the structural order of Au nanoparticles (NPs) in the protection of thiol ligands with different contents. We found that besides protecting Au NPs against aggregation and growth, thiolates can effectively eliminate the dangling bonds of unsaturated Au atoms, and thus increase the structural order. This work enriches our knowledge of Au-S interface interaction and guides the way towards preparing size-controllable nanoparticles with specific physical/chemical properties.

## 1. Introduction

Au NPs are being actively investigated because of their importance in both fundamental researches and technological applications [1-4]. Ultrastable and monodispersed Au nanoparticles have usually been prepared with the protection of thiolates [5, 6]. The interaction of Au-S interface provides a practical and effective strategy to stabilize and functionalize Au NPs. Therefore, it is of great interest to understand and control the influence of thiolates on Au NPs structure.

Recently, significant progress has been made in understanding the structure of ultrasmall (<2 nm) Au NPs [7, 8]. Despite the internal structures of Au NPs with sizes ranging from clusters to nanocrystals (2.0-4.0 nm) have been reported [9], it is not enough for scientific research, partially because of the challenges in synthesizing and crystallizing monodispersed nanoparticles. Moreover, the structural order plays an important role in determining properties of the particles. In such cases, synchrotron-based XAFS technique can provide atomic-scale structural information, including the surface bond length and structural disorder, with crystallographic resolution independent of the sample crystallinity.

Here, we report the preparation of highly monodispersed Au NPs with controllable sizes of ~2.0-4.0 nm, by adjusting the precursor to surfactant ratio in a one-pot chemical solution synthesis. Using synchrotron-based XAFS technique as a sensitive local-structural probe, we present a detailed study on the structural order and bond length of these Au NPs in the protection of thiol ligands.

## 2. Experimental Section

Preparation of dodecanethiol capped Au NPs: The 0.064 g $AuClPPh_3$ was dissolved in 21 mL ethanol at room temperature. Then dodecanethiol was added. After 5 minutes, 0.108 g $C_4H_{14}BN$ was dissolved in the mixture. And then the mixture was kept stirring for 10 hours. The product was collected by centrifugation and washed with ethanol to remove the physical absorbed and residual dodecanethiol.

TEM images were obtained with a JEOL 2010 system. The UV-Vis spectra were recorded on a UV-2501PC/2550 spectrophotometer in the wavelength range of 300–800 nm. XRD patterns were obtained on a Rigaku X-ray diffractometer D/MAX–2200/PC equipped with Cu Kα radiation (40 kV, 20 mA). XAFS measurements were performed at BL14W1 beamline of Shanghai Synchrotron Radiation Facility (SSRF), using ionization chambers with optimized detecting gases in transmission mode. The storage ring of SSRF worked at 3.5 GeV with a maximum current of 210 mA. The samples were taken into a Telfon cell and the thickness of the cell was adjusted to reach the optimum absorption thickness.

## 3. Results and discussio

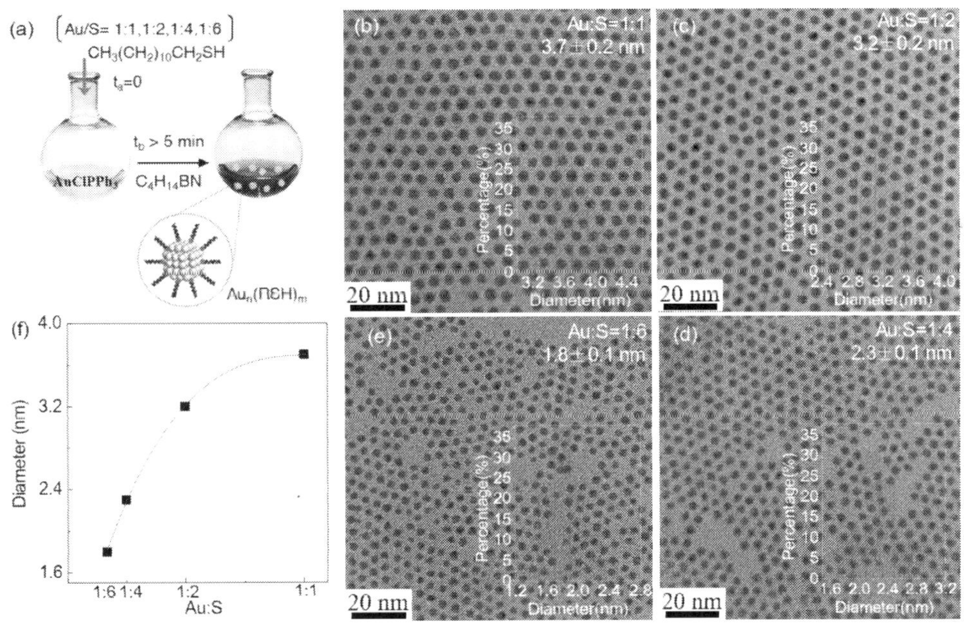


**Figure 1.** Synthetic strategy (a), TEM images (b-e), the insets show the corresponding size distributions. Plot of average size against Au:S ratios (f)

We used ratios (Au:S) of 1:1, 1:2, 1:4, and 1:6 yielding NPs of 3.7, 3.2, 2.3, and 1.8 nm, respectively, as revealed by the TEM images in Figs. 1b-e. These particles are in spherical shape and have a narrow size distribution of ~0.1-0.2 nm, indicating the high monodispersity. For the size distribution, we used sampling methods to estimation the mean value by randomly selecting 100 particles. By decreasing the amount of the thiol, the sizes of NPs gradually increase, indicating the importance of thiol for the preparation of stable and monodispersed Au NPs.

The UV-vis spectra shown in Fig. 2a exhibit one intense surface plasmon band peaked at 510 nm, which is originated from the characteristic plasmon resonance absorbance for spherical Au NPs larger than 2 nm. This absorbance intensifies with increasing Au:S ratios, suggesting the increased NP size. For the NPs synthesized with Au:S ratios of 1:6 and 1:4, the spectral profile is highly structured and displays the fingerprint features of small Au NPs with molecule-like electronic structure, as the sizes are on the edge of metal to semiconductor transition. In Fig. 2b four strong Bragg reflections can be seen, which correspond to the (111), (200), (220) reflections of fcc Au, showing distinct diffraction pattern, indicating better crystalline Au NPs were formed.

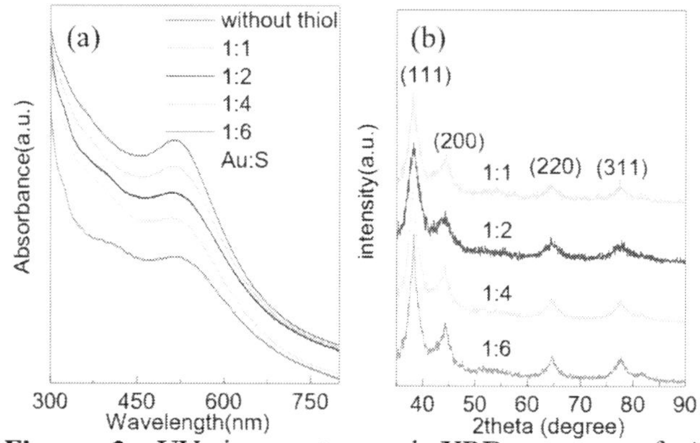


**Figure 2.** UV-vis spectra and XRD spectra of Au nanoparticles at differents Au:S ratios.

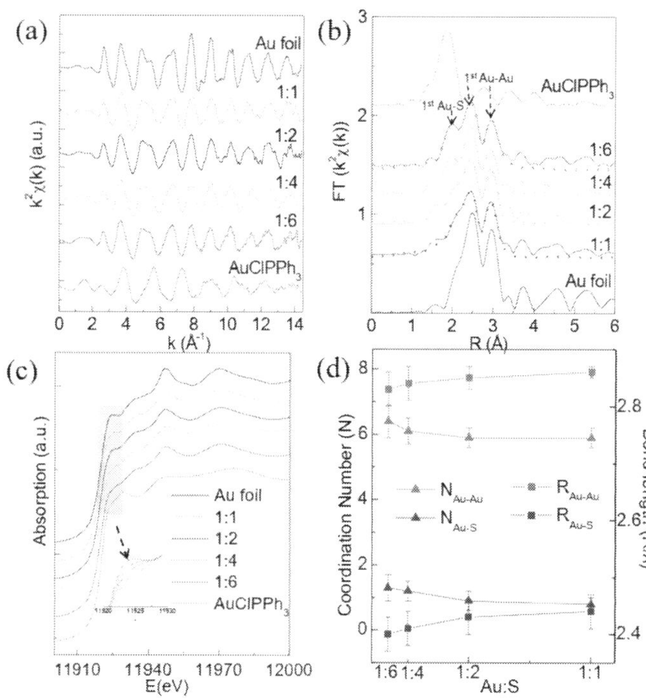


**Figure 3**. Au $L_3$-edge EXAFS [$k^2\chi$ (k)] (a) and their Fourier transforms (b). XANES spectra (c). N and R against Au:S ratios (d).

We plot in Figure 3a,b the EXAFS oscillation $k^2\chi$ (k) and the corresponding Fourier-transformed (FT) spectra. The atomic structures of Au NPs synthesized with different Au:S ratios vary significantly, as reflected by the obviously different EXAFS oscillation shapes. The Au–S coordination peaks at 1.9 Å significantly weaken with the increasing of Au:S ratios, implying the weakened Au-thiols interaction. While the intensity of the Au–Au coordination peaked within 2.15-2.25 Å vary interestingly. The weaker peak gradually reduced relating to the NP size, which can be inferred from the fact that the 2.3 nm and 1.8 nm NPs synthesized with Au:S ratio of 1:4 and 1:6 has the relative lower Au-Au peak than the 3.2 nm and 3.7 nm NPs. However, little evidence of variation is observable in the first stronger peak of Au-Au coordination. In XANES spectra, white-line peak (corresponding to the electronic transition from the $2p_{3/2}$ to the $5d_{5/2,\,3/2}$ states) at around 11926 eV increases in intensity with increasing Au:S ratio. This corresponds to the increased Au→S d-charge transfer, consistent with the enhancement of the interaction between dodecanethiol and Au surfaces. We consider that the tailor of electronic performance was achieved just by control the particle size. Additionally, the variations of the intensities of Au–Au bond peak mainly come from the particle size, without the influence of structure disorder, as consistent with XRD results.

**Table 1**. Structural Parameters of Au nanoparticles

| Sample | Bond | *p* | R (Å) | CN | $\sigma^2(10^{-3}$Å) | ΔE |
|---|---|---|---|---|---|---|
| **Au foil** | Au-Au | | 2.88±0.01 | 12 | 8.0±0.4 | 3.5±1.0 |
| **Au/S=1/6** | Au-S | | 2.40±0.03 | 1.3±0.4 | 3.6±0.3 | 3.3±0.6 |
| | Au-Au (surface) | | 2.83±0.03 | 6.4±0.5 | 9.5±1.1 | 3.7±0.7 |
| | Au-Au (core) | 0.43 | 2.88±0.02 | 12 | 8.7±0.8 | 4.4±0.9 |
| **Au/S=1/4** | Au-S | | 2.41±0.03 | 1.2±0.3 | 4.3±0.3 | 3.2±0.5 |
| | Au-Au (surface) | | 2.84±0.03 | 6.1±0.4 | 9.5±0.9 | 3.7±0.8 |
| | Au-Au (core) | 0.59 | 2.88±0.02 | 12 | 8.7±0.9 | 4.3±0.8 |
| **Au/S=1/2** | Au-S | | 2.43±0.03 | 0.9±0.3 | 3.8±0.2 | 3.6±0.6 |
| | Au-Au (surface) | | 2.85±0.02 | 5.9±0.3 | 9.6±1.0 | 3.6±1.1 |
| | Au-Au (core) | 0.68 | 2.88±0.02 | 12 | 8.7±1.0 | 4.2±0.8 |
| **Au/S=1/1** | Au-S | | 2.44±0.03 | 0.8±0.2 | 3.9±0.2 | 3.5±0.6 |
| | Au-Au (surface) | | 2.86±0.01 | 5.6±0.3 | 9.7±0.8 | 3.6±1.0 |
| | Au-Au (core) | 0.79 | 2.88±0.02 | 12 | 8.7±1.0 | 4.2±0.8 |

The detailed structural information evolution process could be obtained by EXAFS fitting, as played in Table 1. Quantitative least-squares curve-fitting in the R-space with a Fourier transform k-space range of 2.2–12.5 Å$^{-1}$ and a typical R ranges of 1.3-3.2 Å was carried out using module ARTEMIS of programs of IFEFFIT [10]. The contributions of Au atoms at the surface and in the core of a particle were separated in the fitting by defining a parameter p for the fraction of core atoms of the Au NPs. We calculate this value in accordance with the Au atom size and the particle size. The amplitude reduction factor ($S_0^2$) was fixed at 0.86 during the fitting. With reducing the Au:S ratio from 1:1 to 1:6, the statistical factors of reduced chi-square ($\chi_v^2$) were 8.03, 9.42, 8.69, 7.93, and R factor were 0.0019, 0.0018, 0.0019, 0.0020, respectively. And both the surface Au-Au and Au-S coordination number (CN) gradually increased (from 5.6 to 6.4 and 0.8 to 1.3). Generally, unsaturated surface Au atoms have lots of dangling bonds, and hence more structure disorder. Interestingly, the disorder degree $\sigma^2$ of the surface Au-Au bonds (R) in the prepared Au NPs almost remain at about 0.009 Å$^2$. For the Au nanocrystals prepared here, S atoms would occupy the surface hollow sites with a typical tetrahedral configuration formed by three Au surface atoms and a head S atom. Hence, the dodecanethiol coverage can be estimated from the Au-S bond CN using the expression of $CN_{Au-S}/3.0$ [11]. The results indicate that coverage increase from 0.27 to 0.43 with reducing Au:S ratio from 1:1 to 1:6. A larger thiol coverage protect the Au NPs against aggregation and further growth, leading to a small particle size. On the other hand, more dangling bonds residual were replaced by Au-S bonds, resulting enhanced structure order.

## 4. Conclusion

In summary, highly monodispersed plasmonic Au NPs with controllable sizes of ~2.0-4.0 nm were prepared by adjusting the precursor to surfactant in a simple feasible synthesis. Thiolates play two roles in the process of preparing Au NPs. First, the thiol ligands protect Au NPs against aggregation and further growth. On the other hand, the formation of Au-S bonds effectively liminate the dangling bonds on the surface of the Au NPs, and thus increase the structure order of the nanoparticles. This work enriches our knowledge of Au-S interface interaction and guide the way towards preparing size-controllable nanoparticles with specific physical/chemical properties.

**Acknowledgments**
This work was supported by National Science Foundation of China (Grant No. 11175184, 11135008, 11205158, U1332131, and 11305172). The authors are grateful to SSRF for the valuable beamtime.

**References**

[1] Daniel M C and Astruc D 2004 *Chem. Rev.* **104** 293.
[2] Wang L et al 2015 *Nat. Commun.* **6** 6957.
[3] Pyo K et al 2015 *J. Am. Chem. Soc.* **137** 8244.
[4] Yoshida H et al 2012 *Science* **335** 317.
[5] Yin Y and Alivisatos A P 2005 *Nature* **437** 664.
[6] Vericat C et al 2010 *Chem. Soc. Rev.* **39** 1805.
[7] Yao T et al 2010 *J. Am. Chem. Soc.* **132** 7696.
[8] Li Y Y et al 2012 *J. Am. Chem. Soc.* **134** 17997.
[9] Chevrier D M et al 2013 *J. Phys.: Conf. Ser.* **430** 012029.
[10] Ravel B and Newville M 2005 *J. Synchrotron Rad.* ***12*** 537.
[11] Jiang Y et al 2012 *J. Phys, Chem. C* **116** 24999.

# Study of coordination environments around Pd and Pt in a Pd-core Pt-shell nanoparticle during heating

**Y F Nishimura, T Hamaguchi, S Yamaguchi, H Takagi, K Dohmae, T Nonaka and Y Nagai**

Toyota Central R&D Labs., Inc., Aichi 480-1192, Japan

E-mail: yusaku-nishimura@mosk.tytlabs.co.jp

**Abstract.** Local coordination environments around Pd and Pt in a Pd-core Pt-shell nanoparticle (NP) at temperatures ranging from 473 to 873 K was evaluated by utilizing *in situ* XAFS measurement technique to investigate the temperature range in which a core-shell structure is preserved. The core-shell structure was considered to be kept up to 673 K and start to change at about 773 K. Heating to 873 K accelerated atomic mixing in the core-shell NPs. Catalytic properties of the present Pd-core Pt-shell NP are available in the stoichiometric $C_3H_6$-$O_2$ atmosphere at temperatures less than 773 K at most.

## 1. Introduction

Bimetallic nanoparticles (BNPs) possess unique physicochemical properties and functionalities brought by intermetallic interaction and quantum size effect, and thus are expected as heterogeneous catalysts [1,2] and nanomagnets [3]. Regarding the use of BNPs as the former, their catalytic activity is strongly influenced by atomic distribution and electronic structure especially at the surface. In some cases, the BNPs having a core-shell structure (a kind of near surface alloy) exhibit high catalytic activity and selectivity because of the modification of electronic structure by strain and ligand effects [2]. In order to make the most of their unique properties for automotive exhaust catalysts, it is essential to understand not only the temperature range in which a core-shell structure is preserved but also temperature dependence of local coordination environments in a core-shell NP.

Hence in this study, we utilized the *in situ* XAFS measurement technique to investigate temperature dependence of local coordination environments around Pd and Pt in a Pd-core Pt-shell (Pd@Pt) NP in a stoichiometric $C_3H_6$-$O_2$ atmosphere at temperatures ranging from 473 to 873 K.

## 2. Experimental

Monodisperse colloidal Pd@Pt NPs protected by polyvinylpyrrolidone (PVP) were purchased from Tanaka Kikinzoku Kogyo K.K. The molar ratio of Pd to Pt was confirmed to be 1.06 by inductively coupled plasma atomic emission spectroscopy. The colloidal solution was mixed with $\gamma$-$Al_2O_3$ powder, and then dried at 393 K with a hot plate stirrer to form 3 wt% PdPt (1.10 wt% Pd and 1.90 wt% Pt) supported on $Al_2O_3$. Transmission electron microscopy revealed that the size of the Pd@Pt NP was about 3 nm in diameter with one or two layers of Pt shell and the Pd@Pt NP kept a spherical shape on $Al_2O_3$. 200 mg of the PdPt/$Al_2O_3$ powder was pressed to form a disk with the diameter of 10 mm. It was supplied to the *in situ* XAFS measurement.

All the XAFS measurements were performed at the BL33XU (the Toyota beamline) of SPring-8, Japan. EXAFS spectra were recorded at the Pt $L_3$-edge and at the Pd $K$-edge in transmission mode. The PdPt/$Al_2O_3$ disk was firstly treated in an *in situ* cell [4] with diluted $O_2$ gas flow (10 vol% $O_2$ balanced by He, 100 mL min$^{-1}$) at 573 K for 1 h to remove PVP. Oxide species formed at the surface of the PdPt NPs were then reduced in diluted $H_2$ gas flow (2 vol% $H_2$ balanced by He, 100 mL min$^{-1}$) at 573 K for 2 min. Subsequently, the *in situ* XAFS measurement was conducted every 100 degrees from 473 to 873 K in the gas flow comprised of 0.1% $C_3H_6$ and 0.45% $O_2$ balanced by He at a flow rate of 100 mL min$^{-1}$.

Recorded XAFS spectra were normalized with REX2000 [5] using a spline smoothing method based on the criterion proposed by Cook and Sayers [6] and an edge height as a function of energy represented by the Victoreen equation [7]. The extracted EXAFS oscillations $\chi(k)$ were then processed and analysed by using ATHENA and ARTEMIS [8] with FEFF9.05 [9]. Fourier transform of $k^3$-weighted EXAFS oscillation was implemented in the wavenumber $k$ ranges from 27 to 110 nm$^{-1}$ for Pt $L_3$-edge spectra and from 28 to 123 nm$^{-1}$ for Pd $K$-edge spectra. EXAFS curve fitting analysis was carried out in $k$ space for real part of the spectrum brought by Fourier filtering in the range from 0.17 to 0.33 nm for Pt $L_3$-edge and Pd $K$-edge. Both Pt $L_3$-edge and Pd $K$-edge spectra of a sample were simultaneously fitted with EXAFS oscillations calculated for the scattering paths in the first coordination shell of an equimolar PdPt alloy model under the following constraint conditions; average interatomic distance: $R$(Pt–Pd) = $R$(Pd–Pt), XAFS Debye-Waller factor: $\sigma^2$(Pt–Pd) = $\sigma^2$(Pd–Pt), the third cumulant: $C_3$(Pt–Pd) = $C_3$(Pd–Pt), and the product of mole fraction $x$ and average coordination number $N$: $x_{Pd}N$(Pd–Pt) = $x_{Pt}N$(Pt–Pd). The amplitude reduction factors $S_0^2$ of Pt and Pd were experimentally determined to be 0.859 and 0.881 by curve fitting analysis for Pt and Pd foils measured and analysed under the same condition as the PdPt NPs. These values were employed as constants in curve fitting analysis for the Pd@Pt NPs. The goodness of fit was evaluated in the $k$ range where fitting was performed by using R-factor defined as

$$R = \sqrt{\frac{\sum_k \{k^3[\chi_{\text{fitted}}(k) - \chi_{\text{measured}}(k)]\}^2}{\sum_k [k^3 \chi_{\text{measured}}(k)]^2}}$$

where $\chi_{\text{fitted}}(k)$ and $\chi_{\text{measured}}(k)$ stand for fitted and measured EXAFS oscillations, respectively. As a consequence, the R value was less than 0.025 in all curve fitting analyses performed in this study.

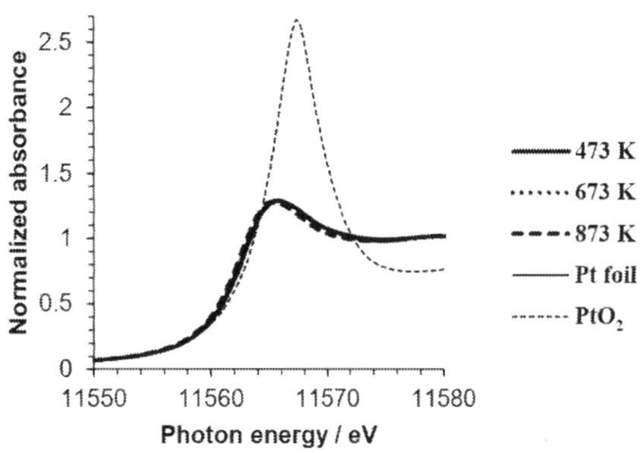


**Figure 1.** Pt $L_3$-edge XANES spectra of the Pd@Pt NPs measured at different temperatures in the 100 mL min$^{-1}$ gas flow consisting of 0.1% $C_3H_6$ and 0.45% $O_2$ balanced by He.

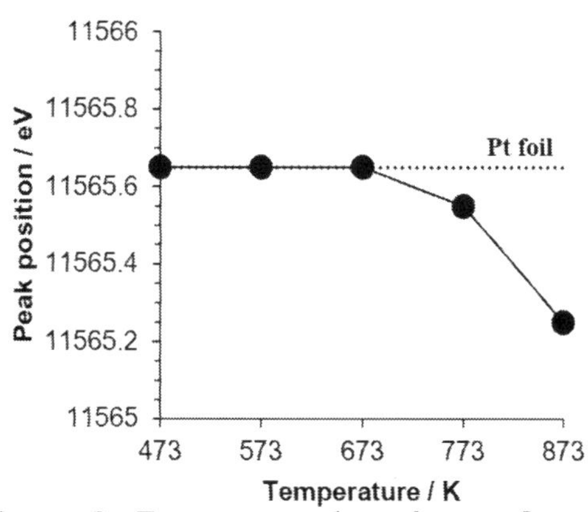


**Figure 2.** Temperature dependence of peak position of the white line at the Pt $L_3$-edge for the Pd@Pt NPs.

## 3. Results and Discussion

Figure 1 shows Pt $L_3$-edge XANES spectra of the Pd@Pt NPs measured at 473, 673, and 873 K. XANES spectra of platinum standards (Pt foil and $PtO_2$ powder) are also indicated in the figure by dotted and dashed lines for comparison, respectively. The peak heights or the integral intensities of the white lines at the Pt $L_3$-edge for the Pd@Pt NPs are almost the same as that for Pt foil. This means that Pt in the Pd@Pt NPs exists as its metal state ($Pt^0$) at temperatures up to 873 K in the present $C_3H_6$-$O_2$ atmosphere. The same tendency was observed for Pd *K*-edge XANES spectra as well. By comparing the Pt $L_3$-edge XANES spectra of the Pd@Pt NPs measured at different temperatures, it is noticed that the white line measured at 873 K appears at slightly lower energy than the others, whose positions are substantially the same. To facilitate the comparison in peak position of the white line, its temperature dependence for the Pd@Pt NPs is shown in figure 2. The unoccupied 5*d* state of Pt starts to change at about 773 K. This change is considered to be accelerated at 873 K.

Figures 3a and 3b compare $k^3$-weighted EXAFS oscillations of the Pd@Pt NPs measured at 473, 673, and 873 K at the Pt $L_3$-edge and at the Pd *K*-edge, respectively. In both figures, it is seen as a general tendency that EXAFS oscillation is damped more strongly as measurement temperature becomes higher due to the increase in XAFS Debye-Waller factor. Careful observation provides the information that the Pt $L_3$-edge EXAFS oscillation measured at 873 K has a different structure at wavenumbers $k \geq 50$ nm$^{-1}$. Besides, the amplitude of the Pd *K*-edge EXAFS oscillation at a low wavenumber (*ca.* 30 nm$^{-1}$), where the damping of EXAFS oscillation is not so strong, decreases at the same temperature. These facts imply the change in coordination environments around Pt and Pd at about 873 K.

EXAFS curve fitting analysis was then performed to discuss local coordination environment around Pd and Pt in the Pd@Pt NPs at different temperatures. Figure 4 depicts temperature dependence of the average coordination numbers for the first coordination shell in the Pd@Pt NPs estimated by the EXAFS curve fitting analysis. Four coordination numbers are almost constant at temperatures at least up to 673 K. These values change slightly at 773 K and significantly at 873 K. This changing trend in coordination numbers, which is consistent with that in the peak position of the white line at the Pt $L_3$-edge, invokes atomic mixing in the Pd@Pt NPs starting at about 773 K. This phenomenon may be explained as follows. In general, above the Tamman temperature, which is approximately half of melting point, atoms in the bulk exhibit mobility with keeping solid state. In the case of bulk materials, the Tamman temperatures of Pd and Pt are 914 K and 1014 K, respectively [10]. In the case of nanomaterials, however, surface energy strongly influences their thermodynamic properties (especially melting point) because of their large surface-to-bulk atomic ratios. Thus, the melting points of Pd and Pt nanoparticles must be low, compared to those of bulk Pd and Pt, respectively. Taking it into consideration, the enhancement of atomic mixing may occur at 873 K.

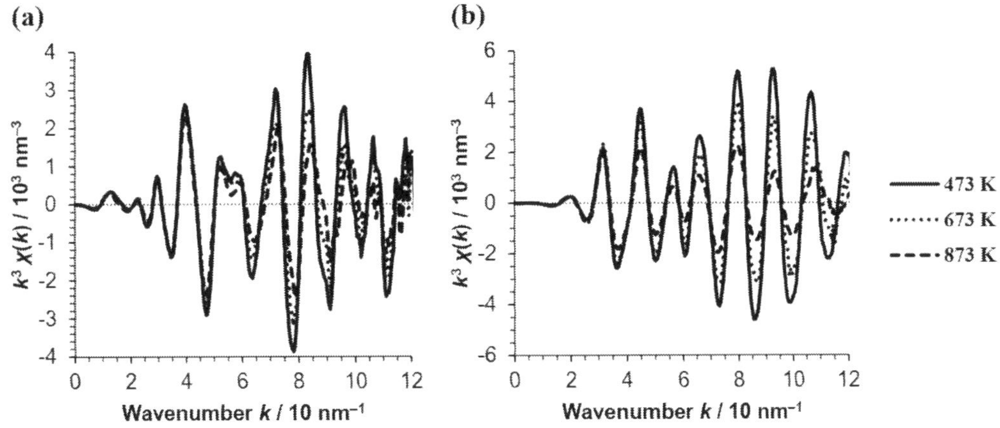


**Figure 3.** $k^3$-weighted EXAFS oscillations of the Pd@Pt NPs measured at different temperatures: (a) Pt $L_3$-edge and (b) Pd *K*-edge.

## 4. Conclusion

Local coordination environments around Pd and Pt in the Pd@Pt NP during heating up to 873 K was investigated by means of the *in situ* XAFS measurement technique. Under the present conditions, the Pd-core Pt-shell structure was considered to be preserved up to 673 K and start to change at about 773 K. Heating to 873 K accelerated atomic mixing in the Pd@Pt NP. The Pd@Pt NPs as heterogeneous catalysts can be employed to make the most of their unique catalytic properties at temperatures less than 773 K at most.

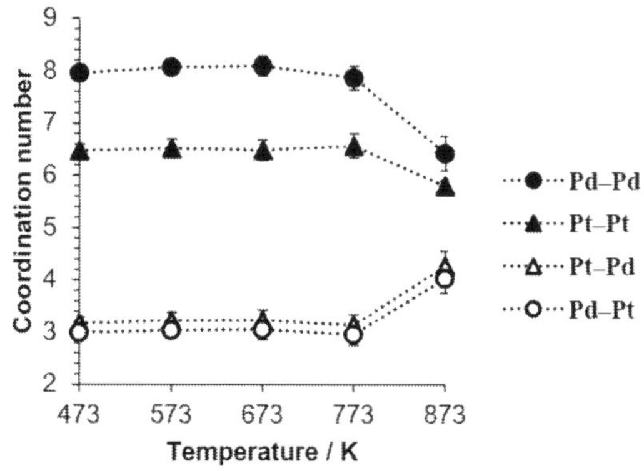


**Figure 4.** Temperature dependence of average coordination numbers for the Pd@Pt NPs estimated by EXAFS curve fitting analysis.

## Acknowledgements

The XAFS measurements were performed at the BL33XU of SPring-8 with the approval of the Japan Synchrotron Radiation Research Institute (JASRI) (Proposal No. 2014B7027 and 2015A7027). This work was financially supported by Japan Society for the Promotion of Science (JSPS) KAKENHI Grant-in-Aid for Young Scientists (B) No. 26810127.

## References

[1] (For example) Toshima N and Yonezawa T 1998 *New J. Chem.* **22** 1179; Guczi L 2005 *Catal. Today* **101** 53; Stamenkovic V R, Mun B S, Arenz M, Mayrhofer K J J, Lucas C A, Wang G, Ross P N and Markovic N M 2007 *Nat. Mater.* **6** 241; Ghosh R and Paria S 2012 *Chem. Rev.* **112** 2373; Graham G W, Jen H-W, Ezekoye O, Kudla R J, Chun W, Pan X Q and McCabe R W 2007 *Catal. Lett.* **116** 1; Croy J R, Mostafa S, Hickman L, Heinrich H and Cuenya B R 2008 *Appl. Catal. A* **350** 207; Ezekoye O K, Drews A R, Jen H-W, Kudia R J, McCabe R W, Sharma M, Howe J Y, Allard L F, Graham G W and Pan X Q 2011 *J. Catal.* **280** 125; Wiebenga M H, Kim C H, Schmieg S J, Oh S H, Brown D B, Kim D H, Lee J-H and Peden C H F 2012 *Catal. Today* **184** 197; Johns T R, Gaudet J R, Peterson E J, Miller J T, Stach E A, Kim C H, Balogh M P and Datye A K 2013 *ChemCatChem* **5** 2636; Haneda M, Suzuki K, Sasaki M, Hamada H and Ozawa M 2014 *Appl. Catal. A* **475** 109; Herreros J M, Gill S S, Lefort I, Tsolakis A, Millington P and Moss E 2014 *Appl. Catal. B* **147** 835
[2] (For example) Greeley J and Mavrikakis M 2004 *Nat. Mater.* **3** 810; Zhou S, Varughese B, Eichhom B, Jackson G and McIlwrath K 2005 *Angew. Chem. Int. Ed.* **44** 4539; Alayoglu S, Nilekar A U, Mavrikakis M and Eichhorn B 2008 *Nat. Mater.* **7** 333; Tang W and Henkelman G 2009 *J. Chem. Phys.* **130** 194504; Strasser P, Koh S, Anniyev T, Greeley J, More K, Yu C, Liu Z, Kaya S, Nordlund D, Ogasawara H, Toney M F and Nilsson A 2010 *Nat. Chem.* **2** 454; Sasaki K, Naohara H, Cai Y, Choi Y M, Liu P, Vukmirovic M B, Wang J X and Adzic R R 2010 *Angew. Chem. Int. Ed.* **49** 8602
[3] (For example) Alloyeau D, Ricolleau C, Mottet C, Oikawa T, Langlois C, Le Bouar Y, Braidy N and Loiseau A 2009 *Nat. Mater.* **8** 940; Antoniak C, Gruner M E, Spasova M, Trunova A V, Römer F M, Warland A, Krumme B, Fauth K, Sun S, Entel P, Farle M and Wende H 2011 *Nat. Commun.* **2** 528; Balasubramanian B, Skomski R, Li X, Valloppilly S R, Shield J E, Hadjipanayis G C and Sellmyer D J 2011 *Nano Lett.* **11** 1747
[4] Nagai Y, Dohmae K, Nishimura Y F, Kato H, Hirata H and Takahashi N 2013 *Phys. Chem. Chem. Phys.* **15** 8461
[5] Taguchi T, Ozawa T and Yashiro H 2005 *Phys. Scr.* **T115** 205
[6] Cook J W and Sayers D E 1981 *J. Appl. Phys.* **52** 5024
[7] Victoreen J A 1949 *J. Appl. Phys.* **20** 1141
[8] Ravel B and Newville M 2005 *J. Synchrotron Rad.* **12** 537
[9] Rehr J J, Kas J J, Vila F D, Prange M P and Jorissen K 2010 *Phys. Chem. Chem. Phys.* **12** 5503
[10] Moulijn J A, van Diepen A E and Kapteijn F 2001 *Appl. Catal. A* **212** 3

Journal of Physics: Conference Series **712** (2016) 012068 doi:10.1088/1742-6596/712/1/012068

# PDMS embedded Ag clusters: Coalescence and cluster-matrix interaction

**S Roese, D Engemann, S Hoffmann, K Latussek, C Sternemann and H Hövel**

Fakultät Physik / DELTA, Technische Universität Dortmund, 44221 Dortmund, Germany

E-mail: stefanie.roese@tu-dortmund.de

**Abstract.** Polydimethylsiloxane (PDMS) has proven to be a suitable embedding medium for silver clusters to prevent aggregation. In order to investigate the influence of the PDMS on the electronic and local atomic structure of the clusters the measurement of x-ray absorption near edge structure (XANES) spectra for different coverages of silver clusters in PDMS and calculations of corresponding XANES spectra have been performed. The coalescence process and the cluster-PDMS interaction were investigated with XANES.

## 1. Motivation

Due to their optical activity [1] and their anti-microbial behavior [2] silver nanoparticles are a recent field of research, which has lately become more and more application-oriented. For the preservation of the size-selective cluster properties the clusters have to be embedded into a medium to prevent them from aggregation. A suitable medium is PDMS, a polymeric substrate, which enables a penetration of clusters with velocities of about $10^3\,\mathrm{ms^{-1}}$ into the PDMS surface [3], where they stay separated up to a certain amount of cluster material. Separating clusters with ligands always has the consequence that the atoms at the cluster surface are in contact with the ligand material, possibly affecting the electronic and structural characteristics of the clusters. The influence of PDMS on the properties of the embedded Ag clusters can be examined with XANES.

## 2. Experimental

### *2.1. Cluster production and sample characterisation*

The Ag clusters with a diameter of $2.0 \pm 0.6$ nm (determined by transmission electron microscopy (TEM) measurements), corresponding to a mean cluster size of 250 atoms, are produced in a supersonic nozzle expansion with a source temperature $\mathrm{T} = 2300\,\mathrm{K}$, resulting in a cluster velocity of about $1.5 \cdot 10^3\ \mathrm{ms^{-1}}$ [4]. The source chamber is connected to the deposition chamber by a heated skimmer. The produced clusters hit a PDMS film of $1.2 \pm 0.2$ µm thickness prepared on a $SiO_2$ silica glass substrate, which is mounted on a manipulator inside the deposition chamber. The cluster coverage was measured with a quartz crystal balance and found to be between 0.09 and 2 ML (cluster-monolayer), where one ML corresponds to a closed packed ML of clusters. UV/Vis transmission measurements on the deposited Ag clusters have been performed in-situ and ex-situ for additional sample characterization [5].

### *2.2. XANES measurements*
XANES with fluorescence detection [6] is a suitable method to investigate the electronic and structural properties of clusters since a high signal to noise ratio is needed due to the low amount of cluster material in the PDMS sample. Changes in the oxidation state or sulfidation are detectable [7]. The experiments were performed at beamline ID26 [8] of the European Synchrotron Radiation Facility (ESRF), Grenoble, France. A Si(111) monochromator (intrinsic resolution of $1.4 \cdot 10^{-4}$) was used and the x-ray beam was focused on the sample with a size of approx. $0.1 \times 0.6\,\mathrm{mm}^2$ at an angle of incidence of 45° with respect to the beam axis. In order to measure XANES spectra in partial fluorescence yield (PFY) an x-ray emission spectrometer with four Ge(220) analyser-crystals was used. Thus, the signal to noise ratio in the XANES measurement was significantly improved and relatively low coverages of matrix embedded Ag clusters could be studied. The analyser energy was set to 3.1505 keV employing the Ge(220) Bragg reflection. Hence, we focused on the Ag $L_2$ edge (3.524 keV) by scanning the incident energy between 3.510 and 3.600 keV. A total energy resolution of 0.4 eV was obtained. Several spectra were recorded at each coverage. No radiation damage was observed. The single spectra were summed up and normalized to the intensity of the impinging beam. After background subtraction the spectra were normalized to their area below the entire spectrum.

## 3. FEFF calculations
The measured XANES spectra are compared to theoretical calculations performed with *FEFF9* [9]. The FEFF parameters were adjusted, so that the calculated fcc Ag bulk spectrum fits to the measurement of the Ag reference. A full multiple scattering radius of 8 Å, the Hedin-Lundqvist self-energy, a 4 eV shift of the entire spectrum and a reduction of the lifetime broadening by 1 eV was used. The construction of coordinate sets describing the atomic structure of the clusters was necessary to calculate the FEFF spectra of separated clusters. It is known, that small free silver metal clusters use to exhibit an icosahedral structure [10, 11]. Clusters with ligand shells (e.g. $Au_{55}$) often are identified with a cuboctahedral structure [12], which is basically a segment of the fcc crystal structure, even if recent studies showed that the geometries might be more complicated [13]. For the FEFF calculation shown here, corresponding to the mean cluster size of 250 atoms, cuboctrahedra with 147 or 309 Ag atoms with bulk next neighbor distances (2.89 Å) are chosen. These numbers are geometrically magic and due to their closed shells they are considered to be distinctly stable [14].

## 4. Results
### *4.1. XANES measurements of samples with different cluster coverages*
The measured XANES spectra for different Ag cluster coverages in PDMS and an Ag reference are shown in fig. 1. All error bars for ≥ 0.20 ML are within the line thickness. The UV/Vis transmission measurements revealed that for the lowest coverage separated cluster with a mean size of 250 atoms can be assumed [5]. Here, about half of the atoms are located at the cluster surface and are therefore in contact with the PDMS. The general structure of the spectra particularly for higher energies hardly varies, but for low coverages (0.09 ML and 0.20 ML) a significant change of the spectra in the energy range of 3.53 - 3.55 keV is observed. With increasing coverage

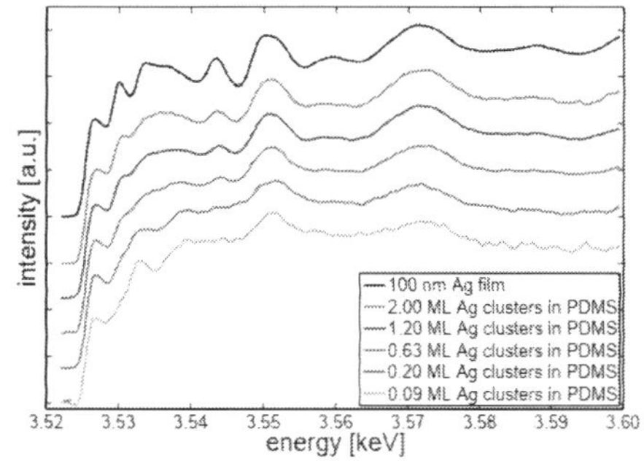


**Figure 1.** XANES spectra for different coverages of Ag clusters in PDMS and corresponding Ag reference.

the spectra approximate progressively the Ag bulk spectrum, which is indicative of cluster coalescence to larger silver islands in agreement with results of UV/Vis spectroscopy [5].

Fixed cross-over points in all spectra indicate that the spectra for cluster coverages of 0.20, 0.63, 1.20 and 2.00 ML might be modeled by a superposition of the absorption spectrum of the silver reference sample (100 nm Ag film) and the XANES signal measured for the separated clusters (lowest coverage of 0.09 ML).

Hence, we modeled the spectra of intermediate coverages by linear combinations of the XANES spectrum of the bulk measurement and that of the separated clusters via a least squares fit as shown in fig. 2. Obviously, the change of the XANES spectra due to cluster coalescence at higher coverage can be described by a decrease of the surface-to-volume ratio of clusters due to their larger mean size. However, superposition also leads to structural features, that do not occur in the measured spectra (see region around 3.53 - 3.55 keV). This leads to the assumption, that these features might be a fingerprint of the PDMS-cluster interaction, which only becomes visible at low coverages as in this case 50% of all atoms within the clusters are in contact with the PDMS. To distinguish between the effect of coalescence and cluster matrix interaction FEFF calculations will be discussed as follows.

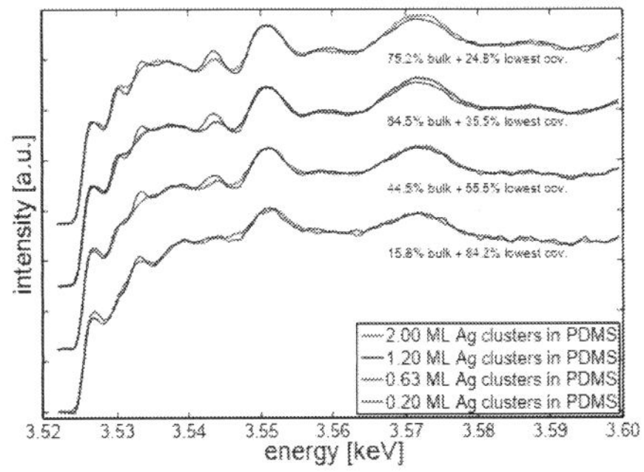


**Figure 2.** Linear superposition of the XANES spectra of a 100 nm Ag film and lowest coverage of Ag clusters in PDMS. The fractions were obtained by fitting.

*4.2. Comparison to FEFF calculation*

Fig. 3 shows the comparison of the measured spectra to the results of the FEFF calculations. The comparison of the measured 100 nm Ag film and the corresponding FEFF calculations shows a good overall agreement. This is valid also with respect to the FEFF calculation of the 309 atoms Ag cluster which is compared to an intermediate coverage (0.63 ML).

The smearing out of features for energies smaller than 3.55 keV by going from bulk to smaller cluster is clearly visible, even if at this coverage already particles larger than the mean size of 250 atoms are formed, as UV/Vis spectroscopy indicates [5].

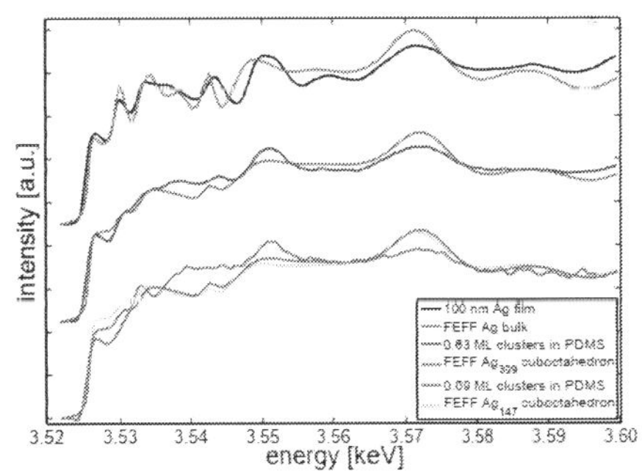


**Figure 3.** Comparison of XANES measurements of different cluster coverages to corresponding FEFF calculations.

However, the measured spectra for coverages below 0.63 ML are not resembled by the FEFF calculation neither of the 309 nor the 147 atoms cluster. This is indicative for a strong interaction between the surface atoms of the cluster and the embedding PDMS matrix, strongly affecting the local geometric and electronic structure. This interaction will have to be characterized in a next step using structural models of embedded clusters, i.e. considering the local environment due to the PDMS matrix.

## 5. Summary

In the present study the influence of PDMS on embedded silver clusters was examined with Ag $L_2$ edge XANES and complementary FEFF calculations. Silver clusters with a mean size of 2 nm deposited and embedded into PDMS were investigated with XANES. The XANES spectra of different cluster coverages vary significantly from each other and from the corresponding Ag bulk measurement in the near edge region. With a linear superposition of the XANES spectra of bulk and the lowest coverage it is possible to reproduce the structural features in the spectra of already aggregated clusters slightly larger than the original 2 nm size. Appropriate FEFF calculations for cuboctahedra consisting of 147 and 309 atoms also describe spectra for this intermediate coverage, but do not resemble some of the structural features, which can be found in the XANES spectrum for the lowest coverage, i.e. for separated matrix embedded Ag clusters. Therefore we draw the conclusion, that these features are tentatively attributed to the interaction between the PDMS and the surface atoms of the well separated clusters. In a next step this effect will be examined for different cluster embedding media.

## Acknowledgments

We are grateful to Kristina Kvashnina and Pieter Glatzel at ESRF for providing assistance in using beamline ID26 at the European Synchrotron Radiation Facility (ESRF), Grenoble, France for the Ag $L_2$ XANES measurements. Samples were precharacterized by XANES measurements at the Ag $L_3$ edge that were previously performed at BL8, DELTA, Dortmund, Germany. The help of Ralph Wagner is gratefully acknowledged. We thank ESRF and DELTA for providing synchrotron radiation and NRW Forschungsschule for financial support.

## References

[1] Kreibig U and Vollmer M 1995 *Optical Properties of Metal Clusters* (Berlin: Springer)
[2] Kim J S et al. 2007 *Nanomedicine* **3** 95-101
[3] Corbelli G, Ghisleri C, Marelli M, Milani P and Ravagnan L 2011 *Adv. Mater.* **23** 4504-08
[4] Hövel H, Fritz S, Hilger A, Kreibig U and Vollmer M 1993 *Phys. Rev.*B **48** 18178-188
[5] Hoffmann S 2012 *Dissertation* TU Dortmund
[6] Bunker G 2010 *Introduction to XAFS* (Cambridge: University Press)
[7] Behrens P 1992 *Solid State Commun.* **81** 235-239
[8] Gauthier C, Sole V A, Signorato R, Goulon J and Moguiline E 1999 *J. Synchrotron Radiat.* **6**, 164-166
[9] Rehr J, Kas J, Vila F D, Prange M P and Jorissen K 2010 *Phys. Chem. Chem. Phys.* **12**, 5503-13
[10] Häkkinen H and Moseler M 2004 *Phys. Rev. Lett.* **93** 093401
[11] Blom M N, Schooss D, Stairs J and Kappes M M 2006 *J. Chem. Phys.* **124** 244308
[12] Wilson N and Johnston R 2002 *Phys. Chem. Chem. Phys.* **4** 4168-71
[13] Walter M, Moseler M, Whetten R L and Häkkinen H 2011 *Chem. Sci.* **2** 1583-87
[14] de Heer W 1993 *Rev. Mod. Phys.* **65** 611-676

# Structural analysis of NiO nanocluster catalysts on $SiO_2$ by using XAFS measurements

**T Sasaki[1], N Ichikuni*[1], T Hara[1] and S Shimazu[1]**

Department of Applied Chemistry and Biotechnology, Graduate School of Engineering, Chiba University,
Yayoi-cho, Inage-ku, Chiba 263-8522, JAPAN

E-mail: ichikuni@faculty.chiba-u.jp

**Abstract**. NiO nanocluster catalysts were prepared on a $SiO_2$ support. Ni K-edge XAFS analysis revealed that the combined structure of NiO and Ni-silicate formed by using $[Ni(NH_3)_6]^{2+}$ as a precursor, whereas pure NiO formed by using Ni-acetylacetonate or Ni-nitrate as a precursor. The obtained NiO nanocluster on Ni-silicate showed the catalytic activity for the oxidative coupling of thiophenol under air atmosphere. The combined structure of NiO and Ni-silicate showed the higher activity than pure NiO nanocluster did.

## 1. Introduction

Support plays important roles to the heterogeneous catalysts. Bulk gold seemed not to have the catalytic activity, however, supported gold nanoclusters enabled to produce the active gold nanocluster catalyst. [1]. As a result, researches on the nanocluster catalysis have become popular. Additionally, we have reported that Ni nanocluster catalysts supported on $Al_2O_3$ showed a Ni cluster size dependence on the WGSR (Water Gas Shift Reaction) [2]. Using a support for the heterogeneous catalysts has following benefits: i) preservation of the cluster size in any condition which causes a wide range of applicability for many reactions, ii) the support effect which can modify the nanocluster catalysis.

In many cases, developing the heterogeneous catalyst uses noble metals such as Pt and Pd because of their higher activity, whereas base metals are not so common. However, developing materials that substitute for noble metals is an urgent issue in terms of a sustainable chemistry. Nickel is one of the elements for this purpose, since it is abundant and cheaper than noble metals and it also has the activities for many reactions such as the hydrogenation of unsaturated compounds [3], the steam reforming of methane or ethane [4, 5], and so on. Moreover, the supported NiO catalyst is active for various reactions such as oxidative dehydrogenation of ethane [6] and that of naphthalene [7]. However, NiO used for oxidative reactions is less effective compared to other catalysts in many instances, and so there are few publications reporting about NiO catalysts. Using the support appropriately can produce the NiO nanocluster and allow us to utilize NiO as a catalyst more effectively.

We have succeeded in a preparation of the supported NiO nanocluster by using an alkoxide stabilizes Ni colloid which was the precursor and in showing the remarkable particle size effect for the oxidative coupling of thiophenol [8]. Our previous report focused on only relation between the NiO particle size and the catalytic activity. In this paper, we attempted to prepare NiO catalysts more simply without using stabilizer and to clarify the origin of this reaction from the viewpoint of the support effect.

In this paper fine surface analysis by XAFS measurements found that Ni-silicate that formed during catalyst preparation had promoting effect for oxidative coupling of thiophenol. The understanding structure-activity relationship is one of the most important factors in terms of development of sustainable heterogeneous catalytic processes. The insight of promoting effect Ni-silicate can lead to the development of new catalytic materials.

## 2. Experimental

### *2.1. Catalyst preparation*

$SiO_2$ (Fuji Silysia Chemical Ltd., CARiACT P-10) supported catalysts were prepared by the impregnation method. NiO(amm)/$SiO_2$-*T*, NiO(ace)/$SiO_2$-*T* and NiO(nit)/$SiO_2$-*T* were obtained by using $[Ni(NH_3)_6](NO_3)_2$, $Ni(acac)_2$ and $Ni(NO_3)_2$ as a precursor, respectively, followed by heat treatment at *T* (K) in air. The heat treatment temperature was varied as 298, 573, 673 and 773 K. Ni loading regulated to 5 wt%. The abbreviation of amm, ace and nit means the precursor complex of ammine, acetylacetonate and nitrate, respectively.

In order to compare the support effect, NiO(amm)/A.C.-*T* prepared by using $[Ni(NH_3)_6]$ $(NO_3)_2$ and active carbon (A.C.) as a precursor and a support, respectively.

A Ni-silicate was prepared by a similar way that M. V. Sivaiah *et al.* [9] reported: Sodium silicate (40 mmol), $NiCl_2$ (30 mmol) and HCl (36 wt%, 20 mmol) were stirred with water (500 mL) in a beaker. This mixture was treated by the hydrothermal synthesis in a Teflon sealed autoclave at 473 K for 96 hours. After the product was cooled to room temperature, it was washed with water and dried overnight. The chemical formula of the Ni silicate was $Si_4Ni_3O_{11}$.

### *2.2. Characterization and catalytic reaction*

Ni K-edge XAFS data were collected at BL-9C of KEK-PF (Proposal No. 2014G575) with Si(111) double crystal monochromator in a transmission mode at room temperature.

Oxidative coupling of thiophenol under air atmosphere was carried out as follows: Thiophenol (5.0 mmol), biphenyl (0.63 mmol) and acetonitrile (25 mL) in Schlenk tube were stirred with the catalyst (0.029 g) in a water bath at 303 K for 0.5 h. The molar ratio of thiophenol/catalyst was 200 based on the Ni amount. The yield was determined by GC (Shimadzu, GC-2025) with a capillary column (Hewlett Packard, Ultra 2, 25 m x 0.20 mm, 0.33 μm) and an FID detector.

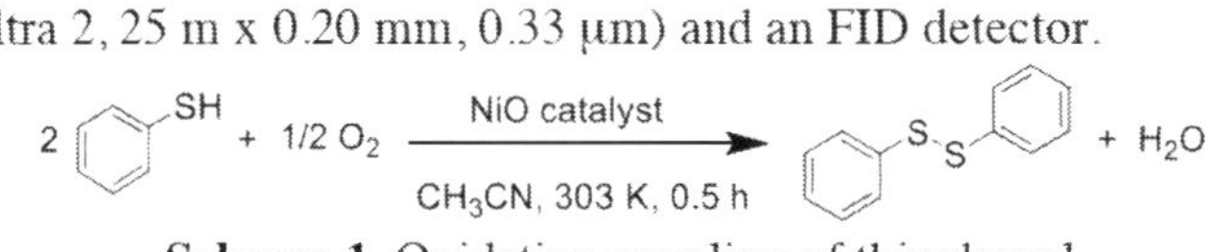


**Scheme 1.** Oxidative coupling of thiophenol

## 3. Results and discussion

Figure 1 shows the Ni K-edge XANES of NiO catalysts and reference compounds. The valence and chemical species of the nickel species on the supported NiO catalysts were determined from these spectra.

XANES shapes of all catalysts were similar to that of bulk NiO (dotted line a), which indicates that nickel species on these catalysts consisted of NiO. However, a presence of other nickel species cannot be excluded. Indeed, there are two shoulders at 8346 eV (α-peak) and 8350 eV (γ-peak) on three types of NiO(amm)/$SiO_2$, besides β-peak which is a characteristic of NiO, and Ni-silicate that is a reference compound also showed two weak peaks at same energy (α- and γ-peak). Furthermore, formation of Ni-silicate was reported when $[Ni(NH_3)_6]^{2+}$ was supported on $SiO_2$ [10, 11]. Therefore, two weak peaks of three kinds of NiO(amm)/$SiO_2$ were assigned to the Ni-silicate, and it can be concluded that Ni-silicate formed with NiO on our NiO(amm)/$SiO_2$ catalysts. Additionally, the similarity of XANES shapes of three types of NiO(amm)/$SiO_2$ indicates that Ni-silicate formed with NiO on all of these and the ratio of these nickel species was mostly the same.

Figure 2 shows the FT of Ni K-edge EXAFS oscillations for NiO catalysts and reference compounds and Table 1 shows the model parameters of reference compounds and the curve fitting results of the nearest Ni-(O)-Ni.

NiO particles on NiO(nit)/$SiO_2$-673 were as large as bulk NiO since the coordination number (*CN*) of Ni-(O)-Ni was 12. On the other hand, *CN*s of the other catalysts were much smaller than that of bulk NiO, which indicates that NiO particles on these catalysts were very small, namely, these were NiO nanoclusters. Also, the peaks of the nearest Ni-(O)-Ni of three types of NiO(amm)/$SiO_2$ were at the same position with that of the Ni-silicate. This fact also indicates that Ni-silicate was on NiO(amm)/$SiO_2$ catalysts. Moreover, the *CN*s of NiO(amm)/$SiO_2$ catalysts were larger than that of Ni-silicate ($CN$ = 6) means that not only Ni-silicate but also NiO was on these catalyst. Additionally, these *CN*s increased with the treatment temperature and it can be said that NiO particles grew with the temperature elevation.

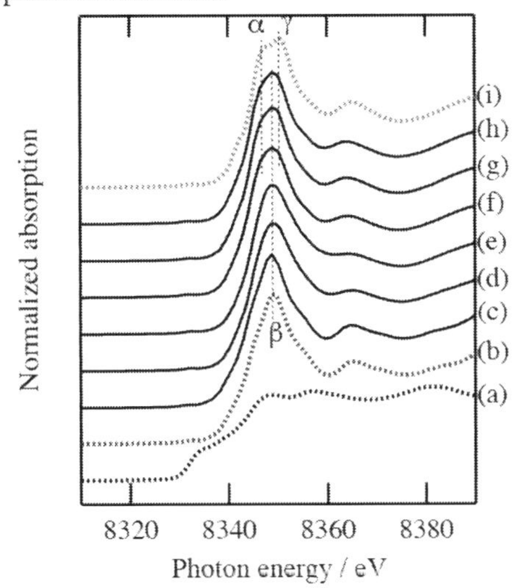


**Figure 1.** Ni K-edge XANES of NiO catalysts and reference compounds: (a) Ni foil, (b) bulk NiO, (c) NiO(nit)/$SiO_2$-673, (d) NiO(ace)/$SiO_2$-673, (e) NiO(amm)/A.C.-573, (f) NiO(amm)/$SiO_2$-773, (g) NiO(amm)/$SiO_2$-573, (h) NiO(amm)/$SiO_2$-298, (i) bulk Ni-silicate.

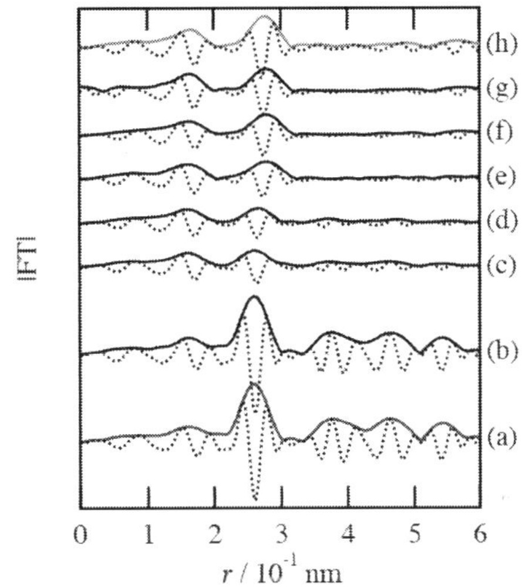


**Figure 2.** FT of Ni K-edge oscillations for NiO catalysts and reference compounds: (a) bulk NiO, (b) NiO(nit)/$SiO_2$-673, (c) NiO(ace)/$SiO_2$-673, (d) NiO(amm)/A.C.-573, (e) NiO(amm)/$SiO_2$-773, (f) NiO(amm)/$SiO_2$-573, (g) NiO(amm)/$SiO_2$-298, (h) bulk Ni-silicate. Imaginary part (dotted lines) and absolute value (solid lines).

**Table 1.** Model parameters of reference compounds and curve fitting results of Ni-(O)-Ni coordination for NiO catalysts and yield of oxidative coupling of thiophenol

| Catalyst | FR[a] (nm) | *CN*[b] | *r*[c] (nm) | d*E*[d] (eV) | *DW*[e] (nm) | *R*[f] (%) | Yield[g] (%) |
|---|---|---|---|---|---|---|---|
| **Ni-silicate (model)**[h] | - | 6 | 0.307 | 0 | 0.006 | - | 0.2 |
| **NiO(amm)/$SiO_2$-298** | 0.233-0.322 | 7.3±1.2 | 0.311 | 4.54 | 0.00730 | 0.012 | 39 |
| **NiO(amm)/$SiO_2$-573** | 0.233-0.322 | 7.4±1.2 | 0.311 | 4.76 | 0.00790 | 0.026 | 22 |
| **NiO(amm)/$SiO_2$-773** | 0.233-0.322 | 8.0±1.3 | 0.310 | 5.33 | 0.00900 | 0.171 | 4.4 |
| **NiO(amm)/A.C.-573** | 0.215-0.307 | 5.0±0.8 | 0.300 | 1.17 | 0.00830 | 0.132 | 2.1 |
| **NiO(ace)/$SiO_2$-673** | 0.212-0.328 | 7.0±1.2 | 0.298 | -1.87 | 0.00950 | 1.449 | 0.2 |
| **NiO(nit)/$SiO_2$-673** | 0.227-0.301 | 12.3±2.1 | 0.295 | 0.459 | 0.00600 | 0.090 | 0.0 |
| **bulk NiO (model)** | - | 12 | 0.295 | 0 | 0.006 | - | 0.0 |

[a] Filtering range, [b] coordination number, [c] bond distance, [d] difference between model compound and experimental threshold energies, [e] Debye-Waller factor, [f] R-factor, [g] diphenyl disulfide yield at 303 K for 0.5 h, [h] reference [11]. FT range: 30-140 $nm^{-1}$.

Table 2 shows the yield of diphenyldisulfide (catalytic activity) and *CN* of Ni-(O)-Ni. Three types of NiO(amm)/$SiO_2$ and NiO(amm)/A.C.-573 showed the catalytic activity for the oxidative coupling of thiophenol and the other catalysts did not. In case of NiO(amm)/A.C.-573, it is considered that the very small size of NiO ($CN = 5.0$) showed the catalytic activity, although this activity was quite low. On the other hand, NiO(nit)/$SiO_2$-673 and NiO(ace)/$SiO_2$-673 did not have the catalytic activity probably due to NiO particles were not small enough. One of the most interesting things is that three types of NiO(amm)/$SiO_2$ showed higher activity than NiO(amm)/A.C.-573 and NiO(ace)/$SiO_2$-673, in spite of the larger *CN*. In light of the discussion of XAFS measurements, it is reasonable to suppose that the coexistence of Ni-silicate with NiO is the key factor in order to produce the activity for this reaction.

The things predicted from these results are that Ni-silicate that has the OH group layer contributed the coverage of SH bond of thiophenol and promote the oxidative coupling of thiophenol or that formation of Ni-silicate caused an unique structural change of NiO particles, which lead to high catalytic activity. The active species for this reaction was the NiO nanocluster and its size affected the catalytic activity.

## Conclusion

The supported NiO catalysts were prepared by using three kinds of nickel species ($[Ni(NH_3)_6]^{2+}$, $Ni(acac)_2$ or $Ni(NO_3)_2$) as a precursor.

Obtained catalysts were used for the oxidative coupling of thiophenol and three types of NiO(amm)/$SiO_2$ and NiO(amm)/A.C.-573 showed the catalytic activity for this reaction while the other catalysts did not. Through the results of the XAFS measurements and the catalytic reaction, we concluded that very small NiO particles (NiO nanoclusters) were needed to cause the catalytic activity and the Ni-silicate promoted it.

## Acknowledgement

This study was supported by a Grant-in-Aid for Scientific Research from JSPS (23560928, 26420784).

## References

[1] Haruta M, Kobayashi T, Sano H and Yamada N 1987 *Chem. Lett.* **16** 405-408
[2] Kitagawa H, Ichikuni N, Hara T and Shimazu S 2012 *e-J. Surf. Sci. Nanotech.* **10** 648-650; Kitagawa H, Ichikuni N, Okuno H, Hara T and Shimazu S 2014 *Appl. Catal. A* **478** 66-70s
[3] Franquin D, Monteverdi S, Molina S, Bettahar M M and Fort Y 1999 *J. Mater. Sci.* **34** 4481-4488
[4] Liberatori J W C, Riberio R U, Zanchet D, Noronha F B and Bueno J M C 2007 *Appl. Catal. A* **327** 197-204
[5] Parizotto N V, Rocha K O, Damyanova S, Passos F B, Zanchet D, Marques C M P and Bueno J M C 2007 *Appl. Catal. A* **330** 12-22
[6] Nakamura K, Miyake T, Konishi T and Suzuki T 2006 *J. Mol. Catal. A* **260** 144-151
[7] Kirumakki S R, Shpeizer B G, Sagar G V, Chary K V R and Clearfield A 2006 *J. Catal.* **242** 319-331
[8] Ichikuni N, Tsuchida O, Naganuma J, Hara T, Tsunoyama H, Tsukuda T and Shimazu S 2012 *Trans. Mat. Res. Soc. Japan* **37** 177-180
[9] Sivaiah M V, Petit S, Beaufort M F, Eyidi D, Barrault J, Batiot-Dupeyrat C and Valange S 2011 *Micro. Meso. Mater.* **140** 69-80
[10] Blackmond D G and Ko E I 1984 *Appl. Catal.* **13** 49-68
[11] Che M, Cheng Z X and Louis C 1995 *J. Am. Chem. Soc.* **117** 2008-2018

# Automated analysis of XANES: A feasibility study of Au reference compounds

**S-Y Chang[1,2], L B Molleta[1], S G Booth[1], A Uehara[1,4], J F W Mosselmans[2], K Ignatyev[2], R A W Dryfe[1], S L M Schroeder[1-3]***
[1]School of Chemistry and School of Chemical Engineering and Analytical Science, University of Manchester, M13 9PL Manchester, UK.
[2]Diamond Light Source Ltd, Didcot, OX11 0DE Oxfordshire, UK.
[3]School of Chemical and Process Engineering, University of Leeds, Leeds LS2 9JT, UK.
[4]Division of Nuclear Engineering Science, Research Reactor Institute, Kyoto University, Asashironishi, Kumatori, Osaka, 590-0494, Japan.

*Email: s.l.m.schroeder@leeds.ac.uk

**Abstract** With the advent of high-throughput and imaging core level spectroscopies (including X-ray absorption spectroscopy, XAS, as well as electron energy loss spectroscopy, EELS), automated data processing, visualisation and analytics will become a necessity. As a first step towards these objectives we examined the possibilities and limitations of a simple automated XANES peak fitting procedure written in MATLAB, for the parametrisation of XANES features, including ionisation potentials as well as the energies and intensities of electronic transitions. Using a series of Au $L_3$-edge XANES reference spectra we show that most of the relevant information can be captured through a small number of rules applied to constrain the fits. Uncertainty in this strategy arises mostly when the ionisation potential (IP) overlaps with weak electronic transitions or features in the continuum beyond the IP, which can result in ambiguity through multiple equally good fits.

## 1. Introduction

Core level spectroscopies such as electron energy loss spectroscopy (EELS) and X-ray absorption spectroscopy (XAS) provide element specific chemical information. The robustness of these spectroscopies has enabled their use across a broad range of applications, including for example oil, gas, nuclear fuel and waste processing research, materials science, electrochemistry, crystallisation, biological and geochemical speciation as well as conservation and restoration of heritage objects. A current trend is to develop high-throughput functionality, which will enable exploration of more applications for these spectroscopies. Some imaging [3-5] and quick scanning EXAFS facilities [6-8] are already providing hundreds and thousands of XANES spectra within minutes to hours. As such, it would be useful to have access to algorithms that permit an initial assessment of such datasets by automated parametrisation for subsequent classification. This would also pave the way to a fuller interpretation of chemical contrast in XANES images.

The fine-structure features in the near-edge region of EELS and XAS (inclusive of XANES and NEXAFS) arise from electronic transitions to bound states and thus contain rich information about the local structure and the chemical state of the probed element. XANES spectra of transition metals are often used for fingerprinting and to determine the oxidation state and coordination geometry (e.g., octahedral vs. tetrahedral) from known spectral characteristics [9]. We have already shown that through deconvolution (peak fitting) of spectral features and subsequent assignment of bound state transitions to these features the interactions of small organic molecules such as protonation, intermolecular solvent-solute hydrogen bonding and solute-solute interaction can be studied in detail [10, 11]. In fact, peak fitting is the only way to determine energies, intensities and widths of the most prominent absorption features.

Functional peak fitting programs for XANES analysis are currently available [12-16], but to enable automated fitting of large numbers of spectra reliable principles allowing both flexibility and robustness are required. The increasing need to process larger numbers of spectra has led us to explore the use of an automated peak fitting procedure based on a simple heuristic approach to obtain the energetic positions and intensities of XANES absorption features by use of a MATLAB script. It builds on the use of a script we recently reported as part of a time-resolved quick EXAFS study of Pd nucleation [17]. In this work we discuss the possibilities and limitations of such an approach by reference to a set of Au $L_3$-edge XANES spectra from various reference compounds. We would like to stress that the aim of this work was not to provide a definite general XANES fitting tool, but to explore the possibilities and limitations of using a few heuristic rules to constrain an automated fitting procedure.

## 2. The Peak Fitting Algorithm

We focused on the parametrisation of the XANES region based on the determination of energies, intensities and widths of the most prominent features. For this we chose to simulate any absorption feature below the ionisation potential (IP) with pseudo-Voigt functions [18, 19], while the absorption edge jumps/IPs were modelled using error functions [15]. This represented a pragmatic and heuristic choice based on our finding that this combination fitted most reference spectra well. In the energy region from just below to slightly above the IP, absorption features due to electronic transitions overlap with EXAFS and multiple scattering features such as shape resonances. Fine structure beyond the IP and in the continuum was modelled using Gaussian peak functions, but we stress that this was done only to remove the EXAFS background which extends into the XANES region, to increase the reliability of the fits of interest in the XANES region, and not to analyse or parametrise the EXAFS part of the spectrum. These Gaussian components of the fits have no physical meaning.

To optimise the fit a MATLAB default trust-region-reflective algorithm was applied, as this method allows boundary limits to be specified. The variables to be optimised during fitting were the number of absorption features to model and then for each fitted function the full width at half maximum (FWHM), the centroid energy, the area and for pseudo-Voigt functions, the Lorentzian/Gaussian ratio η. Using an arctangent instead of an error function resulted in slightly different but still consistent results for IP positions and the absorption features related to transitions to bound states (not shown), albeit with overall worse fit quality. Energy boundaries and constraints were imposed to ensure that the automated fitting results retained physical meaning.

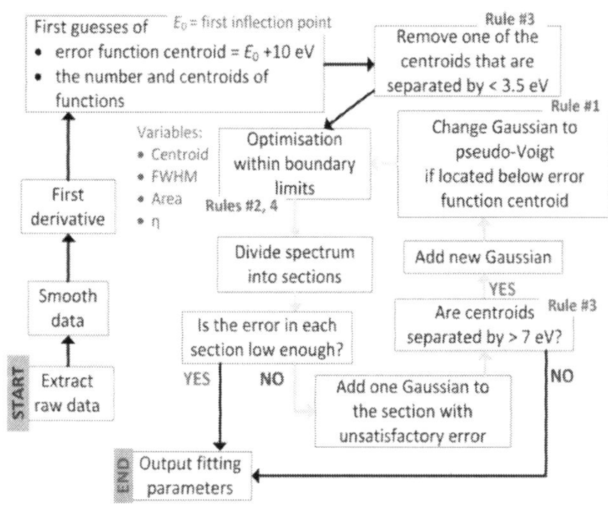


**Figure 1.** The peak fitting algorithm

In other words, by constraining the fits with a few heuristic 'rules', the parameter landscape available for an automated fit was reduced considerably, reflecting prior knowledge about relationships between XANES absorption features, chemical state and local structure.

To start with, we assumed that any absorption maximum located at a lower energy than the step function modelling the IP must correspond to the excitation of a core electron to an unoccupied bound electronic state (Rule #1). In order to determine the FWHM for the error function and the pseudo Voigt functions an initial guess and lower boundary limit was obtained from tabulated natural line width data [20]. These are independent of the electronic structure in the valence region (Rule #2). By way of example, in the samples examined here the natural line width for Au $L_3$ is 5.41 eV therefore the allowable range in the subsequent fitting error function was constrained from 5.41 eV to 10 eV and for the pseudo-Voigt functions from 5.41 eV to 15.41 eV. Rule #3 imposes that electronic transitions must be separated from each other by a threshold value, to prevent overfitting, e.g. 3.5 eV for the initial guess (grey arrows, figure 1) and 7 eV for the subsequent guess (blue arrows, figure 1) gave good results for all Au $L_3$-edge spectra. Aside from the FWHM limit in Rule #2, the height of the IP error functions was fixed at 0.8 for all the Au $L_3$-edge spectra to enable the use of the non-negative Gaussian functions to fit oscillatory fine structure in the continuum range (Rule #4). For the set of data discussed here the FWHM of the electronic transition features tends to increase with photon energy due to increased lifetime broadening [21]. If necessary in a specific data set this trend could also be imposed as Rule #5.

The resulting algorithm is summarised in figure 1. In a first optimisation loop, the number of peaks to be fitted and their centroids are guessed from the first derivative of the smoothed experimental spectrum. A peak centroid is then assigned to each maximum in the spectrum. The initial guess for the IP is the edge inflection point of the experimental spectrum ($E_0$) + 10 eV. In the subsequent minimisation loop (blue arrows in figure 1) the difference between the experimental spectrum and the fitted line shapes is examined, and Gaussians are added where the deviation between experiment and fit is larger than a threshold value, while maintaining Rules #1 and #3.

## 3. Results and Discussion

The peak fitting algorithm was tested on a series of Au $L_3$-edge reference spectra. Fitting each spectrum using the code took at most a few seconds on a computer with a first generation Intel i3 processor. For comparison, we also generated manual (non-automatic) best fits for all spectra using Athena in the Demeter software package [13]. The full experimental methodology, results and the MATLAB code are given in the supporting information (SI).

For gold foil, $[AuCl_2]^-$ and $Au_2S$, good agreement was achieved between automatic and manual best fits (figure 2) producing similar error and pseudo-Voigt function parameters in the XANES region. For $[Au(CN)_2]^-$, $[Au(CN)_4]^-$ and $[AuCl_4]^-$, similarly good agreement in the XANES region can be achieved, although with more than one way to fit the data. For $[Au(CN)_4]^-$, both the automatic fit with three pseudo-Voigt functions and the manual fit with two pseudo-Voigt functions seem equally reasonable. For $[Au(CN)_2]^-$ and $[AuCl_4]^-$, the manual best fits model the regions around 11,930 and 11,926 eV, respectively, better because a weak secondary pseudo-Voigt function improves the fit. The contribution of this transition was too low to be picked up by the automated algorithm. The parameters determined for the strong white line and the IP in these examples are still in good enough agreement to be of analytical value. A similar situation was encountered for $[AuBr_4]^-$ (SI B). A similar but more severe case is $[Au(OH)_4]^-$, which has a secondary and even what appears to be weak tertiary contribution to the white line just below the IP, which are not correctly identified by the automated algorithm.

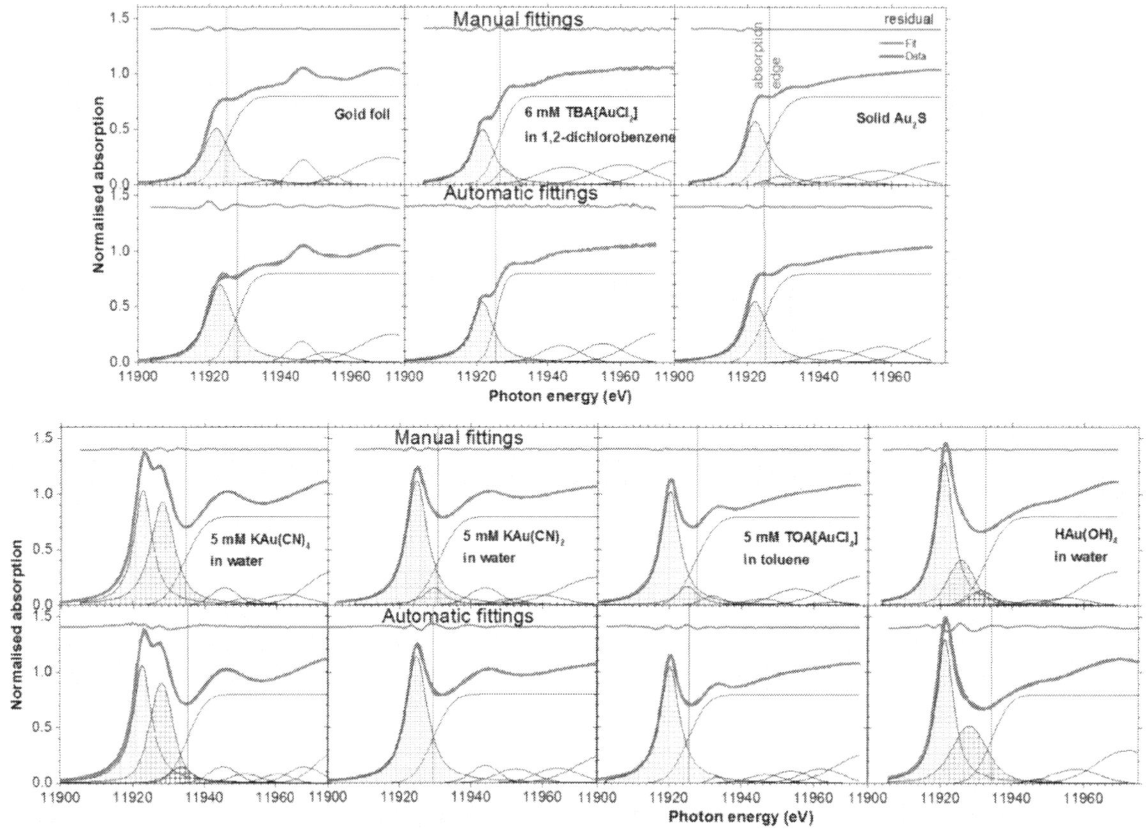


**Figure 2.** The automatic fits in comparison with the manual fits. The pseudo-Voigt functions were shaded in different colours to contrast between the first (green), second (orange) and third (blue) functions. $Au_2S$, $[AuCl_2]^-$ and $[Au(OH)_4]^-$ spectra were obtained from references [1, 2].

## 4. Conclusions

The heuristic automated peak fitting procedure parametrises most of the essential electronic structure and chemical state information at the Au $L_3$-edge with minimal user input. For the test set of Au $L_3$-edge XANES spectra the fits were excellent for most of the compounds while for others some further refinement is required through inclusion of secondary white line features. The method has the flexibility to be applied to different edges through the use of the necessary edge-specific boundary limits. We plan to implement the method in a more open language such as Python. We believe that such simple automation procedures will already be useful for the initial stages of data analysis for large bodies of spectra, e.g. filtering and classification of datasets or interpreting chemical contrast in imaging data more quantitatively. Through further advancement of this technique the methodology could ultimately also be linked with molecular modelling and simulation for deeper understanding of the variations in peak positions and intensities in the XANES region.

## Acknowledgements

All data supporting this study are provided either in the results section of this paper or in the supplementary information accompanying it. The authors acknowledge the Diamond Light Source beamtime on I18 awarded under proposal SP-8861. RAWD and SLMS acknowledge support through an EPSRC-NSF "Materials World Network" grant (EP/H047786/1). SLMS thanks EPSRC for financial support under the critical mass grant EP/I013563/1. AU thanks Kyoto University foundation for supporting a sabbatical visit to UoM. SYC gratefully acknowledges UoM, and Mr and Mrs Clews for the Robert Clews Presidential sponsorship, as well as the Royal Society of Chemistry and XAFS16 for conference travel grants.

**References**


[1] Lengke M F, Ravel B, Fleet M E, Wanger G, Gordon R A and Southam G 2007 *Can. J. Chem.* **85** 651-59

[2] Chang S-Y, Uehara A, Booth S G, Ignatyev K, Mosselmans J F W, Dryfe R A W and Schroeder S L M 2015 *RSC Adv.* **5** 6912-18

[3] Scott R A, Shokes J E, Cosper N J, Jenney F E and Adams M W W 2005 *J. Synchrotron Rad.* **12** 19-22

[4] Fischer D, Jaye C, Scammon K, Sobol P and Principe E 2010 *Microsc. Microanal.* **16** 380-81

[5] Cueva P, Hovden R, Mundy J A, Xin H L and Muller D A 2012 *Microsc. Microanal.* **18** 667-75

[6] Baudelet F, Kong Q, Nataf L, Cafun J D, Congeduti A, Monza A, Chagnot S and Itié J P 2011 *High Pressure Res.* **31** 136-39

[7] Frahm R, Nachtegaal M, Stötzel J, Harfouche M, van Bokhoven J A and Grunwaldt J-D 2010 The Dedicated QEXAFS Facility at the SLS: Performance and Scientific Opportunities *10th International Conference on Radiation Instrumentation*: *AIP Conf. Proc.* (*Melbourne, Australia, 27 September–2 October 2009)* ed R. Garrett, I. Gentle, K. Nugent, and S. Wilkins pp 251-55

[8] Belin S, Briois V, Traverse A, Idir M, Moreno T and Ribbens M 2005 *Phys. Scripta.* **2005** 980

[9] Yamamoto T 2008 *X-Ray Spectrom.* **37** 572-84

[10] Thomason M J, Seabourne C R, Sattelle B M, Hembury G A, Stevens J S, Scott A J, Aziz E F and Schroeder S L M 2015 *Faraday Discuss.* **179** 269-89

[11] Stevens J S, Seabourne C R, Jaye C, Fischer D A, Scott A J and Schroeder S L M 2014 *J. Phys. Chem. B.* **118** 12121-29

[12] Wojdyr M 2010 *J. Appl. Crystallogr.* **43** 1126-28

[13] Ravel B and Newville M 2005 *J. Synchrotron Rad.* **12** 537-41

[14] Newville M 2013 Larch: An Analysis Package for XAFS and Related Spectroscopies *15th International Conference on X-ray Absorption Fine Structure (XAFS15)*: *J Phys. Conf. Ser.* (*Beijing, China, 22-28 July 2012)* ed Z. Wu pp 012007

[15] Boubnov A, Lichtenberg H, Mangold S and Grunwaldt J-D 2015 *J. Synchrotron Rad.* **22** 410-26

[16] Delgado-Jaime M U, Mewis C P and Kennepohl P 2010 *J. Synchrotron Rad.* **17** 132-37

[17] Chang S-Y, Gründer Y, Molleta L B, Uehara A, Mosselmans J F W, Cibin G, Dryfe R A W and Schroeder S L M 2016 *CrystEngComm.* **18** 674-82

[18] Amemiya K, Kitajima Y, Yonamoto Y, Terada S, Tsukabayashi H, Yokoyama T and Ohta T 1999 *Phys. Rev. B.* **59** 2307-12

[19] Stohr J 2003 *NEXAFS Spectroscopy* Springer series in surface sciences vol. 25, ed R. Gomer (Heidelberg: Springer-Verlag)

[20] Krause M O and Oliver J H 1979 *J. Phys. Chem. Ref. Data.* **8** 329-38

[21] Bunker G 2010 *Introduction to XAFS: A Practical Guide to X-ray Absorption Fine Structure Spectroscopy* (New York: Cambridge University Press)

# HERFD-XANES and XES as complementary *operando* tools for monitoring the structure of Cu-based zeolite catalysts during $NO_x$-removal by ammonia SCR

**T Günter[1], D E Doronkin[1,2], H W P Carvalho[1,3], M Casapu[1] and J-D Grunwaldt[1,2]**

[1] Institute for Chemical Technology and Polymer Chemistry, Karlsruhe Institute of Technology (KIT), Kaiserstr. 12, D-76131 Karlsruhe, Germany
[2] Institute of Catalysis Research and Technology, KIT, Hermann-von-Helmholtz-Platz 1, D-76344 Eggenstein-Leopoldshafen, Germany
[3] present address: Centro de Energia Nuclear na Agricultura, Av. Centenário, nº 303, 13400-970, Piracicaba - São Paulo, Brasil

E-mail: grunwaldt@kit.edu

**Abstract.** In this article, we demonstrate the potential of hard X-ray techniques to characterize catalysts under working conditions. *Operando* high energy resolution fluorescence detected (HERFD) XANES and valence to core (vtc) X-ray emission spectroscopy (XES) have been used in a spatially-resolved manner to study Cu-zeolite catalysts during the standard-SCR reaction and related model conditions. The results show a gradient in Cu oxidation state and coordination along the catalyst bed as the reactants are consumed. Vtc-XES gives complementary information on the direct adsorption of ammonia at the Cu sites. The structural information on the catalyst shows the suitability of X-ray techniques to understand catalytic reactions and to facilitate catalyst optimization.

## 1. Introduction

Over the past years, X-ray absorption spectroscopy (XAS) has been successfully applied to elucidate the structure of metal sites in a variety of solid compounds and liquid phase complexes. The application of high energy resolution fluorescence detected - X-ray Absorption Near Edge Structure (HERFD-XANES) spectroscopy offers the possibility to unravel additional spectral features and thus more information on the sample compared to conventional XAS [1,2]. This high resolution is achieved by measuring the fluorescence signal of the Kß-line (3p → 1s). The 3p orbitals interact more with the valence 3d and 4p orbitals and therefore show a stronger influence on the environment than the less pronounced mixing of 2p and 3d orbitals, which results in the Kα-line [3]. By selecting a Kß-line as a single fluorescence channel, a higher energy resolution is achieved as the core-hole life-time during the X-ray absorption process can be circumvented [4].

An additional technique to characterize metal sites is X-ray emission spectroscopy (XES), where electrons from an occupied higher orbital are transferred to an empty lower orbital. Information on occupied states is therefore gained by XES, whereas XANES is a long established complementary technique that provides information on the electron transfer to unoccupied states [5]. Particularly the transfer of electrons from valence orbitals provides new insights into the interaction between metal and its ligands. The weak transitions from the valence level to the 1s orbital are the $K\beta_{2,5}$-line and its satellite feature Kß''. They are strongly sensitive to the coordination number and geometry. As the ligand 2s orbitals interact with the metal valence orbitals, their energy affects the position of the Kß''-line and offers the possibility to distinguish between different ligands with close atomic numbers like oxygen or nitrogen atoms [6], which is otherwise not possible using conventional EXAFS.

So far, these techniques have been applied to study mainly model compounds. Particularly for catalytic reactions, the structure of the catalyst is of interest not only *ex situ* but especially *in situ* or *operando*, i.e. while measuring the catalytic performance. The interaction of reactants with the catalyst can give more insights into the reaction pathways. For the removal of nitrogen oxides from exhaust gas, the selective catalytic reduction (SCR) with ammonia is commonly used. Giordanino et al. [7] recorded HERFD-XANES and XES *in situ* while dosing ammonia over the catalyst Cu-SSZ-13. The interaction with ammonia led to the reduction of $Cu^{2+}$ to linearly coordinated $Cu^{+}$. The adsorption of ammonia on the Cu-sites was additionally supported by the XES data, where a shift of the Kß''-line proves nitrogen atoms (from ammonia) coordinated to Cu. These findings show the valuable application of these spectroscopic methods under *in situ* conditions, giving hints to understand reaction mechanisms.

The logical consequence is the combination of *operando* conditions with the spectroscopic tools. Ideally, the catalyst should be studied during operation. As this is usually not possible for commercial processes, a setup and reaction conditions which reflect as close as possible the real process are used. For studying heterogeneous catalysts, the microreactor setup based on quartz capillaries has shown a good comparability with real fixed bed reactors [8]. The microreactor can be heated to relevant reaction temperatures and various gas mixtures can be dosed and monitored by conventional gas analyzers (e.g. MS, IR etc.). Depending on the X-ray beam size, XAS and XES spectra of the catalyst can be recorded in a spatially-resolved manner, allowing observation of gradients along the catalyst bed, which can give essential information to understand the reaction mechanism [9]. This is particularly important at high temperatures where the reaction occurs mainly at the inlet of the catalyst bed, as demonstrated for the SCR process by Luo et al. [10].

Here, we show the potential of spatially-resolved HERFD-XANES and XES to study the SCR-mechanism on Cu-SSZ-13 under operating conditions.

## 2. Experimental section

The zeolite SSZ-13 was prepared in a similar manner as described by Deka [11] and Zones [12]. It was ion-exchanged with $Cu(OAc)_2$ to 1.2 wt.% Cu-SSZ-13 [13]. HERFD-XANES and XES were measured in fluorescence mode with a beam size of 0.2 x 1 mm at the ID26 beamline at ESRF (European Synchrotron Radiation Facility). The HERFD-XANES was recorded by scanning the incident beam energy with the monochromator (Si(111) double crystal monochromator), while the energy of the fluorescence detector was kept fixed at the maximum of the $K\beta_{1,3}$-emission line (8903.6 eV). In turn, for the XES measurements, the incident beam energy was kept fixed with the monochromator above the K edge at 9100 eV and the fluorescence detector scanned the photons emitted from the sample. A sieved fraction of the catalyst sample (100 - 200 µm, 8 - 10 mm bed length) was loaded into a 1 mm quartz capillary (20 µm wall thickness), which served as a plug-flow reactor, and mounted above a hot air gas blower (FMB Oxford). Gases were dosed with individual mass flow controllers to obtain a gas mixture of 0-1000 ppm NO, 0-1000 ppm $NH_3$, 10% $O_2$ and ~1.5% $H_2O$ in He and analysed by an FTIR spectrometer (MKS 2030). Further experimental details are given elsewhere [13].

## 3. Results and discussion

In a first step, the local structure is typically analysed by a comparison of its spectrum with reference spectra of well-characterised materials, an approach that is used in a similar manner for catalysts under reaction conditions. For Cu-SSZ-13, there are several XAS and XRD studies combined with computational modelling that suggest Cu in positions close to six or eight membered rings in the zeolite pores [14,15]. Under reaction conditions though, reactive gases are expected to adsorb at the Cu sites and therefore change coordination number, geometry and oxidation state of these sites. This makes it necessary to additionally study the influence of each gas separately and in combination with other gases.

Three reference spectra are given in Figure 1, showing the effect on the Cu sites of oxygen, ammonia and their combination. As discussed in more detail earlier [13], Cu is fully oxidized under oxygen flow, but becomes reduced in the presence of ammonia, as the decreased pre-edge feature (specific for $Cu^{2+}$, $3d^9$) confirms. The intense feature in the rising edge at 8982.7 eV is related to $Cu^+$ in linear coordination [7] and is even more pronounced without oxygen. These observations under model gas compositions already show, that the catalytic centres are affected by the single reactants and change their oxidation state and local structure, confirming the *in situ* results by Giordanino et al. [7]. Under reaction conditions, where gases adsorb, react and desorb from the active centres, the reaction progresses along the catalyst bed, resulting in concentration gradients and thus catalytic sites in different states. As essential information on the reaction kinetics can be derived by monitoring the gas concentrations along the reactor [16], also monitoring the catalyst structural dynamics at different positions is equally relevant for understanding the reaction mechanism. The HERFD-XANES spectra obtained at three positions (beginning, middle and end) of the catalyst bed under SCR reaction conditions are depicted in Figure 1, showing clear differences and demonstrating the importance of spatially-resolved compared to integral studies. The decrease in intensity of the feature in the rising edge indicates less linearly coordinated $Cu^+$, which is in line with the ammonia consumption along the reactor. The intensity of the pre-edge feature suggests a mixture of $Cu^+$ and $Cu^{2+}$, with almost no variation under these conditions. By increasing the temperature and thus the reaction rate, a stronger gradient is observed for both features [17].

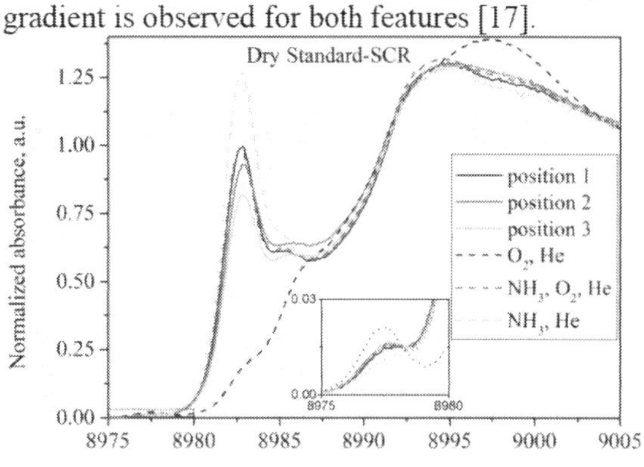


**Figure 1.** Spatially-resolved HERFD-XANES spectra of Cu-SSZ-13 under model gas feeds and at three positions along the catalyst bed under Dry Standard-SCR reaction conditions (1000 ppm NO, 1000 ppm $NH_3$, 10% $O_2$, He; GHSV = 200,000 $h^{-1}$) at 200 °C.

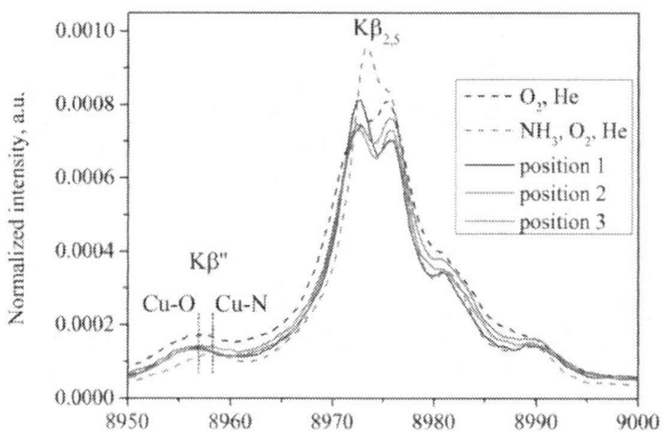


**Figure 2.** Spatially-resolved X-ray Emission Spectra of Cu-SSZ-13 at three positions along the catalyst bed under Standard-SCR reaction conditions (1000 ppm NO, 1000 ppm $NH_3$, 10% $O_2$, 1.5% $H_2O$, He; GHSV = 200,000 $h^{-1}$) at 400 °C and reference spectra in $O_2$, He and $NH_3$, $O_2$, He recorded at 200 °C.

A similar effect is visible in the XES results (Figure 2), where the *operando* spectra vary along the catalyst bed. Especially the $K\beta_{2,5}$-line shows a decrease of the feature at 8972.6 eV and increase of the feature at 8975.8 eV from the inlet towards the outlet of the catalyst bed. As an additional benefit for catalytic reactions, typically information on the ligand nature can be derived from Kß'' region. Also for Cu-SSZ-13 coordination of N or O atoms is clearly resolved at 200°C when adding $NH_3$ to the $O_2$-He stream. However, at 400°C due to a higher reaction rate but also due to a different interaction with $H_2O$ and $NH_3$ molecules the variation of the Kß'' feature is less pronounces. Only at position 3 of the catalyst bed a slight broadening was observed which could be caused by the interaction with the zeolite framework, as indicated by similar spatially resolved EXAFS studies (not shown). However, DFT simulated XES/XANES spectra and additional measurements at high temperatures under model conditions are necessary to obtain a better understanding of this phenomenon, as not only the ligands but also the coordination geometry have an influence on the spectrum. This is currently carried out by our team and a more comprehensive analysis will be reported in our next paper.

## 4. Conclusion

The application of HERFD-XANES and XES to study a catalyst in a spatially-resolved manner under reaction conditions has been demonstrated. Significant structural gradients along the catalyst bed could be observed with both methods, showing less adsorbed ammonia on the Cu-sites towards the end of the reactor. This experiment has illustrated the possibility of describing coordination sphere of metal sites of working catalysts and, thus, the suitability of XAS and XES to give insights into reaction mechanisms. As these methods are not limited to a specific reaction, they could be applied also to other catalytic reactions as understanding a catalyst is the best way for its further optimisation.

## Acknowledgements

We acknowledge funding from KIT, the BMBF (projects 05K10VKB and 05K13VK2) and DBU (T. Günter). We thank ESRF for providing beamtime at the ID26 beamline and financial support at ESRF. Dr. Pieter Glatzel is gratefully acknowledged for the help during the beamtime and for discussions.

## References

[1] Heijboer WM, Glatzel P, Sawant KR, Lobo RF, Bergmann U, Barrea RA, Koningsberger DC, Weckhuysen BM and de Groot FMF 2004 *J Phys Chem B* **108** (28) 10002-11
[2] Glatzel P, Weng T-C, Kvashnina K, Swarbrick J, Sikora M, Gallo E, Smolentsev N and Mori RA 2013 *J Electron Spectrosc* **188** 17-25
[3] Glatzel P and Bergmann U 2005 *Coord Chem Rev* **249** (1–2) 65-95
[4] Hämäläinen K, Siddons DP, Hastings JB and Berman LE 1991 *Phys Rev Lett* **67** (20) 2850-53
[5] Bergmann U and Glatzel P 2009 *Photosynth Res* **102** (2-3) 255-66
[6] Safonov VA, Vykhodtseva LN, Polukarov YM, Safonova OV, Smolentsev G, Sikora M, Eeckhout SG and Glatzel P 2006 *J Phys Chem B* **110** (46) 23192-96
[7] Giordanino F, Borfecchia E, Lomachenko KA, Lazzarini A, Agostini G, Gallo E, Soldatov AV, Beato P, Bordiga S and Lamberti C 2014 *J Phys Chem Lett* **5** (9) 1552-59
[8] Grunwaldt JD, Caravati M, Hannemann S and Baiker A 2004 *Phys Chem Chem Phys* **6** (11) 3037-47
[9] Doronkin DE, Casapu M, Günter T, Müller O, Frahm R and Grunwaldt J-D 2014 *J Phys Chem C* **118** (19) 10204-12
[10] Luo J-Y, Hou X, Wijayakoon P, Schmieg SJ, Li W and Epling WS 2011 *Appl Catal B: Environ* **102** (1–2) 110-19
[11] Deka U, Lezcano-Gonzalez I, Warrender SJ, Lorena Picone A, Wright PA, Weckhuysen BM and Beale AM 2013 *Micropor Mesopor Mat* **166** (0) 144-52
[12] Zones SI (1985) US Patent 4544538
[13] Günter T, Carvalho HWP, Doronkin DE, Sheppard T, Glatzel P, Atkins AJ, Rudolph J, Jacob CR, Casapu M and Grunwaldt J-D 2015 *Chem Commun* **51** (44) 9227-30
[14] McEwen JS, Anggara T, Schneider WF, Kispersky VF, Miller JT, Delgass WN and Ribeiro FH 2012 *Catal Today* **184** (1) 129-44
[15] Andersen CW, Bremholm M, Vennestrom PNR, Blichfeld AB, Lundegaard LF and Iversen BB 2014 *IUCrJ* **1** (6) 382-86
[16] Luo J-Y, Oh H, Henry C and Epling W 2012 *Appl Catal B: Environ* **123–124** (0) 296-305
[17] Günter T, Doronkin D, Boubnov A, Carvalho H, Casapu M and Grunwaldt JD 2015 *Top Catal* submitted

Journal of Physics: Conference Series **712** (2016) 012072 doi:10.1088/1742-6596/712/1/012072

# A Microfluidic Device for the Investigation of Rapid Gold Nanoparticle Formation in Continuous Turbulent Flow

**G Hofmann[1], G Tofighi[1], G Rinke[2], S Baier[1], A Ewinger[2], A Urban[2], A Wenka[2], S Heideker[2], A Jahn[3], R Dittmeyer[2], J-D Grunwaldt[1,4]**

[1] Karlsruhe Institute of Technology, Institute for Chemical Technology and Polymer Chemistry (ITCP), Karlsruhe, Germany
[2] Karlsruhe Institute of Technology, Institute for Micro Process Engineering (IMVT), Karlsruhe, Germany
[3] TU Dresden, Institute of Semiconductors and Microsystems, Dresden, Germany
[4] Karlsruhe Institute of Technology, Institute of Catalysis Research and Technology (IKFT), Karlsruhe, Germany

E-mail: grunwaldt@kit.edu

**Abstract**. A new setup with an integrated microfluidic chip with small dead time, high time resolution and compatibility with *in situ* X-ray absorption (XAS) measurements is presented. It can also be combined with a free liquid jet. By using the microfluidic chip the short reaction times from 2 to 20 milliseconds can be observed, beyond that an external cyclone mixer for extended observation times was applied. The reduction of gold ions with tetrakis(hydroxy-methyl)phosphonium (THPC) has been investigated in the microfluidic setup to monitor this reaction yielding small gold nanoparticles, requiring preferentially a free liquid jet.

## 1. Introduction

Gold nanoparticles (AuNPs) have found various applications in a broad range of scientific fields like medicine and catalysis. Their versatility and suitability in many different applications is due to the fact that gold particles of different size and shape can be prepared [1]. The need for defined AuNPs calls for a better control of the process, knowledge of the reaction mechanisms and kinetics, especially in the early phase of nucleation [2, 3]. For this purpose *in situ* studies are inevitable and microfluidic devices offer the opportunity to provide rapid and thorough mixing [4-6] and to shine X-rays through them. The formation of AuNPs was successfully studied using X-ray absorption spectroscopy (XAS), small angle X-ray scattering (SAXS), and UV-VIS for citrate and $BH_4$ based synthesis routes [7-9]. Notably, a fast gold reduction was observed within the first 200 ms. However, observations of the first 100 ms of reaction were not possible due to the experimental setup or acquisition time of the spectroscopic method used in the stopped flow apparatus, respectively. Therefore, it is required to aim at a fast data acquisition in case of stopped flow or measurements in continuous flow to enhance time resolution at the very early stage of the reaction. This would allow covering the initial growth kinetics or correlating time and measurement position to enhance resolution.

In this work, a first setup with very small dead time and high time resolution is presented. Next, the reaction process can be evaluated by XAS measurements with an option to also use SAXS as complementary technique at a later stage. The reduction of $HAuCl_4$ with tetrakis(hydroxymethyl)phosphonium chloride (THPC) was chosen [10] to understand this rapid reaction yielding small Au nanoparticles suitable for catalysis. Note that up to now mainly the late growth phase of AuNPs from THPC was studied yet from 10 to 700 minutes [1,11,12].

## 2. Experimental Principle and Setup

A high time resolution of the reaction process requires either a very fast acquisition after start of the reaction, which needs a high X-ray flux in order to achieve a good signal to noise ratio, or a method to correlate the result of a "slow" measurement to a specific reaction time. Here, a continous flow microreactor setup was used and the location in the channel or its length corresponds to a certain reaction time [13]. The basic setup is depicted in Figure 1. Both good mixing is and a stable and predictable residence time are important. For this purpose, a micromixer based on a previous design [4] was used and a high flow rate was applied ( Reynolds number above 2300 in the channel) to establish a plug flow with a flat flow velocity profile; the rapid mixing ensures that the reaction kinetics are not determined by the diffusion. In this way, a measurement of the reaction mixture at a given position in the channel corresponds to a specific time of the reaction after mixing; on the other hand the fluid can be monitored at the outlet. The experiments were conducted at beamline P06 (PETRA III) using a fluorescence detector for measuring the fluorescence XAS data [13].

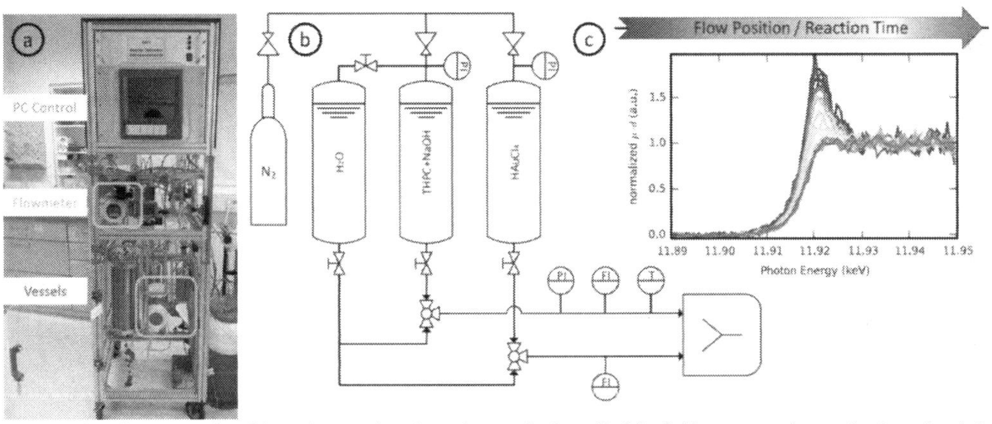


**Figure 1.** a) Photograph, b) schematic drawing of the fluid delivery rack and c) principle for monitoring gold reduction. The rack is grouped into three parts: Vessels, flowmeters, and PC control. Fluids are stored in vessels pressurized by nitrogen gas and are delivered to a microfluidic mixer to be rapidly and thoroughly mixed. The flow to the microfluidic mixer can be adjusted for each branch. The continuous product flow is then measured at several positions, which are mapped to reaction time (not shown). Symbols: PI= pressure indicator, FI= flow indicator, and T= temperature detection.

Sufficiently high and pulsation-free flow rates were provided by a home-built fluidic rack (Fig. 1a). The vessels and all fluidic parts are designed to be pressure resistant up to 15 bars, which is required to achieve a suitably high flow rate. In addition, the vessel, flowmeter, and the branch for the gold precursor solution is kept fully corrosion resistant and chemically inert. The microfluidic mixer as the next key part of the setup was optimized and designed based on computational fluid dynamics (CFD) simulations to obtain a perfect mixing in a short time with little residence time. In Figure 2 a) one simulation is shown for the resulting triple cyclone mixer, which is based on developments at IMVT-KIT [6]. In order to measure the liquid as soon as possible after mixing, the mixer was built into a Si-based microfluidic chip, which features observation lines directly attached to the mixer outlet. A schematic drawing of the microfluidic chip is shown in Figure 2 b). The chip to monitor liquid phase reactions was fabricated in collaboration with GeSiM GmbH and IHM TUD. The chip with a thickness of 1500 μm was obtained by Si fusion bonding of two 750 μm wafers with a diameter of 4 inch. This chip contained the integrated micro fluidic mixer and a 300×300 μm fluidic channel (called "observation line"). This fluidic part was fabricated by structuring the Si wafer by deep reactive ion etching.

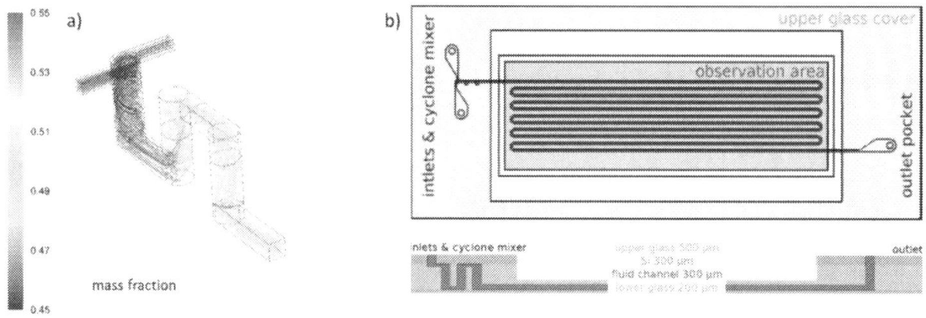


**Figure 2.** a) CFD simulation of the mass fractions passing through the triple cyclone mixer, resulting in a perfect mixing at the outlet. The color scale gives the mixing ratio, where 0.5 corresponds to a 1:1 mixture. b) Simplified construction drawing of the microfluidic chip with integrated triple cyclone mixer and observation lines (present meander design contains 3 straight sections).

The upper side of the chip was partially covered by 500 µm thick glass (blue shades, Fig. 2 b) and hosts inlet and outlet ports. In the central area of the chip the observation area is located, on the opposite side of the meander structure, which was especially thinned down by Si etching (conducted at IHM TUD) to allow experiments in transmission mode. Hence, the X-rays have to pass 300 µm of Si in addition to 200 µm of the lower glass layer and 300 µm of sample in the fluid channel. The bottom side of the chip is completely covered by a further 200 µm thick glass layer and acts as lid for the fluid channel (300 µm × 300 µm) in the meander. Holes in the top glass layer for the inlet and outlet ports and the cylinders for the cyclone mixer in the silicon chip were fabricated by laser drilling. Anodic bonding was used for the glass-Si parts and fusion bonding for the two 750 µm Si wafers. The present configuration is designed for a total flow of 2.5 kg $h^{-1}$ (Reynolds number of 2300), a pressure drop of 12 bar, a dead time of 1 ms and an observation time up to 20 ms (time resolution < 1 µs along the channel due to high flow and small beam). To access longer reaction times, a capillary tube can be attached to the outlet, extending the observation time. Measurements can then also be conducted on the liquid as it is released from the capillary into a free liquid jet. An additional pressure drop of 2 bar for the fluidic rack is still within the operating conditions.

## 3. Results and Discussion

For experiments up to 20 ms observation time the silicon microreactor chip (schematic Figure 2 b) was used alone. The reaction of gold with THPC was investigated. After a certain equilibration time both in the chip and at the outlet (operating it as free liquid jet) oxidized gold was observed. An example spectrum at the outlet of the free liquid jet (residence time 20 ms) is shown in Figure 3 a.

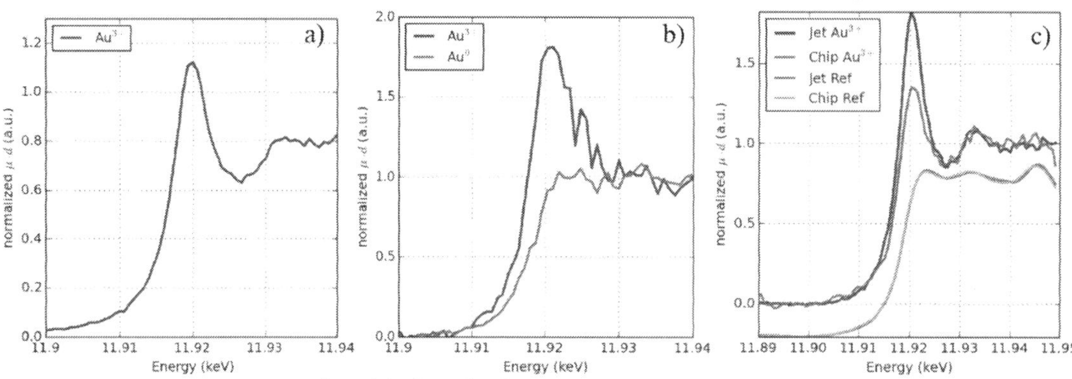


**Figure 3.** XANES spectra after (a) the Si-mixer (20 ms), (b) the capillary (3s, with (green) and without (blue) THPC), and (c) inside the microfluidic chip (Chip) and in the free liquid jet (Jet) for references of gold precursor solution ($Au^{3+}$) and metallic gold foil ($Au^0$).

In the chip itself fast deposition of metallic gold occurred at the position where the X-ray beam was hitting the Si chip despite of the high Reynolds number. Hence, $HAuCl_4$ was reduced by the X-ray beam, possibly enhanced by defect sites at the silicon wall (e.g. seeds for the reduced gold). As the residence time with 20 ms is rather short, we thus exploited the length of the reactor and additionally investigated the reduction using a longer capillary behind the microfluidic mixer. In this case, reduced gold was found (green spectrum in Fig. 3b, compared to reference samples in Fig. 3c). Hence, two different setups need to be used to detect the reduction kinetics in nanoparticle formation under the applied reaction conditions. In case the metal does not interact with the wall the X-ray kinetics of nanoparticle formation could be determined by scanning along the reactor. In case of gold reduction with THPC a free jet can be used, but under the present reaction conditions mostly $Au^{3+}$ is found after 20 ms, however, $Au^0$ nanoparticles coud be found after 3 s. For gaining a full insight, more investigations are required using different precursor solutions and concentrations, different gold precursor/reduction agent ratios, defined equilibration time and different residence times in future.

## 4. Conclusion and Outlook

We have presented a microfluidic device, which allows investigation of reactions with millisecond resolution and does not depend on equally rapid acquisition of data as the spectra can be either taken along the microfluidic reactor or at its outlet (free liquid jet) for X-ray sensitive reactions as in the case of gold. A mixer (< 1 ms dead time) directly integrated on the chip was used to achieve rapid and thorough mixing. The fluidic rack is able to deliver the liquids with high volumetric flow rates as demanded by the experimental approach (turbulent mixing, rather homogeneous flow profile). Pressure driven fluid delivery worked very smooth and no pulsation or discontinuities were observed. This enables the investigation of very early reaction times (2 ms) as well as an external cyclone mixer for extended observation times (>20 ms). Measurements in the free liquid jet are promising, since no depositions can occur and little X-ray absorbing material is present. However, it is more tedious to resolve the temporal evolution of the reaction. This requires external delays (different channel length).

## 5. Acknowledgement

We thank the BMBF projects "X-ray microscopy" (05K10VK1) and "Nanoscopy" (05K13VK2), the virtual institute VI-403 "In-situ Nano Imaging of Biological and Chemical Processes" and the Helmholtz Research Program "Storage and Cross-linked Infrastructures" (SCI) and "Science and Technology of Nanosystems" (STN) for financial support. We also appreciate DESY for allocation of beamtime at P06 and the beamline team in particular Dr. Ulrike Bösenberg and Dr. Gerald Falkenberg.

## References

[1] Grunwaldt J-D, Kiener C, Wögerbauer C and Baiker A 1999 *J.Catal.* **181** 223-232
[2] Wang F, Richards V N, Shields S B, and Buhro W E 2014 *Chem. Mater.* **26** 5–21
[3] Thanh N T K, Maclean N and Mahiddine S 2014 *Chem. Rev.* **114** 7610–7630
[4] Schubert K, Brandner J, Fichtner M, et al., 2001 *Microscale Thermophys.* Eng. **5** 17-39
[5] Fräulin C, Rinke G, Dittmeyer R 2014 *Microfluid Nanofluidics* **16** 149-157
[6] Kölbl A, Kraut M and Wenka A 2011 *Chem. Eng. J.* **167** 444– 454
[7] Wagner J, Tshikhudo T and Kohler J 2008 *Chem. Eng. J.* **135** 104–109
[8] Polte J, Tuaev X, Wuithschick M, Fischer A, Thuenemann A F, Rademann K, Kraehnert R and Emmerling F 2012 *ACS Nano* **6** 5791–5802
[9] Abecassis B, Testard F, Kong Q Y, Francois B and Spalla O 2010 *Langmuir* **26** 13847–13854
[10] Duff D G, Baiker A and Edwards P P 1993 *Langmuir* **9** 2301–2309
[11] Biswas K, Varghese N and Rao C N R 2008 *Small* **4** 649–655
[12] Hueso J L, Sebastian V, Mayoral A, et al. 2013 *RSC Adv.* **3** 10427
[13] Hofmann G, 2015 Dissertation, Karlsruhe Institute of Technology (KIT), DOI(KIT): 10.5445/IR/1000049565.

# *In-situ* XAFS study for calcination process of Cr catalyst supported on $\gamma$-$Al_2O_3$ and $SiO_2$

**T Watanabe, K Ikeda, M Katayama* and Y Inada**

Department of Applied Chemistry, Ritsumeikan University, Kusatsu, Shiga 525-8577, Japan

E-mail: katayama@fc.ritsumei.ac.jp

**Abstract**. The catalytic performance is largely affected by the oxidation state of supported Cr species, and its control changes the activity of Cr catalysts and the selectivity of products. In this study, the calcination process of the supported Cr catalysts on $\gamma$-$Al_2O_3$ and $SiO_2$ was investigated by *in-situ* XAFS spectroscopy. The hydrate species was first supported by the impregnation method and was converted to $CrO_3$ via $Cr_2O_3$ during the calcination process on both supporting materials. It was found that the temperature to complete the oxidation from $Cr_2O_3$ to $CrO_3$ on $SiO_2$ was higher than that on $\gamma$-$Al_2O_3$. The similarity of the interatomic distance between the surface oxygen atoms of the intermediate $Cr_2O_3$ species to that of $SiO_2$ contributes to the stabilization of $Cr_2O_3$ on $SiO_2$ during the calcination process.

## 1. Introduction

The Cr catalyst supported on $SiO_2$ is used for the various reactions. For example, ethylene polymerization, dehydrogenation of alkanes, selective reduction of NO, and hydrodesulfurization which are important processes in modern industrial chemistry. Groppo *et al.* have reviewed about the properties of the Cr/$SiO_2$ catalyst [1], and the XAFS analysis has been published for the Cr catalyst after the polymerization reaction [2]. The supported Cr catalysts are conventionally prepared by the impregnation method using chromium(II) acetate, chromium(II) chloride, bis(triphenylsilyl)chromate(VI), and chromium(VI) oxide as the precursor materials. It has been reported that the composition of Cr(VI) and Cr(III) species is dependent on the used source materials and that the catalytic activity for the polymerization reaction is enhanced by the existence of the Cr(VI) species on $SiO_2$ [3]. In addition, a reported speciation for the Cr catalyst using XPS has revealed that the calcination at 300 °C mainly generates the Cr(III) species while the treatment at higher temperature than 500 °C converts the oxidation state to Cr(VI) [4]. The speciation of Cr catalysts of using $CrO_3$ as a precursor was carried out in calcination process by means of diffuse reflectance spectroscopy, Raman spectra, XAFS, and TPR [5]. The analysis of the oxidation state for the active Cr species is thus fundamental to understand the catalytic performance and to improve its activity. In this study, the *in-situ* XAFS measurements have been applied to the calcination process of the Cr catalyst to identify directly the chemical state of the Cr species on both $\gamma$-$Al_2O_3$ and $SiO_2$. A common precursor of chromium(III) nitrate has been used to compare the chemical conversion of the supported Cr species between two supporting materials.

Journal of Physics: Conference Series **712** (2016) 012073 doi:10.1088/1742-6596/712/1/012073

## 2. Experimental

The powders of γ-$Al_2O_3$ (180 $m^2/g$) or $SiO_2$ (192 $m^2/g$) distributed by the Catalysis Society of Japan were used as the supporting materials, and were suspended in an acidic aqueous solution of $Cr(NO_3)_3 \cdot 9H_2O$. The suspensions were dried for 72 h at 50 °C to prepare the precursor samples for the calcination process. The drying at low temperature prevents the chemical state change by the reaction with air for the supported Cr species before the calcination. The Cr loading was set to 5 wt%.

The *in-situ* XAFS measurements were carried out at the BL-12C station of Photon Factory (KEK, Japan) at the Cr K edge in the transmission mode. The Si(111) double-crystal monochromator was used, and the higher-order harmonics were removed by the Ni-coated double-mirror system. The catalyst precursor powder was set in a flow-type *in-situ* cell, and the temperature was changed from 25 to 600 °C with the increasing rate of 10 °C/min. During the calcination process, the dilute $O_2$ gas (10 vol% balanced by He) was flowed with the total flow rate of 200 $cm^3$/min as the model condition of the calcination in air. The quick-scanning XAFS measurements were repeated in the X-ray energy range from 5.49 to 7.09 keV with the interval time of 2 min. The background of measured XAFS spectra in pre-edge region was fitted by the Victoreen equation and subtracted from whole region. Then, they were normalized to unit edge step. The composition analysis was carried out for the observed XANES spectrum by the linear combination fitting using those of reference samples with various oxidation states.

## 3. Results and discussion

The observed XANES spectral change during the calcination process is shown in figure 1 for the Cr species supported on γ-$Al_2O_3$ and $SiO_2$. The XANES spectra of an aqueous solution of $Cr(NO_3)_3 \cdot 9H_2O$, a $Cr_2O_3$ powder, and a $CrO_3$ powder measured independently are given in figure 1. The initial spectra on both γ-$Al_2O_3$ and $SiO_2$ at 25 °C are identical to that of the aqueous solution of $Cr(NO_3)_3$. The existence of $[Cr(OH_2)_6]^{3+}$ is well established in acidic aqueous solution [6], and the observed XANES spectrum thus confirms that the $[Cr(OH_2)_6]^{3+}$ species adsorbs on the supporting particle without the modification of the first coordination sphere of the Cr(III) center. The absorption band at the pre-edge region observed in the final XANES spectra demonstrates the oxidation to the Cr(VI) state during the calcination process, and the final spectra are consistent with that of $CrO_3$. The present *in-situ* XAFS measurements clearly reveal that the initially supported Cr(III) hydrate is converted to the Cr(VI) oxide by heating under the dilute $O_2$ atmosphere. As clearly shown in figure 1, the XANES spectral changes without the isosbestic points suggest the existence of the intermediate species between the Cr(III) hydrate and $CrO_3$.

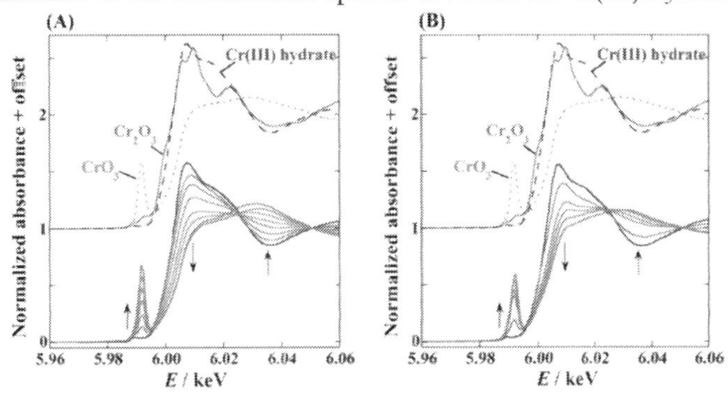


**Figure 1.** Part (A): Evolution of the XANES spectra of Cr species on γ-$Al_2O_3$ along the calcination process from $T = 25$ °C, blue curve, to $T = 600$ °C, red curve (bottom), intermediate spectra every 20 °C. The top part reports, vertically shifted for clarity, the XANES spectra of $CrO_3$, $Cr_2O_3$ and Cr(III) hydrate model compounds. Part (B): as part (A) for Cr species on $SiO_2$.

The composition analysis based on the XANES spectrum was carried out by the linear combination of three reference species, the Cr(III) hydrate, $Cr_2O_3$, and $CrO_3$. The fitting extent of the calculated spectra is represented by R factor:

$$R = \frac{\sqrt{\sum(\mu_{\text{data}}(E) - \mu_{\text{fit}}(E))^2}}{\sqrt{\sum(\mu_{\text{data}}(E))^2}}$$

where $\mu_{\text{data}}(E)$ and $\mu_{\text{fit}}(E)$ are the measured and calculated spectrum, respectively. The observed spectra at 160 °C are compared in figure 2 with the calculated spectra depicted using the estimated mole fraction, and the calculated values of mole fraction are plotted in figure 3 as a function of temperature. The satisfactory reproducibility of the XANES spectrum supports the existence of $Cr_2O_3$ as the intermediate species on both supporting materials (see figure 2). Figure 3 shows that the Cr(III) hydrate is started to convert to $Cr_2O_3$ at around 100 °C and that its composition reaches to *ca.* 40 % at 150 °C. This changeover accompanying the dissociation of bound water molecules is independent of the supporting material. Further increase in temperature promotes the oxidation from $Cr_2O_3$ to $CrO_3$ on both supporting materials. The composition of $CrO_3$ begins to increase at 150 °C and the oxidation is completed up to 250 °C on γ-$Al_2O_3$ under the present experimental condition. It is marked contrast that the completion temperature of the oxidation to $CrO_3$ on $SiO_2$ is much higher than that on γ-$Al_2O_3$, although the formation temperature of *ca.* 150 °C is in accordance with that on γ-$Al_2O_3$. The composition of $Cr_2O_3$ is kept almost constant (*ca.* 25 %) in the temperature range from 200 to 400 °C on $SiO_2$ as shown in figure 3(B). The comparison between γ-$Al_2O_3$ and $SiO_2$ clearly indicates the relative stability of the $Cr_2O_3$ species on $SiO_2$ rather than on γ-$Al_2O_3$.

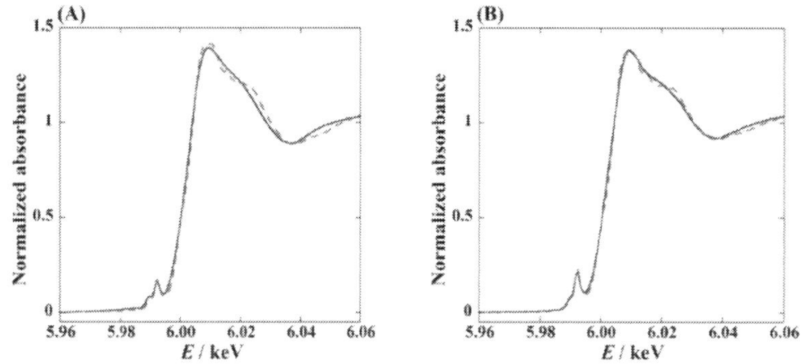


**Figure 2.** The observed (solid line) and the calculated (broken line) XANES spectrum at 160 °C on γ-$Al_2O_3$ (A) and $SiO_2$ (B). The calculated spectra are depicted using the determined mole fractions of Cr(III) hydrate, $Cr_2O_3$, and $CrO_3$.

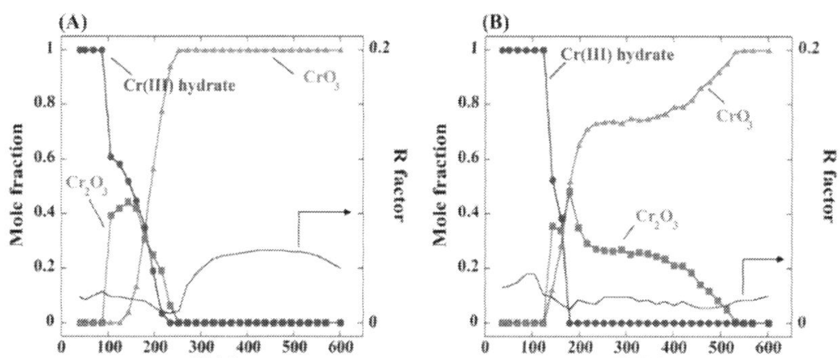


**Figure 3.** The mole fraction of Cr(III) hydrate, $Cr_2O_3$, and $CrO_3$ as a function of temperature on γ-$Al_2O_3$ (A) and $SiO_2$ (B).

The crystal structures of $\gamma$-$Al_2O_3$ and $SiO_2$ are compared in figure 4 with that of the intermediate $Cr_2O_3$ species. The interatomic distance between the surface oxygen atoms is 2.79 Å for $\gamma$-$Al_2O_3$ with the defective spinel structure [7], while that of $SiO_2$ is 2.62 Å [8]. The corundum structure of $Cr_2O_3$ has two different distances of 2.62 and 2.99 Å as shown in figure 4 [9]. Because it is considered that the supported Cr oxides interact with the supporting material via the oxygen atoms, the similarity of the arrangement of the surface oxygen atoms may contribute to the stabilization of supported Cr species. The stabilization of $Cr_2O_3$ species is explained by that both $SiO_2$ and $Cr_2O_3$ have the interatomic distance of 2.62 Å. In the case of $\gamma$-$Al_2O_3$, it is considered that the mismatch of the interatomic distance leads to the easy conversion to $CrO_3$.

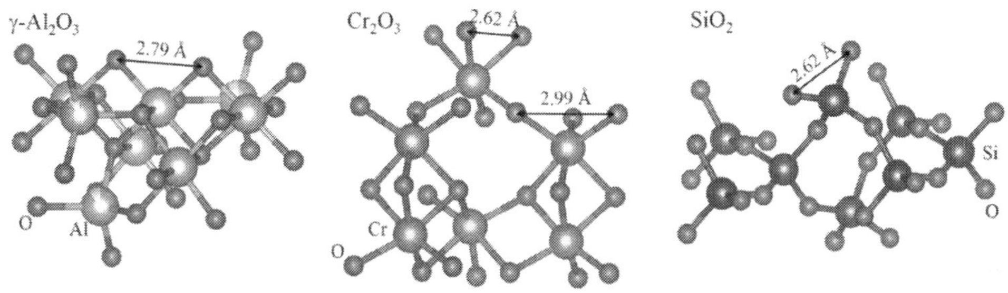


**Figure 4.** The crystal structures of $\gamma$-$Al_2O_3$, $Cr_2O_3$, and $SiO_2$.

## 4. Conclusions

The present *in-situ* XAFS study for the calcination process of the supported Cr catalysts revealed the chemical state changeover from the initial Cr(III) hydrate to the final $CrO_3$ species via the $Cr_2O_3$ intermediate on $\gamma$-$Al_2O_3$ and $SiO_2$. It was evaluated that the stability of $Cr_2O_3$ was significantly affected by the atomic arrangement of surface oxygen atoms, and the similar interatomic O–O distance between $Cr_2O_3$ and $SiO_2$ contributed to the relative stability of $Cr_2O_3$ on $SiO_2$ in the temperature range from 200 to 400 °C. A selection of calcination temperature and supporting materials may change the oxidation state of the supported Cr species. Therefore, the Cr catalyst can be designed with the required oxidation state for a catalytic reaction.

## Acknowledgement

The XAFS measurements have been performed under the approval of the Photon Factory Program Advisory Committee (Proposal Nos. 2011G531 and 2013G596).

## References

[1] Groppo E, Lamberti C, Bordiga S, Spoto G and Zecchina A 2005 *Chem. Rev.* **105** 115
[2] Groppo E, Prestipino C, Cesano F, Bonino F, Bordiga S, Lamberti C, Thüne P C, Niemantsverdriet J W and Zecchina A 2005 *J. Catal.* **230** 98
[3] Gaspar A B and Dieguez L C 2002 *Appl. Catal. A: Gen.* **227** 241
[4] Al-Zahrani S M, Elbashir N O, Abasaeed A E and Abdulwahed M 2001 *Ind. Eng. Chem. Res.* **40** 781
[5] Weckhuysen B M, Schoonheydt R A, Jehng J-M, Wachs I E, Cho S J, Ryoo R, Kijlstra S and Poles E 1995 *J. Chem. Soc. Faraday Trans.* **91** 3245
[6] Ohtaki H and Radnai T 1993 *Chem. Rev.* **93** 1157
[7] Ionescu A, Allouche A, Aycard J P, Rajzmann M and Hutschka E 2002 *J. Phys. Chem.* B **106** 9359
[8] Tucker M G, Keen D A and Dove M T 2001 *Mineral. Mag.* **65** 489
[9] Finger L W and Hazen R M 1980 *J. Appl. Phys.* **51** 5362

# Ag K- and $L_3$-edge XAFS study on Ag species in Ag/$Ga_2O_3$ photocatalysts

**M Yamamoto[1], T Yoshida[2,3], N Yamamoto[1], T Nomoto[4], A Yamamoto[5,6], H Yoshida[5,6] and S Yagi[3]**
[1] Graduate School of Engineering, Nagoya University, Nagoya 464-8603, Japan
[2] Advanced Research Institute for Natural Science and Technology, Osaka City University, Osaka 558-8585, Japan
[3] EcoTopia Science Institute, Nagoya University, Nagoya 464-8603, Japan
[4] Aichi Synchrotron Radiation Center, Seto 489-0965, Japan
[5] Graduate School of Human and Environmental Studies, Kyoto University, Kyoto 606-8501, Japan
[6] Element Strategy Initiative for Catalysts and Batteries (ESICB), Kyoto University, Kyoto 615-8520, Japan

E-mail: yamamoto.muneaki@f.mbox.nagoya-u.ac.jp, tyoshida@ocarina.osaka-cu.ac.jp

**Abstract.** Ag loaded $Ga_2O_3$ (Ag/$Ga_2O_3$) shows photocatalytic activity for reduction of $CO_2$ with water. Ag $L_3$-edge XANES and K-edge EXAFS spectra were measured for various Ag/$Ga_2O_3$ samples, which suggested that structural and chemical states of Ag species varied with the loading amount of Ag and the preparation method. The Ag species were metallic Ag particles with an $AgGaO_2$-like interface structure in the sample with high loading amount of Ag while predominantly Ag metal clusters in the sample with low loading amount of Ag. The XANES feature just above the edge represented the interaction between the Ag species and the $Ga_2O_3$ surface, showing that the Ag metal clusters had more electrons in the *d*-orbitals by interacting with the $Ga_2O_3$ surface, which would contribute the high photocatalytic activity.

## 1. Introduction

An increasing concentration of atmospheric $CO_2$ has become a concern for our planet. One of the solutions for this problem, photocatalytic reduction of $CO_2$ to other useful chemicals has attracted much attention [1–4]. For the $CO_2$ reduction, an external energy and an appropriate catalyst are required, since $CO_2$ is one of the most stable molecules. Recently, it has been reported that Ag loaded $Ga_2O_3$ (Ag/$Ga_2O_3$) shows the photocatalytic activity for reduction of $CO_2$ with water to produce CO [5–7]. In our previous study [8], we found that the chemical state of Ag species in as-prepared Ag/$Ga_2O_3$ samples affected the formation process of the reaction intermediate, which were closely related to the CO production efficiency. However, it is still unclear how the structural and chemical states of the Ag species as well as Ag–$Ga_2O_3$ interaction change with the loading amount of Ag. In the present study, Ag $L_3$-edge and K-edge XAFS analysis was applied to investigate the structural and chemical states of the Ag species, which provided us meaningful information for understanding the photocatalysis of Ag/$Ga_2O_3$ samples.

## 2. Experimental

$Ag/Ga_2O_3$ samples were prepared by an impregnation (IMP) method and a photodeposition (PD) method. Generally, it is expected that IMP method provides Ag ions well interacting with the $Ga_2O_3$ support while PD method is suited to the formation of metallic Ag species. For IMP method, a mixture of $Ga_2O_3$ powders and an aqueous solution of $AgNO_3$ was magnetically stirred and dried, followed by calcination at 673 K for 2 h. The loading amounts of Ag were 0.1, 0.2, 0.5, 1.0 and 5.0 wt%. For PD method, a mixture of a $Ga_2O_3$ powder and an aqueous solution of $AgNO_3$ with ethanol was magnetically stirred and photoirradiated for 3 h, followed by filtration. The loading amounts of Ag were 0.1 and 1.0 wt%. Ag $L_3$-edge XAFS measurements were carried out in a fluorescent X-ray yield mode and in a total electron yield mode at the beam line 6N1 at the Aichi Synchrotron Radiation Center (Proposal No. 2503042). Ag K-edge XAFS spectra were measured at the NW10A station at the Photon Factory in the High Energy Accelerator Research Organization in the fluorescence mode with a 19 element solid state Ge detector and in a transmission mode (Proposal No. 2014G547).

## 3. Results and discussion

In our previous study [8], TEM and HAADF-STEM measurements revealed that around 1 nm sized Ag clusters were formed predominantly in a 0.1 wt% $Ag/Ga_2O_3$ (IMP) photocatalytic active sample while Ag particles with the size of several–several tens nm in a 1.0 wt% $Ag/Ga_2O_3$ (IMP) sample. In a 5.0 wt% $Ag/Ga_2O_3$ (IMP) sample and a 1.0 wt% $Ag/Ga_2O_3$ (PD) sample, large Ag particles with the size of over 30 nm were observed. After heating the 1.0 wt% $Ag/Ga_2O_3$ (PD) sample at 673 K for 2 h, the average Ag particle size became smaller as 1–2 nm and they were highly dispersed on the $Ga_2O_3$ surface.

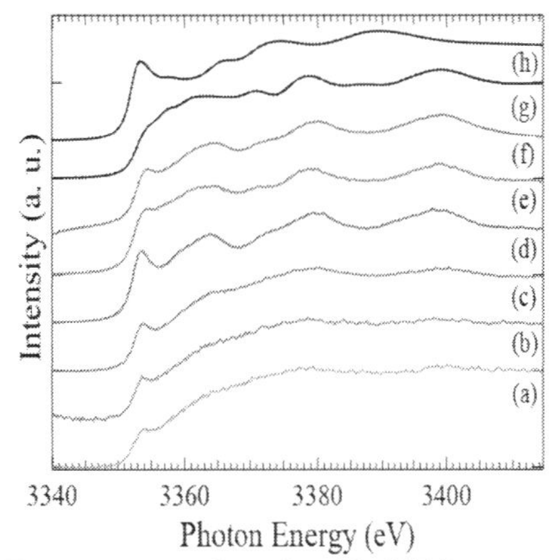


**Figure 1.** Ag $L_3$-edge XANES spectra of the $Ag/Ga_2O_3$ (IMP) samples (a–f), Ag foil (g) and $Ag_2O$ (h). Ag loading (wt%): 0.1 (a), 0.2 (b), 0.5 (c), 1.0 (d), 5.0 (FY) (e) and 5.0 (TEY) (f).

In the present study, the Ag species in the $Ag/Ga_2O_3$ samples were investigated in detail by Ag $L_3$- and K-edge XAFS spectroscopy. Figure 1 shows the spectra of the $Ag/Ga_2O_3$ samples together with Ag foil and $Ag_2O$ as references. We tried to simulate the XANES spectra of the $Ag/Ga_2O_3$ samples by using the reference spectra, but the fitting was not sufficient, indicating that the Ag species in the $Ag/Ga_2O_3$ samples were not a simple mixture of metal and oxides. As is evident from figure 1, the XANES spectrum of the 0.1 wt% $Ag/Ga_2O_3$ (IMP) sample has no distinct peaks, which originates from highly dispersed Ag species such as 1 nm sized Ag clusters [8]. With increasing the loading amount of Ag, the feature of XANES changed gradually to the characteristic one of an Ag bulk.

Note that the peak at 3353 eV grew with the loading amount of Ag up to 1.0 wt%, and then decreased with 5.0 wt% loading. This peak has been assigned to the transition from 2p to unoccupied 4d states [9]. The peak was clearly shown in the spectrum of $Ag_2O$. Therefore, in the $Ag/Ga_2O_3$ samples with higher loading amount of Ag up to 1.0 wt%, most of Ag species would be partially oxidized. As for the 5.0 wt% $Ag/Ga_2O_3$, XANES spectra could be measured both in a fluorescent X-ray yield mode (FY) and a surface sensitive total electron yield mode (TEY) as shown in figures 1e and 1f. The former spectrum (FY) was similar to that of an Ag bulk, suggesting that Ag species was metallic. Nevertheless, a small peak at 3353 eV evidenced the existence of oxides. In the latter spectrum (TEY), this peak was clearly seen and more thorough observation suggested that double peaks at 3364 and 3380 eV became intense, which were similar to the characteristic feature of 1.0 wt% $Ag/Ga_2O_3$. This result indicates that the Ag species in the 5.0 wt% $Ag/Ga_2O_3$ sample is almost metallic Ag like an Ag bulk, but the surface is slightly oxidized. In addition, it is also suggested that the Ag species in 1.0 wt% $Ag/Ga_2O_3$ would be partially oxidized Ag particles or a mixture of the metallic and oxidized species, which will be discussed below.

Figure S1 in the supporting information shows the Ag $L_3$-edge XANES spectra of the 0.1 and 1.0 wt% $Ag/Ga_2O_3$ (PD) samples. Since both the spectra were very similar to that of an Ag bulk, the Ag species in the 0.1 and 1.0 wt% $Ag/Ga_2O_3$ (PD) samples were metallic Ag. It was also confirmed that the broad feature shown in the XANES of the 0.1 wt% $Ag/Ga_2O_3$ (IMP) sample was not due to the low loading amount of Ag. Figure S1, c–f shows the Ag $L_3$-edge XANES spectra of the 1.0 wt% $Ag/Ga_2O_3$ samples prepared by the PD method, followed by heating in air for 2 h at various temperatures, referred to as $Ag/Ga_2O_3$ (PD-H). A small shoulder around 3353 eV would indicate some oxidized moiety in the samples. With increasing the heating temperature, the XANES feature became broad gradually and at more than 573 K, the spectra were similar to that of the 0.1 wt% $Ag/Ga_2O_3$ (IMP) sample. Considering the TEM results in our previous study [8], the spectral change is assignable to the decrease of the Ag particle size by heating at more than 573 K. The decrease of the practice size suggests the strong interaction between the Ag particle and the $Ga_2O_3$ surface.

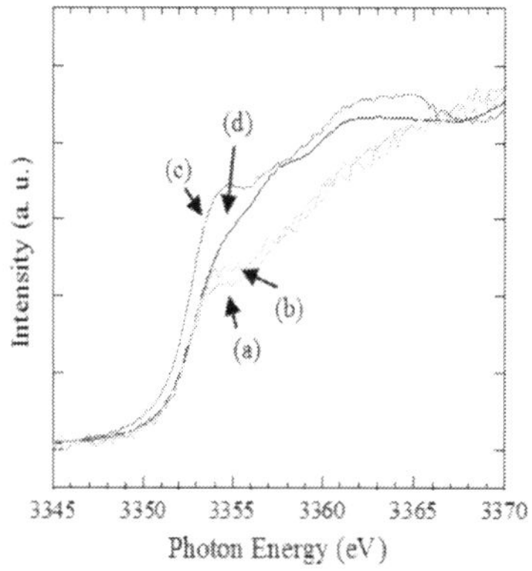


**Figure 2.** Ag $L_3$-edge XANES spectra of the 0.1 wt% $Ag/Ga_2O_3$ (IMP) sample (a), the 1.0 wt% $Ag/Ga_2O_3$ (PD-H) sample (b), the 5.0 wt% $Ag/Ga_2O_3$ (IMP) sample (c) and Ag foil (d).

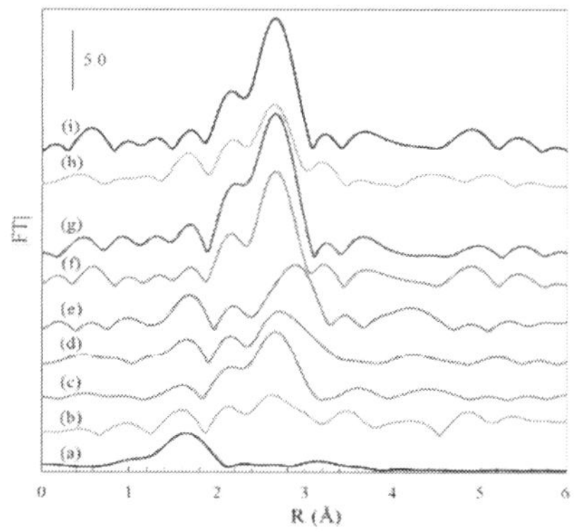


**Figure 3.** Radial structural functions of $Ag_2O$ (a), the $Ag/Ga_2O_3$ (IMP) samples (b–f), the $Ag/Ga_2O_3$ (PD, PD-H) samples (g, h) and Ag foil (i). Ag loading (wt%): 0.1 (b), 0.2 (c), 0.5 (d), 1.0 (e), 5.0 (f), 1.0 (g) and 1.0 (h).

In Figure 2, the characteristic XANES feature just above the edge would represent the interaction between the Ag species and the $Ga_2O_3$ surface. The $Ag/Ga_2O_3$ samples showed lower absorption around 3350–3365 eV than that of an Ag bulk. This feature was most significant for the 0.1 wt% $Ag/Ga_2O_3$ (IMP) sample and the 1.0 wt% $Ag/Ga_2O_3$ heated at 623 K (PD-H) sample. Similar features around 3350–3365 eV were reported for Au–Ag alloys and Ag clusters on Au rods by some researchers [10–12]. They concluded that the electron density in the *d*-orbital of the Ag atom increased by transferring electrons from Au to Ag. Therefore, in the present 0.1 wt% $Ag/Ga_2O_3$ (IMP) sample and the 1.0 wt% $Ag/Ga_2O_3$ (PD-H) sample, the Ag species would probably accept more electrons in the *d*-orbitals from the $Ga_2O_3$ surface as the result of the strong interaction.

Figure 3 shows radial structure functions (RSF) obtained by Fourier transform of $k^3$-weighted Ag K-edge EXAFS (Figure S2 in supporting information) in the region of $k = 3.15–11.80$ Å$^{-1}$. The small peak around 1.7 Å was assigned to the Ag–O pair and/or side lobes, and this peak was larger for the 0.5 wt% $Ag/Ga_2O_3$ (IMP), the 1.0 wt% $Ag/Ga_2O_3$ (IMP) and the 1.0 wt% $Ag/Ga_2O_3$ (PD-H) samples. The peak around 2.5–3 Å observed for all the $Ag/Ga_2O_3$ samples should correspond to the Ag–Ag pair. As for the $Ag/Ga_2O_3$ (IMP) samples, this peak grew with the loading amount of Ag, although it became broader and shifted to longer R region for the 0.5 and 1.0 wt% $Ag/Ga_2O_3$ (IMP) samples.

We also performed curve-fitting analysis on the back Fourier-transformed EXAFS from the first and the second coordination shells (R = 1.35–3.35 Å) as summarized in Table S1 in supporting information. The average Ag–Ag distance was estimated to be ca. 2.89 Å for the 0.1, 0.2 and 5.0 wt% $Ag/Ga_2O_3$ (IMP) samples, which was in good agreement with that of the Ag foil. Note that the photocatalytically active Ag clusters formed predominately in the 0.1 wt% $Ag/Ga_2O_3$ (IMP) sample were metallic. The average coordination number for the Ag–Ag shell ($CN_{Ag}$) for 5.0 wt% $Ag/Ga_2O_3$ (IMP) was close to that of Ag foil, suggesting the large Ag metal particles, while the $CN_{Ag}$ values for the 0.1 and 0.2 wt% $Ag/Ga_2O_3$ (IMP) samples were 4.3 and 6.5, respectively, which were significantly smaller than that of the Ag foil. The geometrical calculation for the 0.1 wt% Ag/$Ga_2O_3$ (IMP) sample demonstrated that the Ag species with the $CN_{Ag}$ values of 4.3 would be smaller than 1 nm, such as metallic cluster, which was quite consistent with the HAADF-STEM result of this sample [8].

On the other hand, the 0.5 and 1.0 wt% Ag/$Ga_2O_3$ (IMP) samples had two types of short (2.89 Å) and long (3.05 Å) Ag–Ag pairs. The latter was longer than that of Ag metal and similar to the Ag–Ag distance (3.03 Å) in $AgGaO_2$ [13]. The Ag–O pair with 2.10–2.12 Å in length which was longer than those of $Ag_2O$ and AgO and similar to the Ag–O distance (2.10 Å) in $AgGaO_2$ [13]. These suggest that a part of Ag atoms at the interface between the Ag metal particle and the $Ga_2O_3$ surface would be in an $AgGaO_2$-like structure as a result of well interaction with the $Ga_2O_3$ surface and other part would form Ag metal particles. In the cases of the 1.0 wt% Ag/$Ga_2O_3$ (PD) and the 1.0 wt% Ag/$Ga_2O_3$ (PD-H) samples, the average Ag–Ag distance was estimated to be 2.89 Å for both samples and $CN_{Ag}$ for the former and the latter samples to 11.9 and 7.3, demonstrating that the large Ag metal particles changed to small ones by heating. It is also noteworthy that the Ag–O pair was found only for the 1.0 wt% Ag/$Ga_2O_3$ (PD-H) sample and the 0.5 and 1.0 wt% Ag/$Ga_2O_3$ (IMP) samples. Since another Ag–Ag pair of long distance was not recognized for the 1.0 wt% Ag/$Ga_2O_3$ (PD-H) sample, this Ag–O bond might be formed by the interaction of small Ag metal particles with $Ga_2O_3$ support.

## 4. Conclusion

We investigated the structural and chemical states of the Ag species as well as its interaction with the $Ga_2O_3$ surface by measuring Ag $L_3$ and K-edge XAFS of the Ag/$Ga_2O_3$ samples. There were some kinds of Ag species such as around 1 nm sized Ag metal cluster in the Ag low loading samples and metallic Ag particles in the Ag high loading samples. The $AgGaO_2$-like interface structure would be formed between metallic Ag particles and the $Ga_2O_3$ surface, observed in the 0.5 and 1.0 wt% Ag/$Ga_2O_3$ (IMP) samples. In the 0.1 wt% Ag/$Ga_2O_3$ sample that was photocatalytically active sample, the Ag metal clusters with the size of around 1 nm were formed predominantly, and they probably accepted more electrons in the *d*-orbitals by the interaction with the $Ga_2O_3$ surface. These differences in structural and chemical states of the Ag species would affect the photocatalytic activity of $CO_2$ reduction with water.

**Acknowledgments**

This work was supported by a Grant-in-Aid for Scientific Research on Innovative Areas "All Nippon Artificial Photosynthesis Project for Living Earth (AnApple)" (No. 25107515 and 15H00871) from the Japan Society for the Promotion of Science (JSPS) and the Program for Element Strategy Initiative for Catalysts & Batteries (ESICB) commissioned by the MEXT of Japan.

**References**

[1] Kubacka A, Fernández-García M and Colón G 2012 *Chem. Rev.* **112** 1555
[2] Habisreutinger S N, Schmidt-Mende L and Stolarczyk J K 2013 *Angew. Chem. Int. Ed.* **52** 7372
[3] Chen D, Zhang X and Lee A F 2015 *J. Mater. Chem. A* **3** 14487
[4] Ola O, Maroto-Valer M M 2015 *J. Photochem. Photobiol. C* **24** 16
[5] Teramura K, Wang Z, Hosokawa S, Sakata Y and Tanaka T 2014 *Chem. Eur. J.* **20** 9906
[6] Yamamoto N, Yoshida T, Yagi S, Like Z, Mizutani T, Ogawa S, Nameki H and Yoshida H 2014 *J. Surf. Sci. Nanotechnol.* **12** 264
[7] Yamamoto M, Yoshida T, Yamamoto N, Yoshida H and Yagi S 2014 *J. Surf. Sci. Nanotechnol.* **12** 299
[8] Yamamoto M, Yoshida T, Yamamoto N, Nomoto T, Yamamoto Y, Yagi S and Yoshida H 2015 *J. Mater. Chem. A* **3** 16810.
[9] Behrens P 1992 *Solid State Commun.* **81** 235
[10] Bzowski A, Yiu Y M and Sham T K 1993 *Jpn. J. Appl. Phys.* **32** 691
[11] Bzowski A, Sham T K and Yiu Y M 1994 *Phys. Rev. B* **49** 13776
[12] Attia Y A, Buceta D, Requejo F G, Giovanetti L J and López Q M A 2015 *Nanoscale* **7** 11273
[13] Kandpal H C and Seshadri R 2002 *Solid State Sci.* **4** 1045

# Oxidation property of $SiO_2$-supported small nickel particle prepared by the sol-gel method

**Y Yamamoto, S Yamashita, N Afiza, M Katayama, and Y Inada***

Department of Applied Chemistry, Ritsumeikan University, Kusatsu, Shiga 525-8577, Japan

E-mail: yinada@fc.ritsumei.ac.jp

**Abstract.** The oxidation property of $SiO_2$-supported small Ni particle has been studied by means of the *in-situ* XAFS method. The Ni particle with the average diameter of 4 nm supported on $SiO_2$ was prepared by the sol-gel method. The XANES spectrum of the small metallic Ni particle was clearly different from that of bulk Ni. The exposure of diluted $O_2$ gas at room temperature promoted the surface oxidation of Ni(0) particle. During the temperature programmed oxidation process, the supported Ni(0) particle was quantitatively oxidized to NiO, and the oxidation temperature was lower by *ca.* 200 °C than that of the $SiO_2$-supported Ni particle with the larger particle radius of 17 nm prepared by the impregnation method.

## 1. Introduction

The supported Ni catalyst shows the high activity for $CO_2$ reforming reaction of methane, and its improvements are continued to achieve the high selectivity and the long lifetime. The reduction of the particle size of active Ni species is one of the important approaches to control the catalytic performance. The fine particle of the active Ni species is expected to have high catalytic activity because of the large surface area. Under the actual reaction conditions, the Ni particles are exposed to the reactant and product gases, and thus the environmental atmosphere must affect to the chemical state of the Ni species. Furthermore, the possible aggregation and dispersion of the supported Ni particle modify the catalytic activity. The knowledge about the chemical state of the supported active Ni species in relation to the particle size is thus necessary to understand the catalysis performance and to construct the guiding design for the high performance Ni catalyst. In the present study, the clarification of the chemical state changeover of the $SiO_2$-supported Ni catalyst have been performed for the small Ni particles prepared by the sol-gel (SG) method during its oxidation process at elevated temperatures. The results will be compared with those for the corresponding Ni catalysts prepared by the conventional impregnation (IMP) method [1].

## 2. Experimental

The supported Ni catalyst was prepared by the SG method according to the reported procedure [2]. The Ni loading was set to be 5 wt%. Nickel(II) nitrate hexahydrate, citric acid, tetraethyl orthosilicate, and deionized water were dissolved in ethanol. The vaporization of the solution at 80 °C formed the solid gel, and it was calcined at 600 °C in air. The obtained Ni catalyst was reduced at 700 °C under $H_2$ gas flow diluted by Ar. The corresponding sample was also prepared by the IMP method [1]. The XRD patterns were recorded using Ultima IV (Rigaku) with the Cu Kα radiation. The transmission electron microscopy (TEM) observation was performed with JEOL2010 microscope (JEOL).

*In-situ* XAFS measurements during the temperature-programmed oxidation (TPO) process were carried out at the BL-12C station of Photon Factory (KEK, Japan) in transmission mode. The higher-order harmonics were removed by detuning the Si(111) double-crystal monochromator. Before the TPO measurements, the Ni catalyst was reduced at 700 °C under dilute $H_2$ gas flow (10 vol.%, 200 $cm^3$/min) in the observation cell to ensure the quantitative reduction. After that, the gas flow was switched to $O_2$ diluted by He (10 vol.%) with the total flow rate of 200 $cm^3$/min at room temperature. The sample was heated up to 700 °C with the increasing rate of 10 °C/min. The measurement time of one XAFS spectrum was 1 min, and the XAFS spectra were recorded with the interval time of 2 min. The EXAFS analysis was conducted by using the FEFFIT program. Fourier transformation was applied to $k^3$-weigted EXAFS data in the $k$ range from 2 to 12 Å$^{-1}$, and the curve fitting analysis was performed in the $R$ range from 1.0 to 3.2 Å. The amplitude reduction factor $S_0^2$ was set to 0.84 and 1.05 for metallic Ni and NiO, respectively. The edge shift energy $\Delta E_0$ was estimated to be 0.8 eV (Ni metal) and –0.2 eV (NiO). The statistical error ($R$ factor) for all fittings was less than 1 %.

## 3. Results and discussion

The observed XRD patterns are given in figure 1(A). The reduced samples show diffraction lines assigned to Ni(111) at 44.5° and Ni(200) at 51.9°, and it is confirmed that the reduction treatment at 700 °C generates the metallic Ni(0) particle. The diffraction lines corresponding to NiO(111) at 37.3°, NiO(200) at 43.2°, and NiO(220) at 62.9° were observed for the oxidized samples after the TPO process, and they clearly demonstrate the existence of NiO. It is noted that the diffraction lines of the SG sample are broader than those for samples prepared by the IMP method. The crystallite size of the supported Ni species for the SG sample is found to be smaller than the IMP sample. Figure 1(B) shows the size distribution of the supported Ni particles estimated by TEM after the reduction treatment. The average diameter of the Ni(0) particle for the SG and IMP sample was 4 nm and 17 nm with the size distribution range of 2–10 nm and 3–40 nm, respectively. It is clarified that the small and uniform Ni particles are generated in the SG sample.

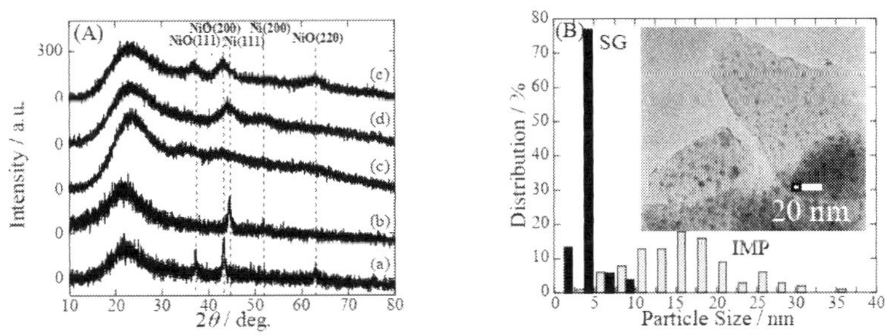


**Figure 1.** XRD patterns (A) of the calcined IMP sample (a), the reduced IMP sample (b), the calcined SG sample (c), the reduced SG sample (d), and the oxidized SG sample (e), and the particle size distribution of the reduced sample (B) estimated by the TEM images. A typical TEM image of the SG sample is inserted.

Figure 2 shows the XANES spectrum and the Fourier transform function of the reduced SG sample measured at room temperature under the $H_2$ atmosphere. The absorption edge energy of the reduced SG sample is in agreement with that of Ni foil. Furthermore, the Fourier transform function indicates that the Ni species with other valence states do not exist. These results indicates that the supported Ni species after the reduction treatment quantitatively reduced to Ni(0). However, the XANES spectrum of the supported Ni(0) particle is clearly different from that of bulk Ni(0). The difference spectrum given in figure 2(A) is obviously indicates that the origin of the spectral difference is not the partial oxidation of the Ni(0) particle, because the spectral difference is not consistent with the difference spectrum between Ni(0) and NiO (figure 2(C)). It is evident from these results that the supported small

Ni(0) particles show the specific XANES spectrum as shown in figure 2(A). The 4p band structure can be affected by the particle size.

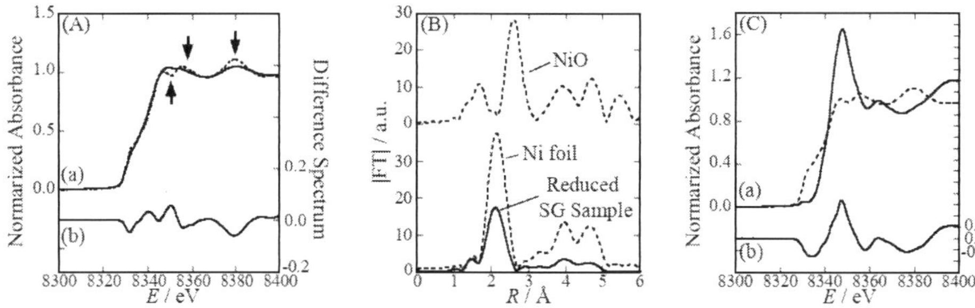


**Figure 2.** XANES spectrum (A) and Fourier transform function (B) of the reduced SG sample measured at room temperature. In A, the XANES spectra of the SG sample (solid line) and bulk Ni foil (broken line) are compared in (a), and the difference spectrum is given in (b). In C, the XANES spectrum of bulk NiO (solid line) and Ni foil (broken line) are compared in (a), and the difference spectrum is given in (b).

The XANES spectral change and the Fourier transform functions are shown in figure 3 for the TPO process. A rapid spectral change was observed just after the switch of the atmosphere gas from $H_2$ to $O_2$ at room temperature. The structure parameters determined by the curve-fitting procedure of the EXAFS data are summarized in table 1. The spectral change observed at room temperature resembles to the difference between Ni(0) and NiO given in figure 2(C). In addition, a weak interaction peak was generated at *ca.* 1.6 Å (see figure 3(B)) after the gas switch, and the average coordination number for the Ni–Ni interaction was decreased from 8.8 to 6.5. The composition analysis based on the linear combination of XANES spectra revealed that *ca.* 40 percent of the Ni species was oxidized to NiO. The similar spectral change was also observed at the beginning of the TPO process for the IMP sample, whereas the fraction of the partial oxidation (*ca.* 10 %) was small because of the larger particle size. The composition difference is quantitatively ascribed to the difference in the surface area of the supported Ni(0) particle. These results indicate that the surface oxidation of the Ni(0) particle proceeded at room temperature by the exposure to $O_2$.

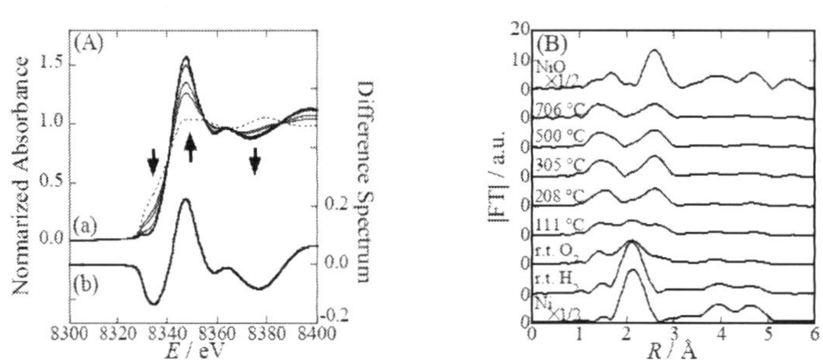


**Figure 3.** The XANES spectral change (A) and the Fourier transform functions (B) for the TPO process. In A, a broken line means the XANES spectrum before the gas switch to $O_2$, and the spectral difference at the gas change is shown in (b).

Figure 4 shows the temperature change of the X-ray absorbance, the composition change estimated by the linear combination fitting of XANES spectrum, and the average coordination number determined by the EXAFS analysis. The oxidation of small Ni(0) particles prepared by the SG method completed at 300 °C. It is lower by *ca.* 200 °C than the corresponding temperature for the larger Ni particles in the IMP sample. The temperature change of the average coordination number supports the lower shift of the oxidation temperature for the SG sample. The temperature difference indicates the relative stability of the smaller NiO particles under the $O_2$ atmosphere and is interpreted by the easy oxygen migration in the small Ni(0) particle. In addition, the lower shift of oxidation temperature suggests the increased surface active sites for the smaller Ni(0) particles.

**Table 1.** Structure parameters of the Ni species during the TPO process.

| Sample | $T$ / °C | Gas | Interaction | $N$ | $R$ / pm | $\sigma^2$ / 10 pm$^2$ |
|---|---|---|---|---|---|---|
| Ni foil | rt | Air | Ni–Ni | 12 (fixed) | 250 ± 1 | 6.3 ± 0.2 |
| NiO | rt | Air | Ni–O | 6 (fixed) | 206 ± 3 | 6.5 ± 1.5 |
| SG | rt | $H_2$ | Ni–Ni | 8.8 ± 0.4 | 250 ± 1 | 9.5 ± 0.4 |
| | rt | $O_2$ | Ni–Ni | 6.5 ± 0.6 | 253 ± 4 | 13.7 ± 0.8 |
| | | | Ni–O | 2.7 ± 0.3 | 202 ± 6 | 14.1 ± 1.9 |
| | 111 | $O_2$ | Ni–Ni | 5.4 ± 0.9 | 257 ± 8 | 16.8 ± 1.5 |
| | | | Ni–O | 3.7 ± 0.3 | 203 ± 5 | 17.8 ± 1.9 |
| | 208 | $O_2$ | Ni–Ni | 0.3 ± 0.5 | 264 ± 15 | 4.2 ± 11.0 |
| | | | Ni–O | 4.2 ± 0.9 | 203 ± 6 | 12.0 ± 4.0 |
| | 305 | $O_2$ | Ni–O | 5.9 ± 0.9 | 204 ± 4 | 15.8 ± 2.6 |
| | 706 | $O_2$ | Ni–O | 6.1 ± 0.7 | 205 ± 4 | 23.8 ± 2.6 |

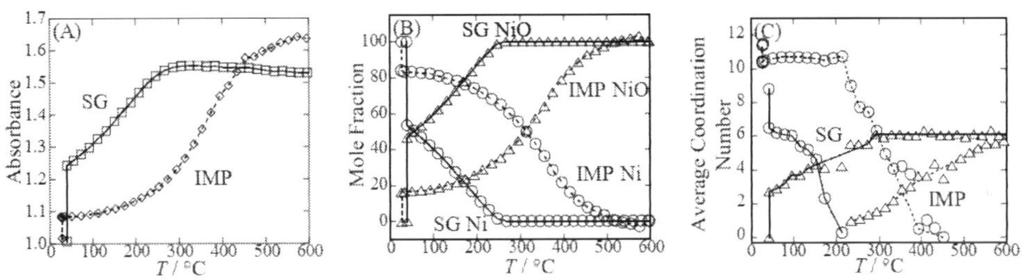


**Figure 4.** The temperature change of the X-ray absorbance (A) at the white line peak of NiO (8347 eV), the composition change (B), and the average coordination number (C) for the Ni–O interaction (triangles) in NiO and the Ni–Ni interaction (circles) in metallic Ni(0) during the TPO process. The solid and broken line means the change of the SG and IMP sample, respectively.

## 4. Conclusions

The solid-phase oxidation of the $SiO_2$-supported small Ni(0) particle prepared by the SG method has been clarified by means of the *in-situ* XAFS method. The small Ni(0) particles show the specific XANES spectrum, and the *in-situ* XAFS measurements demonstrated the surface oxidation at room temperature by the exposure to $O_2$. The temperature of the bulk oxidation becomes lower in comparison to that of the larger Ni(0) particle prepared by the conventional IMP method.

## Acknowledgement

The XAFS measurements have been performed under the approval of the Photon Factory Program Advisory Committee (Proposal Nos. 2011G531 and 2013G596).

## References

[1] Yamashita S, Katayama M and Inada Y 2013 *J. Phys. Conf. Ser.* **430** 012051
[2] Wu C and Williams P T 2012 *Environ. Technol.* **33** 631

# XAFS study on a photodeposition process of Pt nanoparticles on $TiO_2$ photocatalyst

**T. Yoshida*[1,2], Y. Minoura[3], Y. Nakano[3], M. Yamamoto[3], S. Yagi[1], H. Yoshida[4,5]**
[1]EcoTopia Science Institute, Nagoya University, Nagoya 464-8603, Japan,
[2]The OCU Advanced Research Institute for Natural Science and Technology, Osaka City University, Osaka 558-8585, Japan
[3]Graduate School of Engineering, Nagoya University, Nagoya 464-8603, Japan,
[4]Kyoto University, Graduate School of Human and Environmental Studies, Kyoto 606–8501, Japan
[5]Elements Strategy Initiative for Catalysts and Batteries, Kyoto University, Kyoto, 606-8501, Japan

E-mail: tyoshida@ocarina.osaka-cu.ac.jp

**Abstract.** A photodeposition process of Pt metal particles on anatase $TiO_2$ in the aqueous solution of $H_2PtCl_6$ (precursor) and methanol (reductant) was studied using transmission electron microscopy, UV-vis spectroscopy and X-ray absorption fine structure spectroscopy. These analyses proposed the photodeposition mechanism of Pt on $TiO_2$ by the photogenerated electrons: the photo-assisted adsorption of $Pt^{4+}$ complexes on $TiO_2$ through the ligand exchange at the initial stage, followed by the successive rapid reduction of $Pt^{4+}$ to $Pt^0$ to grow the Pt metal particles, which gives small and almost uniform size of Pt metal nanoparticles.

## 1. Introduction

Photoreduction is one of the most popular methods for recovering noble metals, removing metal cations from aqueous effluents and preparing metal-loaded photocatalysts [1-5]. The UV irradiation on a semiconductor material such as $TiO_2$ photocatalyst can form photoexcited electrons on the surface, which can reduce some kinds of metal cations having appropriate redox potential to produce metal particles on the surface. The interaction between the semiconductor and the metal could play an important role for the photocatalytic activity, since the support-metal junction could influence the electronic properties of them, the electron transfer between them and the morphology of the metal particles [6,7]. However, there are only a few studies dealing with the mechanism of the photodeposition process and the structure of the photodeposited metal species [8-10].

Recently, Ohyama *et al.*[8] studied the photodeposition process of Rh metal particles on $TiO_2$ using in situ time-resolved energy dispersive X-ray absorption fine structure spectroscopy (DXAFS) and reported that the $RuCl_3$ precursor was reduced to form Rh metal particles where the coordination number of the reduced Rh atom increased proportionally with the photoirradiation time, suggesting that fine Rh metal particles with uniform size appeared one after another. In the present study, we examined the photodeposition processes of Pt metal particles on anatase $TiO_2$ by means of XAFS, together with UV-vis spectroscopy and transmission electron microscopy.

## 2. Experimental

Photodeposition of Pt metal particles on $TiO_2$ was carried out as follows: $H_2PtCl_6 \cdot 6H_2O$ powder of 13.4 mg was dissolved in a methanol aqueous solution of 60 mL (the mixture of 50 mL of distilled water and 10 mL of methanol). 1.0 g of an anatase $TiO_2$ sample (Kishida Kagaku Co., BET surface area was 5.3 $m^2/g$) calcined at 673 K in air was suspended in the aqueous solution of the Pt precursor. The suspension was irradiated for various periods (0–180 min) with a 300 W Xe lamp equipped with a band path filter (ca. 340±30 nm, 30 $mW/cm^2$), and then filtered and washed with purified water.
UV-vis diffuse reflectance spectra were measured by a JASCO V-670 spectrometer. TEM images were obtained with a JEOL JEM-2100M transmission electron microscope operating at an accelerating voltage of 200 kV. TEM samples were prepared by depositing drops of methanol suspension containing small amounts of the sample powders onto a carbon-coated copper grid and allowing the methanol to evaporate in air. The XAFS measurements at the Pt $L_3$-edge were carried out at BL-9C [11] of the Photon Factory at the High Energy Accelerator Research Organization (KEK, Tsukuba Japan, proposal number 2011G575 and 2014G548) with a Si(111) double-crystal monochromator in the fluorescence mode by using a Lytle detector filled with an Ar(100%) flow with a Ga filter ($\mu t = 6$). Ion chamber of the $I_0$ detector was filled with a $N_2$(85%)/Ar(15%) flow. The above UV-vis, TEM and XAFS measurements were conducted ex-situ for the Pt species deposited/adsorbed on $TiO_2$ samples.

## 3. Results and discussion

Difference UV-vis spectra of the Pt/$TiO_2$ samples are shown in Figure 1. The difference spectrum of each sample was obtained by subtracting UV-vis spectrum of the pristine $TiO_2$ sample. As shown in Figure 1, the spectra of the Pt/$TiO_2$ samples prepared by photodeposition for 5, 10, 15 min showed a band around 400 nm, which would be attributed to the Pt precursors and possibly increasing Pt metallic clusters on the $TiO_2$ surface. Note that the band became broader and flat feature for the Pt/$TiO_2$ samples prepared by photodeposition for 20 min, which would originate from metal nanoparticles. This means that the aggregation of Pt atoms occurred under photoirradiation. After that, the band intensity increased with the further photoirradiation whereas the flat spectral profile did not change. These results suggest that the number or the particles size of the Pt nanoparticles on the surface increased with photoirradiation time.

The size of the Pt nanoparticles was observed by TEM measurements. Although the Pt particles were not observed in the Pt/$TiO_2$ samples prepared by photodeposition for less than 20 min, the Pt nanoparticles were observed and the average size increased from ca. 3.3 to 4 nm with the increase of the photoirradiation time more than 25 min as shown in Figure 2. As mentioned, the absorbance clearly became large as shown in Figure 1, though the size did not varied so drastically as shown in Figure 2. These results suggest that the number of the Pt nanoparticles with similar size of ca. 3–4 nm increased during the photoirradiation after 20 min.

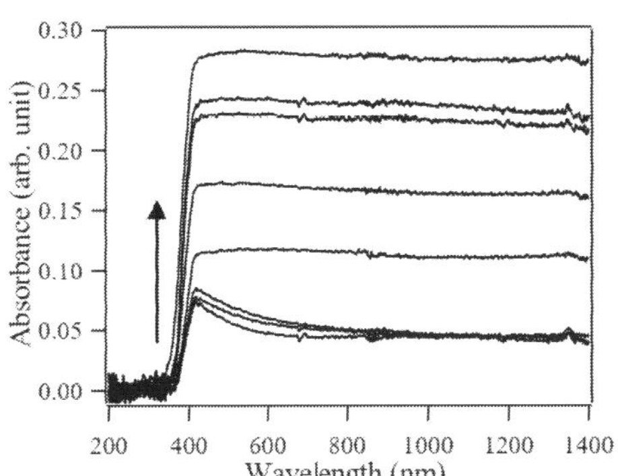


**Figure 1.** Difference diffuse reflectance UV-Vis spectra of a series of Pt species on $TiO_2$ by photodeposition for 5, 10, 15, 20, 25, 30, 60 and 180 minutes.

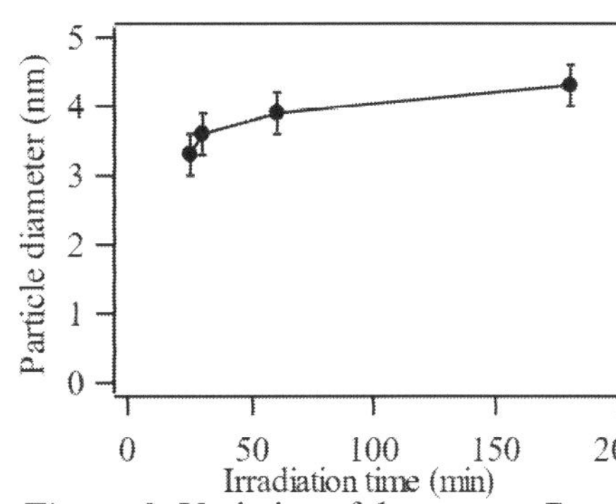


**Figure 2.** Variation of the average Pt particle size with photoirradiation time. The sizes were obtained from TEM images.

We also investigated the photodeposition process of the Pt metal nanoparticles by XAFS spectroscopy. Figure 3(A) shows normalized Pt $L_3$-edge XANES spectra of the Pt/$TiO_2$ samples together with Pt foil and $PtO_2$ as references. Note that these XANES spectra were normalized by the jump of Pt $L_3$-edge, i.e., normalized by a Pt atom on the $TiO_2$ surface. The sharp and narrow absorption band at an absorption edge, which is called white line [12], corresponds to the electronic transition from $2p_{3/2}$ core level states to 6s and 5d orbitals of Pt atoms mainly. The large area of $L_3$ edge absorption around 11565 eV for $PtO_2$ results from the vacancy in the 5d orbital of Pt atoms. As shown in Figure 3(A), the feature of XANES spectrum of the $TiO_2$ sample adsorbing the Pt precursor before photoirradiation was almost similar to that of the reference $PtO_2$ sample, meaning that the adsorbed Pt species also had similar octahedral structure as $PtO_2$. Only the difference between the two spectra was the very small peak around 11.545 keV, which is assignable to $L_2$ edge absorption (11.54 keV) of tungsten as an impurity in the $TiO_2$ support. The small peak was clearly observed for the Pt/$TiO_2$ samples prepared by the photodeposition for 0, 5, 10 and 15 min, suggesting that the amount of $Pt^{4+}$ precursors stabilized on $TiO_2$ surface were still small in the initial stage of the photodeposition.

The energy position of the main peak of XANES shifted to lower X-ray energy with the elongation of photoirradiation time, indicating a decrease in the ratio of $Pt^{4+}$ ions by reduction. Figure 3(B) shows the variation of the spectral shape in this process, where the XANES spectra changed from that of $Pt^{4+}$ ion to that of $Pt^0$ with an isosbestic point, indicating that $Pt^{4+}$ ions were reduced to $Pt^0$ without through any stable intermediates. This means that the photoreduction from the adsorbed Pt species to metallic Pt nanoparticles take place drastically.

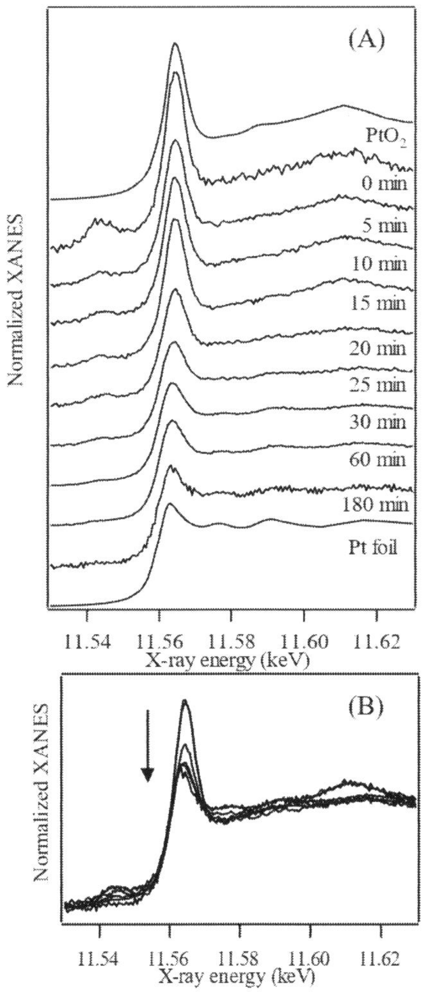


**Figure 3.** Pt $L_3$-edge XANES spectra of Pt Pt/$TiO_2$ samples under photoirradiation for 0, 5, 10, 15, 20, 25, 30, 60 and 180 minutes together with a Pt foil and $PtO_2$.

Actually, all XANES spectra of the Pt/$TiO_2$ samples were reproduced with the linear combination of two XANES spectra of $PtO_2$ and Pt metal, and the fractions of the $Pt^{4+}$ ion and the $Pt^0$ atom for all the Pt/$TiO_2$ samples were directly evaluated as shown in Figure 4. $Pt^{4+}$ was the main component in the Pt/$TiO_2$ samples prepared by photodeposition for 0, 5, 10, and 15 min while the reduction of $Pt^{4+}$ to $Pt^0$ occurred drastically after 15 min and almost completed after 30 min. Here, we also evaluated the height of edge jump in the raw XAFS spectra of the Pt/$TiO_2$ samples recorded in the fluorescence mode in order to compare the amount of the Pt species deposited on these samples. Figure 5 shows the variation of the relative height of the edge jump with the photodeposition time. As is evident from Figures 4 and 5, in the initial stage before 15 min, the valence of the adsorbed Pt species was almost tetravalent and the amount of the Pt species increased. It is clear that ca. 20 % of $Pt^{4+}$ complexes were adsorbed on the $TiO_2$ before photodeposition (0 min) and the $Pt^{4+}$ complexes increased to ca. 50 % by photo-assisted adsorption after 15 min. In the successive period for 20–30 min, the amount of the adsorbed Pt species drastically increased and the reduction of $Pt^{4+}$ to $Pt^0$ drastically occurred on the photoirradiated $TiO_2$. After that, the faction of $Pt^0$ atoms gradually increased by further photoreduction.

These results lead us to speculate the Pt deposition mechanism on the $TiO_2$ surface. On the whole, photoexcited electrons reduce the Pt precursor ($[Pt(IV)Cl_6]^{2-} + 4\ e^- \rightarrow Pt(0) + 6\ Cl^-$), while positive holes oxidize methanol in the solution. The details for the photoreduction would be described as follows: (i) during the induction period, the photoexcited electron would stabilize the precursor with ligand-exchange at the surface ($[Pt(IV)Cl_6]^{2-}$ + $-OH_s^-$ + $e^-$ + $H^+ \rightarrow$ $[Pt(IV)Cl_5O_s]_{ad}^{2-}$ + $Cl^-$ + $H_2$), where the initial amount of the adsorbed Pt complex would be limited by the amount of the surface adsorption sites. (ii) the surface $Pt^{4+}$ complex would be further reduced by the photoexcited electrons to form the Pt metal seeds ($[Pt(IV)Cl_5O_s]_{ad}^{2-} + 3\ e^- \rightarrow Pt(0)O_s + 5\ Cl^-$), and (iii) further photoexcited electrons would drastically reduce the residual $Pt^{4+}$ complexes over the Pt metal seeds to grow itself and to form the Pt metal particles, where the photoexcited electron could be effectively lead to the Pt metallic particles to accelerate the reduction rate of the residual $Pt^{4+}$ complex. The growth of the Pt particles could be limited by the amount of the Pt complexes in the solution.

We have started to investigate the effects of the amount of the Pt precursors and the kind of $TiO_2$ structural phase on the photodeposition process as well as the morphology of the Pt particles on $TiO_2$.

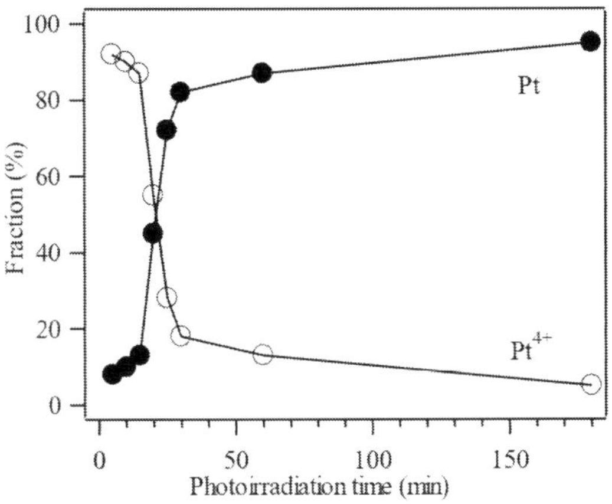


**Figure 4.** Dependence of the fractions of Pt metal and $Pt^{4+}$ ions on the photoirradiation time.

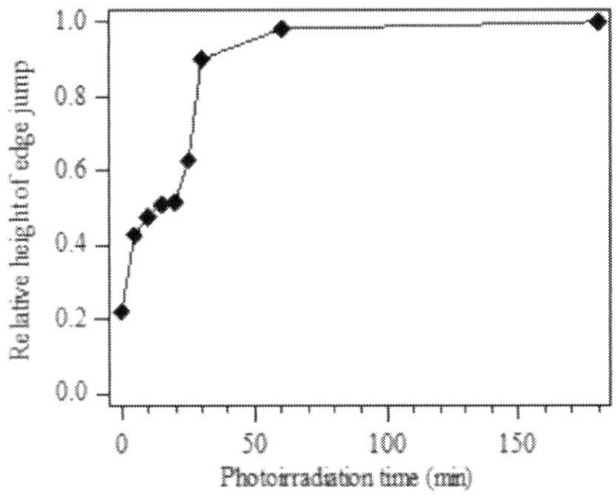


**Figure 5.** Dependence of the relative height of the edge jump on the photoirradiation time.

## 4. Conclusion

We studied photodeposition process of Pt particles on an anatase $TiO_2$ sample. The $Pt^{4+}$ precursors were found to be photo-adsorbed on the $TiO_2$ surface through ligand exchange at the initial stage. After the formation of the metallic Pt seeds, they were drastically reduced to form the Pt metal particles with almost similar size of 3–4 nm in the present condition.

## References

[1] Kraeutler B and Bard A. J, *J. Am. Chem. Soc.* 1977 **99** 7729.
[2] Kraeutler B and Bard A. J *J. Am. Chem. Soc.* 1978 **100** 4317.
[3] Duonghong D, Borgarello E and Graetzel M, *J. Am. Chem. Soc.* 1981 **103** 4685.
[4] Courbon H, Herrmann J M and Pichat P, *J. Phys. Chem.* 1984 **88** 5210.
[5] Goren Z, Willner I, Nelson A J and Frank A J, *J. Phys. Chem.* 1990 **94** 3784.
[6] Bond GC 1982 Metal-support and metal-additive effect in catalysis, (Amsterdam, Elsevier) p 1.
[7] Thomson SJ, *J Chem Soc Faraday Trans Part I* **83** 1893.
[8] Ohyama J, Teramura K, Okuoka K, Yamazoe S, Kato K, Shishido T and Tanaka T 2010 *Langmuir* **26** 13907.
[9] Ohyama J, Yamamoto A, Teramura K, Shishido T and Tanaka T, 2011 *ASC Catal.* **1** 187.
[10] Tanaka T, Ohyama J, Teramura T and Hitomi Y, *Catal. Today*, 2012 **183** 108.
[11] http://pfxafs.kek.jp/beamline/bl9c
[12] Brown M, Peiels RE and Stern EA 1977 *Phys Rev B* **15** 738.

# *In situ* observation of reduction kinetics and 2D mapping of chemical state for heterogeneous reduction in iron-ore sinters

**M Kimura[1,2,3], R Murao[3], N Ohta[4], K Noami[4], Y Uemura[1], Y Niwa[1], K Kimijima[1], Y Takeichi[1,2], H Nitani[1,2]**

[1]Photon Factory, Institute of Materials Structure Science, High Energy Accelerator Research Organization, Tsukuba, Ibaraki 305-0801, Japan

[2]Department of Materials Structure Science, School of High Energy Accelerator Science, SOKENDAI (The Graduate University for Advanced Studies), Tsukuba, Ibaraki 305-0801, Japan

[3]Advanced Technology Research Laboratories, Nippon Steel & Sumitomo Metal Co., Chiba 293-8511, Japan

[4]Nippon Steel & Sumikin Technology Co. Ltd., Chiba 293-0011, Japan

E-mail: masao.kimura@kek.jp

**Abstract.** Iron-ore sinters constitute the major component of the iron-bearing burden in blast furnaces, and the mechanism of their reduction is one of the key processes in iron making. The heterogeneous reduction of sintered oxides was investigated by the combination of X-ray fluorescence and absorption fine structure, X-ray diffraction, and computed tomography. Two-dimensional mapping of the chemical states (CSs) was performed. The iron CSs $Fe^{III}$, $Fe^{II}$, and $Fe^{0}$ exhibited a heterogeneous distribution in a reduced sinter. The reduction started near micro pores, at iron-oxide grains rather than calcium-ferrite ones. The heterogeneous reduction among grains in a sinter may cause the formation of micro cracks. These results provide fundamental insights into heterogeneous reduction schemes for iron-ore sinters.

## 1. Introduction

Iron-ore sinters constitute the major component of the iron-bearing burden in blast furnaces in most countries in the Asia-Pacific region. They are formed by liquid-sintering at $T > 1{,}500$ K and are composed of an iron-oxide phase (mainly $\alpha$-$Fe_2O_3$ and partially $Fe_3O_4$) and calcium ferrites (CFs) [1,2]. Iron-oxide grains with a diameter ranging from several tens to hundreds of μm are bonded by CFs such as $CaFe_2O_4$ and $Ca_2Fe_2O_5$. CFs contain many pores with diameters ranging from micrometers to millimeters (Figure 1). In industrial processes, the speed of reduction must be optimized, and the degradation due to the mechanical strength during the reduction must be minimized, as iron-ore sinters are piled up as high as 30–40 m and reduced in blast furnaces. Therefore, it is very important to understand where the reduction starts and how it progresses among sinters. We investigated the reduction process with respect to the iron chemical-state (CS) changes ($Fe^{III}$, $Fe^{II}$, and $Fe^{0}$) and crack formation among sinters by using X-ray fluorescence (XF), X-ray absorption fine structure (XAFS), and X-ray computed tomography (X-CT) on the scale of a few tens of micrometers.

## 2. Experimental

The iron-oxide ores were mixed with limestone flux and coke breeze and heated by the combustion of the coke breeze, yielding heating up to temperatures of 1,450–1,600 K (above the eutectic temperature of CaO–$Fe_2O_3$: 1,478 K). The sinter was heated and reduced at $T$ = 773 K for 1 h in a gas with a molar ratio of CO:$CO_2$ ~ 1:1. To understand the intrinsic kinetics of the reduction of each phase, *in situ* quick XAFS (QXAFS) measurements were conducted for a powder specimen at BL-9A and BL-9C of the Photon Factory (PF), IMSS, KEK. Iron *K*-edge and calcium *K*-edge XANES spectra were collected at 1,173 K in a He-20 vol. % $H_2$ atmosphere in the transmission mode with a typical time resolution of 15–30 s, and a reaction cell was used for the *in situ* QXAFS [3,4].
Chemical mapping was performed at BL-15A1 [5] at the PF. XAFS and X-ray diffraction (XRD) were conducted for the sinter surface using an X-ray beam ~20 μm in diameter by scanning the specimen at room temperature. XAFS spectra were measured using a silicon drift detector (SDD) in the fluorescence geometry (Figure 2). To estimate the fractions of the iron chemical states (CSs)—$Fe^{III}$, $Fe^{II}$, and $Fe^0$—the fluorescence intensities ($I$) were measured at each point of the specimen, and the ratio of $CS_{Fe} = (I_{at\ 7112eV} - I_{at\ 7100eV})/ (I_{at\ 7200eV} - I_{at\ 7100eV})$ was used as an indicator of the CSs, as the iron *K*-edge energy shifted by ~5 eV around 7,112 eV among $Fe^{III}$, $Fe^{II}$, and $Fe^0$.
X-CT measurements were performed for a sinter 2 × 2 × 10 $mm^3$ in size using an in-house source with an X-ray energy of 70 keV. A charge-coupled device camera with a scintillator was used as a detector. The transmission image was measured with a step of 1 or 2° in a 360° rotation. The minimum special resolution was 0.7 μm.

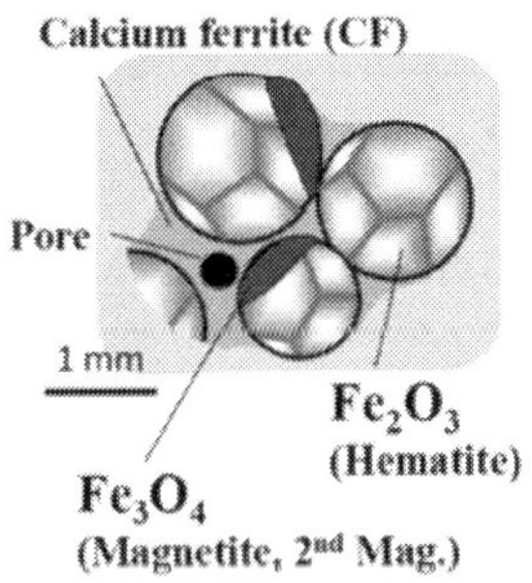


**Figure 1.** Schematic showing the typical microstructure of iron-ore sinters.

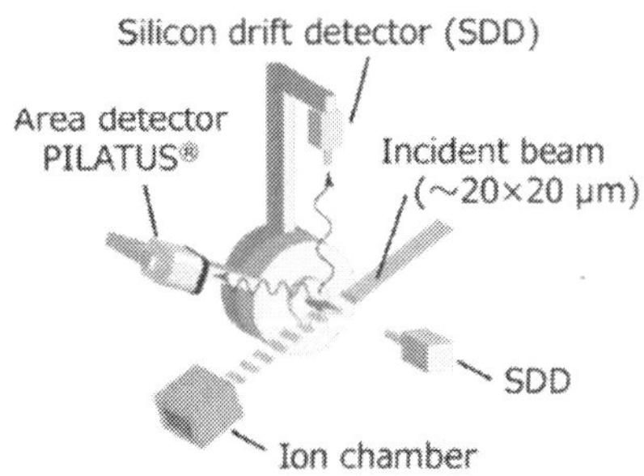


**Figure 2.** Outline of the experimental setup at BL-15A1. We can locate the SDD at two locations.

## 3. Results and discussion

### 3.1. In situ XAFS observation for intrinsic kinetics of reduction

The change in the X-ray absorption near edge structure (XANES) spectra around the iron *K*-edge energies for pellets of $Fe_2O_3$, $Fe_3O_4$, and $CaFe_2O_4$ at 1,173 K with a gas flow of $H_2$/He = 20/200 ccm was obtained by *in situ* QXAFS at the time $t$ after the gas was switched from $O_2$/He to $H_2$/He. A detailed analysis, including a linear combination, was performed for each XANES spectrum, revealing that the total periods of the reduction ($Fe^{III} \rightarrow Fe^0$) of $Fe_2O_3$ ($RedTime_{Fe2O3}$), $Fe_3O_4$ ($RedTime_{Fe3O4}$), and $CaFe_2O_4$ ($RedTime_{CaFe2O4}$) had the following relationship: $RedTime_{Fe2O3} < RedTime_{Fe3O4} \approx RedTime_{CaFe2O4}$. As the results of the simple analysis of the change in the iron *K*-edge energies are consistent with the results of the detailed analysis, the change in the iron *K*-edge energies at the maximum of the first derivation for the specimens is shown in Figure 3. According to the detailed analysis, the periods corresponding to $Fe^{III}$ + $Fe^{II}$ and $Fe^{II}$ + $Fe^{0}$ are shown in pink and green, respectively, in Figure 3. The period of $Fe^{III} \rightarrow Fe^{II}$ in $Fe_2O_3$ was ~5 times faster than those of $Fe_3O_4$ and $CaFe_2O_4$. In $Fe_3O_4$, $Fe^{III}$ atoms were located at the centers of $Fe^{III}$-$O_4$ tetrahedrons and ($Fe^{II}$, $Fe^{III}$)-$O_6$ octahedrons, and $Fe^{II}$ atoms that were not reduced in the first step existed between $Fe^{III}$ atoms that were to be reduced. In $CaFe_2O_4$, the stacks of $Fe^{III}$-$O_6$ octahedrons were adjacent to the stacks of $Ca^{II}$-$O_6$ octahedrons. The existence of other atoms that were not reduced at the first step may explain why it took longer for $Fe^{III}$ to be reduced in $Fe_3O_4$ than in $Fe_2O_3$.

No induction period was observed in $CaFe_2O_4$ by the calcium *K*-edge XANES measurements for $CaFe_2O_4$. Figure 4 shows the change in the absorption intensity at 4,047 eV, corresponding to the white line, at the time *t*. No induction period, as is observed in the iron *K*-edge XANES in Figure 3(c), was observed at the calcium *K*-edge. This suggests that the structure around calcium changed but the iron CS did not change in the period of $t$ = 0 to 2 min. A detailed analysis of the XAFS spectra after the reduction and the *in situ* XRD measurements during the reduction (to be submitted elsewhere) suggested that the following reaction occurred: 2 $CaFe_2O_4 \rightarrow 2FeO + Ca_2Fe_2O_5 + 1/2O_2 \rightarrow 4Fe + 2CaO + 3O_2$.

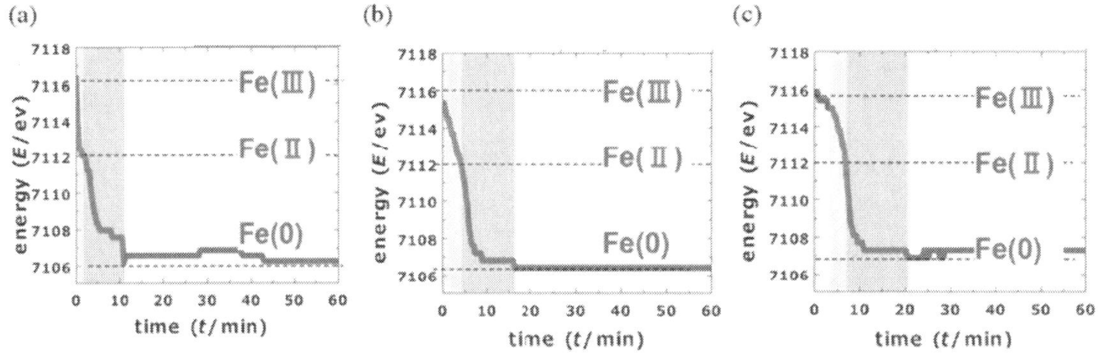


**Figure 3.** Change in the iron *K*-edge energies of pellets at 1,173 K with a gas flow of $H_2$/He = 20/200 ccm for (a) $Fe_2O_3$, (b) $Fe_3O_4$, and (c) $CaFe_2O_4$. The dotted lines show the iron *K*-edge energies at the maximum of the first derivation for the reference materials, i.e., $Fe_2O_3$, FeO, and Fe (foil).

### *3.2. Ex situ two-dimensional mapping of chemical states of iron and X-CT measurements*

Two-dimensional (2D) mapping by XAFS and XRD showed that the prepared sinter was composed of α-$Fe_2O_3$ (~55 vol. %), $Fe_3O_4$ (~20 vol. %), $CaFe_2O_4$ (~20 vol. %), and $Ca_2Fe_2O_5$ (~5 vol. %) over an area of 3 × 3 $mm^2$. Identical 2D mapping was performed for the reduced sinter. We attempted a linear-combination analysis for all the points in the 2D mapping, but this was rather difficult because the XANES spectra were measured with the fluorescence geometry in spite of the high iron concentration. Therefore, we introduced the chemical states (CSs) index of $CS_{Fe}$, which corresponds to the shift of the iron *K*-edge energies at the maximum of the first derivation, between $Fe^{III}$ ($CS_{Fe}$=0.0) and $Fe^{II}$ ($CS_{Fe}$ = 1.0). The CS of $CS_{Fe}$ = 0.5 corresponds to $Fe_3O_4$.

Figure 5 shows a typical example of the 2D mapping at the iron *K*-edge, showing the CS index of $CS_{Fe}$ for the cross section of the reduced sinter. The observed area (Figure 5) corresponds to an interface area where $Fe_2O_3$ grains were sintered to each other with CFs (bonding phases). The CS index of $CS_{Fe}$ was ~0.5, corresponding to $Fe_3O_4$, in most parts of the specimen, which is consistent with the macroscopic XRD measured over an area of 3 × 3 $mm^2$ at $E$ = 7,080 eV. The attenuation lengths in these measurements were ~20 and 8 μm at energies lower and higher than the iron *K*-edge, respectively. Microscopic XRD measurements with an X-ray beam of ~20 μm were also performed, but the type of the diffraction pattern largely depended on the grain size, and it was difficult to obtain the 2D mapping of the crystal structure, which was available in the CS mapping.

Among the region corresponding to $Fe_3O_4$, as shown in Figure 5, the $CS_{Fe}$ values were higher in certain regions, such as those indicated by the blue bold lines, and lower in others, such as those indicated by the pink dotted lines. The regions with lower $CS_{Fe}$ values, where the reduction progressed at a relatively slow speed, may correspond to $CaFe_2O_4$ grains in the as-prepared state (i.e., before reduction) because the intrinsic kinetics of reduction are high in $Fe_2O_3$ and low in $CaFe_2O_4$, as described in Section 3.1. The formation of the regions with higher $CS_{Fe}$ values, where the reduction progressed at a relatively high speed, is attributed to the pores and/or micro cracks, which may have enhanced the diffusion of the reductive gas. Thus, a microstructural analysis using X-CT was performed.

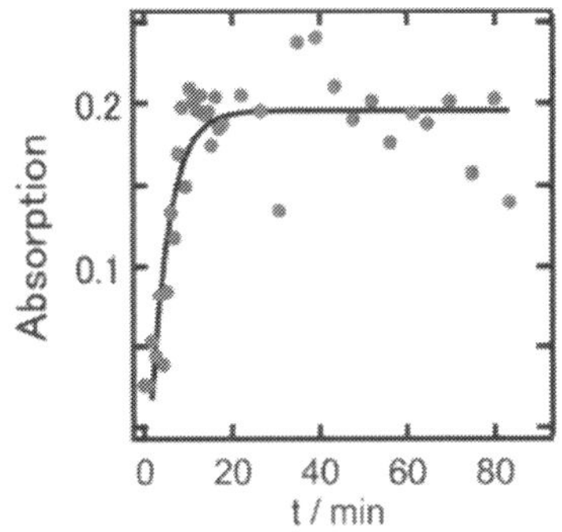


**Figure 4.** Change in the absorption intensity at 4,047 eV for $CaFe_2O_4$. The line is a visual guide.

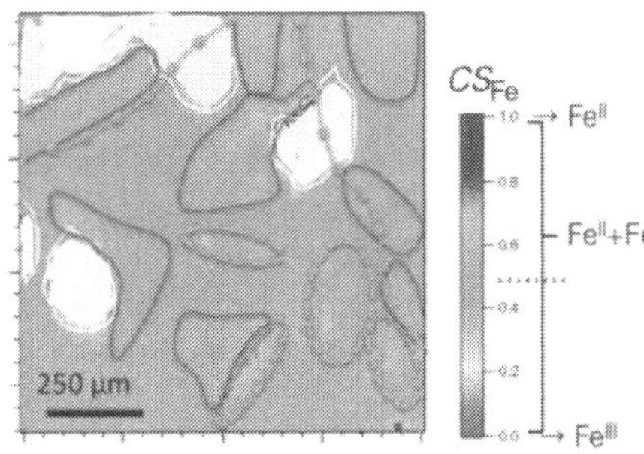


**Figure 5.** CS mapping of iron, where the $CS_{Fe}$ value is indicated by a color scale. The pores are colored white for simplicity, according to optical microscopy measurements. The gray lines with dots show the positions of the cracks observed in Figure 6.

This kind of heterogeneous reduction in the bulk sinters is expected to easily cause micro cracks because the reduction yields a large volume decrease [6]. The three-dimensional microstructure of the other side of the specimen used in the 2D mapping (Figure 5) was investigated by X-CT (Figure 6). Because of the cutting process, the results shown in Figures 5 and 6 do not correspond exactly to the same surface but rather nearby regions. The cracks shown in Figure 6 were located near the boundaries of the high $CS_{Fe}$ regions and/or pores. This suggests that the interface between the reduction-advanced and the reduction-delayed areas and that between the reduction-advanced area and the pre-existing pore are possible sites for cracking. The heterogeneous progress of the reduction may trigger cracking, and further investigation of the correlation of the heterogeneity and the crack initiation is in progress.

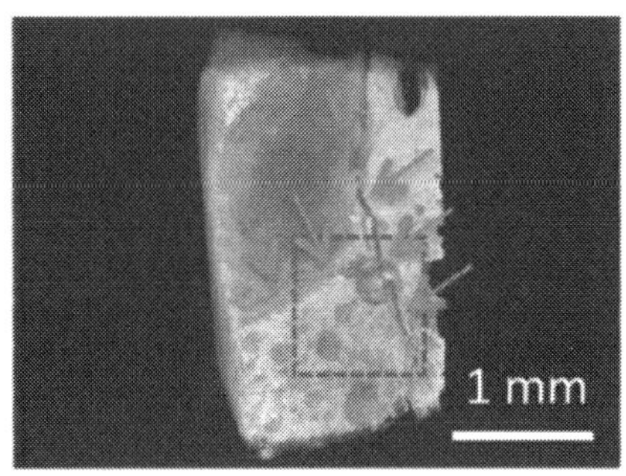


**Figure 6.** X-CT image for the reduced sinter. Cracks are indicated by arrows. The blue square shows the corresponding area in the 2D mapping (Figure 5).

## 4. Conclusion

The heterogeneous reduction reaction of iron-ore sinters was investigated by the combination of XF and XAFS, XRD, and X-CT. To estimate the fractions of iron chemical states (CSs) —$Fe^{III}$, $Fe^{II}$, and

$Fe^0$—we performed 2D mapping of the ratio of the fluorescence intensities at different energies and successfully showed the heterogeneous distribution of the CSs within the reduced sinter. It is suggested that the reduction proceeded at iron-oxide grains rather than CF ones, which is consistent with the intrinsic reduction-kinetics of the individual phase that were obtained by *in situ* QXAFS. The heterogeneous reduction can be also attributed to the existence of pores and/or micro cracks, which may have enhanced the diffusion of the reductive gas. The behavior of the heterogeneous reduction in sinters may be deeply related to the reduction speed and the degradation of the mechanical properties. The approach shown in this study provides fundamental insights for designing an ideal microstructure for iron-ore sinters, which may save energy and natural resources.

**Acknowledgement**

A part of this work, performed by M. Kimura, Y. Niwa, K. Kimijima, Y. Takeichi, and H. Nitani, was supported by the Cross-Ministerial Strategic Innovation Promotion Program (SIP, unit D66) operated by the cabinet office. Experiments at the PF were performed under the approval of the PF Program Advisory Committee (Proposal No. 2014G707 and 2015S2-002). The cooperative research between Nippon Steel & Sumitomo Metal Co. and KEK is highly appreciated. We thank Ms. M Ito for his support with the experiments.

**References**

[1] Ahsan S N, Mukherjee T and Whiteman J A 1983 *Ironmak. Steelmak.* **10** 54.
[2] Lu L, Holmes R J and Manuel J R 2007 *ISIJ Int.* **47** 349-58.
[3] Kimura M, Uemura K, Nagai T, Niwa Y, Inada Y and Nomura M 2009 *J. Phys: Conf. Ser.* **190** 012163.
[4] Kimura M, Uemura Y, Takayama T, Murao R, Asakura K and Nomura M 2013 *J. Phys: Conf. Ser.* **430** 012074.
[5] Igarashi N, Shimizu N, Koyama A, Mori T, Ohta H, Niwa Y, Nitani H, Abe H, Nomura M and Shioya T 2013 *J. Phys. Conf. Ser.* **425** 072016.
[6] Turkdogan E, Olsson R and Vinters J 1971 *Metall. Mater. Trans. B* **2** 3189.

# Probing changes in Hg(II) coordination during its bacterial uptake

**Sara Anne Thomas**[1] **Qing Ma**[2] **and Jean-François Gaillard**[1]

[1]Department of Civil and Environmental Engineering, Northwestern University, 2145 Sheridan Road, Evanston, IL, 60208-3109, USA

[2]DND-CAT Synchrotron Research Center, APS/ANL Building 432A, 9700 S. Cass Ave., Argonne, IL 60439-4857, USA

E-mail: sarathomas2007@u.northwestern.edu, jf-gaillard@northwestern.edu

**Abstract.** We present XAFS data collected at the Hg $L_{III}$-edge for bacterial cells of *Escherichia coli* that have been exposed to 500 and 50 nano-molar $Hg^{2+}$ in aqueous solution, which corresponds to ~30 and ~3 $\mu$g Hg per g cells (wet weight). These concentrations are respectively 1 and 2 orders of magnitude lower than what has been previously reported for Hg-bacteria XAFS experiments. The cells were metabolically active while exposed to Hg(II), providing coordination information that can be directly compared to Hg(II) biouptake experiments. At these amounts of total dissolved metal, Hg(II) binds primarily to thiol moieties that are either present at the cell membrane or localized in the cytoplasm. We show that in this case the Hg binding environment is a mixture of 2- and 4-fold coordination to thiols. This information can be inferred from XANES spectra but the EXAFS provides a more quantitative answer.

## 1. Introduction

Mercury (Hg) is a potent neurotoxin that is affecting human and ecological health worldwide [1]. In its gaseous phase, elemental mercury – $Hg^{o}$ – can travel far distances through the atmosphere and enters aquatic systems after photochemical oxidation to $Hg^{2+}$ [2,3]. In suboxic and anoxic zones Hg(II) is converted to mono-methyl mercury – $CH_3Hg^+$ – after Hg(II) biouptake by methylating microorganisms [3]. This chemical form of Hg is prone to bioaccumulation and biomagnifies through aquatic food webs, causing more fish consumption advisories than any other pollutant [1].

The increasing release of anthropogenic Hg into the environment, primarily as a result of the burning of coal [4,5], has prompted research to better understand the processes that control the microbial biouptake of Hg(II) [6]. For this purpose, we are employing a whole cell *E. coli* biosensor to assess the role of various ligands in controlling Hg(II) transfer through the bacterial cell membrane (*i.e.* biouptake). Recently, we have shown using XANES spectroscopy, that when aminopolycarboxylate ligands, such as EDTA, bind Hg(II) in solution, a ligand exchange reaction takes place between EDTA and thiol moieties at the cell membrane [7]. These XAS experiments were however performed at a higher Hg concentration than those in biouptake experiments and were limited to the collection of the XANES. The purpose of this report is to show that one can probe the coordination environment of Hg(II) in bacterial cells at relevant Hg(II) concentrations (50 nM Hg in solution) and cell densities used in biouptake experiments.

## 2. Material and Methods

### *2.1. Sample preparation*

Bacterial cells of *E. coli* – strain ATCC® 25922™ – were first grown in Lysogeny broth (LB) overnight at 37°C. One mL of the cell suspension was subsequently collected from the LB medium, washed once with a minimal salt medium (MSM), and resuspended in an equivalent volume of MSM. A small aliquot (10-20 $\mu$L) was transferred to 400 mL MSM, and the cells were grown in MSM for ~24 hours until early exponential growth phase characterized by an $OD_{600}$ of 0.2 (~$2\times10^8$ cells per mL). At that stage, the cell suspension was washed twice with and resuspended in an equivalent volume of a third minimally complexing medium (MCM) containing glucose as a carbon source. After adding $Hg(NO_3)_2$ to achieve final concentrations of 50 and 500 nM Hg(II) in cell suspensions (~3 and ~30 $\mu$g Hg per g cells, wet weight), the microbial biouptake of Hg(II) takes place, as is evidenced from our previous results with an *E. coli* based whole cell biosensor [7]. After 3 hours of Hg(II) exposure, the cells were washed twice in 0.1 M $NaClO_4$ and collected as a cell pellet on a 2.5 cm diameter cellulose nitrate filter that was sealed between 2 strips of Kapton tape. The samples were prepared just prior to beamtime, and the bacterial activity was quenched by flash freezing in $LN_2$. The samples remained frozen at a temperature of −80°C prior to XAS data collection. Reference spectra for Hg-thiol standards were prepared following the protocol presented in Jalilevhand et al. [8].

### *2.2. XAFS experiments*

XAFS experiments were performed at the Sector 5 of the Advanced Photon Source using DND-CAT bending magnet beamlime equipped with a Si(111) double crystal monochromator. Harmonics were rejected by detuning by about 40%. Data were collected at the Hg $L_{III}$-edge, $I_0$ was monitored with Oxford ion chambers of 16 cm path-length and the fluorescence signals were collected using a Vortex ME4 silicon drift detector. The frozen cell samples were rapidly positioned in a THMS600 freezing stage – Linkam Scientific Instruments Ltd., UK – maintained at −50°C. Up to 40 scans of 40 minute duration were collected for the samples exposed to the lowest concentrations of dissolved Hg and the data were reduced and interpreted with Athena [9]. Reference spectra were acquired in transmission as well as in fluorescence mode, processed with Athena and fit with Artemis [9] to verify their quality.

## 3. Results

XANES spectra are shown in Figure 1 and the EXAFS spectra in Figure 2. Among the various Hg-thiol reference compounds that were prepared, two explain best the spectra obtained for the cells in exponential growth phase. These standards are the following:

(i) the two-fold thiol coordinated Hg – $Hg(SR)_2$ – a precipitate that forms in a solution of $Hg(NO_3)_2$ and excess cysteine at acidic pH with the stoichiometry: $Hg(Cys)_{2_{(s)}}$ and,

(ii) the four-fold thiol coordinated Hg – $Hg(SR)_4$ – obtained from the reaction of 72 mM Hg(II) with 635 mM cysteine at alkaline pH: $Hg(Cys)_{4_{(aq)}}$.

In the XANES spectra of the Hg-thiol references, two minor spectral features distinguish $Hg(SR)_2$ from $Hg(SR)_4$. The two-fold coordinated Hg-thiol presents two shoulders located at about 12,290 eV and 12,300 eV, whereas the four-fold does not. In addition, the derivative of the normalized XANES of $Hg(SR)_2$ contains a peak at around 12,295 eV that is clearly not present in $Hg(SR)_4$. Linear combination fitting of the XANES derivative of the Hg-cell samples suggest that these samples contain mixtures of both reference coordination environment (data not shown). The XANES derivative fit provides 70% $Hg(SR)_2$ and 30% $Hg(SR)_4$ for cells exposed to 50 nM Hg and 54% $Hg(SR)_2$ and 46% $Hg(SR)_4$ for cells exposed to 500 nM Hg. However, the lack of pronounced edge features in the XANES makes the fit results very sensitive to data processing. Thus, the EXAFS were analyzed as well for comparison.

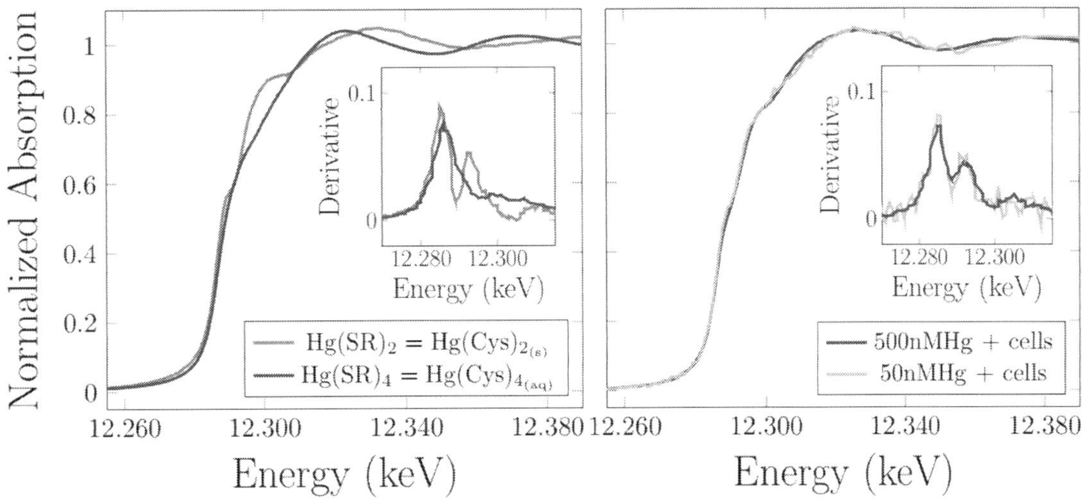


**Figure 1.** XANES spectra of the reference compounds as well as the spectra from exponential phase cells exposed to 500 nM Hg and 50 nM Hg in solution. The derivative of the normalized XANES is shown in the inset.

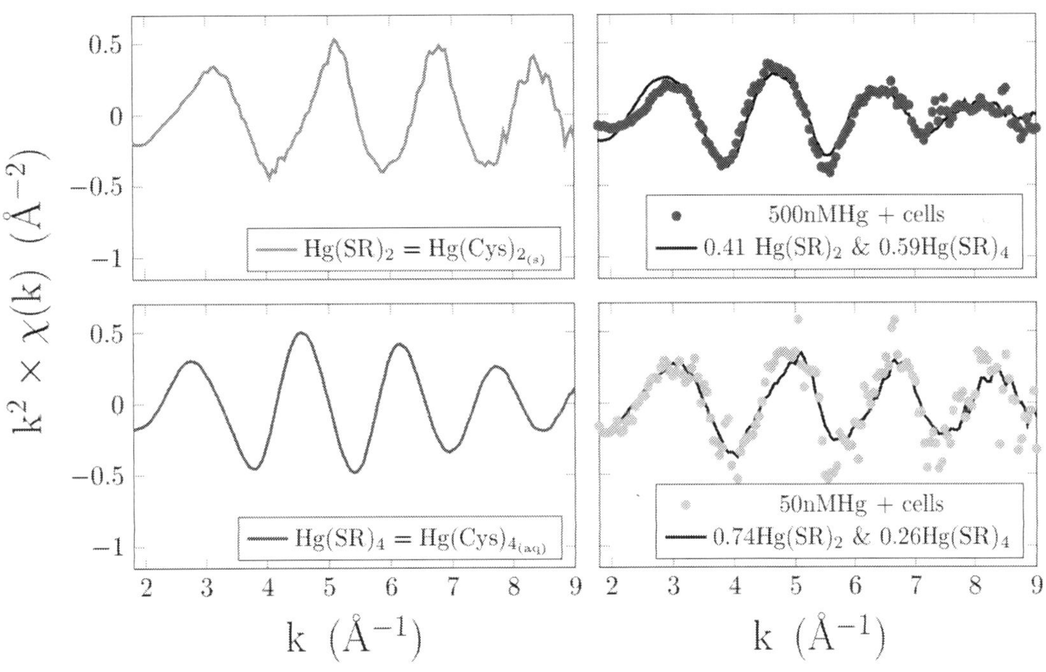


**Figure 2.** EXAFS spectra of the reference compounds as well as the spectra from exponential phase cells exposed to 500 nM Hg and 50 nM Hg in solution. Spectral decomposition was performed using the procedure implemented in Athena [9], after verifying that no other Hg reference compound present in our library could be used.

The EXAFS of the $Hg(SR)_2$ and $Hg(SR)_4$ references show clear differences in the frequency of oscillation, which is dominated by the first sulfur shell. From EXAFS fitting, the Hg-S bond length is $2.35 \pm 0.04$ Å for $Hg(SR)_2$ and $2.50 \pm 0.01$ Å for $Hg(SR)_4$, which agrees with literature values of 2.32 to 2.36 Å and 2.50 to 2.61 Å for $Hg(SR)_2$ and $Hg(SR)_4$, respectively [10]. Principal component analysis was performed on EXAFS spectra of 6 cell samples with similar Hg concentrations (including the 2 samples in this study) and revealed 2 components. Thus, we conclude that there are 2 coordination environments in our cell samples. These EXAFS spectra are reproduced well as a linear combination of the $Hg(SR)_2$ and $Hg(SR)_4$ references. The speciation of Hg in cells exposed to 500 nM of Hg(II) is explained by considering that 41% of Hg is 2-fold coordinated to thiols and 59% is 4-fold. Whereas for the cells exposed to 50 nM Hg(II), we find that 74% of Hg is 2-fold coordinated to thiols while 26% is 4-fold.

## 4. Discussion and Conclusions

Previous XAS studies of Hg(II) interactions with bacterial species have focused on exploring the processes that control its adsorption to cell surfaces [11-13]. In these studies, the lowest concentration assessed was 500 $\mu$g Hg per g cells (wet weight), and only "non-metabolizing" cells were utilized. The XANES and EXAFS spectra from these investigations showed that Hg(II) is coordinated predominantly to sulphydryl groups at low Hg loadings and predominantly carboxylic groups at higher Hg loadings. Another recent study investigated the oxidation of Hg(0) to Hg(II) by anaerobic bacteria and used XANES spectroscopy to determine the redox state as well as the coordination environment of Hg associated with metabolically active cells [14]. It concluded that the speciation of Hg associated with the cell pellet was a mixture of $Hg(SR)_2$ and Hg(0), but the authors lacked $Hg(SR)_4$ as a reference compound. In the studies where the EXAFS is used to determine Hg(II) coordination to thiols, analysis primarily reveals bond distances and coordination numbers characteristic of the formation of $Hg(SR)_2$ [15].

In our study we have obtained Hg $L_{III}$-edge XANES and EXAFS of cells that were actively metabolizing during Hg(II) exposure, *i.e.* in exponential growth phase. The XANES results suggest the presence of both $Hg(SR)_2$ and $Hg(SR)_4$ coordination environments in cells, which is confirmed with the EXAFS spectra. As the existence of $Hg(SR)_4$ in bacteria has yet to be reported, we suggest that the formation of this $Hg(SR)_4$ complex is linked to metabolically active cells and potentially the internalization of Hg(II).

## Acknowledgments

This work is supported by the National Science Foundation under grant CHE-1308504. Portions of this work were performed at the DND-CAT Synchrotron Research Center located at Sector 5 of the APS. DND-CAT is supported by the E.I. DuPont de Nemours & Co., The Dow Chemical Company, the U.S. National Science Foundation through Grant DMR-9304725, and the State of Illinois through the Department of Commerce and the Board of Higher Education Grant IBHE HECA NWU 96.

[1] Mergler D, Anderson H A, Chan L H M, Mahaffey K R, Murray M, Sakamoto M et al. 2007 *Ambio* **36** 3
[2] Morel F M M, Kraepiel A M and Amyot M 1998 *Annu. Rev. Ecol. Syst.* B **29** 543
[3] Sunderland E M, Krabbenhoft D P, Moreau J M, Strode S A, Landing W M 2009. *Global Biogeochem. Cycles* **23** GB2010
[4] Pacyna EG, Pacyna JM, Steenhuisen F, Wilson S 2006 *Atmos. Environ.* **40** 4048
[5] Mason R P, Fitzgerald W F and Morel F M M 1994 *Geochim. Cosmochim. Acta* **58** 3191
[6] Schaefer J K and Morel F M M 2009 *Nat. Geosci.* **2**, 123
[7] Thomas S A, Tong T and Gaillard J-F 2014 *Metallomics* **6** 2213
[8] Jalilehvand F, Leung B O, Izadifard M and Damian E 2006 *Inorg. Chem.* **45** 66
[9] Ravel B and Newville M 2005 *J. Synchrotron Rad.* **12** 537
[10] Manceau A and Nagy K L 2008 *Dalton T.* **11** 1421
[11] Dunham-Cheatham S, Farrell B, Mishra B, Myneni S and Fein J B 2014 *Chem. Geol.* **373** 106
[12] Dunham-Cheatham S, Mishra B, Myneni S and Fein J B 2015 *Geochim. Cosmochim. Acta* **150** 1
[13] Mishra B, O'Loughlin E J, Boyanov M B and Kemner K M 2011 *Environ. Sci. Technol.* **45** 9597
[14] Colombo M J, Ha J, Reinfelder J R, Barkay T and Yee N 2014 *Chem. Geol.* **363** 334
[15] Skylberg U, Bloom P R, Qian J, Lin C-M and Bleam W F 2006 *Environ. Sci. Technol.* **33** 1418

# An x-ray absorption spectroscopy study of Cd binding onto a halophilic archaeon

**Allison R Showalter[1], Jennifer E S Szymanowski[2], Jeremy B Fein[2] and Bruce A Bunker[1]**

[1]Department of Physics, University of Notre Dame, Notre Dame, IN 46566, USA
[2]Department of Civil and Environmental Engineering and Earth Sciences, University of Notre Dame, Notre Dame, IN 46566, USA

Email: ashowal1@nd.edu

**Abstract**. X-ray absorption spectroscopy (XAS) and cadmium (Cd) isotherm experiments determine how Cd adsorbs to the surface of halophilic archaeon *Halobacterium noricense*. This archaeon, isolated from the Waste Isolation Pilot Plant (WIPP) near Carlsbad, New Mexico could be involved with the transport of toxic metals stored in the transuranic waste in the salt mine. The isotherm experiments show that adsorption is relatively constant across the tolerable pH range for *H. noricense*. The XAS results indicate that Cd adsorption occurs predominately via a sulfur site, most likely sulfhydryl, with the same site dominating all measured pH values.

## 1. Introduction

Nuclear activity from energy and weapons production generates radioactive waste, which must be stored for thousands of years with storage in deep geological formations as the current method of disposal. These waste products may result in environmental contamination [1]. The U.S. Department of Energy commissioned the Waste Isolation Pilot Plant (WIPP) in a salt mine near Carlsbad, New Mexico for the U. S. to store transuranic (TRU) waste, made up of items such as tools or debris contaminated with radioactive elements higher than uranium. Microorganisms can increase the mobility of metals, since metal adsorption onto biological surfaces ties transport of the metal to transport of the microorganism [2]. This research project seeks to characterize heavy metal adsorption at high ionic strength on halophilic archaeon *Halobacterium noricense* isolated from the WIPP halite.

Determination of the binding environment of cadmium (Cd) to *H. noricense* will characterize the transport capacity of the archaea. Cd is a toxic metal pollutant present in the TRU waste. Furthermore, Cd adsorption to bacteria has been well-studied, for comparison of Cd adsorption to environmental microorganisms [3]. Previous studies showed that manganese, neptunium and uranium adsorption onto different halophilic archaea or bacteria was rapid (within 120 minutes) and depended on temperature, pH and salinity of the solution [4]. Other studies on thermophiles exhibit similar properties, with Cd adsorption depending on biomass concentration and pH [5]. For this study, isotherm experiments and x-ray absorption spectroscopy (XAS) were used to determine the behavior of Cd adsorption onto *H. noricense*. XAS provides the identity and distance of elements in the local environment surrounding a central atom, can differentiate between elements, requires no long range order and is sensitive to small concentrations of the element of interest [6], making it an ideal technique for this system.

## 2. Materials and methods

The *H. noricense* archaeon was isolated from the halite at the WIPP following standard biological identification, preparation, storage and culture methods [7]. All experiments were performed using 100 g archaea/L in 3 M NaCl solution with varying Cd concentrations. Kinetics experiments used 7.5 ppm Cd and isotherm experiments ranged from 7.5 ppm-100 ppm Cd with a pH range of 5.5-7.5. Analysis made use of inductively coupled plasma-optical emission spectroscopy (ICP-OES) with matrix-matched Cd standards. XAS measurements used 10 or 20 ppm Cd. Standards for the XAS measurements include commercial powders and aqueous standards for including 10 mM Cd in a 100 mM solution of the complexation ligand (perchlorate, acetate, or EDTA) performed in both pure water and in 3 M NaCl solution. Cd K-edge XAS measurements were performed at the MRCAT Sector 10-ID [8] beam line located at the Advanced Photon Source of Argonne National Laboratory (USA). Data analysis was completed using the methodology of the UWXAFS package. Theoretical models were generated using FEFF6 and data analysis was completed using the ARTEMIS graphical interface [9]. Simultaneous $k$, $k^2$, and $k^3$ weighting was applied over a fitting range of 1.2-3.0 Å with a Fourier transform range of 2.1-8.6 Å$^{-1}$. Solid metal oxide powders (CdO and $CdCO_3$) were analyzed to determine the passive electron reduction factor, $S_0^2$, with the resulting empirical $S_0^2$ value of 1.1 ± 0.1.

## 3. Results and discussion

### 3.1. Kinetics experiments

Kinetics experiments were performed in order to determine the equilibration time. The results, shown in figure 1, indicate that full adsorption occurs after about 8 hours. This is in contrast to other systems, including Cd-bacterial systems, which achieve equilibrium in 2 hours or less [3, 4].

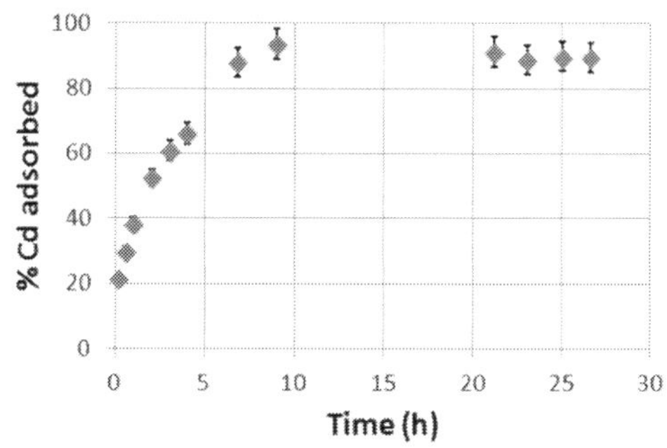


**Figure 1:** Kinetics experimental data for 100 g/L archaea in 3 M NaCl with 7.5 ppm Cd at pH 6.7.

### 3.2. Isotherm experiments

Isotherm experiments determined adsorption dependence on such factors as biomass-to-Cd concentration ratio and pH. The isotherm results show that adsorption seems to be relatively independent of the starting cadmium concentration or pH, as shown in figure 2. The *H. noricense* has a small pH tolerance window of about 5.5-9.7 and all isotherm measurements were performed from 5.5-7.5. This would indicate that adsorption occurs at select sites available in limited quantities on the archaeal surface.

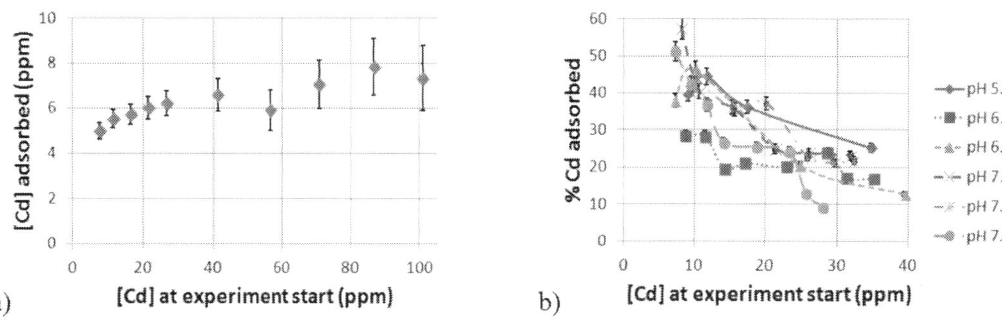


**Figure 2:** Isotherm experimental data for 100 g/L archaea in 3 M NaCl with a) varying biomass-to-Cd concentration ratio at pH 6.7 and b) varying pH. Note for figure 2a that the amount of Cd adsorbed plateaus, indicating a limited number of Cd adsorption sites for a given amount of biomass. In figure 2b, for all pH values the % of Cd adsorbed vs. the starting Cd concentration shows similar trending.

### *3.3. X-ray absorption spectroscopy*

The x-ray absorption near edge spectroscopy (XANES) data show significant differences between the standards used for comparison, but strong similarity between the archaeal samples. Figure 3 shows a comparison between selected standards and the Cd-archaea data. A qualitative assessment of the data indicates that the Cd-archaea adsorption is most similar to the CdS powder standard, indicating a primary binding environment of Cd to sulfur on the archaea surface.

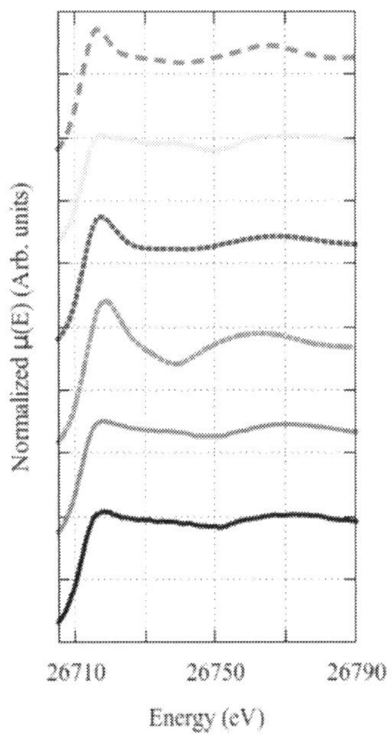


**Figure 3:** XANES data of 10 ppm Cd on archaea and 20 ppm Cd on archaea with standards. Note that the Cd ion in water is different than the Cd ion in 3 M NaCl solution and that there are differences between the S environment standards and Cl environment standards. Both Cd on archaea samples show similar characteristics to the CdS powder standard.

The extended x-ray absorption fine structure (EXAFS) spectroscopy data show similar results to the XANES data, that the Cd-archaea samples most closely resemble the CdS standard. However, in the EXAFS region, the samples also show some similarity to the hydrated $Cd^{+2}$ ion in 3 M NaCl, as seen in figure 4a. This would indicate that though the Cd-S bond is the primary attachment to the surface, there may be some shorter Cd-O or Cd-Cl bonds present, indicating some other minor surface site on the archaeal surface or a Cd-Cl or Cd-O complex being bound to the surface. In order to determine whether the Cd-S bond is the result of precipitation or a sulfhydryl bond, the CdS powder standard is closely compared with the Cd-archaea sample, as shown in figure 4b. Since the Cd-archaea sample is lacking the Cd-Cd peak at about 4.0 Å that is seen in the CdS powder standard, the Cd-S bond in the sample is most likely a sulfhydryl bond.

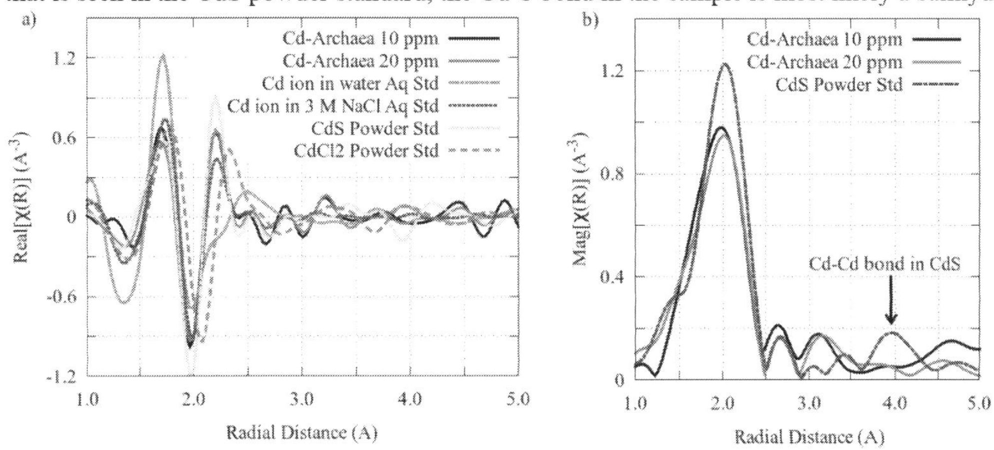


**Figure 4:** The EXAFS data for Cd-archaea samples showing a) the real part with many standards for comparison and b) the magnitude comparing just the Cd-archaea samples with the CdS standard.

## 4. Conclusions

Our results for Cd binding on archaeon *H. noricense* show small variations in adsorption for the moderate pH range where archaea can survive and that the amount of cadmium adsorbed plateaus, indicating adsorption is site-limited and only occurs at selected sites. Both near edge and extended fine structure XAS results show most similarity to the cadmium sulfide standard, indicating a sulfhydryl bond or cadmium sulfide precipitation. The analysis is inconsistent with a Cd-Cd bond at the correct distance for a cadmium sulfide precipitate, so cadmium binding to the archaeal surface is most likely sulfhydryl. This differs from cadmium adsorption to bacteria in that cadmium adsorption to bacteria shows strong pH dependence with a variety of sites, including sulfhydryl, carboxyl, and phosphoryl, with sulfhydryl only appearing in x-ray absorption spectroscopy fitting at low cadmium adsorption. Cadmium adsorption on archaea shows only a sulfhydryl bond at all measured concentrations. Sulfhydryl bonds are rather strong, but site limited on the archaea, so cadmium ions could attach to the archaea in limited quantities at the transuranic waste storage site.

**Acknowledgements**
MRCAT is supported by the U.S. Department of Energy and the member institutions. Use of the APS was supported by the U. S. Department of Energy, Office of Science, Office of Basic Energy Sciences, under Contract No. DE-AC02-06CH11357. A.R.S. wishes to thank the staff at MR-CAT and K. Werellapatha for beamline support and useful conversations. Some experiments and analyses used instrumentation at the Notre Dame Integrated Imaging Facility (NDIIF) and the Center for Environmental Science and Technology (CEST) at the University of Notre Dame. This research was supported, in part, by a CEST/Bayer Predoctoral Research Fellowship to A.R.S.

**References**
[1] Hu Q-H, Weng J-Q and Wang J-S 2010 *J. Environ. Radioact.* **101** 426-37
[2] Yee N and Fein J B 2002 *Chem. Geo.* **185** 303-19
Vijayaraghavan K and Yun Y-S 2008 *Biotech. Adv.* **26** 266-91
[3] Fein J B, Daughney C J, Yee N and Davis T A 1997 *Geochim. Cosmochim. Acta* **61** 3319-28
Burnett P-G G, Daughney C J and Peak D 2006 *Geochim. Cosmochim. Acta* **70** 5253-69
Mishra B, Boyanov M, Bunker B A, Kelly S D, Kemner, K M and Fein J B 2010 *Geochim. Cosmochim. Acta* **74** 4219-33
[4] Naik S and Furtado I 2014 *Geomicrobiology Journal* **31** 708-15
Ams D A, Swanson J S, Szymanowski J E S, Fein J B, Richmann M and Reed D T 2013 *Geochim. Cosmochim. Acta* **110** 45-57
Francis A J, Gillow J B, Dodge C J, Harris R, Beveridge T J and Papenguth H W 2004 *Radiochim. Acta* **92** 481-8
[5] He Z, Yang Y, Zhou S, Zhong H and Sun W 2013 *Colloids Surf., B* **102** 667-673
Daughney C J, Hetzer A, Heinrich H T M, Burnett P-G G, Weerts M, Morgan H, Bremer P J and McQuillan A J 2010 *Chem. Geo.* **273** 82-90
[6] Bunker G 2010 *Introduction to XAFS: A practical guide to x-ray absorption fine structure spectroscopy* (New York: Cambridge University Press)
[7] Swanson J, Simmons K, Norden D and Reed D 2013 *Microbial characterization of halite and groundwater samples from the WIPP* Report for Los Alamos National Laboratory
[8] Segre C U, et al.2000 *AIP Conf. Proc.* **521** 419-22
[9] Stern E A, Newville M, Ravel B, Yacoby Y and Haskel D 1995 *Physica B* 117-20
Newville M, Livins P, Yacoby Y, Rehr J J and Stern E A 1993 *Phys. Rev. B* **47** 14126-31
Ravel B and Newville M 2005 *J. Synchrotron Rad.* **12** 537-41
Zabinsky S I, Rehr J J, Ankudinov A, Albers R C and Eller M J 1995 *Phys. Rev. B* **52** 2995-3009

# Absorption and Mobility of Cr and Zn in Soil in the Vicinity of Jordan River

**M. Harfouche[1], A. M. Ghrair[2], D. M. M. Jaradat[3], G. Aquilanti[4], R. Jaber[5], A. Aldrabee[6] and N. Sawai[1]**

[1]Synchrotron-light for Experimental Science and Applications in the Middle East (SESAME), Allan 19252, Jordan,
[2]Royal Scientific Society, Applied Scientific Research commission, Amman, Jordan,
[3] Department of Chemistry, Faculty of Science, Al-Balqa' Applied Univ., Al-Salt, Jordan
[4]Elettra – Sincrotrone Trieste S.C.p.A.s.s., 34149 Basovizza, Trieste, Italy
[5]University of Jordan, Amman 11941, Department of Civil Engineering, Jordan
[6] Jordan Atomic Energy Commission (JAEC), 11934 Amman, Jordan

messaoud.harfouche@sesame.org.jo

**Abstract**. Sediments from the basin of Jordan River are to some extent significantly contaminated by heavy metals. The goal of this study was to examine the nature of the heavy metals, their mobility, and their binding in the soil matrix in order to understand the accumulation mechanism. Exhaustive mineralogical and geochemical study on the heavy metals in the vicinity of the river will allow knowing the dominant forms of emitted metals. The structural local order around Cr and Zn in soil samples were investigated by analyzing XAFS data. Our investigations show that the coordination sphere, around both Cr and Zn, is constituted of oxygen atoms at an interatomic distance of ~2 Å which is close to those in clay minerals. It is also shown that Cr(III) is the most dominating redox state in soil samples. Moreover, the next backscattering neighbors around Zn and Cr are either carbon or sulfur atoms which reveal a multiple and heterogeneous speciation of Cr and Zn with organic matter.

## 1. Introduction

The Jordan River and its major tributary, flow through the Jordan valley, where many important agricultural activities are located. The study of heavy metal pollutants in the sediments of Jordan River is of interest for two reasons. On one hand the flows through ever-increasingly agricultural and urban areas [1]; on the other hand the flow rate of river has decreased drastically, due to the upstream Dam, the increase of the agricultural surfaces as and the climate changes, thus the potential influence of external sources on the river discharge and chemical composition became significant [2].

Pollution of the natural environment by heavy metals is a worldwide problem because these metals are indestructible and most of them have toxic effects on living organisms, when they exceed a certain concentration [3]. Heavy metals accumulate in the sediments through complex physical and chemical adsorption mechanisms depending on the nature of the sediment matrix and the properties of the adsorbed compounds [4].

The degree of metals bioavailability, and thus metals toxicity, is by large determined by the nature of their complication to sorbent surfaces in soils. Therefore, the objective of this study was to explore the speciation of Cr and Zn in soil system and to investigate the local structure in order to understand the binding system around these heavy metals in soils. This is an essential step to understand the structural behavior around toxic metals in soil and, for future, in the agricultural product.

## 2. Material and methods

### *2.1. Study region*

The Jordan valley is a part of the geological fault which extends from Syria down to the Red Sea. The Lower Jordan River stretches between the Sea of Galilee and the Dead Sea (~ 417 m below sea level). The river is about 105 km meandering distance. The investigated area (Figure 1) is occupied by rural settlements on east side of the river (Jordan) and the majority of the land is used for agriculture (e.g., field crops and fruit trees).

**Figure 1.** Location of the sampling region in the vicinity of the Jordan River showing the exact position of the sampling points.

### *2.2. Sampling, preparation, characterization and Analysis*

Soil samples were collected from 3 different locations using a backhoe sampler from the east bank of the Jordan River (Figure 1). The sampling points were recorded and identified using the latitude and longitude coordinates. In order to consider the representativeness of the samples, we were keen that samples will be collected from different soil types ranging from the sediment at the river bank (JR-0m) to the agricultural zone (JR-200m). For each zone samples were collected at different depths with 5 cm sampling thickness. Therefore, sampling depths were at 0-5 cm, 15-20 cm, 30-35 cm and 45-50 cm from the surface down.

Part of samples were dried in the oven at 100 °C and, then, were subject to series of tests in order, to estimate the levels of total organic matter (TOM) using loss on ignition at 555 °C, to quantify the elemental content of heavy metals and trace elements using ICP-MS and to identify different crystalline structures forming soil through conventional XRD technique.

Natural samples (non-dried) were subject to X-ray Absorption Fine Structure (XAFS) analysis. Data were collected at Cr (5989 eV) and Zn (9659 eV) edges on XAFS beamline at Elettra [5] and mySpot at BessyII [6]. In both beamlines, XAFS data were collected in fluorescence mode using silicon drift detector. XAFS data processing and analysis were performed using IFEFFIT software package following standard procedures [7]. Using different model compounds, linear combination fitting (LCF) has been applied to the X-ray Absorption Near-Edge Structure (XANES) data to estimate the different species that are forming the samples. After normalizing the short-Extend XAFS spectra, EXAFS signals $\chi(k)$ were $k^3$-weighted and Fourier Transformed (FT) using a Bessel window function. Theoretical scattering paths, used for modeling experimental EXAFS, were calculated with FEFF [8].

A multi-shell approach was used for data fitting. In order to reduce the number of free parameters, the amplitude reduction factor ( $S_0^2$ ) was fixed at 0.9. In addition, the shift in the threshold energy ($E_0$) was linked for all the shells. Floating parameters of further shells were the coordination number (N), bond length ($R$) and the Debye–Waller factor ($\sigma^2$). Furthermore, Continuous Cauchy Wavelet (CCW) analyses were performed on the XAFS data collected at the Cr and Zn K-edges [9]. Although the short k range of the EXAFS data (k=~ 7 Å$^{-1}$), combining CCW and EXAFS fitting techniques, lots of information were extracted such as the type of backscattering atoms and their interatomic distances.

## 3. Results and discussion

While XRD patterns didn't reveal more than usual standard soil phases, the elemental content chemical analysis showed that most of heavy metals are in the tolerant contamination range. Exceptionally Cr was found in the moderately contamination range (150 ppm) which can appear to be harmful especially if occurring in its hexavalent form. Knowing the local structure around some of the heavy metals in soil will reveal good information about the soil system in the region.

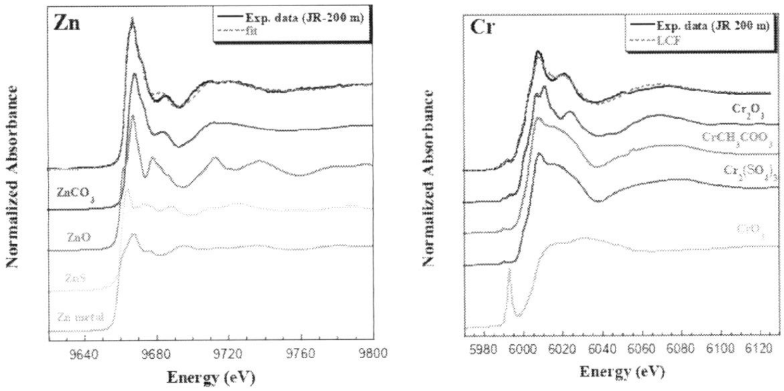


**Figure 2.** Linear combination fitting of the XANES data collected at Zn (left) and Cr (right) K-edges in soil samples from the vicinity of Jordan River

XANES data collected at the Zn and Cr K-edges were subject to LCF using different experimental model compounds. Results show that Zn is present in a form similar to a smithsonite ($ZnCO_3$-like). Moreover Zn species were not influenced by the agricultural processes (at 200 m from the bank) although the use of fertilizers. This can be an indication that Zn may be present in an organic structure form since it is bind to carbon. It also shows that Zn is not in a sulfurous structure.

In contrast to Zn, the Cr shows a large difference as a function of the distance from the river bank. On the zero distance (at the river bank) Cr occurs in inorganic forms as a chromium (III) oxide ($Cr_2O_3$) and in chromium sulfate ($CrSO_4$) distributed equally to ~45% for each phase. In the agricultural region (200 m from the bank) Cr is mainly occurring in a chromium (III) acetate form $Cr(C_2H_3O_2)_3$ (~55%). Inorganic phase of Cr still exists in agricultural soil and occurs in sulfate form [$Cr(SO_4)_3$] but with a smaller content (~30%) compared to the river bank in addition to the oxide phase ($Cr_2O_3$) which is present with a lower concentration to about 20%.

Results derived from EXAFS data analysis are presented in table 1. Due to the short EXAFS range, the number of backscattering atoms is not accurate and thus is not reported. On the contrary, the interatomic distances are accurate, although we speculate that their accuracy decreases with the increase of the distances.

EXAFS analysis show that Cr as well as Zn are surrounded by light elements at short distances which appeared to be oxygen atoms sitting at the interatomic distance of ~2.0 Å for both structures (Table 1). Further EXAFS results to the next nearest neighbors for semi-quantitative analysis, due to the short range, show presence of carbon backscattering atoms around Zn at an interatomic distance of ~3.2 Å which is in agreement with LCF results. Differences in the local structure around Cr atom in soil samples are dependent on the soil type with respect to the distance from river bank. At a zero distance (river bank), Cr is bound to sulfur at an interatomic distance of ~3.4 Å or in the agricultural soil (200m from the river bank) the S bond disappeared and Cr is bound to carbon atoms at a longer distance (~3.8 Å).

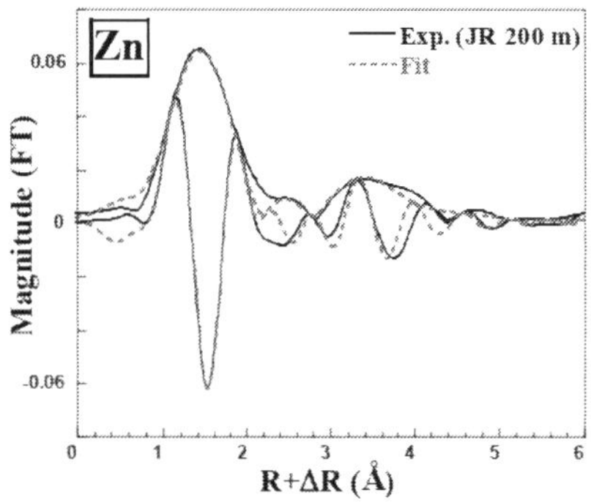


**Figure 3.** Modelling EXAFS data collected at the Zn edge in soil sample at 200 m from the river bank.

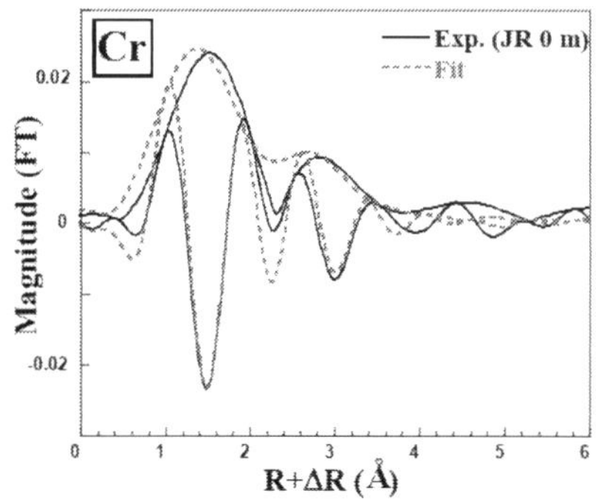


**Figure 4.** Modelling EXAFS data collected at the Cr edge in soil sample from the river bank.

**Table 1.** Structural parameters derived from EXAFS data analysis

| Sample | Bond | CN | Distance (Å) | $\sigma^2$ ($Å^2$) | ΔE (eV) |
|---|---|---|---|---|---|
| Zn in Jordan River 0 m | Zn-O | 5.2 ± 2.0 | 1.99(4) | 0. 004(6) | |
| | Zn-C | - | 3.22(4) | 0.005(5) | 0.57(5) |
| Zn in Jordan River 200 m | Zn-O | 4.6 ± 2.0 | 1.94(5) | 0.002 (6) | |
| | Zn-C | - | 3.21(4) | 0.005 (8) | -1.37(6) |
| Cr in Jordan River 0 m | Cr-O | 4.5 ± 2.5 | 2.02(2) | 0.010(6) | |
| | Cr-S | - | 3.41(4) | 0.045(7) | -2.12(4) |
| Cr in Jordan River 200 m | Cr-O | 3.7 ± 2.5 | 1.97(3) | 0.01(6) | |
| | Cr-C | - | 3.79(2) | 0.08(8) | 0.58(5) |

## 4. Conclusion
Soils from the east bank in the vicinity of the Jordan River, in contrast to speculations of pollutions, seem to be in tolerant range from the heavy metals content side except for the Cr case in studied soil samples. Zn occurs in a stable carbonate structure in the region independent of the soil specifications. In the contrary, the local structure around Cr changes between carbonate-like structures, in agricultural soil, to sulphates in non-exploitable regions.

## 5. Acknowledgment
The authors thank the IAEA for the financial support through the CRP, BessyII and Elettra for allocating beamtime and also the ICTP for financial support for beamtime at Elettra, Italy.

## References
[1] Howari F M and Banat K M 2001, *Journal of Environmental Hydrology* **9** Paper 20
[2] Holtzman R, Shavit U, Segal-Rozenhaimer M, Gavrieli I 2005 *J. Environ. Qual.* **34** 897–906
[3] MacFarlane G R, Burchett M D 2000. *Vierh. Aquat. Bot.* **68** 45–59.
[4] Leivouri M 1998 *Chemosphere* **36** 43–59.
[5] Di Cicco A, Aquilanti G, Minicucci M, Principi E, Novello N, Cognigni A and Olivi L 2009 *J. Phys.: Conf. Ser.* **190** 012043
[6] Paris O, Li C, Siegel S, Weseloh G, Emmerling F, Riesemeier H, Erko A and Fratzl P 2007 *J. Appl. Cryst.* **40** (Supplement) s466-s470
[7] Ressler T 1998 *J. Synch. Rad.* **5** 118
[8] Rehr J J, Albers R C 2000 *Rev. Mod. Phys.* **72** 621.
[9] Muñoz M., Argoul P. and Farges F. 2003 *American Mineralogist* **88** 694-700.

# Speciation of neptunium during sorption and diffusion in natural clay

**T Reich[1,3], S Amayri[1], P J B Börner[1], J Drebert[1], D R Fröhlich[1,4], D Grolimund[2] and U Kaplan[1,5]**

[1] Institute of Nuclear Chemistry, Johannes Gutenberg University Mainz, 55099 Mainz, Germany
[2] Swiss Light Source, Paul Scherrer Institute, 5232 Villigen PSI, Switzerland

E-mail: tobias.reich@uni-mainz.de

**Abstract**. In argillaceous rocks, which are considered as a potential host rock for nuclear waste repositories, sorption and diffusion processes govern the migration behaviour of actinides like neptunium. For the safety analysis of such a repository, a molecular-level understanding of the transport and retardation phenomena of radioactive contaminants in the host rock is mandatory. The speciation of Np during sorption and diffusion in Opalinus Clay was studied at near neutral pH using a combination of spatially resolved synchrotron radiation techniques. During the sorption and diffusion experiments, the interaction of 8 $\mu$M Np(V) solutions with the clay lead to the formation of spots at the clay-water interface with increased Np concentrations as determined by $\mu$-XRF. Several of these spots are correlated with areas of increased Fe concentration. Np $L_3$-edge $\mu$-XANES spectra revealed that up to 85% of the initial Np(V) was reduced to Np(IV). Pyrite could be identified by $\mu$-XRD as a redox-active mineral phase responsible for the formation of Np(IV). The analysis of the diffusion profile within the clay matrix after an in-diffusion experiment for two months showed that Np(V) is progressively reduced with diffusion distance, i.e. Np(IV) amounted to ≈12% and ≈26% at 30 $\mu$m and 525 $\mu$m, respectively.

## 1. Introduction

Argillaceous rock formations are under consideration as a potential host rock for the construction of high-level nuclear waste repositories. Np(IV) and Np(V) are the most stable oxidation states of $^{237}$Np ($t_{1/2} = 2.1 \times 10^6$ a) under environmental conditions. We have investigated the sorption and diffusion of the more mobile Np(V) in Opalinus Clay (OPA), a natural clay from the Mont Terri rock laboratory in Switzerland [1-3]. OPA, which is present also in Southern Germany, possesses a micro-scale heterogeneity and is composed of several types of clay minerals, but also of calcite, quartz and Fe(II)-bearing minerals. In our previous diffusion [1] and anaerobic sorption experiments [2], we observed higher distribution coefficients, $K_d$, than expected from batch experiments performed in air, suggesting

[3] Author to whom any correspondence should be addressed.
[4] Present address: Institute of Physical Chemistry, Universität Heidelberg, 69120 Heidelberg, Germany
[5] Present address: Institute for Nuclear Waste Disposal, Karlsruher Institut für Technologie, 67021 Karlsruhe, Germany

that a partial reduction of Np(V) to Np(IV) had occurred. To verify this hypothesis, different sorption and diffusion samples with Np(V) were prepared at near neutral pH for spatially resolved, molecular-level investigations at the microXAS beamline of the Swiss Light Source (SLS).

## 2. Experimental

The sorption sample was prepared by contacting a 20-30 $\mu$m thick thin section of OPA with 8 $\mu$M $^{237}$Np(V) solution (Millipore water, pH ≈ 7) under anaerobic conditions in an argon glove box. After a contact time of five days, the aqueous phase was removed [4].

For the preparation of the diffusion sample, an intact OPA cylinder (diameter 25.4 mm, thickness 11 mm) was mounted into a diffusion cell as described in [1]. The clay cylinder was oriented so that the bedding was parallel to the direction of transport (diffusion). The high-concentration reservoir of the diffusion setup contained 8 $\mu$M $^{237}$Np(V) solution (artificial OPA pore water, pH = 7.6 [5]). The low-concentration reservoir contained only artificial OPA pore water without Np. The diffusion experiment was performed at 25 °C under ambient-air conditions and lasted for 61 days. During this time the $E_h$ in the high-concentration reservoir was approximately equal to 480 mV (SHE). After two months, the diffusion cell was opened and the clay cylinder was cleaved parallel to the bedding. Smaller segments of the clay cylinder were mounted into the sample holder.

Both samples were measured at the microXAS beamline at the SLS using the Kirkpatrick-Baez mirror microfocusing system and the Si(111) double-crystal monochromator. X-ray fluorescence maps of several elements including Si, S, K, Ca, Ti, Fe, Sr and Np were recorded using a single-element Si detector (KETEK GmbH, Germany). The excitation energy was 17.62 or 17.70 keV, which is above the Np $L_3$-edge. The size of the focused X-ray beam was approximately 2 × 4 $\mu$m (height × width). X-ray diffraction patterns of the OPA thin section on high-purity quartz glass (Heraeus Holding GmbH, Germany) were recorded with a 100k Pilatus detector (DECTRIS Ltd., Switzerland). Np $L_3$-edge XANES spectra were recorded in fluorescence mode at spots where $\mu$-XRF maps had shown areas with higher Np concentrations. A Zr foil ($E_0$ = 17.998 keV) was used for energy calibration. $\mu$-XRD mappings (20 × 20 $\mu$m, 1 $\mu$m step size) were collected for the sorption sample (thin section) at 17.70 keV. The angle between the incoming X-ray beam and the fluorescence detector was 45° or 20° (by simultaneous use of XRD detector) for the sorption sample and equal to 30° for the diffusion sample, respectively. The analysis of the $\mu$-XRF, $\mu$-XANES and $\mu$-XRD data using several software packages is described in [4].

## 3. Results and discussion

### *3.1. Sorption sample*

The $\mu$-XRF mappings of elements contained in OPA (e.g. Si, S, K, Ca, Fe) reflect the mineralogical heterogeneity of the natural clay. For example, the 2 × 2 mm $\mu$-XRF map of Fe in figure 1 (a) shows a large area of increased Fe concentration. As can be seen from the corresponding Np XRF map in figure 1 (a), several spots of increased Np concentration are clearly correlated with the Fe concentration. In the region marked with a white cross (figure 1 (a)), the measured Np $L_3$-edge XANES spectrum is composed of 85 ± 5% Np(IV) and 15 ± 5% Np(V) (figure 1 (b)), indicating that the initial Np(V) has been significantly reduced to Np(IV). $\mu$-XRD measurements in this area (figure 1 (c)) showed that Np was accumulated on pyrite ($FeS_2$). Therefore, pyrite is at least one of the redox-active mineral phases of OPA responsible for the reduction of Np(V). It should be noted that Np spots with considerable fractions of tetravalent Np could be found also in other OPA sorption samples, even when the experiments were performed under ambient-air conditions [4].

### *3.2. Diffusion sample*

In the long-term diffusion experiment using 8 $\mu$M Np(V) in OPA pore water (pH 7.6), we were able to derive the Np diffusion profile along a 3 mm long diffusion path parallel to the bedding of the clay using $\mu$-XRF maps. As can be seen from figure 2 (a), the amount of Np decreases steadily with diffusion distance and reaches the background level after approximately 2.5 mm. Several Np $L_3$-edge $\mu$-XANES spectra were recorded along this diffusion path. The linear combination fit of these spectra using those of Np(IV) and Np(V) aquo ions as references showed that at a distance of ca. 30 $\mu$m from the OPA-water interface 12 ± 5% of the initial Np(V) has been reduced to Np(IV) (figure 2 (b)). The degree of Np reduction increases with distance and reaches 26 ± 5% at 525 $\mu$m (figure 2 (c)). An even higher degree of Np reduction of approximately 80 ± 5% was observed near a spot with increased Fe concentration at a diffusion distance of 450 $\mu$m in a similar long-term diffusion experiment (sample not shown here). This further corroborates the role of the Fe(II)-bearing minerals as described in section 3.1.

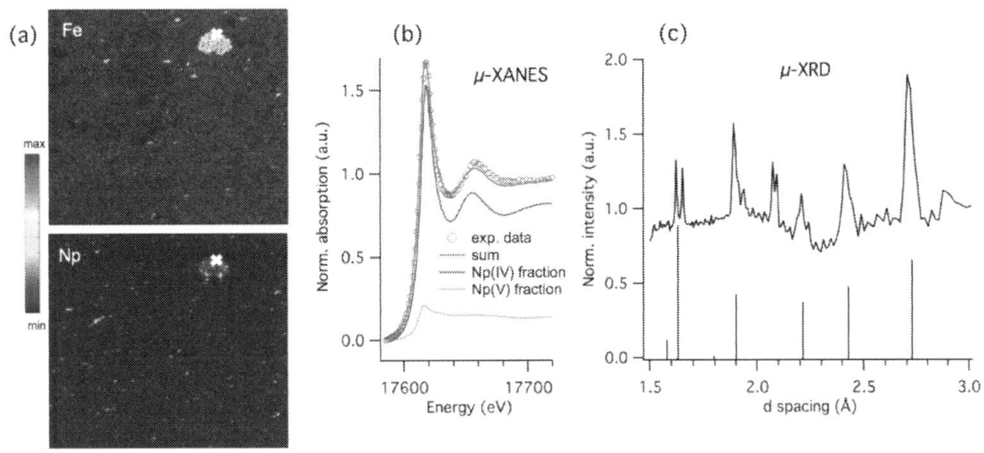


**Figure 1.** (a) $\mu$-XRF elemental distribution maps of Fe and Np (2 × 2 mm, step size 10 $\mu$m). (b) Normalized Np $L_3$-edge $\mu$-XANES spectrum measured on the spot marked with a white cross in the XRF map together with the fractions of Np(IV) and Np(V) reference spectra. (c) Intensity vs. d-spacing plot obtained by $\mu$-XRD mapping of the same spot (20 × 20 $\mu$m, step size 1 $\mu$m). The vertical bars are calculated reflections of pyrite (ICSD database, collection code 43716).

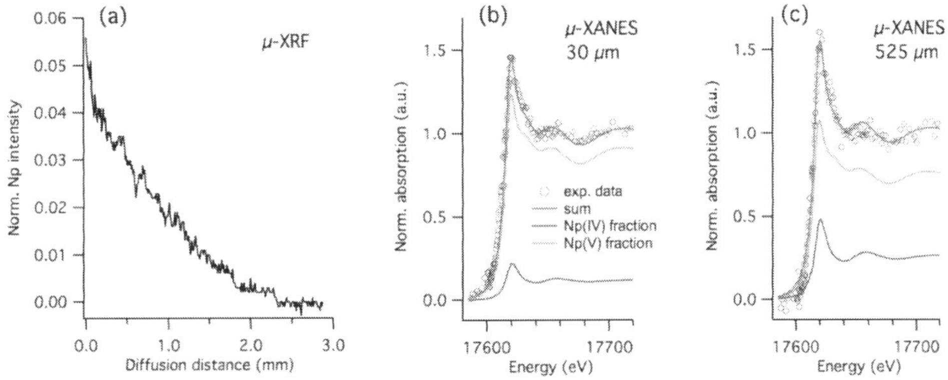


**Figure 2.** (a) Normalized neptunium intensity vs. diffusion distance derived from $\mu$-XRF (3.0 × 1.0 mm, step size 10 $\mu$m). (b) and (c) Normalized Np $L_3$-edge $\mu$-XANES spectra measured at distances of 30 $\mu$m and 525 $\mu$m together with the fractions of Np(IV) and Np(V) reference spectra.

## 4. Conclusions

Although wet-chemistry experiments are important for the determination of parameters quantifying the sorption and diffusion of Np in natural clay, these experiments cannot provide the molecular-level understanding that is needed to support the safety analysis of future high-level nuclear waste repositories. The combination of spatially resolved X-ray microprobe methods (e.g. $\mu$-XRF, $\mu$-XANES, $\mu$-XRD) is a powerful tool for determining the distribution and speciation of actinides in heterogeneous systems like natural clays.

After sorption of 8 $\mu$M Np(V) on Opalinus Clay at pH 7 under anaerobic conditions, $\mu$m-size spots were detected where tetravalent Np is the dominating species. The Fe(II)-bearing mineral pyrite acts as an important redox-active mineral phase. In the long-term diffusion experiment with 8 $\mu$M Np(V) and Opalinus Clay pore water (pH 7.6) as mobile phase, the speciation of Np could be mapped along the diffusion path parallel to the bedding of the clay. These Np $L_3$-edge $\mu$-XANES measurements showed that mobile Np(V) is immobilized in Opalinus Clay as Np(IV) by progressive reduction along its diffusion path, further consolidating the suitability of argillaceous rocks with regard to the long-term storage of Np-containing nuclear waste.

## Acknowledgement

This work was financed by the Federal Ministry for Economic Affairs and Energy (contract no. 02E10166, 02E10981) and TALISMAN (contract no. TALI-C01-15, TALI-C03-09). D. R. Fröhlich was supported by a fellowship of DFG-GRK 826. We are grateful to C. Borka and V. A. Samson for their support of the synchrotron measurements at the microXAS beamline at the Swiss Light Source, Paul Scherrer Institut, Switzerland. We thank Ch. Marquardt (Institute for Nuclear Waste Disposal, Karlsruher Institut für Technologie) for providing OPA bore core samples.

## References

[1] Wu T, Amayri S, Drebert J, Van Loon L R and Reich T 2009 Neptunium(V) sorption and diffusion in Opalinus Clay *Environ. Sci. Technol.* **43** 6567–71

[2] Fröhlich D R, Amayri S, Drebert J and Reich T 2011 Sorption of neptunium(V) on Opalinus Clay under aerobic/anaerobic conditions *Radiochim. Acta* **99** 71–7

[3] Fröhlich D R, Amayri S, Drebert J and Reich T 2012 Influence of temperature and background electrolyte on the sorption of neptunium(V) on Opalinus Clay *Appl. Clay Sci.* **69** 43–9

[4] Fröhlich D R, Amayri S, Drebert J, Grolimund D, Huth J, Kaplan U, Krause J and Reich T 2012 Speciation of Np(V) uptake by Opalinus Clay using synchrotron microbeam techniques *Anal. Bioanal. Chem.* **404** 2151–62

[5] Van Loon L R, Soler J M and Bradbury M H 2003 Diffusion of HTO, $^{36}Cl^-$ and $^{125}I^-$ in Opalinus Clay samples from Mont Terri – Effect of confining pressure *J. Contam. Hydrol.* **61** 73–83

# Metal (Hydr)oxides for the removal of Cr(VI) from drinking water: a XAFS study

**F Pinakidou*, E Kaprara*, M Katsikini, E C Paloura, K Simeonidis, M Mitrakas***

*Aristotle University of Thessaloniki, School of Chemical Engineering, Analytical Chemistry Laboratory, 54124, Thessaloniki, Greece

Aristotle University of Thessaloniki, School of Physics, Department of Solid State Physics, 54124, Thessaloniki, Greece,

E-mail: fpina@physics.auth.gr

**Abstract.** The reduction mechanism and adsorption behaviour of Cr(VI) onto Tin(II) oxy-hydroxides are investigated using Cr-K edge X-ray absorption fine structure (XAFS) spectroscopies. The synthesis of the Sn oxy-hydroxide proceeds via hydrolysis of $SnSO_4$. The successful reduction of Cr(VI) was identified by the analysis of the Cr-*K* edge XANES spectra where only Cr(III) species in octahedral sites are detected. According to the Cr-K edge EXAFS analysis results, Cr(III) forms bidentate inner sphere (binuclear ($^2C$) and mononuclear ($^2E$)) complexes. However, the surface coverage of Cr affects the type of Cr(III)-complexes formed: as the Cr-loading increases, Cr(III)-oxy-anions preferentially sorb in a "combined" $^2C$ and $^1V$ configuration, at the expense of the existing $^2E$ sorption geometry.

## 1. Introduction

Chromium is a naturally occurring element, commonly present in the environment, as trivalent Cr(III) and hexavalent Cr(VI). Cr(VI) is a well-established carcinogen and can enter groundwater from both anthropogenic activities and natural sources, e.g. by underground water contact with ultramafic rocks [1]. Considering the severe health effects of Cr(VI) and its natural occurrence in groundwater, it becomes crucial to develop technologies that can both efficiently remove Cr(VI) from water to concentrations of low μg/L and efficiently scale-up to large-scale treatment. Among the several methods developed to remove Cr(VI) from water, the use of materials that can combine reductive and adsorptive capacity became a very promising solution. For example, solid materials such as Cu-Zn alloys [2], $Fe_3O_4$ nanoparticles [3], zero-valent Fe [4] have been extensively used in water treatment technology, because they act as electron donors that reduce the highly mobile Cr(VI) to Cr(III) which exhibits significantly low water solubility. Recently, the feasibility of using Sn(IV)-compounds for heavy metal-removal from wastewater has been under investigation, mainly due to their low cost and safe disposal. [5,6] However, the effectiveness of Sn(II)-compounds for reduction/adsorption of Cr(VI) from drinking water has not been evaluated yet.

The understanding of chromium adsorption under realistic water treatment can promote the design of optimum adsorbents in respect of their efficiency and environmental impact after their disposal. Therefore, this study attempts to explain the reduction and adsorption behavior of Cr(VI) onto Sn(II) oxy-hydroxides by means of Extended (EXAFS) and Near Edge (XANES) X-ray Absorption Structure spectroscopies, since these techniques are sensitive to the bonding environment of Cr, yielding direct information on the oxidation state, distances and number of atoms bonded to Cr-oxy-anions.

## 2. Experimental details

A Sn(II) oxy-hydroxide ($Sn_6O_4(OH)_4$) was prepared by the hydrolysis of $SnSO_4$ at neutral pH. The collected solid suspension was washed, centrifuged, dried at 100 °C, ground and sieved to obtain either a fine powder form (<45 μm) used for batch adsorption experiments or granules (0.25-0.50 mm) for column tests. For batch adsorption experiments, 250 mg of fine powder were dispersed in 1L of 5 mg/L Cr(VI) solution in NSF water matrix and stirred at 20 °C for 24 h. Adsorption tests were performed at pH-values 5, 7 and 7.5 and at the end of the experiments the suspensions were filtered through a 0.45 μm pore-size membrane. Rapid small scale column tests (RSSCTs) were designed for an empty bed contact time (EBCT) 2 min and a particle size 0.25-0.50 mm, using glass columns of 1.1 cm diameter with PTFE valves and caps and a glass frit in the bottom, which was fed from the top with 100 μg/L Cr(VI) in NSF water matrix at pH=7. Initial and residual Cr(VI) concentration was determined by the diphenyl-carbazide method. Details on the Cr(VI) adsorption experiments are listed in table 1.

XAFS measurements were conducted at the BAM beamline at the synchrotron radiation facility BESSY-II in Berlin.[7] Both EXAFS and XANES spectra were recorded in the fluorescence yield mode using an energy dispersive (Bruker X-Flash) detector. The Cr-*K* edge XANES data were normalized with the intensity of the impinging beam ($I_0$) recorded with an ionization chamber positioned in front of the sample and were subjected to linear background subtraction and normalization to the edge jump. In the EXAFS spectra, after subtraction of the atomic absorption, the theoretical phase and amplitude functions for the scattering paths were calculated by FEFF8.2 [8] and curve fitting was carried out using FEFFIT.[9] The spectra of $Cr(OH)_3$ and $K_2CrO_4$ powder samples were used as references and were recorded in the transmission mode.

**Table 1.** Cr-adsorption configuration, sorption pH and Cr content

| Configuration | Sorption pH | Cr (wt%) |
|---|---|---|
| Batch | 5 | 0.42 |
| Batch | 5 | 1.24 |
| Batch | 7.5 | 1.17 |
| Column | 7 | 0.77 |

## 3. Results and Discussion

### *3.1. Cr-K edge XANES*

The effective reduction of Cr(VI) to Cr(III), using $Sn_6O_4(OH)_4$ at batch and dynamic conditions and after exposure to different Cr-loading was investigated using Cr-*K* edge XANES spectroscopy. As in all 3d-trasistion metals, a pre-edge absorption exists in the Cr-*K* edge XANES spectra, the characteristics of which are modulated by the Cr-valence: a prominent pre-edge peak is present in Cr(VI) compounds, which is related to 1s transitions to empty 3d states. This transition is dipole forbidden in octahedral Cr(III); nevertheless, a weak pre-edge peak is detected, due to the loss in centrosymmetry in $Cr(III)O_6$ compounds, which is attributed to 1s transitions to $3d(t_{2g})$ and $3d(e_g)$ electronic states, respectively [10]. The characteristics (energy position and area) of the pre-edge peak were determined using Voigt functions (blue lines at the inset of Figure 1(a)) and the position of the absorption edge ($E_{abs}$) was determined using a sigmoidal function which simulates transitions to the continuum (red line at the inset of figure 1(a)). In the XANES spectra of all studied samples, shown in Figure 1(a), a weak pre-edge absorption is detected and its characteristics (intensity and energy position) are identical to the respective in reference $Cr(OH)_3$: the two Voigt functions used are located at 5989.5-5989.8 and 5991.9-5992.1 (±0.25) eV respectively, as in $Cr(OH)_3$ (5989.8 and 5992.1

(±0.25) eV), where Cr(III) forms $CrO_6$ octahedra. This finding verifies the presence of Cr(III) in octahedral sites while Cr(VI) is not detected. Thus the reduction of Cr(VI) to Cr(III) at the tin oxy-hydroxide surface takes place successfully and independently of Cr-loading. The presence of Cr(III) species is further supported by the position of the $E_{abs}$; in all studied samples, it is located at 6001.1-6001.5 eV (±0.25), i.e. as in the case of reference $Cr(OH)_3$ (6001.3 eV).

### *3.2. Cr-K edgeEXAFS*

In an effort to determine whether Cr(III) is adsorbed onto the tin oxy-hydroxide surface or precipitates as $Cr(OH)_3$, Cr-*K* edge EXAFS spectroscopy was applied and the Fourier Transforms (FT) of the $k^2 \times \chi(k)$ EXAFS spectra are shown in figure 1(b). The unsuccessful fitting attempts to simulate the EXAFS spectra assuming the formation of Cr(III)-precipitates suggests that Cr(III) must form inner sphere surface complexes. Thus, the EXAFS spectra were fitted assuming that Cr forms mononuclear and binuclear inner sphere complexes. More specifically, the fitting proceeds assuming bidentate mononuclear ($^2$E) and binuclear ($^1$C) contributions of the Cr(III) oxy-anions. In the 1st nearest neighbor (nn) shell, the coordination number (CN) was kept fixed to six while the Debye-Waller ($\sigma^2$) factor was iterated. In the next, Sn-comprised, nn shells, the $\sigma^2$ factors were constrained to be equal, although allowed to vary during the fitting. Finally, a Cr-Sn-O multiple-scattering (MS) path, accounting for scattering at four adjacent Sn atoms around the $CrO_6$ octahedron in the $^2$C geometry, was used. The fitting analysis results are listed in table 2.

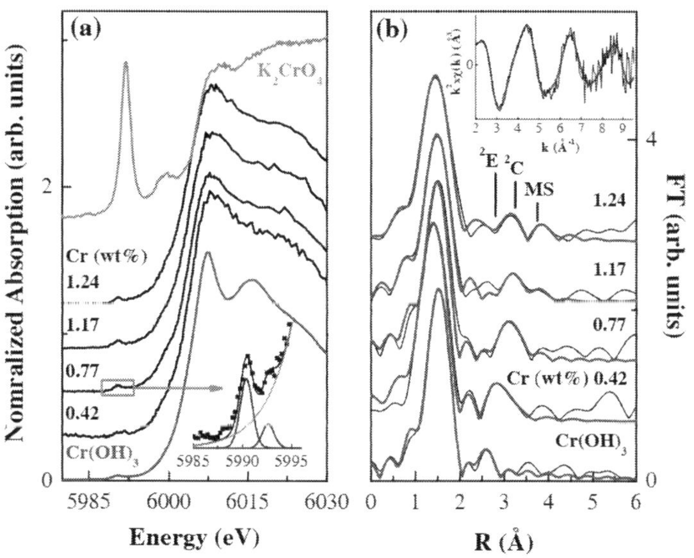


**Figure 1.** (a) Cr-*K* edge XANES spectra and (b) Fourier Transforms (FT) of the $k^2$-weighted $\chi(k)$ Cr-*K* edge EXAFS spectra, of the studied $Sn_6O_4(OH)_4$ and reference $Cr(OH)_3$ and $K_2CrO_4$. The experimental data and the fitting are shown in black and colored lines, respectively. The pre-edge absorption region and the functions used for the simulation of the XANES spectra are shown in the inset of figure 1(a).

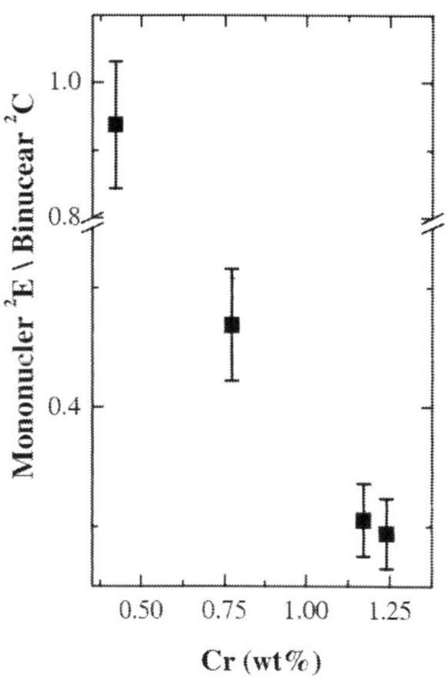


**Figure 2.** The ratio of bidentate $^2$E vs $^2$C complexes as a function of the weight percentage of Cr uptake.

The EXAFS analysis reveal that the Cr-O bond-length is equal to 1.94-1.97Å (±0.01), i.e. characteristic of the presence of $Cr(III)O_6$ species [11]. The Cr-Sn-O length in the MS path is found equal to 3.72-3.75 Å (± 0.05 Å) in all studied samples. Furthermore, as listed in table 2, even though the total number of Cr(III) that participate in the $^2$E and $^2$C configurations does not change, we detect a difference in the type of sorption geometry as a function of the Cr content. More specifically, the total number of inner sphere complexes (sum of $^2$E and $^2$C) is approximately equal to 3.0 – 4.7 (±0.8-1.3). However, as shown in Figure 2, the number of the Cr(III) complexes in the $^2$E geometry versus the number in the $^2$C configuration ($^2$E/$^2$C ratio) decreases as the Cr-loading increases. Thus, it can be stated that the Cr-sorption mechanism changes as a result of increasing surface coverage: as the Cr-uptake increases more Cr(III)-oxy-anions prefer to link via corners with the Sn-octahedra of $SnO_2$. As Cr(VI) approaches $Sn_6O_4(OH)_4$, the latter oxidizes to $SnO_2$ through the hexavalent chromium reduction to Cr(III).

**Table 2.** Cr-*K* edge EXAFS results of the studied samples.

| | 1$^{st}$ nearest neigbor shell | | | $^2$E complexes | | | $^2$C complexes | | |
|---|---|---|---|---|---|---|---|---|---|
| Cr (wt%) | O | $R_{Cr-O}$ (Å) (±0.01) | $\sigma^2$ (×10$^{-3}$Å$^2$) | Sn (±20%) | $R_{Cr-Sn}$ (Å) (±0.02) | $\sigma^2$ (×10$^{-3}$Å$^2$) | Sn (±30%) | $R_{Cr-Sn}$ (Å) (±0.03) | $\sigma^2$ (×10$^{-3}$Å$^2$) |
| 0.42 | 6* | 1.96 | 2.2 | 1.5 | 3.06 | 3.4 | 1.6 | 3.60 | 3.4* |
| 0.77 | 6* | 1.94 | 2.3 | 1.5 | 3.06 | 2.8 | 3.2 | 3.55 | 2.9* |
| 1.17 | 6* | 1.97 | 2.9 | 1.1 | 3.10 | 3.2 | 3.6 | 3.53 | 3.2* |
| 1.24 | 6* | 1.97 | 2.0 | 1.0 | 3.10 | 3.6 | 3.4 | 3.54 | 3.6* |

R is the nearest neighbour distance and $\sigma^2$ is the Debye-Waller factor
* Values kept fixed during the fitting

## 4. Conclusions

The feasibility of implementing $Sn_6O_4(OH)_4$ for the reduction and adsorption of Cr(VI) in water treatment was investigated using XAFS spectroscopies at the Cr-*K*-edge. The analysis of the Cr-*K* edge XANES and EXAFS spectra confirmed the effective use of $Sn_6O_4(OH)_4$ for Cr(VI) removal from drinking water through its oxidation to $SnO_2$. Indeed, only Cr(III) species that form $CrO_6$ octahedra are detected. No $Cr(OH)_3$ precipitates are formed, on the contrary Cr(III)-oxy-anions sorb onto the $SnO_2$ surface. However, changes in the type of inner sphere complexes formed is observed due to increased surface coverage: Cr(III)-oxy-anions preferentially sorb in the $^2$C geometry at the expense of the present $^2$E inner sphere complexes.

## Acknowledgments

The measurements at BESSY were funded from the European Community's Seventh Framework Program (FP7/2007-2013) under grant agreement n. 312284. F. Pinakidou acknowledges financial support from IKY Fellowships of Excellence for Postgraduate Studies in Greece – Siemens Program.

## References

[1] Kaprara E, Kazakis N, Simeonidis K, Coles S, Zouboulis A I, Samaras P and Mitrakas M 2015 *J. Hazard. Mater.* **281** 2
[2] Kaprara E, Seridou P, Tsiamili V, Mitrakas M, Vourlias G, Tsiaoussis I. Kaimakamis G, Pavlidou E, Andritsos N and Simeonidis K, 2013 *J. Hazard. Mater.* **262** 606.
[3] Simeonidis K, Kaprara E, Samaras T, Angelakeris M, Pliatsikas N, Vourlias G, Mitrakas M and Andritsos N, 2015 *Sc. Total Environ.* **535** 61
[4] Gheju M, 2011 *Water Air Soil Poll.* **222** 103.
[5] Rivas B L, Aguirre M D C 2010 *Water Res.* **44** 5730.
[6] Kaur M, Singh M, Mukhopadhyay S S, Singh D, Gupta M 2015 *J. Alloys Comp.* **653** 202.
[7] Görner M, Hentschel M P , Muller B R, Riesemeier H, Krumrey M, Ulm G, Diete W, Klein U, Frahm R, 2001 Nucl. Instrum. Meth. A, **467-468** 703
[8] Newville M. 2001 *J. Synchrotron Rad.* **8** 322
[9] Rehr J J, Kas J J, Vila F D,. Prange M P, Jorissen K, 2010 *Phys. Chem. Chem. Phys.* **12** 5503
[10] Peterson M L, Brown G E Jr, Parks G A, Stein C L 1997 *Geochim. Cosmochim. Acta* **61** 3399
[11] Manceau A and Charlet L 1992 *J. Colloid Interface Sci.* **148** 425

Journal of Physics: Conference Series **712** (2016) 012083 doi:10.1088/1742-6596/712/1/012083

# Structural properties of sodium-rich carbonate-silicate melts: An *in-situ* high-pressure EXAFS study on Y and Sr

**J Pohlenz[1], S Pascarelli[2], O Mathon[2], S Belin[3], A Shiryaev[4], O Safonov[5], A Veligzhanin[6], V Murzin[7], T Irifune[8] and M Wilke[1]**

[1]Helmholtz Centre German Research Centre for Geosciences, Potsdam, Germany, [2]ESRF European Synchrotron Radiation Facility, Grenoble, France, [3]Synchrotron SOLEIL, GIF-sur-YVETTE, France, [4]IPCE, Moscow, Russia, [5]Institute of Experimental Mineralogy RAS, Chernogolovka, Russia, [6]Kurchatov Institute, Moscow, Russia, [7]DESY, Hamburg, Germany, [8]Ehime University, Matsuyama, Japan.

E-Mail: julia.pohlenz@gfz-potsdam.de

**Abstract**. *In-situ* EXAFS combined with a Paris-Edinburgh press (PEP) is an outstanding tool to investigate the local environment of trace elements in melts at high pressure and temperature. A novel design of the pressure assembly ensures a highly stable experimental setup (reaching temperatures of up to 2000 K at 2.5 GPa) while permitting the necessary level of X-ray transmission. This study focuses on the structural incorporation of the geochemically important trace elements Y and Sr in sodium-rich silicate-carbonate melts. Y and Sr K edge EXAFS were collected in transmission mode of the melt (at ~2.5 GPa, 1600 K) and its respective quench products. Distinct changes in the XANES region suggest a change in site symmetry during the cooling process.

## 1. Introduction

Carbonate-bearing silicate and carbonate melts possess unique physical properties. Especially the latter is characterized by extremely low melt viscosities and densities, high surface tensions and electrical conductivities. Of special interest are the distinct geochemical affinities of a wide range of trace elements in carbonate versus silicate melts. The processes of carbonatite genesis and evolution as well as the interaction between carbonate-bearing reservoirs are still unresolved. The only active carbonatitic volcano is the Oldoinyo Lengai, Tanzania, which generates sodium-rich carbonatites in close association with phonolites and nephelinites [1]. Additionally, carbonate-bearing melts evidently play a crucial role during mantle melting, in diamond formation and as metasomatic agents [2]. Understanding the structural properties of carbonate-bearing melts is fundamental to explaining their chemical and physical behavior as well as modeling their role in geochemical processes in the deep Earth. Applying *in-situ* Extended X-ray Absorption Fine Structure (EXAFS) spectroscopy with a large volume Paris-Edinburgh press (PEP) allows for investigating the local structure around geochemically important trance elements in melts at high pressure (*HP*) and high temperature (*HT*) conditions, as found in the Earth's crust and upper mantle.

## 2. *In-situ* EXAFS using a Paris-Edinburgh press

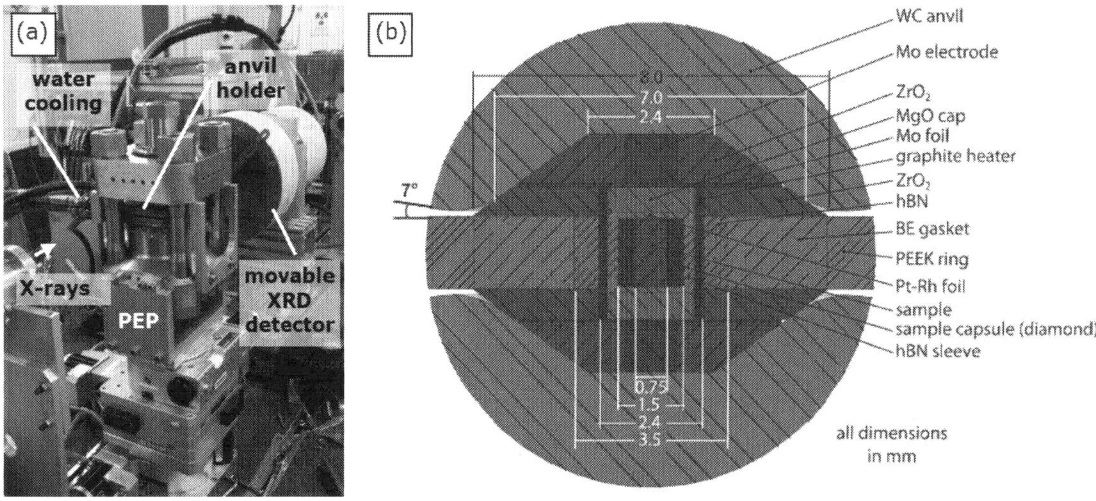


**Figure 1.** (a) Photograph of the Paris-Edinburgh press (PEP) at beam line BM23, ESRF with the XRD detector on the right hand side of the picture. (b) Pressure assembly used in the Paris-Edinburgh press based on zircon dioxide ($ZrO_2$), boron epoxy (BE) and hexagonal boron nitride (hBN) as pressure media. In order to heat, the molybdenum (Mo) parts transfer the current from the wolfram carbide (WC) core of the anvil to the graphite heater. The outer ring made of polyether ether ketone (PEEK) ensures the stability of the setup.

Originally, the Paris-Edinburgh press (Figure 1a) was developed for *HP* experiments on neutron sources [3], but was progressively adapted to XRD and X-ray absorption experiments [e.g. 4]. The press is operated with oil pressure, which is transmitted to the sample-containing pressure assembly through a set of anvils with tungsten-carbide cores. The newly introduced pressure assembly (Figure 1b) used in these *in-situ* experiments is a refined version of the pressure assembly used by Yamada et al. [5] for XRD experiments. The layered setup of the novel assembly firstly assures high stability at high *P* and *T* conditions due to robust and low heat-conducting zircon dioxide parts at the top and the bottom of the assembly. Secondly it assures a sufficient level of X-ray transmission at respective energies due to the radiation penetrated parts consisting of the light materials of boron nitride (as thermal insulator), boron-epoxy (as pressure medium), and polyether ether ketone (for mechanical stabilization). The center of the pressure assembly contains a graphite heater around a nano-polycrystalline diamond, which serves as an inert pressure capsule for the sample itself.

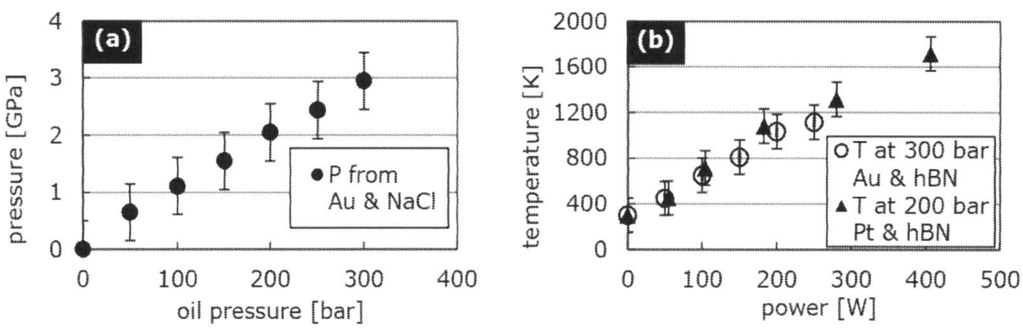


**Figure 2.** Pressure (a) and temperature (b) calibration as determined by the equation of states of reference materials (gold (Au), NaCl, hexagonal boron nitrite (hBN), and platinum (Pt)). Volume changes at *HT/HP* are detected by XRD. Temperature calibration was conducted at ~2 GPa and ~3 GPa sample pressure.

Heating is accomplished by applying high current at low voltage through the anvils, which is transmitted to the graphite heater by central Mo rods and foils at the top and the bottom of the assembly. To avoid high temperature induced damage, the anvils are connected to a water cooling system. With this setup, we were able to reach temperatures of 2000 K at ~2.5 GPa, while retaining a stable gap between the anvils of ca. 300 μm. Pressure and temperature determination is based on the double isochore technique [6] applied to reference materials (e.g. hexagonal boron nitride (hBN), magnesium oxide (MgO), Au, Pt) in calibration runs. The volume changes of these materials are detected with XRD and used in their equation of states, which's intersections indicate the prevailing pressure and temperature conditions.

## 3. Y and Sr incorporation in carbonated silicate melts

This study focuses on unraveling the influence of carbonate concentration on the structural incorporation of the geochemically important trace elements Y (representing rare earth elements) and Sr (representing alkali earths) in silicate and carbonate melts in the system $Na_2O$-$CaO$-$Al_2O_3$-$SiO_2$-$CO_2$ (Figure 3), which resembles a simplified composition of the rocks found at the Oldoinyo Lengai. Starting materials were prepared by mixing carbonate powders ($Na_2CO_3$ + $CaCO_3$) and grinded silicate glass ($Na_2O$-$CaO$-$Al_2O_3$-$SiO_2$ produced by melting the oxide powders) in the respective proportions and then introduced into the diamond capsule.

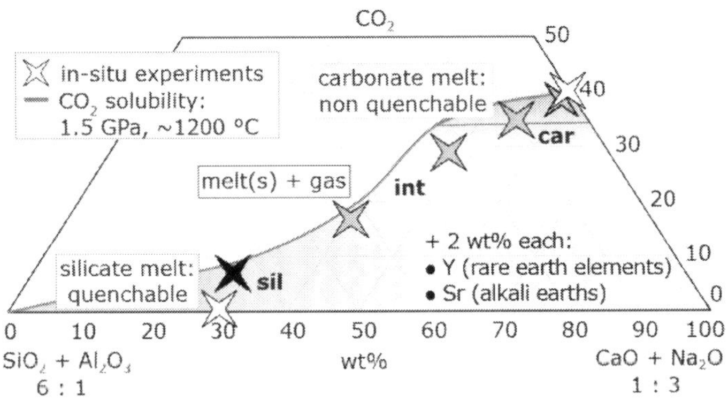


**Figure 3.** Chemical composition of investigated melt samples ranging from silicate (sil) over intermediate (int) carbonate bearing silicate to carbonate (car) compositions. White stars indicate the end members of carbonate free silicate and pure carbonate composition. Grey stars indicate samples prepared out of mixtures between them. The black star indicates the sample of which EXAFS data are presented in this paper.

In order to obtain EXAFS spectra in transmission, the samples were doped with approximately 2 wt% Y and Sr. The *in-situ* experiments were conducted using a Paris-Edinburgh press, installed at BM23, ESRF [7]. $CO_2$-bearing silicate compositions were investigated *in-situ* at 2.2 GPa and 1600 K as well as at the quenched glass after the run. The quench was accomplished by cutting the power supply to the graphite heater. During this process the pressure drops to 1.8 GPa. Pure carbonate melts with $CO_2$ contents between 35 and 42 wt% do not quench to a glass and were investigated only *in-situ* at 2.6 GPa and 1000 K. We determined the liquid state by the absence of crystalline phases as indicated by *in-situ* XRD. The EXAFS measurements were carried out using a Si(111) monochromator. The beam was vertically focused to 50 μm and horizontally collimated to 0.1 mm.

Figure 4 presents Y and Sr K edge EXAFS of a silicate composition containing 5 wt% $CO_2$. The XANES region, especially of Y, shows distinct differences between the measurements taken *in-situ* at *HP/HT* and the quenched glass at *HT*, which might indicate a change in site symmetry during the cooling process. The most prominent difference between EXAFS at *in-situ* conditions and on the quenched sample is the reduction in amplitude. This results in a considerable shortening of the EXAFS spectrum range. It is related to thermally induced vibrations of atoms around their equilibrium position or even migration in the liquid state. This leads to an anharmonic reduction in amplitude and shift in frequency. In order to analyse these data and determine structural parameters of the Y-O and Sr-O pair-distribution an asymmetric pair distribution model needs to be applied. Evaluation of *in-situ* data is still in progress.

In conclusion, we show the feasibility of detecting differences in structural incorporation of elements only being present in low amounts in glasses and melts by transmission EXAFS at high pressure and temperature using the large volume Paris-Edinburgh press.

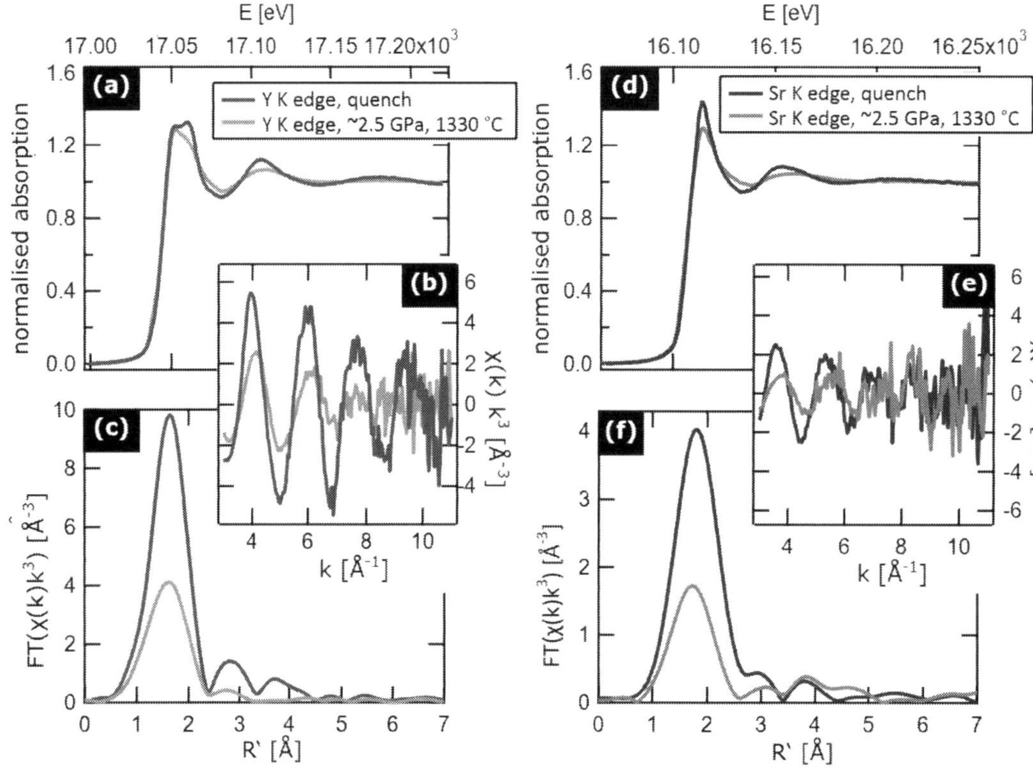


**Figure 4.** Normalised XANES and reduced EXAFS (χ) of Y (a, b) and Sr (d, e) K edge and their Fourier Transforms (FT) (c, f) of *in-situ* (bright) and quenched (dark) samples. Note the reduced amplitude of *in-situ* compared to quench measurements due to enhanced anharmonicity of atomic vibrations. Data was processed with the software xafsX [8].

**References**

[1] Klaudius J and Keller J 2006 Peralkaline silicate lavas at Oldoinyo Lengai, Tanzania *Lithos* **91** 173–90
[2] Jones A P, Genge M and Carmody L 2013 Carbonate Melts and Carbonatites *Rev. Mineral. Geochem.* **75** 289–322
[3] Besson J M, Nelmes R J, Hamel G, Loveday J S, Weill G and Hull S 1992 Neutron powder diffraction above 10 GPa *Phys. B Condens. Matter* **180–181, Part 2** 907–10
[4] Morard G *et al.* 2007 Optimization of Paris–Edinburgh press cell assemblies for in situ monochromatic X-ray diffraction and X-ray absorption *High Press. Res.* **27** 223–33
[5] Yamada A, Wang Y, Inoue T, Yang W, Park C, Yu T and Shen G 2011 High-pressure x-ray diffraction studies on the structure of liquid silicate using a Paris–Edinburgh type large volume press *Rev. Sci. Instrum.* **82** 015103
[6] ZhaoY, Dreele R B V, Weidner D J and Schiferl D 1997 P-V-T Data of hexagonal boron nitride h BN and determination of pressure and temperature using thermoelastic equations of state of multiple phases *High Press. Res.* **15,** 369–386
[7] Mathon O *et al.* 2015 The time-resolved and extreme conditions XAS (TEXAS) facility at the European Synchrotron Radiation Facility: the general-purpose EXAFS bending-magnet beamline BM23 *J Synchrotron Radiat.* **22** 1548–54
[8] Winterer M 1997 XAFS - A Data Analysis Program for Materials Science *J. Phys. IV* **7** C2-243–244

# XAFS study of copper(II) diethylenetriamine complexes having different coordination geometries

**A Gaur[a]*, W Klysubun[b], S K Joshi[c], Balram Soni[d], B. D. Shrivastava[d], J. Prasad[e] and K. Srivastava[e]**

[a]Institute for Chemical Technology and Polymer Chemistry, Karlsruhe Institute of Technology (KIT), Engesserstrasse 20, D-79131 Karlsruhe, Germany
[b]Synchrotron Light Research Institute, 111 University Ave. PO Box 93, Nakhon Ratchasima 30000 Thailand
[c]Physics Department, Government P G Arts and Science College, Ratlam-457001, India
[d]School of Studies in Physics, Vikram University, Ujjain-456010, India
[e]Department of Chemistry, University of Allahabad, Allahabad-211002, India

email: abhijeetgaur9@gmail.com, abhijeet.gaur@kit.edu

**Abstract**. XAFS of three Cu(II) diethylenetriamine complexes (in crystalline form) having different coordination geometries have been investigated. First complex has distorted tetragonal pyramidal, second has distorted square planar and third has distorted square pyramidal geometry. The difference in coordination geometries has been inferred from the differences in pre-edge peak, rising part of edge and in shape of white line, which are seen clearly in the derivative XANES spectra. The distortion in geometry has been correlated with the intensity of peaks in derivative spectra. These inferences have been corroborated from EXAFS analysis where the different paths have been used in the theoretical fits in R space to show contributions of different scatterers at different distances.

## 1. Introduction

The properties of copper (II) sites, catalytic, redox and so forth, generally depend upon the particular copper(II) geometrical arrangement [1]. It is well known that copper(II) exhibits plasticity, i.e., it is able to adapt itself to the ligand coordination constraints, thus assuming different coordination geometries in its complexes [1]. For the present study we have chosen three copper (II) mixed ligand complexes with diethylenetriamine (dien) as one of the ligands. Diethylenetriamine (dien) has varied applications, e.g., in removal of transition metal ions in waste waters and in measuring capacities of clays to exchange cation. It easily forms complexes with metals in aqueous solution. These complexes are stable in air and moisture and hence it is easy to obtain information about the coordination geometries using their spectroscopic signatures. The study of these complexes is interesting since they are stereochemically flexible and can have wide range of coordination geometries [2].

In the present work, X-ray absorption fine structure (XAFS) at the Cu K-edge of the copper (II) diethylenetriamine complexes, $Cu_2(dien)_2Br_2(ClO_4)_2$ (**1**), $Cu_2(dien)_2(na)_2(ClO_4)_3MeOH$ (**2**), $Cu_2(dien)_2(na)_4$ (**3**) (where na = nicotinate), has been investigated. Complex **1** has distorted tetragonal pyramidal geometry [3], complex **2** has distorted square planar geometry [4] and Complex **3** has distorted square pyramidal geometry [4]. The aim of the present work is to study variation in the

XANES features with the change in coordination environment and to study the contribution of different scatterers present at different distances to the EXAFS data of these complexes.

## 2. Experimental

The copper complexes have been prepared and characterized by following standard methods [3,4]. The complexes were in powder form (solid state) and the absorption screens were in the form of pellets. XANES and EXAFS spectra have been recorded (in transmission mode at RT) at beamline BL-8 of the Synchrotron Light Research Institute, Thailand [5]. For calibration, K-edge XANES spectra of pure copper foil were collected simultaneously. The K-edge energy of copper metal was taken as 8,979 eV. The XAFS data have been analyzed by using *Athena* and *Artemis* [6].

## 3. Results and discussions

### 3.1 XANES

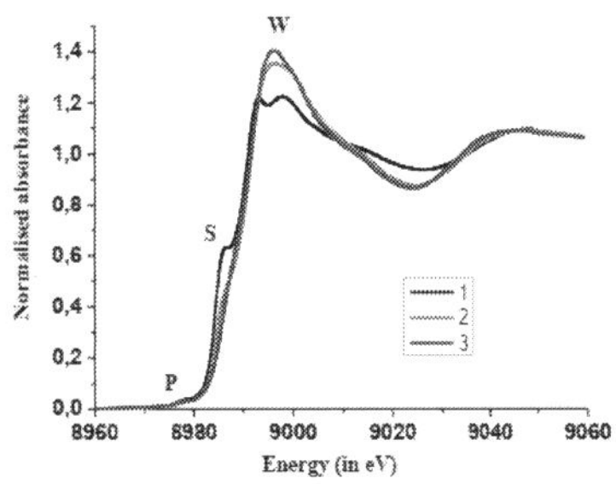


Fig 1. $\mu(E)$ vs. E spectrum

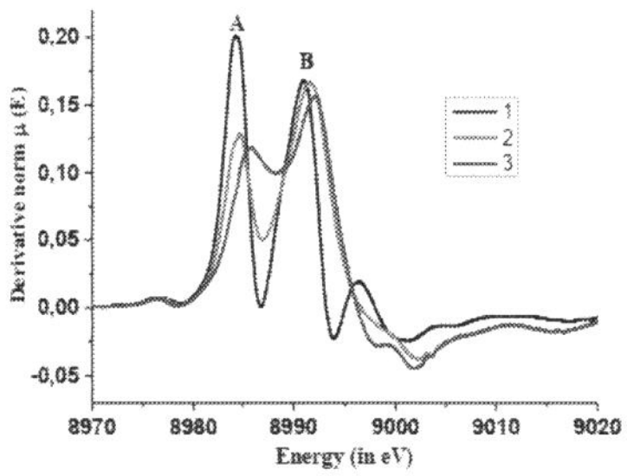


Fig 2. Derivative $\mu(E)$ vs. E

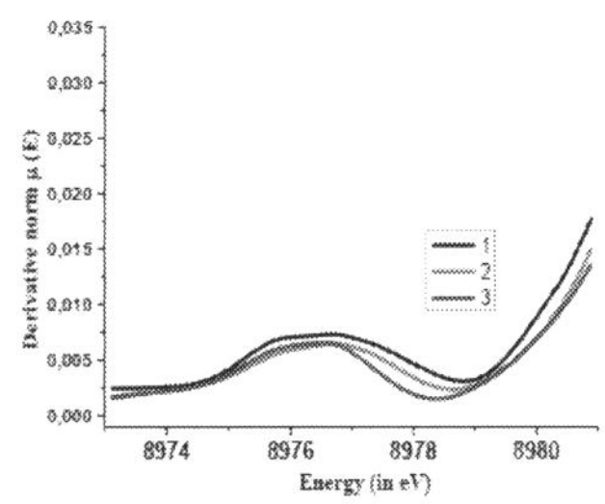


Fig 3. Peak P in the pre-edge

Fig. 1 shows XANES spectra of the studied complexes. The values of chemical shifts are found to be 5.1, 5.8 and 6.8 eV for complexes **1**, **2** and **3**, respectively, showing presence of copper(II). There are three features observed in the absorption edge region of the complexes which reflect coordination geometry of the complexes [7]. First is the pre-edge feature marked P at ~ 8976 eV due to the 1s→3d quadrupole transition. Second is the shoulder S on the rising part of the edge attributed to a 1s→4p transition and the third is the white line W due to the 1s→continuum transitions [8]. The edge portions below S and above S appear as two peaks A and B in the derivative spectra (fig. 2) because of the arctangent nature of the edge portions. The intensity of peak A is found to vary with change in coordination geometry. It has been shown earlier that this peak is characteristic of tetragonal distortion [7]. In the present study, the intensity of this peak is larger for **1** than for **2** and **3** showing higher distortion in **1**. The intensity is comparatively high in **2** than **3**. Thus, it can be inferred that the order of distortion is **1** (tetragonal pyramidal) > **2**(square planar) > **3**(square pyramidal).

The pre-edge peak P has been shown separately in the derivative spectra in fig. 3. It has been observed that the pre-edge intensity of one structural type is distinct from other structural types [9]. Earlier workers have quantified the intensity of the pre-edge peak by the area under the peak and correlated it with 4p mixing. A linear relationship between the 4p character in the ground state and the total pre-edge intensity has been observed which led to a general rule: $I_{OCTAHEDRAL} < I_{5\text{-}COORDINATE} < I_{TETRAHEDRAL}$ (I = intensity), which has been applied to different systems containing Mn, Cu, Ni and Co [8]. In the present work, we have determined the intensity of this feature by calculating the area under the peak by Gaussian fitting. The order of peak intensity as determined from Gaussian fitting is **1** (0.0210±0.0043) > **2** (0.0122±0.0008) ≈ **3** (0.0119±0.0010). In case of complex **1**, the high intensity may be attributed to stronger interaction of Cu orbitals with bromo orbitals in such distorted geometry. Thus, complex **1** having tetragonal pyramidal geometry has maximum contribution from 4p in ground state. However, in case of **2** and **3**, little difference is observed between their peak intensity showing similar amount of 3d-4p mixing in these two complexes.

In fig. 1, the white line peak W is found to have lower intensity in **1** and also the rising edge is shifted to lower energy in **1** with respect to **2** and **3**. In case of Fe, Ni, Mn and Cu complexes, it has been observed that the rising-edge shifts to lower energy for the ligands having higher metal–ligand covalency than with ligands containing lighter atoms such as O and N [8]. This higher covalency decreases the effective nuclear charge and leads to delocalization of metal 4p orbitals. Thus, the rising-edge shifts to lower energies and also the intensity of the main edge transition also decreases. Therefore, in case of complex **1**, the less intense peak W and shifting of the rising edge to lower energy points out towards the presence of ligand with higher covalency, i.e., Br. Presence of Br can thus be inferred from the intensity and position of white line W.

## 3.2 EXAFS

The inferences obtained from XANES analysis have been further investigated by EXAFS analysis. The procedure for EXAFS analysis was similar to that described earlier by the authors [7]. The fitting of the theoretical model to the experimental data of the complexes are described below:

### *3.2.1 Complex 1*

The crystal structure of this complex is available [3] and it has been used to generate the theoretical model by putting input parameters from ref. [3] in *Artemis*. The theoretical model so generated has been fitted to the experimental data of the complex in R-space. A single value of $\Delta E_0$ has been defined for all the paths in the fitting but different values of $\Delta R$ have been defined for different paths. Hence, for complex **1**, one $\Delta R$ has been defined for the three Cu-N paths and two different $\Delta R$ values have been defined for the two Cu-Br paths. Another $\Delta R$ has been defined for the Cu-C paths. Different Debye-Waller factor ($\sigma^2$) values have been defined for each of the scattering path.

For the analysis of the EXAFS data, the input parameter $R_{bkg}$, was set to 1.05 Å. Fourier transform was performed over k-range: $k_{min}$= 2.69 Å$^{-1}$, $k_{max}$= 8.37 Å$^{-1}$. Theoretically modeled data was fitted in the R-space to the experimental data using $k_w$= 3. Fitting was performed for the coordination shells in the R range of 1.0-4.0 Å. Fig. 4(a) shows the fitting in R-space along with the contributions of different paths. The local structure parameters obtained from the analysis are given in table 1. Amplitude reduction factor $S_0^2$ as determined from fitting is 1.03 ± 0.14. The value of goodness-of-fit parameter, i.e., reduced chi-square ($\chi_v^2$) obtained is 42. $\Delta E_0$ value is also reasonable, i.e., 5.00 eV ± 1.40. The three Cu-N distances have been found be 2.00 Å along with the Br atom at 2.39Å, forming the basal plane. The apical Br atom has been found at 2.84 Å. Also, there are C atoms at 2.74, 2.75 and 2.79 Å.

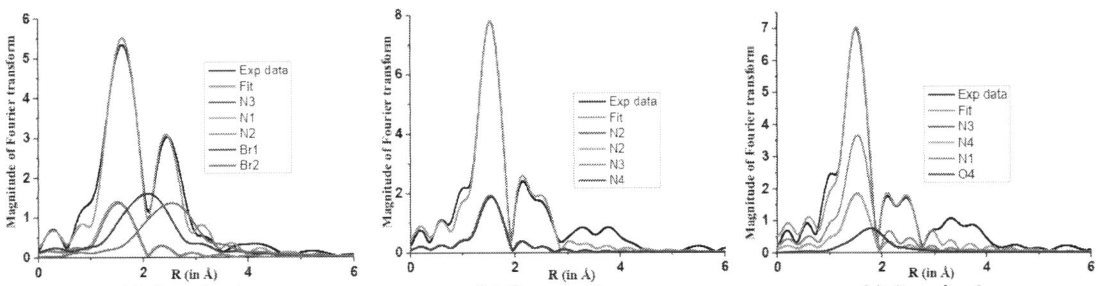

(a) Complex 1 (b) Complex 2 (c) Complex 3

Fig. 4. EXAFS fittings of the complexes in R-space. The experimental curve (black line) and the theoretical fit (red line) along with contributions of different paths are shown in the figure.

### *3.2.2 Complex 2 and 3*

Theoretical models for these complexes have been generated using their own crystal structures [4]. The input parameters used as input in *Artemis* are taken from ref. [4]. For the analysis of the EXAFS data, the input parameter $R_{bkg}$ was set to 1.0 Å. Fourier transform was performed over k-range: $k_{min}$= 2.69 Å$^{-1}$, $k_{max}$= 10.71 Å$^{-1}$. Theoretically modeled data was fitted in the R-space to the experimental data using $k_w$= 3. Fitting was performed for the coordination shells in the R range of 1.0 - 4.0 Å.

In case of **2**, one ΔR has been defined for the four Cu-N paths and another ΔR has been defined for the Cu-C paths. Fig. 4(b) shows the fitting in R-space along with the contributions of different paths. The results obtained from fitting are given in table 1. The value of $\chi_v^2$ obtained is 73 and $\Delta E_0$ value, 5.29 eV ± 1.22, is also reasonable. The value of $S_0^2$ is 1.03 ± 0.15. The four N atoms have been found at a distance of 1.98 Å. Further, there are C atoms at 2.81, 2.82, 2.85 and 2.89 Å.

In case of **3**, One ΔR has been defined for the four Cu-N paths, another for the Cu-O path and separate ΔR has been defined for the Cu-C paths. The fitting in R-space is shown in fig. 4(c) and the results are given in table 1. $\chi_v^2$ obtained is 135, $\Delta E_0$ value is 4.41 eV ± 2.35, and $S_0^2$ is 1.09 ± 0.30. The four N atoms have been found at a distance of 1.99 Å and the Cu-O distance has been found to be 2.24 Å. Further, there are two C atoms at 2.90 Å.

Table 1. The EXAFS fitting results for the complexes

| **Path** | Complex 1 | | | Complex 2 | | | Complex 3 | | |
|---|---|---|---|---|---|---|---|---|---|
| | **N** | **R(Å)** | **$\sigma^2$(Å$^{-2}$)** | **N** | **R(Å)** | **$\sigma^2$(Å$^{-2}$)** | **N** | **R(Å)** | **$\sigma^2$(Å$^{-2}$)** |
| Cu-N | 3 | 2.00 ± 0.01 | 0.0068 ± 0.0027 | 4 | 1.98 ± 0.02 | 0.0051 ± 0.0015 | 4 | 1.99 ± 0.02 | 0.0062 ± 0.0025 |
| Cu-Br /O | 1 | 2.39* ±0.03 | 0.0069 ± 0.0020 | - | - | - | 1 | 2.24** ± 0.02 | 0.0170 ± 0.0085 |
| Cu-Br | 1 | 2.84 ±0.05 | 0.0041 ± 0.0029 | - | - | - | - | - | - |

*For Cu-Br ** For Cu-O

## 4. Conclusion

In the present study, XAFS analysis of three copper(II) diethylenetriamine complexes having different coordination geometries ,i.e., tetragonal pyramidal, square planar and square pyramidal have been performed. The intensity of peak A is larger in **1** than **2** and **3** showing that distortion is maximum in tetragonal pyramidal geometry. Also, the pre-edge peak P intensity as determined from Gaussian fitting has the order **1** > **2** ≈ **3**, which shows that complex **1** has maximum 3d-4p mixing in the ground state. Further, the less intense peak W and shifting of the rising edge to lower energy in **1** indicate the presence of a ligand having higher covalency, i.e., Br.

These inferences have been corroborated from EXAFS analysis where the different paths have been used in the theoretical fits in R space to show contributions of different scatterers at different distances. The contributions of different paths to the experimental data are also shown in the figures. In case of **1**, three N atoms are present at 2.00 Å, one Br atom at 2.39 Å and another Br at 2.84 Å. In **2**, Cu is bound to four N atoms with bond lengths 1.98 Å. In **3**, there are four N atoms at 1.99 Å and one O atom at 2.24 Å. Thus, EXAFS analysis also confirms the presence of different scatterers at different distances in these complexes which corresponds to different coordination geometries.

Thus, the pre-edge and XANES features have been used to infer the presence of different coordination geometries and then inferences were confirmed by EXAFS analysis. Using similar analysis, these edge features can be useful to investigate the local geometry around a metal ion in complicated systems such as metal-enzyme systems and catalysts.

## References

[1] G Tabbi, A. Giuffrida and R. Bonomo J. Inorg. Biochem. **128** (2013) 137.
[2] E. J. Ukpong and I. A. Udoetok, *Journal of Atoms and Molecules* **2** (2012) 312
[3] D K. Towle *et al.*, *Inorg. Chem.* **24** (1985) 4393
[4] Chi Chen, Zhi-Kai Chan, Chun-Wei Yeh and Jhy-Der Chen, *Struct. Chem.* **19** (2008) 87.
[5] W. Klysubun, Y. Thongkam, S. Pongkrapan, K Won-in. J T-Thienprasert and P. Dararutana *Anal. Bioanal. Chem.* **399** (2011) 3033.
[6] Ravel B and Newville M J 2005 *Synchrotron Rad.* **12** 537.
[7] A Gaur, B. D. Shrivastava, K. Srivastava and J. Prasad *X-Ray Spectrom.* **43** (2014) 238.
[8] R Sarangi, Coordination Chemistry Reviews **257** (2013) 459.
[9] M. Sano , S. Komorita , H. Yamatera, *Inorg. Chem.* **31** (1992) 459.

# XAS signatures of Am(III) adsorbed onto magnetite and maghemite

**N Finck[1], L Radulescu[1], D Schild[1], M Rothmeier[1], F Huber[1], J Lützenkirchen[1], T Rabung[1], F Heberling[1], M L Schlegel[2], K Dideriksen[3], S Nedel[3], H Geckeis[1]**

[1] Institute for Nuclear Waste Disposal (INE), Karlsruhe Institute of Technology (KIT), P O Box 3640, D-76021 Karlsruhe, Germany.
[2] CEA, DEN, DPC/SEARS/LISL, Bât. 391, F-91191 Gif-sur-Yvette, France.
[3] Nano-Science Center, Department of Chemistry, University of Copenhagen, Universitetsparken 5, DK-2100 Copenhagen, Denmark.

E-mail: nicolas.finck@kit.edu

**Abstract**. Trivalent americium was adsorbed on magnetite and maghemite under similar chemical conditions and the local environment probed by EXAFS spectroscopy. In both samples, partially hydrated Am(III) binds the surface but slightly different surface complexes were identified. On $Fe_3O_4$, Am(III) forms monomeric tridentate surface complexes similar to that reported for Pu(III) at the (111) surface. In contrast, the lower number of detected Fe atoms may suggest that Am(III) forms monomeric bidentate surface complexes on $\gamma$-$Fe_2O_3$. Alternatively, the lower Fe coordination number can also be due to the presence of vacancies in maghemite. XPS data imply very similar binding environments for Am at both Fe oxide surfaces.

## 1. Introduction

Deep geological disposal of high level nuclear waste (HLW) confined in steel canisters is under consideration in many countries. Over geological time scales, groundwater may reach the canisters and corrode steel. The Fe corrosion products may in turn interact with radionuclides leached out of the HLW matrix, and this process is of considerable interest. Earlier steel corrosion experiments performed in the laboratory under simulated repository conditions showed the formation of various iron (hydr)oxide phases as corrosion products. Among them, magnetite ($Fe_3O_4$) has been frequently detected, and upon partial oxidation was found to partly transform into isostructural maghemite ($\gamma$-$Fe_2O_3$) [1]. The neoformation of such secondary phases represents a significant retention potential for the long lived and radiotoxic actinides (An). Structural incorporation and surface adsorption are relevant long term retention modes.

As reducing conditions are expected in deep geological HLW repositories, notably due to the very low redox potential established by steel corrosion, An may prevail in lower oxidation state(s). Several sorption studies of An on magnetite have been reported. U(VI) [2] contacting pre-formed $Fe_3O_4$ is (partly) reductively immobilized whereas only little U(VI) reduction occurs upon interaction with $\gamma$-$Fe_2O_3$. Also, irrespective of the addition as Pu(V) or Pu(III), only highly specific Pu(III) surface complexes form on magnetite [3]. However, no comparative sorption study of trivalent actinides with both $Fe_3O_4$ and $\gamma$-$Fe_2O_3$ has been reported so far. The goal of this study is to elucidate the adsorption mechanism of trivalent An, using Am(III) as representative, on either magnetite or maghemite under comparable chemical conditions by probing the Am $L_3$-edge by X-ray absorption spectroscopy (XAS).

## 2. Experimental

All experiments were performed under anoxic conditions (Ar glove box). Magnetite and maghemite were prepared as described elsewhere [2]. The sorption samples were prepared by adding Am(III) to either magnetite (sample AmAdsMagn) or maghemite (sample AmAdsMagh) under constant stirring, and similar chemical conditions were considered: $[Am]_{initial}$ = 20 µmol/L, 2.5 g/L $Fe_3O_4$ or γ-$Fe_2O_3$, 0.1 mol/L NaCl. After 2 days, pH and $E_h$ (*vs* S.H.E.) were measured in suspension: pH = 9.7±0.1 and $E_h$ = -0.071±30 mV for AmAdsMagn and pH = 9.9±0.1 and $E_h$ = +0.301±30 mV for AmAdsMagh. The Am and Fe concentrations in the supernatants were determined by HR-ICP-MS (Thermo Element XR) after ultra-centrifugation for 1 hour at 694,000 g (Beckman XL-90). In both samples the Am concentration was < $5.10^{-11}$ mol/L and the Fe concentration was < $2.10^{-8}$ mol/L.

The Fe oxides were characterized by X-ray diffraction (XRD) with a D8 Advance diffractometer using Cu-Kα radiation and an energy-dispersive (Sol-X) detector (Bruker). Fits (Rietveld method) to the XRD data were provided by the TOPAS 4.2 software (Bruker). The solids were further analyzed by scanning electron microscopy (SEM) with an ESEM Quanta 650 FEG (FEI) and the specific surface area was determined by the Brunauer-Emmett-Teller (BET) method [4] using $N_2$ adsorption with an AUTOSORB-1 (Quantachrome Corporation). Information on the valence states of Fe and adsorbed Am in the sorption samples was provided by X-ray photoelectron spectroscopy (XPS) analysis with a VersaProbe II (ULVAC-PHI).

Am $L_3$-edge EXAFS spectra were collected at the INE-Beamline [5] for actinide science at the ANKA Synchrotron Radiation Facility (Karlsruhe, Germany). The energy calibration was done by assigning the first inflection point of a Zr foil at 17998 eV and this reference was measured in parallel with the samples. Data were collected in fluorescence mode using a 5-element low energy germanium solid state detector (Canberra-Eurisys). Data analysis was performed by using Athena and Artemis interfaces to the Ifeffit software [6] following standard procedures [7], with an amplitude reduction factor of $S_0^2$ = 0.88.

## 3. Results and discussion

### *3.1. Solid phases characterization*

Scanning electron micrographs showed that $Fe_3O_4$ particles have octahedral shapes and sizes ranging from ~50 to 400 nm (Figure 1). Maghemite was obtained by oxidation of magnetite and has similar size and shape as the precursor, in agreement with [1]. Thus, Fe(II) oxidation to form γ-$Fe_2O_3$ only marginally affected particle morphology and size. According to the XRD analysis (Figure 2) magnetite (cubic, *Fd3m*) contains marginal amounts (~2 %) of cubic maghemite even though it was stored under anoxic conditions as a suspension (prepared well in advance). Note that the very low amount of maghemite is close to the XRD detection limit and it may likely originate from Fe(II) leaching during magnetite purification by dialysis (several weeks). This minor amount is not expected to significantly affect the sorption studies. Maghemite consists of 78±2 % cubic and 22±2 % tetragonal γ-$Fe_2O_3$, meaning that the cubic magnetite structure is preserved upon oxidation, with vacancies created by Fe(II) release. Vacancies ordering leads to a superstructure that can be indexed as tetragonal [1].

The Fe $2p_{3/2}$ XPS spectrum of maghemite (γ-$Fe^{III}{}_2O_3$), which was obtained by oxidation of $Fe_3O_4$ ($Fe^{II}Fe^{III}{}_2O_4$), almost overlays that of hematite (α-$Fe^{III}{}_2O_3$), meaning that the Fe(II) content in γ-$Fe_2O_3$ is marginal (~1.5±1.5 %) (Figure 3). For the $Fe_3O_4$ used in the sorption study, the shoulder at the low energy side of the Fe $2p_{3/2}$ line, assigned to Fe(II), is of slightly lower intensity than in the fresh sample, indicating some surface oxidation, consistent with XRD data. Accordingly, the Fe(II) content (28±3 %) is slightly lower than the theoretical value of 33 %. Both samples have a similar specific surface area: 13.1±0.5 $m^2/g$ for magnetite and 15.3±0.5 $m^2/g$ for maghemite, consistent with earlier studies [2].

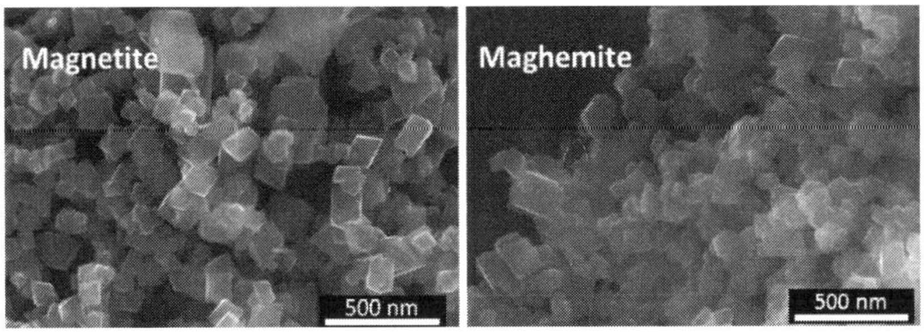


**Figure 1.** Scanning electron micrographs of the magnetite and maghemite substrates.

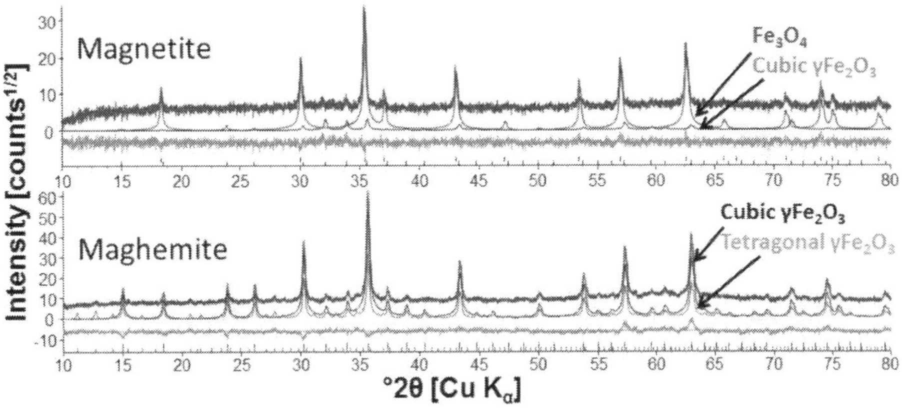


**Figure 2.** Experimental (blue line) and modeled (red line) XRD data with contribution of each single phase (the grey line represents the residual).

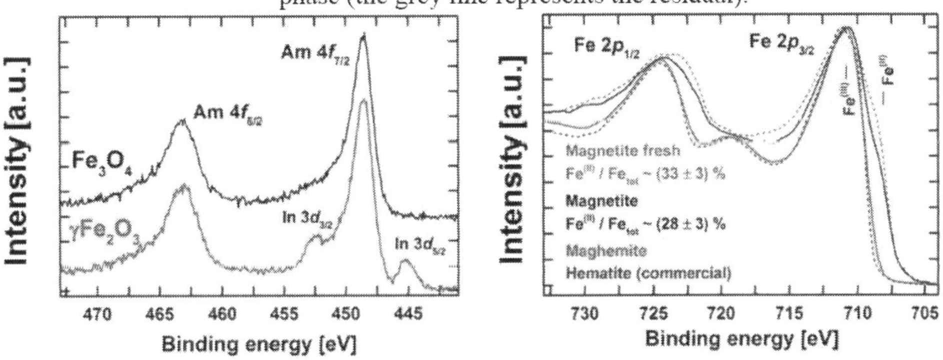


**Figure 3.** Am 4*f* (left) and Fe 2*p* (right) XPS spectra, together with Fe reference compounds (freshly prepared $Fe_3O_4$ and hematite); spectra are charge referenced to O 1*s* at 530.0 eV (the In 3*d* lines originate from the substrate underneath).

### *3.2. Characterization of Am surface species*

According to ICP-MS data, Am is quantitatively retained at the surface of the substrates. XPS was first used to probe the adsorbed actinide (Figure 3). The Am $4f_{7/2}$ binding energy is identical (448.5 eV) in both samples, suggesting similar binding environments, and is not affected by the difference in substrate composition, i.e., the presence or absence of Fe(II).

EXAFS data of the reference Am(III) aquo ions (in perchloric acid) were recorded first (Figure 4) and could be modeled with one O shell containing $N_{O1}$ = 9.0(4) atoms at $R_{Am\text{-}O1}$ = 2.47(1) Å (Table 1), which agrees with earlier findings [7]. After contact with magnetite or maghemite, the Am environment changed as indicated by the differences in EXAFS spectra at $k$ > 6 Å$^{-1}$. In the Fourier transforms (FTs) the first peak corresponding to bound O atoms is of lower intensity in the sorption samples, and contributions from neighboring shells are visible at $R + \Delta R$ ~3 Å and additionally at ~2.5

Å in AmAdsMagh. In both samples, the data could be well fit considering two O subshells in the Am first coordination sphere and higher distance Fe backscatterer(s) (Table 1). No neighboring Am could be detected in any sample, ruling out the presence of surface precipitates or polymers.

**Table 1.** Quantitative EXAFS analysis of the samples

| Sample | FT range[a] | Fit range[b] | Path | $N$ | $R$ [Å] | $\sigma^2$ [Å$^2$] | $R_f$ ($\times 10^3$) |
|---|---|---|---|---|---|---|---|
| $Am(III)_{aq}$ | 3.2-10.2 | 1.7-2.6 | Am-O1 | 9.0(2) | 2.47(1) | 0.008 | 3.1 |
| AmAdsMagn | 3.3-9.0 | 1.6-4.1 | Am-O1 | 5.4(3) | 2.41(1) | 0.003 | 9.0 |
| | | | Am-O2 | 2.5(6) | 2.59(1) | 0.003 | |
| | | | Am-Fe1 | 2.7(3) | 3.45(1) | 0.006 | |
| | | | Am-Fe2 | 0.9(2) | 4.09(4) | 0.003 | |
| AmAdsMagh | 3.3-9.0 | 1.6-3.7 | Am-O1 | 6.0(2) | 2.43(1) | 0.004 | 2.2 |
| | | | Am-O2 | 3.0(3) | 2.67(1) | 0.004 | |
| | | | Am-Fe1 | 1.1(3) | 3.54(2) | 0.006 | |

[a]Fourier transformed range, [b]$R + \Delta R$ interval for the fit. $N$ is the coordination number, $R$ is the interatomic distance, $\sigma^2$ is the mean square displacement ("Debye-Waller" term), $R_f$ is the figure of merit of the fit as defined in [6].

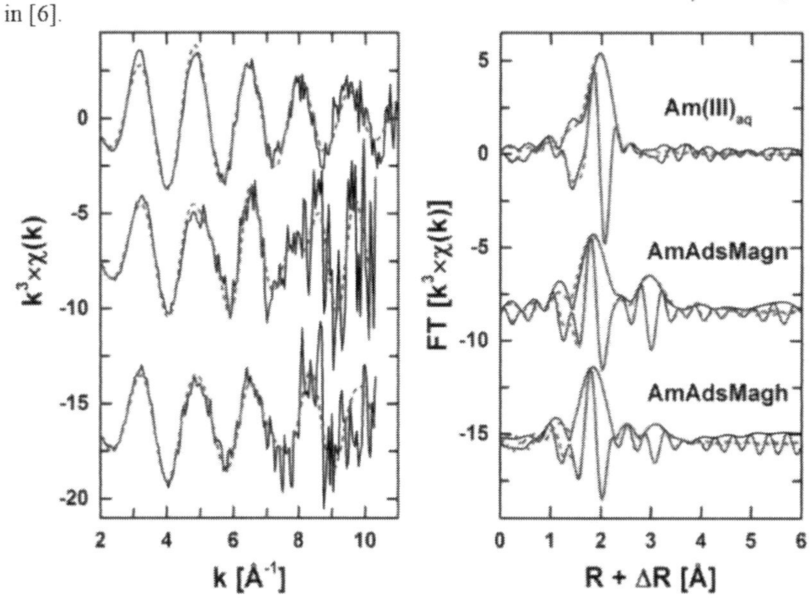


**Figure 1.** Experimental (solid black line) and modeled (dashed red line) EXAFS spectra (left) and corresponding Fourier transforms (right) of the Am(III) aquo ions and of the sorption samples.

In both samples Am bound to the surface (O2 shell at 2.59-2.67 Å) retains part of its hydration sphere (O1 shell at ~2.42 Å). The detection of 3 Fe atoms at ~3.5 Å in AmAdsMagn is consistent with earlier Pu(III) studies [3], where monomeric tridentate inner-sphere surface complexes at $Fe_3O_4$ (111) have been reported. Americium thus forms similarly stable complexes on magnetite. SEM evidenced crystals of octahedral shapes where the (111) face predominates and thus sorption at this face certainly dominates, but Am binds also other crystallographic planes in our powder sample. Compared to $Fe_3O_4$, Am is located at slightly larger distance on $\gamma$-$Fe_2O_3$ (O2 shell at 2.67 Å) and only ~1 Fe atom was detected, hinting at a slightly different binding environment. The AmAdsMagh data can be explained by the formation of monomeric bidentate inner-sphere surface complexes on $\gamma$-$Fe_2O_3$, whereby Am binds two O atoms at the edge of $FeO_6$ octahedra, even though the O2 shell at 2.67 Å contains 3 atoms. In that configuration, the number of detected Fe is limited, in agreement with fit results. Alternatively, the low number of detected Fe can also be explained by the presence of vacancies in $\gamma$-$Fe_2O_3$.

## 4. Conclusions

Am(III) was adsorbed onto magnetite and maghemite under similar chemical conditions and the binding environment probed by XAS. Partially hydrated Am(III) is adsorbed on $Fe_3O_4$ and $\gamma$-$Fe_2O_3$, but dissimilar neighboring Fe shells point to slightly different binding sites. For magnetite, the data are consistent with earlier Pu(III) findings, i.e., the formation of monomeric tridentate Am(III) surface complexes. For maghemite, the lower number of Fe atoms detected at larger distances may indicate the formation of monomeric bidentate complexes at the surface or can be due to vacancies in the $\gamma$-$Fe_2O_3$ lattice.

**Acknowledgements**

We thank E. Soballa (KIT-INE) for SEM analysis, M. Lagos and F. Geyer for ICP-MS measurements and K. Dardenne (KIT-INE) for assistance at the INE-Beamline. ANKA (Karlsruhe, Germany) is acknowledged for providing beamtime. This work has been supported by the European FP7 TALISMAn (**T**ransnational **A**ccess to **L**arge **I**nfrastructure for a **S**afe **M**anagement of **A**cti**N**ides) project under the grant agreement #323300, JRP no TALI-C01-13.

**References**

[1] Cornell R M and Schwertmann U 1996 *The Iron Oxides. Structure, Properties, Reactions, Occurrence and Uses*. VCH, Weinheim, Germany.
[2] Huber F, Schild D, Vitova T, Rothe J, Kirsch R, Schäfer T 2012 *Geochim. Cosmochim. Acta* **96** 154
[3] Kirsch R, Fellhauer D, Altmaier M, Neck V, Rossberg A, Fanghänel T, Charlet L, Scheinost A C 2011 *Environ. Sci. Technol.* **45** 7267
[4] Brunauer S, Emmett H, Teller E 1938 *J. Amer. Chem. Soc.* **60** 309.
[5] Rothe J, Butorin S, Dardenne K, Denecke M A, Kienzler B, Löble M, Metz V, Seibert A, Steppert M, Vitova T, Walther C, Geckeis H 2012 *Rev. Sci. Instrum.* **83**, 043105.
[6] Ravel B, Newville M 2005 *J. Synchrotron Radiat.* **12** 537.
[7] Finck N, Dardenne K, Geckeis H 2015 *Chem. Geol.* **409** 12.

# Aqueous U(VI) interaction with magnetite nanoparticles in a mixed flow reactor system: HR-XANES study

**I Pidchenko[1], F Heberling[1], KO Kvashnina[2], N Finck[1], D Schild[1], E Bohnert[1], T Schäfer[1], J Rothe[1], H Geckeis[1] and T Vitova[1]**
[1] Institute for Nuclear Waste Disposal (INE), Karlsruhe Institute of Technology, PO 3640, 76021 Karlsruhe – Germany
[2] European Synchrotron Radiation Facility (ESRF), 6 Rue Jules Horowitz, BP 220, 38043 Grenoble Cedex - France

E-mail: ivan.pidchenko@kit.edu

**Abstract**. The redox variations and changes in local atomic environment of uranium (U) interacted with the magnetite nanoparticles were studied in a proof of principle experiment by the U $L_3$ and $M_4$ edges high energy resolution X-ray absorption near edge structure (HR-XANES) technique. We designed and applied a mixed flow reactor (MFR) set-up to maintain dynamic flow conditions during U-magnetite interactions. Formation of hydrolyzed, bi- and poly-nuclear U species were excluded by slow continuous injection of U(VI) ($10^{-6}$ M) and pH control integrated in the MFR set-up. The applied U HR-XANES technique is more sensitive to minor changes in the U redox states and bonding compared to the conventional XANES method. Major U(VI) contribution in uranyl type of bonding is found in the magnetite nanoparticles after three days operation time of the MFR. Indications for shortening of the U-$O_{axial}$ bond length for the magnetite compared to the maghemite system are present too.

## 1. Introduction

The safety assessment of long-term repositories for high-level radioactive waste (HLW) motivates speciation studies of actinides (An) and fission products in relevant systems. The element under particular interest is uranium (U), which occurs in the environment mainly as highly mobile U(VI) and sparingly soluble U(IV) form. Over long periods of time ground water might potentially enter the HLW repository, accelerating corrosion processes of steel containers. To model the U interaction mechanisms with one of the main iron corrosion products, magnetite, various laboratory experiments were performed mainly in static conditions. Such static experiments facilitate formation of secondary phases and might not be representative for potential dynamic conditions in a HLW repository. Up to date only one study is available on the interaction of U with magnetite in dynamic conditions coupled with conventional X-ray absorption fine structure (XAFS) investigations [1]. There are still open questions which need to be addressed: 1) What is the speciation of U after continuous redox and phase transformation processes induced by the interaction of U with magnetite; 2) Are the reduced U species long term stable; 3) Are these species incorporated in the vacancies created by the removal of Fe(II), which is favourable at about pH 5.5?

We present benchmark experiment with a mixed flow reactor (MFR) set-up designed for U(VI) interaction with magnetite nanoparticles in aqueous media. For the first time a combination of the U $L_3$ and U $M_4$ edge HR-XANES investigations is used to characterize the U oxidation states and the local atomic environment of U in such a system. Reduced core-hole lifetime broadening effects result in better resolved spectral features in the HR-XANES compared to conventional XANES spectra [2]. The method facilitates the detection of minor contributions of U redox states in mixtures. For example the U $M_4$ edge HR-XANES was successfully applied to characterize U(V) in $U_4O_9$ [3].

## 2. Experimental part

### *2.1. MFR experiment*

A MFR set-up for continuous interaction of U(VI) with magnetite nanoparticles in aqueous suspension employed a 80 mL Plexiglas reactor equipped with a magnetic stirrer and 0.4 μm Millipore filter to prevent loss of particles from suspension (see figure 1). 2 g/L magnetite suspension with pH 7.3 and ionic strength of 0.01 M NaCl was used. Three liters of U(VI) containing aqueous solution ([U]=$1\cdot10^{-6}$ M, pH=5.5) were pumped continuously through the MFR during three days with the help of a Peristaltic pump. pH values were recorded before and after the experiment. Aliquots of aqueous solution after the reactor were sampled every day and the total U concentration was analyzed. The PHREEQC 2.18 code and the Nagra/PSI thermodynamic database [4] for geochemical modelling was used to estimate the U species in the initial U solution. After three days operational time of the MFR, an aliquot of the magnetite suspension was taken and dried in Ar atmosphere. The sample and the references were placed in a Plexiglas holder with two separated Polyethylene windows each with 10 μm thickness. The samples were placed into a gas tight aluminum cylinder for transportation in anoxic conditions and opened shortly prior the HR-XANES measurements.

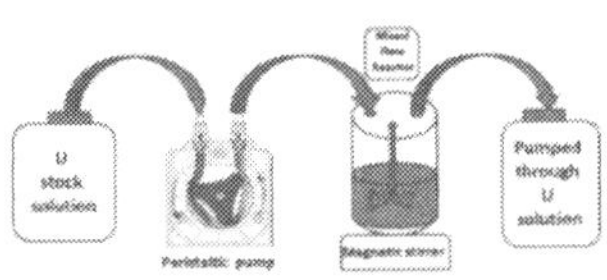


**Figure 1.** Scheme of the MFR set-up.

### *2.2. U $L_3$/$M_4$ HR-XANES experiments*

The U $L_3$/$M_4$ edge HR-XANES spectra of U(VI) interacted with magnetite after three days of operation of the MFR (MFR-3d), $U_4O_9$ and U(VI) adsorbed on maghemite during 55 days (U(VI)-magh) samples were measured at the ID26 beamline at the European Synchrotron Radiation Facility (ESRF) [5]. The incident energy was monochromatized by a Si(111) double crystal monochromator (DCM) (U $L_3$ 17166 eV, $M_4$ 3728 eV). Rejection of higher harmonics was achieved by three Si mirrors at an angle of 3.5 mrad (for the U $M_4$ edge experiment) and three Pd mirrors at an angle of 2.5 mrad (for the U $L_3$ edge experiment) relative to the incident beam. The beam size was focused to ~0.150 mm vertical and 0.450 mm horizontal size for the measurements at both absorption edges. XANES spectra were measured in high energy resolution fluorescence detection mode using an X-ray emission spectrometer [6]. The sample, analyzer crystal and silicon drift diode (SDD) were arranged in a vertical Rowland geometry. The U HR-XANES spectra at the $M_4$/$L_3$ edges were obtained by recording the maximum intensity of the U $L_\alpha$/$M_\beta$ emission lines. Ge(777) (U $L_{\alpha 1}$ 13618 eV) or the Si(220) (U $M_\beta$ 3337 eV) diffraction order of five spherically bent Ge/Si crystal analyzers with 1 m bending radius were used. The crystals were aligned at 75° and at 77° Bragg angles for U $M_4$ and U $L_3$ edges, respectively. The experimental energy resolution was ~2.6 eV for the U $L_3$ and ~0.7 eV for the U $M_4$ edge and was determined by measuring the full width at half maximum (FWHM) of the elastic peak. The paths of the incident and emitted X-rays through air were minimized in order to avoid losses in intensity due to absorption. Depending on the edge, from ten to forty four scans were measured for each sample. Samples were checked for the radiation damage by measuring a very fast scan (1 sec) with and without 200 μm Al foil attenuating the beam intensity. For the U $M_4$ edge changes in the spectral shape were detected after 60 seconds of measuring time, therefore each XANES spectrum has been collected at the fresh sample spot; all spectra were normalized to the maximum absorption intensity.

## 3. Results and discussion

Before interaction with magnetite 90% of the aqueous U is in the form of not hydrolyzed U(VI) uranyl; less than 10% of hydrolyzed species and minor bi- and poly-nuclear hydroxide complexes contribute to the U aqueous solution as well, as predicted by PHREEQC calculations. The U concentration in the outgoing solution was $5 \cdot 10^{-7}$ M, pH~5.8 validating that about 50% of the initial U was removed by the magnetite nanoparticles. It was demonstrated that at pH = 5.5 Fe(II) can be leached from the magnetite and can reduce U(VI) to U(IV)/U(V). These U(IV)/U(V) species might potentially incorporate in the octahedral Fe(II) vacancies in the magnetite structure [7]. Compared to a static batch sorption setup, the MFR set-up allows for better control of the pH, thereby precipitation of U phases is prevented. In addition, formation of secondary Fe(II) phases is avoided due to the fast Fe(II) removal from the MFR volume.

U $M_4/L_3$ edge HR-XANES spectra of MFR-3d, U(VI)-magh and $U_4O_9$ are plotted in figures 2a, a' and 2b, b'. The spectrum of $U_4O_9$ contains two peaks with energy positions characteristic for U(IV) (line A) and U(V) (line B) [3]. The U(VI)-magh spectrum is used as U(VI) reference as there are no U redox changes occurring during U(VI) interaction with maghemite, having all Fe as Fe(III) but the same spinel ferrite structure as magnetite. The shape of the MFR-3d U $M_4$ edge HR-XANES spectrum resembles closely that of the U(VI)-magh spectrum (figure 2a, 2a'). The main peak C in figure 2 has the same energy position in both spectra. The energy positions of peaks D and E shift to higher energies in the MFR-3d spectrum (see table 1). The peaks have been assigned to electronic transitions of $3d_{3/2}$ electrons to $5f\delta/5f\phi$ line C), $5f\pi$ (line D) and $5f\sigma$ (line E) unoccupied valence orbitals of U(VI) in the form of $UO_2^{2+}$ (uranyl) [8]. The $5f\sigma$ mainly mixture of U and O valence orbitals [8]. Therefore it is expected that peak E will be sensitive to variations of the $U\text{-}O_{axial}$ bond length correlated to the level of hybridization of those valence orbitals.

The position of peak E is shifted by ~0.2 eV ± 0.05 eV to higher energies compared to U(VI)-magh implying shorter average $U\text{-}O_{axial}$ bond distance in the MFR-3d sample, compared to U(VI)-magh [9]. A likely explanation is that the shorter $U\text{-}O_{axial}$ bond length results from a stronger bonding of uranyl to the magnetite surface compared to maghemite. This is consistent with the observations made for An(III) adsorption on maghemite and magnetite reporting shorter bond distances for the An(III)-magnetite interactions compared to An-maghemite [10].

The U $L_3$ edge HR-XANES MFR-3d spectrum resembles clearly the spectrum of U(VI)-magh confirming the major U(VI) contribution in the MFR-3d sample (figure 2b, 2b'). The energy position of the multiple scattering H resonance is also sensitive to the $U\text{-}O_{axial}$ bond distance. But it does not change significantly for the two spectra. This is a clear indication that the U $L_3$ edge has lower sensitivity to small variations in the $U\text{-}O_{axial}$ bond length compared to the U $M_4$ edge HR-XANES spectrum.

**Table 1.** Energy positions of the spectral peaks.

| **Edge** | | U $M_4$ (± 0.05 eV) | | | | | |
|---|---|---|---|---|---|---|---|
| **Feature** | A | B | C | D | E' | E | (E-E') |
| [eV] | 3725.3 | 3726.4 | 3726.9 | 3728.8 | 3731.9 | 3732.1 | 0.2 |
| **Edge** | | U $L_3$ (± 0.05 eV) | | | | | |
| **Feature** | F | G | H | | | | |
| [eV] | 17174.8 | 17176.8 | 17188.7 | | | | |

## 4. Conclusions

A combination of U $M_4$ and $L_3$ edge HR-XANES studies is valuable for evaluating minor changes in the redox states and local coordination environment of U in dynamic systems where two or more redox active couples are influencing the reaction process. The combination of spectroscopy studies with these advanced techniques with precise pH and $E_h$ measurements is a powerful tool for investigations of redox processes. We have shown that the set-up is functional and can be applied for longer interaction time studies. One necessary improvement is the insertion of online $E_h$-measurements. Extended X-ray Absorption Fine Structure (EXAFS) investigations will be beneficial as they can potentially help to distinguish between surface adsorbed and/or incorporated U species in magnetite. No shift of the main spectral features in U $M_4$ and U $L_3$ edges HR-XANES spectra indicate no observable change in U redox state after three days of the interaction time with magnetite. Planned additional experiments including longer interaction times and/or different experimental conditions (pH, flow rate) might lead to U redox changes in the system and provide insights about the long term stability of the reduced U species.

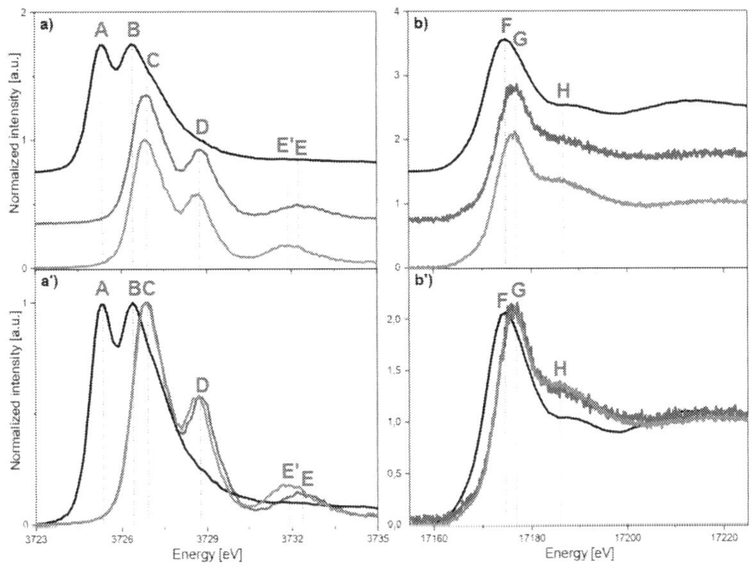


**Figure 2.** U $M_4$ (a, a') and U $L_3$ (b, b') edge HR-XANES spectra of MFR-3d (violet), $U_4O_9$ (black) and U(VI)-magh (red).

**Acknowledgements**

We thank KIT and the Helmholtz Association of German Research Centres for the Helmholtz Young Investigators Group grant (VH-NG-734). We acknowledge ESRF for the provided beamtime. We thank Philippe Martin (CEA-Cadarache, France) for providing the $U_4O_9$ reference spectra.

**References**

[1] Nico NS, Stewart BD and Fendorf S 2009 *Envr. Sci. Tech.* **46** 7391-7396.
[2] Vitova T, Denecke MA, Göttlicher J, Jorissen K, Kas JJ, Kvashnina K, Prüßmann T, Rehr JJ, Rothe J 2013 *J. Phys.: Conf. Ser.* **430** art. no. 012117.
[3] Kvashnina KO, Butorin SM, Martin P and Glatzel P 2013 *Phys. Rev. Let.* **111** 253002.
[4] Hummel W, Berner U, Curti E, Pearson FJ, Thoenen T, 2002 *Radiochim. A.* **90** 805-813.
[5] Gauthier C, Sole VA, Signorato R, Goulon J and Moguiline E 1999 *J. Synch. Rad.* **6** 164-166.
[6] Glatzel P and Bergmann U 2005 *Coordin. Chem. Rev.* **249** 65-95.
[7] Huber F, Schild D, Vitova T, Rothe J, Kirsch R and Schäfer T 2012 *Geoch. Cosm A.* **96** 154-173.
[8] Denning RG 2007 *J. Phys. Chem. A* **111** 4125-4143.
[9] Podkovyrina Y, Pidchenko I, Prüßmann T, Bahl S, Soldatov A,Vitova T 2015 *J. Phys.: Conf. Ser.* submitted.
[10] Finck N, Radulescu L, Schild D, Rothmeier M, Huber F, Lützenkirchen J, Rabung T, Heberling F, Schlegel ML, Dideriksen K, Nedel S, Geckeis H 2015 *J. Phys.: Conf. Ser.* submitted.

# Fe, Ni and Zn speciation in airborne particulate matter

**Bridinette Thiodjio Sendja[1], Giuliana Aquilanti[1], Ivano Vassura[2], Marco Giorgetti[2]**

[1]Elettra - Sincrotrone Trieste, Basovizza, Italy
[2]University of Bologna, Department of Industrial Chemistry, Bologna, Italy

Email: giuliana.aquilanti@elettra.eu

**Abstract**. The study of elemental speciation in atmospheric particulate matter is important for the assessment of the source of the particle as well for the evaluation of its toxicity. XANES data at Fe, Ni, and Zn K-edges are recorded on a sample of urban dust (from the Rimini area of Emilia Romagna region, Italy) deposited on a filter and on the NIST standard reference material 1648. Using linear combination fitting we give an indication of the chemical species of the three metals present in the samples.

## 1. Introduction

Atmospheric particulate matter (PM) is a complex mixture of different chemical components which exist as a suspension of solid or liquid particles in air, with a wide dimension range (10 nm to 100 µm). It has been widely demonstrated that there is an association between increased PM concentrations and adverse health effects [1]. Atmospheric aerosol toxicity is connected especially with its chemical composition and its capacity to adsorb toxic substances on the surface. As these particles may be harmful to humans, the assessment of their level and chemical composition are significant from an environmental health perspective. On a global scale, PM is mainly constituted by sulphates, nitrates and ammonium, organic and elemental carbon and in small extent by heavy metals.

Some heavy metals, such as Fe, Co, Zn, Mn, Ni and Cr, are defined essential, since they are present in human beings and they are indispensable, because of their role in some metabolic processes. If these metals are present in the organism at concentrations higher than what is needed for the correct execution of metabolic processes, they provoke toxic effects as well as non-essential heavy metals do. For these reasons, heavy metals emissions in the environment must be controlled. In addition to this, heavy metals are not involved in decomposition processes, and thus they persist in a certain environmental compartment until they are transported in another environmental matrix by a chemical, biological or physical mechanism.

Heavy metals are present in the atmosphere as airborne PM and thus the determination of their concentration in this matrix is fundamental from an environmental health perspective. In addition to the determination, the chemical speciation of such heavy metals, i.e., how the metal is distributed among different chemical states, is fundamental for the understanding of the complex phenomena related to airborne PM. In this context, the use of an elemental specific technique such as the x-ray absorption spectroscopy (XAS) is one of the most suitable synchrotron radiation technique for these studies. In particular the x-ray absorption near edge structure (XANES) part of the spectra can reveal different oxidation and chemical state of a specific metal in the particulate matter and it probes the bulk of a material.

This contribution deals with the XANES analysis of a sample of urban dust (from the Rimini area of Emilia Romagna region, Italy) deposited on a cellulose filter together with the NIST standard reference material 1648 for comparison. A linear combination fitting allowed us to retrieve information about the chemical speciation of Fe, Ni and Zn present in the sample.

## 2. Experimental

X-ray absorption near edge structure (XANES) spectra are recorded at the XAFS beamline of Elettra-Sincrotrone Trieste (Basovizza, Italy) [2]. The storage ring is operated at 2.0 GeV in top up mode with a typical current of 310 mA. The data are recorded at the Fe K-edge, Ni K-edge and Zn K-edge. The white beam is monochromatised using a fixed exit monochromator equipped with a pair of Si (111) crystals. Harmonics are rejected by detuning the second crystal. Internal reference of Fe, Ni, and Zn foils are used for energy calibration in each scan. The first inflection points for Fe, Ni and Zn are set at 7112 eV, 8333 eV and 9659 eV respectively.

Two samples have been measured. The first sample referred as 29c3 has been collected in urban area of Rimini in the Emilia Romagna Region on 10 cm diameter quartz membrane by means of an ECHOHiVol TCRTECORA sampler equipped with a PM2.5 sampling head and operating at the flow rate of 200 l/min. The sampled volume is 394.4 $m^3$, and the PM2.5 filter density is 0.502 $\mu g/cm^2$. More information about sample 29c3 is reported in [3]. The second sample is a powder of NIST standard reference material 1648. For the XANES measurements the latter is uniformly distributed on a kapton tape. The XANES spectra are collected in fluorescence mode with the sample positioned at 45° with respect to the beam. The total fluorescence yield is measured as a function of the x-ray energy using a silicon drift detector. More than one spectrum was collected per each sample. The spectra are then averaged and normalized according to the standard procedure using the ATHENA program [4].

## 3. Results and discussion

Figure 1 shows the normalized XANES spectra at Fe (panel a)), Ni (panel b)) and Zn K edge (panel c)) of the NIST standard reference material (black line) and of sample 29c3 (red line). For the three edges the data of the samples 29c3 show overall the same features of those of the NIST sample. This indicates that the local environment of the three metals in the 29c3 sample is similar to that in the NIST sample. Nonetheless a few differences can be observed.

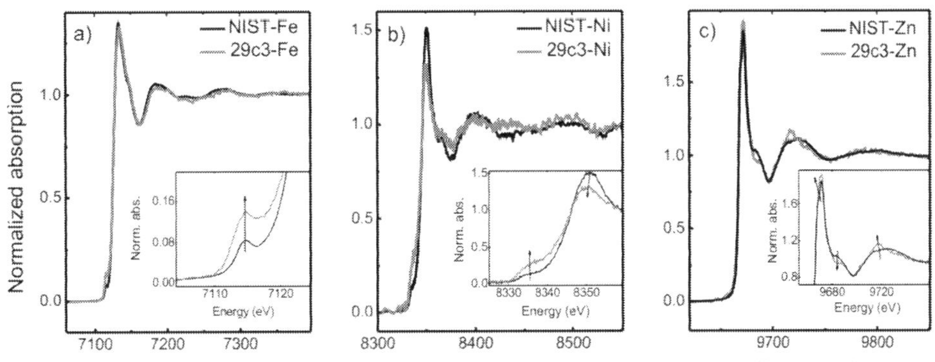


**Figure 1.** Superimposed x-ray absorption spectra around Fe a), Ni b) and Zn c) K edges for NIST (black line) and 29c3 (red line) samples. The inset highlights the differences between the spectra of the two sample per each edge.

For the iron data the centroids of the pre-peak is at 7114.6(1) for both samples. This indicates that the iron species in the two samples contain predominantly ferric iron [5] and, if a fraction of ferrous iron exists, the $Fe^{2+}/Fe^{3+}$ ratio does not vary in a significant way. However, the inset of Fig.1a) shows that the intensity of the pre-peak is higher in the case of the 29c3 sample indicating that an increased fraction of iron possesses a geometry allowing a higher probability of d-p orbital hybridization, i.e. either a tetrahedral geometry or, at least, an octahedral geometry more distorted than that possessed by iron in the NIST sample [6]. The data at nickel K-edge are affected by a higher noise level because of the low concentration of the Ni species. Nonetheless, there are two remarkable differences between the spectra of the two samples (inset of Fig.2b)). The intensity of the shoulder at 8335 eV increases from the NIST to the 29c3 sample. We observe an opposite behavior for the main resonance at 8351 eV that decreases in intensity form the NIST sample to the 29ce sample together with a shift to lower energies. The data at Zn K-edge are shown in Fig.1c). The absence of a pre-peak is expected because of the $d^{10}$ electronic configuration of Zn. The spectral features of the two spectra are compatible with zinc in six-fold coordination by oxygen/nitrogen anions.

In order to quantify the chemical species in the samples that are composed by multicomponent mixture we use a linear combination of spectra of various reference samples. The procedure, implemented on ATHENA [4], uses a least-squares algorithm to refine the sum of a given number of reference spectra to an experimental spectrum. The choice of proper reference spectra is critical in such kind of analysis and must be done gathering as much information as possible about the system under study. Figure 2 shows the XANES part of the normalized spectra together with the linear combination fitting curves for the NIST samples at the three edges. The quality of the fit, which can be quantified by the R-factor (reported in Table 1), indicates the proper choice of the reference compounds.

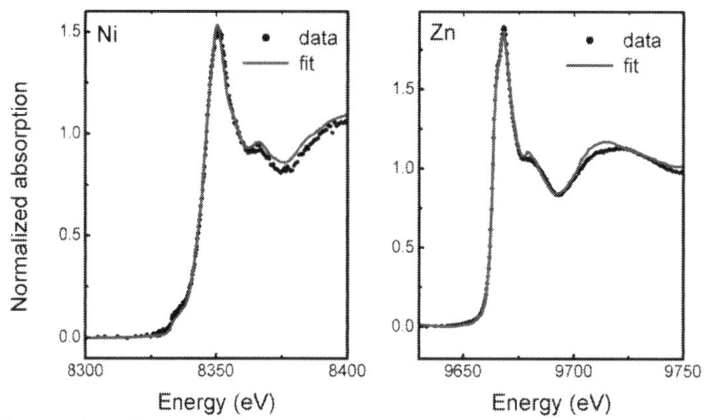


**Figure 2**. Comparison between the normalized spectra of the NIST sample at Ni and Zn K-edge (black dots) and the linear combination fit (blue line).

Table 1 summarizes the results obtained for Ni and Zn by using linear combination fitting method through Athena package.

**Table 1**. Relative amounts of the different chemical species obtained using linear combination fitting for Ni and Zn species.

| **Ni K-edge** | | | | |
|---|---|---|---|---|
| | ***Ni metal*** | ***$NiSO_4 \cdot 6H_2O$*** | ***$NiFe_2O_4$*** | ***$R_{factor}$*** |
| *NIST* | 0.25(1) | 0.33(2) | 0.45(3) | 0.0024 |
| *29c3* | 0.63(2) | 0.19(2) | 0.23(3) | 0.0024 |
| **Zn K-edge** | | | | |
| | ***$ZnCO_3$*** | ***$ZnSO_4.7H_2O$*** | ***$Zn(NO_3)_2$*** | ***$R_{factor}$*** |

| | | | | |
|---|---|---|---|---|
| *NIST* | 0.20(1) | 0.39(3) | 0.48(3) | 0.0012 |
| *29c3* | 0.11(3) | 0.09(8) | 0.8(1) | 0.0092 |

For iron K-edge, we consider the same species reported in [7], as well as Fe(II)$SO_4$· $5H_2O$ and $Fe_2$(III)$(SO_4)_3$· $7H_2O$ without obtaining a satisfactory fit for NIST nor for 29c3 sample. This indicates the presence of a different Fe containing compound for which no reference spectra are available.

For the data at Ni K-edge the chemical species are compatible to what is reported in [8]. The presence of metallic nickel is not reported though. However, the shoulder at 8335 eV highlighted in the inset of Fig. 1b) can be reproduced only if the spectrum of metallic nickel is used in the linear combination fit. The fitting procedure indicates that the fraction of metallic Ni increases significantly from 25% to 63% from the NIST sample to the 29c3 sample.

For Zn K-edge we use the three reference spectra according to [9]. The quality of the linear combination fitting decreases from the NIST sample to the 29c3 indicating that the choice of the reference spectra is not ideal and probably some components are missing. However in both samples the local environment is compatible with a Zn coordination of 6 O/N atoms.

To conclude, we perform a preliminary XANES study at Fe, Ni and Zn K-edges of the NIST reference sample 1648 and of an airborne PM sample deposited on a cellulose filter to get information on the chemical species present. This is important for the assessment of the health effects of aerosols. In fact these depend not only on the size distribution, but also on the chemical composition of the particles. In addition, knowledge on the chemical speciation of metals in PM is able to shed some light on the specific sources (natural, anthropogenic) of the metals in different chemical forms.

**Acknowledgements**
B. Thiodjio Sendja acknowledges the Training and Research in Italian Laboratories (TRIL) of the International Center for Theoretical Physics (ICTP) (Trieste, Italy) for providing financial support and Elettra – Sincrotrone Trieste (Basovizza, Trieste, Italy) for training on XAFS experiments.

**References**

[1] Wiseman C L S and Zereini F 2009 *Science of the Total Environment* 407 2493-2500
[2] Di Cicco A, Aquilanti G, Minicucci M, Principi E, Novello N, Cognigni A and Olivi L 2009 *J Phys: Conf. Series* 190 012043
[3] Venturini E, Vassura I, Raffo S, Ferroni L, Bernardi L and Passarini F 2014 *Environmental Science and Pollution Research* 21 11634-11648
[4] Ravel B and Newville M 2005 *J. of Synchr. Rad.* 12 537–541
[5] Wilke M, Farges F, Petit P E and Brown G E 2001 *Am Min.* 86 714–730
[6] Westre T E, Kennepohl P, de Witt J, Hedman B, Hodgson K O and Solomon E I 1997 *J Am Chem Soc* 119 6297
[7] Bardelli F, Cattaruzza E, Gonnella F, Rampazzo G and Valotto G 2009 *J Phys: Conf. Series* 190 012192
[8] Tirez K, Silversmith G, Vincze L, Servaes K, Vanhoof C, Mertens M, Bleux N and Berghmans B 2011 *J Anal At Spectrom* 26 517-527
[9] Osán J, Meirer F, Groma V, Török S, Ingerle D, Streli C and Pepponi G 2010 *Spectrochimica Acta Part B* 65 1008-1013

# Chemical reactions of As complexation by glutathione: an XAFS study

**M W Franco[1], I F Vasconcelos[2], L V Modolo[3], F A R Barbosa[1]**
[1] Laboratory of Limnology, Ecotoxicology and Aquatic Ecology-LIMNEA, Universidade Federal de Minas Gerais, Belo Horizonte, MG, 31.270-901, Brazil. e-mail: maionewf@hotmail.com
[2]Department of Metallurgical and Materials Engineering, Universidade Federal do Ceará, Campus do Pici Bloco 714, Fortaleza, CE 60455-760, Brazil. e-mail: ifvasco@ufc.br
[3]Department of Botany, Universidade Federal de Minas Gerais. Belo Horizonte, MG, 31.270-901, Brazil

**Abstract.** In this study, the chemical reactions between As(III) and As(V) with glutathione, which is a target compound in As biochemistry due to its primordial role in As immobilization and intracellular reduction, in various molar ratios were investigated using As K-edge XAFS spectroscopy. Results showed a gradual substitution of As-O bonds in the coordination of aqueous As(III) and As(V) for three As-S bonds in the As+GSH complex. Moreover, the data showed reduction of As(V) to As(III) prior or concomitant to the As+GSH complex formation.

## 1. Introduction

Inorganic As oxyanions are known for their high toxicity and represent an environmental concern in many countries [1]. Chemical reactions between organic compounds and inorganic arsenic have been under investigation in an attempt to better understand the mechanisms by which arsenic may be intracellulary immobilized by living organisms or their environment. Once absorbed by living cells, As can be immobilized in peptides containing sulfhydryl groups or methylated to produce organic As compounds. The reduced tripeptide glutathione (γ-glutamylcysteinylglycine; GSH) plays a critical role in the cell homeostasis as the most abundant reducing agents in various biochemical pathways [2]. Complexation of As with thiol functional groups like those in GSH has been studied using various techniques [3-8] but, to our knowledge, a systematic XAFS study is still lacking.

X-ray absorption fine structure (XAFS) spectroscopy is element specific and allows to discriminate the oxidation state of a target atom in combination with its local structure (coordination). This study focused on the investigation of the chemical reaction between arsenite [As(III)] and arsenate [As(V)] and GSH using XAFS. Contributions to understand As behavior in covalent bond with sulfhydryl groups are also provided in this study.

## 2. Materials and Methods

As(III)+GSH and As(V)+GSH solutions (pH = 4.3) were prepared in 1:0.5, 1:1, 1:2, 1:4 (for As(III)) and 1:0.5, 1:2, 1:4, 1:6 (for As(V)) molar ratios, with GSH concentration of 0.16 M. Arsenic K-edge (11868 eV) XAFS data were collected at the National Synchrotron Light Laboratory (LNLS) in

Campinas, Brazil, in transmission mode using a double crystal monochromator equipped with Si (111) crystals and ionization chambers operating with air at atmospheric pressure. Energy calibration was monitored by acquiring simultaneous reference spectra from an Au foil at the Au L1-edge (11918 eV). Theoretical phase shift and scattering amplitude functions were calculated by FEFF 6.0 [9] and the data were analyzed as described elsewhere [10] using software (Athena and Artemis) included in the IFEFFIT computer package [11]. Theoretical scattering paths were calculated based on the following crystalline compounds: $As^{III}$-O path from $NaAsO_2$, $As^{III}$-S path from $As_2S_3$ and $As^V$-O from $As_2O_5$. The theoretical paths were tested by fitting data obtained from aqueous As(III) and As(V) and crystalline standards. A value of $0.87 \pm 0.05$ for $S_0^2$ was obtained from fits to crystalline $NaAsO_2$ and used in all fits. A k-range of 3.0 to 14.5 $Å^{-1}$ and a R-range of 1.0 to 3.0 Å were used for Fourier transform and fit the data, respectively. XANES (X-ray absorption near-edge spectroscopy) spectra for aqueous As(III) and As(V) and As(III)+GSH in the 1:4 ratio were used as basis functions for linear combination fits.

## 3. Results and discussion

The As(III) XANES spectra (Fig. 1A) show a shift in absorption edge as the GSH concentration increased. This is consistent with formation of As bonds with the less electronegative S replacing O in the aqueous complex. As(V) data (Fig. 1B) show the appearance of what looks like a 2nd edge positioned at As(III) energy values, indicating reduction of As(V) to As(III) prior or concomitant to As-S bond formation. The reactions saturate at 1:4 and 1:6 molar rations for As(III) and As(V), respectively, indicating that extra GSH is necessary to reduce As(V).

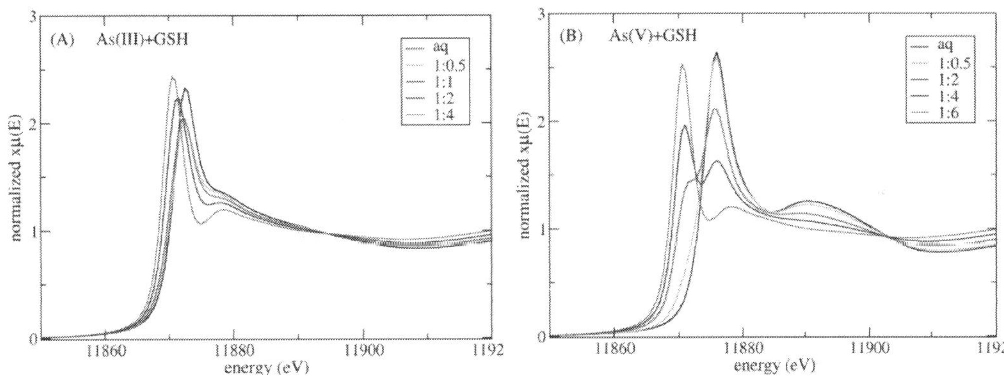


**Figure 1**. Normalized As K-edge XANES spectra for (A) As(III)+GSH and (B) As(V)+GSH at decreasing As:GSH molar ratios. Aqueous As(III) and As(V) are represented by the label aq.

Linear combination fits were applied to the XANES data (Fig. 2) to determine the chemical species present in the samples with intermediate As:GSH ratios (in which saturation was not reached.) The fits show that decreasing As(III):GSH ratio leads to an increase in the proportion of $As^{III}$-S bonds at the expense of $As^{III}$-O bonds. Fits for the As(V)+GSH system required the inclusion of three possible bonds $As^V$-O, $As^{III}$-O, and $As^{III}$-S. The $As^{III}$-S bond was observed in the 1:2 and 1:4 samples but not in the 1:0.5 sample. These samples show the presence of the three possible bonds in different proportions in a fashion consistent with the proposed mechanism of As(V) reduction to As(III) prior or concomitant to the formation of As-S bonds in the $As(GS)_3$ complex.

The changes in As coordination are confirmed by the magnitude of the Fourier transform of the EXAFS (extended X-ray absorption fine structure) oscillations (Fig. 3). The As-O signal gradually decreases accompanied by a gradual increase of the As-S signal (see arrows). The signal of the $As^{III}$-S bond is observed from 1:0.5 up to 1:4 for As(III) and from 1:2 up to 1:6 for As(V). The formation of $As^{III}$-S from $As^V$-O at 1:2 and above is consistent with the consumption of GSH in the As(V) to As(III) reduction process prior or concomitant to $As(GS)_3$ formation. Indeed, this reduction has previously been demonstrated by $^{1}H$ and $^{13}C$ NMR from studies with bench-prepared solutions [4]. Our findings are also consistent with a recent report [5] which showed that the reduction of organoarsenic(V) compounds in the presence of GSH is slower than the complexation. Thus, the reduction step is rate-limiting. Here, the As(V) reduction was complete at a As(V):GSH ratio of 1:6.

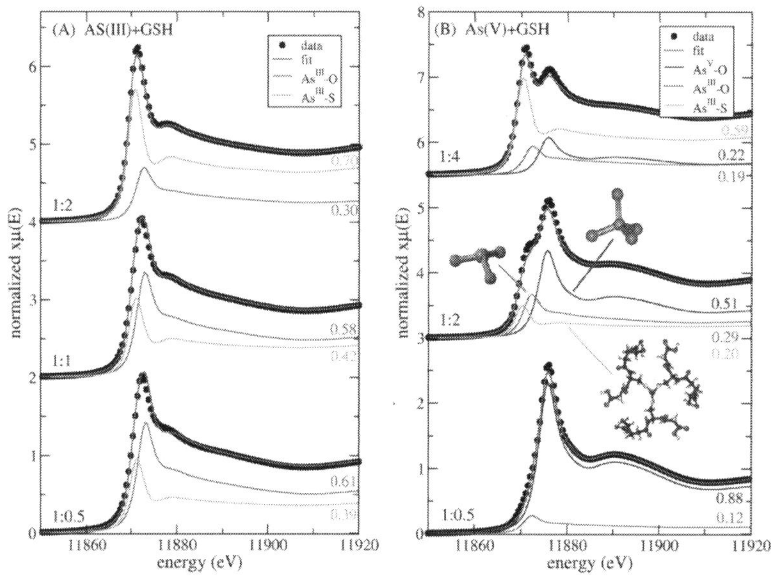


**Figure 2.** Linear combination fits of XANES spectra obtained for (A) As(III)+GSH and (B) As(V)+GSH at decreasing As:GSH molar ratios. Fractions of each As coordination are shown in matching colors with the legend.

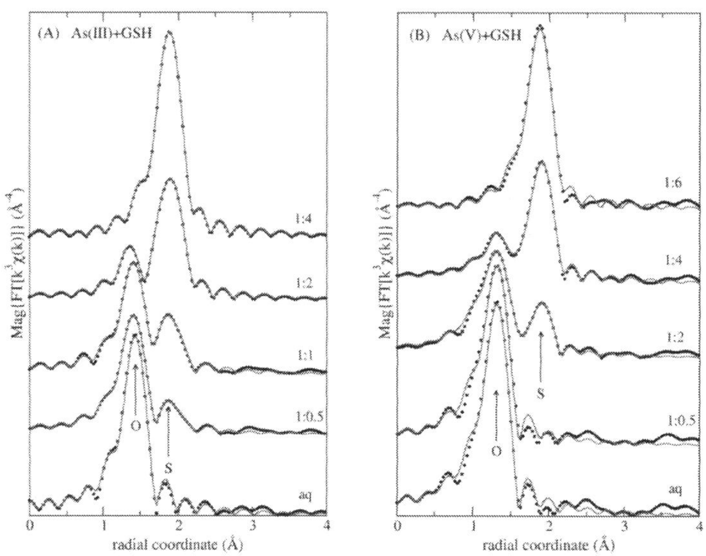


**Figure 3.** Magnitude of Fourier-transformed EXAFS data for (A) As(III)+GSH and (B) As(V)+GSH at decreasing As:GSH molar ratios.

The best-fit results to the EXAFS data show the gradual substitution of As-O bonds in the coordination of As(III) and As(V), formed respectively by approximately 3 to 4 oxygen atoms, towards the formation of the three As-S bonds in the GSH complex (Table 1). As-O bond lengths and $\sigma^2$ values for the As(V)+GSH samples confirm the presence of both $As^{III}$-O and $As^{V}$-O bonds in the intermediate reaction steps.

These fit results are consistent with previous studies on aqueous As(III) characterization [12] and $As(GS)_3$ complex formation [13]. It is important to note that EXAFS data provide an averaged picture of the local coordination of As and cannot rule out the possibility of oxygen and sulfur binding simultaneously to As ions. However, the formation of organometallic complex bearing oxygen and sulfur bound simultaneously to As in living organisms is still to be demonstrated [14,15].

**Table 1.** Best-fit results to EXAFS data ± uncertainty values. $N$ = coordination number; $R$ = bond length; $\sigma^2$ = Debye-Waller factor; relative misfit (R-factor) <= 0.7% (for all fits).

| Samples | As-O | | | As-S | | |
|---|---|---|---|---|---|---|
| | $N$ | $R$ (Å) | $\sigma^2$ ($10^{-3}$ Å$^2$) | $N$ | $R$ (Å) | $\sigma^2$ ($10^{-3}$ Å$^2$) |
| As(III) aqueous | 3.1±0.1 | 1.791±0.007 | 2.1±0.4 | - | - | - |
| As(III)+GSH 1:0.5 | 2.4±0.1 | 1.789±0.002 | 2.9±0.2 | 0.3±0.1 | 2.282±0.011 | 2.3±1.2 |
| As(III)+GSH 1:1 | 2.0±0.1 | 1.785±0.004 | 1.7±0.4 | 0.7±0.1 | 2.272±0.007 | 2.2±0.9 |
| As(III)+GSH 1:2 | 0.8±0.1 | 1.790±0.012 | 1.5±0.9 | 1.9±0.2 | 2.266±0.008 | 2.7±0.8 |
| As(III)+GSH 1:4 | - | - | - | 3.1±0.2 | 2.254±0.007 | 2.8±0.7 |
| As(V) aqueous | 4.2±0.3 | 1.691±0.006 | 3.4±0.7 | - | - | - |
| As(V)+GSH 1:0.5 | 3.9±0.3 | 1.695±0.005 | 3.6±0.8 | - | - | - |
| As(V)+GSH 1:2 | 3.3±0.2 | 1.713±0.006 | 6.0±0.8 | 1.2±0.2 | 2.244±0.009 | 4.4±1.3 |
| As(V)+GSH 1:4 | 1.4±0.2 | 1.728±0.010 | 4.9±1.5 | 2.1±0.2 | 2.258±0.005 | 3.1±0.6 |
| As(V)+GSH 1:6 | - | - | - | 3.2±0.2 | 2.254±0.006 | 3.1±0.7 |

## 4. Conclusion

Arsenic complexation by GSH was studied by XAFS spectroscopy. It has been shown that in the presence of GSH, As-O bonds from aqueous arsenate and arsenite are replaced by As-S bonds to form As+GSH complexes. If arsenic is initially in arsenate form (As(V)), then it is reduced into arsenite (As(III)) prior to the formation of the As+GSH complex. Information about the molecular mechanisms of As bonding can help clarify the complex biochemistry of this element providing means to develop efficient remediation processes and methods

**Acknowledgments:** Financial support was provided by the National Institute of Science and Technology on Mineral Resources, Water and Biodiversity (INCT-Acqua). The Synchrotron Light Laboratory (LNLS) in Campinas, Brazil, provided lodging and logistic support to the first two authors for the measurements.

[1] Smedley PL, Kinniburgh DG 2002 Appl. Geochem. 17 517
[2] Bennett MS, Guan Z, Laurberg M, Su XD 2001 P. Natl. Acad. Sci. USA 98 13577
[3] Scott N Hatlelid KM MacKenzie NE Carter DE 1993 Chem. Res. Toxicol. 6 102
[4] Delnomdedieu M, Basti MM, Otvos JD, Thomas DJ 1994 Chem. Biol. Interact. 90 139
[5] Kretzschmar J, Brendler E, Wagler J, Schmidt A 2014 J. Hazard. Mater. 280 734
[6] Gailer J, Lindner W 1998 J. Chromatogr. Biomed. Sci. Appl. 716 83
[7] Park S, Butcher DJ 2010 Microchem. J. 95 57
[8] Han MJ, Meng X, Lippincott L 2007 Toxicol. Let. 175 57
[9] Zabinsky SI, Rehr JJ, Ankudinov A, Albers RC, Eller MJ 1995 Phys. Rev. B. 52 2995
[10] Vasconcelos IF, Haack EA, Maurice PA, Bunker BA 2008 Chem. Geol. 249 237
[11] Ravel B, Newville M 2005 J. Synchrotron Radiat. 12 537
[12] Ramírez-Solís A, Mukopadhyay R, Rosen BP, Stemmler TL 2004 Inorg. Chem. 43 2954
[13] Smith PG, Koch I, Gordon R, Mandoli DF et al. 2005 Environ. Sci. Technol. 39 248
[14] Páez-Espino D, Tamames J, Lorenzo V, Canovas D 2009 Biometals 22 117
[15] Wang P, Sun G, Jia Y, Meharg AA, Zhu Y 2014 J. Environ. Sci. 26 371

Journal of Physics: Conference Series **712** (2016) 012089 doi:10.1088/1742-6596/712/1/012089

# The Ti environment in natural hibonite: XANES spectroscopy and computer modelling

**Antonina N Kravtsova**[1], **Alexander V Soldatov**[1], **Andrew M Walker**[2] **and Andrew J Berry**[3]

[1] International Research Center "Smart materials", Southern Federal University, 5 Sorge Street, Rostov-on-Don 344090, Russia

[2] School of Earth and Environment, University of Leeds, Leeds LS2 9JT, UK

[3] Research School of Earth Sciences, Australian National University, Canberra ACT 2601, Australia

E-mail: akravtsova@sfedu.ru

**Abstract.** The local atomic structure around Ti in Ti-bearing hibonite ($CaAl_{12}O_{19}$) was studied using X-ray absorption near-edge structure (XANES) spectroscopy and computer modelling. Structural models of the direct substitution of Al by $Ti^{3+}$, Al by $Ti^{4+}$ charge balanced by the coupled substitution of $Mg^{2+}$ for Al, and small Ti clusters were considered. The Ti *K*-XANES spectra of natural hibonite with different Ti concentration were recorded. Theoretical Ti *K*-XANES spectra for structural models of hibonite were calculated. It was shown that the theoretical Ti *K*-XANES spectra for a model with Ti at the five-coordinated M2 site are in agreement with the experimental XANES spectra of hibonite with low concentrations of Ti, while the theoretical spectra for a structural model of clustered Ti are in agreement with the experimental spectra of hibonite with higher Ti contents.

## 1. Introduction

Hibonite ($CaAl_{12}O_{19}$) is a mineral found in calcium-aluminum-rich inclusions (CAIs) of chondritic meteorites. CAIs are among the oldest materials in the Solar System [1], so, hibonite has the potential to record conditions that were present when the Solar nebula condensed to form the Solar System. In particular, hibonite has potential to be used to study the oxygen fugacity ($fO_2$) due to the incorporation of polyvalent elements such as Fe, V, Cr and Ti in its structure. The $Fe^{2+}/Fe^{3+}$ ratio can be used to determine the $fO_2$ of terrestrial materials, but it is not suitable for CAIs because the highly reduced conditions of formation result in the presence of metallic iron. In the case of CAIs, the $Ti^{3+}/Ti^{4+}$ ratio may be an alternative means of determining $fO_2$ (e.g. [2]).

The goal of the present study was to investigate the local atomic structure around Ti in Ti-bearing hibonite.

X-ray absorption near-edge structure (XANES) spectroscopy allows to study the electronic structure and 3D atomic geometry (bond lengths and bond angles) around an absorbing atom at a defect site (i.e. without long-range order) [3]. Recently, XANES spectroscopy was used successfully to investigate the local environment of Ti at defect sites in minerals [4-5]. But the method of extracting information from experimental XANES spectra is not direct and requires computer modelling.

Computer modelling of different Ti defects in Ti-bearing hibonite was recently undertaken using density functional theory (DFT) [6]. In this investigation we used Ti *K*-edge XANES spectroscopy to verify the structural models of Ti-bearing hibonite obtained on the basis of computer modelling. We present the experimental Ti *K*-XANES spectra of natural hibonite with different Ti concentration and compare these to theoretical spectra simulated for the first time.

## 2. Experiment and methods of calculation

Ti *K*-edge XANES spectra of natural and synthetic hibonite with different Ti concentrations were recorded at I18 of Diamond Light Source (United Kingdom). Natural hibonite samples with low and high Ti concentrations contain a mixture of $Ti^{3+}$ and $Ti^{4+}$. In this paper we present Ti *K*-edge XANES spectra of natural hibonite; the spectra of synthetic samples will be presented elsewhere.

Hibonite has a hexagonal structure (e.g. [7]) that is dominated by polyhedral layers perpendicular to the c-axis. Ca occupies a 12-coordinated polyhedron and Al occupies five M sites. The cation site multiplicity and their coordination numbers can be described as $^{[12]}Ca^{[6]}M1^{[5]}M2^{[4]}M3_2{}^{[6]}M4_2{}^{[6]}M5_6O_{19}$ [6]. Here the multiplicity of the site is subscripted and the coordination numbers are superscripted in brackets.

Modelling of Ti defects in hibonite was undertaken using DFT implemented in the SIESTA code, which uses a strictly localized atom-like basis set and pseudopotentials to achieve good computational performance for large systems. The structure of $CaAl_{12}O_{19}$ determined in [7] was used as the starting point for modelling the Ti defects. The structural models of direct substitution of different Al positions by $Ti^{3+}$, the substitution of Al sites by $Ti^{4+}$ coupled to a charge balancing substitution of Mg for Al, and small Ti clusters were considered [6].

Theoretical Ti *K*-XANES spectra for the structural models were calculated on the basis of two approaches.

First, self-consistent real-space full multiple scattering (FMS) theory, within the muffin-tin approximation for the potential shape, implemented into the electronic structure code FEFF9 [8-9] was used. XANES spectra were calculated for atomic clusters of 7.0 Å around absorbing Ti atoms. The influence of different types of exchange-correlation potentials on the Ti *K*-XANES was checked. For the calculation of spectra of hibonite with low Ti concentrations, the exchange-correlation potential of Dirac-Hara with Hedin-Lundqvist imaginary part was used. To calculate the XANES spectra of hibonite with high Ti concentrations, the Hedin-Lundqvist exchange-correlation potential was used.

Second, the full-potential finite difference method (FDM) to solve the Schrödinger equation with the Hedin–Lundqvist exchange-correlation potential realized in the FDMNES code [10] was applied. The main advantage of this method is the possibility of a totally free potential shape, thus avoiding the limitations of the muffin-tin approach. Because of the higher computational demands of FDM, the Ti *K*-edge XANES spectra were simulated for atomic clusters of 5.6 Å around the absorbing Ti atom.

All Ti *K*-XANES spectra of hibonite were calculated taking into account the core hole created by the electronic transition. Spectra for the structural models of small Ti clusters (a cluster of two $Ti^{3+}$ defects on adjacent face sharing six coordinate M4 sites, and a cluster of two $Ti^{4+}$ defects charge balanced by two $Mg^{2+}$ defects) were obtained as a sum of the spectra calculated for the two Ti positions.

## 3. Results and discussion

Experimental Ti *K*-XANES spectra of natural hibonite samples with low (0.1 Ti per formula unit, pfu) and high Ti (0.55 Ti pfu) concentration are shown in Figures 1 and 2, respectively. The main difference in the experimental spectra is the intensity of the pre-edge feature marked as *A*. The pre-edge feature is more intense in the spectrum of the sample with the lower concentration of Ti, which suggests a lower coordination number [11].

In the first instance, the modelling of Ti defects in hibonite was guided by the results of DFT. It was found that the model with Ti on the five-coordinated M2 site is the lowest energy for substitution as an isolated cation on a M site [6]. For higher Ti concentrations, the lowest energy model was a small Ti cluster (two Ti atoms on adjacent face sharing six-coordinate M4 sites) [6]. The cases of $Ti^{3+}$

and $Ti^{4+}$ defects (the latter charge balanced by $Mg^{2+}$ defects) were both considered. Upon optimization an increase in the Ti-O distances was observed.

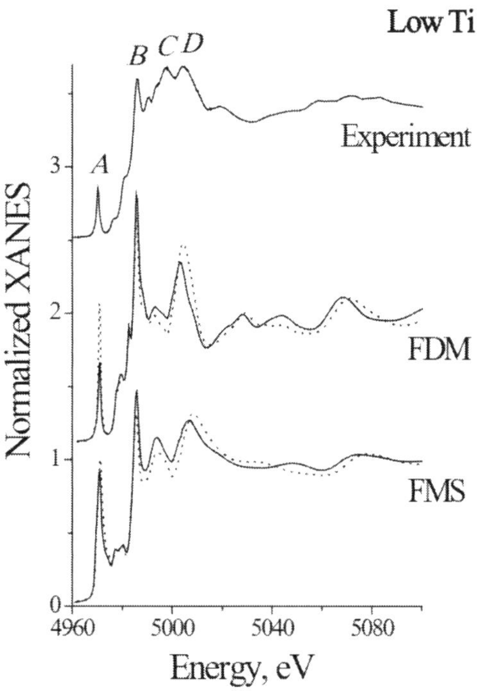


**Figure 1.** Comparison of the experimental Ti *K*-XANES spectrum of natural hibonite with low Ti concentration (0.1 Ti pfu) and theoretical spectra calculated for a structural model with Ti on the M2 site. Theoretical curves are the sum of spectra for a model of $Ti^{3+}$ on the M2 site (20 %) and a model of $Ti^{4+}$ on the M2 site charge balanced by $Mg^{2+}$ on the M3 site (80 %). Theoretical spectra are shown for the non-optimized structure (dotted lines) and after structure optimization (solid lines). Theoretical spectra were simulated using the FMS (FEFF9 code) and FDM (FDMNES code) methods.

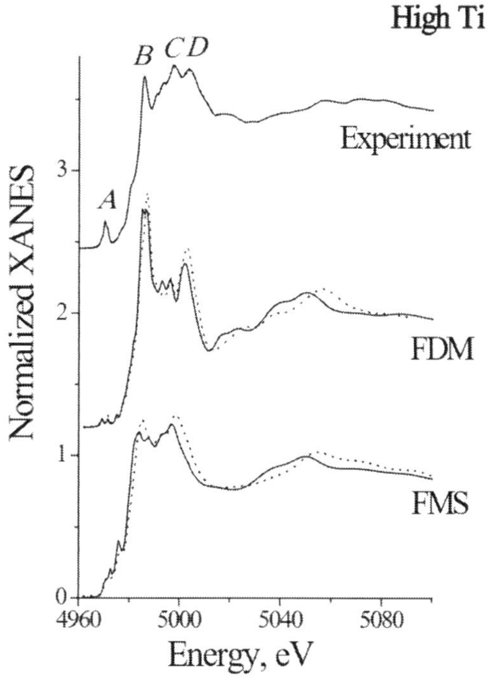


**Figure 2.** Comparison of the experimental Ti *K*-XANES spectrum of natural hibonite with high Ti concentration (0.55 Ti pfu) and theoretical spectra calculated for a structural model with 20% of clustered $Ti^{3+}$ on adjacent face sharing M4 sites and 80 % of a small cluster of two $Ti^{4+}$ defects charge balanced by two $Mg^{2+}$ defects. Theoretical spectra are shown for the non-optimized structure (dotted lines) and after structure optimization (solid lines). Theoretical spectra were simulated using the FMS (FEFF9 code) and FDM (FDMNES code) methods.

The theoretical Ti *K*-XANES spectra were calculated for these structural models using the FMS and FDM methods implemented in the FEFF9 and FDMNES codes, respectively.

As some natural (meteoritic) hibonite contains about 80% of Ti as $Ti^{4+}$ and 20% as $Ti^{3+}$ [12], the sum of the theoretical spectra for the models of a single $Ti^{3+}$ defect on the five-coordinated M2 site and a single $Ti^{4+}$ on the M2 site, charged balanced by $Mg^{2+}$ on the M3 site, with weights of 0.2 and 0.8, respectively, was used to compare to the experimental spectrum of natural hibonite with low Ti concentration (see Figure 1). The sum of the theoretical spectra for the models of small clusters of two $Ti^{3+}$ on adjacent face sharing six coordinate M4 sites and two $Ti^{4+}$ defects on M4 sites charge balanced by two $Mg^{2+}$ defects (with weights of 0.2 and 0.8) was used for comparison with the experimental spectrum of hibonite with higher Ti concentrations (see Figure 2).

Figures 1 and 2 show that the theoretical spectra are in good agreement with the experimental ones. Both FDM and FMS result in adequate predictions of the shape of the experimental spectra, but FDM gives higher intensity of the peak marked as *B* for both samples. Relaxation of the atomic geometries of the Ti defects results in a decrease in the energy difference between spectral maxima (e.g. between B and D) as well as a decrease in the intensity of the pre-edge feature for the sample with low Ti concentration. The decrease in the intensity of pre-edge feature, A, with increasing Ti concentration indicates a change in Ti coordination that is attributed to clustering [11].

## 4. Conclusions

Ti *K*-edge XANES spectroscopy and computer modelling were used to study the atomic structure around Ti in Ti-bearing hibonite. Structural models of direct substitution of Al by Ti ($Ti^{3+}$ and $Ti^{4+}$ charge balanced by substitution of $Mg^{2+}$ for Al) and small Ti clusters were considered. Modelling of Ti defects in hibonite was performed using density functional theory implemented in the SIESTA code. Experimental Ti *K*-XANES spectra of natural samples of hibonite were recorded. Theoretical analysis of the Ti *K*-XANES spectra was done using FMS and FDM theory, which both give results that are in good agreement with the experimental spectra. We conclude that Ti on the five-coordinated M2 position is the most probable for low concentrations of Ti in hibonite, while clustered Ti on the six-coordinated M4 site is most probable for higher Ti concentrations.

## Acknowledgments

A.N.K. is grateful for a grant from the Russian Foundation for Basic Research RFBR 14-05-00580 "Nanodiagnostics of microelements environment in geological materials: X-ray investigation and computer modeling". A.M.W. is supported by an Independent Research Fellowship from the Natural Environment Research Council (NE/K008803/1). A.J.B. thanks the Australian Research Council for the award of a Future Fellowship. T.R. Ireland provided the samples of natural hibonite and P.F. Schofield, P.M. Doyle and J.F.W. Mosselmans assisted with the collection of the experimental spectra.

## Rerferences

[1] Amelin Y, Krot A N, Ulyanov A A and Hutcheon I D 2002 *Science* **297** 1678-1683
[2] Anderson Jr A T, Crewe A V, Goldsmith J R, Moor P B, Newton J C, Olsen E J, Smith J V and Wyllie P J 1970 *Science* **167** 587-590
[3] Bunker G 2010 *Introduction to XAFS. A Practical Guide to X-ray Absorption Fine Structure Spectroscopy* (Cambridge University Press)
[4] Rodina I S, Kravtsova A N, Soldatov A V and Berry A 2011 *Optics and Spectroscopy* **111** 936-939
[5] Tailby N D, Walker A M, Berry A J Hermann J, Evans K A, Mavrogenes J A, O'Neill H St. C, Rodina I S, Soldatov A V, Rubatto D and Sutton S R 2011 *Geochimica et Cosmochimica Acta* **75** 905-921
[6] Doyle P M, Schofield P F, Berry A J, Walker A M and Knight K S 2014 *American Mineralogist* **99** 1369-1382
[7] Bermanec V V, Holtstam D, Sturman D, Criddle A J, Back M E and Scavnicar S 1996 *The Canadian Mineralogist* **34** 1287-1297
[8] Rehr J J, Kas J J, Vila F D, Prange M P and Jorissen K 2010 *Physical Chemistry Chemical Physics* **12** 5503-5513
[9] Rehr J J, Kas J J, Prange M P, Sorini A P, Takimoto Y, Vila F D 2009 *Comptes Rendus Physique* **10** (6) 548-559
[10] Joly Y 2001 *Physical Review B* **63** 125120
[11] Berry A J, Walker A M, Hermann J, O'Neill H St C, Foran G J and Gale J D 2007 *Chemical Geology* **242** 176-186
[12] Beckett J R, Live D, Tsay F-D, Grossman L. and Stolper E 1988 *Geochimica et Cosmochimica Acta* **52** 1479-1495

# Characterization of fossil remains using XRF, XPS and XAFS spectroscopies

**I M Zougrou[1], M Katsikini[1*], F Pinakidou[1], M Brzhezinskaya[2], L Papadopoulou[3], E Vlachos[3], E Tsoukala[3] and E C Paloura[1]**

[1] School of Physics, Aristotle University of Thessaloniki, 54124 Thessaloniki, Greece.
[2] Main Department Scientific-Technical Infrastructure II, Helmholtz Zentrum Berlin, 12489 Berlin, Germany.
[3] School of Geology, Aristotle University of Thessaloniki, 54124 Thessaloniki, Greece.

*E-mail: katsiki@auth.gr

**Abstract.** Synchrotron radiation micro-X-Ray Fluorescence (μ-XRF), X-ray photoelectron (XPS) and X-ray Absorption Fine Structure (XAFS) spectroscopies are applied for the study of paleontological findings. More specifically the costal plate of a gigantic terrestrial turtle *Titanochelon bacharidisi* and a fossilized coprolite of the cave spotted hyena *Crocuta crocuta spelaea* are studied. Ca $L_{2,3}$-edge NEXAFS and Ca 2p XPS are applied for the identification and quantification of apatite and Ca containing minerals. XRF mapping and XAFS are employed for the study of the spatial distribution and speciation of the minerals related to the deposition environment.

## 1. Introduction

The investigation of fossil remains provides valuable information on the evolution of living species, the fossilization conditions and the palaeoenvironment. The aim of this work is to demonstrate the potential of X-ray spectroscopies for the investigation of specimens of paleontological interest. More specifically, Ca $L_{2,3}$-edge near edge X-ray absorption fine structure (NEXAFS) and X-ray photoelectron (XPS) spectroscopies are applied in combination with extended X-ray absorption fine structure (EXAFS) spectroscopy, X-ray fluorescence (XRF) mapping and more commonly used techniques such as optical microscopy (OM) and scanning electron microscopy (SEM) for the assessment of diagenesis of two different types of paleontological remains: a hyena coprolite and a giant tortoise costal plate. The Ca $L_{2,3}$-edge NEXAFS and Ca 2p photoelectron spectra provide quantitative information concerning the bone apatite and Ca-rich secondary phases [1-2]. OM and SEM are used to assess the morphology and chemical composition of the coprolite matrix and inclusions, whereas XRF maps were recorded to acquire complementary information on the distribution of P, Ca, Mn and Fe in the specimens [3-4]. Finally, Fe *K*-edge EXAFS was used to investigate metal-rich inclusions in the coprolite matrix.

## 2. Materials and methods

The studied fossils are a costal plate of the gigantic Pliocene tortoise *Titanochelon bacharidisi* from the New Michaniona, Thessaloniki, Greece, [5] excavation site found in open air as a result of sea erosion and a fossilized coprolite (SKG914) of the Late Pleistocene cave spotted hyena *Crocuta*

*crocuta spelaea* from the Agios Georgios cave in Kilkis, Greece [6]. The XRF, XPS and XAFS measurements were conducted at the BESSY-II Synchrotron Radiation Storage Ring of the Helmholtz Zentrum Berlin. The XRF maps and the Fe *K*-edge XAFS spectra were recorded on the samples' cross sections at the KMC-II beamline using an energy dispersive silicon detector positioned on the horizontal plane at right angle to the incident beam that was impinging at 45° to the sample surface. For the acquisition of the XRF maps, the beam diameter was reduced to 4.8 μm using an elliptical capillary and the energy of the excitation beam was set equal to 7.2 keV. The maps were corrected for the escape peak contributions. The X-ray photoelectron and Ca $L_{2,3}$-edge NEXAFS spectra were recorded at the Optics beamline (PM-4) using the SURICAT end-station and at the Russian German beamline (RGBL) under ultrahigh vacuum conditions. The fossil remains were powdered and then rubbed onto a scratched pure In foil to minimize charging effects. The beam with spot size of 0.1×0.5 mm was impinging at 45° to the sample. The X-ray photoelectron spectra were recorded with excitation photon energy of 1030 eV using a hemispherical capacitor electron energy analyzer (Scienta SES 100) with energy resolution of 160 meV. Energy shifts due to charging effects were corrected by setting the C 1s peak at 284.6 eV. The Ca $L_{2,3}$-edge NEXAFS spectra were recorded in the total electron yield mode measuring the drain current from the sample.

## 3. Results and discussion

Representative OM, SEM-back-scattering electron (BSE) images, and XRF maps of the coprolite specimen are shown in figure 1. SEM- energy dispersive X-ray spectra (EDS) were used for the determination of the chemical composition. The dominant phase was identified as a non-apatitic calcium phosphate matrix with average Ca/P atomic ratio of 1.55±0.11. Furthermore, bone inclusions with well-preserved Haversian bone histology with visible osteocyte lacunae as well as grains of quartz, green hornblende $[(K,Na)_{0-1}Ca_2(Mg,Fe,Al)_5(Si,Al)_8O_{22}(OH)_2]$, biotite $[K,(Mg,Fe)_3AlSi_3O_{10}(OH)_2]$, almadine/glossular garnet $[(Ca,Fe)_3Al_2Si_3O_{12}]$ and microcline ($KAlSi_3O_8$) were also identified throughout the specimen. The latter inclusions are related to minerals introduced from the deposition environment. The primary carbonated hydroxylapatite (HA) of the bone is found enriched in K, Na, Rb, Sr and partially transformed to chlorapatite. The average Ca/P atomic ratio of bone inclusions is similar to HA and equal to 1.65±0.11. In the Ca and P XRF maps, local maxima related to the bone inclusions are observed. However, in areas of low Ca and P intensity, metal-rich (Fe, Mn) inclusions originating from the deposition environment are also present.

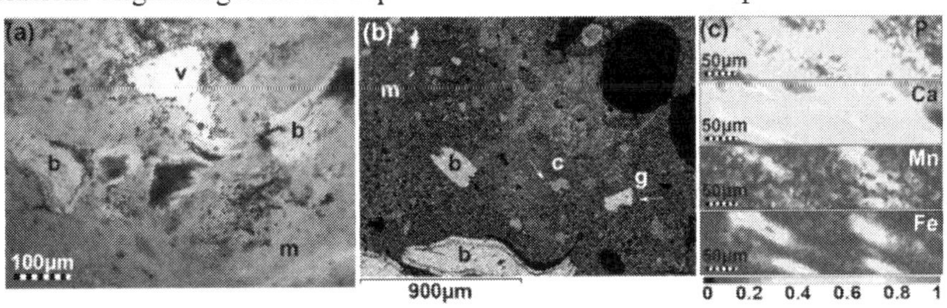


**Figure 1.** Coprolite specimen: a) OM image recorded using plane polarized light that shows digested bone with visible osteocyte lacunae, b) SEM-BSE image revealing grain and bone inclusions in the spotted cave hyena coprolite matrix (SGK 914), c) P, Ca, Mn and Fe XRF maps identify elongated metal-rich inclusions in areas of low Ca and P concentration. *Notation*: b: bone, m: matrix, v: vesicle, c: microcline, g: glossular garnet.

The Fe *K*-edge XANES spectrum recorded in the cross section of the coprolite, reveals that Fe is in the trivalent state, as it is deduced from the position of the absorption edge. The amplitude of the Fourier transform of the $k^3$-weighted $\chi(k)$ Fe *K*-edge EXAFS spectrum is shown in figure 2 along with the $\chi(k)$ spectrum shown in the inset. Fitting of the spectrum using the FEFF8 package [7] reveals that

Fe participates in the formation of the akaganeite mineral. More specifically, 6 O atoms comprise the first nearest neighbouring shell at 1.99 Å, 4 and 4 Fe atoms, at distances 3.09 (edge-sharing) and 3.42 (corner-sharing) Å, respectively, constitute the second shell while the 4[th] nearest neighbouring shell consists of 6 O atoms located at 3.70 Å.

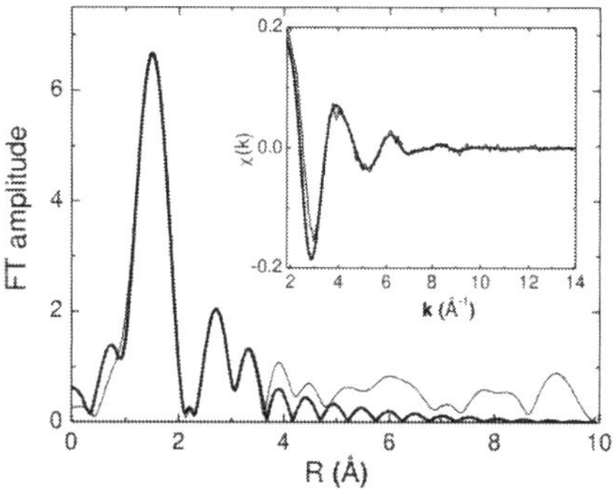


**Figure 2.** Fe *K*-edge $\chi(k)$ (inset) and corresponding amplitude of the Fourier Transform of the $k^3$-weighted $\chi(k)$ spectra of the hyena coprolite. The experimental spectra and the fitting curves are shown in thin and thick lines, respectively.

Ca $L_{2,3}$-edge NEXAFS and Ca 2p X-ray photoelectron spectra were used to investigate the chemical composition of the gigantic tortoise external cortex and cancellous bone as well as the hyena coprolite. To obtain quantitative information, the Ca $L_{2,3}$-edge NEXAFS spectra were fitted using Sigmoidal and Voigt functions as shown in figure 3. A photograph of the gigantic tortoise sample cross section is shown in the inset. The corresponding spectra of reference HA, chlorapatite (ChAp) and calcite samples are also included in figure 3. The fitting results are listed in table 1 along with HA and calcite results from Benzerara *et al.* (2004) [8]. The energy splitting of the ($a_2$-$a_1$) and ($b_2$-$b_1$) peaks is used for the identification of the Ca-containing minerals. The giant tortoise external cortex consists of HA whereas the cancellous bone is comprised of 65% HA and 35% calcite. The coprolite spectrum shows the presence of chlorapatite.

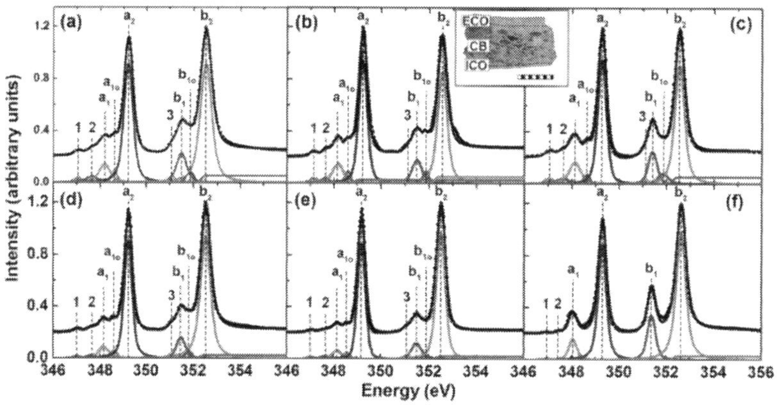


**Figure 3.** Ca $L_{2,3}$-edge NEXAFS spectra (dotted lines) and fitting curves (solid lines) using sigmoidal steps and Voigt peaks of the: (a) hyena coprolite, (b) giant tortoise external cortex, (c) giant tortoise cancellous bone and reference (d) hydroxylapatite, (e) Norway chlorapatite and (f) calcite samples. The experimental and fitting lines are shifted along the y-axis for clarity. A photograph of the giant tortoise costal plate cross section is shown in the inset. The external cortex (ECO), cancellous bone (CB) and internal cortex (ICO) are indicated. *Scale bar:* 1 cm.

Similar to the case of the Ca $L_{2,3}$-edge NEXAFS spectra, Ca 2p XPS spectra are also sensitive to the chemical state of Ca. The spectrum of the giant tortoise cancellous bone is shown in figure 4. It is fitted using two doublets, each one for HA and calcite contributions, with full width at half maximum of 2.20 and 3.27 eV, respectively. The relative contributions of the HA and calcite phases is found equal to 64% and 36%, respectively, in accordance with the NEXAFS results.

**Table 1.** Fitting results of the Ca $L_{2,3}$-edge NEXAFS spectra of the giant tortoise compact and cancellous bone and the hyena coprolite. Fitting results of the reference spectra of HA, geologic Norway chlorapatite (ChAp) and calcite as well as the peak positions of a mixed 65% HA and 35% calcite model are also included. The error in the determination of the peak positions is ±0.03 eV.

| Peak | Tortoise compact | Tortoise cancellous | Hyena Coprolite | HA | HA[8] | Norway ChAp | Calcite | Calcite[8] | mixed model |
|---|---|---|---|---|---|---|---|---|---|
| **1** | 347.13 | 347.03 | 347.05 | 347.02 | 347.1 | 347.02 | 346.92 | 346.9 | 346.99 |
| **2** | 347.63 | 347.60 | 347.64 | 347.62 | 347.7 | 347.63 | 347.45 | 347.4 | 347.56 |
| $\mathbf{a_1}$ | 348.16 | 348.11 | 348.18 | 348.15 | 348.2 | 348.14 | 348.05 | 348.0 | 348.12 |
| $\mathbf{a_{1o}}$ | 348.61 | 348.63 | 348.59 | 348.58 | 348.6 | 348.57 | | | 348.58 |
| $\mathbf{a_2}$ | 349.23 | 349.25 | 349.21 | 349.21 | 349.3 | 349.19 | 349.29 | 349.3 | 349.24 |
| **3** | 351.10 | 351.08 | 351.08 | 351.07 | | 351.07 | | | 351.07 |
| $\mathbf{b_1}$ | 351.48 | 351.39 | 351.50 | 351.48 | 351.6 | 351.48 | 351.37 | 351.4 | 351.44 |
| $\mathbf{b_{1o}}$ | 351.85 | 351.84 | 351.85 | 351.81 | 351.8 | 351.86 | | | 351.81 |
| $\mathbf{b_2}$ | 352.53 | 352.54 | 352.51 | 352.50 | 352.5 | 352.49 | 352.63 | 352.6 | 352.55 |
| $\mathbf{a_2-a_1}$ | 1.07 | 1.13 | 1.03 | 1.06 | 1.1 | 1.05 | 1.24 | 1.3 | 1.12 |
| $\mathbf{b_2-b_1}$ | 1.05 | 1.15 | 1.01 | 1.02 | 0.9 | 1.01 | 1.26 | 1.2 | 1.10 |
| $\mathbf{b_2-a_2}$ | 3.30 | 3.29 | 3.30 | 3.29 | 3.2 | 3.30 | 3.34 | 3.3 | 3.31 |

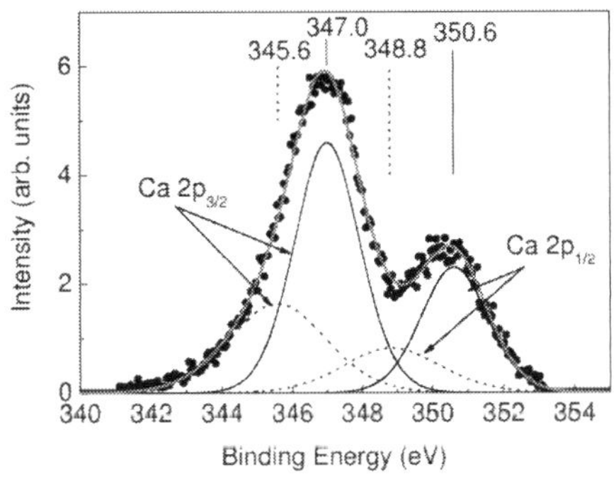


**Figure 4.** High resolution Ca 2p X-ray photoelectron spectrum of cancellous bone from the gigantic tortoise sample. Fitting using peaks that correspond to HA and calcite contributions are plotted in solid and dashed lines, respectively. The experimental and fitting curves are plotted with solid circles and thick solid (red) line, respectively.

## 4. Conclusions

Ca $L_{2,3}$-edge NEXAFS and XPS spectroscopies can provide quantitative information on the Ca chemical state and thus are useful tools for addressing questions of taphonomic interest on bone alterations. OM, SEM, XRF and EXAFS can provide complementary information on the morphology and chemical composition of fossil remains that is strongly affected by the deposition environment.

**Acknowledgements**

The measurements at the electron storage ring BESSY II (HZB) were financially supported from the European Community's Seventh Framework Program (FP7/2007-2013) under grant agreement n.°312284. Dr. R. Ovsyannikov is also greatly acknowledged for his support at the Optics beamline.

**References**

[1] Cosmidis *et al.* 2015 *Acta Biomater.* **12** 260-269.
[2] Nakada *et al.* 2012 *International Journal of Biomaterials* **2012** ID 615018
[3] Zougrou I M *et al.* 2014 *J. Synchr. Rad.* **21** 149–160.
[4] Zougrou I M *et al.* 2014 *J. Phys. Conf. Ser.* **499** 012015.
[5] Vlachos E *et al.* 2014 *J. Vertebrate Paleontology* **34** 560.
[6] Tsoukala E S 1992 *Geobios*, **25** 415.
[7] Ankudinov A L *et al.*1998 *Phys. Rev.* B **58** 7565–7576
[8] Benzerara K *et al.*, 2004, *Geobiology* **2** 249-259.

# Interpretation of the U $L_3$-edge EXAFS in uranium dioxide using molecular dynamics and density functional theory simulations

**Dmitry Bocharov[1,2], Melanie Chollet[1], Matthias Krack[1], Johannes Bertsch[1], Daniel Grolimund[1], Matthias Martin[1], Alexei Kuzmin[2], Juris Purans[2], and Eugene Kotomin[2]**

[1]Paul Scherrer Institute, CH-5232 Villigen PSI, Switzerland
[2]Institute of Solid State Physics, University of Latvia, Kengaraga Str. 8, LV-1063, Riga, Latvia

E-mail: bocharov@latnet.lv, matthias.krack@psi.ch, a.kuzmin@cfi.lu.lv

**Abstract.** X-ray absorption spectroscopy is employed to study the local structure of pure and Cr-doped $UO_2$ at 300 K. The U $L_3$-edge EXAFS spectrum is interpreted within the multiple-scattering (MS) theory using the results of the classical and ab initio molecular dynamics simulations, allowing us to validate the accuracy of theoretical models. The Cr K-edge XANES is simulated within the full-multiple-scattering formalism considering a substitutional model (Cr at U site). It is shown that both unrelaxed and relaxed structures, produced by ab initio density functional theory (DFT) calculations, fail to describe the experiment.

## 1. Introduction

Uranium dioxide ($UO_2$) is employed as a fuel material in most nuclear reactors world-wide. Its doping with small amounts of chromium sesquioxide ($Cr_2O_3$) is technically applied to obtain a larger average grain size after the fuel sintering process [1]. The change in grain structure results in a higher plasticity of the fuel material. The retention of fission gases inside the fuel matrix is also improved due to the larger grain size.

However, the accurate mechanism of Cr cation incorporation into the $UO_2$ matrix is still not fully identified. Indeed, $Cr^{3+}$ ions could replace $U^{4+}$ atom, with a charge compensation by oxygen vacancies. It could be feasible through creation of O ion vacancies. A Coulomb attraction between O vacancies and dopant cations could result in formation of defect complexes [2]. On the other side, the cubic face centered structure of $UO_2$ provides enough space to accommodate interstitial $Cr^{3+}$ ions. In this study, we address this problem using EXAFS spectroscopy, which is well suited for the local structure investigations (e.g. first neighbour bond distances and coordination numbers) of $UO_2$ fuel matrix and Cr dopant.

The interpretation of the X-ray absorption spectra is a non-trivial task, especially for such complicated systems like doped fuel. Here we will address this challenging problem using two approaches based on molecular dynamics (MD) [3] and density functional theory (DFT) simulations [4]. The first approach [3], which combines both classical and ab initio MD using the CP2K code [5, 6] with ab initio MS EXAFS theory [7], is used to interpret the behaviour of pure $UO_2$. Alternatively, DFT simulations are performed to determine the influence of the chromium doping on the fuel matrix. The atomic structure around the chromium impurity is relaxed, and the resulting structures are used to calculate the Cr K-edge X-ray absorption near edge (XANES) spectra within the ab initio full-multiple-scattering (FMS) formalism [7]. The comparison of the simulated XANES spectra derived from DFT results with the experimental one allows the identification of valid atomic configurations.

## 2. Experimental

In this study two X-ray absorption spectroscopy experiments were performed on Cr-doped and undoped $UO_2$ samples at 300 K using the radioactive microprobe end-station at the microXAS beamline (undulator beam line X05LA [8] at the Swiss Light Source facility, Paul Scherrer Institute, Switzerland). The Cr K-edge [9] and U $L_3$-edge [10] X-ray absorption spectra were measured for an irradiated Cr-doped and undoped $UO_2$ samples to retrieve information on the local chemical environment of chromium atoms within the $UO_2$ matrix. Energy is tuned with a Si(111) double crystal monochromator. In case of the irradiated, active materials, spectra are collected in the fluorescence mode using an energy dispersive solid state silicon drift detector. Additionally, pure non-irradiated $UO_2$ was studied as a reference. The low-activity of the non-irradiated $UO_2$ allows us to use an ionization chamber as detector in the transmission mode.

Finally, we also used the U $L_3$-edge X-ray absorption spectrum of $UO_2$, measured at 300 K in transmission mode at the ESRF ROBL BM-20 beamline in [11]. Three scans were collected from each sample then averaged.

## 3. Computational details

Classical [12, 13] and ab initio molecular dynamics (CMD and AIMD) simulations as well as static DFT calculations were performed using the CP2K code [5].

A set of atomic configurations was generated by MD runs within the canonical ensemble (NVT) with periodic boundary conditions at $T$=300 K using supercells of $6a_0 \times 6a_0 \times 6a_0$ (2592 atoms) and $3a_0 \times 3a_0 \times 3a_0$ (324 atoms) for the CMD and AIMD runs, respectively ($a_0$ is the $UO_2$ lattice parameter). A set of 4000 atomic configurations was collected during MD runs of 20 ps and 15 ps in classical and AIMD simulations, respectively. Two force-field models, the core shell potential of Meis et al. [14] ($a_0$=5.4682 Å) and the rigid-ion potential of Karakasidis et al. [15] ($a_0$=5.4657 Å), were used in the CMD runs. The exchange-correlation functional PBE [16] and a fixed lattice constant of $a_0$=5.4682 Å were employed in the AIMD runs.

The configuration-averaged U $L_3$-edge EXAFS spectra were calculated using the scheme described in [3] by averaging spectra calculated independently for each atomic configuration using the ab initio real-space MS code FEFF9.64 [7, 17]. The amplitude of all theoretical EXAFS spectra was multiplied by the scale factor $S_0^2$=0.83, estimated from the prior analysis of the first shell U–O EXAFS signal. The obtained configuration-averaged EXAFS spectra and their Fourier transforms (FTs), calculated employing a 10%-Gaussian window function in the $k$-space range 0.5–18.0 Å$^{-1}$, are reported in Fig. 1. Substitution of uranium with chromium in the $UO_2$ host lattice was studied using the DFT module QUICKSTEP of the CP2K code [6]. Geometry optimised atomic configurations for Cr at a regular U site without and with structure relaxation were used to perform static FMS Cr K-edge XANES calculations by the FEFF9.64 code [7, 17]. The cluster radius around Cr was equal to 8 Å to guarantee the convergence of the FMS series. Complex Hedin-Lundqvist exchange-correlation potential was used to describe inelastic effects in both MS and FMS calculations [7]. The obtained results are compared with the experimental Cr K-edge XANES spectrum of Cr-doped $UO_2$ [10] in Fig. 2.

## 4. Results and discussions

A comparison of the experimental and calculated U $L_3$-edge EXAFS spectra (Fig. 1) indicates rather good overall agreement between them. All three MD simulations describe well the

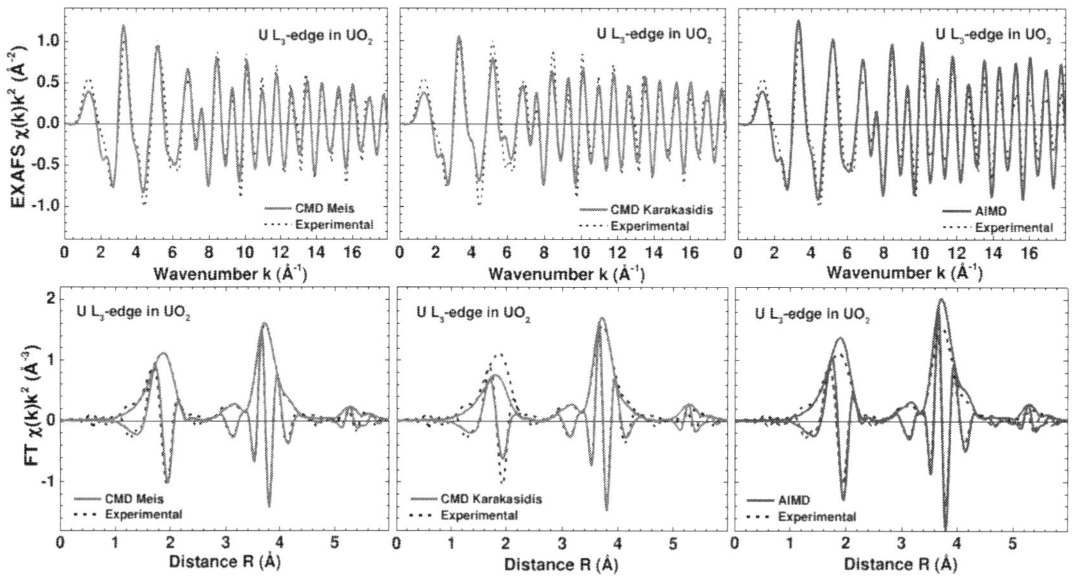

**Figure 1.** Comparison of the experimental (dotted lines) and calculated (solid lines) U $L_3$-edge EXAFS spectra and their Fourier transforms (FTs) (both modulus and imaginary parts are shown) for $UO_2$ at $T$=300 K. The experimental EXAFS spectrum was taken from [11].

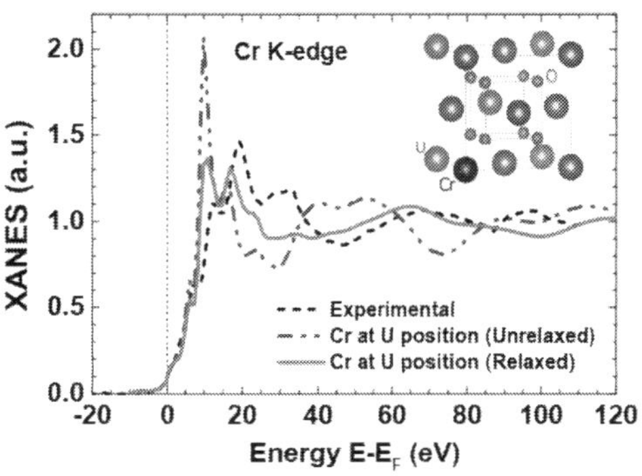

**Figure 2.** Comparison between the experimental [10] and FMS Cr K-edge XANES spectra of Cr-doped $UO_2$. The calculations correspond to chromium atom substituting uranium atom in $UO_2$ matrix without and with structure relaxation. The substitutional model (Cr at U site) is also shown as inset. The energy scale is relative to the Fermi level $E_F$.

structure of $UO_2$ at 300 K, thus reproducing well the phase of the EXAFS spectrum. However, some differences are clearly observed in the EXAFS amplitude due to a simplicity of interaction models employed in the MD simulations that resulted in an inaccurate description of thermal disorder. The best result was obtained for the core-shell potential of Meis et al. [14], which gives good agreement with the experimental data in both $k$ and $R$ spaces. The rigid-ion potential of Karakasidis et al. [15] underestimates the strength of the U–O interactions, resulting in a significantly reduced amplitude of the first coordination shell peak at $\sim$1.9 Å. On the contrary, the strength of the nearest interatomic interactions is slightly overestimated in the AIMD simulation, leading to larger amplitude of the first peak in FT. The values of the mean-squared relative displacement (MSRD) for the U–O atomic pair in the first shell are 0.0063 Å$^2$ for the Meis potential, 0.0120 Å$^2$ for the Karakasidis potential and 0.0050 Å$^2$ for the AIMD compared to the MSRD value 0.0060±0.0005 Å$^2$ determined from the best-fit analysis of the experimental EXAFS spectrum.

The simplest model for Cr-doped $UO_2$ is the substitutional one, when Cr atoms are located at U sites. Our FMS Cr K-edge XANES calculations (Fig. 2) show that such model agrees with the experiment only within the first 6-8 eV above the Fermi level and fails to describe the fine structure at higher energies. For unrelaxed structure, the difference is dramatic, since the theoretical and experimental XANES are nearly out of phase due to overestimated Cr–O bond lengths. Upon structure relaxation, the difference decreases mainly due to a shortening of the nearest Cr–O bonds. However, also in this case, the local environment around Cr atoms is still far from the reality, that is observed as the absence of peaks at ~30 eV and ~100 eV. Future work will have to address alternative more complicated environments for the embedding of Cr atoms into the $UO_2$ matrix, e.g as proposed in the work by Desgranges et al. [18].

## 5. Conclusions

The experimental U $L_3$-edge EXAFS spectra of $UO_2$ were interpreted up to 6 Å in $R$-space using the results of both CMD and AIMD simulations taking into account the MS effects. The accuracy of the CMD method relies on the used force-field model, therefore EXAFS spectra can be employed for its validation. For $UO_2$ a very good agreement between theoretical and experimental EXAFS spectra was found for the Meis force-field [14], whereas the Karakasidis force-field [15] fails to describe correctly mainly U–O interactions. The advantage of the AIMD method is the absence of the adjustable parameters, except for the choice of the functional model. The exchange-correlation functional PBE [16] employed in this study predicts slightly too rigid interactions.

The analysis of the Cr K-edge XANES spectrum of Cr-doped $UO_2$ suggests that the substitutional model (Cr at U site) differs from the experimental data, even when structure relaxation is taken into account. Therefore, further investigations are required to shed light on the accommodation of Cr atoms and precipitates within the $UO_2$ host lattice.

## Acknowledgments

The authors sincerely thank D. Gryaznov, G. Kuri and J. Timoshenko for valuable suggestions and many stimulating discussions. This work was supported by a grant from the Swiss National Supercomputing Centre (CSCS) under the project ID s596. The resource allocation within the PSI share at CSCS and on the PSI compute cluster Merlin4 is also acknowledged. D. B. is grateful for a fellowship within the Sciex-NMS programme. A. K. was supported by Latvian Science Council grant no. 187/2012.

## References

[1] Leenaers A, De Tollenaere L, Delafoy C and Van den Berghe S 2003 *J. Nucl. Mater.* **317** 62–68
[2] Mieszczynski C, Kuri G, Degueldre C, Martin M, Bertsch J, Borca C, Grolimund D, Delafoy C and Simoni E 2014 *J. Nucl. Mater.* **444** 274–282
[3] Kuzmin A and Evarestov R A 2009 *J. Phys. Condens. Matter* **21** 055401
[4] Blokhin E, Kotomin E, Kuzmin A, Purans J, Evarestov R and Maier J 2013 *Appl. Phys. Lett.* **102** 112913
[5] CP2K developers group 2000–2015 CP2K code `http://www.cp2k.org`
[6] VandeVondele J, Krack M, Mohamed F, Parrinello M, Chassaing T and Hutter J 2005 *Comput. Phys. Commun.* **167** 103–128
[7] Rehr J J and Albers R C 2000 *Rev. Mod. Phys.* **72** 621–654
[8] Borca C N, Grolimund D, Willimann M, Meyer B, Jefimovs K, Vila-Comamala J and David C 2009 *J. Phys. Conf. Ser.* **186** 012003
[9] Kuri G, Mieszczynski C, Martin M, Bertsch J, Borca C and Delafoy C 2014 *J. Nucl. Mater.* **449** 158–167
[10] Mieszczynski C, Kuri G, Bertsch J, Martin M, Borca C N, Delafoy C and Simoni E 2014 *J. Phys.: Condens. Matter* **26** 355009
[11] Heisbourg G, Hubert S, Dacheux N and Purans J 2004 *J. Nucl. Mater.* **335** 5–13
[12] Devynck F, Iannuzzi M and Krack M 2012 *Phys. Rev. B* **85** 184103
[13] Krack M 2012 *Mater. Res. Soc. Symp. Proc.* **1383** mrsf11-1383-a03-11
[14] Meis C and Chartier A 2005 *J. Nucl. Mater.* **341** 25–30
[15] Karakasidis T and Lindan P 1994 *J. Phys. Condens. Matter* **6** 2965
[16] Perdew J P, Burke K and Ernzerhof M 1996 *Phys. Rev. Lett.* **77** 3865–3868
[17] Rehr J J, Kas J J, Vila F D, Prange M P and Jorissen K 2010 *Phys. Chem. Chem. Phys.* **12** 5503–5513
[18] Desgranges L, Simon P, Martin P, Guimbretiere G and Baldinozzi G 2014 *JOM* **66** 2546–2552

Journal of Physics: Conference Series **712** (2016) 012092 doi:10.1088/1742-6596/712/1/012092

# Probing Covalency in the $UO_3$ Polymorphs by U $M_4$ edge HR-XANES

**Y Podkovyrina[1], I Pidchenko[2], T Prüβmann[2], S Bahl [2], J. Göttlicher[3], A Soldatov[1], T Vitova[2]**

[1]Internatonal Research Center "Smart Materials", Southern Federal University, ul. Zorge 5, 344090 Rostov-on-Don, Russia
[2]Karlsruhe Institute of Technology, Insitute for Nuclear Waste Disposal, P.O> Box 3640, D-76021 Karlsruhe, Germany
[3]Karlsruhe Institute of Technology, ANKA Synchrotron Radiation Facility, P.O> Box 3640, D-76021 Karlsruhe, Germany

E-mail: podkovyrina@sfedu.ru

**Abstract**. Local atomic and electronic structure investigations of uranium trioxide ($UO_3$) crystalline phases performed by the U $M_4$ edge HR-XANES technique is presented. The experimental U $M_4$ edge HR-XANES spectra of $\alpha$-$UO_3$, $\beta$-$UO_3$ and $\gamma$-$UO_3$ polymorphic phases are compared with spectra of uranate ($CaU_2O_7$) and uranyl ($UO_3$•1-2($H_2O$)) compounds. We describe a finger print approach valuable for characterization of variations of U-O axial bond lengths. Theoretical calculations of spectra using full-multiple-scattering theory (FEFF9.6 code) are performed. We have tested and selected input parameters, which provide best agreement between experimental and calculated spectra.

## 1. Introduction

Both structural and electronic properties of the uranium oxides are of fundamental and practical interest primarily due to the role of $UO_2$ in the nuclear fuel cycle [1, 2]. $UO_2$ is commonly exposed to oxidizing conditions and other binary oxides like for example $UO_3$ and $U_3O_8$ form at different stages of the nuclear fuel cycle [3]. $UO_3$ has the highest oxygen content among the uranium oxides. The chemical and physical properties of its polymorphs are also important to mining, milling, refinement and conversion processes that precede isotope enrichment within a nuclear fuel cycle [4]. X-ray absorption fine structure (XAFS) spectroscopy is successfully used for electronic and geometric structural investigations of uranium bearing materials [5]. The high energy resolution XAFS technique has gained popularity in the last years for structural investigations of broad range of materials [6]. The high energy resolution X-ray absorption near edge structure (HR-XANES) spectra at the U $M_4$ edge ($3d_{3/2} \rightarrow 5f$) have spectral features with reduced broadening compared to the conventionally measured XANES [7-9]. Therefore this advanced technique provides relative energy positions of valence unoccupied actinide (An) f orbitals not accessible by the conventional XANES method [7-9]. In this study we discuss a finger print approach for detection of changes in the bond lengths between U and the two axial O atoms in uranyl type of bonding, i.e. U(VI) forms short (< 1.8 Å) covalent axial bonding with two O atoms (U-Oax). We present results of theoretical *«ab-initio»* full-multiple-scattering (FMS) HR-XANES calculations for the $\alpha$-$UO_3$, $\beta$-$UO_3$ and $\gamma$-$UO_3$ phases by the FEFF 9.6 code [10].

## 2. Experiments and calculations

The α-$UO_3$ and β-$UO_3$ phases were synthesized as reported in [3]; γ-$UO_3$ is a commercial sample (Cameco Corp.); metaschoepite is prepared as reported in [11]. $CaU_2O_7$ was synthesized by recrystallizing $UO_3$•1-2($H_2O$) in concentrated $CaCl_2$ aqueous solution. The structures of the studied compounds were confirmed by powder XRD measurements performed at the SUL-X beamline at ANKA (estimated purity > 95%). The obtained XRD patterns were compared with patterns from the PDF database (ICDD, [12]) using the DIFFRAC.EVA, Bruker software. Each compound with about 10 wt % of U was mixed with cellulose powder and pressed into a pellet. The U $M_4$ edge HR-XANES spectra of $UO_3$•1-2($H_2O$), $CaU_2O_7$, α-$UO_3$, β-$UO_3$ and γ-$UO_3$ were measured using Johann type X-ray emission spectrometer recently installed at the INE-Beamline for actinide research at the ANKA synchrotron radiation facility, Karlsruhe; Germany [13,14]. The incident energy was monochromatized by a Si(111) double crystal monochromator (DCM). The sample, analyzer crystals and a silicon drift detector (KETEK) were arranged in a vertical Rowland geometry. The U HR-XANES spectra at the $M_4$-edge were obtained by recording the maximum intensity of the $M_\beta$ emission line (U $M_\beta$ 3337 eV) with five spherically bent Si(220) crystal analyzers with 1 m bending radius. The crystals were aligned at 75° Bragg angle. The DCM was calibrated by assigning 3725.5 eV eV to the maximum of the WL of a U $M_4$ edge HR-XANES spectrum of a $UO_2$ sample, which was repeatedly measured during the experiments.

The U $M_4$ HR-XANES spectra were calculated with the FEFF9.6 *«ab-initio»* quantum chemical code based on the full-multiple-scattering theory (FMS). The algorithm for the FMS method has been described elsewhere [15]. Phase shifts of the photoelectron were calculated in the framework of the self-consistent crystal muffin-tin (MT) potential scheme with 15% overlapping MT spheres. The spectra have been simulated using several types of exchange potentials: non-local, Dirac-Fock, Hedin-Lundquist or Dirac-Hara potentials. The best agreement with experiment has been achieved for the spectra calculated with the Hedin-Lundquist potential in Final State Rule (FSR) approximation for core-hole, reducing the $3d_{3/2}$ core-hole life-time broadening (3.5 eV) to 2 eV and correcting the Fermi energy for $UO_3$•1-2($H_2O$) and $CaU_2O_7$ by 0.5 eV. The atomic potentials were calculated self consistently for cluster sizes of about 7.0 Å around the absorber (including 101 atoms), while FMS calculations of U $M_4$ HR-XANES were performed for cluster of 10 Å radii (285 atoms). For the simulations of the HR-XANES spectra we have used the crystallographic data presented in Table 1.

Table 1. Crystallographic data used for calculations of the U $M_4$ HR-XANES spectra.

| Phase | Space group | Crystal system | Lattice parameters | #ICSD |
|---|---|---|---|---|
| α-$UO_3$ | P-3m1 (#164 ) | Trigonal | a=b=3.97 c=4.16<br>α=β= 90° γ=120° | 31628 |
| β-$UO_3$ | P121 1 (#4 ) | Monoclinic | a=3.91 b=14.33 c=10.34<br>α=β= γ=90° | 14314 |
| γ-$UO_3$ | I41/amd (#141) | Tetragonal | a=b=6.90 c=19.97<br>α=β=γ=90° | 1093 |
| $UO_3$•1-2($H_2O$) | Pbcn (#60) | Orthorhombic | a=14.68 b=14.02 c=16.71<br>α=β=γ=90° | 31631 |
| $CaU_2O_7$ | R-3m (#166) | Trigonal | a=b=c=6.26<br>α=36.32, β= γ= 90° | 156714 |

## 3. Results and discussion

The U $M_4$ HR-XANES spectra of $UO_3$ polymorphs are compared with the spectra of $UO_3$•1-2($H_2O$) and $CaU_2O_7$ (fig.1a). All U $M_4$ HR-XANES spectra have three distinct spectral features marked with A (~ 3727 eV), B (~ 3729 eV) and C (~ 3732 eV) in figure 1a. The A, B and C peaks have been assigned to electronic transitions of $3d_{3/2}$ electrons to $5f\delta/5f\phi$ (A), $5f\pi$ (B) and $5f\sigma$ (C) unoccupied valence orbitals of U(VI) in the form of $UO_2^{2+}$ (uranyl) [8]. The hybridized U 5f and 6p orbitals form sigma covalent bonds mainly with 2p orbitals of the Oax atoms therefore the energy position of peak C is essentially influenced by variations of the U-Oax bond length [9].

U forms uranyl and long > 1.8 Å (uranate) U-Oax bonds in metaschoepite (1.78 Å) and $CaU_2O_7$ (1.94 Å), respectively. These differences in bond lengths are reflected by the energy positions of peaks A, B and C. Peak A is shifted to higher energies (+ 0.2 eV), whereas peaks B (- 0.2 eV) and particularly C (- 1.1 eV) are shifted to lower energies for $CaU_2O_7$ compared to metaschoepite ($UO_3$•1-2($H_2O$)). The elongation of the U-Oax in $CaU_2O_7$ leads to less electronic density in the vicinity of U, i.e. increased ionicity of the U-Oax bond [16]; as a result, due to worse screening of the core-hole, the main peak A is shifted to higher energies compare to this peak for the metaschoepite spectrum. The energy position of peak B can be in principle influenced by variations of the U-Oax but also by changes in bonding distances between U and the equatorial ligands.

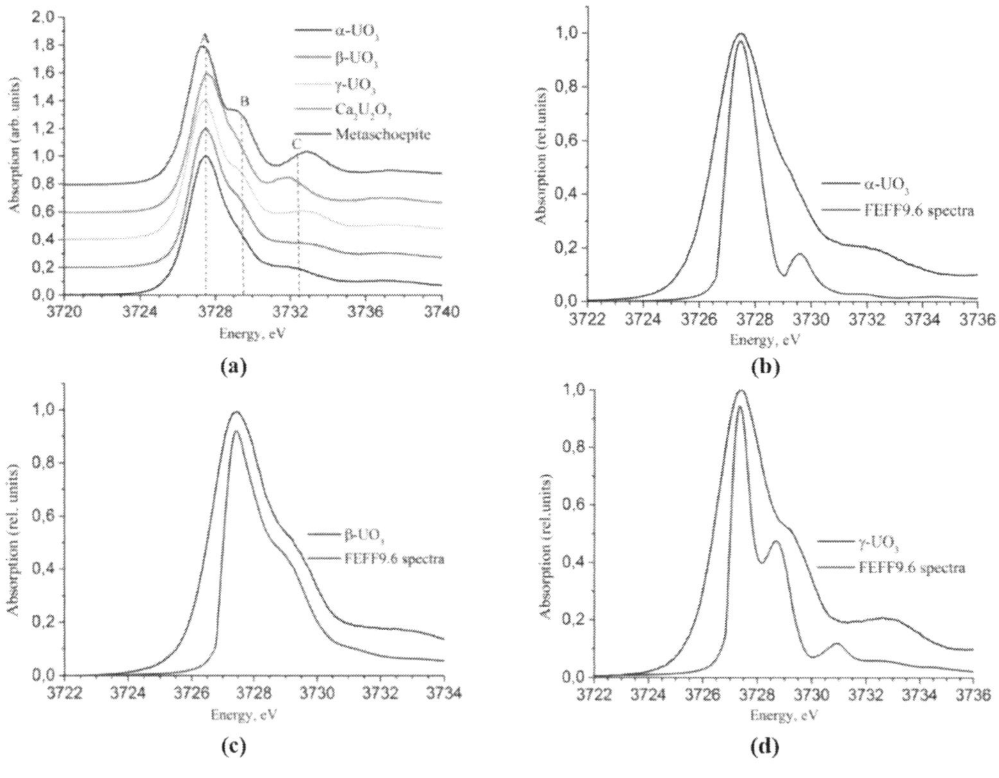


**Figure 1.** U $M_4$ HR-XANES spectra of $UO_3$•1-2($H_2O$), $CaU_2O_7$, α-$UO_3$, β-$UO_3$ and γ-$UO_3$ (a), experimental and calculated spectra for α-$UO_3$ (b), β-$UO_3$ (c) and γ-$UO_3$ (d) phases.

Up to several different U sites with variable U-Oax bond lengths are present in α-$UO_3$, β-$UO_3$ and γ-$UO_3$. α-$UO_3$ has two non-equivalent U positions; U forms bonds with the Oax atoms with two different average bond lengths, i.e. U(1)-$Oax_1$=2.08 Å and U(2)-$Oax_2$=2.39 Å. The uranium atoms in β-$UO_3$ can be divided into three groups: U(1) and U(2) have seven oxygen neighbors at distances varying between 1.69 Å and 2.72 Å, U(3) is coordinated by six oxygen atoms U(3)-$Oax_1$=1.79 Å, U(3)-$Oax_2$=2.17 Å which form a deformed octahedron, U(4) and U(5) have six oxygen neighbors and form uranyl type of bonding [17]. In γ-$UO_3$ all uranium atoms are surrounded by six oxygen atoms in distorted octahedral environment; the average U-Oax bond length is ≈ 1.78 Å [18]. The energy positions of features A, B and C are related to the average U-Oax distances for the three $UO_3$ compounds. The spectra of α- and γ-$UO_3$ are similar to the spectra of $CaU_2O_7$ and $UO_3$•1-2($H_2O$), respectively. Therefore we propose that in average the U-Oax bonds have more covalent character in γ-$UO_3$ compared to α-$UO_3$.

For each $UO_3$ phase the theoretical spectrum is a sum of weighted spectra obtained by placing the absorbing U atom at each non-equivalent crystallographic site. The FEFF9.6 code reproduces all spectral features at the correct energy positions for α-$UO_3$ and β-$UO_3$ (fig. 1 b, c); some intensity differences are present. For γ-$UO_3$ (fig. 1d) the distances between features A and B (A-B), and B and C (B-C) are larger for the experimental (A-B ≈ 2 eV, B-C ≈ 3.5 eV) compared to the calculated (A-B ≈ 1.7 eV, B-C ≈ 2.2 eV) spectra.

## 4. Conclusions

We demonstrated that a fingerprint approach using U $M_4$ HR-XANES can be effectively applied for characterization of "uranyl" and "uranate" type of U-Oax bonding. Using this approach we showed that U-Oax bonds have likely more covalent character in γ-$UO_3$ compared to α-$UO_3$. The U $M_4$ HR-XANES spectra of the $UO_3$ polymorphs were performed within the full-multiple-scattering (FMS) formalism. We varied the input parameters to obtain best agreement between theory and experiment. The FEFF9.6 code emerges as an useful tool for calculation of U $M_4$ HR-XANES spectra, as it successfully reproduced all spectral features for some of the studied compounds.

## Acknowledgements

This work is partially funded by the Russian Foundation for Basic Research (project №14-05-31488 mol_a), the joint program "Mikhail Lomonosov" of the Ministry of Education and Science of the Russian Federation and the Deutscher Akademischer Austauschdienst. We acknowledge KIT and the Helmholtz Association of German Research Centres for the Helmholtz Young Investigators Group grant (VH-NG-734) and thank ANKA for the provided beamtime. We also acknowledge Dario Manara (JRC-ITU), Xavier Gaona (INE-KIT) and Robert Baker (Trinity College Dublin) for providing part of the samples.

## References

[1] Alberman K B , Anderson J S 1949 *Angew.Chem.*, **61**, 416-416.
[2] Hill D J 2008, *Nat.Mater.* **7**, 680.
[3] Hoekstra HR, Siegel S J 1961 *Inorg.Nucl.Chem.*, **18**, 154-165.
[4] Benedict M, Pigford TH, Levi HW 1981 *Nuclear Chemical Engineering, 2nd ed.*; McGraw Hill:New York.
[5] Denecke 2006 *Coordin Chem Rev*, **250**, 730-754.
[6] Bergmann U, Glatzel P 2009 *Photosynth Res*, **102**, 255-266.
[7] Vitova T , Green J C, Denning R G, Loble M, Kvashnina K, Kas J J, Jorissen K, Rehr J J, Malcherek T, Denecke 2015 *Inorg Chem*, **54**, 174-182.
[8] Vitova T., Denecke M A, Göttlicher J, Jorissen K, Kas J, Kvashnina K, Prüßmann T, Rehr J, Rothe J 2013 *Journal of Physics: Conference Series*, **430**, 012117.
[9] Kvashnina K O, Butorin S M, Martin P, Glatzel P 2013 *Phys Rev Lett*, **111.**
[10] Rehr J J, Kas J J, Vila F D, Prange M P, Jorissen K 2010 *Phys. Chem. Chem. Phys.*, , **12**, 5503-5513.
[11] Gorman-Lewis D, Fein J B, Burns P C, Szymanowski J E S, Converse J 2008 J. Chem. Thermodyn. 40 980-990
[12] PDF-2 database, ver. 2.1402; ICDD The International Centre for Diffraction Data (www.icdd.com)
[13] Rothe J, Butorin S, Dardenne K, Denecke M A, Kienzler B, Loble M, Metz V, Seibert A, Steppert M, Vitova T, Walther C, Geckeis H 2012 *Rev Sci Instrum*, **83**.
[14] Walshe A, Prussmann T, Vitova T, Baker R J 2014, *Dalton Trans.*, **43** (11), 4400−4407.
[15] Rehr J J, Kas J J, Prange M P, Sorini A P, Takimoto Y, Vila F D 2009, *Comptes Rendus Physique*, **(6)** 548-559.
[16] Denning R G et al. 2002 *J. of Chem. Phys.*, **117**,17,8008- 8020.
[17] Debets P C 1966, *Acta Cryst*, , **21**, 589.
[18] Engmann R, .de Wolff P M 1963, *Acta Cryst,* **16**, 993.

# Determining the Sulfur species in the dispersants Corexit 9500A and 9527A applying S K-edge XANES spectroscopy

**Lisa Bovenkamp-Langlois[1] and Amitava Roy[1]**

[1]J. Bennett Johnston, Sr., Center for Advanced Microstructures and Devices (CAMD), Louisiana State University, Baton Rouge, LA 70806. USA

Email: lbovenkamp@lsu.edu

**Abstract**: The dispersants Corexit 9500A and 9527A were used extensively during the Deepwater Horizon oil spill in the Gulf of Mexico. Sulfur K-edge X-ray absorption near edge structure (XANES) spectroscopy was used to investigate the dispersants for the sulfur based components. The main sulfur containing component should be dioctyl sodium sulfosuccinate (DOSS). S K-edge XANES analysis shows that indeed the major sulfur species in both kinds of Corexit (9500A and 9527A) is sulfonic acid which is a part of DOSS. In addition some fraction of sulfone was detected.

## 1. Introduction

Following the April 20, 2010, explosion on the Deepwater Horizon oil rig in the Gulf of Mexico, approximately 5 million barrels (185 million gallons) of crude oil gushed out of the Macondo Canyon 252 (MC252) well for over four months [1]. The methods that were used to remediate the oil spill included direct collection of the crude oil from the riser pipe, collection by booms, and burning at the surface. Additionally, a significant fraction evaporated or dissolved. However, these methods were not rapid enough to collect all the oil. Therefore, dispersants, such as Corexit 9500A and 9527A (Nalco Environmental Solutions), were used at the wellhead (in ca. 5000 feet depth) and on the surface (about 2 million gallons altogether [1]) to disperse the oil more efficiently. This method presumably inhibits large volumes of oil from reaching the surface or being transported landward [2].

An oil spill dispersant breaks down the oil into very fine droplets and inhibits the formation of large volumes of emulsified oil. Smaller droplets also enhance biodegradation [3]. A dispersant typically consists of a surfactant in one or more solvents, used to deliver the surfactant to the oil-water interface. The surfactant reduces the surface tension of hydrophobic oil and disperses it into seawater as smaller droplets. Current dispersants, such as Corexit are considered less toxic than earlier mixtures [4]. However, their effects on marine life include acute toxicity of the dispersant itself and toxic oil components that have an increased solubility due to the dispersant [1]. The surfactant in Corexit (9500A and 9527A) is an organic sulfonic salt, dioctyl sodium sulfosuccinate (DOSS, other name: butanedioic acid, 2-sulfo-, 1,4-bis(2-ethylhexyl, ester, sodium salt, cf. Figure 1). Additional compounds are TWEEN 80 (other name: Sorbitan, mono-(9Z)-9-octadecenoate, poly(oxy-1,2-ethanediyl) derivatives), SPAN 80 (other name: sorbitan monooctadecanoate), di(propylene glycol) butyl ether, and some kind of kerosene (http://www.nalcoesllc.com/nes/1602.htm). The exact formulation of Corexit was divulged only after the Deepwater Horizon oil spill.

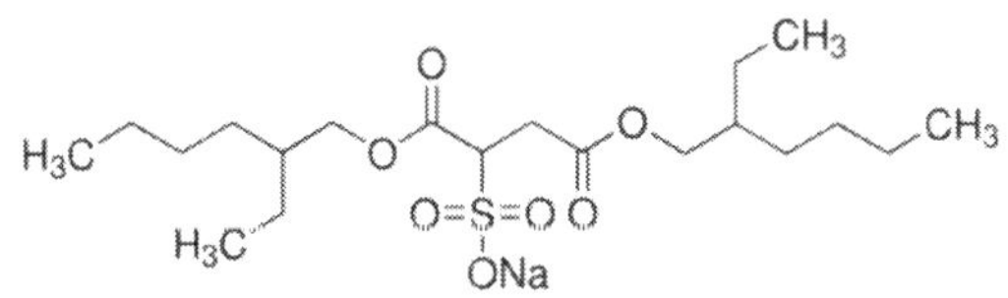


**Figure 1:** Structure of dioctyl sulfosuccinate sodium salt (DOSS)

Generally, sulfur species in crude oil and its derivatives are determined by some type chromatographic separation followed by elemental determination of sulfur, for example, by gas chromatography – inductively coupled plasma mass spectrometry [5]. Such techniques do not allow determination of all the chemical species at once and the preparation process itself may alter the speciation. X-ray absorption near edge structure (XANES) spectroscopy has been successfully used for sulfur speciation determination in crude oil and its derivatives for quite some time [6]. Since the toxicity and persistence of dispersants is of interest to a wide scientific community, the detection of DOSS without any chemical separation by XANES is important.

## 2. Materials and Methods

### 2.1 *Materials*

Corexit 9500A and 9527A were obtained from the Gulf of Mexico Research Initiative (GoMRI) consortium. They were stored in dark brown glass bottles at room temperature, exposed to fluorescent lighting. Reference compounds, such as Pentane-sulfonic acid sodium salt (PSSA), Diphenyl sulfone, and Zinc sulfate were purchased from Sigma Aldrich in high purity.

### *2.2 X-Ray-Absorption Near Edge Structure (XANES) Spectroscopy*

S K-edge XANES measurements were performed at the electron storage ring J. Bennett Johnston, Sr., Center for Advanced Microstructures and Devices (CAMD), Louisiana State University, Baton Rouge, Louisiana. The ring is operating at 1.3 GeV. Data were collected at the DCM beamline. The monochromator, of Lemonnier-type with design modifications made at Bonn University, Germany, was equipped with InSb(111) crystals and calibrated to the maximum of the white line peak of zinc sulfate to 2481.4 eV [7]. For low energy measurements in fluorescence mode a Si drift Vortex detector was used. The pressure inside the sample chamber was reduced to 10 Torr (fluorescence) and 42 Torr (transmission) to accommodate the shorter absorption length of photons in air. Corexit samples were measured in fluorescence mode and prepared as follows. About 50 µl of the fluid was dropped on filter paper which was supported by Kapton tape. The samples were covered with mylar film. Reference compounds (Pentane-sulfonic acid sodium salt (PSSA), Diphenyl sulfone, Zinc sulfate) were measured in transmission mode. The parameters for the measurements were: 2440-2468 eV with 0.5 eV steps, 2468-2485 eV with 0.1 eV steps, and 2485-2520 with 0.3 eV steps. The integration time in all cases was 1 sec and each scan was repeated once to ensure reproducibility. The data were processed (background subtracted, normalized, averaged) and analyzed using the ATHENA software [8].

## 3. Results and Discussion

Sulfur K-edge XANES spectra of Corexit 9500A and 9527A together with reference compounds are shown in Figure 2. These samples were analyzed over a period of three years. The spectra of both samples of Corexit 9500A and 9527A are the same (apart from the variability in sample preparation). This indicates that DOSS is not unstable when exposed to light. Since the liquid was stored in brown bottles in fluorescent indoor lighting this is not contradictory to what was reported earlier [9]. The chemical shift due to the different oxidation states can be observed directly as energy shift in the S K-edge XANES spectra [10, 11]. Reference compounds that are adequate for this analysis of the chemical environment of sulfur in both types of Corexit are diphenyl sulfoxide, diphenyl sulfone, pentane sulfonic acid sodium salt (PSSA), and zinc sulfate. These or similar reference spectra have been published previously in most cases with a different calibration scale [6, 10, 12-14]. Using this fingerprinting method the S species in Corexit (9500A and 9527A) easily can be identified as predominantly sulfonic acid because the White line maximum at 2480.7 eV and also the Shape resonances at ~2486 and ~2496 eV have the same shape and energy position. The shoulder at about 2478.9 is not part of the PSSA spectrum. This indicates that there is some contribution of sulfone. Since the local environment of sulfur in DOSS is just like in PSSA this shoulder must be a contribution from one of the other components. The only candidate is kerosene (one of petroleum distillates). It is known that petroleum products contain sulfur in different forms such as sulfides, thiophenes, sulfoxides, and sulfones [6]..

The shoulder at 2478.9 eV is related to a sulfone which is evident from the literature [6, 12, 13]. However, for the present measurements only diphenol sulfone was available but the White Line (2479.4 eV) is not exactly at the same energetic position as the shoulder.

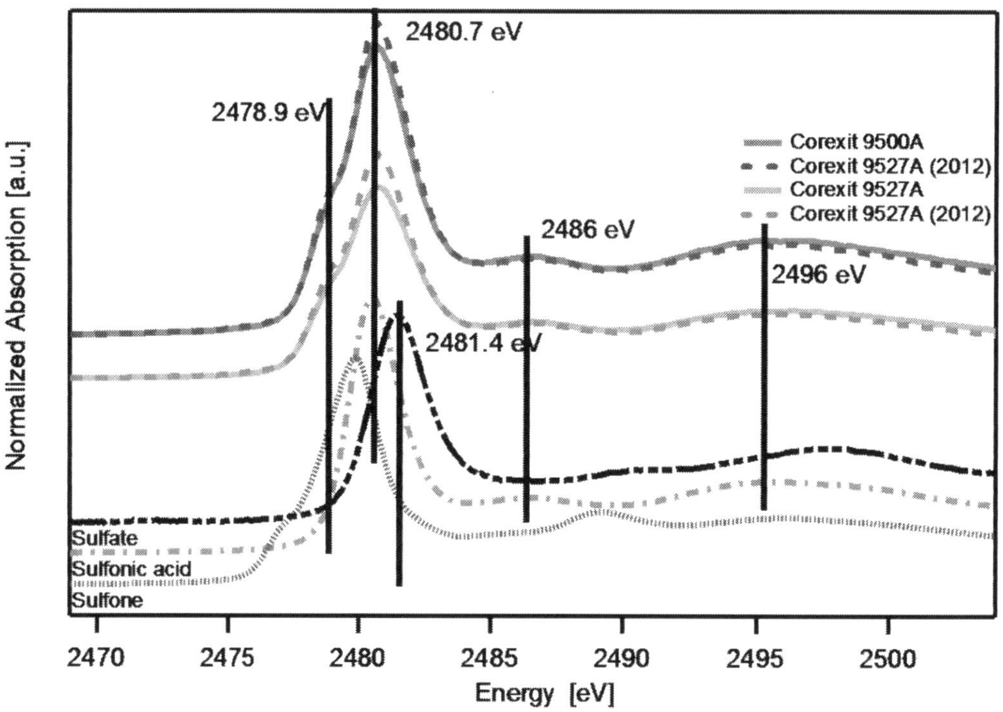


**Figure 2:** S K-edge XANES spectra of Corexit 9500A (red), 9527A (green) and the reference compounds: PSSA (orange), Diphenyl sulfone (blue), and Zinc sulfate (black)

Corexit 9500A and 9527A were used at an unprecedented scale during the Deepwater horizon spill, often deep in the subsurface. These dispersants have also been found to be toxic at certain levels to various organisms. S K-edge XANES of the dispersants confirm that sulfonic acid is indeed an important component. Small contributions from other sulfur containing compounds can be found as well. These are likely related to the petroleum distillates that make up the other components of the dispersants. The ultimate challenge will be the detection of Corexit in other oil spill related samples in the presence of the sulfur species from crude oil and seawater.

**Acknowledgement**
This project was supported by a grant from the Gulf of Mexico Research Initiative (GoMRI), as part of the Consortium for the Molecular Engineering of Dispersant Systems (C-MEDS, http://dispersant.tulane.edu/).

**References**
[1] Gray, J.L., Kanagy, L.K., Furlong, E.T., Kanagy, C.J., McCoy, J.W., Mason A., Lauensteine, G. 2014. *Chemosphere.* 95(2014): p. 124-130.
[2] Baker, J.M., 1995. In: Int. *Oil Spill Conf. Proc. February 1995*, vol. 1, pp. 611–614.
[3] Prince, R. C., McFarlin, K. M. Butler, J. D. Febbo, E. J., Wang, FCY, Nedwed, T. J. 2013. *Chemosphere* 90(2): 521-526.
[4] Anonymous, ExxonMobil 2008 *Oil Spill Dispersant Guidelines* , pp. 202.
[5] Caumette, G., Lienemann, C.P., Merdrignac, I., Bouyssiere, B., Lobinski, R. 2009 *J Anal Atom Spec*, 24 263-276.
[6] Waldo, G.S. Carlson, RMK, Moldowan, J.M., Peters, K.E., Penner-Hahn, J.E. 1991. *Geochim. Cosmochim. Acta* 55(3): p. 801-814.
[7] Chauvistre, R. 1992 Dissertation, *Bonn University*, Bonn, Germany, 1992.
[8] Ravel, B. and M. Newville 2005. *J Synch. Rad.* 12: 537-541.
[9] Batchu, S.R. 2013 *Florida International University Electronic Theses and Dissertations. Paper 859* http://digitalcommons.fiu.edu/etd/859_2013.
[10] Myneni, S.C.B. 2000 in: Alpers, C.N., Jambor, J.L, Nordstrom, D.K. (Eds.) Sulfate Minerals - Crystallography, Geochemistry and Environmental Significance, *Mineral Soc Am*, Washington, D.C., 2000, pp. 113-172.
[11] Modrow, H. 2003 *Habilitation Treatise Bonn University*, Bonn, Germany, 2003.
[12] George, G. N, and Gorbaty, M. L. 1989. *J Am. Chem. Soc.* 111 (9), 3182-3186, DOI: 10.1021/ja00191a012
[13] Franz, B., Gehrke, T., Lichtenberg, H., Hormes, J., Dahl, C., Prange, A. 2009. *Microbiol.*, 155: 2766-2774, DOI: 10.1099/mic.0.027904-0.
[14] Pomerantz, A. E., Seifert, D. J., Bake, K. D., Craddock, P. R., Mullins, O. C., Kodalen, B. G., Mitra-Kirtley, S., Bolin, T. B. 2013. *Energy Fuels* 2013 27 (8), 4604-4608 DOI: 10.1021/ef400773f.

# Sorption mechanisms of metals to graphene oxide

**Allison R Showalter[1], Thomas A Duster[2], Jennifer E S Szymanowski[2], Chongzheng Na[2], Jeremy B Fein[2] and Bruce A Bunker[1]**

[1]Department of Physics, University of Notre Dame, Notre Dame, IN 46556, USA
[2]Department of Civil and Environmental Engineering and Earth Sciences, University of Notre Dame, Notre Dame, IN 46556, USA

Email: ashowal1@nd.edu

**Abstract**. Environmental toxic metal contamination remediation and prevention is an ongoing issue. Graphene oxide is highly sorptive for many heavy metals over a wide pH range under different ionic strength conditions. We present x-ray absorption fine structure (XAFS) spectroscopy results investigating the binding environment of Pb(II), Cd(II) and U(VI) ions onto multi-layered graphene oxide (MLGO). Analysis indicates that the dominant sorption mechanism of Pb to MLGO changes as a function of pH, with increasing inner sphere contribution as pH increases. In contrast, the sorption mechanism of Cd to MLGO remains constant under the studied pH range. This adsorption mechanism is an electrostatic attraction between the hydrated $Cd^{+2}$ ion and the MLGO surface. The U(VI), present as the uranyl ion, changes only subtly as a function of pH and is bound to the surface via an inner sphere bond. Knowledge of the binding mechanism for each metal is necessary to help in optimizing environmental remediation or prevention in filtration systems.

## 1. Introduction

Graphene nanosheets are unique, 2-D systems with physical and chemical properties generating applications in a wide range of fields. One variation is graphene oxide, where oxidation of graphene's aromatic six-member carbon rings results in covalently bonded functional groups on both the basal planes and along the nanosheet edges, possibly including basal sites of epoxide (red), ketone (green), and hydroxyl (black) functional groups and edge site ester or lactol (blue) functional groups, including carboxylic/phenolic groups as shown in figure 1 [1]. The high surface area-to-mass ratio makes graphene oxide an ideal candidate for sorbing heavy metal ions from solution. In fact, multi-layered graphene oxide (MLGO) has exceptionally high reported sorption capacity for several toxic metals including lead (842 mg/g) [2], cadmium (106.3 mg/g) [3] and uranium (97.5 mg/g) [4]. These results also indicate that metal adsorption to MLGO depends strongly on the pH and ionic strength of the system. The high sorption capacity makes graphene oxide a promising new material for filtration systems and for *in situ* or *ex situ* remediation.

Effective use of graphene oxide in this capacity requires knowledge of how and why sorption occurs under environmental conditions. Possibilities for this adsorption include direct covalent inner-sphere bonds formed between de-protonated functional groups on the surface and the metal of interest or electrostatic outer sphere bonds formed between the surface and the metal of interest surrounded by a hydration sphere of water molecules, as shown in figure 1. X-ray absorption fine structure (XAFS)

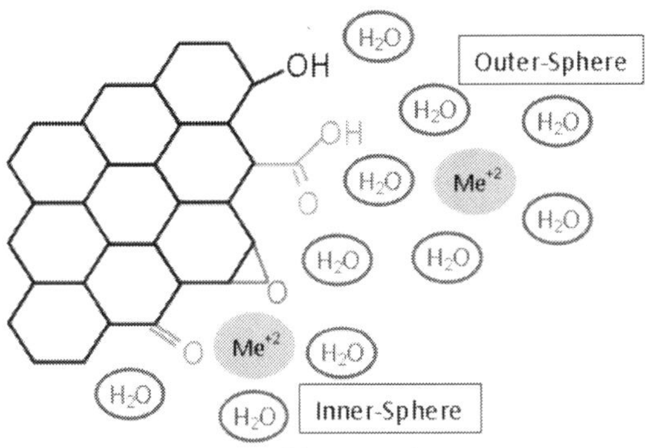


**Figure 1:** Depiction of inner-sphere versus outer sphere adsorption on the MLGO surface. Included at the edge are examples of basal epoxide (red), ketone (green), and hydroxyl (black) functional group sites and edge site ester or lactol (blue) functional groups, including carboxylic/phenolic groups (adapted from Gao et al., 2009 [1]).

spectroscopy provides the identity and distance of elements in the local environment surrounding a central atom [5]. XAFS can differentiate between elements, requires no long range order and is sensitive to small concentrations of the element of interest. These characteristics make XAFS an ideal technique to identify the sorption mechanisms of metals to surfaces in aqueous environments.

In a previous study, Pb adsorption onto MLGO was studied for the pH range of 4.9-8.3. The sorption mechanism of Pb onto MLGO is strongly dependent upon pH, with outer sphere electrostatic attraction being the dominant mechanism at low pH and inner sphere covalent bonding being the dominant mechanism at high pH. In contrast, Cd adsorption to MLGO over the pH range of 4.9-8.1 exhibited only a single mode of sorption. The spectra are all nearly identical to each other. The features of the samples closely resemble the hydrated free $Cd^{+2}$ ions in solution, but do not exactly replicate it. The best fit for the samples used only a Cd-O path with slightly larger Debye-Waller factors than the free $Cd^{+2}$ ions. This would be consistent with an electrostatic outer sphere adsorption since the close presence of the surface to the Cd would increase the disorder compared to the $Cd^{+2}$ ions in solution.

## 2. Materials and methods

The preparation, characterization and experimental methodologies are described in Duster et al. [6]. In brief, for the XAFS analysis, batch metal sorption experiments use 420 mg/L MLGO with 0.1 M $NaClO_4$ in ambient air (carbonate concentration < 0.6 ppm in solution) exposed to 10 ppm U before being centrifuged and the hydrated paste taken for XAFS analysis. Aqueous standards containing 10 mM U and 100 mM perchlorate salt, acetate salt, or ethylenediaminetetraacetate (EDTA) salt at appropriate pH values were also prepared to mimic the hydrated metal ion, bidentate inner spherically bound metal ion, and monodentate inner spherically bound metal ion, respectively. The hydrated paste was sealed in slotted Teflon holders using Kapton tape, and samples and standards were measured at the U L(III)-edge (17,166 eV) at the MRCAT 10-ID and 10-BM beamlines [7] at the Advanced Photon Source (APS) at Argonne National Laboratory near Chicago, IL, USA. Data analysis used the methodology of the UWXAFS package [8] . For the U fitting, a range of 1.15-3.5 Å with a Fourier transform range of 3.5-12.0 $Å^{-1}$ was used for pH values of 4.0, 5.8, 7.2 and 8.5.

## 3. Results and discussion

U(VI) in solution generally appears as the uranyl ion, $UO_2^{+2}$, where the two oxygen atoms from the ion align axially and the bonding atoms coordinate in the equatorial plane. X-ray absorption near edge structure (XANES), which is sensitive to the oxidation state of the uranium, indicates that all uranium in this system remains as U(VI). The U-MLGO XANES data of the samples show only subtle differences from each other, especially when compared with the hydrated uranyl ion standard as seen in figure 2a. In the extended x-ray absorption fine structure (EXAFS) region, the U-MLGO data again shows only slight differences from each other. This suggests that there is one dominant type of adsorption mechanism across the studied pH range, similar to the Cd-MLGO system. In contrast though, the U-MLGO is clearly different from the hydrated uranyl ion and the precipitates formed at high pH, as shown in figure 2b. Subtle differences can be seen between the U-MLGO samples, with these differences trending according to pH. This may be the result of increased disorder associated with varying aqueous speciation of the uranyl ion, as it forms several aqueous complexes with hydroxyl or carbonate ions in solution that change within the studied pH range. However, this effect seems to be small and does not change the dominant mode of sorption. As such, the binding mechanism of the U on the MLGO surface is most likely an inner sphere adsorption.

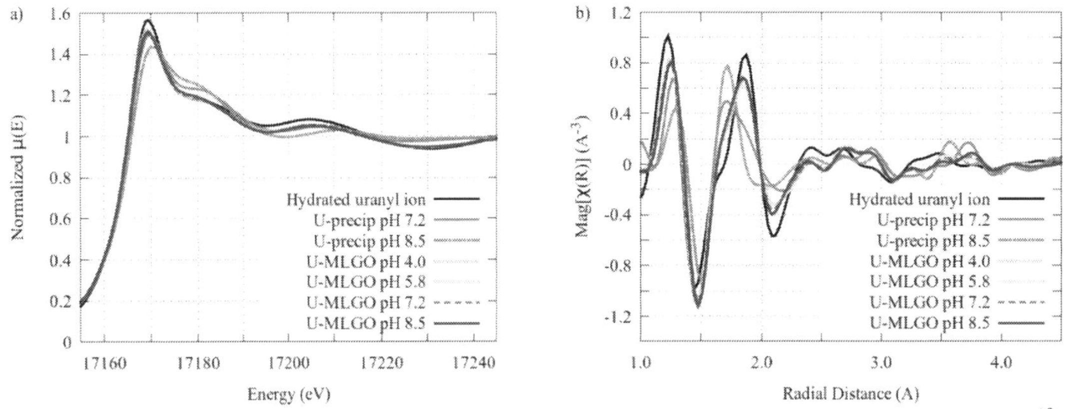


**Figure 2:** XAFS data for U-MLGO, uranyl precipitates and the hydrated uranyl ion ($UO_2^{+2}$) in aqueous solution showing a) the XANES region and b) the EXAFS region. Note that the four samples are only subtly different from each other, but do show significant variation from the hydrated uranyl ions and uranyl precipitates.

A comparison of the U-MLGO samples and precipitates formed by adding concentrated amounts of uranium (10 mM) to solution at pH 7.2 or 8.5 is shown in figure 3. Another study found a uranyl precipitate (similar to the one observed at pH 7.2 in figure 3) forming on a single-layer graphene oxide surface as low as pH 6.1 [9]. However, this study used a uranium concentration of 0.5 mM (119 ppm) on 0.4 g/L graphene oxide, which is a similar graphene oxide concentration to the current study, but an order of magnitude higher uranium concentration. At the lower uranium concentration used in this study, there does not seem to be any uranium precipitation occurring on the MLGO surface. The uranyl precipitate data show prominent differences to the data at low R (1.0-2.5 Å) and features between 3.5-4.0 Å that are lacking in the U-MLGO samples. Fitting is also inconsistent with a U-U bond for the U-MLGO data at this location, so there is no evidence that a significant amount of precipitate is forming on the U-MLGO samples in this study. As such, the sorption mechanism of U to MLGO in this study is determined to be inner sphere adsorption.

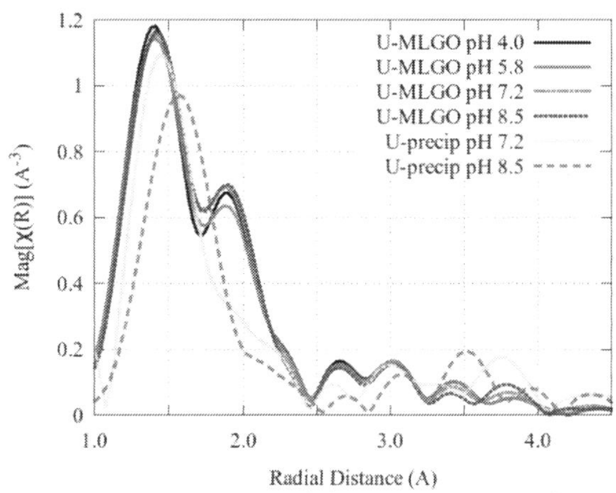


**Figure 3:** Comparison of the U-MLGO samples to uranyl precipitates formed in solution at pH 7.2 or 8.5. Not all U(VI) compounds contain the uranyl moiety [10], explaining the lack of axial signal in the precipitates here.

## 4. Conclusions

We have shown that different heavy metals have different adsorption mechanisms to MLGO in a moderate pH range that would be common in environmental remediation situations or in filtration systems. XAFS has enabled us to identify the dominant adsorption mechanism for Pb, Cd and U to MLGO over this pH range. The Pb(II) adsorption mechanism is strongly dependent on the system pH, with adsorption controlled by a combination of outer sphere electrostatic attraction and inner sphere covalent bonding. At lower system pH, outer sphere adsorption dominates, while inner sphere adsorption takes over as the main sorption mechanism at higher pH values. In contrast, the adsorption mechanism for Cd and U is relatively independent of pH. For Cd, at all measured pH values adsorption is dominated by an electrostatic outer shell attraction between the Cd(II) and its coordinating water molecules and the MLGO surface. The U-MLGO systems show subtle variation across the measured pH range, but adsorption seems to be dominated by an inner sphere bond for all the samples. These slight differences may be attributable to aqueous complexation of the uranyl ion as it changes in pH. However, the U-MLGO samples are clearly unique from the hydrated uranyl ion in solution, indicating that attraction is not electrostatic. Similarly, the samples are inconsistent with appreciable precipitation, indicating that the U(VI) is bound to the MLGO surface via an inner sphere bond. Overall, metal adsorption to MLGO varies with the type of metal, indicating that it might be possible to tailor the material to best adsorb certain heavy metals, making it ideal for environmental remediation or filtration systems.

## Acknowledgements

MRCAT is supported by the U.S. Department of Energy and the member institutions. Use of the APS was supported by the U. S. Department of Energy, Office of Science, Office of Basic Energy Sciences, under Contract No. DE-AC02-06CH11357. A.R.S. wishes to thank the staff at MRCAT, B. Mishra and K. Werellapatha for beamline support and useful conversations. Experiment used instrumentation at the Center for Environmental Science and Technology (CEST) at the University of Notre Dame. This research was supported, in part, by a fellowship to TAD from the Arthur J. Schmitt Foundation. The authors also wish to thank the two anonymous reviewers for their excellent comments.

## References

[1] Szabo T, Berkesi O, Forgo P, Josepovits K, Sanakis Y, Petridis D and Dekany I 2006 *Chem. Mater.* **18** 2740-9
Gao W, Alemany L B, Ci L J and Ajayan P M 2009 *Nature Chem.* **1** 403-8
Lee D W, De Los Santos L, Seo J W, Felix L L, Bustamante A, Cole J M and Bames C H W 2010 *J. Phys. Chem. B* **114** 5723-8
[2] Zhao G, Ren X, Chen C, Huang Y and Wang X 2011 *Dalton Trans.* **40** 10945-52
[3] Zhao G, Li J, Ren X, Chen C and Wang X 2011 *Environ. Sci. Technol.* **45** 10454-62
[4] Zhao G, Wen T, Yang X, Hu J, Shao D and Wang X 2012 *Dalon. Trans.* **41** 6182-8
[5] Bunker G 2010 *Introduction to XAFS: A practical guide to x-ray absorption fine structure spectroscopy* (New York: Cambridge University Press)
[6] Duster T A, Szymanowski J E S, Na C, Showalter A R, Bunker B A and Fein J B 2015 *Colloids Surf., A* **466** 28-39
[7] Segre C U, et al.2000 *AIP Conf. Proc.* **521** 419-22
Kropf A J, et al. 2010 *AIP Conf. Proc.* **1234** 299-302
[8] Stern E A, Newville M, Ravel B, Yacoby Y and Haskel D 1995 *Physica B* **208-209** 117-20
[9] Li Z, Chen F, Yuan L, Liu Y, Zhao Y, Chai Z and Shi W 2012 *Chem. Eng. J.* **210** 539-46
[10] Catalano J G and Brown G E 2004 *Am. Mineral.* **89** 1004-1021

# Local structures of Ca, Ti and Fe in meteorite fusion crusts

**T Tobase*[1], A Yoshiasa[1], T Hiratoko[1], H Hongu[1], H Isobe[1], A Nakatsuka[2], H Arima[3], K Sugiyama[3]**

[1] Graduated School of Sci. and Tech., Kumamoto Univ., Kumamoto, 860-8555, Japan

[2] Graduate School of Science and Engineering, Yamaguchi Univ. Ube 755-8611, Japan

[3] Department of Institute for material research, Tohoku Univ. Sendai 980-8577, Japan

E-mail: 155d9009@st.kumamoto-u.ac.jp

**Abstract.** The local structures of meteorite fusion crusts were studied by Ca, Ti and Fe K-edge XANES and EXAFS spectroscopy. The surface of meteorites were melted and volatilized with extreme high temperature and large temperature gradient when meteorites were rushed into atmosphere. This study indicated that meteorite fusion crusts have unique local structures. The local structures of fusion crusts differ from tektites especially in intensity of the shoulder in the rising flank of the edge in Ca XANES spectra. It is consistent with chemical composition change by the volatilization of Si at fusion during atmospheric entry. The high estimated $Fe^{3+}/(Fe^{2+}+Fe^{3+})$ rat io in meteorite fusion crusts indicates that meteorite fusion crusts are formed into atmospheric oxidation condition. The Ca-O distances in meteorite fusion crusts are 2.61-2.66 Å and are extremely longer than in other natural glasses. The fusion crusts have unique local structure since they experienced extremely high temperature and short quenching time. The XAFS method is effective in distinction of meteorite fusion crusts and classification of natural glass.

## 1. Introduction

All meteorites develop fusion crusts during atmospheric entry due to heating by hypervelocity collisions with air molecules [5]. Meteorite fusion crusts (fusion crusts) have been formed in extremely short times (several seconds) under the Earth's atmosphere conditions [2]. The surface of meteorites were melted and volatilized with extreme high temperature and large temperature gradient when meteorites were rushed into atmosphere. It is assumed that the unique local structures were formed in fusion crust glass under the ultra-high temperature with volatilization and rapid quenching process. Local structural difference can be predicted between fusion crusts and bulk. Local structures are affected by the pressure and temperature conditions during the glass formation and quenching process [7, 9, 16]. Wang et al. [9], Tobase et al. [11] and Wang et al. [13] reported unique Ti, Ca, Fe local structures in meteorite impact-related glasses and tektite. Upon a devastating impact of a giant meteoroid on the Earth, particles of the Earth's surface were melted and catapulted into outer space, where they finally solidified and fell back to the Earth as tektites. Tektites should be passed through several kinds of vitrification processes. Since local structures in glass are strongly related to forming process, comparing local structures of fusion crusts and impact-related glasses are important to reveal the vitrification processes of natural glasses. It is indicated that the XAFS method is useful for classification of natural glasses and distinction of meteorite fusion crusts.

## 2. Sample and Experiment

In order to study the local st ructures around Ca, Ti and Fe atoms in the fusion crusts of L5 chondrite (unnamed, North African), Dar al Gani 456 (L6 chondrite), millbillillie (Eucrite achondrite), Allende (CV3 carbonaceous chondrite) and Odessa (Iron meteorite, IAB) meteorites, we prepared several reference materials such as minerals, tektite glasses, impact glasses and volcanic glasses. The Ca , Ti, and Fe K-edge XAFS measurements were performed at beam line BL-9C, equipped with Si (111) double-crystal monochromater , Cu calibration, fluorescence mode, lytle detector, of the Photon Factory, KEK, Tsukuba, Japan. The EXAFS functions were transformed into the radial distribution functions. In order to obtain information on the detailed structure parameters, we conducted parameter fitting with analytical EXAFS formulae. Analysis of XAFS data was performed by using XAFS93 and MBF93 programs [17].

## 3. Result and discussion

### *3.1 XANES spectra of Ca, Ti and Fe*

The Ca, Ti and Fe XANES spectra are shown in figure 1 and 2. The Ca XANES spectra in fusion crusts are unique and have various pre-edge and shoulder intensity (figure 1A). The pre-edge peak observed in Dar al Gani 456 is the highest intensity. It indicates that site symmetry of Dar al Gani 456 is lower than other meteorites based on the relative intensity of the pre-edge [6]. The pre-edge intensity of L5 chondrite is the lowest and the local symmetry around Ca in L5 chondrite is higher than other fusion crusts. XANES pattern of Ca for tektite is unique and distinguishable from other reference materials such as volcanic glass (obsidian), impact glass (impactite) and mineral (apatite) [11]. All fusion crusts have almost similar spectra which differ from any other reference materials. Ca XANES in fusion crusts differ from tektites especially in shoulder intensity (figure 1B). The silicon contents of calcium aluminosilicate glass has positive correlation to the shoulder intensity [6]. Table 2 shows that chemical composition of fusion crusts and reference materials. Combining the result of silicate content and shoulder intensity, silicon content of fusion crusts also have positive correlation to shoulder intensity. Although the value of $CaO/Al_2O_3$ is lower than 1, the correlation ship of silicon content and shoulder intensity is consistent with that of Neuville et al., 2004 [6]. Lower silicon content of fusion crusts related to chemical composition change by the volatilization of Si at fusion during atmospheric entry. The XANES pattern of Ca indicates that the meteorite fusion crusts experienced high temperature with volatilization.

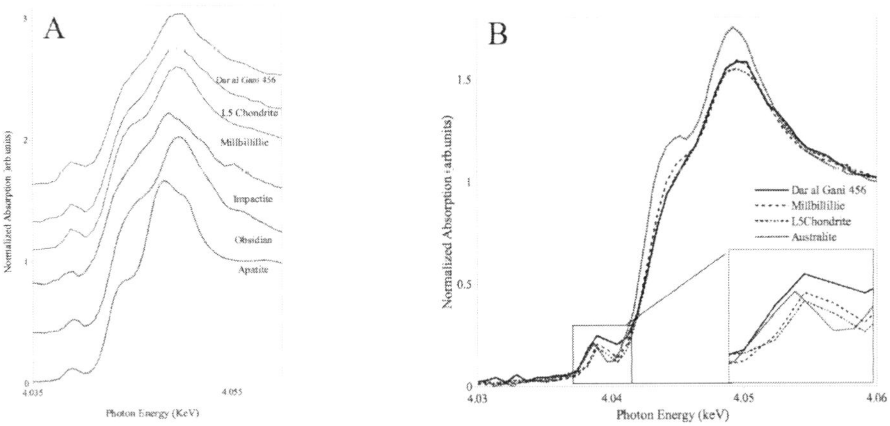


**Figure 1**. (A) Experimental Ca XANES spectra of meteorite fusion crusts and reference materials. (B) Detailed comparison of XANES spectra of Ca in Dar al Gani 456, L5chondrite, Millbillillie and australite tektite.

Comparing to previous study of Wang et al. [12], Ti XANES spectra of fusion crusts are similar to that of moldavite-green tektite (figure 2A). Shape of Ti XANES spectra of fusion crusts in Millbillillie and Allende are similar to moldavite-green. However XANES spectra of fusion crusts are slightly different to that of tektite. The pre-edge peak intensity is related to Ti coordination number. The pre-edge peak intensities of fusion crusts are 22.6-28.7% (the peak top of spectra is defined as 100%) and

Journal of Physics: Conference Series **712** (2016) 012095 doi:10.1088/1742-6596/712/1/012095

are higher than that of moldavite-green tektite. The pre-edge peak intensity of fusion crusts indicates that fusion crusts have mixture of five- and six- coordinated Ti. Since coordination environment in Ti is related to formation temperature, it is presumed that forming temperature of fusion crusts is higher than that of moldavite-green tektite.

The Fe XANES spectra of fusion crusts (figure 2B) are different from tektites and impact glasses [14]. The local structures of Fe in fusion crusts have unique local structures. The threshold energy of Fe is strongly related to oxidation state of Fe ions. Since the threshold energy of moldvite-green tektite is lower than fusion crusts and impact glasses, the total charge of Fe ions in fusion crusts is higher than that in tektites. According to Wang et al. [14], the estimated Fe $^{3+}$/ ($Fe^{2+}$+$Fe^{3+}$) ratio in fusion crusts are 0.34-0.66 and that in tektites are 0.04-0.13. The estimated Fe $^{3+}$/ ($Fe^{2+}$+$Fe^{3+}$) ratio in fusion crusts are higher than that in tektite. Tektites are well known that they were produced in minimal oxidation environment such as space. The high estimated Fe $^{3+}$/ ($Fe^{2+}$+$Fe^{3+}$) ratio in fusion crusts indicates that fusion crusts are formed into atmospheric oxidation condition. The threshold energy of fusion crusts have almost same as suevite impact glass because of meteorite fusion crusts formed in the atmosphere.

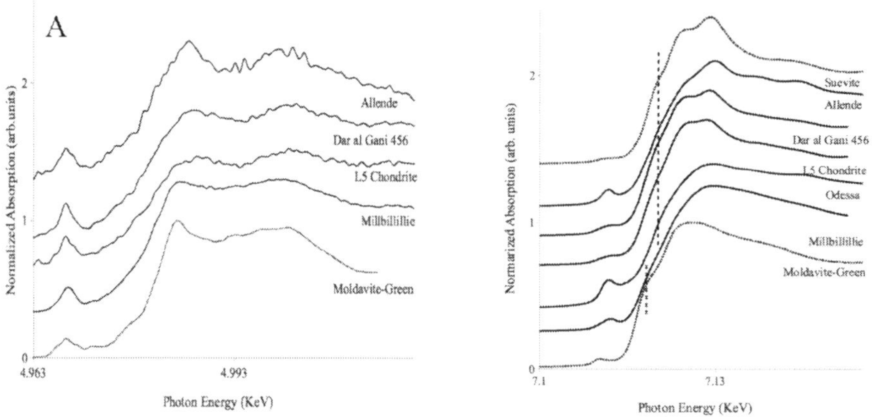


**Figure 2**. (A) Experimental Ti and (B) Fe spectra of meteorite fusion crusts and moldavite tektite. Solid lines are meteorite fusion crusts, dotted line is tektite and impact glass. Vertical dashed lines are threshold energy.

### *3.2 Radial distribution function and bonding distances of Ca, Ti and Fe*

The EXAFS $k^3\chi(k)$ functions were transformed into the radial distribution functions (RDF) for the Ca, Ti and Fe k-edge of fusion crusts, as shown in figure 3. In order to obtain information on the detailed structures parameters, we conducted parameter fitting with analytical EXAFS formulae. The obtained structural parameters of Ca, Ti, Fe are summarized table 1.

The Ca-O distances in fusion crusts are 2.61-2.66 Å and are extremely longer than that in tektites (2.4-2.419 Å) and impact glasses (2.438 Å) [11]. The fusion crusts have unique local structures since they were experienced extremely high temperature and short quenching time. Ca-O distance of Allende is much longer than that of other fusion crusts. Longer Ca-O distance indicated that forming temperature of Allende fusion crusts is higher than that of other fusion crusts.

The Ti-O distances in fusion crusts are 1.90-2.00 Å and are shorter than that in moldavite-green tektite (2.001 Å) . The estimated coordination number of Ti in meteorite fusion crusts are 5- and 6-coordinated Ti and are lower than that in moldavite-green tekite. Since coordination number of Ti has positive correlation to forming temperature, forming temperatures of fusion crusts are higher than that of moldavite-green tektite.

The Fe-O distances in fusion crusts are 2.02- 2.11 Å. The Fe-O distance in Odessa is the longest which is 2.11 Å. Wang et al. [14] reported Fe-O distances in tektites are 2.01-2.06 Å and Fe-O distance in natural glasses are 1.99-2.11 Å. The pre-edge intensity has positive correlation to Fe-O distance. The estimated coordination number in fusion crusts are mixture of 5- and 6- coordinated Fe and are higher than that in tektites.

The local structures of cation in fusion crusts are unique and the XAFS method is effective in distinction of fusion crusts and classification of natural glasses.

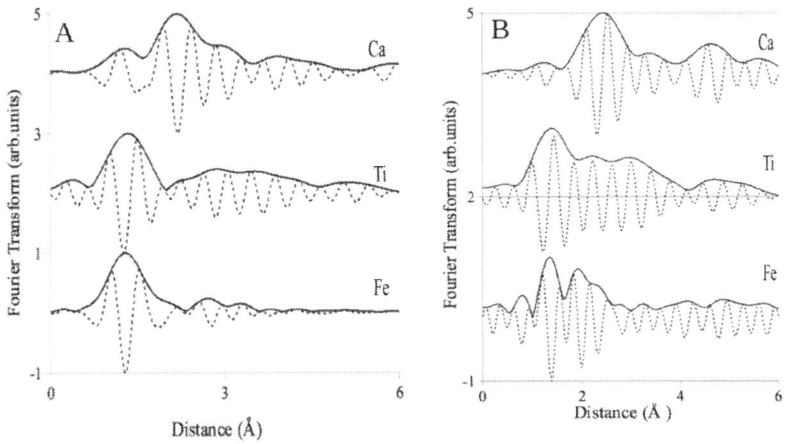


**Figure 3.** Fourier Transform of the Ca, Ti and Fe K-edge EXAFS oscillation function $k^3\chi(k)$ of Millbillillie (A) and L5 chondrite (B).

Table 1. The structural parameters determined by XAFS.

| Sample | Ca-O(Å) | Ca R-factor (%) | Ti-O (Å) | Ti R-factor (%) | Fe-O (Å) | Fe R-factor (%) |
|---|---|---|---|---|---|---|
| Allende | 2.780 (7) | 11.6 | 2.008 (3) | 5.8 | 2.021 (2) | 4.9 |
| Dar al Gani 456 | 2.665 (9) | 1.9 | 1.903 (1) | 2 | 2.016 (1) | 0.67 |
| L5 chondrite | 2.639 (11) | 6.7 | 1.906 (1) | 3.8 | 2.086 (1) | 9.4 |
| Millbillillie | 2.609 (4) | 0.6 | 1.921 (4) | 2.5 | 2.057 (1) | 0.53 |

Table2. Chemical composition of bulk meteorites and tektite (by EPMA). Chemical composition of Dar al Gani 456 (L6 chondrite) is not published. As a reference, we used the chemical composition of other L6 chondrite.

| Sample | | $SiO_2$ (wt%) | CaO (wt%) | $TiO_2$ (wt%) | $Al_2O_3$(wt%) | FeO (wt%) | Reference |
|---|---|---|---|---|---|---|---|
| Allende | | 34.2 | 2.71 | 0.16 | 3.18 | 22.5 | [1] |
| Dar al Gani 456 | Feldspar | 64.8 | 2.35 | - | 20.5 | 0.60 | [11] |
| | Hollandite | 63.8 | 2.92 | - | 20.7 | 1.01 | |
| L5 chondrite | Oligoclase | 63.4 | 2.41 | - | 22.4 | - | [8] |
| | Chrysolite | 36.0 | - | - | - | 24.6 | |
| Millbillillie | Pyroxene | 49.1-50.2 | 0.72-2.18 | 0.18-0.45 | 0.18-0.33 | 34.4-35.1 | [14] |
| | Plagioclase | 45.0-48.5 | 14.5-19.1 | 0.02-0.58 | 32.8-36.1 | 0.09-1.76 | |

| | | | | | | |
|---|---|---|---|---|---|---|
| Odessa | - | - | - | - | 60.6-62.4 | [3] |
| Australite | 71.4-72.2 | 3.27-3.42 | 0.62-0.86 | 12.75-13.27 | 4.73-5.04 | Measured |
| Impactite | 65.3-68.3 | 0.01-0.20 | 19.1-20.5 | 0.01-0.02 | 0.06-0.84 | [4] |
| Obsidian | 68.16-69.1 | 2.36-3.32 | 14.94-15.32 | 0.7-0.81 | 2.64-3.23 | Measured |

### *4. Acknowledgment*
*This study was performed under the auspices of the Photon Factory (PAC No. 2014G535). The synchrotron radiation XAFS experimental is performed at the beam line BL9C at the Photon Factory.*

**References**

[1] Elliott C J 1987 *Smithsonian Contributions of the Earth Science* **27** 49
[2] El Goresy A and Fechtig H 1967 *Astrophys.* **11** 391
[3] El Goresy 1967 *Geochemica et Cosmochimica Acta* **31** 1667
[4] French B M, Koeberl C, Gilmour I, Shirey B, Dons J A and Naterstad J, 1997 *Geochimica et Cosmochimica Acta* **61** 873
[5] Genge M J and Grady M M 1999 *Meteoritics and Planetary Science* **34** 341
[6] Neuville D R, Cormier L, Flank A, Briois V and Missiot D 2004 *Chemical Geology* **213** 153
[7] Mysen B and Neuville D 1995 *Geochimica et Cosmochimica Acta* **59** 325
[8] Paris E, Romano C and Wu Z 1995 *Physica B* **208** 351
[9] Seixas T M, Almeida B, Mendes M H, Salgueiro da Silva M A, Santos J F, Almeida F, Diogo Z, Lopes A B and Vieira JM 2013 *Materials Science Forum* **730-732** 170
[10] Stebbins J F and McMillan P 1989 *American Mineralogist* **74** 965
[11] Tobase T, Wang L, Yoshiasa A, Okube M, Nakatani T, Hayasaka Y and Isobe H 2013 *J. Phy. Conference Series* **430** 012070
[12] Tomioka N, Mori H and Fujino K 2000 *Geophysical Research Letters* **27** 3997
[13] Wang L, Yoshiasa A, Okube M and Takeda T 2011 *J. Synchrotron Radiation* **18** 855
[14] Wang L, Yoshiasa A, Hiratoko T, Hu Y, Arima H and Sugiyama K 2013 *J. Mineralogical and Petrological Sciences* **108** 288
[15] Yamaguchi A, Takeda H, Bogard D D and Garrison D 1994 *Meteoritics and Planetary Science* **29** 237
[16] Yarger J L, Diefenbacher J, Smith K H, Wolf G H, Poe B and McMillan P F 1995 *Science* **270** 1964
[17] Maeda H 1987 *Journal of Physical Society of Japan* **56** 2777

# Valence determination of rare earth elements in lanthanide silicates by $L_3$-XANES spectroscopy

**Antonina N Kravtsova[1], Alexander A Guda[1], Joerg Goettlicher[2], Alexander V Soldatov[1], Vladimir K Taroev[3], Anvar A Kashaev[4], Lyudmila F Suvorova[3], Vladimir L Tauson[3]**
[1] International Research Center "Smart materials", Southern Federal University, 5 Sorge Street, Rostov-on-Don 344090, Russia
[2] Karlsruhe Institute of Technology, ANKA Synchrotron Radiation Facility, Karlsruhe, Germany
[3] Vinogradov Institute of Geochemistry, Siberian Branch of R.A.S., Irkutsk, Russia
[4] Institute of the Earth's Crust, Siberian Branch of R.A.S., Irkutsk, Russia,

E-mail: akravtsova@sfedu.ru

**Abstract.** Lanthanide silicates have been hydrothermally synthesized using Cu and Ni containers. Chemical formulae of the synthesized compounds correspond to $K_3Eu[Si_6O_{15}]\cdot 2H_2O$, $HK_6Eu[Si_{10}O_{25}]$, $K_7Sm_3[Si_{12}O_{32}]$, $K_2Sm[AlSi_4O_{12}]\cdot 0.375H_2O$, $K_4Yb_2[Si_8O_{21}]$, $K_4Ce_2[Al_2Si_8O_{24}]$. The oxidation state of lanthanides (Eu, Ce, Tb, Sm, Yb) in these silicates has been determined using XANES spectroscopy at the Eu, Ce, Tb, Sm, Yb, $L_3$-edges. The experimental XANES spectra were recorded using the synchrotron radiation source ANKA (Karlsruhe Institute of Technology) and the X-ray laboratory spectrometer Rigaku R-XAS. By comparing the absorption edge energies and white line intensities of the silicates with the ones of reference spectra the oxidation state of lanthanides Eu, Ce, Tb, Sm, Yb has been found to be equal to +3 in all investigated silicates except of the Ce-containing silicate from the run in Cu container where the cerium oxidation state ranges from +3 (Ce in silicate apatite and in a KCe silicate with $Si_{12}O_{32}$ layers) to +4 (starting $CeO_2$ or oxidized $Ce_2O_3$).

## 1. Introduction

Synthesis and investigation of new silicates containing rare earth elements (REE) are of particular interest as they have a potential to determine the environmental conditions of mineral formation. From geochemical point of view such silicates may have significance as indicators of hydrothermal conditions of rare earth mineralization. REE-bearing silicates are also multi-functional group of novel compounds with increasing application areas. They have potential in photoluminescence, ion-exchange properties and electronics.

In the present investigation the synthesis of single crystals (up to 3 mm length) of REE-containing (REE = Eu, Ce, Tb, Sm, Yb) silicates were done under controlled conditions of formation [1-3]. Analysis of the chemical compositions and chemical structures of most of the synthesized silicates has already been performed [1-3].

In the paper we present the determination of the oxidation state of rare earth elements in the synthesized silicates on the basis of X-ray absorption near-edge structure (XANES) spectroscopy. XANES spectroscopy [4] is an effective method which allows obtaining information on electronic and local atomic structure around the absorbing type of atoms in materials in condensed state even without long-range order in atomic distribution [5]. In our research the oxidation state of lanthanides (Eu, Ce,

Tb, Sm, Yb) was determined by comparing the absorption edge energies and white line intensities of the $L_3$-edge XANES spectra of the studied silicates (with an unknown oxidation state of the lanthanides) and reference samples (with well known lanthanide oxidation state). This technique was shown to be very effective for the oxidation state of ions determination in complex solids [6]. These data coupled with the crystal structures characteristics may be useful for diagnostics of mineral formation processes.

## 2. Experimental

Lanthanide (Eu, Ce, Tb, Sm, Yb) containing phases have been hydrothermally synthesized in the Vinogradov Institute of Geochemistry (Irkutsk, Russia). Synthesis of Eu, Ce, Sm, Yb-containing silicates was performed in Ni containers where oxygen fugacity ($fO_2$) was controlled by buffer association Ni-NiO (nickel-bunsenite, NB). For comparison Tb, Ce-containing silicates were synthesized in Cu containers where oxygen fugacity is controlled by a Cu-$Cu_2O$ buffer. Experimental temperature was 500 ± 5 °C, the pressure in autoclaves was about 100 MPa. The duration of hydrothermal synthesis was 45 days. For Ce silicate synthesis in Cu-container three runs with differences in starting materials and in one case without $Cu_2O$ have been performed (Table 1). Chemical formulae of the compounds synthesized in Ni containers correspond to $K_3Eu[Si_6O_{15}] \cdot 2H_2O$, $HK_6Eu[Si_{10}O_{25}]$, $K_7Sm_3[Si_{12}O_{32}]$, $K_2Sm[AlSi_4O_{12}] \cdot 0.375H_2O$, $K_4Yb_2[Si_8O_{21}]$, $K_4Ce_2[Al_2Si_8O_{24}]$. Electron microprobe analyses of the Ce run products from syntheses in Cu-container have not yet been carried out but powder synchrotron X-ray diffraction (XRD) gives first information about formed crystal phases.

**Table 1.** Experimental conditions for the three Ce silicate syntheses in Cu containers.

| Experiment | Starting solution KOH [wt. %] | $SiO_2$ [g] | $Al_2O_3$ [g] | $Cu_2O$ [g] | Ce-oxide [g] | $CeF_3$ [g] | Buffer after run |
|---|---|---|---|---|---|---|---|
| Ce-Cu1 | 15.25 | 12.48 | 1.90 | 5.0 | 1.0 ($CeO_2$) | -- | Cu, $Cu_2O$ |
| Ce-Cu2 | 15.25 | 9.61 | 1.6 | 5.5 | 8.0 ($Ce_2O_3$) | -- | Cu only |
| Ce-Cu3 | 15.25 | 9.61 | 1,.6 | – | 8.02 ($Ce_2O_3$) | 0.5 | Cu only |

$L_3$-XANES spectroscopy at the SUL-X beamline of the synchrotron radiation source ANKA (Karlsruhe Institute of Technology (KIT)) has been applied to obtain valence state of Ce and Tb in the run products from REE (aluminium) silicate syntheses. The SUL-X beamline uses a wiggler as radiation source. A Si(111) crystal pair installed in a fixed exit double crystal monochromator provided the monochromatic radiation. ANKA was operating at 2.5 GeV electron energy at electron current of 100 – 150 mA. X-ray energy was calibrated for the Ce $L_3$-edge with a Cr and for Tb $L_3$-edge with a Co metal foil to 5989 eV and 7009 eV, respectively (first inflection point). Calibration spectra were measured simultaneously with the sample spectra. Spot size at sample positon was about 0.8 mm x 0.8 mm (KB mirrors collimated). For small aggregates the beam has been focused to about 0.15 mm (hor.) x 0.1 mm (vert.) at the sample position.

Ce containing silicates have been selected as single crystals from the run in Ni container ($K_4Ce_2[Al_2Si_8O_{24}]$) [2] and as single crystals and polycrystalline aggregates from the runs in Cu containers. Samples were pressed to pellets when sufficient material could be selected or fixed on Kapton tape and measured in transmission using ionization chambers as detectors and at the same time in fluorescence mode (beam – sample – detector angles: each 45°) with a 7 element Si(Li) solid state detector (SGX Sensortech, former Gresham). Transmission data were preferred when concentrations of the absorbing element were high like for Ce sample in Ni-container and the Tb sample, whereas for some Ce containing single crystals and polycrystalline aggregates of the run in Cu-containers fluorescence data gave better signal to noise ratio. Gas fillings of the first, second and third ionization chamber were - for the Ce $L_3$-edge 600 mbar $N_2$, 370 and 630 mbar Ar20%N80%, and for the Tb $L_3$-edge 200 and 630 mbar Ar20%N80% and 266 mbar Ar. For measurements in fluorescence mode Ce and Tb $L\alpha$ emission intensities have been recorded between 4440 and 5160 eV, and 5910 and 6550 eV, respectively. Pre-edge background have been measured 150 eV before the edge with 5 and 2 eV energy step widths, across the edge (-20 to +20 eV) with 0.3 eV step width (all counting times 1s per step) and above the edge up to $k$ 10.8 for Ce and 13.9 for Tb with square root of $k$ as counting time. Edge energies for Ce $L_3$ and Tb $L_3$ were set to 5.723 and 7.514 keV, respectively. Pre-edge, post-edge corrections and edge jump normalization have been performed with the ATHENA code [7]. Data quality of the spectra was not sufficient for EXAFS analysis because focus was put on screening the XANES region of many aggregates. Hence, measurement parameters were optimized for fast data accumulation giving sufficient signal to noise ratio just in the XANES region. Moreover, EXAFS data evaluation is not possible in the cases of polycrystalline aggregates that consist of at least two Ce containing phases, and is difficult in the cases where measurements had to be done on single crystals directly due to orientation effects because sufficient material was not available for grinding and preparing pellets.

Eu $L_3$-, Ce $L_3$-, Sm $L_3$-, Yb $L_3$-XANES spectra of investigated silicates and reference samples were also recorded using X-ray absorption laboratory spectrometer Rigaku R-XAS. XANES spectra were measured in the transmission mode. A tungsten cathode and a Ge (311) crystal monochromator were used. X-ray tube voltage was 20 kV and current was 70 mA. A gas-filled proportional counter (argon gas at a pressure of 300 mbar) was used to register the intensity of the incident radiation; while a scintillation detector was used to register transmitted radiation.

Additionally, powder synchrotron X-ray diffraction (XRD) at the SUL-X beamline (Si(111) crystal pair as monochromator) was used to identify crystal phases from the Ce silicate syntheses in Cu containers where electron microprobe analysis have not yet been carried out. A CCD detector (XDR VHR-2 150, Photonic Science) have been used in symmetric transmission geometry in 16 bit fusion mode with 2 x 2 binning with a beam focused to about 150 µm (hor.) x 100 µm (vert.) at the sample position. Lattice plane distances have been calibrated with $LaB_6$ (NIST 660b), and CCD diffraction images were radially integrated to 1D XRD pattern using the Fit2D software [8]. With the given energy (17 keV, about 0.73 Å) and sample detector distance (about 106 mm) the lower limit of d-values range down to about 1.2 Å. The upper limit of d values was about 15 Å due to the primary beam stop. Further data evaluation for crystal phase identification has been performed with the DIFFRAC.EVA software (V 4.1.1, Bruker) [9] and the ICDD powder diffraction database.

## 3. Results and discussion

The positions of the white line of the Ce $L_3$ spectrum of the single crystals from the runs in Ni and Cu containers are close to the white line position of the $Ce(III)(NO_3)_3$ reference spectrum whereas the polycrystalline aggregates from the runs in Cu containers contain Ce(IV) besides Ce(III) in different fractions, even in the runs with $Ce_2O_3$ and $CeF_3$ as starting material (Figure 1b,c). Single crystals in general show only Ce(III). Ce(IV) in the fine and coarse grained material represent unreacted or dissolved and re-precipitated $Ce(IV)O_2$ verified by XRD for the run Ce-Cu1 or is due to oxidized $Ce_2O_3$ starting material for the runs Ce-Cu2 and Ce-Cu3 (Table 1, Figure 1b,c,d). Ce(III) in the fine and coarse grained material as well as in single crystals (Figure 1b,c) are most probably incorporated into silicate structures.

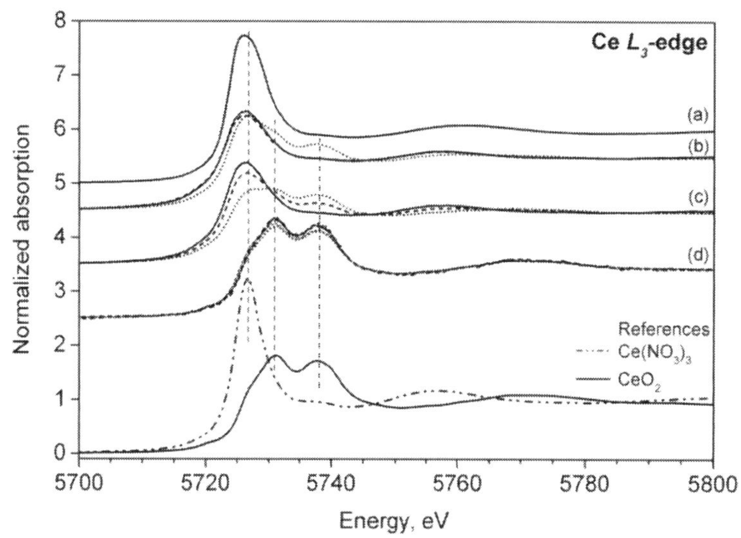


**Figure 1.** Ce $L_3$-XANES spectra of run products from synthesis in Ni-container ($K_4Ce_2[Al_2Si_8O_{24}$ (a)) and from runs in Cu-containers Ce-Cu3 (b), Ce-Cu2 (c), and Ce-Cu1 (d). Spectra of single crystals (solid lines), of coarse polycrystalline (dashed lines) and fine polycrystalline aggregates (dotted lines) are compared with Ce(III)$(NO_3)_3$ (dash-double dotted line) and Ce(IV)$O_2$ (solid line) reference spectra.

XRD confirms $CeO_2$ for the sample fractions where Ce(IV) has been detected in the Ce $L_3$-XANES spectra (Figure 2). Other Bragg reflexions matches with KAl feldspar ($KAlSi_3O_8$, most probably as the Al, Si disordered sanidine phase), and sometimes with $KAlSiO_4$ (kalsilite). The latter is not shown here. The remaining patterns of the Bragg peaks of the run products from Cu containers (Ce-Cu2, Ce-Cu3) are very similar to $Ce_{9.33}(SiO_4)_6O_2$ which has apatite structure [10]. Due to the presence of K during synthesis also a silicate apatite with $KCe_9(SiO_4)_6O_2$ composition analogous to $KPr_9(SiO_4)_6O_2$ [11] is under discussion, and has to be verified by electron microprobe analysis. A KCe silicate apatite has not been described so far. Ce is 3+ in the apatite structure which is confirmed also by the Ce(III) whiteline increases in the Ce $L_3$-XANES spectra of the sample fractions were Bragg reflexion of the silicate apatite are intense and reflexions of the $CeO_2$ are low in the X-ray diffractograms (Figure 2). The apatite structure is also supported by the hexagonal habitus of some of the single crystals detected in the run products. Single crystal structure investigation has been started for verification. $KPr_9(SiO_4)_6O_2$ is mentioned as a potential oxide ion conductor [11]. Recently silicate apatites have been synthesized with Na, Ca and Eu, Gd, Sm, and also K and Pr [12], and Li and Eu, Ce, Tb [13]. Such compounds are discussed as new optical materials, e.g., as phosphors for application in solid-state lighting. Our investigations show that REE silicate apatites can also form under hydrothermal conditions (500 °C, 1 kbar) apart from the solid state and flux procedures described in [12,13].

The XRD of one aggregate of the Ce-Cu2 run is similar to the pattern of $K_8Nd_3(Si_{12}O_{32})(OH)$ (ICDD, pdf No. 49-53 orr 80-2178) (Figure 3). Such type of K-REE silicates with $Si_{12}O_{32}$ layers have recently been described also for K-Sm and K-Tb silicates ($(K_{7.814}Sm_3Si_{12}O_{32}(OH)_{0.814} \cdot 0.77H_2O$, $K_{7.04}Tb_3Si_{12}O_{32.02} \cdot 1.36H_2O$) [14]. The new compound has probably the composition $K_7Ce_3(Si_{12}O_{32}) \cdot nH_2O$ [this study, and personal communication, Sergey Aksenov]. The corresponding Ce $L_3$-XANES spectrum of the polycrystalline aggregate shows some Ce(IV) because $CeO_2$ is present besides the KCe silicate with $Si_{12}O_{32}$ layers where Ce is +3 (Figure 3).

Only Ce(IV) have been found in the run Cu-Cu1 (Figure 1d) where few large fragments do not show crystal faces and are gel like amorphous products. XRD of the aggregates that are the main fraction in this run shows $CeO_2$ and a KAl feldspar ($KAlSi_3O_8$). This run was the only one in Cu-containers with $CeO_2$ as starting material and where $Cu_2O$ added as buffer was still present after the run. High oxygen fugacity may have prevented Ce(IV) from reduction and hence from incorporation into silicates, and just $KAlSi_3O_8$ feldspar had formed.

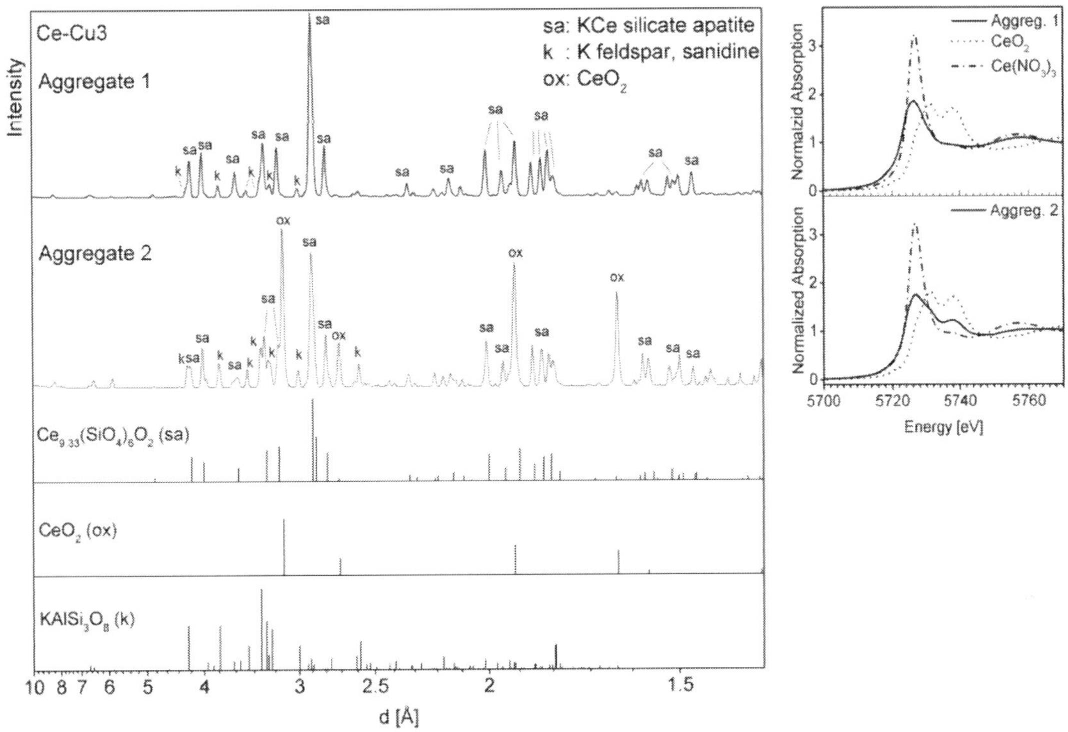


**Figure 2.** Powder synchrotron X-ray diffractograms and Ce $L_3$-XANES spectra of products from Ce silicate synthesis in Cu container (Ce-Cu3) resulting in $CeO_2$, K-feldspar and a Ce or KCe silicate apatite and Ce(IV) and Ce(III) fractions for polycrystalline aggregate 2; and K-feldspar and (K)Ce silicate apatite and only Ce(III) for aggregate 1. For the latter, the absence of Ce(IV) in the XANES data is in agreement with the absence of $CeO_2$ in the XRD and proves that Ce is +3 in the (K)Ce silicate apatite. Reference line patterns for XRD are $Ce_{9.33}(SiO_4)_6O_2$ (ICDD pdf No. 43-443), $CeO_2$ (34-394)and $KAlSi_3O_8$ (sanidine) (25-618); and for the Ce $L_3$-XANES spectroscopy $CeO_2$ and $Ce(NO_3)_3$ are references for Ce(IV) and Ce(III), respectively.

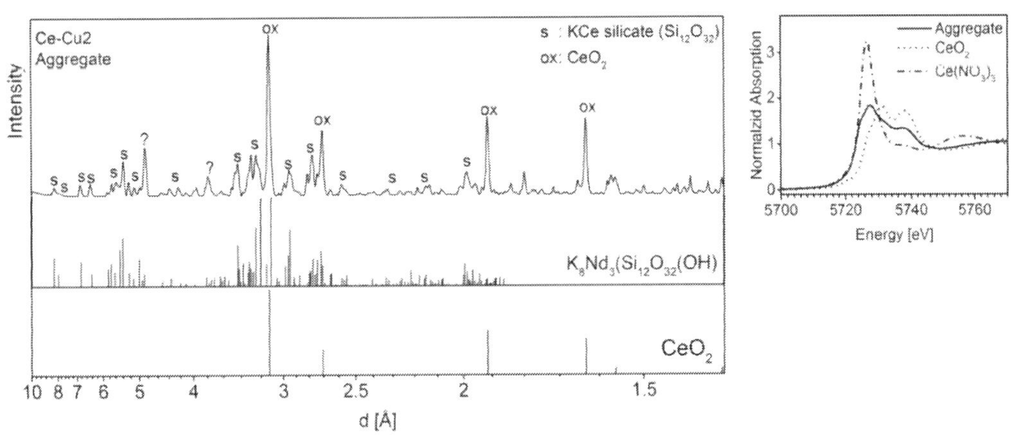


**Figure 3.** Powder synchrotron X-ray diffractogram and corresponding Ce $L_3$-XANES spectrum of an aggregate from Ce silicate synthesis in Cu container (Ce-Cu2). The X-ray diffractogram shows $CeO_2$ and a pattern similar to the one of $K_8Nd(Si_{12}O_{32})OH$ (ICDD, pdf No. 49-53 or 80-2178, here 80-2178). The Ce $L_3$-XANES spectrum confirms the Ce $^{4+}$ of $CeO_2$. From Ce $L_3$-XANES analysis and the structure of such silicates Ce is assumed to be 3+.

Tb ion oxidation state has been determined to be equal to +3 because the white line of the spectrum of the Tb polycrystalline aggregates from the run in Cu-container is close to the lower energy position of the two white lines of a Tb(III,IV) mixed oxide (Figure 4; [15]). This is a similar result as for Ce in some of the coarse aggregates and single crystals from runs Ce-Cu2 and Ce-Cu3 (Figure 1b,c). The Tb silicate has already been determined to $K_7Tb_3Si_{12}O_{32} \cdot 1.35\ H_2O$ in [14].

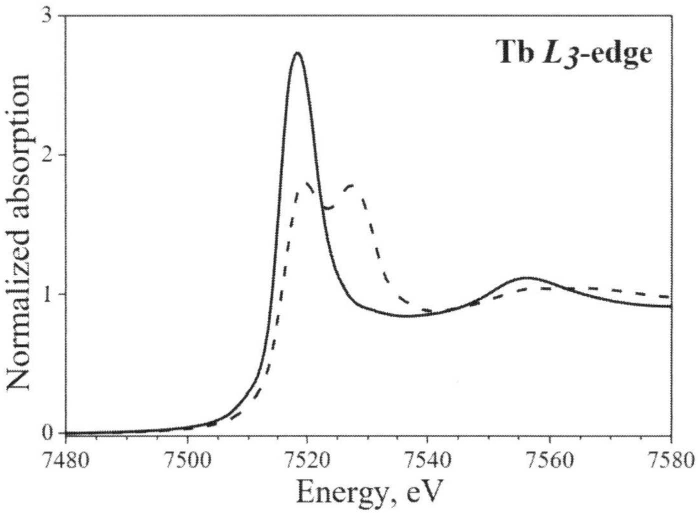


**Figure 4.** Tb $L_3$-XANES spectra: Run product from synthesis in Cu-container ($K_7Tb_3Si_{12}O_{32} \cdot 1.35\ H_2O$) [14] (solid line), $TbO_x$ (x=1.5-2) reference (dashed line).

In Figure 5 we present a comparison of the experimental Eu $L_3$-XANES spectra of $K_3Eu[Si_6O_{15}] \cdot \cdot 2H_2O$, $HK_6Eu[Si_{10}O_{25}]$ and the ones of reference substances, $Eu^{2+}S$ with oxidation state +2 and $Eu^{3+}F_3$, $Eu^{3+}{}_2O_3$ with oxidation state +3. It can be seen that the energy of the main maximum of Eu $L_3$-XANES spectra is ~6983.2 eV for $EuF_3$, $Eu_2O_3$ (Eu oxidation state is +3 ) and ~6975.2 for EuS (Eu oxidation state is +2). The main maximum energies of Eu $L_3$-XANES spectra of $K_3Eu[Si_6O_{15}] \cdot 2H_2O$ and $HK_6Eu[Si_{10}O_{25}]$ silicates are consistent with the main maximum energy of the $Eu^{3+}{}_2O_3$. So, it is possible to conclude that Eu in the $K_3Eu[Si_6O_{15}] \cdot 2H_2O$, $HK_6Eu[Si_{10}O_{25}]$ synthesized silicates has the oxidation state +3.

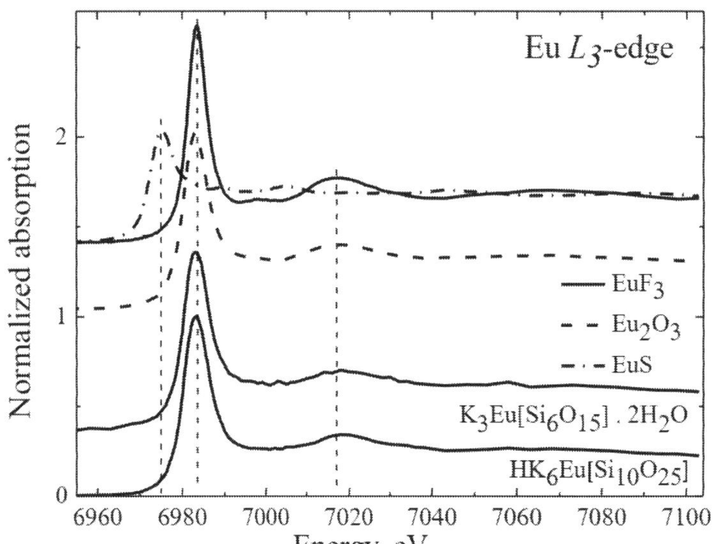


**Figure 5.** Comparison of the experimental Eu $L_3$-XANES spectra of $K_3Eu[Si_6O_{15}] \cdot 2H_2O$, $HK_6Eu[Si_{10}O_{25}]$ silicates and $Eu^{3+}F_3$ (upper solid line), $Eu^{3+}{}_2O_3$ (dashed line), $Eu^{2+}S$ (dash-dotted line) reference spectra.

On the basis of the analysis of the edge energies of the Sm $L_3$-, Yb $L_3$-XANES spectra it was found that the oxidation state of Sm and Yb in $K_7Sm^{3+}{}_3[Si_{12}O_{32}]$, $K_2Sm^{3+}[AlSi_4O_{12}] \cdot 0.375H_2O$, $K_4Yb^{3+}{}_2[Si_8O_{21}]$ silicates is equal to 3+ [16].

## 4 Conclusions

The oxidation state of rare-earth elements (REE=Eu, Ce, Tb, Sm, Yb) in new synthesized REE-containing silicates has been determined using XANES spectroscopy near REE $L_3$-edges. Comparison of the absorption edge energies of the silicates with the ones of reference substances allowed to determine the oxidation state of Eu, Ce, Tb, Sm, Yb in all investigated silicates equal to +3 except for the Ce-containing silicate synthesized in Cu containers where Ce was found also in oxidation state +4 most probably due to non-reacted $CeO_2$ or oxidized $Ce_2O_3$ starting material that has not been incorporated into silicates. A Ce or KCe silicate apatite has been discovered from two runs in Cu-containers.

We are incredibly sad to report that Vladimir Taroev died suddenly on January 28, 2016. He was an experimental mineralogist and petrologist by heart with his specialty of hydrothermal synthesis of feldspars and new REE silicates.

## Acknowledgments

The work is supported by the grant of the Russian Foundation for Basic Research No. 14-05-00580 and No. 16-05-00392.

## References

[1] Aksenov S M, Rastsvetaeva R K, Rassylov V A, Bolotina N B, Taroev V K and Tauson V L (2013) *Micropor. Mesopor. Mater.* **182** 95-110
[2] Aksenov S M, Rassylov V A, Rastsvetaeva R K and Taroev V K (2013) *Crystallography Reports* **58**, N 6 835-41
[3] Rastsvetaeva R K, Aksenov S M and Taroev V K (2010) *Crystallography Reports* **55**, 6 1041-49
[4] Bunker G 2010 *Introduction to XAFS. A Practical Guide to X-ray Absorption Fine Structure Spectroscopy* (Cambridge University Press)
[5] Mansour A N, A. Marcelli A, Cibin G, G. Yalovega G, Sevastyanova T and Soldatov A V (2002) *Phys Rev B* **65** 134207
[6] Vitova T, K. Kvashnina K O, Nocton G, Sukharina G, Denecke M A, Butorin S M, Mazzanti M, Caciuffo R, Soldatov A, Behrends T and Geckeis H (2010) *Phys Rev B* **82** 235118
[7] Ravel B, Newville M (2005) *J. Synchrotron Rad.* **12** 537–41
[8] Hammersley A P, Svensson S O, Hanfland M, Fitch A N, Hausermann D (1996) *High Pressure Res.* **14**, 235-48
[9] BRUKER (2015): DIFFRAC.EVA. V 4.1.1 - https://www.bruker.com/products/x-ray-diffraction-and-elemental-analysis/x-ray-diffraction/xrd-software/eva/overview.html
[10] Belokoneva E L, Petrova T L, Simonov M A, Belov N V (1972) *Soviet physics, Crystallography* **17** 429-31
[11] Werner F, Kubel F. (2005) *Materials Letters*, **59**, 3660–5
[12] Latshaw A M, Hughey K D, Smith M D, Yeon J, zur Loye H-C (2015) *Inorg. Chem.* **54**(3) 876–84
[13] Kim D, Park D, Oh N, Kim J, Jeong E D, Kim S-J, Kim S, Park J-C (2015) *Inorg. Chem.* **54** 1325–36
[14] Taroev V K, Kashaev A A, Malcherec T, Goettlicher J, Kaneva E V, Vasiljev A D, Suvorova L F, Suvorova D S and Tauson V L (2015) *Solid State Chemistry* **227** 196-203
[15] Potdevin A, Chadeyron G, Briois V, Leroux F and Mahiou R (2010) *Dalton Trans.* **39** 8718–24
[16] Kravtsova A N, Guda A A, Soldatov A V, Goettlicher J, Taroev V K, Kashaev A A, Suvorova L F and Tauson V L (2015) *Optics and Spectroscopy* **119**, N 6 982-86

Journal of Physics: Conference Series **712** (2016) 012097 doi:10.1088/1742-6596/712/1/012097

# Weathering and precipitation after meteorite impact of Ni, Cr, Fe, Ca and Mn in K-T boundary clays from Stevns Klint

**Yumiko Miyano*[1], Akira Yoshiasa[1], Tsubasa Tobase[1], Hiroshi Isobe[1], Hidetomo Hongu[1], Maki Okube[2], Akihiko Nakatsuka[3], Kazumasa Sugiyama[4]**

[1] Graduated School of Science and Technology, Kumamoto Univ., Kumamoto, Japan
[2] IMR, Tokyo Institute of Technology, Yokohama, Japan
[3] Graduate School of Science and Engineering, Yamaguchi Univ. Ube, Japan
[4] Department of Institute for material research, Tohoku Univ. Sendai, Japan

E-mail: yoshiasa@sci.kumamoto-u.ac.jp

**Abstract**. Ni, Cr, Fe, Ca and Mn K-edge XANES and EXAFS spectra were measured on K-T boundary clays from Stevns Klint in Denmark. According to XANES spectra and EXAFS analyses, the local structures of Ni, Cr and Fe in K-T boundary clays is similar to $Ni(OH)_2$, $Cr_2O_3$ and FeOOH, respectively. It is assumed that the Ni, Cr and Fe elements in impact related glasses is changing into stable hydrate and oxide by the weathering and diagenesis at the surface of the Earth. Ca in K-T boundary clays maintains the diopside-like structure. Local structure of Ca in K-T clays seems to keep information on the condition at meteorite impact. Mn has a local structure like $MnCO_3$ with divalent state. It is assumed that the origin on low abundant of Mn in the Fe-group element in K-T clays was the consumption by life activity and the diffusion to other parts.

## 1. Introduction

K-T boundary clays were formed in between the Cretaceous (Kreide) and Tertiary geological times. As they are composed of black and reddish brown clays with different chemical composition compared with surrounding sedimentary rocks, they stand out clearly in many localities. These clays contain relatively high Pt-group element concentration including iridium [1] and Fe-group element concentration except for Mn [2]. The Fe-group elements increased extremely are believed to originate from the meteorite itself and from impact ejecta, but the accumulation of Mn isn't admitted in meteorite impact origin sediments. Additionally, the abnormal concentration of As and Zn is reported [2]. Sakai et al., [3] and Okube et al., [4] presumed sedimentation process and environment after meteorite impact based on the local structure analyses around As and Zn in K-T clays by XAFS experiments, respectively. Metastable compounds and glasses formed by impact will be weathered. Elements ionized by weathering precipitate as stable compounds. Precipitates should also be affected by diagenesis process. Records of various unique formation processes are sometimes left in the local structure around minor and trace elements. The thermal quenching history at meteorite impact was kept preserved as the local structure of Zr in K–T clays [5]. Local structure is affected by several conditions during the formation process [6-9]. We have determined the local structure of Ni, Cr, Fe, Ca and Mn contained in K-T boundary sediments by using X-ray absorption spectroscopy in order to understand their origin.

## 2. Experimental

The specimen of K-T boundary clays is from Stevns Klint in Denmark. Analytical point was the same layer as that studied by Okube et al., [4]. The layer consists of high transition metals concentration including Pt-group elements, Cr and Ni. In order to study the local structure around Ni, Cr, Fe, Ca and Mn atoms in K-T boundary clays, we prepared several reference materials NiAsS, NiO, $Ni(OH)_2$, $NiF_2$ (related structure to $Ni(OH)_2$), chromite, $Cr_2O_3$, FeO, $Fe_3O_4$, $\alpha$-$Fe_2O_3$ (hematite), $\alpha$-FeOOH (goethite), CaO, $CaTiO_3$, $CaMgSi_2O_6$ (diopside), calcite, aragonite, vaterite, apatite, $MnCO_3$, $Mn_2O_3$ and $MnO_2$. The Ni, Cr, Fe, Ca and Mn K-edge XAFS measurements were performed at beam line BL-9C, equipped with Si (111) double-crystal monochromator, of the Photon Factory, KEK, Tsukuba, Japan. Spectra were taken in transmission mode and fluorescence mode using Lytle-type detector (e.g. Fe K-edge from 6709.8 to 8109.5 eV with step from 0.855 to 7.604 eV and 1-s to 5-s counting time). X-ray energy calibration was performed by setting the copper metal pre-edge absorption peak to 8978.8 eV. Analysis of XAFS data was performed by using XAFS93 programs [10], whose details are given by Yoshiasa et al. [11].The radial structural function was obtained by the Fourier transform over the k range of $2.5 < k < 9.0$-$10.5$ Å$^{-1}$ for the K-T clay sample. We carried out the Fourier filtering technique and a non-linear least-squares structure parameters fitting method with an analytical EXAFS formula [10]. The single-shell fitting was carried out for first nearest neighbor distance (Table 1).

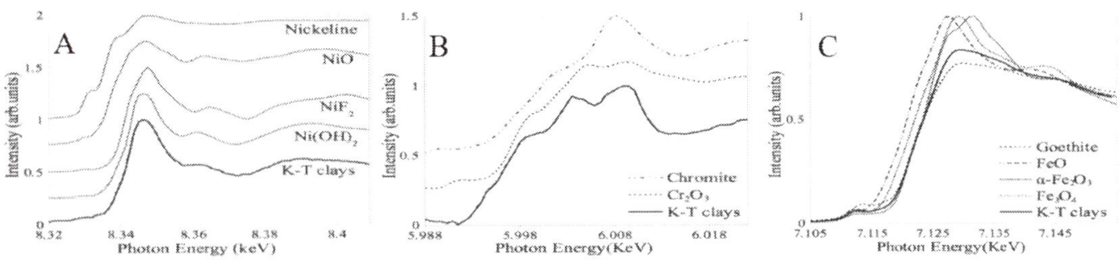


**Figure 1.** (A) Experimental Ni, (B) Cr and (C) Fe spectra in KT-clays and reference materials.

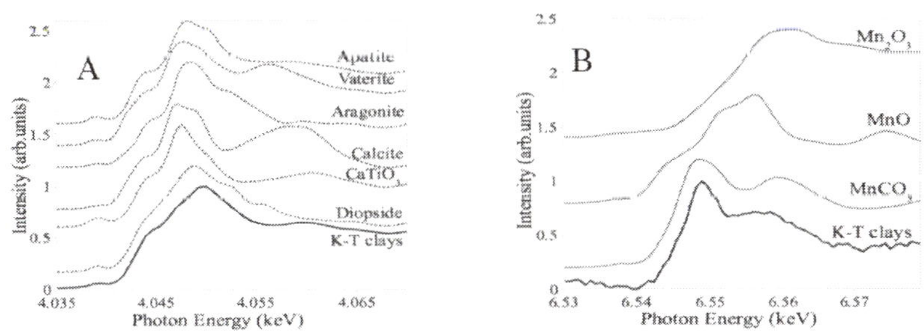


**Figure 2.** (A) Experimental Ca and (B) Mn spectra in KT-clays and reference materials.

## 3. Results and discussion

### *3.1 XANES spectra of Ni, Cr, Fe, Ca and Mn*

The Ni and Cr K-edge XANES spectra are shown in Figs 1A and 1B. The Ni K XANES spectrum in K-T clays is in good agreement with $Ni(OH)_2$. The Cr XANES spectrum in K-T clays is most similar to $Cr_2O_3$. The Fe K-edge XANES spectra are shown in Fig 1C. The Fe XANES spectrum in K-T clays is most similar to goethite. The XANES pattern, threshold energy and pre-edge peak heights of K-T clays is middle of goethite and hematite. Fe ions were oxidized completely up to trivalent.

The Ca XANES spectrum in K-T clays (Fig 2A) is different from any single phase natural glass [12]. Calcium is included in several kinds of minerals and non-crystalline materials. The Ca XANES spectrum in K-T clays is most similar to diopside which is found in igneous and metamorphic rocks. It is formed under high temperature and weathers under the Earth's surface conditions. The detailed spectrum may be able to explain by a mixture of several minerals such as diopside and $CaTiO_3$

perovskite. The XAFS method is effective in structure analyses of non- and low-crystalline materials in particular. We consider that large parts of Ca local structure in K-T clays maintain diopside-like structure and keep information on before weathering process. The Ca local structure in K-T clays is different from constituent mineral in the lower and upper deposited sediments.

The Mn XANES spectrum in K-T clays (Fig 2B) is most similar to $MnCO_3$. We consider that Mn in K-T clays was ionized by weathering of metastable materials formed under high temperature and precipitated as $MnCO_3$. The valence of Mn is divalent and different from highly oxidised Fe. It should be indicated that Mn is easily incorporated into life body as with calcium carbonate.

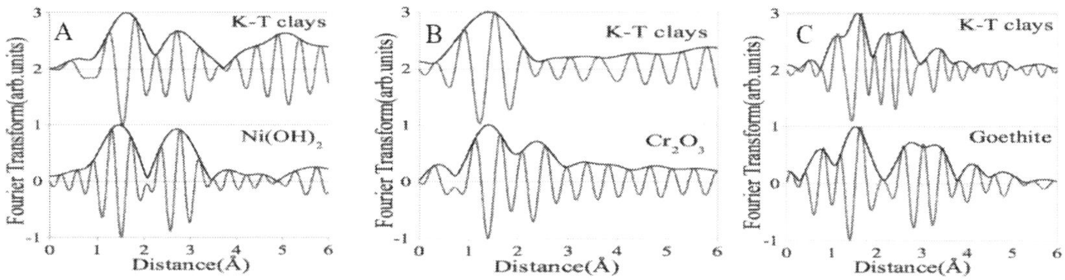


**Figure 3**. Fourier Transform of the (A) Ni, (B) Cr and (C) Fe K-edge EXAFS oscillation function $k^3\chi(k)$ of K-T clays and reference materials.

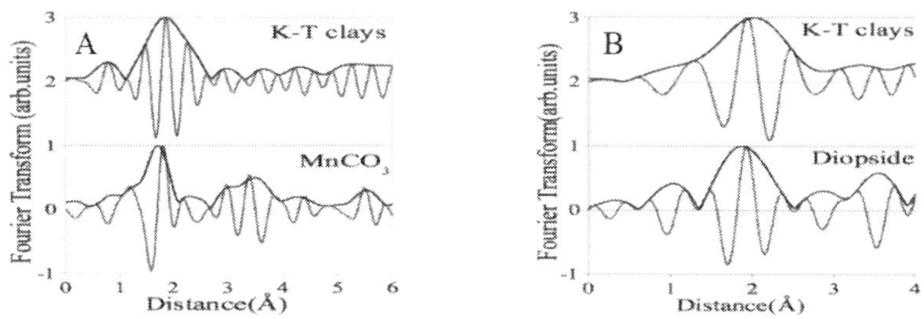


**Figure 4**. Fourier Transform of the (A) Mn and (B) Ca K-edge EXAFS oscillation function $k^3\chi(k)$ of K-T clays and reference materials.

Table.1 The Cation-O bonding distances determined by EXAFS.

| | Ni-O (Å) | Cr-O (Å) | Fe-O (Å) | Ca-O (Å) | Mn-O (Å) |
|---|---|---|---|---|---|
| K-T clays | 2.126(5) | 1.957(1) | 2.015(13) | 2.459 (11) | 2.145 (2) |
| reference | $Ni(OH)_2$ | $Cr_2O_3$ | goethite | $CaMgSi_2O_6$ | $MnCO_3$ |
| minerals | 2.065(4) | 1.981(6) | 2.009(2) | 2.425(11) | 2.186 (10) |

*3.2 Radial distribution function and bonding distance of Ni, Cr, Fe, Ca and Mn*

Figures 3 and 4 show the radial distribution functions for the Ni, Cr, Fe, Ca and Mn K-edge of K-T clays and reference materials. The obtained first nearest neighbor distances for each cation are summarized with reference materials in Table 1. The Ni-O (2.126(5) Å) and Cr-O (1.957(1) Å) distances in K-T clays are similar to those in $Ni(OH)_2$ and $Cr_2O_3$, respectively. The Fe-O distance in K-T clays is 2.015(13) Å and is consist with that in goethite (2.004(15) Å by EXAFS) which has two crystallographically non-equivalent Fe-O distances (2.15 and 1.91 Å). The Fe-O distance in K-T clays is similar to typical octahedral coordinated $Fe^{3+}$-O bonding distance. The Ca-O distance in K-T clays (2.459(11) Å) is same distance in $CaMgSi_2O_6$ diopside. The Mn-O distance in K-T clays (2.145(2) Å) is similar to 2.186(10) Å in $MnCO_3$.

*3.3 Discussion*

According to XANES spectra and EXAFS analyses, the local structures of Ni and Cr in K-T boundary clays are similar to $Ni(OH)_2$ and $Cr_2O_3$, respectively. Most of Ni and Cr is a meteorite origin [2]. It is guessed that the Ni, Cr, Fe elements is changing into stable hydrate and oxide by the weathering and diagenesis near the surface of the Earth. The chromium hydrate $Cr(OH)_3$ is unstable under conditions of the Earth's surface and changed into $Cr_2O_3$ in K-T clays. When a meteorite impact occurs on the surface of the Earth, the Earth's crust rock and meteorite are melted under extremely high temperature and the melt blow off in the sky and space. The solidified melt under reduced conditions [13] flows on the surface of the Earth as a micro-tektite and ash with glass structure. Fe-group elements in glass structure react to water and oxygen by weathering and diagenesis in the Earth and changes into a hydrate and oxide. These are a usual process at the surface of the Earth. The record of the unique local structure by meteorite impact is erased by the weathering and precipitation processes. Sedimentation process and environment after impact were inferred from local structure of As and Zn by XAFS methods. The concentration of As has happened by the high oxidation state of As and co-precipitate with Fe ion [3]. Tobase et al. [5] showed that the thermal quenching history at meteorite impact was left in the local structure of Zr in K–T clays. The local structure of Ca in K-T clays seems to keep information on the situation at meteorite impact. It was shown that the existence of stable phase, sedimentation process (weathering, precipitation and diagenesis) and life activity affect to local structure of guest and minor elements.

We found out that Mn in K-T boundary clays have a local structure like manganese carbonate with divalent state. The extreme accumulation of Mn is not admitted in K-T clays. We guess that the manganese local structure in K-T clays was formed by activities at the Earth's surface (it may be precipitation by life activity) via weathering of metastable solids formed by impact and ionization by dissolution from solids. Mn is one of the indispensable elements for life while there is toxicity in Ni and Cr. Manganese can also be co-precipitated with calcium ion as a minor impurity element. Concentration of the high Fe-group element like Ni, Cr and Fe would be the meteorite origin. Mn is also included of course. It can assumable that manganese will be consumed by life activity and scattered to other parts.

## References

[1] Alvarez, L. W., Alvarez, W., Asaro, F. and Michel, H., V., 1980 *Science* **208** 1095.

[2] Strong, C.P., Brooks, R.R., Wilson, S.M., Reeves, R.D., Orth, C.J., Mao, X.-Y., Quintana, L.R., Anders, E., 1987 *Geochimica et Cosmochimica Acta* **51** 2769

[3] Sakai, S., Yoshiasa, A., Arima, H., Okube, M., Numako, C. and Sato., T., 2007 *AIP Conference Proceedings* **882** 274.

[4] Okube, M. Sasaki, S., Yoshiasa, A., Wnag, L., Nakatani, T., Hongu, H., Murai, K. Nakatsuka, A. and Miyawaki, R., 2012 *Journal of Mineralogical and Petrological Sciences* **107** 192.

[5] Tobase, T. Yoshiasa, A., Wang, L. Hiraroko, T., Hongu, H., Okube, M. and Sugiyama, K., 2015 *Journal of Mineralogical and Petrological Sciences* **110** 88

[6] Stebbins, J. F. and McMillan, P., 1989 *American Mineralogist* **74** 965

[7] Paris, E. Dingwell, D.B., Seifert, F.A., Mottana, A. and Romano, C., 1994 *Physics and Chemistry of Minerals* **21** 520

[8] Mysen, B. and Neuville, D., 1998 *Geochimica et Cosmochimica Acta* **59** 325

[9] Yarger, J.L., Diefenbacher, J., Smith, K.H., Wolf, G.H., Poe, B., and McMillan, P.F., 1995 *Science* **270** 1964

[10] Maeda, H., 1987 *Journal of the Physical Society of Japan* **56** 2777.

[11] Yoshiasa, A., Nagai, T., Ohtaka, O., Kamishima, O. and Shimomura, O., 1999 *J. Sychrotron Rad.* **6** 43.

[12] Tobase, T., Wang, L., Yoshiasa, A., Okube, M., Nakatani, T., Hayasaka, Y. and Isobe, H., 2013 *Journal of Physics: Conference Series* **430** 012070 (4pp).

[13] Wang, L. Yoshiasa, A., Okube, M., Hiratoko, T., Hu,Y., Isobe, H., Arima, H. and Sugiyama, K., 2013 *Journal of Mineralogical and Petrological Sciences* **108** 288

# Fission products behaviour in $UO_2$ submitted to nuclear severe accident conditions

**E Geiger[1], R Bès[2], P Martin[3], Y Pontillon[1], P L Solari[4] and M Salome[5]**
[1] CEA/DEN/CAD/DEC/SA3C/LAMIR, Saint Paul lez Durance, France
[2] Department of Applied Physics, Aalto University, Espoo, Finland
[3] CEA/DEN/MAR/ DTEC/SECA/LCC, Bagnols sur Cèze, France
[4] Soleil Synchrotron, Paris, France
[5] ESRF Synchrotron, Grenoble, France
Corresponding author: ernesto.geiger@cea.fr

**Abstract.** The objective of this work was to study the molybdenum chemistry in $UO_2$ based materials, known as SIMFUELS. These materials could be used as an alternative to irradiated nuclear fuels in the study of fission products behaviour during a nuclear severe accident. $UO_2$ samples doped with 12 stable isotopes of fission products were submitted to annealing tests in conditions representative to intermediate steps of severe accidents. Samples were characterized by SEM-EDS and XAS. It was found that Mo chemistry seems to be more complex than what is normally estimated by thermodynamic calculations: XAS spectra indicate the presence of Mo species such as metallic Mo, $MoO_2$, $MoO_3$ and $Cs_2MoO_4$.

## 1. Introduction

Severe Accidents in nuclear power plants may involve the release of radioactive elements such as Fission Products (FP) into the biosphere. Many integral and analytical international research programs, such as PHEBUS FP, HI/VI, VEGA, VERCORS, etc. [1], have been carried out aiming at understanding the behaviour of FP during a Nuclear Severe Accident, like that of Fukushima in 2011. Most of them used irradiated nuclear fuel from Pressurized Water Reactors or Boiling Water Reactors. Another possibility would be to use model materials, known as SIMFUEL, consisting in $UO_2$ samples doped during fabrication with stable FP (Ba, Ce, Cs, La, Mo, Nd, Pd, Rh, Ru, Y and Zr) in concentrations representative of a given burn-up. Despite the limitations of SIMFUELS, such as the difficulty to reproduce some phenomena directly linked to irradiation, these materials could be used to study FP behaviour such as precipitation, release and species formation in severe accident conditions. These conditions can be recreated in dedicated annealing test under controlled atmosphere. Besides, one of the main advantages of SIMFUELS is that they lack of radioactivity, other than that of natural uranium, reducing radioprotection constraints and enabling to use powerful characterisation techniques such as XAS. Among the many FP, molybdenum presents a particular interest. The $Mo/MoO_2$ redox couple is believed to act as a oxygen potential buffer, affecting the chemical state of other FP and their release [2]. In order to study Mo chemistry in oxidizing conditions, three SIMFUELS samples doped with 12 stable FP in concentration representative of a 76 $GWd.t^{-1}$ irradiated fuel were submitted to annealing tests under flowing argon + 1% $O_2$. Mo behaviour has been characterized by SEM-EDS and XAS.

## 2. Methods and materials

### *2.1. Samples characteristics*

SIMFUEL samples were acquired from the Chalk River Laboratories, Canada. Samples fabrication process is described in [3] and were latter implanted with Cs. The concentration of each FP, presented in **Table 1**, matches that of a 76 $GWd.t^{-1}$ Burn-up $UO_2$ irradiated fuel. Samples are 500 μm thick, polished quarter discs made from 10 mm diameter discs.

| Samples composition (at. %) | | | | | | | | | | | |
|---|---|---|---|---|---|---|---|---|---|---|---|
| Ba | Ce | Cs | La | Mo | Sr | Y | Zr | Rh | Pd | Ru | Nd |
| 0.26 | 0.61 | 0.35 | 0.20 | 0.61 | 0.13 | 0.06 | 0.60 | 0.03 | 0.42 | 0.64 | 0.91 |

Table 1: FP concentration in SIMFUEL samples

### *2.2. Experimental loop and test conditions*

The annealing tests were performed in the DURANCE experimental loop, very similar to the MERARG one [4], located at the *Bernard François laboratory* (CEA Cadarache, France). The $PO_2$ in the inlet gas was monitored thanks to a MicroPoas probe (GENAIR), allowing to calculate the oxygen potential for each test ($\mu O_2$). Three annealing sequences were performed under flowing argon containing 1% $O_2$. Sample A was annealed at 300°C for 2 h, Sample B at 300°C for 2 h and then at 1200 °C for 30 min, and finally Sample C at 1700°C for 10 min.

### *2.3. Characterization methods*

2.3.1. *Scanning Electron Microscopy.* Scanning Electron Microscopy (SEM) was performed using a Philips XL30 FEG instrument, equipped with an EDAX-EDS system allowing us to determine the distribution and content of Mo using Energy Dispersive X-ray Spectroscopy (EDS). Acceleration tension was set to 15 kV. Mo $L\alpha 1$ X-ray lines were used to analyse samples.

2.3.2. *X-ray Absorption Spectroscopy.* Measurements were carried out at the MARS beam-line, SOLEIL synchrotron radiation facility (France) [5], under top-up 430 mA ring mode, in the (17.6 - 22.4) keV energy range. A Si(220) double-crystal monochromator (DCM) was used. The monochromatic beam was focused to 350 x 350 $\mu m^2$. XAS spectra were collected in fluorescence mode, with a Vortex-90-EX silicon drift detector. The acquired spectra were processed using the ATHENA software [6]. Pre-edge removal and normalization was achieved using linear functions. The post-edge line for XANES spectra was taken using the position 30 eV and 150 eV relatively to edge position $E_0$. The $E_0$ and white line positions were taken as the first inflection point and the first zero-crossing of its first derivative, respectively. Linear combination fits were performed in the (-10, +50) eV range relative to the absorption edge.

## 3. Results

### *3.1. SEM-EDS*

SEM-EDS images and Mo X-ray maps of samples A, B and C are presented in **Figure 1**. Two main precipitates are observed in sample A: Mo precipitates which contain Ru, Rh and Pd, with a Mo/Ru ratio equal to 0.87 and the second type containing Ba, Sr, Zr, Ce, Y, U and O. Three different precipitates were observed in sample B, two of them containing Mo and Ba along with Sr, Ce and Zr, with Mo/Ba ratios equal to 1.2 and 7.0, respectively. The third type contains also Pd, Rh and Ru, with a Mo/Ru ratio equal to 0.1. Three types of precipitates containing Mo were observed in Sample C: the first ones contain also Ba, with a Mo/Ba ratio equal to 8.4 while the other two were constituted by Mo, Ru, Rh and Pd, with different Mo/Ru ratios: 4.5 and 1.0, respectively. No Cs precipitates were observed in the three samples. A predominance diagram, calculated using the FactSage software/ SGPS database, is presented in **Figure 2**. The calculated oxygen potentials were: -350, -210 and -230 $kJ.mol^{-1}$ for samples A, B and C, respectively. According to **Figure 2**, the stable species in samples A and B would be $BaMoO_4$, $MoO_2$ and $Cs_2MoO_4$ (zone 3), while for sample C they would be $BaZrO_3$, $MoO_2$ and $Cs_2MoO_4$ (zone 2).

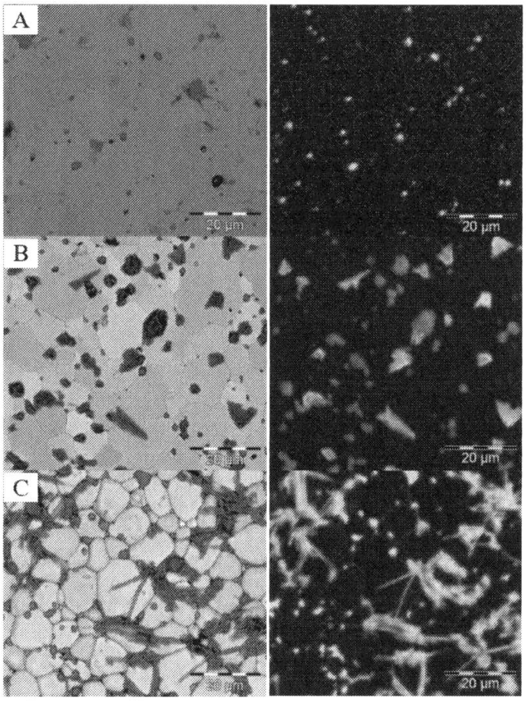


Figure 1: SEM images (left) of samples A, B and C, along with the respective Mo X-ray maps (right)

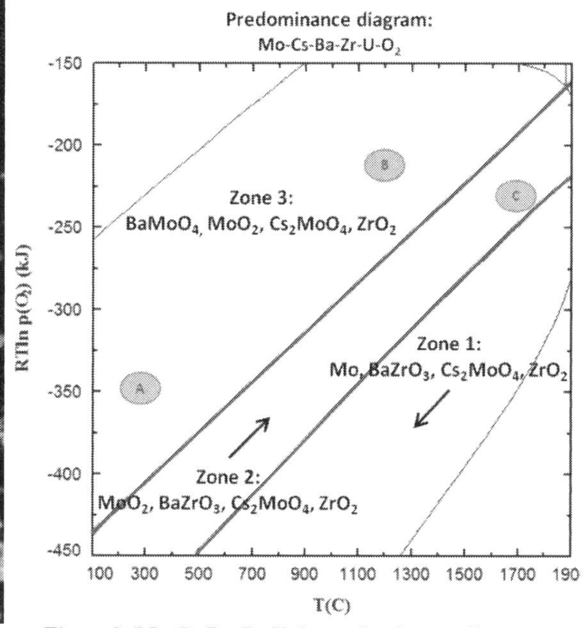


Figure 2: Mo-Cs-Ba-Zr-U-$O_2$ predominance diagram

*3.2. XANES*

Mo XANES spectra obtained for the three samples and four reference compounds spectra: metallic Mo, $MoO_2$, $MoO_3$ and $Cs_2MoO_4$, are presented in **Figure 3**. The $E_0$ and white line energies for each acquired spectra are presented in **Table 2**. As observed, these values were modified as consequence of the annealing tests: sample B presents an $E_0$ energy shift of 2.9 eV regarding to sample A, while the $E_0$ position of sample C presents an energy shift of -0.2 and -3.1 compared to Samples A and B, respectively. Linear combination fits results are presented in **Table 3**.

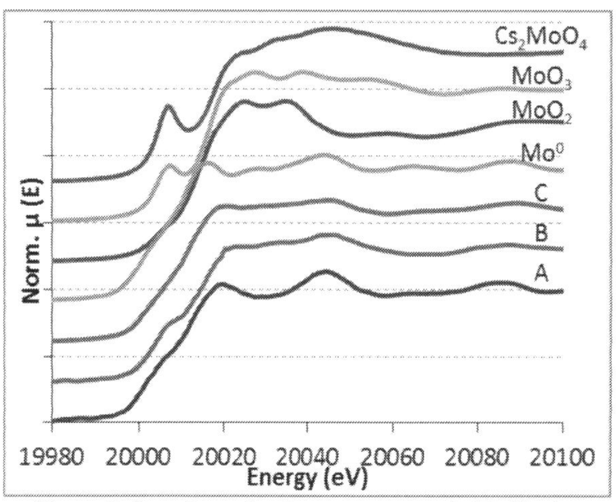


Figure 3: Samples and reference compounds Mo K-edge XANES spectra

| Sample | $E_0$ (eV) | White line (eV) |
|---|---|---|
| A | 20001.4 | 20019.3 |
| B | 20004.3 | 20023.0 |
| C | 20001.2 | 20019.1 |
| Mo | 20000.0 | 20016.6 |
| $MoO_2$ | 20005.0 | 20025.0 |
| $MoO_3$ | 20004.6 | 20026.8 |
| $Cs_2MoO_4$ | 20003.8 | 20045.5 |

Table 2: Samples and references compounds spectra $E_0$ and white line position

| Sample | Reference weight | | | | R factor |
|---|---|---|---|---|---|
| | Mo | $MoO_2$ | $MoO_3$ | $Cs_2MoO_4$ | |
| A | 0.729 | 0.093 | - | 0.158 | 0.002367 |
| B | 0.317 | 0.099 | 0.427 | 0.144 | 0.000833 |
| C | 0.592 | 0.251 | 0.136 | 0 | 0.000720 |

Table 3: Mo spectra fit results

## 4. Discussion

SEM-EDS and XANES results highlight the evolution of the chemical phases as consequence of the annealing tests under oxidizing atmosphere. Regarding to sample A, SEM-EDS results show that Mo is found only in precipitates containing Pd, Rh and Ru. These precipitates have already been observed in irradiated nuclear fuels and are referred to as white inclusions [2]. Though these precipitates are supposed to be metallic, XANES results indicate that Mo oxidized species such as $Mo^{4+}$ and $Mo^{6+}$ are present as well. Based in thermodynamic estimations (**Figure 2**), the existing species would be $MoO_2$ and $BaMoO_4$. Regarding to sample B, SEM-EDS results indicate the presence of three different Mo precipitates: the first ones containing Mo, Pd, Rh and Ru, the two others containing Ba, Ce, Mo, Sr, Y and Zr, with an important difference in their Mo/Ba ratio. As described in bibliography, Ba is normally found in an oxide phase with perovskite structure, described as (Ba, Cs, Sr)(Mo, Zr, U, RE)$O_3$ (RE = Rare Earths) [2]. XANES results confirm the presence of Mo oxidized species. The presence of two different types of Mo/Ba precipitates would imply the existence of Mo species other than $BaMoO_3$ or $BaMoO_4$, not predicted by thermodynamic estimations. Regarding to sample C, SEM-EDS results show the presence of big sized precipitates, with an equivalent circle diameter superior to 40 μm, containing Ba, Sr and Ce, but no Zr. Sample C Mo XANES spectrum presents an energy shift to lower energy compared to that of sample B, which would imply an important presence of metallic Mo. SEM-EDS results confirm this, since more precipitates containing Mo, Pd, Rh and Ru are observed in this sample. The presence of $Mo^0$ in the three samples would indicate that chemical equilibrium was not attained during tests.

## 5. Conclusion

Molybdenum chemistry in oxidizing conditions was studied in model materials representative of high burn-up irradiated nuclear fuels. Three samples were studied, which were annealed up to 300, 1200 and 1700°C. Samples were characterized by SEM-EDS and XAS. It was that found that Mo, initially present in metallic precipitates, was oxidized and interacted with other FP such as Ba, Sr, Ce and Cs. Though there is a rather good agreement between experimental observations and thermodynamic estimations, an unidentified species other than $BaMoO_3$ or $BaMoO_4$ seems to be created at 1200°C.

## 6. Acknowledgements

Authors wish to thank the CEA Cadarache LAMIR, LCU, LEMCI and LLCC staff for their contribution to this project. We acknowledge SOLEIL for provision of synchrotron radiation facilities and Sandrine Schlutig and Hervé Hermange for assistance in using MARS beam-line.

## 7. References

[1] Lewis B J, Dickson R, Iglesias F C, Ducros G and Kudo T 2008 *J. Nucl. Mater.* **380** 126–143

[2] Kleykamp H, 1985 *J. Nucl. Mater.* **131** 221–246

[3] Lucuta P G, 1991 *J. Nucl. Mater.* **178** 48–60

[4] Pontillon Y, Bonnin J, Addes E, Cochaud J J, Piquemal J, Pontillon M, Roure C and Volle G 2005 European Working Group on Hot Laboratories and Remote handling Plenary Meeting, Petten, Netherlands

[5] Llorens I, Solari P L, Sitaud B, Bès R, Camelli S, Hermange H, Othmane G, Safi S, Moisy P, Wahu S, Bresson C, Schlegel M L, Menut D, Bechade J L, Martin P, Hazemann J L, Proux O and Den Auwer C 2014 *Radiochim Acta* **102** 957–972

[6] Ravel B and Newville M 2005 *J. Synchrotron Radiat.* **12** 537–541

Journal of Physics: Conference Series **712** (2016) 012099 doi:10.1088/1742-6596/712/1/012099

# Micro and conventional XAFS study of incinerated Cr-rich tannery sludge

**F Pinakidou, M Katsikini, S Varitis, P Kavouras, E C Paloura**

Aristotle University of Thessaloniki, School of Physics, Department of Solid State Physics, 54124, Thessaloniki, Greece

E-mail: fpina@physics.auth.gr

**Abstract**. μ-XRF, XAFS and μ-XAFS spectroscopies were implemented in order to investigate the effect of thermal annealing on the stability of a vitrified tannery waste. The two dimensional μ-XRF scans disclose the presence of Cr-rich and depleted regions where both Cr(III) and Cr(VI) are present. However the polymerization of the Cr-O-Cr chains is different in these two distinct regions. More specifically, the μ-XANES and μ-EXAFS results reveal that in the Cr-rich inclusions, the Cr(III)$O_6$ belong to $Cr_2O_3$, while in the Cr-depleted regions, different linkages among the Cr(III)$O_6$ octahedra are detected. The percentage of the different type of polymeric structure around Cr(III) in the two distinct regions was assessed using Cr-*K*-EXAFS and XANES spectroscopies. It is revealed that upon thermal annealing the number of Cr(VI) ions is preserved and participate in the formation of the glass matrix. Even though Cr(III) is a glass modifier, the local coordination of the Cr(III)-octahedra changes upon thermal treatment. The presence of Cr(VI) suggests the limited capacity of the vitrification process to inhibit Cr(III) oxidation, however, Cr(VI) is safely trapped into both the glass and glass-ceramic matrix.

## 1. Introduction

The common practice of land-based disposal of tannery wastes, where trivalent chromium [Cr(III)] is the dominant species, is responsible for the contamination of soils. Furthermore, the presence of naturally occurring minerals (e.g. Mn oxides) can promote the oxidation of Cr(III) to the more toxic Cr(VI), thus water pollution with Cr(VI) can also be related to the current disposal criteria for tannery wastes [1]. Chrome tanning produces two types of sludge, hair-burn and chrome: the former is produced during the mechanically scrapping and soaking of salt-cured hides and is mainly organic, while the latter is a chrome sludge, produced after sequentially cleansing and soaking of hair in a NaCl-$H_2SO_4$ solution, bathed for several hours in a basic $Cr(SO_4)_3$ solution and neutralized with $NaHCO_3$ [2].

Incineration of the Cr-sludge at extremely high temperatures is a process that minimizes the need for additional land disposal; during incineration, the toxic components are converted to harmless or less harmful compounds whereas the volume of waste is significantly reduced. However, the wastes produced during pre-treatment and in the combustion unit (ashes and other residues) may still be toxic and must be disposed in an environmentally sound manner. Stabilization via vitrification of incinerated tannery waste is a promising and cost effective method that leads to the formation of glass and vitroceramic products, capable of immobilizing the Cr(III) and Cr(VI) species [3]. The aim of this work is to investigate the feasibility of preparing glass and vitroceramic materials that immobilize safely chromate and chromite, whether the oxidation of Cr(III) has been provoked and if the presence of Cr(III) and Cr(VI) species are involved in the building of the glass matrix (glass forming role of Cr(VI) and network modifying role of Cr(III)).

## 2. Experimental details

The studied tannery sludge originates from the industrial zone of Thessaloniki in northern Greece and was incinerated for 1.5 h at 500 °C in an oxygen-starved atmosphere under anoxic conditions. In this process Cr(III) is not oxidized to the highly toxic and mobile Cr(VI) and also nearly most of organic carbon is removed. The resulting chromium-loaded ash was rich mainly in Ca (% 16 mass%) and Cr (% 15.5 mass%) with few remaining organic content [4]. The Cr-loaded ash was then mixed with glass-forming oxides ($SiO_2$) and network modifiers ($CaCO_3$ and $Na_2CO_3$ in the form of $CaCO_3$ and $Na_2CO_3$ respectively) and melted in a platinum crucible at 1400 C for 2 h. The melt was then quenched in air on a refractory steel plate. The chemical composition of the studied glass is 10 wt% ash, 20 wt% CaO and a $SiO_2/Na_2O$ ratio equal to 3.66. In an effort to produce vitroceramic products by controlled crystallization of the initial vitreous products, thermal treatment is applied at 880°C.

XAFS measurements were conducted at the KMC2 beamline at the synchrotron radiation facility BESSY in Berlin. The beamline is equipped with a double-crystal monochromator and capillary optics that allow the reduction of the beam diameter down to 4.8μm. Two dimensional μ-XRF scans allows for the detection of variations in the distribution of Cr in samples that developed annealing-induced inhomogeneities. Thus, combination of μ-XRF with μ-XAFS spectra recorded from selected sample spots permit the investigation of the local bonding environment of Cr, as a function of the Cr-varying distribution. Both XAFS and μ-XAFS spectra were recorded in the fluorescence yield mode using an energy dispersive (Röntec) detector and $Cr_2O_3$, $Cr(OH)_3$ and $CrO_3$ powders were used as references.

## 3. Results and Discussion

### *3.1. μ-XRF*

A representative 300 $\mu m^2$ 2D-map of the studied thermally treated glass depicting the varying distribution of Cr is shown in Fig. 1(a). It is characteristic that in the Cr-rich regions, the Cr-content is 5 to 10 times higher compared to the Cr-depleted regions. The analysis of the μ-XRF spectra (figure 1(b)) recorded from the two distinct spots shown by arrows in Fig.1(a) reveal that the Cr/Ca concentration ratio differs: in the Cr-rich (Cr-high) inclusions the Cr/Ca ratio is three times higher than the respective in the Cr-poor (Cr-low) regions (0.35 and 0.11, respectively).

### *3.2. Cr-K-μ-XANES and Cr-K-XANES*

In an effort to correlate modifications in the oxidation state and the bonding environment of Cr as a result of its inhomogeneous distribution, Cr-*K*-μ-XANES spectra were recorded from spots high- and low-Cr concentration. Similar to all 3d-trasistion metals, a pre-edge absorption occurs in the XANES spectra from both regions (shown in figure 2(a)), the characteristics of which are modulated by the Cr-valence: a prominent pre-edge peak occurs for Cr(VI) species, caused by transitions from an 1s state to empty d-orbital and though forbidden for octahedral Cr(III), a small pre-edge peak is present due to $3d(t_{2g})$ and $3d(e_g)$ electronic transitions, respectively. As shown in figure 2(a), both the intensity and energy position of the pre-edge peak, which is a measure of the proportion of Cr(VI) in mixed valence samples, is not significantly altered [5]. However, the structure above the absorption edge ($E_{abs}$) is modified between the two regions: in the Cr-rich region, the spectrum resembles the respective in $Cr_2O_3$ while weak structure exists in the Cr-depleted region. The same conclusion can be drawn for the XANES spectra of the glassy and annealed sample: the intensity and energy position of the pre-edge peak are similar; however the changes in the structure above the $E_{abs}$ are prominent. These findings imply the presence of both Cr(III) and Cr(VI) species and thus linear combination fitting (LCF) was applied in order to evaluate the changes in the Cr(III)/Cr(VI) ratio as a function of both Cr-concentration and annealing process.

The application of LCF using the experimental spectra of the $Cr_2O_3$ and $CrO_3$ reference samples does not yield satisfactory results in most of the samples. Nevertheless, if the experimental spectrum of a $Cr(OH)_3$ sample is used, the fitting quality is significantly improved (colored dotted lines in figure 2(a)). Even though in both $Cr_2O_3$ and $Cr(OH)_3$ compounds Cr is trivalent, the $Cr(III)O_6$ octahedra link in different modes: in the former case they share one face, two edges and three corners while in the latter case, two Cr-octahedra share two edges and link to four adjacent polyhedra. Thus, the use of $Cr(OH)_3$ in LCF does not indicate the presence of Cr-oxyhydroxides but rather the modification in the polymerization of the Cr-O-Cr chains. As listed in table 1, Cr(VI) species are always detected: the same amount of Cr(VI) exists in both the Cr-depleted and Cr-rich regions, while upon annealing of the glassy sample, a change in the percentage of Cr(VI) is not detected. However, as listed in table 1, Cr(III) belongs only to $Cr_2O_3$ inclusions in the Cr-rich region. In the Cr-depleted region, 53.1 % of the Cr(III) ions connect in a network similar to the structure of $Cr_2O_3$ while 29.5 % of $Cr(III)O_6$ follow the polymeric structure of $Cr(OH)_3$. In the glassy sample, the Cr-octahedra connect via edges and corners as in $Cr(OH)_3$, while upon annealing the polymerization of the Cr-O-Cr chains changes completely. Thus, it is concluded that the Cr(VI) concentration is not affected by the varying distribution of Cr in the annealed sample, however changes in the type of linkage among the Cr-octahedra are induced. Upon thermal treatment of the glass, reduction of Cr(VI) is not detected while the Cr(III)-groups change their polymerization

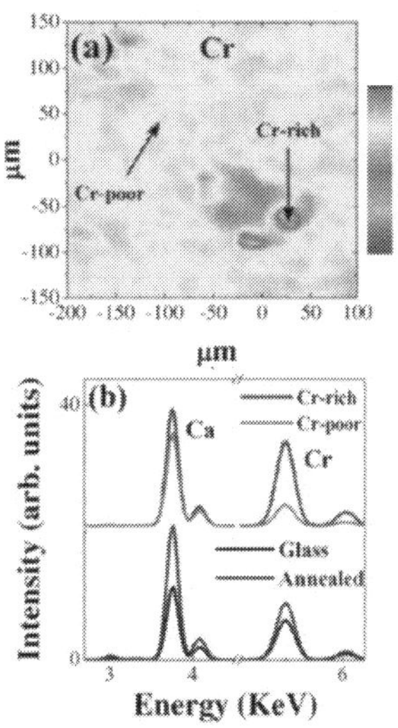


**Figure 1.** (a) 2D μ-XRF scans of the annealed glass showing the inhomogeneous distribution of Cr. The color bar represents the normalized intensity. (b) μ-XRF spectra recorded from the Cr-rich and Cr-depleted regions and XRF spectra of the glassy and annealed sample.

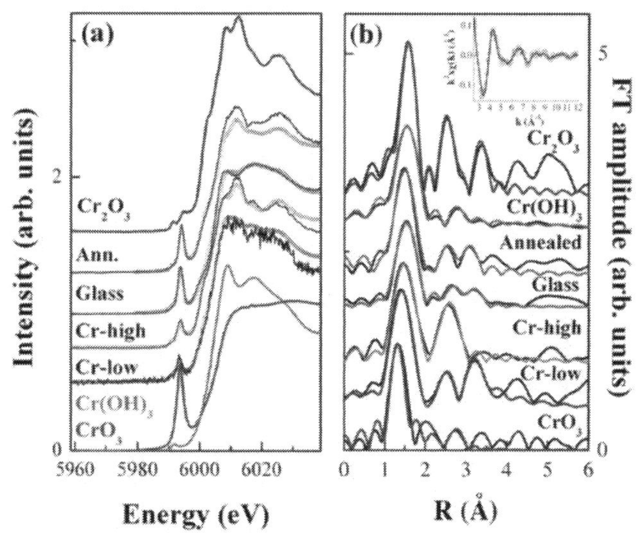


**Figure 2.** (a) Cr-*K*-μ-XANES spectra recorded from the Cr-rich and Cr-depleted regions of the annealed sample and Cr-*K*-XANES spectra of the glassy and annealed sample. The results of the linear combination fit (LCF) are shown in colored dotted lines. (b) Fourier Transforms (FT) of the $k^2 \times \chi(k)$ Cr-*K*-μ-EXAFS spectra recorded from the Cr-rich and Cr-poor regions of the annealed sample and Cr-*K*-EXAFS spectra of the glassy and annealed samples. A representative $k^2 \times \chi(k)$ EXAFS spectrum and the fitting is shown in circles and in a black line, respectively.

### *3.3. Cr-K-μ-EXAFS and Cr-K-EXAFS*

Taking into consideration the results from the μ-XANES analysis, the Cr-*K*-μ-EXAFS spectra were fitted assuming that in the Cr-rich and Cr-depleted regions, the Cr(III)- and Cr(VI)-polyhedra link via edges and corners, i.e. form Cr(III)$O_6$ octahedra that share edges and corners and Cr(VI)$O_4$ tetrahedra that share only corners. Curve fitting was carried out simultaneously in the R- and k-space using FEFF. The Fourier Transforms (FTs) of the $k^2$-weighted Cr-*K*-μ-EXAFS spectra recorded from the high- and low-concentration-Cr-regions are shown in figure 2(b) and the analysis results are listed in table 1. In very good agreement with the Cr-*K*-μ-XANES results, it is found that approximately 25 (±8) % of Cr exists as Cr(VI) in both the Cr-rich and Cr-depleted regions where it is tetrahedrally coordinated and the Cr-O bondlenght is equal to 1.60-1.62Å. However, even though the number of the Cr(III)$O_6$ is equal (within the error bar) in both regions, they link to the glass matrix in a different manner. Cr(III) belongs to $Cr_2O_3$ inclusions in the Cr-rich regions while in the Cr-depleted regions the polymerization of the Cr-O-O chains is modified: approximately 75% of Cr-polyhedra connect to adjacent polyhedra as in $Cr_2O_3$ while the rest link only via edges and corners, in a less compact structure, similar to $Cr(OH)_3$. The analysis of Cr-*K*-EXAFS spectra of the glassy and annealed sample reveal that upon annealing, the percentage of Cr(VI) does not change. In analogy with the μ-XAFS results, the polymerization of the Cr(III)$O_6$ chains is modified due to thermal treatment: Cr(III) links to adjacent Cr-polyhedra as in $Cr(OH)_3$ in the vitreous sample while after annealing, the Cr(III) are mainly involved in a linkage as in $Cr_2O_3$.

**Table 1.** LCF results of the μ- & conventional Cr-*K*-XANES of all studies samples. μ- & conventional Cr-*K*-EXAFS results of all studies samples.

| | XANES LCF (%) (±5) | | | EXAFS | | | | |
|---|---|---|---|---|---|---|---|---|
| | $CrO_3$ | $Cr_2O_3$ | $Cr(OH)_3$ | Cr(III)$O_6$ (%) (±8 %) | $Cr_2O_3$ (%) (±12 %) | $R_{Edge}$ Cr(III) (±0.02 Å)* | $R_{corner}$ Cr(VI) (±0.03 Å)* | $R_{corner}$ Cr(III) (±0.04 Å)* |
| **Cr-rich** | 11.2 | 88.8 | - | 73.1 | 75.5 | 2.90 | 3.31 | 3.46 |
| **Cr-poor** | 17.4 | 53.1 | 29.5 | 74.2 | 69.2 | 2.86 & 2.94 | - | 3.41 & 3.53 |
| **Glass** | 49.4 | - | 51.6 | 58.4 | - | 3.01 | 3.28 | 3.51 |
| **Annealed** | 37.9 | 49.5 | 17.6 | 49.6 | 72.6 | 2.86 & 2.95 | - | 3.45 & 3.52 |
| **$CrO_3$** | | | | 0 | - | | 3.34 | - |
| **$Cr(OH)_3$** | | | | 100 | - | 2.98 | - | 3.50 |
| **$Cr_2O_3$** | | | | 100 | 100 | 2.90 | - | 3.44 |

*$R_{egde}$ *and* $R_{corner}$ *are the inter-atomic distances in the higher shells.*

## 4. Conclusions

The effective stabilization via vitrification of Cr-rich tannery waste was investigated using micro- and conventional XAFS spectroscopies. Cr(VI) is present before and after thermal treatment, however it acts as glass former, and thus is safely immobilized into the glass matrix. The percentage of Cr(VI) is not affected by either Cr-concentration or thermal treatment. The Cr-inclusions formed after thermal treatment are rich in $Cr_2O_3$ while in the Cr-depleted regions, the bonding geometry of Cr(III) changes due to the different configuration of the Cr(III)$O_6$ polyhedra in space. In the vitreous sample, the polymerization of the Cr(III)-chains resembles the respective in $Cr(OH)_3$ while upon annealing the existing Cr(III)$O_6$ units preferentially link as in $Cr_2O_3$.

**Acknowledgments**
The measurements at BESSY were funded from the European Community's Seventh Framework Program (FP7/2007-2013) under grant agreement n. 312284.

**References**

[1] Saleh F Y, Parkerton T F, Lewis R V, Huang J H, Dickson K L 1989 *Sci Total Environ* **86** 25
[2] Driess S J, 1986 *Ground Water* **24** 312
[3] Basegio T, et al. P, 2009 *J. Hazard. Mater.* **165** 604
[4] Varitis S, et al. *J Therm. Anal. Calor.* **121** 203
[5] Peterson M L, Brown G E Jr, Parks G A, Stein C L 1997 *Geochim. Cosmochim. Acta* **61** 3399

Journal of Physics: Conference Series **712** (2016) 012100 doi:10.1088/1742-6596/712/1/012100

# Structural and magnetic properties of nanoclusters formed in III-V semiconductors

**Krystyna Lawniczak-Jablonska[1*], Anna Wolska[1] and Marcin T. Klepka[1]**
[1]Institute of Physics, Polish Academy of Science, Al. Lotnikow 32/46, 02-668 Warsaw, Poland
*E-mail: jablo@ifpan.edu.pl

**Abstract**. Studies of X-ray magnetic circular dichroism (XMCD) were performed for a set of GaMnAs films with different Mn concentrations priory and after high temperature annealing (500 and 600 °C). After thermal treatment, GaMnAs samples with zinc blende structure and MnAs hexagonal nano-clusters were formed. In most of the samples, both types of clusters were detected by EXAFS studies. Dependence of the orbital and the spin moments on magnetic field were calculated from XMCD data by applying the sum rule. It was shown that both moments were much larger for MnAs nano-clusters. When these inclusions are formed even in a small amount, they dominate the XMCD signal. Interestingly, in some of samples the zinc blende GaMnAs nano–clusters were observed at a surface while in the bulk of hexagonal MnAs. Therefore, the location of magnetic ions in the host matrix is crucial for their magnetic properties. This unique information can be provided by XAS and XMCD.

## 1. Introduction

Magnetic semiconductors have attracted considerable attention due to expectation that manipulation of an electron spin, as an alternative to manipulation of an electron charge, can be used for the storage of information in semiconducting devices. In spite of the efforts of many technologists and scientists, until now the magnetic semiconductors with uniformly distributed magnetic ions do not exhibit ferromagnetic properties at room temperature. The highest solubility limit reported for $Ga_{1-x}Mn_xAs$ is around x=0.1 [1].Therefore, there is an increasing interest in granular materials. Presence of magnetically active metallic nanoclusters in GaMnAs leads to enhanced magnetotransport [2] and magneto-optical [3] properties over a wide spectral range. This phenomenon opens an opportunity for various applications of such hybrid systems. Location of magnetic ions in the host matrix is crucial for magnetic properties and is correlated with all other important physical properties. Moreover, an understanding of the local environment of magnetic ions in semiconductors and its valence state has an important impact on optimization of material properties. Information about the atom location in the crystal lattice is provided by only a few techniques. Among them, EXAFS is a well-established tool which has been proven to be useful as a chemically sensitive local probe for identification of the site and the valence state of magnetic ions in DMSs [4]. The calculated XANES and EXAFS spectra for models with different location of magnetic ions in the crystal structure, under assumption that all ions are located only in the given position, show significant differences [5-8]. The quantitative comparison of the calculated model spectra with the measured spectra, is a procedure for a quick check of homogeneity. It is known that, in the as-grown MBE samples of the GaAs matrix, the Mn atoms may occupy substitutional and interstitial positions, but the abundance of each position is not easy to find. The number of Mn atoms in the interstitial positions is increasing with the Mn concentration. In our earlier paper we demonstrated, how the proportion between these two locations can be estimated applying EXAFS and XANES [8].

Two kinds of nanoclusters existing in the GaMnAs samples have been reported: cubic zinc blende (ZB) and hexagonal MnAs phases [e.g. 9]. The formation of nanoclusters in the MBE as grown GaMnAs layers started during the annealing above 400 °C. Presence of these nanoclusters changes drastically the magnetic properties of layers increasing additionally the $T_C$ above RT. Creation of small and exclusively cubic MnAs clusters was assumed by annealing at 500 °C and larger hexagonal by annealing at 600 °C and above. Recently performed detailed EXAFS studies on a set of samples with different Mn concentration unambiguously demonstrated that such assumption is not justified [10, 11]. The problem is more complicated and depends on the location of Mn in as grown sample (interstitial or substitutional). Study of the atomic order around Mn atoms showed, that the ZB MnAs clusters do not exist because 12 Mn atoms in the second coordination sphere cannot be fitted to the experimental spectra. Instead, the study revealed that small cubic GaMnAs clusters are formed with much higher content of Mn than ever produced in GaMnAs layers with randomly distributed Mn atoms (about 20%) [10]. Clusters with the size larger than 8 nm have had already the MnAs hexagonal structure. In most of studied samples, the clusters with both structures were present. The proportion of Mn in each structure estimated by EXAFS was in perfect agreement with the size distribution of clusters found from TEM analysis [11]. Knowing the position of Mn ions in crystal structure it is very interesting to exam their magnetic properties. The x-ray magnetic circular dichroism (XMCD) is well established technique to perform such studies. In the present paper, the results of XMCD studies on a set of samples discussed in papers [10, 11] are reported.

## 2. Experimental

Sets of $Ga_{1-x}Mn_xAs$ layers with nominal Mn contents x = 0.015, 0.06, 0.07, and 0.08 were grown by MBE on the (001)-oriented GaAs substrates at the temperature of 230 °C. Nominal Mn concentrations were estimated from Mn/Ga flux ration. The thickness of layers was around 1 μm. After deposition, each sample was cut into three pieces: one of them was left untreated (as grown), the remaining two were reintroduced into the MBE growth chamber and annealed under $As_2$ flux for 30 min at T = 500 °C and at T = 600 °C, respectively. Details of samples growth and their characterisation are reported in refs [10, 11]. In table 1, results of EXAFS studies at the set of samples under consideration are collected. In most of samples, both kind of nanoclusters were found: GaMnAs ZB and hexagonal MnAs. In the sample with smallest Mn concentration (x=0.015), only the GaMnAs ZB nanoclusters were detected.

**Table 1.** The percentage of Mn atoms in GaMnAs ZB clusters and in hexagonal MnAs clusters as estimated by EXAFS analysis.

| **Sample** | **Annealed at 500 °C** | | **Annealed at 600 °C** | |
|---|---|---|---|---|
| Clusters type | ZB | Hexagonal | ZB | Hexagonal |
| $x$ = 0.015 | 100% | 0 | 100% | 0% |
| $x$ = 0.06 | 72(2)% | 28% | 40(5) % | 60% |
| $x$ = 0.07 | 53(5)% | 47% | 15(5)% | 85% |
| $x$ = 0.08 | 69 (2)% | 31% | 50(5)% | 50% |

XMCD studies were performed at ESRF ID08 beamline on the $L_{2,3}$ Mn edge in the function of the temperature and magnetic field. Although, the best mode of XMCD detection is the transmission, the total electron yield (TEY) and the total fluorescence yield (TFY) were simultaneously applied for spectra recording because samples were thin films deposited on a thick substrate. One should be aware that TEY method can be sensitive to variable magnetic field and changes in electron detecting efficiency (photocurrent). The TFY is insensitive to the applied field, but the yield is intrinsically not proportional to the absorption cross section because the radiative to non-radiative relative core-hole

decay probability depends strongly on symmetry and spin polarization of the XAS final states. These two detection modes differ also in the examined depth (TEY surface, TFY bulk) and can provide different result in the case of inhomogeneous samples.
As an example in the figure 1a XMCD spectra for the sample with nominal content of Mn=0.08 annealed at 600 °C is shown. Measurements were performed at room temperature and in 5T magnetic field, simultaneously with above two detection modes. The shape of the XMCD is very different for each of modes. This is an additional confirmation that two kinds of Mn locations in this sample exist as was already reported from EXAFS studies [10] and is shown in the table 1. The XMCD signal registered by TEY is characteristic for the cubic ZB atomic order around Mn as was reported in several papers [e.g. 12].The XMCD recorded by TFY is characteristic for octahedral symmetry of atomic arrangement as in hexagonal MnAs clusters [13 and references therein]. This indicates that Mn atoms have cubic arrangement at the surface and can be located in substitutional position either in the GaAs matrix or cubic $Ga_{1-x}Mn_xAs$ clusters and in the bulk of sample hexagonal MnAs clusters dominate. It is worth notice that XMCD for hexagonal clusters is almost one order of magnitude larger than for cubic clusters. In the case of the sample with x=0.015 annealed at 500 °C (figure 1b), where only cubic nano-clusters were reported from EXAFS analysis (see table 1), no significant difference between both modes of detection was observed. Therefore, performing the XMCD analysis one should pay special attention to the used mode of detection. For detailed analysis of XMCD, the TFY measurements were taken because bulk properties of samples were of interest.

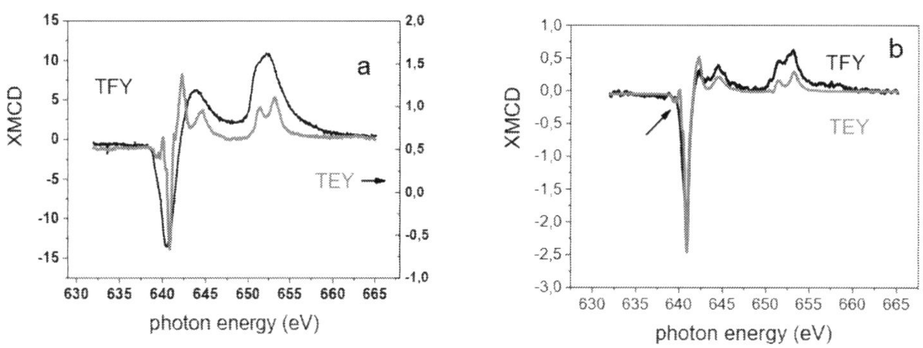


**Figure 1.** XMCD spectra measured with TEY (red) and TFY (black) detection mode, for samples $Ga_{1-x}Mn_xAs$ with a) x=0.08 and b) x=0.015 Mn content. The measurements performed at room temperature and 5T magnetic field.

The quantitative evaluation of magnetic moments, separated into the spin and the orbital contributions, was performed from integrals over XMCD spectra by applying the so-called sum rules [14,15]. In the numerical evaluation of the orbital and the spin moments in the case of transition metals and particularly Mn atoms, two main sources of errors are expected. First is the coupling between $L_2$ and $L_3$ edges (so called branching ratio) which influences spin moment. It was predicted that the intermixing of the $L_2$ and $L_3$ absorption edges is mainly present in the early transition metals [16] where the electron core hole interaction increases while the spin-orbit splitting decreases. For such metals (e.g. Mn) the application of the spin sum rule can produce an error up to 30%. Error at this level is indicated in numerical values presented in table 2 and treated as confidence interval for reveling the trends in the changes of magnetic orbital and spin moment. Second possible source of errors in applying the spin sum rule is the energy dependent radial matrix elements [17]. Although the sum rules are normalized to the isotropic absorption cross-section, which is proportional to the number of holes in the d-shell, also the continuum states contribute. In the performed calculation, the number of holes in the d-shell was set equal to 4.8, similarly as in numerous other papers, reported XMCD for $Ga_{1-x}Mn_xAs$ [e.g. 12].To eliminate contribution from continuum states, a double step function was fitted to the absorption spectrum and subtracted. This procedure is likely to introduce a systematic error in the determination of a number of holes in the final state. Therefore, not the numerical values but the observed relative changes of the orbital and the spin magnetic moments in the function of magnetic field and the structure of samples will be rather discussed.

## 3. Results and discussion

Quantitative results of the XMCD spectra analysis for samples with x=0.015 are shown in the figure 2 for as grown and annealed in 500 °C samples. Spectra were measured at 150 K as a function of magnetic field. Without magnetic field at this temperature, no XMCD signal was detected. In the as grown sample similar as in other reports [e.g. 12], the small orbital moment was detected, whose absolute value increased with the magnetic field from -0.0012 $\mu_B$ at 0.5T to -0.02 $\mu_B$ at 5T and T=150 K. The orbital moment was not changing noticeably, when cubic clusters were formed after 500 °C annealing (table 2 and figure 2). Starting from 2T, the orbital moment was found not to depend significantly on the magnetic field as can be seen in figure 2b and table 2. Interestingly, the orientation of the orbital to the spin moment changes in these two samples depending on the magnetic field. For the as grown sample, the parallel orientation was observed at 1 and 2 T, but for 500 °C annealed sample only for 1 T magnetic field. For all other cases, antiparallel orientation was found. The reason for the observed changes needs more detailed studies.

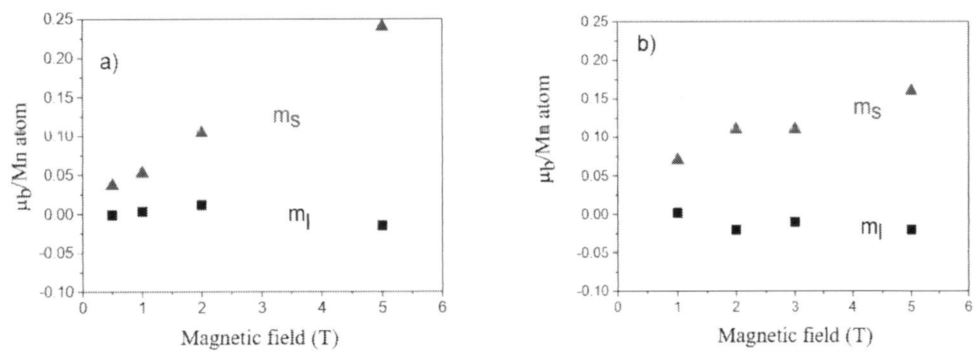


**Figure 2.** The field dependency of the spin (triangles) and orbital (squares) of the magnetic moments estimated from the XMCD measurements for $Ga_{1-x}Mn_xAs$ for x=0.015 a) as grown samples; b) after annealing at 500 °C.

The value of the spin moment changes with the magnetic field depending on the kind of samples. In the as grown sample, similarly as it was reported for solid solution of Mn in GaAs [12], the spin moment increases linearly with the magnetic field (figure 2a). For the annealed film (500 °C) with small (~5nm) cubic nano-clusters, the XMCD signal initially increases rapidly and the spin moment increases by 57% on increasing the field from 1 to 2 T, then remains constant in the limit of error (saturates) indicating the magnetic coupling of Mn ions (figure 2b table 2). This effect is more pronounced in the sample with still small but already hexagonal MnAs nano-clusters (figure 3, table 2). When the hexagonal MnAs nano-clusters are formed, the strong magnetic coupling between Mn ions can be observed already at 150 K. The saturation is reached at 0.5 T and no change of the orbital and spin moment with the increase of magnetic field is observed. Both moments are antiparallel coupled in the whole range of magnetic field. The XMCD is observed already at 0.05T with quite large orbital and spin moments.

**Table 2.** The value of the orbital $m_l$ and the spin $m_s$ magnetic moment for $Ga_{1-x}Mn_xAs$ samples with x=0.015 after different thermal treatment measured at 150 K in the function of magnetic field.

| Sample/ moment | 0.05T | 0.1T | 0.5T | 1T | 2T | 3T | 5T |
|---|---|---|---|---|---|---|---|
| as grown | | | | | | | |
| $m_l$ ($\mu_b$) | | | -0.0012(4) | 0.0030(6) | 0.012(4) | | -0.020(6) |
| $m_s$ ($\mu_b$) | | | 0.04(1) | 0.05(1) | 0.11(4) | | 0.24(7) |
| an. 500 °C | | | | | | | |
| $m_l$ ($\mu_b$) | | | | 0.0020(6) | -0.020(6) | -0.013(4) | -0.020(6) |
| $m_s$ ($\mu_b$) | | | | 0.07(2) | 0.11(4) | 0.11(4) | 0.16(5) |
| an. 600 °C | | | | | | | |
| $m_l$ ($\mu_b$) | -0.020(6) | -0.09(3) | -0.20(6) | -0.25(7) | -0.26(7) | | -0.20(6) |
| $m_s$ ($\mu_b$) | 0.15(4) | 0.28(9) | 0.64(21) | 0.77(25) | 0.79(26) | | 0.82 (27) |

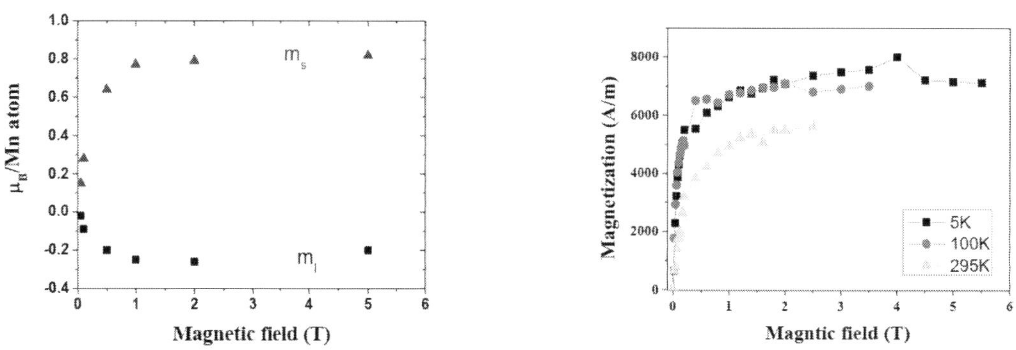


**Figure 3.** The field dependence of the spin (triangles) and the orbital (squares) magnetic moments estimated from the XMCD measurements for $Ga_{1-x}Mn_xAs$ film with x=0.015 after annealing at 600 °C (left). The field dependence of magnetization measured by SQUID performed for the same sample at different temperatures (adapted from paper [11] right).

Measurements of the superconducting quantum interference device (SQUID) magnetization (figure 3 right) performed for the same sample, after the correction on diamagnetic compound typical for GaAs, show also typical ferromagnetic-like behavior (i.e. very fast saturation with magnetic field) even at 295 K, in agreement with magnetic properties of bulk MnAs. These both finding confirmed that when MnAs hexagonal nano-cluster are formed even in small amount which was not detected by EXAFS (table1), XRD or TEM, they dominate the magnetic signal.

**Acknowledgments:**

Financial support from the EU FP7 EAgLE project under the grant agreement REGPOT-CT-2013-316014 is gratefully acknowledged. The authors would like to thank Violetta Sessi for help during the measurements at ESRF.

**References**

[1] Mack S, Myers R C, Heron J T, Gossard A C, and Awschalom D D 2008 *Appl. Phys. Let.* **92** 192502
[2] Ye S, Klar P J, Hartmann Th, Heimbrodt W, Lampalzer M, nau S, Torunski T, Stolz W, Kurz T, Krug von Nidda H-A, Loidi A 2003 *Appl. Phys. Lett.* **83** 3927
[3] Yokoyama M, Yamaguchi H, Ogawa T and Tanaka M 2005 *J. Appl. Phys.* **97,** 10D317
[4] Lawniczak-Jablonska K, 2015 Chapter 15, Magnetic Ions in Group III–V Semiconductors, Springer-Verlag Berlin Heidelberg 2015, C.S. Schnohr and M.C. Ridgway (eds.), *X-Ray Absorption Spectroscopy of Semiconductors*, Springer Series in Optical Sciences 190
[5] Shioda R, Ando K, Hayashi T, and Tanaka M 1998 *Phys. Rev.* B **58**, 1100
[6] Bacewicz R, Twarog A, Malinowska A, Wojtowicz T, Liu X, Furdyna J K 2005 J. *Phys. Chem. Solids* **66** 2004
[7] Demchenko I N, Lawniczak-Jablonska K, Story T, Osinniy V, Jakiela R, Domagala J Z, Sadowski J, Klepka M, Wolska A, Chernyshova M, 2007 J. *Phys: Condens. Mat*ter **19** 496205
[8] Lawniczak-Jablonska K, Libera J, Wolska A, Klepka M T, Jakiela R, J. Sadowski 2009 *Rad. Phys. Chem.***78**, S80
[9] Kwiatkowski A, Wasik D, Kaminska M, Bozek R, Szczytko J, Twardowski A, Borysiuk J, Sadowski J, Gosk J 2007 *J. Appl. Phys.* **101** 113912
[10] Lawniczak-Jablonska K, Libera J, Wolska A, Klepka M T, Dluzewski P, Sadowski J, Wasik D, Twardowski A, Kwiatkowski A and Sato K 2011 *Phys. Stat. Sol.* RRL **5**, 62
[11] Lawniczak-Jablonska K, Bak-Misiuk J, Dynowska E, Romanowski P, Domagala J Z K, Libera J, Wolska A, Klepka M T, Dluzewski P, Sadowski J, Barcz A, Wasik D, Twardowski A, Kwiatkowski A 2011 *J. Solid State Chem.* **184,** 1530
[12] Edmonds K W, Farley N R S, Johal T K, van der Laan G, Campion R P, Gallagher B L, and Foxon C T 2005 *Phys. Rev.* **B 71**, 064418
[13] Lawniczak-Jablonska K, Wolska A, Klepka M T, Kret S, Gosk J, Twardowsk A, Wasik D, Kwiatkowski A, KurowskaB, Kowalski B J, Sadowski J 2011 *J. Appl. Phys.* **109**, 074308
[14] Thole B T, Carra P, Sette F, van der Laan G 1992 *Phys. Rev. Lett.* **68**, 1943
[15] Carra P, Thole B T, Altarelli M, Wang X 1993 *Phys. Rev. Lett.***70**, 694
[16] Schwitalla J and Ebert H 2005 comment *in Phys. Rev.* Lett. **80**, 4586
[17] Wu R, Wang D and Freeman A J 1993 *Phys. Rev. Lett.* **71**, 3581

# Local dynamics and phase transition in quantum paraelectric $SrTiO_3$ studied by Ti $K$-edge x-ray absorption spectroscopy

**Andris Anspoks[1], Janis Timoshenko[1], Juris Purans[1], Francesco Rocca[2], Vladimir Trepakov[3,4], Alexander Dejneka[4] and Mitsuru Itoh[5]**

1 Institute of Solid State Physics, University of Latvia
2 IFN-CNR, Institute for Photonics and Nanotechnologies, Unit FBK-Photonics, Trento, Italy
3 Institute of Physics, AS CR, Prague, Czech Republic
4 Ioffe Physical-Technical Institute RAS, St-Petersburg, Russia
5 Tokyo Institute of Technology, Japan

E-mail: andris.anspoks@cfi.lu.lv

**Abstract.** Strontium titanate is a model quantum paraelectric in which, in the region of dominating quantum statistics, the ferroelectric instability is inhibited due to nearly complete compensation of the harmonic contribution into ferroelectric soft mode frequency by the zero-point motion contribution. The enhancement of atomic masses by the substitution of $^{16}O$ with $^{18}O$ decreases the zero-point atomic motion, and low-T ferroelectricity in $SrTi^{18}O_3$ is realized. In this study we report on the local structure of Ti in $SrTi^{16}O_3$ and $SrTi^{18}O_3$ investigated by Extended X-ray Absorption Fine Structure measurements in the temperature range 6 – 300 K.

## 1. Introduction

Strontium titanate ($SrTiO_3$, STO) is a well-known and thoroughly studied member of the family of $ABO_3$ oxides, which at room temperature has the cubic $Pm3m$ perovskite-type structure. Upon cooling, STO undergoes a second order $Pm3m$ to $I4/mcm$ antiferrodistorsive (AFD) phase transition (PT) at $T_C = 105$ K, with condensation of the $R_{25}$ phonon mode at the Brillouin zone boundary [1]. STO is called quantum paraelectric [2] because at low temperatures, where quantum statistics dominates, the ferroelectric ordering is suppressed and the rising of dielectric permittivity saturates. This leads to the idea that ferroelectricity in STO can be induced by suppression of the quantum fluctuations. Experimentally, this was observed using isotope exchange ($^{16}O$ with $^{18}O$) [3, 4].

The long debate about the structure of the polar phase and the mechanism of the phase transition in $SrTi^{18}O_3$ is still open [4, 5]. The structure of STO has been studied by different experiments as X-ray diffraction (XRD) [6], neutron diffraction [7] and X-ray absorption spectroscopy [8, 9], but a clear picture and consistence of opinions have not yet been achieved.

In this work the local atomic structure around Ti in $SrTi^{16}O_3$ (STO16) and $SrTi^{18}O_3$ (STO18) is studied in the temperature range 6 – 300 K with x-ray absorption spectroscopy methods, namely Extended x-ray Absorption Fine Structure (EXAFS), using radial distribution function reconstruction [10] and advanced Reverse Monte Carlo (RMC) and Evolutionary Algorithm (EA) techniques for EXAFS data analysis (RMC/EA-EXAFS) [11].

## 2. Samples and experiment

In our study we used powders of standard STO16 and STO18 where 96 % of oxygen atoms were substituted by $^{18}O$ as described in [4].

The Ti K-edge x-ray absorption spectra were measured in transmission mode at the ESRF bending-magnet beamline BM23 and at the HASYLAB/DESY A1 bending-magnet beamline in the temperature range from 6 K to 300 K. The x-ray radiation was monochromatized by a 40 % detuned Si(111) double-crystal monochromator, and the beam intensity was measured using two ionization chambers filled with argon and krypton gases. The Oxford Instruments liquid helium flow cryostat was used to maintain the required sample temperature. To achieve the best sample homogeneity and thickness at the Ti K-edge, necessary for transmission mode experiments, the proper amount of the STO powder was deposited on polytetraflouroethylene Millipore membranes from suspensions in methyl alcohol.

## 3. Results

The extended x-ray absorption fine structure (EXAFS) oscillations $\chi(k)$ and X-ray absorption near edge structure (XANES) data were extracted following the conventional procedure [12, 13] using the EDA software package [14]. The theoretical spectra, scattering amplitudes and phase shift functions were calculated by ab initio real-space multiple-scattering code FEFF8.5 [15] using the complex exchange-correlation Hedin-Lundqvist potential in the muffin-tin (MT) self-consistent-field approximation with the default values of MT radii.

First results and analyses including detailed studies of the x-ray absorption near edge structure (XANES) pre-peak structure [16] show that spectra of STO18 and STO16 look very similar because the difference in the structure between STO18 and STO16 is very small [3]. XANES analysis was done on the second pre-peak ("B"), because its intensity is proportional to the squared displacement of the Ti [17]. This peak is originating from transitions of Ti 1s electron to the p-d hybrid orbitals (titanium 3d orbital mixing with the oxygen 2p orbitals due to the Ti displacement from the center of the octahedra) [17]. Our data [16] revealed noticeable anomaly of the temperature dependence for STO18 pre-peak "B" intensity in the low temperature region: STO18 has higher intensity of the pre-peak "B" than STO16. Normally one should expect the opposite: that with the increase of the oxygen mass the oscillation amplitudes should decrease.

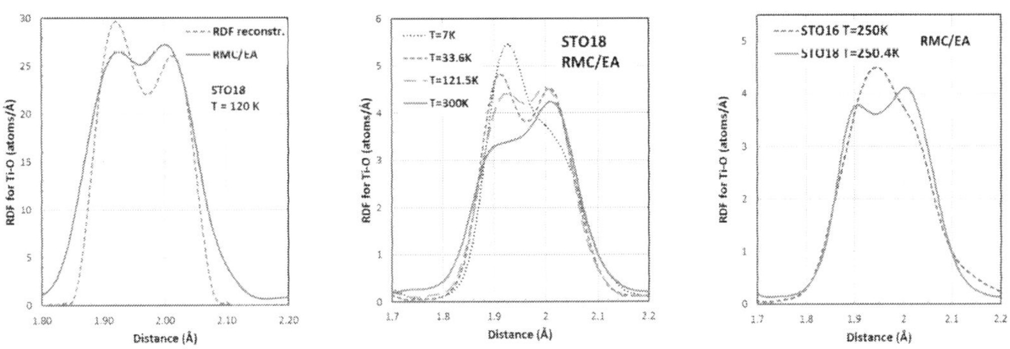


**Figure 1.** RDF for the first coordination shell of Ti (Ti–$O_1$).

### *3.1. Radial distribution function (RDF) reconstruction method*

As next step we started with the analysis for the first coordination shell Ti–$O_1$ using a model-independent approach [10], which allowed one to reconstruct the true radial distribution function (RDF). The Ti $K$-edge EXAFS contribution $\chi(k)k^2$ for the first coordination shell was isolated using Fourier filtering procedure and best-fitted in 3 – 16 Å$^{-1}$ of the k-space using EDARDF code [14]. The obtained results (left panel of figure 1) show a clear non-Gaussian character of the RDFs for the first coordination shell of Ti even for T = 120 K which is a cubic phase according to the crystallographic data.

### *3.2. RDF analysis from RMC/EA method*

To reveal the local atomic structure beyond the first coordination shell we applied our newly developed reverse Monte Carlo (RMC) and evolutionary algorithm (EA) technique for EXAFS data analysis (RMC/EA-EAXFS) [11]. The method does not restrict the geometry of the system and takes into account all multiple scattering effects, which are very sensitive to the bonding angles and distances, e.g. in Ti-O-Ti chains and triangles.

In the presented RMC/EA-EXAFS simulations we used a set of 128 supercells, where each supercell has size $4a \times 4b \times 4c$ and contains 1280 atoms. Lattice constants $a$, $b$ and $c$ and initial positions of the atoms were taken from diffraction studies [7]. Comparison of the experimental and theoretical EXAFS spectra has been performed using a Morlet wavelet transform (WT) [18] in 3 – 16 Å$^{-1}$ of the k-space, and 0.9 – 4.5 Å of the R-space range. The WT allows one to obtain a two-dimensional representation of the periodic signal with simultaneous localization in k and R spaces, WT allows to better separate the contribution to the total EXAFS from the noise due to imperfections in background subtraction procedure [18]. Scattering paths with the half-path length up to 5.5 Å and the multiple-scattering contributions up to the third order were considered in our simulations. The calculation of the cluster potential was performed only once for the average atom configuration, corresponding to the STO crystallographic structure. The complex exchange-correlation Hedin-Lundqvist potential and default values of muffin-tin radii, as provided within the FEFF8.5 code [15], were employed. As the output of RMC/EA-EXAFS analysis we have obtained a set of atomic coordinates, which can be used to calculate RDF, average distance between atoms, mean-square relative displacement (MSRD), and other values. MSRD values here are calculated as the second moment for the corresponding path.

The obtained RDFs for Ti–O$_1$ (figure 1) are consistent with the RDF reconstruction method and confirm the non-Gaussian character of the RDFs. We have found that there are two well resolved groups of oxygen atoms separated by $\sim$ 0.1 Å in the first coordination shell of Ti in STO18, in the whole range of investigated temperatures (figure 1). The same applies to STO16, but the corresponding distance between two groups of oxygen atoms is smaller. The variations of the amplitudes seen in figure 1 most probably are due to the too low statistics, because they vary with the changes of the simulation parameters. At the same time, in all simulations there are two well resolved groups of oxygen atoms.

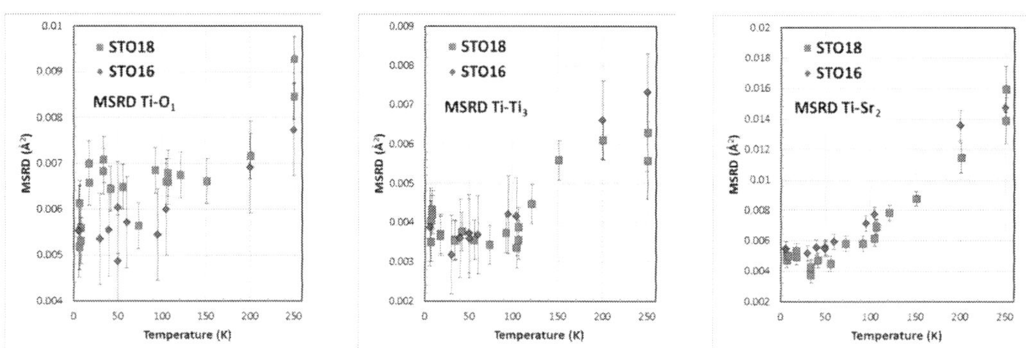


**Figure 2.** MSRD values obtained with RMC/EA-EXAFS for Ti–O$_1$, Ti–Sr$_2$, and Ti–Ti$_3$.

### *3.3. MSRD analysis from RMC/EA method*

The absolute values of MSRD for the first coordination shell of Ti (Ti–O) (left panel of figure 2) in the low temperature region for STO18 are larger than those for STO16 which is in-line with XANES analysis results. Normally one should expect opposite effect where with the increase of the oxygen mass the MSRD values should decrease as it was observed in the case of germanium [19].

The MSRD values for the first coordination shell of Ti in STO18 have an anomaly in the vicinity of the low-T phase transition into polar phase (25 K).

There is a strong correlation between nearest neighboring Ti atoms for STO in the whole range of the temperatures, as shown by the low MSRD value for Ti third coordination shell (Ti–Ti) (middle panel of figure 2). At the same time, Ti–Ti MSRD values show a noticeable increase around AFD PT in the 105 – 150 K region.

MSRD values for the second coordination shell of Ti (Ti–Sr) (right panel of figure 2) have the strongest temperature dependence, with standard Einstein-like temperature behavior in the high temperature region, corresponding to the paraelectric phase. Ti–Sr MSRD for STO18 shows a jump-like discontinuity at 25 K where the MSRD value increases going to lower T. No anomalies are observed for Ti–Sr MSRD for STO16 in the low temperature region.

The nature of the observed effects can be static or dynamic. It is important to note that our measurements revealed the presence of some static and dynamic disorders also in STO16, but less pronounced than in STO18. This can be connected to precursor effect widely discussed today [5].

## Acknowledgments

This work has been supported by the Latvian Science Council grant no. 402/2012. The work was supported in part from large infrastructure SAFMAT Project CZ.2.13/3.1.00/22132, P108/12/1941 of GACR. Authors wish to thank Dr. Olivier Mathon from ESRF and Dr. Edmund Welter from DESY for assistance during the beamtimes, and Dr. Alexei Kuzmin for fruitful discussions and feedback.

## References

[1] Shirane G, and Yamada Y 1969 *Phys. Rev.* **177** 858–863
[2] Muller K A and Burkard H 1979 *Phys. Rev.* B **19** 3593–3602
[3] Itoh M, Wang R, Inaguma Y, Yamaguchi T, Shan Y-J and Nakamura T 1999 *Phys. Rev. Lett.* **82** 3540–3543
[4] Taniguchi H, Itoh M, and Yagi T 2007 *Phys. Rev. Lett.* **99** 017602
[5] Bussmann-Holder A and Bishop A R 2006 *Eur. Phys. J.* B **53** 279-282
[6] Lytle F W 1964 *J. Appl. Phys.* **35** 2212–2215
[7] Hui Q, Tucker M G, Dove M T, Wells S A, and Keen D A 2005 *J. Phys.: Condens. Matter* **17** S111–S124
[8] Fischer M, Lahmar A, Maglione M, San Miguel A, Itie J P, and Polian A 1994 *Phys. Rev.* B **49** 12451–12456
[9] Ravel B and Stern E A 1995 *Physica* B **208/209** 316–318
[10] Kuzmin A and Purans J 2000 *J. Phys.: Condens. Matter* **12** 1959–1970
[11] Timoshenko J, Kuzmin A, Purans J 2014 *J. Phys.: Condens. Matter* **26** 055401.
[12] Rehr J J and Albers R C 2000 *Rev. Mod. Phys.* **72** 621
[13] Aksenov V L, Kovalchuk M V, Kuzmin A, Purans J, and Tyutyunnikov S I 2006 *Crystallogr. Rep.* **51** 908
[14] Kuzmin A 1995 *Physica B* **208-209** 175
[15] Ankudinov A L, Ravel B, Rehr J J and Conradson S D 1998 *Phys. Rev.* B **58** 7565–7576
[16] Anspoks A, Bocharov D, Purans J, Rocca F, Sarakovskis A, Trepakov V, Dejneka A, and Itoh M 2014 *Phys. Scripta* **89** 044002
[17] Vedrinskii R V, Kraizman V L, Novakovich A A, Demekhin Ph V, and Urazhdin S V 1998 *J. Phys.: Condens. Mat.* **10** 9561–9580
[18] Timoshenko J and Kuzmin A 2009 *Comp. Phys. Commun.* **180** 920–925
[19] Purans J, Afify N D, Dalba G, Grisenti R, De Panfilis S, Kuzmin A, Ozhogin V I, Rocca F, Sanson A, Tiutiunnikov S I, Fornasini P 2008 *Phys. Rev. Lett.* **100** 055901:1-4

# EXAFS study of the structural properties of In and In + C implanted Ge

**R Feng [1], F Kremer [1], D J Sprouster [2], S Mirzaei [1], S Decoster [3], C J Glover [4], S A Medling [1], S P Russo [5] and M C Ridgway [1]**

[1] Department of Electronic Materials Engineering, Research School of Physics and Engineering, Australian National University, Canberra, ACT 0200, Australia
[2] Nuclear Science and Technology Department, Brookhaven National Laboratory, Upton, NY, 11973, USA
[3] Instituut voor Kern-en Stralingsfysica, KU Leuven, 3001 Leuven, Belgium
[4] Australian Synchrotron, 800 Blackburn Road, Clayton, Australia.
[5] Applied Physics, School Applied Sciences, RMIT University, Melbourne 3001, Australia

E-mail: ruixing.feng@anu.edu.au

**Abstract.** The structural configurations of In implanted Ge have been studied via x-ray absorption spectroscopy with and without the codoping of C. In the case of In singly implanted Ge, while the In atoms occupy an substitutional site in Ge ($InGe_4$) at low In concentration ($\leq$ 0.2 at. %), they precipitate into a metallic phase (In metal) and form complexes composed of one vacancy and three Ge atoms ($InVGe_3$) at concentration $\geq$ 0.6 at. %. This behaviour can be suppressed by the addition of C leading to In-C pairing to form $InCGe_3$ complexes. This cluster enables In atoms to recover a four-fold coordinated structure and has the potential to improve the electrical activation of In atoms in Ge.

## 1. Introduction

It is widely accepted that an alternative channel material replacing Si is required to maintain CMOS (complementary metal-oxide semiconductor) performance improvements with a scaling beyond the 14-nm technology generation [1]. Ge has been considered as a promising candidate as it offers superior electron (~ 2×) and hole (~ 4×) mobilites compared to Si [2]. The low activation energy (0.012 eV) and low diffusion coefficient of In in Ge makes it an ideal *p*-type dopant, which benefits both device performance and scaling [3]. Since the lattice structure significantly influences the electrical properties, a detailed study of the In-doped Ge atomic configuration is necessary to understand the effects of doping concentration and defect environments that ensue with processing.

Recent lattice site location studies employing the emission channeling technique concluded that at low In implantation fluences ($2.9\times10^{12}$ In atoms/$cm^2$), In preferentially occupies Ge bond-centered sites before annealing and then redistributes to substitutional lattice sites after 300 °C annealing [4]. In-vacancy ($In_mV_n$) clusters within the Ge lattice were found to be energetically favourable by density functional theory (DFT) calculations [5]. In the case of In + C co-implanted Ge, substitutional In-C pairs in the Ge lattice were detected by perturbed angular correlation (PAC), and C co-doping was thus considered to be an approach to improve the electrical properties by compensating the strain caused by oversized In atoms in the Ge lattice [6]. Complementary atomic information such as coordination numbers (CN), nearest neighbour (NN) type and bond length are still undetermined for In-rich Ge and In + C co-implanted Ge, a knowledge of which could potentially enable correlation of the atomic environment to the electrical properties as a function of In concentration.

In this work, we employed extended x-ray absorption spectroscopy (EXAFS) to characterize the atomic environment of In-doped Ge as a function of In concentration and C co-doping, examining the configurations of the In absorber with Ge, V and C.

## 2. Experimental Methods

In ions were implanted into undoped 1.8 μm (100)-Ge on Si substrates at 250 °C with In concentrations ranging from 0.02 to 1.2 at. %. C was co-implanted before the In in selected samples to the same concentrations. The In concentrations for the two sample sets were measured by Rutherford backscattering spectrometry. A three step annealing process was applied, at temperatures of 550, 450 and 350 °C for 0.5, 1, 2 h, respectively, in $N_2$, to initially activate In dopants and reduce lattice disorder. Fluorescence mode XAS measurements were performed at the XAS beamline of the Australian Synchrotron at the In *K* edge, with a Si (3 1 1) double crystal monochromator and a 10 × 10 pixel-array Ge solid state detector [7]. Data analysis was carried out using the IFEFFIT package based on the FEFF9 code [8]. To aid the simulations and fits of the EXAFS spectra, four different models were proposed as in Figure 2, after structural optimizations within DFT calculations performed with the VASP code [9].

## 3. Results and Discussion

Figure 1(a) shows the Fourier transformed (FT) EXAFS spectra of the In-doped Ge samples. It is clear that the 0.02, 0.06 and 0.2 at. % samples share a similar spectrum, while the 0.6 and 1.2 at. % samples are comparable to the In standard. The fits of the data confirm this argument. For samples with In concentration ≤ 0.2 at. %, In atoms occupy a substitutional site in a highly ordered Ge matrix, consistent with model (a) in Figure 2 and in agreement with the results from emission channeling [4].

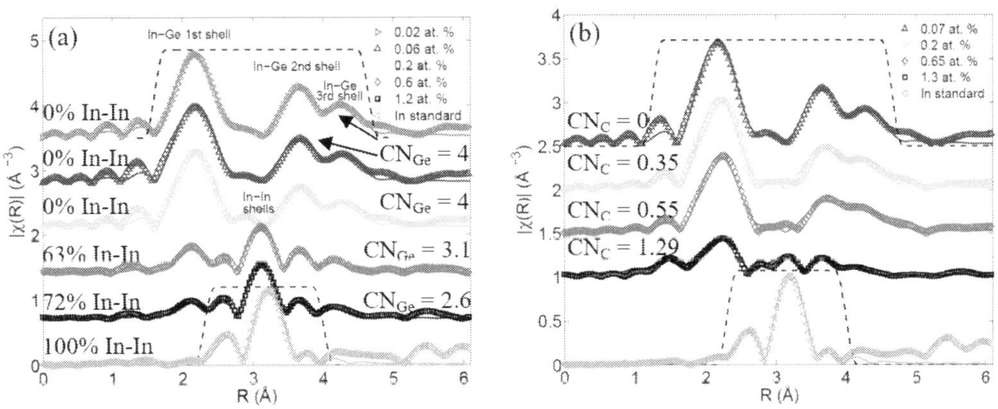


**Figure 1.** The magnitude of Fourier transform of $k^2$ weighted EXAFS spectra in radial distance (*R*-range), markers represent the data points while the solid lines are the best fits, where (a) is the result for In implanted Ge samples and (b) for In + C implanted Ge samples (Ref. [10]). The In metal fraction and coordination number of Ge in the first shell (in (a)), and the coordination number of C in the first shell (in (b)) are indicated above the spectrum. Reprinted with permission from J. Appl. Phys. **118**, 165701 (2015). Copyright 2015, AIP Publishing LLC.

For the 0.6 and 1.2 at. % samples, the FT EXAFS spectra appear to be comprised of substitutional In and metallic In. However, a combination of models (a) and (b) in Figure 2, where substitutional In atoms have a first shell of four Ge atoms, yielded non-physical Debye-Waller factors (DWFs) in the EXAFS analysis. As a consequence, model (c) in Figure 2 was invoked after a structural relaxation using a DFT calculation, where the Ge coordination number for In atoms in the non-metallic environment was allowed to vary with $CN_{Ge} \leq 4$. The resultant fit qualities were much improved using a combination of models (b) and (c), attesting to the validity of this approach. The fits of 0.6 and 1.2 at. % samples yielded 63 % metal fraction with $CN_{Ge} = 3.1$ and 72 % metal fraction with $CN_{Ge} = 2.6$, respectively. The precipitation of In was confirmed by transmission electron microscopy (TEM) and

XANES analysis [10], and the $InVGe_3$ cluster configuration used here was supported by the earlier DFT studies [5]. A more comprehensive investigation about these samples was provided in our previous work [10].

In our earlier report, the In metal precipitation and In-V clusters were found to be electrically inactive in Ge as expected [10]. Aiming to suppress the formation of these electrically inactive structures and enhance the In atom active fraction at high concentration, C was introduced as a co-dopant.

For an In concentration of 0.07 at. %, the EXAFS spectra (Figure 1(b)) are comparable with and without C co-implantation. The high quality fit of the co-implanted sample with model (a) confirmed that the In atoms are in substitutional sites of Ge without In-C pairing. The low concentrations of both In and C yields a low probability of In-C pair formation.

In-C pairs were found in the 0.2 at. % co-implanted sample. Though it has an EXAFS spectrum similar to that without C (Figure 1) and the first scattering peak in the FT spectra shows In atoms are substitutional, the smearing of the second and third peaks implies more disorder in the In surroundings. Indeed, model (a) was not able to provide a good fit for the second and third peaks. A model of a substitutional In atom coordinated with a C atom (model (d)) was thus proposed. After a DFT geometry optimization, the relaxed model shows that the introduction of a nearest neighbour C atom separates the second and third Ge atom shells into six different shells (with distances of 3.51, 4.10, 4.38, 4.45, 4.82 and 5.09 Å). The paired C component was allowed to vary given not all In atoms are expected to form pairs. An improved quality fit was obtained using model (d). The number of NN C atoms was 0.35 for the 0.2 at. % In+C implanted Ge sample. In-C pair formation starts before the In precipitation concentration is reached.

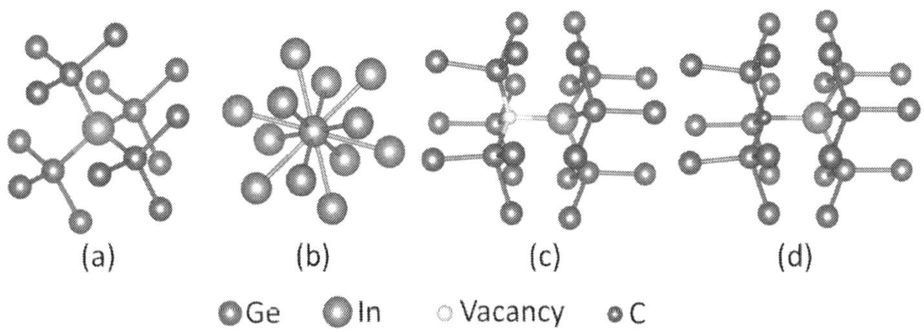


**Figure 2.** Schematics of models of (a) In atom in the substitutional site of Ge lattice ($InGe_4$), (b) In metal, (c) In atom coordinating with a vacancy in the substitutional sites of Ge lattice ($InVGe_3$) and (d) In atom coordinating with a C atom in the substitutional sites of Ge lattice ($InCGe_3$)

For In and C concentrations $\geq$ 0.65 at. %, Figure 1(b) shows that the amplitude of the In metal peaks were dramatically reduced for co-implanted samples with the spectra more similar to that of substitutional In than metallic In. The In metallic precipitation at high doping concentration was obviously suppressed when C was co-implanted as supported by TEM and XANES analysis (not shown). The reduction of In metal precipitation is a consequence of the formation of substitutional In–C pair ($InCGe_3$ cluster) in the Ge lattice. For the 0.65 at. % In+C doped sample, a considerable smearing of the second and third peaks is visible in the EXAFS spectrum, which is an indication of In-C pair as described above. The fit of the spectrum using model (d) gave the NN C atoms coordination number to In of 0.55.

For the 1.3 at. % sample, a good fit was also obtained using the same approach, with the number of NN C atoms being 1.29. The peak at ~3 Å in FT spectrum of the 1.3 at. % sample was due to the Ge atoms in the 3.51 Å shell, which was not observed in the 0.2 and 0.7 at. % samples due to their low fraction of In–C pairs. A low spectral amplitude was visible, which means the In atoms are in a highly disordered environment. In fact, the Ge lattice was polycrystalline as observed by TEM due to the high implanted concentration of both In and C atoms.

The bond lengths calculated from the fits of the EXAFS spectra were in good agreement with those from the DFT calculation for both In and In + C implanted samples. Alternative models were tested in the fits of the EXAFS spectra, but yielded low fitting quality or non-physical DWFs. A binding energy of -0.46 eV was calculated for an In-C pair in Ge by Chroneos *et al.* [11] using DFT, indicating that pairing is energetically favourable as experimentally demonstrated here. Our results were also supported by the PAC measurements of Tessema *et al.* [6] who observed In-C pairing when the annealing temperature was below 650 °C. Not only the In metal precipitation was suppressed by co-doping C, evidence of In-V cluster was also not found in the In+C co-implanted Ge samples according to the XAS analysis. With the introduction of C, an improvement of In atom activation was achieved with the apparent reduction of both electrical inactive In complexes and In atoms recover into a four-fold coordinated structure ($InCGe_3$).

## 4. Conclusion

In conclusion, the structural properties of In implanted and In+C co-implanted Ge have been analysed by XAS. When In was singly implanted in Ge, the In atoms precipitate from the substitutional sites of Ge and form In metal and In-V clusters when the In concentration is $\geq$ 0.6 at. %. The formation of these two electrical inactive In related clusters was successfully suppressed by co-implanting C, which leads to In-C pairs occupying substitutional sites of Ge. The reduction of electrically inactive complex formation and the recovery of In atoms back to the four-fold coordinated structure enhanced the In atom activation in Ge.

## Acknowledgements

We acknowledge access to NCRIS and AMMRF infrastructure at the Australian National University including the Australian National Fabrication Facility, the Heavy Ion Accelerator Capability and the Center for Advanced Microscopy. We also thank the Australian Research Council and the Australian Synchrotron for support.

## References

[1] Riddet C, Watling J R, Chan K, Parker E H C, Whall T E, Leadley D R and Asenov A 2012 *IEEE Trans. Electron Devices* **59** 1878
[2] Yang Y J, Ho W S, Huang C F, Chang S T and Liu C W 2007 *Appl. Phys. Lett.* **91** 102103
[3] Geballe T H and Morin F J 1954 *Phys. Rev.* **95** 1085
[4] Decoster S, De Vries B, Wahl U, Correia J G and Vantomme A 2009 *J. Appl. Phys.* **105** 083522
[5] Chroneos A, Bracht H, Grimes R W and Uberuaga B P 2008 *Appl. Phys. Lett.* **92** 172103
[6] Tessema G and Vianden R 2003 *J. Phys.: Condens. Matter* **15** 5297
[7] https://www.synchrotron.org.au/images/stories/beamline/spectrum/srms_282.pdf
[8] Rehr J J, Kas J J, Vila F D, Prange M P and Jorissen K 2010 *Phys. Chem. Chem. Phys.* **12** 5503
[9] Kresse G and Furthmüller J 1996 *Phys. Rev. B* **54** 11169
[10] Feng R, Kremer F, Sprouster D J, Mirzaei S, Decoster S, Glover C J, Medling S A, Russo S P and Ridgway M C 2015 *J. Appl. Phys.* **118** 165701
[11] Chroneos A 2011 *Semicond. Sci. Technol.* **26** 095017

Journal of Physics: Conference Series **712** (2016) 012103 doi:10.1088/1742-6596/712/1/012103

# Subtle local structural variations in oxygen deficient niobium germanate thin film glasses as revealed by x-ray absorption spectroscopy

**M A Sahiner[1], A Nabizadeh[1], D Rivella[1], L Cerqueira[1], J Hachlica[1], R Morea[2], J Gonzalo[2], J C Woicik[3]**

[1]Seton Hall University, Physics Department, South Orange, NJ 07079, USA
[2]Laser Processing Group, Instituto de Optica, CSIC, Serrano 121, 28006 Madrid Spain
[3]National Institute of Standards and Technology, Gaithersburg, Maryland 20899, USA

E-mail: mehmet.sahiner@shu.edu

**Abstract**. The local electronic and crystal structure of niobium-lead-germanate, $Nb_2O_5$-PbO-$GeO_2$ (NPG), glass thin films on silicon substrates were probed by XANES and EXAFS. NPG glasses are promising candidates for applications in nonlinear optical devices because they exhibit interesting optical characteristics such as high nonlinear third order optical susceptibility. In this work NPG glasses were prepared with pulsed laser deposition method with varying oxygen partial pressure to induce thin films with different oxygen stoichiometry. Previously, it was shown that oxygen stoichiometry has a very important effect to produce unusual high optical susceptibility. Detailed EXAFS and XANES analyses in a series of NPG thin films revealed the subtle variations in the local environment around Nb atoms and the Nb oxidation states caused by oxygen deficiencies.

## 1. Introduction

Multicomponent oxide film glasses containing large fractions of transition and post transition metals such as Nb and Pb are promising candidates for integrated photonic devices due to their excellent optical properties. They possess a broad transparence range ($0.4 < \lambda < 8$ μm), low phonon energies (~ 800 cm-1) and finally, they present large linear refractive indices ($\geq 2$) and nonlinear optical susceptibilities ($\chi(3)$) if compared to silica glasses.[1,2]

However, the practical use of these materials requires in many cases the production of high optical quality thin film glasses. Different thin film preparation methods have been attempted, such as sol-gel, sputtering or pulsed laser deposition (PLD) [3,4,5]. In particular, PLD allows the synthesis of multicomponent oxide thin films, although a careful choice of the deposition parameters is required to obtain smooth films with good optical quality [6]. Moreover, in the specific case of glasses having large contents of transition and post-transition metal oxides, PLD offers unprecedented possibilities to tailor their optical response. This characteristic is of extraordinary relevance in the case of glasses containing $Nb_2O_5$, since there exists a strong relationship between the $Nb_2O_5$ content and the magnitude of their nonlinear optical response [7,8]. This behavior has been related to the formation of three dimensional arrangements of $[NbO_6]$ octahedra, which enhance the hyperpolarizability of the glass.

However, conventional structural analysis techniques such as Raman spectroscopy and X-ray photoelectron spectroscopy did not allow to provide an absolute proof of the reduction of Nb and the modification of its local environment in the glasses that were proposed to be responsible for the strong enhancement of the nonlinear response [10]. In this work, we have performed X-ray absorption spectroscopy analysis (XAS) to obtain an exhaustive picture of the local environment of Nb and its oxidation state in the film glasses. In order to study the effect of oxygen deficiency, a series of thin films has been produced at different oxygen pressures. XAS analysis demonstrates that oxygen deficient films have significant fraction of $Nb^{4+}$ that can be as high as 100% for films deposited in vacuum to decrease to values close to 20% for films grown at 5 Pa.

## 2. Experimental

Niobium-lead-germanate (NPG) thin film glasses were produced by PLD using an ArF excimer laser ($\lambda$=193nm, $\tau$=12ns FWHM). The laser beam was focused on a partially crystallized opal bulk NPG target with a nominal composition of 55 $Nb_2O5$ - 35 PbO - 10 $GeO_2$ (mol %), that was prepared using standard melting methods as described elsewhere [4]. The laser energy density at the target surface was in all cases $\approx$2.5 J cm$^{-2}$. The vacuum chamber was evacuated to a residual pressure of $1 \times 10^{-4}$ Pa prior film deposition. Films were grown in a broad $O_2$ dynamic pressure range ($10^{-4}$ Pa $\leq P_{O2} \leq$ 10 Pa) on Si substrates held at room temperature and placed at 3 cm from the target surface. The film thicknesses varied between 100-150nm.

In a previous work [6] we demonstrated that NPG films grown at $P_{O2} \leq 1$ Pa were absorbent. Films deposited in vacuum presented a strong oxygen deficiency (> 50%), while the oxygen content in the films increased with $P_{O2}$ up to a value that was $\approx$80% that of the bulk target at $P_{O2}$= 5 Pa. Only films grown above this pressure presented an excess of oxygen, but this worsened their optical response [6]. Thus, NPG films grown at 5 Pa were selected for further optical characterization [9,10].

Details of the optical characterization techniques can be found elsewhere [4,6,9].

In the present work we have performed glancing incidence EXAFS and XANES measurements at the X23A2 beamline of National Synchrotron Light Source at Brookhaven National Laboratory to obtain an exhaustive picture of the local environment of Nb and its oxidation state in the film glasses. The Nb K absorption edge (18986 eV) was used in the EXAFS data acquisition in the fluorescence detection mode with a 0.5° angle of x-ray incidence. Nb metal foil was used for energy calibration. University of Washington's Feffit EXAFS analysis and fitting code was used in for data analyses and EXAFS fittings. The EXAFS functions, $\chi(k)$'s were obtained by subtracting the atomic absorption background using the AUTOBK code [11]. The resultant functions were Fourier Transformed (FT) using a Gaussian window set for 2.0 – 12.0 Å$^{-1}$ k-range. The theoretical EXAFS functions were calculated using University of Washington's FEFF8.4 code [12].

## 3. Results

In Figure 1 Fourier Transformed (FT) EXAFS data for all the PLD prepared films on silicon substrates at different partial oxygen pressures were shown. Also shown in the figure are the FT data for the $NbO_2$ and $Nb_2O_5$ polycrystalline samples for comparison purposes. Vacuum means that there was no oxygen flooding during the deposition of the thin film. The first shell below is dominated by Nb-O scattering paths so the first shell peak is sensitive to oxygen coordination around the Nb atoms. In Figure 1, the first shell peaks below 2 Å for the NPG thin films exhibit an evolution from $NbO_2$ to $Nb_2O_5$ structure increasing partial oxygen pressure set during the depositions. Increasing the oxygen partial pressure during the pulsed laser deposition can dramatically decrease the oxygen deficiency and could also induce structural changes in the synthesized thin films. With increasing oxygen pressure, the oxygen deficiency in these thins films diminishes and furthermore the local structure around the Nb atoms is pushed from $NbO_2$ (tetrahedrally coordinated) to $Nb_2O_5$ (octahedrally coordinated) symmetry.

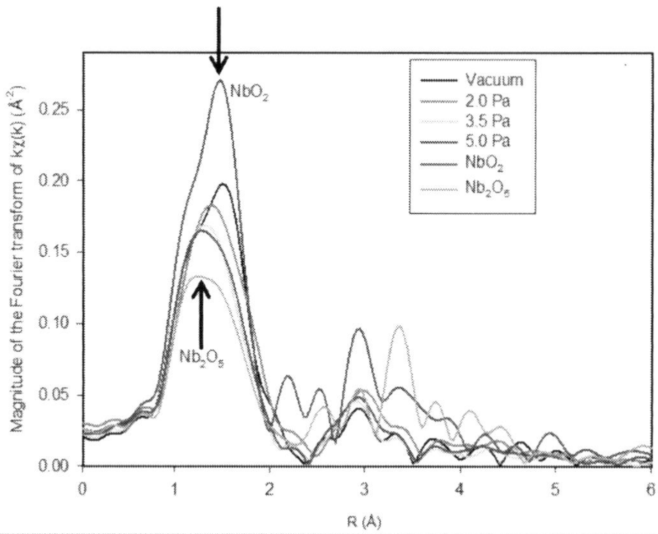


**Figure 1.** FT EXAFS data for the Nb K edge for $Nb_2O_5$-PbO-$GeO_2$ (NPG), glass thin films on silicon substrates NPG thin films with deposited with $O_2$ varying partial pressure. Also shown in the figure FT EXAFS data for $NbO_2$ and $Nb_2O_5$.

In Figure 2, Nb K-edge x-ray absorption near-edge spectroscopy (XANES) results are shown for the vacuum and the 5 Pa sample. $NbO_2$ and $Nb_2O_5$ XANES data are also plotted for comparison. $NbO_2$ does not exhibit a pre-edge peak as well as the vacuum sample whereas, $Nb_2O_5$ and 5 Pa samples have pre-edge peaks indicating a similar local symmetry. The XANES post-edge features also align well for the $NbO_2$ and the NPG film deposited with no oxygen flooding (vacuum). Similarly, XANES post-edge features 5 Pa film is in better alignment with $Nb_2O_5$. Therefore, qualitative XANES analyses support the EXAFS observations that oxygen deficiency pushes the local structure towards $NbO_2$ crystal symmetry.

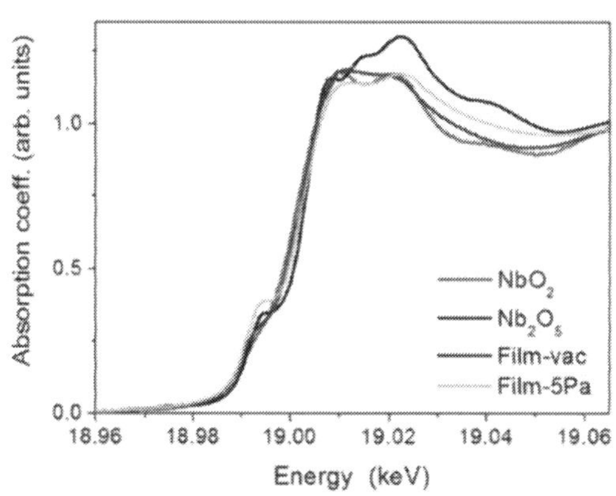


**Figure 2.** XANES data for the Nb K edge for $Nb_2O_5$-PbO-$GeO_2$ (NPG), glass thin films on silicon substrates NPG thin films with deposited under vacuum and with 5 Pa oxygen partial pressure. Also shown in the figure FT EXAFS data for $NbO_2$ and $Nb_2O_5$.

In Figure 3, background subtracted $\chi(k)$'s for the NPG thin films and corresponding first-shell EXAFS fits are shown. The EXAFS fits were performed using FEFF8.4 calculated $\chi(k)$'s for $NbO_2$ and $Nb_2O_5$ with $Nb_2O_5/NbO_2$ ratios are floated. Similar analyses to our previous Hf based high-k dielectric thin films in determining the mixed non-equilibrium crystal phases by using weighted EXAFS analyses of the existent structures were used in these thin films.[13-15]. The FT data overlays and EXAFS fits in R-space are shown in Figure 4. The fits were performed on the first-shell since the first-shell peak is well resolved and it is the determining factor in distinguishing the crystal symmetry between $NbO_2$ versus $Nb_2O_5$ environment.

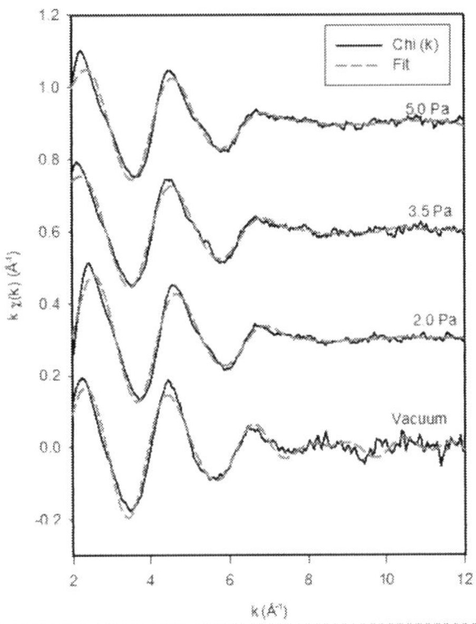


**Figure 3.** EXAFS functions, $\chi(k)$'s (solid-black lines) and EXAFS fits (dashed-red lines) for the PLD deposited NPG thin films under different partial $O_2$ pressures.

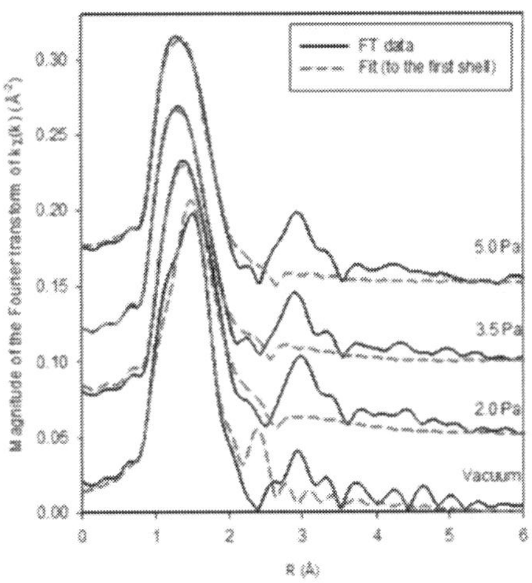


**Figure 4.** Overlays of the FT EXAFS data (solid black lines) and EXAFS fits (dashed-red lines) for the PLD deposited NPG thin films under different partial $O_2$ pressures.

The ratio of the $NbO_2$ to $Nb_2O_5$ for all the films are listed in Table I. The vacuum deposited thin film has no $Nb_2O_5$ contribution and increasing $P_{O2}$ pulls the symmetry towards octahedral $Nb_2O_5$ structure.

**Table 1.** The ratio of $Nb_2O_5$ to $NbO_2$ content as determined by the weighted EXAFS fits.

| Sample | $Nb_2O_5/NbO_2$ (%) |
|---|---|
| Vacuum | 0.0 |
| 2.0 Pa | 60 ±3 |
| 3.5 Pa | 70 ±3 |
| 5.0 Pa | 80 ±4 |

The EXAFS fit results for near-neighbor distances, coordination numbers, and Debye-Waller factors for the first-shell are listed in the Table 2. In these fits, the coordination numbers are not varied but used from the crystal symmetry values. The Debye-Waller factors for the first shell are varied but constrained to be equal to each other for all Nb-O near neighbor distances to limit the floating parameters in the fits. The R values are listed in a range in Table 2. The differences in Nb-O near-neighbor distance for $NbO_2$ and $Nb_2O_5$ from fit results are noted.

**Table 2.** The ranges of Nb-O near-neighbor distances, Nb-O coordination numbers, and Debye-Waller factors from to the first-shell EXAFS fits of the NPG thin films.

| First-Shell Fits | | | | | | |
|---|---|---|---|---|---|---|
| | $NbO_2$ (I 41/a :1) | | | $Nb_2O_5$ (C 1 2 1) | | |
| Sample ID | R(Å) (±0.03) | $N_{Nb-O}$ | $\sigma^2$(Å$^2$) | R(Å) (±0.03) | $N_{Nb-O}$ | $\sigma^2$(Å$^2$) |
| **Vacuum** | 1.93-2.29 | 6 | 0.006±0.001 | | | |
| **2.0 Pa** | 1.95-2.24 | 6 | 0.007±0.001 | 1.56-1.98 | 6 | 0.006±0.001 |
| **3.5 Pa** | 1.95-2.21 | 6 | 0.006±0.001 | 1.53-1.96 | 6 | 0.007±0.001 |
| **5.0 Pa** | 1.96-2.21 | 6 | 0.008±0.001 | 1.52-1.96 | 6 | 0.008±0.001 |

## 4. Conclusion

EXAFS and XANES analyses on the pulsed laser deposited niobium-lead-germanate films showed that the partial oxygen pressure during the deposition was crucial in establishing in the final crystal symmetry around the Nb atoms. EXAFS fits determined the ratio of $Nb_2O_5/NbO_2$ for the films deposed in various partial oxygen pressures. NPG film deposited with no oxygen flooding exhibits $NbO_2$ symmetry whereas, for the NPG film deposited at 5 Pa the $Nb_2O_5$ to $NbO_2$ ratio is 80%.

## References

[1] Lines M E 1991, *J. Appl. Phys.* **69** 6876
[2] Vogel E M, Weber M J and Krol D M 1991 *Phys. Chem. Glasses* **32** 231
[3] Hashimoto T and Yoko T 1995 *Appl. Opt.* **34** 2941
[4] Gonzalo, Sanz O, Perea A, Fernandez-Navarro J M, Afonso C N and Garcia-Lopez J, 2003 *Appl. Phys. A.* **76** 743
[5] Intyushin E B and Novikov V A 2008 *Thin Solid Films* **516** 4194
[6] Sanz O, Gonzalo J, Perea A, Fernandez-Navarro J M, Afonso C N and Garcia-Lopez J 2004 *Appl. Phys. A.* **79** 1907
[7] Flambard A, Videau J, Delevoye J L, Cardinal T, Labrugre C, Rivero C A, Couzi M and Montagne L, 2008 *J. Non Cryst. Sol.* **354** 3540
[8] Royon A, Canioni L, Bousquet B, Rodriguez V, Couzi M, Rivero C, Cardinal T, Fargin E, Richardson M and Richardson K, 2007 *Phys. Rev. B* **75** 104207
[9] Gonzalo J, Fernandez H, Solis J, Munoz-Martin D, Fernandez-Navarro J M, Afonso C N and J. Fierro J L G 2007 *Appl. Phys. Lett.* **90** 251907
[10] Munoz-Martin D, Ruiz de La Cruz A, Fernandez-Navarro J M, Domingo C, Solis J and Gonzalo J 2011 *J. Appl. Phys* **110** 023522
[11] Ankudinov L, Ravel B, Rehr J J and Conradson S D 1998 *Phys. Rev.* B **58** 7565.
[12] Newville M, Livins P, Yacoby Y, Rehr J J, and Stern E A 1993 *Phys. Rev.* B **47** 14126
[13] Lysaght P, Woicik J C, Sahiner M A, Lee B-H, Jammy R 2007 *Appl. Phys. Lett.* **91** 122910
[14] Sahiner M A, Lysaght P S, Woicik J C, Park C S, Huang J, Bersuker G, Taylor W, Kirsch P D and Jammy R 2012 *Phys. Status Solidi* (a) **209** 679
[15] Sahiner M A, Lysaght P S, Price J, Kirsch P D, Woicik J C, Klump A, Reehil C, Manners W A, and Nabizadeh A 2014 *Appl. Phys.* A **117** 93

Journal of Physics: Conference Series **712** (2016) 012104 doi:10.1088/1742-6596/712/1/012104

# Soft x-ray absorption spectroscopy on Co doped ZnO: structural distortions and electronic structure

**I A Kowalik**[1], **E Guziewicz**[1], **M Godlewski**[1] **and D Arvanitis**[2]

[1] Inst. of Physics, Polish Academy of Sciences, Al. Lotnikow 32/46, 02-668 Warsaw, Poland
[2] Department of Physics and Astronomy, Uppsala University, Box 516, Uppsala 75120, Sweden

E-mail: `ikowalik@ifpan.edu.pl`

**Abstract.** We present soft x-ray absorption spectra from a series of Co doped ZnO films. We discuss systematic variations of the Co L-edge white line intensity and multiplet features for this series of samples. We document sizeable differences in the electronic state of the Co ionic cores, as well as in the local environment of the host lattice atoms, characterised by means of x-ray absorption spectra at the O K-edge and Zn L-edges. Model calculations allow to correlate the observed effects to small structural distortions of the ZnO lattice.

## 1. Introduction

Within the area of magnetism applications, ZnO:Co is an important model material for spintronics applications [1]. A series of ZnO:Co thin film samples, with Co diluted in the 10% range, are grown by means of the Atomic Layer Deposition (ALD) technique on Si substrates at temperatures 160 – 300 °C. Table 1 summarises some of the growth parameters. The samples are characterised by means of Atomic Force Microscopy (AFM), Scanning Electron Microscopy (SEM), X-Ray Diffraction (XRD), Secondary Ion Mass Spectrometry (SIMS), X-ray Photoelectron Spectroscopy (XPS) and SQUID magnetometry [2, 3]. The absence of stable ferromagnetism in the case of ZnO:Co has been linked in the literature with the possible existence of Co rich clusters within the ZnO matrix [4]. Given previous work documenting strong intensity variations in the Ni L-edge white line intensity and in the O K-edge spectral features, upon the controlled oxidation of ultra thin Ni films [5], it appears appropriate to study the XAS of the Co L-edges, O K-edge, Zn L-edges to correlate the absence of ferromagnetism in ZnO:Co samples with the electronic state and local environment of the Co, O and Zn atoms. The existence of small Co rich clusters, can lead to local strain within the ZnO lattice.

## 2. Experimental

We discuss XAS measurements taken at the I1011 beam line at the MAX IV synchrotron radiation facility in Lund, Sweden [6, 7]. We focus here on measurements performed in the Total Electron Yield (TEY) and characterise the near surface region of the samples. In figure 1, and figure 2 TEY results using linearly polarised x-rays are shown at normal x-ray incidence. All XAS spectra are normalised to the atomic continuum on a *per atom* basis to allow for a meaningful comparison. No measurable X-ray Magnetic Circular Dichroism (XMCD) was found in the TEY mode, for magnetic fields up to 0.5 T at 300K. The absence of ferromagnetism at 300K for the near surface region is in agreement with previous observations [3, 8]. To allow for

| Sample | $T_g$ [°C] | ZnO:CoO $m$:$n$ | t [nm] | $x_{Co}$ [%] |
|---|---|---|---|---|
| F213 | 300 | 8:1 | 130 | 8.8 |
| F254 | 160 | 8:1 | 70 | 18-36 |
| F279 | 300 | 8:1 | 1280 | 5.9 |
| F307 | 200 | 2:1 | 250 | 8 |
| F328 | 200 | 2:1 | 60 | 11.4 |

Table 1: List of the samples investigated in this study. We indicate the growth temperature, the ZnO to CoO cycles' ratio, thicknesses, and the Co concentrations obtained from SIMS.

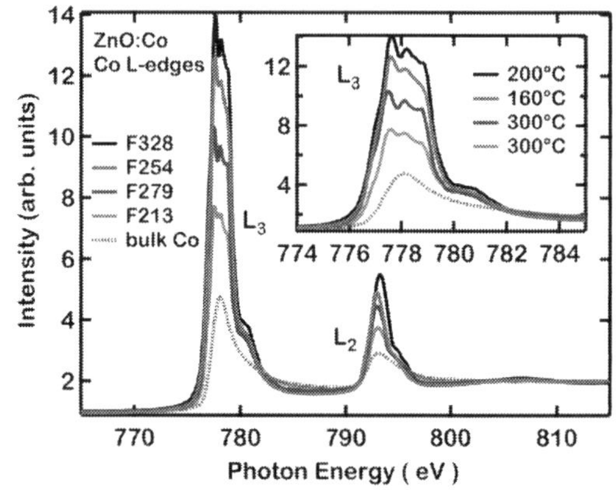


Figure 1: Co L-edge XAS spectra. White line intensity variations are found as a function of the growth parameters. A correlation with the growth temperature (see insert) is found.

a determination of the electronic state of the Co atoms, spectra were taken also for a reference metallic Co sample using a thin film grown and characterised *in situ* [7].

## 3. Results

The Co L-edges of the ZnO:Co films do indeed exhibit the typical multiplet shapes observed earlier on ZnO:Co magnetically dilute samples [8, 10]. We denominate the samples using a growth sequence number, such as F328 (Table 1, Fig. 1). We observe in the data presented in figure 1 a strong increase of the white lines intensity, versus the Co bulk. The white line area is proportional to the number of empty Co(3d) final states. We observe also a decrease of intensity in the spectral regions following the white lines (782-790 eV and 796-802 eV). These spectral areas correspond to Co(4s) final states. The data of Table 1 indicate that the spectral area variations correlate best with the growth temperature. Growth at low temperatures leads to higher white line intensities. Furthermore, in the data of figure 1, it can be observed that small relative intensity variations within the multiplet fine structure components take place. Interestingly, even in cases where the Co L-edges indicate only small variations in the few percent range in terms of overall intensity, clear variations in the O K-edge and Zn L-edges can be observed, as shown in the TEY data of figure 2. The variations observed at the Co L-edge are of the order of several 10% of the XAS intensity, at the Zn L-edge they are of order few %, in particular in the region close to the edge. Even if in terms of absolute intensity these can be considered small, their relative intensity variation cannot be neglected. The data of figure 2 indicate that intensity variations at the O K-edge can be very strong close to the edge, comparable in strength to the intensity variations of the Co data (around 530 eV).

## 4. Discussion

In figure 1 the XAS intensity variations at the Co L-edges traces the existence of more empty Co(3d) states and of less Co(4s) empty states, versus metallic Co. A similar result is found for the case of controlled oxidation of ultra thin Ni films for the Ni(3d) and Ni(4s) states [5]. For the Ni films these intensity variations correlate with the decrease of the ferromagnetic response [5]. As each Ni atom is losing first neighbour Ni atoms and acquiring as first neighbour O atoms, the alignment of the magnetic moment of the Ni atoms tends to switch from parallel to antiparallel versus the closest Ni neighbour atoms. A similar effect can be expected here with the Co atoms. The strong variations of spectral intensities for the Co atoms can be understood assuming the existence of Co rich clusters within the ZnO lattice, as suggested earlier [4]. Stronger L-edge white lines allow to conclude, assuming Co clusters are present within the ZnO lattice, either that they are less numerous leading to more substitutional Co, or smaller in size, or possibly incorporating more O atoms. Given the fact that Co is diluted in the ZnO lattice in the 10% range, the variations in intensity at the Zn and O edges link the stronger variations observed at the Co-edges, to small structural distortions relating with the ZnO lattice. Strain is indeed expected within the ZnO lattice if Co clusters are present, with Co not fully substituting the Zn atoms. It can be seen from the data of figure 1, and figure 2 that in cases where the Co white lines are strongest such as in the case of samples F254 and F328, still sizeable intensity variations are observed at the Zn L-edge and O K-edge. It is furthermore observed that the multiplet fine structure at the Co L-edge presents small differences in energy and relative intensities of the features observed, indicating that the Co atoms are at least partly located in an environment of different crystal symmetry. To elucidate the origin of the spectral variations we observe to structural distortions in the ZnO lattice, we perform model calculations at the O K-edge and Zn $L_3$-edge using the FEFF code, and compare model spectra with our experimental results.

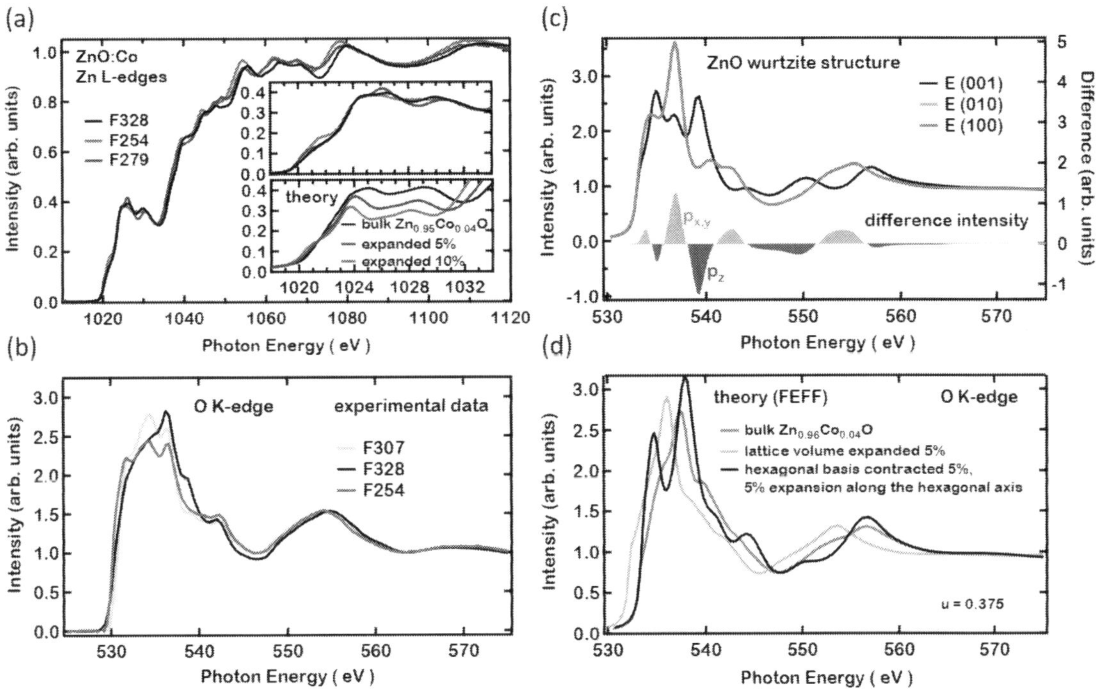


Figure 2: (a) XAS spectra for the Zn L-edges. In the insert the experimental spectra are shown, as well as two model calculations. (b) XAS O K-edge experimental spectra. (c) and (d) theoretical spectra for ZnO:Co thin film samples. For theory the Co content is 4%.

To calculate the XAS spectra we use the FEFF9 version of the *ab initio* FEFF code [11, 12]. We have used earlier the FEFF code to calculate the N K-edge XAS of GaN, and our results for GaN are consistent with earlier FEFF calculations and experimental data of c-axis oriented GaN films [13]. For the present calculations a ZnO cluster with a radius of 12Å (608 atoms) is used, around the photo-excited O atom, in the wurtzite crystal structure, with lattice constants of 3.249Å (x,y axis of the hexagonal unit cell in the hexagonal plane), 5.207Å (z axis along the hexagonal axis) and the parameter $u = 0.375\mathring{A}$. A Hedin-Lundqvist energy dependent exchange correlation potential with a constant imaginary part of 0.25 eV is used here. The calculations of figure 2 are performed with unpolarized light given the random granular growth. Linearly polarized light is used for calculations shown in figure 2, to explore the angle dependence in the wurtzite structure and assign the various spectral features. The Co atoms occupy Zn sites and are randomly distributed. Several different Co atom configurations have been considered, the results being similar to better than 2%. We use a $8\mathring{A}$ radius for self-consistency and full multiple scattering. This value is sufficient to assure the convergence of the results versus cluster size.

It is found that it is possible to reproduce the order of magnitude of the difference of spectral shapes of samples F307 and F254 versus F328 by considering small structural distortions of order of 5% of the lattice constant. We are able to reproduce in particular the strong variations at the edge, around 530 eV in the experimental data. The samples exhibit granular growth, with a typical grain size of order 4-5 nm and it is probable that a broadening of the absorption edge should be observed, as can be reproduced by the calculations, due to small structural distortions for each grain as presented in figure 2. figure 2 shows the symmetry assignment of the various spectral features at the O K-edge. The hexagonal z-axis is probed by orienting the electric field along the (001) direction. Focusing on the Zn $L_3$-edge data of figure 2, we also find that it is possible to reproduce the observed intensity variations with a ZnO lattice expansion of the 5-10% range. The existence of structural distortions for ZnO:Co constitutes an important result for magnetism applications. It has been proposed that structural distortions in ZnO:Co affect the magnetic properties. A decrease in the lattice constant of the order of 5% contribute to an increase in the Curie temperature of the order of 100 K in theoretical models [1].

## 5. Conclusions

We document intensity variations at the core level edges of all constituent atoms for ALD grown ZnO:Co samples, correlating with the sample growth temperature. The observed effects at the Co L-edges support that small Co rich clusters exist within the ZnO lattice. Co clusters are expected to induce local strain within the host lattice. Modelling of the O K-edge and Zn L-edges, using the FEFF *ab initio* code supports that small structural variations of the ZnO lattice constitute a possible explanation for the observed spectral intensity variations. We conclude that it is possible to use ALD growth, in particular focusing on the growth temperature in combination with XAS, to establish specific properties of the samples, such as the ferromagnetism of ZnO:Co.

## 6. Acknowledgements

We acknowledge M I Łukasiewicz for the sample growth. We acknowledge the support of the Polish National Science Centre under Grant No. 2011/03/D/ST3/02654, the Swedish VR Baltic Science Link project, and the Carl Tryggers foundation for Science Research (No. CTS 14:32).

[1] Pearton S J, Norton D P, Ivill M P et al. 2007 *IEEE Trans. Elec. Dev.* **54** 1040
[2] Godlewski M, Guziewicz E, Lukasiewicz M I et al. 2011 *Phys. Status Solidi B* **248** 1596
[3] Sawicki M, Guziewicz E, Lukasiewicz M I et al. 2013 *Phys. Rev. B* **88** 085204
[4] Iusan D, Knut R, Sanyal B et al. 2008 *Phys. Rev. B* **78** 085319
[5] May F, Tischer M, Arvanitis D et al. 1996 *Phys. Rev. B* **53** 1076
[6] Kowalik I A, Ohrwall G., Jensen B N et al. 2010 *J. Phys.: Conf. Series* **211** 012030
[7] Kowalik I A 2015 *Acta Phys. Pol. A* **127** 831
[8] Kobayashi M, Ishida Y, Hwang J I et al. 2010 *Phys. Rev. B* **81** 075204
[9] Mulders A M, Loosvelt H, Fraile Rodriguez A et al. 2009 *J. Phys.: Cond. Matter* **21** 124211
[10] Wi S C, Kang J-S, Kim J H et al. 2004 *Appl. Phys. Lett.* **84** 4233
[11] Rehr J J, Kas J J, Prange M P et al. 2009 *Comptes Rendus Physique* **10 (6)** 548
[12] Rehr J J, Kas J J , Vila F D et al. 2010 *Phys. Chem. Chem. Phys.* **12** 5503
[13] Kowalik I A, Persson A, Nino M A et al. 2012 *Phys. Rev. B* **85** 184411

Journal of Physics: Conference Series **712** (2016) 012105 doi:10.1088/1742-6596/712/1/012105

# Structural characterisation of $Fe_2O_3$ nanoparticles

**Michael Hagelstein[1], Dorothée Vinga Szabó[2], Sabine Schlabach[2], Paolo Masala[3], Marco Scavini[3], Mauro Coduri[4], Claudio Ferrero[5]**
[1]Karlsruhe Institute of Technology, ANKA, Karlsruhe, Germany,
[2]Karlsruhe Institute of Technology, Institute for Applied Materials (IAM) and Karlsruhe Nano-Micro-Facility (KNMF), Karlsruhe, Germany,
[3]University of Milan, Department of Chemistry, Italy,
[4]CNR-IENI Institute for Energetics and Interphases, Lecco, Italy,
[5]ESRF-The European Synchrotron, Grenoble, France

E-mail: michael.hagelstein@kit.edu

**Abstract**. The structure of nano-crystalline $Fe_2O_3$ particles, synthesized using the microwave plasma technique, has been analysed using synchrotron based X-ray absorption spectroscopy and X-ray powder diffraction, as well as transmission electron microscopy. Furthermore, magnetic properties, the crystal structure, and the microstructures are compared and the potential model character of the samples for structure simulations is discussed.

## 1. Introduction

There is growing interest in the application of monodisperse magnetic nanoparticles in magnetic suspensions, ferrofluids, or for drug delivery. In general, magnetic properties of nanoparticles are observed to be size dependent, independent from the method of synthesis [1]. However, our studies with microwave plasma synthesized $Fe_2O_3$ nanoparticles [2] show that the magnetic properties do not always correlate with the particle sizes. Although transmission electron microscopy studies reveal quite similar particle sizes of 5 nm, the saturation magnetization differs significantly (see [3]). The reason for this discrepancy is not yet fully clear. We believe that it is intimately related to the atomic structure of the nanoparticles coupled with possible structural disorder induced by the particle size. In addition to the antiparallel coupling of the spin and the orbital moment in the maghemite structure which has been observed in $Fe_2O_3/ZrO_2$ core/shell particles earlier [4], competing magnetic interactions of disordered crystal sub-lattices, strain as well as surface effects may lead to frustration and

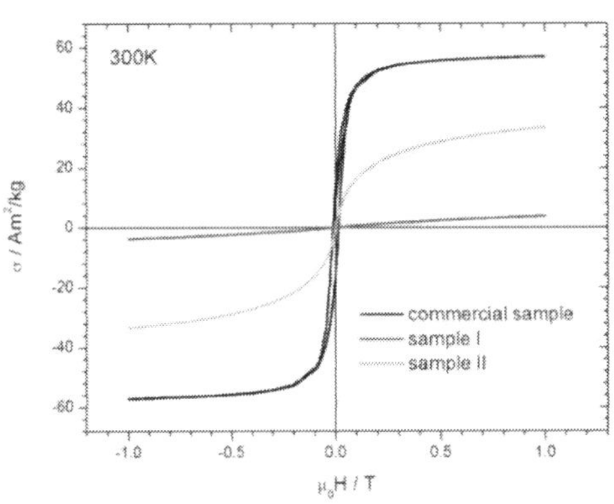


**Figure 1.** Magnetic moment per unit mass $\sigma$ as a function of applied field $\mu_0H$ of the samples measured at 300K.

disorder of the spin lattices. It is the aim of this paper to study the discrepancy between observed magnetization and particle size in more depth.

## 2. Results and discussion

We have determined the magnetic properties and performed an accurate structural and microstructural investigation of two nanocrystalline, microwave plasma synthesized $Fe_2O_3$ samples with similar particle size deduced by TEM and one commercial nano-$Fe_2O_3$ as a reference. Special emphasis is laid on X-ray absorption fine structure spectroscopy as well as high resolution X-ray powder diffraction.

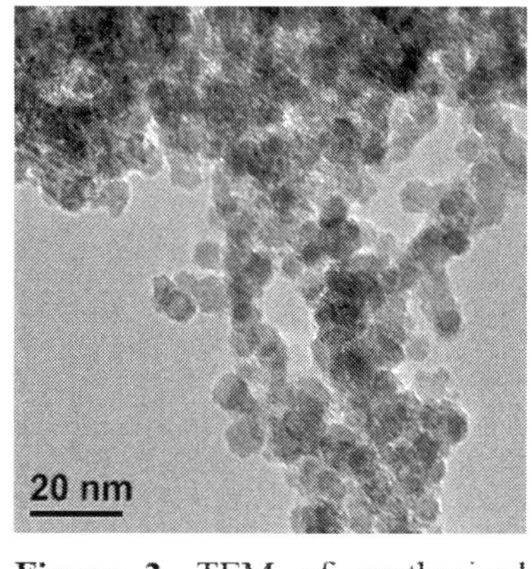


**Figure 2.** TEM of synthesised nanoparticles (sample II). A typical particle size of 5 nm is determined.

### *2.1. Magnetic properties and transmission electron microscopy*

The magnetic moment $\sigma$ as a function of applied magnetic field H has been determined with a SQUID magnetometer at room temperature in the range from 1 T to -1 T. The investigated $Fe_2O_3$ nanoparticles are superparamagnetic and are showing different saturation magnetization. The commercial material in contrast exhibit hysteresis and a significantly higher saturation magnetization (see figure 1). A typical TEM image from microwave-plasma synthesized $Fe_2O_3$ is displayed in figure 2.

### *2.2. X-ray powder diffraction*

X-ray powder diffraction measurements on samples I and II and on a $\gamma$-$Fe_2O_3$ reference sample were performed at room temperature at the ID31 beamline of the ESRF (now ID22). The samples were investigated at $\lambda$=0.400 Å covering a wave-vector $Q$ region up to $Q_{max} \sim 25.7$ Å$^{-1}$. The experimental data have been analysed using: i) the Williamson-Hall (W-H) method [5] in order to establish the crystallite size and strain, ii) the Rietveld method to determine the average crystal structure and iii) the total pair distribution function G($r$) to unveil the possible structural disorder. The last one is obtained by Fourier transforming the total scattering functions S($Q$) which, in turn, is obtained by the experimental data after suitable corrections and normalizations [6].

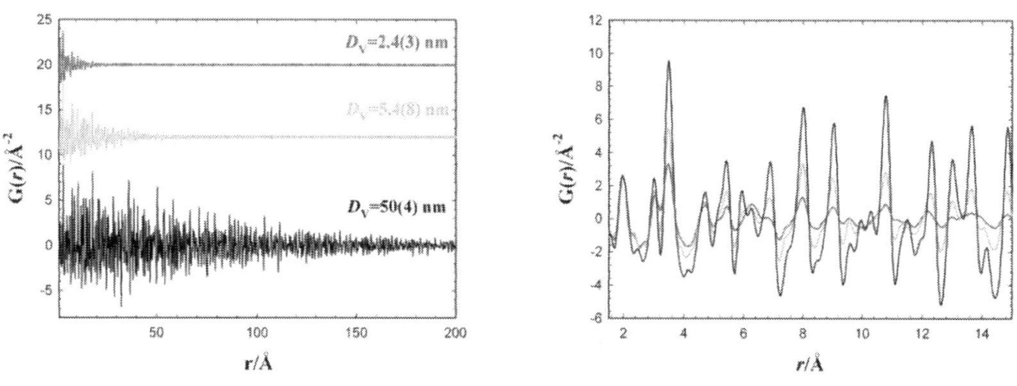


**Figure 3.** G($r$) data collected on the $\gamma$-$Fe_2O_3$ samples at the ESRF. The left hand panel presents data up to 200 Å, while the right hand panel highlights the low $r$ region. $\gamma$-$Fe_2O_3$ reference (black curves), sample I (red) and sample II (green).

Selected results of the Rietveld analysis are displayed in table 1. The most abundant phase in all the investigated $\gamma$-$Fe_2O_3$ samples is maghemite with a secondary hematite phase. The $\gamma$-$Fe_2O_3$ reference sample had to be refined in a tetragonal space group with a tripling of the c-axis because of ordering of iron vacancies. Figure 3 (left) shows the G($r$) data points on the two samples and reference sample in a large interatomic distance range.

The damping of the G($r$) functions is related to the finite dimension of the nanoparticles and can be mapped by a spherical envelope function which supplies alternative values of the crystallite dimension (see numbers in Figure 3, left). The damping of G($r$) is related to the average volume-weighted diameters ($D_V$) of the different sample crystallite sizes. On the right are depicted the same G($r$) curves. The widths of G($r$) peaks pertinent to the nearest-neighbors Fe-O distances at about 2 Å look quite similar for all the samples; conversely, for greater distances a broadening of the G($r$) peaks is apparent for samples I and II, as compared to the reference sample. See e.g. the Fe-Fe distances between 3 and 4 Å. This suggests a progressive disordering of interatomic distances, by decreasing the crystallite diameter, i.e. increasing the surface/volume ratio. This finding is in agreement with the high strain parameters detected for samples I and II by the W-H analysis compared to the $\gamma$-$Fe_2O_3$ reference sample (see table 1).

**Table 1.** Crystallographic data for the $\gamma$-$Fe_2O_3$ reference sample and the two nano-crystalline samples obtained by the Rietveld method. Size and strain data obtained by Williamson-Hall and PDF analysis.

| | $\gamma$-$Fe_2O_3$ | | | | | | | $\alpha$-$Fe_2O_3$ |
|---|---|---|---|---|---|---|---|---|
| | **Space group** | **a (Å)** | **c (Å)** | **Fraction (w%)** | **$D_V$ (W-H)** | **$D_V$ (PDF)** | **$\varepsilon$** | **Fraction (w%)** |
| $\gamma$-$Fe_2O_3$ reference | $P4_12_12$ | 8.3463(1) | 25.0269(1) | 96.0(0.1) | 55(5) | 50(4) | 0.0022(5) | 4.0(0.4) |
| Sample I | $P4_332$ | 8.3464(9) | -------- | 80.8(6) | 2.3(1) | 2.4(3) | 0.0095(9) | 19.2(9) |
| Sample II | $P4_332$ | 8.3439(8) | -------- | 87.2(1) | 3.8(1) | 5.4(8) | 0.0084(6) | 12.8(7) |

### *2.4. X-ray absorption spectroscopy*

XAFS spectra have been collected at the Fe-K edge in transmission mode. Experiments were performed at the SUL-X beamline at ANKA. This beamline delivers a focused, monochromatic beam by using a wiggler source, focusing mirrors and a cryogenically cooled Si(111) double crystal monochromator. The data analysis followed a standard procedure using the code WINXAS. For the

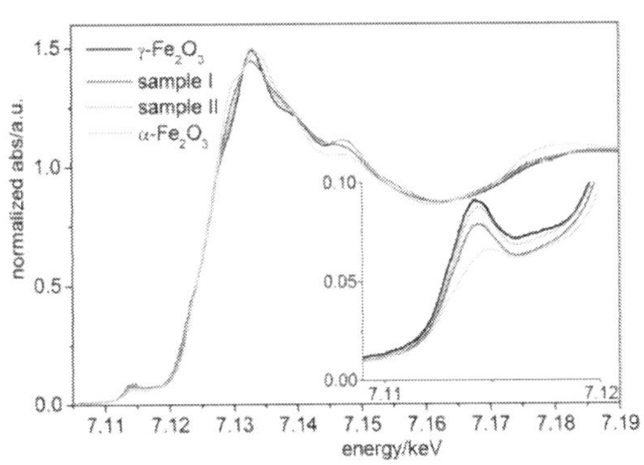


**Figure 4.** The near edge region of iron oxide nanoparticles (sample I, sample II) together with $\gamma$-$Fe_2O_3$ reference sample and $\alpha$-$Fe_2O_3$.

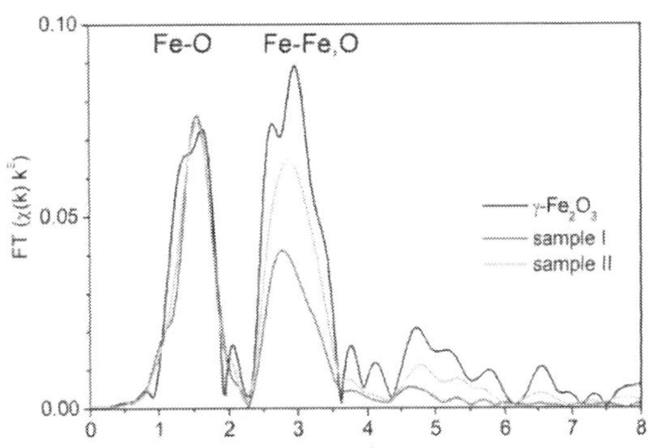


**Figure 5.** Fourier transforms of the XAFS (not phase shift corrected) of the sample I and II together with the $\gamma$-$Fe_2O_3$ reference sample.

XAFS analysis, a $k$-range of 2.4 - 14.5 $Å^{-1}$ was used for calculating the Fourier transform. The near edge structure and the Fourier transform of the XAFS are displayed in figure 4 and 5, respectively.

The XANES region with the pre edge peak assigned to a 1s to 3d transition provides valuable information on the relative occupation of tetrahedral and octahedral sites [7]. In particular, the decrease of the pre-edge peak intensity of the nanomaterials with respect to the reference sample suggests a smaller $Fe_{tetra}/Fe_{oct}$ ratio.

XAFS analysis using $\chi(k)$ $k^3$ weighting with FEFF theoretical functions [8] have been calculated for the first Fe-O shell and two next nearest Fe-Fe shells using structural data from Shmakov et al. [9] and Petkov et al. [10]. Higher Fe-O coordination shells and multiple scattering paths are not included. The bond lengths $R_i$ and coordination numbers $CN_i$ for the coordination shells are summarized in table 2.

**Table 2.** Radial distances $R_i$ (±0.01 Å) and coordination numbers $CN_i$ (error ±10%) determined from the XAFS data (the many body amplitude reduction factor $S_0^2$ fixed at 0.9). The Debye-Waller factors for the next nearest Fe shells $\sigma_2^2$ and $\sigma_3^2$ have been kept fixed at 0.01 $Å^2$ during refinement of the three shells.

| | $R_1$ [Å] Fe-O | $\sigma_1^2$ [$Å^2$] | $CN_1$ Fe-O | $R_2$ [Å] Fe-Fe | $CN_2$ Fe-Fe | $R_3$ [Å] Fe-Fe | $CN_3$ Fe-Fe |
|---|---|---|---|---|---|---|---|
| $\gamma$-$Fe_2O_3$ reference | 1.92 | 0.012 | 6.2 | 2.99 | 3.1 | 3.42 | 6.9 |
| Sample I | 1.95 | 0.010 | 5.2 | 3.01 | 1.5 | 3.41 | 5.7 |
| Sample II | 1.94 | 0.011 | 5.3 | 3.02 | 3.5 | 3.43 | 6.5 |

## 3. Conclusions

This investigation reveals major structural characteristics of the synthesized samples. The samples are predominantly maghemite but with considerable fractions of hematite. Lattice parameters are comparable to the bulk; the coordination of Fe ions at the surface appears to be distorted. The site occupation ratio $Fe_{tetra}/Fe_{oct}$ is smaller than for bulk maghemite which is also corroborated by the longer average bond length for the first Fe-O shell determined by XAFS. Although particle sizes seem similar by TEM investigations, the crystallite sizes are significantly different. At decreasing the particle size, disorder increases both allowing atomic relaxations and enhancing the ratio of tetrahedral to octahedral coordinated iron. These differences may explain the differences in magnetic properties.

Still many unanswered questions further remain, i.e. the structural differences of surface and bulk of nanoparticles are to be analysed and discussed in greater detail, whether the surface contains glassy states which would explain magnetic properties, and how particles interact with each other. We hope to reveal such issues by comparing single particle structure simulations with our experimental data.

**Acknowledgments**

We are particularly indebted to the ANKA and ESRF synchrotron user facilities. The scientists at ESRF, Andy Fitch and Yves Watier, as well as at ANKA, Tao Liu, Stefan Mangold, Ralph Steininger, and Stephen Doyle helped us collecting and analyzing the data.

**References**

[1] Szabó D V, Schlabach S 2014 *Inorganics* **2(3)** 468
[2] Vollath D, Szabó D V, Taylor R D and Willis J O J 1997 *Mater. Res.* **12** 2175
[3] Nadeem K, Krenn H, Traussnig T, Würschum R, Szabó D V, Letofsky-Pabst I 2011 *J. Magn. Magn. Mat.* **323** 1998
[4] Pellegrin E, Hagelstein M, Doyle S, Moser H O, Fuchs J, Vollath D, Schuppler S, James M A, Saxena S S, Niesen L, Rogojanu O, Sawatzky G A, Ferrero C, Borowski M, Tjernberg O, and Brookes N B 1999 *phys. stat. sol. (b)* **215** 797
[5] Williamson G K, Hall W H 1953 *Acta Metall.* **1** 22
[6] Egami T, Billinge S J L, *Underneath the Bragg peaks: Structural Analysis of Complex Materials* 2003 **V. 7** Pergamon Materials Series: Oxford
[7] Meneghini C, Boscherini F, Pasquini L, Renevier H 2009 *J. Appl. Cryst.* **42** 642
[8] Rehr J J, Kas J J, Vila F D, Prange M P, Jorissen K 2010 *Phys. Chem. Chem. Phys.* **12** 5503
[9] Shmakov A N, Kryukova G N, Tsybulya S V, Chuvilin A L, Solov'eva L P 1995 *J. Appl. Cryst.* **28** 141
[10] Petkov V, Cozzoli P D, Buonsanti R, Cingolani R, Ren Yang 2009 *J. Am. Chem. Soc.* **131(40)** 1426

Journal of Physics: Conference Series **712** (2016) 012106 doi:10.1088/1742-6596/712/1/012106

# Structure analyses of Cu nanoclusters in the soft magnetic $Fe_{85.2}Si_1B_9P_4Cu_{0.8}$ alloy by XAFS and fcc cluster model

**M Matsuura[1], M Nishijima[1], K Konno[2], H Ofuchi[3], K Takenaka[1], A Makino[1]**

[1] Research and Development Center for Ultra High Efficiency Nano-crystalline Soft Magnetic Material, Institute for Materials Research, Tohoku University, Sendai 980-8577, Japan

[2] Sendai National College of Technology, Medeshima, Natori, Miyagi 981-1239, Japan

[3] Industrial Application Division, Japan Synchrotron Radiation Research Institute (JASRI), Kouto, Sayo, Hyogo 679-5198, Japan

E-mail:m_matsuura@imr.tohoku.ac.jp

**Abstract.** Size of the clusters and structure details of fcc Cu clusters in nanocrystalline soft magnetic alloy of $Fe_{85-86}Si_{1-2}B_8P_4Cu_1$ (NANOMET) are investigated. A linear combination fitting of XAFS data indicates that about 30% of Cu atoms are partitioned in the fcc clusters and the rest in the amorphous matrix. EXAFS of the fcc Cu nanocluster embedded in amorphous matrix is calculated on the basis of a simple fcc structure model using FEFF9. Surface effect of the nanoclusters is considered by counting a fraction of the nearest neighbour atoms in amorphous matrix. Good agreement with the experimental result is obtained for the fcc nanocluster with 9 coordination shells which consists of total 177 atoms within 1.5 nm in a diameter.

## 1. Introduction

Nanocrystalline soft magnetic alloys have been attracted much attention because of their high saturation magnetic flux density ($B_s$), low coercivity ($H_c$) and low magnetic core loss ($W$). Those soft magnets, however, exhibit relatively low $B_s$ compared with a conventional silicon steel (Fe-3.5wt. % Si) with 2 T. Recently Makino et al. have developed a new Fe-based alloy of $Fe_{85-86}Si_{1-2}B_8P_4Cu_1$ (NANOMET) which exhibits high $B_s$ (1.82~1.85 T) together with low $H_c$ ( 2.6~5.8 A/m) and considerably low $W$ [1-2]. These superior soft magnetic properties of NANOMET originate from the precipitated α-Fe nanocrystallites embedded in a residual amorphous matrix. NANOMET contains a small amount of Cu, ~1 at%, which effectively reduces the grain size of the precipitated α-Fe crystallites to a nanometer scale [3]. It is not, however, well understood about the role of Cu in the nanocrystallization of precipitated α-Fe crystallites. Because of a dilute content of Cu (<1.0 at%), conventional structural analyses, XRD and TEM, are inappropriate, whereas XAFS, especially fluorescence XAFS, is an advantageous tool for studying the local structure of a dilute element. In the previous work the authors observed the local structure changes around Cu in NANOMET during nanocrystallization by XAFS [4-5]. These XAFS results and TEM observations [6] conform that fcc Cu clusters are formed prior to the α-Fe precipitation like as the case of FINEMET [7-9]. But unlike

FINEMET fcc Cu clusters in NANOMET transform into bcc structure in the early stage of α-Fe precipitation. In this paper we calculated EXAFS of the fcc Cu nanoclusters based on a simple fcc structure model and estimated the size of the cluster.

## 2. EXAFS calculation of the fcc Cu nanoclusters embedded in amorphous matrix

Cu atoms are distributed both in fcc Cu clusters and amorphous matrix. So at first, the distribution fractions of Cu in the two phases are evaluated from the experimental data. Secondly, EXAFS functions $\chi(k)_{fcc\ cluster}$ of the model fcc Cu clusters are calculated using FEFF9. Thirdly, fraction of the nearest neighbour atoms outside clusters $x$ is calculated. Contribution of Cu in amorphous matrix surrounding fcc clusters is estimated from the fraction $x$ and $\chi(k)_{amor}$ deduced from the experimental data.

### 2.1. Fraction of Cu partitioned into fcc Cu clusters.

Figure 1 shows results of the Cu K-edge EXAFS in terms of radial structure function (RSF) of the as-quenched, annealed at 340°C and 700°C for $Fe_{85.2}Si_1B_9P_4Cu_{0.8}$ alloy together with that of a Cu foil. XAFS measurements were carried out at BL14B2 in SPring-8 by a fluorescence mode using 19-elements SSD at 10-20K. Experimental details were described in the previous works [4-5]. The RSF of the as-quenched ribbon shows a typical of amorphous structure and those of the annealed at 340°C and 700°C show an fcc structure. In the previous papers, authors proved that Cu atoms form fcc clusters prior to α-Fe precipitation and they segregate above the second crystallization temperature ($T_{x2}$=571°C) [4-5]. Here we try to analyze the fcc Cu clusters formed during nanocrystallization of $Fe_{85.2}Si_1B_9P_4Cu_{0.8}$ alloy. Cu atoms in the sample annealed at 340°C are partitioned both into the fcc Cu clusters and amorphous matrix. The EXAFS function of the annealed at 340°C, $\chi(k)_{340℃}$, can be described as

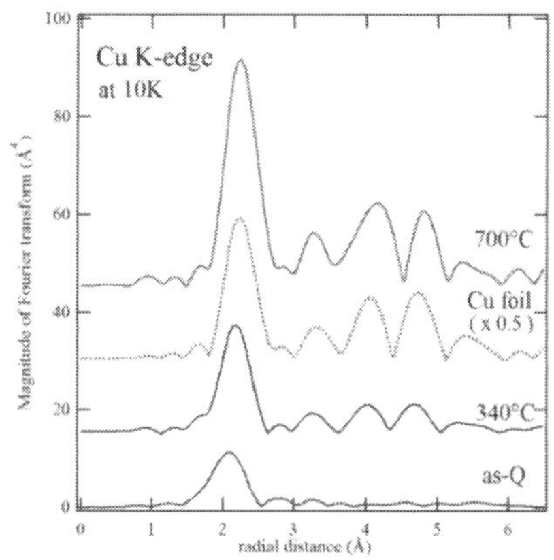


**Figure 1.** Radial structure functions of the Cu K-edge for $Fe_{85.2}Si_1B_9P_4Cu_{0.8}$ alloy: as-quenched, annealed at 340°C and 700°C together with a Cu foil.

$$\chi(k)_{340℃} = \alpha\cdot\chi(k)_{fcc\ cluster} + (1-\alpha)\cdot\chi(k)_{amor.} \quad (1)$$

, where $\alpha$ is a fraction of Cu atoms included in the fcc clusters, $(1-\alpha)$ a fraction of those in the amorphous matrix, $\chi(k)_{fcc\ cluster}$ an EXAFS function of the fcc Cu clusters and $\chi(k)_{amor.}$ that of the amorphous matrix. α is estimated by fitting the $\chi(k)_{340℃}$ to a linear combination of these data of $\chi(k)_{700℃}$ and $\chi(k)_{as\text{-}quench}$ which are considered to correspond to $\chi(k)_{fcc\ cluster}$ and $\chi(k)_{amor.}$, respectively in equation (1). The estimated value of α is 0.29, i.e. around 30% of Cu atoms contribute to form fcc clusters and the rest are in the amorphous matrix. We thus can evaluate $\chi(k)_{fcc\ cluster}$ in equation (1) to which the calculation is compared.

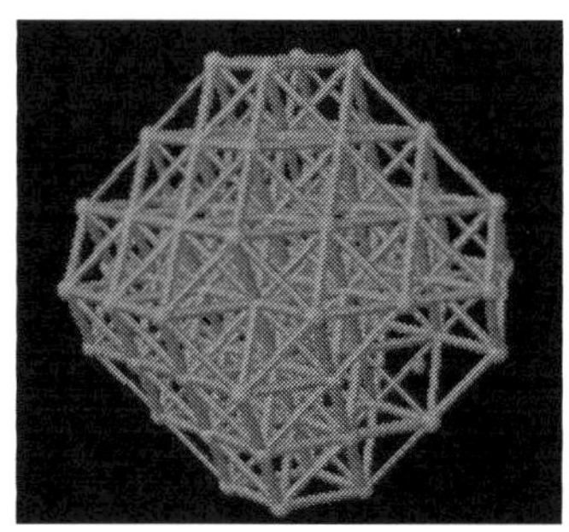

**Figure 2.** Scattering atoms within the clusters when an absorbing atom (shown as pink color) locates in the third coordination shell of the *5-th* cluster.

### 2.2. Structure model of an fcc Cu nanocluster.

The fcc Cu clusters in NANOMET are assumed to consist exclusively of Cu atoms, i.e. pure Cu, and a size distribution of the fcc clusters is neglected. Coordinates, (x,y,z)'s, are assigned to all atoms in the cluster of which center is (0,0,0). We call the "*n-th* cluster" which has the outermost coordination shell of $n$. Figure 2 shows all the scattering atoms of the clusters when an absorbing atom (colored with pink) locates in the third coordination shell of the *5-th* cluster. When the specific coordination shell $j$ of the *n-th* cluster consists of $N_j$ atoms, the EXAFS functions, $\chi(k)_{n,j}$, of those atoms in the $j$-th shell are same because they have same symmetry. The $\chi(k)_{n,j}$'s are calculated for all coordinate shell, $j$=1, 2, , ,$n$, supposing an isolated fcc Cu cluster, i.e. nothing is surrounded around the cluster, using ab initio calculation code of FEFF9 [10]. Finally the EXAFS function of the isolated *n-th* cluster $\chi(k)_{n,i}$, is calculated by averaging over all atoms $N(=1+\sum_{j=1}^{n}N_j)$ in the cluster using equation (1):

$$\chi(\mathrm{k})_{n,i} = \frac{\sum_{j=1}^{n} N_j \chi(k)_{n,j}}{N} \quad (1)$$

*2.3. Contribution of the amorphous matrix: surface effects of the nanocluster.*

Because the Cu clusters in NANOMET are embedded in the amorphous matrix, photo electrons emitted from atoms close to the surface of the cluster are scattered by amorphous matrix. Contribution from amorphous matrix to $\chi(k)$ is not easy to evaluate exactly but as indicated in Figure 1 the RSF of the as-quenched one has only one small peak of the nearest neighbor shell. When an absorbing atom near the surface has $p$ nearest neighbors within the cluster, then we assume that the nearest neighbor atoms in the amorphous matrix are *12-p* atoms. A fraction of the nearest neighbor atoms, $x$, outside the cluster to the total nearest neighbor atoms, i.e. $N_{nn}(=12\times N)$, is evaluated by following equation:

$$x = \frac{\sum_{j=0}^{n} N_{j,o}}{N_{nn}} \quad (2)$$

, where $N_{j,o}$ is number of the nearest neighbor atoms in the amorphous matrix around an atom in the *j-th* coordinate shell. EXAFS of the *n-th* cluster $\chi(k)_n$ can be described as:

$$\chi(k)_n = (1-x)\chi(k)_{n,i} + x\cdot\chi(k)_{n,o} \quad (3)$$

, where $\chi(k)_{n,o}$ is $\chi(k)$ contributed from atoms outside the cluster which is substituted by $\chi(k)_{as-quench}$.

*2.4. Calculated results of EXAFS for the fcc Cu nanocluster.*

In the calculation a coordination shell around a center of the cluster is changed from $n$=1 to 12, which corresponds to 0.512 nm to 1.77 nm in diameter of the clusters in which $N$=13 to 249 atoms are included. In the calculation an interatomic distance is 0.250 nm, which leads to fit the first peak of the RSF to the experimental data. Figure 3 shows calculated results of the RSF of the fcc Cu clusters with a different coordination shell ($n$=3, 9 and 12), together with the $\chi(k)_{fcc\ cluster}$ reduced from the experimental data. Cluster diameters are 0.88, 1.53, 1.77 nm for $n$=3, 9, 12. As a matter of course, surface effect of a cluster increases with decrease in a cluster size; the $3^{rd}$-cluster, has the lowest RSF peaks and besides they attenuate rapidly with increase in a radial distance and the RSF curve of the *12-th* cluster exhibits the most similar features to that of a Cu foil among other cases. Figure 3 shows that the result of the $n$=9 exhibits the best agreement with that of the $\chi(k)_{fcc\ cluster}$. Below and above $n$=9 the agreement becomes poorer than that of $n$=9. From these results it can be said that about 30% of Cu atoms in $Fe_{85.2}Si_1B_9P_4Cu_{0.8}$ alloy are partitioned in the fcc

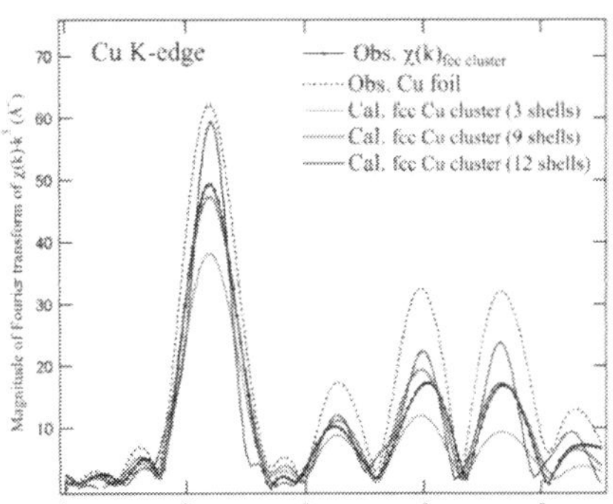


**Figure 3.** Calculated RSFs of the fcc Cu clusters with different coordination shells: n=3, 9, and 12, together with the experimental results of the $\chi(k)_{fcc\ cluster}$ and a Cu foil.

Cu clusters and size of the fcc nanocluster is about 1.8 nm and 249 atoms are included in it. The value of 1.8 nm in size of the cluster agrees with a recent TEM observation [6]

## 3. Discussion

The Cu K-edge EXAFS of $Fe_{85.2}Si_1B_9P_4Cu_{0.8}$ alloy shows that local structure around Cu changes in a sequence of "amorphous→fcc→bcc→fcc" in the cause of nanocrystallization [4-5]. Here, we focus our attention to the structure of the fcc clusters formed prior to the α-Fe precipitation. A fraction of Cu atoms α which form fcc clusters embedded in amorphous matrix is determined by a liner combination fitting (LCF). In the LCF procedures data of the $\chi(k)_{700℃}$ is applied rather than that of a Cu foil because the segregated Cu crystals are more similar to the fcc Cu clusters rather than a Cu foil: concentration and surface effects. Interatomic distance which can fit to the experiments in the calculation is 0.250 nm. This value is a little bit shorter than that of a Cu foil, 0.2556 nm. Probably it stems from that in the present calculation the nanoclusters are assumed to consist only from Cu but actually some Fe atoms are also included in it. A whole future of the EXAFS function of the present fcc cluster alters little if some fraction of Cu atoms are replaced by Fe atoms. Although the present fcc cluster model can fit the experimental EXAFS result, it cannot explain how fcc Cu clusters contribute nanocrystallization of precipitated α-Fe crystallites. Previous XAFS work [4] indicates that before the onset of α-Fe precipitation ($T_{x1}$=432°C), fcc structure of Cu clusters transforms into bcc and TEM observation shows that nanometer size Cu rich particles are observed mostly at around peripheral of α-Fe nanocrystallites [6]. From these facts α-Fe precipitation occurs preferentially nearby fcc Cu clusters and surface energy at interfaces between the fcc Cu clusters and bcc Fe crystals may cause to transform structure of Cu clusters from fcc to bcc. The Cu K-edge EXAFS data indicates that the sample annealed at 340°C shows fcc structure and bcc at 380°C, and it shows mixture of fcc and bcc at 365°C. It is necessary to examine whether this structure model of the fcc clusters can explain such a structure transformation, which will be presented somewhere else in near future.

## 4. Acknowledgements

The synchrotron radiation experiments were performed at the BL14B2 of SPring-8 with the approval of the Japan Synchrotron Radiation Research Institute (JASRI) (Proposal Nos. 2013A1641, 2013B1715, 2014A1705, 2014B1794 and 2015A1850). This work was financially supported by Ministry of Education, Culture, Sports, Science and Technology (MEXT), Japan under "Tohoku Innovative Materials Technology Initiatives for Reconstruction" project, "Ultra-low Core Loss Magnetic Material Technology Area."

## References

[1] Makino A, Kubota T, Yubuta K, Inoue A, Urata A, Matsumoto K and Yoshida S 2011 *J. Appl. Phys.* **109** 07A302-1.
[2] Makino A, Men H, Kubota T, Yubuta K, and Inoue A 2009 *Mater. Trans.* **50** 204.
[3] Cui L, Men H, Makino A, Kubota T, Yubuta K, Qi M and Inoue A 2009 *Mater. Trans.* **50** 2515.
[4] Nishijima M, Matsuura M, Takenaka K, Takeuchi A, Ofuchi H and Makino A 2014 *AIP Adv.* **4** 057129.
[5] Matsuura M, Nishijima M, Takenaka K, Takeuchi A., Ofuchi H and Makino A 2015 *J. Appl. Phys.* **117** 17A324.
[6] Nishijima M, Matsuura M, Zhang Y and A Makino 2015 *Phil. Mag. Let.* 2015 **95** 277.
[7] Yoshizawa Y, Oguma S and Yamauchi K 1988 *J. Appl. Phys.* **64** 6044.
[8] Kim S H, Matsuura M, Sakurai M and Suzuki K 1993 *Jpn. J. Appl. Phys.* Part 1 **32**(Suppl. 32-2) 676.
[9] Ayers J D, Harris V G, Sprague J A, Elam W T and Jones H N 1998 *Acta Mater.* **46** 1861.
[10] Rehr J J, Kas J J,Vila F D, Prange M P and Jorissen K 2010 *Phys. Chem. Chem. Phys.* **12** 5503-5513.

# XAFS study on the temperature-dependent occupation sites of Co codopants in (Co, Cu)-codoped ZnO films

**Fengchun Hu†, Shibao Zhang†, Zhiyun Pan, Wensheng Yan, Qinghua Liu*, Tao Yao, and Shiqiang Wei**

National Synchrotron Radiation Laboratory, University of Science and Technology of China, Hefei, Anhui 230029, P. R. China

*Corresponding author. E-mail: qhliu@ustc.edu.cn
†These authors contributed equally to this work.

**Abstract.** Elemental codoping has been an effective way to regulate the structural and electronic properties of semiconductors. By using x-ray diffraction and x-ray absorption fine structure spectroscopy, we investigate the local structure and spatial occupations of Co dopants in Cu-doped ZnO thin films prepared by pulsed-laser deposition method. It is revealed that the Co dopants are substantially incorporated into ZnO matrix when the deposition temperature is increased up to 650 °C, although the preferential orientation of ZnO film is changed. The results provide experimental guidance in the synthesis of the co-doped ZnO based dilute magnetic semiconductors.

## 1. Introduction

Zinc oxide (ZnO) semiconductor has drawn intense scientific interest because of its potential application in photoelectronics, blue/UV light luminescence, and spintronics [1-3]. Until now, the most effective method to improve the properties of ZnO is to introduce impurity atoms [2, 4]. Doping of magnetic atoms into the ZnO lattice could achieve ferromagnetism for spintronic devices [5]. As a promising candidate for practical use, high mobility of electrons and enough carrier density are demanded, which could be resolved by codoping of another dopant into the single-doped materials [6]. Although a great number of studies have been performed to investigate the single- or co-doped ZnO systems, the relationship between the atomic occupation and electronic structure of the dopants in the ZnO matrix is still unclear.

The processes of the crystal formation are sensitive to the environmental conditions of samples preparation. For example, Chakraborti *et al.* have studied the effect of substrate temperature on the properties and structure of the (Co, Cu):ZnO thin films deposited on sapphire c-plane by pulsed-laser deposition (PLD). The magnetization of ZnO thin films prepared at 500 °C decreased evidently compared with that of the films fabricated at 600 °C, which would be attributed to the formation of Cu nanoclusters in the ZnO lattice [6]. Our group has reported that the growth orientation of ZnO and the dopant occupation sites would be changed by varying the substrate temperature [7]. All of these studies have clearly implied that the atomic occupation, electronic properties, and crystal structure of doped ZnO systems can be affected by the growth temperature. In order to rationally control the structure and functional properties of ZnO-based semiconductors, it would be excited to seeking a comprehensive knowledge on the atomic and electronic interactions of the co-dopants in ZnO thin films under different growth temperatures.

In this study, a series of (Co, Cu)-co-doped ZnO thin films have been prepared with the same composition under different substrate temperatures by pulsed-laser deposition method. The x-ray absorption fine structure (XAFS) spectroscopy and x-ray diffraction (XRD) were used to probe the local structure and spatial occupations of Co dopants in the (Co, Cu)-codoped ZnO thin films. These results broad our understanding on the modulation of co-dopants for modifying the atomic and electronic properties of ZnO-based thin films.

## 2. Experimental Section

The $Zn_{0.93}Co_{0.05}Cu_{0.02}O$ thin films were prepared on Si (100) substrates by pulsed laser deposition (PLD) (248 nm wavelength, 5 Hz repetition frequency) under various substrate temperatures (RT, 200, 400 and 650 °C). The $Zn_{0.93}Co_{0.05}Cu_{0.02}O$ targets were synthesized by modified solid-state reaction method and the detailed procedures were described in our previous work [8]. The target and substrate were kept rotation during the experiment process. The structural properties of the (Co, Cu)-codoped ZnO thin films were characterized by X-ray Diffractometer. The Co *K*-edge XAFS spectra were used to ascertain the local structure around the Co ions performed in the fluorescence mode at the U7C beamline in the National Synchrotron Radiation Laboratory (NSRL), China.

## 3. Results and discussion

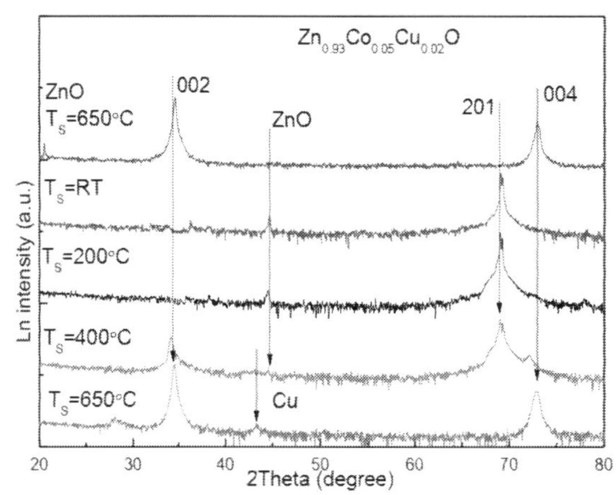


**Figure 1**. XRD patterns of (Co, Cu)-codoped ZnO films grown at different temperatures and the pristine ZnO deposited at 650 °C as comparison.

Figure 1 shows the XRD patterns of (Co, Cu)-codoped ZnO thin films deposited at different substrate temperatures. To highlight the dopant-related weak peaks, the peak intensity is shown at a logarithmic scale. The (201) diffraction peak of wurtzite ZnO located at 69° is predominant for the films grown at Ts≤400 °C. As the substrate temperature up to 400 °C, there are two additional peaks corresponding to the (002) and (004) planes of ZnO abruptly arising. Finally, the (201) peak vanishes entirely for the film deposited at 650 °C. These results indicate that the crystalline orientation of the codoped ZnO films transforms from the [201] direction at low substrate temperature to the c-axial orientation when the temperature is up to 400 °C. Furthermore, a low intensity peak corresponding to metallic Cu appears at ca. 43° for the films grown at Ts≥400 °C. The characteristic peaks of second phases related to Co or Cu do not emerge for all the films grown at various substrate temperatures.

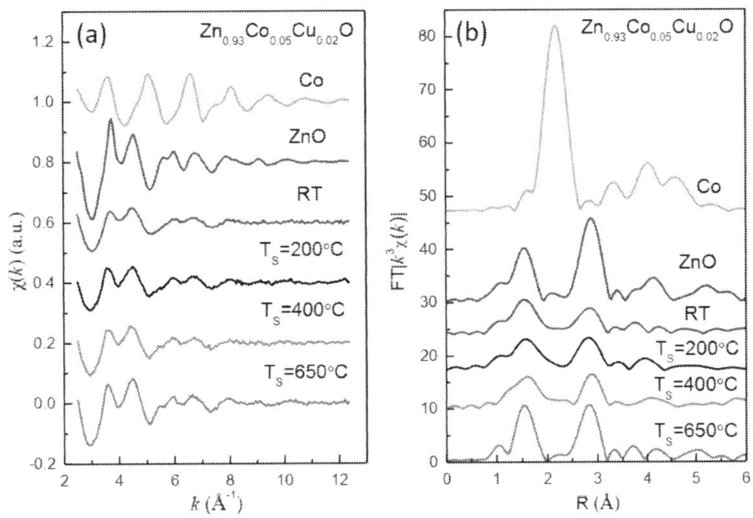


**Figure 2.** (a) Co $K$-edge EXAFS oscillation functions $\chi(k)$ and (b) FT spectra of $k^3\chi(k)$ of $Zn_{0.93}Co_{0.05}Cu_{0.02}O$ thin films deposited at different temperatures of RT, 200 °C, 400 °C, and 650 °C, and the pristine ZnO deposited at 650 °C and Co foil for references.

In order to clarify the atomic structures of the doped ions in the thin films, the extended x-ray absorption fine structure (EXAFS) technique is used to detect the local structures of Co dopants in ZnO lattice prepared at various substrate temperatures. The EXAFS oscillation $\chi(k)$ functions and the Fourier transform (FT) curves of the $k^3$-weighted $\chi(k)$ at Co $K$-edges for the films grown at different substrate temperatures are plotted in Figure 2. In the FT spectra for the samples, two strong peaks corresponding to the first-nearest Co–O and the second-nearest Co–Zn neighbors are located at around 1.55 and 2.88 Å, respectively [9]. For all the (Co, Cu) codoped ZnO thin films, there is no peak corresponding to the first-nearest Co–Co coordination of Co metal appears in the FT curve [10], which indicates that the Co ions are still substituted for the Zn sites in the ZnO matrix.

Summarizing above XRD and XAFS results, we conclude that temperature plays an important role on the interrelations between the atomic occupation of (Co, Cu) dopants and ZnO crystal orientations. Additionally, because the ionization energy of Co (17.3 eV) is lower than that of Zn (17.9 eV) and the solubility of Co in ZnO is relatively large, Co ions could be successfully incorporated into the ZnO lattice and substituted for the Zn sites [11]. In contrast, the ionization energy of Cu (20.3 eV) is larger than that of Zn. The Cu atoms are energetically inert to be oxidized and substituted for the Zn sites. Consequently, the Co dopants are precipitated out as metallic species even under the relatively mild growth environment of low growth temperature [7]. Although the transformation of ZnO matrix preferential orientation would affect the dopants interactions and electronic exchange, the low concentration of Cu could not afford stronger electrostatic forces and enough electrons to reduce the Co ions. Therefore, the Co atoms were substituted into ZnO matrix for the samples prepared at all substrate temperatures.

## 4. Conclusion

The $Zn_{0.93}Co_{0.05}Cu_{0.02}O$ thin films were prepared by PLD on Si (100) substrate at different substrate temperatures. The XRD results suggest that the orientation of the codoped ZnO films was changed and the Cu dopants were easily separated from the ZnO matrix with increasing the substrate temperature. In-depth analysis of the Co *K*-edge EXAFS data shows that the Co atoms were substituted into ZnO matrix for the samples prepared at all substrate temperatures. The results provide some hints for understanding the interaction between the dopant ions and the ZnO lattice in the transition metal-doped ZnO thin films fabricated at different substrate temperatures.

## Acknowledgments

This work was supported by National Natural Science Foundation of China (Grant No. 11175184, 11135008, U1332131, and 11305172). The authors are grateful to NSRL for the synchrotron radiation beamtime.

## References

[1] Ohno H 1998 *Science* **281** 951-956

[2] Ozgur U, Alivov Y I, Liu C, Teke A, Reshchikov M A, Dogan S, Avrutin V, Cho S J and Morkoc H 2005 *J. Appl. Phys.* **98** 041301

[3] McCluskey M D and Jokela S J 2009 *J. Appl. Phys.* **106** 071101

[4] Schmidt-Mende L and MacManus-Driscoll J L 2007 *Mater. Today* **10** 40-48

[5] Coey J M D, Venkatesan M and Fitzgerald C B 2005 *Nat. Mater.* **4** 173-179

[6] Chakraborti D, Ramachandran S, Trichy G, Narayan J and Prater J T 2007 *J. Appl. Phys.* **101** 052918

[7] Zhang S B, Hu F C, Hu J F, Chen W R, Liu Q H, Jiang Y, Pan Z Y, Yan W S, Sun Z H and Wei S Q 2013 *J. Phys. Chem. C* **117** 24913-24919

[8] Hu F C, Pan Z Y, Yao T, Oyanagi H, Jiang Y, Yan W S, He B, Ye J and Wei S Q 2009 *J. Phys.D: Appl. Phys.* **42** 125408

[9] Pan Z Y, Hu F C, He S, Liu Q H, Sun Z H, Yao T, Xie Y, Oyanagi H, Xie Z, Jiang Y, Yan W S and Wei S Q 2012 *J. Phys. Chem. C* **116** 4855-4861

[10] Zhang H, Yao T, Sun Z H, Li Y Y, Liu Q H, Hu F C, Pan Z Y, He B, Xie Z and Wei S Q 2010 *J. Phys. Chem. C* **114** 13596-13600

[11] Sharpe A G 1981Inorganic Chemistry Longman Publishing Group: London

# Combined spectroscopic study on the growth mechanism of Diphosphine-stabilized Gold Nanoclusters

**Jie Bao, Lina Yang, Wei Liu, Yuanyuan Huang, Ting Huang, Yuanjie Cao, Tao Yao*, Zhihu Sun*, and Shiqiang Wei**

National Synchrotron Radiation Laboratory, University of Science and Technology of China, Hefei 230029, P.R. China.

*Corresponding author. E-mail: yaot@ustc.edu.cn , zhsun@ustc.edu.cn

**ABSTRACT**: The formation process of 1,5-Bis(diphenylphosphino) pentane ($L^5$)-protected Au nanocluster through the reduction of precursor $Au_2L^5Cl_2$ by borane-tert-butylamine (TBAB) is traced by a combination of time-dependent x-ray/UV-vis absorption spectroscopies and mass spectrometry. It is revealed that the initial generation of dual-core basic unit $Au_4L^5{}_2Cl$ is a critical step, which allows for a subsequent size-growth process via incorporation of the existing Au monomer (Au(I)-Cl) or $Au_2L^5$ to form $Au_5L^5{}_2$, $Au_8L^5{}_3Cl$, $Au_{10}L^5{}_4Cl$ and finally main $Au_{11}L^5{}_4Cl_2$. This work advances one step further toward understanding the mechanism of formation and growth of diphosphine ligands-protected Au NCs.

## 1. Introduction

Small metal clusters composed of a few to tens of core atoms have size-specific unusual physical and chemical properties that differ from those of larger nanocrystals, which have potential applications in fields such as catalysis, optical devices and imaging [1-3]. Particularly, major advances in the chemical solution synthesis of Au nanoclusters has been achieved, and some well-defined thiolate-protected nanoclusters such as $Au_{25}$ and $Au_{38}$ have been reported [4, 5].

Recently, diphosphine ligands have shown promise for facilitating the formation of smaller monodispersed Au clusters [6-8]. Understanding the growth mechanism of the gold-phosphine clusters is necessary for designing synthetic protocols to produce specific nuclearity clusters, and has recently attracted considerable interest [9, 10]. Moreover, it is highly desirable to develop synthetic strategies that permit the precise and efficient synthesis of nanoclusters for fundamental studies as well as practical applications. This requires a profound understanding of the underlying mechanism governing the formation of the nanoclusters.

In this work, we combine the quick X-ray absorption fine structure (XAFS) and UV−vis absorption spectroscopies, as well as matrix-assisted laser desorption ionization mass spectrometry (MALDI-MS) to probe the formation process of the 1,5-bis(diphenylphosphino) pentane ($L^5$)-protected Au nanocluster.

Journal of Physics: Conference Series **712** (2016) 012108 doi:10.1088/1742-6596/712/1/012108

## 2. Experimental

Chlorogold(I) diphosphine complexes of $Au_2L^5Cl_2$ were prepared according to the literature for other Au(I)-diphosphine compounds [11]. Then 36.5 mg (0.04 mmol) $Au_2L^5Cl_2$ was disolved in 4 mL ethanol solution, to which an ethanol solution (2 mL) of borane-tert-butylamine (TBAB, 3.5 mg, 0.04 mmol) was quickly added at room temperature under magnetic stirring. At different reaction intervals, aliquotes of the reacting solution were quickly taken from the flask and subjected to characterization. Au $L_3$-edge XAFS spectra were measured in the quick-XAFS mode (30 sec per scan) at the 1W1B beamline of Beijing Synchrotron Radiation Facility (BSRF), China. The electron storage ring was operated at 2.5 GeV with a maximum current of 250 mA. A double Si (111) crystal monochromator was used. The measurements were performed in the tramsission mode, by using a 10-mm-thick Teflon cell. As a reference, the initial precursor solution prior to TBAB addition was also measured and labeled as 0 min. The UV-vis absorption data were acquired with a Purkinje General TU-9001 spectrometer in the wavelength range of 250–800 nm. Matrix-assisted laser desorption ionization mass-spectrometry (MALDI-MS) measurements were performed with a Bruker Autoflex III mass spectrometer in positive linear mode.

## 3. Results and discussion

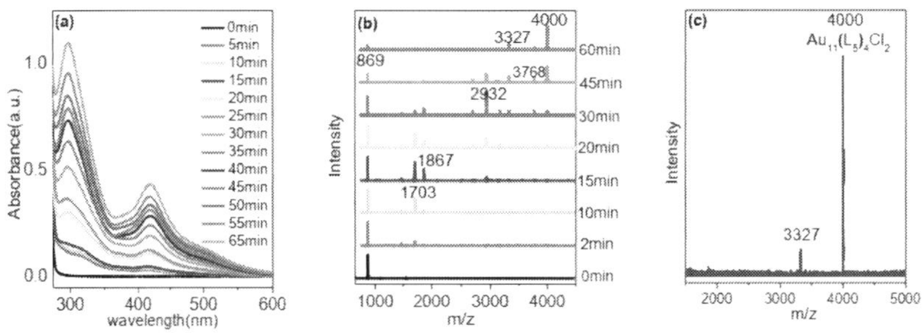


**Figure 1.** (a) Time-dependent UV-vis absorption spectra and (b) MALDI-MS during the Au cluster formation in ethanol. (c) MALDI-MS of the final product after 24 h of reaction.

Shown in Figure 1(a) are the time-dependent UV-vis absorption spectra. In the initial 15 min, two subtle absorption peaks at about 317 and 420 nm appears, and the first peak broadens. At 20 min, the peak at 317 nm blue shifts to 297 nm, and as the reaction going on, these two peaks at 297 and 420 nm become progressively intensified, while their positions and spectral shape remain unchanged. It is reported that the 420 nm band mainly comprise the contributions from members of the phosphine-ligated $Au_x$ ($8 \leq x \leq 12$) [9, 12], where the similar charge density transfer between the metal core and the phenyl group and phosphine group result in the similar shape but different intensity [13]. Therefore, the gradually pronounced absorption peaks in the UV-vis spectra, along with the unchanged spectral shape, suggest the growth toward $Au_8$-$Au_{12}$ clusters after 20 min of reactions. To identify the intermediate species, time-dependent MALDI-MS were measured as displayed in Figure 1(b). In the first 10 min, the formation of $Au_4L^5{}_2Cl$ intermediate cluster is indicated by the peak at $m/z$=1703 Da, followed by the appearance of another peak at $m/z$=1867 Da corresponding to $Au_5L^5{}_2$ at 15 min. After that, $Au_4L^5{}_2Cl$ and $Au_5L^5{}_2$ gradually disappeared, and the $Au_7L^5{}_3$ (m/z=2700 Da) and $Au_8L^5{}_3Cl$ (m/z=2932 Da) species emerged simultaneously. At further prolonged reaction times >30 min, end product of $Au_{10}$-$Au_{11}$ clusters ($m/z$=3327-4000 Da) with the dominant $Au_{11}L^5{}_4Cl_2$ ($m/z$ =4000 Da) were formed (Fig. 1(c)), which coincides with the UV results.

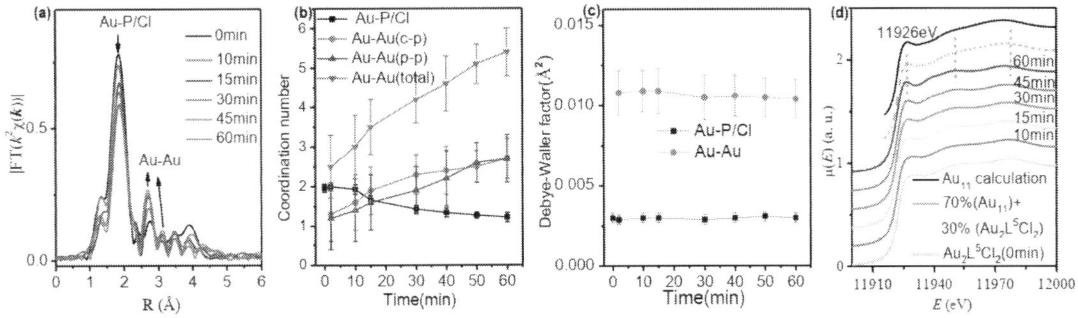


**Figure 2.** (a) Magnitudes of Foruier-transformed EXAFS $k^2\chi(k)$ functions at typical reaction times. Temporal evolution of coordination number CNs (b), and Debye-Waller factor $\sigma^2$ (c) for the Au–P/Cl and Au–Au coordination pairs extracted from EXAFS curve-fitting. (d) XANES spectra at typical reaction times. The calculated XANES spectrum for a $Au_{11}P_8Cl_2$ cluster is also displayed, along with a linear combination of the spectra of $Au_{11}P_8Cl_2$ (70%) and $Au_2L^5Cl_2$ (30%).

To obtain structure and composition information in the reaction process, we performed time-dependent Au $L_3$-edge XAFS measurements. The EXAFS data processing and curve-fitting were done using the ATHENA and ARTEMIS modules implemented in the IFEFFIT software packages[14], respectively. The quantitative curve-fittings were carried out in the *R*-space with a Fourier transform *k*-space range of 3.0–12.0 $Å^{-1}$. The backscattering amplitudes $F(k)$ and phase-shifts $\Phi(k)$ of the Au–P and Au–Au pairs were calculated for a $Au_{11}P_8Cl_2$ cluster [15, 16] by using FEFF8.0 code [17]. In the incomplete icosahedron of $Au_{11}$, the central-peripheral and peripheral-peripheral Au–Au bonds are quite different (2.60–2.72 Å *vs* 2.84–3.19 Å )[16], hence two Au–Au pairs labelled as Au–Au (c-p) and Au–Au (p-p) were considered. During the curve-fitting, the overall amplitude reduction factor $S_0^2$ was fixed to the best-fit value of 0.90 determined from fitting the data of Au foil. For each pair, four parameters of coordination number (CN) $N$, interatomic distance $R$, Debye–Waller factor $\sigma^2$, and energy shift $\Delta E_0$ were allowed to vary. To reduce the number of adjustable parameters, a common $\sigma^2$ and a common $\Delta E_0$ were used for the two Au-Au pairs.

The magnitudes of the Fourier transformed (FT) EXAFS $k^2\chi(k)$ functions at typical reaction times are shown in Figure 2(a). Evidently, with the increase of reaction time, the Au-ligand (Au–P or Au–Cl, undistinguishable) peak at 1.9 Å is weakened gradually, while the Au-Au peaks (2.7 and 3.1 Å) are significantly intensified, suggesting the gradual growth of the Au clusters. More quantitative results of the cluster growth could be obtained from the least-squares EXAFS curve-fittings. The obtained $N$ and $\sigma^2$ as functions of reaction time are shown in Figure 2(b) and (c), respectively. At 30 min, the total $N_{Au-Au}$ (4.2) is very close to the nominal $N_{Au-Au}$ (4) of $Au_8L^5{}_3Cl$ [18], and the end product with total $N_{Au-Au}$ of 5.4 is very close to the $N_{Au-Au}$ (5.5) of $Au_{11}L^5{}_4Cl_2$ [15, 16], consistent with the MS results that demonstrate the main products of $Au_8L^5{}_3Cl$ and $Au_{11}L^5{}_4Cl_2$ at 30 and 60 min, respectively. Moreover, at 60 min the Au-ligand CN (1.22) is between the nominal values in the end product of $Au_{11}L^5{}_4Cl_2$ (0.91) and the precursor $Au_2L^5Cl_2$ or analogues (2.0)[15, 16, 19], suggesting the coexistence of both species. The real Au-ligand CN could be roughly expressed as a weighted-sum of the CNs in $Au_{11}L^5{}_4Cl_2$ and $Au_2L^5Cl_2$ *via* $0.91\times p+2\times(1-p)=1.22$, which gives the proportion $p$ of Au atoms in $Au_{11}L^5{}_4Cl_2$ as $p{\sim}0.7$. This point is also supported by the XANES spectra (Fig. 2(d)). The starting material $Au_2L^5Cl_2$ shows a double-peak structure in the white-line region of XANES. With reaction going on, the double-peak disappears and a peak at 11926 eV is gradually intensified. Using the FEFF8.0 code [17], we calculated the spectrum of the $Au_{11}P_8Cl_2$ cluster. A linear combination of the spectrum of $Au_{11}P_8Cl_2$ (70%) and $Au_2L^5Cl_2$ (30%) could well reproduce the main experimental features at 60 min, as labelled by the vertical bars in Fig. 2(d). These results lead us to conclude that the majority (~70%) of Au atoms are in the form of $Au_{11}$ cluster after 60 min, with a small part (~30%) existing as $Au_2L^5Cl_2$ or other analogous species.

Finally, we discuss briefly the formation mechanism of Au nanoclusters through the reduction of the precursor $Au_2L^5Cl_2$. Yao et al. have reported an initial nucleation of gold nanocrystals by formation of intermediate $Cl^{3-}$Au-Au$Cl^{3-}$dimer [20]. Analogously, we speculate that two precursor molecules of $Au_2L^5Cl_2$ may form a dual-core basic unit $Au_4L^5_2$ by connecting two $Au_2L^5$ units, due to the breaking of the Au-Cl bonds under the reduction reaction of TBAB. Then a Au monomer (Au(I)-Cl) which could be formed from $Au_2L^5Cl_2$ by breaking the Au-P bond (and Au-Cl bond) could be linked to the dual-core basic unit to form $Au_5L^5_2$. After that the larger clusters such as $Au_7L^5_3$, $Au_8L^5_3Cl$, $Au_{10}L^5_4Cl$ and $Au_{11}L^5_4Cl_2$ are formed on the basis of the $Au_5L^5_2$, via incorporation of the existing Au monomer (Au(I)-Cl) or $Au_2L^5$.

## 4. Conclusions

In summary, using a combination of time-dependent UV-vis/ XAFS and mass spectra, the formation process of $L^5$-protected Au nanocluster through the reduction of precursor $Au_2L^5Cl_2$ by TBAB is investigated. It is found that the cluster formation is achieved in a nucleating/growth manner including two distinct reaction steps. The initial generation of dual-core basic units of $Au_4L^5_2$ and $Au_5L^5_2$ is a critical step allowing for subsequent size-growth process. This work deepens understanding of the mechanism of formation and growth of diphosphine ligands-protected Au nanoclusters and may be helpful for further controllable synthesis of metal nanoclusters.

**Acknowledgments**

This work was supported by the National Natural Science Foundation of China (Grant No. 11175184, 11135008, 11475176, 11422547, 11435012, 21471143 and U1332131). The authors are grateful to BSRF for the valuable synchrotron radiation XAFS beamtime.

**References**

[1] Spivey J J, Krishna K S and Kumar C 2014 *J.Phys.Chem.C* **118** 20043-69
[2] Lu Y Z and Chen W 2012 *Chem.Soc.Rev.* **41** 3594-623
[3] Jin R C 2010 *Nanoscale* **2** 343-62
[4] Qian H F, Zhu Y and Jin R C 2009 *Acs Nano* **3** 3795-803
[5] Wu Z, Suhan J and Jin R C 2009 *J. Mat.Chem.* **19** 622-6
[6] Shichibu Y, Suzuki K and Konishi K 2012 *Nanoscale* **4** 4125-9
[7] Shichibu Y and Konishi K 2010 *Small* **6** 1216-20
[8] Chen J, Zhang Q F and Bonaccorso T 2014 *J.Am.Chem.Soc.* **136** 92-5
[9] Hudgens J W, Pettibone J M and Senftle T P 2011 *Inorg.Chem.* **50** 10178-89
[10] Guidez E B, Hadley A and Aikens C M 2011 *J.Phys.Chem.C* **115** 6305-16
[11] Brandys M C, Jennings M C and Puddephatt R J 2000 *J.Chem.Soc.Dalton* 4601-6
[12] Pettibone J M and Hudgens J W 2010 *J.Phys.Chem.Lett.* **1** 2536-40
[13] Wu Z K and Jin R C 2013 *Chemistry* **19** 12259-63
[14] Newville M 2001 J.Synchrotron Rad. **8** 96-100
[15] McKenzie L C, Zaikova T O and Hutchison J E 2014 *J.Am.Chem.Soc.* **136** 13426-35
[16] Hall K P and Mingos D M P 1984 Prog. Inorg. Chem.**32** 237-325
[17] Ankudinov A L, Ravel B, Rehr J J and Conradson S D 1998 *Phys.Rev.B* **58** 7565-76.
[18] Vandervelden J W A, Bour J J, Bosman W P and Noordik J H 1983 *Inorg.Chem.* **22** 1913-8
[19] Pettibone J M and Hudgens J W 2012 Small **8** 715-25
[20] Yao T, Sun Z H , Li Y Y and Wei S Q 2010 *J.Am.Chem.Soc.* **132** 7696-701

# Investigation of Prussian Blue Analogs by XMCD at the K-edge of transition metals

**A BORDAGE[1], L NATAF[2], F BAUDELET[2] and A BLEUZEN[1]**
[1]ICMMO, Univ. Paris Sud, UMR CNRS 8182, Univ. Paris Saclay, Rue du doyen G. Poitou, 91405 Orsay Cedex, France, [2] Synchrotron SOLEIL, L'Orme les merisiers Saint-Aubin, BP48, 91192 Gif sur Yvette Cedex, France

E-mail: amelie.bordage@u-psud.fr

**Abstract.** Despite transition metal (TM) K-edge x-ray magnetic circular dichroism (XMCD) seems an interesting tool to get magnetic and structural information at the atomic scale, the effects originating this signal are still poorly understood. We thus initiated a deep investigation of the TM K-edge XMCD using Prussian Blue analogs (PBA) as model-compounds. In a recent study of the NiFe PBA family, we demonstrated that the XMCD signals at the TM K-edges strongly vary with external (mechanical) or internal (chemical) pressure and so that they are highly sensitive to small structural distortions. Following these first results, we extended this approach to the MnFe and CoFe families to evaluate the effect of electronic parameters (number of unpaired electrons of the $M^{II}$ TM) on the XMCD signal. All the results set milestones in the disentanglement of the components originating the XMCD signals at the K-edge of TM and will eventually help in a better understanding of the photomagnetic properties of PBAs.

## 1. Introduction

Transition metals (TM) $L_{2,3}$-edges X-ray Magnetic Circular Dichroism (XMCD) is a well-known and widely used technique to characterize magnetic compounds [1, 2, 3, 4, 5, 6] but only very few studies were conducted at the TM K-edges [7, 8, 9]. This technique has been used for instance to characterize under pressure Co [10], magnetite [11] or the structural $\alpha$-$\epsilon$ phase transition in iron [12, 13]. In the case of molecular magnets, TM K-edge XMCD was used in two cases. Verdaguer et al. first investigated three molecule-based magnets from the Prussian Blue Analogs (PBA) family ($Cs^{I}Mn^{II}[Cr^{III}(CN)_6]$, $Cs^{I}Ni^{II}[Cr^{III}(CN)_6]$ and $Ni_3^{II}[Cr^{III}(CN)_6]_2$) and observed at each metal edge (Ni, Mn and Cr) a dichroic signal in the energy range corresponding to the allowed transitions to the metal $4p$ levels [14]. Champion et al. then concentrated on the $Rb_{1.8}Co_4[Fe(CN)_6]_{3.3}$ PBA at both the Co and Fe K-edges for the $\{Co^{II}Fe^{III}\}$ photoexcited states [15]. They characterized the relative orientation of the local magnetic moments of the metallic ions in the photoinduced metastable state and proposed the first direct experimental evidence of the ferrimagnetic behavior of the photoinduced metastable state in this PBA.

However, TM K-edge XMCD is still poorly understood and the effects giving rise to a signal and their influence on its spectral shape have still to be disentangled. PBA are ideal candidates as model-compounds for such an investigation. Their general formula is $C_xA_y[B(CN)_6]_y\square_{(4-y)}$, where A and B are transition metals, C an alkali cation and $\square$ $[B(CN)_6]$ unit vacancies. The nature of the alkali cation and of the two transition metals as well as the number of alkali cation can indeed be varied while most often conserving a face-centered cubic structure.

These characteristics offer the unique possibility to independently vary chemical, structural and electronic parameters and then evaluate their impact on the XMCD signal at the TM K-edge.

We present here the latest results obtained by TM K-edge XMCD on several PBAs (NiFe, CoFe and MnFe families). All the measurements were performed on the ODE beamline (SOLEIL synchrotron, Saint Aubin, France) [16]. The XMCD signals were recorded for a 1.3T magnetic field, at 10K for the NiFe PBAs and at 4.2K for the CoFe and MnFe PBAs in order to be below the Curie temperature of the compounds.

## 2. XMCD and structural deformation

The NiFe PBAs family was the first investigated by TM K-edge XMCD to understand the impact of structural deformation on the XMCD signals, since they never present a pressure-induced charge transfer between the $Ni^{II}$ and $Fe^{III}$ ions. Four NiFe PBAs were investigated: $Ni_4[Fe(CN)_6]_{2.7}$ (called **NiFe**), $CsNi_4[Fe(CN)_6]_3$ (called **CsNiFe**), $Cs_2Ni_4[Fe(CN)_6]_{3.3}$ (called **Cs2NiFe**) and $Rb_2Ni_4[Fe(CN)_6]_{3.3}$ (called **Rb2NiFe**). The synthesis of these compounds and the experimental details are given in Ref [17], and we remind here the main results of this study.

The Ni and Fe K-edges XMCD signals of **NiFe** for variable pressure (0-7 GPa) are presented in figure 1. At ambient pressure, both signals are mainly characterized by an intense peak corresponding to the absorption edge (at 7128 eV in the case of Fe and at 8348 eV in the case of Ni). The relative sign of the XMCD signals at both edges indicates a ferromagnetic 3d-3d exchange interaction between the magnetic moments borne by the Ni and Fe ions. When applying a pressure (up to 7 GPa), the sign and shape of the XMCD signals remain unchanged but a significant variation of the intensity is observed : it first increases and then decreases until the signal completely disappears at 7 GPa (resp. 1.6 GPa) in the case of Fe (resp. Ni). Consistently with the variations observed on the x-ray absorption spectra for the different pressures, this pressure dependence of the XMCD signal indicates that its intensity is strongly related to the 3d-4p and 4p-4p orbital overlapping and exchange interactions, as well as to symmetry changes of the TM sites (progressive tilt of the rigid [Fe(CN)] entities around the $Fe^{III}$ ions crystallographic positions or/and the departure of the Ni coordination polyhedra from the $O_h$ symmetry), as explained in detail in Ref. [17].

The insertion of alkali cation inside the structure enables to induce an internal chemical pressure on the structure, and consequently better understand the results obtained in the opposite case of an external mechanical pressure. Figure 2 thus compares the Fe K-edge XMCD signals for the four NiFe PBAs (**NiFe**, **CsNiFe**, **Cs2NiFe** and **Rb2NiFe**). Here also, only the intensity of the peak corresponding to the absorption edge (7128 eV) varies with the number of alkali cation present inside the PBA unit cell. The highest intensity is observed for **NiFe**, which presents the structure with the highest number of vacancies and distortion since no alkali cation is present. The intensity then decreases for **CsNiFe** by almost a factor of 2 and the two compounds with two cations per unit cell (**Cs2NiFe** and **Rb2NiFe**) display the lowest intensity. These changes are consistent with those observed for **NiFe** for low pressures below 1 GPa (Figure 1): the higher the pressure, the more the Ni coordination polyhedra departs from the $O_h$ symmetry and the higher the intensity of the XMCD signal at the TM K-edge. It is to be noted that the nature of the alkali cation seems to be of little influence, since the intensity is similar for **Cs2NiFe** and **Rb2NiFe** despite their quite different ionic radii (1.72Å in the case of $Rb^+$ and 1.88Å in the case of $Cs^+$). This may suggest that for a given number of alkali cation in the unit cell, the size of the alkali cation induces only a weak dilatation/contraction of the structure.

## 3. Influence of the number of unpaired 3d electrons at the divalent site

We present now the first results on the $K_{0.1}Co_4[Fe(CN)_6]_{2.7}$ and $Mn_4[Fe(CN)_6]_{2.7}$ PBAs (respectvely called **CoFe** and **MnFe**). If including **NiFe**, this Mn-Co-Ni serie illustrates the

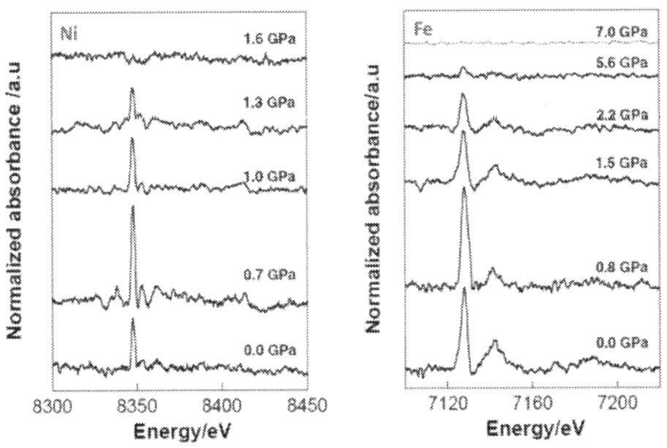


**Figure 1.** Variable pressure normalized XMCD signals at the Ni (left) and Fe (right) K-edges of **NiFe** (recorded at 10K).

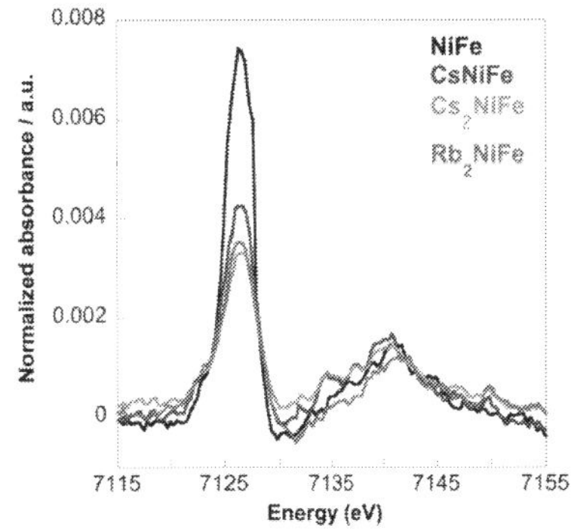


**Figure 2.** Normalized XMCD signals at the Fe K-edge of a serie of NiFe PBA at ambient pressure (recorded at 10K).

effect of the number of unpaired 3d electrons of the divalent transition metal on the XMCD signal at the corresponding TM K-edge. Indeed, **MnFe** has a spin $S=5/2$, **CoFe** $S=1/2$ and **NiFe** $S=1$.

The normalized XMCD signals at the Fe K-edge and the $M^{II}$ (M=Mn, Co, Ni) edges are presented in figure 3. The relative sign of the Fe K-edge signal and the $M^{II}$ K-edge one are consistent with the Fe-$M^{II}$ 3d-3d exchange interactions. In the case of **CoFe** (resp. **NiFe**), the opposite (resp. identical) sign of the signals is the signature of a ferrimagnetic (resp. ferromagnetic) coupling of the $Co^{2+}$ (resp. $Ni^{2+}$) and $Fe^{3+}$ sublattices. The case of **MnFe** is more complicated, since the XMCD signal at the Mn K-edge presents both a negative and positive contributions; the positive contribution clearly relates to the ferrimagnetic coupling between the $Mn^{2+}$ and $Fe^{3+}$ sublattices, but the origin of the negative contribution is still unclear. This difference between $Co^{2+}$ and $Ni^{2+}$ on the one side and $Mn^{2+}$ on the other side probably originates from the different filling of the 3d orbitals between these three $M^{II}$ cations (insets in figure 3). Indeed, in the cases of $Co^{2+}$ and $Ni^{2+}$, each $e_g$ orbital contains an unpaired electron and only one (resp. zero) unpaired electron is present in the $t_{2g}$ orbital in the case of $Co^{2+}$ (resp. $Ni^{2+}$). The opposite situation is observed in the case of $Mn^{2+}$ : three unpaired electrons fill the $t_{2g}$ orbital while only two in the $e_g$ orbital. Finally, at the Fe K-edge, the less intense broader peak at 7145.9 eV observed in the case of **NiFe** was attributed to the multiple-scattering contribution [17]. Despite a lower signal-to-noise ratio, this feature is also visible in the case of **MnFe**; we can also expect this feature to be present for **CoFe** but measurements with a better statistics are still required to confirm it.

## 4. Conclusion

We summarized here recent results obtained by XMCD at the TM K-edges of several PBAs. We undertook the comparison of the XMCD signals at different metallic edges for three $M^{II}Fe^{III}$ PBAs (M = Mn, Co, Ni). We could conclude from these first preliminary results that the number of unpaired 3d electrons of the divalent transition metal in a $M^{II}Fe^{III}$ PBA and the repartition of these unpaired electrons between the $t_{2g}$ and $e_g$ orbitals have a strong influence on the XMCD signal at the $M^{II}$ K-edge, whereas at the Fe K-edge, the dominant influence is the 3d-3d exchange interaction. The investigation of NiFe PBAs also demonstrated the potential for the XMCD signals at the TM K-edges as a sensitive probe of small structural distortions, induced either by an external or internal (via the insertion of alkali cation) pressure. All these

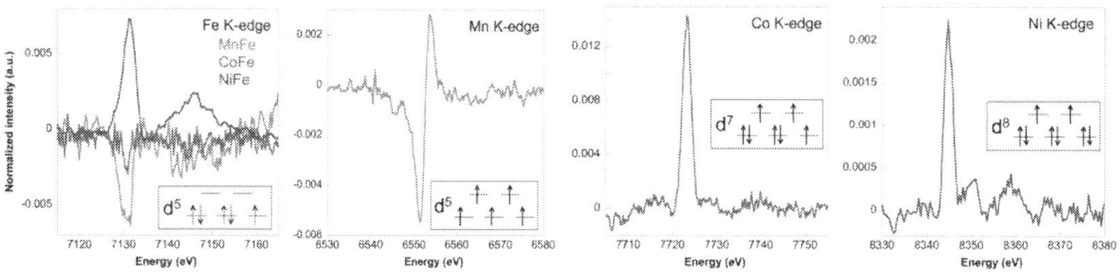


**Figure 3.** Normalized XMCD signals of the $M^{II}$-$Fe^{III}$ (M = Mn, Co, Ni) PBA series at the Fe and M K-edges recorded at 4K (**MnFe** and **CoFe**) or 10K (**NiFe**) The electronic configuration of each cation ($Fe^{3+}$ low-spin, $Mn^{2+}$/$Co^{2+}$/$Ni^{2+}$ high spin) are illustrated in the insets.

results eventually indicate that the deep investigation of PBA by XMCD at the TM K-edge will bring new insights into the versatile properties of PBAs, in particular photomagnetism [18, 19, 20], and therefore shed a new light on these compounds for their use in new high-density data-storage systems.

*Acknowledgments*

The authors are thankful to J. Coquet (ODE beamline, SOLEIL), A. Juhin and N. Mas (IMPMC, Paris, France), A. Beauvois, R. Moulin and G. Fornasieri (ICMMO,Orsay, France) for help during the experiments.

*References*

[1] Van der Laan G and Figueroa A I 2014 *Coord. Chem. Rev.* **277-278** 95
[2] De Groot F and Kotani A 2008 *Core Level Spectroscopy of Solids* (Taylor & Francis, New York)
[3] Funk T, Deb A, George S J, Wang H and Cramer S P 2005 *Coord. Chem. Rev.* **249** 3
[4] Goering E, Will J, Geissler J, Justen M, Weigand F and Schuetz G 2001 *Journal of Alloys and Compounds* **328** 14
[5] Thole T, Carra P, Sette F and Van der Laan G 1992 *Physical Review Letters* **68** 1943
[6] Brouder Ch and Hikam M 1991 *Physical review B* **43** 3809
[7] Torchio R, Mathon O and Pascarelli S 2014 *Coord. Chem. Rev.* **277-278** 80
[8] Brouder Ch, Alouani M and Bennemann K H 1996 *Physical review B* **54** 7334
[9] Guo G Y 1996 *J. Phys.: Condens. Matter* **8** L747
[10] Torchio R, Monza A, Baudelet F, Pascarelli S, Mathon O, Pugh E, Antonangeli D and Itié J-P 2011 *Physical Review B* **84** 060403(R)
[11] Baudelet F, Pascarelli S, Mathon O, Itié J-P, Polian A and Chervin J-C 2010 *Physical review B* **82** 140412(R)
[12] Baudelet F, Pascarelli S, Mathon O, Itié J-P, Polian A, d'Astuto M and Chervin J-C 2005 *J. Phys.:Condens. Matter* **17** S957
[13] Mathon O, Baudelet F, Itié J-P, Polian A, d'Astuto M, Chervin J-C and Pascarelli S 2004 *Physical review Letters* **93** 255503
[14] Verdaguer M, Mallah T, Helary C, L'Hermite F, Sainctavit Ph, Arrio M-A, Babel D, Baudelet F, Dartyge E and Fontaine 1995 *Physica B* **208-209** 765
[15] Champion G, Escax V, Cartier dit Moulin C, Bleuzen A, Villain F, Baudelet F, Dartyge E and Verdaguer M 2001 *J. Am. Chem. Soc.* **123** 12544
[16] Baudelet F, Kong Q, Nataf L, Cafun J-D, Congeduti A, Monza A, Chagnot S and Itié J-P 2011 *High Pressure Research* **31** 136
[17] Cafun J-D, Lejeune J, Itié J-P, Baudelet F and Bleuzen A 2013 *The Journal of Physical Chemistry C* **117** 19645
[18] Sato O, Iyoda T, Fujishima A, Hasimoto K 1996 *Science* **272** 704
[19] Ohkoshi S, Tokoro H, Hashimoto K 2005 *Coord. Chem. Rev.* **249** 1830
[20] Cafun J-D, Lejeune J, Baudelet F, Dumas P, Itié J-P and Bleuzen A 2012 *Angew. Chem. Int. Ed.* **51** 9146

# Silver nanoparticles in silicate glass prepared by UV laser irradiation: dependences of size and atomic structure of particles upon irradiation parameters

**M Dubiel[1,5], M Heinz[1], V V Srabionyan[2], V V Pryadchenko[2], L A Avakyan[2], Ya V Zubavichus[3], J Meinertz[4], J Ihlemann[4] and L A Bugaev[2]**
[1]Martin Luther University Halle-Wittenberg, Halle (Saale), Germany
[2]Southern Federal University, Rostov-on-Don, Russia
[3]Kurchatov Institute, Moscow, Russia
[4]Laser-Laboratorium Göttingen e.V., Göttingen, Germany

E-mail: manfred.dubiel@physik.uni-halle.de

**Abstract.** Plasmonic Ag nanostructures were generated in glass surfaces by means of UV laser irradiation (193 nm). The aim of the work was to identify the correlations between the atomic structure of such nanoparticles and their optical properties. The structural characterization of the samples, prepared with different number of laser pulses was performed by Ag *K*-edge EXAFS. The processing of the spectra by the Fourier-transform analysis gave the values of Ag-Ag and Ag-O interactions averaged over ionic and neutral states of silver in the sample. The consistent analysis of the behavior of features in optical spectra and the obtained dependencies of Ag-Ag and Ag-O structural parameters upon the number of laser pulses enabled to suggest the mechanism of plasmonic Ag nanoparticles formation in silicate glass, which is influenced by the interaction with polyvalent ions (for example $Fe^{2+}$) and by defect centers (for example HTC and NBOHC centers) generated by UV irradiation.

## 1. Introduction

The formation of metal nanoparticles, especially silver particles in dielectric matrixes, is extensively studied. These nanocomposite materials exhibit linear and non-linear optical properties, which are determined by the surface plasmon resonance (SPR) of the nanoparticles [1]. Generally, silver nanoparticles in sodium silicate glasses are prepared by the $Ag^+ \leftrightarrow Na^+$ ion exchange process and subsequent thermal treatments [2, 3]. However, this method does not allow creating nanoscaled arrays of particles localized in a certain way. In our previous work [4] to obtain spatially specific precipitation of nanoparticles (line structures) in $Ag^+ \leftrightarrow Na^+$ ion exchanged glass surfaces the laser irradiation below the ablation threshold of the glass was implemented. By means of UV laser irradiation high resolution patterning could be realized. However, the UV laser irradiation, depending on the laser parameters, can lead to the reduction of the size or to the dissolution of nanoparticles [5] and to their instability even at room temperature. These processes can occur since the UV irradiation of glass induces the creation of various defect centers [6], which results both in the enhancement of particles formation and in the recombination processes giving a decrease in the particles concentration. In some cases it was also shown that the formation of stable nanoparticles can be obtained without

[5] To whom any correspondence should be addressed.

additional thermal treatments [7]. Therefore, to obtain stable nanoparticles with certain characteristics of SPR, for example peak position and full width at half maximum (FWHM), it is necessary to reveal the dependences of these characteristics upon the parameters of the laser irradiation and to obtain the values of these parameters that provide the stability of the nanoparticles.

In this paper the formation of stable silver nanoparticles in $Ag^+ \leftrightarrow Na^+$ ion exchanged glass surfaces by means of ArF-excimer laser irradiation (193 nm) without additional thermal treatments is succeeded. Experimental methods to characterize the nanoparticles are optical spectroscopy to determine the size of the nanoparticles using Mie theory, and Ag *K*-edge extended X-ray absorption fine structure (EXAFS) to reveal the dependences of the structural parameters of Ag-Ag and Ag-O interactions upon the number of laser pulses.

## 2. Experimental results

The basic glass was soda-lime silicate glass (float glass). The $Ag^+ \leftrightarrow Na^+$ ion exchange has been performed by immersing the glass slides into a molten $NaNO_3$ / 5 wt.-% $AgNO_3$ salt bath at a temperature of 330 °C for 20 minutes. Then the ion exchanged glass slides have been irradiated by the ArF-excimer laser (Lambda Physik LPX 315; wavelength 193 nm, 20 ns pulse duration) at the tin bath side. The laser fluence was approximately 100 mJ/cm$^2$ and the repetition rate was 10 Hz. The number of laser pulses has been varied from 2 to 5000. The size of the laser spots was 5 to 10 mm. For simplicity of notations, the sample obtained using 2 laser pulses will be referred to as Sample 2, the other one using 10 laser pulses will be referred to as Sample 10 and so on.

Extinction spectra have been measured at a UV/Vis/NIR spectrometer (Perkin Elmer, Lambda 900). Figure 1 shows the extinction spectra of Samples 10, 100, 500, 1000 and 5000 (here the extinction coefficient is represented). It can be seen, that up to 100 pulses (Samples 10 and 100) the spectra are characterized by a wide absorption band with a maximum at ~ 450 nm. In the spectra of Samples 500, 1000 and 5000 (more than 500 pulses), this absorption band takes the form of a SPR band with a maximum shifted from ~ 450 nm in Sample 10 to ~ 435 nm in Sample 5000. Concerning the SPR bands Mie theory simulations have been performed (dashed lines in Figure 1) assuming that the absorption in the visible region occurs only owing to the SPR of silver nanoparticles which are described by a log-normal size distribution. This results in average particle sizes of $D = 28$, 20 and 14 nm in Samples 500, 1000 and 5000, respectively. Required equations for the simulations are taken from [8]. Deviations in the UV region can be explained by atomic and ionic silver in the glass, which were not taken into account by the simulations. Probably, the calculated average sizes of nanoparticles are not exactly determined because of the neglect of absorption contributions of any other Ag centers. The concentration of such structural units is not known.

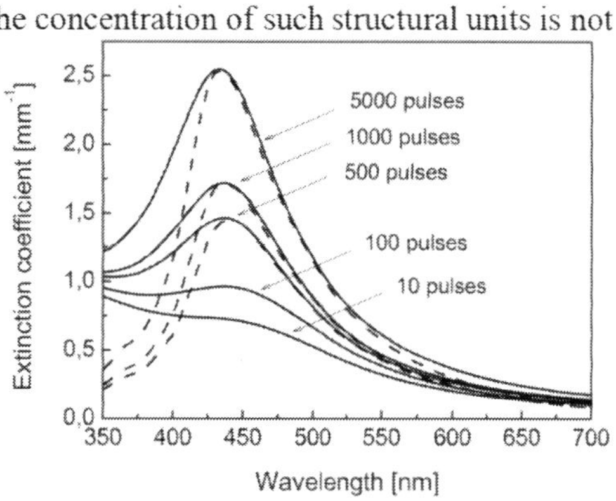


**Figure 1.** Extinction spectra of samples 10, 100, 500, 1000 and 5000; the solid lines represent the experimentally obtained spectra and the dashed lines refer to the Mie theory simulations.

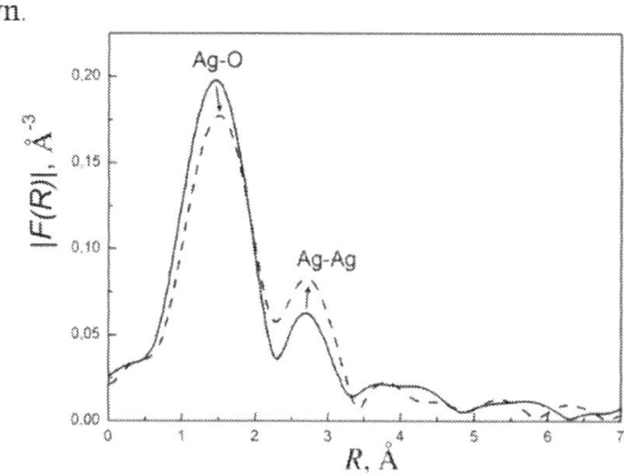


**Figure 2.** FT magnitudes $|F(R)|$ of experimental Ag *K*-edge EXAFS in Sample 20 (solid curve) and in Sample 5000 (dashed curve).

Journal of Physics: Conference Series **712** (2016) 012110 doi:10.1088/1742-6596/712/1/012110

In Figure 2 the magnitudes of the Fourier-transforms $|F(R)|$ of experimental Ag *K*-edge EXAFS in Sample 20 (low number of pulses) and in Sample 5000 (largest number of pulses) are compared. The magnitude of the first peak, which corresponds to the contribution of Ag-O interactions, decreases with the increasing of the number of laser pulses, while the magnitude of the second peak, attributed to the contribution of Ag-Ag bonds, increases. To confirm these qualitative conclusions, experimental Ag *K*-edge EXAFS in Samples 2 – 5000 were processed by the technique of [9], which enables to reduce the effect of correlations among fitting parameters on the determined values of structural parameters by using interdependencies between all correlating parameters of a fitting model, which were obtained at different weights $k^n$ for oscillatory part $\chi^{exp}(k)$ and at different $\Delta k$ intervals for FT. The fit of $F(R)$ of EXAFS spectrum can be performed using the following expression for $\chi(k)$:

$$\chi\ (k) = A_{\text{Ag-Ag}} \cdot \chi_{\text{Ag-Ag}} + A_{\text{Ag-O}} \cdot \chi_{\text{Ag-O}} + \widetilde{\chi}_{\text{Ag}} \qquad (1)$$

where $\chi_{\text{Ag-Ag}}$ – is the contribution of one atomic pair Ag-Ag without the reduction factor $S_0^2$(Ag-Ag) [10]; $\chi_{\text{Ag-O}}$ – is the contribution of one atomic pair Ag-O without $S_0^2$(Ag-O). These reduction factors are included into the amplitudes $A_{\text{Ag-Ag}} = C_{\text{Ag-Ag}} \cdot N_{\text{Ag-Ag}} \cdot S_0^2(\text{Ag-Ag})$ and $A_{\text{Ag-O}} = C_{\text{Ag-O}} \cdot N_{\text{Ag-O}} \cdot S_0^2(\text{Ag-O})$, where $C_{\text{Ag-Ag}}$ and $C_{\text{Ag-O}}$ are the fractions of Ag atoms in the sample bonded respectively with Ag and O atoms; $N_{\text{Ag-Ag}}$ and $N_{\text{Ag-O}}$ are respectively the number of Ag and O atoms neighboring to the absorbing Ag atom. $\widetilde{\chi}_{\text{Ag}}$ in (1) is the contribution of photoelectron scattering processes on the second and more distant shells of the absorbing Ag atom. Determining the structural parameters for the nearest neighbors of Ag atoms using FT of EXAFS enables to neglect the contribution of the term $\widetilde{\chi}_{\text{Ag}}$ in (1). This can be done because of the strong spread of Ag-Ag and Ag-O distances for the second neighbors of the absorbing Ag atoms in glass and hence, in the reduction of their contributions in relation to the nearest neighbors one. The fit of $F(R)$ of Ag *K*-edge EXAFS in the samples was performed by such a "nearest neighbors" approximation of (1). The functions $\chi(k)$ of the Samples 2 – 5000 were obtained in the *k*-scales with the zero at the inflection point $E_0$ = 25513 eV. The fit gave the average values of structural and non-structural parameters: amplitudes $A_{\text{Ag-Ag}}$, $A_{\text{Ag-O}}$ and corresponding Debye-Waller (DW) parameters $\sigma^2_{\text{Ag-Ag}}$, $\sigma^2_{\text{Ag-O}}$, interatomic distances $R_{\text{Ag-Ag}}$, $R_{\text{Ag-O}}$, and energy-shift parameters $e_0$(Ag-Ag), $e_0$(Ag-O) [10].

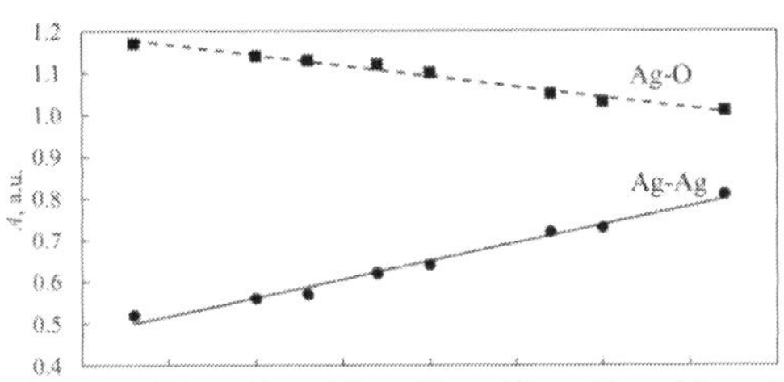


**Figure 3.** Dependences of amplitudes ($A$) of Ag-Ag (solid curve) and Ag-O (dashed curve) interactions upon the number $N$ of laser pulses.

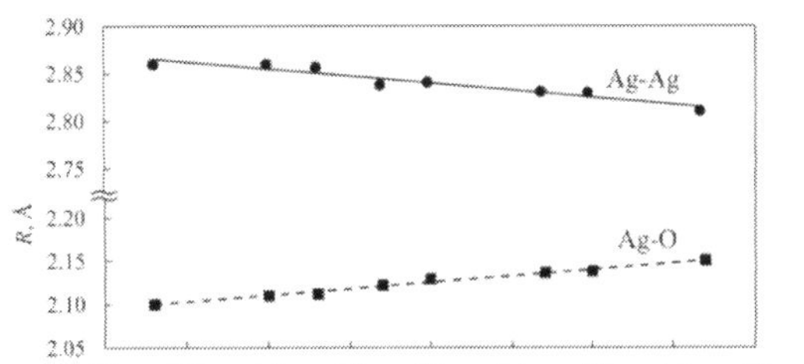


**Figure 4.** Dependences of interatomic distances ($R$) of Ag-Ag (solid curve) and Ag-O (dashed curve) bonds upon the number $N$ of laser pulses.

Figure 3 shows the dependences of amplitudes $A_{\text{Ag-Ag}}$ and $A_{\text{Ag-O}}$ upon the number $N$ of laser pulses. Both curves are approximately linear functions of lg($N$), however, $A_{\text{Ag-Ag}}$ increases with $N$, while $A_{\text{Ag-O}}$

decreases. In Figure 4 the dependences of interatomic distances $R_{Ag\text{-}Ag}$ and $R_{Ag\text{-}O}$ upon the number $N$ of laser radiation pulses are presented.

## 3. Discussion and conclusions

According to the optical spectra in Figure 1, the samples 10 and 100 exhibit a wide absorption band with a maximum at ~ 450 nm, which can be attributed to different structural units containing silver atoms (e. g. color centers) absorbing in the visible range at $\lambda$ ~ 420 – 460 nm. The most suitable constructions are the well-known hole-trap centers (HTC) [11,12] and small clusters of neutral silver atoms $Ag^0_n$ ($n$ = 2 – 15 atoms). The increasing number of laser pulses from 500 to 5000 leads to the formation of the distinct SPR line of silver nanoparticles, shifted respectively from ~ 440 nm to ~ 435 nm.

The decrease of $A_{Ag\text{-}O}$ and the increase of $A_{Ag\text{-}Ag}$ amplitudes in Figure 3 agree with the behavior of Ag-O and Ag-Ag peaks of $|F(R)|$ in Figure 2, and reflect the decreasing amount of silver ions in the glass matrix and the simultaneously increasing number of agglomerations of silver atoms (for example nanoparticles) upon the number $N$ of laser pulses. In the glass samples after $Ag^+ \leftrightarrow Na^+$ ion exchange and irradiation with small number of laser pulses one can expect the predominance of the Ag-Ag interactions, which can be attributed to the following plausible color centers in glass: $Ag^+$–$Ag^+$, $Ag^+$–$Ag^0$ (dimers) and HTC. These centers are characterized by different values of $R_{Ag\text{-}Ag}$ distributed from ~ 2.9 to 3.3 Å [13], which are distinctly longer than $R_{Ag\text{-}Ag}$ between the neighboring neutral $Ag^0$ atoms. The dependences of $R_{Ag\text{-}Ag}$ and $R_{Ag\text{-}O}$ upon the number $N$ of laser radiation pulses in Figure 4 reflects the decrease of amount of initially dominating silver ions owing to their reduction into $Ag^0$ atoms according to the reaction $Ag^+ + e^- \rightarrow Ag^0$. Thus, considering the behavior of $R_{Ag\text{-}O}$ upon $N$, one must take into account that the value $R_{Ag\text{-}O}$ = 2.1 Å in Sample 10 ($N$ = 10 pulses) is typical for the bonds $Ag^+$–O and $Ag^{3+}$–O in different silver oxides [14]. Meanwhile, the value $R_{Ag\text{-}O}$ = 2.15 Å in Sample 5000 is closer to $Ag^0$–O distances (2.16 Å). The decrease of $R_{Ag\text{-}Ag}$ with the increasing number $N$ of laser pulses can be also explained as the decreasing of contribution from the number of silver ions constructions (like HTC centers) due to the reduction of silver ions, simultaneously with the increasing of contribution of silver nanoparticles, whose size decreases with the increasing number of pulses.

The work was supported by DFG-RFBR project (DFG № DU 214/14-1, RFBR № 14-02-91334)

## 4. References

[1] Stockman M I 2012 *Opt. Express* **19** 22029–50
[2] Dubiel M, Haug J, Kruth H, Hofmeister H and Schicke K D 2008 *Mater.Sci.Eng.* B **149** 146-51
[3] Berger A 1992 *J. Non-Cryst. Solids* **151** 88–94
[4] Dubiel M, Heinz M, Stiebing M, Meinertz J, Ihlemann J and Rainer Th 2014 *Proc. SPIE* **9163** 91631M-1 – 91631M-6
[5] Stepanov A L, Hole D E and Townsend P D 1999 *Nucl.Instrum.Methods Phys.Res.* B **149** 89-98
[6] Hosono H, Kajihara K, Suzuki T, Ikuta Y, Skuja L and Hirano M 2002 *Solid State Commun.* **122** 117–120
[7] Zhang J, Dong W, Sheng J, Zheng J, Li J, Qiao L and Jiang L 2008 *J. Cryst. Growth* **310** 234–9
[8] Kreibig U and Vollmer M 1995 *Optical Properties of Metal Clusters* vol 25 (Berlin: Springer)
[9] Srabionyan V V, Bugaev A L, Pryadchenko V V, Makhiboroda A V, Rusakova E B, Avakyan L A, Schneider R, Dubiel M and Bugaev L A 2013 *J. Non-Cryst. Solids* **382** 24–31
[10] Koningsberger D C, Mojet B L, van Dorssen G E and Ramaker D E 2000 *Top.Catal.* **10** 143–55
[11] Wang Z and Toshio K 2011 *J. Lumin.* **131** 36–40
[12] Shimotsuma Y, Hirao K, Kazansky P G and Qiu J 2005 *Jpn. J. Appl. Phys.* **44** 4735–48
[13] Allen J P, Scanlone D O and Watson G W 2011 *Phys. Rev.* B **84** 115141-1
[14] McMillan J A 1960 *J. Inorg. Nucl. Chem.* **13** 28

Journal of Physics: Conference Series **712** (2016) 012111 doi:10.1088/1742-6596/712/1/012111

# Co *K*-edge magnetic circular dichroism across the spin state transition in $LaCoO_3$ single crystal

**V Efimov[1], A Ignatov[2], I O Troyanchuk[3], V V Sikolenko[1,4], A. Rogalev[5], F. Wilhelm[5], E Efimova[1], S I Tiutiunnikov[1], D Karpinsky[3], V Kriventsov[6], E Yakimchuk[6], S Molodtsov[7], P Sainctavit[8] and D Prabhakaran[9]**

[1] Joint Institute for Nuclear Research, 141980 Dubna, Russia
[2] Oak Ridge National Laboratory, Oak Ridge, TN 37831, USA
[3] Scientific-Practical Material Research Center NAS Belarus, 220072 Minsk, Belarus
[4] REC "Functional nanomaterials" Immanuel Kant Baltic Federal University, 236041 Kaliningrad, Russia
[5] European Synchrotron Radiation Facility, BP 220, 38043 Grenoble, France
[6] Boreskov Institute of Catalysis SB RAS, 630090 Novosibirsk, Russia
[7] European XFEL Gmbh, Albert-Einstein-Ring 19, 22761 Hamburg, Germany
[8] Institut de minéralogie et de physique des milieux condensés CNRS F-75012, France
[9] Department of Physics, University of Oxford, OX1 3PU, Oxford, Great Britain

E-mail: efimovvv2006@mail.ru

**Abstract.** We report on Co *K*-edge x-ray magnetic circular dichroism (XMCD) measurements of $LaCoO_3$ single crystal in temperature range from 5 to 300 K and external magnetic field of 17 T. The response consists of pre-edge (at 7712 eV) and bi-polar peak (up at 7727, down at 7731 eV) with amplitudes, respectively, less than $10^{-3}$ and $10^{-2}$ of the Co *K*-edge jump. Using the sum rule the orbital magnetic moment of 4*p* Co is evaluated. Its temperature dependence reaches a maximum of $(2.7 \pm 0.9) \times 10^{-3}\ \mu_B$ at 120 K, following the trend for the total magnetic moment on the Co obtained from the superconducting quantum interference device measurements. However, on warming from 25 to 120 K, the orbital magnetic moment of the 4*p* Co doubles while total magnetic moment of Co increases 10 times. First principle calculations are in order to relate the Co *K*-edge XMCD results to the orbital and spin moment of 3*d* Co.

## 1. Introduction

A perovskite $LaCoO_3$ exhibits variety of physical properties [1-11]. At low temperature the $Co^{3+}$ ion is in the low-spin state (LS; $t^6_{2g}e^0_g$, S = 0). Thermal activation of LS state into intermediate-spin state (IS; $t^5_{2g}e_{1g}$, S = 1) or high-spin state (HS; $t^4_{2g}e_{2g}$, S = 2) is observed in the temperature interval from 40 to 120 K [3-10]. Korotin *et. al.* [2] argued that IS state is stabilized by strong 3*d*(Co)-2*p*(O) hybridization. Currently, there are plenty experimental works supporting either LS-IS [2-6] or LS-HS [7-9] scenario. Co $L_{2,3}$-edges x-ray magnetic circular dichroism (XMCD) accompanied by theoretical calculations for the $CoO_6$ cluster [9] have shown that the spin-state transition in $LaCoO_3$ can be well described by thermal excitation from LS ground state into a triply degenerated HS state. However, bonds breaking, strains, and oxygen vacancies on the surface can modify spin state of the Co ion [10, 11]. The present experimental work is undertaken to provide an essentially bulk probe of Co spin states by using a hard X-ray XMCD at the Co *K*-edge.

Journal of Physics: Conference Series **712** (2016) 012111 doi:10.1088/1742-6596/712/1/012111

## 2. Experimental

$LaCoO_3$ single crystal (~ 10 mm in diameter and ~10 cm long) were grown by the floating zone method. The sample were cut in pieces and characterized using x-ray diffraction, EXAFS [12], and superconducting quantum interference device (SQUID) magnetometry. The X-ray absorption near edge spectra (XANES) and X-ray magnetic circular dichroism (XMCD) measurements were performed on beamline ID12 of the European Synchrotron Radiation Facility at Grenoble. All Co K-edge absorption spectra were recorded in total fluorescence yield (TFY) measurements for normal incidence (a backscattering geometry) of the x-ray beam using 98% circularly polarized X-rays. XMCD measurements were performed using 98% circularly polarized x-rays. The signal was recorded from the difference in absorption for parallel and antiparallel alignments of the X-ray helicity pseudovector with respect to an external magnetic field of 17 T, applied along the beam direction. The x-ray helicity and the external magnetic field direction were alternately flipped to minimize experimental artifacts. Six to eight measurements were averaged in order to improve the signal-to-noise ratio.

## 3. Results and discussion

The Co *K* edge absorption and XMCD spectra measured at 5, 25, 60, 120, 150, and 300 K are shown in Fig. 1. The XANES consists of two areas: the pre-edge (7709-7716 eV) due to a quadruple transition from 1*s* to unoccupied 3*d*, and the main edge (7716-7740 eV) due to a dipole transition from 1*s* to unoccupied 4*p* -states. The 3*d* of Co are split into $t_{2g}$ and $e_g$ by the crystal field that clearly seen as a doublet P1- P2. Peaks A, B, and C (at ~7720, 7727, and 7738 eV, respectively) are the best described in the formalism of multiple scattering (MS) of the photoelectron in the cluster of 40-60 atoms centered on the Co absorption site: The major peak B is formed by scattering from the first coordination shell comprised of six oxygen atoms. The peaks A and C are shaped by MS of the photoelectron from more distant coordination shells.

The Co *K*-edge XMCD is driven by orbital polarization of unoccupied 4*p* states coupled to spin polarization of the same 4*p* by spin-orbit interaction, $\xi_{4p}$ . The spin polarization of 4*p* occurs due to hybridization of those states with essentially spin-polarized 3*d* band. The XMCD manifests itself as a week pre-edge doublet and a bi-polar feature (peak "up" at ~7727 eV followed by peak "down" at ~7731 eV) located in vicinity of the major absorption peak B. In the external magnetic field of 17 T the amplitudes of the pre-edge and the bi-polar future do not exceed $10^{-3}$ and $10^{-2}$ of the Co *K*-edge jump. XMCD signal has been successfully measured due to high signal-to-noise ratio, good stability of the beamline, and accurate energy reference. Contrary to ferromagnetics, metal Co, for instance, where the XMCD signal is given by single peak, the observed bi-polar structure is typical for paramagnetic Co, i.e. Co doped ZnO [13] and antiferromagnetic Co, i.e. Co doped $LaMnO_3$ [14].

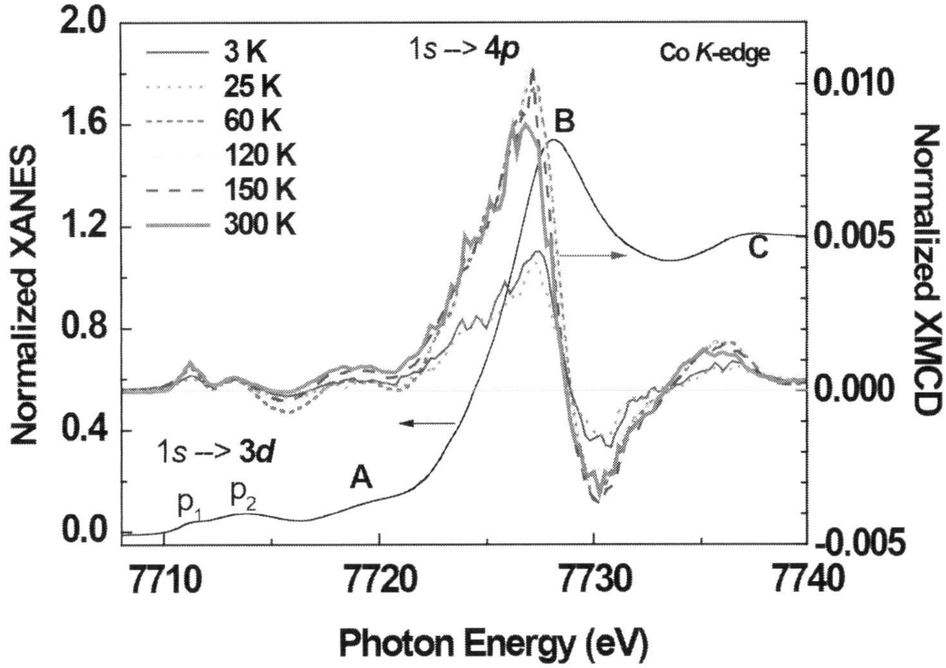


**Fig. 1:** Co *K*-edge absorption and temperature-dependent XMCD spectra of $LaCoO_3$ single crystal, measured under external magnetic field of 17 Tesla applied along the incident beam (that is also the *c*-axis). The background was subtracted and the absorption spectrum was normalized to a unit jump at 50-60 eV above the edge. The XMCD is given as a fraction of the absorption edge jump.

By application of a sum rule [15] the integrated *K*-edge XMCD intensity can be related to the orbital magnetic moment of 4*p*, $<L_z>_{4p}$ as per one Co atom:

$$<L_z>_{4p} = -(6 - n_{4p})\cdot(2I_{XMCD} / 3I_{XANES})$$

where $n_{4p}$ is the electron count in the 4*p*, $I_{XMCD}$ and $I_{XANES}$ are the integrated over the range from 7716 to 7750 eV XMCD and XANES signals. The low limit cut off negligible contribution from *p*-states over the pre-edge. The upper limit aims at separating atomic from continuum contributions. Uncertainties in selection of $n_{4p}$ that were taken equal to 1.0 and selection of the upper integration limit propagates to 20-30% uncertainty in the evaluated $<L_z>_{4p}$

The obtained orbital moments for all measured temperatures are shown in Fig 2 along with total magnetic moments per Co ion derived from our SQUID magnetization measurements, though in slightly smaller applied field. Clearly, the $<L_z>_{4p}$ following the trend for the total magnetic moment on the Co ion. However, on warming from 25 to 120 K, the orbital magnetic moment of the 4*p* Co doubles while total magnetic moment of Co increases 10 times. First principle calculations are in order to relate the Co *K*-edge XMCD results to the orbital and spin moment of 3*d* Co.

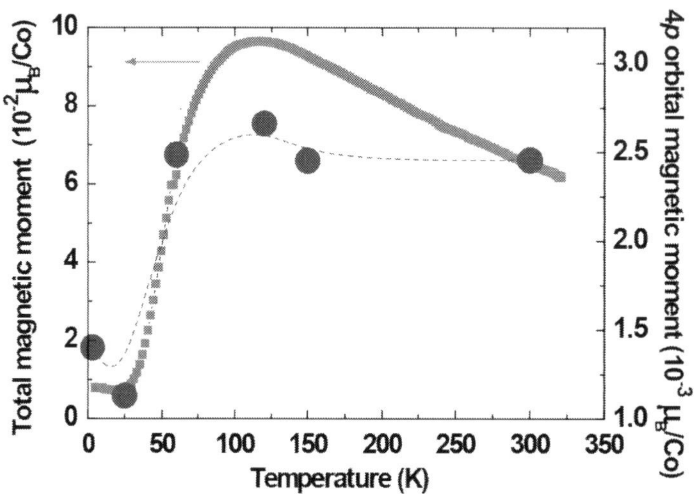


**Fig. 2.** Comparison of 4*p* orbital magnetic moment per Co derived from XMCD sum rule with total magnetic moment per Co ion obtained from SQUID magnetization measurements as a function of the temperature. The XMCD and SQUID data were collected in applied magnetic fields of 17 and 14 T, respectively. Uncertainties for 4*p* orbital magnetic moment are ~ 20-30% (refer to the text for details).

**Acknowledgments**
This work was supported by RFBR Grants No. 15-32-50367_mol_np and No. 13-02-00699_a.

**References**

[1] J. B. Goodenough, J. Phys. Chem. Solids 6, 287 (1958)
[2] M. A. Korotin, S. Y. Ezhov, I. V. Solovyev, V. I. Anisimov, D. I. Khomskii, and G. A. Sawatzky, Phys. Rev. B 54, 5309 (1996)
[3] P. G. Radaelli and S.-W. Cheong, Phys. Rev. B 66, 094408 (2002).
[4] K. Asai, A. Yoneda, O. Yokokura, J. M. Tranquada, G. Shirane, and K. Kohn, J. Phys. Soc. Jpn. 67, 290 (1998)
[5] C. Zobel, M. Kriener, D. Bruns, J. Baier, M. Gruninger, T. Lorenz, P. Reutler, and A. Revcolevschi, Phys. Rev. B 66, 020402 (2002)
[6] R. F. Klie, J. C. Zheng, Y. Zhu, M. Varela, J. Wu, and C. Leighton, Phys. Rev. Lett. 99, 047203 (2007)
[7] A. Podlesnyak, S. Streule, J. Mesot, M. Medarde, E. Pomjakushina, K. Conder, A. Tanaka, M. W. Haverkort, and D. I. Khomskii, Phys. Rev. Lett. 97, 247208 (2006)
[8] S. Noguchi, S. Kawamata, K. Okuda, H. Nojiri, M. Motokawa, Phys. Rev. B 66, 094404 (2002)
[9] M. W. Haverkort, Z. Hu, J. C. Cezar, T. Burnus, H. Hartmann, M. Reuther, C. Zobel, T. Lorenz, A. Tanaka, N. B. Brookes, H. H. Hsieh, H.-J. Lin, C. T. Chen, and L. H. Tjeng, Phys. Rev. Lett. 97, 176405 (2006)
[10] J.-Q. Yan, J.-S. Zhou, and J. B. Goodenough, Phys. Rev. B 70, 014402 (2004)
[11] I. O. Troyanchuk, M. V. Bushinsky, and L. S. Lobanovsky, J. Appl. Phys. 114, 213910 (2013).
[12] Sikolenko V, Troyanchuk I, Efimov V, Efimova E, Karpinsky D, Sakura P, Zaharko O, Ignatov A, Aquilanti D, Selutin A, Shmakov A and Prabhakaran D, J.Phys. Conference Series, in press.
[13] Ney A, Ollefs K, Ye S, Kammermeier T, Ney V, Kaspar T C, Chambers S A, Wilhelm A F, and Rogalev A, 2008 Phys. Rev. Lett. 100, 157201
[14] Burnus T, Hu Z, Hsieh H H, Joly V L J, Joy P A, Haverkort M W, Hua Wu, A. Tanaka A, Lin H J, Chen C T, and Tjeng L H, 2008 Phys. Rev. B 77 125124
[15] Guo G P, 1998 Phys. Rev. B 57 10295.

Journal of Physics: Conference Series **712** (2016) 012112 doi:10.1088/1742-6596/712/1/012112

# Pressure-induced electronic phase transition in compound $EuCu_2Ge_2$

**A Y Geondzhian[1], A A Yaroslavtsev[1,2], P A Alekseev[1,3], R V Chernikov[5], B R Gaynanov[1], F Baudelet[6], L Nataf[6] and A P Menushenkov[1]**

[1] National Research Nuclear University MEPhI (Moscow Engineering Physics Institute), Kashirskoe sh. 31, Moscow 115409, Russia
[2] European XFEL GmbH, Albert-Einstein-Ring 19, 22761 Hamburg, Germany
[3] National Research Centre "Kurchatov Institute", Kurchatov sq., 123182 Moscow, Russia
[4] DESY Photon Science, Notkestr., 85, D-22607 Hamburg, German
[6] Synchrotron SOLEIL, l'Orme des Merisiers, 91190 Saint-Aubin, France

Email: a.y.geondzhian@gmail.com

**Abstract**. We report the high-pressure XANES study of the electronic phase transition from $4f^7$ to $4f^6$ configuration of europium in the rare-earth compound $EuCu_2Ge_2$. The hydrostatic pressure dependence of the europium valence was obtained in a wide pressure range (1-30 GPa) at room temperature. It was found that upon the pressure increase above 20 GPa the europium valence does not reach the integer value +3 but stabilizes at 2.87. The experimental results were supported by the band structure calculations in the framework of DFT, which allowed us to discuss the features of $3d$-$4f$ hybridization in this system. The study also compares the mechanisms of external and "chemical" pressure by the Si substitution in Ge site in series $EuCu_2(Si_xGe_{1-x})_2$.

## 1. Introduction

Compounds $EuCu_2X_2$ (X =Si, Ge) crystallize in the $ThCr_2Si_2$-type structure (tetragonal space group I4/mmm) which is famous for a variety of fascinating phenomena such as valence fluctuations, "heavy-fermion superconductivity", non-Fermi liquid states, Kondo insulators, etc. [1-4]. These compounds are usually highly sensitive to the impacts like electron/hole doping or pressure, which lead to the modification of their electronic structure near Fermi level and eventually to various phase transitions. That is how the superconductivity was recently found in $EuFe_2As_2$ under compression [5]. The compound $EuCu_2Ge_2$ is antiferromagnetic ($T_N \approx 15$ K) at ambient conditions and at the same time shows the intermediate valence state of europium with a value of 2.15 [6]. We talk about electronic phase transition in terms of transition between $4f$ electronic configurations of rare-earth element or the valence instability. Recently it has been reported in [6] that silicon substitution in Ge site in series $EuCu_2(Si_xGe_{1-x})_2$ leads to a chemical compression and induces the transition from the magnetically ordered state ($x < 0.65$) to a heavy-fermion state ($0.65 < x < 0.85$) and then to a nonmagnetic strong valence fluctuating state which is accompanied by the Eu valence increase from 2.15 in $EuCu_2Ge_2$ to 2.67 in $EuCu_2Si_2$. The suppression of the magnetic ordering may be due to the predominance of the contribution from non-magnetic $Eu^{3+}$ ($4f^6$, $J = S = 0$). Thus, the europium valence might be an indicator of certain electronic and magnetic properties in this system. In this way we decided to investigate the valence in pure parent compound $EuCu_2Ge_2$ under the high hydrostatic pressure and to compare the results with the chemical pressure effects observed in $EuCu_2(Si_xGe_{1-x})_2$.

## 2. Experiment

The X-ray absorption experiment was carried out at the hard X-ray energy dispersive beamline ODE of synchrotron SOLEIL (France). The high pressure was generated by the diamond anvil cell (DAC) and controlled using the ruby luminescence. The XANES spectra were measured at Eu-$L_2$ absorption edge in a wide pressure range 1-30 GPa.

The separation of the spectral components arising from different europium valence states was done by fitting the experimental XANES, corrected for a polynomial background, with combination of Lorentzian (representing the absorption resonance) and arctangent (describing the transitions to the continuum) curves of constrained widths and energy positions (Fig. 1). In this approach the amount of Eu ions in each valence state is assumed to be proportional to the area under the corresponding Lorentzian curve or, equivalently, to its weight in the fitting formula. Europium valence state at ambient pressure in $EuCu_2(Si_xGe_{1-x})_2$ was explored in our earlier XANES experiment at the Eu-$L_3$ absorption edge [7].

## 3. Results and Discussion

The resulting pressure dependence of europium valence is shown in Fig. 2. We can point out the three characteristic regions. In the first region from 1 to 6 GPa the valence changes very slightly with the pressure increase. Probably it is caused by the exchange-spin interaction in the presence of the long-range magnetic order. The transition between the first and the second regions (with the value of valence around 2.3) corresponds to the suppression of magnetic order in series $EuCu_2(Si_xGe_{1-x})_2$. After this the strong magnetic interaction is replaced by the pure valence fluctuations and in the second region the europium valence begins to grow rapidly. However, the most interesting is the region above 20 GPa in which the valence begins to stabilize. The highest valence value (~ 2.87) obtained under the pressure of 30 GPa in compound $EuCu_2Ge_2$ is very close to the europium valence in compound $EuCu_2Si_2$ at low temperature (4 K) [3,6-7]. In such a way the following question arises: does the europium valence in these compounds reach the integer value +3 at all?

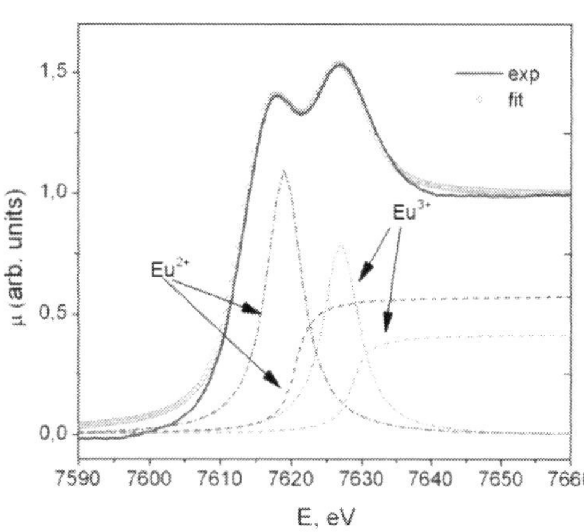


**Figure 1.** $L_2$ absorption edge of Eu in $EuCu_2Ge_2$ at room temperature and pressure 10 GPa. The fitting with $Eu^{2+}$ and $Eu^{3+}$ contributions is shown.

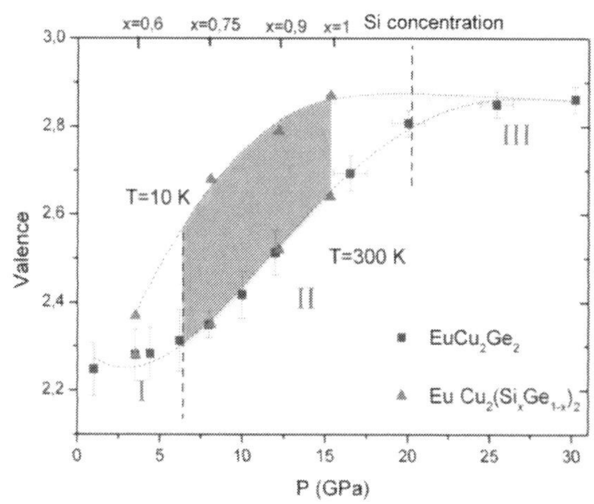


**Figure 2.** The diagram of Eu valence behaviour in the system $EuCu_2(Si_xGe_{1-x})_2$ vs. temperature, Si concentration and external pressure applied to the pure $EuCu_2Ge_2$, obtained in our X-ray absorption experiments.

Extrapolation of the saturation region shows that the valence may reach the integer value at pressure ~ 80 GPa. It is also possible that above 30 GPa the $EuCu_2Ge_2$ experiences a sharp transition to the stable phase with europium valence +3 as it was observed for cerium [8]. However, the situation when the valence transition stops at the intermediate phase is also known [9], and thus we do an attempt to propose the mechanism for the valence stabilization in $EuCu_2Ge_2$ using the theoretical calculations of electronic structure.

*3.1 DFT calculations*

*Ab initio* electronic structure calculations were performed within the density-functional theory (DFT) using a full-potential linearized augmented plane-wave (FP-LAPW) approach using the EXCITING code [10]. The Perdew, Burke, Ernzerhof (PBE) gradient corrected local spin density approximation (LSDA-GGA) for the exchange correlation (XC) potential has been used. To account for the Coulomb correlation interaction within the Eu 4*f* shell, the PBE XC potential corrected according to GGA+U method was additionally considered. For europium the $U$ and $J$ parameters values of 2.7 and 1.1 eV have been taken, respectively. The pressure was modeled in the approximation of uniform compression by variation of the elementary cell volume, which was chosen because of the absence of exact information on the lattice deformation in $EuCu_2Ge_2$ under high pressure.

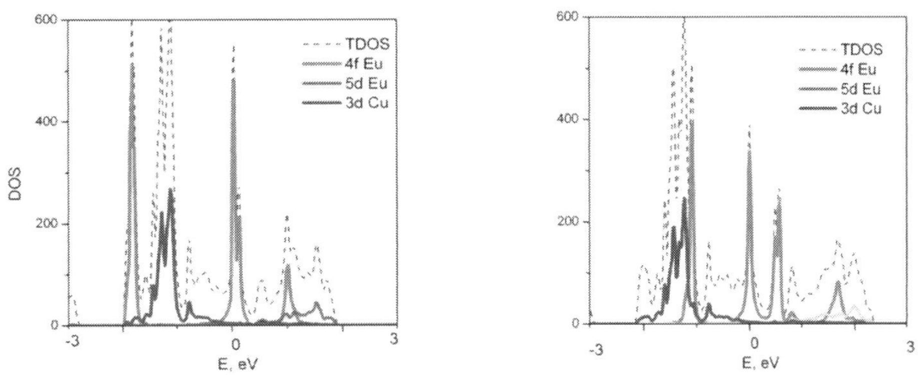


**Figure 3.** Density of states for the different elementary cell volumes in $EuCu_2Ge_2$ (from left to right: cell's scale =100%, 95 %)

Based on the results of DFT calculation we propose a qualitative model for the formation of europium intermediate valance state by the peculiar hybridization between Cu 3*d* and Eu 4*f* electronic levels. In the consequence of Coulomb interaction between localized electrons the localized 4*f* level of Eu splits into few components (Fig. 3) and one of them overlaps with Cu 3*d* subband under pressure. The saturation region with the maximum value of the europium valence at high pressure is probably due to the saturation of the 4*f*-3*d* overlap when the narrow 4*f* level reaches the middle of the wide 3*d* subband and the overlapping area becomes almost independent of levels position. According to this model, the value +3 of europium valence in compound $EuCu_2Ge_2$ is unreachable.

*3.2 "Chemical' vs. hydrostatic pressure*

As Fig. 2 shows the "chemical" (top horizontal axis) and external (lower horizontal axis) pressures have a nonlinear relation and, consequently, a different effect on the europium valence. This fact has been also partially confirmed by the DFT calculations of $EuCu_2Ge_2$ and $EuCu_2Si_2$ compounds electronic structure.

At ambient pressure the Cu 3*d* DOS maximum in compound $EuCu_2Si_2$ is situated higher in energy than in compound $EuCu_2Ge_2$ and with increasing pressure it slightly moves down from the Fermi level. Thus $EuCu_2Ge_2$ under pressure and $EuCu_2Si_2$ have different energy position of Cu 3*d* levels. But in spite of that the saturation of europium valence under high pressure was obtained in compound $EuCu_2Si_2$ as well in the previous inelastic neutron scattering experiment [11].

It was expected that the application of external pressure of the order of 2 GPa would lead to a noticeable valence change (quantitative evaluation of the relation between valence and pressure up to 5 GPa for $EuCu_2Si_2$ was given in [1]), which is experimentally manifested in renormalization of the magnetic spectrum. However, external pressure did not produce a significant change in the structure of the magnetic response nor in its temperature dependence. This fact can be explained in the way of compensation of two main reasons for renormalization of the magnetic spectrum. On one hand, it is a high hybridization of *f*-electrons with the conduction band states, and on another, it is an opposite renormalization expected due to the shift of the average valence to +3. Therefore it is possible that the compound $EuCu_2Si_2$ has only quantitative difference but similar (with $EuCu_2Ge_2$) physical reasons for the stabilization of the europium valence under high pressure.

## 4. Conclusion

The pressure dependence of europium valence in $EuCu_2Ge_2$ has been obtained in a wide pressure range using the XANES spectroscopy. It was found that the mechanisms of the "chemical" and external pressure most likely affect the europium valence in a different way. However, the stabilization of europium valence near the similar value of 2.9 was obtained for both Si-substituted and pure compound. Based on the DFT calculations the qualitative model of hybridization between 4*f* level and 3*d* subband has been proposed, which explains the stabilization of valence at high pressure and predict that even at a higher compression it will stay below +3.

## 5. Acknowledgements

This work is supported by Russian Science Foundation (grant 14-22-00098).

## References

[1] Alekseev P A, Nemkovski K S, Mignot Mignot J-M, Nemkovski K S, Lazukov V N, Nefeodova E V, Menushenkov A P, Kuznetsov A V, Bewley R I and Gribanov A V 2007 *JETP* **105** 14

[2] Hossain Z, Geibel C, Senthilkumaran N, Deppe M, Baenitz M, Schiller F and Molodtsov S 2004 *Phys. Rev. B* **69** 014422

[3] Fukuda Sh, Nakanuma Y, Sakurai J, Mitsuda A, Isikawa Y, Ishikawa F, Goto T, Yamamoto T 2003 *J. Phys. Soc. Jap.* **72**(12) 3189

[4] Alekseev PA, Nemkovski K S, Mignot J-M, Lazukov V N, Nikonov A A, Menushenkov A P, Yaroslavtsev A A, Bewley R I, Stewart J R and Gribanov A V 2012 *J. Phys.: Condens Matter* **24** 375601

[5] Miclea C F, Nicklas M, Jeevan H S, Kasinathan D, Hossain Z, Rosner H, Gegenwart P, Geibel C and Steglich F 2009 *Phys. Rev. B* **79** 212509

[6] Banik S, Bendounan A, Thamizhavel A, Arya A, Risterucci P, Sirotti F, Sinha A K, Dhar S K, and Deb S K 2012 *Phys. Rev. B* **86** 085134

[7] Alekseev P A, Nemkovski K S, Kozlenko D P, Menushenkov A P, Yaroslavtsev A A, Gribanov A V, Clementyev E S, Pantalei C, Klobes B, Hermann R P 2014 *JETP Lett.* **99** 164

[8] Franceschi E and Olcese G L 1969 *Phys. Rev. Lett* **22** 1299

[9] Khomskiy D I 1979 *UFN* **129** 443

[10] Gulans A, Kontur S, Meisenbichler C, Nabok D, Pavone P, Rigamonti S, Sagmeister S, Werner U, Draxl C 2014 *J. Phys.: Condens. Matter* **26**, 363202

[11] Alekseev P A, Mignot J-M, Nemkovski K S, Nefeodova E V, Lazukov V N, Karpunin D Yu, Bewley R I, Gribanov A V 2008 *Physica B* **403**, 864

# High local disorder in $Tb_2Hf_2O_7$ pyrochlore oxide nanocrystals

**V A Kabanova[1], V V Popov[1], Ya V Zubavichus[2], E S Kulik[2], A A Yaroslavtsev[1,3,4], R V Chernikov[5], A P Menushenkov[1]**

[1]National Research Nuclear University "MEPhI", Kashirskoe sh. 31, 115409 Moscow, Russia
[2]NRC "Kurchatov Institute", Akademika Kurhatova sq. 1, 123182 Moscow, Russia
[3]European XFEL GmbH, Albert-Einstein-Ring 19, 22761 Hamburg, Germany
[4]I. Institut für Theoretische Physik, Universität Hamburg, Jungiusstrasse 9, 20355 Hamburg, Germany
[5]DESY Photon Science, Notkestrasse 85, 22607 Hamburg, Germany

E-mail: victoria.kabanova@gmail.com

**Abstract**. The process of $Tb_2Hf_2O_7$ nanocrystals formation upon annealing to 1600°C was investigated by means of X-ray absorption fine structure (XAFS) spectroscopy combined with X-ray diffraction (XRD) and pair distribution function (PDF) analysis. The structure ordering and the growth of nanocrystals upon annealing were estimated independently from XRD patterns and PDF. The probable content of $Tb^{4+}$ ions in $Tb_2Hf_2O_7$ was estimated from XANES. All studies indicate a high disorder and a large number of local structure defects in $Tb_2Hf_2O_7$ pyrochlore oxide.

## 1. Introduction

Double oxides of the form $Ln_2M_2O_7$ (where Ln = lanthanide, M = Ti, Zr, Hf) have a variety of potential applications as thermal barrier coatings [1], ionic conductors [2], scintillators [3], semiconductors [4] and matrices for high level nuclear wastes immobilization [5]. A number of these applications depend upon the structure of these compounds, which are able to undergo an order – disorder phase transition from pyrochlore (*Fd-3m* space group) to defect fluorite (*Fm-3m* space group) structure [6,7]. When defect fluorite structure is formed, both cations local surrounding is identical, containing 8 oxygen atoms in the cube vertices. In case of pyrochlore structure, $M^{4+}$ oxygen surrounding changes to octahedral (CN=6), while cubic $Ln^{4+}$ surrounding splits into 2 shorter and 6 longer Ln-O distances. Stability of pyrochlore structure is generally related to the ratio of $Ln^{3+}$ and $M^{4+}$ ionic radii, and is normally observed for $R_{Ln^{3+}}/R_{M^{4+}}$=1.46-1.78, while smaller values lead to defect fluorite structure formation [6,8].

However, the question concerning phase transition boundary is controversial [7-12], and the most interesting are compounds with the ratio $R_{Ln^{3+}}/R_{M^{4+}}$~1.46, in which both pyrochlore and defect fluorite structures might form depending on the synthesis conditions [13]. The aim of the current work was to study the local crystalline structure formation in one of these "boundary" compounds, $Tb_2Hf_2O_7$ ($R_{Tb^{3+}}/R_{Hf^{4+}}$=1.46), upon isothermal annealing of an amorphous precursor.

## 2. Experimental

The mixed Tb-Hf hydroxide was prepared by co-precipitation from $Tb(NO_3)_3 \cdot 5H_2O$ (99.99% pure), $HfOCl_2 \cdot 8H_2O$ (98+% pure) and $NH_3 \cdot H_2O$ (analytical grade). The $Tb_2Hf_2O_7$ powders were obtained by isothermal annealing of the resulting $Tb_2O_3 \cdot 2HfO_2 \cdot 10.5H_2O$ precursor at 600-1600°C for 3 h. A more detailed description of the preparation technique may be found in [14].

The powder XRD patterns were measured at beamline "Structural Material Science" of Kurchatov Synchrotron Radiation Source (Moscow, Russia) (λ=0.68886 Å, sample to detector distance 150 mm) and processed by Rietveld method. The powder XRD patterns for pair distribution function (PDF) analysis were collected at beamline P02.1 of PETRA-III storage ring (DESY, Hamburg) using high-energy photons of 59.82 keV (λ=0.207270 Å) to get a high $q$-range up to 21 $Å^{-1}$. PDFgetX2 [15] was used to calculate full structure factors and pair distribution functions from experimental XRD patterns. Modeling of the obtained experimental PDFs was performed using PDFgui [16].

The X-ray absorption spectra were measured at I811 beamline of MAX-lab (Lund, Sweden) and at beamline "Structural Material Science" of Kurchatov Synchrotron Radiation Source in transmission geometry above the $L_3$-Tb (7514 eV) and $L_3$-Hf (9561 eV) absorption edges at room temperature. VIPER software pack [17] was used for extraction and fitting of EXAFS-functions. Amplitudes and phases of photoelectron back scattering were calculated with FEFF code [18] using the parameters of pyrochlore structure and diffraction data. Modeling of the near-edge region of spectra was performed using XANDA software pack [19].

## 3. Results and discussion

XRD study has shown that isothermal annealing of X-ray amorphous mixed hydroxide at 600-700°C leads to the formation of nanocrystalline powders with the defect fluorite structure. A further annealing temperature increase leads to an increase in coherent scattering lengths and a decrease in microstrain values. Heat treatment at temperature ≥1400°C initiates nucleation and growth of nanodomains with pyrochlore-type superstructural ordering of cations inside a microcrystalline matrix with the defect fluorite structure.

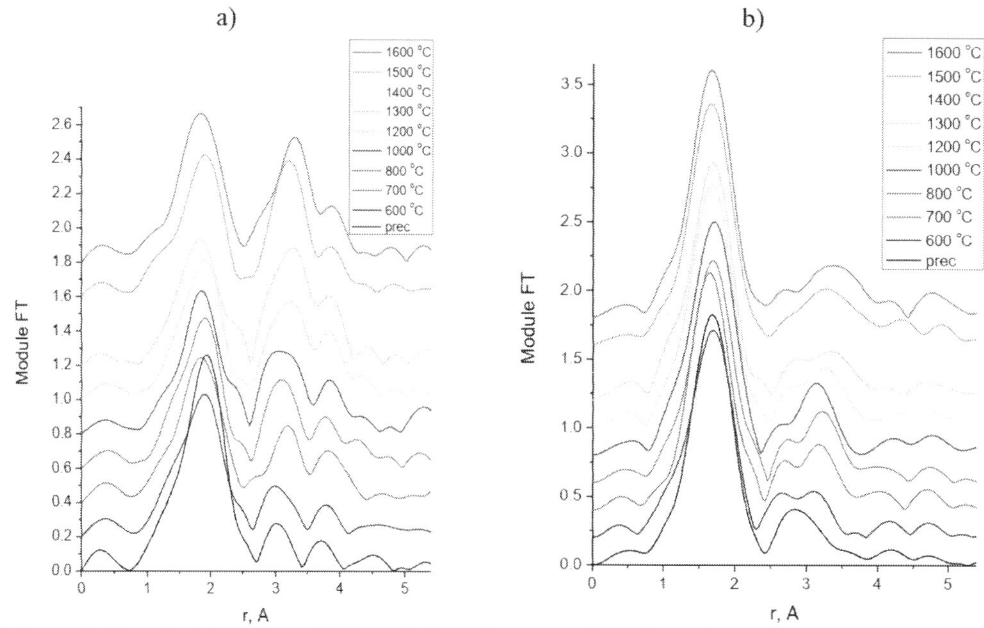


**Figure 1.** Fourier-Transform (FT) modules of EXAFS-functions above $L_3$-Tb (a) and $L_3$-Hf (b) edges.

According to the qualitative EXAFS analysis (figure 1), there is a slight shift of the EXAFS Fourier transform peaks corresponding to the 1st coordination shell, which indicates Tb-O and Hf-O bond lengths contraction upon annealing. The next maxima corresponding to the metal-metal bonds change their shape and intensities, signifying the structure ordering in the course of annealing.

Modeling of EXAFS spectra was performed using both pyrochlore and defect fluorite models with fixed coordination numbers for all bonds. As a clearly pronounced non-equivalence of cations local surroundings was obvious (figure 1) and as there was a significant discrepancy of experimental data and a defect fluorite model curves, the pyrochlore model was chosen. The splitting of the 1$^{st}$ coordination shell Tb-O into two components was present for the samples in the whole range of investigated annealing temperatures including the precursor sample. The refined values of bond lengths and Debye-Waller factors for the 1$^{st}$ coordination shell are shown in figure 2. The fitting results indicate high disorder and a large number of structure defects.

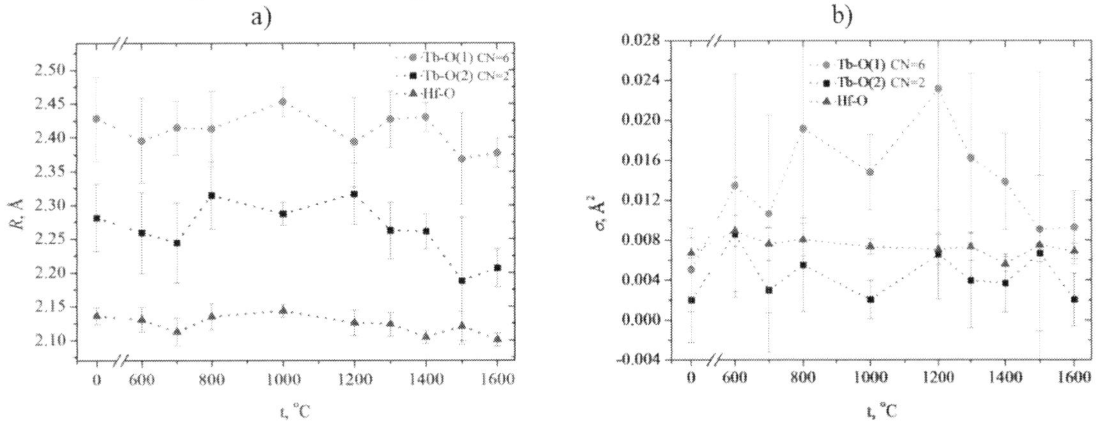


**Figure 2**. Bond lengths (a) and Debye-Waller factors (b) evolution with an annealing temperature increase.

It was found that X-ray absorption near edge structure (XANES) of $Tb_2Hf_2O_7$ spectra contains two contributions, which probably correspond to the $Tb^{3+}$ and $Tb^{4+}$ oxidation states. $Tb^{4+}$ content was estimated by modeling of XANES spectra with two lorentzian + arctangent functions. The highest amount was found to be ~7-8% for the sample annealed at 1000°C. It was assumed that, according to the charge, $Tb^{4+}$ ions might partially occupy the $Hf^{4+}$ positions, which leads to an increase in structure disorder and an obstruction of the pyrochlore-type ordering.

Atomic pair distribution function (PDF) analysis shows that annealing causes an increase of interatomic distance range, in which pair correlation function G(r) maxima are still distinguishable. This corresponds to the nanocrystallites size growth upon annealing as the PDF oscillations for more calcined samples are present at longer distances. This qualitatively correlates with the XRD results for coherent-scattered regions.

**Table 1.** $Tb_2Hf_2O_7$ XRD and pair distribution function modeling results.

| Annealing temperature | Latt. param., Å (XRD results) | Latt.param., Å (PDF analysis) | $U_{iso}(Tb^{3+}/Hf^{4+})$, Å$^2$ |
|---|---|---|---|
| **1000°C** | 5.21439 | 5.2107 | 0.01942 |
| **1200°C** | 5.2181 | 5.20443 | 0.01613 |
| **1300°C** | 5.22182 | 5.21037 | 0.01559 |
| **1400°C** | 5.22443 | 5.21242 | 0.01462 |
| **1500°C** | 5.22974 | 5.21641 | 0.01444 |
| **1600°C** | 5.23358 | 5.2186 | 0.01457 |

Modeling of the experimental pair distribution functions was performed using a defect fluorite model. A good convergence of experimental and refined curves was present for the long range order, but in the range of short interatomic distances a significant discrepancy was observed. This indicates pyrochlore-type local structure of samples, while the average structure is well described by the defect fluorite model. The refined lattice parameter values demonstrate increasing dependency from an annealing temperature, which is in agreement with diffraction results (table 1). It was also found that cations anisotropic displacement parameters $U_{iso}$ decrease upon calcination (table 1), corresponding to structure ordering.

## 4. Conclusion

The evolution of $Tb_2Hf_2O_7$ crystalline, local and electronic structure parameters upon annealing of initial mixed hydroxide has been studied by a complex of diffraction and absorption synchrotron methods. It has been shown that isothermal annealing of the X-ray amorphous precursor at temperatures above 600-700°C first leads to the formation of oxide nanocrystalline powders with a defect fluorite structure. Also, calcination temperature increase reveals an increase in coherent scattering lengths and a decrease in microstrain values. Heat treatment at temperature ≥1400°C initiates nucleation and growth of nanodomains with pyrochlore-type superstructural ordering of cations inside a microcrystalline fluorite matrix. Investigation of the local structure of prepared samples showed that all synthesized $Tb_2Hf_2O_7$ powders had clearly pronounced nonequivalence of the $Tb^{3+}$ and $Hf^{4+}$ cations local environment. It was assumed that the observed local structure disorder of $Tb_2Hf_2O_7$ samples may partially occur due to the presence of $Tb^{4+}$ cations.

## Acknowledgements

The authors thank J. Bednarcik (DESY Photon Science), R.V. Chernikov (DESY Photon Science) and S. Carlson (MAX IV lab.) for their help in experiments and the Russian Science Foundation (grant 14-22-00098) for financial support.

## References

[1] Cao X Q, Vassen R and Stoever D 2004 *J. Eur. Ceram. Soc.* **24** 1-10
[2] Yamamura H, Nishino H, Kakinuma K and Nomura K 2003 *Solid State Ionics* **158** 359-65
[3] Hansel R A, Desai S K, Allison S W, Heyes A and Walker D G 2010 *J. Appl. Phys.* **107** 016101
[4] Wei F, Tu H, Wang Y, Yue S and Du J 2009 *J. Phys.: Conf. Ser.* **152** 012003
[5] Ewing R C, Weber W J and Lian J 2004 *J. Appl. Phys.* **95** 5949-71
[6] Subramanian M A, Aravamudan G and Rao G V S 1983 *Prog. Solid State Chem.* **15** 55-143
[7] Blanchard P E R, Liu S, Kennedy B J, Ling C D, Zhang Z, Avdeev M, Cowie B C C,Thomsen L and Jang L Y 2013 *J. Phys. Chem. C.* **117** 2266-73
[8] Popov V V, Menushenkov A P, Zubavichus Ya V, Yaroslastsev A A and Leshchev D S 2013 *Russ. J. Inorg. Chem.* **58** 1400-7
[9] Stanek C R 2003 Ph. D. Thesis. Imperial College. London. UK
[10] Mandal B P, Garg N and Sharma S M 2006 *J. Solid State Chem.* **179** 1990-4
[11] Zu X T, Li N and Gao F 2008 *J. Appl. Phys.* **104** 043517
[12] Popov V V, Zubavichus Ya V, Menushenkov A P, Yaroslavtsev A A, Kulik E S, Pisarev A A and Kolyshkin N A 2015 *Russ. J. Inorg. Chem.* **60** 16-22
[13] Popov V V, Zubavichus Ya V, Menushenkov A P, Yaroslavtsev A A and Kulik E S 2014 *Russ. J. Inorg. Chem.* **59** 279-85
[14] Popov V V, Zubavichus Y V, Petrunin V F, Menushenkov A P, Kashurnikova O V, Korovin S A, Chernikov R V, Yaroslavtsev A A 2011 *Glass Phys. Chem.* **37** 512-20
[15] Qiu X, Thompson J W and Billinge S J L 2004 *J. Appl. Crystal* **37** 678
[16] Farrow C L, Juhas P, Liu J W, Bryndin D, Bozin E S, Bloch J, Proffen Th and Billinge S J L 2007 *J. Phys.: Condens. Matter.* **19** 335219
[17] Klementev K V 2001 *J. Phys. D: Appl. Phys.* **34** 209
[18] Rehr J J and Albers R C 2000 *Rev. Mod. Phys. B* **72** 621-54
[19] Klementiev K V, XANES dactyloscope for Windows, freeware http://www.cells.es/old/Beamlines/CLAESS/software/xanda.html

# Thermally activated decomposition of (Ga,Mn)As thin layer at medium temperature post growth annealing

**Y Melikhov[1], P Konstantynov[1], J Domagala[1], J Sadowski[1,2], M Chernyshova[3], T Wojciechowski[1], Y Syryanyy[1], I N Demchenko[1]**

[1] Institute of Physics, Polish Academy of Sciences, Warsaw, Poland
[2] MAX IV Laboratoriet, Lund, Sweden
[3] Institute of Plasma Physics and Laser Microfusion, Warsaw, Poland

E-mail: melikhov@ifpan.edu.pl

**Abstract**. The redistribution of Mn atoms in $Ga_{1-x}Mn_xAs$ layer during medium-temperature annealing, 250-450 °C, by Mn K-edge X-ray absorption fine structure (XAFS) recorded at ALBA facility, was studied. For this purpose $Ga_{1-x}Mn_xAs$ thin layer with x=0.01 was grown on AlAs buffer layer deposited on GaAs(100) substrate by molecular beam epitaxy (MBE) followed by annealing. The examined layer was detached from the substrate using a "lift-off" procedure in order to eliminate elastic scattering in XAFS spectra. Fourier transform analysis of experimentally obtained EXAFS spectra allowed to propose a model which describes a redistribution/diffusion of Mn atoms in the host matrix. Theoretical XANES spectra, simulated using multiple scattering formalism (FEFF code) with the support of density functional theory (WIEN2k code), qualitatively describe the features observed in the experimental fine structure.

## 1. Introduction

A potential simultaneous use of the ferromagnetic and semiconductors properties in spintronics has attracted a great interest to diluted magnetic semiconductors (DMS) with the most studied material being (Ga,Mn)As. With the optimized MBE growth and post growth annealing procedures nowadays (Ga,Mn)As layers have achieved the Curie temperature, $T_C$, as high as about 200 K. This is remarkably high as for DMS, but still too low in view of potential application in spintronics devices. From physics point of view, an indirect coupling between localized spins mediated by charge carriers is of paramount importance for the possibility of magnetic ordering in DMS [1]. However, the Ruderman-Kittel-Kasuya-Yosida (RKKY) interaction mechanism, which is used to describe the properties of DMS systems, has limitations predicting an increase of $T_C$ with increasing concentration of Mn impurities, the situation which is not always experimentally observed. Further studies not only on the influence of microstructure and its inhomogeneities upon material's properties but also on transformation processes (including formation and migration of point defects) which occur in (Ga,Mn)As during growth and post growth annealing should lead to an improved understanding of the microstructure evolution and could potentially lead to a further progress in reaching larger $T_C$ in (Ga,Mn)As.

The paper aims to check the effectiveness of x-ray absorption fine structure (XAFS) as a probe of thermally activated decomposition of DMS, namely, (Ga,Mn)As after medium temperature post growth annealing.

## 2. Sample Preparation and Experimental Details

The $Ga_{1-x}Mn_xAs$ layer was grown in a SVT Associates MBE system [2] on GaAs (100) substrate. The test measurements of the energy dependence of the X-ray absorption coefficient of Mn atoms embedded in a single GaAs crystal matrix on bulk GaAs substrate were disturbed by elastic scattering from the substrate. To overcome this problem, a possibility of a physical separation of the studied layers from the substrate was investigated. For this purpose, thin AlAs buffer layer was deposited on GaAs substrate prior to the $Ga_{1-x}Mn_xAs$ growth. Finally, about 1000 Å amorphous As capping layer was deposited on $Ga_{1-x}Mn_xAs$. After removing from the MBE setup, the film was cleaved into four pieces: one piece was left intact to serve as a reference sample (referred further as "as-grown") and the other three were annealed at temperatures 250 °C, 350 °C and 450 °C. The annealing temperature was calibrated by the temperature of desorption of As capping layer and by the thermocouple placed at the vicinity of the MBE substrate holder.

The content of Mn in the $Ga_{1-x}Mn_xAs$ layer was estimated to be at the level of 1 at. % by Energy-dispersive X-ray spectroscopy (EDX) using Auriga 40 FIB-SEM workstation. High resolution diffraction (HR-XRD) studies were performed using a Philips X'Pert-MRD diffractometer equipped with a parabolic x-ray mirror, a four-bounce Ga 220 monochromator at the incident beam, and a three-bounce Ge analyzer at the diffracted beam. After conventional initial analysis, all samples were subjected to the separation of the $Ga_{1-x}Mn_xAs$ layer from the GaAs substrate by chemical etching, a so-called "lift-off" procedure [3]. Its schematic representation is shown in Figure 1. The local atomic and electronic structure around Mn atoms was studied by XAFS spectroscopy with the spectra collected at the BL22 CLÆSS beamline at ALBA light source [4] at room temperature.

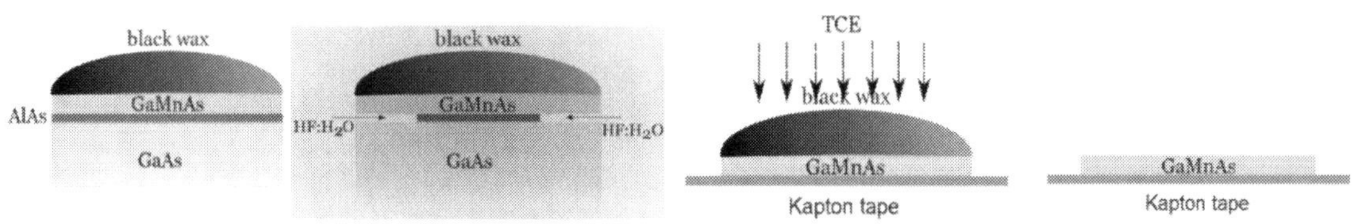


**Figure 1.** Schematic representation of the steps during the "lift-off" procedure.

## 3. Results and Discussion

Figure 2(a) shows the ω/2θ scans obtained for the symmetric (004) reflection for all samples. The "as-grown" and 250 °C annealed samples revealed high crystalline perfection of the $Ga_{1-x}Mn_xAs$ layer. A satellite peak close to the dominant GaAs (004) peak is present in the XRD data for both samples. These satellite peaks are centered at lower angle with respect to the GaAs (004) with slight moving to lower angles when the annealing temperature is increased to 250 °C. A substantial change in the satellite peak position occurs when the annealing temperature is increased to 350 °C (generally, peak is not visible), where, according to literature data, the remanent magnetization and Curie temperature are appreciably lower than in the "as-grown" sample. After annealing at 450 °C, the satellite peak is dramatically shifted towards the GaAs (004) peak (revealed as a shoulder) accompanied, according to literature reports, by the total disappearance of remanent magnetization. Consequently, the lattice constant of the investigated layer decreases with the increasing annealing temperature reaching, or even exceeding, the lattice constant of the GaAs substrate. This decrease in itself is an indicator that some sort of reorganization occurs, possibly a formation of other Mn-containing phases, e.g., a precursor phase for Mn-rich (Ga,Mn)As clusters.

In order to determine the local atomic and electronic structure around Mn atoms the XAFS spectra at the K-edge of Mn were collected at room temperature on samples detached from the substrate (success in elimination of the elastic scattering is shown in Figure 2(b)). Figure 3(a) presents normalized XANES spectra around Mn K edge, whereas, an inset shows modulus of Fourier transforms, |FT(R)|, of the EXAFS function for all samples.

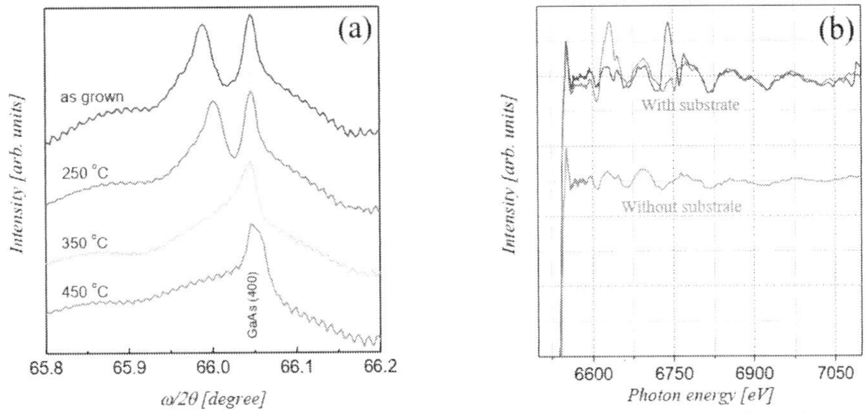


**Figure 2.** (a) XRD ω/2θ scans for the 004 reflection from "as-grown" and annealed (Ga,Mn)As layers (i.e. before the "lift-off" procedure). (b) Comparison of raw XAFS spectra for samples with (two angles of incidence are chosen) and without substrate.

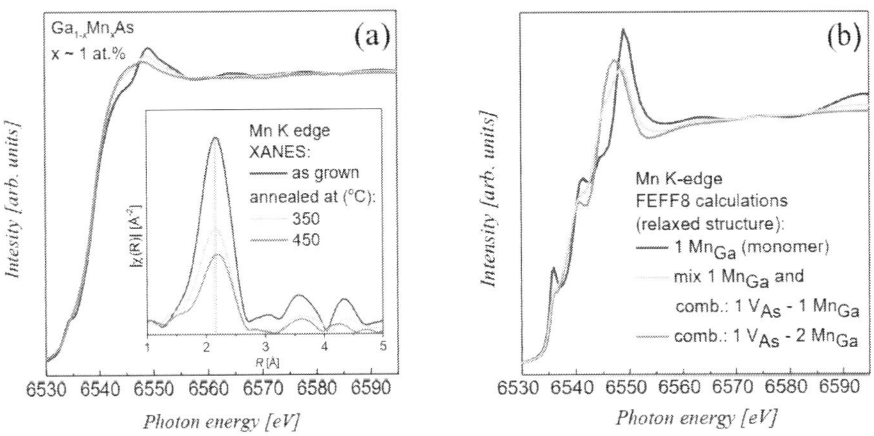


**Figure 3.** (a) Normalized Mn K-edge XANES spectra of the (Ga,Mn)As layers (i.e. after the "lift-off" procedure) (The inset shows Fourier transforms of the k-weighted Mn K-edge EXAFS χ function). (b) Mn K-edge XANES spectra calculated by FEFF code using models formed based on structure optimization performed by WIEN2k.

The analysis of XANES spectrum of the "as-grown" sample indicates that Mn atoms most likely substitute Ga in the GaAs matrix, which is also supported by our own *ab initio* simulation results. The behavior of "as grown" sample is consistent with a small amount of structural disorder of the cubic structure of GaAs. An increase of annealing temperature to 350 °C leads to the slow broadening and suppression of main feature around 6549 eV along with increase of electronic states around 6544 eV of the absorption profiles (see Figure 3 (a)). With an increase of annealing temperature to 450 °C, the XANES spectrum is transformed into a featureless profile. This behavior is consistent with the loss of long-range periodicity of Mn atoms into investigated layer what is supported by HR-XRD observations.

Analysis of the modulus of Fourier transforms, |FT(R)|, of the EXAFS function show a dramatic decrease of the |FT(R)| amplitude with increase of the annealing temperature: decrease by a factor of two is observed (see inset of Figure 3 (a)). Generally, such decrease can be modeled either via (i) an increase of the Debye-Waller factor (which corresponds to high structural disorder around Mn atoms) and/or (ii) a decrease of neighboring atoms number around Mn (which is a result of structural transformation caused by Mn atoms migration in GaAs matrix before formation of Mn-rich inclusions in GaAs matrix). The latter can take place only when As/Ga vacancies are considered as well.

If the second possibility is considered, the following working hypothesis can be proposed: starting from the randomly distributed substitutional Mn (referred as monomers) and point defects like As/Ga vacancies formation of non-regular distributed areas of "N-mers" (dimer, trimer and so on combined through As) could occur. Theoretical support of this working hypothesis comes from our own *ab initio* calculations performed by WIEN2k code. In order to determine the possibility to form a specific type of point defect including, for instance, As vacancy, substitutional Mn, and their combination, their formation energies were calculated by introducing these defects within the structure and by relaxing the entire structure to minimize the forces on the atoms. The defect formation energy is then calculated as the difference between the total energy of the non-defected structure and the relaxed defected structure at constant volume. It was found that despite the fact that creation of As vacancy itself is unfavorable to the system, the lowest energy formation is developed when one (or more) substitutional Mn and one (or more) As vacancy are placed close to each other. Theoretical study by Raebiger *et al.* [5] also confirms the preference of Mn atoms to form "N-mers" (As vacancy/ies were not considered in this case, however). In addition to that, the quantitative analysis of the XANES spectra of annealed samples by FEFF 8 code (where models were formed based on results of EXAFS fitting, combined with structure optimization performed by WIEN2k) shows that the signal is a combination of signals coming from both Mn monomers and "N-mers" (see Figure 3(b)).

The proposed working hypothesis is further supported, though indirectly, if one considers annealing at high enough temperatures (up to about 600 °C): such annealing eventually leads to formation of the separated secondary phases consisting of Mn-rich phase and voids [6]. Annealing at much higher temperatures (above 650 °C) or for longer times will eventually lead to formation of separated hexagonal MnAs clusters [7].

## 4. Conclusions

The redistribution of substitutional Mn in $Ga_{1-x}Mn_xAs$ thin layer with x=0.01 during medium-temperature annealing, 250-450 °C, was studied using XAFS measurements at Mn K-edge. Analysis of experimental spectra allowed to propose a theoretical model where substitutional Mn and point defects like As/Ga vacancies form non-regular distributed areas of "N-mers". Theoretical XANES spectra, simulated using multiple scattering formalism (FEFF code) with the support of density functional theory (WIEN2k code), qualitatively describe the features observed in the experimental spectra. To test the proposed model positron annihilation measurements are planned as this technique directly probes vacancies in sample and may identify a presence, type and amount of vacancies.

## Acknowledgments

We acknowledge the support from the Baltic Science Link project coordinated by the Swedish Research Council, EAgLE project (Project Number: 316014), CALIPSO program and PL-Grid and ICM infrastructures. The XAFS experiments were performed at CLÆSS beamline at ALBA Synchrotron Light Facility with the collaboration of ALBA staff.

## References

[1] Ohno H *et al* 1996 *Appl. Phys. Lett.* **69** 363
[2] Sadowski J and Domagala J Z 2012 *Advanced Functional materials: A perspective from theory to experiment* eds B Sanyal and O Eriksson (Elsevier) chapter 2 pp 21-44
[3] Decoster S *et al* 2013 *Synchrotron Rad.* **20** 426
[4] BL22 CLÆSS beamline: Technical description, ALBA light source, Spain, http://www.cells.es/en/beamlines/bl22-claess
[5] Raebiger H *et al* 2006 *Appl. Phys. Lett.* **89** 012505
[6] Kovacs A *et al* 2013 *J. Phys. D: Appl. Phys.* **46** 145309
[7] Demchenko I N *et al* 2007 *J. Phys.: Condens. Matter* **19** 496205

# XAFS study on the impact of local structure on electrochemical performance for $Co_3O_4$ nanowire arrays

**Shan Jiang, Weiren Cheng, Jingfu He, Junheng Huang, Qinghua Liu*, Yong Jiang, Shiqiang Wei***

National Synchrotron Radiation Laboratory, University of Science and Technology of China, Hefei, Anhui 230029, P. R. China

*Email: qhliu@ustc.edu.cn, sqwei@ustc.edu.cn

**Abstract.** Determining the local structure of catalyst materials is critical for understanding the mechanism of enhanced electrochemical activity in semiconductor electrode. Here, using X-ray absorption fine structure (XAFS) spectra, we reveal that the local disorder structure is formed for the mixed-phase $Co_3O_4$ nanowire arrays due to the interaction between the phases of $Co_3O_4$ and $Co_2(OH)_2CO_3$. Comparing to pure $Co_3O_4$ nanowire arrays, the mixed phase sample is richer in $Co^{2+}$ and the electronic structure is changed by the local structure, which are demonstrated by the X-ray absorption near-edge structure (XANES) spectra. It is deduced that the mixed-phase $Co_3O_4$ nanowire arrays with abundant $Co^{2+}$ sites provide more redox centres in electrochemical reaction than the pure $Co_3O_4$ nanowire arrays.

## 1. Introduction

In order to solve the energy crisis and environment problems, exploring high capacity energy storage device is one of the primary areas of research [1]. Spinel oxide $Co_3O_4$ as an important p-type versatile semiconductor has been widely utilized in energy storage devices. Among various $Co_3O_4$ electrodes, nanowire arrays electrode [2, 3] is an eminent candidate because of its large surface area per unit volume, high utilization of the active sites, and excellent mass transport property.

In recent years, there have been numerous investigations on the influence of material substrate [4] and annealing temperature [5] on the performance of $Co_3O_4$ nanowire arrays electrode. These results have suggested that the various preparation conditions of electrode may change the formed electrode structure and corresponding properties. However, the deep-understanding of the structure-function relationship of $Co_3O_4$ nanowire arrays electrode remains unknown. To this end, using a local structure sensitive measurement to correlate the structure and electrochemical properties of $Co_3O_4$ nanowire arrays electrodes is essential.

Herein, utilizing extended X-ray absorption fine structure (EXAFS) spectra, we reveal that the local disorder structure of $Co_3O_4$ nanowire arrays is formed due to the interaction between the phases of $Co_3O_4$ and $Co_2(OH)_2CO_3$. Moreover, X-ray absorption near-edge structure (XANES) spectra show that the mixed phase sample contains richer $Co^{2+}$ than that of $Co_3O_4$ nanowire arrays. These $Co^{2+}$ ions provide redox centers in electrochemical reaction and enhance the electrochemical performance.

## 2. Experimental

The precursor $Co_2(OH)_2CO_3$ nanowire arrays were prepared according to the reported hydrothermal method [6]. Then, the precursor was annealed at 350 °C in air for 2 h, leading to the formation of $Co_3O_4$. And the precursor was annealed at 350 °C in nitrogen mixed with 5% air for 2 h leading to the formation of mixed phase of $Co_3O_4$ and $Co_2(OH)_2CO_3$. The Co K-edges X-ray absorption fine structure (XAFS) spectra were recorded in the fluorescence mode at BL14W1 station in SSRF.

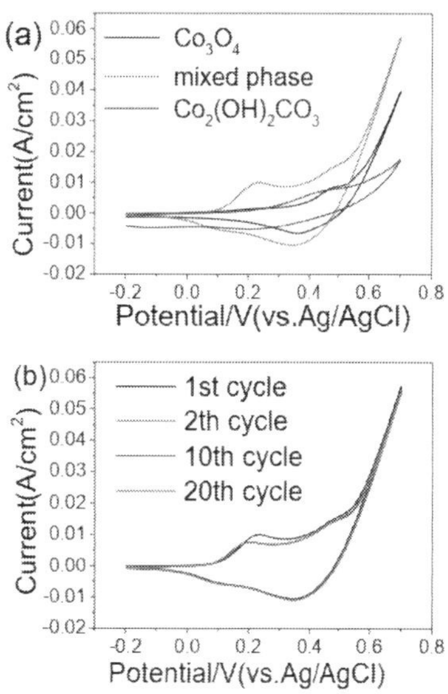


**Figure 1**. (a) CV of $Co_3O_4$ nanowire arrays (black line), mixed phase (red line), and $Co_2(OH)_2CO_3$ (blue line). (b) The cycle stability test of the mixed phase.

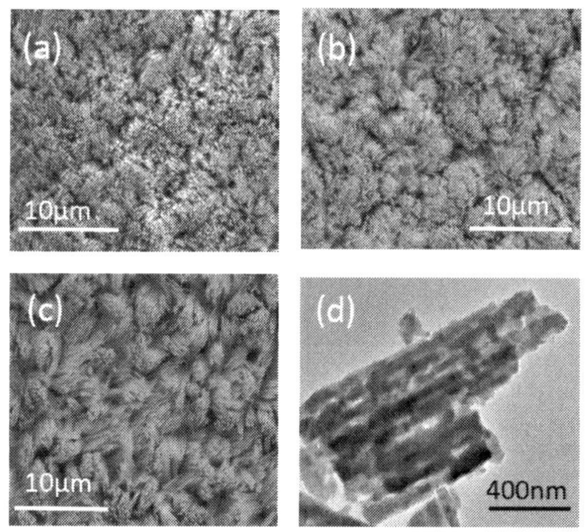


**Figure 2**. (a) SEM images of $Co_2(OH)_2CO_3$, (b) mixed phase, and (c) $Co_3O_4$ nanowire arrays electrode. (d) TEM image of a bunch of the nanowire arrays scratched from the Ti foam.

## 3. Results and discussion

We performed the electrochemical test and compared the electrochemical performance of the electrode materials annealed under different conditions. As seen from Figure 1(a), two pairs of redox peaks were observed, corresponding to the conversion between different cobalt oxidation states according to literature [7]. From these results we can see that $Co_3O_4$ electrode has a better performance than the precursor $Co_2(OH)_2CO_3$ electrode. While the sample annealed in nitrogen has an even better performance than the $Co_3O_4$ electrode. As shown in Figure 1(b), the discharge curve (the sample annealed in nitrogen) did not change after 20 cycles, indicating that our electrode had relatively good electrochemical stability.

SEM and TEM images of the $Co_2(OH)_2CO_3$ precursor and the electrode annealed under different conditions are presented in Figure 2. The $Co_2(OH)_2CO_3$ precursor film showed 1D nanowire arrays architecture (Figure 2(a)). The morphology of the sample before and after annealing treatment did not change much, maintaining the nanowire arrays structure (Figure 2(b) and (c)). TEM image (Figure 2 (d)) was a bunch of the nanowire arrays scratched from the Ti foam.

Figure 3(a) provided the XRD patterns of the precursor film, the sample annealed in nitrogen and in air. All the reflection peaks of the precursor film could be well indexed to crystalline orthorhombic basic cobalt carbonate hydroxide $Co_2(OH)_2CO_3$ (JCPDS 38-0547). After heat treatment in air, all the diffraction peaks were attributed to spinel $Co_3O_4$ phase (JCPDS 42-1467). This result indicates that the crystalline $Co_3O_4$ has been formed after annealing treatment, also supported by the color change of the product from the original pink to black. The precursor annealed in nitrogen mixed with a little amount of air resulted in a different phase with that annealed in air. There are several peaks corresponding to $Co_3O_4$ and $Co_2(OH)_2CO_3$, displaying a mixed phase. This is probably because the nitrogen atmosphere has a reduction effect.

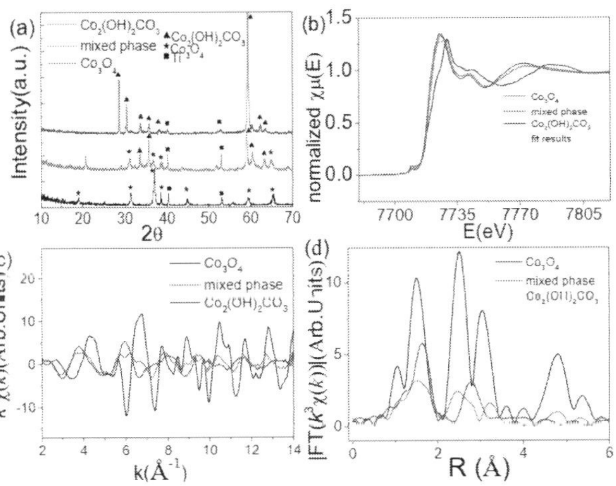


**Figure 3**. (a) XRD patterns, (b) Co K-edge XANES spectra, (c) oscillation function $k^3\chi(k)$ at the Co K-edge, and (d) FTs curves of $Co_2(OH)_2CO_3$ precursor film (blue line), mixed phase (red line), and $Co_3O_4$ (black line).

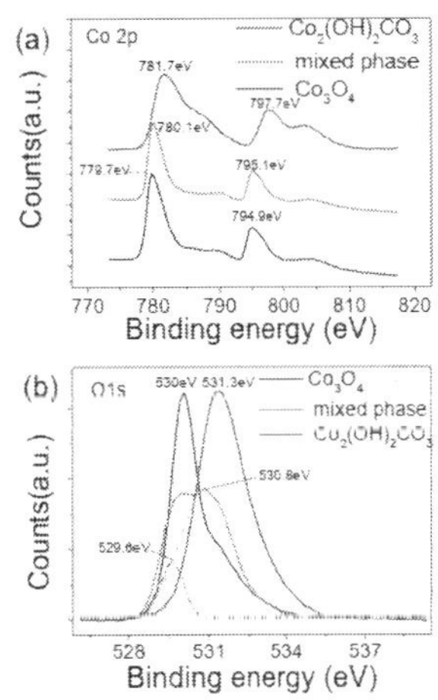


**Figure 4**. XPS patterns of Co 2p (a), O1s (b), ($Co_2(OH)_2CO_3$ (blue line), mixed phase (red line) and $Co_3O_4$ (black line)).

To clarify the atomic structure of the Co atom in mixed phase, X-ray absorption fine structure (XAFS) technique was used as a sensitive local structure probe [8]. The XANES spectra, extended-XAFS (EXAFS) $k^3\chi$ (k) functions and their Fourier transforms (FTs) at Co K-edge of annealed and pristine samples are shown in Figure 3(b)–(d). We further performed XANES linear combination fitting. The results provided the exact content of each phase of 0.175 and 0.825 for $Co_3O_4$ and $Co_2(OH)_2CO_3$, respectively. EXAFS spectra in k space of the mixed phase displayed a similar oscillation with the precursor $Co_2(OH)_2CO_3$. An increase of the degree of disorder and a reduction of the coordination number were deduced from the FTs curve (Figure 3(d)) for the sample annealed in nitrogen. They may result from a distortion of the surface structure.

The composition of samples synthesized under different annealing conditions was further identified by XPS characterization. The typical core spectra of Co and O elements were measured and are shown in Figure 4(a) and (b). The peaks near 779.7 and 790.3 eV, and 794.9 and 804.5 eV in the Co core spectra can be assigned to Co 2p3/2 and Co 2p1/2, respectively. Co 2p peak binding energy of $Co_2(OH)_2CO_3$ is higher than those in $Co_3O_4$ [9].O1s spectra of the mixed phase sample displayed two peaks, which appeared in $Co_3O_4$ (530 eV) and $Co_2(OH)_2CO_3$ (531.3 eV). We can see that the Co 2p peaks of the mixed phase located near that of $Co_3O_4$, and it may be caused by the increased ratio of $Co_3O_4$ on the mixed phase surface. Therefore, combining the XAFS results, it can be deduced that the mixed phase had a richer $Co^{2+}$ content compared with $Co_3O_4$. The mixed-phase $Co_3O_4$ nanowire arrays electrode with abundant $Co^{2+}$ sites provide more redox centers in electrochemical reaction than the pure $Co_3O_4$ nanowire arrays electrode [10].

## 4. Conclusions

In summary, we have synthesized the vertically oriented $Co_3O_4$ nanowire arrays by hydrothermal route. SEM results show that the nanowire morphology did not change after annealing treatment, but XRD results show that the composition is a mixed phase of $Co_3O_4$ and $Co_2(OH)_2CO_3$. The XAFS analysis confirmed that the mixed phase has local structure distortion and richer $Co^{2+}$ which result in an improvement of electrochemical performance. These results may provide some guidance for improving cobalt-based catalysts as electrochemical electrode materials.

## Acknowledgments

This work was supported by the Scientific Research Grant of Hefei Science Center of CAS (2015SRG-HSC027) and the National Natural Science Foundation of China (Grants No. 21533007, U1332111, U1532265, and 11305174).

## References

[1] Liu C, Li F, Ma L P and Cheng H M 2010 *Adv. Mater.* **22(8)** E28-E62.
[2] Hochbaum A I and Yang P 2009 *Chem. Rev.* **110(1)** 527-46.
[3] Jiang J, Liu J P, Huang X T, Li Y Y, Ding R M, Ji X X, Hu Y Y, Chi Q B and Zhu Z H 2009 *Cryst. Growth Des.* **10(1)** 70-75.
[4] Gao Y, Chen S, Cao D, Wang G, and Yin J 2010 *J. Power Sources* **195(6)** 1757-60.
[5] Sinha A K, Gupta R K and Deb S K 2012 *Appl. Phys. A* **108(3)** 607-13.
[6] Xia X H, Tu J P, Zhang Y Q, Mai Y J, Wang X L, Gu C D and Zhao X B 2012 *RSC Adv.* **2(5)** 1835-41.
[7] Xia X H, Tu J P, Wang X L, Gu C D and Zhao X B 2011 *Chem. Commun.* **47(20)** 5786-88.
[8] Yao T, Zhang X D, Sun Z H, Liu S J, Huang Y Y, Xie Y, Wu C Z, Yuan X, Zhang WQ, Wu Z Y, Pan G Q, Hu F C, Wu L H, Liu Q H and Wei S Q 2010 *Phys. Rev. Lett.* **105** 226405.
[9] Yang J J, Cheng H and Frost R L 2011 *Spectrochim. Acta A* **78(1)** 420-8.
[10] Simon P and Gogotsi Y 2008 *Nat. Mater.* **7(11)** 845-54.

# Grazing incidence X-ray absorption characterization of amorphous Zn-Sn-O thin film

S L Moffitt,[1] Q Ma,[2] D B Buchholz,[1] R P H Chang,[1] M J Bedzyk[1] and T O Mason[1]
[1]*Department of Materials Science and Engineering, Northwestern University, Evanston, Illinois 60208, USA*
[2]*DND-CAT, Northwestern Synchrotron Research Center at Advanced Photon Source, Argonne, Illinois 60439, USA*

E-mail: q-ma@northwestern.edu

**Abstract**. We report a surface structure study of an amorphous Zn-Sn-O (*a*-ZTO) transparent conducting film using the grazing incidence X-ray absorption spectroscopy technique. By setting the measuring angles far below the critical angle at which the total external reflection occurs, the details of the surface structure of a film or bulk can be successfully accessed. The results show that unlike in the film where Zn is severely under coordinated ($N < 4$), it is fully coordinated ($N = 4$) near the surface while the coordination number around Sn is slightly smaller near the surface than in the film. Despite a 30% Zn doping, the local structure in the film is rutile-like.

## 1. Introduction

Crystalline transparent conducting oxides (TCOs) have been widely used as transparent electrodes in flat panel displays, solar cells, light emitting diodes, etc [1]. There is, however, an increasing interest in amorphous (*a*-) TCOs in device applications. They have comparable electrical and optical properties to their crystalline counterparts [2]. In addition, amorphous films can be deposited at low temperature on flexible substrates, have good mechanical stabilities under stress [3] and smoother surface morphologies [4]. These are some of the prerequisites for flexible electronics.

Amorphous Zn-Sn-O (*a*-ZTO) films, besides being indium-free, have shown good electrical properties. A 10 at.% Zn-doped $SnO_2$ film, deposited by radio frequency (RF) magnetron sputtering, has shown conductivity of ~1000 S/cm after vacuum annealing at 300°C [5]. To understand the structural origin of their physical properties, X-ray absorption spectroscopy techniques are often used [2, 6]. However, surface structures are rarely addressed. Here, we report some preliminary results of grazing incidence X-ray absorption measurements on *a*-ZTO film with 30% Zn as an example.

## 2. Sample preparation and characterization

The *a*-ZTO thin film of 529 nm thick with 30 at.% Zn was deposited on $2\times1$ cm$^2$ fused quartz substrate at room temperature by a pulsed laser deposition (PLD) system employing a 248 nm KrF excimer laser. The laser beam was focused using a fused silica lens to a $1\times2$ mm$^2$ spot size on the targets, which were rotated at a rate of 5 rpm to prevent localized heating. A dual-target deposition method was employed, where the constituent oxides (ZnO and $SnO_2$) were used as the targets. The oxygen partial pressure was maintained at 18 mTorr. The film composition was controlled by the deposition time of each target and was later confirmed by energy dispersive X-ray analysis. The film thickness was measured by optical ellipsometry. Carrier concentration ($-7.2\times10^{18}$ cm$^{-3}$), Hall mobility (10.2 cm$^2$/V·s), and conductivity (12 S/cm) of the film were measured by Hall effect in van der Pauw configuration with a 0.58 Tesla field.

X-ray absorption measurements were carried out at the 5-BM-D beamline of the DND-CAT located at Sector 5 of the Advanced Photon Source (Argonne, IL). A Si(111) double-crystal monochromator was used for energy scan ($\Delta E/E = 1.4\times10^{-4}$). The incident X-ray intensity was 60% detuned for harmonic rejection and monitored by an ion chamber filled with gasses absorbing 10% of the incident X rays. The EXAFS data were all collected in fluorescence mode using 4-element Si-drifted solid state detectors (ME4). For the film EXAFS, the sample was held vertically with the X-ray incidence angle $\alpha=54^\circ$ and the data were collected with a ME4 detector at 90° to the X-ray beam. For the surface EXAFS, the sample was placed on a two-circle Huber stage with its surface normal upward (or 90° to the synchrotron X-ray electric vector) and the data were collected with $\alpha <$ the critical angle $\theta_c$ using two ME4 detectors. The beam size was set using slits to $0.05\times8$ mm$^2$ for Zn K edge and to $0.02\times8$ mm$^2$ for Sn K edge. A second slit set was placed behind the sample for alignment. Direct and reflected beams were measured by an ion chamber with gasses absorbing 20% of the X rays. Figure 1 shows reflected and fluorescence intensities vs. θ measured 50 eV above Zn and Sn K edges, respectively. For energy scan to 600 eV above $E_0$, Changes in $\theta_c$ are negligibly small (<0.003° for Sn K edge and <0.01° for Zn K edge).

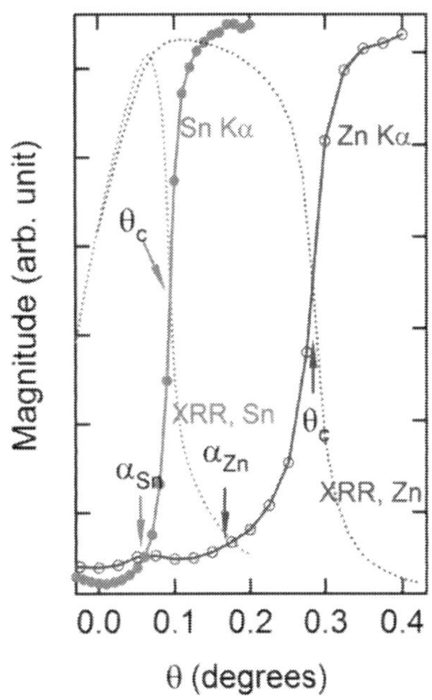


Figure 1. X-ray reflectivity (XRR) and Zn and Sn Kα fluorescence intensities. EXAFS are collected at $\alpha_{Sn}$ and $\alpha_{Zn}$.

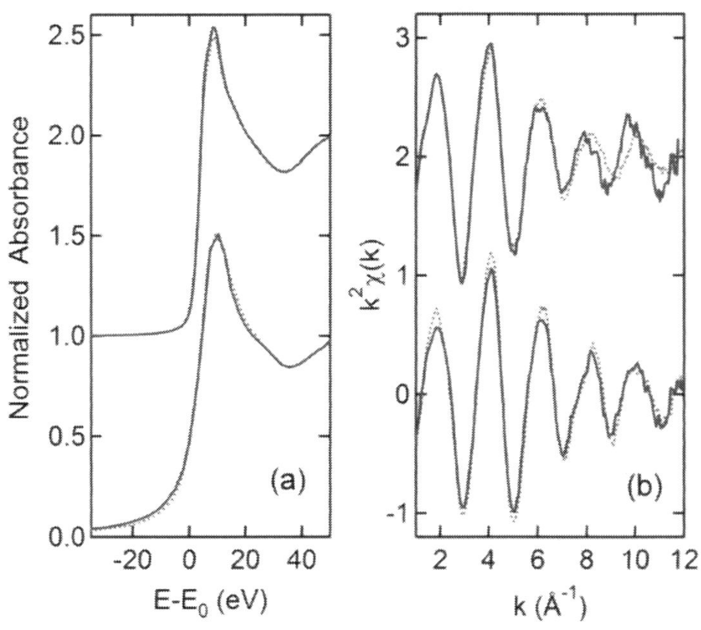


Figure 2. The Zn K edge (upper) and the Sn K edge (lower) X-ray absorption measurements of the *a*-ZTO film: dot lines – film and full lines – surface; (a) XANES; (b) EXAFS.

## 3. Results and discussion

In a grazing incidence geometry, the EXAFS may be measured through collecting specular reflected intensities, i.e., so called ReflEXAFS, or fluorescence intensities, as the X-ray energy is scanned around an absorption edge $E_0$ of the element of interest [7]. Yet, the strong correlation between the real δ(E) and imaginary β(E) increments of the refraction index n ($= 1 - \delta - i\beta$) makes it cumbersome to extract the structures from the data. Fortunately, under $\alpha < \theta_c$ conditions, the measured signals from both channels contain the same fine structure oscillations proportional to the absorption coefficients [8], and X-ray probing depth reduces to a few nanometers, providing a means to access the top surface of a film.

To avoid the self-absorption effect, the $\alpha$-dependence of the EXAFS was monitored. It was found to be absent when $\alpha \leq \frac{2}{3}\theta_c$. Figure 2 compares (a) the X-ray absorption near edge structures (XANES) and

(b) the EXAFS, of the film (dots) and film surface (full lines) of the *a*-ZTO film. The surface EXAFS were measured at $\alpha = 0.16^{\circ}$ ($< \theta_c = 0.28^{\circ}$) and $\alpha = 0.06^{\circ}$ ($< \theta_c = 0.1^{\circ}$) for the Zn $K_{\alpha}$ and Sn $K_{\alpha}$, respectively. Four and ten scans were merged to obtain the data presented in figure 2, respectively. Note that the self-absorption effect is largely absent in the surface data. In fact, the whiteline of the Zn K edge surface data is higher than that of the film, which may indicate a larger coordination number (N).

As seen in figure 2, the surface EXAFS are similar to the film ones up to k~8 $Å^{-1}$, beyond which differences in phases are seen. The Zn-K edge EXAFS is much more so. The origin of differences is unclear, except that they are reproducible experimentally and not an artifact of data processing. Given $\alpha << \theta_c$, these differences are unlikely due to a $\delta(E)$ effect. The Fourier transforms (FT) including data > 8.3 $Å^{-1}$ produce a prominent shoulder or satellite peak at a location smaller than the first main peak. This is rather peculiar in that it occurs to the backscattering amplitude of a light element, i.e., oxygen.

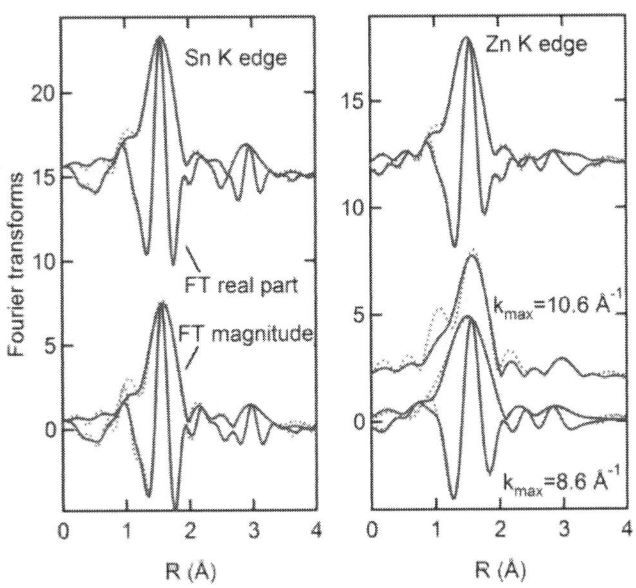


Figure 3. The best fits of Sn K edge and Zn K edge EXAFS spectra of $a$-$Zn_{0.3}Sn_{0.7}O_2$ film (upper) and film surface (lower). Dotted lines: experimental data and full line: curving fittings. For Sn K edge, the result with k = 3.4-11.9 $Å^{-1}$ and R = 1.0-3.6 Å is shown. For Zn K edge: the result with k = 2.3-10.6 $Å^{-1}$ and R = 1.0-3.6 Å is shown for the film EXAFS, and the results with two k ranges are shown (R=1.0-3.6 Å) for the surface EXAFS. Table 1 shows the numerical results. In total, 9 variables are fitted and the independent points ( $= 2\Delta r\Delta k/\pi$) are 11 and 14 for $k_{max}$ = 8.6 and 10.6 $Å^{-1}$ for the Zn K edge, respectively, and 14 for the Sn K edge.

The data were quantified by the FEFF8 analyses based on rutile (*r*-) $SnO_2$ structure using the Artemis software [9]. The amplitude reduction factors $S_0^2$ were determined to be 0.9 and 1.0 from the EXAFS's of wutzite (*w*-) ZnO and *r*-$SnO_2$, respectively. For the Zn-O bond, the distance R = 1.96 Å and the bond disorder $\sigma^2 = 4.3\times10^{-3}$ $Å^2$, and 2.06 Å and $3.0\times10^{-3}$ $Å^2$ for the Sn-O bond. Figure 3 shows the best fits. Clearly, the satellite features are not accounted for. This is more apparent for the Zn K edge. The results using two different k ranges are presented to show this effect. Throughout the analyses, the results are weakly dependent on the $k_{max}$ value for the film. This is due to the fact that the oxygen backscattering amplitude is located mainly in the low k region. However, the results of the film surface did show a somewhat larger $k_{max}$ dependence due to the satellite features, whose effect is included in the error bars. Table 1 summarizes the local structures around Zn and Sn. The structures can be well interpreted with single scattering paths of Zn-O, Sn-O and Zn-Sn. Attempts to use Zn-Zn or mix of Zn-Zn and Zn-Sn paths in the fitting yielded no physically meaningful results, indicating no measurable phase separation in the film. For a better fit, it appears that the third cumulant term $C_3$ in the EXAFS cumulant expansion is needed [10] to account for none Gaussian bonding distribution in this amorphous film.

As seen in Table 1 the local structure around Sn is similar to that of *r*-$SnO_2$ in which $N_1/N_2 = 6/2$ and $R_1/R_2 = 2.06/3.19$ Å, despite high Zn doping level. The structure around Zn is defective. The oxygen vacancies around Zn do not contribute to carrier concentration and are believed to form the neutral vacancies [11]. The short Zn-O bond and the large $\sigma^2$ around Zn in the film is a strong indication of how the local structure around Zn is geometrically frustrated which induces strain in the film. E.g., $\sigma^2 = 0.023$ $Å^2$ for the Zn-Sn interaction is equivalent to a spread of the Zn-Sn interaction of ±0.15 Å over its mean.

Therefore, it may not be surprising that this is no longer the case at the surface where the number of this interaction has decreased, resulting a smaller $\sigma^2$ for the Zn-Sn interaction at a shorter distance (see Table 1). For a disordered structure whose $\sigma^2$ ($= \sigma_S^2 + \sigma_T^2$) is dominated by $\sigma_S^2$, such as the one present in this film, the observed N-$\sigma^2$ correlation may be largely inherent. Near the surface Zn is fully coordinated (N = 4) likely through oxygen uptake from ambient. Unlike Zn, Sn may have a slightly lower N near the surface and the changes in its local structure are much subdued.

Table 1. Local structures of the film surface and the film (bracketed)

| Bonding | *N* | *R* (Å)[§] | $\sigma^2 \times 10^3$ | $C_3 \times 10^4$ |
|---|---|---|---|---|
| Zn → O | 3.9±0.3 (3.4±0.1) | 1.99±0.02 (2.01±0.01) | 8.9 (6.8) | 0.0 (5.4) |
| Zn → Sn | 1.0±0.1 (2.0±0.2) | 3.18±0.03 (3.24±0.01) | 11.0 (23.0) | 30.5 (25.0) |
| Sn → O | 5.7±0.4 (6.0±0.1) | 2.06±0.02 (2.06±0.01) | 6.2 (5.8) | 3.1 (6.4) |
| Sn → Sn | 2.6±0.2 (2.5±0.1) | 3.27±0.03 (3.30±0.01) | 12.0 (9.9) | 2.2 (8.4) |

[§]Energy shifts ΔE= 3.6 eV - 6 eV.

## 3. Summary

We report a surface structure study of an amorphous Zn-Sn-O transparent conducting film using the grazing incidence X-ray absorption technique. By setting the measuring angles far below the critical angle, the details of the surface structure of a film or bulk can be successfully accessed. The measurements show that when $\alpha \leq \frac{2}{3}\theta_c$ the self-absorption effect disappear for systems such as ZnO and $SnO_2$, which makes this technique suitable for the surface studies of films or bulk. The results show that unlike in the film where Zn is severely under coordinated (N < 4), it is fully coordinated (N=4) near the surface while the coordination number around Sn is slightly less near the surface than in the film. Despite the high Zn doping (30 at. %), the local structure in the film is rutile-like, while some deviation from it is seen around Zn near the film surface. Analyses to link these structural results with the film's physical properties are underway.

**ACKNOWLEDGMENT**

Film growth and electrical characterization were supported by the NSF funded Materials Research Science and Engineering Center (MRSEC) at Northwestern under Grant No. DMR-1121262. Optical characterization and structural analysis was supported by the ANSER Center, an Energy Frontier Research Center funded by the U.S. Department of Energy, Office of Basic Energy Sciences, under Award No. DE-SC0001059. X-ray absorption and scattering measurements were conducted at the DND-CAT at the Advanced Photon Source (APS). DND-CAT is supported by the NU Office of Research, E.I. DuPont de Nemours & Co., and The Dow Chemical Company. Use of the APS was supported by the U. S. Department of Energy, Office of Science, Office of Basic Energy Sciences, under Contract No. DEAC02-06CH11357.

[1] Ginley D S, Hosono H and Paine D C, in *Handbook of Transparent Conductors*, edited by David S. Ginley (Springer, New York, 2010).
[2] Hoel C A, Mason T O, Gaillard J F and Poeppelmeier K R 2010 *Chem. Mater.* **22** (12) 3569
[3] Kamiya T and Hosono H 2010 *NPG Asia Materials* **2** (1) 15.
[4] Hara H, Hanada T, Shiro T and T Yatabe 2004 *J. Vac. Sci. & Tech. A* **22** (4) 1726.
[5] Ko J H, Kim I H, Kim D, Lee K S, Lee T S, Jeong J H, Cheong B, Baik Y J and Kim W M 2006 *Thin Solid Films* 494 (1-2) 42.
[6] Buchholz D B, Ma Q, Alducin D, Ponce A, Jose-Yacaman M, Khanal R, Medvedeva J E and Chang R P H 2014 *Chem. Mater.* 26 5401
[7] Heald S M, Chen H and Tranquada J M 1988 *Phys. Rev.* B38 1016.
[8] Martens G and Rabe P 1980 *Phys. Stat. Sol.* (a) 58 415.
*[9]* *http://bruceravel.github.io/demeter/*
[10] Rehr J J and Albers R C 2000 *Rev. Mod. Phys.* 72(3) 621.
[11] Zhu Q, Ma Q, Buchholz D B, Chang R P H, Bedzyk M J and Mason T O 2014 *J Appl. Phys.* 115 033512.

# Importance of local structural distortions in magnetocaloric effect in Mn based antiperovskites

**K R Priolkar[1,*], E T Dias[1], G Aquilanti[2], Ö Çakir[3], M Acet[4], A K Nigam[5]**

[1]Department of Physics, Goa University, Taleigao Plateau, Goa 403206 India
[2]Elettra-Sincrotrone Trieste S.C.p.A., s.s. 14, km 163.5 I-34149 Basovizza, Trieste, Italy
[3]Physics Department, Yildiz Technical University, TR-34220 Esenler, Istanbul, Turkey
[4]Faculty of Physics and CENIDE, Universitat Duisburg-Essen, D-47048 Duisburg, Germany
[5]Tata Institute of Fundamental Research, Dr. Homi Bhabha Road, Colaba, Mumbai 400005, India

E-mail: krp@unigoa.ac.in

**Abstract.** $Mn_3GaC$ and $Mn_3SnC$ are two antiperovskites which exhibit a large magnetocaloric effect associated with a magnetostructural transformation. Although structurally similar, the magnetic transition and the nature of magnetocaloric effect in the two compounds is completely different. While $Mn_3GaC$ transforms from a ferromagnetic to an antiferromagnetic state at $T_t$ = 178K with the antiferromagnetic propagation vector along $k = \frac{1}{2}, \frac{1}{2}, \frac{1}{2}$, in $Mn_3SnC$ the transformation occurs at $T_t$ = 280K from a paramagnetic state to an antiferromagnetic state with $k = \frac{1}{2}, \frac{1}{2}, 0$. Mn K EXAFS has been employed to show that these differences in magnetic properties are related to local structural distortions in $Mn_6C$ octahedra which are quite different in the two compounds.

## 1. Introduction

A class of materials that have attracted attention as candidates for ferroic cooling applications are the Mn based antiperovskite materials. Amongst these $Mn_3GaC$ undergoes a volume discontinuous first order transition from a ferromagnetic (FM) to an antiferromagnetic (AFM) ground state around 160K [1, 2] accompanied by a large inverse table like magnetocaloric effect (MCE) in relatively low fields especially preferred in designing a practical refrigerant unit [3, 4]. On the other hand, isostructural $Mn_3SnC$ exhibits a sharp first order change from a paramagnetic (PM) state to a non-collinear antiferromagnetic state at 279K accompanied by a conventional magnetic entropy change (80.69 mJ/cm$^3$-K under a magnetic field of 2T) [2, 5].

Although both the samples undergo a cubic-cubic phase transformation, neutron diffraction studies have shown that the antiferromagnetic propagation vector $k$ in the two compounds is different. In $Mn_3GaC$, $k$ is along $\frac{1}{2}, \frac{1}{2}, \frac{1}{2}$ direction [6] while $Mn_3SnC$ orders with $k = \frac{1}{2}, \frac{1}{2}, 0$ which makes two of the three Mn atoms to order antiferromagnetically while the third one has a small ferromagnetic component along the $z$ direction [7]. There are however, no signatures of any structural disorder that could explain this difference in magnetic behavior. In order to check for any local structural disorder, Mn K edge EXAFS studies were performed across the first order transition on $Mn_3GaC$ and $Mn_3SnC$ compounds. Our study reveals presence of such

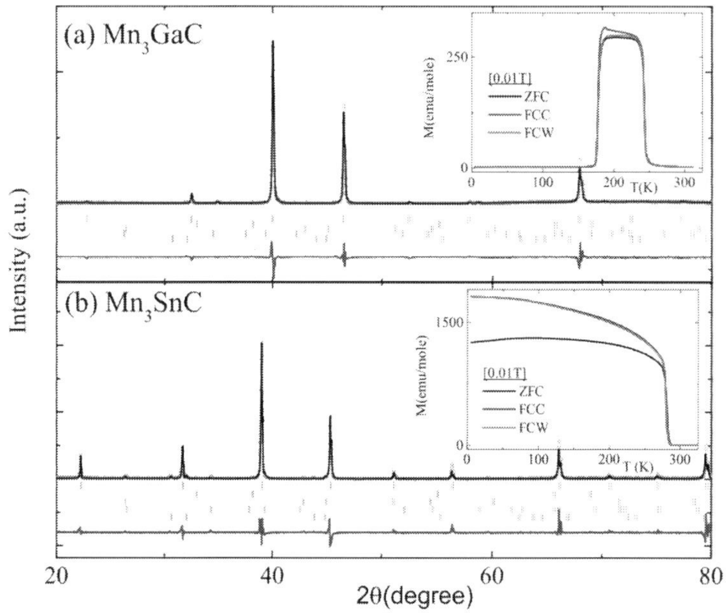


**Figure 1.** Rietveld refined X-ray diffraction patterns recorded for (a) $Mn_3GaC$ and (b) $Mn_3SnC$ at room temperature. Insets in the respective panels show temperature evolution of magnetization recorded during ZFC, FCC and FCW cycles.

a disorder that could be related to the observed differences in the magnetic properties of the two compounds.

## 2. Experimental

The compounds used here were prepared by solid state reaction method, details of which are given in Ref. [8] and characterized by x-ray diffraction (XRD) and magnetization as a function of temperature during the zero field cooling (ZFC), field cooled cooling (FCC) and field cooled warming (FCW) cycles. Extended x-ray absorption fine structure (EXAFS) data at the Mn K edge (6539 eV) was collected in transmission mode at 300K (RT) and 80K (LT) in the range from -200 to 1300 eV with respect to the Mn K edge at the XAFS beamline at Elettra, Trieste [9]. The data was processed using Demeter program [10].

## 3. Results and Conclusions

Rietveld refined XRD patterns recorded in the angular ($2\theta$) range of 20° to 100° in steps of 0.02° using Cu $K_\alpha$ radiation are presented in Figure 1 and show the formation of cubic antiperovskite phase for both the compounds along with minor impurity phases ($\sim$ 1%).

Magnetization (M) of $Mn_3GaC$, recorded as a function of temperature under ZFC, FCC and FCW conditions at an applied field of 0.01T (see inset of Figure 1a) exhibits a PM to FM transition at $T_C$ = 242K followed by a FM to AFM first order transition at $T_t$ = 178K [11]. $Mn_3SnC$ on the other hand exhibits a single transition depicted by a sharp increase in magnetization at $T_t$ = 280 K (inset of Figure 1b) which corresponds to a first order transition from a high temperature low volume PM state to a low temperature high volume magnetically ordered phase with competing FM and AFM interactions [7].

k weighted Mn K edge EXAFS data in R space and back transformed k space recorded

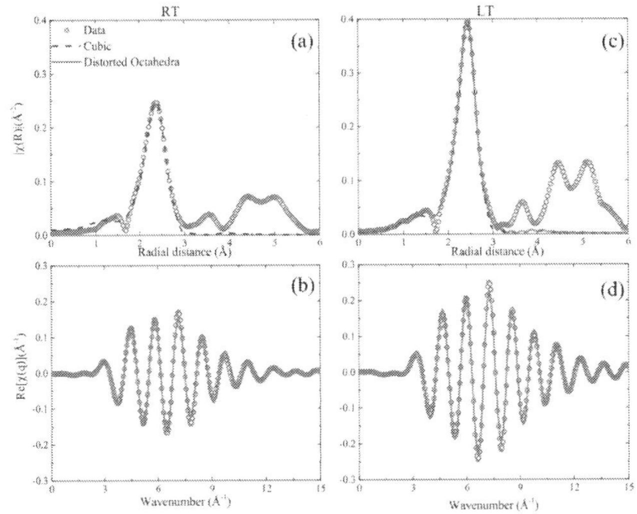


| Bond | $R_{\_XRD}$ | $R_{\_EXAFS}$ |
|---|---|---|
| **RT** | | |
| Mn-C | 1.949 | 1.923(4) |
| Mn-Mn | 2.757 | 2.720(6) |
| Mn-Ga | 2.757 | 2.720(6) |
| **LT** | | |
| Mn-C | 1.943 | 1.937(3) |
| Mn-Mn | 2.748 | 2.779(5) |
| Mn-Ga | 2.748 | 2.735(2) |

**Figure 2.** EXAFS spectra in R space ((a) and (c)) and Fourier filtered k space ((b) and (d)) at RT and LT in $Mn_3GaC$. The dashed lines and the solid lines indicate best fit for cubic and distorted octahedra model respectively. The table gives the best fitted values of bond lengths from XRD and EXAFS.

at RT and LT for $Mn_3GaC$ and $Mn_3SnC$ are presented in Figures 2 and 3 respectively. While $Mn_3GaC$ EXAFS data at RT could be well fitted to a model incorporating structural constraints (shown as a dashed line (cubic model) in Figure 2(a)), satisfactory fits could not be obtained for $Mn_3SnC$ EXAFS data recorded at both, RT and LT using this model (Figure 3(a) and (c)). Similarly, the cubic model, though fitted well (dashed line in Figure 2(c)) to the LT data of $Mn_3GaC$, gave some unphysical parameters (mean square relative displacements, $\sigma^2 < 0$). Therefore, the structural constraints were relaxed and bond distances and $\sigma^2$ for each scattering path were varied independently. The resulting fit is shown as a solid line (distorted octahedra) in Figure 2(c). The fitted values of bond distances are listed in a table alongside the Figure 2. It can be seen that, at LT, the Mn–C and Mn–Ga distances slightly shorten while Mn–Mn bonds elongate. This situation corresponds to a rhombohedral distortion of the cubic unit cell in the transformed phase. The AFM ground state of $Mn_3GaC$ is believed to be due to increased Mn–C hybridization. The shorter Mn–C bond distance obtained here agrees well with this hypothesis. The RT and LT data in the Fourier filtered space along with best fits are presented in Figures 2(b) and (d) respectively.

In the case of $Mn_3SnC$, it was found that local structure of Mn consists of a distribution of shorter and longer bond distances (distorted octahedra model). While about 2/3 of Mn–C and Mn–Mn bonds shorten and the rest grow longer. In the case of Mn–Sn distances the distribution was exactly reversed. Such a local structural distortion though preserves the overall cubic symmetry, distort the $Mn_6C$ octahedra so as to elongate along one direction and compress along other two. The best fits to the data in R space are shown as solid lines in Figure 3(a) and (c) and the back transformed data at RT and LT along with the best fits are shown in Figure 3(b) and (d) respectively. The best values of bond distances obtained are listed in a Table alongside the Figure 3.

These distortions of $Mn_6C$ octahedra agree well with the observed propagation vector $k$ and

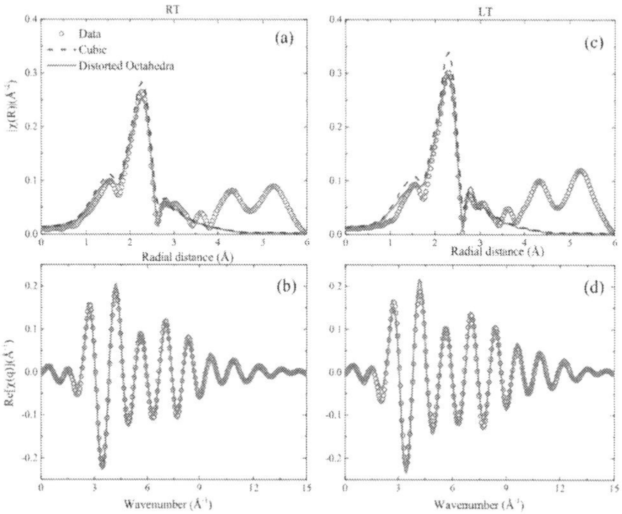


| Bond | $R_{XRD}$ | $R_{EXAFS}$ |
|---|---|---|
| RT | | |
| Mn-C | 1.998 | 2.032(2)x0.33<br>1.958(2)x0.67 |
| Mn-Mn | 2.826 | 2.874(2)x0.33<br>2.769(2)x0.67 |
| Mn-Sn | 2.826 | 2.874(2)x0.67<br>2.769(2)x0.33 |
| LT | | |
| Mn-C | 1.997 | 2.029(3)x0.33<br>1.961(3)x0.67 |
| Mn-Mn | 2.824 | 2.869(3)x0.33<br>2.773(3)x0.67 |
| Mn-Sn | 2.824 | 2.869(3)x0.67<br>2.773(3)x0.33 |

**Figure 3.** R space ((a) and (c)) and back transformed k space ((b and (d)) EXAFS spectra recorded at RT and LT for $Mn_3SnC$. The dashed lines and the solid lines indicates best fit with cubic and distorted octahedra model respectively. The table gives the best fitted values of bond lengths from XRD and EXAFS.

the observe MCE. In $Mn_3GaC$, the octahedra are distorted along [111] direction so also is the antiferromagnetic alignment of Mn spins resulting in an inverse MCE [6]. While, in $Mn_3SnC$ the elongation of octahedra along one direction is perhaps responsible for the antiferromagnetic alignment of two of the three Mn atoms and a pure ferromagnetic moment on the third [7]. This pure ferromagnetic moment explains the conventional MCE in $Mn_3SnC$.

## Acknowledgments
Financial support by Board of Research in Nuclear Sciences (BRNS) under the project 2011/37P/06 is acknowledged.

## References
[1] Bouchaud J, Fruchart R, Pauthenet R, Guillot M, Bartholin H and Chaiss F 1966 *J. Appl. Phys.* **37** 971–972 URL `http://scitation.aip.org/content/aip/journal/jap/37/3/10.1063/1.1708544`
[2] Fruchart D and F Bertaut E 1978 *Journal of the Physical Society of Japan* **44** 781–791 URL `http://dx.doi.org/10.1143/JPSJ.44.781`
[3] García J, Bartolomé J, González D, Navarro R and Fruchart D 1983 *J. Chem. Thermodyn.* **15** 1059 – 1069 URL `http://www.sciencedirect.com/science/article/pii/0021961483900319`
[4] Tohei T, Wada H and Kanomata T 2003 *J. Appl. Phys.* **94** 1800–1802 URL `http://scitation.aip.org/content/aip/journal/jap/94/3/10.1063/1.1587265`
[5] Wang B S, Tong P, Sun Y P, Luo X, Zhu X B, Li G, Zhu X D, Zhang S B, Yang Z R, Song W H and Dai J M 2009 *Europhys. Lett.* **85** 47004
[6] Ö Çakir, Acet M, Farle M and Senyshyn A 2014 *J. Appl. Phys.* **115** 043913 URL `http://scitation.aip.org/content/aip/journal/jap/115/4/10.1063/1.4862903`
[7] Dias E T, Priolkar K R, Das A, Aquilanti G, akir, Acet M and Nigam A K 2015 *Journal of Physics D: Applied Physics* **48** 295001 URL `http://stacks.iop.org/0022-3727/48/i=29/a=295001`
[8] Dias E T, Priolkar K R and Nigam A K 2014 *Mater. Res. Express* **1** 026106 URL `http://stacks.iop.org/2053-1591/1/i=2/a=026106`

[9] Cicco A D, Aquilanti G, Minicucci M, Principi E, Novello N, Cognigni A and Olivi L 2009 *J Phys Conf Ser* **190** 012043
[10] Ravel B and Newville M 2005 *J Synchrotron Radiat* **12** 537–541
[11] Dias E, Priolkar K and Nigam A 2014 *J. Magn. Magn. Mater.* **363** 140 – 144 URL `http://www.sciencedirect.com/science/article/pii/S0304885314002807`

# EXAFS and X-ray diffraction study of $LaCoO_3$ across the spin-state transition

**V V Sikolenko**[1,9], **I O Troyanchuk**[2], **V V Efimov**[1], **E A Efimova**[1], **S I Tiutiunnikov**[1], **D V Karpinsky**[2], **S Pascarelli**[3], **O Zaharko**[4], **A Ignatov**[5], **D Aquilanti**[6], **A G Selutin**[7], **A N Shmakov**[7] **and D Prabhakaran**[8]

[1] Joint Institute for Nuclear Research, 141980 Dubna, Russia
[2] Scientific-Practical Material Research Center NAS Belarus, 220072 Minsk, Belarus
[3] European Synchrotron Radiation Facility, BP 220, 38043 Grenoble, France
[4] Paul Scherrer Institute, Laboratory for Neutron Scattering and Imaging, CH-5232 Villigen PSI, Switzerland
[5] Oak Ridge National Laboratory, Oak Ridge, TN 37831, USA
[6] Elettra synchrotron Trieste, 34149 Trieste, Italy
[7] Boreskov Institute of Catalysis SB RAS, 630090 Novosibirsk, Russia
[8] Department of Physics, University of Oxford, OX1 3PU, Oxford, Great Britain
[9] REC "Functional nanomaterials" Immanuel Kant Baltic Federal University, 236041 Kaliningrad, Russia

E-mail: vadim.sikolenko@jinr.ru

**Abstract.** A combined high-resolution Co *K*-edge extended x-ray absorption fine structure (EXAFS) and high-resolution X-ray powder diffraction (XRD) study has been performed to clarify the detail of anomalous behavior of temperature-dependent magnetic susceptibility curve on the $LaCoO_3$ across the spin-state (~120 K) transition. According to XRD analysis, the Debye-Waller factor of Co-O bond exhibit rapid growth below 20 K whereas the temperature dependence of the average Co-O bond length shows linear behavior from 10 K to 400 K. The EXAFS data show an anomalous decrease of the Co–O bond lengths with respect to those obtained by XRD. No local distortion of $CoO_6$ octahedral as temperature increases up to 400 K has been detected.

## 1. Introduction

Perovskite-like lanthanum cobalt oxide $LaCoO_3$ is a fascinating material studied since the 1950's with many controversial explanations of its peculiar structural [1,2], transport [3] and magnetic [4,5] properties. With temperature increase a maximum of the magnetic susceptibility [4] and thermal expansion coefficient [3] was observed near 120 K whereas a second anomaly [3] followed by a plateau at 500 K ÷ 520 K is associated with the metal-insulator transition. Goodenough et al. [6] originally interpreted these magnetic transitions as spin-state transitions of $Co^{3+}$ ions from the nonmagnetic ground state low-spin state (LS; $t^6_{2g}e^0_g$, $S = 0$) to a high-spin state (HS; $t^4_{2g}e^2_g$, $S = 2$) due to the close values of the intra-atomic exchange energy ($J_H$) and the crystal field splitting ($10Dq$) at the $Co^{3+}$ sites. These spin-state transitions are further manifested by observable changes in the crystal structure since the HS $Co^{3+}$ has a much larger radius (0.61 Å) than the LS state (0.54 Å) and intermediate IS (0.55 Å) [7].


Published under licence by IOP Publishing Ltd

Potze *et al* introduced the concept of intermediate-spin state (IS; $t^5_{2g} e^1_g$, $S = 1$) [8] and showed that the spin-state transition near 120 K is associated with the thermal excitation of $Co^{3+}$ ions from the LS ground state to an IS [8]. Moreover, this point of view was supported by Korotin and coworkers [9] who performed the LDA+U band structure calculations assuming that temperature effects and spin state transition can be simulated by expanding the lattice parameter. According to the Korotin and coworkers [9] the stabilization of the IS state due to the large hybridization between the Co-$e_g$ and O-2p levels. Due to the partially filled $e_g$ level, the IS state is Jahn–Teller (JT) active. The LS-IS scenario became widely used in interpretation of numerous experimental data [1-6, 8, 9]. However, in resent works, including EXAFS [10], magnetic circular dichroism (XMCD) [11], inelastic neutron scattering [12] as well as recent theoretical work on GGA+*U* (GGA-generalized gradient approximation) calculations [13] and theoretical study by Pandey *et al.* [13] has attempted to revive the LS-HS scenario.
Here we present the analysis result of EXAFS and XRD experiments on $LaCoO_3$ powder.

## 2. Experimental

The powder sample $LaCoO_3$ was synthesized using the method described in [1]. XRD experiments were carried out at the synchrotron facility HASYLAB/DESY (Hamburg, Germany) using the powder diffractometer at the B2 beamline with the temperature range 10 – 290 K and the single crystal diffractometer at the BM01A beamline at ESRF in the temperature range 80 – 400 K. The obtained X-ray powder diffraction data were analyzed by a Rietveld method using FullProf program.
EXAFS experiments have been performed at the beam line BM29 of ESRF (Grenoble, France). The EXAFS spectra were measured at the Co K-edge in the energy range 7400 eV – 9500 eV in standard transmission mode simultaneously with reference sample (cobalt thickness 9 μm metallic foil – in order to fix $E_0$ not bigger 0.005 Å) in the temperature range 20 – 400 K. Each temperature point was measured 3 times with a count rate of 2.5 seconds per point. To reduce the harmonic content in the x-ray beam, we detuned the monochromator crystals 40% at 7900 eV. The $LaCoO_3$ powder sample were deposited on the millipore cellulose membranes with thicknesses specially selected to obtain an X-ray absorption edge jump $\Delta\mu \cdot x \sim 1$ at the Co *K*-edge.
A curve fitting procedure by the EDA software package [14] was used to determine the average *R*(Co–O) distance and the parallel mean square relative displacement ($MSRD_{\|}$) $\Delta\sigma^2_{Co-O}$ (or EXAFS Debye-Waller factor). The energy position $E_0$, used in the definition of the photoelectron wave number $k = [(2m_e/\hbar^2)(E - E_0)]^{1/2}$, was set at the threshold energy $E_0 = 7714$ eV. The Fourier transforms (FTs) of the EXAFS $\chi(k)k^2$ spectra were calculated in the wave number interval up to $k = 1.0 \div 20$ Å$^{-1}$ with a 10% Gaussian-type window function. At low temperatures, the signal to noise is very good out up to 20 Å$^{-1}$; however, it deteriorates at high $k$ as the temperature increases and is poor beyond about 18.0 Å$^{-1}$ at 300 K. Consequently, in all the fits the upper end of the FT $k$ range is restricted to 17.5 Å$^{-1}$.
Experimental scattering amplitude and phase shift functions for the Co-O atom pair were used in the EXAFS analysis. They were obtained from the EXAFS spectra of a reference Co-foil sample at $T$ = 20 K. We assumed that in these conditions, anharmonicity effects in the dynamics of the $CoO_6$ octahedron can be neglected, and the sample is composed of regular $CoO_6$ octahedra. The cobalt coordination number and the Co-O distance were set respectively equal to $N_{ref} = 6$ and $R_{ref} = 1.925$ Å, according to the results of the Rietveld refinement of our X-ray powder diffraction data on the same $LaCoO_3$ sample.

## 3. Results and discussion

All observed Bragg peaks for $LaCoO_3$ in the temperature range from 10 K to 400 K were indexed in the frame of the rhombohedral *R*-3*c* space group. The temperature dependence of $MSRD_{\|}$ is the set of the contributions of all normal modes and can be well approximated by the correlated Einstein model [15].
The temperature dependence of the $MSRD_{\|}$ and diffraction Debye-Waller factor or uncorrelated mean squared displacement (MSD) for Co and O atoms in $LaCoO_3$ is shown in Fig. 1. Our experimental

$MSRD_{||}$ $\Delta\sigma^2$(Co–O) values for $LaCoO_3$ are in agreement with the previous EXAFS data [10] which are limited 300 K. The $MSRD_{||}$ values in the range from 20 K to 400 K are reasonably well fitted by the Einstein model [15] with a characteristic temperature of 550 K. The DCF (i.e. difference between $MSRD_{||}$ and MSD), reflecting the correlation in atomic motion of distant atoms (cobalt and oxygen), grows gradually with temperature (Fig. 1). Such growth of the interaction strength between atoms in the Co-O pairs can be associated to a gradual transition from HS $Co^{3+}$ ions to an high-hybridized IS spin state [9].
The absence of growth $MSRD_{||}$ in temperature range from 5 K to 20 K is probably associated to a high correlation of both LS and of HS $Co^{3+}$ ions motion with the motion of oxygen.

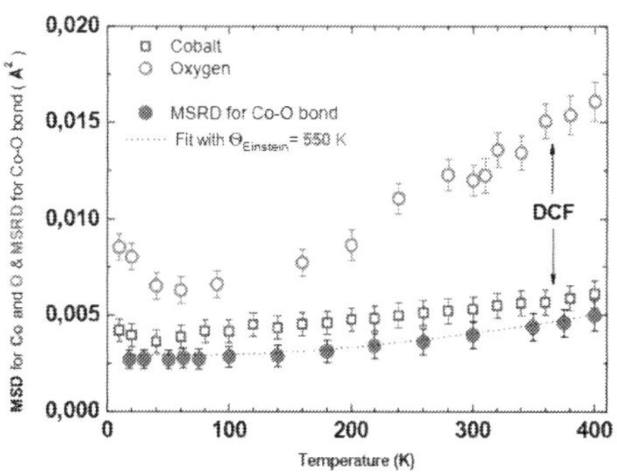


**Figure 1.** Temperature dependence of MSD (calculated from XRD) for cobalt (squeres) and oxygen (empty circles) and correlated $MSRD_{||}$ (full circles) for Co-O bond in $LaCoO_3$. The dot line is Einstein model best-fitting the $MSRD_{||}$ ($\Theta_E$ = 550 K).

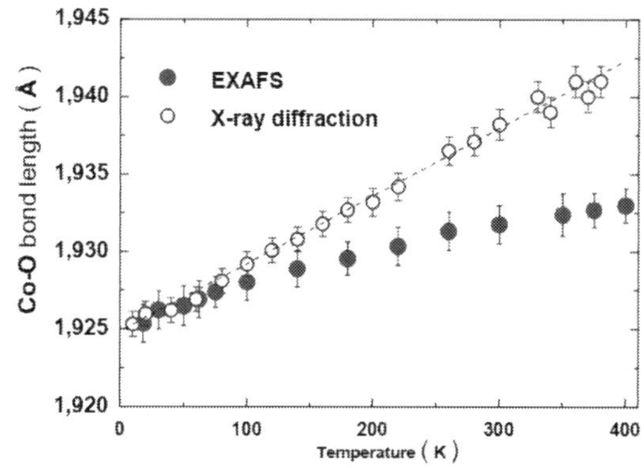


**Figure 2.** The temperature dependence of the Co–O bond lengths for $LaCoO_3$ obtained by EXAFS and XRD

Fig. 2 shows the result of the temperature-dependant Co–O bond lengths obtained by the EXAFS fitting procedure and the Rietveld refinement of diffraction data. Note that the local interatomic distance $\langle r_{Co\text{-}O}\rangle = \langle | r_O - r_{Co} | \rangle$ probed by EXAFS is usually larger than the equilibrium crystallographic distance $R_{Co\text{-}O} = | \langle r_O\rangle - \langle r_{Co}\rangle |$ measured by diffraction. The difference between $\langle r_{Co\text{-}O}\rangle$ and $R_{Co\text{-}O}$ is associated with the influence of the *perpendicular* $MSRD_{\perp}$ $\langle\Delta\sigma^2_{\perp}$(Co–O)$\rangle$, i.e. the thermal atomic displacement in the direction perpendicular to the Co-O bond. However, the Co–O bond lengths determined from the EXAFS analysis in our experiment are always shorter with respect to those obtained from the diffraction, and this deviation increases with temperature. We interpret this deviation to the $Co^{3+}$ spin-state transition from HS to IS at ~80 K, leading to important difference Co–O bond lengths measured by EXAFS and XRD. The observed anomalous decrease of EXAFS Co–O bond lengths with respect to diffraction Co–O bond lengths above ~80 K could be explained in the frame of a model based on the gradual growth of IS/HS ratio $Co^{3+}$ fraction in a LS basic matrix. Such result can be explained by similar values of small ion radiuses of LS and IS with respect to the HS ion radius [7], i.e. the HS Co–O distance essentially bigger LS and IS Co-O distance. The existing of significant fraction of HS states in LS basic matrix at low temperature can be also confirmed by an increase of Debye-Waller factor at low temperatures (Fig. 1).

## 4. Conclusions

We have observed an anomalous temperature behaviour of the Co–O bond length in $LaCoO_3$ by a combined analysis of EXAFS and diffraction results which is well described by a thermally induced spin-state transition from HS to the IS in LS matrix in both cases [4, 10-13, 17-19]. An increase of

Debye-Waller factor obtained from XRD at 10-20 K can be explained by the contribution of HS $Co^{3+}$ stabilized by influence of structural defects due to oxygen vacancies and scraps of chemical bonds at the boundary of powder grains [16].
The main results our study are supported by the existing magnetic susceptibility [3], the MCD measurements [11], inelastic neutron data [12] as well as by structural properties under high pressure [17-19].

**Acknowledgments**
This work was supported by RFBR Grants No.13-02-00699_a and No. 15-32-50378_mol_np.
Authors are indebted to Dr. Dmitry Chernyshov (SNBL at ESRF, France) for his help with diffraction experiments and data analysis, to Dr. Alexei Bossak for his assistance with sample preparation and numerous discussions.

**References**
[1] Radaelli P G and Cheong S W 2002 *Phys.Rev.*B **66** 094408
[2] Maris G, Ren Y, Volotchaev V, Zobel C, Lorenz T and Palstra T, 2003 *Phys. Rev. B* **67**, 224423
[3] Knízek K, Jirak Z, Hejtmanek J, Veverka M, Marysko M, Maris G, 2005 *Eur.Phys.J.* B **47** 213
[4] Zobel C, Kriener M, Bruns D, Baier J, Grüninger M, Lorenz T, Reutler P and Revcolevschi A, 2002 *Phys.Rev.*B **66** R020402
[5] Sazonov A P, Troyanchuk I O, Gamari-Seale H, Sikolenko V V, Stefanopoulos K L, Nicolaides G K, Atanassova Y K 2009 *J.Phys.:Cond.Mat* **21** 156004
[6] Senaris-Rodriguez M A and Goodenough J B 1995 *J. Solid State Chem.* **118** 323
[7] Shannon R D 1976 *Acta Crystallogr. Sect.A* **32** 751
[8] Potze R H, Sawatzky G A and Abbate M 1995 *Phys. Rev.* B **51** 11501
[9] Korotin M A, Anisimov V I, Khomskii D I, S.Y. Ezhov, I.V. Solovyev, D.I. Khomskii, and G.A. Sawatzky,1996 *Phys. Rev.* B **54**, 5309
[10] Sundaram N, Jiang Z, Anderson I E, Belanger D P, Booth C H, Bridges F, Mitchell J F, Proffen T, Zheng H, 2009 *Phys. Rev. Lett.* **102**, 026401
[11] Haverkort M V, Hu Z, Cezar J C, Burnus T, Hartmann H, Reuther M, Zobel C, Lorenz T, Tanaka A, Brookes N B, Hsieh H H, Lin H J, Chen C T, and Tjeng L H. 2006 *Phys. Rev. Lett.* **97** 176405
[12] Podlesnyak A, Streule S, Mesot J, Medarde M, Pomjakushina E, Conder K, Tanaka A, Haverkort M V, Khomskii D I, 2006 *Phys. Rev. Lett.* **97** 247208
[13] Pandey S, Kumar A, Prabhakaran D, 2008 *Phys. Rev. B* **77** 045123
[14] Kuzmin A, 1995 *Physica **B*** 208/209 175
[15] Fornasini P, Beccara G, Dalba R, Grisenti A, Sanson M, Vaccari M, 2004 *Phys. Rev. B* **70**, 174301
[16] Herklotz A, Rata A D, Schultz L, Doerr K 2009 *Phys. Rev. B* **79**, 092409
[17] Troyanchuk I, Balagurov A, Sikolenko V, Efimov V, Sheptyakov D, 2013 *J. Appl.Phys.* **113** 053909
[18] Troyanchuk I, Bushinsky M, Sikolenko V, Efmov V, Ritter C, Hansen T and Többens D M 2013 *Eur. Phys. J. B* **86** 435
[19] Troyanchuk I O, Bushinsky M V, and Lobanovsky L S 2013 *J. Appl. Phys.* **114** 213910

# Dependence of the Jahn-Teller distortion in $LaMn_{1-x}Sc_xO_3$ on the isovalent Mn-site substitution

**G Subías[1], V Cuartero[2], J Blasco[1], J García[1], C Meneghini[3] and G Aquilanti[4]**

[1]Instituto de Ciencia de Materiales de Aragón, Departamento de Física de la Materia Condensada, CSIC-Universidad de Zaragoza, C/ Pedro Cerbuna 12, 50009 Zaragoza (Spain)
[2]ESRF-The European Synchrotron, 71, Avenue des Martyrs, Grenoble (France)
[3]Dipartimento di Scienze, Università Roma TRE, Via della Vasca Navale 84 – 00146, Roma, (Italy)
[4]Elettra-Sincrottrone Trieste, S.S. 14, Km 163.5, 34149 Basovizza, Trieste (Italy)

E-mail: gloria@unizar.es

**Abstract.** We investigated the relative importance of removing the $Mn^{3+}$ Jahn-Teller distortion in driving ferromagnetism in $LaMn_{1-x}Sc_xO_3$ combining x-ray powder diffraction and x-ray absorption spectroscopy at the Mn and Sc K-edges. By increasing the Sc content, the orthorhombic distortion of the *Pbnm* cell in $LaMnO_3$ decreases but the unit-cell remains slightly distorted in $LaScO_3$. Besides, the nearly tetragonal-distorted $MO_6$ in $LaMnO_3$ continuously evolves into a nearly regular one in $LaScO_3$. On the other hand, x-ray absorption spectra show that the $MnO_6$ octahedron remains Jahn-Teller distorted and the $ScO_6$ octahedron is nearly regular along the whole series. Moreover, the ordering of the $Mn^{3+}$ Jahn-Teller distortion is not disrupted in the ab plane for any Sc concentration. This contrasts with the Ga-substituted compounds, where a regular $MnO_6$ is found for x>0.5. However, both $LaMn_{0.5}Sc_{0.5}O_3$ and $LaMn_{0.5}Ga_{0.5}O_3$ show ferromagnetic behavior independently of the presence (or not) of Jahn-Teller distorted $Mn^{3+}$. Thus, our results point to the Mn-sublattice dilution as the main effect in driving ferromagnetism in these manganites over local structure effects previously proposed by the spin flipping or the vibronic superexchange models.

## 1. Introduction

Doped manganites have been extensively studied owing to their colossal magnetoresistance property and its potential applications to magnetic devices. In particular, $LaMnO_3$ is considered as the prototype of the cooperative Jahn-Teller (JT) systems and orbital-ordered state [1]. It also develops long-range antiferromagnetic (AFM) ordering of type A below $T_N$=140 K. The substitution at the Mn dramatically modifies the magnetic and electronic properties. It is known that the A-type AFM orbitally ordered state of $LaMnO_3$ is easy destroyed to give way to ferromagnetism (FM) by introducing various cations on the Mn site [2-5]. However, this ferromagnetic behavior derives from a complex interplay between several mechanisms: double-exchange and super-exchange interactions, orbital order and JT distortions. The substitution of Mn with non-magnetic isovalent ions such as $Ga^{3+}$ ($3d^{10}$) or $Sc^{3+}$ ($3d^0$) weakens the $Mn^{3+}$-$Mn^{3+}$ super-exchange interactions but surprisingly favors the appearance of a FM ground state for $LaMn_{0.5}Ga_{0.5}O_3$ [3,4] and $LaMn_{0.5}Sc_{0.5}O_3$ [5]. Two models have been reported to account for this property: the spin flipping of Mn $e_g$-orbital in JT-distorted $Mn^{3+}$ atoms surrounded by

Ga (Sc) atoms [6] and the FM vibronic superexchange interaction coming from disordered and fluctuating JT-distorted $Mn^{3+}$ atoms [7]. Both models assume that the change from A-type AFM to FM ordering is originated by the orbital direction adopted by the $Mn^{3+}$ ion and it is similar for Ga and Sc non-magnetic substitutions, so the atomic $e_g$ orbital would be perturbed in the same way for the two series.

These models contrast with previous x-ray absorption spectroscopy (XAS) studies [8-10]. A XAS study of the $LaMn_{1-x}Ga_xO_3$ series have revealed that the tetragonal distortion of $MnO_6$ octahedron continuously decreases with the gallium content in such a way that for x>0.5 $MnO_6$ distortion disappears and becomes regular [8,9], simultaneously to the appearance of long-range FM. Moreover, a XAS study of the isostructural $TbMn_{1-x}Sc_xO_3$ series showed that the $MnO_6$ octahedra remain tetragonally distorted along the whole dilution range, independently of the Sc content [10].

Here, we have studied the crystallographic structure by x-ray powder diffraction (XRD) and the local structure by XAS at the Mn and Sc K-edges in the $LaMn_{1-x}Sc_xO_3$ series in order to clarify the importance of removing the Jahn-Teller distortion in driving the FM behavior.

## 2. Experimental details

Powder $LaMn_{1-x}Sc_xO_3$ (x= 0, 0.1, 0.2, 0.3, 0.4, 0.5, 0.6, 0.7, 0.9, 1) samples were synthesized following the same ceramic procedure as previously outlined for $LaMn_{1-x}Ga_xO_3$ [3]. XRD patterns were collected for the whole series in the 2-theta range between 18° and 135° at room temperature by using a D-max Rigaku system with a rotating anode and selecting the Cu Kα radiation. The crystal structures were refined by the Rietveld method using the program FULLPROF [11]

The Mn K-edge XAS spectra were recorded at the BM23 beamline of the European Synchrotron Radiation Facility. Data were recorded at a fixed T~80 K in transmission mode for Sc concentration up to x=0.6 while for higher dilutions fluorescence detection was used. The Sc K-edge XAS measurements were carried out at fixed temperature at the CLAESS beamline of the ALBA synchrotron for Sc concentration x≥0.3 whereas temperature-dependent XAS spectra of the $LaMn_{0.4}Sc_{0.6}O_3$ sample were collected at the XAFS beamline of the Elettra Synchrotron. Data for x>0.4 were recorded in transmission mode while for lower Sc concentration fluorescence detection was used. XANES spectra were normalized to unity edge jump using the Athena software from the Demeter package [12]. The extraction of the $\chi(k)$ signals were also performed using Athena program, and R-space EXAFS spectra were obtained by calculating the Fourier Transform (FT) of the $k^2\chi(k)$ signals in the (2.5-14) $Å^{-1}$ k-range. The EXAFS structural analysis was performed using theoretical phases and amplitudes calculated by the FEFF-6 code [13] and fits to the experimental data were carried out in R-space with the Artemis program of the Demeter package [12].

## 3. Results and conclusions

The XRD results show that the tetragonal distortion of the $MO_6$ octahedron continuously decreases with the Sc content (figure 1). There are two possibilities that could explain this fact: either both $MnO_6$ and $ScO_6$ octahedron are equally distorted, and the magnitude of the correlated distortion decreases with x, or the observed distortion of $MO_6$ by XRD corresponds to the average of different $MnO_6$ and $ScO_6$ octahedra. In figure 1, we have compared the M-O distances from XRD with the weighted addition of the Mn-O and Sc-O distances taken from XRD for the end-member compounds of the series, $LaMnO_3$ and $LaScO_3$. The same comparison was also made with the weighted addition of the Mn-O and Ga-O distances for the $LaMn_{1-x}Ga_xO_3$ series as reported in the inset of figure 1. The agreement is very good, which means that the geometrical orientation of the $MnO_6$ tetragonal distortion is preserved in the *ab* plane in both, Sc-substituted and Ga-substituted compounds. In the Ga-substituted samples, different $MnO_6$ and $GaO_6$ octahedra are deduced by XAS where the JT-distortion of the $Mn^{3+}O_6$ decreases with the Ga substitution and the appearing of FM concurs with the presence of regular $Mn^{3+}O_6$ octahedron (x>0.5). The occurrence of Mn tetragonal distortions, which remain ordered even up to high Sc content, has been also reported for the $TbMn_{1-x}Sc_xO_3$ series [10].

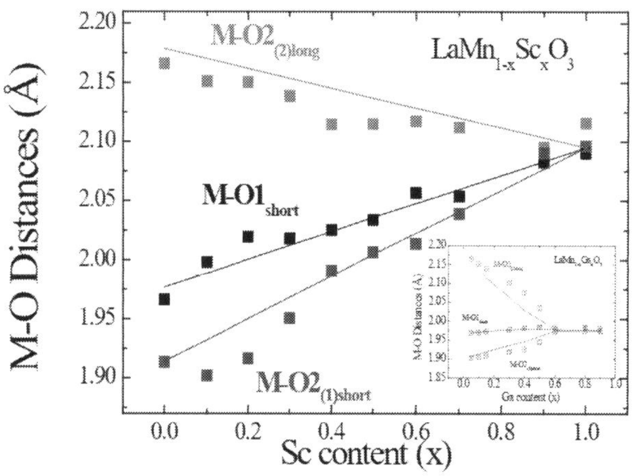


**Figure 1.** Evolution with the Sc content of the $MO_6$ octahedron (closed symbols) compared to the weighted addition of the M-O distances of the two end-member $LaMnO_3$ and $LaScO_3$ (lines) from XRD. **Inset:** same comparison for the $LaMn_{1-x}Ga_xO_3$ series.

No edge energy shift is observed between the $LaMn_{1-x}Sc_xO_3$ samples and either the $LaMnO_3$ sample at the Mn K-edge or the $LaScO_3$ sample at the Sc K-edge (figure 2) within the precision of the experiments (0.4 eV). Thus, $Mn^{3+}$ formal valence keeps constant along the dilution.

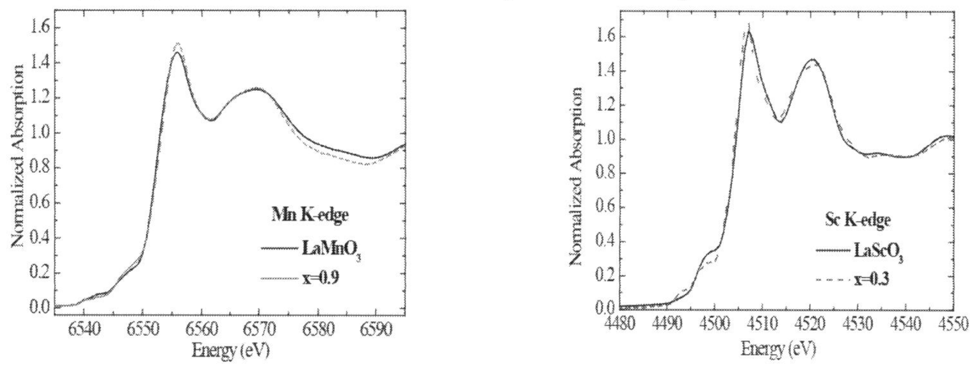


**Figure 2. Left.** Normalized Mn K-edge XANES spectra of $LaMn_{1-x}Sc_xO_3$ (x=0, 0.9). **Right.** Normalized Sc K-edge XANES spectra of $LaMn_{1-x}Sc_xO_3$ (x=0.3, 1).

Figure 3 shows the modulus of the FT of the $k^2$-weighted EXAFS signals of selected concentrations of Sc. All samples show a main peak at ~1.5 Å (Mn K-edge) and ~1.6 Å (Sc K-edge) corresponding to the first (Mn-O or Sc-O) coordination shell without phase shift correction. The intensity and shape of this peak is practically independent of the dilution indicating that the oxygen coordination geometry around both, Mn and Sc atoms, remains almost unaltered along the whole series. The structural analysis at the Mn K-edge was performed considering the same shift on the equatorial plane (Mn-$O1_{short}$ and Mn-$O2_{short}$, $delr_1$) and a different one for the apical distances (Mn-$O2_{long}$, $delr_2$) of the $MnO_6$ octahedron. The relevant structural parameters for the first oxygen coordination shell of all samples are shown in the inset of figure 3(left). The same tetragonal distorted (JT) $MnO_6$ as in $LaMnO_3$ is found for all Sc substitutions. At the Sc K-edge, we have performed the fits considering the three pairs of Sc-O distances slightly distorted of $LaScO_3$ but they converge to a unique distance. As shown in the inset of figure 3 (right), the same nearly regular $ScO_6$ octahedron as in $LaScO_3$ is found for all Sc concentrations. The temperature-dependent EXAFS spectra of $LaMn_{0.4}Sc_{0.6}O_3$ at the Mn and Sc K-edges between 80 and 300 K confirm the robustness of $MnO_6$ tetragonal distortion and the lack of significant static distortion of $ScO_6$ for $LaMn_{0.4}Sc_{0.6}O_3$ along this temperature range.

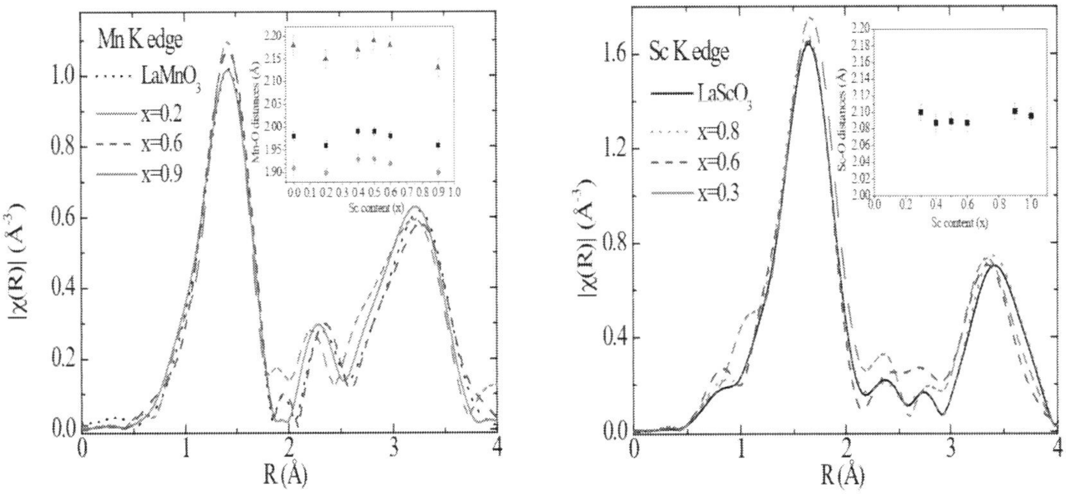


**Figure 3. Left.** Modulus of the FT at the Mn K-edge for selected $LaMn_{1-x}Sc_xO_3$ samples. **Inset:** Evolution of the fitted Mn-O distances with the Sc content. **Right.** Modulus of the FT at the Sc K-edge for selected $LaMn_{1-x}Sc_xO_3$ samples. **Inset:** Evolution of the fitted Sc-O distances with the Sc content.

The present results show a marked difference in the local structure for $LaMn_{1-x}Ga_xO_3$ and $LaMn_{1-x}Sc_xO_3$. $MnO_6$ is hardly distorted in an environment composed by regular $GaO_6$ octahedra whereas this distortion results more feasible when $MnO_6$ is surrounded by $ScO_6$ octahedra because of the larger tilts of the $ScO_6$ octahedron in the orthorhombic *Pbnm* cell. Therefore, the occurrence of the tetragonal distortion of the $MnO_6$ octahedron strongly depends on the lattice in which Mn atom is allocated. Despite the different structural behaviour found for Ga and Sc substitutions, the magnetic response for $x \geq 0.5$ is similar in both cases. The two proposed models [6,7] to account for the change from A-type AFM to FM cannot be supported by either the present structural study or that in the Ga-substituted samples [8,9] so the Mn-sublattice dilution effect may be the main responsible for the magnetic behaviour of these manganites.

## Acknowledgments

Authors acknowledge the financial support from Mineco/FEDER MAT2012-38213-C02-01 and DGA E69 projects. They also thank Alba, ESRF, and Elettra for granting beam time.

## References

[1] Ritter C et al. 1997 *Phys. Rev. B* **56** 8902
[2] Hébert S et al. 2002 *Phys. Rev. B* **65** 104420
[3] Blasco J et al. 2002 *Phys. Rev. B* **66** 174431
[4] Goodenough J B et al. 2002 *Sol. State Sci.* **4** 297
[5] Blasco J et al. 2012 *J. Phys.:Condens. Matter* **24** 076006
[6] Farrell J and Ghering G A 2004 *New J. Phys.* **6** 168
[7] Zhou J-S and Goodenough J B 2003 *Phys. Rev. B* **68** 144406
[8] Sánchez M C et al. 2004 *Phys. Rev. B* **69** 184415
[9] Sánchez M C et al. 2006 *Phys. Rev. B* **73** 094416
[10] Cuartero V et al. 2010 *Phys. Rev. B* **81** 224117
[11] Rodríguez-Carvajal J and Roisnel T, http://www-llb.cea.fr/ fullweb/fullprof.98/fp98.htm
[12] Ravel B and Newville M 2005 *J. Synchrotron Radiat.* **12** 537
[13] Rehr J J and Albers R C 2000 *Rev. Mod. Phys.* **72** 621

Journal of Physics: Conference Series **712** (2016) 012120 doi:10.1088/1742-6596/712/1/012120

# Oxidation-resistive copper nanoparticles: photoreduction synthesis and their oxidation state measurements by XAFS and HRTEM

**H Tanaka[1], T Aoki[1], M Yonemura[1], M Miyagawa[1] and K Okumura[2]**

[1]Department of Applied Chemistry, Faculty of Science and Engineering, Chuo University, 1-13-27 Kasuga Bunkyo-ku 112-8551 Tokyo, JAPAN

[2]Department of Applied Chemistry, Kogakuin University, Tokyo, JAPAN

E-mail: htanaka@kc.chuo-u.ac.jp

**Abstract**. We examined the synthesis of copper nanoparticles by photoreduction, and characterized them using optical spectroscopy, XAFS measurement, and electron microscopy. Ethanol solution of copper acetate with $TiO_2$ nanoparticles was photoirradated. Optical absorption observation indicated that copper nanoparticles were formed in the solution. XAFS measurement indicated that the nanoparticles were metallic, not oxidized. Electron microscopy observation exhibited that the nanoparticles kept metallic even under exposure to air.

## 1. Introduction

Copper nanoparticles have attracted much attention because they have a potential for substitution of precious and worth metal nanoparticles such as gold nanoparticles [1-2]. In fact, various synthesis methods have been studied, but they usually use radical and hazardous reagents such as hydrazine in order to reduce copper ions which are more difficult to reduce than noble metal ions [3-4]. In addition, they are usually obtained more or less oxidized because of their high reactivity [5]. In these points of view, it is important to establish the environmental-friendly synthesis method and to confirm the oxidation state of produced nanoparticles.

In this study, we have examined synthesis of copper nanoparticles by environmental-friendly photoreduction method, which has been developed for synthesis of silver nanoplates [6]. The synthesized nanoparticles were characterized by optical absorption and electron microscopy. Additionally, oxidation state of the synthesized nanoparticles was measured by XAFS measurements.

## 2. Experimental

The synthesis of copper nanoparticles was carried out using a similar method to that reported previously [6-7]. In brief, copper acetate ($(CH_3COO)_2Cu{\cdot}H_2O$, Wako Pure Chemical Industries) was dissolved in ethanol solution which was diluted 10-fold with water, and its concentration was adjusted to 8 mmolL$^{-1}$. $TiO_2$ nanoparticles dispersion was added to the copper acetate solution without any degassing processes, and was sealed. The mixed solution was photoirradiated using a high-pressure Hg lamp (USH-500SC, Ushio) equipped with a wavelength-selective mirror (UV365, Ushio) for 24 h at room temperature.

The obtained dispersion was analyzed by optical absorption, and Cu K-edge XAFS (JASRI, Hyogo, Japan). The Cu K-edge XAFS spectra were measured at beam line BL01B1 with fluorescence mode. The XAFS spectra were analyzed by the program package REX2000 (Rigaku Co.). Model parameters for curve-fitting analysis (back scattering amplitude and phase shift) was extended from an EXAFS oscillations observed for Cu foil. Additionally, the dispersion was deposited onto a collodion-coated Cu grid under air, and was observed by high resolution TEM (spherical aberration corrected HD-2700 Type A, HITACHI High-Technologies Co.) operated at 200 kV equipped with EELS (electron energy-loss spectroscopy) analyzer.

## 3. Results and discussion

### *3.1. Optical spectroscopy measurements*
When the initial dispersion started to be photoirradiated by Hg lamp, its color changed light blue to dark red rapidly. In its optical absorption spectrum, a peak ascribed to surface plasmon band for Cu nanoparticles was observed. This photosynthesis process of Cu nanoparticles is probably considered to be proceeded by two-step reaction as follows: pairs of electron and hole were induced by photons emitted from Hg lamp by help of $TiO_2$, and $Cu^{2+}$ ions are reduced into Cu metal by the electrons. On the other hand, when the photoirradated dispersion was exposed to air, its color changed dark red to light blue gradually. Futhermore, this light-blue dispersion was photoirradiated again, its color turned to dark red. These observations were almost the same as those previously reported [7], and this similarity indicates that reproducible copper nanoparticles were synthesized by the present method.

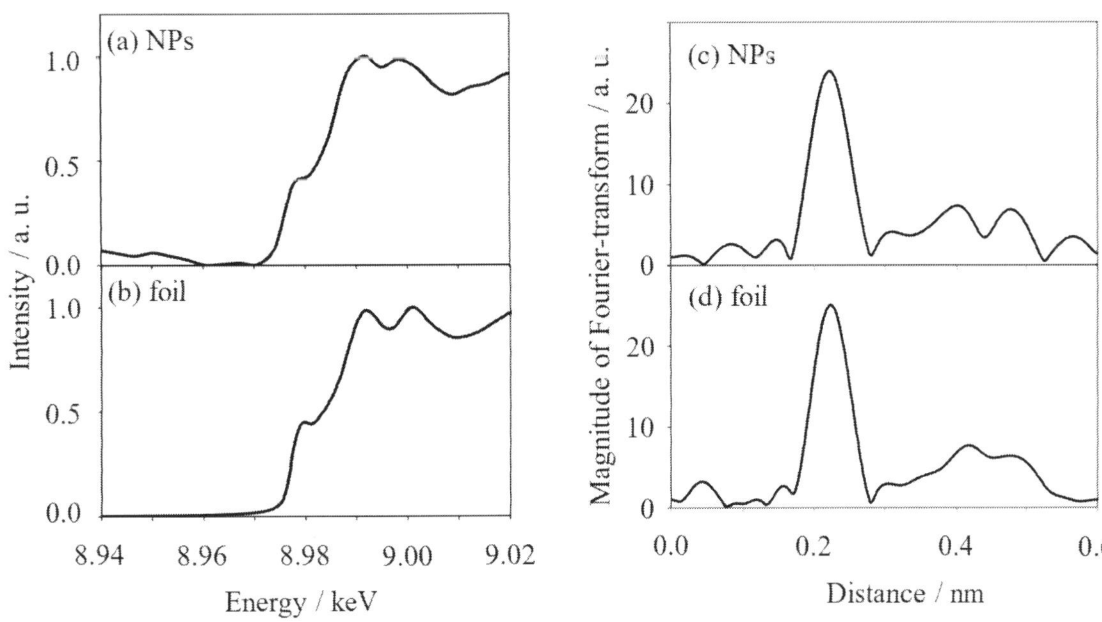


**Figure 1.** XANES spectrum for (a) Cu nanoparticles and (b) Cu foil. Fourier-transformed EXAFS spectrum for (c) Cu nanoparticles and (d) Cu foil.

### *3.2. XAFS measurements*

Figure 1 shows XANES and Fourier-transformed EXAFS spectra for Cu nanoparticles and Cu foil. Since absorption edge and peak profile in the XANES spectrum are found to be almost coincident, the oxidation state of the Cu nanoparticles is considered to be metallic. In the EXAFS spectrum, as the highest peak is distributed in the same range for both, this peak can be assigned to the scattering from the first nearest-neighbored Cu atoms.

Table 1 shows summary of curve-fitting analysis for Cu nanoparticles. The coordination number of 11.9 for the Cu nanoparticles agrees well with that for the Cu foil, and this suggests that the Cu atoms in the nanoparticles are fully coordinated with the first nearest-neighbored Cu atoms as the fcc structure. In addition, the coordination distance of 0.259 for the Cu nanoparticles agrees well with that for the Cu foil. These results indicate that the Cu nanoparticles are composed of metallic Cu in solution. In addition, since the energy difference of 3 eV and $R$ factor of 0.38 % are small enough, the curve-fitting process was considered to be well conducted.

**Table 1**. Summary of curve-fitting analysis for Cu NPs using the values listed in the row of Cu foil.

| Samples | Coordination | Number | Distance (nm) | $dE^a$ (eV) | $\sigma^b$ (nm) | $R^c$ (%) |
|---|---|---|---|---|---|---|
| Cu NPs | Cu-Cu | 11.9±0.9 | 0.259±0.001 | 3±2 | 0.0060 | 0.38 |
| Cu foil | Cu-Cu | 12.0 | 0.256 | 0 | 0.0060 | - |

[a] Energy difference in the absorption threshold between the model compounds and Cu NPs.
[b] Debye-Waller factor (fixed).
[c] $R$ factor.

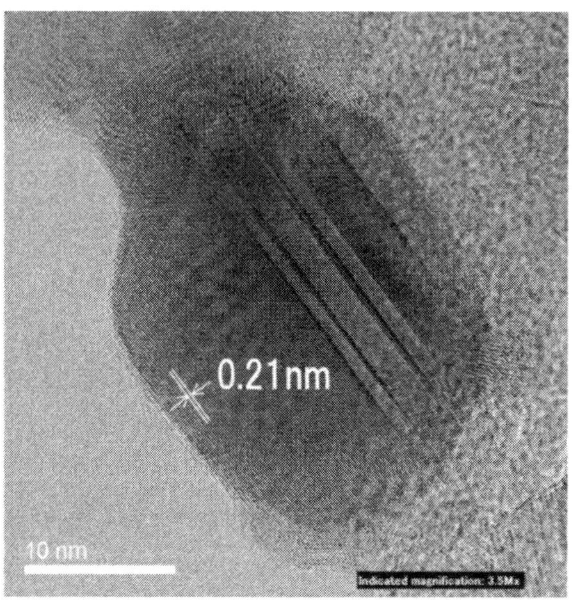


**Figure 2.** High-resolution TEM image for Cu NPs.

### *3.3. HRTEM observations*

Figure 2 shows the HRTEM image for the Cu nanoparticles. Spacing of the lattice fringes is 0.21 nm, and this corresponds well with that for Cu metal. This can be confirmed by EELS spectrum shown in Figure 3(a). This profile is almost the same for metallic Cu, neither CuO nor $Cu_2O$ [8]. In addition, even in the surface of nanoparticle in Figure 2, no oxides are observed. In fact, as shown in Figure 3(b), no distribution indicates that there are no substances relevant to oxygen. Considering the treatment of the specimen under exposure to air, the Cu nanoparticles synthesized by the present method are oxidation resistive.

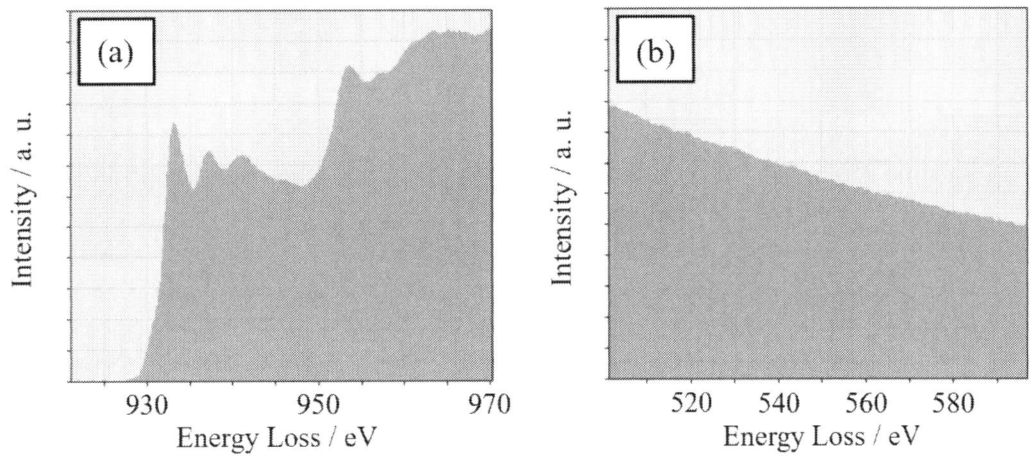


**Figure 3.** Electron energy loss spectra for Cu NPs. (a) Cu *L* and (b) O *K* absorption edge.

**Acknowledgements**

The XAFS measurements were performed at the BL01B1 of Spring-8 with the approval of the Japan Synchrotron Radiation Research Institute (JASRI) (Proposal No. 2015A1134). We thank Mr. Shohei Arai at HITACHI High-Technologies Co. for observation of HRTEM and EELS. This work was supported by MEXT KAKENHI Grant Number 25248004 and by JSPS KAKENHI Grant Number 25410028.

**References**

[1] Dold B 2008 *Rev. Environ. Sci. Biotechnol.* **7** 275
[2] Toshima N and Yonezawa T 1998 *New J. Chem.* **22** 1179
[3] Yonezawa T, Nishida N and Hyono A 2010 *Chem. Lett.* **39** 548
[4] Huang X, Wu H, Liao X and Shi B 2010 *Green Chem.* **12** 395
[5] Nishida N, Miyashita A, Tsukuda T and Tanaka H 2013 *Chem. Lett.* **42** 168
[6] Murayama H, Hashimoto N and Tanaka H 2009 *Chem. Phys. Lett.* **482** 291
[7] Nishida N, Miyashita A, Hashimoto N, Murayama H and Tanaka H 2011 *Eur. Phys. J. D* **63** 307
[8] Leapman R D, Grunes L A and Fejes P L 1982 *Phys. Rev. B* **26** 614

# Temperature dependence of Zr and Ti K-edge XANES spectra for para- and ferro-electric perovskite-type PbZrO3, PbTiO3 and BaTiO3

**A Yoshiasa*[1], T Nakatani[1], T Hiratoko[1], T Tobase[1], A Nakatsuka[2], M Okube[3], H Arima[4] and K Sugiyama[4]**

[1] Graduated School of Sci. and Tech., Kumamoto Univ., Kumamoto, 860-8555, Japan

[2] Graduate School of Science and Engineering, Yamaguchi Univ. Ube 755-8611, Japan

[3] IMR, Tokyo Institute of Technology, Yokohama 226-8503, Japan

[4] Department of Institute for material research, Tohoku Univ. Sendai 980-8577, Japan

E-mail: yoshiasa@sci.kumamoto-u.ac.jp

**Abstract**. Zr and Ti K-edge XANES spectra of $PbZrO_3$, $PbTiO_3$ and $BaTiO_3$ perovskite-type compounds were measured in the temperature range from 10K to 850K. Quantitative comparisons for the near-edge spectra were performed in a wide temperature range using the absorption intensity invariant point (AIIP) standardization. Clear temperature dependence for pre-edge shoulder is identified by the calculating the temperature difference of the XANES spectrum intensity. Decrease of pre-edge shoulder and peak intensity is observed only in the para- and ferro-electric phases and draw curves, not straight lines. The gradients for shoulder and pre-edge peak intensity are rich in a variety. The decrease in absorption of pre-edge peak and shoulder is speculated due to the shift from the off-centre position of the Zr atom with respect to the oxygen octahedron to center position. The Zr ion in the $PbZrO_3$ para-electric phase has same temperature behaviors of Ti ions in the ferroelectric perovskite.

## 1. Introduction

Characterization of pre-edge peak structure in K-edge XANES spectra of transition metals is useful for understanding materials in the fields of physics, chemistry and earth sciences. The pre-edge features are sensitive to the electronic and geometrical structure around the X-ray absorbing central atom. Temperature dependence on the intensity of pre-edge peaks were observed in various compounds [1-5]. Hashimoto et al. [3] measured the Ti K-edge XANES spectra of several titanates at various temperatures. They have revealed that the pre-edge peak intensities are normally increase with temperature but ferroelectric phases shows different type temperature dependences. Hiratoko et al. [5] succeeded in the quantitative comparative study of temperature dependence over a wide temperature range of pre-edge peaks intensity in various titanium compounds using the absorption intensity invariant point (AIIP) standardization. Large temperature dependence of pre-edge peaks originates in the quadrupole transition and the large density of d states below the p states. Temperature dependence is not only due to a phase transition or to a local geometrical distortion of titanium site but also to thermal vibration amplitude of the absorbing and scattering atoms [5]. The aspect of temperature

Journal of Physics: Conference Series **712** (2016) 012121 doi:10.1088/1742-6596/712/1/012121

change in X-ray absorptivity at each peak position is fluctuated by the hybridized orbital proportion and the site geometry.

$PbTiO_3$ and $BaTiO_3$ perovskites have been intensively studied, because of ferroelectric properties at ambient condition and being very useful as functional materials. Several XAFS studies were carried out to investigate the mechanism of ferroelectric-dielectric phase transition for these materials [3,6-8]. Miyanaga et al. [8] proposed that the Ti ion occupies an off-center position of the distorted $TiO_6$ octahedral site and the geometrical distortion in tetragonal phase decreases with increasing temperature and A2 pre-edge peak intensity decreases. These results showed that the A2 pre-edge peak intensity depends on the degree of distortion around the Ti atoms. $PbZrO_3$ has a para-electric property. $PbZrO_3$ has the same perovskite type structure and phase transitions as $BaTiO_3$ and $PbTiO_3$. There are little researches of comparing them on detailed temperature dependence of XANES patterns because there is no clear pre-edge peak in Zr K-edge XANES spectrum.

A sufficient comparative study and the detail interpretation of the temperature dependence of pre-edge and near-edge shoulder intensities have not been made in ferro- and para-electric perovskite-type compounds in wide temperature range. We present a quantitative comparison for the pre-edge shoulders in $PbZrO_3$, $PbTiO_3$ and $BaTiO_3$ perovskite-type compounds in a wide temperature range using the AIIP standardization and investigated how intensity of the pre-edge shoulders changed with temperature.

## 2. Experiment

The specimens of $PbZrO_3$ perovskite (99.99%), $PbTiO_3$ perovskite (99.99%) and $BaTiO_3$ perovskite (99.99%) were commercial chemicals (Furu-uchi Chemical Co. Ltd) and were examined by X-ray diffraction methods at beam line BL-10A of the Photon Factory in the High Energy Accelerator Research Organization (KEK), Tsukuba, Japan (PAC No. 2011G611 and 2013G542). Appropriate amounts of fine-powder sample and boron nitride powder were mixed and pressed into pellets of <0.2 mm in thickness and 10.0 mm in diameter. All samples had edge-jumps of 0.7 ($\Delta\mu d$), where $\mu$ is the linear absorption coefficient and d is the thickness. XAFS spectra near the Zr and Ti K-edges were measured in transmission mode at beamlines NW10A and BL-9C of KEK, Tsukuba, Japan. Synchrotron radiation was monochromatized by Si (311) or Si (111) double-crystal monochromator. X-ray energy calibration was performed by setting the copper metal pre-edge absorption peak to 8978.8 eV. X-ray absorption measurements in the temperature range 10–850 K were performed under a helium atmosphere. Details of the measurements are given by Yoshiasa et al. [9] and Hiratoko et al. [5]. Analysis of XAFS data was performed by using XAFS93 programs [10].

## 3. Results and discussion

### *3.1 AIIP-based normalization and temperature dependence of pre-edge peak intensity*

Figs. 1-3 show the Zr and Ti K-edge XANES spectra for $PbZrO_3$, $PbTiO_3$ and $BaTiO_3$ perovskite-type compounds at various temperatures. In order to clarify the temperature dependence, the difference spectrum between the low-temperature spectra and those at higher temperatures is plotted and shown on the right side. The absorption intensity invariant point (AIIP) [5], i.e. an energy point where the intensity does not change with temperature, was observed near each main peak. Thick arrows in the figures indicate the AIIP. The physical meaning of AIIP will be regarded as extension of EXAFS baseline. Using the AIIP peak intensity as a standard point, it is possible to present a quantitative comparison for the pre-edge peaks in various compounds in a wide temperature range. We normalized the spectrum intensity for the samples in a way that the intensity at AIIP is 1.0 and compared the temperature dependence of the XANES pre-edge peaks among the samples. The AIIP-based normalization allows the elimination of errors caused by the reference point itself changing with temperature. XANES amplitudes at energies higher than the AIIP damp owing to Debye–Waller type effects with increasing temperature [5]. Amplitudes increase linearly with temperature at energies lower than the AIIP usually except for ferroelectric $PbTiO_3$ and $BaTiO_3$ phases. The height of pre-edge peak for ferro-electric perovskite increases with temperature [5,11].

Decreases in Ti K-edge peak A2 intensities were observed mainly in the ferroelectric tetragonal phases. The observations are consistent with previous results [3,6-8,12]. The pre-edge feature of ferroelectric perovskite tends to decrease toward Curie point (Tc) with increasing temperature. This decrease in absorption of pre-edge peak will be speculated due to the shift from the off-centre position of the Ti

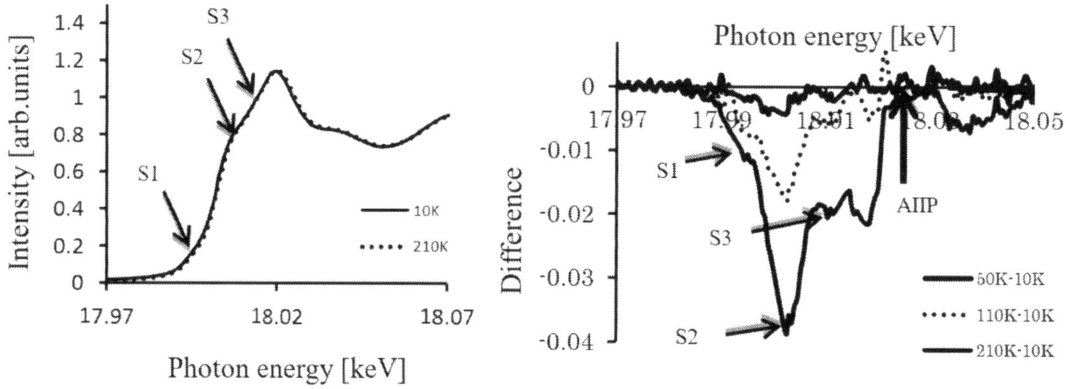


**Figure 1**. (A) Zr K-edge XANES spectrum of $PbZrO_3$ at 10 and 210 K (AIIP: 18.025). (B) The difference spectra for $PbZrO_3$ that is the difference between the spectrum at 10K and those of higher temperatures (50K - 210K).

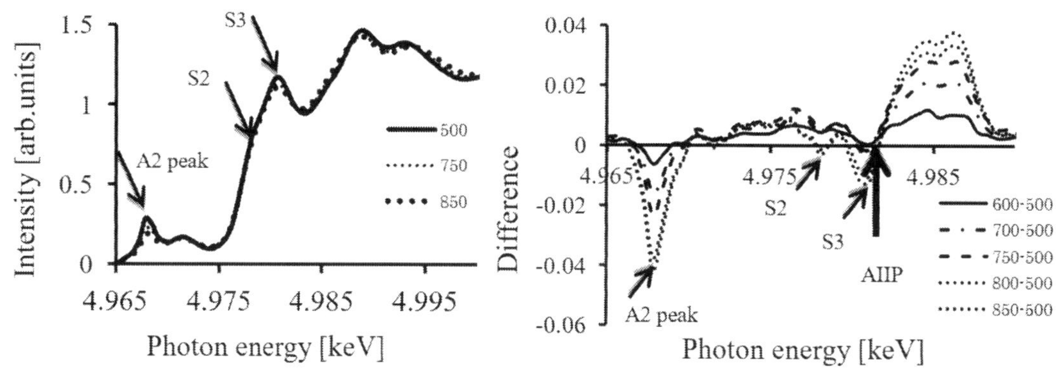


**Figure 2**. (A) Ti K-edge XANES spectrum of $PbTiO_3$ at 500 and 750K K (AIIP: 4.982) . (B) The difference spectra for $PbTiO_3$ that is the difference between the spectrum at 500K and those of higher temperatures (600K - 850K).

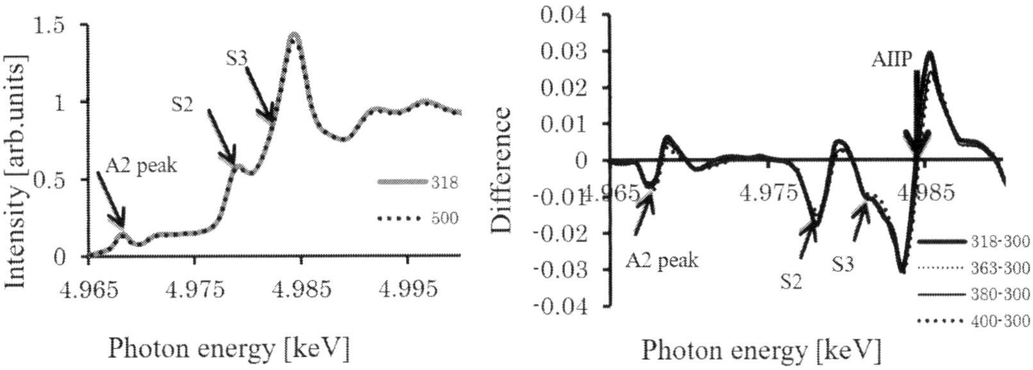


**Figure 3.** (A) Ti K-edge XANES spectrum of $BaTiO_3$ at 318 and 500 K (AIIP: 4.985). (B) The difference spectra for $BaTiO_3$ that is the difference between the spectrum at 300K and those of higher temperatures (318K - 400K).

atom with respect to the oxygen octahedron to center position. It should be also concluded that the decreasing in intensities observed in the para- and ferro-electric materials arise from the atomic movement from off-center to center position [11].

*3.2 Decrease of pre-edge shoulder intensity and gradient against temperature.*

A clear pre-edge peak isn't observed in Zr K-edge for $PbZrO_3$ and some shoulders appear on the low energy side of the main peak and AIIP position. These shoulders were named S1–S3. The assignments for electron transitions near the pre-edge and shoulders position were proposed by theoretical calculations [13,14]. The hybridizations between cation and oxygen are important in each compound. In the difference spectra between the low temperature and higher temperature spectra (Figs 1-3), clear negative and positive peaks can be distinguish according to the respective temperature dependences. Three negative peaks or shoulders are identified and observed only in the Ti and Zr K-edges for para- and ferro-electric $PbZrO_3$, $PbTiO_3$ and $BaTiO_3$ by thin arrows in Figures. These peaks in $PbZrO_3$ correspond to S1 – S3 shoulders. In para- and ferro-electric perovskite-type compounds, some pre-edge peak and shoulder intensities decrease with increasing temperature. The validity of observing the temperature dependence of XANES was indicated by the difference spectra between the low temperature and higher temperature spectra.

Figure 4 shows the comparison of temperature dependence of the S2 shoulder intensity for $PbZrO_3$ and $BaTiO_3$ and the A2 pre-edge peak intensity observed in $BaTiO_3$ and $PbTiO_3$. The intensities of shoulder S1 and peak A2 decrease clearly with increasing temperature and draw curves, not straight lines. The gradients for S1 shoulder and A2 pre-edge peak intensity are rich in a variety. It is known that the pre-edge peak intensity increase usually linearly with increasing temperature by harmonic thermal vibration and that the gradient of the pre-edge A2 peak intensity against temperature has two tendencies of rutile- and anatase-types [5]. It is important that the Zr ion in the $PbZrO_3$ para-electric phase has similar temperature behaviors of Ti ions in the ferroelectric $PbTiO_3$ and $BaTiO_3$ perovskite. We guess that the decrease in pre-edge shoulder intensity occur by the same factor as the Ti K-edge peak A2 intensities in the ferroelectric phases. The decrease in absorption of pre-edge shoulder is speculated due to the shift from the off-center position of the Zr and Ti atom with respect to the oxygen octahedron to center position.

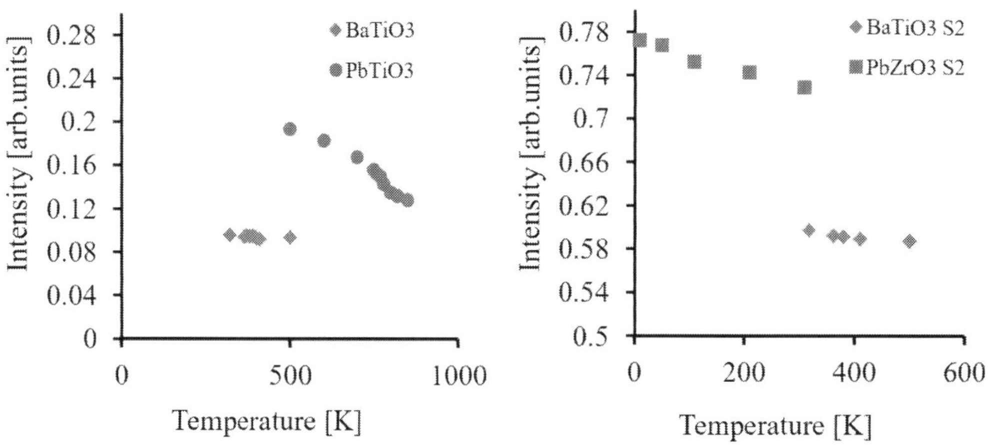


**Figure 4**. (A) Temperature dependence of normalized pre-edge peak A2 intensity of Ti K-edge for PbTiO3 and BaTiO3. (B) Temperature dependence of normalized pre-edge peak S2 intensity of Zr and Ti K-edges for PbZrO3 and BaTiO3 perovskite.

**References**

[1] Durmeyer O, Kappler J P, Beaurepaire E, Heintz J M and Drillon M 1990 *J. Phys. Condens. Matter* **2** 6127
[2] Nozawa S, Iwazumi T and Osawa H2005 *Phys. Rev.* **B72**, 121101
[3] Hashimoto T, Yoshiasa A, Okube M, Okudera H and Nakatsuka A 2007 *AIP Conf. Proc.* **882** 428
[4] Durmeyer O, Beaurepaire E, Kappler J P, Brouder C and Baudelet F 2010 *J. Phys. Condens. Matter* **22** 125504
[5] Hiratoko T, Yoshiasa A, Nakatani T, Okube M, Nakatsuka A and Sugiyama K 2013 *Journal of Synchrotron Radiation* **20** 641
[6] Sicron N, Ravel B, Yacoby Y, Stern E A, Dogan F and Rehr J J1994 *Physical Review* **B50** 168
[7] Ravel B, Stern E A, Vedrinski R I and Kraizman V 1998 *Ferroelectics* **206** 407
[8] Miyanaga T, Diop D, Ikeda S I and Kon H 2002 *Ferroelectics* **274** 41
[9] Yoshiasa A, Tamura T, Murai K, Ogata K and Mori H 1999 *J. Sychrotron Radiation* **6** 43
[10] Yoshiasa A, Koto K, Maeda H and Ishii T 1997 *Jpn. Appl.Phys* **36** 781
[11] Nakatani T, Hiratoko T, Okube M, Murai K and Yoshiasa A 2011 *Acta Cryst.* **A67**, C584
[12] Revel B and Stern E A 1995 *Physica B* **208-209** 316
[13] Vedrinskii R V, Kraizman V L, Novakovich A A, Demekhin P V and Urazhdin S V 1998 *J. Phys: Condens. Matter* **10** 9561
[14] Fujita M, Nakamatsu H,Sugihara S, Aihara J and Sekine R 2004 *J. Comput. Chem. Jpn.* **3** 21

# Pressure-induced insulator-to-metal transition in $\alpha$-$SnWO_4$

**Alexei Kuzmin[1], Andris Anspoks[1], Aleksandr Kalinko[2,3], Janis Timoshenko[1], Robert Kalendarev[1], Lucie Nataf[4], François Baudelet[4], Tetsuo Irifune[5], Pascale Roy[4]**

[1] Institute of Solid State Physics, University of Latvia, Riga, Latvia
[2] University of Paderborn, Faculty of Sciences, Department of Chemistry, Germany
[3] DESY Photon Science, Hamburg, Germany
[4] Synchrotron SOLEIL, l'Orme des Merisiers, Saint-Aubin, Gif-sur-Yvette, France
[5] Geodynamics Research Center, Ehime University, Matsuyama, Ehime, Japan

E-mail: a.kuzmin@cfi.lu.lv

**Abstract.** In-situ high-pressure W $L_1$ and $L_3$ edges x-ray absorption and mid-infrared spectroscopies complemented by first-principles calculations suggest the existence of pressure-induced insulator-to-metal transition in $\alpha$-$SnWO_4$ in the range of 5-7 GPa. Its origin is explained by a symmetrization of metal-oxygen octahedra due to a strong interaction of Sn 5s, W 5d and O 2p states along the $b$-axis direction, leading to a collapse of the band gap.

## 1. Introduction

Metal-insulator transitions in oxides are of considerable interest already for more than half a century [1, 2, 3]. The phenomenon finds applications in novel electronic and photonic devices, thus stimulating the discovery of new functional materials [4, 5]. In this study we present an overview of recent experimental and theoretical findings on novel pressure-induced insulator-to-metal transition in low-temperature $\alpha$-phase of tin tungstate ($\alpha$-$SnWO_4$).

Tin tungstate has two polymorphs, the low-temperature orthorhombic $\alpha$-phase (space group $Pnna$, no. 52) and high-temperature cubic $\beta$-phase (space group $P2_13$, no. 198), which transform into each other by a diffusion-controlled phase transition mechanism at about 670°C [6]. The two phases have narrow band gaps ($E_g$=1.7 eV in $\alpha$-$SnWO_4$ and 2.6 eV in $\beta$-$SnWO_4$) and unique band structures, strongly influenced by the charge transfer phenomena between oxygen and tungsten atoms and the presence of the lone pair of Sn 5s electrons [7, 8].

The crystal structure of $\alpha$-$SnWO_4$ (Fig. 1) is composed of two-dimensional sheets of $WO_6$ octahedra, separated by layers of $Sn^{2+}$ ions, which are also six-fold coordinated by oxygen atoms [6]. The $WO_6$ octahedra within a sheet are joined by four vertices, forming a 2D perovskite-like layer, and are distorted due to the second-order Jahn-Teller (SOJT) effect [9] because of the $W^{6+}(5d^0)$ electronic configuration. This distortion is associated with an off-center displacement of $W^{6+}$ ions in the direction of the octahedron edge, so that there are three distinct W–O bonds. The six-fold coordination of $Sn^{2+}(5s^26p^0)$ ions is distorted even more, leading also to three different Sn–O bonds. The pair of stereochemically active Sn 5s electrons is responsible for the lone pair distortion of the $SnO_6$ octahedron via the SOJT effect [7, 8].

Journal of Physics: Conference Series **712** (2016) 012122 doi:10.1088/1742-6596/712/1/012122

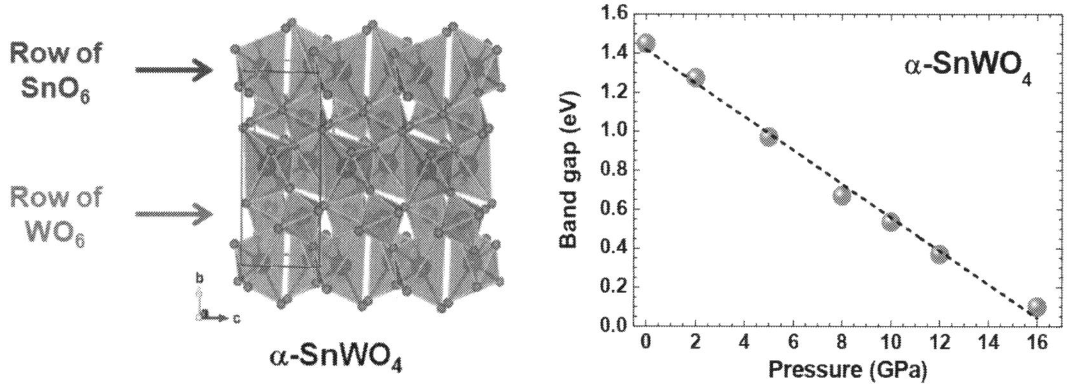


**Figure 1.** Crystal structure of $\alpha$-$SnWO_4$ [6] and pressure dependence of band gap according to first-principles linear combination of atomic orbital (LCAO) calculations [10].

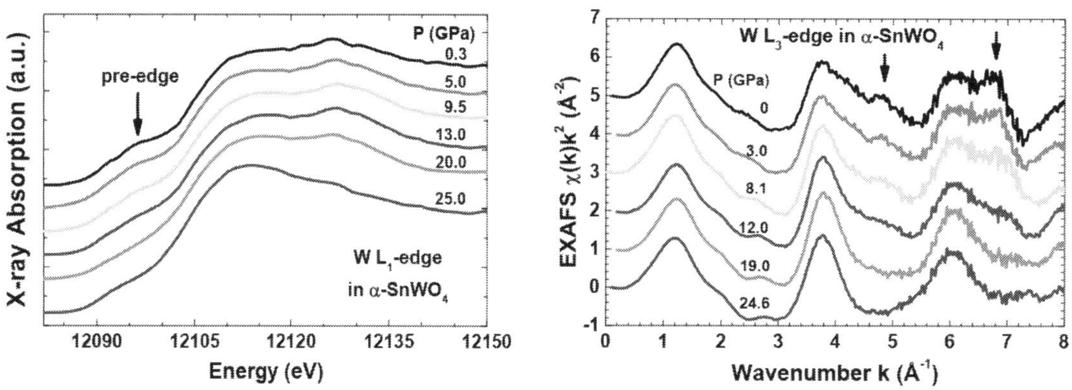


**Figure 2.** Pressure dependence of the W $L_1$-edge XANES and W $L_3$-edge EXAFS in $\alpha$-$SnWO_4$.

Narrow band gap and easily deformable metal-oxygen octahedra make $\alpha$-$SnWO_4$ a good candidate for pressure-induced insulator-to-metal transition. Here we have tackled this issue by in-situ high-pressure x-ray absorption and infrared spectroscopies complemented by first-principles calculations.

## 2. Experimental and calculation details

The synthesis of polycrystalline $\alpha$-$SnWO_4$ was performed by solid-state reaction: an equimolar amount of SnO (99.99%) and $WO_3$ (99.9%) powders was placed in a sealed silica ampoule under a vacuum, annealed at 600°C for 8 hours and slowly cooled down to room temperature.

Room-temperature pressure-dependent x-ray absorption spectroscopy study was performed in the pressure range from 0 to 25 GPa at the W $L_1$ and $L_3$ edges using the dispersive set-up of the bending-magnet SOLEIL ODE beamline [11] and high-pressure nano-polycrystalline diamond anvil cell (DAC) [12]. The use of nano-polycrystalline DAC allowed us to eliminate spurious contributions from the Bragg peaks into the x-ray absorption spectra. Mid-infrared (550-8000 $cm^{-1}$) absorption measurements were performed in transmission mode using Bruker IFS 125MR spectrometer combined with recently developed high-pressure setup based on the DAC at the SOLEIL AILES beamline [13]. The sample pressure was monitored by the ruby

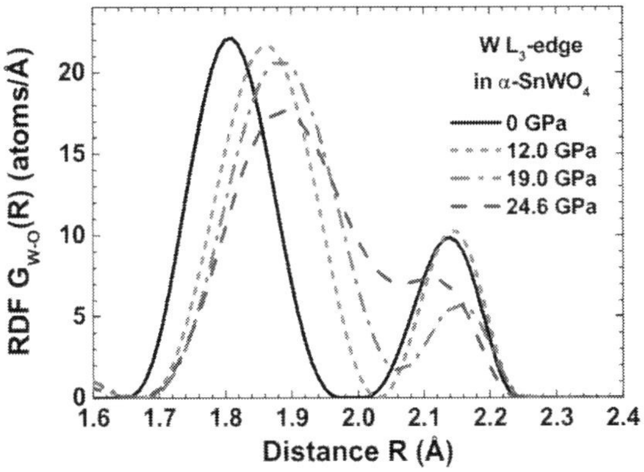


**Figure 3.** Pressure dependence of radial distribution functions (RDFs) $G_{W-O}(R)$ for the first coordination shell of tungsten in $\alpha$-$SnWO_4$.

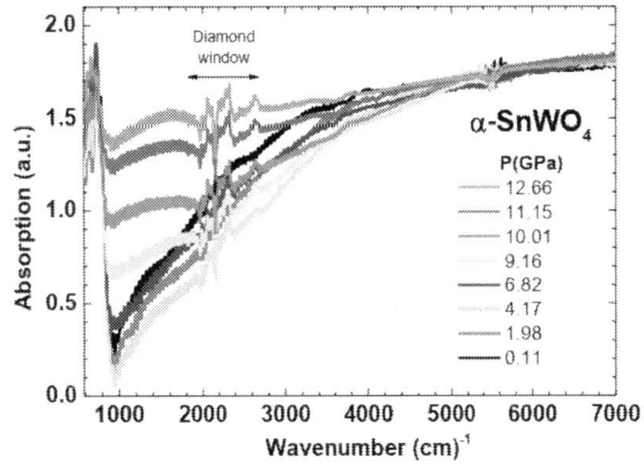


**Figure 4.** Pressure dependence of mid-infrared spectra of $\alpha$-$SnWO_4$. The features between ~1830 and 2700 $cm^{-1}$ are due to a diamond window.

luminescence in both experiments.

The W $L_1$-edge x-ray absorption near edge structure (XANES) and $L_3$-edge extended x-ray absorption fine structure (EXAFS) were extracted following the conventional procedure [14] using the EDA package [15] and are shown in Fig. 2. The radial distribution functions (RDFs) within the first coordination shell of tungsten (Fig. 3) were reconstructed from EXAFS by the regularization-like method [16, 17], which was successfully employed by us in the analysis of several tungstates [18, 19, 20]. Theoretical scattering amplitude and phase shift functions, calculated by the *ab initio* FEFF8 code [21] using the complex exchange-correlation Hedin-Lundqvist potential [22], were used in all simulations (see [8] for more details).

## 3. Results and discussion

Our first-principles linear combination of atomic orbital (LCAO) calculations [10] predict the existence of the insulator-to-metal transition in $\alpha$-$SnWO_4$ (Fig. 1). Its origin is explained by a symmetrization of metal-oxygen octahedra due to a strong interaction of Sn 5s, W 5d and O 2p states, leading to a collapse of the band gap at about 16 GPa [10]. The evidence of this transition was experimentally observed in the mid-infrared region (Fig. 4). Here the contribution of lattice vibrations is responsible for the absorption bands below ~850 $cm^{-1}$. An abrupt increase of the absorption, which is associated with optical indirect transition across the band gap [8, 10], was detected above ~850 $cm^{-1}$ upon increasing pressure above 5-7 GPa. Lower value of the experimental transition pressure compared to the predicted one [10] can be attributed to possible non-stoichiometry of the sample.

The structural origin of the insulator-to-metal transition can be probed directly by studying the W $L_{1,3}$-edges absorption spectra. The analysis of the W $L_1$-edge XANES indicates that upon increasing pressure $WO_6$ octahedra become less distorted, as is evidenced from a decrease of the pre-edge peak due to 2s(W)→5d(W)+2p(O) transition [23] (left panel in Fig. 2). This conclusion is well supported by the analysis of the W $L_3$-edge EXAFS spectra (right panel in Fig. 2).

Pressure dependence of the radial distribution function (RDF) around tungsten $G_{W-O}(R)$ was reconstructed and is shown in Fig. 3. The RDF at 0 GPa consists of two well defined peaks located at 1.80 and 2.14 Å, which correspond to four nearest and two distant oxygen atoms of the $WO_6$ octahedra distorted due to the SOJT effect [9]. Upon increasing pressure a distortion of the $WO_6$ octahedra decreases, affecting mainly the group of four nearest oxygen atoms: the average value of their W–O bond length becomes longer by ~0.09 Å. At the same time, the group of two distant oxygen atoms at 2.14 Å is less influenced by pressure, however their average W–O bond length decreases by ~0.03 Å above 19 GPa. Such pressure-induced reduction of the $WO_6$ octahedra distortion is compatible with a displacement of tungsten atoms towards the center of the $WO_6$ octahedra by about 0.2 Å at highest pressure. The observed symmetrization of the $WO_6$ octahedra correlates well with our results of first-principles LCAO calculations [10], which predict such behaviour for both $WO_6$ and $SnO_6$ octahedra upon increasing pressure.

## 4. Conclusions

X-ray absorption spectroscopy at the W $L_1$ and $L_3$-edges and infrared absorption spectroscopy were successfully applied to in-situ investigation of the pressure effect on the local atomic and electronic structure in $\alpha$-$SnWO_4$. It is shown that a displacement of tungsten atoms by about 0.2 Å towards the center of the $WO_6$ octahedra occurs in $\alpha$-$SnWO_4$ upon increasing pressure up to 25 GPa. The strong modification of $\alpha$-$SnWO_4$ lattice is accompanied by the insulator-to-metal phase transition caused, according to the first-principles LCAO calculations [10], by interaction of Sn 5s, W 5d and O 2p states along the $b$-axis direction.

## Acknowledgments

The authors are grateful to Professor Alain Polian for providing the NDAC cell. This work was supported by Latvian Government Research grant no. 187/2012. The experiment at the SOLEIL synchrotron radiation facility has received funding from the European Community's Seventh Framework Programme (FP7/2007-2013) CALIPSO under grant agreement no. 312284.

## References

[1] Mott N F 1968 *Rev. Mod. Phys.* **40** 677
[2] Honig J M and Zandt L L V 1975 *Ann. Rev. Mater. Sci.* **5** 225
[3] Imada M, Fujimori A and Tokura Y 1998 *Rev. Mod. Phys.* **70** 1040
[4] Yang Z, Ko C and Ramanathan S 2011 *Annu. Rev. Mater. Res.* **41** 337
[5] Granqvist C G 2014 *J. Vac. Sci. Technol. B* **32** 060801
[6] Jeitschko W and Sleight A W 1974 *Acta Cryst. B* **30** 2088
[7] Cho I S, Kwak C H, Kim D W, Lee S and Hong K S 2009 *J. Phys. Chem. C* **113** 10647
[8] Kuzmin A, Anspoks A, Kalinko A, Timoshenko J and Kalendarev R 2014 *Phys. Scr.* **89** 044005
[9] Kunz M and Brown I D 1995 *J. Solid State Chem.* **115** 395
[10] Kuzmin A, Anspoks A, Kalinko A, Timoshenko J and Kalendarev R 2015 *Sol. Energy Sol. Mater. Cells* **143** 627
[11] Baudelet F, Kong Q, Nataf L, Cafun J D, Congeduti A, Monza A, Chagnot S and Itié J P 2011 *High Press. Res.* **31** 136
[12] Tetsuo I, Ayako K, Shizue S, Toru I and Hitoshi S 2003 *Nature* **421** 599
[13] Bernardina S D, Alabarse F G, Kalinko A, Roy P, Chapuis M, Vita N, Hienerwadel R, Berthomieu C, Judeinstein P, Zanotti J M, Bantignies J L, Haines J, Catafesta J, Creff G, Manceron L and Brubach J B 2014 *Vib. Spectrosc.* **75** 154
[14] Aksenov V L, Kuzmin A Y, Purans J and Tyutyunnikov S I 2001 *Phys. Part. Nucl.* **32** 675
[15] Kuzmin A 1995 *Physica* B **208-209** 175
[16] Kuzmin A 1997 *J. Physique IV (France)* **7** C2-213
[17] Kuzmin A and Purans J 2000 *J. Phys.: Condens. Matter* **12** 1959
[18] Kalinko A and Kuzmin A 2011 *J. Non-Cryst. Solids* **357** 2595
[19] Kuzmin A, Kalinko A and Evarestov R A 2013 *Acta Mater.* **61** 371
[20] Kuzmin A, Anspoks A, Kalinko A and Timoshenko J 2013 *J. Phys.: Conf. Ser.* **430** 012109
[21] Ankudinov A L, Ravel B, Rehr J J and Conradson S D 1998 *Phys. Rev.* B **58** 7565
[22] Rehr J J and Albers R C 2000 *Rev. Mod. Phys.* **72** 621
[23] Kuzmin A and Purans J 1993 *J. Phys.: Condens. Matter* **5** 9423

# Polarization-dependent DANES study on vertically-aligned ZnO nanorods

Chengjun Sun[1], Chang-In Park[2], Zhenlan Jin[2], In-Hui Hwang[2], S. M. Heald[1], and Sang-Wook Han[2*]
[1]X-ray Science Division, Advanced Photon Source, Argonne National Laboratory, Argonne, IL 60440, USA
[2]Department of Physics Education and Institute of Fusion Science, Jeonbuk(Chonbuk) National University, Jeonju 54896, Korea

[*]Email: shan@jbnu.ac.kr

**Abstract**. The local structural and local density of states of vertically-aligned ZnO nanorods are examined by using polarization-dependent diffraction anomalous near edge structure (DANES) measurements from *c*-oriented ZnO nanorods at the Zn K edge at the geometry of the incident x-ray electric field parallel and perpendicular to the x-ray momentum transfer direction. Orientation-dependent local structures determined by DANES are comparable with polarization-dependent EXAFS results. Unlike other techniques, polarization-dependent DANES can uniquely describe the orientation-dependent local structural properties and the local density of states of a selected element in selected-phased crystals of compounds or mixed-phased structures.

## 1. Introduction

Anomalous x-ray scattering techniques have been widely used to determine local structural properties in the crystalline regions of matters, particularly, including phase-mixture structures and textured structures [1-4]. Polarization-dependent diffraction anomalous fine structure (DAFS) can describe the orientation-dependent local structural and the chemical properties of a probing atom because the atomic form factor square of $|f_0 + f' + if''|^2$ depends on x-ray electric field directions. The Thomson scattering factor of $f_0$ is independent of the orientation of crystalline structures, whereas the resonant anomalous scattering amplitude of $f' + if''$ is contributed by crystalline orientations due to the interaction of incident x-ray electric field and electron-hole dipoles [1-4]. Polarization-dependent diffraction anomalous near edge structure (DANES) can uniquely describe the chemical valence state and the local electron density of states of probing atoms within selected-phased crystals in a specimen. Previous studies have mostly focused on DAFS compared to EXAFS [1-4]. In this paper, we demonstrate polarization-dependent DANES measurements from vertically-aligned ZnO nanorods, comparing with *simultaneously*-measured XANES. The direct comparison of DANES and XANES measured from single-crystal ZnO nanorods will prove that DANES alone can probe the chemical valence state and the local density of state of a selected element within selected-phased crystals.

Wurtzite-phase ZnO nanorods have been intensively studied as building blocks for nanodevice applications, including light emitting diodes, solar cells, piezoelectric devices, sensors, and actuators [5]. The observation of quantum confinement effects from ZnO nanorods perpendicular to the nanorod length [6] suggested that the behavior of electrons in nanorods depends on the direction. Furthermore, previous studies with polarization-dependent XAFS measurements from ZnO nanorods demonstrated that the bond lengths of ZnO pairs parallel and perpendicular to the nanorod length direction were elongated and shrunken, respectively, compared to those of ZnO powder [8]. Since XAFS detects an average structural properties of a specimen, it cannot determine the local structural properties of selected-phased crystals. The orientation-dependent local structural properties and the local density of states in crystalline regions are often critical in understanding of fundamental properties of materials, including superconductivity and orientation-dependent magnetization [9].

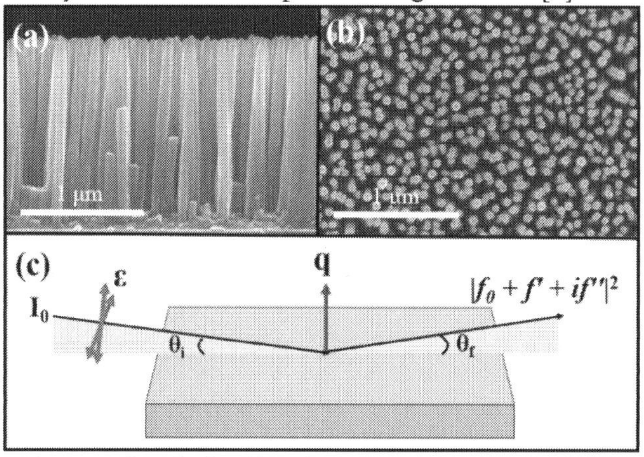


**Figure 1**. (a) and (b) SEM images of vertically-aligned ZnO nanorods and (c) schematic of polarization-dependent DANES geometry.

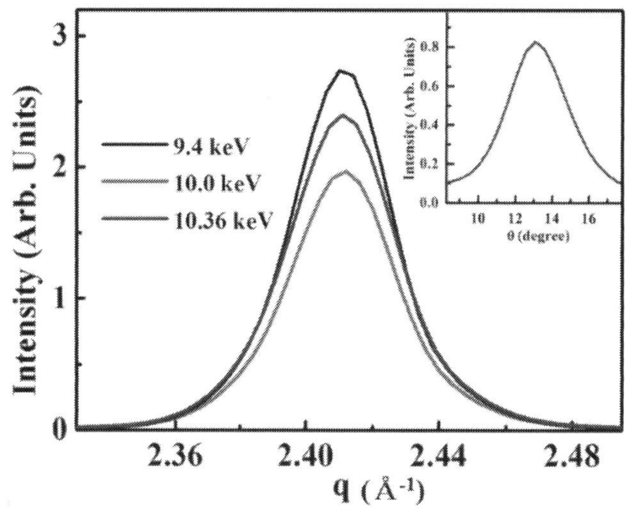


**Figure 2**. XRD from the ZnO(0002) planes as a function of momentum transfer, q, at the electric field of incident x-rays perpendicular to the nanorod length direction at the incident x-ray energies of 9.4, 10.0, and 10.36 keV, respectively. The inset shows the θ-rocking curve at the ZnO(0002) diffraction peak at the incident x-ray energy of 9.4 keV.

## 2. Experiment

For polarization-dependent DANES measurements, wurtzite-phase ZnO nanorods are synthesized on $\alpha$-$Al_2O_3$ substrates using a catalyst-free metal organic chemical vapor deposition(MOCVD). The details of ZnO nanorod growth by using MOCVD are described elsewhere [5]. The structural properties of ZnO nanorods are examined by using field-emission scanning electron microscopy (SEM, S-5500, Hitachi) and x-ray diffraction(XRD) measurements. SEM images demonstrate that ZnO nanorods are vertically-well oriented with a uniform distribution on the substrate surface, as shown in Figs. 1 (a) and (b). From the SEM images, the mean length and diameter of ZnO nanorods are determined to be 1.4 μm and 80 nm, respectively. High-resolution XRD measurements (data not shown) from the nanorods using a conventional tube x-rays with the Cu $K_{\alpha 1}$ radiation show the ZnO lattice constant $c$ of 5.2098 Å with the residual strain of 0.038 Å. The mosaicity of the nanorods are determined to be $3.70 \pm 0.06°$ by θ-rocking at the ZnO(0002) diffraction, as shown in the inset of Fig. 2. SEM images from the nanorods shown in Fig. 1 (a) suggest that the mosaicity is mainly contributed by the orientation of the nanorods. DANES measurements at the ZnO(0002) diffraction peak from the ZnO nanorods are performed with the electric field of incident x-rays parallel ($\hat{e} \parallel \hat{c}$) and perpendicular ($\hat{e} \perp \hat{c}$) to the ZnO nanorod length at the Zn K edge, as shown in Fig. 1 (c). The incident x-ray energy is selected by the Si(111) double crystal monochromator at the 20BM beamline of the Advanced Photon Source(APS). Near the ZnO(0002) peak, the two-dimensional measurements of θ-2θ and energy scans are performed. XANES data are *simultaneously* taken with the DANES measurements to directly compare the two techniques. For the *ab*-and *c*-polarized DANES measurements, the *c*-axis of ZnO nanorods is vertically and horizontally placed by rotating the sample to 90°, respectively, as shown in Fig. 1 (c) because the x-ray is horizontally polarized. The polarization-dependent XANES measurements are performed with a fluoresce mode.

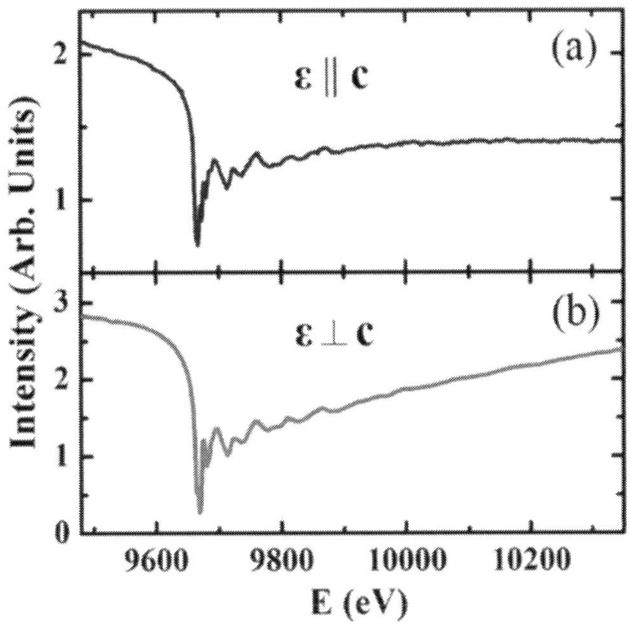


**Figure 3**. Polarization-dependent DAFS from ZnO nanorods at the Zn K edge at the incident x-ray electric fields (a) parallel and (b) perpendicular to the nanorod length.

## 3. Results and Discussion

When the angle between the incident and the out-going x-rays with a parallel geometry ($\hat{e} \parallel \hat{c}$) approaches to the Brewster angle of 90°, the intensity of diffraction decreases to zero by an angle dependent factor of $\cos2(\theta_i + \theta_f)$. The lattice constant *c* of ZnO is ~5.21 Å corresponding to the x-ray momentum transfer of q ~ 2.41 $Å^{-1}$ and to the incident angle of ~14.26° at the x-ray energy of 9659 eV. At the Zn K edge (9659 eV), the angle dependent factors are estimated to be 0.76 and 0.8 at 9400 eV and 10360 eV, respectively, for the parallel geometry. The angle dependent factor does not affect the resonant anomalous scattering intensity because it gradually increases from 0.76 to 0.8 when incident x-ray energy increases from 9400 eV to 10360 eV. The diffraction angle changes from ~13.3° to ~14.7° within the x-ray energy range and the change of the diffraction angle does not affect XANES spectra much. Figure 2 shows XRD from the ZnO(0002) planes at different x-ray energies of 9400, 10000, and 10360 eV for below, just above, and far above the Zn K edge, respectively. XRD obviously shows the contribution of resonant anomalous scattering, $f' + if''$, near the absorption edge. Polarization-dependent XRD measurements as a function of incident x-ray energy near the absorption edge can determine the orientation-dependent local structural properties around a probing atom due to the absorption factor of $f''$. DANES spectra are obtained by taking the total intensity of the XRD peaks as a function of the incident x-ray energy.

Figure 3 shows polarization-dependent DANES on the ZnO(0002) diffraction peak as a function of incident x-ray energy near the Zn K edge. The peak intensities dramatically decrease at the absorption edge and gradually increased above the edge for both polarizations, due to the anomalous scattering factor of $f'$. The polarization-dependent intensities agree well with the theoretical prediction of DAFS [1-4]. The small oscillations above the absorption edges are owing to the absorption terms of $f''$ which is dependent of the angle between the incident x-ray electric field and the direction of a neighboring atom [8]. The oscillations in DAFS above an absorption edge are originated from the local structures around probing atoms in crystalline sites, similar to EXAFS which detects average local structural properties of a whole specimen [1-4]. From the absorption contribution to diffraction intensity near the absorption edge, the chemical valence states and the local density of states can be determined because $f''$ corresponds to the electron transitions from a core shell to the empty outermost orbitals of a probing atom. Thus, DANES can describe the local density of states in the outermost orbitals.

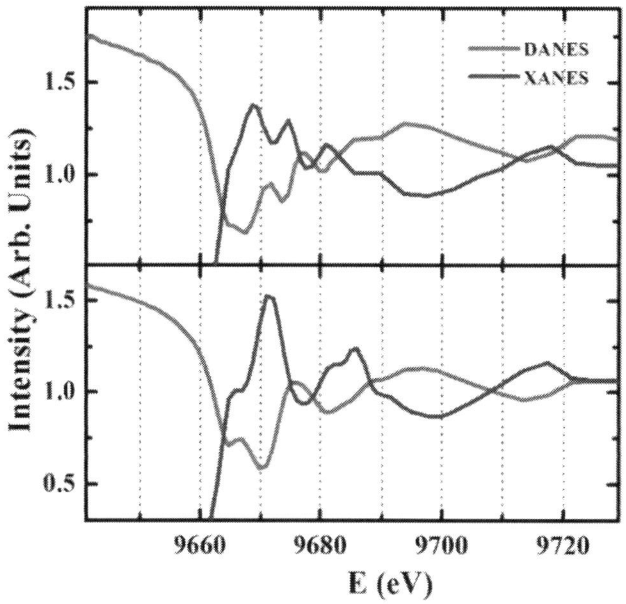


**Figure 4**. Polarization-dependent DANES from ZnO nanorods at the Zn K edge with polarization-dependent XANES at the incident x-ray electric field (a) parallel and (b) perpendicular to the nanorod length.

Figures 4 (a) and (b) show a direct comparison of polarization-dependent DANES and XANES. The DANES is considerably similar to the XANES. However, we observe a shift of DANES from XANES. The minimum position of DANES is shifted by ~2.0 eV from the maximum position of XANES for both polarization. The shift is ascribed to the Kramers-Kronig relation of $f'(E)$ and $f''(E)$. Polarization-dependent XANES data from vertically-aligned ZnO nanorods shows the typical XANES from a ZnO film [10] and vertically-aligned ZnO nanorods [7]. XANES from ZnO with *c*-polarization ($\hat{e} \parallel \hat{c}$) is quite different that with *ab*-polarization ($\hat{e} \perp \hat{c}$) owing to oxygen positions in the Zn-O tetrahedron; one oxygen in the *c*-axis and the other three oxygen atoms in nearly the *ab*-plane. The polarization-dependent XANES data corresponds to the average distribution of the local density of states of Zn atoms in ZnO nanorods, whereas polarization-dependent DANES corresponds the orientation-dependent local density of states of Zn atoms in ZnO crystals. Since the ZnO nanorods are single crystals, the DANES is nearly identical to the XANES that corresponds to the local density of states and the chemical valence state of a probing atom. This result demonstrates that polarization-dependent DANES can determine the orientation-dependent local density of states and the chemical valence state of probing atoms in selected-phased crystalline sites.

## 4. Conclusions

We demonstrate that polarization-dependent DANES measurements from vertically-aligned ZnO nanorods at the Zn K edge can be a probe to determine the orientation-dependent local density of states and the chemical properties of probing atoms in crystalline sites, comparing to simultaneously-measured XANES. Polarization-dependent DANES can detect orientation-dependent local structure and density of states around a selected element in selected-phased crystals, whereas polarization-dependent XANES can provide average information of a whole specimen. The direct comparison of polarization-dependent DANES and XANES measurements on single-phase ZnO nanorods demonstrates that the DANES will be considerably useful to determine the local structural and the local density of states of a selected element in complex-phase structures and mixed-phase nanostructures.

## Acknowledgements

This study was supported by the 2013 X-ray Division Visitor Program of APS, Argonne National Laboratory(ANL) and the Basic Science Research Program through the National Research Foundation of Korea government(NRF) grant funded by the Ministry of Science and ICT and Future Planning(MSIP) (No. 2015M2B2A4032132), the PAL-XFEL project, and KRISS funds for 2015 basic science research projects in Korea. The DANES and XAFS data were taken at the 20BM beamline of APS, ANL. Sector 20 facilities at APS, and research at these facilities, are supported by the US Department of Energy - Basic Energy Sciences, the Canadian Light Source and its funding partners, the University of Washington, and APS. Use of the APS, an Office of Science User Facility operated for the U.S. Department of Energy(DOE) Office of Science by ANL, was supported by the U.S. DOE under Contract No. DE-AC02-06CH11357. The SEM measurements were conducted at the Jeonju Center of the Korean Basic Science Institute in Korea.

## References

[1] Stragier H, Cross J O, Rehr J J, Sorensen L B, Bouldin C E, and Woicik J C 1992 *Phys. Rev. Lett.* **69**, 3064.
[2] L. B. Sorensen, Cross J O, Newville M, Ravel B, Rehr J J, Stragier H, Bouldin C E, and Woicik J C 1994 *Resonant Anomalous X-Ray Scattering: Theory and Applications* edited by G. Materlik, C. J. Sparks, and K. Fischer, Elsevier Science, North-Holland, 389-420.
[3] Ravel B, Bouldin C E, Renevier H, Hodeau J-L, and Berar J-F 1999 *Phys. Rev.* B **60**, 778.
[4] Woicik J C, Cross J O, Bouldin C E, Ravel B, Pellegrino J G, and Steiner B, Bompadre S G, Sorensen L B, Miyano K E, and Kirkland J P 1998 *Phys. Rev.* B **58**, R4215.
[5] Park Sun-Hong, Kim Seon-Hyo, and Han Sang-Wook 2007 *Nanotechnology* **18**, 055608.
[6] Park W I, Jun Y H, Jung S W, and Yi Gyu-Chul 2003 *Appl. Phys. Lett.* **82**, 964.
[7] Han S-W, Yoo H-J, An Sung Jin, Yoo Jinkyoung, and Yi Gyu-Chul 2005 *Appl. Phys. Lett.* **86**, 21917.
[8] Han Sang-Wook 2006 *Int. J. Nanotechnology* **3**, 396.
[9] Renevier H, Hodeau J L, Wolfers P, Andrieu S, Weigelt J, and Frahm R 1997 *Phys. Rev. Lett.* **78**, 2775.
[10] Chiou J W, Jan J C, Tsai H M, Bao C W, Pong W F, Tsai M-H, Hong L-H, Klauser R, Lee J F, Wu J J, and Liu S C 2004 *Appl. Phys. Lett.* **84**, 3462.

# *In situ* XANES & XRD Study of interphasial reaction between uncharged $Li_2FeSiO_4$ cathode and $LiPF_6$-based electrolyte

Z Arthur[1], H C Chiu[2], X Lu[2], N Chen[3], V Emond[1], G P Demopoulos[2], D T Jiang[1,*]

[1]Department of Physics, University of Guelph, Guelph, ON N1G 2W1, Canada
[2]Department of Materials Engineering, McGill University, Montreal, QC H3A 0C5, Canada
[3]Canadian Light Source, Saskatoon, SK S7N 2V3, Canada

*E-mail: djiang@uoguelph.ca

**Abstract.** *In situ* synchrotron radiation XANES and XRD have been carried out on $Li_2FeSiO_4$ cathode material in a lithium-ion-battery (LIB) cell. The evolution of the long range lattice structure and the local iron oxidation state has been observed at a charging rate of C/20 for the formation cycle for one Lithium extraction; additional *ex situ* measurements of the pristine cathode material were taken for comparison. The observed spontaneous interaction between the cathode and the fluorinated electrolyte and the impact of subsequent cycling are discussed.

## 1. Introduction

The ubiquity of Lithium Ion Batteries (LIBs) has brought with it a surge in rapid battery prototyping. Synchrotron radiation techniques have been widely applied in this field to gain insight into the electrochemical/solid state phase transition processes that accompany the formation cycle of electrode materials. These processes are believed to be a key aspect of designing high-performance lithium ion batteries [1-3]. A quick perusal of the current literature reveals that *in situ* studies using both XRD and XANES abound. However, due to various limitations, these two techniques are usually performed independently on electrochemical cells of the same prototype. In such cases, the hope is that enough controls have been put in place that the results from the two techniques are comparable. In an effort to alleviate concerns over sample reproducibility, a setup for the rapid measurement of $Li_2FeSiO_4$ (LFS) cathode material within a single active cell by both XANES and XRD has been constructed. The two measurements can be performed sequentially on the same sampling area, with less than 100-micron spatial uncertainty, in less than 30 minutes. In this study, the pristine material, uncharged constructed cell and cycled cell have each been measured using this combined technique (the latter two *in situ*). The results of our observations are reported here.

## 2. Experimental

XANES and XRD data were collected at 06ID-1 Hard X-ray Micro-Analysis (HXMA) beamline [4] at the Canadian Lightsource (CLS). The beamline optics were optimized to allow rapid movement between 7112 eV (nominal Fe K-edge) and 17 keV without sample re-alignment. The setup within the experimental hutch is shown in Figure 1. The reproducibility of the beam location was confirmed by scanning the center of X-ray slits with a 100 micron opening both vertically and horizontally. The setup (Figure 1) consists of a fast photon shutter, followed by a He beam-pass, at the end of the He pass JJ X-ray® slits (beam defining; ca. 0.5x 0.5 $mm^2$ aperture under He flow), and connected downstream to this aperture a (independent) He gas flowed ion chamber. In an effort to further improve the diffraction signal, a 2 mm W pinhole collimator was placed downstream of the ion chamber, directly adjacent to the LIB *in situ* cell. Downstream of the *in situ* LIB cell, a photo-diode detector served a dual purpose, both as a transmitted XANES detector and as a beam stop for the mar345® image plate detector positioned at 252.58(8) mm downstream of the sample (Figure 1). The XRD patterns were collected in the Debye-Scherrer geometry at 17 keV, and the Fe K-edge XANES were collected using the photo-diode and ion chamber in transmission mode. When detector linearity was not achievable (i.e. crystal glitches not normalizing out), a short (4") ion chamber was used in lieu of the photo-diode (not pictured in Figure 1).

Journal of Physics: Conference Series **712** (2016) 012124 doi:10.1088/1742-6596/712/1/012124

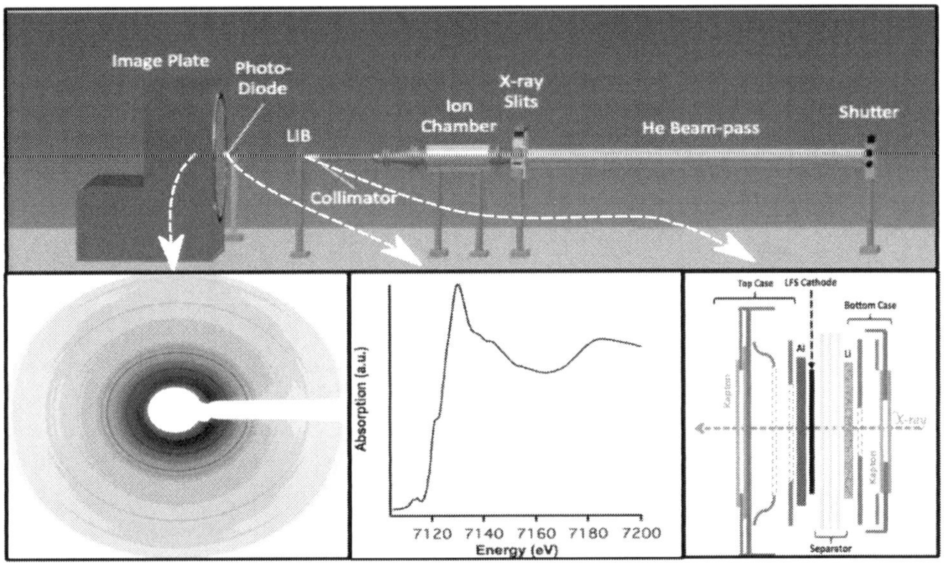


**Figure 1.** [top] *In situ* experimental setup with labeled components (beam goes right to left), allowing for rapid sequential measurement of XANES & XRD at the same sample location (within 100 micron); [bottom left] Debye-Scherrer pattern of LFS pristine material taken with mar345® image plate; [bottom middle] transmission Fe K-edge XANES spectrum of pristine LFS material; [bottom right] Stack view of the modified CR2032 cell with: LFS cathode rolled onto Al foil; 25 μm Separators (Celgard® 2325); Li anode; and two Kapton® windows for transmitted X-rays.

The *in situ* battery sample cells were modified CR2032 battery cells, having 10 mm through holes punched on the beam exiting side and 4 mm holes on the beam incident side of the cell casing. Both holes were covered with Kapton® tape from inside and outside the cell. The assembled cells were sealed inside a glove box under an argon atmosphere. The LFS cathode consisted of $Li_2FeSiO_4$ nanoparticles (80% LFS, 10% acetylene black and 10% polyvinylidene fluoride) and were prepared via an organic-assisted hydrothermal-annealing (180°C and 400°C, respectively) synthesis method [5]. The electrochemical *in situ* cell consisted of $Li_2FeSiO_4$ nanoparticle aggregates rolled onto Al foil; 25 μm Celgard® 2325 separators; a lithium anode; and 1M of $LiPF_6$ in ethylene carbonate and dimethyl carbonate (EC/DMC) electrolyte (Figure 1: bottom right). The *ex situ* battery cathode was prepared by removing the cathode from an assembled cell in a glove box under high purity $N_2$ atmosphere, after being carefully washed using EC/DMC solvent they were hermetically sealed until the XANES and XRD measurements.

XANES data reduction was performed using Athena software [6]. The XANES data were normalized to the edge jump size as determined by using the post-edge data from approximately 70 eV up to 250 eV. The photon energy calibration was carried out by defining the Fe metal foil inflection point at 7112 eV. Calibration of the mar345® detector to sample distance was completed using a NIST $LaB_6$ sample mounted inside a CR2032 cell casing and FIT2D's calibration routine [7]. XRD two-dimensional patterns were processed using FIT2D, and integrated into 2θ diffraction patterns [8]. The Debye-Scherrer ring pattern for the pristine LFS material can be seen in Figure 1 (bottom left). The pristine XANES spectrum for the same material at the same location is provided in Figure 1 (bottom middle).

## 3. Results and Discussions

*In situ* XANES and XRD measurements of the cathode were performed on the initially uncharged battery 96 hours after the coin cell was assembled. From the time of assembly LFS remained in contact with 1M of $LiPF_6$ in EC/DMC electrolytic solution. Measurements were also carried out after cycling the cell at a rate of C/20 for a single Li extraction (i.e. the formation cycle). These cells will be referred to as the "uncharged" cell and the "cycled" cell, respectively. The *ex situ* measurement is directly comparable to the "uncharged" *in situ* measurement in terms of state-of-charge (SOC) and interaction with the electrolyte for ca. 90 hours before disassembly.

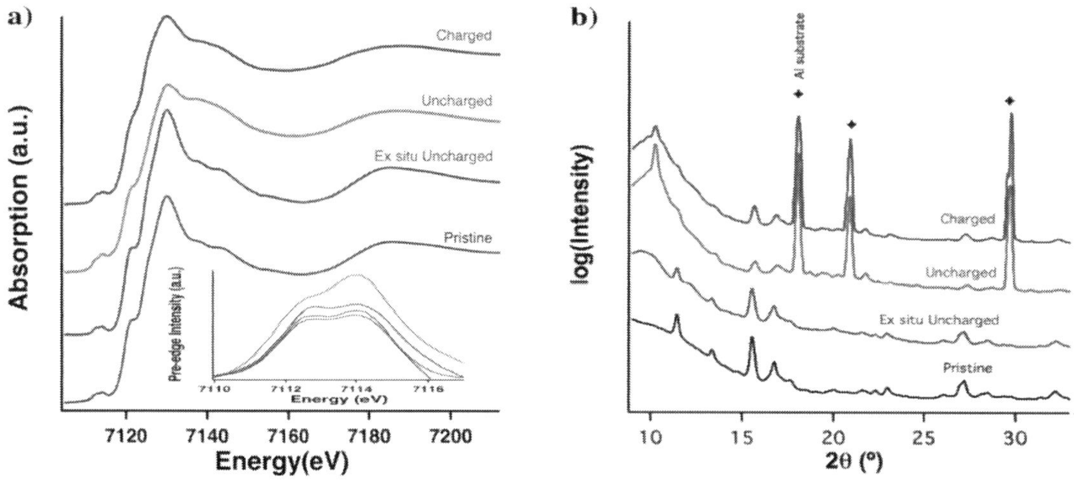


**Figure 2.** (a) Normalized Fe K-edge XANES from (black) pristine, (purple) *ex situ* uncharged, (red) uncharged, and (blue) cycled; (inset) background removed pre-edge region. (b) XRD patterns for pristine, cycled, and both uncharged cathode states. The three aluminum substrate peaks are indicated by the stars on top.

Figure 2a depicts the three LFS XANES spectra, capturing the pristine material state (black), the uncharged state (red) and the cycled state (blue). From comparison between the uncharged and pristine states, it is apparent that the 96-hour interaction with the fluorinated electrolyte had already caused changes to the iron site local atomic structure; in the region around 7140 eV of the *in situ* uncharged spectrum (red; Figure 2a), there is broadening of features and an intensity increase as compared to the pristine spectrum (black). This is consistent with an increase in short-range disorder following an interphasial reaction between the cathode and the electrolyte. Evidence of this spontaneous interphasial reaction is almost completely lost from the *ex situ* uncharged spectrum (purple), in fact upon washing it appears that the *ex situ* cathode is indistinguishable from the pristine. More distinct spectral changes related to this interphasial reaction can be observed in the comparison of the background removed pre-edge spectra (inset; Figure 2a). Here, the uncharged spectrum (red trace) is of larger intensity than that of either the pristine or cycled traces. This is an indicator of larger Jahn-Teller distortions of the Fe tetrahedra [9]. The Fe K pre-edge features in the tetrahedral ferrous and ferric iron complexes have been identified [10]. They include *primarily* dipole transitions to: $^4T_2$ and $^4T_1$ (both low & high energy states) in the ferrous case; and to $^5T_2$ in the ferric case [10]. This means for a ferrous iron tetrahedral complex the pre-edge possesses an apparent "two-peak" pattern; while for a ferric iron tetrahedral complex the pre-edge has a single peak [10]. However, the energy location of the ferrous pre-edge $^4T_1$ (high energy state) feature coincides with that of the ferric $^5T_2$ feature. As a result, for a system with both ferrous and ferric contents the overall pre-edge appears to still have an apparent "two-peak" pattern as shown in Figure 2a(insert). Therefore when comparing the *in situ* uncharged (red) to the pristine (black) (Figure 2a insert), the observed change in the apparent "two-peak" intensity ratio indicates a change in the ratio of underlying components, i.e. the ferrous and ferric content of the sample changes during the 96 hours interaction between the cathode and the electrolyte. This phenomenon is again almost invisible in the relaxed *ex situ* uncharged state (purple). Also shown in Figure 2a(inset): after the formation cycle (blue), the aforementioned ratio between the apparent "two peaks" is largely restored to that similar to the pristine case (black).

The *in situ* XRD patterns depicted in Figure 2b reveal the long-range structural differences among the pristine, "uncharged" and "cycled" states. The as-prepared pristine material was found to be a mixture of approximately 3:2 of monoclinic $P2_1/n$ ($\gamma_s$) and the "low temperature" orthorhombic $Pmn2_1$ ($\beta_{II}$) phases, determined respectively through Rietveld refinement [12, 13]. Upon reaction with the electrolyte in the uncharged cell, a new phase was formed. This can be observed in the birth of a new peak around 10°, and the increase in peak intensity around 22°. These indicators of phase change are not present in the *ex situ* uncharged XRD patterns (Figure 2b; purple). The XRD pattern for the uncharged cell also exhibits broadening in the features associated with $P2_1/n$ ($\gamma_s$) and $Pmn2_1$ ($\beta_{II}$) phases, as well as decreased intensity. In terms of peak intensity and broadening, once again the cycled state has been partially restored to the pristine state. However, the new spontaneous phase formation that occurred in the uncharged state persisted in the cycled cathode. Additionally, in the multi-phase mixture the relative ratio of $P2_1/n$ ($\gamma_s$) to $Pmn2_1$ ($\beta_{II}$) may not be assumed constant in this case.

## 3. Conclusions

A rapid XANES and XRD mode-switching setup has been implemented at the HXMA Beamline at the CLS, employing *in situ* spectroscopy and diffraction studies of same-sample areas of lithium-ion-batteries. Use of both *in situ* XANES & XRD in this way has allowed a spontaneous reaction to be observed between the $Li_2FeSiO_4$ cathode and $LiPF_6$-based electrolyte in the uncharged lithium-ion-battery state which was not otherwise observable in the *ex situ* measurements. This reaction results in partial oxidation of the ferrous content in the cathode material; however, for the cycled battery state (formation cycle), the ferrous content is then partially restored. This local structural phenomenon is accompanied by the emergence of a new crystalline phase as indicated by the *in situ* XRD.

## 4. Acknowledgements

The authors acknowledge assistance provided during data acquisition by Dr. W. Chen and Dr. C.-Y. Kim. Further acknowledgement and gratitude is extended to Dr. L. Zhang (University of Nebraska-Lincoln) for providing the reference compound data. DTJ acknowledges Canadian NSERC support for the work done at the University of Guelph. SR research described in this paper was performed at the CLS, which is supported by CFI, NSERC of Canada, University of Saskatchewan, Government of Saskatchewan, Western Economic Diversification Canada, NRC Canada, and CIHR.

**References:**

[1] Goodenough J B and Park K S 2013 The Li-Ion Rechargeable Battery: A Perspective *J. Am. Chem. Soc.* **135** 1167-1176

[2] Xu K 2014 Electrolytes and Interphases in Li-Ion Batteries and Beyond *Chem. Rev.* **114** 11503-11618

[3] Tarascon J M and Armand M 2001 Issues and challenges facing rechargeable lithium batteries *Nature* **414** 359-367

[4] Jiang D T, Chen N and Sheng W 2007 CLS 06ID-1: A Wiggler-based Hard X-ray Spectroscopy Beamline *AIP Proc.* **879** 800-803.

[5] Lu X, Wei H, Chiu H -C, Gauvin R, Hovington P, Guerfi A, Zaghib K and Demopoulos G P 2015 Rate-dependent phase transitions in Li2FeSiO4 cathode nanocrystals *Sci. Rep.* **5** 8599

[6] Ravel B and Newville M 2005 ATHENA, ARTEMIS, HEPHAESTUS *J. Synchrotron Rad.* **12** 537-541

[7] Hammersley A P, Svensson S O and Thompson A 1994 Calibration and correction of spatial distortions in 2D detector systems *Nucl. Instr. Meth. Phys. Res. Sec. A: Accelerators, Spectrometers, Detectors and Associated Equipment* **346** 312-321

[8] Hammersley A P, Svensson S O, Hanfland M, Fitch A N and Hausermann D 1996 Two-dimensional detector software: From real detector to idealised image or two-theta scan *High Pressure Res.* **14** 235-248

[9] Frank de G, György V and Pieter G 2009 The 1s x-ray absorption pre-edge structures in transition metal oxides *J. Phys.: Condens. Matter* **21** 104207

[10] Westre T E, Kennepohl P, DeWitt J G, Hedman B, Hodgson K O and Solomon E I 1997 A Multiplet Analysis of Fe K-Edge 1s → 3d Pre-Edge Features of Iron Complexes *J. Am. Chem. Soc.* **119** 6297-6314

[11] Zhang L 2015 XANES spectra of reference compounds tetrahedral ferrous and ferric sofides (unpublished)

[12] Eames C, Armstrong A, Bruce P and Islam M 2012 Insights into changes in voltage and structure of Li2FeSiO4 polymorphs for lithium-ion batteries *Chem. Mater.* **24** 2155-2161

[13] Rodríguez-Carvajal J 1993 Recent advances in magnetic structure determination by neutron powder diffraction *Phys. B: Condens. Matter* **192** 55-69

# A XAFS study of the local environment and reactivity of Pt-sites in functionalized UiO-67 MOFs

**E Borfecchia**[*,1], **S Øien**[2], **S Svelle**[2], **L Mino**[1], **L Braglia,**[1,4] **G Agostini**[1,3], **E Gallo**[1,3], **K A Lomachenko**[1,4], **S Bordiga**[1], **A A Guda**[4], **M A Soldatov**[4], **A V Soldatov**[4], **U Olsbye**[2], **K P Lillerud**[2], **C Lamberti**[4,5]

[1]Dept. of Chemistry, NIS and INSTM Reference Centers, University of Turin, Turin (Italy);
[2]inGAP Centre for Research Based Innovation, Dept. of Chemistry, University of Oslo, Oslo (Norway);
[3]European Synchrotron Radiation Facility (ESRF), Grenoble (France);
[4]Southern Federal University, Rostov-on-Don (Russia);
[5]Dept. of Chemistry, CrisDi Centre for Crystallography, University of Turin, Turin (Italy)

E-mail: elisa.borfecchia@unito.it

**Abstract.** We synthesized UiO-67 Metal Organic Frameworks (MOFs) functionalized with bpydcPt(II)$Cl_2$ and bpydcPt(IV)$Cl_4$ complexes (bpydc = bipyridine-dicarboxylate), as attractive candidates for the heterogenization of homogeneous catalytic reactions. Pt $L_3$-edge XAFS experiments allowed us to thoroughly characterize these materials, in the local environment of the Pt centers. XAFS studies evidenced the rich reactivity of UiO-67-Pt(II) MOFs, including reduction to bpydcPt(0) under $H_2$ flow in the 600−700 K range, room-temperature oxidation to bpydcPt(IV)$Br_4$ through oxidative addition of liquid $Br_2$ and ligand exchange between 2 $Cl^-$ and even bulky ligands such as toluene-3,4-dithiol. Preliminary XANES simulations with ADF code provide additional information on the oxidation state of Pt sites.

## 1. Introduction

Metal-organic frameworks (MOFs) are crystalline, porous solids consisting of metal ions or clusters, coordinated with organic linkers [1]. The recently discovered UiO-66 and UiO-67 classes of iso-structural MOFs are obtained connecting $Zr_6O_4(OH)_4$ inorganic cornerstones with 1,4-benzene-dicarboxylate or 4,4′ biphenyl-dicarboxylate linkers, for the UiO-66 and UiO-67 MOFs, respectively [2]. Due to their outstanding stability at high temperatures, high pressures and in presence of different solvents, these materials are among the few MOFs already commercialized for applications in the fields of catalysis, $H_2$ storage, and gas purification [1,2]. We are currently exploring the possibility to enhance the capabilities of the UiO-67 MOF by grafting to the framework an additional catalytically-active Pt center, by chelating bipyridine-dicarboxylate (bpydc) linkers [3]. The resulting metal-functionalized MOFs are attractive candidates for industrial applications aiming to heterogenization of homogeneous catalytic reactions. Due to the local character of the functionalization process, XAFS has played a key role in clarifying the local structural and electronic properties of the grafted metal center [4, 5].

## 2. Experimental

Pt-functionalized UiO-67 materials, hereafter UiO-67-Pt(II) and UiO-67-Pt(IV), were synthesized by the standard solvothermal method, by reacting $ZrCl_4$ with a mixture of $H_2bpdc$ and ($H_2bpydc$)/$PtCl_x(H_2bpydc)$ (x = 2 or 4; the ratio between the linkers being 9:1, and equal molar quantity of $ZrCl_4$ and linker) in a solution of DMF [3].

Pt $L_3$-edge (11564 eV) XAFS data were collected at the I811 beamline @ Max Lab II (Lund, Sweden). The white beam produced by a liquid He-cooled superconducting wiggler was monochromatized by an horizontally sagitally focused double-crystal Si(111) monochromator, detuned to 20% to minimize the third harmonic. Spectra were collected in transmission mode using 30 cm ionization chambers for $I_0$ and $I_1$ and by a photodiode for $I_2$. MOF samples were measured inside a home-made cell allowing sample activation and gas dosage under *in situ* or *operando* conditions [6]. The EXAFS spectra were extracted and analyzed with Athena and Arthemis codes [7]. Geometry relaxation, electronic structure and Pt $L_3$ XANES calculations were performed by means of molecular orbital density functional theory implemented in ADF-2014 software [8]. B3LYP exchange correlation functional and large QZ4P basis set were used in all simulations.

## 3. Results and discussion

In this contribution we present results on UiO-67 functionalized with bpydcPt(II)$Cl_2$ and bpydcPt(IV)$Cl_4$ coordination complexes acting as linkers in the MOF framework and prepared following three different synthesis methods: (i) one-pot synthesis (OPS) where $ZrCl_4$ and $PtCl_x$ (x = 2 or 4) precursors salts react with biphenyl and bipyridine linkers; (ii) premade linker synthesis (PMLS), where a previously prepared $PtCl_x(H_2bpydc)$ linker reacts with biphenyl linkers and $ZrCl_4$; and (iii) post-synthesis functionalization (PSF), where a premade UiO-67-bpy MOF is suspended in a solution of precursor $PtCl_x$ salt [3].

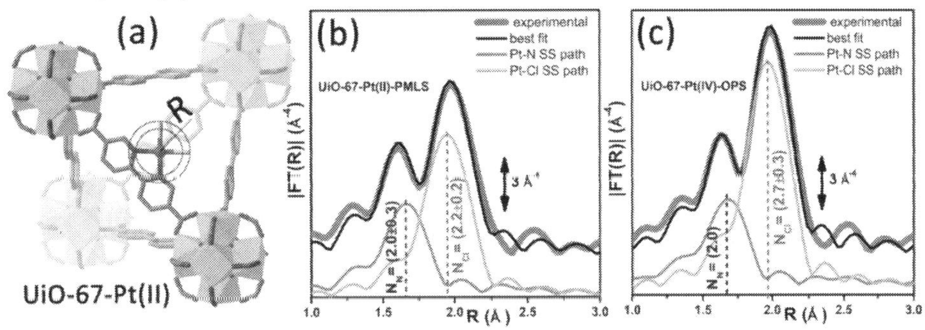


**Figure 1.** (a) Three-dimensional representation of UiO-67-Pt(II) MOF, evidencing the structure of the isolated $H_2bpydcPtCl_2$ center inserted in the MOF structure, with 2 N and 2 Cl in the first coordination shell of Pt(II) (atoms color code: Pt: dark cyan, N: blue, Cl: light green, C: gray, O: red, Zr cornerstones: light cyan). (b,c) Experimental FT of the $k^3$-weighted EXAFS spectra collected for a typical (b) UiO-67-Pt(II) sample (PMLS synthesis) and (c) UiO-67-Pt(IV) sample (OPS synthesis) and their corresponding best fits; moduli of Pt−N and Pt−Cl single scattering paths are also reported, vertically shifted for the sake of clarity, together with the respective coordination numbers refined in the EXAFS fit.

With the only exception of the UiO-67-Pt(II)-OPS, that crystallizes together with a fraction of about 40% of amorphous (nonporous) phase, the remaining syntheses resulted in high crystalline materials with porosity close to the target ideal structure. XRPD and Pt $L_3$-edge XAFS studies proved that the three synthesis methods are equivalent and for Pt(II) samples fully comply with the target structure (figure 1a) on both long-range (ordered MOF framework) and short-range (local environment of Pt sites probed by XAFS) scales. The last point has also been supported by Pt $L_3$ valence-to-core RIXS maps [3]. XAFS also revealed that the incorporation of Pt(IV) sites into the UiO-67 MOF is more critical. Although the synthesis of stable Pt(IV) linkers was successful (see figure 3a for XANES spectra of $H_2bpydcPtCl_2$ and $H_2bpydcPtCl_4$ linkers), during the preparation of the UiO-67-Pt(IV) MOFs a significant fraction of bpydcPt(IV)$Cl_4$ sites is reduced to bpydcPt(II)$Cl_2$, resulting in a final average Pt-coordination number $N_{Cl} = (2.7 \pm 0.3)$, lower than the target value $N_{Cl} = 4$ (figure 1c).

UiO-67-Pt(II) systems were further tested toward accessibility and reactivity to molecules of small- (e.g. $H_2$), medium- (e.g. $Br_2$), and large-size (e.g. thiol), resulting in the reactive paths represented in figure 2. $H_2$-temperature programmed reduction (TPR) treatments have been followed under *operando* conditions by Pt $L_3$-edge EXAFS, showing that Cl ligands can be selectively removed as HCl molecules in the 600–700 K temperature range, resulting in bpydcPt(0) complexes linked to the MOF framework (reduction path, figure 2a). This result was derived from a sophisticated parametric analysis of the $H_2$-TPR datasets (figure 2b), where all the EXAFS spectra collected in the 300–750 K range have been simultaneously refined, adopting the Einstein model for the Pt–N and Pt–Cl Debye-Waller factors [3]. These findings were also supported by FTIR spectroscopy, which evidences the high coordinative unsaturation of the reduced Pt centers, able to form a variety of Pt monocarbonyl complexes and also bpydcPt(0)$(CO)_2$ dicarbonyl complexes upon CO adsorption. The formation of EXAFS-silent Pt(0)-hydrogen species in these conditions is currently under investigation, as well as the effect of higher $H_2$ concentration during $H_2$-TPR, possibly inducing aggregation of the isolated Pt(0) sites into highly reactive metal nanoclusters.

The large pore size of UiO-67 allows for ligand exchange between 2 $Cl^-$ and even bulky ligands such as toluene-3,4-dithiol ($H_2$tdt, ligand exchange path, figure 2c). Framework bpydcPt(II)$Cl_2$ complexes can also be oxidized at room temperature to bpydcPt(IV)$Br_4$ through oxidative addition of liquid $Br_2$ (oxidation path, figure 2d). With this respect, EXAFS monitored the ligand exchange in the first coordination shell of Pt (figure 2), while XANES spectroscopy was used to monitor the changes in the Pt oxidation state along the observed reactions (figure 3).

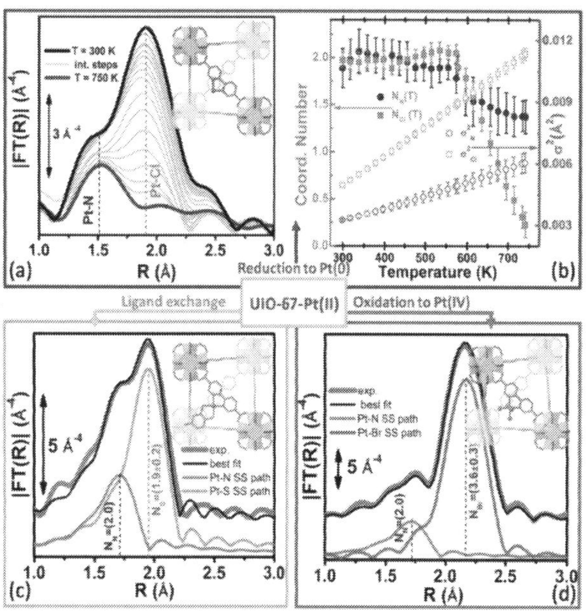


**Figure 2.** Reactivity of Pt(II) species in functionalized UiO-67-Pt MOFs as evidenced by EXAFS in *operando* and static conditions. (a,b) Reduction path: (a) FT of the $k^3$-weighted Pt $L_3$-edge EXAFS spectra collected during *operando* $H_2$-TPR experiments on UiO-67-Pt(II)_PMLS MOFs, simultaneously refined adopting the Einstein model for the Pt–N and Pt–Cl Debye-Waller factors, as shown in part (b). (c,d) Experimental FT of the $k^3$-weighted Pt $L_3$-edge EXAFS spectra and corresponding best fit curves for UiO-67-Pt(II) after interaction with (c) $H_2$tdt (ligand exchange path) and (d) $Br_2$ (oxidation path). Single scattering contributions to first shell EXAFS signal in the two cases (Pt–N and Pt–S in part (c) and Pt–N and Pt–Br in part (d)) are also reported, vertically shifted for the sake of clarity, with indicated the correspondent coordination numbers refined in the fit.

Simulated XANES spectra, calculated with ADF code (figure 3b,d), were able to reproduce correctly the variation of the white line intensity for UiO-67-Pt(II) and -Pt(IV) linkers (figure 3a), and UiO-67-Pt(II) MOF before and after the interaction with $H_2$tdt and $Br_2$ (figure 3c). Indeed, white line intensity in $L_3$-edge XANES is directly proportional to the density of the unoccupied 5d states and consequently it provides information on the formal charge (or valence) of Pt species [3]. In particular, XANES simulations confirmed that after ligand exchange with $H_2$tdt the pristine Pt(II) oxidation state is conserved, whereas interaction with $Br_2$ results in oxidized $H_2bpydcPt(IV)Br_4$ complexes. In a successive work, we plan to use the ADF calculated unoccupied molecular orbitals for improving the simulation of the XANES spectra in dipole one-electron approximation along with finite difference method with an upgraded version of the FDMNES code [9].

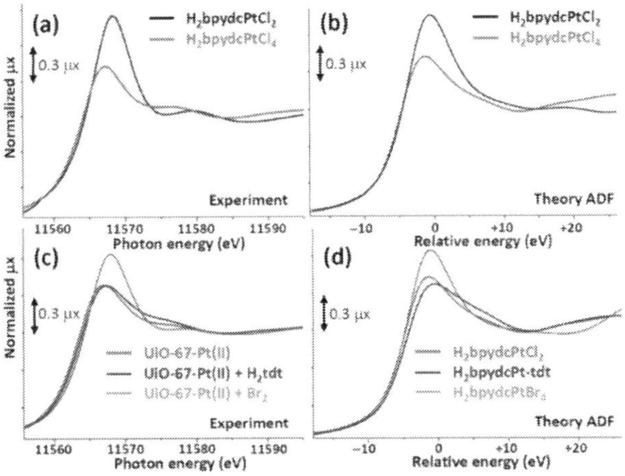


**Figure 3.** (a) Experimental XANES spectra of the $H_2bpydcPtCl_2$ and $H_2bpydcPtCl_4$ linkers. (b) Corresponding theoretical spectra computed with ADF code. (c) Experimental XANES spectra of UiO-67-Pt(II) MOF before (red curve) and after interaction from the liquid phase with $H_2tdt$ and $Br_2$. (d) Theoretical spectra computed with ADF code of the $H_2bpydcPtCl_2$, $H_2bpydcPt(tdt)$ and $H_2bpydcPtBr_4$ molecular fragments.

In conclusion, XAFS allowed us to thoroughly characterize the local coordination environment and oxidation states in a series of Pt(II)- and Pt(IV)-functionalized UiO-67 MOFs, synthesized by different methods. By combing EXAFS in static and *operando* conditions with DFT-assisted XANES simulations we also explored the rich reactivity of Pt(II) sites, undergoing reduction, oxidation and ligand exchange reactions. These results pave the way to further studies aiming to assess the performance of these systems for the heterogenization of specific homogeneous catalytic reactions.

**Acknowledgments**
C L, K A L, A A G and A V S acknowledge the megagrant of the Russian Federation Government to support scientific research at the Southern Federal University, no. 14.Y26.31.0001.

**References**

[1] Ferey G 2008 *Chem. Soc. Rev.* **37** 191; Long J.R and Yaghi O M 2009 *Chem. Soc. Rev.* **38** 1213; Bordiga S, Bonino F, Lillerud K-P and Lamberti C 2010 *Chem. Soc. Rev.* **39** 4885; Butova V V, Soldatov M A, Guda A A, Lomachenko K A and Lamberti C, 2016 *Russ. Chem. Rev.*, **85** doi:10.1070/RCR4554.

[2] Cavka J H, Jakobsen S, Olsbye U, Guillou N, Lamberti C, Bordiga S, Lillerud K-P 2008 *J. Am. Chem. Soc.* **130** 13850; Valenzano L, Civalleri B, Chavan S, Bordiga S, Nilsen M H, Jakobsen S, Lillerud K-P, Lamberti C. 2011 *Chem. Mater.* **23** 1700; Chavan S, Vitillo J G, Gianolio D, Zavorotynska O, Civalleri B, Jakobsen S, Nilsen M H, Valenzano L, Lamberti C, Lillerud K-P, Bordiga S. 2012 *Phys. Chem. Chem. Phys.* **14** 1614.

[3] Øien S, Agostini G, Svelle S, Borfecchia E, Lomachenko K A, Mino L, Gallo E, Bordiga S, Olsbye U, Lillerud K-P and Lamberti C 2015 *Chem. Mater.* **27** 1042.

[4] Bordiga S, Groppo E, Agostini G, van Bokhoven J A and Lamberti C. 2013 *Chem. Rev.* **113** 1736; Mino L, Agostini G, Borfecchia E, Gianolio D, Piovano A, Gallo E and Lamberti C. 2013 *J. Phys. D: Appl. Phys.* **46** 423001; Garino C, Borfecchia E, Gobetto R, van Bokhoven J A and Lamberti C. 2014 *Coord. Chem. Rev.* **277–278** 130.

[5] van Bokhoven J A, and Lamberti C, 2016 *X-Ray Absorption and X-ray Emission Spectroscopy: Theory and Applications*, Eds. Vol. 1,2 (Chichester: John Wiley & Sons).

[6] Lamberti C, Prestipino C, Bordiga S, Berlier G, et al. 2003 *Nucl. Instr. Meth. B,* **200** 196.

[7] Ravel B and Newville M 2005 *J. Synchrotron Radiat.* **12** 537.

[8] Fonseca Guerra C, Snijders J G, te Velde G, Baerends E J, 1998 *Theor. Chem. Acc.* **99**, 391; te Velde, G, Bickelhaupt F M, Baerends E J, Guerra C F, et al. 2001 *Comput. Chem.* **22** 931.

[9] Guda S A, Guda A A, Soldatov M A, Lomachenko K A, Bugaev A L, Lamberti C, Gawelda W, Bressler C, Smolentsev G, Soldatov A V, Joly Y, 2015 *J. Chem. Theory Comput.* **11** 4512.; *J. Phys.: Conf. Ser.*, these proceedings.

# Simulation of the EXAFS and Raman spectra of $In_xGa_{1-x}N$ utilizing the equation of motion routine of FEFF8.

**M Katsikini[1*], F Pinakidou[1], E C Paloura[1], J Arvanitidis[1], S Ves[1], U Reinholz[2], E Papadomanolaki[3] and E Iliopoulos[3]**

[1] Aristotle University of Thessaloniki, School of Physics, 54124 Thessaloniki, Greece
[2] BAM Federal Institute for Materials Research and Testing, 12200 Berlin, Germany
[3] University of Crete, Department of Physics and IESL-FORTH, 70013 Heraklion-Crete, Greece

* E-mail: katsiki@auth.gr

**Abstract**. A combined analysis of EXAFS and Raman spectra is applied for the study of $In_xGa_{1-x}N$ alloys with 0.3<x<0.5. Alloying causes relaxation of the selection rules resulting in Raman spectra that resemble the vibrational density of states. On the other hand, theoretical simulation of the Raman spectra using the Equation of Motion routine of FEFF8 provides the vibrational component of the Debye-Waller factor (DWF). The static disorder component of the DWFs was obtained by fitting the Ga and In *K*-edge EXAFS spectra. The analysis revealed that the nearest neighbor distances of the 1st and 2nd shell deviate from the values predicted by the law of Vegard and the virtual crystal approximation. The static disorder in the first nearest neighboring shell (In-N and Ga-N) is null whereas in the cation-cation neighboring shells the static component is generally smaller than the vibrational.

## 1. Introduction

The tunability of the direct band gap of $In_xGa_{1-x}N$ alloys from near-IR to UV, by varying x, renders them important materials for photonic and optoelectronic applications as well as for efficient solar energy conversion devices [1,2]. InN, GaN and their alloys crystallize in the wurtzite structure where a cation (Ga, In) and a N *hcp* sublattice interpenetrate so as both the N and the cations are tetrahedrally coordinated. Although the lattice constants determined by X-ray diffraction (XRD) obey the law of Vegard, it has been theoretically predicted [3,4] and experimentally verified [5-8] that the In-N and Ga-N bonds retain their length close to the corresponding values in the binary compounds. As a result, the virtual crystal approximation (VCA), which assumes that all atoms occupy the average lattice positions defined by the lattice constants as determined by XRD [9], breaks down since the In and Ga atoms are displaced from their ideal positions and three different second nearest neighbor (*nn*) distances are detected in the cation sublattice (i.e. In-In, Ga-Ga, In-Ga) [10]. These local distortions affect the periodic lattice potential and consequently the carrier mobility [11]. Local fluctuations in the periodic potential reduce strongly the mean phonon free path, causing relaxation of the selection rules in the Raman spectra which become proportional to the vibrational density of states (VDOS) [12,13]. Extended X-ray absorption fine structure (EXAFS) spectroscopy is well suited for the study of local variations in the bonding geometry of the absorbing atom (for example In and Ga). The ratio of In/Ga atoms that populate the second *nn* shell is also obtained along with information on the static and thermal disorder described by the Debye-Waller factor (DWF). Assuming harmonic thermal vibrations and symmetric atom pair distribution functions [14], the latter can be written as a sum of a vibrational and a static component i.e. $\sigma^2 = \sigma^2_{vib} + \sigma^2_{stat}$. In this work, we benefit from the equation of motion routine [15,16] of the FEFF8 program [17] in order to simulate the Raman spectra of $In_xGa_{1-x}N$ alloys with $0.3<x<0.5$ and to obtain the $\sigma^2_{vib}$. The $\sigma^2_{stat}$ component is determined from the analysis of the Ga and In *K*-edge EXAFS spectra of the same samples.

## 2. Growth conditions and experimental details

The $In_xGa_{1-x}N$ layers with x values 0.32, 0.37, 0.42 and 0.48 were grown by plasma-assisted molecular beam epitaxy on $Al_2O_3$ (0001) substrates using AlN buffer layers. The In/Ga flux ratio was kept the same for all the samples whereas the growth temperature was varied, from 520 to 400 °C, in order to control the In content [18]. The *a* and *c* lattice constants were determined from XRD measurements. The Ga and In *K*-edge EXAFS spectra were recorded in the fluorescence yield mode, at room temperature at the KMC-II and BAM*line* beamlines, respectively, of the BESSY II storage ring at the Helmholtz Zentrum, Berlin. The Raman spectra were recorded in the backscattering geometry using a micro-Raman system and the 457.9 nm line of an Ar+ laser for the excitation.

## 3. Results and discussion

A representative Raman spectrum of the sample $In_{0.48}Ga_{0.52}N$ is shown in figure 1. Strong enhancement in the intensity around the longitudinal optical (LO) modes is observed due to resonance effects as the energy gap of the studied samples approaches the energy of the laser line. The lattice constants were used for the construction of a cluster of approximately 200 atoms taking into account the In/Ga ratio. The force constants, k, listed in table 1 for the cation-cation, cation-N and N-N atom pairs were used for the simulation of the Raman spectra as shown in figure 1. The contribution in the VDOS of various scattering paths is also shown in figure 1. As expected, cation-N and cation-cation shells contribute in the high- and low-frequency part of the spectrum, due to the small and large reduced mass of the atom pair, respectively. The $\sigma^2_{vib}$ values listed in table 1 were determined after fitting the projected VDOS for each path [15,16].

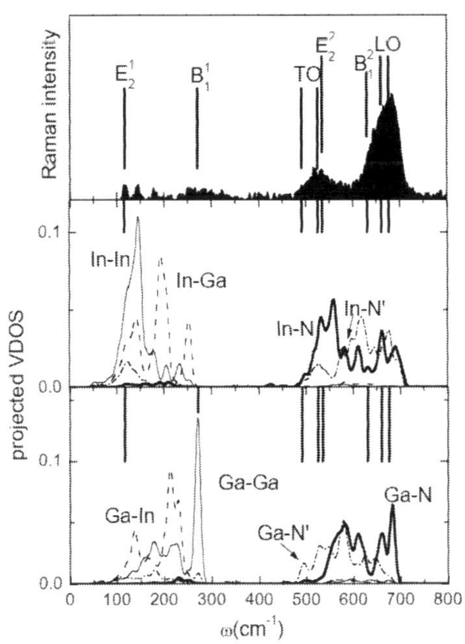


**Figure 1.** *Top:* Raman spectrum of the $In_{0.48}Ga_{52}N$ sample; *Middle:* Calculated VDOS, using the equation of motion routine, projected in the first (In-N), second (In-In and In-Ga) and third (In-N´) *nn* shells of In; *Bottom:* Calculated VDOS projected in the first (Ga-N), second (Ga-Ga and Ga-In) and third (Ga-N´) *nn* shells of Ga; Vertical lines denote the Raman frequencies of the vibrational transverse optical (TO), longitudinal optical (LO) and $E_1$, $E_2$ and $B_1$ modes of $In_{0.5}Ga_{0.5}N$.

**Table 1.** Atom-pair force constants, k, used for the calculation of the projected VDOS and $\sigma^2_{vib}$, obtained using the equation of motion method for the paths G1-G6 (Ga-N, Ga-Ga, Ga-In, Ga-N´, Ga-Ga-N and Ga-In-N, respectively) and I1-I6 (In-N, In-In, In-Ga, In-N´, In-In-N and In-Ga-N, respectively). x=1 and 0 correspond to InN and GaN binary compounds.

| | k (N/m) | | | | | | $\sigma^2_{vib}\times10^{-3}$ (Å$^2$) | | | | | | | | | | | |
|---|---|---|---|---|---|---|---|---|---|---|---|---|---|---|---|---|---|---|
| **x** | **Ga-N** | **In-N** | **Ga-Ga** | **In-In** | **Ga-In** | **N-N** | **G1** | **G2** | **G3** | **G4** | **G5** | **G6** | **I1** | **I2** | **I3** | **I4** | **I5** | **I6** |
| 1 | - | 100 | - | 27 | - | 20 | - | - | - | - | - | - | 3.9 | 5.4 | - | 8.3 | 4.0 | - |
| 0.48 | 135 | 112 | 23 | 8 | 12 | 32 | 3.5 | 5.3 | 6.4 | 8.1 | 3.9 | 4.2 | 3.5 | 6.9 | 6.9 | 8.3 | 4.5 | 4.9 |
| 0.42 | 130 | 118 | 25 | 15 | 18 | 35 | 3.2 | 5.1 | 7.1 | 7.5 | 3.6 | 5.0 | 3.2 | 5.2 | 5.0 | 6.9 | 3.8 | 3.9 |
| 0.37 | 132 | 120 | 28 | 20 | 17 | 35 | 3.2 | 4.7 | 4.9 | 7.1 | 4.0 | 3.9 | 3.2 | 5.2 | 5.1 | 6.8 | 3.9 | 3.8 |
| 0.32 | 135 | 125 | 30 | 20 | 25 | 35 | 3.6 | 4.5 | 4.9 | 6.7 | 3.6 | 3.5 | 3.4 | 5.0 | 4.7 | 6.9 | 3.5 | 3.6 |
| 0 | 135 | - | 40 | - | - | 30 | 3.0 | 4.1 | - | 6.0 | 3.0 | - | - | - | - | - | - | - |

The EXAFS spectra recorded at the Ga and In *K*-edges were fitted using scattering paths that were constructed using the experimentally determined lattice constants. The $\sigma^2_{vib}$ component of the DWF was set equal to the values listed in table 1. The fitting was performed for the 1$^{st}$ (4 N atoms), 2$^{nd}$ [12·x In and 12·(1-x) Ga atoms] and 3$^{rd}$ (9 N atoms) *nn* shells. The triangular scattering cation-cation-N paths were also considered with their path-length fixed according to the *nn* distances of the 1$^{st}$ and 2$^{nd}$ shell. The $\sigma^2_{stat}$ component of the DWF was set equal to the value of the corresponding cation-cation path since the cation-N path is characterized by null $\sigma^2_{stat}$. Fitting of the Ga and In *K*-edge spectra in the k- and R-space is shown in figure 2 and the fitting parameters are listed in table 2.

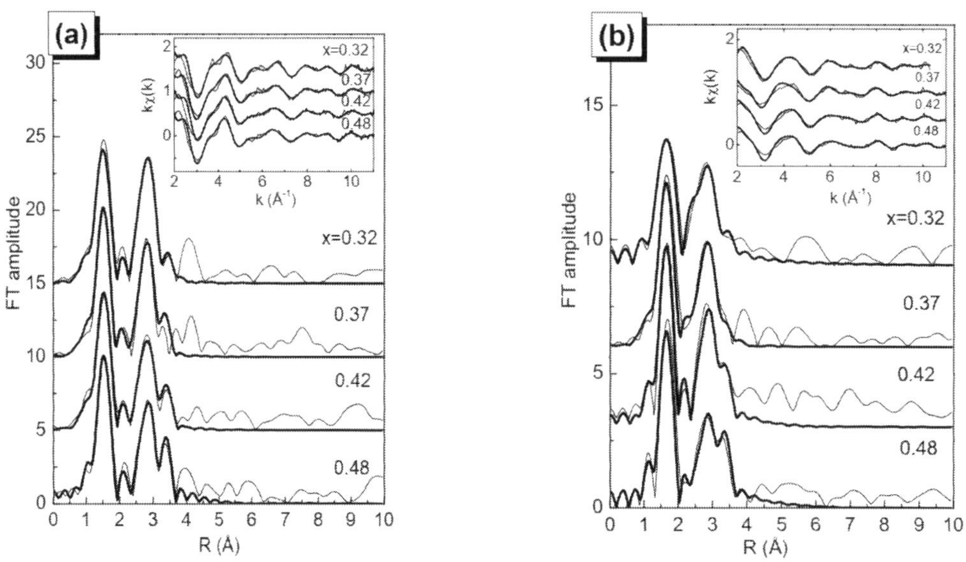


**Figure 2.** Fourier transform amplitude of the k$^3$-weighted χ(k) EXAFS spectra recorded at the (a) Ga and (b) In *K*-edge. The k$^3$-weighted χ(k) spectra are shown in the inset. The experimental and fitting curves are shown in thin and thick solid lines, respectively.

The results verify safely the violation of the VCA, as it is deduced from different Ga-N and In-N distances. Consequently, the local strain induced by alloying is mainly accommodated by cation–N–cation bond bending rather than cation–N bond stretching. Splitting is also observed in the 2$^{nd}$ *nn* shell. Regarding the DWF, the $\sigma^2_{stat}$ contribution is null for the 1$^{st}$ *nn* shell. In the 2$^{nd}$ *nn* shell the values of $\sigma^2_{stat}$ are generally smaller compared to the $\sigma^2_{vib}$.

**Table 2.** Fitting results of the In and Ga *K*-edge EXAFS spectra of the $In_xGa_{1-x}N$ samples for the 1st (cation-N) and 2nd (cation-cation) *nn* shells. $x_{XRD}$ and $x_{EXAFS}$ is the In molar fraction determined by XRD and EXAFS, respectively, R (Å) is the *nn* distance and $\sigma^2_{stat} \times 10^{-3}$ (Å$^2$) is the static component of the DWF.

| | | In-N | | Ga-N | | In-In | | Ga-Ga | | In-Ga | |
|---|---|---|---|---|---|---|---|---|---|---|---|
| $x_{XRD}$ | $x_{EXAFS}$ ±15% | R± 0.01 | $\sigma^2_{stat}$ | R± 0.01 | $\sigma^2_{stat}$ | R± 0.02 | $\sigma^2_{stat}$± 30% | R± 0.02 | $\sigma^2_{stat}$± 30% | R± 0.02 | $\sigma^2_{stat}$ ± 30% |
| 0.48 | 0.49 | 2.13 | 0.0 | 1.97 | 0.0 | 3.39 | 1.6 | 3.28 | 0.9 | 3.33 | 3.3 |
| 0.42 | 0.40 | 2.12 | 0.0 | 1.96 | 0.1 | 3.38 | 3.7 | 3.27 | 2.6 | 3.33 | 2.8 |
| 0.37 | 0.36 | 2.13 | 0.0 | 1.95 | 0.0 | 3.36 | 5.2 | 3.24 | 2.8 | 3.30 | 5.5 |
| 0.32 | 0.33 | 2.12 | 0.2 | 1.96 | 0.0 | 3.36 | 2.9 | 3.24 | 1.9 | 3.29 | 5.2 |

## 4. Conclusions

The vibrational and static disorder component of the DWF of the Ga and In *K*-edge EXAFS spectra of $In_xGa_{1-x}N$ were determined utilizing the equation of motion routine of the FEFF8 software and the VDOS obtained by means of Raman spectroscopy. The static disorder component is found null for the cation-N pairs whereas in the 2nd *nn* cation-cation shells the static disorder component was found generally smaller compared to the thermal component. Finally, splitting in the distances in the 1st and 2nd *nn* shell imposes that local distortions caused by alloying are accommodated mainly by bond bending

## Acknowledgements

This work is supported by the European Union (European Social Fund – ESF) and Greek national funds through the Operational Program "Education and Lifelong Learning" of the National Strategic Reference Framework (NSRF)-Research Funding Program: THALES, project NitPhoto. The measurements at BESSY II were funded from the European Community's Seventh Framework Programme (FP7/2007-2013) under grant agreement no.312284.

## References

[1] Kamiyama S, Amano H and Akasaki I 2005 The Evolution of Nitride Semiconductors *Optoelectronic Devices: III Nitrides* ed. M Razeghi and M Henini (Oxford: Elsevier) chapter 3 pp 23–38.
[2] McLaughlin D V P and Pearce J M 2013 *J Met. Mater. Trans.* **A44** 1947.
[3] Ferhat M and Bechstedt F 2002 *Phys. Rev.* **B44** 07523.
[4] Saito T and Arakawa Y 1999 *Phys. Rev.* **B60** 1701.
[5] O' Donnell K P *et al.* 2001 *J. Phys. Cond. Mat.* **13** 6977.
[6] Katsikini M *et al.* 2003 *Nucl. Instrum. Meth.* **B200**, 114.
[7] Kachkanov V, O'Donnel K P, Pereire S and Martin R W 2007 *Philos. Mag.* **87** 1999.
[8] Katsikini M *et al.* 2008 *Phys. Status Sol.* **A205** 2593.
[9] Mikkelsen Jr. J C and Boyce J B 1982 *Phys. Rev. Lett.* **49** 1412.
[10] Waychunas G A, Dollase W A and Ross II C R 1994 *Am. Mineral.* **79** 274.
[11] Adachi A 1994 *GaAs and related materials* (Singapore: World Scientific Publishing) chapter 14 p 556.
[12] Hernández S *et al.* 2005 *J. Appl. Phys.* **98** 013511.
[13] Katsikini M *et al.* 2010 *Phys. Status Solidi* **C7** 76.
[14] Mattila T and Zunger A 1999 *Phys. Rev.* **B85** 160.
[15] Poiarkova A V and Rehr J J 1999 *Phys. Rev.* **B59** 948.
[16] Poiarkova A V 1999 *X-ray absorption fine structure Debye-Waller factors* PhD Thesis, University of Wasington.
[17] Ankudinov A L, Ravel B, Rehr J J and Conradson S D 1998 *Phys. Rev.* **B58** 7565.
[18] Iliopoulos E *et al.* 2006 *Phys. Stat. Sol.* **A203** 102.

# Structural and electronic studies of metal hexacyanoferrates based cathodes for Li rechargeable batteries

**Marco Giorgetti*[1], Adriana Mignani[1], Giuliana Aquilanti[2], Paolo Conti[3], Marcus Fehse[4], Lorenzo Stievano[4]**

[1]Department of Industrial Chemistry, University of Bologna, 40136 Bologna, Italy
[2]Elettra - Sincrotrone Trieste S.C.p.A. s.s 14 km 163.5, 39149 Basovizza, Trieste Italy
[3]School of Science and Technology, Chemistry Division, University of Camerino, 62032 Camerino, Italy
[4]ICGM (UMR 5253) AIME Université Montpellier, 34095 Montpellier, France.

E-mail: marco.giorgetti@unibo.it

**Abstract**. *Operando* XANES and EXAFS spectra on the newly prepared Fe hexacyanocobaltate active material for positive electrodes in lithium batteries have been recorded at the XAFS beamline of Elettra using a suitable *in situ* cell. In this way, it was possible to follow in detail the main structural and electronic changes during the charge and discharge processes of the battery. The use of a chemometric approach for data analysis is also underlined.

## 1. Introduction

The development of low-cost, efficient storage of off-peak electric power and of electrical energy generated by energy sources other than fossil fuels is a global priority. Rechargeable batteries offers efficient electrochemical energy storage and batteries based on lithium technology has been used in several devices [1]. At present, the searching for new materials with sufficiently large interstitial spaces for the fast intercalation/release process of $Li^+$ and other cations is one of the most interesting chemical challenges. Metal hexacyanoferrates (M*hcf*) represent an interesting class of intercalation materials, with high void space, which can be a viable and low-cost alternative to conventional electrodes based on metal oxides. Manganese hexacyanoferrate has been used in Na batteries, delivering a reversible capacity of 120-130 mAh/g for in the first 30 cycles [2]. Wessels *et al.* proposed application of M*hcf* as electrode material in aqueous batteries [3], the copper analogues has been proposed by Asakura for lithium batteries, investigating the specific role of the Fe and Cu site during the lithium insertion [4]. Aqueous zinc-ion battery based hexacyanoferrate has been recently reported [5], whereas Kurihara and Moritomo studied the structural and electronic properties of a Mn and Co hexacyanoferrate [6] family of cathodes for Li batteries. This paper deals with the study by X-ray absorption spectroscopy (XAS) [7] of the modification of the structural and electronic properties of the iron(II) hexacyanocobaltate analogues while used as positive electrode materials for lithium rechargeable battery.

## 2. Experimental

The sample was prepared by co-precipitation method at 40 °C [8] resulting in a stoichiometric $K_{0.44}Fe_{1.56}Co(CN)_6$ material. The electrode was obtained by tape casting a slurry containing the pure active material (80%), 10% polyvinylidene fluoride (PVDF) and 10% carbon black in NMP (N-methylpyrrolidone) on a flat teflon surface. The mass loading of the electrode was about 2.5 mg/cm$^2$ of active materials. *Operando* X-ray absorption (XAS) spectra were recorded at Elettra (Basovizza, Italy) at the XAFS beamline [9]. A suitable electrochemical cell for the *operando* measurements, which work in transmission geometry, was charged and successively discharged at C/30 rate (insertion of a mole of Li per mole of active material in 30 hours). The storage ring was operated at 2.0 GeV in top-up mode with a typical current of 300 mA. The data were recorded alternatively at the Fe K-edge (7112 eV) and Co K-edge (7709 eV) in transmission mode using an ionization chamber filled with a mixture of Ar, $N_2$ and He in order to have 10%, 70% and 95% of absorption in the $I_0$, $I_1$ and $I_2$ chambers, respectively. The white beam was monochromatized using a fixed exit monochromator equipped with a pair of Si (111) crystals. Harmonics were rejected by detuning one of the two crystal of the monochromator of 30% with respect to the maximum of the rocking curve. Internal reference of Fe and Co foils were used for energy calibration in each scan. Spectra were collected with a constant k-step of 0.03 Å$^{-1}$ with 3s/point acquisition time from 6850 to 7700eV (Fe K-edge) and from 7550 to 8320eV (Co K-edge).

XAS spectra were calibrated using the Athena program [10]. The pre-edge background was removed by subtraction of a linear function extrapolated from the pre-edge region, and the XANES spectra were normalized at the unity by extrapolation of the atomic background. The EXAFS analysis has been conducted by using the GNXAS software package [11]. $E_0$ values were found to be displaced by several eV respect to the edge inflection point. $S_0^2$ values were found to be 0.70 and 0.74 for Fe and Co, respectively.

## 3. Results and Discussion

From the structural viewpoint iron hexacyanocobaltate (Fe*hcc*) materials, as per the complete series of metal hexacyanoferrates, are based on three dimensional -Co-C-N-Fe- network, *Fm3m* (No.225), where both metals occupy the octahedral 4a Wyckoff position and water molecules are present at interstitial sites (see the sketch inside Figure 1). Alkali-metals, such as $Li^+$ of $Na^+$, are placed at the 8c site to guarantee neutrality, and of course can be removed and inserted during an electrochemical reaction process. Figure 1 displays the charge and discharge curve as obtained during the *operando* XAS experiment. During the charge the $K^+$ cation is extracted from the hexacyanocobaltate structure making feasible the $Li^+$ insertion in the successive discharge. Overall we have recorded 73 spectra at both Fe and Co K-edge, 42 during the charge and 28 in the subsequent discharge, this number enables a very close monitoring of the local charge and structure modification upon the potassium release and the lithium insertion process. The charge curve displays a curved-shape behavior during the registration of the first 15 spectra which may indicate an activation process of the active material, most probably related to the irreversible dehydration of the starting material.

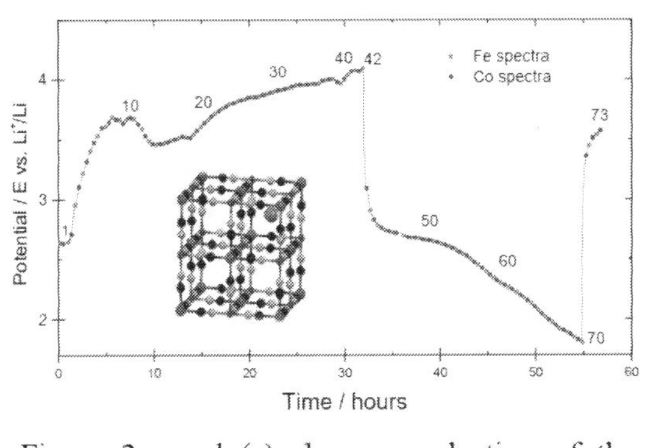


**Figure 1.** Voltage profile of the charge and discharge curves of Fe*hcc* cathode, as obtained during the *operando* XAS experiment. The battery has delivered 77 and 59 mAh/g during charge and discharge, respectively. Numbers indicate the i$^{th}$ spectrum during the experiment. 42 spectra during charge and 28 during discharge have been recorded. Reference and counter electrode: Li. electrolyte: 1 M $LiPF_6$ in EC:PC:3DMC (ethylene carbonate, propylene carbonate, dimetlylcarbonate)

Figure 2 panel (a) shows a selection of the XANES spectra during the first charge and the successively discharge. A dramatic change of the spectra in the rising portion of the edge, the edge and on the main peak are observed. This suggests a relevant electronic and local structure modification of the Fe environment during the first charge (spectra 1-42). An almost opposite behavior is observed during the discharge process, spectra 42-70. Figure 2(b) compares the pristine (black), the charged electrode (blue) and the discharged (red) XANES curves. The curve moves towards higher energy during charge and therefore $Fe^{2+}$ is oxidized and, successively, reduced in the subsequent discharge. This process is not fully reversible. In fact, even though the threshold energy is set back to the initial point after the discharge, other characteristics such as the edge feature B, the main peak C and the resonance D are different. Overall, this experiment points out that the Fe is electroactive but a different local atomic arrangement is present after one full redox cycle. A complete different behavior is observed at the Co site, as shown in the panel (c): the XANES spectrum of the pristine material ($1^{st}$) appears the same of the charges ($42^{th}$ spectrum) and discharged ($70^{rd}$) electrode. The local and electronic structure of cobalt does not change during the electrochemical process, and is thus completely inert in this range of investigated electrochemical potentials.

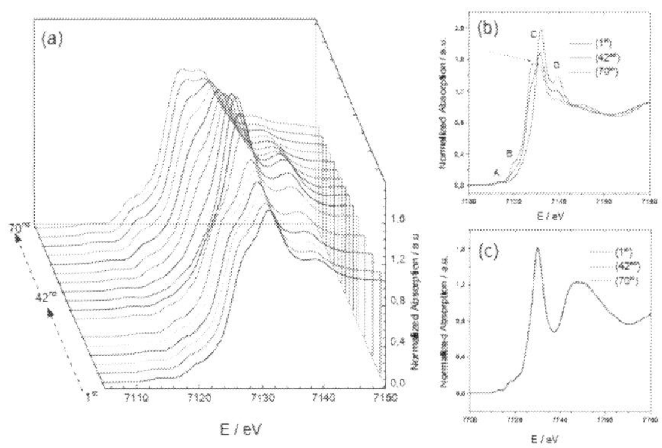


**Figure 2.** (a) Selection of Fe K-edge XANES spectra obtained during the *operando* XAS experiment. They are indicated spectra # 1, 4, 9, 13, 18, 22, 27, 31, 36, 42, 46, 50, 54, 58, 63, 67, 70. (b) Comparison of the normalized XANES spectra at the Fe K-edge for the pristine, the charged and discharged electrode. The arrow indicates the presence of an isosbestic point. (c) Corresponding normalized XANES spectra at the Co K-edge.

A chemometric approach can be used to obtain a more complete understanding of the cell dynamic during the electrochemical process. Among some different chemometric techniques, the application of the Multicurve Resolution Analysis (MRC) has been demonstrated to be very useful in the study of XAS spectra on battery systems [12]. Following this approach not only the number of species but also the existence range of the various species involved during the electrochemical approach can be revealed. Fig 3a shows the concentration profiles obtained by this analysis. They support the presence of three species during the overall electrochemical cycle (charge and discharge). Generally speaking, species 1 converts to species 2 during the first charge and species 2 converts to species 3 after the subsequent discharge. In addition, there is a sub-range where each species exists almost exclusively. This is the case of species 2, which is present after the first charge (spectrum 42th), and species 3 which forms after the subsequent discharge (spectrum 70th). Interestingly, the pristine electrode does not completely matches with species 1, but only to about 80%. Also, the situation concerning the first 15 spectra is quite stable and different to the other ranges where, typically, there is a progressive conversion of one species to the other. This is in agreement to the behaviour of the first 15 points of the charge curve presented in Fig. 1.

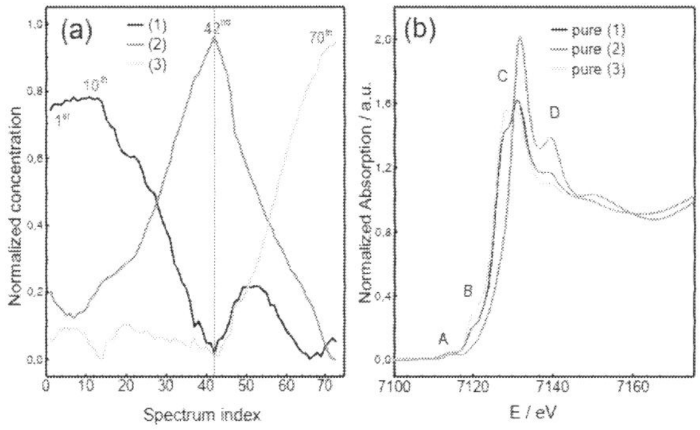


**Figure 3.** MRC Analysis computed using XANES data at the Fe K-edge. (a) concentration profiles and (b) corresponding pure spectra. Three species can be observed during the overall electrochemical process (charge and discharge).

Figure 3b displays the XANES spectra of the pure species computed by the MRC analysis. The curves are rather similar to the experimental ones displayed in the Figure 2b, especially for species 2 ad 3, which agree to the $42^{th}$ and $70^{th}$ spectra, respectively. In addition, the pure species 1 is slightly different to the $1^{st}$ experimental spectrum: the rising edge feature B and the peak maximum C are more intense.

The shape of the curve of Fig. 3a also suggests that when a relative concentration of two species is the same, like the crossing points $28^{th}$ and $56^{th}$, an isosbestic point should be present in the experimental spectra. Indeed, Fig. 2b confirms the appearance of such a condition at about 7130 eV, marked by an arrow. In summary, this underlines the capability of the present chemometric analysis to study *operando* XAS experiments.

A preliminary EXAFS data analysis computed by multiple-edge approach [8] on the experimental raw data at both Fe and Co K-edges has evidenced some structural variation of the Fe-N-C-Co structural unit. The analysis was conducted on a selected number of samples, indicated by the MRC analysis. These fits suggest that the Fe-N distance is almost the same during the first discharge, from 2.06(1) to 2.04(1) Å and sets back to the value of 2.05(1) Å in the $70^{th}$ spectrum. On the contrary, the corresponding Fe-N Debye-Waller factor is strongly influenced by the electrochemical process. A higher value is observed for spectrum $10^{th}$, indicating higher structural disorder, but a progressive decrease is then observed from 0.008(2) to 0.003(1) Å$^2$ during the full charge ($K^+$ removal, spectrum $42^{th}$). The subsequent lithium insertion takes place with an increase of the structural disorder, and the value of 0.008 Å$^2$ is restored. At the Co site, the Co-C bond length is the same in all investigated spectra, and the same holds true for the corresponding Debye-Waller factor. Concerning the Fe-N-C-Co structural unit, both angles Fe-N-C and N-C-Co are subject of small deviation from the perfect linearity during charge. The latter could be studied because of the strong four-body Multiple Scattering effects usually observed in these materials [13].

## 4. Conclusions

The experiment carried out in this work concerned the structural and electronic characterisation of the iron hexacyanocobaltate as positive electrode material for secondary Li battery. It has revealed that Fe is the electroactive metal contributing to the charge balance, by giving activity of the $Fe^{3+}/Fe^{2+}$ couple, whereas the Co site is electrochemically inert. Emphasis was also devoted to the combined used of chemometry to the *operando* XAS experiment at the Fe K-edge.

## 5. Acknowledgements

Experiments at Elettra have been performed through the project # 20130225.

## References

[1] Goodenough JB 2012 *J. Solid State Electrochem.* **16** 2019
[2] Wang L, Lu Y, Lu J, Xu M, Cheng J, Zhang D, Goodenough JB 2013 *Angew. Chem. Int. Ed.* **52** 1964
[3] Wessels CD, Peddada SV, McDowell MT, Huggins RA, Cui Y, 2012 *J. Electrochem. Soc.* **159** A98
[4] Asakura D, Li CH, Miruno Y, Okubo M, Zhou H, Talham DR 2013 *J. Am. Chem. Soc.* **135** 2793
[5] Zhang L, Chen L, Zhou X, Zhaoping L 2015 *Adv. Energy Mater.* **5** doi:10.1002/aenm.201400930
[6] Kurihara Y, Moritomo Y 2014 *Jpn. J. Appl. Phys.* **53** 067101
[7] Giorgetti M 2013 *ISRN Materials Science* **2013**, Art. No. 938625, 22 pages
[8] Giorgetti M, Berrettoni M 2008 *Inorg. Chem.* **47** 6001.
[9] Di Cicco A, Aquilanti G, Minicucci M, Principi E, Novello N, Cognigni A, Olivi L 2009 *Journal of Physics: Conference Series* **190** 012043
[10] Ravel B, Newville, M 2005 *J. Synchrotron Radiat.* **12**, 537

[11] Filipponi A, Di Cicco A 1995 *Phys. Rev. B* **52** 15135.
[12] Conti P, Zamponi S, Giorgetti M, Berrettoni M, Smyrl WH 2010 *Anal. Chem.* **82** 3629
[13] Giorgetti M, Berrettoni M, Filipponi A, Kulesza PJ, Marassi R 1997 *Chem. Phys. Lett.* **275** 108

# XAFS study on structure-activity correlations of $\alpha$-$Co(OH)_2$ nanosheets water oxidation catalysts

**Junheng Huang, Qinghua Liu, Tao Yao, Zhiyun Pan*, and Shiqiang Wei***

National Synchrotron Radiation Laboratory, University of Science and Technology of China, Hefei, Anhui 230029, P. R. China

*Corresponding author. E-mail: zhypan@ustc.edu.cn, sqwei@ustc.edu.cn

**Abstract**. Understanding the structure-activity of the metal hydroxide materials is critical to the design of an efficient oxygen evolution reaction (OER) catalyst for water oxidation. A challenge is to identify and collect surface active site relative to bulk. Here, we have prepared an ultrathin $\alpha$-$Co(OH)_2$ nanosheet with large exposed surface Co sites as an high-efficiency $O_2$ evolution catalyst. Using XAFS technique, we have investigated the oxidation state and local structural evolutions of the $\alpha$-$Co(OH)_2$ nanosheets catalyst. A coordination-miss Co sites ($CoO_{6-x}$) with oxidation state of +3.3 on the nanosheet surface is revealed after the OER procedure, indicating the in situ formation of $\gamma$-CoOOH nanosheet is a key factor leading to an enhanced water oxidation performance.

## 1. Introduction

The worldwide environment and energy problems have made it urgently important to change the traditional energy production style and exploit renewable energy, such as solar-driven water splitting to $H_2$ and $O_2$ [1]. The oxygen evolution reaction (OER) is kinetically sluggish, which decrease the efficiency of the overall photoelectric water splitting [2]. Accordingly, a number of OER catalyst have been developed to accelerate the charge carrier transfers and utility and to lower the required overpotential.

Transition metal oxides or oxyhydroxides are attractive OER catalysts candidates due to their low cost, excellent OER activity and stability, such like Co and Ni based oxides [3-5]. Therefore, it is a very meaningful work to study the structure-function relationship for this highly active Co-, Ni-catalyst. On the other hand, X-ray absorption fine structure (XAFS) spectroscopy as a local atomic structure sensitive measurement is widely applied in catalysis research. However, relating to the bulk catalyst, it is difficult to identify the detection signal of the surface active site since the proportion of the rest metal atomic is large.

Here, we report that 2-dimensional $\alpha$-$Co(OH)_2$ nanosheets are successfully synthesized, which behave only two $CoO_6$ octahedron slabs thickness and show superior OER activity than the state-of-the-art $IrO_2$ catalyst. Due to the exposed high density active Co site in this nanosheet catalyst, we can reasonably utilize XANES and EXAFS to track the changes of the valency and the local atomic structure during water oxidation reaction.

## 2. Experimental

Sample synthesis: $\alpha$-$Co(OH)_2$ nanosheet was prepared by a hydrothermal method [5,6]. 0.0952 g $CoCl_2{\cdot}6H_2O$ was dispersed in 40 ml DI-water /ethylene glycol (1:3) solution and adjusted the PH value to about 9 by addition of Ammonia solution (28 wt%). Then these solution were transfer to a Teflon-lined stainless steel autoclave and heated at 130 ℃ for 24 h. The $\alpha$-$Co(OH)_2$ and $\gamma$-CoOOH model compound were synthesized by solution means according to the previous report [7,8].

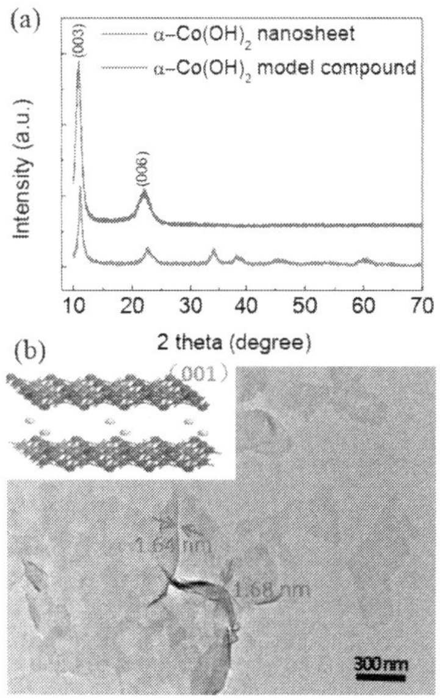


**Figure 1.** (a) XRD pattern of synthesized α-$Co(OH)_2$ nanosheet and α-$Co(OH)_2$ model compound. (b) TEM image for α-$Co(OH)_2$ nanosheet, (inset) shows the crystal structure of α-$Co(OH)_2$ nanosheet with two $CoO_6$ slabs thickness, displaying cobalt (blue), oxygen (red), and chlorine (green) ions.

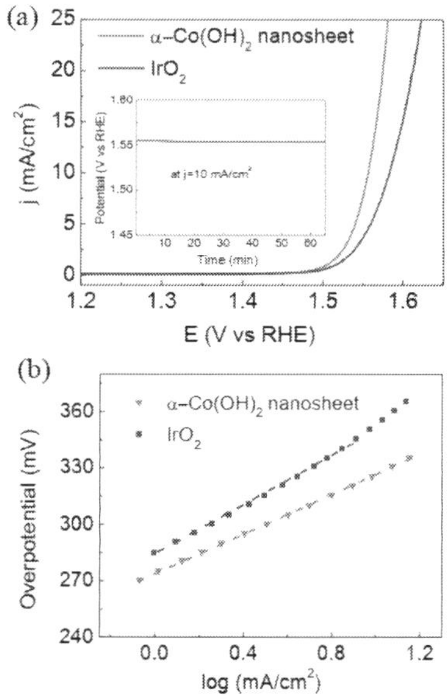


**Figure 2.** (a) Polarization curves and (b) Tafel polts of α-$Co(OH)_2$ nanosheet and $IrO_2$ catalyst coating on Au/ITO electrode in 1 M KOH medium. The inset in (a) shows Chronopotentiometric measurements at $J$ = 10 mA/cm$^2$ for α-$Co(OH)_2$ nanosheet.

Electrochemical measurements were performed in an electrochemical workstation with a three-electrode system. Electrocatalyst powder and 30 μl Nafion solution were dissolved in 1 ml of a 3:1 mixture of DI-water and ethanol to give the final concentration of 2 mg/ml. Then this dispersion was loaded onto Au/ITO substrate to form a thin film as OER working electrode, platinum mesh as the counter electrode, and saturated Ag/AgCl as reference electrode.

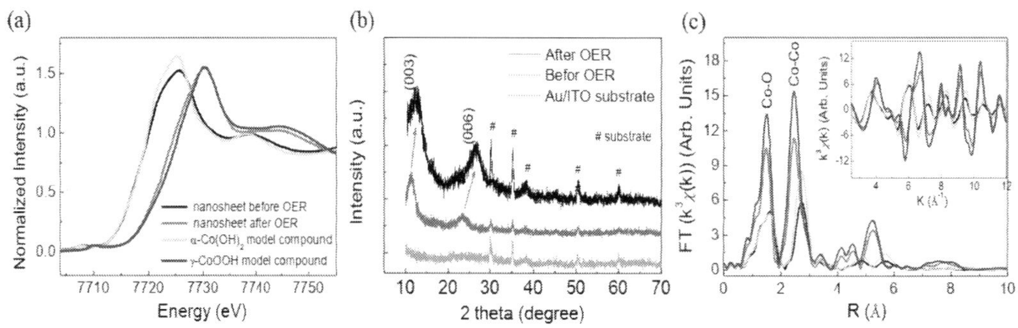


**Figure 3.** (a) XANES spectra, (c) EXAFS spectra and (inset) oscillation function $k^3\chi(k)$ of α-$Co(OH)_2$ nanosheet before OER (black), α-$Co(OH)_2$ nanosheet after OER (red), α-$Co(OH)_2$ model compound (green) and γ-CoOOH model compound (blue) respectively. (b) XRD pattern for α-$Co(OH)_2$ nanosheet before and after OER procedure at 10 mA cm$^{-2}$.

XAFS measurements at Co K-edge were carried out in fluorescence mode at beamline U7C of the National Synchrotron Radiation Laboratory (NSRL), beamline BL14W1 of the Shanghai Synchrotron Radiation Facility (SSRF). The storage rings of NSRL, SSRF were operated at 0.8 GeV with the current of 250 mA, at 3.5 GeV with the current of 300 mA, respectively. The X-ray energy was calibrated by using a Co foil. All the acquired XAFS data were processed by the IFEFFIT software packages.

## 3. Result and discussion

X-ray diffraction pattern (XRD) in figure 1a suggest that comparing with the α-$Co(OH)_2$ model compound, α-$Co(OH)_2$ nanosheet possesses a highly (001) preferred orientation. Figure 1b displays the transmission electron microscopy (TEM) image of α-$Co(OH)_2$ nanosheet, which are several hundred nanometers in size with the average thickness of only about 1.6 nm. These results demonstrate that the ultrathin α-$Co(OH)_2$ nanosheet with largely exposed (001) face were successfully fabricated (see inset of figure 1b).

To determine the water oxidation activities of α-$Co(OH)_2$ nanosheet, the linear sweep voltammetry was carried out at a scanning rate of 2 mV $s^{-1}$ in 1 M KOH electrolytes. The iR-corrected polarization curves (figure 2a) recorded with the α-$Co(OH)_2$ nanosheet reveal a markedly small OER overpotential of 320 mV at current density of 10 mA $cm^{-2}$, which is obviously lower than the state-of-the-art $IrO_2$ electrocatalyst ( overpotential = 340 mV at 10 mA $cm^{-2}$ ). Tafel curves shown in figure 2b were derived from the OER polarization curves at the current density from 1-10 mA $cm^{-2}$. The characteristic of catalytic activity for the α-$Co(OH)_2$ nanosheet is also reflected by its Tafel slope of 48 mV $dec^{-1}$, even lower than that of $IrO_2$ about 59 mV $dec^{-1}$. The durability measurement was carried out for more than 60 min in alkaline environment. Inset image of Figure 2a shows the α-$Co(OH)_2$ nanosheet have good electrochemical stability as the overpotential remained nearly constant at OER current of 10 mA $cm^{-2}$. These results demonstrate that the α-$Co(OH)_2$ nanosheet exhibit an excellent oxygen evolution activity, even higher than the precious metal $IrO_2$ electrocatalyst.

In order to give an in-depth understanding of the structure-activity correlation of the α-$Co(OH)_2$ nanosheet, oxidation state and structure before and after the OER procedure were characterized by X-ray absorption fine structure (XAFS) measurement (including XANES). Figure 3a presents the XANES spectra of α-$Co(OH)_2$ nanosheet before and after the OER procedure. The α-$Co(OH)_2$ model compound and γ-CoOOH model compound synthesized by solution means have oxidation state of +2[7] and +3.3[8] respectively. With increasing formal oxidation state, the edge position of the compound will shift to higher energy. The edge shape and position of α-$Co(OH)_2$ nanosheet (7720.2 eV) is similar to that of α-$Co(OH)_2$ model compound (7720.1 eV), however, after bearing an OER current for more than 60 min, the edge shape and position of α-$Co(OH)_2$ nanosheet (7724.0 eV) are obviously changed and match well with that of γ-CoOOH model compound (7724.8 eV), indicating an increase average oxidation state of Co centers about +3.3 and proportion of Co site contain a formal oxidation state of Co (IV). Similar phenomenon that anodization leads to the higher oxidation state active metal centers has been also observed for NiOOH compound [3,9].

To detect the phase and local atomic structure transformation of α-$Co(OH)_2$ nanosheet, XRD pattern and EXAFS data were obtained. XRD pattern in figure 3b clearly exhibits a small decrease in interplanar spacing of (003) and (006) face, it can be explained by the phase change from α-$Co(OH)_2$ nanosheet to γ-CoOOH nanosheet [5,7-8]. The XANES and EXAFS results showing in figure 3c display that the position of peaks in the FT EXAFS spectrum of α-$Co(OH)_2$ nanosheet after OER is similar to those of γ-CoOOH model compound, and the shortening of Co-O distances (see Table 1) is most likely due to the formation of Co-O dangling bonds in the surface Co octahedron which in turn favors local structural distortions and a concomitant decrease of the interplanar spacing. In addition, the Co-O coordination number decrease from 5.9 for the α-$Co(OH)_2$ nanosheet before OER to 5.0 for the nanosheet catalyst after OER, revealing the presence of O vacancies in the surface Co octahedron ($CoO_{6-x}$) during the water oxidation reaction. The coordination-miss Co sites ($CoO_{6-x}$) with oxidation state of +4 can facilitate the formation of hydroperoxy (OOH) species [3,9]. Based on the discussion above, it can be suggested that the outstanding catalytic activity of α-$Co(OH)_2$ nanosheet originated in the formation of γ-CoOOH nanosheet with the highly oxidized Co site and $CoO_{6-x}$ structure.

## 4. Conclusions

In summary, we have prepared an ultrathin α-$Co(OH)_2$ nanosheet as an high-efficiency $O_2$ evolution catalyst. The XANES and XRD results reveal that the γ-CoOOH nanosheet of the valency about +3.3 in Co site was in situ formed when water oxidation process occurs. Further detected by EXAFS indicate that the Co-O neighbor number reduce from 5.9 to 5.0 leading to a coordination-miss $CoO_{6-x}$ octahedron structure, which may facilitate the oxygen evolution and enhance the catalytic activity. Our results will provide helpful design insights for high performance catalysts.

**Table 1.** EXAFS fitting parameters[a]

| Sample | Path | N[b] | R (Å) | $\sigma^2$($Å^{-2}$) | $\Delta E^0$ (eV) |
|---|---|---|---|---|---|
| α-$Co(OH)_2$ nanosheet before OER | Co-O | 5.9 (0.2) | 2.09 (0.01) | 0.009 (0.001) | -1.8 |
| | Co-Co | 5.0 (0.2) | 3.12 (0.01) | 0.007 (0.001) | |
| α-$Co(OH)_2$ nanosheet after OER | Co-O | 5.0 (0.2) | 1.89 (0.01) | 0.005 (0.001) | -3.2 |
| | Co-Co | 5.1 (0.2) | 2.84 (0.01) | 0.006 (0.001) | |
| α-$Co(OH)_2$ standard sample | Co-O | 6.0 | 2.10 (0.01) | 0.009 (0.001) | -1.2 |
| | Co-Co | 6.0 | 3.12 (0.01) | 0.007 (0.001) | |
| γ-CoOOH standard sample | Co-O | 6.0 | 1.90 (0.01) | 0.004 (0.001) | -2.8 |
| | Co-Co | 6.0 | 2.85 (0.01) | 0.005 (0.001) | |

[a] Fitting region $1 \leq R \leq 3.2$, $2.5 \leq k \leq 12$. Values in parenthesis indicate uncertainties.

[b] Coordination number of standard sample were fixed to the theoretical values.

**Acknowledgments**

This work was supported by the National Natural Science Foundation of China (Grants No. 11179004, U1332111, 11305174 and 11135008).

**References**

[1] McKone J R, Lewis N S and Gray H B 2014 *Chemistry of Materials* **26** 407-14
[2] Smith R D, Prevot M S, Fagan R D, Zhang Z, Sedach P A, Siu M K, Trudel S and Berlinguette C P 2013 *Science* **340** 60-3
[3] Gao M, Sheng W, Zhuang Z, Fang Q, Gu S, Jiang J and Yan Y 2014 *J. Am. Chem. Soc.* **136** 7077-84
[4] Song F and Hu X 2014 *Nat. Commun.* **5** 4477
[5] Huang J H, Chen J T, Yao T, He J F, Jiang S, Sun Z H, Liu Q H, Cheng W R, Hu F C, Jiang Y, Pan Z Y and Wei S Q 2015 *Angew. Chem. Int. Ed.* **54** 8722-7
[6] Zhu J, Bai L, Sun Y, Zhang X, Li Q, Cao B, Yan W and Xie Y 2013 *Nanoscale* **5** 5241-6
[7] Ma R Z, Liu Z P, Takada K, Fukuda K, Ebina Y, Bando Y and Sasaki 2006 *Inorganic Chemistry* **45** 3964-9
[8] Bardé F, Palacin M R, Beaudoin B, Delahaye-Vidal A and Tarascon J M 2004 *Chemistry of Materials* **16** 299-306
[9] Bediako D K, Lassalle-Kaiser B, Surendranath Y, Yano J, Yachandra V K and Nocera D G 2012 *J. Am. Chem. Soc.* **134** 6801-9

# Investigation on Electrochemical Property of AB/$Al_2O_3$-coated Li-excess Mn-based Layered Oxides

**H Kobayashi[1], K Takada[2], Y Arachi[2], T Okumura[1], M Shikano[1], and H Nitani[3]**

[1]National Institute of Advanced Industrial Science and Technology (AIST), 1-8-31, Midorigaoka, Ikeda, Osaka 563-8577, Japan
[2]Kansai University, 3-3-35 Yamate-chuo, Suita, Osaka 564-8680 Japan
[3]High Energy Accelerator Research Organization (KEK), 1-1 Oho, Tsukuba 305-0801, Japan

E-mail: hironori-kobayashi@aist.go.jp

**Abstract**. The AB and/or $Al_2O_3$-coated $Li_{1.20}Mn_{0.55}Ni_{0.16}Co_{0.09}O_2$ were prepared by the mechanochemical reaction. The optimal condition for sample preparation was selected to be the rotation speed of 2000 rpm and the reaction time of 5 minutes by SEM and XANES measurements. Surface analysis using SEM and XANES data demonstrated that all the samples were rather uniformly covered with AB and/or nano-$Al_2O_3$ particles. The $Al_2O_3$-coated samples showed less discharge capacity and C-rate performance compared with the pristine sample, while the AB/$Al_2O_3$-coated sample showed almost the same discharge capacity and C-rate performance and had better cycle performance at 318 K compared with the pristine sample. These results demonstrate that the mechanochemical AB/$Al_2O_3$-coating process is an effective way of improving the cycle performance at high temperature without degrading the discharge capacity and C-rate performance.

## 1. Introduction

The layered $Li_2MnO_3$-Li$MO_2$ ($M$ = transition metal) materials are one of the promising positive electrode materials for lithium secondary battery [1]. Especially, it has been reported that $Li[Mn_{0.56}Ni_{0.17}Co_{0.07}Li_{0.2}]O_2$ displayed an initial discharge capacity of c.a. 280 mAh/g in the voltage range of 2.5 to 4.8 V and maintained a reversible capacity of c.a. 250 mAh/g after 50 cycles [2]. On the other hand, the decomposition of the electrolyte proceeded on the positive electrode material above 4.5 V is a big problem to be solved for realizing long cycle life. Recently several efforts have been made to improve the cycle performance through surface modification of positive electrode materials. We investigated the structural change with Li de-intercalation from $Li[Mn_{0.56}Ni_{0.17}Co_{0.07}Li_{0.2}]O_2$ and reported the possibility of removing oxygen from the lattice during the 1$^{st}$ charge process using synchrotron radiation technique [3]. In addition, limited efforts have been made to improve the cycle performance through surface modification of Li-excess Mn-based layered materials. Recently, we reported that the mechanochemical $Al_2O_3$-coating process is an effective way of improving the cycle performance of $Li_{1.20}Mn_{0.55}Ni_{0.16}Co_{0.09}O_2$ at high temperature [4]. In this paper, the effect of acetylene black(AB)/$Al_2O_3$-coating on electrochemical properties was studied in the same chemical composition of $Li_{1.20}Mn_{0.55}Ni_{0.16}Co_{0.09}O_2$.

## 2. Experimental

We used the $Li_{1.20}Mn_{0.55}Ni_{0.16}Co_{0.09}O_2$ powder (Toda Industrial Corp), the AB powder (Denki Kagaku Kogyo Kabushiki Kaisha), and the nano-$Al_2O_3$ powder with the order of 10-20 nm (TECNAN) in this study. The coated $Li_{1.20}Mn_{0.55}Ni_{0.16}Co_{0.09}O_2$ were prepared by mechanochemical reaction with the nano-powder (AB and $Al_2O_3$) using NOB-MINI (Hosokawa Micron Co.) [4]. The particle morphology was measured by SEM (JSM-6700FV, JEOL). Crystal and electronic structures were investigated by synchrotron XRD (BL19B2 at SPring-8 / Proposal No. 2012B1574) and XAFS (BL9C at PF / Proposal No. 2010G599 and BL4B at UVSOR / Proposal No. 25-521) measurements. The crystal and electronic structures were determined using the analysis programs RIETAN-FP [5] and REX2000. Electrochemical property was measured using coin-type cells with Li/1M $LiPF_6$ in EC:DMC(1:2)/samples. The stepwise pre-cycling treatment was carried out by increasing the upper voltage limit by 0.1 V every two cycles from 4.5 V to 4.8 V [3].

## 3. Results and Discussion

The mechanochemical reaction was used to prepare the 5 wt% AB-coated, 5 wt% $Al_2O_3$-coated, and 5 wt% AB and 5 wt% $Al_2O_3$-coated $Li_{1.20}Mn_{0.55}Ni_{0.16}Co_{0.09}O_2$ and the 2000 rpm of the rotation speed and 5 minutes of the reaction time were selected as the optimal preparation condition. Figure 1 shows the SEM images of the AB and/or $Al_2O_3$-coated $Li_{1.20}Mn_{0.55}Ni_{0.16}Co_{0.09}O_2$ together with the pristine sample. The SEM images of all the coated samples showed that these particles of AB and $Al_2O_3$ were uniformly distributed over the surface of $Li_{1.20}Mn_{0.55}Ni_{0.16}Co_{0.09}O_2$ particles. The XRD patterns of the pristine and coated $Li_{1.20}Mn_{0.55}Ni_{0.16}Co_{0.09}O_2$ showed that the coated samples were well defined layered phases without additional impurity peak. The crystal structure was refined using the synchrotron XRD data, the initial model being that of $LiNi_{1/2}Mn_{1/2}O_2$ [6], and lattice parameters for all the samples were almost the same values of $a$ = 2.855 Å and $c$ = 14.245 Å. In order to obtain information on the bulk structure of the coated $Li_{1.20}Mn_{0.55}Ni_{0.16}Co_{0.09}O_2$, we collected the Ni, Co, and Mn *K*-edge XAFS data. The Ni, Co, and Mn *K*-edge XANES spectrum essentially showed no deviation from the valence state of $Ni^{(2-\alpha)+}$, $Co^{(3+\beta)+}$, and $Mn^{4+}$, respectively, after mechanochemical reaction. These XRD and XANES

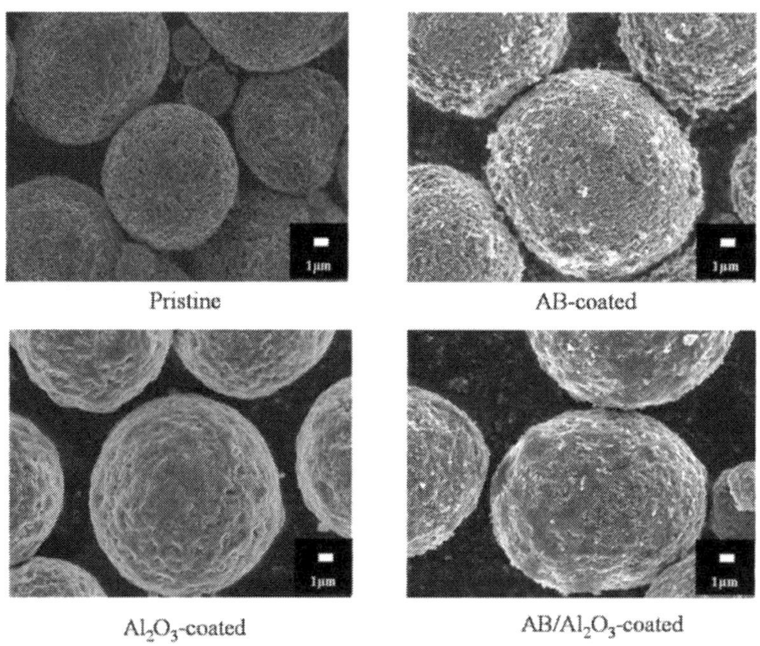


**Figure 1.** SEM images of the AB and/or $Al_2O_3$-coated $Li_{1.20}Mn_{0.55}Ni_{0.16}Co_{0.09}O_2$ together with the pristine sample

results clarified that the bulk structure and the valance state of AB and/or $Al_2O_3$-coated samples were essentially same as those of the pristine sample.

In order to obtain information on the surface of the coated $Li_{1.20}Mn_{0.55}Ni_{0.16}Co_{0.09}O_2$, we collected the O *K*-edge and Mn *L*-edge XANES spectra in TEY mode. Figure 2 shows the O *K*-edge and Mn *L*-edge XANES spectra for the pristine and coated $Li_{1.20}Mn_{0.55}Ni_{0.16}Co_{0.09}O_2$. The O *K*-edge spectra contained two pre-edge peaks of A and B originating from the layered structures [3,4] and the Mn $L_{III}$-edge spectra contained two peaks of C and D originating from the layered structured [3,4]. For the O *K*-edge XANES spectra, the intensity of peaks A and B clearly decreased for $Al_2O_3$ and AB/$Al_2O_3$-coated samples, as shown in Fig. 2(a). On the other hands, the new peaks originated from AB were observed for AB-coated sample. For the Mn *L*-edge XANES spectra, the intensity of peaks C and D clearly decreased for $Al_2O_3$ and AB/$Al_2O_3$-coated samples, as shown in Fig. 2(b). On the other hands, the intensity of peaks C and D was strongly observed for AB-coated sample because the AB has only six electrons and X-ray penetrates AB easily. These results indicated that the nano-$Al_2O_3$ particles were rather uniformly distributed over the surface of $Li_{1.20}Mn_{0.55}Ni_{0.16}Co_{0.09}O_2$ particles, even if AB particles were mixed with nano-$Al_2O_3$ particles, because a decrease in the intensity of the peaks A and B in the O *K*-edge spectra and peaks C and D in the Mn *L*-edge spectra corresponding to the layered structure means that the surface of $Li_{1.20}Mn_{0.55}Ni_{0.16}Co_{0.09}O_2$ particles was well covered with nano-$Al_2O_3$ powder.

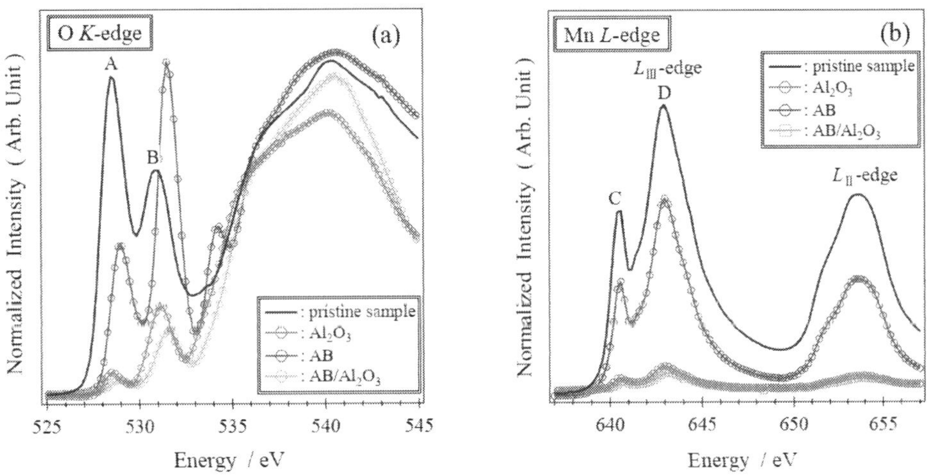


**Figure 2.** (a) O *K*-edge and (b) Mn *L*-edge XANES spectra for the pristine and coated $Li_{1.20}Mn_{0.55}Ni_{0.16}Co_{0.09}O_2$ (UVSOR/BL4B).

Figure 3(a) shows the C-rate dependence of discharge capacity retention of the Li/pristine and coated $Li_{1.20}Mn_{0.55}Ni_{0.16}Co_{0.09}O_2$ cells after stepwise pre-cycling treatment at 298 K. The $Al_2O_3$-coated sample showed the smaller discharge capacity retention with C-rate when it was compared with the pristine sample. $Al_2O_3$ is an insulating material without Li-ion conductivity, therefore, it is thought that the diffusion of Li-ion on the surface of $Li_{1.20}Mn_{0.55}Ni_{0.16}Co_{0.09}O_2$ was barred by $Al_2O_3$ layer. On the other hand, the AB and AB/$Al_2O_3$-coated samples showed almost the same discharge capacity retention. Figure 3(b) shows the cycle dependence of discharge capacity of the Li/pristine and coated $Li_{1.20}Mn_{0.55}Ni_{0.16}Co_{0.09}O_2$ cells at 318 K. The $Al_2O_3$-coated sample showed the smaller initial discharge capacity compared with that of the pristine sample. With cycles, the discharge capacity of the Li/pristine and AB-coated samples cells decreased gradually, while the discharge capacity of the Li/$Al_2O_3$ and AB/$Al_2O_3$-coated samples cells were almost constant value after 35 cycles. In this case, it is thought that the cycle performance at high temperature was improved since the chemical reaction between $Li_{1.20}Mn_{0.55}Ni_{0.16}Co_{0.09}O_2$ and electrolyte was suppressed by the existence of oxide-layer.

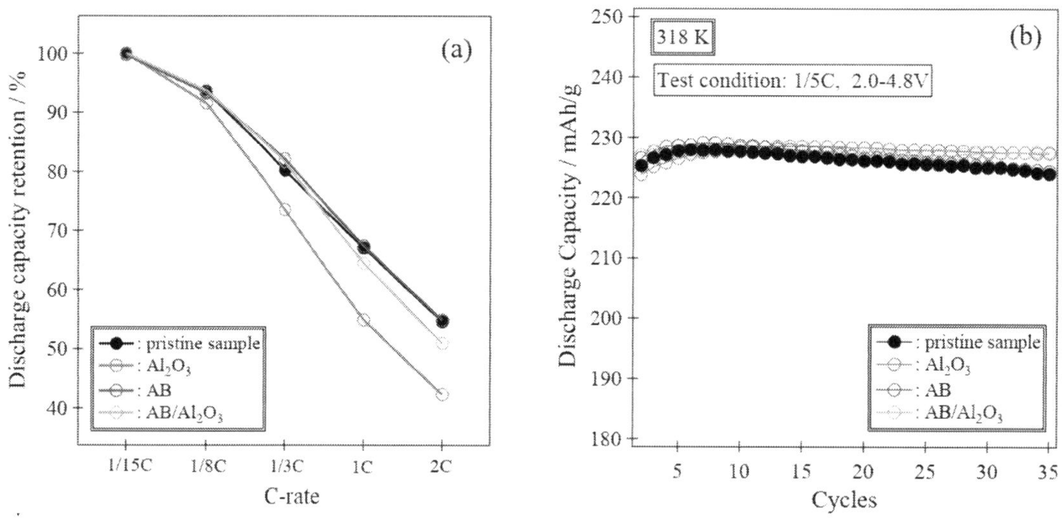


**Figure 3.** (a) C-rate dependence of discharge capacity retention of the Li/pristine and coated $Li_{1.20}Mn_{0.55}Ni_{0.16}Co_{0.09}O_2$ cell after stepwise pre-cycling treatment at 298 K. (b) Cycle dependence of discharge capacity of the Li/pristine and coated $Li_{1.20}Mn_{0.55}Ni_{0.16}Co_{0.09}O_2$ cells at 318 K with a current density of 1/5C and cut-off voltages of 2.0 and 4.8 V.

## 4. Summary

In this study, AB and/or $Al_2O_3$-coated $Li_{1.20}Mn_{0.55}Ni_{0.16}Co_{0.09}O_2$ were prepared using mechanochemical reaction. The optimal preparation condition of coating to $Li_{1.20}Mn_{0.55}Ni_{0.16}Co_{0.09}O_2$ was investigated by SEM, XRD, and XANES measurements. In electrochemical tests, the AB/$Al_2O_3$-coated sample showed the better cycle performance at 318 K, compared with that of the Li/pristine $Li_{1.20}Mn_{0.55}Ni_{0.16}Co_{0.09}O_2$ cell, without degrading the discharge capacity and C-rate performance. These results demonstrate that the mechanochemical AB/$Al_2O_3$-coating process is an effective way of improving the electrochemical performance.

## References

[1] Ito A, Osawa Y and Sato Y, 2008 *J. Power Sources* **183** 344.
[2] Ito A, Li D, Sato Y, Arao M, Watanabe M, Hatano M, Horie H and Ohsawa Y, 2010 *J. Power Sources* **195** 567.
[3] Kobayashi H, Takenaka Y, Arachi Y, Okumura T, Shikano M, Kageyama H and Nitani H, 2012 *Solid State Ionics* **225** 580.
[4] Kobayashi H, Okumura T, Shikano M, Takada K, Arachi Y and Nitani H, 2014 *Solid State Ionics* **262** 43.
[5] Izumi F and Momma K, 2007 *Solid State Phenom.* **130** 15.
[6] Kobayashi H, Sakaebe H, Kageyama H, Tatsumi K, Arachi Y and Kamiyama T, 2003 *J. Mater. Chem.* **13** 590.

Journal of Physics: Conference Series **712** (2016) 012130 doi:10.1088/1742-6596/712/1/012130

# Local structure modification in lithium rich layered Li-Mn-O cathode material

**Marco Giorgetti*[1], Diandian Wang[1], Giuliana Aquilanti[2], Daniel Buchholz[3], Stefano Passerini[3]**

[1]Department of Industrial Chemistry, University of Bologna, 40136 Bologna, Italy
[2]ELETTRA, Sincrotrone Trieste S.C.p.A. s.s 14 km 163.5, 39149 Basovizza, Italy
[3]Helmholtz Institute Ulm (HIU), 89081 Ulm, Germany.

E-mail: marco.giorgetti@unibo.it

**Abstract**. X-ray absorption spectroscopy (XAS) is applied to study the local geometry of Co, Ni, and Mn sites in a new high voltage cathode for lithium batteries. The material is a solid solution between $Li_2MnO_3$ and $Li_{(x)}Mn_{0.4}Ni_{0.4}Co_{0.2}O_2$. The XAS technique has permitted to check the local atomic structure and charge associated with the metals in a series of electrodes with different lithium concentration x, obtained during the first charge operation, and compared to the first discharge and a successive charge. The ex-situ XAS investigation on the initial activation of the cathode material (first charge) can be described by two separated reaction of $LiMO_2$ (M = Ni and Co) and $Li_2MnO_3$. The strength and limitations of the EXAFS approach in these materials is underlined.

## 1. Introduction

Batteries are currently being developed to power an increasingly diverse range of application, from cars to microchip [1]. Among the cathode materials, layered transition metal oxides have been investigated extensively. $LiCoO_2$ is the most important commercial material because of its very good electrochemical performance, despite its major drawbacks such as high cost, toxicity and safety problems [2]. Solid solution of layered cathode materials such as the combination of $Li_2MnO_3$ and $LiMO_2$ (M= Mn, Co, Ni..) have been shown as promising candidates for cheaper, higher capacity and safer cathodes for lithium batteries [3]. The presence of cobalt reduces the electrode polarization and improves the activation of the $Li_2MnO_3$ component, even if present in small amount, but certainly represents an issue with regard to cost. It has been reported that $Li[Li_{0.2}Mn_{0.54}Ni_{0.13}Co_{0.13}]O_2$ , which can be indicated as 0.6 $Li_2MnO_3 \cdot 0.4LiMn_{1/3}Ni_{1/3}Co_{1/3}O_2$, delivers an initial capacity of about 250 mAh/g within the potential range extending from 2.0 to 4.8V [4]. A new high voltage layered $Li[Li_{0.2}Ni_{0.16}Mn_{0.56}Co_{0.08}]O_2$ cathode material, synthesized by co-precipitation method followed by high temperature annealing at 900 °C, displays a capacity of 270 mAh/g (first cycle) at even high rate of charge with a reduction of the use of cobalt in the active material [5]. We present here the ex-situ XAS investigation [6] of the lithium-rich cathode material, $Li[Li_{0.2}Ni_{0.16}Mn_{0.56}Co_{0.08}]O_2$ at different state of charge (SOC), in order to gain an understanding of the role of the three metal sites in the electrochemical redox process occurring during the initial activation. An extensive XANES and a detailed pre-edge study of such electrodes [7] during the first charge has revealed that the manganese is not taking part in the electrochemical oxidation process and a complete $Ni^{2+}/Ni^{4+}$ and a partial $Co^{3+}/Co^{4+}$ redox processes occur. The present paper focuses on the EXAFS characterization of the same electrode at Mn, Ni and Co K-edges. We anticipate that the results of the EXAFS fittings has been found to be correlated to the XANES analysis.

## 2. Experimental

X-ray absorption Spectroscopy (XAS) spectra have been recorded at Elettra (Basovizza, Italy) at the XAFS beamline [8]. The storage ring was operated at 2.0 GeV in top up mode with a typical current of 300 mA. The data were recorded at the Mn K-edge (6539 eV), Co K-edge (7709 eV) and Ni K-edge (8333 eV) in transmission mode using an ionization chamber filled with a mixture of Ar, $N_2$ and He in order to have 10%, 70% and 95% of absorption in the $I_0$, $I_1$ and $I_2$ chambers, respectively. The white beam was monochromatized using a fixed exit monochromator equipped with a pair of Si (111) crystals. Harmonics were rejected by detuning one of the two crystal of the monochromator of 30% with respect to the maximum of the rocking curve. Internal reference of Mn, Co, and Ni foils were used for energy calibration in each scan. Spectra were collected with a constant k-step of 0.03 $Å^{-1}$ with 3s/point acquisition time from 6330 to 7700 eV (Mn K-edge); from 7550 to 8320 eV (Co K-edge) and from 8150 to 9500eV (Ni K-edge).

The preparation of $Li[Li_{0.2}Ni_{0.16}Mn_{0.56}Co_{0.08}]O_2$ electrodes for the ex-situ XAS measurements is as follows: 10 pellets of Li-half cell were prepared (d=13mm) and cycled with C/10 (20 mA/g) to a certain potential, *e.g.*, certain capacity. The cells were then rested for a fixed amount of time before open circuit voltage (OCV) of each cell was checked. Afterwards they were disassembled, the electrodes washed with dimethyl carbonate (DMC), dried and sealed.

The EXAFS analysis has been conducted by using the GNXAS package [9]. $E_0$ values have been found to be displaced by several eV respect to the edge inflection point. The utilized $S_0^2$ values were 0.69, 0.76 and 0.84 for Mn, Co, and Ni, respectively according to the literature [10].

## 3. Results and Discussions

Figure 1 shows the voltage profile of the cell during charge-discharge operation. The numbered points in the curve indicates predetermined states of charge (SOC) at which cells were prepared for the XAS measurements. Point 1 represents the fresh, unchanged electrode, point 2 to 4 corresponds to the extraction of ~83, 103, 200 mAh/g capacity, and point 5 represents a fully charged electrode at 4.8V corresponding to the full ~347 mAh/g practical capacity of the electrode. Point 8 represents a fully discharged electrode at 2.6 V after one complete cycle. Point 9 and 10 were taken on the second charge process.

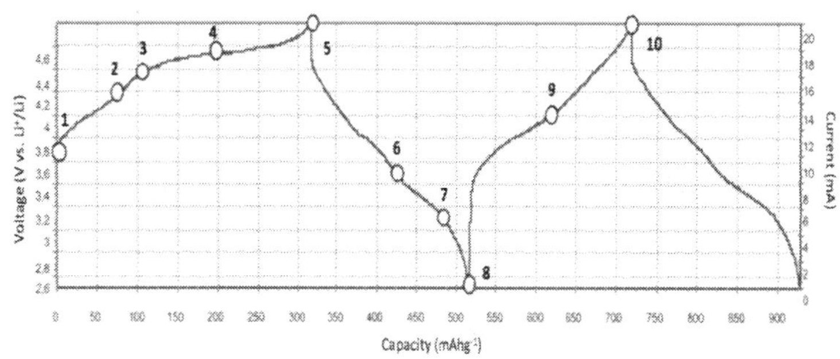


**Figure 1.** Voltage profile of two successive charge and discharge curves of Li-rich NCM at 20mA/g. Representative points of 1-10 in the process of XAS measurements are indicated. Reference and counter electrode: Li. electrolyte: 1 M $LiPF_6$ in EC/DMC.

The active material, $Li[Li_{0.2}Ni_{0.16}Mn_{0.56}Co_{0.08}]O_2$, is an equimolar solid solution of monoclinic *(C2/m)* $Li_2MnO_3$ and hexagonal layered *(R3m)* $LiMn_{0.4}Ni_{0.4}Co_{0.2}O_2$. Structurally speaking the presence of two phases can make the EXAFS analysis difficult. A structural model based on the hexagonal layered structure works well for Ni and Co, whereas a mixed phase should be considered for the Mn K-edge data. Because of the observed $Li_2MnO_3$ activation step [7] during the first charge (sample 1 to 5) with concomitant conversion of the monoclinic component to the hexagonal one, we have used the hexagonal structural model for the EXAFS analysis at the Ni, Co K-edges (all samples) and Mn (samples 5 to 10), whereas modifications for samples 1-4 in the case of Mn K-edge data were applied.

The structural differences between the two phases, at the local domain probed by EXAFS, arises in the Mn-Mn interactions of the second shell (three in the monoclinic vs. six in the hexagonal layered structure). Figure 2 displays an example of the EXAFS data analysis at the Mn K-edge of sample 6. The quality of the fits is good. The only frequency missing, and not added in the fitting procedure, is the one ascribable to the Mn-Mn pair at distance higher than 5 Å (Fig 2b).

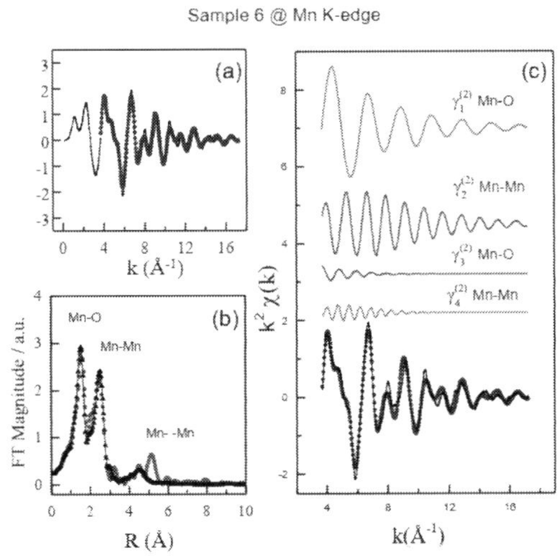


**Figure 2.** Best fit for sample 6 at the Mn K-edge as an example, in terms of EXAFS signal (a), the corresponding FT (b), and details of the single EXAFS contribution (c).

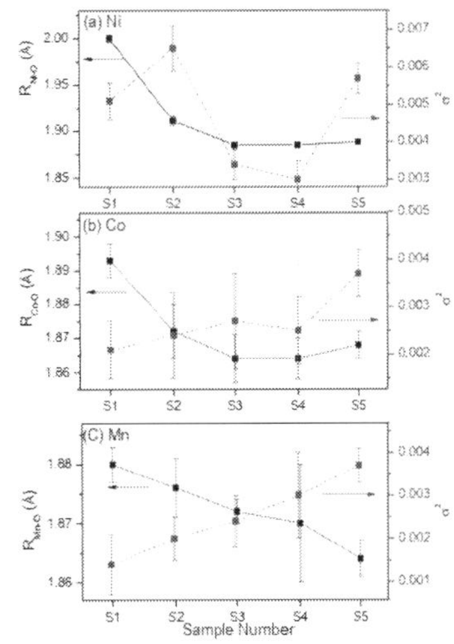


**Figure 3.** M-O first shell best fit results and corresponding Debye Waller. The error bars are also indicated. They have been calculated by the CONTOUR plots.

Figure 3 reports the M-O first shell best fit results and corresponding Debye Waller at the three metal edges, for the first charge. The first shell of Ni shortens dramatically (by about 0.12 Å for the Ni- O distance) and also for Co (by about 0.03 Å, coordinated by six oxygen atoms). The corresponding Debye Waller factors generally increase during the lithium release process (first charge), indicating an increase of the structural disorder around the Ni and Co site. The Mn-O first shell distance exhibited only minor changes (about 0.02 Å). The corresponding EXAFS Debye Waller factor has revealed a large increase of the structural disorder. This phenomenon is in good agreement with the pre-edge analysis [7]. Overall these findings agree well with the proposed mechanism of initial $Li_2MnO_3$-like domain activation to a hexagonal phase, which is retained during subsequent cycling.

EXAFS confirms that the manganese is not fully redox active in this material but, instead, assists only to a low extent in the overall redox reaction. As seen, a large decrease of the Ni-O first shell distance was observed. This bond distance is about 1.88 Å in sample 3, 4, and 5, which agrees with the one observed in similar electrodes [9] for $Ni^{4+}$-O. This confirms that $Ni^{2+}/Ni^{4+}$ is the most reacting electrochemical couple during the first activation process. Data for samples 6 to 10 (not reported in figure 3) agrees with a reversible structural modification for both $Co^{2+}/Co^{3+}$ and $Ni^{2+}/Ni^{4+}$ electrochemical couple.

The use of Contour plots in the parameter error determination, depicted in Figure 4, suggests a further consideration. Among the few examples shown in the figure for samples 5 and 10, where parameters have been selected among those having strong correlation, in order to reflect the highest error, the last two at the bottom (in blue) involving the coordination number of the Mn--Mn in the second shell, deserve a particular attention. This parameter is highly relevant because it allows a direct comparison of the two possible phases of the lithium manganese oxide cathode. This number is expected to be three in the monoclinic case and six in the hexagonal layered structure case. As seen from the figure, this relevant parameter has a strong correlation when considered against the EXAFS Debye Waller. This means that the two fitting models, namely A (hexagonal like) and B (monoclinic plus hexagonal) in the figure, are statistically of the same quality and both possible. In our case we have kept the coordination number fix to 6. In another test, both parameters were allowed to float, and the results indicated a coordination number close to 6. In other cases [11] there has been a speculation to the ability of the EXAFS to discriminate among the two different phases. The work presented here suggests to take into account about these pitfalls while investigating the structures of the lithium based manganese oxide cathode materials.

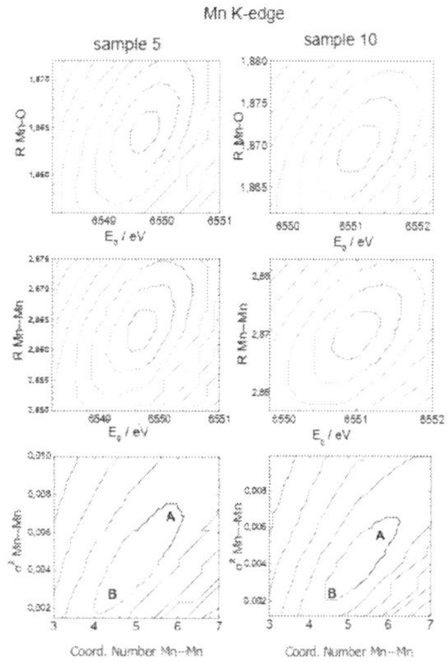


**Figure 4.** Examples of two-dimensional section of the parameter space (contour plots) for the error parameter determination, for Mn K-edge EXAFS data (sample 5 and 10). These plots were selected among the parameters having strong correlation to reflect the highest error. This evaluation provides only statistical errors on EXAFS refined parameters. The inner elliptical contour corresponds to the 95% confidence level. The parameters errors have been indicated in Figure 3. The two plots at the bottom have been obtained by left floating both $\sigma^2_{Mn-Mn}$ and $CN_{Mn-Mn}$. (third shell). It is seen that at the 95% confidence level two structural configuration indicated as A ($\sigma^2_{Mn-Mn}$= 0.003 and $CN_{Mn-Mn}$ = 4.5) and B ($\sigma^2_{Mn-Mn}$= 0.007 and $CN_{Mn-Mn}$ = 6) can be considered equivalent at the 95% confidence level.

## 4. Conclusions

EXAFS data at Mn, Ni, and Co K-edges of the lithium rich, cobalt poor, $Li[Li_{0.2}Ni_{0.16}Mn_{0.56}Co_{0.08}]O_2$ layered material at different state of charge provided new insights on the mechanism of the lithium release and insertion in their first stages of battery life.

## 5. Acknowledgements

Experiments at Elettra have been performed through the proposal # 20110095. D. Wang wishes to thank the EC for providing the 2-years scholarship through the "Advanced Spectroscopy in Chemistry" Erasmus Mundus European Master Programme.

## References

[1] Guo YG, Hu JS, Wan LJ 2008 *Adv Mater* **20** 2878
[2] Zhang XY, Jiang WJ, Maguer A, Quilu F, Gendron F, Julien CM 2010 *J Power Sources* **195** 1292
[3] Li J, Zheng JM, Yang Y 2007 *J Electrochem Soc* **154** A427
[4] Gao J, Kim J, Manthiram A 2009 *Electrochem Commun* **11** 84
[5] Li J, Kloepsch R, Stan MC, Novak S, Kunze M, Winter M, Passerini S, 2011 *J Power Sources* **196** 4821
[6] Giorgetti M 2013 *ISRN Materials Science* **2013**, Art. No. 938625, 22 pages
[7] Buchholz D, Passerini S, Aquilanti G, Wang D, Giorgetti M 2015 *ChemElectroChem* **2** 85
[8] Di Cicco A, Aquilanti G, Minicucci M, Principi E, Novello N, Cognigni A, Olivi L **2009** *Journal of Physics: Conference Series* 190 012043
[9] Filipponi A, Di Cicco A 1995 *Phys Rev B* **52** 15135
[10] Bareno J, Balamasubramian M, Kang SH, Wen JG, et al. 2011 *Chem Mater* **23** 2039
[11] Rana J, Stan M, Kloepsch R, Li J et al 2014 *Adv Energy Mater*, **4**, Art. No. 1300998

# Fe local structure in Pt-free nitrogen-modified carbon based electrocatalysts: XAFS study

**Agnieszka Witkowska**[1], **Gabriele Giuli**[2], **Marco Renzi**[3], **Stefania Marzorati**[4], **Wubulikasimu Yiming**[2], **Francesco Nobili**[3], **Mariangela Longhi**[4]

[1]Gdansk University of Technology, Department of Solid State Physics, Poland
[2]University of Camerino, Geology Division, Camerino, Italy
[3]University of Camerino, Chemistry Division, Camerino, Italy
[4]University of Milano, Chemistry Department, Milano, Italy

E-mail: agnieszka@mif.pg.gda.pl

**Abstract.** The paper presents a new results on the bonding environment (coordination number and geometry) and on oxidation states of Fe in nitrogen-modified Fe/C composites used as Pt-free catalysts for oxygen reduction in Direct Hydrogen Fuel Cells. Starting from glucose or fructose, two catalysts displaying different electrochemical performance were prepared and studied in the form of pristine powder and thin catalytic layer of electrode by Fe K-edge XAFS spectroscopy. The results show how the Fe local structure varies as a function of different synthesis conditions and how changes in the structural properties of the catalysts are related to fuel cell electrochemical performance increase during a cell activation period.

## 1. Introduction

Nowadays, Direct Hydrogen Fuel Cells (DHFCs) represent one of the most promising alternatives for the development of low temperature power devices. However, one of the most severe limitations to a large-scale application of this technology is the high cost of both polymeric membranes and of catalytic electrode materials, commonly requiring noble metals as Pt due to the intrinsic difficulty of the cathodic Oxygen Reduction Reaction (ORR). In this context, non-precious metal catalysts are actively searched for, such as to meet already established operational benchmarks for conventional platinum low temperature FC vehicular requirements, with the additional benefit of significant cost reduction and stability increase.

Among various materials, nitrogen-modified carbons containing non-precious transition metals are considered interesting because of their optimized morphology (in terms of porosity and specific surface area), ORR activity (closely approaching those of commercial Pt catalysts) and reliability as confirmed by results from independent groups [1, 2, 3]. This work is dedicated to nitrogen-modified carbons containing Fe centres. Innovation of considered catalysts involves the use of organic compounds, fructose and glucose, as a precursor of carbon. Very promising electrochemical performance (clearly revealed by rotating disk electrode measurements), electrochemical stability (depending on the synthesis pocedure and working conditions in real FC) and a very scarce data about the nano and atomic-level structure of these catalysts became the motivation of the present study. Scaning electron microscopy (SEM), X-ray diffraction (XRD) and X-ray Absorption Fine Structure (XAFS) measurements and analysis were performed to define the location and local structure of metal centres and their chemical states in the catalyst at various stages of the electrode preparation and to determine if and how changes in the iron local environment affect the catalyst stability.

## 2. Materials

Among several preparation approaches, thermal treatment of mixed precursors from a carbon and nitrogen source added with transition metal salts [4] seems promising due to the ease and flexibility of the synthetic route. Thus, nitrogen-modified carbons containing Fe centres (about 1-2 wt.%) were prepared by high-temperature pyrolysis in $N_2$ atmosphere of a gel formed by organic precursors of C (fructose or glucose), N (guanidine acetate) and Fe (Fe(II) acetate) [3]. For the structural study pristine powders and conditioned in real FC electrodes prepared on the base of these powders were used. Catalytic inks and electrodes for FC test were prepared in a standard way. Resulting Fe loading in the catalytic layer was about 0.1-0.2 mg/cm$^2$. Electrode conditioning was realized passing fully humidified oxygen and hydrogen at cathode and anode side, respectively, at $T_{O_2}/T_{H_2}/T_{cell} = 70/70/70^oC$. Activation process was conducted at 0.6 V by about 3 h, until stable polarization curve was obtained. The obtained samples are hereafter called Fe-FAG and Fe-GAG when fructose and glucose, respectively, was used as a C source.

## 3. Results and discussion

X-ray diffraction data of Fe-FAG and Fe-GAG pristine powders are shown in Figure 1. Obtained patterns exhibit apparent differences between Fe-FAG and Fe-GAG samples. In Fe-FAG pristine powder, it can be easly observed the presence of ordered carbon or carbon-iron phases (peaks indicated by full circles in Figure 1). The broad nature of these reflections in Fe-GAG pristine powder pattern indicate poor ordering of the carbon structure. Moreover, it should be underlined that additional diffraction peaks (indicated by crosses in Figure 1), related to the presence of small amount of metallic Fe, can also be noticed, but only in the pattern of Fe-FAG pristine powder. At the same time, SEM images show the presence of small (nanometric) spherules in both cases (see Figure 2). However, the spherules are more numerous in Fe-FAG powder than in Fe-GAG one, thus explaining the absence of metallic iron diffraction peaks in Fe-GAG pattern.

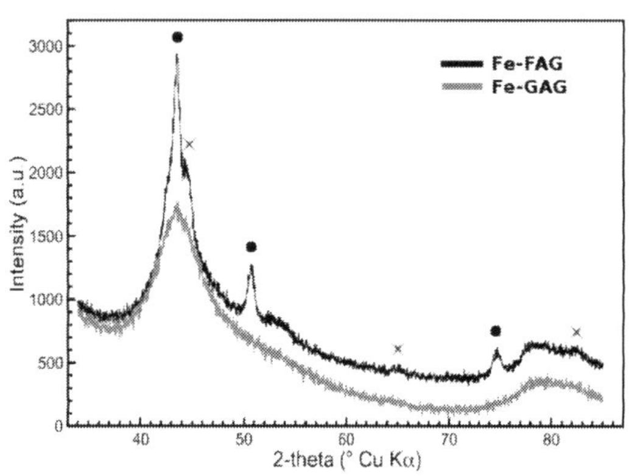


**Figure 1.** X-ray diffraction patterns of Fe-FAG and Fe-GAG pristine powders. Full circles – poorly ordered carbon or carbon-iron phases, crosses – metallic Fe.

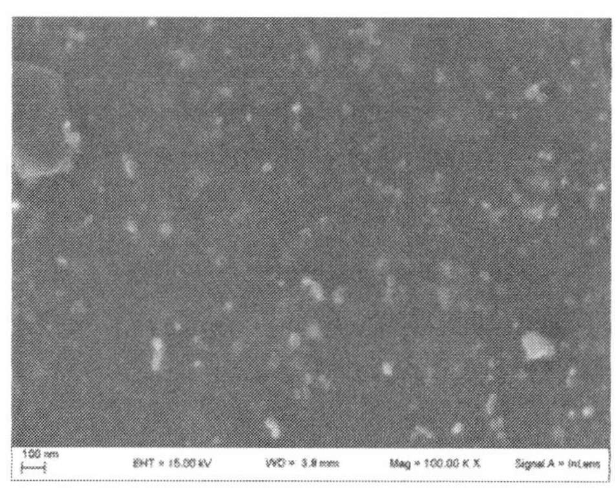

**Figure 2.** SEM image of the Fe-FAG powder displaying nanometric spherules (possibly agglomerates of metallic Fe or carbon-iron compounds).

Fe K-edge XAFS measurements in both near-edge (XANES) and extended (EXAFS) range were also performed. Spectra of pristine powders, conditioned electrodes and a series of reference samples were recorded at ELETTRA 11.1 XAFS station. All XAFS measurements were realized ex situ at RT using transmission geometry. Coarse comparison of the XANES spectra (Figure 3) and Fourier transforms (FTs) of the total EXAFS spectra (Figure 4) reveals a strong difference in the Fe local structure for catalysts prepared in the various synthesis conditions (compare pristine powders of Fe-FAG and Fe-GAG). Then, it should be noted that in both catalysts electrode preparation and activation process realized in a real FC changed Fe local structure. However, the most pronounced structural modification occurred in the case of Fe-GAG. Thus, resulted XAFS spectra of both conditioned electrocatalysts are rather similar (differences are observed especially in the near-edge region, see inset in Figure 3).

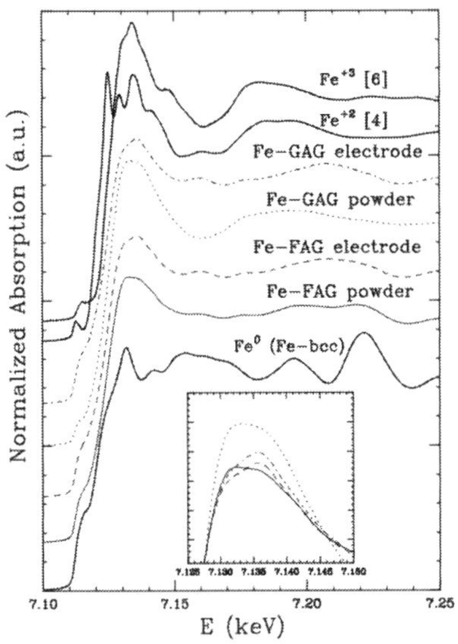


**Figure 3.** Fe K-edge normalized XANES spectra of Fe-FAG, Fe-GAG samples and of some reference bulk samples (thick solid lines). Inset: Fe-FAG and Fe-GAG white line intensity and shape comparison.

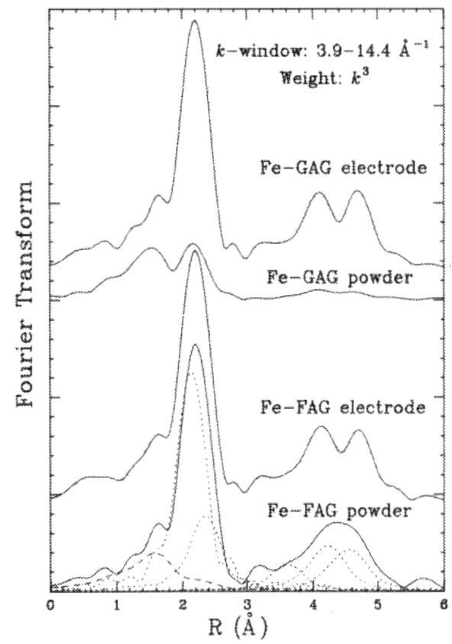


**Figure 4.** Fourier transforms of the EXAFS experimental signals of Fe-FAG and Fe-GAG samples. Theoretical partial $\gamma^{(n)}$ signals used in GNXAS analysis are also shown: dashed line – Fe-C; dotted lines – nano Fe-bcc.

Qualitative structural analysis of studied samples was started by performing reference spectra linear combination fitting in the XANES region. However, it should be underlined that fit result for any spectra combination was not satisfactory. Thus, only Fe metallic phase contents was properly estimated. Obtained values (over 50% for Fe-FAG powder, about 20% for Fe-GAG powder and over 60% for both activated electrodes) fully confirm XRD and SEM observations.

Then, structural models for detailed EXAFS data-analysis performed using the GNXAS method [5] were constructed taking into account the above mentioned results. Thus, the calculated signals were modeled based on the $Fe_3C$ structure (only first shell Fe-C two-body contribution) and Fe-bcc structure (two- and three-body contributions, up to 5th coordination shell) – see FT of Fe-FAG pristine powder spectra in Figure 4. Best fit results for the first-shell structural parameters are shown in Table 1. First of all, it should be mentioned that in all studied samples Fe-Fe configurations (up to 5th coordination shell) precisely reconstruct metallic Fe-bcc structure. However, reduction of coordination numbers and higher value of disorder parameters in respect to bulk metallic iron, indicate the presence of nanostructured Fe. Moreover, it is interesting to note that obtained Fe local structure parameters are different for Fe-FAG and Fe-GAG conditioned electrodes, especially these which describe metallic phase structure. In Fe-FAG electrode, larger Fe nanoparticles with more disordered Fe-Fe distributions can be detected.

**Table 1.** The best GNXAS fit results for the first shell Fe-C and Fe-Fe configurations obtained for all considered samples: $R$ [Å] – the average inter-atomic distance, $\sigma^2$ [$10^{-3}$Å$^2$] – the standard deviation of distance, $\beta$ – asymmetry parameter and $N$ – the coordination number.

| Sample | $R_{\mathrm{Fe-C}}$ | $\sigma^2_{\mathrm{Fe-C}}$ | $\beta_{\mathrm{Fe-C}}$ | $N_{\mathrm{Fe-C}}$ | $R_{\mathrm{Fe-Fe}}$ | $\sigma^2_{\mathrm{Fe-Fe}}$ | $N_{\mathrm{Fe-Fe}}$ |
|---|---|---|---|---|---|---|---|
| Fe-FAG powder | 1.97(1) | 12(1) | 0.1 | 6.1(6) | 2.475(5) | 6.0(6) | 5.1(2) |
| Fe-FAG electrode | 1.96(1) | 9(1) | 0.05 | 4.2(4) | 2.488(5) | 8.0(8) | 7.0(3) |
| Fe-GAG powder | 1.94(1) | 8(1) | 0.0 | 4.8(5) | 2.47(1) | 8(1) | 1.4(1) |
| Fe-GAG electrode | 1.96(1) | 10(1) | 0.05 | 4.4(4) | 2.489(5) | 6.9(7) | 6.3(3) |

These differences may cause different catalyst activity and stability observed when durability tests are performed. At the same time, in all samples Fe-C distribution is broad and sligtly asymmetric. Only in the case of Fe-GAG powder the distribution is a little better defined, giving in EXAFS signal comparable contribution with the Fe metallic phase. However, there can not be clearly concluded whether Fe-C contribution is related to carbon-iron compounds or Fe nanoparticle attached to carbon support. Therefore, detailed XANES spectra analysis based on the theoretical calculations realized by means of the MXAN code [6] were performed. MXAN preliminary results, obtained using the model structure constructed on the base of Fe-C and Fe-Fe distances and multiplicities determined by EXAFS data analysis (Table 1), show the possible existence of small Fe clusters attached to a grafen-like layer. In particular, model structure with a single Fe atom attached to that layer produces theoretical spectra similar to that of Fe-GAG pristine powder, whereas larger Fe cluster supported on the grafen-like structure gives theoretical spectra more similar to the spectra measured for Fe-FAG powder and conditioned electrodes. However, more advanced and sophisticated calculations are necessary to define precisely the coordination geometry of Fe ions and location of Fe nanoparticles.

## 4. Conclusions

XANES and EXAFS results combined with XRD and SEM analysis allowed us to construct model of Fe environment in nitrogen-modified carbon based electrocatalysts in which organic compounds, fructose and glucose, were used as a carbon precursor. Detected subtle structural differences and changes occurred on the atomic level during the catalysts preparation and activation in real FC were used as a possible explanation of a different catalysts stability and their electrochemical activity observed during the fuel cell operation. In order to improve our knowledge about Pt-free catalysts degradation mechanisms, detailed structural analysis of considered catalysts subjected to durability tests (various protocols were applied) are in progress.

## Acknowledgments

We acknowledge the support of the SLL ELETTRA in providing SR facilities (XASF11.1 station) for 2035398 experiment. This research has been carried out within the PRIN Project: "Advanced nanocomposite membranes and innovative electrocatalysts for durable PEM FCs".

## References

[1] Lefèvre M, Proietti E, Jaouen F and Dodelet J-P 2009 *Science* **324** 71
[2] Garsuch A, MacIntyre K, Michaud X, Stevens D A and Dahn J R 2008 *J. Electrochem. Soc.* **155** B953
[3] Galbiati I, Bianchi C L, Longhi M, Carrà A and Formaro L 2010 *Fuel Cells* **10** 251
[4] Maruyama J, Fukui N, Kawaguchi M, Abe I 2008 *J. Power Sources* **182** 489
[5] Filipponi A et al., 2009 *GNXAS extended suite of programs for advanced x-ray absorption data-analysis: methodology and practice* (Gdansk, Poland: TASK Publishing)
[6] Benfatto M, Congiu-Castellano A, Daniele A and DellaLonga S 2001 *J. Synchr. Rad.* **8** 267

# Structural phase transitions in ionic conductor $Bi_2O_3$ by temperature dependent XPD and XAS

**Yingcai Zhu,** [1ξ]**Pengfei An,** [1ξ]**Meijuan Yu,** [1*]**Augusto Marcelli,** [2] **Yong Liu,**[3] **Tiandou Hu,** [1] **and Wei Xu**[1*]

[1]Beijing Synchrotron Radiation Facility, Institute of High Energy Physics, Chinese Academy of Sciences, China

[2]INFN – Laboratori Nazionali di Frascati, Via E. Fermi 40, 00044 Frascati, Rome, Italy

[3]AVIC, Beijing Institute of Aeronautical Materials, Beijing 100095, China.

[ξ] These two authors contributed equally

E-mail: yumj@mail.ihep.ac.cn and xuw@mail.ihep.ac.cn

**Abstract**. The superionic behavior of cubic δ-phase $Bi_2O_3$, a metastable phase at high temperature, is of great interests from both scientific and technological perspectives. With the highest ionic conductivity among all known compounds, the δ-phase $Bi_2O_3$ possesses promising applications in solid-oxide fuel cells. Previous investigations pointed out the α to δ-phase transition occurs during the heating process, as supported by the X-ray and Neutron diffraction experiments. Through in situ measurements of the long-range order structure and the local structure by X-ray powder diffraction and X-ray absorption spectroscopy, we investigated the evolution of the structures under different temperatures. Both techniques provided ample evidence that the existence of meta-stable β-phase are crucial for forming the defective fluorite cubic δ phase. Our finding suggested that the phase transition from tetragonal β-phase to δ-phase is an influencing factor for the generation of the oxygen-ion pathways.

## 1. Introduction

The demand of energy is triggering large investments in energy related researches spanning from physics, chemistry, materials sciences to biology. The main goal of these researches is essentially that to resolve bottleneck issues regarding storage, conversion and transport of energy. Among the promising candidates for energy storage applications, the solid-oxide fuel cells (SOFCs) possess unique merits such as high efficiency, great stability, low cost and a low environmental impact. However, the mechanical/physical compatibility of components hamper the performance of SOFCs because these structural phase transitions occur at high temperature. Investigation of the structural phase transitions involving SOFCs components is mandatory, being a real critical issue that focus interests from both a fundamental and a technological perspective.

Journal of Physics: Conference Series **712** (2016) 012132 doi:10.1088/1742-6596/712/1/012132

The ionic conductor $Bi_2O_3$ is a promising electrolyte for SOFC because it exhibits an extraordinary high ionic conductivity at high temperature. Actually, at high temperature, $Bi_2O_3$ exhibits a complicated phase transition diagram involving four polymorphs (Fig.S1).[1] By heating up the material from room temperature to 700° C, $Bi_2O_3$ experience a monoclinic α-phase to cubic δ-phase transition, whereas several other phases (tetragonal β-phase or bcc γ-phase) appear as the material is cooled down to low temperature.[1] The ionic conductivity in the high temperature cubic phase is four orders of magnitude larger than that at room temperature in the monoclinic phase,[2] pointing out a structural origin for this ionic conductivity. The cubic δ-phase exhibits a defective fluorite structure with 25% of oxygen sites empty. Intriguingly, theoretical studies pointed out that the ionic conduction mechanism is closely related with 3D configurations of oxygen vacancies.[3] Configurations (Figure. S1)derived as (100), (110) and (111), respectively, are probably correlated to the lone electron pairs (LEP) of $Bi^{3+}$ ions.[4] Moreover, due to the "soft" cation lattice, ionic diffusions are supported by the formation of transport paths associated to oxygen vacancies.[5] Furthermore, as previously showed the atomic disorder is ubiquitous at high temperature while the average symmetry is cubic. In spite of the available experimental data, a detailed understanding of the phase transition path from the α- to δ-phase while heating and cooling, which may explain the principles of the SOFC degradation in $Bi_2O_3$ is lacking. Furthermore, the understanding of nonequilibrium phenomena such as the temperature-dependent phase transitions is of fundamental significance in condensed matter physics.

It is well known that X-ray powder diffraction and X-ray absorption spectroscopy are complementary techniques in resolving the average long-range ordered atomic structure as well as the average local atomic structure. Owing to the unique element and site selectivity, the X-ray Absorption spectroscopy probes the local geometric and electronic structure of the system.[6] Previously, the temperature dependent structural phase transitions in $Bi_2O_3$ were investigated by X-ray diffraction, [1] neutron diffraction[7]and Raman spectroscopy,[8] revealing the α- to δ-phase transition. Recently, the short-range order structure were investigated by analyzing the diffuse scattering part of neutron total scattering data.[9] Interestingly, the short range ion-ion correlation[9] in disordered δ phase $Bi_2O_3$ are different from that deduced from the long-range order structure. [6-8]

In this article, we investigate the phase transformation path of $Bi_2O_3$ polymorphs during a cycle of heating and cooling by resolving the variations in local structure as well as the evolution of the average long-rang ordered structure. By employing the *in situ* X-ray powder diffraction and the *in situ* X-ray absorption spectroscopy, the structural phase transition can be deciphered readily. Our findings suggest the formation of β-phase nanoclusters on the track of phase transition from α- to δ-phase, implying the meta-stable β -phase is crucial for in forming the defective fluorite superionic state in disordered cubic δ phase.

## 2. Experimental and calculation method

Temperature dependent X-ray Absorption Fine structure Spectroscopy (T-XAFS) experiments were performed at the 1W2B endstation of the Beijing Synchrotron Radiation Facility. Bi2O3 pellets were housed in a home-made furnace.[10] The samples were heated from RT to 750°C then cooled down to RT. The heating rate was about 10°/min from room temperature to 750°C when performing the QXAFS; the cooling rate was controlled as 10°/min from 750°C to 500°C; then the sample was left to naturally cooling down to room temperature within 10 minutes. During the heating process, we collected QXAFS spectra in the transmission mode; while fine scan XAFS spectra were measured at separate temperature.

The temperature dependent X-ray powder diffraction (T-XPD) experiments were performed at the 4B9A endstation of the Beijing Synchrotron Radiation Facility. The $Bi_2O_3$ powder were brushed upon a Pt heater in the commercial furnace. The XPD experiments were performed in both quick and fine scan mode while heating from RT to 800°C. The heating rate for XPD was set the same as for XAFS.

XAFS spectra are normalized after subtracting the atomic absorption background using the Demeter and IFEFFIT package.[11] Fourier Transforms of $k^3$ weighted EXAFS $\chi(k)$ were performed with a Hanning type window over the $k$ range of 2 to 13 $Å^{-1}$. To perform a consistent fitting for all temperatures, the k range was fixed to [2.2, 10.5] Å$^{-1}$ and the fitted R range lied in [1, 4] Å. The simulation of near-edge region of the X-ray Absorption Spectra (XAS) were carried out using the full multiple scattering (FMS) theory as implemented in FEFF9.6 code.[12] The muffin-tin potential and Hedin-Lundqvist exchange-correlation potential were adopted to calculate the atomic potential in self-consistent field (SCF). To achieve good convergence, in our calculations the cluster radius for SCF potential and full multiple scattering is 5 and 8 Å, respectively.

## 3. Results and Discussions

The temperature dependent X-ray powder diffraction profiles (Fig.1(a) and Fig.S4) can be used to determine the phase qualitatively. As shown in Fig.1 (a), the XPD profiles of $Bi_2O_3$ from room temperature to 800 °C clearly show phase transformation during the heating process. The room temperature phase can be purely indexed to the monoclinic α-phase ( PDF#41-1449). The relative intensity ratio between the (002) and (120) plane are slightly different from the PDF card, indicating preference of certain orientations in the powder samples in which the particle orientations are anisotropic. The peaks at high angles above 60° remain for all measured temperature, which is systematic background not related with the sample. After heating up to 400 °C, the intensity of (002) peak decreases but the α-phase is still maintained. Above 600°C, the peak at 28° becomes more pronounced indicating the emergence of β-phase. Therefore, a mixture of α and β is shown in temperatures above 600°C. Both δ-and α-phase show a strong line at 26°; so we may resort to the line at 54° to identify the δ-phase. The (002) peak disappears at 800°C, indicating that the corresponding preferred orientation was broken due to large temperature-induced disordering occurring at high temperature. The peak at 54° emerges in 800°C that is likely related to the δ-phase. Hence it is concluded that phase stable temperature window for α- and β-phase is 400 to 600°C and 600°C to 700°C, respectively. On the other hand, since the $Bi_2O_3$ melts quickly about 830°C, the phase stable window for δ-phase is quite narrow (from 800 to 830°C). Meanwhile, we observed that the phase transition is not sharp since the mixed α-, β-, and δ- phase are indexed even at 800°C.

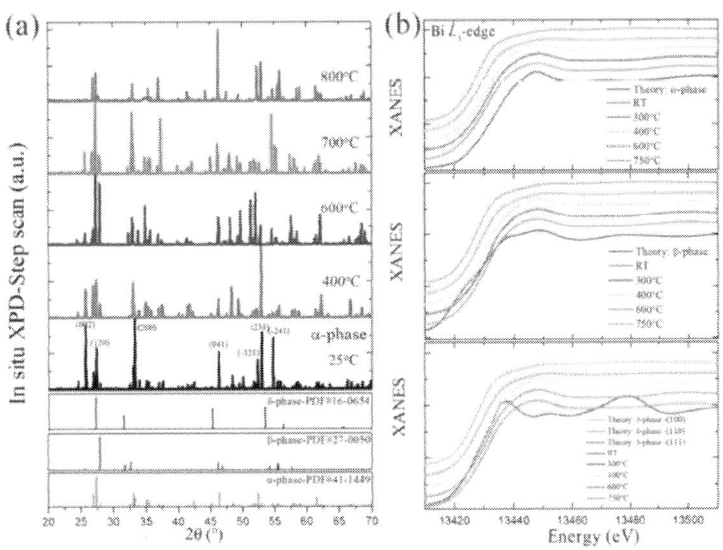


**Figure 1.** (a)Temperature dependent X-ray Powder diffraction and the PDF cards corresponding to $Bi_2O_3$ polymorphs; (b) T-XAS versus the theoretical calculations of all $Bi_2O_3$ polymorphs.

Next, we focus on the temperature induced local structure transition. As shown in Fig.1 (b), the XANES spectra at Bi $L_3$-edge not only reflect the electronic transition from Bi 2p core electrons to Bi 5d empty states but also the multiple scattering of the photon excited electrons by coordinate atoms.

Owing to the lack of empty d shells and longer core-hole width (~5.98 eV), the spectra at Bi $L_3$-edge are flat and seemingly featureless. The feature becomes more broadened at high temperature due to thermal atomic disordering. To identify the phase, we performed theoretical simulations of the XANES spectra by employing Full Multiple Scattering Theory. Firstly, comparison of the experimental XANES spectra of $Bi_2O_3$ and theoretical XANES spectra based on different atomic models suggest that the mixture phase appears at high temperatures above 600 °C. Different configurations of oxygen vacancies are adopted to simulate the defective fluorite cubic δ-phase. However, the theoretical spectra deviate from the experimental spectra, suggesting that δ-phase does not exist even at 750°C. Furthermore, the EXAFS fitting (Figure.S3, and Table. S1) of the phases using α-and β-phase models suggested that the mixture of both phases at high temperatures above 600 °C while heating. The β-phase is stabilized while cooling down to 500°C. As shown in Fig.1(a), there is evidence for δ-phase at 800°C from XPD. But we cannot identify δ-phase from XAFS.

## 4. Conclusions

The *in situ* X-ray powder diffraction and *in situ* X-ray absorption spectroscopy were employed to investigate the polymorphs of $Bi_2O_3$ under temperature. Upon heating from room temperature to 400°C, $Bi_2O_3$ are well crystallized in α-phase; while mixed α-and β-phase are identified above 400°C. The δ-phase exists in very narrow window (800°C-830°C). Unlike the previous reports for sharp phase transition from α- to δ-phase,[1] our XAFS and XPD data indicate a sluggish phase transition from α- to δ-phase mediated by β-phase. The formation of β-phase in heating process participate in the formation of oxygen ionic indicates an alternative thermodynamic pathways form forming vacancies as required for δ-phase. Upon cooling down to 500°C, the intermediate tetragonal β-phase formed in heating process can be stabilized.

Moreover, the soft oxygen sub-lattice and the $6s^2$ lone electron pairs of $Bi^{3+}$ plays an essential role in the temperature induced phase transitions owing to the highly modulated potential energy surface.[13] The formation of β-phase prior to the δ-phase may induce drastic variations of the potential energy surface; hence, the phase transition from α to δ can be either hampered or facilitated depending on the thermodynamic pathways, remaining an open question for further investigations. The findings provide an insight to the understanding of superionic conductivity in $Bi_2O_3$ at high temperature as applied in solid-oxide fuel cells.

## Acknowledgments

The authors acknowledge the financial support from the NSF of China under Grant No. 11205186, No. U1532128, and No. 11105172. W. X acknowledge the financial support from the INFN-IHEP collaborative framework. This work is dedicated to the late Prof. Dr. Ziyu Wu.

## References

[1] Harwig, H. A.; Gerards, A. G. J. Solid State Chem. 1978, 26, (3), 265-274
[2] Stephen, H. Rep. Prog. Phys. 2004, 67, (7), 1233
[3] Mohn, C. E.; Stølen, S.; Norberg, S. T.; Hull, S. Phys. Rev. Lett. 2009, 102, (15), 155502
[4] Gabuda, S.; Kozlova, S.; Érenburg, S.; Bausk, N.; Davidovich, R.; Zyryanov, V.; Yukhin, Y. JETP Lett. 2002, 76, (1), 50-52
[5] Mohn, C. E.; Stølen, S.; Norberg, S. T.; Hull, S. Phys. Rev. B 2009, 80, (2), 024205
[6] Rehr, J. J.; Albers, R. C. Rev. Mod. Phys. 2000, 72, (3), 621-654
[7] Yashima, M.; Ishimura, D. Chem.Phys. Lett 2003, 378, (3–4), 395-399
[8] Schröder, F.; Bagdassarov, N.; Ritter, F.; Bayarjargal, L. Phase.Transit. 2010, 83, (5), 311-325
[9] Norberg, S. T.; Eriksson, S. G.; Hull, S. Solid State Ionics. 2011, 192, (1), 409-412
[10] An, P.; Hong, C.; Zhang, J.; Xu, W.; Hu, T. J.Synchrotron.Rad. 2014, 21, (1), 165-9
[11] Ravel, B. and Newville, M. J.Synchrotron.Rad., 2005, 12, 537-541.
[12] Rehr, J.J.; Kas, J.J.; Vila, F.D.; Prange, M.P.; Jorissen, K. Phys. Chem. Chem. Phys., 2010, 12 5503.
[13] Mohn, C. E.; Stølen, S.; Norberg, S. T.; Hull, S. Phys.Rev. B 2009, 80, (2), 024205

# NEXAFS and XPS of p-Aminobenzoic Acid Polymorphs: The Influence of Local Environment

**J S Stevens[1], A Gainar[1], C Jaye[2], D A Fischer[2], S L M Schroeder[3,4]**

[1]School of Chemical Engineering and Analytical Science, The University of Manchester, Oxford Road, Manchester, M13 9PL, UK
[2]Material Measurement Laboratory, National Institute of Standards and Technology, Gaithersburg, MD, 20899, USA
[3]School of Chemical and Process Engineering, University of Leeds, Leeds, LS2 9JT, UK
[4]DIAMOND Light Source Ltd., Chilton, Didcot, OX11 0QX, UK

E-mail: joanna.stevens@manchester.ac.uk; s.l.m.schroeder@leeds.ac.uk

**Abstract**. Nitrogen K-edge XPS and NEXAFS of the two polymorphic forms of para-aminobenzoic acid (PABA) are significantly different reflecting variation in hydrogen bonding. Alteration in hydrogen bonding at the amino group leads to a shift to high energy for both the XPS N 1s core level and the 3π* NEXAFS resonance with β-PABA. Participation of the amine group in the aromatic system causes the 1π* resonance to be sensitive to the nature of the intermolecular bonding at the *para*-carboxylic acid group, and a shift to low energy for α-PABA is observed due to hydrogen-bonded carboxylic acid dimer formation. FEFF calculations also successfully reproduce both the energy and intensity variations observed for the σ* shape resonance associated with the C–N bond, with the majority of the decrease in energy observed for β-PABA arising from the longer C–N bond.

## 1. Introduction

The local 3-dimensional environment and interactions of a molecule in the solid-state can have a substantial influence on physical and chemical properties, such as stability, solubility, melting temperature, and reactivity [1]. The ability to characterize and identify such forms (polymorphs) with identical chemical composition but variations in crystal structure (structural packing, interactions, and bonding) is thus extremely important. Understanding or insight into the electronic nature of these materials could provide the basis for investigating how they form in solution, and therefore ultimately how to control the outcome of crystallization [2]. Near-edge X-ray absorption fine structure (NEXAFS) spectroscopy and X-ray photoelectron spectroscopy (XPS) are ideal for this purpose, as they intrinsically probe the local environment of X-ray absorbing atoms. The effects of the local environment and interactions were initially investigated for *para*-aminobenzoic acid (PABA), an intermediate for folic acid synthesis with pharmaceutical applications, by nitrogen NEXAFS and *ab initio* density functional theory (DFT) [3]. Comparison of the crystalline α-form and an isolated molecule revealed substantial differences for the extreme case of removing all effects of the surroundings, and this was especially notable for the calculated N K-edge NEXAFS [3]. XPS and NEXAFS of the nitrogen moiety in PABA have therefore been collected for the two different crystalline polymorphs (α- and β-, Figure 1) [4,5] to investigate their sensitivity to more subtle variations in local environment and intermolecular interactions. This is complemented by investigation into the influence of relatively minor alteration in bond length with FEFF calculations.

## 2. Methods

### *2.1. Materials*
α-PABA was used as supplied (>99% purity, Sigma-Aldrich UK) and β-PABA was formed through an aqueous slurry of α-PABA at 5°C. Powder samples were used for analysis.

### *2.2. X-ray Photoelectron Spectroscopy (XPS)*
N 1s XP spectra were recorded with a Kratos Axis Ultra instrument, employing a monochromatic Al $K_\alpha$ source (1486.69 eV) operating at 180 W (15 kV and 12 mA) [3]. Samples were fixed using double-sided tape and measurements were obtained in constant analysis energy mode below $10^{-8}$ mbar, with a pass energy of 20 eV, 0.1 eV steps, and 500 ms dwell time per data point. Analysis of the data was carried out with CasaXPS software using a Shirley background and GL(30) lineshape [6]. Samples were referenced to the C 1s photoemission C=C at 284.8 eV.

### *2.3. Near Edge X-ray Absorption Fine Structure (NEXAFS) Spectroscopy*
N K-edge partial electron yield (PEY) NEXAFS measurements were performed on the U7a beamline of the National Synchrotron Light Source (NSLS) at Brookhaven National Laboratory, NY with the sample at 54.7°, a 600 l/mm monochromator grating, and an entrance grid bias of -150 V [3]. Spectra were normalized by the simultaneously recorded $I_0$, and the monochromator energy scale was calibrated using the 400.6 eV 1$^{st}$ $\pi^*$ of a titanium nitride grid using the Athena software [7,8]. Repeatability of resonance positions is <0.1 eV.

### *2.4. FEFF calculations*
FEFF8.2 [9] was used to simulate the $\sigma^*$ shape resonance in the N K-edge NEXAFS for a molecule from the crystal structure of α- and β-PABA following geometry optimization of the crystal structure with CASTEP [3]. The SCF (7.0), XANES (4.0), and FMS (10.0) cards were used, and the calculated data were rigid shifted by +2.3 eV. To identify the influence of the C–N bond length, calculations were also run in which the average C–N bond length of α-PABA was increased by 0.035Å relative to the equilibrium value in the crystal structure so that it matched that of β-PABA (1.399 Å).

## 3. Results and Discussion
The nitrogen XPS and NEXAFS for the α- and β-PABA polymorphs yield characteristic spectra (Figures 1 and 2), with NEXAFS clearly differentiating between the two forms. For N 1s XPS, a small yet reproducible increase of around 0.2 eV in binding energy is observed for β-PABA (Figure 1). This follows the change in intermolecular hydrogen bonding with the additional hydrogen-acceptor effect at the amine nitrogen in β-PABA (Figure 1), reducing the electron density on nitrogen.

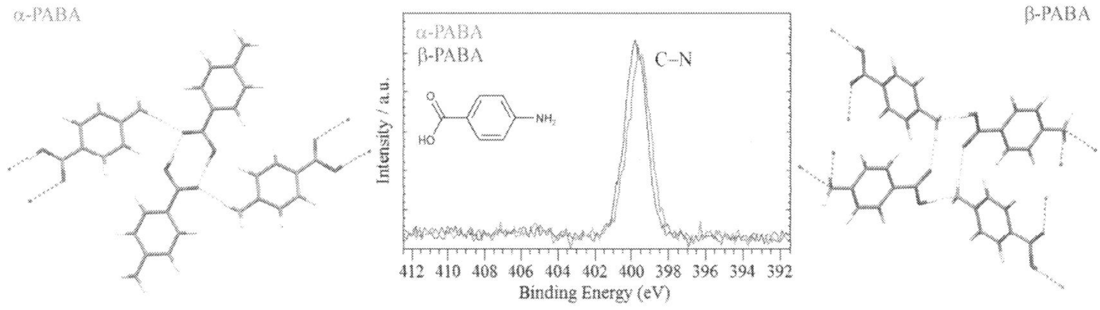


**Figure 1.** Nitrogen 1s XPS and intermolecular interactions of the α- and β-PABA polymorphs.

In the N K-edge NEXAFS spectra, both energy and intensity shifts occur for each of the peaks (Figure 2), reflecting the influence of variations in the local environment (intermolecular H-bonding, electron density distribution) and geometry (such as bond lengths) between the polymorphs. Visualization of the associated unoccupied π* molecular orbitals (MOs) shows the first energy resonance (1π*) involves promotion from the core N 1s level into the lowest unoccupied molecular orbital (LUMO), which is delocalized across the amine, aromatic ring and *para*-carboxylic acid group (Figure 2) [3]. The chemical state and interactions at the opposite end of the ring to nitrogen appear to influence the position of this 1π* resonance, with successively increasing absorption energy reflecting the carboxylic acid dimer in α-PABA, hydrogen bonding to the amino group of another molecule in β-PABA (Figure 1), and the carboxylate $COO^-$ in the anionic form [10]. In contrast, the second resonance (3π*) involves the LUMO+2 (Figure 2), which is much more concentrated around the nitrogen atom [3]. This leads to it being more sensitive to the nature of the interactions at nitrogen, with the additional hydrogen-acceptor effect at the amine nitrogen in β-PABA (Figure 1) increasing the energy of this 3π* MO in addition to that of the N 1s core level (Table 1). The O K-edge π* resonances are correspondingly decreased in energy with the accompanying additional hydrogen-donor at oxygen.

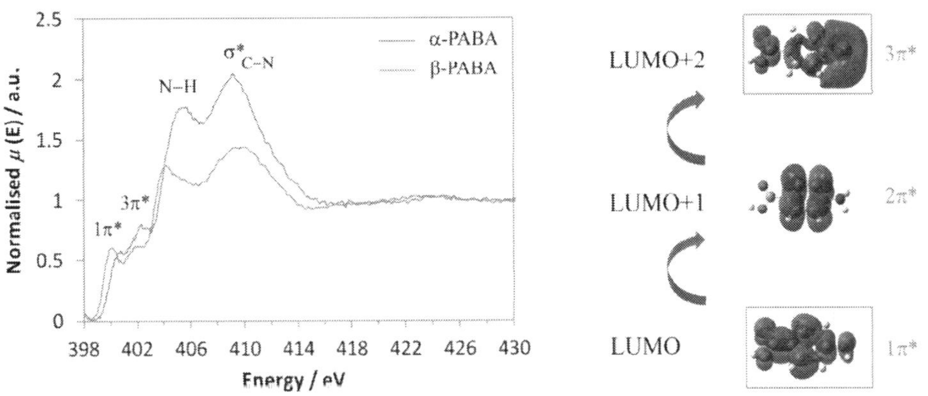


**Figure 2.** Nitrogen K-edge NEXAFS for the α- and β-PABA polymorphs, and unoccupied MOs involved in the first two resonances (outlined with boxes). Significant promotion to LUMO+1 (2π*) is not expected as it has no density at N.

**Table 1.** Nitrogen K-edge NEXAFS features

| NEXAFS features | α-PABA / eV | β-PABA / eV |
|---|---|---|
| IP (aromatic $NH_2$) | 403.30 | 403.60 |
| N 1s → 1π* | 400.00 | 400.58 |
| N 1s → 3π* | 401.70 | 402.30 |
| N 1s → N–H | 404.09 | 405.11 |
| N 1s → $\sigma^*_{C-N}$ | 409.61 | 409.09 |

Following the edge step (ionization potential, IP) and third resonance associated with N–H, the broader post-edge fourth resonance is a σ* shape resonance for the C–N bond [3,10]. This $\sigma^*_{C-N}$ peak shows a shift to lower energy for β-PABA, in contrast to the increased π* and IP positions (Table 1). This is investigated for the two polymorphs using the FEFF code (as it predicts the post-continuum resonances) [9]. The results for α- and β-PABA show significant differences in both energy and intensity for the two polymorphs (Figure 3). Comparison of the FEFF output with the experimental data (Figure 3) shows excellent correlation, successfully predicting both the relative energy and intensity variations.

The dependence of this σ* resonance on the C–N bond was further investigated through variation of the C–N bond length for α-PABA. Lengthening of the C–N bond to that for β-PABA (+0.035 Å), leads to a decrease in σ* energy relative to the IP, accounting for the majority of the energy and intensity shifts between α- and β-PABA (Figure 3).

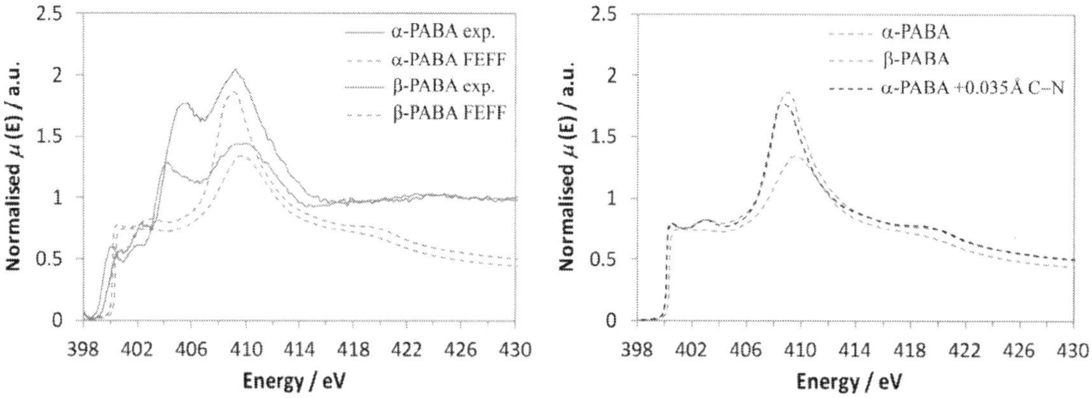


**Figure 3.** Comparison of experimental (solid line) and FEFF-calculated (dashed line) nitrogen K-edge NEXAFS for the two polymorphs of PABA, and the impact of variation in C–N bond length.

## 4. Conclusions

Insights into the electronic and structural nature of the two polymorphs of PABA are obtained with nitrogen K-edge NEXAFS and XPS. A small shift in core level N 1s energy is attributed to the variation in intermolecular hydrogen bonding. Significantly different, highly characteristic NEXAFS spectra arise from the changes in local surroundings and bonding, with the π* shifts related to the nature of the MOs and local environment, and the σ* shape resonance primarily influenced by the C–N bond length. This demonstrates the sensitivity of XPS and NEXAFS to the relatively minor changes in local environment despite their identical chemical composition.

## Acknowledgements

Use of the NSLS at BNL was supported by U.S. Department of Energy under Contract No. DE-AC02-98CH10886. JSS, AG and SLMS acknowledge support by EPSRC Critical Mass Grant EP/I013563/1.

## References

[1] Cartensen J T *Pharmaceutical Principles of Solid Dosage Forms* 1993 (Lancaster, PA: Technomic Publishing Co.)
[2] Davey R. J., Schroeder S L M, and ter Horst J H 2103 *Angew. Chem., Int. Ed.* **52** 2166-2179
[3] Stevens J S, Seabourne C R, Jaye C, Fischer D A, Scott A J and Schroeder S L M 2014 *J. Phys. Chem. B* **118** 12121-12128
[4] Lai T F and Marsh R E 1967 *Acta Crystallogr.* **22** 885-893
[5] Gracin S and Fischer A 2005 *Acta Crystallogr., Sect. E: Struct. Rep. Online* **61** o1242.
[6] Fairley N and Carrick A 2005 *The Casa Cookbook - Part 1: Recipes for XPS Data Processing. Acolyte Science* (Knutsford, Cheshire: Acolyte Science)
[7] Ravel B and Newville M 2005 *J. Synchrotron Radiat.* **12** 537-541
[8] Newville M 2001 *J. Synchrotron Radiat.* **8** 322-324
[9] Ankudinov A L, Ravel B, Rehr J J and Conradson S D, 1998 *Phys. Rev. B: Condens. Matter Mater. Phys.* **58** 7565-7576
[10] Stevens J S, Gainar A, Suljoti E, Xiao J, Golnak R, Aziz E F and Schroeder S L M 2015 *Chem. - Eur. J.* **21** 7256-7263

Journal of Physics: Conference Series 712 (2016) 012134 doi:10.1088/1742-6596/712/1/012134

# Biomimetic mono- and dinuclear Ni(I) and Ni(II) complexes studied by X-ray absorption and emission spectroscopy and quantum chemical calculations

**N Schuth[1], S Mebs[1], H Gehring[2], B Horn[2], P Holze[2], R Kositzki[1], P Schrapers[1], C Limberg[2] and M Haumann[1]***

[1] Department of Physics, Freie Universität Berlin, 14195 Berlin, Germany
[2] Department of Chemistry, Humboldt-Universität zu Berlin, 12489 Berlin, Germany

*E-mail: michael.haumann@fu-berlin.de

**Abstract**. Five biomimetic mono- or dinuclear nickel complexes featuring Ni(I) or Ni(II) sites were studied by X-ray absorption and emission spectroscopy and DFT calculations. Ni K-edge XANES spectra and Kß main and satellite emission lines were collected on powder samples. The pre-edge absorption transitions (core-to-valence excitation) and $Kß_{2,5}$ emission transitions (valence-to-core decay) were calculated using DFT (TPSSh/TZVP) on crystal structures. This yielded theoretical ctv and vtc spectra in near-quantitative agreement with the experiment, showing the adequacy of the DFT approach for electronic structure description, emphasizing the sensitivity of the XAS/XES spectra for ligation/redox changes at nickel, and revealing the configuration of unoccupied and occupied valence levels, as well as the spin-coupling modes in the dinuclear complexes. XAS/XES-DFT is valuable for molecular and electronic structure analysis of synthetic complexes and of nickel centers in $H_2$ or $CO_x$ converting metalloenzymes.

## 1. Introduction

Nickel ions play important roles in many biological enzymes involved in small molecule activation catalysis. Prominent examples are hydrogen-converting nickel-iron hydrogenases and acetyl-CoA synthase (ACS) and carbon monoxide dehydrogenase (CODH) involved in carbon oxide ($CO_x$) chemistry [1-2]. These enzymes integrate mono- or dinuclear nickel centers in mixed ligand environments, combining H, C, N, or O ligands from hydride groups, $CO/CO_2$ or water species, and amino acids with S ligands from cysteine residues or metal-bridging sulfido groups. ACS, for example, holds a dinickel center, in which one Ni ion is ligated by two nitrogen ligands from the protein backbone and by two cysteine thiolate groups, which bridge between the two Ni ions. The nickel centers in the proteins often undergo redox changes in the catalytic cycle, so that, e.g., Ni(I) and Ni(II) species are formed. However, the redox and spin state of the nickel sites, as well as the detailed metal ligation environment may not always be accessible by protein crystallography and many spectroscopic methods. Knowing these properties is essential for unraveling the reaction mechanisms.

Synthetic model chemistry has been extensively used to mimic the metal coordinations and valence states in the proteins, aiming at a better understanding of their reactivity. In the present study, five crystallographically characterized mono- or dinuclear complexes containing formal Ni(I) or Ni(II) sites were studied (figure 1), with metal coordinations including either only nitrogen donors or additional S/O/H ligands, thus resembling the biological centers. X-ray absorption and emission spectroscopy (XAS/XES) is a tool, by which all oxidation and spin states of metal centers, as well as their molecular and electronic configurations can be studied in proteins and synthetic systems. In particular the pre-edge absorption region of the K-edge, reflecting resonant core-level (1s) electron excitation into unoccupied valence orbitals (core-to-valence transitions, ctv), and the Kß emission region, due to electronic 3p→1s decay (Kß main lines, $Kß_{1,3}$ and Kß´) or decay from occupied valence levels in the $Kß_{2,5}$ region (valence-to-core transitions, vtc), bear valuable information on the electronic structure [3-7]. The respective absorption or emission features can be calculated using density functional theory (DFT), for electronic transition assignment and generation of model structures. Here, we used Ni K-edge XAS and Kß emission spectroscopy and DFT calculation of the ctv and vtc spectral features to access the electronic structures of the nickel complexes.

## 2. Materials and Methods

Nickel complexes were synthesized as previously described [8-12] and powder samples were prepared by grounding with boron nitride (1:10 w/w). XAS/XES experiments were carried out at beamline ID26 of the ESRF (Grenoble, France) using a Si[311] double-crystal monochromator for excitation energy tuning and a vertical-plane Rowland spectrometer with 5 Si[551] analyzer crystals and an avalanche photodiode detector for narrow-band emission detection and a scintillation detector for total-fluorescence detection. Samples were held in a liquid-He cryostat at 20 K. Single-point DFT calculations on crystal structures of the complexes were performed with the program package ORCA [13] (TPSSh exchange-correlation functional, triple-zeta-valence-plus-polarization basis set). Calculated spectra were 183 eV shifted on the energy axis and stick spectra broadened by Gaussians (ctv, FWHM 1.2 eV) or Lorentzians (vtc, FWHM 3.0 eV) for comparison with experimental data.

## 3. Results

Figure 1 shows the crystal structures of the studied nickel complexes **1** ($L(^tBu)Ni^IOEt_2$ ($L(^tBu/Me)$ = $[HC(C(^tBu/Me)NC_6H_3(^iPr)_2)_2]^-$ [8]), **2** ($L(^tBu)Ni^{II}SEt$ [9]), **3** ($[L(^tBu)Ni^I]_2N_2$ [8, 10]), **4** $(PYR)Ni^{II}HNi^I$ ($PYR^{2-}$ = $[\{NC(Me)C(H)C(Me)\text{-}NC_6H_3(iPr)_2\}_2(C_5H_3N)]^{2-}$ [11]), and **5** ($[L(Me)Ni^{II}H]_2$ [11, 12]). Both mononuclear Ni complexes show a planar-trigonal coordination site with a bidentate β-diketiminate ligand ("nacnac") and respective third O/S ligand, resembling the distal nickel ion in the ACS protein. **1** is a Ni(I) S = ½ ($d^9$) system and **2** contains a high-spin ($d^8$) Ni(II) ion. The dinuclear complexes represent increasing oxidation states. Complex **3** comprises a bridging $N_2$ ligand and two weakly coupled Ni(I) centers (each S = ½). **4** has a distorted trigonal Ni(I) site and a square-planar Ni(II) site, the hydride is bound only to the Ni(II), and the complex shows a doublet state with a low-spin Ni(II). **5** shows two hydride bridges and presumably antiferromagnetically coupled high-spin Ni(II) centers.

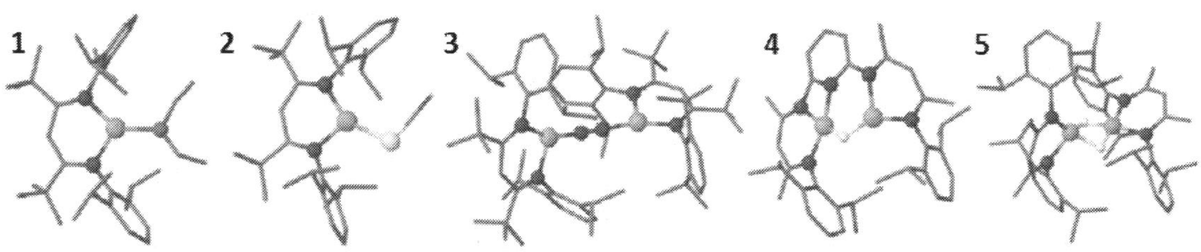


**Figure 1.** Crystal structure of complex **1**, **2**, **3**, **4**, and **5** [8-12]. Color code: Ni, green; N, blue; O, red; S, yellow; C, grey; H, white. The Ni ions and their first-sphere ligands are highlighted as spheres, non-bridging hydrogen atoms are omitted for clarity.

Ni K-edge XAS and Kß XES spectra of the complexes are shown in figure 2. The overall shape changes in the XANES reflect the varying first-sphere coordination geometries of the nickel centers, featuring distorted 3- or 4-coordinated Ni sites, which cause pronounced 1s→4p transition features around 8337 eV in particular in **3** and **5**. The pre-edge absorption was relatively weak in all complexes, showing a single resolved feature around 8330-8332 eV. The Kß main line emission was broad and rather similar for all complexes, showing only small variations, in response to the

Ni(I)/Ni(II) and ligand variations, in the intensity ratio of the $K\beta_{1,3}$ (~8265 eV) and unresolved Kß´ (~8255 eV) spin-polarization features. The isolated ctv excitation transitions (figure 2C) varied mostly in their center energy, with **1** and **3** containing Ni(I) showing a ~1 eV lower ctv maximum than **2** containing Ni(II). **4** with a Ni(I)Ni(II) site showed a ctv feature at low energies, whereas **5** with a Ni(II)Ni(II) center showed a ~0.5 eV up-shifted feature. Pronounced variations in the $K\beta_{2,5}$ emission line shapes reflected the N/S/O/H ligand variations in the five complexes (figure 2D).

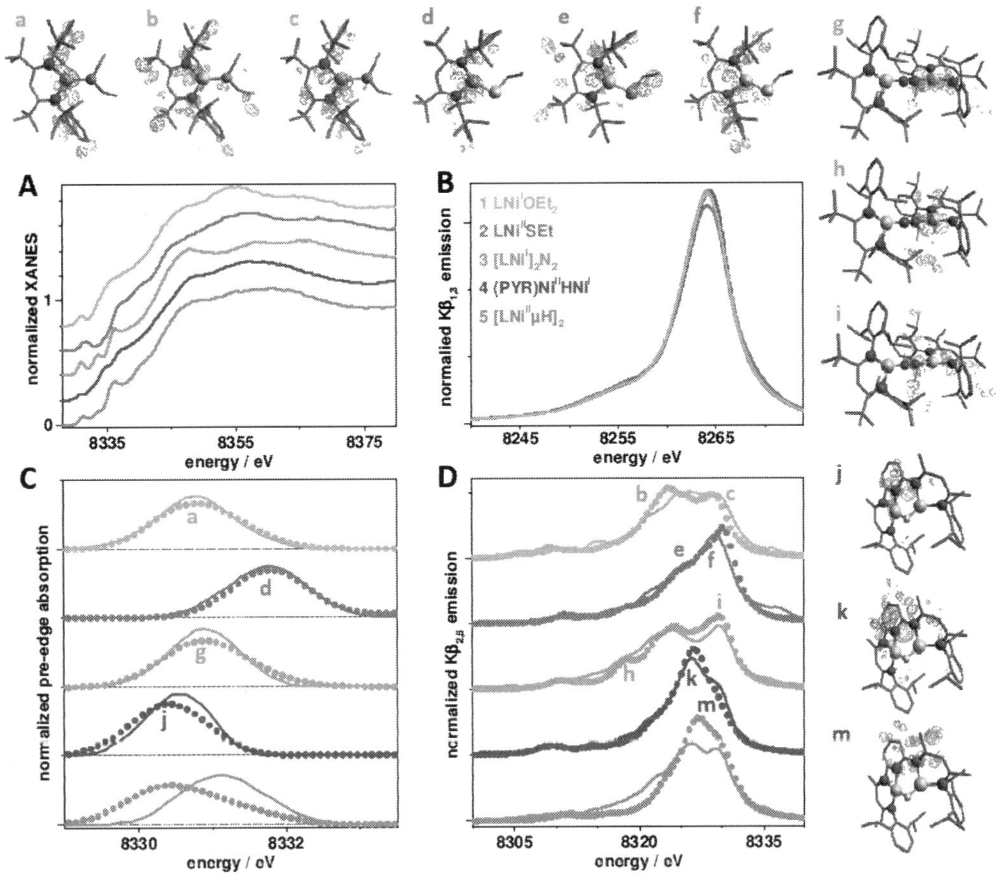


**Figure 2.** XAS/XES spectra of five nickel complexes. (A) Ni XANES spectra, (B) Kß main line ($K\beta_{1,3}$, Kß´) emission spectra, (C) core-to-valence (pre-edge) absorption spectra, (D) valence-to-core ($K\beta_{2,5}$) emission spectra. Emission spectra were non-resonantly excited at 8800 eV; Kß main lines and DFT-calculated (dots) and experimental (lines) ctv and vtc spectra after background removal were normalized to unity area. Structures labelled a-m show metal- or ligand-centered molecular orbitals corresponding to the most intense excitation or decay transitions in the respective ctv or vtc regions.

The DFT-calculated and experimental ctv and vtc spectra are compared in figure 2 (C and D). We used the valence, spin, and multiplicity assignments from earlier studies [8-12] or alternative spin multiplicities (M = 2S+1) in the calculations. For complexes **1-4**, the calculated spectra essentially reproduced the single resolved ctv feature as well as the pronounced transition intensity and energy variations in the vtc spectra for M = 2 (**1**), M = 3 (**2**), M = 3 (**3**), and M = 2 (**4**) states, confirming the earlier spin state assignments such as the weak coupling of the two Ni(I) centers in **3** and the S = ½ Ni(I) and low-spin Ni(II) sites in **4**. Using ferromagnetic coupling of the two high-spin Ni(II) ions in **5** yielded ctv and vtc spectra, which clearly deviated from the experimental data, thereby supporting the alternative antiferromagnetic coupling assignment based on magnetic moment data [10, 12]. Further calculations on **5** will be reported in a forthcoming publication.

## 4. Discussion

We report XAS/XES spectra for five structurally characterized nickel complexes in the solid state, which mimic biological nickel centers [8-12]. The XANES spectral shape is pronouncedly affected by the first-sphere ligation variations causing altered 3d/4p mixing. However, The Kß main line emission is rather insensitive to the Ni(I)/Ni(II) and ligand variations in the Ni complexes, presumably due to core-hole lifetime spectral broadening leading to unresolved spin-polarization features and few unpaired Fe(d) spins in the closed-shell systems, in contrast to earlier first-row transition metals [3-7].

The ctv (pre-edge absorption) and vtc ($Kß_{2,5}$) emission features clearly are more sensitive to the relatively moderate ligation and valence state variations in the complexes. DFT at the TPSSh/TZVP level reproduced the experimental ctv and vtc spectra near-quantitatively, thus facilitating electronic transition and molecular orbital (MO) configuration assignment. The energy shifts of the ctv features are explained by energy variations of unoccupied valence MOs and the peak features of the vtc spectra reflect decay contributions from occupied valence MOs dominated by the respective ligand species. Our results confirm the coordination geometries and spin states of the nickel sites as deduced from earlier studies by other methods [8-12]. The results emphasize the sensitivity and selectivity of the XAS/XES-DFT approach for detection of metal ligation variations such as O/S exchange and metal-hydride bond formation [4-5], as well as specific spin state assignment even for multinuclear metal centers. This approach is highly valuable for investigations of nickel complexes in proteins and expected to provide new insights into their unique mechanisms of small molecule catalysis.

**Acknowledgments**

M.H. and C.L. gratefully acknowledge support within Unicat (Cluster of Excellence Berlin). M.H. thanks the Deutsche Forschungsgemeinschaft for a Heisenberg Fellowship and for funding (grant Ha3265/6-1), the Bundesministerium für Bildung und Forschung for financial support within the Röntgen-Angström Cluster (grant 05K14KE1), and the ESRF for a Long Term Project. We thank the group of P. Glatzel at ID26 of ESRF for excellent technical support.

**References**

[1] Appel A M., Bercaw J E, Bocarsly A B, Dobbek H, DuBois D L, Dupuis M, Ferry J G, Fujita E, Hille, R, Kenis P J, Kerfeld C A, Morris R H, Peden C H, Portis A R, Ragsdale S W, Rauchfuss T B, Reek J N, Seefeldt L C, Thauer R K and Waldrop G L 2013 *Chem. Rev.* **113** 6621
[2] Haumann M 2015 In *Biohydrogen* (Rögner M, Ed) DeGruyter, Berlin, 97
[3] Lambertz C, Chernev P, Klingan K, Leidel N, Siegfridsson K G V, Happe T and Haumann M 2014 *Chem. Sci.* **5** 1187
[4] Chernev P, Lambertz C, Brünje A, Leidel N, Sigfridsson K G V, Kositzki R, Hsieh C H, Yao S, Schiwon R, Driess M, Limberg C, Happe T and Haumann M 2014 *Inorg. Chem.* **53** 12164
[5] Leidel N, Hsieh C H, Chernev P, Sigfridsson K G V, Darensbourg M Y and Haumann M 2013 *Dalton Trans.* **42** 7539
[6] Leidel N, Chernev P, Havelius K G V, Schwartz L, Ott S and Haumann M 2012 *J. Am. Chem. Soc.* **134** 14142
[7] Leidel N, Chernev P, Havelius K G V, Ezzaher S, Ott S and Haumann M 2012 *Inorg. Chem.* **51** 4546
[8] Pfirrmann S, Limberg C, Herwig C, Stosser R and Ziemer B 2009 *Angew. Chem. Int. Ed.* **48** 3357
[9] Horn B, Limberg C, Herwig C and Braun B 2014 *Inorg. Chem.* **53** 6867
[10] Horn B, Pfirrmann S, Limberg C, Herwig C, Braun B, Mebs S and Metzinger R 2011 *Z. Anorg. Allg. Chem.* **637** 1169
[11] Gehring H, Metzinger R, Herwig C, Intemann J, Harder S and Limberg C 2013 *Chem. Eur. J.* **19** 1629
[12] Pfirrmann S, Limberg C and Ziemer B (2008) *Dalton Trans.* **47** 6689
[13] Neese F 2012 *Wiley Interdisc. Rev. Comput. Mol. Sci.* **2** 73

Journal of Physics: Conference Series **712** (2016) 012135 doi:10.1088/1742-6596/712/1/012135

# Unveiling the complex network of interactions in Ionic Liquids: a combined EXAFS and Molecular Dynamics approach

**A Serva[1], V Migliorati[1], A Lapi[1] and P D'Angelo[1]**

[1] Dipartimento di Chimica, Università di Roma "La Sapienza", P.le A. Moro 5, 00185 Roma, Italy

E-mail: alessandra.serva@uniroma1.it

**Abstract.** The structural properties of geminal dicationic ionic liquids ($[C_n(mim)_2]Br_2$)/water mixtures have been investigated by means of extended X-ray absorption fine structure (EXAFS) spectroscopy and Molecular Dynamics (MD) simulations. This synergic approach allowed us to assess the reliability of the MD results and to provide accurate structural information about the first coordination shell of the $Br^-$ ion. We found that the local environment around the anion changes as a function of the water concentration, while it is the same independently from the length of the bridge-alkyl chain. Moreover, as regards the long-range structural organization, no tail-tail aggregation occurs with increasing alkyl chain length.

## 1. Introduction

Ionic liquids (ILs) are an emerging class of solvents, showing a melting point near room temperature. Recently, a special class of "high stability" ILs, geminal dicationic ionic liquids (DILs), has been obtained [1]. DILs represent a very interesting variation of the cationic partner with various advantages over the traditional monocationic ILs when used as lubricants, catalysts, solvents, and separation media. Up to now, only few experimental and theoretical studies of pure DILs [2, 3] and mixtures of DILs and water [4] have been reported in the literature, thus the complex molecular organization taking place in DIL/water solutions is still lacking.

Here, we present a combined Molecular Dynamics (MD) and extended X-ray absorption fine structure (EXAFS) study of the structural properties of 1,3-bis[3-methylimidazolium-1-yl]alkane bromide ($[C_n(mim)_2]Br_2$)/water mixtures (see Figure 1). In particular, we investigated $[C_3(mim)_2]Br_2$/water mixtures as a function of water concentration (DIL/water molar ratios of 1:16, 1:70 and 1:400) and $[C_n(mim)_2]Br_2$/water mixtures, obtained by changing the alkyl-bridge chain length (n= 3, 6, 9) with constant water concentration (DIL/water molar ratio of 1:70). The synergic use of EXAFS and MD is particularly well suited to provide a global structural picture of DIL/water solutions both in proximity of the anion, and in the longer distance range.

## 2. Methods

The $[C_n(mim)_2]Br_2$/water mixtures were prepared by adding the proper amount of bidistilled water to pure $[C_n(mim)_2]Br_2$. X-ray absorption spectra above the Br K-edge were collected at room-temperature in transmission mode at the Elettra Synchrotron (Trieste, Italy) on the

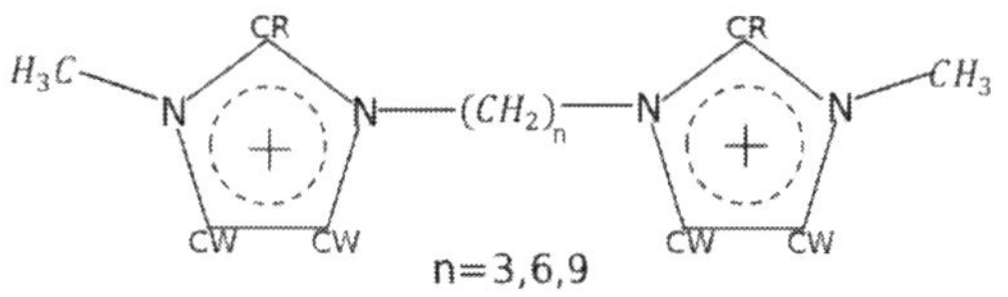


**Figure 1.** Molecular structure of 1,3-bis[3-methylimidazolium-1-yl]alkane dication, $[C_n(mim)_2]^{2+}$.

beamline 11.1. The storage ring was operating at 2 GeV with an optimal storage beam current between 300 and 30 mA. EXAFS data have been analyzed with the GNXAS program. Five MD simulations of $[C_n(mim)_2]Br_2$/water mixtures have been carried out using the DL_POLY package [5]. The ions and the water molecules were placed in a cubic box, with periodic boundary conditions. The number of ion pairs used in each simulation depends on the mixture and computational cost. The force field used for $[C_n(mim)_2]Br_2$ was taken from Lopes and Pádua [6, 7], with a minor modification introduced by Yeganegi et al. for the $[C_3(mim)_2]^{2+}$ cation [2], while the SPC/E forcefield was employed for water [8]. As regards the $Br^-$-water interaction, the Lennard-Jones parameters were taken from the OPLS force field [9], that in a combined EXAFS-MD study has been shown to provide a reliable description of the structural properties of a $Br^-$ aqueous solution [10]. Each system was first equilibrated in the NPT ensemble for 1 ns to obtain an almost constant density and subsequently the production runs were carried out for 6 ns in the NVT ensemble.

## 3. Results

The EXAFS spectra of the $[C_3(mim)_2]Br_2$/water mixtures with molar ratio of 1:16, 1:70 and 1:400 are shown in Figure 2 (a), together with the experimental data of pure $[C_3(mim)_2]Br_2$ and the corresponding Fourier transform calculated in the interval k=0.7-12.0 Å$^{-1}$ (b). The first evidence is the difference between the EXAFS spectra of the $[C_3(mim)_2]Br_2$/water mixtures and that of pure $[C_3(mim)_2]Br_2$: the experimental spectrum of pure $[C_3(mim)_2]Br_2$ is characterized by a lower amplitude as compared to the mixtures. These results suggest that water molecules have a high affinity with the anion and enter its coordination sphere, also for the most diluted solution. Moreover, the EXAFS and FT signals change as a function of the water concentration: going from the DIL/water mixture with molar ratio of 1:16 to 1:400, the EXAFS amplitudes become higher and there is an increase in intensity of the FT signals, indicating a progressive crowding of the $Br^-$ ion coordination environment. After studying the effect of the water concentration on the $[C_3(mim)_2]Br_2$/water mixtures, we then analyzed the structural arrangement around the $Br^-$ ion as a function of the alkyl chain length linking the two imidazolium rings. In this case the water concentration was kept fixed. Figure 2 (d, e) shows the EXAFS spectra of the 1:70 $[C_n(mim)_2]Br_2$/water mixtures (with n=3, 6, 9) and the corresponding Fourier transform calculated in the interval k=0.7-12.0 Å$^{-1}$. Both the EXAFS and FT signals are almost identical for the three members of the series, thus meaning that the local coordination around the bromide ion is the same independently from the length of the alkyl-bridge chain. To better understand these results, we calculated the Br-O and Br-H radial distribution functions, g(r), from the MD simulations of the DIL/water mixtures under investigation (O and H are the oxygen and hydrogen atoms of the water molecules, respectively.) Figure 3 shows the g(r)s, that have been multiplied by the numerical density of the observed atoms, $\rho$, in order to compare DIL/water mixtures with different densities. In all cases the Br-O and Br-H g(r)s show a first distinct and sharp peak at 3.33 and 2.35 Å, respectively. Moreover, as regards the Br-H g(r)s, two distinct peaks, at shorter and longer distances as compared to the Br-O first peak, can be observed: only one hydrogen atom of the water molecules belonging to the first shell is oriented toward the $Br^-$ ion, due to the ability of the bromide ion to make hydrogen bonds [11]. When considering the effect of the water concentration (Figure 3 (a,c)) on the $[C_3(mim)_2]Br_2$/water mixtures, the position of the g(r) first peak does not change with

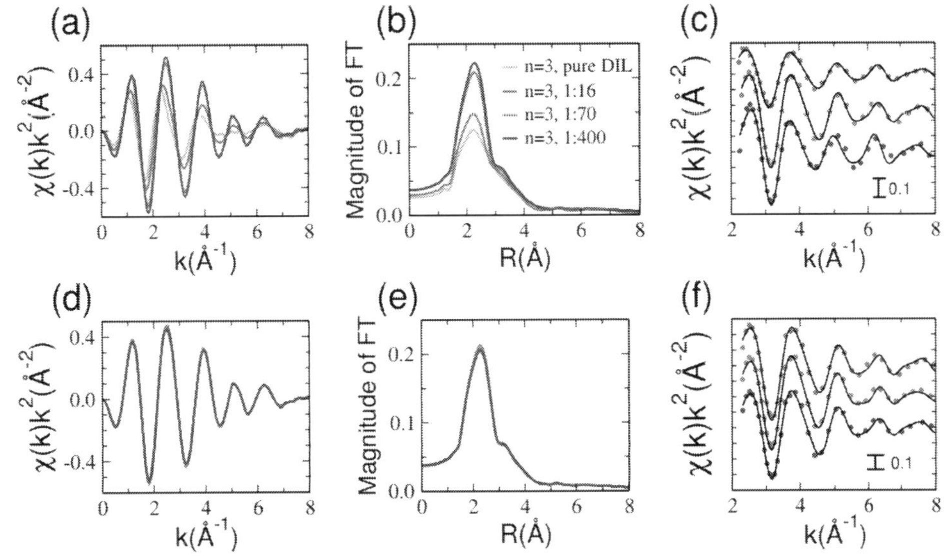


**Figure 2.** a,d: Experimental EXAFS spectra of the $[C_3(mim)_2]Br_2$/water mixtures with different molar ratios and of the $[C_n(mim)_2]Br_2$/water mixtures with different alkyl-bridge chain lengths, respectively. b,e: Non-phase shifted corrected Fourier tranforms of the experimental data. c,f: Comparison between the experimental spectra (dotted line) and the theoretical signals (solid black line) calculated from the simulations.

increasing water content, while the intensity of the peak becomes higher, meaning that the bromide coordination sphere becomes more populated. Note that the $Br^-$ hydration number is 4.4, 6.0 and 6.5 for the 1:16, 1:70 and 1:400 $[C_3(mim)_2]Br_2$/water mixtures, respectively. Conversely, if we look at the Br-O and Br-H g(r)s obtained by fixing the amount of water and changing the alkyl chain length from n=3 to n=9, they are almost identical with a Br-O coordination number of 6.0. These results are in agreement with the EXAFS experimental spectra and confirm that the local structural environment seen by the $Br^-$ ion is not influenced by the alkyl-bridge chain length of the imidazolium dications. The Br-O first-shell bond lengths and the coordination numbers obtained in our work are comparable to that found in an MD-EXAFS study of monocationic $[C_4mim]Br$/water mixtures with different molar ratios [10]. Moreover, for the most diluted solution (DIL/water molar ratio of 1:400), the Br-O coordination number is similar to that obtained in theoretical studies of bromide aqueous solutions [10, 12].

The Br-O and Br-H g(r)s, obtained from the MD trajectories for all of the DIL/water mixtures under investigation, have been used to calculate $\chi(k)$ theoretical signals. Comparison of the EXAFS experimental spectra with the MD results allows one on the one hand to carry out a reliable EXAFS analysis, that represent a non trivial task when the EXAFS technique is applied to the study of disordered systems. On the other hand it allows us to prove that the force field used is able to correctly describe the structural properties of the investigated systems. The comparison between the experimental EXAFS spectra and the corresponding theoretical signals for all of the $[C_n(mim)_2]Br_2$/water mixtures is shown in Figure 2 (c,f). In all cases, a very good agreement between the theoretical and experimental data has been obtained. It is important to stress that non-polarizable force fields like those used here usually provide a good description of the structural properties of ILs, while failing to predict dynamic properties, such as diffusion coefficients and conductivities. However, this is outside the scope of this work.

From the analysis of the MD trajectories it is also possible to explore the long-range structural

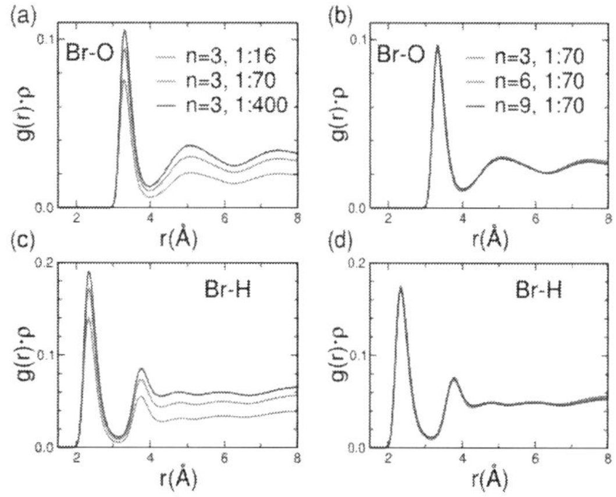


**Figure 3.** Br-O and Br-H g(r)s, multiplied by the numerical density $\rho$ for the $[C_n(mim)_2]Br_2$/water mixtures.

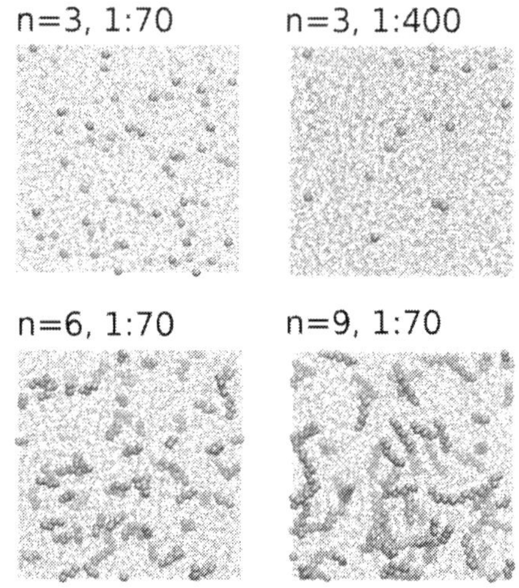


**Figure 4.** MD snapshots of the DIL/water mixtures, with the polar regions in red and the alkyl-bridge chains in cyan.

properties of the investigated mixtures. Figure 4 shows some MD snapshots of the simulation boxes, where the polar regions (imidazolium cations, water molecules and $Br^-$ ions) and the alkyl-bridge chains are colored red and cyan, respectively. In particular, no evidence of the aggregation of linkage alkyl chains has been found with increasing alkyl chain length. These results are in good agreement with a previous study of $[C_n(mim)_2](Tf_2N)_2$, where aggregation occurs only when n is 12 [3]. This behaviour differs from that of monocationic ILs: several theoretical and experimental studies on $[C_n mim]Br$/water mixtures show the segregation of the alkyl tails and the formation of polar and apolar domains occurs with long enough alkyl chains ($n \geq 5$) [13].

## 4. Conclusions

In this paper we presented a detailed investigation of $[C_n(mim)_2]Br_2$/water mixtures as a function of the water concentration and of the alkyl-bridge chain length. Our combined EXAFS-MD approach allowed us to provide direct structural information of the systems both in proximity of the anion and in the longer distance range.

## References

[1] Anderson J L, Ding R, Ellern A and Armstrong D W 2005 *J. Am. Chem. Soc.* **127** 593-604
[2] Yeganegi S, Soltanabadi A and Farmanzadeh D 2012 *J. Phys. Chem.* B **116** 11517-26
[3] Li S, Bañuelos J L, Zhang P, Feng G, Dai S, Rother G and Cumming P T 2014 *Soft. Matter.* **10** 9193-9200
[4] Barghava B L and Klein M L 2010 *J. Chem. Theory Comput.* **6** 873-79
[5] Smith W and Forester T 1996 *J. Mol. Graphics* **14** 136-41
[6] Canongia Lopes J N, Deschamps J and Pádua A H H 2004 *J. Phys. Chem.* B **108** 2038-47
[7] Canongia Lopes J N and Pádua A H H 2006 *J. Phys. Chem.* B **110** 19586-92
[8] Berendsen H J C, Grigera J R and Straatsma T P 1987 *J. Phys. Chem.* **91** 6269-71
[9] Jorgensen W L, Ulmschneider J P and Tirado-Rives J 2004 *J. Phys. Chem.* B **108** 16264-70
[10] D'Angelo P, Zitolo A, Aquilanti G and Migliorati V 2013 *J. Phys. Chem.* B **117** 12516-24
[11] D'Angelo P, Migliorati V and Guidoni L 2010 *Inorg. Chem.* **49** 4224-31
[12] Migliorati V, Sessa F, Aquilanti G and D'Angelo P 2014 *J. Chem. Phys.* **141** 044509
[13] Zhao Y, Gao S, Wang J and Tang J 2008 *J. Phys. Chem.* B **112** 2031-39

# NEXAFS Chemical State and Bond Lengths of p-Aminobenzoic Acid in Solution and Solid State

**J S Stevens[1], A Gainar[1], E Suljoti[2], J Xiao[2], R Golnak[2], E F Aziz[2,3], S L M Schroeder[4,5]**

[1]School of Chemical Engineering and Analytical Science, The University of Manchester, Oxford Road, Manchester, M13 9PL, UK
[2]Joint Ultrafast Dynamics Lab in Solutions and at Interfaces (JULiq), Helmholtz-Zentrum Berlin für Materialien und Energie, Albert Einstein Strasse 15, 12489 Berlin, Germany
[3]Freie Universität Berlin, Arnimallee 14, 14195 Berlin, Germany
[4]School of Chemical and Process Engineering, University of Leeds, Leeds, LS2 9JT, UK
[5]DIAMOND Light Source Ltd., Chilton, Didcot, OX11 0QX, UK

E-mail: joanna.stevens@manchester.ac.uk; s.l.m..schroeder@leeds.ac.uk

**Abstract**. Solid-state and solution pH-dependent NEXAFS studies allow direct observation of the electronic state of para-aminobenzoic acid (PABA) as a function of its chemical environment, revealing the chemical state and bonding of the chemical species. Variations in the ionization potential (IP) and 1s→π* resonances unequivocally identify the chemical species (neutral, cationic, or anionic) present and the varying local environment. Shifts in σ* shape resonances relative to the IP in the NEXAFS spectra vary with C–N bond length, and the important effect of minor alterations in bond length is confirmed with nitrogen FEFF calculations, leading to the possibility of bond length determination in solution.

## 1. Introduction

Through X-ray absorption and emission spectroscopies, the chemical, electronic and structural properties of organic species can be investigated. The nature of solution species and their interactions can play an important role in the activity of biological systems [1] and in the structural outcome and therefore properties of crystallization products [2]. Near-edge X-ray absorption fine structure (NEXAFS) measurements allow direct observation of the electronic state of *para*-aminobenzoic acid (PABA) as a function of its chemical environment [3]. While the IP and π* resonances are characteristic of the species and related to their chemical and electronic nature in the solution and solid state, the dominating influence of bond length on the σ* shape resonance proposed from experimental data is here investigated with FEFF calculations, exploring the ability and sensitivity of measuring bond lengths.

## 2. Methods

### *2.1. N K-edge Near Edge X-ray Absorption Fine Structure (NEXAFS) Spectroscopy*

Solid-state partial electron yield (PEY) NEXAFS measurements were performed on the U7a beamline of the National Synchrotron Light Source (NSLS) at Brookhaven National Laboratory, NY with the sample at 54.7°, a 600 l/mm monochromator grating, and an entrance grid bias of -150 V [3]. Spectra were normalized by the simultaneously recorded total electron yield signal from a gold-coated, 90% transmission grid in the incident beam ($I_0$), and the monochromator energy scale was calibrated using the 400.6 eV 1st $\pi^*$ of a titanium nitride grid. Solution-state partial fluorescence yield (PFY) spectra were recorded using the liquid microjet of the LiXEdrom endstation at the U41-PGM beamline of the BESSY II synchrotron at Helmholtz Zentrum Berlin (HZB) using an 18 μm diameter glass nozzle with 0.6 mL min$^{-1}$ flow rate [3]. Beamline energy calibration was performed with $N_2$ gas (total electron yield X-ray absorption spectroscopy 1s→2p $\pi$ transition). Peak fitting and normalization were performed using the Athena software [4,5], with arctan steps for the ionization potentials (IPs, edge steps) and Gaussian functions for the peaks.

### *2.2. FEFF calculations*

FEFF8.2 [6] was used to simulate the $\sigma^*$ shape resonance in the N K-edge NEXAFS for a geometry optimized monomer of non-ionic PABA. The SCF (7.0), XANES (4.0), and FMS (10.0) cards were used, and the calculated data were rigid shifted by +2.3 eV. To identify the influence of the C–N bond length, the C–N bond length of α-PABA was increased by 0.002-0.1 Å relative to the equilibrium value for the non-ionic monomer.

## 3. Results and Discussion

Nitrogen K-edge NEXAFS measurements reveal variations for the PABA species in both the solid state and in solution (Figure 1). Characteristic spectra are obtained for the different chemical species (non-ionic, cationic, anionic) present in solution with variation in pH, with a distinctive electronic signature for the cationic form following the change in nitrogen protonation state ($NH_2$ to $NH_3^+$). Shifts in the ionization potential (IP, edge step) and 1s→$\pi^*$ resonances unequivocally identify the chemical and electronic nature of the species present [3,7]. Further slight shifts and intensity variations are observed for the impact of differing local environments and interactions for the three non-ionic species investigated (α- and β-crystalline forms and the solutions species in methanol).

The energy of the $\sigma^*$ resonance associated with the C–N bond relative to the IP also varies across the series (term value $\delta_{C\text{-}N}$, Figure 1). The decrease in $\delta_{C\text{-}N}$ appears to correlate well with successively longer C–N bond lengths reported by X-ray diffraction (XRD) for the solid-state (Table 1, $y = -39.664x + 61.133$) [3]. This relationship can then be applied to the NEXAFS data, using it to provide an indicator of bond lengths for PABA species in solution.

**Table 1.** N K-edge NEXAFS features and bond lengths [3].

| | $\delta_{C\text{-}N}$ / eV | $\Delta\delta_{C\text{-}N}$ / eV | XRD C–N / Å | XRD lengthening C–N / Å | NEXAFS C–N / Å[b] |
|---|---|---|---|---|---|
| Non-ionic α-PABA | 6.31 | 0 | 1.379(2) | 0 | 1.382 |
| Non-ionic β-PABA | 5.49 | -0.82 | 1.408(3) | 0.029 | 1.403 |
| Anionic pH 11 | 5.40 | -0.91 | 1.410(2)[a] | 0.031 | 1.405 |
| Non-ionic Methanol | 5.19 | -1.12 | / | / | 1.411 |
| Cationic pH 1 | 3.16 | -3.15 | / | / | 1.462 |
| Cationic HCl salt | 3.05 | -3.26 | 1.4626(5) | 0.0836 | 1.464 |

[a] C–N bond length from the hydrated Na salt shown for comparison
[b] $y = -39.664x + 61.133$ for linear relationship between $\delta_{C\text{-}N}$ and XRD C–N/Å [3].

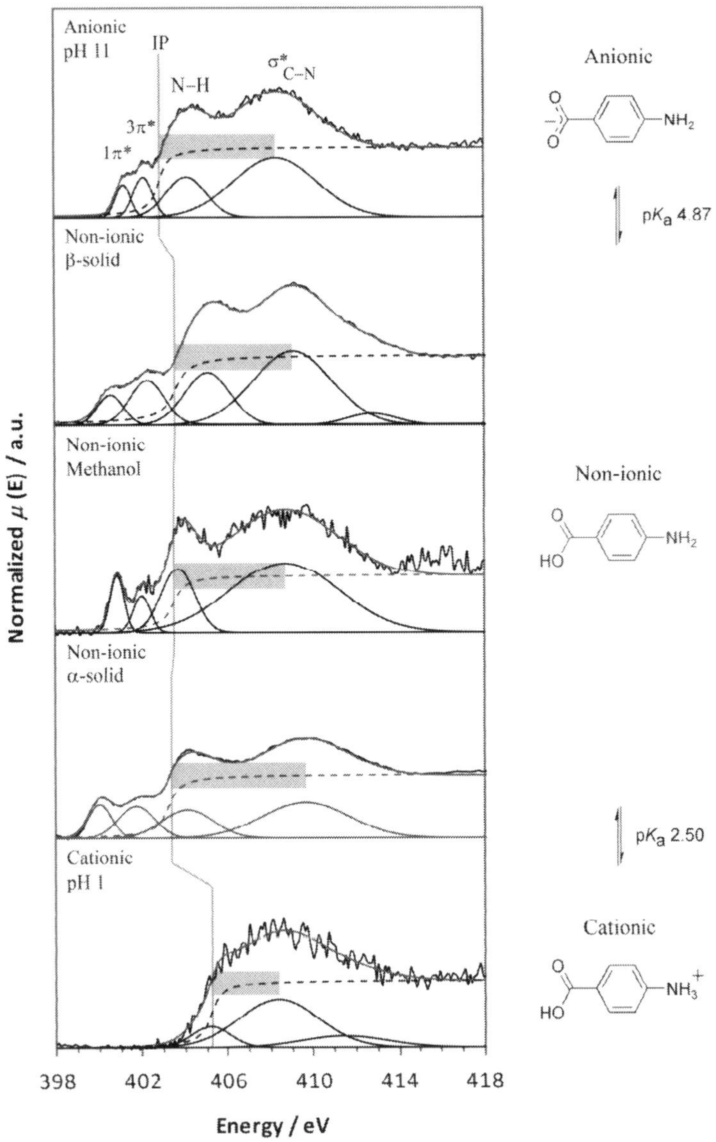


**Figure 1.** Nitrogen K-edge NEXAFS for anionic, non-ionic, and cationic PABA species in the solid and solution state, with the IPs marked by grey lines and the magnitude of the term value $\delta_{C\text{-}N}$ highlighted by grey boxes (adapted under CC BY license [3,7]).

To further investigate the impact of the C–N bond of PABA on the nitrogen NEXAFS, FEFF calculations were performed. Using an optimized non-ionic monomer as input, only the C–N length was varied, keeping all other parameters constant. This enables the effect of the C–N bond length to be isolated from other changes occurring between the species.

The C–N length was initially extended by 0.1 Ångstroms (Å), where a clear alteration and decrease in σ* energy of almost 2.5 eV is observed relative to the IP ($\Delta\delta_{C\text{-}N}$, Figure 2). The changes in C–N bond length in practice are smaller than 0.1 Å, ranging from 0.002 to 0.08 Å for the different PABA species (Table 1). Smaller increments were therefore tested, with lengthening by 0.03 Å (to mimic that occurring on moving from non-ionic α-PABA to β- or anionic PABA) resulting in a decrease in energy by around 0.8 eV (Figure 2). This compares favourably with experimental NEXAFS, where a 0.82-0.91 eV change occurs (for cationic species, underestimation of FEFF $\Delta\delta_{C-N}$ values may be related to the alteration in electronic structure and geometry accompanying the positive charge on N, bent $NH_2$ → tetrahedral $NH_3^+$). The extreme sensitivity is further demonstrated for the case with just 0.005 Å longer C–N (Figure 2), and even the small shift occurring with 0.002 Å lengthening is observable, showing that even relatively subtle variations have a noticeable influence. The results suggest that bond length variations should be quantifiable to a precision of approximately 0.01 Å.

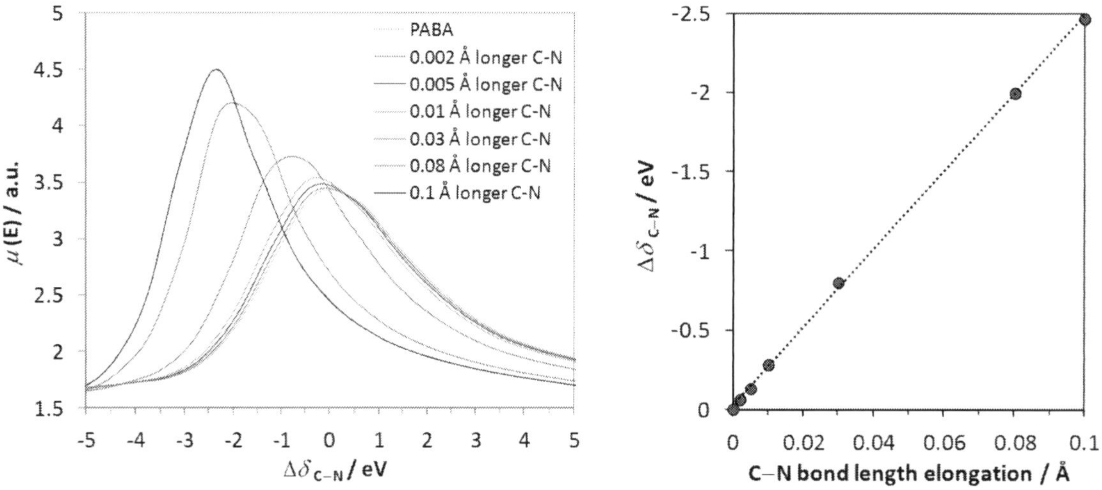


**Figure 2.** FEFF calculations for the nitrogen K-edge NEXAFS of PABA, demonstrating the effect of lengthening the C–N bond on the σ* energy relative to IP ($\delta^*_{C-N}$).

## 4. Conclusions

Nitrogen NEXAFS acts as a chemical and structural indicator for PABA, identifying the chemical species present in both the solid-state and in solution as well as providing a measure of C–N bond length. The ability to detect even minor variations in structure is verified by the FEFF calculations through incremental variation of bond length, with changes from as little as 0.002 Å alteration visible. NEXAFS can then potentially provide bond lengths for solution (or amorphous) species, such as demonstrated with PABA, and the combination of experiment and calculations is particularly strong.

## Acknowledgements

Use of the NSLS at BNL was supported by U.S. Department of Energy Contract No. DE-AC02-98CH10886, and BESSY II at HZB by European Community's Seventh Framework Programme, grant no. 312284. JSS, AG and SLMS acknowledge support by EPSRC Critical Mass Grant EP/I013563/1.

## References

[1] Chatterjea M N and Shinde R. 2012 *Textbook of Medical Biochemistry* (New Delhi: Jaypee Brothers Medical Publishers (P) Ltd.)
[2] Davey R. J., Schroeder S L M, and ter Horst J H 2013 *Angew. Chem., Int. Ed.* **52** 2166-2179
[3] Stevens J S, Gainar A, Suljoti E, Xiao J, Golnak R, Aziz E F and Schroeder S L M 2015 *Chem. - Eur. J.* **21** 7256-7263
[4] Ravel B and Newville M 2005 *J. Synchrotron Radiat.* **12** 537-541
[5] Newville M 2001 *J. Synchrotron Radiat.* **8** 322-324
[6] Ankudinov A L, Ravel B, Rehr J J and Conradson S D, 1998 *Phys. Rev. B: Condens. Matter Mater. Phys.* **58** 7565-7576
[7] Stevens J S, Seabourne C R, Jaye C, Fischer D A, Scott A J and Schroeder S L M 2014 *J. Phys. Chem. B* **118** 12121-12128

# EXAFS characterisation of metal bonding in highly luminescent, UV stable, water-soluble and biocompatible lanthanide complexes

**A. Kalyakina**[1,4], **V. Utochnikova**[4,5], **A. Trigub**[3], **Y. Zubavichus**[3], **N. Kuzmina**[3], **S. Bräse**[1,2]

[1]Institute of Organic Chemistry, Karlsruhe Institute of Technology, Karlsruhe, Germany

[2]Institute of Toxicology and Genetics, Karlsruhe Institute of Technology, Karlsruhe, Germany

[3]NRC «Kurchatov Institute», Moscow, Russia

[4]Department of Chemistry, Lomonosov Moscow State University, Moscow, Russia

[5]P. N. Lebedev Physical Institute, Russian Academy of Sciences, Moscow, Russia

E-mail: alena.kalyakina@kit.edu

**Abstract.** The combination of X-ray diffraction with EXAFS was employed to assess the co-ordination environment of lanthanide complexes in solutions. This method is based on the assumption that the local structure of lanthanide complexes in solution combines elements of the crystal structure of the complex in the solid state (single- or polycrystalline) and the elements of the local structure of a lanthanide salt, completely dissociated in the solvent (usually chlorides). The success of this approach is demonstrated with the lanthanide (III) 2,3,4,5,6-pentafluorobenzoate complexes, where the local structure in aqueous and methanol solutions were estimated. Moreover, the dissociation degree of the complexes in aqueous and methanol solutions was evaluated.

## 1. Introduction

Over the past decades EXAFS has been widely used for structural analyses of various compounds and materials, particularly when X-ray diffraction methods are not applicable or their application is relatively limited. For instance, EXAFS has proved to be a useful tool for analyses of metal complexes in solutions. Here we present a comprehensive approach for analysis of coordination environment of lanthanide complexes with pentafluorobenzoates (pfb) in solution with EXAFS by comparison with structural data obtained by conventional X-ray diffraction techniques in single- and polycrystalline samples. Previously it was shown, that combination of X-ray diffraction and EXAFS is a powerful tool in structural elucidation [1, 2].

The main idea underpinning the present study is the finding that pfb ligands possess different coordination modes depending on the metal used, and even complexes with the same metal can give rise to different molecular structures. Such a versatile structural chemistry observed in the complexes with this ligand urged us to investigate their molecular structures in solution. Given the fact that different ligating modes can be easily distinguished due to a large difference in Ln…C contact lengths, the presence of different scattering pathways in EXAFS spectra, corresponding to these contacts in different modes, can be successfully utilized for discrimination between them. Thereby, it is possible to evaluate the composition of coordination spheres of these complexes in solution.

The significance of knowing the coordination environment of these complexes in solution is further augmented by their possible application as bio-imaging probes for cellular studies due to their high photophysical performance, water solubility and low toxicity [3].

## 2. Experiment

Lanthanide complexes $Ln(pfb)_3(H_2O)_n$ were synthesized under the same conditions: concentrated aqueous ammonia was added to aqueous $LnCl_3$ (Ln=Pr-Lu except of Pm) followed by adding a suspension of Hpfb in acetone/methanol (3:1), then the solution was evaporated (Fig. 1). The solid product was dried on air. All the solutions for XAFS measurements were the same concentration of 0.05M and adjusted to pH=7.0. The local structure of **δ-Ln, «ε»-Ln**, **$H_2O$-Ln**, **$CH_3OH$-Ln**, **$H_2O$-$LnCl_3$**, **$CH_3OH$-$LnCl_3$**, has been elucidated using Ln $L_{III}$-edge X-ray absorption spectroscopy. The spectra were measured at the Structural Materials Science beamline of the Kurchatov Centre for Synchrotron Radiation (National Research Centre "Kurchatov Institute", Moscow, Russia) in transmission mode using two ionization chambers filled with appropriate $N_2$/Ar mixtures. Experimental data reduction and analysis were performed using the IFEFFIT software suite [4] with FEFF [5] ab initio photoelectron phase and amplitude functions.

## 3. Results and discussion

In the present study crystal structures of polycrystalline samples of $Ln(pfb)_3(H_2O)_n$ (Fig. 1) designated as **δ-Ln**, Ln = Eu, Gd, Tb, Dy, Ho, Er, and of $Ln(pfb)_3$ designated as **«ε»-Ln**, Ln = Tm, Yb, Lu) were determined by X-ray diffraction, and their EXAFS spectra were measured at the Ln $L_{III}$-edge. To estimate the composition of coordination sphere around the lanthanide ion, EXAFS spectra were also obtained for aqueous (**$H_2O$-Ln**) and methanol solutions (**$CH_3OH$-Ln**). The applicability of the proposed method was evaluated using the **δ-Ln** complexes with known crystal structures. The first co-ordination sphere of **δ-Ln** consists of 9 oxygen atoms at the distances of 2.24-2.55 Å, belonging either to water or carboxyl groups of the ligand. These atoms were also revealed in EXAFS as Ln···O scattering path in the first coordination sphere. EXAFS spectroscopy allows us to distinguish them by the appearance of the Ln···C scattering path in the second shell. Indeed, according to the determined distance d(Ln···C), the $\kappa^1$ (R = 3.43-3.61 Å) or $\kappa^2$ (R = 2.68-2.73 Å) ligand these co-ordination modes can be easily discriminated (Fig. 2). It goes without saying that different co-ordination modes of the ligand affect the Ln-O distance as well, but the deviation in the average distance between Ln and all 9 oxygen atoms was beyond the sensitivity of the experiment, and remained the same regardless of the co-ordination mode of the ligands used in the fitting model.

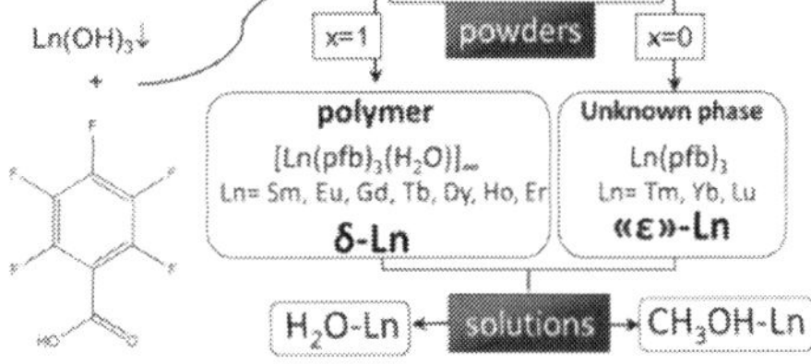


**Figure 1.** Schematic representation of samples studied in the present work.

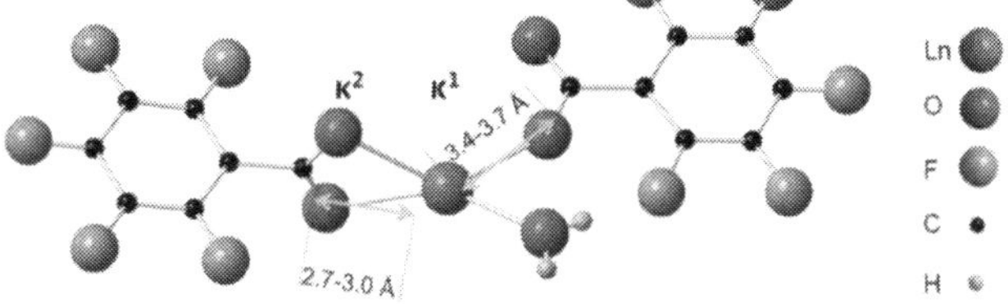


**Figure 2.** Two types $\kappa^1$ and $\kappa^2$ of Ln ion co-ordination by perfluorinated aromatic carboxylates.

The second co-ordination sphere of **δ-Ln** (Fig. 3) contained 2 carbon atoms of $\kappa^2$-carboxylic group (R=2.68-2.73 Å) and 4 carbon atoms of $\kappa^1$ carboxylic group (R=3.43-3.61 Å). The contribution of the Ln···Ln contacts is apparent for all the Ln complexes in powder state, corresponding to the polymeric structure of **δ-Ln**. The decrease of the Ln···Ln distances from **δ-Eu** to **δ-Er** is in good agreement with the lanthanide contraction (Fig. 4). The obtained EXAFS spectroscopy data are consistent with the XRD results and confirm the same composition for all the complexes in the **δ-Ln** group, similar composition in the case of **«ε»-Tm** and **«ε»-Yb**, and different co-ordination modes of the ligands in **«ε»-Lu**. Moreover, high accuracy in the determination of the Ln···Ln distance in the EXAFS experiment was employed to optimise this interatomic distance during the Rietveld-refinement in X-ray analysis. At the same time, though the goodness of fit was satisfactory for all the complexes, a considerable deviation was observed in some cases.

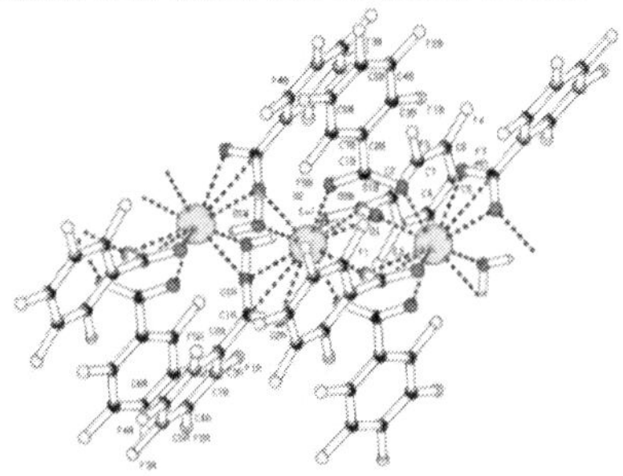

**Figure 3.** The structure of **δ-Eu**

To evaluate the dissociation degree of the complexes in aqueous and methanol solutions, their EXAFS spectra were compared with the spectra of corresponding powder samples and spectra of $LnCl_3$ aqueous solutions (**$H_2O$-$LnCl_3$**). The comparison between solutes and powder samples can provide an estimate for the number of co-ordinated ligands, along with their co-ordination modes. Using crystal structures of powder samples as a starting model, EXAFS spectra of complexes in water were analysed. The presence and the quantity of the Ln···C scattering paths should characterise the number and type of carboxylic-groups, and therefore the number of co-ordinated ligands, as was previously demonstrated for solid-state samples. The FTs of the experimental EXAFS spectra and fitting curves for **δ-Ln**, **«ε»-Lu**, **$H_2O$-Ln** and **$H_2O$-$LnCl_3$** are shown in Fig. 4 for Ln = Eu, Er (Table 1). The EXAFS spectra of **$H_2O$-$LnCl_3$** revealed two co-ordination spheres with the radii of 2.45Å (oxygen atoms of the water molecules) and 3.28 Å (chloride atoms). It is in a good agreement with the previously reported results of MD simulation [6], which also claim that the second shell of water molecules appears at a distance of more than 4.5 Å.

**Table 1.** The data obtained from EXAFS spectra for **Eu-series** .

| Sample | | **δ-Eu** | | **$H_2O$-Eu** | | **$H_2O$-$EuCl_3$** | |
|---|---|---|---|---|---|---|---|
| Shell number | Path | **N** | **R, Å** | **N** | **R, Å** | **N** | **R, Å** |
| 1 | Eu···O | 9 | 2.38 | 10 | 2.46 | 10 | 2.45 |
| 2 | Eu···C ($\kappa^2$) | 2 | 2.68 | 2 | 2.76 | | |
| 2 | Eu···C ($\kappa^1$) | 4 | 3.43 | | | | |
| 2 | Eu···Cl | | | | | 1.5 | 3.28 |
| 3 | Eu···Eu | 2 | 4.10 | | | | |

In contrast to **$H_2O$-$EuCl_3$**, the EXAFS spectra of **$H_2O$-Eu** reveal the presence of the co-ordination sphere with the radius of 2.8 Å, which corresponds to the two $\kappa^2$-carboxylic carbon atoms.

Therefore the EXAFS spectrum of **$H_2O$-Eu** shows the partial dissociation of the complex, since only two ligands are co-ordinated, and $Eu^{3+}$ co-ordination sphere is further supplemented by water molecules (Fig. 3). In contrast, in **$H_2O$-Er** three $\kappa^1$-carboxylic carbon atoms are present according to the EXAFS spectrum (Fig. 4), which suggests that erbium pentafluorobenzoate does not dissociate in water. The same behaviour was found for **$H_2O$-Ho**, **$H_2O$-Tm** and **$H_2O$-Yb**, while the second co-ordination shell of **$H_2O$-Lu** contains two $\kappa^2$ and one $\kappa^1$-carbon atom as in the «ε»-**Lu** powder.

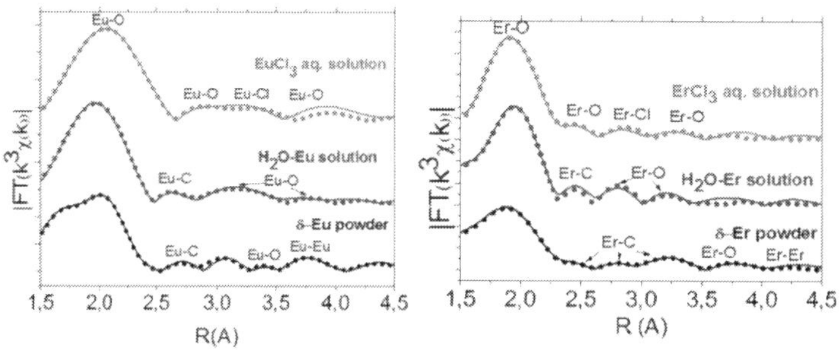


**Figure 4.** The comparison of the EXAFS spectra of **δ-Ln**, **$H_2O$-Ln (0.05M)** and **$H_2O$-$LnCl_3$ (0.05M)**: Ln = Eu (left) and Ln = Er (right), lines – experiment, dots - theory

Thus, water solutions of $Ln(pfb)_3(H_2O)_x$ (Ln=Eu, Gd, Tb, Dy, Ho) partially dissociate, while the lanthanides from the end of the row (Ln = Er, Tm, Yb, Lu) do not undergo dissociation, which is consistent with the solubility data [3]. The similar findings were obtained for the methanol solutions of the complexes. The major difference was that all the complexes had the same molecular structure with only one scattering path for Ln···C in the second shell, which is in line with $^{19}F$ NMR spectroscopy data and mirrors lower co-ordination ability of methanol compared to water [3].

## 4. Conclusions

In the present study we have demonstrated, how convenient can be the synergy of X-ray diffraction and EXAFS for elucidation of coordination environment of lanthanide complexes in crystals and solutions. The X-ray diffraction of polycrystalline samples $Ln(pfb)_3(H_2O)_x$ revealed the presence of several phases, depending on the size of the Ln atom. Following the increase of charge density on $Ln^{3+}$, rearrangements in the co-ordination sphere were observed, as water molecules competed with $pfb^-$ ligands. This structural change was also observed in EXAFS spectroscopy by the presence of two scattering paths in the second shell. These crystal structures were used as starting models for analysis of molecular structures solutes. In the latter case, we observed the alteration of co-ordination mode in **$H_2O$-Ho**, which appeared in the change of corresponding scattering path Ln···C, while in the case of **$H_2O$-Lu** both scattering paths were revealed, similarly to the powder sample, following the lanthanide contraction.

## Acknowledgments

A. Kalyakina and V. Utochnikova gratefully acknowledge the financial support of the state Baden-Württemberg (exchange ZOIV program) and RFBR (Grant Nos. 13-03-12453 and 14-03-32052).

## References

[1] Bordiga S., Bonino F., Lillerud K.P. and Lamberti C., *Chem. Soc. Rev.*, **2010**, *39*, 4885–4927
[2] Shuvaev S., L. Marciniak, A. Freidzon, et. al., *Dalton Trans.*, **2014**, *43*, 3121-3136
[3] A. Kalyakina, V. Utochnikova, S. Bräse, et. al. *Chem. Eur. J.* **2015**, *21*, 17921-17932
[4] B. Ravel, M. Newville, *J. Synchrotron Rad.*, **2005**, *12*, 537–541.
[5] M. Newville, B. Ravel, D. Haskel, Y. Yacoby et. al., *Physica B*, **1995**, *209*, 154–156
[6] C. Beuchat, D. Hagberg, R. Spezia, L. Gagliardi, *J. Phys. Chem. B* **2010**, *114*, 15590–15597

# Local atomic structure and oxidation processes of Cu(I) binding site in amyloid beta peptide: XAS Study

**M A Kremennaya[1], M A Soldatov[1], V A Stretsov[2] and A V Soldatov[1]**

[1]International Research Center "Smart materials", Southern Federal University, Zorge Str. 5, 344090 Rostov-on-Don, Russia
[2]CSIRO Manufacturing Flagship, Parkville, VIC 3052, Australia

E-mail: kremennayamariya@gmail.com

**Abstract**. There are two different motifs of X-ray absorption spectra for Cu(I) K-edge in amyloid-β peptide which could be due to two different configurations of local Cu(I) environment. Two or three histidine ligands can coordinate copper ion in varying conformations. On the other hand, oxidation of amyloid-β peptide could play an additional role in local copper environment. In order to explore the peculiarities of local atomic and electronic structure of Cu(I) binding sites in amyloid-β peptide the x-ray absorption spectra were simulated for various Cu(I) environments including oxidized amyloid-β and compared with experimental data.

## 1. Inroduction

Alzheimer's disease is a complex multifactorial neurodegenerative disease and this creates difficulties in accurate identification of its etiology [1, 2]. The main morphological feature of Alzheimer's disease is the formation of insoluble deposition of amyloid-β peptide into amyloid plaques [4]. It is also known that the brains of patients with Alzheimer's disease exhibit high levels of oxidative stress and high levels of redox active transition metals such as copper and iron [3]. Metal ions play a key role in amyloid beta aggregation processes [5].

The copper and iron ions can be involved in production of reactive oxygen species (ROS) via Fenton and Haber-Weiss reactions [6]. Under aerobic conditions, oxidation of copper/iron ion leads to the formation of hydrogen peroxide $H_2O_2$, $O_2^-$ superoxide and hydroxyl radical OH. The formation of ROS causes the oxidation of amyloid-β peptide (oxidation of amyloid-β peptide occurs through oxidation of individual amino acids, such as methionine and histidine) as well as other biomolecules in the vicinity [6, 7].

## 2. Methods and experiment

We have prepared *Aβ*(1-16)-Cu complexes in phosphate buffer saline (PBS) as described [8, 9] and mixed them with 10-fold molar ratio of ascorbate dissolved in water. *Aβ*(1-16) was obtained from Auspep Pty Ltd. Immediately after preparation the samples were injected into solution cells and rapidly frozen. The $Cu^{2+}$ concentration in solution cells was up to 2.2mM. A series of Cu *K*-edge (8980.4 eV) XAS scans were obtained from samples in a fluorescence mode at 15-20K using a helium displex cryostat. The experiments were conducted at the PNC-CAT 20BM bending magnet beamline at the Advanced Photon Source (APS), USA. The experimental set up and data processing were identical to those reported [8]. Energy calibration was accomplished using the first inflection point of a Cu foil spectrum measured simultaneously with each scan. Up to 15-40 min scans were measured for each sample. No X-radiation reduction of $Cu^{2+}$ to $Cu^{1+}$ was detected by comparing edge spectra for consecutive scans. Each scan was collected from freshly exposed region of the sample by moving the sample stepwise within the cell window area. The XANES regions were extracted from the experimentally measured absorption coefficient using background subtraction and normalization methods implemented in the program ATHENA [10], an interface to IFEFFIT [11].

Atomic structure of Cu(I) binding site and oxidation processes in amyloid-β peptide were studied by combination of quantum chemical calculations and advanced theoretical XANES analysis (X-ray Absorption Near Edge Structure). Theoretical models of Cu(I) binding site in amyloid-β peptide were calculated using the ADF (Amsterdam Density Functional) package [12]. The calculation of theoretical X-ray absorption spectra was carried out using an accelerated version of the FDMNES code [13]. The acceleration code for FDMNES was developed by International Research Center "Smart materials" [14].

## 3. Results and discussions

There are two different motifs of X-ray absorption spectra for Cu(I) *K*-edge in amyloid-β peptide which corresponds to different structures of Cu(I) binding site in amyloid beta peptide. On one hand, this can be explained by two different configurations of local Cu(I) environment: two or three histidine ligands coordinate copper ion in varying conformations [15, 16]. On the other hand, oxidation of amyloid-β peptide could play an important role in local copper environment as well. In order to explore this hypothesis the X-ray absorption spectra were simulated for various Cu(I) environments including oxidized amyloid-β and compared with experimental data [17].

Two different motifs of X-ray absorption spectra for Cu(I) *K*-edge in amyloid-β peptide are shown in Fig. 1. For comparison X-ray absorption spectrum of Cu(II) *K*-edges in amyloid-β peptide is also shown in black in Fig.1. The first group of spectra for Cu(I) *K*-edge is characterized by two main spectral features: pre-edge A and edge B. The second group of experimental spectra is characterized by three main spectral features: pre-edge A, edge B and peak C.

A B

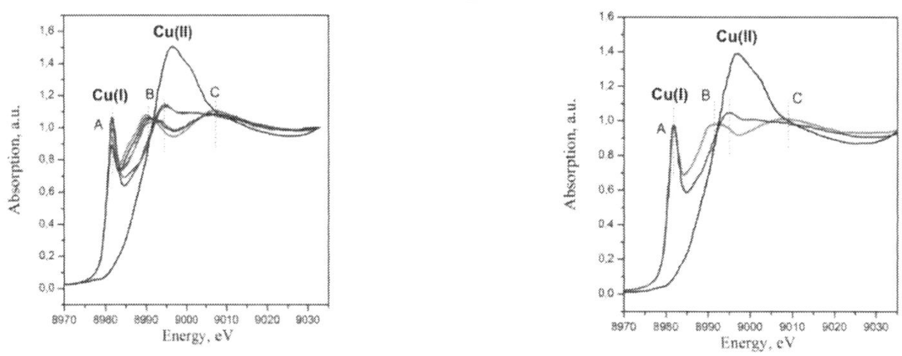


**Figure 1.** Experimental X-ray absorption spectra for Cu(I) and Cu(II) *K*-edge in amyloid-β peptide (A). Selected experimental motifs for Cu(I) – "Blue" and "Pink", Cu(II) – "Black"

There are two hypotheses that can be proposed to explain the existence of two experimental motifs. The first hypothesis is related to mixture of Cu(I) and Cu(II) species due to incomplete reduction of Cu(II) to Cu(I) with ascorbate and the second is related to the existence of two different structures of Cu(I) binding site in amyloid-β peptide.

To test the first hypothesis a linear combination two groups of experimental spectra for Cu(I) *K*-edge and Cu(II) *K*-edge were constructed (Figure 2). As can be seen from Fig. 2, this hypothesis is somewhat true in one case (Fig. 2A) showing possible 15% contribution from Cu(II) and the hypothesis is false in the seconds case (Fig. 2B).

To test second hypothesis that there exist two different structures of Cu(I) binding site in amyloid-β peptide, we utilized two proposed geometries of copper (I) binding site in amyloid-β peptide [11]:

Linear (two-coordinate structure)

*His – Cu – His*

Three-coordinated

*His – Cu – His*
|
*His*

Journal of Physics: Conference Series **712** (2016) 012138 doi:10.1088/1742-6596/712/1/012138

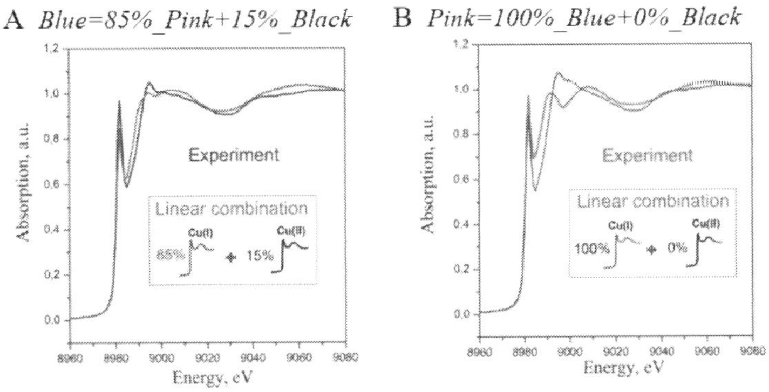


**Figure 2.** Linear combination spectra for three main motifs of Cu(I) and Cu(II) experimental XANES spectra in amyloid-β peptide. A – Linear combination for experimental motif "Blue", R-factor=0.014, B - Linear combination for experimental motif "Pink", R-factor=0.029

For the basis of linear model (two-coordinate structure) was used a simple structure consisting of a copper ion coordinated by two imidazole rings. The three-coordinated model of Cu(I) binding site in amyloid beta peptide is consists of copper coordinated by three imidazole rings forming a T-shaped structure. Geometric optimization of local structure was carried out by minimizing the total energy of system by using the exchange-correlation potential in generalized gradient approximation Becke-Perdew and basis set DZ (Double Z) and +1 net charge. For linear model and three-coordinated model the structural parameters were obtained (Table 1) and theoretical XANES spectra were calculated for linear model and three-coordinated model, respectively (Figure 3 A and Figure 3 B, red line).

**Table 1.** Linear and three-coordinated Cu(I) binding sites in amyloid-β-Cu(I) complexes.

| | Linear models | Three-coordinated models |
|---|---|---|
| | 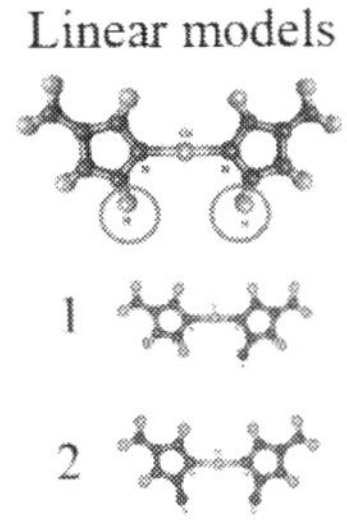 | 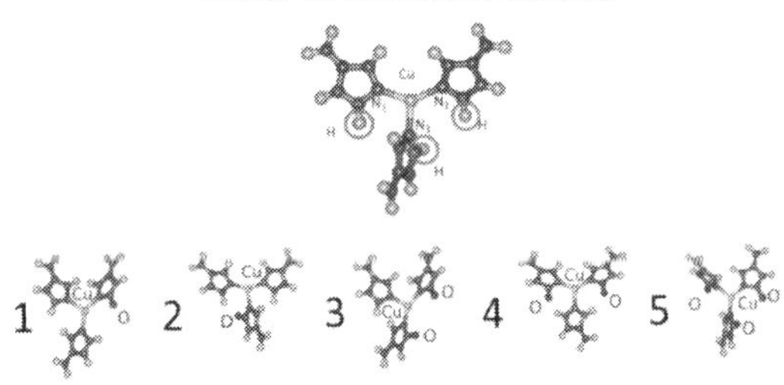 |
| Bond length (Non-oxidized models) | Cu-N = 1.88 ± 0.01 Å | N1 – Cu = 1.96Å ± 0.01Å, N2 – Cu = 1.97 Å ± 0.01Å, N3 – Cu = 2.04Å ± 0.01Å |
| Angles (Non-oxidized models) | N-Cu-N = 178.4° ± 0.7 ° | N1 – Cu – N2 = 136.1° ± 0.2°, N1 – Cu – N3 = 117.49° ± 0.2°, N2 – Cu – N3 = 106.21° ± 0.2° |

Modelling oxidation of amyloid-β near the Cu(I) binding site for linear and three-coordinate binding sites was based on histidine oxidation [17]. Theoretical X-ray absorption spectra were calculated for all amyloid-β oxidized models of Cu(I) binding site shown in Figure 3 A, B.

## 4. Conclusions

The main structural parameters such as interatomic distances and bond angles were obtained by means of *in vacuo* density functional theory for both: histidine two-coordinated linear and histidine three-coordinated copper (I) binding site configurations of amyloid-β. The X-ray absorption spectra are particularly sensitive to the changes in the nearest environment such as replacement of hydrogen atoms on oxygen atoms. It was found that linear two-coordinated model copper (I) binding site with its oxidized histidines version is better suited for the description of one experimental motifs. Possible oxidation of Cu(I) to Cu(II) (or incomplete reduction of Cu(II) to Cu(I)) may also contribute to the final spectra.

A B

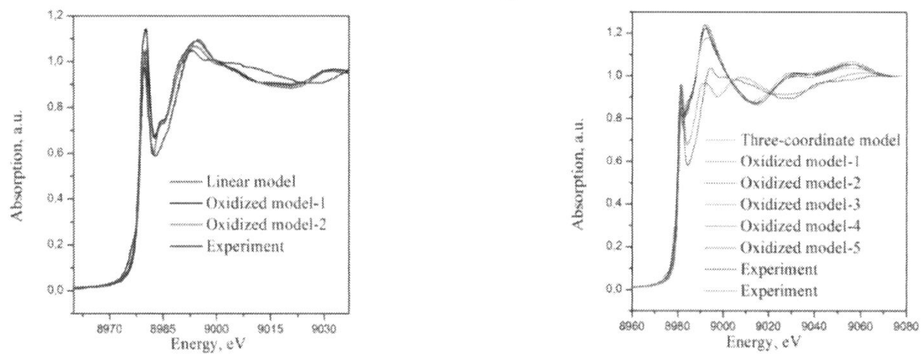


**Figure 3.** Comparison theoretical and experimental X-ray absorption spectra for Cu(I) linear non-oxidized and oxidized models (1,2) and three-coordinated non-oxidized and oxidized models (1-5)

## 4. Funding

This study was supported by RFBR, research project No. 14-02-31790/15. Part of the work was financially supported by the Ministry of Education and Science of Russia Scholarship SP-2373.2013.5 (M.S. author).

## References

[1] T. Kowalik-Jankowska, M. Ruta, K. Wiśniewska, L. Łankiewicz, and M. Dyba, 2004 *J. Inorg. Biochem.* **98**, 940.
[2] M. G. Savelieff, Y. Liu, R. R. P. Senthamarai, K. J. Korshavn, H. J. Lee, A. Ramamoorthy, and M. H. Lim, 2014 *Chem. Commun.* **50**, 5301.
[3] C. Schoneich, 2000 *J. Pharm. Biomed. Anal.* **21**,1093.
[4] R. Vassar, B. D. Bennett, S. Babu-Khan, S. Kahn, E. A. Mendiaz, P. Denis, D. B. Teplow, S. Ross, P. Amarante, R. Loeloff, Y. Luo, S. Fisher, J. Fuller, S. Edenson, J. Lile, M. A. Jarosinski, A. L. Biere, E. Curran, T. Burgess, J.-C. Louis, F. Collins, J. Treanor, G. Rogers, and M. Citron, 1999 *Science* **286**, 735.
[5] A. Abelein, A. Gräslund, and J. Danielsson, 2015 *Proc. Natl. Acad. Sci. USA* **112** , 5407.
[6] R. C. Nadal, S. E. J. Rigby, and J. H. Viles, 2008 *Biochemistry* **47**, 11653.
[7] C. Schöneich, 2004 *Ann. N.Y. Acad. Sci.* **1012**, 164.
[8] V.A. Streltsov, S.J. Titmuss, V.C. Epa, K.J. Barnham, C.L. Masters, J.N. Varghese, A Combined X-ray Absorption Spectroscopy and Computational Study of Amyloid beta-peptide Copper II Complexes, (in preparation, unpublished work).
[9] V. A. Streltsov and J. N. Varghese, 2008 *Chem. Commun.* 3169.
[10] B. Ravel and M. Newville, 2005 *J Synchrotron Radiat.* **12**, 537.
[11] M. Newville, 2001 *J Synchrotron Radiat.* **8**, 96.
[12] G. te Velde, F. M. Bickelhaupt, E. J. Baerends, C. Fonseca Guerra, S. J. A. van Gisbergen, J. G. Snijders, and T. Ziegler, 2001 *J. Comput. Chem.* **22**, 931.
[13] Y. Joly, 2001 *Phys. Rev. B* **63**, 125120.
[14] S. A. Guda, A. A. Guda, M. A. Soldatov, K. A. Lomachenko, A. L. Bugaev, C. Lamberti, W. Gawelda, C. Bressler, G. Smolentsev, A. V. Soldatov, and Y. Joly, 2015 *J Chem Theory Comput.*
[15] C. Hureau, V. Balland, Y. Coppel, P. L. Solari, E. Fonda, and P. Faller, 2009 *JBIC J. Biol. Inorg. Chem.* **14**, 995.
[16] C. Hureau and P. Dorlet, 2012 *Coord. Chem. Rev.* **256**, 2175.
[17] C. Schöneich and T. D. Williams, 2002 *Chem ResToxicol.* **15**, 717.

# Time resolved XANES illustrates a substrate-mediated redox process in Prussian blue cultural heritage materials

**Claire Gervais[1*], Marie-Angélique Lanquille[2], Giulia Moretti[1], Solenn Réguer[3]**

[1] Bern University of the Arts, Fellerstrasse 11, 3027 Bern, Switzerland, [2] Centre de recherche sur la conservation (CRC, USR 3224), Sorbonne Universités, Muséum national d'Histoire naturelle, Ministère de la Culture et de la Communication, CNRS Paris, France and IPANEMA (CNRS, MCC, USR 3461), Saint-Aubin BP 48, 91192 Gif-sur-Yvette, France, [3] Synchrotron SOLEIL, Saint-Aubin BP 48, 91192 Gif-sur-Yvette, France

E-mail: claire.gervais@bfh.ch

**Abstract.** The pigment Prussian blue is studied in heritage science because of its capricious fading behavior under light exposure. We show here that XANES can be used to study the photosensitivity of Prussian blue heritage materials despite X-ray radiation damage. We used an original approach based on X-ray photochemistry to investigate in depth the redox process of Prussian blue when it is associated with a cellulosic substrate, as in cyanotypes and watercolors. By modifying cation and proton contents of the paper substrate, we could tune both rate and extent of Prussian blue reduction. These results demonstrate that the photoreduction and fading of Prussian blue is principally mediated by the substrate and its interaction with the oxygen of the environment.

## 1. Introduction

Prussian blue (PB, iron(III) hexacyanoferrate(II)) is better known as electrochemically synthesized thin films, or as the original compound of a series of analogues where iron is substituted by other transitional metal atoms. However, before becoming a star compound in materials sciences, Prussian blue got an equal success as a pigment and dye. Thanks to its high tinting strength and its cheap and straightforward synthesis, it was used in various heritage artefacts, such as oil paintings, textiles, early-photographic processes or watercolors [1].

Despite this wide use, Prussian blue is still depicted as a problematic and mysterious pigment because of its capricious fading behavior under light exposure, which depends a lot on the object investigated. The fading is due to the photoreduction of Prussian blue, a process where the substrate, the environment, impurities, as well as the structure and chemical composition of Prussian blue play a role [2, 3]. Much remains to be done to understand the exact redox mechanism behind fading. For instance, the migration pathways of cations and electrons within the structure upon Prussian blue reduction remains unknown. In previous studies, we showed that X-ray absorption spectroscopies are particularly well dedicated to investigate the photoreduction of Prussian blue [2, 4, 5]. However, X-ray radiation damages were observed for Prussian blue embedded in a cellulosic paper, a model used to study cyanotypes and watercolors, cultural heritage artefacts known to be very photosensitive [6]. X-ray radiation involves a reduction of the pigment, in a similar way than visible light does. We could recently put into profit this analogy to study by X-ray photochemistry the influence of the cellulosic substrate on the photosensitivity of Prussian blue in paper heritage materials [4].

Here, we further investigate the redox properties of those materials by investigating the impact of anoxia on the X-ray photoreduction kinetics of PB cellulosic model samples. These findings help us refining our understanding of the substrate-mediated redox process responsible for Prussian blue fading in paper heritage materials.

## 2. Experimental details

Prussian-blue paper model samples were prepared by immersion of a 100% cellulosic paper (Whatman) in a Prussian blue sol (synthesized as in [2]) and let dry in open air. KCl and HCl variantes were prepared by preliminary immersing the paper in KCl (1M) and HCl (1M) respectively. A section of approximately $0.5\,\mathrm{cm}\times0.5\,\mathrm{cm}$ was cut from the samples and inserted into a controlled atmosphere flow cell designed specifically for X-ray measurements in transmission geometry and to perform in-situ light fading of the samples [5]. A gas flow of nitrogen humidified at 50% RH was injected into the cell and kept flowing during the entire experiment, ensuring stable anoxic and relative humidity conditions around the sample ($N_2$-50%RH).

The Fe K-edge XAS spectra of the PB-paper samples were acquired by energy-dispersive X-ray absorption spectroscopy at the ODE beamline, SOLEIL synchrotron, Paris [7]. The X-ray beam was around $30\mu m\times35\mu m$ FWHM in size and the photon flux hitting the sample was $1.1\times10^{11}$ photons/s. An oscillating movement of the sample over $200\mu m$ in horizontal and vertical was performed during acquisition to take into account sample intrinsic heterogeneity (distribution of PB particles within fibers). XAS spectra were recorded over a range of 425 eV (7050–7475 eV). For each final XAS spectrum, up to 150 successive exposures (frames) were saved and averaged, with an exposure time to the CCD camera for each frame chosen from 5 to 25 ms depending on the pigment concentration into the sample. Each sample was continuously exposed to the beam during approximately one hour. The $I_0$ was regularly recorded by exposing the direct beam onto the CCD camera. Energy calibration was based on the inflection point of the absorption edge at 7112 eV of a metallic iron foil regularly measured during the session.

Data treatment was performed using the Iffefit 3.0.3 package with the computational code Larch [8]. The pre-edge range background was removed using an empirical polynomial function, and the absorption background was removed using a cubic spline with a normalization range between 40 and 220 eV after the edge.

Structural information was extracted from the XANES spectra by focusing on two features: On one hand, the maximum of the derivative of the XANES signal was chosen as the absorption edge and its variation upon time was monitored to estimate changes in the oxidation state of Prussian blue. On the other hand, the shape of the XANES derivatives was also investigated to estimate the amount of structural disorder generated during the redox process. The complete methodology is described in [4].

## 3. Results

During exposure of the Prussian blue paper samples to the X-ray beam, an absorption edge shift towards lower energies was systematically observed. This shift can be attributed to a reduction of iron(III) into iron(II) in Prussian blue and was associated with a visual discoloration of the pigment.

**Photochemistry of Prussian blue in air.** This X-ray photochemistry was recently put into profit to refine our understanding of the redox process between Prussian blue and a cellulosic

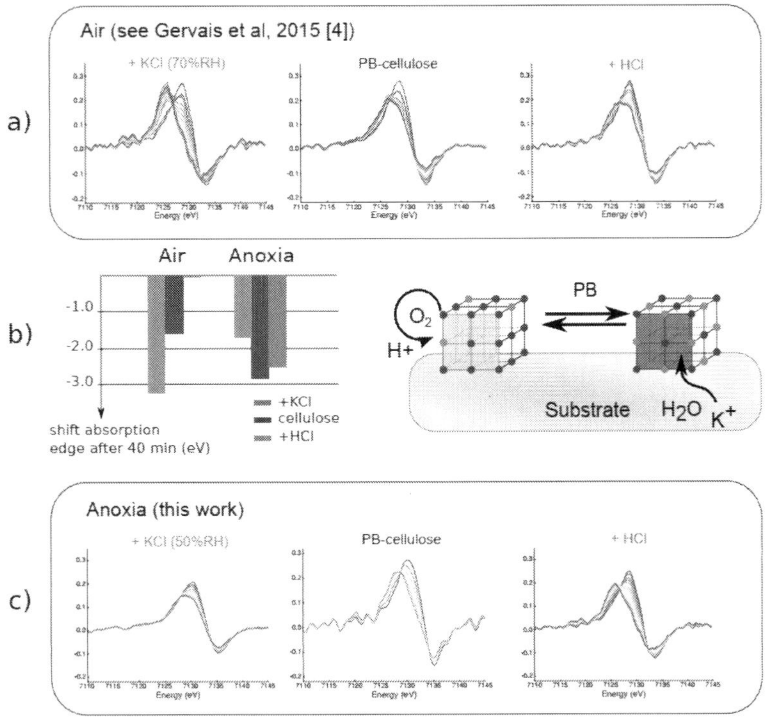


**Figure 1.** X-ray photochemistry of Prussian blue paper samples followed by Fe K-edge XANES. a) XANES derivatives in air in function of time (from [4]). (b) Left: Shift of the absorption edge after 40 minutes of continuous X-ray exposure. Right: Schematic overview of the redox process in these materials. (c) XANES derivatives in anoxia in function of time. In a) and c), colors of the spectra follow a rainbow code (from blue t = 0, to red t = 40 min).

paper. The pH and alkali cation content of the paper substrate was modified and their influence on the X-ray photoreduction kinetics in air studied in [4]. Results are recalled Figure 1a).

Addition of KCl to the cellulosic paper and high relative humidity (70 %RH) promotes the reduction of Prussian blue and results in a clear and greater shift of the absorption edge in the XANES spectra (-3.2 eV compared to -1.6 eV for Prussian blue on a pure cellulosic substrate). These findings can be interpreted by considering that upon reduction, solvated cations are migrating into the Prussian blue crystal lattice in view of compensating the increase of charges induced by iron reduction.

To the contrary, acidification of the cellulosic paper leads to a very small shift of the absorption edge (almost 0 eV) as well as broader and less intense XANES derivatives. Protons brought into the system, together with the X-ray beam reductive effect, are reducing the oxygen contained in the air, which in turn leads to a re-oxidation of the photoreduced Prussian blue (Figure 1b, right). This dynamic, competitive redox process lead to defects in Prussian blue and a photoreduction which is ill-defined and small in extent.

**Photochemistry of Prussian blue in anoxia.** To further refine the role of the environment and particularly of the oxygen, we investigated in this work the X-ray photoreduction of the same three PB cellulosic samples, but this time in anoxia instead of air. Anoxic conditions were obtained with a flow of nitrogen humidified at 50% RH. Measurements in both works were performed during the same beamtime session on the same samples. Therefore, the photoreduction kinetics can be compared.

As shown Figure 1c, the photoreduction differs in amount and quality for the three samples. In general, anoxia increases the photoreduction and the redox process seems to be less sensitive to the substrate (see Figure 1b: -1.7 eV, -2.8 eV and -2.5 eV for KCl, cellulose and HCl samples respectively). In agreement with previous studies [5, 9], these results confirm that anoxic conditions are the driving factor of the reduction kinetics and not anymore the nature of the substrate. This latter plays a smaller role but leads nevertheless to interesting spectral variations.

For the acidified sample, anoxia induces a drastic increase of photoreduction compared to air. This can be explained by the fact that no re-oxidation of the photoreduced Prussian blue is possible. Protons added to the system are thus not anymore participating directly to the redox process, and the photoreduction looks similar in quality and extent to that of the Prussian blue pure cellulosic sample in anoxia.

For the sample loaded in cations, the photoreduction is still consequent in anoxia but lower than in air (Figure 1b: -3.2 eV in air and -1.7 eV in anoxia). The difference lies mainly in the shape of the XANES derivatives. While a clear shift of the absorption edge occurs in air, XANES derivatives are broad and loosely defined in anoxia, suggesting the creation of disorder and many oxidation regions in the photoreduced Prussian blue structure. One possible explanation lies in the multiple roles played by water. Water is necessary to solvate cations and allow them to migrate through the zeolite-like channels of the lattice, where the water molecular network acts as a cation carrier. The photoreduction kinetics in anoxia were recorded in a lower relative humidity environment (50%RH) than in air (70%RH), so that cations might be less readily solvated. As a consequence, the photoreduction might become limited by the low diffusion coefficients of the cations rather than by the anoxic environment. Another explanation lies in the water network present in the Prussian blue lattice, which might change in anoxic conditions.

## 4. Conclusion

Thanks to an original approach based on X-ray photochemistry monitored by energy-dispersive XAS, we could refine the redox process of the Prussian blue in paper support and show that in fine, this one is controlled by the substrate content, structure and photosensitivity as well as its close interaction with the oxygen environment. Applied to Prussian blue in the context of heritage materials, the present approach could be interesting to investigate the photochemistry of novel multifunctional materials based on Prussian blue or Prussian blue analogues [10].

## Acknowledgments

Claire Gervais acknowledges the Swiss National Science Foundation for the grant no. 138986.

## References

[1] Berrie B H 1997 Prussian blue *Artist's pigments: A Handbook of Their History and Characteristics. Volume 3* ed Feller R L and FitzHugh E W chap 7, pp 191–217

[2] Gervais C, Languille M A, Réguer S, Gillet M, Pelletier S, Garnier C, Vicenzi E P and Bertrand L 2013 *Journal of Analytical Atomic Spectrometry* **28** 1600–1609

[3] Samain L, Gilbert B, Grandjean F, Long G J and Strivay D 2013 *Journal of Analytical Atomic Spectrometry* **28** 524–535

[4] Gervais C, Languille M A, Moretti G and Reguer S 2015 *Langmuir* **31** 8168–8175

[5] Gervais C, Languille M A, Reguer S, Gillet M, Vicenzi E P, Chagnot S, Baudelet F and Bertrand L 2013 *Applied Physics A* **111** 15–22

[6] Ware M 1999 *Cyanotype: The History, Science, and Art of Photographic Printing in Prussian Blue* (Bradford, UK: National Museum of Photography, Film, and Television)

[7] Baudelet F, Kong Q, Nataf L, Cafun J D, Congeduti A, Monza A, Chagnot S and Itie J P 2011 *High Pressure Research* **31** 136–139

[8] Newville M 2001 *Journal of Synchrotron Radiation* **8** 322–324

[9] Gervais C, Languille M A, Reguer S, Garnier C and Gillet M 2014 *Heritage Science* **2** 26

[10] Pajerowski D M, Ravel B, Li C H, Dumont M F and Talham D R 2014 *Chemistry of Materials* **26** 2586–2594

# A microchannel confocal examination of arsenic speciation and distribution in *Bufo americanus*

**Michelle M. Nearing[1], Iris Koch[1], Robert A. Gordon[2], Kenneth J. Reimer[1]**

[1]Environmental Sciences Group, Royal Military College, Kingston, ON, Canada
[2]Moyie Institute and Dept. of Physics, Simon Fraser University, Port Coquitlam, BC, Canada

E-mail: ragordon@alumni.sfu.ca

**Abstract**. We have used confocal methods to examine the distribution and speciation of arsenic within amphibian tissue (*Bufo americanus*) from a contaminated mine site. The use of new microchannel technology permits a confocal, and energy-independent, examination of a given voxel within the amphibian tissue without the need for sectioning. We observe differences in arsenic concentration and speciation depending on tissue type, with the Eberth-Katschenko layer in particular containing Mn, Fe, Cu and Zn in addition to Ca and pentavalent arsenic.

## 1. Introduction

Amphibians such as *Bufo americanus*, whose life-cycles comprise both aquatic and terrestrial environments, can be exposed to arsenic through soil, food or water sources, allowing these species to serve as indicators of overall environmental quality. Understanding the potential uptake and biotransformation pathways for arsenic in these indicator species is important for assessing risk to ecological systems. A previous study has examined arsenic speciation in frog and toad species from an arsenic contaminated site using high-performance-liquid-chromatography – inductively-coupled-plasma-mass-spectrometry (HPLC-ICPMS) and complementary X-ray absorption spectroscopy (XAS) [1]. Although the XAS techniques used in this study allowed for the speciation of all arsenic in these samples, the spatial resolution of the arsenic in the skin and legs tissues was not determined due to difficulties in obtaining suitable tissue sections.

Confocal methods can provide an alternate means of obtaining distribution and speciation through fluorescence imaging and XAS without obtaining a thin section. A confocal set-up is comprised of both a focusing element and solid-angle-limiting detection element, and where these two elements overlap focus and detection, there is a defined volume (voxel) of investigation [2-4]. Polycapillary glass optics are commonly used in such x-ray fluorescence imaging set-ups, and, when used as the detection optic, the solid-angle defining the detection region is dependent on the critical angle of the glass, which leads to an energy-dependence [4] of the voxel size, with lower energy fluorescence being detectable from a larger volume than higher energies. This is problematic when one is seeking to correlate the presence of lighter and heavier atoms.

Here we make use of a germanium lithographically-fabricated microchannel optic for the detection aspect of the measurement [5]. While the polycapillary optic functions like a two-dimensional spatial filter [6], albeit energy-dependent, the microchannel optic functions similarly to Soller slits [7], constraining the detection of emitted x-rays along the beam, while the focusing constrains the other two dimensions to give a defined voxel without the energy-dependence of the glass polycapillary. By using this method, we have determined elemental distributions and measured arsenic speciation at select locations in a bulk piece of *Bufo americanus* skin.

## 2. Experimental

A patch of tissue from the back of a eastern American toad (*Bufo americanus*) was obtained from a freeze-dried specimen collected from a contaminated mine site near Upper Seal Harbour, Nova Scotia, as per Moriarty *et al.*[1]. The irregularly-textured patch, approximately 20 $mm^2$ in area, was sealed in kapton tape for measurement at room temperature. The region of examination is indicated in Figure 1.

**Figure 1.** (Colour online) Patch of skin tissue indicating region of examination (green, middle), approximate location on a representative (not actual before freeze-drying, left) toad back and overhead view (right) of measurement showing 1mm working distance of optic to sample.

X-ray measurements were carried out using the microprobe end-station of beamline 20ID at the Advanced Photon Source [8]. The undulator beam through the LN2-cooled Si(111) monochromator was focused to a cross-section of 2 × 2 $\mu m^2$ using Kirkpatrick-Baez mirrors, with the beam incident on the sample stage at a 38° angle. Samples were affixed by tape to a plastic holder mounted on an xy-stage (Melles-Griot) capable of submicron stepping, further mounted on an xyz stage (ADC) with approximately 0.5 um resolution as per the typical operating set-up at 20ID [8]. For finer resolution depth-scanning, a high-resolution stage (Newport) was added. Incident beam intensity was monitored using a He-filled transmission ion chamber (incident intensity ~ $9 \times 10^{10}$ ph/s at 12 keV). A germanium microchannel optic [5], lithographically fabricated at Cornell University, was mounted in front of a single-element Vortex silicon drift detector and optimized, using similar motor stages, at 1mm working distance from the focal spot of the beam with nominal 2 μm selectivity. The tissue sample was scanned either in horizontal position and depth or in horizontal and vertical positions with the full fluorescence energy spectrum collected at each voxel at an excitation energy of 12 keV. X-ray Absorption Near Edge Spectra (XANES) at the arsenic K-edge were collected at select voxels after mapping. Three to five XANES scans (10 minute duration) were averaged before background removal and normalization to edge jump. While some changes in amplitude were evident, possibly due to some positional drift (not observed optically, but within limits of instrumentation and thermal stability of the experimental set-up and a sample heterogeneous in three dimensions) during or between scans, no change in white line energy was observed from first to last scan, indicating the beam was not changing the arsenic speciation during the XANES measurements.

## 3. Results and Discussion

To locate the irregular surface, an initial coarse (10 μm step) scan of the horizontal position and depth (xz) was conducted. Upon doing so, the existence of a calcium-rich sub-surface layer, some 10 – 30 μm thick, became apparent, as shown in Figure 2. For *Bufo americanus*, this likely corresponds to the Eberth-Katschenko (EK) layer [9] observed in *Bufo itericus*. The EK layer helps the amphibian maintain hydric balance on land, but can also regulate or inhibit transport of other ionic species between the external environment and the interior of the toad. Arsenic is present in the EK layer (Figure 2) as are Mn, Fe, and Zn, with a trace of copper as well. A comparison of x-ray fluorescence spectra confirms these metals are substantially less evident below the EK layer. Note, with the beam incident from lower right and fluorescence detection to lower left in Figure 2, no correction for matrix effects have been applied to these spectra – i.e. at the EK position, the incident beam must pass through more tissue (~ 500 μm of 1g/ml low density carbon, or a 6 – 7% additional incident attenuation), while at the sub-EK position, fluorescence must pass through an extra ~80 μm of tissue, making the Ca signal about 1/3 less than if the escape depths were equivalent. Such a correction here would be small in comparison to the apparent differences. The presence of arsenic on the surface may indicate some residual contamination from the mine site, where the majority species of arsenic is in arsenate form [10].

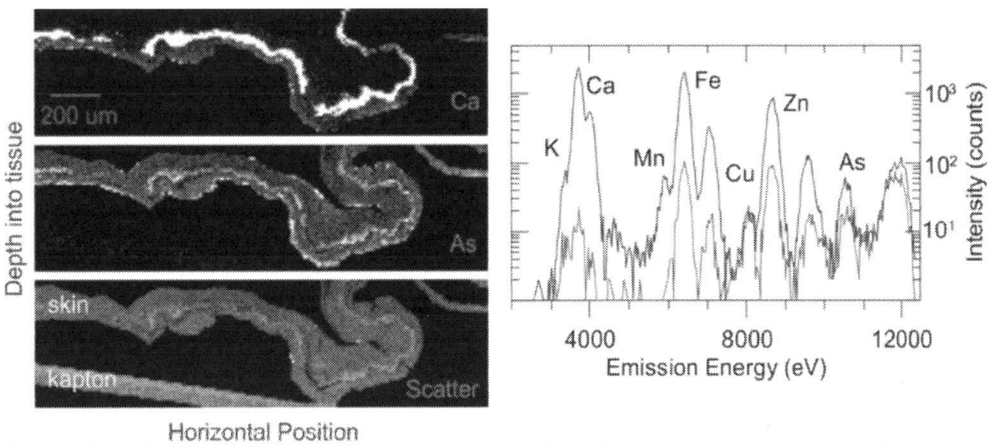


**Figure 2.** (colour online) X-ray fluorescence maps for calcium, arsenic and scattered intensity (left) and sample XRF spectra (right) for two regions: one in the Eberth-Katschenko layer (blue) and one below the E-K layer (green) illustrating the change in metal content. Spectra were obtained from binning 12 pixel-spectra in the image as indicated, with 0.4s collection time per pixel.

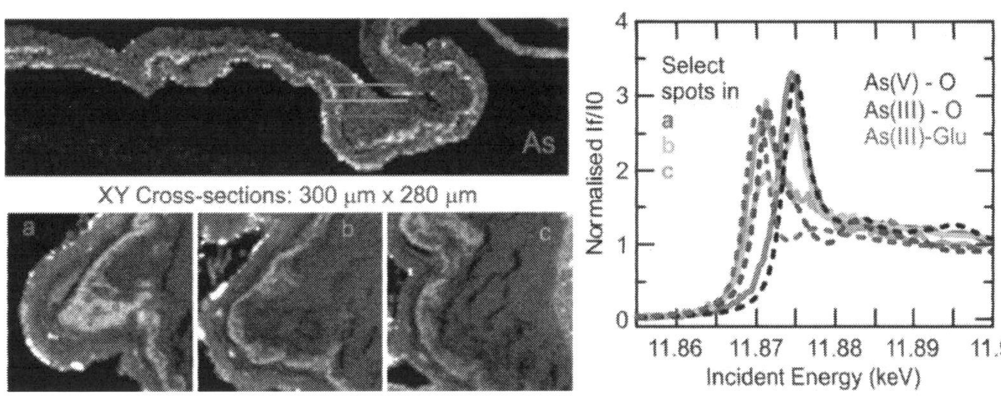


**Figure 3**. (colour online) Arsenic cross-section maps and XANES data from select locations showing different species of arsenic present.

A higher-resolution examination in cross-section (horizontal – vertical, xy) was conducted at three different depths of the protruding portion in Figure 2 (not centered on the x-depth section map).

While presenting a different view, such xy mapping does not permit one to follow the path of the beam through the tissue as the xz does. The arsenic maps at these different depths and XANES spectra from select points in them are presented in Figure 3. Spot b, near the surface of the tissue, exhibits two distinct valence (3,5+) white line features. Spot a, from within the EK layer, only exhibits pentavalent arsenic, while spot c, further within the tissue (away from epidermis), is dominated by trivalent species, closer to $As^{3+}$- O than $As^{3+}$-S. Both $As^{3+}$-O and $As^{3+}$-S were identified in bulk analysis of the dried toad leg [1]. Since spot b is bordering the surface, and though the locale favoured $As^{5+}$-O species [10], if the 3+ species observed interior is part of the glandular secretion used to coat the skin (which is the function of the glands on the skin examined), then it is possible that some of the trivalent arsenic is due to remnant surface coating, but further tests would be needed to confirm this.

There are two possible routes for arsenic into the body of the toad – ingestion and absorption. The presence of $As^{5+}$ within the EK layer suggests that the absorption route is occurring, however, further within the tissue $As^{3+}$ is observed. Additional control studies would be needed to determine if the EK layer is an effective barrier against arsenic migration further into the toad, or if some reduction occurs on transiting through the EK layer.

## 4. Conclusions

We have used novel microchannel confocal methods to examine skin tissue from the back of a *Bufo americanus* (eastern American toad). The Eberth-Katschenko layer within the dermal tissue, in addition to a strong calcium signature, ragordoncontains arsenic and additional metals, likely obtained from absorption. These metals are less evident in lower tissue. Evidence for reduced arsenic species was found within the skin tissue, but the EK layer was dominated by pentavalent arsenic. Further control studies are needed to identify the EK layer barrier function.

## Acknowledgements

The RMC-ESG work was supported by NSERC of Canada. Sector 20 facilities at the Advanced Photon Source, and research at these facilities, are supported by the US Department of Energy - Basic Energy Sciences, the Canadian Light Source and its funding partners, the University of Washington, and the Advanced Photon Source. Use of the Advanced Photon Source, an Office of Science User Facility operated for the U.S. Department of Energy (DOE) Office of Science by Argonne National Laboratory, was supported by the U.S. DOE under Contract No. DE-AC02-06CH11357. We thank A.R. Woll and D. Agyeman-Budu of CHESS/Cornell U. for providing the microchannel optic.

## References

[1] Moriarty M M, Koch I and Reimer K J 2013 *Environmental Science Processes & Impacts* **15** 1520.
[2] Janssens K, Proost K and Falkenberg G 2004 *Spectrochiminica Acta* B **59** 1637.
[3] Woll A, Mass J,Bisulca C, Huang R, Bilderback D H, Gruner S and Gao N 2006 *Appl. Phys.* A **83** 235.
[4] Nakazawa T and Tsuji K 2013 *X-ray Spectrometry* **42** 374.
[5] Woll A, Agyeman-Budu D, Choudhury S, Coulthard I, Finnefrock A C, Gordon R, Hallin E and Mass J 2014 *J. Phys.: Conf. Series* **493** 012028.
[6] Behne E A, Feng Y and Seidler G T 2001 *Rev. Sci. Instrum.* **72** 3908.
[7] Bewer B 2012 *J. Synch. Rad.* **19** 185.
[8] Heald S M, Cross J O, Brewe D L and Gordon R A 2007 *Nucl. Instrum. Methods Phys.* A **582** 215.
[9] Azevedo R A, Pelli A A, Ferreira-Pereira A, Santana A S J, Felsemburgh and de Brito-Gitirana L 2005 *Micron* **36** 61 and 385 (erratum).
[10] Meunier L, Koch, I and Reimer K J 2011 *Sci. Tot. Environ.* **409** 2233.

# New data evaluation procedure including advanced background subtraction for radiography using the example of insect mandibles

**Stefan Mangold*[1], Thomas van de Kamp[2,3], Ralph Steininger[1]**

[1] Karlsruhe Institute of Technology, ANKA/IBPT (KIT), P.O. Box 3640, D-76021 Karlsruhe, Germany
[2] Karlsruhe Institute of Technology, Institute for Photon Science and Synchrotron Radiation (KIT-ANKA-IPS), P.O. Box 3640, D-76021 Karlsruhe, Germany
[3]Karlsruhe Institute of Technology, Laboratory for Applications of Synchrotron Radiation, P.O. Box 3640, D-76021 Karlsruhe, Germany

E-mail: stefan.mangold@kit.edu

**Abstract**. *The usefulness of full field transmission spectroscopy is shown using the example of mandible of the stick insect Peruphasma schultei. An advanced data evaluation tool chain with an energy drift correction and highly reproducible automatic background correction is presented. The results show significant difference between the top and the bottom of the mandible of an adult stick insect.*

## 1. Introduction

The usefulness of full field transmission spectroscopy can be attributed to the fact that spectral information is collected in parallel, thus avoiding the time-consuming scanning of the sample through a micro-focussed X-ray beam. Therefore this technique is more and more used to gain insight into complex and large sample systems in catalysis, environmental science and biology. Pioneering work in X-ray Absorption Near Edge Structure (XANES) imaging of catalyst has been reported in [1]. During this work a correction of the (vertical) energy drift over the detector and a background subtraction of the spectra were omitted. A formula to calculate the energy drift dependent on the divergence of the x-ray beam and type of crystal in the monochromator is described here [2]. A significantly enhanced data evaluation chain was presented in combination with measurements of an x-radia microscope set-up here [3]. Because this group still omitted the energy calibration/inspection and used a very simple background subtraction, we aimed to develop our own data evaluation chain.

Zinc plays a critical role in the mechanical properties of invertebrate jaws and mandibles [4]. The zinc species distribution is important to understand the exceptional high hardness and Young's modulus. Volume renderings of the head of the stick insect Peruphasma schultei [5] based on tomographic white beam measurements showed the suitability of this mandibles for full field XANES measurements [6]. A large and homogenous beam is needed to take full advantage of the large available detector area. As a bulk bending magnet beamline, ANKA-XAS is therefore very well suited. Due to the absence of any focusing mirrors the beamline delivers a beam of very uniform intensity across 11 x 1 mm$^2$ and is therefore well-suited for radiography of laterally extended samples.

## 2. Experimental

The difference to the standard XAFS set-up was the installation of the camera set-up between **M-850.50** Hexapod 6 Axis micro positioning system from PI and the second ionisation chamber. The high-resolution X-ray imaging detector system consists of a YAG fluorescence screen a vertical reflecting mirror, Mitutoyu long working distance microscope objective and a PCO-edge camera [7]. The effective pixel size on the sample is 5 by 5 $\mu m^2$ and therefore the set-up generates more than 500.000 Absorption Near Edge Structure (XANES) data sets in each radiogram. Even with the reduction of the measurement time per point by a factor of 10 to earlier measurements this systems offers a dramatically improved signal to noise ratio. Since a full field XANES measurement relies on a significant absorption jump per pixel or voxel (tomographic measurements) and therefore many samples have to be thicker than 50 $\mu$m in the medium X-ray energy range, the setup uses a flux-optimized optic with a well suitable lateral resolution of 5 $\mu$m.

## 3. Data evaluation

The image data are cached on advanced RAID-0 system and uploaded regular to a data cruncher with RAID5 data storages connected by Thunderbolt (MacPro, 1Gb/s SSD, 64 Gb RAM, 600 Mb/s read/write on storages). The single steps of the advanced data evaluation tool chain based on IgorPro [8] are described in table 1. Similar data evaluation procedures have been reported by other groups [9].

**Table 1:** Data evaluation steps

| Data evaluation step | Description | Time needed per core (standard intel CPU) |
|---|---|---|
| Calculation of spectra | Calculation of spectra by ln(measurement without sample/ measurement with sample) for each detector pixel (ln = natural logarithm) | < 6 min per 4Gb |
| Calculation of the vertical energy shift | Used for correction of the vertical energy drift over the image | 5h (500.000 spectra); only once per edge / core |
| Correction of the vertical energy shift | Used to correct the beamline dependent vertical energy shift | < 3 min (500.000 spectra) |
| Edge filtering | Filter out all pixels, which contains no reasonable spectrum | < 2min |
| Background subtraction | Independent background subtraction for each raw XAFS spectra in each pixel | ~100 spectra / core / s |
| PCA analysis | Detection of components (PCA = principal component analysis) | 10s - 15 min |
| Linear combination | Calculate the loadings of the different components | 10 min – 2h / core |

Unique steps of this data evaluation are the correction of the small but notable energy drift over the detector due to the illumination of the flat monochromator crystal with a divergent beam and the highly reproducible automatic background correction. The energy correction is done by a measurement of the respective metal foil (or homogenous sample), the fit of the energy-position and a scaling factor of the spectrum in each pixel and the usage of this shift values for each sample data set. The latter obtains its stability due a combination of a sigmoidal fit and a long distance pre- and post-edge fit. Due to this mechanism it can operate also at very low edge jumps down to 0.05. Because the fits are based on highly optimized C-code within IgorPro math libraries, it enables near real time data evaluation. The planned parallelisation of the code will enable faster data evaluation than acquisition. The energy correction and the background subtraction are needed to ensure, that the input data for the PCA contains only background subtracted, normalised data with reasonable edge jump. Without the clean-up steps, the PCA would be overrun by not useful information.

## 4. Results

Figure 1 shows the transmission image of the sample and differences between the absorption below and above the Zn absorption edge. Parts of the mandible of the stick insect *Peruphasma schultei* inherits an edge jump of nearly 1. This step is used to find area of interest. While the edge jump on the mandibles of the youngest insects is in the range of 0.20 (figure 3 A) and rather constant, the older insects have a significant stronger absorption on the tip of the mandible.

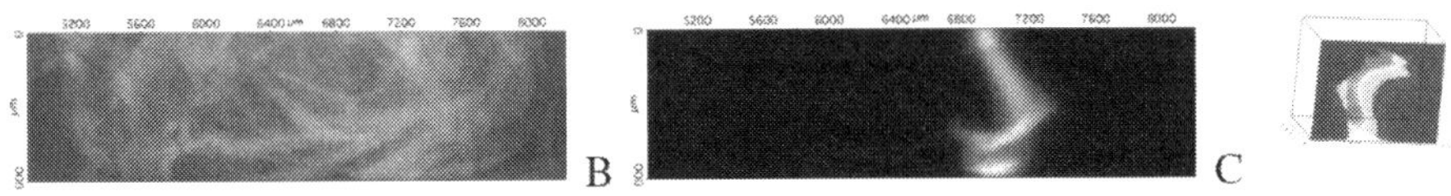

**Figure 1.** A) Transmission image of the sample; B) difference between after and before absorption edge. C) Zinc concentration over sample area, all plots show the mandible of an adult insect

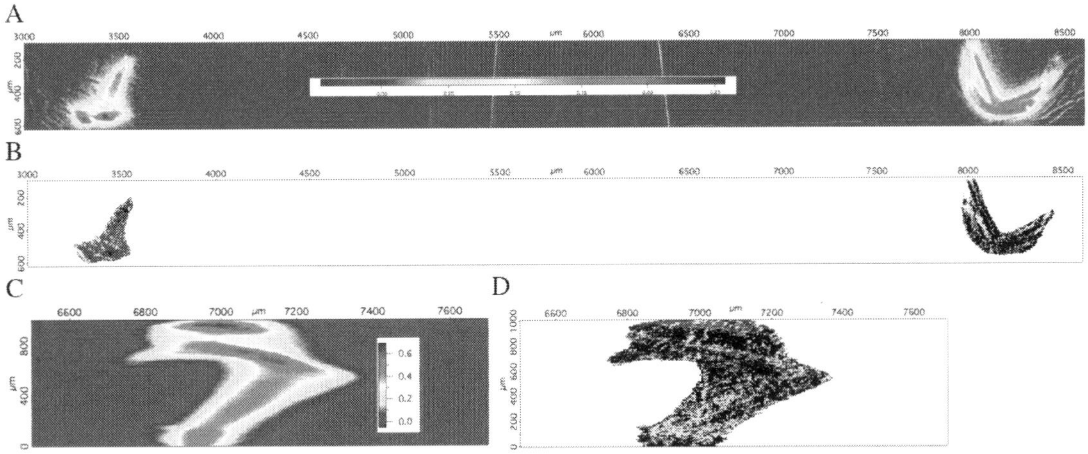

**Figure 2. A)** Absorption image plot shows the difference between after and before absorption edge. **B)** Distribution of the 3 PCA components over the mandibles, right side shows a young hatchling (larval stage 1) before it started feeding, left side an older larva of the same stage. C) Absorption image plot of the mandible of adult insect (difference between after and before absorption edge). D) Distribution of the 3 PCA components over the same mandible

Figure 2A illustrates the results of the linear combination fit of the 3 PCA-components over the mandibles in the youngest stick insects. There is a significant difference between the right (hatchling before it started feeding) and left mandible, but no remarkable difference over the mandible itself. The absorption for the mandibles of the adult insect at the zinc edge is larger than the juvenile ones (2C). The difference of the extracted PCA components are also significantly larger. While there is only a small change between inner and outer exoskeleton, there are significant differences between basis (bottom) and cutting edge (top) of the mandible.

The spectra presented in figure 3 indicate the small but distinct difference over the height of the mandible of an adult insect, which should be responsible for different hardness of the structures. The difference also between the mandibles of the younger insects (hatchling and older larval stage 1) is even smaller (figure 3). First literature studies indicate the presences of two different types of zinc-proteins [10]. Two mixtures (75/60 % and 25/40 %) of zinc-histidine and zinc cysteine are added to the plots for comparison.

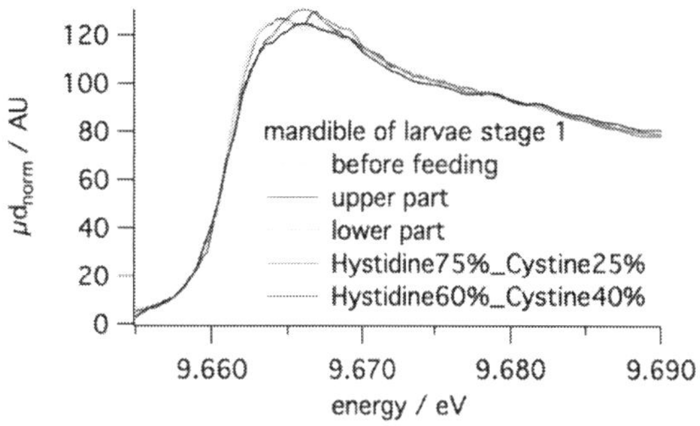


**Figure 3.** Distribution of the 3 PCA components over the mandible of hatchling and older larval stage 1

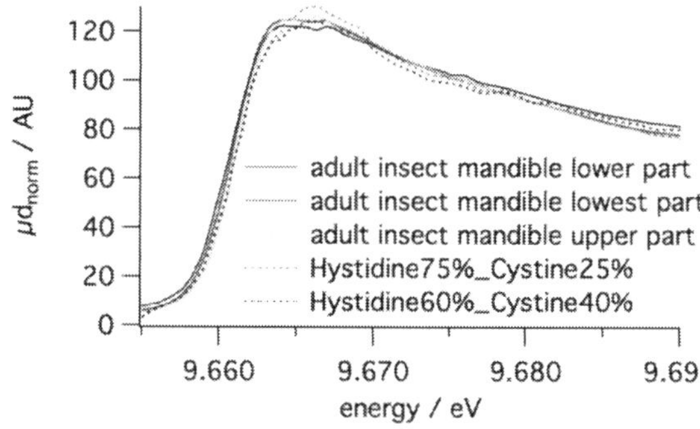


**Figure 4.** Distribution of the 3 PCA components over the mandible of the adult insect.

## 5. Conclusion

The improvement of the signal to noise ratio of the measurement and the improved data reduction tool chain shows the progress of a prototype experiment to a reliable analytical tool, which can be used for a large set of different sample like batteries, catalytic systems, art and mineral samples. The quality of the data evaluation is needed to detect the small difference between species in the mandibles. With the growth of the insect also the distribution of the species is changing.

## Acknowledgement

We acknowledge the Synchrotron Light Source ANKA for provision of beamtime.

## References

[1] Grunwaldt J-D, Hannemann S, Schroer C G and Baiker A 2006 *J. Phys. Chem.* B **110** 8674-8680
[2] Katayama M, Sumiwaka K, Hayashi K, Ozutsumi K, Otha T and Inada Y 2012 *J. Synchrotron Rad.* **19** 717–721
[3] Meirer F, Cabana J, Liu Y, Mehta A, Andrews J C and Pianettac P 2011 *J. Synchrotron Rad.* **18** 773–781
[4] Broomell C C, Mattoni M A, Zok F W and Waite J H 2006 *J. Exp. Biol.* **209** 3219-3225
[5] Conle O V and Hennemann F H 2005 *Zootaxa* **1068** 59-68
[6] Van de Kamp T, Ershov A, Dos Santos Rolo T, Riedel A and Baumbach T 2013 *Entomologie heute* **25** 147-160
[7] Cecilia A, Rack A, Douissard P A, Martin T, dos Santos Rolo T, Vagovič P, Hamann E, van de Kamp T, Riedel A, Fiederle M and Baumbach T 2011 *Nucl. Instr. Methods. A* **648** Supplement 1 321-323
[8] IgorPro - http://www.wavemetrics.com/ 10.10.2015
[9] Liu Y, Meirer F, Williams P A, Wang J, Andrews J C and Pianetta P 2012, J. Synchrotron Rad. **19** 281-287
[10] Terzano R, Chami Z A, Vekemans B, Janssens K, Miano T and Ruggiero P 2008, *J. Agric. Food Chem.* **56** 3222–3231

# SPring-8 BL36XU: Catalytic Reaction Dynamics for Fuel Cells

**O Sekizawa[1], T Uruga[1,2], Y Takagi[3], K Nitta[2], K Kato[2], H Tanida[2], K Uesugi[2], M Hoshino[2], E Ikenaga[2], K Takeshita[2,4], S Takahashi[2,4], M Sano[2], H Aoyagi[2], A Watanabe[2], N Nariyama[2,4], H Ohashi[2,4], H Yumoto[2], T Koyama[2], Y Senba[2], T Takeuchi[2], Y Furukawa[2,4], T Ohata[2,4], T Matsushita[2,4], Y Ishizawa[2], T Kudo[2,4], H Kimura[2,4], H Yamazaki[2], T Tanaka[2,4], T Bizen[2,4], T Seike[2,4], S Goto[2,4], H Ohno[2], M Takata[2,4], H Kitamura[2,4], T Ishikawa[2,4], M Tada[3,5], T Yokoyama[3] and Y Iwasawa[1]**

[1] Innovation Research Cnter for Fuel cells, The University of Electro-Communications, Chofu, Tokyo 182-8585, Japan
[2] Japan Synchrotron Radiation Research Institute, SPring-8, Sayo, Hyogo 679-5198, Japan
[3] Department of Molecular Structure, Institute for Molecular Science, Okazaki Aichi 444-8585, Japan
[4] RIKEN SPring-8 Center, Sayo, Hyogo 679-5148, Japan
[5] Research Center for Materials Science, Nagoya University, Nagoya, Aichi 464-8602, Japan

E-mail: iwasawa@pc.uec.ac.jp

**Abstract**. A tapered undulator beamline BL36XU was constructed at SPring-8 to conduct structural and electronic analysis of dynamic events on polymer electrolyte fuel cell (PEFC) cathode catalysts for the development of next-generation PEFCs. BL36XU provides various time and spatially resolved XAFS techniques in an energy range from 4.5 to 35 keV for investigating PEFCs under the operating conditions. In addition, we developed in-situ complementary measurement systems, such as in-situ time-resolved XAFS/XRD and ambient pressure HAXPES systems. This report describes the performance and present status of the BL36XU.

## 1. Introduction

Polymer electrolyte fuel cells (PEFCs) have extensively been studied as a next-generation clean energy supply source. Stationary fuel-cell cogeneration systems and fuel-cell vehicles have been launched on the market in Japan, but various problems have to be solved to spread them widely. Crucial issues for widely spread commercialization of fuel-cell cars are an increase of activity, much improvement of durability and reduction in the cost. To solve these problems and to elucidate the oxygen reduction reaction mechanism and the degradation process of cathode catalysts, we constructed a high time- and spatially resolved XAFS beamline BL36XU at SPring-8. The time scale of milliseconds to seconds is necessary to study the structural kinetics of the catalytic reactions in PEFCs, and the spatial resolution of 100 nm is necessary to observe the spatial distribution of catalyst in PEFCs during the deterioration process.

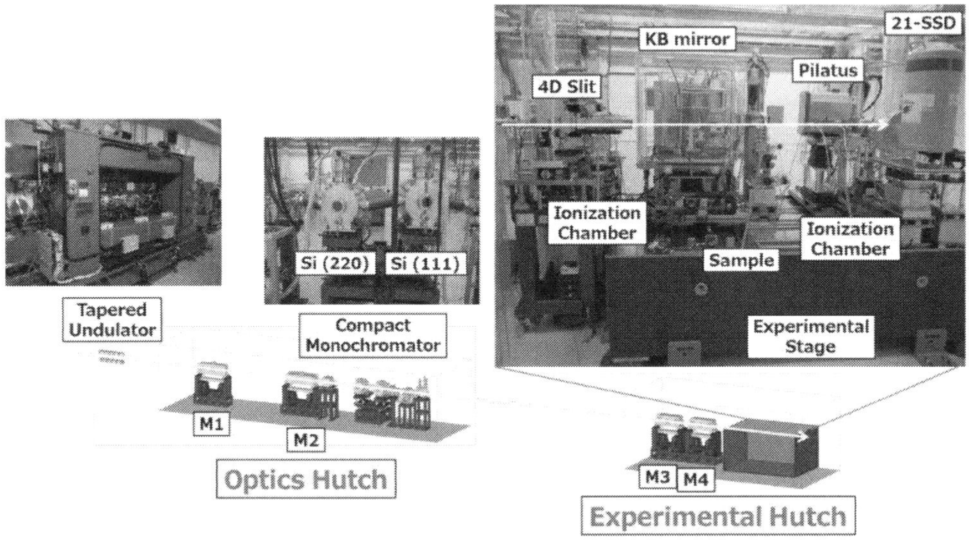


**Figure 1.** Layout of BL36XU.

The BL36XU construction was completed in November 2012 and made available to users in January 2013. The outline and design of the BL36XU are described in a previous report [1]. Figure 1 shows a schematic layout of BL36XU. It consists of a SPring-8 standard tapered undulator, front-end section, one optics hutch, and one experimental hutch. The main X-ray optics consists of two servomotor-driven monochromators and four mirrors. The main measurement system and performance specifications are described below.

## 2. Time-resolved measurements

Three types of time resolved XAFS systems were developed and installed at BL36XU corresponding to the time resolution, sample, and measurement condition using quick XAFS (QXAFS) and energy dispersive XAFS (DXAFS) techniques.

### *2.1. Quick XAFS measurement system*

A QXAFS system using direct-drive servomotor-driven Si channel-cut crystal monochromators with a minimum time-resolution of 10 ms [2] is a base system for XAFS measurements at BL36XU. Two monochromators consisting of Si(111) and Si(220) crystals were installed for covering an energy range from 4.5 to 35 keV. This system has been used for 20-ms time-resolved QXAFS measurements conducted for PEFCs under various operating conditions [3-5]. Figure 2 shows the Pt $L_3$-edge EXAFS spectra of a PEFC Pt/C catalyst under the operating conditions at the BL36XU. The minimum Pt concentrations to measure good quality XANES and EXAFS spectra with 20-ms time resolution is

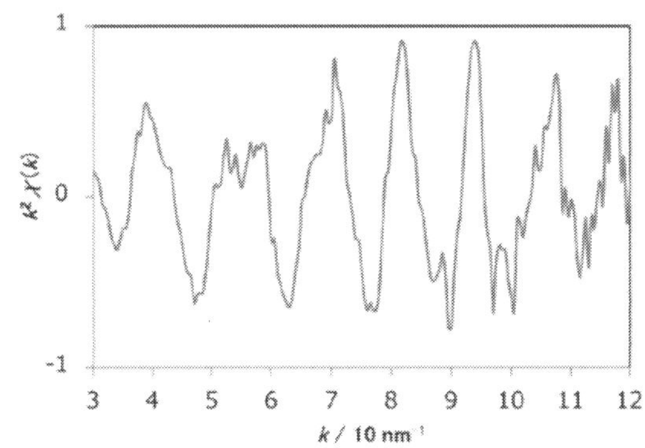


**Figure 2.** $k^2$-weighted Pt $L_3$-edge EXAFS oscillation of PEFC (0.5 $mg_{-Pt}/cm^2$) by 20 ms time-resolved QXAFS.

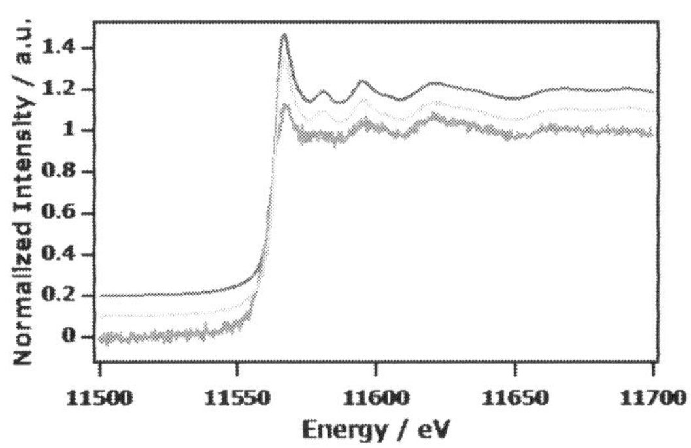


**Figure 3.** Pt $L_3$-edge XANES spectrum of Pt foil measured by step scan (blue), QXAFS (green) and DXAFS (red) techniques. Measurement times; 12 min, 20 ms and 100 μs, respectively.

ca. 0.1 and 0.5 $mg_{-Pt}/cm^2$, respectively

Another faster QXAFS system consists of a galvano motor driven monochromator [6], which enables a time resolution of milliseconds and is used to observe fast elementary steps in real PEFCs.

### *2.2. Energy dispersive XAFS measurement system*

The fastest measurement system is a DXAFS system with a minimum time resolution of 100 μs, which is especially used for model samples of PEFCs with high catalyst concentrations to obtain information on the faster elementary steps in a cathode catalytic reaction. Figure 3 shows the Pt $L_3$-edge DXANES spectra of Pt foil together with step scan XANES and QXANES spectra.

## 3. Spatially resolved XAFS

Membrane electrode assembly (MEA) of PEFCs has multi-layered structures consisting of anode catalyst layer, polymer electrolyte membrane and cathode catalyst layer. The electrochemical reactions occur at the cathode catalyst layer with spatially heterogeneous distribution, so spatial distribution and morphology of cathode nanoparticles must be investigated. Two types of two-dimensional imaging XAFS methods were installed in BL36XU depending on the catalyst concentration in MEAs: a fast scanning microscopic XAFS system and an imaging QXAFS system. An in-situ X-ray computed laminography (XCL) XAFS measurement system was developed and installed as a three-dimensional microscopic XAFS.

### *3.1. Fast scanning microscopic XAFS systems*

A fast scanning microscopic XAFS system [7] using Kirkpatrick-Baez (KB) mirrors (JTEC Co.) was developed with a minimum special resolution of 100 nm. Two sets of KB mirror systems were installed to cover an energy range from 4.5 to 35 keV: one has Rh-coated mirrors, and the other has Pt-coated mirrors. Both mirror systems can be automatically switched using motorized stages. Figure 4 shows focused beam profiles. For example, the X-ray flux is $2\times10^{10}$ photons/s with a beam size of 200 nm × 200 nm at 12 keV. This system has been used for 2D chemical state imaging of dilute samples, such as a sliced MEA sample [8] and single-catalyst particle [9].

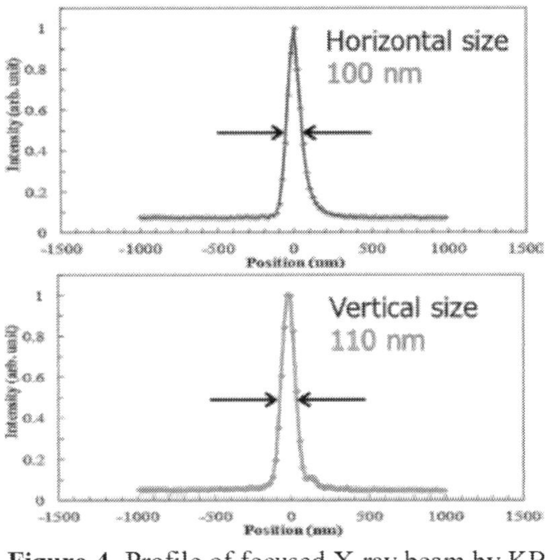


**Figure 4.** Profile of focused X-ray beam by KB mirror at X-ray energy of 25 keV.

### *3.2. Imaging QXAFS measurement system*

Imaging QXAFS enables time- and spatially resolved QXAFS by measuring transmitted X-ray images by a high-resolution X-ray imaging detector consisting of an imaging unit (Hamamatsu Photonics K.K., AA50) and a low noise CMOS camera (Hamamatsu photonics K.K., ORCA-Flash 4.0). The field of view depends on magnification of lens (266 μm ~ 1 mm), and maximum spatially resolution is 1 μm. This system is used for high concentration samples. The time-resolved imaging QXAFS spectra can be measured with a minimum time resolution of one minute.

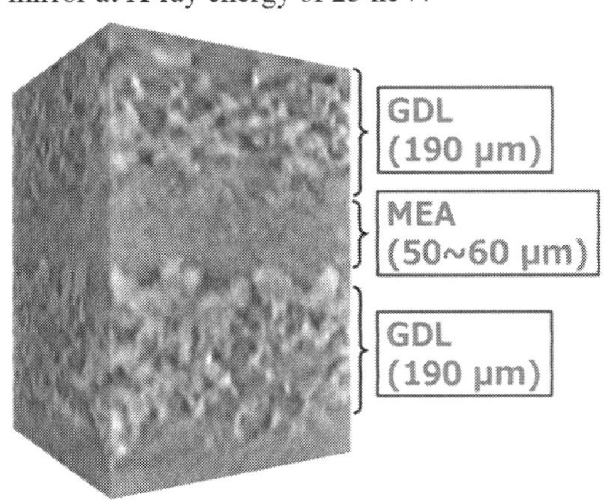


**Figure 5.** Reconstructed image of PEFCs.

### *3.3. Laminography XAFS measurement system*

An in-situ 3D imaging XAFS system was developed using XCL, which is suitable for 3D imaging of laterally extended planar objects such as PEFCs [10, 11]. The inclination angle of the rotational axis is set to 30 degrees and measures transmitted X-ray images by using a position sensitive detector. Figure 5 shows a reconstructed image of PEFC by XCL. 3D reconstruction was conducted using a filtered back projection method considering the inclination of the rotational axis. Each part of the MEA and gas diffusion layer (GDL) can be clearly seen.

## 4. Complementary measurement systems

### *4.1. Time-resolved X-ray diffraction*

To obtain complementary structural information of the crystalline part in catalyst particles, an in-situ 20 ms time-resolved X-ray diffraction (XRD) measurement system was installed. Diffracted X-rays are detected using a two-dimensional pixel detector (DECTRIS Ltd., PILATUS 300K-W). This system can be coupled with a time-resolved XAFS measurement system using the same PEFC, which enables sequential time-resolved QXAFS/XRD measurement with time resolution of tens milliseconds by fast control of the QXAFS monochromators.

### *4.2. Ambient pressure hard X-ray photoelectron spectroscopy*

An ambient pressure hard X-ray photoelectron spectroscopy (AP-HAXPES) system (VG Scienta Ltd., R4000 HiPP-2) was developed for investigating the electronic states of metal and light elements in catalysts under a maximum pressure of 5,000 Pa. This system has been applied for measuring PEFC cathode catalysts under operating conditions [12].

## 5. Conclusion

BL36XU is a high time- and spatially resolved XAFS measurement system under PEFC operating conditions. We have proceeded to elucidate the oxygen reduction reaction mechanism and degradation process of the cathode catalysts by complementary uses of XAFS, XRD, and HAXPES techniques.

## 6. Acknowledgement

This work was supported by the New Energy and Industrial Technology Development Organization (NEDO), Japan.

## References

[1] Sekizawa O, Uruga T, Tada M, Nitta K, Kato K, *et al.* 2013 *J. Phys.: Conf. Ser.* **430** 012020
[2] Nonaka T, Dohmae K, Araki T, Hayashi Y, Hirose Y, *et al.* 2012 *Rev. Sci. Instrum.* **83** 083112
[3] Ishiguro N, Uruga T, Sekizawa O, Tada M, *et al.* 2013 *Phys. Chem. Chem. Phys.* **15** 18827-34
[4] Samjeské G, Higashi K, Sekizawa O, Iwasawa Y, *et al.* 2014 *ECS Trans.* **64** 113-19
[5] Ishiguro N, Kityakarn S, Sekizawa O, Tada M, *et al.* 2014 *J. Phys. Chem. C* **118** 15874-83
[6] Uruga T, Tanida H, Inoue K, Yamazaki H and Irie T 2007 *AIP Conf. Proc.* **882** 914
[7] Tsuji T, Uruga T, Sekizawa O, Ishiguro N, Tada M, *et al.* 2013 *J. Phys.: Conf. Ser.* **430** 012019
[8] Takao S, Sekizawa O, Uruga T, Iwasawa Y, *et al.* 2014 *Angew. Chem. Int. Ed.* **53** 14110-14
[9] Ishiguro N, Uruga T, Sekizawa O, Tada M, *et al.* 2014 *ChemPhysChem.* **15** 1563-68
[10] Saida T, Sekizawa O, Uruga T, Tada M, *et al.* 2012 *Angew. Chem. Int. Ed.* **51** 10311-14
[11] Sekizawa O, *et al.* 2014 *X-Ray Microscopy 2014 Conference Program Handbook* p 287
[12] Takagi Y, Wang H, Sekizawa O, Tada M, Iwasawa Y, *et al.* 2014 *Appl. Phys. Lett.* **105** 131602

Journal of Physics: Conference Series **712** (2016) 012143 doi:10.1088/1742-6596/712/1/012143

# Inhomogeneous distribution of chemical species in lithium nickel oxide cathode of lithium ion battery

**T Uenoyama, R Miyahara, M Katayama*, and Y Inada**

Department of Applied Chemistry, Ritsumeikan University, Kusatsu 525-8577, Japan

E-mail: katayama@fc.ritsumei.ac.jp

**Abstract.** The spatial distribution of the oxidation state for the nickel species in the $LiNiO_2$ cathode was analyzed by means of the *in-situ* XAFS imaging technique during the charging and discharging processes. The inhomogeneous reaction for the $LiNiO_2$ cathode was observed under the operating condition. The pattern in the 2-dimensional map of the oxidation state of the active material in the discharging process was similar to that in the charging process. It means that the areas preceding the discharging reaction agree with the areas delaying the charging reaction. It was suggested that the diffusion of $Li^+$ was restricted by the surface product of the $LiNiO_2$ electrode, and the concentration gradient of $Li^+$ delayed the charging reaction at the reaction channel of the discharging reaction.

## 1. Introduction

The inhomogeneous distribution of the chemical state for the active material in the electrode of lithium ion secondary battery (LIB) should be understood to prevent unexpected accidents, because it causes the local excess current and the local deactivation of the electrode. The control of such inhomogeneous distribution contributes to improve the battery performance for the stabilized utilization. The LIB electrodes are generally made as a composite of the active material, the conductive additive, and the binder. In such composite electrodes, the pathway of electrons and $Li^+$ ions for the electrochemical reaction include various interfaces. If the interfacial resistance differs by each position in the electrode, the inhomogeneous reaction may be caused during the charging and discharging processes. The analysis of the spatial distribution of the state of charge in the operating condition is essential to understand the spatial distribution of the resistance in the electrode. X-Ray absorption fine structure (XAFS) is the most powerful technique to observe the electrochemical reaction in operating battery cells. In the LIB cathode, the charging and discharging process involves the redox reaction of the transition metal species in the active material. The oxidation state of such metal species can be determined by the XANES spectrum. The full field XAFS imaging technique using a 2-dimensional detector was developed by Grunwaldt *et al.* in the investigation on $Rh/Al_2O_3$ catalyst [1]. We applied the XAFS imaging to analyze the progress of the electrode reaction with the spatial resolution in the order of 10 $\mu$m [2].

Lithium nickel oxide ($LiNiO_2$) has been expected as a promising active material of the next generation LIB [3]. However, the practical use of $LiNiO_2$ is not achieved due to the synthetic difficulty and the structural instability [4]. The partial substitution of Ni is thus attempted to improve the cycle performance [5,6] and the stability [7]. The knowledge of the spatial distribution for the charging/discharging reaction in the $LiNiO_2$ cathode is fundamental to understand the reaction of the partially substituted $LiNiO_2$ electrodes. In this research, we clarified the inhomogeneous reaction of the $LiNiO_2$ cathode under the operating conditions by means of the 2-dimensional XAFS imaging technique.

## 2. Experimental

Lithium nickel oxide ($LiNiO_2$) was synthesized by the sol-gel method [8]. The adipic acid solution was added to an aqueous solution of nickel acetate and lithium acetate (Li:Ni = 1:1) at 333 K and was evaporated to form the sol at 363 K in air. It was further dried at 363 K in vacuum for 24 h and was calcined at 973 K for 14 h under the $O_2$/He gas flow. The obtained material was identified as $LiNiO_2$ by the X-ray diffraction pattern. A cathode sheet was made as the composite of $LiNiO_2$, acetylene black, and a polymer binder applied on an aluminum foil as a current collector. The test LIB cell was assembled with the cathode sheet, two sheets of polypropylene as the separator, Li foil as the anode, and 1 mol $dm^{-3}$ $LiPF_6$ solution as the electrolyte solution. The solvent of the electrolyte solution was the 3:7 mixture of ethylene carbonate and ethyl methyl carbonate. An Al-laminated polymer bag was used as the outer shell of the battery for the *in-situ* XAFS experiments. The battery capacity was 86 mA h $g^{-1}$, and the charging/discharging processes were performed with the constant current of 39 mA $g^{-1}$.

The 2-dimensional XAFS imaging measurements were performed during the charging/discharging processes at the NW2A station of Photon Factory Advanced Ring (KEK, Japan). The Si(111) double-crystal monochromator was used without the focusing optics. The higher-order harmonics were removed by the Rh-coated double-mirror system. The incident X-ray intensity was monitored by an ionization chamber. A CMOS image sensor (ORCA Flash 2.8) coupled with the $Gd_2O_2S$:Tb phosphor was used as the transmitted X-ray detector. The measurement area was 4 mm × 3 mm. The XANES measurement was carried out at the photon energy from 8300 eV to 8400 eV around the Ni K edge, and the total measurement time for one XANES image was 10 min. The energy difference caused by the vertical dispersion of the beam and its incident angle to the monochromator was corrected by the previously reported procedure [2]. A chemical state map was drawn by judging the X-ray energy of the white line peak at each detector element.

## 3. Results and discussion

The chemical state maps of the $LiNiO_2$ cathode were shown in figure 1 together with the charging and discharging curves. The XAFS image shown in figure 1 were obtained for the third charging/discharging cycle. The similar patterns were observed for the further cycles. The chemical state of the nickel species in figure 1(a) and (k) shows the uncharged $LiNiO_2$ and charged $Li_{1-x}NiO_2$ state, respectively. Similar to the reported $LiFePO_4$ cathode [9], the inhomogeneous distribution was observed also in the case of the $LiNiO_2$ cathode, although the $LiFePO_4$ cathode consists of only two phases of $LiFePO_4$ and $FePO_4$

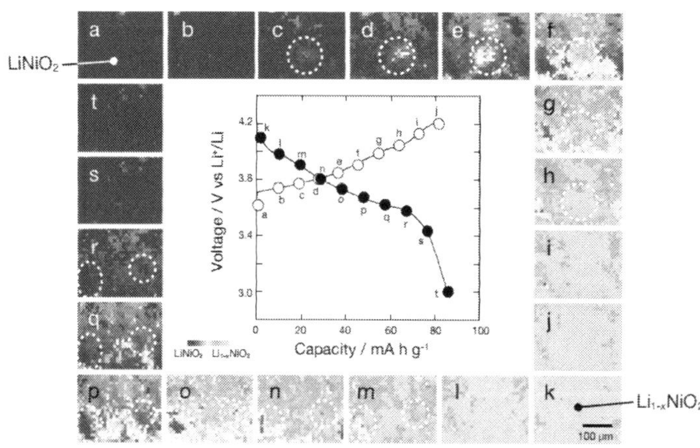


**Figure 1.** The chemical state maps of the $LiNiO_2$ cathode for the charging (from a to j) and discharging (from k to t) processes. The dashed circle shows the preceding area for the chemical state change of the Ni species during the electrode reactions.

during the charging process in contrast to the non-stoichiometric $Li_{1-x}NiO_2$ phase. The preceding area during the charging process was observed at almost the center in the presented maps (dashed circle in figure 1(c)-(h)). In the discharging process, the preceding area (dashed circle in figure 1(m)-(r)) was not identical to that in the charging process.

The images at the 50 % charged and discharged states are compared in figure 2 between the $LiNiO_2$ and $LiFePO_4$ cathodes. The scale of images is different between figure 2(a,b) and (c,d) according as the scale of inhomogeneity. The size of the distribution for the $LiNiO_2$ cathode is close to the spatial resolution of the present XAFS imaging analysis. The size of the preceding area of $LiNiO_2$ was *ca.* 100 $\mu$m, while that of $LiFePO_4$ was the order of 1 mm. The different scale of the spatial distribution indicates that the formation mechanism of the inhomogeneous distribution differs between two active materials. The reverse pattern was observed in the maps of $LiFePO_4$ between the charging and discharging processes (figure 2(c) and (d)), which means that the common reaction channel exists in both processes. The inhomogeneous distribution of $LiFePO_4$ becomes more uniform with increasing the carbon content in the cathode composite [9]. It has been suggested that the electronic conductance due to the carbon network controls the spatially inhomogeneous electrode reaction. In contrast, the maps of $LiNiO_2$ (figure 2(a) and (b)) show a similar pattern between both processes. It means that the charging reaction is delayed at the preceding area for the discharging process in the case of the $LiNiO_2$ electrode.

The disagreement of the preceding area at each process is explained in considering the accumulation of surface products, such as the polycarbonate species [10], on the $LiNiO_2$ cathode. The SEM images of the $LiNiO_2$ cathode were shown in figure 3 before and after the fifteen charging/discharging cycles. The porous shape of the pristine electrode, which can supply a lot of pathways for the $Li^+$ diffusion, disappeared after the redox cycles as shown in figure 3(b). This surface products formed during the aging process can restrict the $Li^+$ diffusion between the inside of the electrode composite and the bulk electrolyte solution. Because some cracks are found in the surface product, it is considered that the $Li^+$ diffusion occur mainly through such remaining pathways.

The observed inhomogeneous distribution of $LiNiO_2$ is thus ascribed by the restricted pathway for the $Li^+$ diffusion. At the discharging process, the pathway of $Li^+$ becomes the reaction channel, because the $Li^+$ supply from the bulk electrolyte solution is necessary to progress the electrode reaction. On the other hand, the following change of the $Li^+$ concentration is expected during the charging process. The charging reaction increases the $Li^+$ concentration in the cathode composite. The diffusion of the $Li^+$ ion occurs from the inside of the cathode composite to the bulk solution through the reaction channel. As

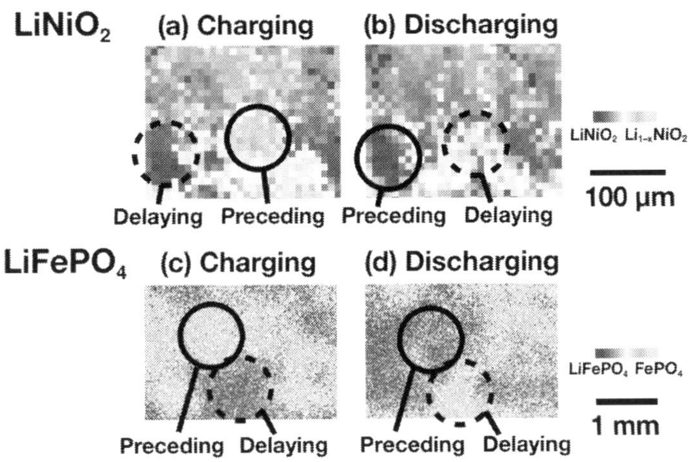


**Figure 2.** The comparison of the spatial distribution of the chemical state of active materials at the 50 % charged and discharged states between the $LiNiO_2$ and $LiFePO_4$ cathodes.

**Figure 3.** The SEM images of the $LiNiO_2$ cathode (a) before and (b) after the fifteen charging/discharging cycles.

the result, the concentration gradient forms from the reaction channel to the covered area in the $LiNiO_2$ cathode. The increase in the $Li^+$ concentration at the reaction channel causes the delay of the charging reaction.

## 4. Conclusions

The spatial distribution of of the oxidation state of the Ni species in the $LiNiO_2$ cathode during the charging and discharging processes was analyzed using the 2-dimensional XAFS imaging technique. The inhomogeneous distribution was observed in the scale of 100 $\mu$m. The disagreement of the preceding area between the charging and discharging processes is explained in considering the $Li^+$ diffusion inside the cathode sheet. It has been clarified that the surface products on the $LiNiO_2$ cathode cause the inhomogeneous reaction by restricting the $Li^+$ diffusion.

## Acknowledgement

The XAFS imaging measurements have been performed under the approval of the Photon Factory Program Advisory Committee (Nos. 2012G020 and 2014G542).

## References

[1] Grunwaldt J D, Hannemann S, Schroer C G, and Baiker A 2006 *J. Phys. Chem.* B **110** 8674
[2] Katayama M, Sumiwaka K, Hayashi K, Ozutsumi K, Ohta T and Inada Y 2012 *J. Synchrotron Rad.* **19** 717
[3] Ohzuku T, Ueda A and Nagayama M 1993 *J. Electrochem. Soc.* **140** 1862
[4] Kalyani P and Kalaiselvi N 2005 *Sci. Technol. Adv. Mater.* **6** 689
[5] Guilmard M, Rougier A, Grüne M, Croguennec L and Delmas C 2003 *J. Power Sources* **115** 305
[6] Mohan P, Kumar K A, Kalaignan G P and Muralidharan V S 2012 *J. Solid State Electrochem.* **16** 3695
[7] Belharouak I, Vissers D and Amine K 2006 *J. Electrochem. Soc.* **153** A2030
[8] Lee Y S, Sun Y K and Nahm K S 1999 *Solid State Ionics* **118** 159
[9] Katayama M, Sumiwaka K, Miyahara R, Yamashige H, Arai H, Uchimoto Y, Ohta T, Inada Y and Ogumi Z 2014 *J. Power Sources* **269** 994
[10] Aurbach D, Gamolsky K, Markovsky B, Salitra G, Gofer Y, Heider U, Oesten R and Schmidt M 2000 *J. Electrochem. Soc* **147** 1322

# Study of the relation between Mg content and dissolution kinetics of natural lime stone using μXRF, μXRD and μXAS

**H S Grunwaldt[1], A Zimina[2,3], J Göttlicher [4], R Steininger[4] and J D Grunwaldt[2,3]**

[1]Faculty of Agriculture, University of Applied Science Kiel, Germany
[2]Institute of Catalysis Research and Technology, Karlsruhe Institute of Technology, Germany
[3]Institute for Chemical Technology and Polymer Chemistry, Karlsruhe Institute of Technology, Germany
[4]ANKA Karlsruhe Institute of Technology, Germany

E-mail: anna.zimina@kit.edu

**Abstract**. The dissolution parameters of calcium carbonates play a key role in agriculture as they regulate the plant nutrients uptake and buffer the pH-value of the soil. The combination of μXRF, μXAS and μXRD mapping in the hard X-ray regime for the examination of both short- and long range order in natural limestones revealed that their dissolution properties depend both on the crystallinity of the mineral phases and their composition. The results particularly show that in many cases the presence of magnesium inhibits dissolution of the phases.

## 1. Introduction

Calcium carbonate as natural limestone plays an important role in agriculture [1,2]. Calcium carbonates regulate the pH-value of the soil, the availability of several plant nutrients like phosphates, and prevent the dissolution of hazardous elements like aluminum and heavy metals. On one hand, calcium deficiency causes damage to plants. Calcium excess in turn causes indirect damage, as it blocks the availability of micronutrients and can furthermore lead to magnesium and potassium deficiency.

Natural limestones consist primarily of calcite and aragonite ($CaCO_3$), but may contain dolomite when the calcium ions are replaced by magnesium ions ($MgCa[CO_3]_2$). It is well known that the crystal structure, crystallinity and Mg content influence the solubility of calcite carbonates [3-5], but there are inconsistencies in liming material evaluation [6,7]. Hence, in this study the particle size, chemical composition and crystallinity of different phases of limestones in their natural form are traced and linked to their effectiveness as fertilizer, judged by acid neutralization capacity measurements [6], a method nowadays applied in Europe (Commission Regulation (EU) 463/2013 : Method 14.2 EN 13971 and Method 14.3 EN 16357). A preliminary study on samples from Northern Germany, USA and the United Kingdom which show different reactivity and crystallinity, already visible by bulk X-ray diffraction and in the optical microscope, hint to the presence of structural selectivity due to size fractionating [3,4,6,8-10].

Detailed information on the distribution of magnesium and calcium in the sample can be obtained by applying X-ray fluorescence (μXRF) mapping on the 50 μm length scale. Combining μXRF with X-ray diffraction (μXRD) and X-ray absorption (μXAS) measurements on the same length scale provides a link between the chemical composition, the microcrystallinity and the calcium coordination of the sample. This helps to clarify the relation between the structure and the dissolution kinetics of natural limestones [11,12]. The crystal size influence on the dissolution behavior of the limestones can be followed by examining different sieve fractions of the same material.

## 2. Experimental and data analysis

The experiment was performed at the SUL-X beamline at ANKA, where it is possible to focus the beam to 50μm at 4.2keV using a cryogenically cooled Si(111) monochromator and Kirkpatrick –Baez mirror setup. μXRD was measured with an area detector, while μXRF and μXAS measurements were performed using a 7-element SiLi detector at the Ca K-edge at the same sample position. The experimental vessel was evacuated to $2x10^{-5}$ mbar to prevent the absorption of low energy photons in air, enabling recording of Mg $K_\alpha$-fluorescence at 1253.6eV. Samples were spread onto Kapton tapes and mapped with respect to Mg and Ca content at 4.2keV. Each point was analyzed by μXRD at 14 keV in transmission mode and μXAS at the Ca K edge at 4038.5eV in fluorescence mode, in order to relate the Mg content to the changes in crystalline structure and in electron density in the vicinity of the Ca atoms.

In this study 6 natural limestones were investigated. Milled calcite, dolomite, and aragonite were used as reference materials. The reactivity measurements were conducted as follows: 200 mg $CaCO_3$-equivalent treated with 10 ml citrate buffer at pH 4.1 (2 N citric acid, 2.5 sodium citrate) 30 min in Scheibler-Apparatus, released $CO_2$ determined volumetrically and converted to % dissolved carbonate. For each sample μXRF mapping was performed inside 1x1 $mm^2$ optically preselected regions of interest with a 50x50$\mu m^2$ beam. μXRD mapping was performed on the same 1x1$mm^2$ area in transmission. A smaller area of 0.24x0.24$mm^2$ was chosen for the μXAS mapping based on μXRF and μXRD, the XANES spectrum up to $k = 8Å^{-1}$ was recorded at each point of the mesh in 40μm steps.

The elemental distributions of Mg and Ca were calculated using the integrated intensity of the relevant $K_\alpha$ fluorescence lines. The XRD pattern was calibrated with the help of $LaB_6$ milled powder and radial and azimuthal integration performed using the IgorPro Nika software package [13].

The natural limestones can strongly deviate from perfect randomly oriented powders, and this induces changes in the relative intensity of the diffraction peaks due to the presence of large crystallites. The presence of specific mineral phases was confirmed if the significant intensity of (104), (2-13) and (202) for calcite/dolomite and (111) and (102) for aragonite diffraction peaks were observed during measurement. The XRD maps were calculated for different components using the intensities of the (104) diffraction lines of calcite and dolomite and (111) of aragonite. Additionally, the crystallinity of each sample was deduced by analyzing the radial intensity distribution of the (104)/(102) diffraction line.

The Ca K-edge XAS data were preliminarily processed using the Athena software [14] and calibrated to calcite and dolomite standards. The XAS spectra of different areas were preliminarily examined using principal component analysis (PCA built-in function in Athena) and observed components were compared to the reference materials. The XAS chemical component maps were calculated based on linear combination analysis (LCA) in the energy range from -20 eV to 100 eV relative to the absorption edge, using the spectra of calcite, dolomite and components from the previous PCA procedure. The unavoidable effect of self-absorption at these photon energies has to be considered. The assignment of the chemical state of the Ca ion is based on the energy position of the characteristic spectral features and not on their relative intensities. In such a way the difference in damping of the post-edge intensity at every single mapping point might be compensated by selection of sample areas with a similar thickness and by identical preparation of the reference materials.

## 3. Results and discussion

In this study 6 natural limestones of three different sieve fractions (350-100μm, 100-50μm and below 50μm) were investigated: Oker, Tilcon, Salzhemmendorf, Columbia River Carbonate Calder (CRC Calder), Columbia River Carbonate Wauconda (CRC Wauconda, 1mm and 100-50μm), Dolomite

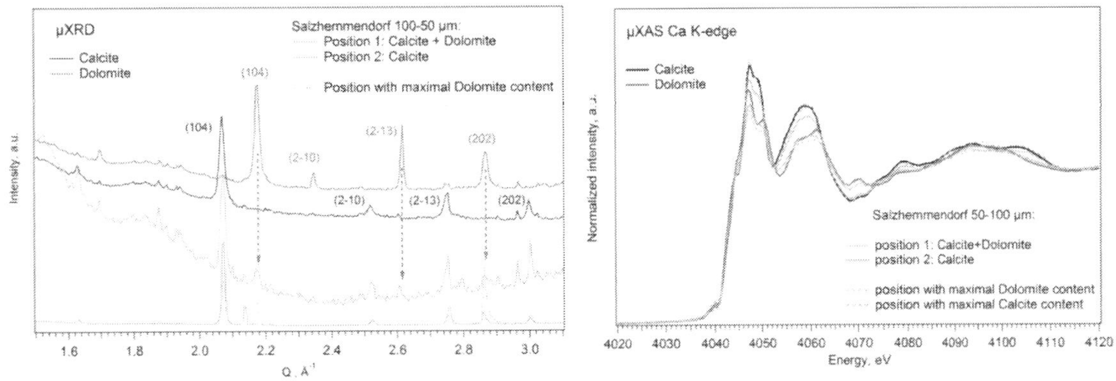


**Figure 1.** XRD pattern (left) and XAS Ca K spectra (right) of representative point of calcitic Salzhemmendorf, 100-50 µm sieve fraction.

**Table 1.** Characteristics of the studied materials deduced from µXRD and µXAS and their reactivity as a fertilizer, the relative phase content is estimated based on statistics of the LCA of µXAS data, with a detection limit of ±5% due to experimental uncertainty.

| *Limestones* | *Main phase* | *Additional phase* | *Reactivity* |
|---|---|---|---|
| Calcite Oker | Calcite, polycrystalline | 10% Dolomite, 40% Mg containing Calcite, 50-100 µm | high |
| Dolomite Tilcon | Dolomite with Mg deficit, monocrystalline, all sizes | >5% Calcite, 350-100 µm | low |
| Calcite Salzhemmendorf | Calcite, polycrystalline, all sizes | 20% Dolomite, 350-100 µm | high |
| CCR Wauconda, | Calcite, monocrystalline, 350-100µm | 30% Mg containing Calcite, polycrystalline | low, increasing with decrease of sieve fraction |
| CCR Calder | Mg containing Calcite, all sizes | 35% Calcite, polycrystalline, all sizes | low, increasing with decrease of sieve fraction |
| Dolomite Jettenberg, | Dolomite, polycrystalline | 10% Calcite, polycrystalline | low |

Jettenberg (100-50µm). As an example the results on the Salzhemmendorf 100-50 µm sieve fraction are shown here in detail.

Figure 1 shows the XRF Mg map normalized to the total Ca and Mg intensity, in order to trace the location of the dolomite phase in the Salzhemmendorf sample (100-50 µm). The XRD component maps of the same area are shown, the smaller area where the µXAS measurements were conducted is marked with white boxes. At the bottom of figure 1 the XRD patterns and XAS spectra of 2 selected points on the maps where the most pronounced difference in XRD signal was measured are shown. The XAS spectra at these 2 points reveal the presence of dolomite and calcite and they are linked to the Mg content plotted as an XRF map. Note furthermore the difference in the XAS component map compared to the XRD phase map. This might be due to the different penetration length of these two methods and insensitivity of the XRD measurements to thin amorphous surface layers. Hence, a combination of XRF, XRD and XAS seems to be beneficial.

All the samples were processed in the same way and detailed thorough analysis of the data was performed to obtain a representative picture of the investigated material and a detailed view of the structure at various sieve fractions. Based on this detailed analysis we concluded that high Mg concentration causes the formation of the dolomite related structure which is supported by XRD and XAS methods. The deficit of Mg (concentration below certain dissolution limit) leads to the formation of calcite-like crystalline structures ("modified calcite", $Mg_xCa_{2-x}(CO_3)_2$ $0<x<1$) with modified electronic structure/bond length as detected by μXAS. At very low Mg concentration bulk calcite was found. Analysis of the data obtained by this comprehensive study provided information on the crystallographic phases, the chemical composition and the crystallinity of the limestone, as summarized in table 1.

From this study we can conclude that the dissolution rate is defined by the concentration of both the fine calcite grains and Mg containing calcite ("modified calcite"). The presence of large crystallites or dolomite phases significantly reduced the reactivity. Therefore for practical applications, the dissolution behavior of material consisting of a pure calcite and Mg containing calcite could be improved by milling. In addition, sieving to the 100-50 μm fraction could help to remove the large dolomite phases for the materials with significant content of fine calcite phases, thereby enhancing their reactivity.

## 4. Conclusion

The combined study demonstrated that high solubility is provided by small calcite crystals and is furthermore hindered due to the presence of large dolomite crystals. The increase in solubility by removing the dolomite phase via sieving can be therefore successful for some natural limestone materials. These studies provided insight into the microscopic mechanism to tune the reactivity, and in this way might help to optimize this process as well as influence the proper choice of limestone deposits in future.

## References

[1] Finck A 2007 *Pflanzenernährung und Düngung* (Baumträger-Verlag, Berlin-Stuttgart)
[2] Eimer C 2011 *Bewertung von kohlensauren Kalkdüngern für die Anwendung in Landwirtschaft, Forst und Gartenbau*Master-Thesis, FH Kiel, FB Agrarwirtschaft
[3] Grunwaldt H-S „Granulierter Düngekalk", 1980 *Bauernblatt Schleswig-Holstein* 787 - 793.
[4] Grunwaldt H-S „Neubewertung der Kohlensauren Düngekalke" 1986, *Bauernblatt Schleswig-Holstein* 3499 - 3500
[5] Morse J W and Mackenzie F T 1990 *Geochemistry of Sedimentary Carbonates* (Amsterdam: Elsevier)
[6] Grunwaldt H-S "Assessment of the Effectiveness of Carbonate Liming Materials, USDA Strategy Meeting, 12.11.2013 Ames/USA, http://reactivelime.org./4.html (12.8.2015).
[7] Hoiberg A "Background and Inconsistencies on Liming Material Evaluation", USDA Strategy Meeting, 12.11.2013 Ames/USA, http://reactivelime.org./4.html (12.8.2015).
[8] Hugenroth P, Meyer B and Steffens G 1979 *Landw. Forsch.* **32** 157-164
[9] Hugenroth P and Meyer B 1974 *Landw. Forsch.* **30/I** 160
[10] Sauerbeck D and Rietz E, 1985 *VDLUFA-Schriften* **16** 431 -438
[11] Chalmin E, Sansot E, Orial G, Bousta F and Reiche I 2008 *X-ray Spectrom.* **37** 424
[12] Brinza L, Schofield P F, Hodson M E, Weller S, Ignatiev K, Geraki K, Quinn P D, Mosselmann J F W 2014 *J. Synchrotron Rad.* **21** 235
[13] Ilavsky J 2012 *J. Appl. Cryst.* **45** 324.
[14] Ravel B and Newville M 2005 *J. Synchrotron Rad.* **12** 537–541

Journal of Physics: Conference Series **712** (2016) 012145 doi:10.1088/1742-6596/712/1/012145

# XAFS characterization of industrial catalysts: *in situ* study of phase transformation of nickel sulfide

**J Wang[1,*], Z Jia[1], Q Wang[2], S Zhao[1], Z Xu[1], W Yang[1] and A I Frenkel[1]**
[1]Sinopec Shanghai Research Institute of Petrochemical Technology, Shanghai 201208, China
[2]Physics Department, Yeshiva University, New York, NY 10016, United States

E-mail: wangjq.sshy@sinopec.com

**Abstract.** The online sulfiding process for nickel-contained catalyst often ends up with a nickel sulfide mixture in refinery plant. To elucidate the local environment of nickel and its corresponding sulfur species, a model catalyst (nickel sulfide) and model thermal process were employed to explore the possibilities for characterization of real catalysts in industrial conditions. The present investigation shows effectiveness of *in situ* XANES and EXAFS measurements for studying the phase stability and phase composition in these systems, which could be used to simulate real sulfiding process in industrial reactions, such as hydrodesulfurizations of oil.

## 1. Introduction

Nickel sulfide materials have a wide industrial application in petrochemical and refinery processes for catalyst [1-5]. Generally, nickel sulfide catalyst is obtained by an *in situ* sulfiding procedure [6] using nickel oxide precursor with mixed gas stream of $H_2$ and $H_2S$ at certain temperature and pressure. The catalyst activity and selectivity for these processes are a function of sulfiding procedure because nickel-contained catalyst can form a variety of nickel sulfides with various phase compositions ($Ni_3S_2$, $Ni_6S_5$, $Ni_7S_6$, $NiS$, $NiS_2$) [7-13]. However, both the structure and chemical composition of the final nickel and sulfur species are generally not known due to the complexity of nickel-sulfur compounds existing in the real catalyst system during and after sulfurization process. Another open question is the extent of stability of nickel sulfur species subjected to hydrogen atmosphere at reaction temperatures due to the occurrence of phase transformation [14] and the sequential loss of sulfur [15].

It is difficult to elucidate the structure of nickel sulfides during the sulfiding process by means of the conventional in-house characterization. For example, X-ray diffraction is a conventional method to study phase transformations but it is limited in this case due to the unavoidable contamination of the expensive *in situ* cell of the diffractometer, and the difficulty to identify, quantify or even isolate the diffraction pattern of nickel sulfide catalyst from that of the support materials. X-ray absorption near-edge structure (XANES) and extended X-ray absorption fine structure (EXAFS) spectroscopies are quite useful for studies of local atomic environments [16]. The sensitivity of XANES spectra to the electronic structure, oxidation state, and geometry of the absorbing site is well known. A combination of the XANES and EXAFS analyses of nickel K-edge is expected to enable determination of the

[1] To whom any correspondence should be addressed.

chemical state and detailed local environment of nickel but their use for industrial catalysts is scarce and not well documented, to the best of our knowledge.

In this investigation, a well characterized nickel sulfide mixture (a reference compound, from a commercial supplier, consisting of 70 wt.% $\alpha$-$Ni_3S_2$ and 30 wt.% $\alpha$-$Ni_7S_6$) was used as a starting material for exploring the local structure changes and phase composition of nickel sulfide by XANES and EXAFS data analysis during the phase transformation of $\alpha$-$Ni_7S_6$ to $\alpha$-$Ni_3S_2$ under hydrogen atmosphere. These results could facilitate characterizations in the sulfiding process of nickel catalyst or nickel sulfide catalysts itself in hydrodesulfurization reaction.

## 2. Experimental

### *2.1. Sample preparation*

A commercial nickel (II) sulfide (Alfa Aesar) was used as a starting material (denoted as $NiS_x$). It was analyzed to be made up of 70 wt.% of $\alpha$-$Ni_3S_2$ and 30 wt.% of $\alpha$-$Ni_7S_6$ by chemical analysis. Sample $NiS_x$ was reduced at different temperatures ($T$) with the pure hydrogen at a flowrate of 50 ml/min for 240 min at most. Finally, the reduced sample was characterized to be pure $\alpha$-$Ni_3S_2$ by XRD (Bruker AXS D8 Advance), chemical analysis (ICP-AES, Varian 725 ES) and X-ray Absorption Spectroscopy.

### *2.2. X-ray absorption measurements*

Sample $NiS_x$ for the nickel K-edge measurements were conducted on the beamline of NSLS X19A at National Synchrotron Light Source (NSLS), Brookhaven National Laboratory, Upton, New York, USA. A pellet made by mixing ground $NiS_x$ and alumina in the mass ratio of 1:3 was prepared for *in situ* nickel K-edge measurements. The sample was mounted in an *in situ* XAS cell (the Nashner-Adler cell) [17] and reduced in 5 vol.% $H_2$ in He at 450°C for 4 h. EXAFS scans before and after reduction were recorded at room temperature, denoted as $NiS_x$ and $NiS_x$-$R$. The pure $\alpha$-$Ni_3S_2$ for nickel K-edge measured at room temperature was also measured. Composition of sample $NiS_x$-$R$ was analyzed by chemical analysis, which consisted of 80 wt.% of $\alpha$-$Ni_3S_2$ and 20 wt.% of $\alpha$-$Ni_7S_6$.

### *2.3. Data Analysis*

Data processing and analysis were performed using the IFEFFIT package [18, 19]. The relative amount of $\alpha$-$Ni_7S_6$ and $\alpha$-$Ni_3S_2$ in the sample $NiS_x$-$R$ was analyzed by linear combination analysis (LCA) on $k^2$-weighted EXAFS spectra using Athena software.

## 3. Results

### *3.1. Phase transformation of $\alpha$-$Ni_7S_6$ to $\alpha$-$Ni_3S_2$ evidenced by XRD*

Figure 1 presents the XRD patterns of $NiS_x$ to a pure $\alpha$-$Ni_3S_2$ measured at room temperature after reduction at different temperature and time. By identification of the structure in the PDF database, the crystal phase for $NiS_x$ shows a mixed phase of $\alpha$-$Ni_7S_6$ (JCPDF#24-1021) and $\alpha$-$Ni_3S_2$ (JCPDF#44-1418). According to the chemical analysis by ICP-AES, the major phase in the raw nickel sulfide material is $\alpha$-$Ni_3S_2$ with a content of 70.0 wt.%, the minor phase for $\alpha$-$Ni_7S_6$ is 30.0 wt.%. XRD analysis demonstrated that $\alpha$-$Ni_7S_6$ could be transformed to $\alpha$-$Ni_3S_2$ when heating sample $NiS_x$ in hydrogen.

### *3.2. Electronic structure changes and phase composition of $NiS_x$ during in situ XAS*

Figure 2 compares the experimental Ni K-edge spectra of $NiS_x$, $NiS_x$-$R$ and $\alpha$-$Ni_3S_2$ in E-space, K-space and R-space measured at 25°C. It can be seen that a slightly shift to higher energy occurred in Figure 2a with the loss of sulfur from $NiS_x$ to $\alpha$-$Ni_3S_2$ via $NiS_x$-$R$, which means the loss of sulfur could be perceptive from XANES shift. However, it is impossible to find the phase composition of $NiS_x$-$R$ from XANES analysis by linear combination method because the features do not change too significantly, and the contrast is weak. EXAFS fitting is also impossible to use for this purpose because of the very disordered structure of $\alpha$-$Ni_7S_6$.

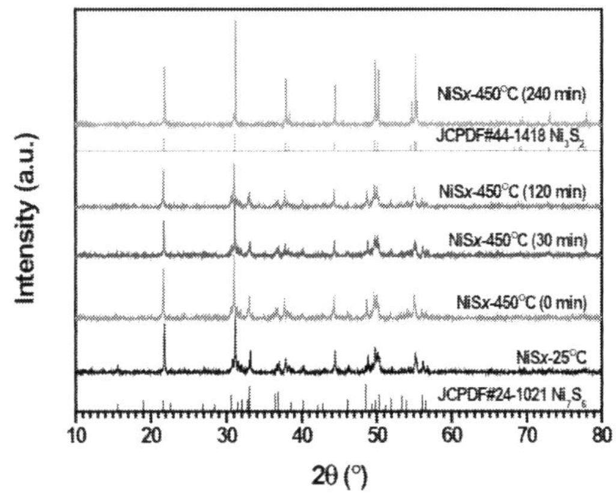


**Figure 1.** XRD patterns of $NiS_x$ measured at room temperature after reduction at different temperatures and durations.

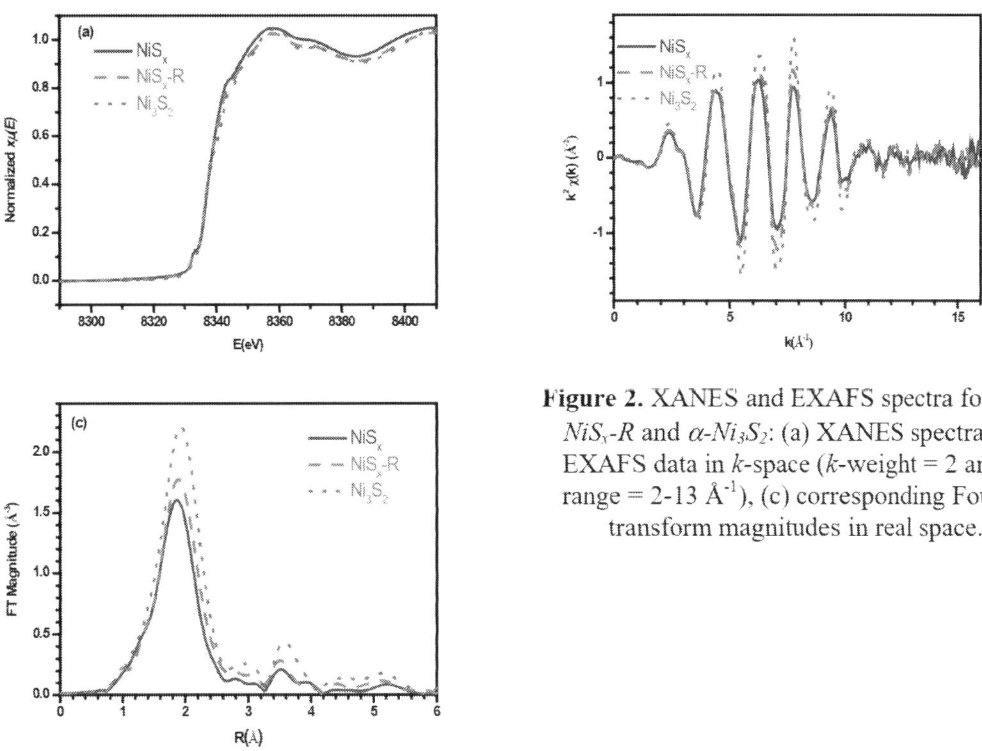


**Figure 2.** XANES and EXAFS spectra for $NiS_x$, $NiS_x$-$R$ and $\alpha$-$Ni_3S_2$: (a) XANES spectra, (b) EXAFS data in $k$-space ($k$-weight = 2 and $k$-range = 2-13 Å$^{-1}$), (c) corresponding Fourier transform magnitudes in real space.

However, EXAFS data do change significantly between the starting sample and the final sample (Figures 2b,c). The changes show presence of isosbestic points in k-space, hence, the unknown composition is a two-phase mixture that can be analyzed by linear combination method. Therefore, we used linear combination analysis for EXAFS, not XANES spectra. The $k$-range for linear combination is 2-13 Å$^{-1}$. Figure 3 shows the results of linear combination fitting of $NiS_x$-$R$ spectrum to the spectra of $NiS_x$ and $\alpha$-$Ni_3S_2$. It was found sample $NiS_x$-$R$ was composed of 19.0% $\alpha$-$Ni_7S_6$ and 81.0% $\alpha$-$Ni_3S_2$, which is in agreement with the phase calculation by nickel content measured from chemical analysis.

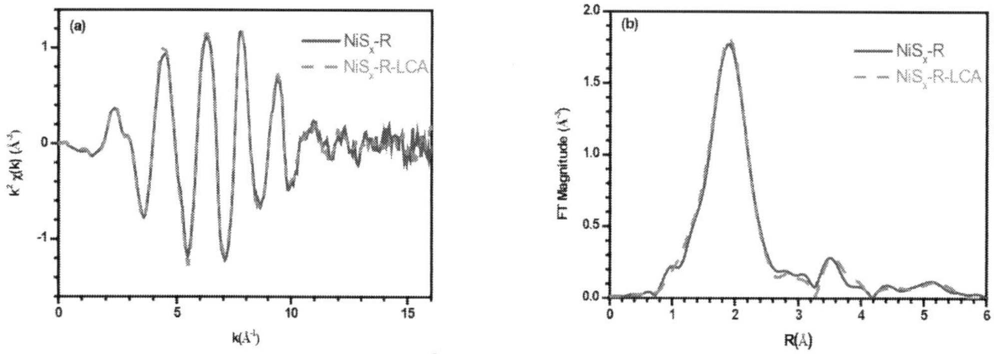


**Figure 3.** Linear combination analysis in k$^2$chi data for $NiS_x$-$R$ spectrum (k-weight = 2 and k-range = 2-13 Å$^{-1}$).

*3.3. Considerations of the in situ methods for investigating the mechanism of sulfiding process for industrial nickel catalysts*

The purpose of this work was to investigate the sensitivity of XANES and EXAFS to structure changes of nickel sulfide. As emphasized in the introduction, this knowledge is important for guiding *in situ* experimental design for investigating the mechanism of sulfiding process for real nickel catalysts.

One analysis demonstrated important complementarity of synchrotron-based characterization by XAS and the lab-based XRD studies. Specifically, laboratory X-ray diffraction turned out to be adequate for detecting the phase change of nickel sulfide but it failed to analyze the phase composition in the real catalyst system due to the diffraction peaks from the catalyst support. XANES analysis showed no sensitivity to the local structure and phase composition changes of nickel sulfide. EXAFS analysis by fitting methods is not suitable for obtaining the coordination number changes during the phase transformation of $\alpha$-$Ni_7S_6$ to $\alpha$-$Ni_3S_2$ because of the large structural disorder and because EXAFS fitting method is not generally applicable for phase speciation in multi-component mixtures [20]. Linear combination analysis of EXAFS data in k-space was a good alternative to the above methods and was used in this work to calculate the mixing fractions of different phases in this model catalyst, hence, show a strong promise as a technique of choice in real catalytic investigations.

## 4. Summary and conclusions

We presented a method, based on in situ XAS studies and ex-situ characterization by XRD and chemical analysis techniques, to measure the phase composition and phase stability of nickel sulfide-based catalysts during an online sulfiding process and evaluate possible sulfur loss or addition. Our results are potentially useful for a large class of nickel sulfide catalysts used in hydrodesulfurization reactions.

## Acknowledgements

The authors acknowledge support from China Petrochemical Technology Company, Ltd and the facilities support provided at the National Synchrotron Light Source at the Brookhaven National Laboratory (US DOE, Contract No. DE-SC0012704) and the Synchrotron Catalysis Consortium (US DOE, Grant No. DE-SC0012335). The authors acknowledge Drs. Nebojsa Marinkovic and Syed Khalid for assistance with synchrotron measurements. The authors are grateful to Dr. Rongwei Wang for useful discussions.

**References**

[1] Licea Y E, Amaya S L, Echavarria A, Bettini J, Eon J G, Palacio L A and Faro A C 2014 *Catal. Sci. Technol.* **4** 1227
[2] Hein J, Hrabar A, Jentys A, Gutiérrez O Y and Lercher J A 2014 *ChemCatChem* **6** 485
[3] Bahzad D, Al-Fadhli J, Al-Dhafeeri A and Abdal A 2010 *Energ. Fuel* **24** 1495
[4] Callejas M A and Martínez M T 1999 *Ind. Eng. Chem. Res.* **38** 3285
[5] Poels E, Van Beek W, Den Hoed W and Visser C 1995 *Fuel* **74** 1800
[6] Scheffer B, Mangnus P J and Moulijn J A 1990 *J. Catal.* **121** 18
[7] Bezverkhyy I, Danot M and Afanasiev P 2003 *Inorg. Chem.* **42** 1764
[8] Tilley R D and Jefferson D A 2002 *J. Phys. Chem. B* **106** 10895
[9] Olivas A, Cruz-Reyes J, Petranovskii V, Avalos M and Fuentes S 1998 *J. Vac. Sci. Technol., A* **16** 3515
[10] Faber M S, Lukowski M A, Ding Q, Kaiser N S and Jin S 2014 *J. Phys. Chem. C* **118** 21347
[11] Maxted E B 1951 *Advances in Catalysis* vol 3, ed W G Frankenburg, V I Komarewsky, E K Rideal, P H Emmett and H S Taylor (New York: Academic Press), p 129
[12] Takeuchi A, Tanaka K-I and Miyahara K 1974 *Chem. Lett.* **3** 171
[13] Dunn J and Kelly C 1977 *J. Therm. Anal. Calorim.* **12** 43
[14] Rosenqvist T 1954 *J. Iron Steel Inst.* **176** 37
[15] Kloprogge J T, Welters W J J, Booy E, de Beer V H J, van Santen R A, Geus J W and Jansen J B H 1993 *Appl. Catal., A* **97** 77
[16] Kim D H, Szanyi J, Kwak J H, Wang X, Hanson J C, Engelhard M and Peden C H F 1993 *J. Phys. Chem. C* **113** 7336
[17] Frenkel A I, Hills C W and Nuzzo R G 2001 *J. Phys. Chem. B* **105** 12689
[18] Newville M 2001 *J. Synchrotron Radiat.* **8** 322
[19] Ravel B and Newville M 2005 *J. Synchrotron Radiat.* **12** 537
[20] Wang Q, Hanson J C and Frenkel A I 2008 *J. Chem. Phys.* **129** 234502

Journal of Physics: Conference Series **712** (2016) 012146 doi:10.1088/1742-6596/712/1/012146

# XAFS beam lines at Aichi Synchrotron Radiation Center dedicated to industrial use

**Yoshikazu Takeda**

Aichi Synchrotron Radiation Center, Aichi Science & Technology Foundation, Seto, Aichi 489-0985, Japan

E-mail: takeda@astf.or.jp

**Abstract.** Aichi Synchrotron Radiation Center was designed for industrial use following five years of discussion among academia, industry and local government in the Aichi area. Among the six beam lines constructed, those that facilitated X-ray absorption fine structure (XAFS) analysis were given first priority. In addition to the hardware, attention was given to the development of operating procedures that were quick and user-friendly. The facility entered public service in March 2013. In the year 2013, 55% of the experiments involved XAFS analysis (hard X-ray, soft X-ray and vacuum ultraviolet regions) and in 2014 it was 57%. The range of research fields is very broad, emphasizing the importance of the XAFS beam lines.

## 1. Introduction

The Aichi Synchrotron Radiation Center (AichiSR) was designed mainly for industrial use and entered public service on March 26, 2013. The facility is owned by Aichi Science & Technology Foundation, which cooperates with industry, academia and the Aichi Prefectural Government. The AichiSR is responsible for maintenance, operation of the facility and user services.

The AichiSR is the 8th SR facility constructed for public use in Japan. Since three general-purpose facilities already exist (Photon Factory, UVSOR and SPring-8), other facilities can be designed for specific purposes. When we first started discussing an SR facility in the Aichi area in 1991, the Photon Factory and UVSOR were already operating, and construction of the SPring-8 had begun. Therefore, our aim was to construct a compact and easily accessible facility. Based on investigations carried out in 2000 and 2001, we learned that there were a large number of active SR users in industry and universities in the Aichi area, and some of them expressed a desire to have an SR facility in this area. Considering the needs of industrial and academic researchers, working mainly in the fields of applied physics and chemistry, in 2003 we proposed a plan called "Photo-Science Nanofactory". In this proposal, a center would be constructed that focused on SR, but also provided facilities for other high-level measurement and analysis methods such as transmission electron microscopy, scanning electron microscopy, secondary ion mass spectroscopy and nuclear magnetic resonance [1]. The technicians, scientists, and researchers working at the facility would fully support the needs of users. Research laboratories, incubation laboratories, and conference and meeting rooms would also be constructed. This plan attracted the attention of both industry and the Aichi Prefectural Government.

Based on opinions expressed by both industry and academia, the design was based on a 1.2-GeV storage ring with a booster ring for top-up operation, and four superconducting bending magnets (superbends). The working group was made up of members of Nagoya University, members of the Aichi Prefectural Government, and power users from industry and brushed up the design. A plan for implementation of the Nanofactory was completed, and the final report of the "Knowledge Hub" project was submitted to Aichi Prefecture. In 2009, Aichi Prefecture approved construction, which then started in 2010. In the summer of 2011 the accelerators were installed, and commissioning began in 2012. The first light was observed in the summer of 2012, and the opening ceremony took place on March 22, 2013. The construction costs for the facility were covered by Aichi Prefecture, donations from industry and the Japanese National Government. Therefore, from the beginning, the AichiSR facility was a common local and national asset, and is basically open to any researcher without restrictions based on region or country.

## 2. Facility overview

*2-1. Accelerators*

The accelerators consist of a 50-MeV LINAC, a 1.2-GeV booster ring and a 1.2-GeV storage ring [2]. The storage energy of 1.2 GeV was determined based on the desire to construct a compact ring (72-m circumference) that could supply hard X-rays (up to about 20 keV) into more than eight beam lines from four superbends with a magnetic field of 5 T [2,3]. The combination of compactness and the ability to produce both hard and soft X-rays was a basic requirement during the design phase. The booster ring has two purposes. First, it removes the need to ramp up the magnetic field in the bending magnets of the storage ring, thus avoiding any possible nonlinear dependence of the magnetic field in the superbends on the driving current. Second, it allows the storage ring to be run in top-up mode. Light with the same quality and brilliance is required for experimental repeatability and for checking the reproducibility of products.

*2-2. Beam lines*

The six beam lines currently available for use are: 1) hard X-ray absorption fine structure (XAFS) and fluorescence X-ray analysis (BL5S1: 5-20 keV), 2) soft X-ray XAFS and photoelectron spectroscopy (BL6N1: 1.75-6 keV), 3) ultra-soft X-ray XAFS, vacuum ultraviolet (VUV) and photoelectron spectroscopy (BL7U: 0.03-0.85 keV), 4) powder X-ray diffraction (BL5S2: 5-23 keV), 5) X-ray reflection, thin film and surface diffraction (BL8S1: 9.3-14.6 keV) and 6) small-angle X-ray scattering (BL8S3: 8.2 and 13.5 keV) [4]. This beam line configuration is based on user demand.

*2-3. Management*

The key factors that were considered with regard to industrial use were: 1) ease of use (low threshold for entry), 2) measurements under normal atmosphere, 3) repeat experiments on same topic, 4) short time from application to execution and 5) nondisclosure of results. To realize 4), applications are accepted every two months. All users pay for beam time, but the charge is much lower for academic use if results are disclosed.

## 3. Industrial use

In 2014, 66% of our users were from industry (including industry-university collaborations), 25% were purely academic and the remainder were from public research laboratories, as shown in Fig. 1. The research fields of these users were very diverse, as shown in Fig. 2. In 2013, the percentage of industrial users was 70%, but the beam time allocated for industrial use increased from 2976 hours in 2013 to 3732 hours in 2014.

The goal of industrial research is not to publish papers, but to produce profitable products. Therefore, evaluation of requests for beam time should not be performed based on scientific importance or impact on novel technology. Repeat experiments are also important in industry. Beam-line and beam-time assignment is carried out through negotiation between

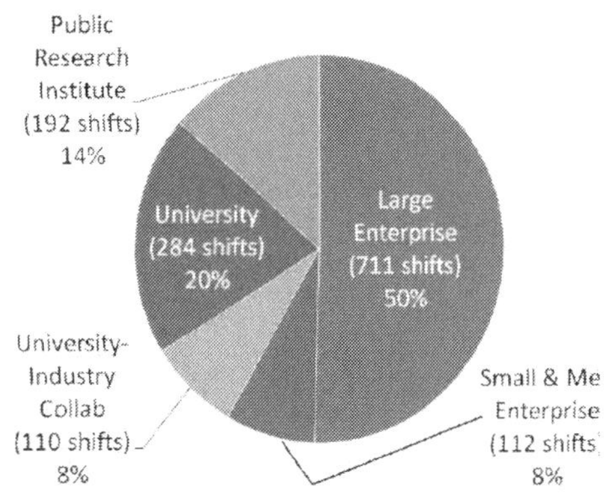


Fig. 1 AichiSR users from industry, universities and public research laboratories in 2014.

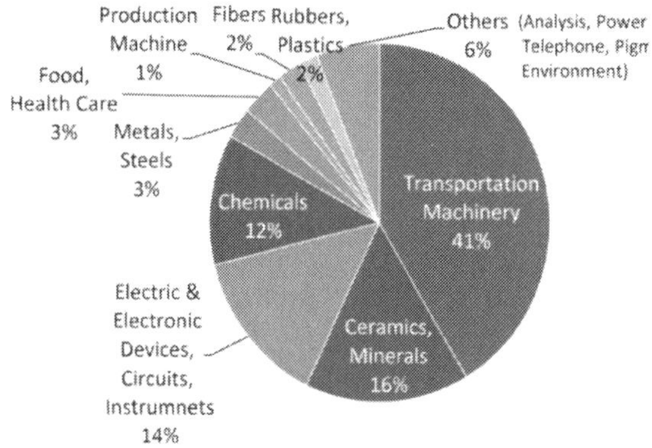


Fig. 2 Diverse research fields of AichiSR users.

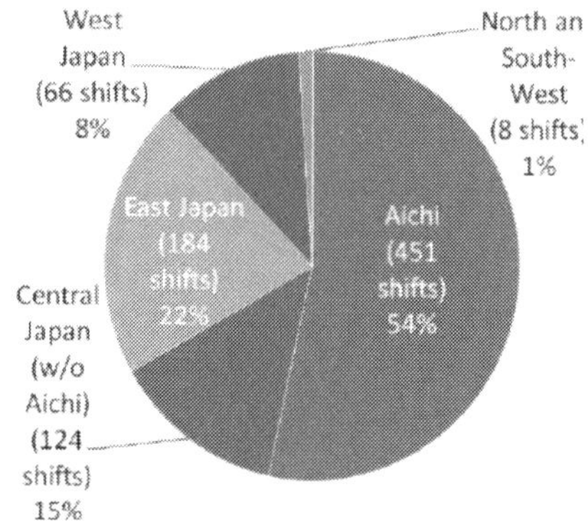


Fig. 3 Location of companies of AichiSR users.

coordinators and users. Use of multiple beam lines for one application is possible, and sometimes we actually suggest this to users.

Fig. 3 shows the locations of AichiSR user companies. As mentioned in the Introduction, many SR users existed even in 2000 and 2001 in the Aichi area, and AichiSR was constructed to meet their demands. The results shown in Fig. 3 demonstrate that a real demand exists.

## 4. Beam lines

*4-1. XAFS beam lines*

Three of the six beam lines are for XAFS measurements in the hard X-ray, soft X-ray and VUV regions. Another XAFS beam line covering the soft X-ray to VUV region is under construction. The most popular beam line is the hard X-ray XAFS beam line, which is often overbooked. For this reason, it has been decided to construct another new beam line for hard X-ray XAFS measurements. In 2013, 55% of the experiments involved XAFS (hard X-ray, soft X-ray and VUV) and this increased to 57% in 2014, as shown in Fig. 4.

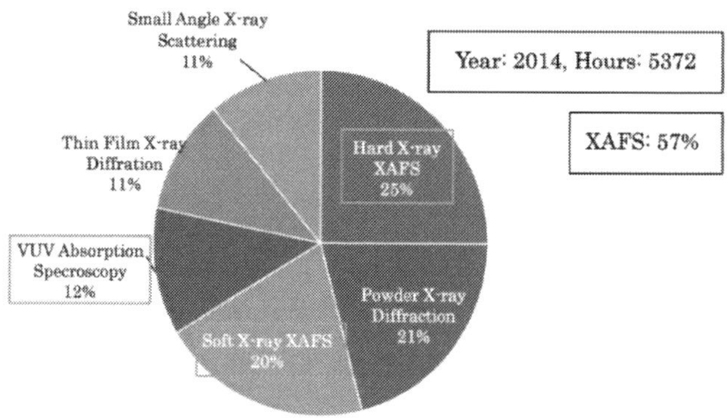


Fig. 4 The three XAFS beam lines use 57% of the beam time at AichiSR.

The most popular beam line BL5S1 is capable of operating in transmission mode, fluorescence yield mode, conversion electron yield mode, and can be used to perform quick XAFS experiments. The most common elements analyzed are Ni, Fe, Mn, Co, Ti, Zn and Cu. For example, Ni, Fe and Mn are typically found in battery electrodes. Analysis of these elements involves K-edge measurements. BL6N1 allows electron and fluorescence yield modes, either under vacuum or in a He atmosphere. The most common elements analyzed are S, Si and Ca in solids and liquids, and in particular S in catalysts and rubber. $L_{III}$-edge measurements of Rh, Pd, and Ag are also commonly performed. At BL7U, the electron yield mode is currently used, typically for elements such as N, C and O. Photoelectron spectroscopy can be performed using the soft X-ray and VUV beam lines.

The percentage of industrial use for BL5S1, BL6N1 and BL7U is 70%, 71% and 47%, respectively. Although this often involves analysis of heavier elements in the hard X-ray region, analysis of lighter elements is becoming more popular due to the desire for lightweight products. Even in the area of structural materials, carbon fiber reinforced polymer is increasing in importance, so that the very light element C has become a major target for measurements. Development of functional organic materials has also increased the need for analysis of C and O in the VUV region. Li in batteries can also be investigated using BL7U. Although there is a strong need for technical and scientific support for work in the VUV region, we currently have a shortage of scientists with the relevant industrial experience.

*4-2. Examples of XAFS measurement*

*4-2-1. XAFS analysis of liquid solution*
Liquid solutions can be analyzed in a He chamber at atmospheric pressure using X-rays with energies above 1 keV. At BL6N1, measurements of S K-edge and Ag $L_{III}$-edge spectra of such solutions have been extensively performed.

In Fig. 5 an example of XAFS signal of Ag nanoparticle dispersed in water is compared with those of Ag metal and $AgNO_3$ powder [5].

*4-2-2. Transfer vessel*
A transfer vessel has been developed to keep samples under a high vacuum while transferring them between high-vacuum chambers at AichiSR and other member facilities of the project "Photon-Beam Platform" for different experiments at different beam lines [6].

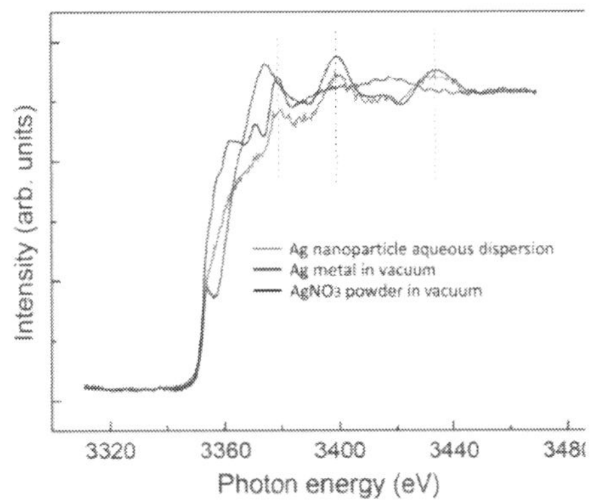


Fig. 5 XAFS signal of Ag nanoparticle in water is compared with those of metal Ag and $AgNO_3$ powder.

## 5. Summary

After five years of discussion and planning by universities, industry, and local government in the Aichi area, the Aichi Synchrotron Radiation Center was designed for industrial use. The choice of accelerators, beam lines, and service management was determined based on discussion and feedback involving industry and academia. Of the six beam lines available, the highest priority was given to XAFS beam lines that allow analysis of elements from Li to U. In addition to the hardware, attention was given to the development of operating procedures that were quick and user-friendly. The facility became available for public use in March 2013. In 2013, 55% of the experiments involved XAFS analysis (hard X-ray, soft X-ray and VUV regions). The beam time allocated to industrial use increased from 2976 hours in 2013 to 3732 hours in 2014. The range of research fields for users of the facility was found to be very broad. Although at present there is an energy gap for the XAFS beam lines, this will be filled by the new BL1N2 beam line, which will allow analysis of Na, Mg and Al.

## References

[1] Takashima Y, Yamane T, Takeda Y, Soda K, Yagi S, Takeuchi T, Akimoto K, Sakata M, Suzuki A, Tanaka K, Nakamura A, Hori M, Morita S, Seki K, Mizutani U, Kobayakawa H, Yamashita K, Katoh M 2007 *AIP Conference Proceedings(American Institute of Physics)* vol 879 pp 75-78
[2] Yamamoto N, Takashima Y, Hosaka M, Morimoto H, Takami K, Hori Y, Sasaki S, Koda S and Katoh M 2010 *Proceedings of IPAC2010 (Kyoto Japan)* pp 2567-2569
[3] Yamamoto N, TakashimaY, Hosaka M, Takami K, Mano A, Morimoto H, Hori Y, Sasaki S and Koda S, Katoh M 2011 *Proceedings of IPAC2011 (San Sebastian, Spain)* pp 2987-2989
[4] Yamamoto N, Takashima Y, Katoh M, Hosaka M, Takami K, Morimoto H, Hori Y, Sasaki S, Koda S, Ito T, Sakurai I, Hara H, Okamoto W, Watanabe N and Takeda Y 2010 *The 10th International Conference on Synchrotron Radiation Instrumentation (Melbourne, Australia, American Institute of Physics)* pp 591-594
[5] http://www.astf-kha.jp/synchrotron/en/about/greeting.html
[6] Nakanishi Y, Yagi S and Ohta T 2010 *IEEJ Electronics, Information and Systems* vol 130 pp 1762-1767 (in Japanese)

# XAFS data acquisition with 2D-detectors: Transmission mode XAFS and grazing incidence EXAFS spectroscopy

**D Lützenkirchen-Hecht*, J-C Gasse, R Bögel, R Wagner, R Frahm**
Bergische Universität Wuppertal, Gaußstr. 20, 42097 Wuppertal, Germany

*E-mail: dirklh@uni-wuppertal.de

**Abstract.** XAFS-experiments in transmission and reflection modes have been performed using a Pilatus 100K pixel detector. Transmission mode XAFS spectra from a Co metal foil and $Co_3O_4$ were recorded to evaluate the data quality offered by this 2D-detector. Furthermore, the pixel detector was also used to measure reflection mode grazing incidence EXAFS data. Using different regions of interest in the collected scattering patterns, we will show that the diffuse scattering can be separated for the different contributing surfaces and interfaces, allowing simultaneous investigations of surfaces and buried interfaces within multi-layered samples.

## 1. Introduction

The quality of transmission mode EXAFS measurements rely on the homogeneity of the investigated samples in the region transmitted by the X-ray beam. Lateral thickness variations, holes or inhomogeneous sample compositions may lead to severe distortions of the EXAFS spectra. So the amplitude of the EXAFS oscillations is reduced compared to a perfect sample, and the determined coordination numbers and Debye-Waller factors are erroneous [1]. A careful sample preparation may prevent such artifacts in the case of ex-situ experiments. For in-situ studies, however, there are many situations where the sample considerably changes, e.g., in the case of catalysts in a reacting environment [2] or a battery material during charge [3]. While ionization chambers measure the average absorption through the sample, a 2D-area pixel detector may provide local absorption spectra for each position within the sample. Recent experiments have demonstrated the capabilities of such a setup for in-situ experiments of battery materials, where local charging / discharging occurs [3]. In this contribution, we will present transmission mode EXAFS data of model samples measured with a large, unfocussed beam of a wiggler source in order to evaluate the data quality offered by pixel detectors also for real samples.

In a second series of experiments, we addressed grazing incidence X-ray absorption spectroscopy measurements with the 2D detector, where the X-rays impinge on the surface of the sample under a small incident angle $\phi_i$. Besides the specular reflected beam (incident angle $\phi_i$ = exit angle $\phi_e$), a second distinct scattering peak (Yoneda-peak) occurs if $\phi_e = \phi_c$, the critical angle of the scattering material [4], with $\phi_c(Au) \approx 0.35°$ and $\phi_c(Bi) \approx 0.25°$. From related XAFS experiments, a depth profiling of the sample is feasible, and structural information of the surface and buried layers may be obtained [5, 6]. If ionization chambers are used as detectors, the EXAFS has to be measured sequentially for the different scattering (exit) angles, which makes Yoneda-XAFS experiments extremely time-consuming. Using a 2D-area detector, the entire scattering pattern is acquired simultaneously for each energy in the EXAFS scan, and a wealth of information is available in a moderate time. Here we will discuss the feasibility of Yoneda-XAFS measurements for the investigation of multilayered samples.

## 2. Experimental details

A Pilatus 100K detector was used for all experiments [7]. Transmission mode studies were conducted at beamline 10 at the DELTA storage ring [8], where incident and transmitted intensities from the sample were monitored by $N_2$-filled ionization chambers, and the 2D-detector was placed behind the second ionization chamber in order to measure the lateral distribution of the transmitted X-rays. The unfocussed incident beam from the Si(111) channel-cut monochromator was defined by two slit systems in front of the sample. Kapton foils downstream the second ionization chamber were used to adjust the photon flux on the Pilatus detector in order to avoid counter overflows and not to exceed the maximum countrate accepted by the detector ($10^6$ counts $s^{-1}$ $pixel^{-1}$). Co metal foils and $Co_3O_4$ powder samples were investigated, with about 0.5 s – 1 s integration time per data point (ca. 25 min/spectrum).

For the conducted grazing incidence X-ray scattering experiments, the 2D-detector was directly placed in the scattered radiation fan without a second ionization chamber. The measurements were done at the MRH endstation of the ROBL beamline BM20 at the ESRF using a Si(111) monochromator and a mirrored, focussed beam. Test sample was a Bi (20 nm)/Au (200 nm)-bilayer on a float glass substrate prepared by vacuum deposition. The entire scattering pattern was evaluated for each energy while scanning across the Au $L_3$ (11919 eV) and the Bi $L_3$ (13419 eV) absorption edges.

## 3. XAFS experiments

### *3.1. Transmission mode EXAFS*

In order to study the feasibility of laterally resolved EXAFS experiments and to evaluate the data quality that can be achieved, first tests were performed using a Co-metal foil. In figure 1(a), a map of the beam behind the sample is shown in the right inset for an acquisition time of 1 s. The illuminated region of the sample consists of ca. 6 pixels vertically, and 40 pixels horizontally. With a pixel size of 172 μm ([7], see inset), this corresponds to a beam size of 1 x 7 $mm^2$. The raw absorption data measured using the ionization chambers are compared to different areas of the pixel detector. Obviously, the data quality of a spectrum calculated from a single pixel is excellent with respect to the reproduction of sharp absorption features close to the edge as well as to the low noise of the $k^3$-weighted $\chi(k)$ data depicted in the inset. If a vertical cross section of the beam, i.e. a line of 6 pixels, is used for the calculation of an absorption spectrum, the data quality slightly improves, and if the entire region of the sample that is illuminated from the beam is taken into consideration (40 lines of 6 pixels = 240 pixels), the obtained spectrum is almost indistinguishable to that from the ionization chamber.

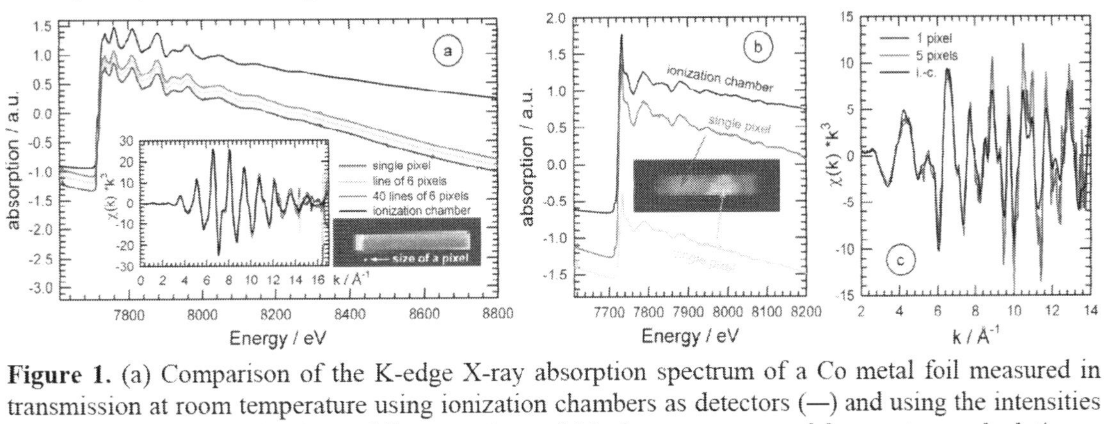


**Figure 1.** (a) Comparison of the K-edge X-ray absorption spectrum of a Co metal foil measured in transmission at room temperature using ionization chambers as detectors (—) and using the intensities on the 2D-detector (1 s / point). Different regions of this detector were used for spectrum calculation: a single pixel (—), a vertical cross section of 6 pixels (—) and the entire illuminated area (40x6 pixel,—). The insets show the $k^3$-weighted EXAFS oscillations $\chi(k)*k^3$ for the different spectra depicted, and the intensity distribution on the Pilatus detector measured at E = 7600 eV. (b) Raw absorption data (0.5 s / point) for an inhomogeneous $Co_3O_4$ powder sample with an intensity map. (c) $\chi(k)*k^3$ data extracted from a single pixel (—) and 5 neighboring pixels (—) compared to the ionization chamber data (—).

While no substantial differences can be seen in the extracted $\chi(k)*k^3$ data up to ca. 12 $Å^{-1}$ for all the different measurements, uncompensated glitches are visible at about 14.4 $Å^{-1}$ and 16 $Å^{-1}$ in the data of the pixel detector, limiting the usable k-range here. The EXAFS of a $Co_3O_4$ powder sample is shown in figure 1(b). The inhomogeneous intensity distribution on the 2D-detector is directly reflected in the variation of the edge jump on two selected positions (arrows in figure 1(b)). Moreover, the EXAFS oscillations measured using the ionization chamber have substantially smaller amplitudes compared to the data of single pixels, especially at larger k-values, see figure 1(c). Thus, it should be promising to measure laterally resolved EXAFS spectra of inhomogeneous samples containing variations of thickness, composition, etc., with a lateral resolution of the pixel size, i.e. ca. 200 μm in the present case.

*3.2. Grazing incidence XAFS*

Some selected scattering patterns obtained from the Bi/Au-bilayer sample are shown in figure 2 for three different energies in the vicinity of the Au $L_3$ edge and $\phi_i$=0.5°, placing the 2D-detector 0.6 m behind the sample. As can clearly be seen, the intensity of the diffuse scattering strongly depends on the X-ray energy and the scattering angle, i.e. the diffuse scattering from the smaller exit angle, i.e. the outer Bi-air surface with smaller density (ROI 2), is not dramatically affected when the energy is scanned through the Au $L_3$-edge. In contrast, the pattern related to the larger exit angle (ROI 2), the inner Bi-Au interface with the larger Au density, dramatically drops down in this energy range (figure 2(c)).

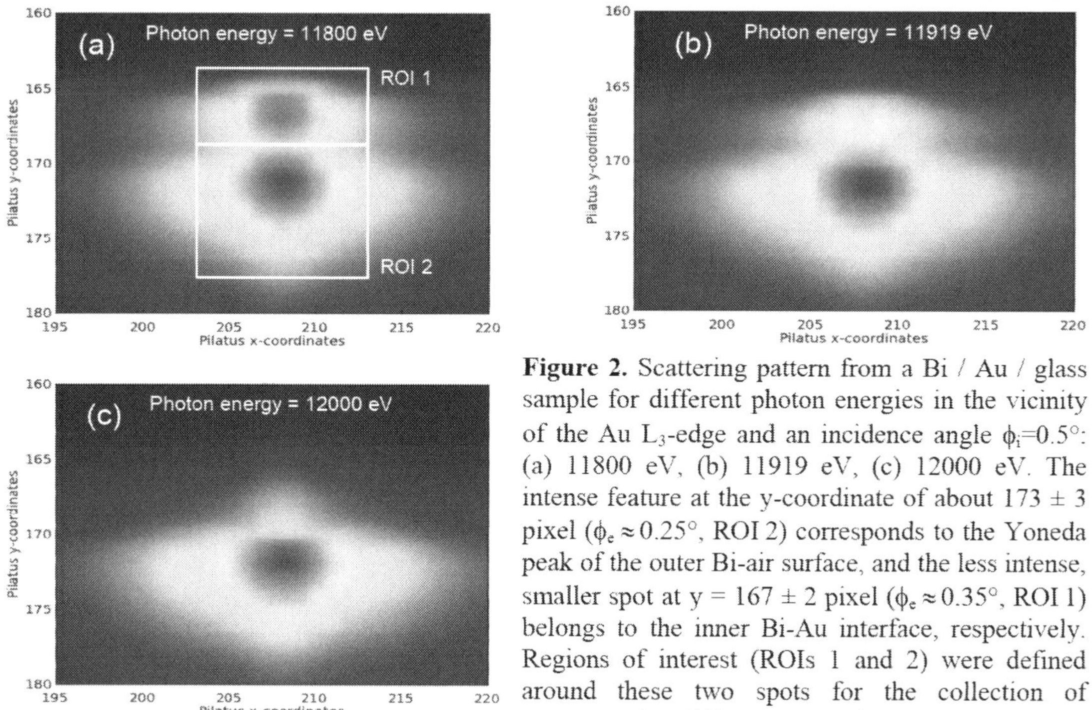


**Figure 2.** Scattering pattern from a Bi / Au / glass sample for different photon energies in the vicinity of the Au $L_3$-edge and an incidence angle $\phi_i$=0.5°: (a) 11800 eV, (b) 11919 eV, (c) 12000 eV. The intense feature at the y-coordinate of about 173 ± 3 pixel ($\phi_e \approx 0.25°$, ROI 2) corresponds to the Yoneda peak of the outer Bi-air surface, and the less intense, smaller spot at y = 167 ± 2 pixel ($\phi_e \approx 0.35°$, ROI 1) belongs to the inner Bi-Au interface, respectively. Regions of interest (ROIs 1 and 2) were defined around these two spots for the collection of representative diffuse EXAFS data.

This behaviour is also observed in the EXAFS spectra calculated from these diffuse X-ray scattering profiles. As can be seen in figure 3, the intensity measured for the Yoneda-peak with the larger $\phi_e$ (smaller y-coordinate in figure 2) is always smaller compared to that related to the smaller $\phi_e$ (larger y-coordinate). While a distinct Au $L_3$-edge can be seen in the first Yoneda peak, no edge can be seen in the scattering related to the second feature. Furthermore, the Bi $L_3$-edge spectrum related to the first Yoneda peak shows a fine structure similar to that of metallic Bi, while that of the second peak resembles that of Bi-oxide $Bi_2O_3$. The qualitative interpretation of these observations is clear: The diffuse scattering related to the larger $\phi_e$ is related to the inner Au-Bi interface, where both elements are present in a metallic form. On the other hand, no gold is present at the outer surface of the bilayer, and thus no Au $L_3$-edge is observed for the smaller $\phi_e$. Due to the chemical reactivity of Bi metal, the latter metal is oxidized at its air-side surface, forming a thin oxide layer of about 1-2 nm thickness [9]. Although the incidence angle is larger than $\phi_c$, the technique appears to be extremely surface sensitive [5]. Thus, the X-ray absorption fine structure in the diffuse scattering reveals features similar to that of $Bi_2O_3$ (see e.g. [10]) due to strong contributions of the outermost surface to the scattered signals [5].

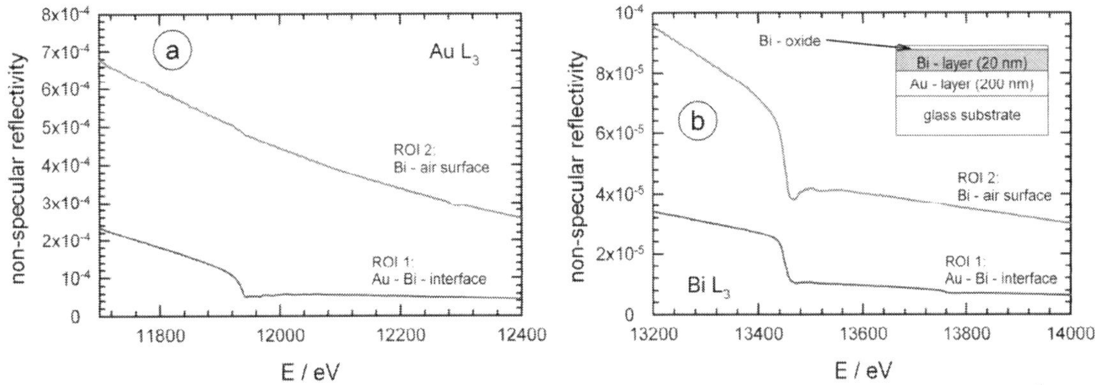


**Figure 3.** Non-specular reflection mode EXAFS experiments at the (a) Au $L_3$- and (b) Bi $L_3$-edges for an incidence angle $\phi_i = 0.5°$ and exit angles according to the Yoneda peaks shown in figure 2, i.e. $\phi_e \approx 0.25°$ (—) and $\phi_e \approx 0.35°$ (—). The inset depicts a model of the layer structure of the sample.

A detailed modelling of the measured data (for calculation details see [5]) is currently in progress in order to determine the thickness of the Bi oxide layer at the surface. Furthermore, the formation of a gold-bismuth alloy at the inner interface can be initiated by an annealing process of the bilayer. Such an alloy formation may result in substantial modifications of the EXAFS detected at the Yoneda peak at larger exit angles, and, thus, a careful analysis may identify these interfacial phases, their thickness and their changes with temperature and time, which will be subject of a forthcoming publication.

## 4. Conclusions

We have investigated the feasibility of transmission and reflection mode EXAFS experiments using a 2D-pixel detector. The results show that acceptable data quality can be achieved even if a short acquisition time and a single detector pixel is used for the evaluation, and the data quality improves if the signals from several pixels are averaged. For surface sensitive grazing incidence X-ray absorption spectroscopy, the pixel detector allows the simultaneous measurement of the entire scattering pattern, thus the structure of different surfaces and interfaces can be investigated in parallel.

## References

[1] Bridges F, Li GG and Wang X 1992 *Nucl. Instrum. Meth. A* **320** 548
[2] Schroer CG and Grunwaldt J-D 2013 *In: In-situ characterization of heterogeneous catalysts. Eds.: J.A. Rodriguez, J.C. Hanson, and P.J. Chupas*. John Wiley & Sons pp. 49 - 73.
[3] Tanida H, Yamashige H, Orikasa Y, Oishi M, Takanashi Y, Fujimoto T, Sato K, Takamatsu D, Murayama H, Arai H, et. al. 2011 *J. Synchrotron Rad.* **18** 919
[4] Yoneda Y 1963 *Phys. Rev.* **131** 2010
[5] Keil P, Frahm R and Lützenkirchen-Hecht D 2010 *Corrosion Sci.* **52** 1305
[6] Lützenkirchen-Hecht D, Keil P and Frahm R 2007 *Surf. Sci.* **601** 4232
[7] Kraft P, Bergamaschi A, Broennimann C, Dinapoli R, Eikenberry EF, Henrich B, Johnson I, Mozzanica A, Schlepütz CM, Willmott PR and Schmitt B 2009 *J. Synchrotron Rad.* **16** 368
[8] Lützenkirchen-Hecht D, Wagner R, Szillat S, Hüsecken AK, Istomin K, Pietsch U and Frahm R 2014 *J. Synchrotron Rad.* **21** 819
[9] Atkinson R and Curran E 1985 *Thin Solid Films* **128** 333
[10] Jiang N and Spence JCH 2006 *J. Phys.: Condens. Matter* **18** 8029

# XAFS Data Interchange: A single spectrum XAFS data file format

**B. Ravel[1] and M. Newville[2]**

[1] National Instituts of Standards and Technology Gaithersburg, MD 20899, USA
[2] Center for Advanced Radiation Studies, University of Chicago, Chicago, IL 60637, USA

E-mail: bruce.ravel@nist.gov

**Abstract.** We propose a standard data format for the interchange of XAFS data. The XAFS Data Interchange (XDI) standard is meant to encapsulate a single spectrum of XAFS along with relevant metadata. XDI is a text-based format with a simple syntax which clearly delineates metadata from the data table in a way that is easily interpreted both by a computer and by a human. The metadata header is inspired by the format of an electronic mail header, representing metadata names and values as an associative array. The data table is represented as columns of numbers. This format can be imported as is into most existing XAFS data analysis, spreadsheet, or data visualization programs. Along with a specification and a dictionary of metadata types, we provide an application-programming interface written in C and bindings for programming dynamic languages.

## 1. The problem of XAFS data representation

A poorly structured and poorly documented file format hinders the use of the data contained in the file by limits its portability between computers, between data analysis and data visualization programs, and over time. What's more, data contained in a poorly formatted file is ill-suited for deposition with a journal as supplemental material.

Virtually every XAFS beamline in the world provides some kind of single-scan file format to its visitors. All such files capture the data table associated with the measurement in some manner. Many capture some metadata relevant to the provenance and measurement conditions of the data. The representation of the metadata and the data table is wildly inconsistent from beamline to beamline. Indeed, there is not even agreement as to a standard representation of the abscissa of the data, with extant examples of the abscissa represented in units of eV, keV, monochromator angle, and photon wavelength.

We propose a solution to the problem of representing a single spectrum of XAFS data called "XAFS Data Interchange" or XDI. Our goals for XDI include:

(1) Establish a common format for transferring data between XAS beamlines, XAS experimenters, data analysis packages, web applications, and anything else that needs to process XAS data.
(2) Increase the relevance and longevity of experimental data by reducing the amount of data archaeology future interpretations of that data will require.
(3) Enhance the user experience by promoting interoperability among data acquisition systems, data analysis packages, and other applications.

(4) Provide a mechanism for extracting and preserving a single XAS-like data set from a related experiment (for example, a resonant diffraction or inelastic scattering measurement) or from a complex data structure (for example, a database or a hierarchical data file used to store a multi-spectral data set).

(5) Provide a representation of an XAS spectrum suitable for deposition with a journal or into a database.

In short, we need to share data across continents, decades, and analysis toolkits.

XDI is designed to capture a $\mu(E)$ spectrum as an ASCII column data file. While it may, in certain situations, be a suitable format for raw data from a beamline, it cannot possibly capture the full complexity of a modern synchrotron measurement. For example, it is not suitable for multi-dimensional or multi-spectral measurements such as a microspectroscopy image. It is not intended to represent a fluorescence measurement made with a large number of energy-discriminating detector elements. To fully capture the data from an energy-discriminating detector requires at least three numbers per channel at each energy point. This would make an unmanageably wide data table when represented as a column text file. It is also not intended to fully encapsulate the path of a measurement through data processing and analysis. There exist hierarchical, relational, and compressed file formats which would be more suitable for any of those chores.

The intent, then, of XDI is to represent a single, processed $\mu(E)$ spectrum. This might be a single scan, or the average of repeated scans on a sample, or the consolidation of the channels of a multi-element detector after dead-time corrections have been applied. In any case, a processed $\mu(E)$ spectrum is the basic unit of currency for an XAFS analysis problem and an XDI file is the encoding of this basic unit of currency.

## 2. The XAFS Data interchange format

Figure 1 shows an example of a transmission XAFS spectrum represented in the XDI format. There is a clear separation between the parts of the file which contain metadata and the data table. The file begins with the metadata portion. Every line of the metadata portion begins with a hash (`#`) character. After the metadata portion, the data table is represented as columns of integer or floating point numbers.

The XDI file format and the version number of the specification used to generate the file are given in line 1. Lines 2–21 show various examples of metadata. A line containing metadata starts with a hash character, followed by a two-part name, followed by a colon (`:`). The two parts of the metadata name represent a semantic grouping, or *family*, and a keyword. For example `Mono.d_spacing` at line 9 is in the family of metadata pertaining to the monochromator and contains the *d*-spacing of the crystal used to monochromate the incident light. The value, in this case 3.13550 Å, follows the colon.

This way of representing the metadata maps naturally onto the concept of an associative array, with the colon separating the key/value pair. The definition of semantic families, like `Mono`, allows a mapping of metadata onto nested associative arrays, useful for programming languages that support nested data structures.

Information about the data table – including number of columns, column labels, and units on the abscissa – are shown in lines 2–4. It is good practice to replicate the column labels in the last header line preceding the data table, as shown at line 28.

Three items of metadata are *required* in any compliant XDI file, `Mono.d_spacing`, `Element.symbol`, and `Element.edge` (seen at lines 5, 6, and 9). The monochromator *d*-spacing is required if a correction needs to be made to the energy axis of the data table. The atomic and edge symbols of the absorbing element are required to unambiguously identify the data. For

```
1  # XDI/1.0 GSE/1.0
2  # Column.1: energy eV
3  # Column.2: mutrans
4  # Column.3: i0
5  # Element.edge: K
6  # Element.symbol: Fe
7  # Scan.edge_energy: 7112.0 eV
8  # Mono.name: Si 111
9  # Mono.d_spacing: 3.13550
10 # Beamline.name: 13-BM-D
11 # Beamline.collimation: none
12 # Beamline.focusing: no
13 # Beamline.harmonic_rejection: detuned 50%
14 # Facility.name: APS
15 # Facility.energy: 7.00 GeV
16 # Facility.xray_source: bending magnet
17 # Scan.start_time: 2002-03-16T17:50:01
18 # Detector.I0: 15cm  N2
19 # Detector.I1: 15cm  N2
20 # Sample.name: FeN
21 # Sample.prep: powder on tape, 5 layers
22 # GSE.extra: some configuration parameter
23 # ///
24 # room temperature
25 # measured at beamline 13-BM-D
26 # vert slits = 2mm (at 45m)
27 #----
28 # energy   mutrans      i0
29   6962.0  -0.39452033  302897.20
30   6972.0  -0.41004339  304569.20
31   6982.0  -0.42398716  304504.20
32   6992.0  -0.43863027  304514.20
33   7002.0  -0.45328045  304407.20
   ⋱
```

**Figure 1.** Listing of an XDI file containing transmission XAFS data measured on iron nitride, FeN. The entire header and the first five data points are shown. The data table has columns for the energy axis, $\mu(E)$, and the signal on the incident intensity detector and it continues to the end of the file.

example, both Cr K and Ba $L_I$ have tabulated edge energies of 5989 eV, while the Se K and Tl $L_{III}$ edges are both at 12658 eV.

Five items of metadata are *recommended* to be included in any XDI file. Four of them – `Beamline.name`, `Facility.name`, `Facility.xray_source`, and `Scan.start_time` (lines 10, 14, 16, and 17) – capture the most essential details of the provenance of the data. `Column.1` (line 2) unambiguously identifies the energy units of the abscissa.

XDI allows the capturing of metadata outside the dictionary through the use of *extension fields*. An example is shown at line 22. The letters "GSE" refer to the data acquisition system at the beamline where these data were measured. Any data acquisition, processing, or analysis program can add metadata to the XDI file header. First the application identifies itself in line 1 after the string which identifies the XDI version number. Using the same identifier string as the family part of the metadata label, the application appends metadata lines using the same syntax as for dictionary metadata lines. In this way, any application which touches the XDI file can encode its metadata into the file.

There is a space for user-supplied comments, as seen at lines 24-26. The metadata section of the header ends at line 23 with text that contains the hash character followed by three or more forward slash characters (/). This is followed by optional, freely-formatted text, then terminated at line 27 with text that contains the hash character followed by three or more dash characters

(-). This comment text is most commonly used at the time of data acquisition by the beamline visitor. This text is free-form and there is no expectation that it should be interpreted by any later application.

Finally, the data table is simply columns of numbers extending to the end of the file. Every entry in the data table must be interpretable as a number. The leftmost column of the data table must contain the abscissa in the units defined in the `Column.1` metadata line. Each row of the data table must have the same number of columns.

## 3. The XDI implementation

All work related to XDI is hosted as a GitHub repository at `https://github.com/XraySpectroscopy/XAS-Data-Interchange`. The documentation at the repository includes a full specification of the XDI format and a dictionary of metadata defining several meaningful concepts related to the provenance and measurement conditions of a spectrum. The specification includes complete details for what constitutes a valid metadata name string or a valid numeric entry in the data table.

The specification is implemented as a library written in the C programming language. This library reads and interprets the contents of the XDI file in the following steps:

(1) Parse the XDI file, stopping only in the case of an unrecoverable error, then
(2) Check that the required metadata, as identified in the specification, are present, then
(3) Check that the recommended metadata, as identified in the specification, are present, then
(4) Validate each individual metadata item against its dictionary definition, if available

The library returns a data structure containing the full content of the XDI file. The entire content of the specification and the dictionary are encoded in the library.

This library is easily bound to dynamic languages. Bindings are provided at this time for Python and Perl. New bindings for other languages can be created using the C library. A less fully-featured implementation is provided in the FORTRAN language.

The distribution also includes a cross-platform build system, documentation, an extensive test suite of compliant and non-compliant data files, and automation for testing the library.

XDI is available for use. Participation in this effort via the GitHub site is encouraged. We hope to see wide adoption of XDI at beamlines, in programs for data processing and analysis, in XAS theory programs, and elsewhere.

Journal of Physics: Conference Series **712** (2016) 012149 doi:10.1088/1742-6596/712/1/012149

# ROCK: the new Quick-EXAFS beamline at SOLEIL

**V Briois**[1], **C La Fontaine, S Belin, L Barthe, Th Moreno, V Pinty, A Carcy, R Girardot, E Fonda**

Synchrotron SOLEIL, L'Orme des Merisiers, BP 48, Saint-Aubin, 91 192 Gif-sur-Yvette, France

E-mail: briois@synchrotron-soleil.fr

**Abstract.** ROCK is a new beamline at SOLEIL dedicated to Quick-EXAFS measurements. The optical layout has been optimized to get full advantage of the monochromators, which were designed at SOLEIL and successfully used at SAMBA from 2009 to 2014. ROCK has started user operations since March 2015. It is mainly employed to monitor fast kinetic processes in materials used in catalysis and energy sciences. A review of the ROCK performances and capabilities is presented. The high automation achieved for fast change of monochromators, optimization of mirrors for harmonic rejection and detectors allows the simultaneous *operando* characterization of different chemical elements present in a material during the same reaction.

## 1. Introduction

X-ray Absorption Spectroscopy is a well-known technique for *in situ* characterization of materials such as catalysts and batteries. With the advent of $3^{rd}$ generation synchrotron radiation facilities allowing a sub-second time resolution in the monitoring of kinetics, the experimentalist can now access to a deeper and more accurate temporal description of the chemical species involved during operative use of these materials. The simultaneous determination of their local order evolution and catalytic or electrochemical properties can be achieved at X-ray Absorption beamlines dedicated to the study of catalysis and materials used for energy storage. The SOLEIL's ROCK (Rocking Optics for Chemical Kinetics) beamline, recently funded by the French *Agence Nationale de la Recherche* in the framework of the national "*Investissements d'Avenir*" program (ANR-10-EQX-45-01), is one of these dedicated infrastructures for the *operando* characterization of materials by time-resolved Quick-EXAFS.

## 2. Beamline Design

The SOLEIL's Quick-EXAFS monochromators installed at ROCK are up-graded versions of the monochromator developed by the Frahm's group [1], which consists of a channel cut crystal mounted on a cam driven tilt table that oscillates around a selected central Bragg angle, $\theta$, with a user selectable angular amplitude, $\Delta\theta$. After a joint work between SOLEIL and Frahm's group for integrating an angular encoder attached to the crystal holder in order to have a direct reading of the crystal position [2], each group worked independently to develop a new flexible monochromator. The SOLEIL's Quick-EXAFS monochromators, fully described in [3], can oscillate over an angular

[1] To whom any correspondence should be addressed.

amplitude $\Delta\theta$ up to 4°. The Si(111) or Si(220) crystals equipping those monochromators can be remotely selected by the user, allowing to work at energies between 4 and 43 keV. These monochromators were transferred from SAMBA to ROCK in 2014 and are now surrounded by optics designed to obtain the highest flux possible delivered by a SOLEIL's bending magnet source. The optical setup is designed to follow *in operando* conditions the reactions of several distinct elements. This "element- or edge-jumping" capability is based on the alternate use of the two Quick-EXAFS monochromators and was already demonstrated at SAMBA to be a powerful skill for unravelling synergetic behaviours between elements during chemical processes [3, 4]. The optical layout of ROCK is shown schematically in Figure 1.

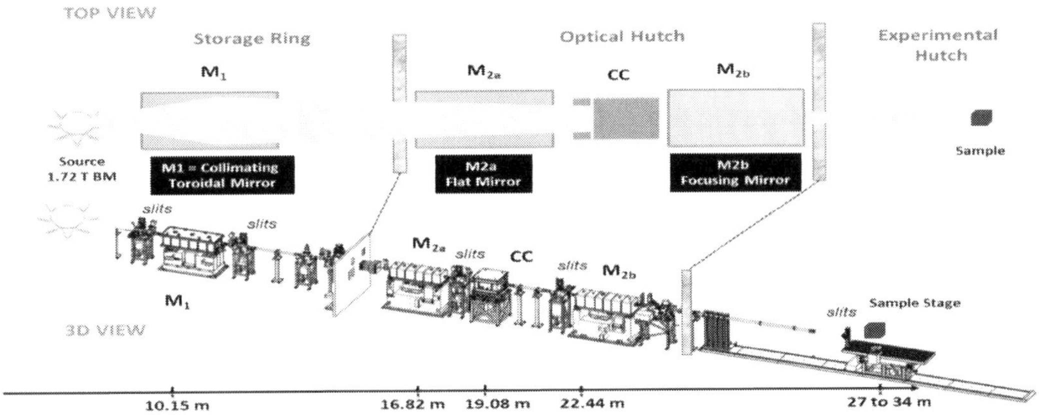


**Figure 1.** Optical layout of the ROCK beamline.

In order to maximize the beamline acceptance defined as 1.5 mrad (H) x 0.22 mrad (V), the first mirror (M1) is installed inside the shield wall of the storage ring at 10.15 m from the bending magnet source (B = 1.72 T, $\sigma_x$ = 60 μm and $\sigma_z$ = 24.5 μm). The beamline is separated from the machine vacuum by a water-cooled 300 μm thick beryllium window. Water-cooled primary horizontal and vertical slits define the X-ray beam before it impinges on the first optic. M1 is a 1.1 meter-long toroidal and cylindrically bendable silicon mirror coated with an Ir layer (50 nm) over a Cr binder layer. The power absorbed by M1 is lower than 100 W allowing the use of In/Ga eutectic cooling channels as heat transmitter to the water-cooling system. M1 deflects upward a horizontally focused beam according to its toroidal shape (fixed sagittal radius = 31.7 mm) with vertical collimation according to its cylindrical shape. Although, the standard working grazing incidence of M1 is 2.25 mrad with a cut-off energy at 37 keV, this angle can be slightly decreased in order to use the beamline up to 43 keV. The curvature of the mirror is adjustable in order to adapt the beam collimation to the machine current (the nominal tangential radius used at 2.25 mrad is 9.02 km).

In the optical hutch, M2a and M2b mirrors surround the Quick-EXAFS channel-cut (CC) monochromators. These mirrors are coated with 3 stripes (Pt, Pd and $B_4C$ stripes) used to eliminate the high order harmonics with a rejection coefficient ranging from $10^4$ to $10^8$ depending on the cut-off energy. Additionally, M2a reduces the heat load on the monochromators, keeping the water-cooling solution of these channel-cuts still efficient. The length of both mirrors is 1.1 m. For each mirror, a double axis goniometer ensures the pitch (from 1.75 to 5.2 mrad) and roll adjustments of the mirrors with respect to the incident beam whereas robust motorized translations are used for vertical alignment of the mirror's center and horizontal stripe exchange. M2a is a flat water-cooled mirror which preserves the beam collimation at the entrance of the monochromator whereas M2b is an adjustable cylindrically bent mirror used to vertically focus the beam at the sample position. M2a and M2b deflect the beam upward and downward, respectively.

Journal of Physics: Conference Series **712** (2016) 012149 doi:10.1088/1742-6596/712/1/012149

Si(111) and Si(220) channel-cuts are installed at a distance of 18.92 and 19.25 m from the source, respectively. As the position of the beam varies vertically depending on the grazing angle used for M2a, two independent 60 mm vertical translations are used to properly align each crystal on the beam. Vertical slits are installed after each optic for cutting diffuse beam while horizontal slits are used after the Quick-EXAFS monochromators in order to cut possible Laue reflections by the channel-cut which may partially overlap the Bragg reflection of interest at the sample position. A SOLEIL's designed pink screen is installed after M2a in order to visualize the beam deflected by either M1 or M2a while a SOLEIL's designed monochromatic screen is installed after M2b in order to image the beam coming from the monochromators, which is either by-passed by M2b or deflected by this mirror.
X-ray shutters are placed before the shield walls of the storage ring and optical hutch in order to allow the access to the optical and experimental hutches, respectively, while keeping a constant heat load on the optics.

## 3. Beamline Performances

The calculated and measured fluxes arriving at the sample are shown in Figure 2. The absolute flux was measured using a calibrated silicon photodiode with 450 mA of stored current in the machine whereas the calculations considered perfect optics with the nominal 500 mA current, already achieved at SOLEIL but not yet used for operations. For these reasons, the measured flux is slightly lower than the calculated one. A flux between $10^{11}$ and $10^{12}$ ph/s is achieved over a large energy range, offering the necessary condition to achieve 25 ms of time resolution with convenient signal-to-noise ratio.

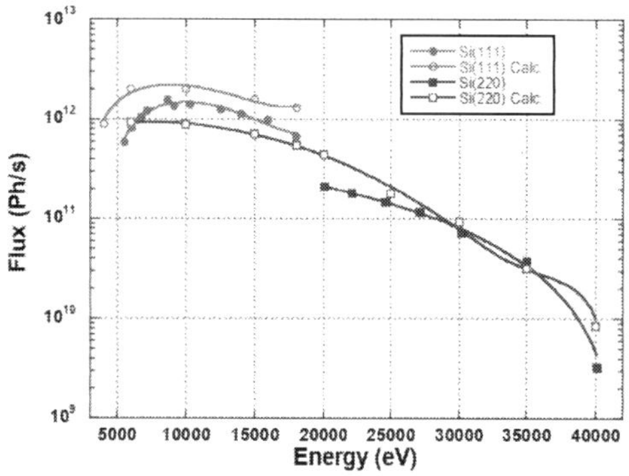


**Figure 2.** Flux of the monochromatic beam measured at the sample position for a full spot size, plotted as a function of X-ray energies for a beam current of 450 mA. The flux calculated for perfect optics and a beam current of 500 mA is reported for comparison purpose.

The beam size achieved in the horizontal direction is invariant with the monochromator in use and depends only on the position of the sample with respect to the M1 focus point, which is located at 32.68 m from the source. At this focus position, the horizontal size of the beam is ca. 380 µm (FWHM). At upstream positions from this focus point, the beam is defocused. Moving the sample from 32.68 to 27 m in the experimental hutch allows for increasing by a factor 10 the horizontal beam size, as shown in Figure 3. The size in the vertical plane is determined by the curvature of the cylindrically bent M2b mirror and can be tuned from defocused to optimized focusing conditions at any sample position in the experimental hutch. A maximum vertical size of ca. 1 mm can be achieved versus ca. 130 µm (FWHM) in optimized vertical focusing condition. The versatility in beam size at the sample position, which can be modified by just moving the sample between 27 and 32.68 m, allows varying the density of photons on the sample while keeping nearly constant the flux and thus the time resolution performance of the beamline. The flexible beam size is important for the study of samples suffering from radiation damage as illustrated in Figure 4 with the decomposition kinetics of a AuCl pellet transformed into Au(0) crystallites under radiation. The lower the photon density, the longer the stability of AuCl.

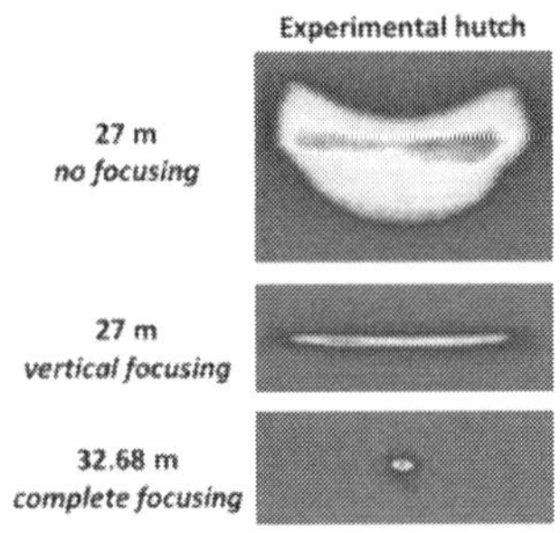


**Figure 3.** Versatile beam size as a function of the position in the experimental hutch.

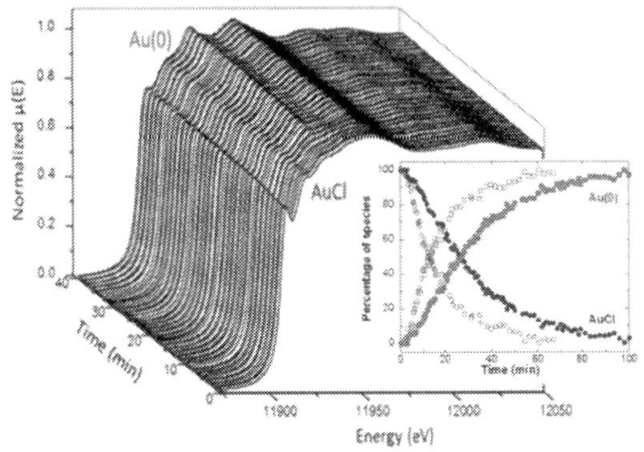


**Figure 4.** Change of decomposition kinetics of a pellet of AuCl with the photon density. The spectra were recorded at the M1 focus point. The speciation curves in the insert compare the decomposition rate at the focus point (open symbols) and at 27 m from the source (closed symbols).

Mechanically, the channel-cut crystals can oscillate with a frequency up to 30 Hz for small oscillation amplitudes (typically 0.3-0.5°), and is reduced to 20 Hz for larger oscillation amplitudes. Those mechanical characteristics fit well with the acceptable damping of the measured signal observed with increasing oscillation frequencies. As shown in Figure 5, the EXAFS signals recorded for a copper foil over an oscillation amplitude of 2° of the Si(111) channel-cut present only small damping effects for the data recorded at 20 Hz compared to the EXAFS signal recorded step-by-step at the SAMBA beamline. Nevertheless, the inspection of the XANES spectra recorded in the same conditions of wide oscillation amplitude clearly indicates that the intrinsic rise time of detectors and amplifiers is no more adapted, it induces a strong damping in this energy range where fine structures are observed. By reducing the oscillation amplitude to 0.3°, the Quick-XANES spectra recorded at the maximum oscillation frequency (30 Hz) are now well comparable to the step-by-step XANES spectrum. The use of gridded ionization chambers could overcome the observed damping at high oscillation frequency, as recently demonstrated for oscillations at 50 Hz achieved with the newly built direct drive torque motor Quick-EXAFS monochromator developed by the Frahm's group [5].

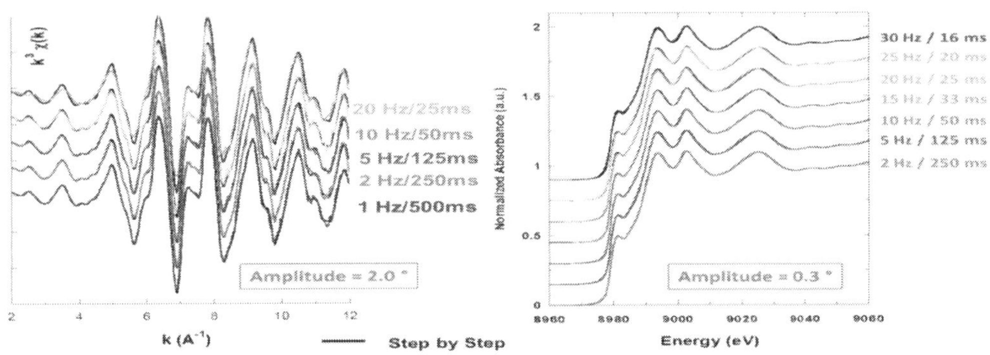


**Figure 5.** Time resolution performances measured on a Cu foil for two oscillation amplitudes. Quick-EXAFS spectra are compared to a step-by-step spectrum (black curve).

## 4. End Station Facilities

The detectors used for transmission experiments are S-1194 and S-1196 ionization chambers provided by Ohyo Koken Kogyo Co. Ltd. They are installed on the 2.5 m long × 1 m wide motorized experimental table, which can move along the beam axis from 27 to 34 m from the source. According to the working position of the table, pipes of variable lengths are installed to keep the beam under a primary vacuum in the experimental hutch up to the slits placed in front of the $I_0$ ionization chamber. Figure 6 displays the experimental table set at the M1 focus position.

A wide range of sample environments can be mounted on the motorized sample stage with 60 mm stroke in both x and z directions. ROCK provides to users dedicated cells for heterogeneous catalysis experiments [4] (together with gas delivery system and a mass spectrometer for activity measurement) and for electrochemical cycling of batteries [6]. Sample environment data such as oven temperature or electrochemical conditions can be associated with the Quick-EXAFS data, making easier data analysis and link between local order and applied constraints. Exhaust gas lines are available allowing the safe handling of toxic gases such as $H_2S$, CO or NO. The beamline also offers the possibility to combine Quick-EXAFS measurements with complementary techniques such as UV-vis. and/or Raman spectroscopies and X-ray Diffraction [4].

Due to the highly demanded "edge jumping" feature offered by ROCK for the *operando* study of bimetallic systems, special supports hosting two identical ionization chambers have been designed at the beamline: each set of ionization chambers is dedicated to one edge, with mixture of gases optimized for the corresponding energy, and is alternatively moved into the beam path (Figure 7).

**Figure 6.** Motorized experimental table located at the M1 focus position. Acquisition electronics are loaded in dedicated spaces in and on the experimental table.

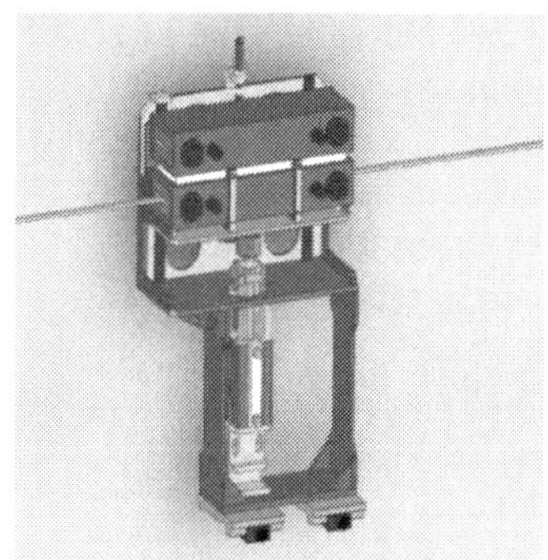

**Figure 7.** View of stacked $I_1$ ionization chambers for "edge jumping" capability.

## 5. Control System, Data Acquisition and Survey, Automation

The beamline is controlled through the Tango software bus. Electronics and software used to operate data acquisition with the Quick-EXAFS monochromators were developed by the SOLEIL's Control Command group. When acquisition is started, 5 s double data buffers continuously collect the channels of a 16-bit analog-to-digital converter (ADC), accounting for the ionization chamber currents amplified by fast Keithley 428 current-to-voltage amplifiers. A scaler card collects synchronously the position of the angular encoder at a user-selectable frequency, usually in the range of 250 kHz, this card provide the trigger line to the ADC card. A dedicated software merges the ADC signals and the encoder values in a single binary nexus file. A visualization interface enables the user to look at the evolution of the raw spectra during the course of a kinetic experiment, in a meaningful absorption versus energy plot (Figure 8). The data continuously acquired can be entirely or partly displayed in real-time by this visualisation interface, named QuickExafsViewer. The former operation mode is particularly useful for the staff during the set-up of the beamline and to check the energy calibration while the latter operation mode is well-suited to follow the evolution of kinetics by picking up selected snapshots representative of the reaction. Reference spectra previously recorded and archived can be also loaded in this visualization interface for comparison purposes with the data in acquisition.

Automation of data acquisition sequences for "edge jumping" experiments is obtained by using the interactive Python scripting environment (IPython 0.13), which is easily interfaced with the Tango software bus. The Python shell can then be easily handled by the user to start and stop acquisitions sequences with repeated changes of beamline configuration, including swap of monochromators, change of M2a and M2b stripes, and exchange of ionization chambers, together with suitable reconfiguration of recording parameters. Typical sequences used for edge jumping are 20 to 30 s of data recording with one monochromator and a period of ca. 30 s required to set-up the beamline in the new configuration before recording data for 20 to 30 s with the second monochromator.

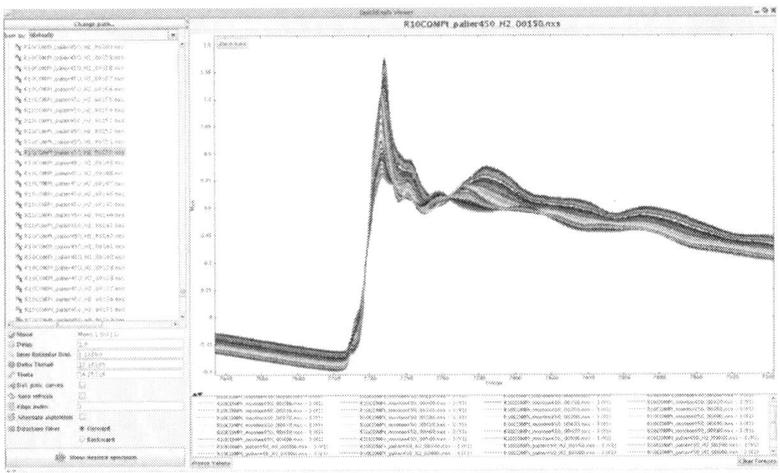

**Figure 8.** Use of the QuickExafsViewer visualization interface for monitoring the *in situ* reduction of a cobalt-based supported catalyst.

## 6. Conclusion

Fully optimized focusing optics in both horizontal and vertical directions built around the Quick-EXAFS monochromators deliver at the sample position a flux ranging between $10^{11}$ and $10^{12}$ ph/s in a large energy range covered by the beamline. At these conditions, 25 ms is a typical achievable time resolution. With such time resolution, users obtain a huge amount of data in a couple of minutes making the common strategy of data evaluation, reduction, and analysis quite inefficient. New tools for data analysis must be proposed to the users to interactively optimize the outcome of an experiment carried out at the beamline. Chemometric tools, such as Multivariate Curve Regression with Alternating Least Square (MCR-ALS) fitting, are emerging as powerful methods for getting more information from XAS spectra of evolving mixtures upon reaction [7]. Therefore, a user-friendly interface is under active development at the beamline to handle this high number of data into normalized matrices of absorption spectra directly suitable as input files for MCR-ALS analysis.

## Acknowledgement

The beamline staff sincerely thanks Didier Dallé (previously SOLEIL and now ESRF) for his contribution to the ROCK design.

## References

[1] Frahm R, Richwin M and Lützenkirchen-Hecht D 2005 *Phys. Scr.* **T115** 974
[2] Stözel J, Lützenkirchen-Hecht D, Fonda E, De Oliveira N, Briois V and Frahm R 2008 *Rev. Sci. Instrum.* **79** 083107
[3] Fonda E, Rochet A, Ribbens M, Barthe L, Belin S and Briois V 2012 *J. Synchrotron Rad.* **19** 417
[4] La Fontaine C, Barthe L, Rochet A and Briois V 2013 *Cat. Today* **205** 148
[5] Müller O, D. Lützenkirchen-Hecht D and Frahm R, 2015 *Rev. Sci. Inst.* **86** 093905
[6] Leriche J B, Hamelet S, Shu J, Morcrette M, Masquelier C, Ouvrard G, Zerrouki M, Soudan P, Belin S, Elkaïm E and Baudelet F 2010 *J. Electrochem. Soc* **157** A606
[7] Cassinelli W H, Martins L, Passos A R, Pulcinelli S H, Santilli C V, Rochet A and Briois V 2014 *Cat. Today* **229** 114